Methods of Soil Analysis

Part 2
Microbiological and Biochemical Properties

Editorial Committee: R. W. Weaver, chair
Scott Angle
Peter Bottomley
David Bezdicek
Scott Smith
Ali Tabatabai
Art Wollum

Managing Editor: S. H. Mickelson
Editor-in-Chief SSSA: J. M. Bigham

Number 5 in the Soil Science Society of America Book Series

Published by: Soil Science Society of America, Inc.

1994

Soil Science Society of America, Inc.
677 South Segoe Road, Madison, Wisconsin 53711 USA

Library of Congress Cataloging-in-Publication Data

Methods of soil analysis. Part 2, Microbiological and biochemical
    properties / editorial committee, R.W. Weaver, chair . . . [et al.].
        p.   cm. — (Soil Science Society of America book series ; no.
    5)
    Includes bibliographical references and index.
    ISBN 0-89118-810-X
    1. Soil microbiology—Methodology.   2. Soils—Analysis.
I. Weaver, R. W. (Richard W.), 1944–   . II. Soil Science Society
of America.   III. Series.
QR111.M34   1994
631.4'17'0287—dc20   94-20752
                                                                              CIP

Printed in the USA

CONTENTS

Chapter 5   Most Probable Number Counts   **59**

PAUL L. WOOMER

Chapter 6   Light Microscopic Methods for Studying Soil Microorganisms   81

PETER J. BOTTOMLEY

Chapter 7   Viruses   **107**

J. SCOTT ANGLE

Chapter 8   Recovery and Enumeration of Viable Bacteria   **119**

DAVID A. ZUBERER

Chapter 13   Anaerobic Bacteria and Processes   223

HEINRICH F. KASPAR AND JAMES M. TIEDJE

Chapter 14   Denitrifiers   245

JAMES M. TIEDJE

Chapter 15   Actinomycetes   269

E. M. H. WELLINGTON AND I. K. TOTH

Chapter 16   *Frankia* and the Actinorhizal Symbiosis   291

DAVID D. MYROLD

Chapter 17   Filamentous Fungi   329

DENNIS PARKINSON

FOREWORD

The methods pertinent to soil microbiology were formerly included in Part 2 of the Agronomy Monograph No. 9, Methods of Soil Analysis. Since the 2nd edition of this document, the number of biochemical and microbiological methods have expanded greatly. In addition because the clientele of scientists engaged in these efforts are primarily soils based, the ASA Board of Directors in 1993 elected to place this document in the SSSA Book Series. It is most refreshing and encouraging to see this stand-alone contribution specifically dedicated to soil microbiological and biochemical methods. This text will be well received by an ever-expanding spectra of biogeoscientists. It is very timely given the rise in public and private interests in soil and water quality, biodiversity, biodegradation, terrestrial ecology, environmental quality protection, sustainability of the biosphere and issues of global climatic change. For too long soil quality has been defined in terms of soil physical and chemical attributes with little or no regard to biological components. Part of this oversight has been a function of techniques available to accurately identify, define and quantify biological health and diversity. Another aspect is the rather recent explosion of interest and awareness among geoscientists in the functionality and import of soil microbiological and biochemical attributes in near-surface earth processes. The methods reported herein are at the cutting edge of science. Analytical techniques range in resolution from whole organisms to molecular fragments. In unravelling the identity and behavior of the complex soil biological system, temporal and dynamic diversity are considered in sampling methods. The authors represent a select spectra in biogeoscience expertise and career development. Such a synergistic assemblage of scientists assures that the methodology presented is current and relevant. The document is comprehensive in scope, interdisciplinary in character, and offers a high probability of acceptance among biologists. These methods will serve as the standard bearer for both professional and practicing biological scientists. It is the goal that common methodology will enhance collaboration and interchange among scientists and generate data sets using similar analytical approaches. We commend the authors and editors for their diligence and genius in bringing this new book to fruition in such a timely manner. This addition to the SSSA Book Series, will be well received and widely used by a growing number of biogeoscientist professionals wishing to document soil microbiological-biochemical attributes in near surface earth systems.

Larry P. Wilding, *president*
Soil Science Society of America

PREFACE

The books, *Methods of Soil Analysis*—Parts 1 and 2, published as Agronomy Monograph No. 9 have been the primary references on analytical methods used by soil scientists and persons in other disciplines involved with making measurements on soils. Part 2 of the second edition covered both methods on soil chemistry and soil microbiology. The need for more extensive coverage in both of these areas resulted in necessity of dividing Part 2 into two new books. One covering the topic of soil chemistry and the second covering soil microbiology and soil biochemistry. Revision was so extensive and involved so many new authors that it seemed best to consider this book a new publication rather than a third edition. It is published as one of the Soil Science of America Book Series.

Division of some subject matter between the book on chemical methods for soil analysis and this book was not always straightforward because some chemical methods are needed in measuring microbiological and biochemical processes. In such cases, a chemical method is provided within chapters of this book but the depth of coverage on theory is not complete nor are alternative methods presented as is the case for the book on soil chemical methods. Our desire was to make it possible to use the methods in this book independently without having to purchase both books.

Early in the book the topics of statistical methods, soil sampling, and measurement of soil moisture tension are covered. These chapters were not covered in the previous editions of Part 2 but are particularly important for investigations in soil microbiology and biochemistry. Several methods are provided on use of molecular techniques that were not in previous editions but are needed in many modern soil microbiology laboratories. The treatment of the material on molecular topics is such that a person would not need extensive training in molecular techniques to take advantage of the methods.

It is hoped that many laboratories outside of soil science will take advantage of the methods contained in this book. They will be particularly relevant and useful to laboratories with interest in environmental microbiology or bioremediation. Analytical methods are essential to progress in science and the methods presented in this book are recognized by soil scientists as being among the best currently available. All chapters were reviewed by persons having expertise on particular methods, by an associate editor, and by the editor. The help of the many reviewers,

efforts and patience of authors, and advice from the editorial board are all gratefully acknowledged. A book such as this one is very much a team effort and is
beyond the capability of any individual or small group of individuals.

R. W. Weaver, *editor*
Texas A&M University
College Station, Texas

J. Scott Angle, *associate editor*
University of Maryland
College Park, Maryland

Peter J. Bottomley, *associate editor*
Oregon State University
Corvallis, Oregon

CONTRIBUTORS

J. Scott Angle

Professor of Agronomy, Agronomy Department, University of Maryland, College Park, MD 20742

Wilfredo Laserna Barraquio

Associate Professor of Microbiology, Institute of Biology, University of the Philippines, Diliman, QL 1101, Philippines

L. W. Belser

School of Science and Computer Studies, Nelson Polytechnic, Nelson, New Zealand

D. F. Bezdicek

Professor of Soils, Department of Crop and Soil Sciences, Washington State University, Pullman, WA 99164-6420

Peter J. Bottomley

Professor of Microbiology and Soil Science, Department of Microbiology, Oregon State University, Corvallis, OR 97331-3802

Thomas W. Boutton

Associate Professor of Ecology, Department of Rangeland Ecology and Management, Texas A&M University, College Station, TX 77843-2126

L. G. Bundy

Professor of Soil Science, Department of Soil Science, University of Wisconsin, Madison, WI 53706

Seth K. A. Danso

Technical Officer, Soil Fertility and Crop Production Section, Joint FAO/IAEA Division, International Atomic Energy Agency, Wagramerstrasse 5, P.O. Box 100, A-1400 Vienna, Austria

Eric A. Davidson

Associate Research Scientist, The Woods Hole Research Center, P.O. Box 296, Woods Hole, MA 02543

D. J. Drahos

Director of Research and Development and Senior Scientist, SBP Technologies, Inc.; Sybron Chemicals, Inc., Salem, VA 24153

B. D. Eardly

Assistant Professor of Biology, Pennsylvania State University, Berks Campus, Reading, PA 19610

Mary K. Firestone

Professor of Soil Microbial Ecology, Department of Soil Science, University of California, 108 Hilgard Hall, Berkeley, CA 94720

Dennis D. Focht

Professor of Soil Microbiology, Department of Soil and Environmental Sciences, University of California, Riverside, CA 92521

Jeffry J. Fuhrmann   Associate Professor of Soil Microbiology, Department of Plant and Soil Sciences, University of Delaware, Newark, DE 19717-1303

William C. Ghiorse   Professor and Chairman, Section of Microbiology, Division of Biological Sciences, Cornell University, Ithaca, NY 14853

Peter H. Graham   Professor, Department of Soil Science, University of Minnesota, St. Paul, MN 55108

Charles Hagedorn   Professor of Soil Microbiology, Department of Crop and Soil Environmental Sciences, Virginia Polytechnic Institute and State University, Blacksburg, VA 24061-0404

Stephen C. Hart   Assistant Professor, School of Forestry, Northern Arizona University, P.O. Box 15018, Flagstaff, AZ 86011-15018

R. D. Hauck   Senior Scientist, Tennessee Valley Authority, NFE-IA, Muscle Shoals, AL 35660

William E. Holben   Research Assistant Professor, Center for Microbial Ecology and Department of Crop and Soil Sciences, Michigan State University, East Lansing, MI 48824. Currently Research Scientist, Environmental Microbiology, The Agouron Institute, La Jolla, CA 92037-4696; Email bholben@vaxkiller.agr.org.

W. R. Horwath   Faculty Research Associate, 3450 S.W. Campus Way, Crop and Soil Science Department, Oregon State University, Corvallis, OR 97331

Elaine R. Ingham   Associate Professor of Soil Ecology, Department of Botany and Plant Pathology, Oregon State University, Corvallis, OR 97331-2902

Russell E. Ingham   Associate Professor, Department of Botany and Plant Pathology, Oregon State University, Corvallis, OR 97331-2902

Heinrich F. Kaspar   Cawthron Institute, P.O. Box 175, Nelson, New Zealand

A. C. Kennedy   Soil Scientist, USDA-ARS, Washington State University, Pullman, WA 99164-6421

Leif Klemedtsson   Ph.D., Swedish Environmental Research Institute (IVL), Gothenburg, Sweden

Roger Knowles   Professor of Microbiology, McGill University, Macdonald Campus, Ste. Anne de Bellevue, PQ H9X, Canada

J. O. Legg   Adjunct Professor, Agronomy Department, University of Arkansas, Fayetteville, AR 72701

K. J. McInnes   Assistant Professor of Environmental Physics, Department of Soil and Crop Sciences, Texas A&M University, College Station, TX 77843-2474

J. J. Meisinger — Soil Scientist, USDA-ARS, BARC-West, Beltsville, MD 20705

F. Blaine Metting, Jr. — Senior Program Manager, Battelle, Pacific Northwest Laboratories, Richland, WA 99352

Andrew R. Moldenke — Research Professor of Entomology, Oregon State University, Corvallis, OR 97331

A. R. Mosier — Research Chemist, USDA-ARS, Fort Collins, CO 80522

R. L. Mulvaney — Professor, Department of Agronomy, University of Illinois, Urbana, IL 61801

David D. Myrold — Associate Professor, Department of Crop and Soil Sciences, Oregon State University, Corvallis, OR 97331-7306

A. V. Ogram — Assistant Professor of Soils, Department of Crop and Soil Science, Washington State University, Pullman, WA 99164-6420

Timothy B. Parkin — Research Microbiologist, USDA-ARS, National Soil Tilth Laboratory, Ames, IA 50011

Dennis Parkinson — Professor, Department of Biological Sciences, University of Calgary, Calgary, Alberta, Canada T2N 1N4

E. A. Paul — Professor, Crop and Soil Sciences, Michigan State University, East Lansing, MI 48824

Ian L. Pepper — Professor of Environmental Microbiology, Department of Soil and Water Science, University of Arizona, Tucson, AZ 85721

Suresh D. Pillai — Assistant Professor of Environmental Microbiology, Texas A&M University Research Center, El Paso, TX 79927

Joseph A. Robinson — Associate Director of Biostatistics and Environmental Research, The Upjohn Company, Kalamazoo, MI 49001

M. J. Sadowsky — Associate Professor of Soil Science and Microbiology, Soil Science Department, University of Minnesota, St. Paul, MN 55108

M. J. Savage — Professor of Agrometeorology, Department of Agronomy, University of Natal, Pietermaritzburg 3201, Republic of South Africa

Edwin L. Schmidt — Professor Emeritus of Soil Science, Department of Soil Science, University of Minnesota, St. Paul, MN 55108

Dipankar Sen — Research Scientist, Department of Soil and Crop Sciences, Texas A&M University, College Station, TX 77843

Horace D. Skipper — Professor, Department of Agronomy and Soils, Clemson University, Clemson, SC 29634-0359

T. E. Staley   Research Microbiologist, USDA-ARS, NAA, ASWCRL,
                        Beckley, WV 25813

John M. Stark   Assistant Professor of Microbial Ecology, Department of
                        Biology and the Ecology Center, Utah State University,
                        Logan, UT 84322-5500

David M. Sylvia   Professor of Soil Microbiology, Soil and Water Science
                        Department, University of Florida, Gainesville, FL 32611-
                        0290

M. A. Tabatabai   Professor of Soil Biochemistry, Department of Agronomy,
                        Iowa State University, Ames, IA 50011

James M. Tiedje   University Distinguished Professor, Center for Microbial
                        Ecology, Michigan State University, East Lansing, MI
                        48824-1325

I. K. Toth   Research Fellow, Biological Sciences, University of War-
                        wick, Coventry, CV4 7AL U.K.

Ronald F. Turco   Professor, Department of Agronomy, 1150 Lilly Hall of
                        Life Sciences, Purdue University, West Lafayette, IN
                        47907-1150

R. W. Weaver   Professor of Soil Microbiology, Department of Soil and
                        Crop Sciences, Texas A&M University, College Station,
                        TX 77843-2474

E. M. H. Wellington   Senior Lecturer, Department of Biological Sciences, Uni-
                        versity of Warwick, Coventry CV4 7AL U.K.

Duane C. Wolf   Professor, Department of Agronomy, University of Ar-
                        kansas, Fayetteville, AR 72701

A. G. Wollum, II   Professor of Soil Microbiology, Department of Soil Sci-
                        ence, North Carolina State University, Raleigh, NC
                        27695-7619

Paul L. Woomer   Programme Officer, Tropical Soil Biology and Fertility
                        Programme, P.O. Box 30592, Nairobi, Kenya

S. F. Wright   Research Scientist, USDA-ARS, Soil Microbial Systems
                        Laboratory, BARC-East, Beltsville, MD 20705

L. M. Zibilske   Associate Professor of Soil Microbiology, Department of
                        Plant, Soil and Environmental Sciences, University of
                        Maine, Orono, ME 04469-5722

David A. Zuberer   Professor of Soil Microbiology, Department of Soil and
                        Crop Sciences, Texas A&M University, College Station,
                        TX 77843-2474

Conversion Factors for SI and non-SI Units

Conversion Factors for SI and non-SI Units

To convert Column 1 into Column 2, multiply by	Column 1 SI Unit	Column 2 non-SI Unit	To convert Column 2 into Column 1, multiply by
Length			
0.621	kilometer, km (10^3 m)	mile, mi	1.609
1.094	meter, m	yard, yd	0.914
3.28	meter, m	foot, ft	0.304
1.0	micrometer, μm (10^{-6} m)	micron, μ	1.0
3.94×10^{-2}	millimeter, mm (10^{-3} m)	inch, in	25.4
10	nanometer, nm (10^{-9} m)	Angstrom, Å	0.1
Area			
2.47	hectare, ha	acre	0.405
247	square kilometer, km^2 (10^3 m)^2	acre	4.05×10^{-3}
0.386	square kilometer, km^2 (10^3 m)^2	square mile, mi^2	2.590
2.47×10^{-4}	square meter, m^2	acre	4.05×10^3
10.76	square meter, m^2	square foot, ft^2	9.29×10^{-2}
1.55×10^{-3}	square millimeter, mm^2 (10^{-3} m)^2	square inch, in^2	645
Volume			
9.73×10^{-3}	cubic meter, m^3	acre-inch	102.8
35.3	cubic meter, m^3	cubic foot, ft^3	2.83×10^{-2}
6.10×10^4	cubic meter, m^3	cubic inch, in^3	1.64×10^{-5}
2.84×10^{-2}	liter, L (10^{-3} m^3)	bushel, bu	35.24
1.057	liter, L (10^{-3} m^3)	quart (liquid), qt	0.946
3.53×10^{-2}	liter, L (10^{-3} m^3)	cubic foot, ft^3	28.3
0.265	liter, L (10^{-3} m^3)	gallon	3.78
33.78	liter, L (10^{-3} m^3)	ounce (fluid), oz	2.96×10^{-2}
2.11	liter, L (10^{-3} m^3)	pint (fluid), pt	0.473

Mass

To convert Column 1 into Column 2, multiply by	Column 1 SI Unit	Column 2 non-SI Unit	To convert Column 2 into Column 1, multiply by
2.20×10^{-3}	gram, g (10^{-3} kg)	pound, lb	454
3.52×10^{-2}	gram, g (10^{-3} kg)	ounce (avdp), oz	28.4
2.205	kilogram, kg	pound, lb	0.454
0.01	kilogram, kg	quintal (metric), q	100
1.10×10^{-3}	kilogram, kg	ton (2000 lb), ton	907
1.102	megagram, Mg (tonne)	ton (U.S.), ton	0.907
1.102	tonne, t	ton (U.S.), ton	0.907

Yield and Rate

To convert Column 1 into Column 2, multiply by	Column 1 SI Unit	Column 2 non-SI Unit	To convert Column 2 into Column 1, multiply by
0.893	kilogram per hectare, kg ha^{-1}	pound per acre, lb acre^{-1}	1.12
7.77×10^{-2}	kilogram per cubic meter, kg m^{-3}	pound per bushel, lb bu^{-1}	12.87
1.49×10^{-2}	kilogram per hectare, kg ha^{-1}	bushel per acre, 60 lb	67.19
1.59×10^{-2}	kilogram per hectare, kg ha^{-1}	bushel per acre, 56 lb	62.71
1.86×10^{-2}	kilogram per hectare, kg ha^{-1}	bushel per acre, 48 lb	53.75
0.107	liter per hectare, L ha^{-1}	gallon per acre	9.35
893	tonnes per hectare, t ha^{-1}	pound per acre, lb acre^{-1}	1.12×10^{-3}
893	megagram per hectare, Mg ha^{-1}	pound per acre, lb acre^{-1}	1.12×10^{-3}
0.446	megagram per hectare, Mg ha^{-1}	ton (2000 lb) per acre, ton acre^{-1}	2.24
2.24	meter per second, m s^{-1}	mile per hour	0.447

Specific Surface

To convert Column 1 into Column 2, multiply by	Column 1 SI Unit	Column 2 non-SI Unit	To convert Column 2 into Column 1, multiply by
10	square meter per kilogram, m^2 kg^{-1}	square centimeter per gram, cm^2 g^{-1}	0.1
1000	square meter per kilogram, m^2 kg^{-1}	square millimeter per gram, mm^2 g^{-1}	0.001

Pressure

To convert Column 1 into Column 2, multiply by	Column 1 SI Unit	Column 2 non-SI Unit	To convert Column 2 into Column 1, multiply by
9.90	megapascal, MPa (10^6 Pa)	atmosphere	0.101
10	megapascal, MPa (10^6 Pa)	bar	0.1
1.00	megagram per cubic meter, Mg m^{-3}	gram per cubic centimeter, g cm^{-3}	1.00
2.09×10^{-2}	pascal, Pa	pound per square foot, lb ft^{-2}	47.9
1.45×10^{-4}	pascal, Pa	pound per square inch, lb in^{-2}	6.90×10^3

(continued on next page)

Conversion Factors for SI and non-SI Units

To convert Column 1 into Column 2, multiply by	Column 1 SI Unit	Column 2 non-SI Unit	To convert Column 2 into Column 1, multiply by
Temperature			
$1.00 (K - 273)$	Kelvin, K	Celsius, °C	$1.00 (°C + 273)$
$(9/5 °C) + 32$	Celsius, °C	Fahrenheit, °F	$5/9 (°F - 32)$
Energy, Work, Quantity of Heat			
9.52×10^{-4}	joule, J	British thermal unit, Btu	1.05×10^{3}
0.239	joule, J	calorie, cal	4.19
10^{7}	joule, J	erg	10^{-7}
0.735	joule, J	foot-pound	1.36
2.387×10^{-5}	joule per square meter, J m^{-2}	calorie per square centimeter (langley)	4.19×10^{4}
10^{5}	newton, N	dyne	10^{-5}
1.43×10^{-3}	watt per square meter, W m^{-2}	calorie per square centimeter minute (irradiance), cal cm^{-2} min^{-1}	698
Transpiration and Photosynthesis			
3.60×10^{-2}	milligram per square meter second, mg m^{-2} s^{-1}	gram per square decimeter hour, g dm^{-2} h^{-1}	27.8
5.56×10^{-3}	milligram (H$_2$O) per square meter second, mg m^{-2} s^{-1}	micromole (H$_2$O) per square centimeter second, μmol cm^{-2} s^{-1}	180
10^{-4}	milligram per square meter second, mg m^{-2} s^{-1}	milligram per square centimeter second, mg cm^{-2} s^{-1}	10^{4}
35.97	milligram per square meter second, mg m^{-2} s^{-1}	milligram per square decimeter hour, mg dm^{-2} h^{-1}	2.78×10^{-2}
Plane Angle			
57.3	radian, rad	degrees (angle), °	1.75×10^{-2}

Electrical Conductivity, Electricity, and Magnetism

To convert Col. 1 into Col. 2, multiply by	Column 1 SI Unit	Column 2 non-SI Unit	To convert Col. 2 into Col. 1, multiply by
10	siemen per meter, S m^{-1}	millimho per centimeter, mmho cm^{-1}	0.1
10^4	tesla, T	gauss, G	10^{-4}

Water Measurement

To convert Col. 1 into Col. 2, multiply by	Column 1 SI Unit	Column 2 non-SI Unit	To convert Col. 2 into Col. 1, multiply by
9.73×10^{-3}	cubic meter, m^3	acre-inches, acre-in	102.8
9.81×10^{-3}	cubic meter per hour, m^3 h^{-1}	cubic feet per second, ft^3 s^{-1}	101.9
4.40	cubic meter per hour, m^3 h^{-1}	U.S. gallons per minute, gal min^{-1}	0.227
8.11	hectare-meters, ha-m	acre-feet, acre-ft	0.123
97.28	hectare-meters, ha-m	acre-inches, acre-in	1.03×10^{-2}
8.1×10^{-2}	hectare-centimeters, ha-cm	acre-feet, acre-ft	12.33

Concentrations

To convert Col. 1 into Col. 2, multiply by	Column 1 SI Unit	Column 2 non-SI Unit	To convert Col. 2 into Col. 1, multiply by
1	centimole per kilogram, cmol kg^{-1}	milliequivalents per 100 grams, meq 100 g^{-1}	1
0.1	gram per kilogram, g kg^{-1}	percent, %	10
1	milligram per kilogram, mg kg^{-1}	parts per million, ppm	1

Radioactivity

To convert Col. 1 into Col. 2, multiply by	Column 1 SI Unit	Column 2 non-SI Unit	To convert Col. 2 into Col. 1, multiply by
2.7×10^{-11}	becquerel, Bq	curie, Ci	3.7×10^{10}
2.7×10^{-2}	becquerel per kilogram, Bq kg^{-1}	picocurie per gram, pCi g^{-1}	37
100	gray, Gy (absorbed dose)	rad, rd	0.01
100	sievert, Sv (equivalent dose)	rem (roentgen equivalent man)	0.01

Plant Nutrient Conversion

To convert Col. 1 into Col. 2, multiply by	Elemental	Oxide	To convert Col. 2 into Col. 1, multiply by
2.29	P	P$_2$O$_5$	0.437
1.20	K	K$_2$O	0.830
1.39	Ca	CaO	0.715
1.66	Mg	MgO	0.602

Chapter 1

Soil Sampling for Microbiological Analysis

A. G. WOLLUM, II, *North Carolina State University, Raleigh, North Carolina*

Soil is a unique medium containing a diverse community of organisms, representing many morphological and physiological types. Attempting to numerically characterize these organisms or their activities requires an understanding of both the spatial and temporal distribution of organisms within the soil environment. The fact that organisms in soils are rarely static in numbers or activity, compounds the problem of characterizing populations or their activities. Many organisms exist at relatively low levels numerically but can have a profound affect on nutrient availability, plant development, or environmental quality. In most cases, it is impossible to equate the numbers of a particular organism with its importance in the soil ecosystem. Further, the enumeration of any population or magnitude of activity represents a point-in-time measurement that is at some dynamic equilibrium governed by the physical, chemical, and biological environment.

Variability is a familiar problem to most scientists, especially those dealing with environmental issues or habitats. Even the novice soil scientist recognizes that soils differ from site to site, based on observable differences such as color, depth, or arrangement of soil horizons. Less obvious is the fact that not only do the gross soil properties vary from site to site, but within a site, significant variation may occur. Intuitively, one might suspect that microbial populations vary by depth, with the surface horizons generally having more organisms and a greater abundance of types than the subsurface horizons. Information provided by Waksman and Starkey (1931) support this contention. For different soils, organisms were always more numerous in the surface horizons as compared to the subsurface horizons. These differences were attributed to the fact that the physical and chemical properties were different for different layers of soil, thus giving rise to heterogenous distributions of microorganisms.

Besides variation within the profile, it is reasonable to expect spatial variation in soil microbiological properties within sampling area smaller

Copyright © 1994 Soil Science Society of America, 677 S. Segoe Rd., Madison, WI 53711, USA. *Methods of Soil Analysis, Part 2. Microbiological and Biochemical Properties*—SSSA Book Series, no. 5.

1

than landscape units. Spatial variation has been documented for a variety of soil physical and chemical properties including hydraulic conductivity (Russo & Bresler, 1981), water content (Vauclin et al., 1982), soil acidity as measured by pH (Yost et al., 1982), and P (Yost et al., 1982). Less well documented are the spatial variabilities of soil microbial populations or microbe activities, although Wollum and Cassel (1984) demonstrated spatial structure for *Bradyrhizobium japonicum* in Coastal Plain soils of North Carolina. Other investigators have correlated various microbial populations or activities to other spatially dependent soil physical or chemical properties (Darmighi et al., 1967; Weaver et al., 1972; Dick, 1984; West et al., 1988; Mpepereki & Wollum, 1991).

Distinct boundaries between soil classification units mapped in a field rarely occur and it is common for units to grade from one to another in a continuum of gradually changing characteristics. Because soil properties vary from location to location and vary with depth or even season of the year, such intrinsic variation must be considered prior to developing sampling plans to determine soil biological or biochemical properties. Within the highly complex soil environment, attempts to study community structure or soil organism activity will be difficult. Therefore, the formulation of study objectives and soil sampling and handling procedures are among the most critical investigative steps the individual scientist undertakes. Before starting, one must have a set of clearly defined study objectives, for without such a framework, data that are difficult to interpret will be collected and will not be as useful as they might otherwise be. It is the overall purpose of this chapter to provide a general overview on sampling for microbiological studies. The specific objectives are to: (i) review some of the problems associated with sampling for soil microbial attributes; (ii) describe some general sampling plans; and (iii) how to handle samples for routine laboratory analyses. To implement statistical interpretations of the various sampling designs, and allocation of resources in sampling, the reader is encouraged to consult any one of several excellent texts, including Green (1979), Gilbert (1987), or Snedecor and Cochran (1980).

1–1 PRINCIPLES

Often the soil microbiologist is confronted with the problem of trying to determine a value for some microbial attribute, for instance, the number of gram-negative bacteria in the surface soil horizon or N mineralization in a soil amended with crop residues or enzyme activity. To obtain a value for the attribute, the soil microbiologist must be concerned about the material used (sample) and the measurement process (methodology used to make the measurement).

While the purpose of this chapter is to emphasize the sampling process, several issues related to the measurement must be addressed. It is assumed that the researcher has a choice of measurement methodologies. Given a choice, the best measurement would be one that accurately measures the selected attribute, giving the same result time after time. Any deviation from the same value, using the same measurement technique on the same sample can be considered a measurement error. For most soil biological attributes repetitive measurements cannot be performed on the same material. Therefore, measurements are made on subsamples taken from the whole. Deviations of measurement among subsamples, not attributable to the measurement technique itself are called the subsampling error. The error of observation on subsamples is the result of both measurement and subsampling error. Within limits, the investigator controls these errors by using the best and most reproducible measurement methodologies and by using subsamples from thoroughly homogenized samples.

The measurement and subsampling error are usually small when compared to experimental error. Experimental error arises from the fact that there is usually variation among the individual observations made on materials receiving the same treatment. This can arise when more than one independent sample is drawn from the same experimental unit or that the investigator is using replications in the experimental design. In a statistical sense, the experimental error is the deviation of an observation from its population mean.

Two additional considerations are important concerning measurement methodologies, bias, and precision. For the best case situation, we assume that determinations accurately represent the population mean. In the event there is a difference between the determined mean and the actual population mean, that difference is referred to as bias (Kempthorne & Allmaras, 1986). Thus to reduce or eliminate bias, particular attention must be paid to measurement technology itself. Every effort must be taken to identify and eliminate those factors leading to bias. The solution to the problem might be as simple as modifying an extraction procedure, lengthening the time of extraction or altering the extracting solution. Conversely, the measurement procedure may be so flawed so as to require an entirely new methodology. Precision on the other hand can be defined as the variation from the observation mean (Kempthorne & Allmaras, 1986). Precision can be improved by increasing the number of repetitions of the measurement. However, increasing the number of repetitions of a measurements will not reduce bias if the measurement methodology itself is flawed.

Apart from the statistical considerations of the measurement regarding bias and precision the investigator is faced with the problem of defining the sampling unit. In some respects the sampling unit is defined by the experimental design. In the laboratory, 10 g of soil in a beaker, perhaps amended with some genetically engineered microorganism might constitute the sampling unit. Utilizing the entire sample in this case would be relatively simple. However, when dealing with a larger sample, for instance a greenhouse pot with 1 to 2 kg of soil, a field plot of even moderate dimensions, or even a geographic area, one is forced to consider using only a portion of the entire soil mass. In these cases, it is necessary to think of the larger soil mass as being composed of an almost infinite number of separate and smaller sampling units. By definition the size of the sampling unit may vary from something that is a few cubic centimeters to thousands of liters.

Since it will be impossible to withdraw the entire soil mass, one will be forced to use a sampling unit. Several important decisions must be made at this point; however, the basic decision being, what constitutes a satisfactory sampling unit. Not only must the investigator address the basic issue of how large the sampling unit should be, a spatula full to a shovel full, but should the subsample be composed of a series of individual samples composited into one or one large sample from the middle of the sampling area. These decisions are often based on pragmatic considerations such as, how much can the investigator handle, is there sufficient storage space for large individual samples, or what are the existing processing facilities? If one routinely processes 10 000 g of soil per day, it doesn't make sense to make the sampling unit 10 000 g.

Lastly, the number of samples will depend on at least two issues. First, what is the desired precision for the measurement? Secondly, what is the magnitude of variation both within the sampling area or among areas treated in a similar fashion? The more heterogeneous the area, the more samples will be required to attain a specific level of precision. Although variation is known to exist within soil biological or biochemical properties, few studies have been conducted to evaluate variation; thus preliminary evaluations may be necessary to estimate variation within areas to provide a basis for the best sampling plan.

Within the limits of the desired precision and the scope of the study objectives, one should have a specific sampling plan. While a great number of plans exist, the best one is that which provides the greatest precision at a fixed cost or a specified precision at the lowest cost. Snedecor and Cochran (1980) provided a good statistical basis for deciding on the allocation of resources between replication and subsampling within replications. Due to the inherent variability of the soil biological or biochemical properties, any plan is bound to be a compromise between the funds available for sampling and analysis and the precision desired for the study.

1–1.1 Judgement Samples

Judgement samples are not recommended for characterizing basic soil microbiological characteristics. Such samples are highly biased, since the individual has deliberately avoided areas not representative of the entire area. While measurements made on samples obtained in this manner would have a lower error associated with the mean as compared with a strictly random sample, conditions for randomness of the sample could not be met and traditional statistical analyses would not be valid. The use of the judgment sample primarily should be associated with those instances when the investigator merely requires soil as a source for microorganisms. To take a judgement sample, the investigator delineates representative areas from which the sample may be withdrawn. Depending on experimental needs, the sample may range in size from small (< 100 g) to large (many kilograms).

1–1.2 Simple Random Samples

To meet the statistical requirements regarding simple random samples, each sample must have an equal opportunity to be selected. Random samples can be obtained in several ways although the establishment of a grid, commonly with two sets of parallel lines at right angles to each other. Pairs of random numbers drawn from a random number table (e.g., Fisher & Yates, 1963) are used to establish intersection points for sampling. Each pair of numbers establishes coordinates along base lines, that when perpendicular lines are drawn the lines will intersect. The intersection is then established as the sampling point. One advantage of this procedure is that sampling points can be established before going to the field, if the shape of the sampling area is known ahead of time. Alternatively, one can use random number pairs to generate a direction, generally in degrees and a distance from a starting point. Either approach is acceptable.

1–1.3 Stratified Random Samples

Stratified random samples are obtained in a similar fashion as the simple random samples except that the area to be sampled is broken into smaller subareas. Then each subarea is sampled following the simple random sampling procedure previously described. These subareas may represent different series or unique subareas that differ because of some easily measured or recognizable physical or chemical property. Collecting random samples by subareas allows the researcher to make statements about each of the subareas separately and greatly increases the precision of estimates over the entire population. Increasing precision must be weighed against the extra time and effort that must be exerted not only to obtain the samples but to run the analyses. However, increasing stratification too much will ultimately decrease the precision and further complicate the statistical analyses.

1–1.4 Systematic Samples

A systematic sampling procedure is used to ensure that the entire area being sampled is well represented by the individual samples. In this procedure, samples are obtained at predetermined points. These points might be at regular intervals along sets of parallel lines or straight lines set by compass bearing. One might further add other sampling criteria, that is, that only those points associated with a particular plant, such as the north side of the plant, a fixed distance from the plant or related to geographic or topographic features. These kinds of stipulations make the sampling process highly selective.

A major problem associated with this kind of sampling relates to the estimation of the sampling error. Peterson and Calvin (1986) suggest three options to determine sampling error. First, one can treat the samples as random. This is a questionable procedure and should not be done without testing for the independence of the samples themselves. A second option, which blocks or stratifies the samples assumes that the variation among areas within a block is due to sampling variation. In this case, the statistical analysis would be similar to the one used for stratified random samples. Finally, one could take several separate systematic samples. After experimental analysis, a subset of observations from the systematic samples could be drawn at random and their respective experimental values used to calculate the sample mean. As an additional alternative, prior to analysis, one could select a subset also drawn at random on which to perform the desired analysis. Using the subset of systematic samples, means could be calculated and compared. The third option involving a randomly selected subset is the only one in which the assumption of unbiased assessment of the sampling error is valid. The advantages of following this approach must be balanced against the fact that using a subset of the data usually reduces the precision of the estimate of the mean. However, this disadvantage must be weighed against the cost saving that will accrue because fewer samples need to be run.

1–1.5 Composite Samples

One possible way to reduce the total cost of running samples in the laboratory, thereby making more efficient use of all possible resources is to use composite samples. For this type of sampling, individual samples are obtained from the area, bulked together and mixed. Overall, this seems to be a simple solution to reducing the analytical load in the laboratory. However, the approach has validity only under certain conditions and with certain key provisions (Parkinson et al., 1971). These conditions are that:

1. An equal number and amount of the individual samples are used for each composite sample, that is, the sampling unit must be the same.
2. No interactions exist among the individual sampling unit that affect the results.
3. The only objective of the study is to obtain an unbiased estimate of the mean.
4. The analysis of subsamples from a composite sample cannot provide an estimation of the variance of the original and individual sampling units. However, one can estimate the variance between similar composite samples.

When there is no estimate of the variability of sample from the same area, compositing should be avoided. However, if an estimate can be obtained by analyzing a subset of individual sampling units, one can determine the number of individual sampling units to use to make up the composite sample to attain a specified level of precision. A procedure to make this calculation is given in the following section.

1–1.6 Number of Samples to Take

From a scientific viewpoint the purpose of sampling is to draw a collection of sampling units from a population on which an observation may be made for the purpose of estimating the population mean without measuring all sampling units in the population. For instance, the observation may represent the activity of a soil enzyme, acid phosphatase in some particular fixed area of a no-till treatment, where the population is represented by the set of all possible observations about the attribute being studied. Estimating the population mean from a subset of observations saves the investigator both time and money, as well as being able to use the information to make statistical inferences about the population, if the sample has been properly obtained. However collecting too much information wastes time and resources, whereas not getting enough may render the entire study useless or lead to improper conclusions. Prior to any study, two important considerations must be addressed. First of all, is the population normally distributed? All of the previous examples about sampling designs have been based on the assumption that the population is normally distributed. If this is not the case the reader is encouraged to consult chapter 2 on log-normal distribution or other statistical texts for information dealing with non-normal distributions. The other issue deals with how close does the determined mean have to be to the population mean? Is it sufficient to be within ±25%, ? ±10%? or must the determined value be within ±5% of the population mean? Depending on which value is chosen, there will be a tremendous impact on the amount of sampling needed. Specifying how close to the population mean one wants to be, will permit the design of a sampling scheme with an adequate number of samples to meet study objectives.

To determine the number of samples needed to be within a certain value of the mean, consider the following: if sampling from a population with a normal distribution, the sample mean can be compared to the theoretical population mean using the t-distribution equation as follows:

$$t-value = \frac{sample\ mean - population\ mean}{\sqrt{sample\ variance/no.\ of\ samples}} \qquad [1]$$

Intuitively as the number of samples goes to infinity the sample mean approaches the theoretical population mean (the difference approaches zero). The denominator of Eq. [1] is also known as the standard error of the mean. If the denominator is multiplied by the appropriate t-value, a confidence interval can be established when estimating the population mean. For example, if the desired level of significance is 0.05, the value X in Eq. [2] represents the 95% confidence interval about the mean.

$$\overline{X} = \pm \text{ (t-value) (SE)} \qquad [2]$$

According to Eckblad (1991), if we know the population mean, the sample variance and specify the t-value for some level of significance, we can infer the approximate number of samples needed to estimate the theoretical population mean, within some prespecified value of the population mean. By rearranging Eq. [1], we get:

$$\text{No. of samples} = \frac{(t-value)^2 \text{ (sample variance)}}{(\text{sample mean} - \text{population mean})^2} \qquad [3]$$

This equation can be approximated in the following manner. Using a pre-study sampling, sample variance and sample mean can be determined. In the subsequent study, assuming that the mean needs to be within 10% of the true population mean, Eq. [3] can be rewritten:

$$\text{No. of samples} = \frac{(t-value)^2 \text{ (sample variance)}}{[(\text{sample mean}) \ (0.1)]^2} \qquad [4]$$

Consider the following information: a presampling mean for fungi-g^{-1} soil, based on eight observations $= 2.47 \times 10^5$ ranging from 1.32×10^5 to 3.8×10^5; sample variance $= 5.11 \times 10^{10}$; and t-value for the 0.05 level of significance $= 2.306$ (with 8 df). Using this information to solve Eq. [4], it would take 45 samples to predict the theoretical population mean within 10%.

1–2 METHODS

1–2.1 Bulk Samples for Isolation

1–2.1.1 Nonrhizosphere

Modest precautions are necessary when one samples and handles soils merely as a source of organisms. Handling samples as described will assure that representative information will be forthcoming.

1–2.1.1.1 Materials
1. Shovel.
2. Auger.
3. Plastic bags.
4. Knife or scissors.
5. Ice pick or trowel.
6. Marking pen.
7. Insulated box, with ice.
8. Plastic pail (20 L).
9. Optional: Sealable containers for anaerobic samples.

1–2.1.1.2 Procedures. Based on prior observation, delineate the boundaries of the sampling area. Traverse the entire area and at random points collect equal volumes of soil material from the same horizon, using either an auger, shovel, trowel, or other suitable instrument. Composite the sample in a large clean plastic pail. Thoroughly mix the material and place the desired quantity into a labelled plastic bag. Do not let the sample dryout or expose it to high temperatures. Transport the sample to the laboratory in an ice chest. To prevent contamination from a second sample area, wash the soil sampling instruments in water, rinse with 95% ethanol and sterilize by evaporating the alcohol by flaming.

1–2.1.2 Rhizosphere/Rhizoplane

Based on the definition of Hiltner (1904), *rhizosphere* represents the portion of the soil that is under the immediate influence of the plant root. Generally, microorganisms are found in greater numbers and diversity in the rhizosphere compared with nonrhizosphere locations. Differences are attributable in part to root exudates, alteration of the partial pressures of O_2-CO_2, coupled with changes in nutrient availability that may be controlled by acidity, plant species, stage of growth or moisture stress. Most soil microbiologists recognize that the rhizosphere does not represent a fixed distance from the root surface but is thought to be a continuum, extending from the root surface (the *rhizoplane*) to a point where the root has no influence on microbial properties. The number of microbes per unit volume decreases as the reference point moves away from the root.

1–2.1.2.1 Materials. Use items listed under section 1–2.1.1.1.

1–2.1.2.2 Procedures. Carefully excavate the root specimens from the soil using a shovel or trowel. Remove the root material from the bulk soil. A small trowel and ice pick often facilitates the removal of roots from the soil. Leave as much soil adhering to the roots as possible. Place the roots with adhering soil into a labelled bag and store in an ice chest for transport to the laboratory. Since large variation may exist from sample to sample, bulk samples should consist of numerous subsamples. Do not dry the roots or expose them to high temperature.

1–2.2 Characterization Studies

Characterization of soil microbiological or biochemical properties within an experimental area or comparisons among different treatments or cropping systems requires the utmost attention to planning details. It is imperative to establish experimental objectives as a prelude to developing the sampling plan. Besides delineating the sampling area and unit, one must decide on the sampling plan. Often the sampling issue is governed by the experimental objectives or hypotheses to be tested.

As a component of the development of the sampling plan, one should begin by evaluating the environment from which samples will come. To accomplish this goal one must determine the salient physical and biological features of the experimental area including slope, aspect, elevation, kind and abundance of different vegetational components as well as preliminary descriptions of the soils within each sampling area. Minimal soil information should include an understanding of major soil boundaries (at least to soil type), the horizonization within each major soil type and the relative amount of each soil type within the sampling area.

1–2.2.1 Materials

Use items listed under section 1–2.1.1.1.

1–2.2.2 Procedures

Having established the sampling areas and the sampling scheme, collect soil or rhizosphere/rhizoplane samples from each horizon as required to meet the objective statement of the study. Package the soil material in a container. Plastic bags will suffice for most samples. Label each bag as to its sample area and sampling date. Place on ice for transport to the laboratory and subsequent use.

The advantage of using plastic bags is that the plastic bags generally are permeable to CO_2 and O_2 and prevent sample drying. Because of this fact, aerobic samples will remain aerobic during transport to the laboratory. Materials should be used as quickly as possible, preferably without storage. Since not all studies involve aerobic samples, special consideration is due anaerobic samples. Those working with anaerobic organisms should consider using sealable glass or rigid and sealable plastic containers for transportation to the lab. Canning jars come in a variety of sizes and with a sealable lid they make excellent containers for anaerobic samples.

1–2.3 Processing Samples for Microbiological Studies

Soil samples used for characterization of numbers or activities of organisms should be used as quickly as possible after collection. While not recommended, samples may be stored in the refrigerator at 4 °C in a field collected condition without further processing if they cannot be used immediately. If working with stored samples, the investigator has several key issues to evaluate: (i) Should samples be air-dried or processed in any way before storage? (ii) What kind of containers should be used if samples must be stored? (iii) At what temperature should the samples be stored? (iv) How long can samples be stored before they no longer possess their original characteristics? Several studies are indicative of what happens when samples are air-dried. Air-drying the soil sample results in a reduction of microbes proportional to the length of the storage period (Sparling & Cheshire, 1979). In addition to the reduction of numbers, there is a differential effect of drying on the composition of the microbial community. In the Sparling and Cheshire (1979) study, yeasts increased in proportion to the total community as the length of storage increased. Some people suggest that to restore the dried sample to its original composition, it should be remoistened and incubated for a period of time. However, Patten et al. (1980) observed that previously dried soil samples had a greater capacity to denitrify nitrate after they had been dried and subsequently remoistened as compared to their original capacity at the time of sampling. Others have noted a literal explosion in microbial numbers following remoistening of air-dried and stored samples (Bartlett & James, 1980).

Storing samples in the refrigerator could promote the development of anaerobic conditions (Gordon, 1988). When plastic bags are used as storage containers, anaerobiosis may be a function in part of the thickness of the bag, coupled with the decreased gas permeability of the plastic bag as temperatures are decreased (Bremner & Douglas, 1971). According to Gordon (1988), the thickness of the bags used should be specified. He recommends 0.025-mm (1 mil) thick bags for incubation studies that might be taken as a standard for sampling containers.

Normal refrigerator temperatures of 4 °C appear to be the temperature of choice if samples have to be stored before use (Wollum, 1982). Freezing should be avoided as the structure of the organic matter seems to be disrupted, although not as much as drying the sample would (Bartlett & James, 1980). An additional factor related to freezing is that desiccation frequently accompanies freezing, compounding the freezing effect.

Storage in the refrigerator at 4 °C even in a field moist condition may result in changes of the microbial community and activities of the organisms (Stotzky et al., 1962). After 3 mo of storage numbers of organisms decreased compared with the zero time evaluation, except for the actinomycetes. Part of this decrease was attributed to sample drying, even though materials were stored in plastic bags. Others have also noted that some drying occurs in plastic bags (Gordon, 1988). To overcome drying, one might double bag the soil materials. Stotzky et al. (1962) also observed that there were greater changes in microbial numbers and activities for soil materials from surface horizons as compared to those from deeper within the soil profile. The difference was attributed to different organic matter contents in the surface and subsurface soil, which allowed microbes to prosper by providing an energy source for cell division in surface soil, but with only enough energy to keep the cell viable in materials from other horizons, but perhaps not enough to permit cell division.

Based on the work of Stotzky et al. (1962) one might conclude that storage of soil materials for microbiological studies is not acceptable. However, they used a 3-mo storage and transportation period as the standard. Perhaps shorter periods of storage might be acceptable. A study of soil microbiological and biochemical properties from low-input sustainable agricultural plots, revealed that most properties remained constant for storage periods from 7 to 21 d (M. J. Kirchner & A.G. Wollum, 1992, unpublished data). For example during 21 d of storage, properties of microbial biomass, available N, and the enzyme activities for acid and alkaline phosphatase and β-glucosidase did not change compared to the values obtained immediately after collection. However, arylsulfatase activities were significantly lower after 21 d of storage than those obtained without storage. For actinomycetes, *Bacillus* spp. and fungi determined on DPY medium (Papavizas & Davey, 1959), there was no significant effect of up to 21 d storage on population numbers. Also, the numbers of fluorescent *Pseudomonas* spp. and gram-negative bacteria remained constant for 7 d of storage, but not for longer periods. Only total bacteria as measured using tryptic soy medium (Difco) failed to remain constant during any length of storage (minimum tested equaled 7 d). These results suggest that investigators should be extremely cautious about storing soil samples for microbiological studies. Until a wider range of soil materials have been tested for the effect of length of storage on microbiological and biochemical properties, the general rule should probably be to avoid storage if at all possible. When samples must be stored, they should be stored for the shortest time possible. Others are encouraged to evaluate the effect of the length of storage on soil microbiological and biochemical properties.

Prior to use, samples frequently must be processed for ease of use. During the preparative processing, do not let the sample dry and do not expose it to high temperatures. Large aggregates may be broken by forcing the soil sample through 2-mm sieves with large rubber stoppers. During this phase, large pieces of organic matter, including root segments should be carefully removed. For rhizosphere samples, smaller roots (approximately < 1 mm) should be severed from larger roots using a scissors. Be careful to leave as much of the adhering soil to the roots as possible. Use samples immediately, do not store.

1–3 SOURCES OF ERROR

Despite the utmost care, relying on samples and measuring some attribute will result in error while trying to estimate the population mean. These sources of error can be ascribed to three different categories. The first relates to sampling error and is due to the inherent variability within different subsets of the original population. Increasing the number of subsets drawn from the population will not decrease sample error, but will improve precision.

Selection error occurs from the tendency to avoid nonrepresentative areas within the population. Designing a good sampling program and following it regardless of the sampling condition will aid considerably in reducing selection error. To some extent there is a tendency for selection errors to cancel one another.

The last source of error is that attributable to measurement. Compartmentalized measurement error is composed of random errors of measurement that tend to be cancelled as the sample size increases. Biases tend to be independent of sample size. Only through constant vigilance can bias be minimized. It is the responsibility of the individual to use the best possible technique and eliminate those aspects of sample handling or preparation that might lead to a measurement error. These precautions might be as simple as assuring that large organic debris or extraneous rocks are removed, large aggregates broken up and that the sample has been thoroughly mixed.

1–4 CONCLUDING REMARKS

Regrettably too little attention has been given to the adequacy of sampling for soil biological and biochemical characterization. It is recognized that variability exists among samples for specific soil microbiological properties. However, what constitutes the best sampling plan or an adequate number of samples to reasonably predict sample means within some predetermined level of confidence have been ignored by many soil microbiologists.

What I have attempted to do in this chapter is to outline some reasonable sampling plans and ways to process samples for microbiological studies. What remains is for individual investigators to evaluate the sampling plans for their own conditions, and establish the validity of the approach through the application of the proper statistical treatment. Information about statistical procedures can be found in numerous texts and the reader is referred to any one of their choice. Some suggested texts include Green (1979), Gilbert (1987), and Snedecor and Cochran (1980). Any of these sources can be used with confidence. It is hoped that the information presented will serve to stimulate others to conduct the kinds of studies that will lead to the selection of the best sampling plans and with the fewest samples to reach a specified level of confidence.

REFERENCES

Bartlett, R., and B. James. 1980. Studying dried, stored soil samples—some pitfalls. Soil Sci. Soc. Am. J. 44:721–724.

Bremner, J.M., and L.A. Douglas. 1971. Use of plastic films for aeration in soil incubation experiments. Soil Biol. Biochem. 3:289–296.

Darmighi, S.M., L.R. Frederick, and J.C. Anderson. 1967. Serogroups of *Rhizobium japonicum* in soybean nodules as affected by soil types. Agron. J. 59:10–12.

Dick, W.A. 1984. Influence of long-term tillage and crop rotation combinations on soil enzyme activities. Soil Sci. Soc. Am. J. 48:569–574.

Eckblad, J.W. 1991. How many samples should be taken? BioScience 41:346–348.

Fisher, R.A., and F. Yates. 1963. Statistical tables for biological, agricultural and medical research. 6th ed. Oliver and Boyd, Edinburgh.

Gilbert, R.O. 1987. Statistical methods for environmental pollution monitoring. Van Nostrand Reinhold Co., New York.

Gordon, A.M. 1988. Use of polyethylene bags and films in soil incubation studies. Soil Sci. Soc. Am. J. 52:1519–1520.

Green, R.H. 1979. Sampling design and statistical methods for environmental biologists. John Wiley and Sons, New York.

Hiltner, A. 1904. Uber neuere erfahrungen und probleme auf dem gebiet der bodenbakteiol-
ogie und unter besonderer berusckichtigung der grundungung und brache. Arb. Dtsch.
Landwirtsch. Ges., Berlin 98:59–78.

Kempthorne, O., and R.R. Allmaras. 1986. Errors and variability of observations. p. 1–31.
In A. Klute (ed.) Methods of soil analysis. Part 1. 2nd ed. Agron. Monogr. 9. ASA and
SSSA, Madison, WI.

Mpepereki, S., and A.G. Wollum, II. 1991. Diversity of indigenous *Bradyrhizobium japonicum* in North Carolina Soils. Biol. Fertil. Soils 11:121–127.

Papavizas, G.C., and C.B. Davey. 1959. Evaluation of various media and antimicrobial
agents for isolation of soil fungi. Soil Sci. 88:112–117.

Parkinson, D., T.R.G. Gray, and S.T. Williams. 1971. Methods for studying the ecology of
soil micro-organisms. IBP Handb. 19. Int. Biol. Program, Blackwell Sci. Publ., Oxford.

Patten, D.K., J.M. Bremner, and A.M. Blackmer. 1980. Effects of drying and air-dry storage
of soils on their capacity for denitrification of nitrate. Soil Sci. Soc. Am. J. 44:67–70.

Peterson, R.G., and L.D. Calvin. 1986. Sampling. p. 33–51. *In* A. Klute (ed.) Methods of soil
analysis. Part 1. 2nd ed. Agron. Monogr. 9. ASA and SSSA, Madison, WI.

Russo, D., and E. Bresler. 1981. Effect of field variability in soil hydraulic properties on
solutions of unsaturated water and salt flows. Soil Sci. Soc. Am. J. 45:675–681.

Snedecor, G.W., and W.G. Cochran. 1980. Statistical methods. 7th ed. Iowa State College
Press, Ames.

Sparling, G.P., and M.V. Cheshire. 1979. Effects of soil drying and storage on subsequent
microbial growth. Soil Biol. Biochem. 11:317–319.

Stotzky, G., R.D. Goos, and M.I. Timonin. 1962. Microbial changes occurring in soil as a
result of storage. Plant Soil 16:1–18.

Vauclin, M., S.R. Vieira, R. Bernard, and J.L. Hatfield. 1982. Spatial variability of surface
temperature along two transects of a bare soil. Water Resour. Res. 18:1677–1686.

Waksman, S.A., and R.L. Starkey. 1931. The soil and the microbe. John Wiley and Sons,
London.

Weaver, R.W., L.R. Frederick, and L.C. Dumenil. 1972. Effect of cropping and soil properties on numbers of *Rhizobium japonicum* in Iowa soils. Soil Sci. 114:137–141.

West, A.W., G.P. Sparling, T.W. Speir, and J.M. Wood. 1988. Comparison of microbial C,
N-flush and ATP and certain enzyme activities of different textured soils subject to
gradual drying. Aust. J. Soil Res. 26:217–229.

Wollum, A.G., II. 1982. Cultural methods for soil microorganisms. p. 781–802. *In* A.L. Page
et al. (ed.) Methods of soil analysis, Part 2. 2nd ed. Agron. Monogr. 9. ASA and SSSA,
Madison, WI.

Wollum, A.G., II, and D.K. Cassel. 1984. Spatial variability of *Rhizobium japonicum* in two
North Carolina soils. Soil Sci. Soc. Am. J. 48:1082–1086.

Yost, R.S., G. Uehara, and R.L. Fox. 1982. Geostatistical analysis of soil chemical properties
of large land areas. I. Variograms. Soil Sci. Soc. Am. J. 46:1028–1032.

Chapter 2

Statistical Treatment of Microbial Data

TIMOTHY B. PARKIN, *USDA-ARS, National Soil Tilth Laboratory, Ames, Iowa*

JOSEPH A. ROBINSON, *The Upjohn Company, Kalamazoo, MI*

Statistics is a dynamic field. Developments in the applied and theoretical areas of statistics are progressing at a rate unprecedented in this field of mathematics, largely as a result of the advent of inexpensive computing power. The majority of these developments have been driven by individuals interested in solving problems in data description and statistical inference, that is, by experimentalists with practical problems to solve in data interpretation. Indeed the field of statistics got its start with persons motivated by the need to summarize groups of data and interpret the probabilistic meaning of differences between experimental groups treated in different ways.

One of the strengths of statistics lies in its generic nature. One set of tools may be applied to practical problems in data interpretation encountered in many fields, from physics through applied agriculture to molecular biology. Soil microbiologists, like other experimentalists, have data interpretation problems that run the gamut from single variable (univariate) to multiple variable (multivariate) investigations. There are many excellent texts and reviews, geared toward the practicing soil microbiologist, that present the mechanics of these univariate and multivariate techniques: this information will not be reviewed in the present chapter. Finally, the physical continuity that exists between samples collected from a given soil often shows up as a correlation in the measurements made on those samples. The field of geostatistics has been championed by many as a source of mathematical tools for describing and interpreting differences in soil variables that are correlated because of their physical proximity. Excellent review articles and texts have been published on geostatistical methods. Thus, we have elected to not include information in this chapter on geostatistics as well. The practical problem for soil microbiologists we have elected to address in this chapter is the challenge of describing, summarizing, and

15

interpreting differences between variables that are highly skewed, and in particular, seem to be described best by the 2-parameter lognormal distribution.

The frequency distributions exhibited by many soil variables are highly skewed. Positively skewed distributions are a consequence of the fact that many soil variables cannot take on negative values and are, hence, constrained by zero. Several skewed probability density models have been used to describe environmental data, including the Poisson, negative binomial, Weibull, gamma, exponential, and lognormal. Of these distributions, the lognormal has been most commonly applied. Because of the relatively widespread application of the lognormal distribution to environmental variables, the following discussion will focus on the analysis of lognormally distributed data. This chapter reviews the features of the lognormal distribution and summarizes the mechanics of using it for summarizing batches of skewed data, and for interpreting differences between batches of numbers that each appear to be best described by the lognormal distribution.

2–1 CHARACTERISTICS OF THE LOGNORMAL DISTRIBUTION

A variable is considered to be lognormally distributed if the logarithm of the variable is normally distributed. In practical terms, it does not make any difference whether $\log_{10}$ or $\log_e$ is used since log values to different bases are linearly related. For all the examples presented in this chapter the natural logarithm (base e) is used.

With symmetric distributions, such as the normal distribution, the mean and median have the same value. However, because the lognormal distribution is skewed, the mean and median have different values. Figure 2–1 shows a lognormal probability density function with the relative positions of the mean and median indicated. From Fig. 2–1, it is evident that the value of the median is less than that of the mean. This fact has several implications in parameter estimation and hypothesis testing. These topics are dealt with in greater detail later in this chapter.

The population mean and median are defined by Eq. [1] and [2].

$$\text{Mean} = \exp(\mu + \sigma^2/2) \qquad [1]$$

$$\text{Median} = \exp(\mu) \qquad [2]$$

where μ and σ^2 are the mean and variance of the ln-transformed variable.

It should be noted that these two expressions for computing the mean and the median are only valid when the entire population is known. Application of these two equations to sample data will result in biased estimates of the population median and mean, with the extent of the bias as a function of the sample size (Parkin et al., 1988; Parkin & Robinson, 1992b). In estimating the population mean and median from sample data,

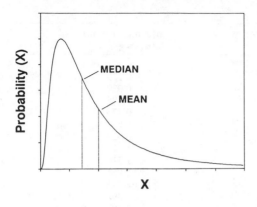

Fig. 2–1. The log-normal probability density function showing the relative locations of the population mean and median.

the uniformly minimum variance unbiased (UMVU) estimators are recommended. A description of the application of these estimators is described in section 2–3.

2–2 DIAGNOSING LOGNORMALITY

2–2.1 Sample Statistics as Indicators of Asymmetry

The predominant characteristic of the lognormal probability model is its asymmetry. Several indications of asymmetry of a given data set can be gleaned from inspection of the descriptive sample statistics. This can be illustrated by considering a sample data set of soil denitrification measurements (Table 2–1). This denitrification data were obtained using a soil core-acetylene inhibition technique. Presented are denitrification rate measurements obtained from 36 individual soil cores using the acetylene inhibition technique. Also shown in Table 2–1 are the sample mean ($\bar{x}$), sample standard deviation (s), coefficient of variation (CV), and the sample geometric mean that is the antilogarithm of the average of the ln-transformed values [$\exp(\bar{y})$]. The wide difference between the sample mean (56.5) and the geometric mean (17.6) indicates that asymmetry exists and a lognormal model may be appropriate. Other clues suggesting that asymmetry may exist are the high coefficient of variation (170%), and the high coefficient of skewness (CS = 2.76). The expected value of the CS for the normal distribution is 0, and it is observed that the CS associated with the ln-transformed data is close to 0.

2–2.2 Graphical Methods

Asymmetry in a data set can also be visualized by graphical techniques. One such technique is to construct a sample histogram. Figure 2–2A shows a sample histogram of the untransformed denitrification data (column 1, Table 2–1). It is evident from the sample histogram that a high

Table 2–1. Denitrification rate data.

Denitrification rate —ng-N g^{-1} d^{-1}—	ln(denitrification) —ln(ng-N g^{-1} d^{-1})—
0.60	−0.511
0.87	−0.139
1.04	0.039
1.47	0.385
1.61	0.476
1.87	0.626
2.30	0.833
4.56	1.517
5.56	1.716
7.07	1.956
9.42	2.243
10.00	2.303
11.10	2.407
12.90	2.557
12.90	2.557
13.20	2.580
14.30	2.660
16.80	2.821
19.10	2.950
20.40	3.016
22.70	3.122
37.20	3.616
41.00	3.714
43.30	3.768
48.20	3.875
51.20	3.936
54.10	3.991
55.80	4.022
70.90	4.261
77.70	4.353
106.00	4.663
114.00	4.736
126.00	4.836
226.00	5.421
358.00	5.881
435.00	6.075

$$\bar{x} = 56.5 \qquad\qquad \bar{y} = 2.868$$
$$s^2 = 9238 \qquad\qquad \hat{\sigma}^2 = 2.922$$
$$s = 96.1 \qquad\qquad \exp(\bar{y}) = 17.6$$
$$\%CV = 170$$
$$CS = 2.76 \qquad\qquad CS = -0.19$$

proportion of samples have low values but a few samples exhibit extreme values. If the data are lognormally distributed then the histogram of the ln-transformed values (column 2, Table 2–1) will exhibit a Gaussian distribution (Fig. 2–2B). Plotting the sample histogram is useful in visualizing skewness but does not necessarily confirm that the sample data are indeed from a lognormal distribution, as the shape of the histogram is influenced, in part, by how the data are distributed into the different size classes during construction of the histogram.

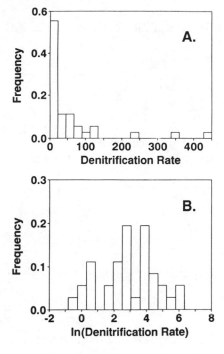

Fig. 2–2. Sample histograms of untransformed (Panel A) and log-transformed (Panel B) denitrification rate measurements of Table 2–1.

Another graphical method with greater diagnostic power is a probit plot. Probit analysis is a special case of a Quantile-Quantile plot (Miller, 1986), where the empirical quantiles of the sample data are plotted against the expected quantiles for a given distribution. A detailed description of probit analysis for diagnosing normality (or lognormality) has been presented elsewhere (Miller, 1986; Warrick & Nielsen, 1980; Parkin & Robinson, 1992b).

2–2.3 Goodness-of-Fit Tests

Graphical methods such as sample histograms and probit plots only provide indications of lognormality. There are several statistical methods available for testing normality. When these tests are performed on ln-transformed data they can be used as tests of lognormality.

For testing normality (or lognormality), Gilbert (1987) recommends the use of the W-test developed by Shapiro and Wilk (1965). This test has good power for detecting departures from normality; however, tables have only been developed for sample sizes ≤ 50. The D'Agostino test (D'Agostino, 1971) is similar to the W-test, and tables of the test statistic have been developed for $50 < n < 1000$. A description of the implementation of the Shapiro-Wilk test for normality is presented below. For details on the implementation of the D'Agostino test see Gilbert (1987) or Parkin and Robinson (1992b).

2–2.3.1 The Shapiro-Wilk W-Test

The ln-transformed denitrification data of Table 2–1 are used as an example in the description of this test.

Step 1. Compute the denominator of the W statistic.

$$d = \sum_{i=1}^{n} x_i^2 - \frac{1}{n} [\sum_{i=1}^{n} x_i]^2$$

d = 102.3 for the ln-transformed denitrification data.

Step 2. Sort the data from smallest to largest (this has already been done in Table 2–1).

Step 3. Compute k.

$k = n/2$ if n is even

$k = (n-1)/2$ if n is odd

For the data in Table 2–1:

$k = (36)/2 = 18$

Step 4. Find the coefficients $a_1, a_2, \ldots a_k$, corresponding to the n samples available (Table 2–A1, Appendix).

For the example data the coefficients are:
$a_1 = 0.4068$ $a_2 = 0.2813$ $a_3 = 0.2415$ $a_4 = 0.2121$ $a_5 = 0.1883$
$a_6 = 0.1678$ $a_7 = 0.1496$ $a_8 = 0.1331$ $a_9 = 0.1179$ $a_{10} = 0.1036$
$a_{11} = 0.0900$ $a_{12} = 0.0770$ $a_{13} = 0.0645$ $a_{14} = 0.0523$ $a_{15} = 0.0404$
$a_{16} = 0.0287$ $a_{17} = 0.0172$ $a_{18} = 0.0057$

Step 5. Compute the W statistic

$$W = \frac{1}{d} [\sum_{i=1}^{k} a_i*(x_{n-i+1}-x_i)]^2$$

For the example data this equation becomes:
$$W = \frac{1}{102.3} [\,(0.4068)\,(6.075+0.511) + (0.2813)\,(5.881+0.139) +$$
$$(0.2415)\,(5.421-0.039) + (0.2121)\,(4.836-0.385) +$$
$$(0.1883)\,(4.736-0.476) + (0.1678)\,(4.663-0.626) +$$
$$(0.1496)\,(4.353-0.833) + (0.1331)\,(4.261-1.517) +$$
$$(0.1179)\,(4.022-1.716) + (0.1036)\,(3.991-1.956) +$$
$$(0.0900)\,(3.936-2.243) + (0.077)\,(3.875-2.303) +$$
$$(0.0645)\,(3.768-2.407) + (0.0523)\,(3.714-2.557) +$$
$$(0.0404)\,(3.616-2.557) + (0.0287)\,(3.122-2.58) +$$
$$(0.0172)\,(3.016-2.660) + (0.0057)\,(2.950-2.821)]^2$$

$W = 0.9699$

Step 6. Reject the null hypothesis if W is less than the quantile given in Table 2–A2 for the given sample n and alpha level.

H_o = The population is normally distributed.
H_A = The population is not normally distributed.

From Table 2–A2 the quantile for alpha = 0.05 and n = 36 is 0.935. Since the W for the ln-transformed denitrification data (0.9699) is > 0.935 we cannot reject the null hypothesis and conclude that the ln-transformed data is normally distributed or, in other words, the untransformed data are lognormal.

2–3 ESTIMATING POPULATION PARAMETERS FROM SAMPLE DATA

2–3.1 General Considerations

In most environmental studies, it is impossible to sample the entire population of the variable of interest. Thus, we are forced to estimate the parameters of the underlying population, such as the mean, median, and variance, from sample data. Several different statistical procedures for estimating these population parameters have been applied to lognormally distributed data, but recent statistical evaluations suggest that, for lognormally distributed data with a CV exceeding 100%, the class of UMVU estimators is preferred over other methods (Parkin et al., 1988; Parkin & Robinson, 1992a). A brief description of these estimators, and examples of their implementation follows.

2–3.2 The Uniformly Minimum Variance Unbiased Estimators

The UMVU estimators of the mean and variance were developed independently by Finney (1941) and Sichel (1952). The UMVU estimator of the median was developed by Bradu and Mundlak (1970) based on Finney's work. These methods for estimating the population mean, median, and variance from sample data are proposed as alternatives to computing the arithmetic average, variance, and sample median according to traditional methods. The advantages of the UMVU techniques are that they are unbiased and have minimum associated variance. Estimators of the population mean, variance, and median are given by Eq. [3], [4], and [5], respectively.

$$\text{Mean} = \exp(\bar{y})\ \psi_n\ (\hat{\sigma}^2/2) \qquad\qquad [3]$$

$$\text{Variance} = \exp(2\bar{y}) \{\psi_n (2\hat{\sigma}^2) - \psi_n [\hat{\sigma}^2(n-2)/(n-1)]\} \qquad [4]$$

$$\text{Median} = \exp(\bar{y})\psi_n \{-\hat{\sigma}^2/[2^*(n-1)]\} \qquad [5]$$

where $\bar{y}$ is the average of the ln-transformed samples
σ^2 is the variance of the ln-transformed samples
ψ_n is the power function presented in Eq. [6].

$$\psi_n(z) = 1 + \frac{z(n-1)}{n} + \frac{z^2(n-1)^3}{n^2(n+1)2!} + \frac{z^3(n-1)^5}{n^3(n+1)(n+3)3!} +$$

$$\frac{z^4(n-1)^7}{n^4(n+1)(n+3)(n+5)4!} + \cdots \qquad [6]$$

where n = the number of samples.
Depending upon n and s^2, 6 to 10 terms of Eq. [6] must be evaluated to achieve stability.

2–3.3 Application of the UMVU Estimators

At first glance, implementation of these estimators may appear to be complex, as in all cases the power series given in Eq. [6] must be evaluated. However, this equation has been converted to machine form (Appendix 2), which makes the implementation of the UMVU estimators straightforward. The calculations that follow illustrate the implementation of the UMVU estimators of the mean, variance, and median when applied to the denitrification data of Table 2–1.

2–3.3.1 Calculating the UMVU Estimator of the Mean

Application of Eq. [3] to calculate the UMVU estimator of the mean first requires calculating the mean and variance of the ln-transformed data. These quantities are given in Table 2–1.

$$\bar{y} = 2.868 \text{ and } \hat{\sigma}^2 = 2.922$$

Next the power series (ψ_n) given in Eq. [6] must be evaluated for $\hat{\sigma}^2/2$, with n = 36.

$$\hat{\sigma}^2/2 = 1.461, \text{ thus } \psi_{36}(1.461) = 3.938$$

Now solving Eq. [3] yields,

$$\exp(\bar{y}) \ \psi_{36}(\hat{\sigma}^2/2) = (17.6)^*(3.938) = 69.3$$

Note that the UMVU estimate of the mean (69.3) is different than the arithmetic average of untransformed sample values ($\bar{x} = 56.5$) given in Table 2–1.

2–3.3.2 Calculating the UMVU Estimate of the Variance

To estimate the population variance Eq. [4] is used. This requires that the psi function (Eq. [6]) be evaluated for both the quantities $2\hat{\sigma}^2$ and $[\hat{\sigma}^2 * (n-2)/(n-1)]$.

$$\psi_{36}(2\hat{\sigma}^2) = \psi_{36}(5.844) = 153.8 \text{ and}$$

$$\psi_{36}[\hat{\sigma}^2 * (n-2)/(n-1)] = \psi_{36}(2.839) = 13.26$$

Subsequent application of Eq. [4] results in:

$$\exp(2\bar{y}) * (153.8 - 13.26) = (309.8) * (140.5) = 43527 \text{ or a SD}$$

$$\text{of } (43527)^{\frac{1}{2}} = 208.6$$

The CV is calculated by dividing the standard deviation by the UMVU estimate of the mean;

$$208.6/69.3 = 3.01, \text{ or } 301\%$$

Note here that the CV computed using the UMVU estimators is different than that observed in Table 2–1. Simulation studies of Parkin et al. (1988) indicate that the UMVU estimate of the CV is the best estimator.

2–3.3.3 Calculating the UMVU Estimate of the Median

The UMVU estimator of the median is given by Eq. [5], which requires the power function ψ_n to be evaluated for the quantity $(-\hat{\sigma}^2/(2*(n-1)))$.

$$\psi_{36}\{-\hat{\sigma}^2/[2*(n-1)]\} = \psi_{36}(-0.0417) = 0.959$$

The estimate of the median from Eq. [5] is then,

$$\exp(\bar{y}) * (0.959) = (17.6) * (0.959) = 16.9$$

2–3.4 Confidence Intervals

Whereas, knowledge of underlying distribution allows for the selection of better estimators for the mean, median, and variance, the real value of identifying the underlying distribution is in selection of the appropriate method for calculating confidence intervals. Since the expected values of the mean and median of a lognormally distributed variable are different, the confidence intervals associated with these two parameters will also be different. Methods for computing exact confidence intervals of the mean and median are presented below.

2–3.4.1 Land's Exact Confidence Interval of the Mean

Several methods have been proposed for calculation of the confidence interval of the mean of a lognormally distributed variable (Parkin et al., 1990; Parkin & Robinson, 1992b). A method for the calculation of exact confidence intervals of the mean of a lognormally distributed variable was developed by Land (1973). This method, in essence involves calculation of two one-sided confidence limits, at the user-defined probability level. Lower and upper confidence limits are calculated according to Eq. [7] and [8].

$$LCI_{Mean} = \exp[\bar{y} + \hat{\sigma}^2/2 + \hat{\sigma}C_l/(n-1)^{\frac{1}{2}}] \qquad [7]$$

$$UCI_{Mean} = \exp[\bar{y} + \hat{\sigma}^2/2 + \hat{\sigma}C_u/(n-1)^{\frac{1}{2}}] \qquad [8]$$

where $\bar{y}$ = the mean of the ln-transformed sample values,
   $\hat{\sigma}^2$ = the variance of the ln-transformed sample values,
   C_l = the tabulated constant (Table 2–A3) for the lower 95% confidence limit, and
   C_u = the tabulated constant (Table 2–A4) for the upper 95% confidence limit.

An illustration of the implementation of Land's (1973) method, using the data of Table 2–1 is presented below.

Using Eq. [7] and [8] to calculate the lower and upper confidence limits about the mean requires $\bar{y}$, $\hat{\sigma}^2$, and the confidence limit factors, C_l and C_u. Values of C_l and C_u are obtained from linear interpolation of Tables 2–A3 and 2–A4 (see Appendix) for n = 36 and $\hat{\sigma}$ = 1.709. These correspond to C_u = 3.443 and C_l = −2.166. Using these values and $\bar{y}$ for the denitrification data of Table 2–1 yields:

$$LCL = \exp[2.868 + 2.922/2 + 1.709 * -2.166/(36-1)^{\frac{1}{2}}]$$

$$= \exp(3.703) = 40.6$$

$$UCL = \exp[2.868 + 2.922/2 + 1.709 * 3.443/(36-1)^{\frac{1}{2}}]$$

$$= \exp(5.324) = 205$$

It is noted that the resulting 90% confidence interval obtained is asymmetric about the estimate of the mean.

2–3.4.2 Exact Confidence Limits About the Median

This technique is the asymptotic or normal theory method, which, when applied to ln-transformed sample data yields exact confidence limits of the median (Parkin & Robinson, 1992a) Lower and upper confidence 95% limits are given by Eq. [9] and [10], respectively.

$$LCL_{med} = exp[\bar{y} - t_{0.95,n-1}(\hat{\sigma}^2/n)^{1/2}] \qquad [9]$$

$$UCL_{med} = exp[\bar{y} + t_{0.95,n-1}(\hat{\sigma}^2/n)^{1/2}] \qquad [10]$$

where t equals the critical value from the Student's t distribution with $n-1$ degrees of freedom for a two-tailed value of $\alpha = 0.10$.

The application of the median confidence interval calculation is illustrated using the denitrification data of Table 2–1. The following example illustrates computation of lower and upper confidence 95% confidence limits of the median.

$$LCL_{med} = exp[2.868 - 1.689 (2.922/36)^{1/2}] = exp(2.387) = 10.9$$

$$UCL_{med} = exp[2.868 + 1.689 (2.922/36)^{1/2}] = exp(3.349) = 28.5$$

The resulting confidence interval is a 90%. To obtain a 95% confidence interval (i.e., 97.5% confidence limits) the two-tailed value for $\alpha = 0.05$ and $n-1$ degrees of freedom is used.

2–4 SELECTING THE APPROPRIATE LOCATION PARAMETER

2–4.1 The Mean vs. the Median: General Considerations

It was pointed out earlier in this chapter that the mean and median of a lognormal distribution have different values. These are the two location parameters that are most often used to summarize lognormal data. It should be recognized that the mean and the median of a lognormal distribution actually convey different information about the distribution. The mean is the center of gravity of the distribution while the median is the center of probability of the distribution. The choice of the appropriate location parameter is critical, as it can affect the conclusions drawn from the data. Unfortunately, few guidelines concerning the validity of focusing on the mean or median as a summary statistic exist in the literature.

It has been recognized that bacterial numbers in nature can be approximated by lognormal distributions, and it has been recommended that "the preferred statistic for summarizing microbiological data is the geometric mean" (Greenberg et al., 1985). This recommendation is based on the assumption that the sample geometric mean is the best indicator of the central tendency of the lognormal distribution (for lognormal distributions the geometric mean is an estimator of the median). However, the mean and median both indicate central tendency (i.e., mean is the central tendency of mass and the median is the central tendency of the individuals in the population). The recommendation given above, that the geometric mean is the best estimator of central tendency for bacterial numbers is somewhat arbitrary, as the definition of "best" depends upon the objective of the specific study being conducted. A major factor that must be considered is the influence of sample volume on the median.

2–4.2 Effects of Sample Volume on the Mean and Median

The Central Limit Theorem predicts that, regardless of the form of the underlying population, the distribution of sample means approaches normality as the number of samples used in computing the mean increases. To illustrate this effect, a computer simulation was performed where 1000 samples randomly drawn from a lognormal population with a mean of 70, median of 31.3, and a SD of 140. This batch of 1000 samples was then randomly bulked (averaged) in groups of 2, 5, 10, 20, and 50 to observe the effect of sample bulking on estimates of the population mean, median, and SD (Table 2–2). The original batch of 1000 samples has a mean of 71.1, a median of 29.4 and a SD of 173. When these 1000 samples are randomly averaged in groups of 2 the mean of the new batch of 500 samples remains the same (71.1), yet the value of the median increases to 37.8, and the SD decreases to 129.

Through the process of averaging, the population of bulked samples approaches symmetry and the value of the median approaches the value of the mean, yet the value of the mean remains unchanged. In natural systems if the variable of interest is randomly dispersed, collecting large samples has the same effect as bulking or pooling small samples. Thus, the value of the median obtained is functionally dependent upon the sample volume collected. Because of this effect, focus on the median or geometric mean may only be appropriate when the samples themselves have some inherent identity and significance. This implies that in terrestrial systems, where the sample volume is arbitrarily defined by the size of the shovel, core, or soil chamber available to the investigator, the median (or the geometric mean) may not be the appropriate population parameter to estimate. The following discussion provides some additional guidelines on when the median or the mean are to be emphasized as the location parameter of choice.

2–4.3 The Median as the Location Parameter of Choice

When is it appropriate to use the median? Use of the median as the summary parameter is desirable in situations where the sample unit has identity and significance as an individual. An excellent example of the appropriate use of the median is presented by the work of Hirano et al. (1982). These researchers found that epiphytic bacterial populations of leaf surfaces were lognormally distributed. The sample units of this study were the individual plant leaves. According to the criterion stated above, the appropriate use of the median as the summary parameter requires that the samples have identity and significance. The significance of considering the bacterial populations on individual leaves is given by Hirano et al. (1982).

Table 2–2. Influence of sample bulking on the mean and median, and standard deviation. Original sample set (n = 1000) was selected from a log-normal distribution with mean of 70 and SD of 140.

Sample size (n)	Bulking size (n)	Mean	Median	Standard deviation
1000	1	71.1	29.4	173
500	2	71.1	37.8	129
200	5	71.1	50.7	84.3
100	10	71.1	53.7	49.9
50	20	71.1	63.9	37.2
20	50	71.1	64.4	31.6

They state, "The quantitative variability of epiphytic bacterial populations on individual leaves may be an expression of the uniqueness of each leaf as an ecosystem with one or more environmental or biological characteristics significantly different from that of a neighboring leaf." It was also observed in this study that the practice of bulking or compositing individual leaves results in overestimates of the population median for individual leaves. From a study on bacterial populations in the rhizosphere, Loper et al. (1984) reached similar conclusions concerning the effects of sample bulking on estimates of the median. These reported effects of sample bulking on the median are directly analogous to the simulation results observed in Table 2–2.

In many cases, the median or the geometric mean is arbitrarily chosen over the mean because of the resistance of these two parameters to the influence of extreme values often observed with lognormal sample data. However, this should not be the only criterion used. Careful consideration of relationship between the sample unit and the questions being asked should play a predominant role in the decision-making process.

2–4.4 The Mean as the Location Parameter of Choice

In the area of soil microbiology, often what is desired is an estimate of the magnitude of a given microbial process in a particular environment. For example, soil denitrification in agricultural systems may be an important mechanism of fertilizer N loss. Measurements of denitrification exhibit highly skewed frequency distributions. If the objective of a given study is to estimate the magnitude of denitrification-N loss from a field, then the mean, which is an estimator of the center of mass of the distribution, is the appropriate location parameter to estimate. Since the individual soil samples, either cores or soil chambers, have no inherent significance as individuals, the median is not the location parameter of choice. Indeed, in a study of the influence of sample volume effects on denitrification, it was found that the value of the median obtained from a given set of samples was significantly influenced by the size of the soil cores used (Parkin, 1991). It was suggested in this study that collection of larger-sized cores had the same effect on the median as bulking of several smaller samples. In addition to being influenced by sample volume, the median will underestimate the total N loss. For situations where an estimate of the total mass of a variable (either a chemical constituent or a microbial process) is required, the mean should be the location estimator of choice.

2–5 HYPOTHESIS TESTING

2–5.1 General Considerations

In previous sections, optimum methods for computing summary statistics of lognormally distributed data have been discussed. However, a typical objective of many studies extends beyond estimation of population parameters from sample data. In many cases, sampling is conducted to evaluate treatment effects. Application of standard analysis of variance procedures require several assumptions concerning the underlying error structure of the data, and among these is the assumption of normality. Violation of the assumption of normality will influence the ability of a statistical test to perform at the stated alpha level, and will also affect the power of a statistical test to detect differences between the batches of samples being compared (Hey, 1938; Cochran, 1947). Two procedures have been recommended when the normality assumption has been violated. These are: (i) first transform for normality or (ii) apply nonparametric statistical methods (Snedecor and Cochran, 1967). This section examines some of the consequences of implementing these two approaches.

2–5.2 Efficacy of Hypothesis Testing Procedures

By now it should be recognized that, when faced with lognormally distributed data, the investigator has a choice of location parameters (the mean and the median), and that the choice of the appropriate location parameter must be consistent with the objectives and methodologies of the problem under study. After the choice of the appropriate location parameter has been made, consideration must be given to the statistical methods used at the hypothesis-testing stage.

A recent study of the efficacy of hypothesis-testing methods applied to lognormal data revealed that some statistical procedures were better for detecting differences in medians, while other techniques were more sensitive to differences in population means (Parkin, 1993). The sensitivities of the statistical tests (Table 2–3) to detect differences in medians was evaluated by applying the tests to samples drawn from lognormal populations whose medians differed by a factor of 2 (Table 2–4). The tests were applied at alpha levels of 0.05. With increasing sample size, the power of tests 1, 2, and 3 to detect differences in the underlying populations increased. However, the power associated with test 4 decreased with increasing sample size. Test 5 was insensitive to changes in sample size and could only detect differences approximately 1% of the time, despite the fact that medians of

Table 2–3. Five statistical tests used in hypothesis testing evaluations.

Test	Description	Reference
1	Two-tailed, unpaired t-test equal sized samples, ln-transformed data	Snedecor & Cochran, 1967
2	Nonparametric signed rank	Wilcoxon, 1945; Snedecor & Cochran, 1967
3	Overlap of confidence limits of the median	Parkin & Robinson, 1992a
4	Two-tailed, unpaired t-test equal sized samples, untransformed data	Snedecor & Cochran, 1967
5	Overlap of Land's exact confidence limits of the mean	Land, 1973; Parkin & Robinson, 1992b

Table 2–4. Power of statistical tests to detect differences between two log-normal populations with differing medians.

Sample size	Test 1	Test 2	Test 3	Test 4	Test 5
—n—	\multicolumn{5}{c}{—frequency at which differences were observed—}				
4	19.4	19.7	4.0	19.4	0.51
8	30.4	28.9	11.2	18.7	0.93
12	41.6	39.9	19.7	17.0	1.36
20	61.0	58.3	37.9	14.8	1.04
60	97.8	96.3	92.2	11.2	1.12
100	99.9	99.7	99.4	9.8	1.28

the underlying populations differed by a factor of 2. Thus, if demonstration of differences in medians is of interest, either tests 1, 2, or 3 should be used, as the latter two methods (tests 4 and 5) are insensitive to differences in the median.

An evaluation of these tests using lognormal populations with equal medians, yet means that differed by a factor of 4, yielded information on the power of the statistical tests to detect differences in means (Table 2–5). Again, all tests were applied at an alpha level of 0.05. In this situation tests 1, 2, and 3 were all insensitive to differences in means, yet tests 4 and 5 were sensitive to differences in the means of the underlying populations. The power of the tests 4 and 5 to detect differences is poor for small sample sizes, but increases with increasing sample sizes. It is also observed that at any given sample size (n > 4) test 5 has greater power to detect differences than test 4.

The five statistical tests were also rated with regard to Type I error rate. Table 2–6 summarizes results from experiments designed to evaluate the actual Type I error rates of the statistical tests. The tests were applied at three alpha levels (0.05, 0.1, and 0.2) to two lognormal populations (CVs of 50 and 200%). With population 1 (CV = 50%), tests 1, 2, and 3 performed at the stated alpha level with regard to Type I error rate. With population 2 (CV = 200%), tests 1 and 2 again performed at the stated alpha level; however, unlike the results obtained for population 1, test 1

Table 2–5. Power of two statistical tests to detect differences between log-normal populations with differing means.

Sample size	Test 1	Test 2	Test 3	Test 4	Test 5
—n—	—frequency at which differences were observed—				
4	8.74	9.51	1.43	5.1	2.6
8	6.42	6.75	1.44	6.5	21.3
12	6.07	7.68	1.58	8.7	39.5
20	5.63	7.55	1.70	20.6	35.2
60	5.26	7.56	1.73	77.5	98.7
100	7.17	7.48	1.54	91.6	99.9

Table 2–6. Realized Type I error rate of hypothesis testing methods applied to sample data from two lognormal populations.

Pop. CV	Test level	Actual Type I error rates observed				
		Test 1	Test 2	Test 3	Test 4	Test 5
	—alpha—			–%–		
	0.05	5.06	4.93	0.53	4.55	0.43
50%	0.10	10.1	10.1	1.78	9.43	1.93
	0.20	20.0	19.9	6.76	19.4	5.92
	0.05	5.02	4.90	0.35	1.46	0.15
200%	0.10	9.94	10.0	1.90	4.79	0.61
	0.20	19.8	19.7	6.86	15.4	2.93

Tests 1, 2, and 3 are median separation tests, while Tests 4 and 5 are mean separation tests.

did not perform at the stated alpha level. The Type 1 error rates of tests 3 and 5 were substantially less than the nominal alpha levels at which the tests were applied.

2–5.3 Recommendations

Based on simulation studies conducted over a range of population variances, the recommended approach to solving the two-sample problem when the mean is of interest involves constructing confidence intervals for the two batches and comparing them with one another (Parkin, 1993). If the 90% confidence limits for the means of two lognormally distributed batches do not overlap, then the hypothesis that the two lognormal means are equal can be rejected. The precise alpha level, or Type I error rate depends, to some degree, upon the variance of the underlying population; however, as indicated in Table 2–6, it will be substantially less than the nominal alpha level at which the confidence limits are calculated. If, however, it has been determined that the median is the location parameter of interest, then either the nonparametric method or the t-test on ln-transformed data can be used.

Discussion in this chapter has focused on the two-sample problem. There is little information in the literature that deals with multiple comparisons procedures appropriate for skewed data. However, when interest is in the median as the location estimator, the simulation results of the two-sample problem indicate that an appropriate procedure would be application of the FLSD test (Carmer & Swanson, 1971) on the ln-transformed data. The FLSD is implemented by first performing an F test, followed by ordinary LSD if the F test indicates significance. We are not aware of any simulation studies that have evaluated procedures for comparing the means of more than two lognormally distributed variables. In such situations, one option may be to perform the FLSD on untransformed data; realizing that the power to detect real differences may be poor. Alternatively, in the spirit of Exploratory Data Analysis (Hoaglin et al., 1983), one could compute a plot of means and confidence intervals and visually observe overlapping ranges between treatments. Due to the general lack of information on this topic, it is not possible to make specific recommendations beyond this point.

2–6 SAMPLE NUMBER REQUIREMENTS

Many of the considerations associated with experimental design and sample allocation have been discussed in chapter 1 by Wollum in this book. This section will address three related topics: (i) sample number requirement to estimate the mean within a given precision, (ii) sample compositing, and (iii) the power of statistical tests.

2–6.1 Estimating the Mean

The question, "How many samples do I need to accurately estimate the mean?" is one that is frequently raised by experimentalists. When the underlying population is normally distributed, the sample size required to obtain an estimate of the population mean within a given precision is easily obtained (see chapter 1 by Wollum in this book). However, this approach will underestimate the sample number requirements if the underlying population is lognormally distributed. With log-normally distributed data, calculation of the number of samples needed to estimate the mean of a lognormal distribution at a given probability level can be done; however, this process is not straightforward. This is accomplished by iteratively searching for the value n in Eq. [7] and [8]. In this process, alpha equals the probability level desired and the upper and lower confidence limits are set equal to the estimate of the mean plus or minus the user specified degree of precision. It should be noted that, since the lognormal confidence interval is asymmetric about the median, the sample number requirement for obtaining an estimate of the mean at the lower precision limit will differ from the number of samples required at the upper limit of precision

2–6.2 Compositing

Bulking or compositing of samples prior to analysis is a procedure that is often used to reduce analytical load. Compositing of samples is also useful in reducing the variance associated with the estimate of the population mean. The dramatic effect of sample bulking on the variability associated with the mean is illustrated in Table 2–2. As bulking size is increased, the standard deviation shows an asymptotic decrease. Thus, if sample identity is not a significant factor, and the mean is the location estimator of choice, then bulking or compositing of samples may be a valuable strategy to reduce variability. However, if samples have identity and significance, and the median is the estimator of choice, then bulking or compositing of samples is not recommended.

2–6.3 Power of Statistical Tests

In addition to influencing the precision with which population parameters are estimated, sample size will also influence the power of hypothesis-testing procedures. The impact of sample size on the power of statistical tests was observed in Tables 2–4 and 2–5. It is apparent that all the tests lack power to detect differences when sample numbers are low. High replication is needed to detect differences between log-normal populations whose location parameters (either the mean or median) differ by a factor of two.

If availability of resources are the limiting factor controlling sample size, power of statistical tests to detect differences can be increased by increasing the alpha level at which the testing procedure is applied. For example, simulation experiments indicate that with sample data ($n = 8$) drawn from two lognormal populations with means that differ by a factor of 2, overlap of 97.5% confidence intervals will only detect differences 6.04% of the time. Yet, if 90% confidence intervals are computed, power is increased to 33.4% (Parkin, 1993). In the latter case, the associated Type I error rate is approximately 5.3%, compared to a Type I error rate of 0.5% when 97.5% confidence interval overlap is performed.

2–7 CONCLUDING REMARKS

The focus of this chapter has been on the statistical analysis of lognormal data. Techniques for diagnosing lognormality have been presented, as well as methods for estimating population parameters and testing hypotheses. When the data are lognormally distributed, then methods described in this chapter should be used. Methods optimal for Gaussian distributed data will be suboptimal for the lognormal parametric family. The lack of power that exists when traditional approaches are applied to positively skewed data must be recognized.

As a cautionary note, it should be stated that the consequences of applying lognormal parametric analysis methods to positively skewed data that are not truly lognormally distributed are largely unknown. Robustness of these UMVU estimators to deviations from true lognormality has only recently been addressed (Sichel, 1987; Meyers & Pepin, 1990). Meyers and Pepin (1990) reported that the UMVU estimator of the mean, when applied to data from contaminated lognormal distributions, is inefficient relative to the arithmetic average. We agree with this recommendation and encourage additional work in this area. What is really needed is for the gap that Miller (1986) refers to, regarding the absence of robust statistical theory for asymmetric distributions, be filled. When these methods are in place, then the soil microbiologist with asymmetric data to analyze will have tools for analyzing positively skewed data that may equal or just approximate the lognormal probability density family.

We would like to encourage additional research in this area. The availability of inexpensive computational power offered by desktop computers offer opportunities for a more detailed analysis of the adequacy of statistical procedures applied to a given data set. The techniques described in this chapter, along with some knowledge of the properties of the underlying distribution derived from the sample data can be used on a case-by-case basis, to evaluate the adequacy of mean or median separation test for a given sample size.

APPENDIX 1: STATISTICAL TABLES

Table 2–A1. Coefficients (a_i) for the Shapiro-Wilk test of normality. (From Shapiro and Wilk, 1965.)

i\n	2	3	4	5	6	7	8	9	10	11
1	0.7071	0.7071	0.6872	0.6646	0.6431	0.6233	0.6052	0.5888	0.5739	0.5601
2	—	—	0.1677	0.2413	0.2806	0.3031	0.3164	0.3244	0.3291	0.3315
3	—	—	—	—	0.0875	0.1401	0.1743	0.1976	0.2141	0.2260
4	—	—	—	—	—	—	0.0561	0.0947	0.1224	0.1429
5	—	—	—	—	—	—	—	—	0.0399	0.0695

i\n	12	13	14	15	16	17	18	19	20	21
1	0.5475	0.5359	0.5251	0.5150	0.5056	0.4968	0.4886	0.4808	0.4734	0.4643
2	0.3325	0.3325	0.3318	0.3306	0.3290	0.3273	0.3253	0.3232	0.3211	0.3185
3	0.2347	0.2412	0.2460	0.2495	0.2521	0.2540	0.2553	0.2561	0.2565	0.2578
4	0.1586	0.1707	0.1802	0.1878	0.1939	0.1988	0.2027	0.2059	0.2085	0.2119
5	0.0922	0.1099	0.1240	0.1353	0.1447	0.1524	0.1587	0.1641	0.1686	0.1736
6	0.0303	0.0539	0.0727	0.0880	0.1005	0.1109	0.1197	0.1271	0.1334	0.1399
7	—	—	0.0240	0.0433	0.0593	0.0725	0.0837	0.0932	0.1013	0.1092
8	—	—	—	—	0.0193	0.0359	0.0496	0.0612	0.0711	0.0804
9	—	—	—	—	—	—	0.0163	0.0303	0.0422	0.0530
10	—	—	—	—	—	—	—	—	0.0140	0.0263

i\n	22	23	24	25	26	27	28	29	30	31
1	0.4590	0.4542	0.4493	0.4450	0.4407	0.4366	0.4328	0.4291	0.4254	0.4220
2	0.3156	0.3126	0.3098	0.3069	0.3043	0.3018	0.2992	0.2968	0.2944	0.2921
3	0.2571	0.2563	0.2554	0.2543	0.2533	0.2522	0.2510	0.2499	0.2487	0.2475
4	0.2131	0.2139	0.2145	0.2148	0.2151	0.2152	0.2151	0.2150	0.2148	0.2145
5	0.1764	0.1787	0.1807	0.1822	0.1836	0.1848	0.1857	0.1864	0.1870	0.1874
6	0.1443	0.1480	0.1512	0.1539	0.1563	0.1584	0.1601	0.1616	0.1630	0.1641
7	0.1150	0.1201	0.1245	0.1283	0.1316	0.1346	0.1372	0.1395	0.1415	0.1433
8	0.0878	0.0941	0.0997	0.1046	0.1089	0.1128	0.1162	0.1192	0.1219	0.1243
9	0.0618	0.0696	0.0764	0.0823	0.0876	0.0923	0.0965	0.1002	0.1036	0.1066
10	0.0368	0.0459	0.0539	0.0610	0.0672	0.0728	0.0778	0.0822	0.0862	0.0899
11	0.0122	0.0228	0.0321	0.0403	0.0476	0.0540	0.0598	0.0650	0.0697	0.0739
12	—	—	0.0107	0.0200	0.0284	0.0358	0.0424	0.0483	0.0537	0.0585
13	—	—	—	—	0.0094	0.0178	0.0253	0.0320	0.0381	0.0435
14	—	—	—	—	—	—	0.0084	0.0159	0.0227	0.0289
15	—	—	—	—	—	—	—	—	0.076	0.0144

i\n	32	33	34	35	36	37	38	39	40	41
1	0.4188	0.4156	0.4127	0.4096	0.4068	0.4040	0.4015	0.3989	0.3964	0.3940
2	0.2898	0.2876	0.2854	0.2834	0.2813	0.2794	0.2774	0.2755	0.2737	0.2719
3	0.2462	0.2451	0.2439	0.2427	0.2415	0.2403	0.2391	0.2380	0.2368	0.2357
4	0.2141	0.2137	0.2132	0.2127	0.2121	0.2116	0.2110	0.2104	0.2098	0.2091
5	0.1878	0.1880	0.1882	0.1883	0.1883	0.1883	0.1881	0.1880	0.1878	0.1876
6	0.1651	0.1660	0.1667	0.1673	0.1678	0.1683	0.1686	0.1689	0.1691	0.1693
7	0.1449	0.1463	0.1475	0.1487	0.1496	0.1505	0.1513	0.1520	0.1526	0.1531
8	0.1265	0.1284	0.1301	0.1317	0.1331	0.1344	0.1356	0.1366	0.1376	0.1384
9	0.1093	0.1118	0.1140	0.1160	0.1179	0.1196	0.1211	0.1225	0.1237	0.1249
10	0.0931	0.0961	0.0988	0.1013	0.1036	0.1056	0.1075	0.1092	0.1108	0.1123
11	0.0777	0.0812	0.0844	0.0873	0.0900	0.0924	0.0947	0.0967	0.0986	0.1004
12	0.0629	0.0669	0.0706	0.0739	0.0770	0.0798	0.0824	0.0848	0.0870	0.0891
13	0.0485	0.0530	0.0572	0.0610	0.0645	0.0677	0.0706	0.0733	0.0759	0.0782
14	0.0344	0.0395	0.0441	0.0484	0.0523	0.0559	0.0592	0.0622	0.0651	0.0677

Table 2–A1. Continued.

15	0.0206	0.0262	0.0314	0.0361	0.0404	0.0444	0.0481	0.0515	0.0546	0.0575
16	0.0068	0.0131	0.0187	0.0239	0.0287	0.0331	0.0372	0.0409	0.0444	0.0476
17	—	—	0.0062	0.0119	0.0172	0.0220	0.0264	0.0305	0.0343	0.0379
18	—	—	—	—	0.0057	0.0110	0.0158	0.0203	0.0244	0.0283
19	—	—	—	—	—	—	0.0053	0.0101	0.0146	0.0188
20	—	—	—	—	—	—	—	—	0.0049	0.0094

i\n	42	43	44	45	46	47	48	49	50
1	0.3917	0.3894	0.3872	0.3850	0.3830	0.3808	0.3789	0.3770	0.3751
2	0.2701	0.2684	0.2667	0.2651	0.2635	0.2620	0.2604	0.2589	0.2574
3	0.2345	0.2334	0.2323	0.2313	0.2302	0.2291	0.2281	0.2271	0.2260
4	0.2085	0.2078	0.2072	0.2065	0.2058	0.2052	0.2045	0.2038	0.2032
5	0.1874	0.1871	0.1868	0.1865	0.1862	0.1859	0.1855	0.1851	0.1847
6	0.1694	0.1695	0.1695	0.1695	0.1695	0.1695	0.1693	0.1692	0.1691
7	0.1535	0.1539	0.1542	0.1545	0.1548	0.1550	0.1551	0.1553	0.1554
8	0.1392	0.1398	0.1405	0.1410	0.1415	0.1420	0.1423	0.1427	0.1430
9	0.1259	0.1269	0.1278	0.1286	0.1293	0.1300	0.1306	0.1312	0.1317
10	0.1136	0.1149	0.1160	0.1170	0.1180	0.1189	0.1197	0.1205	0.1212
11	0.1020	0.1035	0.1049	0.1062	0.1073	0.1085	0.1095	0.1105	0.1113
12	0.0909	0.0927	0.0943	0.0959	0.0972	0.0986	0.0998	0.1010	0.1020
13	0.0804	0.0824	0.0842	0.0860	0.0876	0.0892	0.0906	0.0919	0.0932
14	0.0701	0.0724	0.0745	0.0765	0.0783	0.0801	0.0817	0.0832	0.0846
15	0.0602	0.0628	0.0651	0.0673	0.0694	0.0713	0.0731	0.0478	0.0764
16	0.0506	0.0534	0.0560	0.0584	0.0607	0.0628	0.0648	0.0667	0.0685
17	0.0411	0.0442	0.0471	0.0497	0.0522	0.0546	0.0568	0.0588	0.0608
18	0.0318	0.0352	0.0383	0.0412	0.0439	0.0465	0.0489	0.0511	0.0532
19	0.0227	0.0263	0.0296	0.0328	0.0357	0.0385	0.0411	0.0436	0.0459
20	0.0136	0.0175	0.0211	0.0245	0.0277	0.0307	0.0335	0.0361	0.0386
21	0.0045	0.0087	0.0126	0.0163	0.0197	0.0229	0.0259	0.0288	0.0314
22	—	—	0.0042	0.0081	0.0118	0.0153	0.0185	0.0215	0.0244
23	—	—	—	—	0.0039	0.0076	0.0111	0.0143	0.0174
24	—	—	—	—	—	—	0.0037	0.0071	0.0104
25	—	—	—	—	—	—	—	—	0.0035

Table 2–A2. Quantiles of the Shapiro-Wilk test of normality. (From Shapiro and Wilk, 1965.)

			Probability level			
n	0.010	0.020	0.050	0.100	0.500	0.900
3	0.753	0.756	0.767	0.789	0.959	0.998
4	0.687	0.707	0.748	0.792	0.935	0.987
5	0.686	0.715	0.762	0.806	0.927	0.979
6	0.713	0.743	0.788	0.826	0.927	0.974
7	0.730	0.760	0.803	0.838	0.928	0.972
8	0.749	0.778	0.818	0.851	0.932	0.972
9	0.764	0.791	0.829	0.859	0.935	0.972
10	0.781	0.806	0.842	0.869	0.938	0.972
11	0.792	0.817	0.850	0.876	0.940	0.973
12	0.805	0.828	0.859	0.883	0.943	0.973
13	0.814	0.837	0.866	0.889	0.945	0.974
14	0.825	0.846	0.874	0.895	0.947	0.975
15	0.835	0.855	0.881	0.901	0.950	0.975
16	0.844	0.863	0.887	0.906	0.952	0.976
17	0.851	0.869	0.892	0.910	0.954	0.977
18	0.858	0.874	0.897	0.914	0.956	0.978
19	0.863	0.879	0.901	0.917	0.957	0.978
20	0.868	0.884	0.905	0.920	0.959	0.979
21	0.873	0.888	0.908	0.923	0.960	0.980
22	0.878	0.892	0.911	0.926	0.961	0.980
23	0.881	0.895	0.914	0.928	0.962	0.981
24	0.884	0.898	0.916	0.930	0.963	0.981
25	0.888	0.901	0.918	0.931	0.964	0.981
26	0.891	0.904	0.920	0.933	0.965	0.982
27	0.894	0.906	0.923	0.935	0.965	0.982
28	0.896	0.908	0.924	0.936	0.966	0.982
29	0.898	0.910	0.926	0.937	0.966	0.982
30	0.900	0.912	0.927	0.939	0.067	0.983
31	0.902	0.914	0.929	0.940	0.967	0.983
32	0.904	0.915	0.930	0.941	0.968	0.983
33	0.906	0.917	0.931	0.942	0.968	0.983
34	0.908	0.919	0.933	0.943	0.969	0.983
35	0.910	0.920	0.934	0.944	0.969	0.984
36	0.912	0.922	0.935	0.945	0.970	0.984
37	0.914	0.924	0.936	0.946	0.970	0.984
38	0.916	0.925	0.938	0.947	0.971	0.984
39	0.917	0.927	0.939	0.948	0.971	0.984
40	0.919	0.928	0.940	0.949	0.972	0.985
41	0.920	0.929	0.941	0.950	0.972	0.985
42	0.922	0.930	0.942	0.951	0.972	0.985
43	0.923	0.932	0.943	0.951	0.973	0.985
44	0.924	0.933	0.944	0.952	0.973	0.985
45	0.926	0.934	0.945	0.953	0.973	0.985
46	0.927	0.935	0.945	0.953	0.974	0.985
47	0.928	0.936	0.946	0.954	0.974	0.985
48	0.929	0.937	0.947	0.954	0.974	0.985
49	0.929	0.937	0.947	0.955	0.974	0.985
50	0.930	0.938	0.947	0.955	0.974	0.985

Table 2–A3. Factors for calculating lower 0.95 confidence limit of lognormally distributed data. Interpolated from Land (1973).

$\hat{\sigma}$	Sample number					
	4	6	12	20	40	100
0.10	−2.006	−1.759	−1.679	−1.657	−1.648	−1.643
0.20	−1.865	−1.697	−1.646	−1.637	−1.636	−1.641
0.50	−1.626	−1.592	−1.608	−1.628	−1.651	−1.682
1.00	−1.522	−1.588	−1.677	−1.737	−1.796	−1.873
2.00	−1.686	−1.864	−2.075	−2.211	−2.347	−2.523
5.00	−2.877	−3.357	−3.919	−4.290	−4.659	−5.140
10.00	−5.328	−6.280	−7.416	−8.170	−8.921	−9.904

Table 2–A4. Factors for calculating upper 0.95 confidence limit of lognormally distributed data. Interpolated from Land (1973).

$\hat{\sigma}$	Sample number					
	4	6	12	20	40	100
0.10	2.481	1.942	1.780	1.729	1.698	1.671
0.20	2.886	2.069	1.852	1.780	1.737	1.697
0.50	5.209	2.638	2.157	2.009	1.917	1.831
1.00	10.076	4.128	2.954	2.607	2.401	2.207
2.00	19.994	7.701	5.054	4.243	3.756	3.301
5.00	49.887	18.880	12.010	9.849	8.534	7.291
10.00	99.753	37.660	23.835	19.475	16.815	14.287

APPENDIX 2: COMPUTER PROGRAMS

The following is a listing of a BASIC computer program for evaluating Finney's PSI function (Eq. [6]), used in computing the UMVU estimators of the mean, variance, and median. This function is evaluated until a user specified precision is achieved.

```
10 REM ********************************************************
20 REM ******            PROGRAM 'PSIEVAL'           ******
30 REM ***    THIS PROGRAM EVALUATES FINNEY'S PSI    ***
40 REM ***      FUNCTION UNTIL A USER DEFINED        ***
50 REM ***      PRECISION IS REACHED. DEFAULT        ***
60 REM ***    PRECISION IS 0.01% (TOL=0.0001, LINE 70)   ***
      ********************************************************
70 TOL=0.0001
80 INPUT "INPUT VALUE ",Z
90 INPUT "INPUT NUMBER OF SAMPLES ",N
100 COEF1=1 : COEF=1 : PSI=1 : I=0
110 IF COEF/PSI<TOL THEN 170
120 I=I+1
130 COEF=COEF1*(Z*(N−1)^2)/(N*(N+2*I−3)*I)
140 PSI=PSI+COEF
```

```
150 COEF1 = COEF
160 GOTO 110
170 PRINT "PSI = ",PSI
180 END
```

Computer Programs Available in USDA Technology Transfer Documents

A series of BASIC computer programs have been developed by the authors to perform many of the computations described in this chapter. These programs are supplied, free of charge, upon request.

1. NSTL91-1 PROGEN.BAS Contains routines for reading, printing, and writing ASCII files to disk.

2. NSTL91-2 PROBIT.BAS Performs probit analysis.
   NORMTEST.BAS Performs test of normality using Shapiro-Wolk procedure and D'Agostino's procedure.

3. NSTL91-3 LSTATS.BAS Calculates descriptive sample statistics for log-normally distributed data. UMVU, ML, and MM estimators are calculated along with Land's exact confidence interval method. Algorithms for computing and plotting sample histogram are also included.

4. NSTL91-4 LMCARLO.BAS Performs Monte Carlo sampling from lognormal distribution with user specified mean and standard deviation.

REFERENCES

Bradu, D. and Y. Mundlak. 1970. Estimation in lognormal linear models. J. Am. Stat. Assoc. 65:198–211.

Carmer, S.G., and M.R. Swanson. 1971. Detection of differences between means: A Monte Carlo study of five pairwise multiple comparison procedures. Agron. J. 63:940–945.

Cochran, W.G. 1947. Some consequences when the assumptions for the analysis of variance are not satisfied. Biometrics 3:22–38.

D'Agostino, R.B. 1971. An omnibus test of normality for moderate and large size samples. Biometrika 58:341–348.

Finney, D.J. 1941. On the distribution of a variate whose logarithm is normally distributed. J. Royal Stat. Soc. Suppl. 7:144–161.

Gilbert, R.O. 1987. Statistical methods for environmental pollution monitoring. Van Nostrand Reinhold Co., New York.

Greenberg, A.A., R.R. Trussell, L.S. Clesceri, and M.A.H. Franson. (ed.). 1985. Standard methods for the examination of water and wastewater, 16th ed. Am. Publ. Health Assoc. et al., Washington, DC.

Hey, G.B. 1938. A new method of experimental sampling illustrated on certain non-normal populations. Biometrika 30:68–80.

Hirano, S.S., E.V. Nordheim, D.C. Arny, and C.D. Upper. 1982. Lognormal distribution of epiphytic bacterial populations on leaf surfaces. Appl. Environ. Microbiol. 44:695–700.

Hoaglin, D.C., F. Mosteller, and J.W. Tukey. 1983. Understanding robust and exploratory data analysis. John Wiley and Sons, New York.

Land, C.E. 1973. Standard confidence limits for linear functions of the normal mean and variance. J. Am. Stat. Assoc. 68:960–963.

Loper, J.E., T.V. Suslow, and N.M. Schroth. 1984. Lognormal distribution of bacterial populations in the rhizosphere. Phytopathology 74:1454–1460.

Meyers, R.A., and P. Pepin. 1990. The robustness of lognormal-based estimators of abundance. Biometrics 46:1185–1192

Miller, R.G. 1986. Beyond ANOVA, basics of applied statistics. John Wiley and Sons, New York.

Parkin, T.B. 1991. Characterizing the variability of denitrification. p. 213–228. In N.P. Revsbech and J. Sorensen (ed.) Denitrification in soil and sediment. Plenum Press, New York.

Parkin, T.B. 1993. Evaluation of statistical methods for determining differences between lognormal populations. Agron. J. 85:747–753.

Parkin, T.B., S.T. Chester, and J.A. Robinson. 1990. Evaluation of methods for calculating confidence intervals about the mean of a lognormally distributed variable. Soil Sci. Soc. Am. J. 54:321–326.

Parkin, T.B., J.J. Meisinger, S.T. Chester, J.L. Starr, and J.A. Robinson. 1988. Evaluation of statistical methods for lognormally distributed variables. Soil. Sci. Soc. Am. J. 52:323–329.

Parkin, T.B., and J.A. Robinson. 1992a. Statistical evaluation of median estimators for lognormally distributed variables. Soil Sci. Soc. Am. J. 57:317–323.

Parkin, T.B., and J.A. Robinson. 1992b. Analysis of lognormal data. Adv. Soil Sci. 20:193–235.

Shapiro, S.S., and M.B. Wilk. 1965. An analysis of variance test for normality (complete samples). Biometrika 52:591–611.

Sichel, H.S. 1952. New methods in the statistical evaluation of mine sampling. London Inst. Mining Met. Trans. 61:261–288.

Sichel, H.S. 1987. Some advances in lognormal theory, APCOM 87. p. 3–8. In Proc. 20 Int. Symp. on the Applications of Computers and Mathematics in the Mineral Industries. Vol. 3: Geostatistics. SAIMM, Johannesburg.

Snedecor, G.W., and W.G. Cochran. 1967. Statistical methods. 6th ed. Iowa State Univ., Ames.

Warrick, A.W., and D.R. Nielsen. 1980. Spatial variability of soil physical properties in the field. p. 385. In D. Hillel (ed.) Applications of soil physics. Academic Press, New York.

Wilcoxon, F. 1945. Some uses of statistics in plant pathology. Biometrics 1:41–45.

Chapter 3

Soil Sterilization

DUANE C. WOLF, *University of Arkansas, Fayetteville, Arkansas*

HORACE D. SKIPPER, *Clemson University, Clemson, South Carolina*

Microorganisms are ubiquitous in soil and responsible for numerous and diverse biochemical activities. Elimination of indigenous soil microorganisms may be necessary to separate biological from chemical transformations of specific organic compounds, or to study the growth or metabolic activity of specific microorganisms inoculated into the soil. The study of adsorption or volatilization of labile compounds added to soil also requires elimination of microbial metabolism.

The goal of soil sterilization is to destroy the microbial population with minimum alteration of soil chemical and physical properties. The objective of this chapter is to present laboratory methods of soil sterilization and report their potential impact on soil properties.

3–1 PRINCIPLES

Sterilization requires destruction of both actively growing microorganisms and any resting structures such as spores (Joslyn, 1983; McLaren, 1969). As a probability function, sterilization should reduce the chances of having a survivor to $\leq 10^{-6}$ (Korczynski, 1981).

Historically, the most common method for soil sterilization has been moist heat such as autoclaving. Additional sterilization methods include dry heat; ionizing radiation; or gaseous compounds such as ethylene oxide, propylene oxide, or methyl bromide. Nongaseous chemicals such as mercuric chloride and sodium azide have also been used as microbial inhibitors.

In all soil sterilization procedures, effectiveness of the process must be verified by demonstrating an absence of microbial growth (Beloian, 1983). In addition, special attention must be given to storing, opening, and using the sterilized soil in a manner to prevent contamination. Verification can be accomplished by inoculating a test tube containing 10 mL of nutrient broth (see chapter 8 by Zuberer in this book) with 1 g of sterilized soil and incubating for 7 d at 25 °C. If the broth becomes turbid, the sample was not sterile. Similarly, sterile soil can be added directly to petri dishes containing a general agar-base growth medium such as tryptic soy agar (Martin, 1975). Pour or spread plate techniques can be used. Another approach to evaluate sterilization efficiency is to conduct serial dilutions and use a pour or spread plate technique (see chapter 8 in this book) or a suitable most probable number procedure (see chapter 5 by Woomer in this book). Several media are acceptable and could include tryptic soy agar for bacteria and Martin's medium for fungi (see chapter 17 by Parkinson in this book). However, it should be noted that many soil microorganisms do not grow on conventional media using traditional microbiological techniques.

3–2 MOIST HEAT

3–2.1 Materials

1. Autoclave.
2. Glass, stainless steel, or other heat resistant containers for soil.
3. Aluminum foil or paper to cover the containers.

3–2.2 Procedures

The most commonly used method for soil sterilization involves moist heat. Moist heat is much more effective than dry heat for sterilization. Pressurized steam at 121 °C as supplied by an autoclave is the most commonly used sterilization technique. Before the soil sample is autoclaved, it is advisable to pre-incubate the moist soil for 2 to 3 d at room temperature to stimulate microbial growth. Ideally, the moisture potential of the soil should be adjusted to −0.01 to −0.03 MPa or approximately 60% of the soil's moisture-holding capacity. The moist soil should be placed in a shallow glass or stainless steel container and covered with aluminum foil or paper to prevent subsequent contamination. Steam sterilization of large volumes of soil should be avoided because of uneven heat distribution in the sample. More efficient sterilization can be accomplished by spreading the soil to a depth of ≤2.5 cm in a shallow container for the preincubation and subsequent steam sterilization. Adding ≤10 g of soil to a 20 by 125 mm screw-top culture tube is also acceptable.

The soil should be autoclaved at 0.10 MPa (15 lb/in.^{2}) and 121 °C for 1 h. After the initial autoclaving, sterile distilled water may need to be aseptically added to restore the desired moisture level of the soil. The moist soil should be incubated an additional 2 d and autoclaved a second time for 1 h to eliminate any microorganisms that were not destroyed in the initial autoclaving. Only one autoclaving of air-dry soil is not adequate for sterilization (Wolf et al., 1989). Sterility can be verified by inoculating suitable media and confirming the absence of microbial growth.

3–2.3 Comments

Autoclaving soil has been shown to influence soil chemical properties. Particularly noteworthy is the potentially large increase in the extractable Mn levels that can result from autoclaving soil (Lopes & Wollum, 1976; Martin et al., 1973; Skipper & Westermann, 1973; Wolf et al., 1989). Eno and Popenoe (1964) reported that steam treatment for 6 h at 0.14 MPa (20 lb/in.[2]) resulted in increased levels of extractable N, P, S, and organic matter. Lopes and Wollum (1976) reported that NH_4-N and NO_3-N increased significantly in autoclaved soil. Available P and water-soluble amino acids and carbohydrate levels are increased by autoclaving (Ferriss, 1984; Skipper & Westermann, 1973). This method has also been shown to influence extractable Al and Fe and to decrease trace element levels in soil but generally does not result in significant changes in Ca, Mg, or K levels (Ferriss, 1984). Cation exchange capacity, surface area, and pH are not usually influenced by autoclaving (Wolf et al., 1989).

Peterson (1962) reported that colonization of autoclaved soil by soil microorganisms is inhibited and the duration of inhibition is related to the time of autoclaving and the type of microbe tested. Jenkinson (1966) suggested that autoclaving would make a portion of the nonbiomass organic matter more decomposable in re-inoculation studies. Autoclaving soil destroys the free radical mechanisms involved in abiotic transformations of certain organic compounds (Fletcher & Kaufman, 1980). Others have suggested that autoclaving can change the surface charge of pores in sandstone and reduce the surface area of clays (Jenneman et al., 1986).

3–3 DRY HEAT

3–3.1 Materials

1. Forced-draft oven capable of temperatures ≥ 200 °C.
2. Glass, porcelain, or stainless steel containers for soil.

3–3.2 Procedures

The soil sample in a suitable container is placed in an oven adjusted to 200 °C. A 150-mm diam. glass petri dish will accommodate a 50-g soil sample. Partially offsetting the lid of the petri dish during the heating procedure allows for more rapid sample drying. To assure sterilization, an exposure time of 24 h at 200 °C is recommended. Before the sample is allowed to cool, the petri dish lid should be replaced to prevent contamination of the sterilized sample. To allow rapid and uniform heat convection, it is important not to overload a dry-heat sterilizer (Joslyn, 1983).

3–3.3 Comments

Spores are more resistant than vegetative cells to dry heat sterilization. Preincubation of moist soil should facilitate dry heat sterilization by allowing a portion of the spores to germinate. Exposure temperatures of 160 °C for 3 h (Labeda et al., 1975) or 90 °C for 48 h (Wolf et al., 1989) reduce soil microbial populations substantially but are not adequate for soil sterilization. Kitur and Frye (1983) found that heating the soil to 200 °C substantially decreased the soil organic matter and increased extractable Mn and NH_4-N and electrical conductivity levels in three soils they studied. Plant growth was also suppressed in soil sterilized by dry heat.

3–4 GAMMA IRRADIATION

3–4.1 Materials

1. ^{60}Co or ^{137}Cs irradiation facility.
2. Glass or polyethylene containers for soil.

3–4.2 Procedures

Although cost has limited their numbers, facilities are available that can sterilize soil with ^{60}Co or ^{137}Cs radiation. In general terms, following preincubation, the soil samples are placed in either glass screw-top culture tubes or polyethylene bags and delivered to the irradiation facility. Glass containers will turn amber color after irradiation. The samples are irradiated with 0.03 to 0.06 MGy (3–6 Mrad) (Davis, 1975; Jackson et al., 1967; McLaren, 1969). Soil samples of up to 25 kg can be sterilized in one exposure (Cawse, 1975).

3–4.3 Comments

Irradiation can more easily eliminate the microbial population of moist soil compared to air-dry soil, and fungi are more easily eliminated than bacteria (Jackson et al., 1967). Irradiation has been shown to increase levels of extractable Mn, NH_4-N, and organic N in soil but generally does not influence other chemical or physical properties, and no toxicity problems have been noted in irradiated soils (McLaren, 1969; Wolf et al., 1989). However, irradiation will cause some heating depending upon the radiation dose.

Thompson (1990) recommended 0.02 MGy radiation to eliminate vesicular-arbuscular mycorrhizal (VAM) fungi in soil used for VAM-plant nutritional studies, but Jakobsen (1984) suggested even lower levels. In contrast to autoclaved soils, irradiated soils do not show a marked lag period when inoculated with a mixed microbial population (Peterson,

1962; Powlson & Jenkinson, 1976). Cawse (1975) and Silverman (1983) provide more detailed discussions of the microbiology and biochemistry of irradiated soils.

3–5 MICROWAVE IRRADIATION

Recently, the use of microwave radiation (2450 MHz) has been adapted to use in preparing microbiological growth media (CEM Corp., P.O. Box 200, Matthews, NC 28106). Commercially available microwave ovens have not been successfully used to sterilize soil samples to date, but the potential for using them to selectively eliminate certain plant pathogens has been demonstrated (Ferriss, 1984). Using plate count data, Wainwright et al. (1980) showed that microwave treatment of moist soil eliminated the soil fungal population while having no influence on the heterotrophic bacteria levels. However, Speir et al. (1986) used selective inhibition respiration to show that soil fungi and bacteria were equally susceptible to microwave radiation. The lethal action of microwave radiation has been shown by Vela and Wu (1979) to be due to the high temperatures that result when water absorbs the radiation. This theory would explain why the influence of microwave radiation is greater in moist than dry soils. Microwave radiation has been shown to influence soil chemical properties with large increases in extractable Mn levels noted (Thien et al., 1978).

3–6 GASEOUS COMPOUNDS

3–6.1 Materials

1. Ethylene oxide, or propylene oxide, or methyl bromide (Monobromomethane).
2. Commercial sterilization chambers for ethylene or propylene oxide. Vacuum desiccator can be used for small soil samples.
3. Glass or polyethylene containers for soil.

3–6.2 Procedures

Commercially available sterilization chambers are generally used with ethylene or propylene oxide, and the instructions prepared by the manufacturer should be carefully followed. For sterilization of small soil samples (about 100 g), a vacuum desiccator can be used. Because ethylene and propylene oxides are extremely flammable, *all operations should be carried out in a properly vented fume hood.* The soil should be added to a flask or culture tube, plugged with cotton, pre-incubated under moist conditions, and placed in the vacuum desiccator. A 250-mL beaker is placed in the desiccator, and 50 mL of cold ethylene oxide is added to the beaker. With 50 mL 100 g^{-1} of soil, the application rate is approximately 10 mol kg^{-1} of soil. The fumigant should be cold because the boiling point is 11 °C. The desiccator is quickly sealed and a vacuum applied. After fumigation for 24 h at room temperature, the vacuum is released and the beaker containing any remaining ethylene oxide is removed. The gaseous fumigant is removed from the soil by repeated evacuations with a minimum of six cycles generally being adequate to remove the ether-like odor of the remaining ethylene oxide.

A similar procedure can be followed when using propylene oxide to sterilize soil. A 50-mL volume of propylene oxide 100 g^{-1} of soil is approximately 7 mol kg^{-1} of soil. Because the boiling point for propylene oxide is 35 °C, an exposure time of 48 h is more suitable. Bartlett and Zelazny (1967) modified a pressure cooker such that the unit could be heated, and the exposure time was reduced to 15 min. They also placed the soil in sealed polyethylene freezer bags that were permeable to the gaseous fumigant but prevented subsequent contamination of the fumigated soil. Lopes and Wollum (1976), Skipper and Westermann (1973), and Wolf et al. (1989) added propylene oxide directly to soil in flasks or culture tubes and successfully sterilized soil.

Methyl bromide can also be used to sterilize soil. Because it is a highly poisonous gas, it must be used in a fume hood. Commercial preparations of the fumigant generally contain 2% chloropicrin (teargas) as a warning agent (Van Berkum & Hoestra, 1979). The pressurized fumigant is generally purchased in 454-g (1 lb) cans such as Brom-O-Gas[1] (Great Lakes Chemical Corp., West Lafayette, IN). A special unit marketed as Simplex opener (Soil Fumigants Co., Orlando, FL) can be used to puncture the can and release the fumigant. With a boiling point of 5 °C, the liquid vaporizes rapidly. Eno and Popenoe (1964) sterilized 1-kg units of soil by placing polyethylene bags containing pre-incubated moist soil in a desiccator. Following evacuation of the desiccator, methyl bromide was introduced into the desiccator until the pressure equilibrated. The desiccator was sealed for 24 h and then evacuated until the fumigant was removed. Ferriss (1984) fumigated 45 kg of soil in a polyethylene bag in a 76-L (20-gallon) garbage can into which he released 454 g of methyl bromide-chloropicrin (98%–2%).

3–6.3 Comments

Ethylene oxide, propylene oxide, and methyl bromide are effective soil-sterilizing agents. The length of exposure time for sterilization depends upon the concentration of the sterilant, temperature, and initial microbial population of the sample (Caputo & Odlaug, 1983; Korczynski, 1981). Because of the highly flammable nature of pure ethylene oxide, commercially available sterilants use a combination of ethylene oxide and dichlorodifluoromethane or CO_2 (Pusino et al., 1990; Korczynski, 1981).

---

[1]Trade name included for the reader's benefit and does not indicate preferential endorsement over similar products.

Both ethylene and propylene oxides can result in substantial increases in soil pH. The increase may be as large as 2 pH units in acid soils with high organic matter content (Clark, 1950; Lopes & Wollum, 1976; Skipper & Westermann, 1973; Wolf et al., 1989). The pH increase has been related to the esterification of carboxyl groups in soil organic matter (Bartlett & Zelazny, 1967).

Propylene oxide also interferes with the determination of soil surface area when the ethylene glycol monoethyl ether procedure is used and, thus, results in erroneously low surface area measurements (Wolf et al., 1989). Ethylene oxide has been reported to be converted to ethylene glycol (Greenhalgh, 1978) and polyethylene glycol (Pusino et al., 1990) when used to sterilize soil.

Binding of the oxide and glycol to the soil would be expected to influence adsorption of organic chemicals. Griffiths (1987) reported that the residual effects of ethylene oxide fumigation should be carefully evaluated before the fumigated soil is used in subsequent experiments involving plant growth or inoculation with microorganisms.

Propylene oxide has also been shown to increase extractable Mn levels and decrease extractable Fe levels in most soils. Other macro- and micronutrient levels have been largely unaffected by propylene oxide treatment. However, measured soil organic matter levels are increased by treatment with ethylene or propylene oxides, most likely as the result of adsorption of the oxides by organic and inorganic soil colloids (Allison, 1951; Lopes & Wollum, 1976).

Methyl bromide has been used to selectively eliminate groups of microorganisms and to sterilize soil (Eno & Popenoe, 1964; Thompson, 1990). Because methyl bromide decomposes to methanol and bromide, elevated Br^- levels from fumigation can result in plant phytotoxicity problems (Hoffmann & Malkomes, 1979; Thompson, 1990; Van Berkum & Hoestra, 1979). Fumigation with methyl bromide results in increased extractable organic C in soil (Powlson & Jenkinson, 1976) but does not increase extractable Mn levels (Lopes & Wollum, 1976). Electrical conductivity would also be expected to increase in soils fumigated with methyl bromide.

3–7 NONGASEOUS COMPOUNDS

3–7.1 Mercuric Chloride

3–7.1.1 Materials

1. Mercuric chloride.
2. Glass or polyethylene flask or culture tube.

3–7.1.2 Procedures

The amount of mercuric chloride ($HgCl_2$) needed to inhibit soil microbial activity depends upon the clay and organic matter content. Levels of 500 to 20 000 mg of $HgCl_2$ kg^{-1} of dry soil have been used, but a level of 1.84 mmol (500 mg) $HgCl_2$ kg^{-1} of soil appears adequate (Fletcher & Kaufman, 1980; Rozycki & Bartha, 1981; Wolf et al., 1989). The level of metabolic inhibitor required for a given soil should be evaluated on a case by case basis. The $HgCl_2$ can be added directly to the moist soil as a dry chemical or in a concentrated solution, thoroughly mixed, and incubated 48 h. The soil container should be covered with foil or paper.

3–7.1.3. Comments

In long-term incubation studies such as those often used to separate biological from chemical degradation of organic materials, contamination of previously sterilized soil is a problem. For such studies, it is often advantageous to use a metabolic inhibitor such as $HgCl_2$.

Because of the toxicity of mercury to humans, care should be exercised in using it for soil treatment (Grier, 1983). Safe disposal of the treated soil must be practiced. Under anaerobic conditions, the potential could exist for formation of highly toxic methylated mercury forms. Thus, the use of $HgCl_2$ as a microbial inhibitor in anaerobic studies is not recommended. It is obvious that soil treated with $HgCl_2$ cannot be used in subsequent inoculation or plant growth studies.

Of the methods evaluated, $HgCl_2$ produces the fewest changes in soil chemical and physical properties. Surface area, caption exchange capacity, pH, extractable Mn, Fe, Al, and macro- and micronutrients were not significantly affected (Wolf et al., 1989).

Phenylmercuric acetate has been used as an antimicrobial agent in leather, paper, and wood (Grier, 1983) and has also been used to inhibit soil enzymatic activity (Bremner, 1982).

3–7.2 Azide

3–7.2.1 Materials

1. Sodium or potassium azide.
2. Glass or polyethylene flask or culture tube.

3–7.2.2 Procedures

Azide has been used as an effective microbial inhibitor, but it is not a sterilizing agent. To moist pre-incubated soil, 3 mmol of NaN_3 (200 mg) or KN_3 (250 mg) kg^{-1} soil is added and thoroughly mixed. The soil is covered with foil or paper and incubated at 25 °C for 48 h. **CAUTION:** Azide represents an explosion hazard!

3–7.2.3 Comments

Rozycki and Bartha (1981) concluded that NaN_3 or KN_3 was not a suitable inhibitor of soil respiration because the azide produced erroneous

CO_2 evolution values and increased pH values. Working with three soils, Wolf et al. (1989) showed that NaN_3 significantly increased soil pH and Na content. The pH increase was the result of the conversion of azide to hydrazoic acid (Rozycki & Bartha, 1981), and the pH increase would be determined by the buffering capacity of the soil. Azide treatment would not influence soil cation exchange capacity, surface area, or extractable nutrient levels other than Na or K added with the azide. However, extractable Mn levels have been shown to increase following azide treatment even though the pH increased.

Kaufman et al. (1968) used KN_3 at 250 mg kg^{-1} soil and found that inhibition of microbial activity was nearly complete. Total CO_2 evolution was not significantly different among soil treated with azide, ethylene oxide, or autoclaved.

At 400 or 800 mg NaN_3 kg^{-1} soil, Skipper and Westermann (1973) noted that bacterial plate counts were not reduced, but fungal counts were reduced 60 to 70%. They also reported a pH increase in treated soil.

3–8 CONCLUSIONS

Several methods are available to sterilize soil or inhibit microbial activity. Regardless of the method used, it is necessary to experimentally verify that the soil microbial population was eliminated by the sterilization method. Many of the materials used in soil sterilization represent serious hazards to the researcher and extreme caution must be exercised in using the chemicals.

Most sterilization procedures result in some degree of alteration of soil chemical or physical properties. The researcher should determine which method results in the smallest changes in the major properties of interest and how those changes may impact the results of the specific research to be conducted. In studies not requiring inoculation of microorganisms or plant growth, $HgCl_2$ appears to be the best overall treatment to prevent microbial activity in aerobic soils and results in the smallest changes in soil chemical and physical properties. Where inoculation or plant growth is required, autoclaving, dry heat, irradiation, and gaseous fumigants can be used. The changes in soil properties produced by any of the methods should be clearly recognized and considered in subsequent research.

REFERENCES

Allison, L.E. 1951. Vapor-phase sterilization of soil with ethylene oxide. Soil Sci. 72:341–352.

Bartlett, R.J., and L.W. Zelazny. 1967. A simple technique for preparing and maintaining sterile soils for plant studies. Soil Sci. Soc. Am. Proc. 31:436–437.

Beloian, A. 1983. Methods of testing for sterility: Efficacy of sterilizers, sporicides, and sterilizing processes. p. 885–917. In S.S. Block (ed.) Disinfection, sterilization, and preservation. 3rd ed. Lea and Febiger, Philadelphia.

Bremner, J.M. 1982. Nitrogen-urea. p. 699–709. In A.L. Page et al. (ed.) Methods of soil analysis. Part 2. 2nd ed. Agron. Monogr. 9. ASA and SSSA, Madison, WI.

Caputo, R.A., and T.E. Odlaug. 1983. Sterilization with ethylene oxide and other gases. p. 47–64. In S.S. Block (ed.) Disinfection, sterilization, and preservation. 3rd ed. Lea and Febiger, Philadelphia.

Cawse, P.A. 1975. Microbiology and biochemistry of irradiated soils. p. 213–267. In E.A. Paul and A.D. McLaren (ed.) Soil biochemistry. Vol. 3. Marcel Dekker, New York.

Clark, F.E. 1950. Changes induced in soil by ethylene oxide sterilization. Soil Sci. 70:345–349.

Davis, R.D. 1975. Bacteriostasis in soils sterilized by gamma irradiation and in reinoculated sterilized soils. Can. J. Microbiol. 21:481–484.

Eno, C.F., and H. Popenoe. 1964. Gamma radiation compared with steam and methyl bromide as a soil sterilizing agent. Soil Sci. Soc. Am. Proc. 28:533–535.

Ferriss, R.S. 1984. Effects of microwave oven treatment on microorganisms in soil. Phytopathology 74:121–126.

Fletcher, C.L., and D.D. Kaufman. 1980. Effect of sterilization methods on 3-chloroaniline behavior in soil. J. Agric. Food Chem. 28:667–671.

Greenhalgh, R. 1978. IUPAC commission on terminal pesticide residues. J. Assoc. Official Anal. Chem. 61:841–868.

Grier, N. 1983. Mercurials—Inorganic and organic. p. 346–374. In S.S. Block (ed.) Disinfection, sterilization, and preservation. 3rd ed. Lea and Febiger, Philadelphia.

Griffiths, B.S. 1987. Growth of selected microorganisms and plants in soil sterilized by ethylene oxide or gamma-irradiation. Soil Biol. Biochem. 19:115–116.

Hoffmann, G.M., and H.P. Malkomes. 1979. The fate of fumigants. p. 291–335. In D. Mulder (ed.) Soil disinfestation. Elsevier Sci. Publ. Co., Amsterdam, the Netherlands.

Jackson, N.E., J.C. Corey, L.R. Frederick, and J.C. Picken, Jr. 1967. Gamma irradiation and the microbial population of soils at two water contents. Soil Sci. Soc. Am. Proc. 31:491–494.

Jakobsen, I. 1984. Mycorrhizal infectivity of soils eliminated by low doses of ionizing radiation. Soil Biol. Biochem. 16:281–282.

Jenkinson, D.S. 1966. Studies on the decomposition of plant material in soil. II. Partial sterilization of soil and the soil biomass. J. Soil Sci. 17:280–302.

Jenneman, G.E., M.J. McInerney, M.E. Crocker, and R.M. Knapp. 1986. Effect of sterilization by dry heat or autoclaving on bacterial penetration through Berea sandstone. Appl. Environ. Microbiol. 51:39–43.

Joslyn, L. 1983. Sterilization by heat. p. 3–46. In S.S. Block (ed.) Disinfection, sterilization, and preservation. 3rd ed. Lea and Febiger, Philadelphia.

Kaufman, D.D., J.R. Plimmer, P.C. Kearney, J. Blake, and F.S. Guardia. 1968. Chemical versus microbial decomposition of amitrole in soil. Weed Sci. 16:266–272.

Kitur, B.K., and W.W. Frye. 1983. Effects of heating on soil chemical properties and growth and nutrient composition of corn and millet. Soil Sci. Soc. Am. J. 47:91–94.

Korczynski, M.S. 1981. Sterilization. p. 476–486. In P. Gerhardt (ed.) Manual of methods for general bacteriology. Am. Soc. Microbiol., Washington, DC.

Labeda, D.P., D.L. Balkwill, and L.E. Casida, Jr. 1975. Soil sterilization effects on in situ indigenous microbial cells in soil. Can. J. Microbiol. 21:263–269.

Lopes, A.S., and A.G. Wollum. 1976. Comparative effects of methyl bromide, propylene oxide, and autoclave sterilization on specific soil chemical characteristics. Turrialba 26:351–355.

Martin, J.K. 1975. Comparison of agar media for counts of viable soil bacteria. Soil. Biol. Biochem. 7:401–402.

Martin, J.P., W.J. Farmer, and J.O. Ervin. 1973. Influence of steam treatment and fumigation of soil on growth and elemental composition of avocado seedlings. Soil Sci. Soc. Am. Proc. 37:56–60.

McLaren, A.D. 1969. Radiation as a technique in soil biology and biochemistry. Soil Biol. Biochem. 1:63–73.

Peterson, G.H. 1962. Microbial activity in heat- and electron-sterilized soil seeded with microorganisms. Can. J. Microbiol. 8:519–524.

Powlson, D.S., and D.S. Jenkinson. 1976. The effects of biocidal treatments on metabolism in soil. II. Gamma irradiation, autoclaving, air-drying and fumigation. Soil Biol. Biochem. 8:179–188.

Pusino, A., M. Gennari, A. Premoli, and C. Gessa. 1990. Formation of polyethylene glycol on montmorillonite by sterilization with ethylene oxide. Clays Clay Miner. 38:213–215.

Rozycki, M., and R. Bartha. 1981. Problems associated with the use of azide as an inhibitor of microbial activity in soil. Appl. Environ. Microbiol. 41:833–836.

Silverman, G.J. 1983. Sterilization by ionizing irradiation. p. 89–105. *In* S.S. Block (ed.) Disinfection, sterilization, and preservation. 3rd ed. Lea and Febiger, Philadelphia.

Skipper, H.D., and D.T. Westermann. 1973. Comparative effects of propylene oxide, sodium azide, and autoclaving on selected soil properties. Soil Biol. Biochem. 5:409–414.

Speir, T.W., J.C. Cowling, G.P. Sparling, A.W. West, and D.M. Corderoy. 1986. Effects of microwave radiation on the microbial biomass, phosphatase activity and levels of extractable N and P in a low fertility soil under pasture. Soil. Biol. Biochem. 18:377–382.

Thien, S.J., D.A. Whitney, and D.L. Karlen. 1978. Effect of microwave radiation drying on soil chemical and mineralogical analysis. Commun. Soil Sci. Plant Anal. 9:231–241.

Thompson, J.P. 1990. Soil sterilization methods to show VA-mycorrhizae aid P and Zn nutrition of wheat in vertisols. Soil Biol. Biochem. 22:229–240.

Van Berkum, J.A., and H. Hoestra. 1979. Practical aspects of the chemical control of nematodes in soil. p. 53–134. *In* D. Mulder (ed.) Soil disinfestation. Elsevier Sci. Publ. Co., Amsterdam, the Netherlands.

Vela, G.R., and J.F. Wu. 1979. Mechanism of lethal action of 2,450-MHz radiation on microorganisms. Appl. Environ. Microbiol. 37:550–553.

Wainwright, M., K. Killham, and M.F. Diprose. 1980. Effects of 2450 MHz microwave radiation on nitrification, respiration and S-oxidation in soil. Soil Biol. Biochem. 12:489–493.

Wolf, D.C., T.H. Dao, H.D. Scott, and T.L. Lavy. 1989. Influence of sterilization methods on selected soil microbiological, physical, and chemical properties. J. Environ. Qual. 18:39–44.

Chapter 4

Soil Water Potential

K. J. McINNES AND **R. W. WEAVER,** *Texas A&M University, College Station, Texas*

M. J. SAVAGE, *University of Natal, Pietermaritzburg, South Africa*

Soil biochemists and microbiologists readily agree that it is important to measure soil water potential. The importance of measuring water potential along with water content in describing soil conditions influencing microbial activity was the topic of a symposium that resulted in a Soil Science Society of America Publication (Parr et al., 1981). The ability of soil microorganisms to physiologically adjust to changes in soil water potential largely determines their ability to maintain activity and survive during periods of water stress. Whereas soil water content, aside from its relation to water potential, influences the mobility of microbial cells in soil, and the diffusion of essential or toxic gases and solutes to and from microbial cells.

Commonly used methods for measuring soil water potential require specialized equipment, training and considerable effort. A gravimetric method, however has been available for many years and does not require specialized equipment, training, or considerable effort (Hansen, 1926; Gardner, 1937) but is accurate enough for many applications (Fawcett & Collis-George, 1967; McQueen & Miller, 1968; Al-Khafaf & Hanks, 1974; Hamblin, 1981; Savage et al., 1992). The method entails measuring the water content of filter paper equilibrated with soil. The method is described in this chapter to bring it to the attention of soil microbiologists and biochemists in the hope that it will facilitate more frequent measurement of soil water potential. The method may be used in both the laboratory, as we describe here, and in the field (Savage et al., 1992). Other more accurate but more meticulous methods for measuring soil water potential along with theory are described by Rawlins and Campbell (1986), Campbell and Gee (1986), and Klute (1986).

4–1 PRINCIPLES

The technique is based on the premise that under isothermal conditions, filter paper confined in an air-tight container with soil will equilibrate with the soil and reach the same water potential. The mass of water contained in the soil sample must be large enough so that water transferred between the soil and paper does not significantly alter the soil water potential. After equilibration of the filter paper with the soil, it is relatively straightforward to gravimetrically measure the water content of the filter paper. The water potential is then determined from the water potential-water content relationship of the filter paper.

Variability of the relationship between water content and water potential for filter paper within a batch is reported to be small (Sibley et al., 1990). It would be convenient if the relationship were similar between batches so that a calibration curve would not have to be determined by different researchers. Water potential-water content relationships of ashless Whatman no. 42 filter paper have been determined by several researchers (Fawcett & Collis-George, 1967; Hamblin, 1981; Chandler & Gutierrez, 1986; Greacen et al., 1987; Sibley & Williams, 1990; Savage et al., 1992). For water potentials < -0.03 kJ kg^{-1}, drier than field capacity, little difference existed between water potential-water content relationships from these individual research efforts. Because there are many methods of determining the water potential-water content relationship, all with their own problems, differences between batches may be smaller than differences in methods (Savage et al., 1992). Relationships from Fawcett and Collis-George (1967), Hamblin (1981), and Greacen et al. (1987), more than likely on different batches, were essentially the same. The greatest relative differences in the relationships were for water potentials > -0.03 kJ kg^{-1}, where the larger pores of the paper are most influential. Relationships between water content and water potential have also been determined for Schleicher and Schuell no. 589 White Ribbon filter paper (Gardner, 1937; McQueen & Miller, 1968; Al-Khafaf & Hanks, 1974). When water potential-water content relationships (< -0.03 kJ kg^{-1}) from the literature for Whatman no. 42 are averaged, they produce essentially the same curve as the average for Schleicher and Schuell no. 589. A mean curve from the relationships of the above researchers (from relationships reported by Savage et al., 1992) along with that from Sibley and Williams (1990) is presented (Fig. 4–1).

Other filter papers could be used. Millipore filtration membranes MF 0.05 μm and MF 0.025 μm have increased sensitivities (changes in water content to changes in water potential) between -0.03 J kg^{-1} and -1 kJ kg^{-1}, but this advantage is almost negated by an increased variability in the water content measurement (Sibley & Williams, 1990).

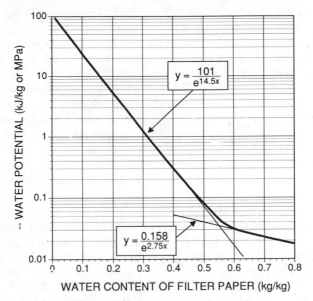

Fig. 4–1. Water potential-water content relationship for ashless Whatman no. 42 or Schleicher and Schuell no. 589 White Ribbon filter paper.

4–2 MATERIALS

1. Whatman no. 42 ashless filter paper or Schleicher and Schuell no. 589 White Ribbon filter paper of 70-mm diam.
2. 80-mm diam. cans with lids (soil-water-content can). The cans should be at least 40 mm deep.
3. Roll of plastic tape used for electrical connections.
4. Balance weighing to 0.1 mg.
5. Tweezers for handling filter paper.
6. Fine-haired brush.
7. Insulated box for incubating samples.
8. Oven for drying samples at 105 °C.

4–3 PROCEDURE

Collect the soil sample and place it in a soil-water-content can before significant water loss occurs. The can should not be completely filled. It is desirable to leave at least 10 mm of space above the soil. Using the tweezers, place a filter paper on top of the soil. Immediately close the container with its lid and seal the joint with plastic electrical tape. It is not necessary or even desirable to sieve the soil before placing it in the container. Sieving field samples would result in water loss and in the case of very wet samples possibly alter the water potential by disruption of the size of soil pores. Place the sample upside down in an insulated container and store at room temperature to allow time for the water in the filter paper to equilibrate with the soil water. Placing the sample upside down ensures good contact between the filter paper and the soil. Placement of samples in an insulated box reduces temperature gradients in the sample containers that may cause condensation and result in inappropriate mass measurements. Storage for one day is adequate for equilibration of relatively wet (-0.1 kJ kg^{-1}) samples but up to 3 d may be needed for relatively dry samples (-20 kJ kg^{-1}).

After storage, open the lid to the insulated box, remove a container, close the box, invert the container, and remove the filter paper using the tweezers. The inclusion of the 10-mm space above the soil in the can at the time of sampling allows the filter paper to fall away from the lid when it is inverted. Do not place the container on a lab bench before removing the filter paper. Temperatures one or two degrees cooler than the insulated box may result in water condensing on the inside walls of the can and absorption by the paper. If the filter paper has some adhering soil, quickly brush the soil off and immediately place the filter paper in an oven-dried can and replace the lid. Only a few seconds may be allowed for this process because water is being lost from the filter paper by evaporation. Weigh the wet filter paper and can combination, remove the lid, and place both in an oven set at 105 °C for a few hours. After drying, replace the lid on the can to prevent water being taken up by the filter and allow the can to cool for a few minutes before weighing. Weigh the dry filter paper and can combination. Calculate the water content of the equilibrated filter paper using Eq. [1].

water content (kg kg^{-1}) =

$$\frac{(\text{wet mass} + \text{can mass}) - (\text{dry mass} + \text{can mass})}{(\text{dry mass} + \text{can mass}) - \text{can mass}} \qquad [1]$$

Because the can may absorb a significant amount of water in relation to the amount of water in the filter paper, it is important that the can mass be determined after it has been oven-dried.

Calculate the water potential of the filter paper using the solid dark line or the appropriate equation in Fig. 4–1. Greater uncertainty exists in the water potential-water content relationship for water contents > 0.6 kg kg^{-1} than < 0.6 kg kg^{-1}. We do not recommend the extrapolation of water potentials from the relationships in Fig. 4–1 for water contents of filter paper > 0.8 kg kg^{-1}. Further, soil water potential associated with water contents wetter than field capacity is not as meaningful to the survival and activity of microorganisms as water content (Parr et al., 1981).

4–4 COMMENTS

We propose the use of a single filter paper, but there is some merit in using two filter papers layered on top of each other. If two filter papers are used, the filter paper next to the soil could be discarded and the "clean" paper used for the water content determination. Disadvantages of using two filter papers are that a longer time would be required for equilibration with the soil and more water would be absorbed into the filter papers from the soil that would reduce the water content of the soil and decrease the water potential. Actually, the situation when two filter papers would be useful is at high soil water content. Under such conditions, equilibration time is rapid and the amount of water absorbed by the filter paper is negligible compared to the amount of soil water.

Another variable in the methodology is to measure only the soil matric potential or the combined matric and osmotic potentials. When the filter paper is in contact with the soil, the measured potential is predominantly the matric potential. If the paper is kept from contacting the soil and water potential of the paper equilibrates with water potential of the soil by vapor diffusion then the measured water potential is a result of both soil matric and osmotic potential (Al-Khafaf & Hanks, 1974). The disadvantage of not allowing the filter paper to contact the soil is that the time for equilibration is greatly extended. For nonsaline soils, the osmotic potential does not contribute enough to the total water potential to obtain significant differences between measurements for filter paper in contact with soil or kept from contacting soil.

It is not advisable to cut the filter papers because this may introduce additional errors. Cutting the paper and undue handling influences its absorption capabilities (Savage et al., 1992). Paper is available in various sizes that preclude the need for cutting. It is important to use a large enough filter paper mass to absorb adequate water to minimize the errors associated with gravimetric determinations (i.e., the accuracy of the balance. A 70-mm diam. piece of Whatman no. 42 filter paper has a mass of approximately 0.36 g and can absorb nearly double that mass in water. To keep errors in water potential from water loss from the soil to the filter paper to a minimum, we suggest using a soil mass to filter paper area ratio of at least 20 kg m^{-2}.

REFERENCES

Al-Khafaf, S., and R.J. Hanks. 1974. Evaluation of the filter paper method for estimating soil water potential. Soil Sci. 117:194–199.

Campbell, G.S, and G.W. Gee. 1986. Water potential: Miscellaneous methods. p. 619–633. *In* A. Klute (ed.) Methods of soil analysis. Part 1. 2nd ed. Agron. Monogr. 9. ASA and SSSA, Madison, WI.

Chandler, R.J., and C.I. Gutierrez. 1986. The filter-paper method of suction measurement. Geotechnique 36:265–268.

Fawcett, R.G., and N. Collis-George. 1967. A filter paper method for determining the moisture characteristic of soil. Aust. J. Exp. Agric. Anim. Husb. 7:162–167.

Gardner, R. 1937. A method of measuring the capillary tension of soil moisture over a wide moisture range. Soil Sci. 43:277–283.

Greacen, E.L., G.R. Walker, and P.G. Cook. 1987. Evaluation of the filter paper method for measuring soil water suction. p. 137–143. *In* International conference on measurement of soil and plant water status. Vol. 1 Utah State Univ., Logan.

Hamblin, A.P. 1981. Filter paper method for routine measurement of field water potential. J. Hydrol. 53:355–360.

Hansen, H.C. 1926. The water-retaining power of soil. J. Ecol. 14:111–119.

Klute, A. 1986. Water retention: Laboratory methods. p. 635–662. *In* A. Klute (ed.) Methods of soil analysis. Part 1. 2nd ed. Agron. Monogr. 9. ASA and SSSA, Madison, WI.

McQueen, I.S., and R.F. Miller. 1968. Calibration and evaluation of a wide-range gravimetric method for measuring moisture stress. Soil Sci. 106:225–231.

Parr, J.F., W.R. Gardner, and L.F. Elliott. 1981. Water potential relations in soil microbiology. Soil Sci. Soc. Am. Spec. Publ. 9. SSSA, Madison, WI.

Rawlins, S.L., and G.S. Campbell. 1986. Water potential; thermocouple psychrometry. p. 597–618. *In* A. Klute (ed.) Methods of soil analysis. Part 1. 2nd ed. Agron. Monogr. 9. ASA and SSSA, Madison, WI.

Savage, M.J., I.N. Khuvutlu, and H. Bohne. 1992. Estimating water potential of porous media using filter paper. South Afr. J. Sci. 88:269–274.

Sibley, J.W., G.K. Smyth, and D.J. Williams. 1990. Suction moisture content calibration of filter papers from different boxes. ASTM Geotech. Test. J. 13:257–262.

Sibley, J.W., and D.J. Williams. 1990. A new filter material for measuring soil suction. ASTM Geotech. Test. J. 13:381–384.

Chapter 5

Most Probable Number Counts

PAUL L. WOOMER, *Tropical Soil Biology and Fertility Programme, Nairobi, Kenya*

The most probable number (MPN) technique is used to estimate microbial population sizes when quantitative assessment of individual cells is not possible. Exact cell numbers of an individual organism frequently cannot be measured in soils due to heterogeneous populations or unavailability of a suitable diagnostic media. The MPN technique relies on the detection of specific qualitative attributes of the microorganism of interest. Population estimates are derived from the pattern of attribute occurrence across a serial dilution from MPN tables that are based on the mathematical approaches of Halvorson and Ziegler (1933). Cochran (1950) later contributed procedures that estimate error and calculate confidence limits. Recently, computer software was developed that combine these operations to increase the flexibility of MPN procedures (Woomer et al., 1990). Others have addressed the acceptability of experimental results and developed tests of experimental technique (Halvorson & Moeglein, 1940; Woodward, 1957; deMan, 1975; Stevens, 1958; Scott & Porter, 1986; Woomer et al., 1988).

An important aspect of the MPN methodology is the ability to estimate a microbial population size based on a process-related attribute. For example, all soil organisms capable of infecting the root system and forming root nodules with a given legume host can be measured when plant infection methodology (Brockwell, 1963) is linked to MPN approaches (Brockwell et al., 1975; Vincent, 1970; Weaver & Frederick, 1972). This is in contrast to direct counts of an individual strain or serotype recovered from a soil via filtration procedures and then enumerated microscopically (Schmidt et al., 1968). A distinct advantage of MPN methodology over filter counts is the more uniform recovery of a microbial population across diverse soil types (Kingsley & Bohlool, 1981; Woomer et al., 1988). Furthermore, the detection of an organism through process-related attributes often results in the recovery of mixed populations with similar functional roles in soils. These populations may then be separated into individual colonies for more detailed study.

This chapter addresses the appropriate use of MPN methodology, assists in the selection of an experimental design offering sufficient resolution to distinguish differences between population estimates, presents a generalized MPN methodological guideline, and provides several tables that allow population estimates to be obtained from experimental results. A unique feature of this discussion is the attention given to the reliability of experimental results, including the criteria for discarding data that do not comply with the principle assumptions that underlie use of this technique. Specific application of MPN methodology may be acquired from other chapters of this volume through the selection of specific culture media that differentiates soil microorganisms based upon their various functional roles in soil.

5–1 PRINCIPLES

5–1.1 Theoretical Assumptions

The MPN technique is a means to estimate microbial population sizes in a liquid substrate. This method does not rely upon the population size within any single dilution or aliquot, but rather on the pattern of positive and negative test results following inoculation of a suitable test medium (Alexander, 1982). These results are then used to derive a population estimate based on the mathematics of Halvorson and Ziegler (1933). These authors describe three different classes of MPN methodology. Two early approaches included the incubation of many subsamples from a single dilution, and serial dilution and incubation of an unreplicated test sample. These two approaches, when combined, result in the third class, which is dilution and incubation of replicated cultures across several serial dilution steps. This combined technique is explained here in detail.

Underlying the mathematical solution for the MPN are two key assumptions (Cochran, 1950). It is assumed that organisms in the initial and all subsequent dilutions are randomly distributed. To satisfy this assumption, the test substrate must be well dispersed throughout the initial and subsequent dilutions. Secondly, it is assumed that one or more organisms contained within an inoculant volume are capable of producing a positive result. Failure to satisfy the second assumption often results in unusual patterns of positive and negative results such as lower dilutions showing some negative results across replicates followed by higher dilutions that are entirely positive. An ancillary assumption underlying the MPN technique is that all test organisms occupy a similar volume.

5–1.2 The Mathematical Solution of the Most Probable Number

The following is the general equation for determining the MPN of organisms in a substrate that has been serially diluted and several units inoculated at each dilution level (Halvorson & Ziegler, 1933):

$$[a_1p_1/(1-e^{-a1x})] + \dots + [a_kp_k/(1-e^{-akx})] = a_1n_1 + \dots + a_kn_k \quad [1]$$

where
  a = the dilution level of each dilution,
  n = the number of inoculated units at each dilutions level,
  p = the number of positive units within each dilution level,
  k = the highest dilution level of the series,
  e = the base of the natural logarithm.

x yields the MPN. A special case solution for a six-step serial dilution based on Brockwell and co-workers (1963, 1975) was modified by Woomer et al. (1990) as follows:

$$[Rp_1/(1-e^{-Rx})] + \dots + \{[(1/R^4)p_6]/(1-e^{(R/4)x})\} =$$

$$N(R+1+R^{-1}+\dots+R^{-4}) \quad [2]$$

where
  R = the dilution ratio,
  N = the number of units per dilution, and
  p_1 through p_6 = the number of positive units in dilutions 1 through 6.

Upon solving for x, the MPN of the first dilution level is R^2x.

Needless to say, the mathematical solution for any given MPN is time consuming. For this reason, MPN tables have been developed that yield estimates for a given experimental design and pattern of positive results. Publications that include various MPN tables are listed in Table 5–1.

5–1.3 Confidence Limits and Population Estimate Separation

The standard error, confidence limit and tests of significance between two population estimates were established by Cochran (1950). The base ratio and the number of units per dilution level are used in the calculation of the confidence factor (CF) as follows:

$$CF_{0.05} = \text{antilog}_{10}\{2 \times 0.55 \times \sqrt{[(log_{10}a)/n]}\} \text{ when } a < 10 \text{ and}$$

$$= \text{antilog}_{10}\{2 \times 0.58 \times \sqrt{[(log_{10}a)/n]}\} \text{ when } a \geq 10 \quad [3]$$

where
  a = the dilution ratio
  n = the number of units per dilution level.

The confidence factor is multiplied by and divided into the population estimate to establish the estimate's upper and lower limits. Decreases in the base dilution ratio or increases in the number of replicate plants per dilution result in a reduction of the confidence factor and greater resolution of the population estimate. Alternatively, increases in the base dilution ratio or decreases in replication result in broader confidence limits and reduced

Table 5–1. Sources of most probable number tables for various dilution series and replication.
(After Woomer et al., 1990.)

Source	Dilution ratio	Replicate no.
Halvorson & Ziegler, 1933	10	10
Woodward, 1957	10	3,5,10
Brockwell, 1963	5	4
Vincent, 1970	2,4,10	2,4
deMan, 1975	10	3,5,10
Brockwell et al., 1975	10	3
Alexander, 1982	10	5
Woomer et al., 1990	12.5	2
Woomer et al., 1990	3	3

ability to significantly distinguish different MPN estimates from one another. The confidence factors derived from different dilution ratio and replicate combinations is presented in Table 5–2 (Cochran, 1950). For two population estimates to be significantly different from one another, the lower limit of the greater population must be larger than the upper limit of the lesser population. This relationship is also useful in planning MPN experimental designs since:

$$\sqrt{(P_G/P_L)} > CF \qquad [4]$$

where P_G and P_L are the greater and lesser population sizes, respectively, and CF is the confidence factor ($P = 0.05$). Alternatively stated, in order for two MPN estimates to differ significantly from one another the **ratio** of the **greater** population estimate to the **lesser** population estimate must be **greater than the square of the confidence factor.** If the ratio of the population estimates for two test samples can be predicted, an experimental design that includes the optimal dilution ratio and the number of replicate units within each dilution level may be identified from Table 5–2. Such a design maximizes the likelihood of distinguishing the two population estimates significantly while minimizing the total number of experimental units.

5–1.4 Reliability of Experimental Results and Tests of Technique

The compliance of an individual set of experimental MPN results with the underlying assumptions of the technique is testable within those results. Not only are some outcomes intuitively more likely than others but every experimental outcome (pattern of results) has a discrete probability. The probability of the occurrence of an MPN result was first identified by Halvorson and Moeglein (1940). Using this approach, researchers can either accept or reject experimental outcomes based on the probability that the particular result would occur when all underlying assumptions of the MPN technique are met. Unfortunately, very few of these probability tables are available, although fairly complete tables for 10-fold dilutions

Table 5-2. Factors for calculating the confidence intervals of most probable number estimates.

Replicates per dilution	Factor for 95% confidence interval at various dilution ratios[†]				
	2	3	4	5	10
1	4.01	5.75	7.14	8.31	14.45
2	2.67	3.45	4.01	4.47	6.61
3	2.23	2.75	3.11	3.40	4.67
3	2.00	2.40	2.67	2.88	3.80
5	1.86	2.19	2.41	2.58	3.30
6	1.76	2.04	2.23	2.37	2.98
7	1.69	1.94	2.10	2.23	2.74
8	1.63	1.86	2.00	2.11	2.57
9	1.59	1.79	1.93	2.03	2.44
10	1.55	1.74	1.86	1.95	2.33

† Confidence factors are multiplied by and divided into the population estimate to establish the upper and lower confidence intervals ($P = 0.05$), respectively. Confidence factors calculated using MPNES software (Woomer et al., 1990). (After Cochran, 1950.)

with 10 units per dilution level were developed by Halvorson and Moeglein (1940) and should be consulted when appropriate.

The probability of an individual experimental outcome's compliance with the underlying assumptions of the MPN technique for fivefold dilutions with four units per dilution are presented in Table 5-3 (Scott & Porter, 1986). Researchers are encouraged to use this table whenever possible as a means of accepting or rejecting lower probability results that are occasionally obtained from MPN investigations. The number of successive, entirely positive dilution levels does not influence the probability of an experimental outcome; rather, the pattern of incompletely positive and entirely negative results provide insight into the reliability of the experimental results. In this way, the experimental outcomes 430100, 443010, and 444301 have the same probability of occurrence ($P = 0.0189$) regardless of their different MPN estimates.

Another useful test of experimental technique was first developed by Stevens (1958). This is the range of transition (ROT). The ROT is the inclusive number of dilution steps between the first not-entirely-positive to the last not-entirely-negative results. This is a direct measure of the experimental compliance with the principle assumptions underlying the MPN procedure. Thus, the smaller the ROT the greater the probability that the data are not significantly different from the theoretical values that would result when all underlying assumptions are met.

The probabilities of the ROT for many dilution series/replicate combinations are presented in Table 5-4 (Stevens, 1958; Scott & Porter, 1986). Stevens (1958) suggests that this test of technique not be applied to individual series until a bulk of results has been examined and that this technique be used to discover and remove procedural deficiencies. Following this, researchers may adopt a rule of rejecting results at $P = 0.001$.

Table 5–3. Expected frequency of relatively common codes for a six-step fivefold dilution series with four replicate units per dilution. (From Scott and Porter, 1986.)

Probability class			
≥0.0005		≥0.01	
Results	Probability	Results	Probability
4 4 4 4 0 0	0.069311	4 4 4 4 0 0	0.069311
4 4 3 4 0 0	0.002614	4 4 4 3 0 0	0.108684
4 4 4 3 0 0	0.108684	4 4 3 3 0 0	0.016419
4 4 3 3 0 0	0.016419	4 4 4 2 0 0	0.134553
4 4 2 3 0 0	0.004858	4 4 3 2 0 0	0.053246
4 4 1 3 0 0	0.001406	4 4 2 2 0 0	0.026264
4 4 4 2 0 0	0.134553	4 4 1 2 0 0	0.011349
4 4 3 2 0 0	0.053246	4 4 4 1 0 0	0.126018
4 3 3 2 0 0	0.001192	4 4 3 1 0 0	0.105657
4 4 2 2 0 0	0.026264	4 4 2 1 0 0	0.083751
4 3 2 2 0 0	0.002314	4 3 2 1 0 0	0.016232
4 4 1 2 0 0	0.011349	4 4 4 1 1 0	0.053458
4 3 1 2 0 0	0.002696	4 4 3 1 1 0	0.024457
4 2 1 2 0 0	0.000914	4 4 2 1 1 0	0.012713
4 4 0 2 0 0	0.003060	4 4 4 0 1 0	0.020544
4 3 0 2 0 0	0.001588	4 4 3 0 1 0	0.018857
4 2 0 2 0 0	0.000867	4 4 2 0 1 0	0.015538
4 4 4 1 0 0	0.126018	4 4 1 0 1 0	0.010145
4 3 4 1 0 0	0.001343		
4 4 3 1 0 0	0.105657	Total frequency	0.907196
4 3 3 1 0 0	0.006520		
4 2 3 1 0 0	0.001213		
4 4 2 1 0 0	0.083751		
4 3 2 1 0 0	0.016232		
4 2 2 1 0 0	0.005155		
4 1 2 1 0 0	0.001549		
4 4 4 1 1 0	0.053458		
4 4 3 1 1 0	0.024457		
4 3 3 1 1 0	0.000602		
4 4 2 1 1 0	0.012713		
4 3 2 1 1 0	0.001181		
4 4 1 1 1 0	0.005657		
4 3 1 1 1 0	0.001389		
4 4 0 1 1 0	0.001554		
4 3 0 1 1 0	0.000824		
4 4 4 0 1 0	0.020544		
4 4 3 0 1 0	0.018857		
4 3 3 0 1 0	0.001245		
4 4 2 0 1 0	0.015538		
4 3 2 0 1 0	0.003131		
4 2 2 0 1 0	0.001007		
4 4 1 0 1 0	0.010145		
4 3 1 0 1 0	0.004755		
4 2 1 0 1 0	0.002496		
4 1 1 0 1 0	0.001116		
4 4 0 0 1 0	0.003957		
4 3 0 0 1 0	0.003690		
4 2 0 0 1 0	0.003063		
4 1 0 0 1 0	0.002008		
4 0 0 0 1 0	0.000786		
Total frequency	0.982946		

Regardless of which test is employed, researchers must be aware that statistical methods allow for tests of experimental technique and, whenever possible, they should assess the quality of their experimental results using these methods. Researchers should establish their own limits for accepting or rejecting experimental results. When $P \leq 0.005$, the probability of a failure in experimental technique and noncompliance with the assumptions of the MPN is $\geq 99.5\%$.

5–1.5 Calibration of MPN Technique

Another important test of experimental technique is by the recovery of known populations. Most probable number estimates can be obtained from pure cultures and compared with the results obtained using other techniques. Also, pure cultures of known population densities can be added to a test substrate and the MPN determined. While pure culture controls are not required for routine MPN determinations, any new MPN procedure should be developed with careful consideration of the experimental results obtained from known population densities.

5–2 METHODOLOGY

5–2.1 Experimental Design

The design of an MPN determination involves: (i) selection of the base dilution ratio (A); (ii) selection of the number of units at each dilution level (N); (iii) determination of an initial dilution and its value (I) and; (iv) selection of the inoculation volume (V). These choices determine the limits of detection of a given assay, the confidence limits of that assay, the required number of units containing diagnostic media, the volume of diluent and mass of test sample necessary, and the need and value of applying correction terms (CT) to the MPN population estimate obtained from the appropriate table. These correction terms are based on the inoculation volumes and whether or not the first dilution is equal to subsequent serial dilutions. Researchers must be careful to design an assay for which reliable tables are available for both population estimates and acceptance of experimental results. All of the MPN estimate tables presented in this chapter (Tables 5–6 to 5–9) assume that 1 mL of inoculant is applied to all test units and the initial dilution is the same as all subsequent dilutions (I = A). The MPN estimate represents the densities in the original test sample.

Researchers are encouraged to array dilution series systematically as there is no strict requirement for randomization of inoculated units. Placing inoculated test units from lowest to highest dilution is the common practice as this reduces the likelihood of contamination and researcher errors. When large numbers of plant infection tests are conducted in a growth chamber or glasshouse, it is recommended that researchers randomize the

Table 5–4. A test of technique of dilution series results: The expected frequency of equalling or exceeding the Range of Transition. (From Stevens, 1957; Scott & Porter, 1986.)

Range†	Probability of compliance with assumptions (test of technique‡)			
	Dilution ratio			
	2	4	5	10
	Two replicates per dilution level			
1	0.930	0.717	0.660	0.525
2	0.820	0.373	0.281	0.114
3	0.625	0.123	0.075	0.013
4	0.415	0.034	0.015	0.001
5	0.246	0.009	0.003	0.0001
6	0.136	0.002	0.0006	0.00001
	Three replicates per dilution level			
1	NA§	0.891	0.851	0.731
2	NA	0.511	0.435	0.193
3	NA	0.209	0.123	0.023
4	NA	0.060	0.027	0.002
5	NA	0.015	0.005	0.0002
6	NA	0.004	0.001	0.00002
	Four replicates per dilution level			
1	NA	0.955	0.931	0.838
2	NA	0.682	0.561	0.271
3	NA	0.294	0.178	0.035
4	NA	0.088	0.040	0.004
5	NA	0.023	0.008	0.0004
6	NA	0.006	0.002	0.00004
	Five replicates per dilution level			
1	NA	0.981	0.967	0.899
2	NA	0.777	0.661	0.340
3	NA	0.372	0.233	0.047
4	NA	0.118	0.054	0.005
5	NA	0.031	0.011	0.0005
6	NA	0.008	0.002	0.00005

† The range of transition is the number of dilution steps between the first not entirely positive dilution step to the last non-negative dilution step.
‡ Frequency distributions for threefold dilution series are not available.
§ NA, frequency distribution not available.

placement of the individual treatments (e.g., different test samples) and place adjacent treatments so that lowest dilutions of different test samples are adjacent. This reduces the likelihood of cross contamination between different experimental units. Uninoculated controls are not required to obtain an MPN estimate but should be inserted among inoculated cultures as an additional test for cross contamination.

5–2.2 Materials

The following materials are required to perform the MPN determination illustrated in Fig. 5–1. This example is a fivefold dilution series rep-

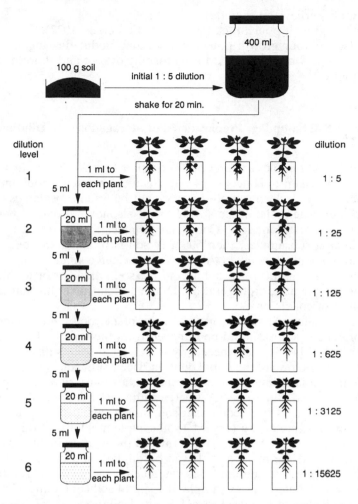

Fig. 5–1. Procedure and example results of an MPN determination with 6 fivefold dilution steps and four units per dilution level.

licated four times at each dilution level and used to inoculate a plant infection assay enumerating soil rhizobia. The exact amount and nature of materials required for other MPN determinations vary with the experimental design and the microorganism of interest.

1. 100 g (dw) soil.
2. One sterile 1 L wide-neck Erlenmeyer flask and stopper containing 400 mL of sterile -N mineral nutrient diluent (0.125 g K_2HPO_4 and 0.05 g $MgSO_4 \cdot 7H_2O$ in 1000 mL of distilled water).
3. One sterile 5-mL wide-mouth pipette.
4. Five sterile 5-mL pipettes.

5. One sterile 1-mL pipette.
6. One growth pouch rack (Weaver & Frederick, 1972).
7. Twenty-four growth pouches containing nodulating legume seed-lings previously selected for uniformity of root and shoot development.

5–2.3 Soil Sampling, Preparation and Storage Prior to Dilution

Collect soil samples as described in chapter 1 by Wollum in this book. Care must be taken to collect soils with clean soil sampling equipment. It is not necessary to clean sampling equipment between subsamples but all sampling equipment should be sprayed with alcohol and flamed between different sites or treatments. Composite samples are prepared by thoroughly mixing subsamples, then 500 g of soil is retained and excess soil discarded. Care must be taken to store soils in clean containers that do not contaminate the samples. Do not air dry the soil sample prior to dilution. Store the sample under refrigeration. Dilute and inoculate the soil sample within 48 h of collection.

Prior to dilution, stones, gravel, large plant roots, and nondecomposed organic residues should be removed from the sample. Crush larger soil aggregates, but it is not necessary to sieve the soil as dispersion will occur during agitation of the initial dilution. Use a sufficiently large volume of soil in the initial dilution to ensure representativeness of the sample. For example, a 10-fold dilution series is better initiated by dilution of 100 g (dry wt.) of soil obtained from a composite soil sample in 900 mL of diluent than by dilution of a single 10 g (dry wt.) soil sample in 90 mL of diluent.

The soil moisture content is determined by oven drying a subsample overnight so that the moist sample can be diluted on a dry weight basis. After the soil moisture content is known, an equivalent volume is removed from the initial diluent. For example, if 135 g of soil containing 35% H_2O (= 100 g soil dry wt.) is used in a 1:5 initial dilution, 35 mL should be removed from the initial diluent (= 135 g soil in 365 mL of diluent). This ensures the consistency between the initial dilution and subsequent dilution steps.

Alternatively, soils of unknown moisture content may be diluted and the moisture content of a subsample later determined. In this case, the final experimental results are corrected by treating the first dilution as an initial dilution (I) with a dilution ratio different from subsequent dilutions. For example, if a soil containing 50% moisture on a dry weight basis (0.333 H_2O and 0.667 dry soil) is diluted 10-fold without correction for the moisture content, the initial dilution (I) is actually 1:15 (= 10/0.667) and the resultant population estimate obtained from the appropriate MPN table should be corrected for this moisture through multiplication by 1.5 (= 15/10). In this way, researchers can subsequently adjust for unknown amounts of moisture in a test sample.

5–2.4 Preparation of the Dilution Series and Culture of Inoculated Test Units

1. Place 100 g (dry wt.) soil into a large-mouthed jar or flask containing 400 mL of sterile diluent. The volume of diluent may be adjusted in advance to account for the moisture contained in the soil sample. Remember that the initial dilution of a 2:1 dilution series contains 100 g soil (dry wt.) and 100 mL of diluent, a 5:1 dilution contains 100 g soil (dry wt.) and 400-mL diluent and a 10:1 dilution contains 100 g soil (dry wt.) and 900 mL diluent. As was stated earlier, do not prepare the initial dilution by placing relatively small amounts of soil into relatively few milliliters of diluent (e.g., 1 g soil into 9 mL diluent for a 10:1 dilution) as small quantities of soil are more likely to be nonrepresentative. Low concentrations of mineral salts are used as diluents instead of distilled water to avoid differences in osmotic potential across the dilution series.

2. Mix the soil and diluent using a wrist action shaker set at a high agitation level for 20 to 25 min. When completed, this is either the first dilution level (A_1) or the initial dilution (I). Further dilution of this initial dilution is acceptable to bring the anticipated MPN estimates within range of the six-step dilution assay.

3. Transfer 5 mL of the initial dilution to a sterile test tube containing 20 mL of diluent using a sterile, wide-mouth pipette. The soil must not settle to the bottom of the vessel before the aliquot is withdrawn. Close the test tube and agitate using a vortex mixer or by rapid hand shaking.

4. Continue the dilution series to 5^{-6} using sterile, narrow-mouth pipettes for each dilution step. Again, take care to avoid sedimentation before withdrawal of the transferred aliquots. These are the second through sixth dilution levels.

5. Apply 1.0 mL directly to the legume root systems in each growth pouch using a sterile 1.0 mL pipette. A single pipette can be used for the entire dilution series of each test substrate if test units are inoculated in sequence from the highest (5^{-6}) to the lowest (5^{-1}) dilution. With each incrementally lower dilution, the pipette must be rinsed twice before transferring the 1.0 mL of inoculant to the culture units by drawing the dilution into the pipette and releasing it back into the test tube before inoculation.

6. Place growth pouches in a clean growth chamber or glasshouse. Check for the appearance of microbial presence (root nodules) daily. In general, results for microbial cultures can be recorded within 7 to 28 d and plant infection counts yield results between 14 and 21 d.

7. Record data as explained in section 5–2.5. Culture units that yield negative results should not be discarded immediately, but should be checked daily to ensure that no delayed growth occurs.

5–2.5 Recording Experimental Results

Prepare a data sheet that allows the entry of each experimental unit as displaying either positive or negative results. A sample data sheet for a

Table 5–5. An example data sheet for recording the results of MPN experimentation.

Dilution level	Replicate				Total
	1	2	3	4	
5^{-1}	+	+	+	+	4
5^{-2}	+	+	+	+	4
5^{-3}	+	+	−	+	3
5^{-4}	−	−	+	−	1
5^{-5}	−	−	−	−	0
5^{-6}	−	−	−	−	0

Base dilution level (A)	5
Number of units per dilution level (N)	4
Experimental results	4-4-3-1-0-0
Tabular MPN (TMPN)	165
Initial dilution (I)†	NA
Inoculation volume (V)	1.0 mL
Probability of experimental outcome‡	0.106
Population estimate	165 cells per gram

† Note that initial dilutions only apply when the first dilution is different from all subsequent dilution steps (see section 5–2.7), NA = not applicable.
‡ In this case, the probability is obtained from Table 5–3.

fivefold dilution with four replicates is given in Table 5–5. Enter the results for each unit after careful inspection for growth of the inoculant. In the example, the results of the plant infection counts in Fig. 5–1 have been recorded to coincide with root nodulation as a positive (+) event.

5–2.6 Assigning Tabular Population Estimates to Results

The population estimates are assigned by locating the experimental results on the appropriate MPN table (Tables 5–6 to 5–9). These tables are organized by replicate number. Tables are presented for 2, 3, 4, and 5 replicate units per dilution level and for 2, 3, 4, 5, and 10-fold dilution series. The most likely results of the dilution series are listed in the first column. When the correct code is located, the researcher obtains a population estimate from the adjacent column that corresponds to the ratio of the serial dilution (2, 3, 4, 5, or 10-fold dilutions).

For the six-step, fivefold serial dilution with four replicates that yield the experimental results 4-4-3-1-0-0 (Table 5–5), first locate the table for four replicates (Table 5–8). Then, identify the correct experimental code (near the center of the experimental results column) and obtain the population estimate for a fivefold dilution (165 cells g^{-1} soil). This population estimate is the density of cells in the original test sample for inoculation volumes of 1.0 mL. For initial dilutions other than fivefold, or for inoculant volumes other than 1.0 mL, additional calculations are required as described in section 5–2.7.

Table 5–6. Most probable number of 2, 3, 4, 5, and 10-fold dilution series replicated twice.†

No. of positive results at dilution level	Population estimate (tabular MPN) ratio of the dilution series				
1-2-3-4-5-6	2	3	4	5	10
0-1-0-0-0-0	0.5	1.0	1.5	2.0	4.6
1-0-0-0-0-0	0.6	1.2	1.8	2.5	6.0
1-0-1-0-0-0	1.2	2.5	3.9	5.2	12
1-1-0-0-0-0	1.3	2.6	4.1	5.5	12
2-0-0-0-0-0	1.4	3.3	5.5	8.0	23
2-0-1-0-0-0	2.3	5.5	9.4	14	49
2-1-0-0-0-0	2.4	6.0	11	16	61
2-1-1-0-0-0	3.6	9.3	17	28	128
2-2-0-0-0-0	3.9	11	23	40	230
2-2-0-1-0-0	5.4	17	38	71	493
2-2-1-0-0-0	5.7	18	43	81	614
2-2-1-1-0-0	7.8	28	70	142	1270
2-2-2-0-0-0	8.4	33	91	202	2305
2-2-2-0-1-0	11	51	152	354	4844
2-2-2-1-0-0	12	56	172	408	5938
2-2-2-1-1-0	17	86	280	713	12704
2-2-2-2-0-0	18	104	372	1016	23054
2-2-2-2-0-1	25	162	638	1797	48442
2-2-2-2-1-0	27	179	688	2109	59379
2-2-2-2-1-1	38	285	1136	3565	120000
2-2-2-2-2-0	44	359	1484	5075	230545
2-2-2-2-2-1	73	703	2752	10545	593790
Confidence factor‡	2.67	3.45	4.01	4.47	6.61

† This is the population density in the original test sample that assumes 1-mL inoculation volume. Table generated using MPNES software (Woomer et al., 1990).
‡ The confidence factor is divided into and multiplied by the population estimate to establish the lower and upper confidence limits ($P = 0.05$), respectively. (After Cochran, 1950.)

5–2.7 Correcting for Initial Dilution and Inoculant Volume

If all dilution steps are equal (the initial dilution is the same as all subsequent dilutions) and if the inoculant volume is 1.0 mL, the population estimate may be obtained directly from the appropriate MPN table as described in section 5–2.6.

If the volume inoculated to each culture unit is not 1.0 mL, then the population estimate must be divided by this amount. For example, if 0.5 mL is inoculated, then the tabular population estimate is divided by 0.5. Similarly, if 2.5 mL is inoculated, then the tabular population estimate is divided by 2.5.

If the initial dilution differs from the subsequent dilution series, and if that initial dilution is not used to inoculate dilution level 1 but is rather further diluted, then the tabular population estimate is multiplied by that dilution value. For example, if a test sample is first diluted 100-fold and if this is not applied as the lowest dilution to the culture units but is rather further diluted, then the tabular value obtained from the MPN table is multiplied by 100 (see section 5–2.6).

Table 5–7. Most probable number of 2, 3, 4, 5, and 10-fold dilution series with three replicate units per dilution level.†

No. of positive results at dilution level	Population estimate (tabular MPN) ratio of the dilution series				
1-2-3-4-5-6	2	3	4	5	10
1-0-0-0-0-0	0.3	0.7	1.1	1.5	3.5
1-0-1-0-0-0	0.7	1.5	2.3	3.1	7.2
1-1-0-0-0-0	0.7	1.5	2.4	3.2	7.3
2-0-0-0-0-0	0.8	1.7	2.7	3.8	9.1
2-1-0-0-0-0	1.3	2.8	4.4	6.1	14
3-0-0-0-0-0	1.4	3.3	5.5	8.0	23
3-0-1-0-0-0	2.0	4.7	8.0	12	38
3-1-0-0-0-0	2.0	5.0	8.6	13	42
3-1-1-0-0-0	2.7	6.7	12	18	74
3-2-0-1-0-0	2.9	7.2	13	20	92
3-2-1-0-0-0	3.7	9.8	19	31	147
3-3-0-0-0-0	4.0	11	23	40	230
3-3-0-1-0-0	4.8	15	32	59	382
3-3-1-0-0-0	5.0	15	34	64	425
3-3-1-1-0-0	6.2	20	48	93	738
3-3-2-0-0-0	6.4	22	53	105	919
3-3-2-1-0-0	7.8	30	75	156	1466
3-3-3-0-0-0	8.4	33	91	203	2305
3-3-3-0-1-0	10	44	129	298	3829
3-3-3-1-0-0	11	47	139	322	4219
3-3-3-1-1-0	13	63	194	465	7188
3-3-3-2-0-0	14	68	216	532	9375
3-3-3-2-1-0	17	91	305	791	14375
3-3-3-3-0-0	18	104	372	1035	23016
3-3-3-3-0-1	22	140	531	1524	38243
3-3-3-3-1-0	23	149	556	1680	42454
3-3-3-3-1-1	29	200	776	2422	73823
3-3-3-3-2-0	31	220	864	2969	93750
3-3-3-3-2-1	40	306	1220	4219	143752
3-3-3-3-2-2	53	422	1677	6250	209064
3-3-3-3-3-0	44	359	1480	6563	230545
3-3-3-3-3-1	60	548	2224	8062	421932
3-3-3-3-3-2	93	956	3456	13311	937506
Confidence factor‡	2.23	2.75	3.11	3.40	4.67

† This is the population density in the original test sample that assumes 1-mL inoculation volume. Table generated using MPNES software (Woomer et al., 1990).

‡ The confidence factor is divided into and multiplied by the population estimate to establish the lower and upper confidence limits ($P = 0.05$), respectively. (After Cochran, 1950.)

If an initial dilution is used as the source of subsequent dilutions and also applied as the inoculant to dilution level 1, then the corrected initial dilution is the ratio of the initial to the serial dilution. For example, if an initial dilution of 1:100 is prepared, and then further diluted 1:5, and if that 1:100 dilution is used as the source of inoculant of the first dilution level, then the corrected initial dilution is 100/5 (or 20). This is multiplied by the MPN estimate from the table.

Table 5–8. Most probable number of 2, 3, 4, 5, and 10-fold dilution series with four replicate units per dilution level.†

No. of positive results at dilution level	Population estimate (tabular MPN) ratio of the dilution series (A)				
1-2-3-4-5-6	2	3	4	5	10
1-0-0-0-0-0	0.2	0.5	0.8	1.1	2.5
1-1-0-0-0-0	0.5	1.1	1.7	2.2	5.1
2-0-0-0-0-0	0.5	1.2	1.8	2.5	5.9
2-1-0-0-0-0	0.9	1.9	2.9	3.9	9.2
3-0-0-0-0-0	1.0	2.0	3.3	4.5	11
3-1-0-0-0-0	1.3	2.9	4.6	6.4	15
3-2-0-0-0-0	1.7	3.9	6.2	8.7	21
4-0-0-0-0-0	1.4	3.3	5.5	8.0	23
4-1-0-0-0-0	1.9	4.5	7.7	11	36
4-1-1-0-0-0	2.4	5.7	10	15	54
4-2-0-0-0-0	2.4	6.0	11	16	61
4-2-1-0-0-0	3.0	7.5	14	22	93
4-3-0-0-0-0	3.0	8.0	15	24	112
4-3-1-0-0-0	3.8	10	20	33	159
4-3-2-0-0-0	4.5	13	25	44	213
4-4-0-0-0-0	3.9	11	23	40	230
4-4-1-0-0-0	4.7	14	31	57	359
4-4-1-1-0-0	5.5	17	40	75	544
4-4-2-0-0-0	5.7	18	43	81	614
4-4-2-1-0-0	6.6	23	55	107	926
4-4-3-0-0-0	6.9	24	60	121	1124
4-4-3-1-0-0	8.1	30	78	165	1593
4-4-3-2-0-0	9.5	38	102	218	2130
4-4-4-0-0-0	8.4	33	91	203	2305
4-4-4-1-0-0	10	43	125	287	3594
4-4-4-1-1-0	12	53	161	380	5469
4-4-4-2-0-0	12	56	172	409	6137
4-4-4-2-1-0	15	70	222	545	9262
4-4-4-3-0-0	14	75	243	611	11239
4-4-4-3-1-0	17	94	319	830	15926
4-4-4-3-2-0	21	119	417	1113	21297
4-4-4-4-0-0	18	104	374	1035	230545
4-4-4-4-1-0	22	136	519	1465	359439
4-4-4-4-1-1	26	170	669	1992	546920
4-4-4-4-2-0	27	179	719	2109	613730
4-4-4-4-2-1	32	226	938	2813	1123930
4-4-4-4-3-0	33	244	972	3281	1592630
4-4-4-4-3-1	41	313	1272	4375	2129690
4-4-4-4-3-2	50	404	1686	5938	3594390
4-4-4-4-4-0	44	355	1496	5175	5469200
4-4-4-4-4-1	55	485	2072	7177	6137300
4-4-4-4-4-2	73	689	2872	10228	9262000
4-4-4-4-4-3	107	1069	3988	15263	11239300
Confidence factor‡	2.00	2.40	2.67	2.88	3.80

† This is the population density in the original test sample assumes 1-mL inoculation volume. Table generated using MPNES software (Woomer et al., 1990).

‡ The confidence factor is divided into and multiplied by the population estimate to establish the lower and upper confidence limits ($P = 0.05$), respectively. (After Cochran, 1950.)

Table 5–9. Most probable number of 2, 3, 4, 5, and 10-fold dilution series with five replicate units per dilution level.†

No. of positive results at dilution level	Population estimate (tabular MPN) ratio of the dilution series (A)				
1-2-3-4-5-6	2	3	4	5	10
1-0-0-0-0-0	0.2	0.4	0.6	0.8	1.9
1-1-0-0-0-0	0.4	0.8	1.3	1.7	4.0
2-0-0-0-0-0	0.4	0.9	1.4	1.9	4.4
2-1-0-0-0-0	0.7	1.4	2.2	2.9	6.8
3-0-0-0-0-0	0.7	1.5	2.3	3.2	7.7
3-1-0-0-0-0	1.0	2.1	3.3	4.5	10
3-2-0-0-0-0	1.3	2.7	4.3	5.8	13
4-0-0-0-0-0	1.0	2.2	3.6	5.1	12
4-1-0-0-0-0	1.3	3.0	4.8	6.7	16
4-2-0-0-0-0	1.7	3.8	6.1	8.5	21
4-3-0-0-0-0	2.0	4.7	7.6	10	27
5-0-0-0-0-0	1.4	3.3	5.5	8.0	23
5-0-1-0-0-0	1.8	4.1	7.0	10	31
5-1-0-0-0-0	1.8	4.2	7.2	11	33
5-1-1-0-0-0	2.2	5.2	8.9	13	45
5-2-0-0-0-0	2.2	5.4	9.3	14	49
5-2-1-0-0-0	2.7	6.4	11	18	69
5-3-0-0-0-0	2.6	6.7	12	19	78
5-3-1-0-0-0	3.1	8.0	15	24	107
5-3-2-0-0-0	3.6	9.6	18	30	138
5-4-0-0-0-0	3.2	8.5	16	26	127
5-4-1-0-0-0	3.8	10	20	34	169
5-4-2-0-0-0	4.4	12	25	43	216
5-4-3-0-0-0	5.0	15	31	54	270
5-5-0-0-0-0	3.9	11	23	40	230
5-5-0-1-0-0	4.5	13	28	51	312
5-5-1-0-0-0	4.5	13	29	53	327
5-5-1-1-0-0	5.2	16	36	67	453
5-5-2-0-0-0	5.3	17	37	70	488
5-5-2-1-0-0	6.0	20	46	88	692
5-5-3-0-0-0	6.1	20	49	94	780
5-5-3-1-0-0	6.9	24	60	119	1070
5-5-3-2-0-0	7.9	29	73	150	1383
5-5-4-0-0-0	7.1	26	65	133	1275
5-5-4-1-0-0	8.1	31	81	170	1690
5-5-4-2-0-0	9.3	37	100	216	2159
5-5-4-3-0-0	11	45	124	269	2716
5-5-5-0-0-0	8.4	33	91	203	2305
5-5-5-0-1-0	9.5	40	113	257	3126
5-5-5-1-0-0	9.7	41	118	268	3282
5-5-5-1-1-0	11	49	144	336	4532
5-5-5-2-0-0	11	50	151	353	4922
5-5-5-2-1-0	13	60	186	442	6918
5-5-5-3-0-0	13	63	197	476	7797
5-5-5-3-1-0	15	75	242	601	10702
5-5-5-3-2-0	17	89	296	757	13826
5-5-5-4-0-0	15	80	263	669	12753
5-5-5-4-1-0	17	96	328	860	16902
5-5-5-4-2-0	20	116	409	1089	21589
5-5-5-4-3-0	23	140	509	1358	27150
5-5-5-5-0-0	18	104	374	1026	23054
5-5-5-5-0-1	21	124	466	1309	31225
5-5-5-5-1-0	21	129	484	1356	32720

Table 5–9. Continued.

No. of positive results at dilution level	Population estimate (tabular MPN) ratio of the dilution series (A)				
1-2-3-4-5-6	2	3	4	5	10
5-5-5-5-1-1	24	154	597	1699	45261
5-5-5-5-2-0	25	160	625	1797	49224
5-5-5-5-2-1	28	192	769	2305	69148
5-5-5-5-3-0	29	202	825	2460	78127
5-5-5-5-3-1	34	244	968	3125	107022
5-5-5-5-3-2	39	295	1167	3906	138269
5-5-5-5-4-0	35	263	1036	3594	127528
5-5-5-5-4-1	41	322	1290	4688	169028
5-5-5-5-4-2	49	396	1602	5938	215899
5-5-5-5-4-3	59	492	1980	7188	271557
5-5-5-5-4-4	73	619	2436	9375	334051
5-5-5-5-5-0	44	357	1463	5781	230546
5-5-5-5-5-1	53	454	1880	6785	328192
5-5-5-5-5-2	65	598	2421	8980	492238
5-5-5-5-5-3	84	816	3146	11906	781272
5-5-5-5-5-4	119	1238	4202	17965	1312535
Confidence factor‡	1.86	2.19	2.41	2.58	3.30

† This is the population density in the original test sample that assumes 1-mL inoculation volume. Table generated using MPNES software (Woomer et al., 1990).
‡ The confidence factor is divided into and multiplied by the population estimate to establish the lower and upper confidence limits ($P = 0.05$), respectively. (After Cochran, 1950.)

5–2.8 Constructing Confidence Limits

The lower confidence limit is calculated by dividing the population estimate by the confidence factor (see section 5–1.3). Similarly, the upper limit results from multiplying the population estimate by the confidence factor. The confidence factors are listed at the bottom of each column of population estimates (Tables 5–6 to 5–9) or can be obtained from Table 5–2. For the fivefold dilution series with four replicates (Table 5–8), the confidence factor is 2.88 ($P = 0.05$). For the experimental results 4-4-3-1-0-0, the table gives an MPN estimate of 165. The lower confidence limit ($P = 0.05$) equals 165/2.88, or 57. The upper limit is 165×2.88 or 475. These results may be expressed as 165 (57–475, $P = 0.05$). Two population estimates differ significantly ($P = 0.05$) when the upper limit of the lesser estimate does not overlap with the lower limit of the greater estimate.

5–2.9 Sample Calculations

5–2.9.1 A Serial Dilution Where the First and Subsequent Dilutions are Equal and is Inoculated with 1 mL/unit

A fourfold serial dilution is prepared and inoculated onto four culture units at each dilution (A = 4, N = 4). Positive results are observed as follows:

Dilution level	Dilution	Positive results
1	4^{-1}	4 of 4
2	4^{-2}	4 of 4
3	4^{-3}	2 of 4
4	4^{-4}	1 of 4
5	4^{-5}	0 of 4
6	4^{-6}	0 of 4

These results are summarized as A = 4, N = 4, 4-4-2-1-0-0. Because there are four replicate units at each dilution level, Table 5–8 is consulted. Once the correct experimental result is identified in the first column of the table, the population estimate is obtained from the column corresponding to the dilution ration A = 4. This value is 55 cells/g of test substrate. Because all dilutions are equal and the inoculant volume is 1.0 mL, the MPN may be obtained directly from the table without further correction. The upper and lower confidence limits ($P = 0.05$) are calculated by multiplying and dividing by the confidence factor in the bottom row of the table. The confidence factor for a fourfold dilution with four units per dilution level is 2.67. The lower and upper confidence levels are 55/2.67 and 55 × 2.67, or 21 and 147. The MPN estimate is expressed as 55 (21 to 147, $P = 0.05$).

The ROT (see section 5–1.4) of these results is 2 (the third and fourth dilution steps, inclusively). Table 5–4 is used by locating the probability that corresponds to four replicates per dilution level, a fourfold dilution series and an ROT of 2. The probability of a failure of experimental technique obtained is 0.682. This result indicates compliance with the assumptions upon which the MPN technique is based.

5–2.9.2 A Serial Dilution Where the First and Subsequent Dilutions are not Equal and the Initial Dilution and Subsequent Dilution Level were Inoculated onto the Culture Units by Applying 2.0 mL of Inoculant

A dilution series is prepared by first diluting 100 g soil (dry wt.) with 900 mL of diluent (1:10) and further diluted at 2:1. Two milliliters of inoculant is applied to each of two culture units at each dilution level. The results follow:

Dilution level	Dilution	Positive results
1	2^0	2
2	2^{-1}	2
3	2^{-2}	0
4	2^{-3}	1
5	2^{-4}	0
6	2^{-5}	0

Remember that in this particular design 2^0 represents the initial dilution prepared at 1:10 and used to inoculate the first dilution level. The results are summarized as A = 2, N = 2, I = 10/2, V = 2, 2-1-0-1-0-0. Table 5–6 is consulted because there were two replicate units per dilution. A population estimate is obtained from the appropriate column. This value is 5.4 but must be corrected for both the initial dilution (1:10) and the inoculation volume (2 mL). Because the initial dilution was used as a source of inoculant for the lowest dilution level (2^0), the corrected initial dilution is 10/2 or 5 and the intermediate corrected MPN estimate is 5.4×5 (or 27.0). However, 2 mL was applied to each culture unit, so the MPN estimate must also be divided by 2. The final corrected MPN is 27.0/2 or 13.5. The confidence limits ($P = 0.05$) are constructed by dividing and multiplying by 2.67 (the confidence factor for a twofold dilution series with two units at each dilution level). The final MPN estimate is expressed as 13.5 (5.0 to 36.0, $P = 0.05$).

5–2.9.3 A Dilution Series Yields Unlikely Results

A 10-fold dilution series is prepared and inoculated onto three culture units at each dilution level (A = 10, N = 3). The experimental results expressed across dilution levels 1 to 6 are recorded as 2-3-2-0-3-0. These results are considered to be unlikely and the ROT test applied. The ROT is 5 consisting of the first to fifth dilutions, inclusively. The probability of compliance with the assumptions of the MPN, obtained from Table 5–4, equals 0.0002. These experimental results are discarded due to a 99.98% probability that the underlying assumptions of the MPN technique have not been satisfied.

5–3 COMMENTS

Despite the emergence of more precise and sophisticated methods of microbial enumeration in soils, water, and agricultural products, the MPN technique continues to remain an important means of assessing microbial populations. A distinct advantage of MPN techniques compared with more direct quantitative procedures is that MPN estimates measure only live and active organisms, thus reducing the confusion between live and dead cells common with microscopic techniques. The MPN technique is less sensitive to the effects of soil mineralogy than direct measurement techniques that require extraction and filtration of soil organisms (Woomer et al., 1988). MPN methodologies provide more realistic estimates of infective propagules of VA mycorrhizae in field soils than does the more conventional method of spore counts (Porter, 1979). MPN procedures continue to be employed as standardized methodology in the quality control of rhizobial inoculants (Scott & Porter, 1986), water (American Public Health Association, 1985), and food (West & Coleman, 1986).

A distinct disadvantage is that MPN procedures tend to require more labor and materials than microscopic procedures. Furthermore, MPN estimates often have a lower order of precision than do well-replicated direct counts. However, the precision of MPN estimates can be greatly improved by conducting preliminary investigations of a test sample using a dilution series with a wide dilution ratio (A = 10) and few units per dilution level (N = 2 or 3) followed by more detailed experimentation employing lower dilution ratios and greater replication (e.g., A = 2 or 3, N = 5). The range of the second MPN determination is centered around the results of the preliminary investigation by adjustment of the initial dilution (I). This approach not only improves precision but increases the accuracy of MPN determinations as it is desirable to design experiments such that the ROT occurs within the middle dilution steps rather than later in the dilution series. If the final dilution step is not entirely negative, it is possible that the MPN has yielded an underestimate. Placement of the ROT within the middle dilution steps is also important because some computer-generated MPN estimates (see Woomer et al., 1990) produce "large number errors" when the ratio of the serial dilution is large (A≥10), the number of units per dilution exceeds three and the bulk of the experimental results are positive (e.g., 5-5-5-5-5-4). However, these errors have been corrected in the MPN tables presented within this chapter.

In conclusion, MPN procedures are best employed when a functional attribute of mixed microbial populations is under study or when researchers can detect a qualitative property but cannot count the organisms responsible for that property directly. For these reasons, the MPN technique will remain an important means of estimating microbial populations in soils, water, and agricultural products.

ACKNOWLEDGMENT

The author is posted with the Tropical Soil Biology and Fertility Programme (TSBF), Nairobi, Kenya, through the Natural Environment Research Council, UK and gratefully acknowledges the support of these organizations and that of the United Nations Educational, Scientific and Cultural Organization (UNESCO), which hosts TSBF staff in Nairobi.

REFERENCES

Alexander, M. 1982. Most probable number method for microbial populations. p. 815–820. *In* A.L. Page et al. (ed.) Methods of soil analysis. Part 2. 2nd ed. Agron. Monogr. 9. ASA and SSSA, Madison, WI.

American Public Health Association. 1985. Estimation of bacterial density. p. 802–804. *In* Standard methods for the examination of water and wastewater. 16th ed. American Public Health Assoc., Washington, DC.

Brockwell, J. 1963. Accuracy of a plant-infection technique for counting populations of *Rhizobium trifolii*. Appl. Microbiol. 11:377–383.

Brockwell, J., A. Diatloff, A.L. Garcia, and A.C. Robinson. 1975. Use of wild soybean (*Glycine ussuriensis* Regel and Maack) as a test plant in dilution-nodulation frequency tests for counting *Rhizobium japonicum*. Soil Biol. Biochem. 7:305–311.

Cochran, W.G. 1950. Estimation of bacterial densities by means of the "most probable number." Biometrics 6:105–116.

deMan, J.C. 1975. The probability of most probable numbers. Eur. J. Appl. Microbiol. 1:67–78.

Halvorson, H.O., and A. Moeglein. 1940. Application of statistics to problems in bacteriology. V. The probability of occurrence of various experimental results. Growth 4:157–168.

Halvorson, H.O., and N.R. Ziegler. 1933. Applications of statistics to problems in bacteriology. I. A means of determining bacterial population by the dilution method. J. Bacteriol. 25:101–121.

Kingsley, M.T., and B.B. Bohlool. 1981. Release of *Rhizobium* spp. from tropical soils and recovery for immunofluorescence enumeration. Appl. Environ. Microbiol. 42:241–248.

Porter, W.M. 1979. The "most probable number" method for enumerating infective propagules of vesicular arbuscular mycorrhizal fungi in soil. Aust. J. Soil Res. 17:515–519.

Schmidt, E.L., R.O. Bankole, and B.B. Bohlool. 1968. Flourescent-antibody approach to study of rhizobia in soil. J. Bacteriol. 95:1987–1992.

Scott, J.M., and F.E. Porter. 1986. An analysis of the accuracy of a plant infection technique for counting rhizobia. Soil Biol. Biochem. 18:355–362.

Stevens, W.L. 1958. Dilution series: A statistical test of technique. J. R. Stat. Soc. Ser. B. 20:205–214.

Vincent, J.M. 1970. A manual for the practical study of the root nodule bacteria. Blackwell Scientific Publ., Oxford and Edinburgh.

Weaver, R.W., and L.R. Frederick. 1972. A new technique for most-probable-number counts of rhizobia. Plant Soil 36:219–222.

West, P.A., and M.R. Coleman. 1986. A tentative national reference procedure for isolation and enumeration of *Escherichia coli* for bivalve molluscan shellfish by most probable number method. J. Appl. Bacteriol. 1:505–516.

Woodward, R.L. 1957. How probable is the most probable number. J. AWWA 49:1060–1068.

Woomer, P., J. Bennet, and R. Yost. 1990. Overcoming the inflexibility of most-probable-number procedures Agron. J. 82:349–353.

Woomer, P., P.W. Singleton, and B.B. Bohlool. 1988. Reliability of the most-probable-number technique for enumerating rhizobia in tropical soils. Appl. Environ. Microbiol. 54:1494–1497.

Chapter 6

Light Microscopic Methods for Studying Soil Microorganisms

PETER J. BOTTOMLEY, *Oregon State University, Corvallis, Oregon*

Microscopy provided some of the earliest insights into the spatial relationships between soil microorganisms and the physical components of their environment (Starkey, 1938). Soil biologists have a long history of using microscopy for identifying and quantifying soil fauna, and studying the autecology of both free-living and symbiotic bacteria and fungi. When the concept was developed that soil microorganisms could be significant sources and sinks for nutrients, researchers used microscopy, along with other methods, to measure the magnitude of the biomass. Concurrent with the interest in measuring soil biomass, microscopy has been used to address the long-standing problems of how to quantify fungal biomass, how to distinguish between dead and viable microbial cells, and how to differentiate between physiologically active and dormant cells.

Despite the inherent limitations to microscopic methods, they continue to play a vital role in the detection and enumeration of specific organisms, and serve as a tool to validate the findings generated with other methods. Unfortunately, soil remains one of the most difficult of natural environments upon which to practice microscopy and the literature describes many "variations-on-a-theme."

In this chapter, I do not plan to discuss the use of electron microscopic techniques for studying soil microorganisms in situ. I draw the reader's attention to the work of R.C. Foster, CSIRO Division of Soils, Adelaide, South Australia, Australia. This work has been summarized as an excellent collection of electron micrographs (Foster et al., 1983).

6–1 SAMPLING OF SOIL FOR MICROSCOPIC OBSERVATION

Although many of the general issues about soil sampling have been covered elsewhere (see chapter 1 by Wollum and chapter 2 by Parkin and Robinson in this book), three important details are worth emphasizing when a researcher is interested in studying soil microbial populations by microscopic means.

6–1.1 Time of Sampling

The density of a soil microbial population can fluctuate during the calendar year. These differences can be pronounced if there is a period when plants are completely absent (e.g., cropping systems), or where seasonal changes result in the plant community existing in either an active, dormant, or dead state.

6–1.2 Depth of Sampling

As mentioned elsewhere (see chapter 8 by Zuberer in this book), the researcher should be aware of the physical and chemical changes that occur in soil with respect to depth (horizon development). In addition, the rooting patterns of the plant community on the study site should also influence the depth to which soil samples are taken.

6–1.3 Rhizosphere vs. Nonrhizosphere Soil

The distance that the rhizosphere extends from the root surface is influenced by the plant species. As a consequence, the proportion of the soil volume that is influenced by plant roots will depend upon the rooting pattern of the particular plant species present.

6–2 MICROSCOPIC ENUMERATION OF TOTAL BACTERIA IN SOIL

Since soil is optically opaque, microorganisms must be extracted and transferred to an optically suitable background before quantitative light microscopic observations can be made. The choice of extraction method depends upon the density of the population of interest, and whether numbers, or microbial activity is the parameter of primary interest. In the case of the latter, composite samples of soil should be brought from the field immediately before the time of experimentation. No attempt should be made to alter soil moisture content unless it is so wet to be physically impossible to sieve (< 2 mm) and create uniform subsamples. When the soil is to be used solely for enumeration purposes, samples can be air-dried to a point where they can be thoroughly mixed and conveniently sieved.

6–2.1 Separation of Bacteria from Soil by Dispersion, Dilution, and Selective Filtration

In this procedure, bacteria are separated from soil structure by mechanical dispersion. Then, the majority of the soil bacteria are separated from organic debris and large mineral particles by consecutive filtration through polycarbonate membranes of 8.0- and 3.0-μm pore size. To facilitate the filtration step, soil is diluted up to 500-fold depending upon the density of the bacterial population. Most of the soil debris is collected on the filters and a high percentage of the soil bacteria pass through both the 8.0 and the 3.0-μm pore size filters. At this stage, the bacterial population can be collected upon 0.45-μm pore size polycarbonate membrane filters for microscopic enumeration, or, incubated in a viability or metabolic competence assay prior to microscopy. The extent of the initial soil dilution or the amount of sample stained and filtered through 0.45 μm filters can be varied depending upon the density of the soil bacterial population. Although a portion (10–20%) of the soil bacteria are retained on the 8.0- and 3.0-μm pore size filters, this procedure produces filtered specimens from which accurate cell counts can be made and, of equal importance, a high percentge of the soil bacteria in the suspension that pass the 3.0-μm filter are respiratorily competent.

6–2.1.1 Dispersion Procedure

1. Suspend quadruplicate 3-g samples of soil in acid-washed milk dilution bottles (160-mL capacity). Each bottle contains a monolayer of acid-washed glass beads of 3-mm diam., and 27 mL of filter-sterilized (0.2-μm pore size) 0.15 M NaCl. The bottle caps should be thoroughly clean and checked for their ability to form a water-tight seal with the bottle. Do not use bottles with chipped rims.

2. Shake the bottles vigorously by hand until soil structure has been completely destroyed. About 10 to 15 min of shaking is required to disrupt aggregates in a well-structured soil to ≤ 1-mm diam. Aggregate disruption can be monitored by periodically assessing the settling soil suspension for the presence and size of intact aggregates through the bottom of the bottle.

3. To facilitate removal of supernatant without disturbing the sediment, place the bottles on a platform at eye level and allow the soil suspensions to settle for exactly 5 min.

4. Using a safety pipette, recover a 10-mL portion from the supernatant above the sediment layer in each bottle and transfer to 490 mL (for a total bacterial count) or 190 mL (for counting specific bacterial types by immunofluorescence) of filter-sterilized (0.2-μm pore size) deionized water. This procedure creates either a 500- or 200-fold soil dilution. The settled sediment must not be disturbed during the sampling.

6–2.1.2 Filtration and Staining Procedures

1. Set up a 47-mm-diam. polycarbonate membrane filter (8.0-μm pore size) in each of four Buchner funnels. Attach each of the funnels to a side-arm flask. Attach the flasks, one at a time, via a moisture trap and a vacuum gauge to a vacuum pump. Filter a 50-mL portion from each of the dispersed and diluted samples of soil through a separate filter under −0.05 MPa of suction.

2. Replace the 8.0-μm filters in the Buchner funnels with 47-mm-diam. polycarbonate filters of 3.0-μm pore size. Pass the 8.0-μm filtrates from step 1 through the 3.-μm filters under −0.05 MPa of suction.

Note: The following steps are designed specifically for the enumeratioin of total bacteria. See section 6–2.2 for the procedure to enumerate specific serotypes by immunofluorescence.

3. Add a 0.1-mL portion of formalin (final concentration, 2% wt/vol) to a 1.9-mL portion of each of the four filtered microbial suspensions (< 3.0 μm).

4. Add to each formalin-fixed sample a 0.2-mL portion of a filter-sterilized (0.2-μm pore size) solution of acridine orange (0.1% wt/vol in 0.1 M citrate buffer, pH 6.6).

5. Incubate in the dark at room temperature for 10 min.

6. To promote uniform dispersion of bacterial cells upon the filter, add an aliquot (0.25-mL) of the acridine orange-treated suspension to 8.0 mL of filter-sterilized 0.15 M NaCl.

7. Transfer each 8.25-mL sample to a filter unit consisting of a glass chimney-filter support combination and a "blackened" 25-mm diam. polycarbonate filter (0.45-μm pore size). Apply a suction of −0.05 MPa and draw each sample through the filter.

6–2.1.3 Filter Blackening Procedure

The filters need to be "blackened" since they will be viewed using short wavelength light and epifluorescence microscopy. I have found the following procedure to produce "blacker" filters than I have been able to produce from commercially available batches of the traditional "blackener" Irgalan Black.

1. Suspend the filters in a 0.3% (wt/vol) solution of Sudan Black (CI 26150, Kodak Chemical Co.) in 70% (vol/vol) ethanol.
2. Rotate slowly on a shaker for 2 d.
3. Separate the filters from dye by pouring slowly through cheese-cloth.
4. Rinse the filters gently with filter-sterilized water.
5. Using blunt-edged forceps, spread the filters on tissue paper to air dry.

6–2.1.4 Comments about Filtration

Two aspects of filtration should be mentioned. First, a vacuum gauge should be inserted in line so that suction pressure can be monitored in the filtering unit. Excess suction should be avoided since it creates nonuniform distribution of cells on the filter and can result in passage of bacteria that are "thin" rods (1.0 × 0.2 μm). Secondly, some of the commercially available filter supports contain a fritted center disc that is set lower than the glass rim of the support. When resting on these units, polycarbonate membrane filters are prone to collapse and tear upon application of a vacuum.

While filter supports manufactured by some companies do not have this problem, it can be circumvented by using 25-mm silver-coated membranes of 8.0-μm pore size (Osmotics, Inc., Minnetonka, MN) underneath the polycarbonate membranes on the filter supports. I have no doubt that others have solved the same problems with equally ingenious methods.

6–2.1.5 Destaining Procedure

Several procedures for destaining AO-treated filters have been described in the literature. We follow the procedure originally described by Meyer-Reil (1978). While the filters are still on the filter supports and under a slight vacuum, 20-mL portions of prefiltered 0.1 M Na citrate buffers, pH 6.6, 5.5, and 4.0 and deionized water are passed sequentially through each filter.

6–2.1.6 Viewing the Filters

Transfer each filter to a microscope slide and place on top of a drop of Cargille immersion oil (Type A or B). Apply the same oil to the upper side of the filter and then apply a cover slip. Specimens are viewed with an epifluorescence microscope using a fluorescein filter set. For a Zeiss microscope, this set consists of an excitation filter, BP 450-490, a FT 510 chromatic beam splitter, and a LP520 cut-off filter as barrier. A planachromat 100X 1.25 oil objective lens provides superior viewing optics without the accompanying eye fatigue caused by field distortions and edge effects of cheaper lenses. It is extremely helpful to have a calibrated "whippledisc" in one of the eye pieces so that a uniform area can be counted in each field of view.

6–2.1.7 Comments upon the Counting Procedure

The procedure described above is sufficient to provide about 50 to 100 bacteria per field of view if the soil initially contains about 1 to 2 $\times$ 10^9 cells per gram of soil recoverable in one dispersion and filtration event. In soils containing a lower density of bacteria the researcher can either (i) filter a larger volume of the acridine orange stained suspension or (ii) reduce the extent of soil dilution prior to passage through 8.0- and 3.0-μm filters. Despite the precautions described above, uniform distribution of cells on the filter is not easy to achieve. The number of bacteria in 25 to 30 fields of view should be counted.

There can be tremendous diversity in the shape and size of the "cells" and in their color (shades of orange and green). Since fluorescence can fade during the time it takes to debate over the nature of a fluorescing particle, my approach is to first make a count of all particles that are of a size and shape similar to bacteria. After counting the appropriate number of fields per filter, I rest my eyes for a few minutes, and then return to the microscope to determine a "correction factor." This involves making an estimate of the proportion of the fluorescing particles that are inert mineral or organic fragments. Although eye fatigue is not described in the literature, it is a serious problem that a researcher should be aware of.

A systematic approach to microscopy is essential if consistent results are to be obtained. In soil samples containing about 10^9 bacteria per gram that are recoverable with one extraction, the reproducibility of population estimates made from replicate volumes recovered from the same soil suspension can be good (e.g., $2.7 \pm 0.5 \times 10^9$ per gram of soil). Similar reproducibility of population estimates can be obtained from replicate soil samples taken from the same composite sample. The following are examples of mean values and standard errors for densities of soil bacteria that I have obtained on different soils with this method: $1.6 \pm 0.3 \times 10^9$, $7.4 \pm 0.9 \times 10^8$, and $4.0 \pm 0.4 \times 10^8$ per gram of oven-dry soil.

6–2.1.8 Calculations

The number of bacteria per gram of oven-dry soil is given by the extended Eq [1] that can be reduced to Eq [2].

$$N = \frac{n \times B \times 2.2 \times 500 \times (27 + X + Sv)}{a \times 0.25 \times 1.9 \times 10 \times Sw} \qquad [1]$$

$$N = \frac{nB1100}{4.75a} \frac{(27 + X + Sv)}{Sw} \qquad [2]$$

Symbol Definitions.

N = the number of bacteria per gram of oven-dry soil;
n = the number of bacteria per field of view;
B = the effective filtering area of the membrane filter (μm^2);
a = the area of the microscopic field of view (μm^2);
0.25, the volume (mL) of the acridine orange-stained sample that was filtered for enumeration;
2.2, the final volume (mL) of the 1.9-mL portion of the 500-fold dilution that was fixed with formalin (0.1 mL) and stained with acridine orange (0.2 mL);
500, the total volume (mL) of the 500-fold dilution;
1.9, the initial volume (mL) of the 500-fold dilution that was fixed for staining;
10, the volume (mL) of the initial soil suspension that was used to make the 500-fold dilution;
X, the volume of water (mL) in the 3-g soil sample;
Sv, the volume (mL) of soil particles (assume a particle density of 2.7 g cm^{-3});
Sw, the weight (g) of oven-dry soil.

6–2.1.9 Use of Alternative Nucleic Acid Stains

There are a few reports in the literature where soil scientists have experimented with other nucleic acid-specific stains. Macdonald (1986b) and Bottomley and Dughri (1989) used the stain diphenylamidino indophenol (DAPI). It has been reported that DAPI is highly specific for DNA (Porter & Feig, 1980) and it has been used in my laboratory to confirm soil bacterial population counts made with AO.

6–2.1.10 Procedure

1. Make a DAPI stock solution (1 mg per mL) in filtered deionized water and keep refrigerated in a dark bottle.
2. Make a working solution of final concentration 10.0 µg per mL by adding a 100-µL portion of the stock to 10 mL of filtered deionized water.
3. Samples are stained, filtered, and destained as described above for acridine orange.

6–2.1.11 Special Requirements for DAPI

Since the excitation and emission characteristics of DAPI are quite distinct from AO, a different filter set is required in the epifluorescence microscope. A BP365/10 excitation filter, FT390 chromatic beam splitter, and a LP395 cut-off barrier filter are required (Zeiss filter set: 47 77 01). Unfortunately, the coating of the optically superior planachromat lens absorbs light in the excitation region of the DAPI filter set; a neofluor (100/1.30) oil objective, without flat-field correction, must be used. The bacterial cells fluoresce a light blue color. The bis-benzimide dye, Hoechst 33258, also requires the same filter set and objective lens combination as described for DAPI (Paul, 1982).

6–2.2 Enumeration of Specific Bacterial Types by Immunofluorescence Microscopy

The details of preparing fluorescein-labeled antibody conjugates have been described elsewhere in this volume (see chapter 28 in this book). I will concentrate upon the details of staining, destaining, and microscopy as they pertain to studying specific organisms within soil populations.

6–2.2.1 Preparation of a Working Solution of a Fluorescein-labeled Immunoglobulin Conjugate (FA)

After calibrating FAs for titer and cross-reactivity, we routinely store them, supplemented with merthiolate (0.01% wt/vol), in sealed ampoules (0.1–0.5 mL) at −60 °C. An ampoule is recovered, thawed, and diluted with 0.02 M sodium phosphate buffer, pH 7.2 by a predetermined amount that gives maximum fluorescence with homologous isolates of a serotype and minimum fluorescence with isolates from other serotypes (typically between 10- and 25-fold). The dilute sample is filtered through a 0.2-µm pore size polycarbonate membrane. All manipulations of a FA should be conducted under dim light, and the working solution kept in a darkened tube on ice, or in a refrigerator at 2 to 4 °C.

6–2.2.2 Separation of Bacteria from Soil

Bacteria are separated from soil by the mechanical dispersion, dilution (200-fold), and filtration procedure described above (sections 6–2.1.1 and 6–2.1.2).

6–2.2.3 Procedure

1. Depending upon the density of the specific bacterial serotype to be enumerated, between 5- and 20-mL portions of the 200-fold soil dilution are filtered onto blackened (see section 6–2.1.3) 25-mm diam. polycarbonate membranes (0.45-µm pore size).

2. Using blunt-edged forceps, place membranes on microscope slides and cover with a few drops of a rhodamine-gelatin conjugate. This treatment reduces the nonspecific binding of FA to "universal acceptors" such as soil colloids (Bohlool & Schmidt, 1968). Place the slide-filter combinations in an oven at 60 °C until the conjugate is barely dry. Do not be concerned that the filters are stuck to the slides at this stage.

3. Under dim light, treat the filters with a few drops of the diluted FA and incubate for 1 h at room temperature in a darkened and humidified chamber. During this procedure the filters become detached from the slide and amenable to handling.

4. Transfer the filters to filter supports and destain each with 50 mL of 0.02 M phospate buffer under −0.05 MPa of suction.

5. Place the membranes upon clean glass microscope slides. To the upper surface of the filter, add a drop of mounting fluid containing the fluorescence enhancer, p-phenylenediamine (100-mg phenylenediamine in 100 mL of glycerol, pH adjusted to 8.0 with 0.5 M carbonate-bicarbonate buffer), and apply a no. 1 coverslip. Apply a drop of immersion oil onto the cover slip and view the specimen under epifluorescent microscopy at 1000X with the fluorescein filter set described earlier. Use a whipple disc in the eyepiece for convenience of counting. Count 25 to 30 fields of view. Increase to 50 fields if fewer than five cells per field are observed.

6–2.2.4 Comments

It is important to separate the issue of the lower limit of population detection with FA from that of accurate determination of population size. Using fluorescent antibodies, an experienced worker can detect a population existing at 10^2 per gram of soil but cannot measure it accurately using this technique. In soil samples where certain serotypes exist between 10^5 and 10^6 per gram, the reproducibility between replicate volumes recovered from the same filtered suspension can be excellent. For example, the following values have been obtained from my laboratory for serotypes of *Rhizobium leguminosarum* bv. *trifolii;* $1.22 \pm 0.06 \times 10^6$ and $2.00 \pm 0.05 \times 10^5$ per gram of soil. When serotype populations exist in the range of 10^3 to 10^4 per gram, the variability between replicate volumes increases; e.g., $1.17 \pm 0.21 \times 10^4$ and $4.82 \pm 2.73 \times 10^3$ per gram. The reproducibility of population density values determined from replicate soil samples taken from the same composite sample can also be excellent. The following mean values and their standard errors are typical of those obtained from my research laboratory for indigenous serotypes of *R. leguminosarum* bv. *trifolii:* $5.1 \pm 0.6 \times 10^5$, $1.7 \pm 0.5 \times 10^5$, and $5.0 \pm 0.6 \times 10^4$ per gram of soil.

6–2.3 Separation of Bacteria from Soil by Extraction and Flocculation

In autecological studies, the population density of a specific microorganism of interest may be too low to tolerate the dilution step described in 6–2.1.1. For example, if an organism exists at 10^4 per gram of soil and if 20-mL of a 1:200 soil suspension are filtered and viewed at 1000X, only one cell per 10 fields of view will be seen. Several procedures have been described whereby soil colloids are flocculated away from the microorganisms suspended in the original 1:10 dilution of soil. Unfortunately, since bacteria are colloidal particles, they are prone to co-flocculating with the soil. The flocculation mixtures invariably contain a calcium salt (sometimes alone, or in combination with an insoluble colloid such as $MgCO_3$), and either an ammonium salt with an anion that will form an insoluble precipitate with Ca [e.g., NH_4OH, $(NH_4)_2$ oxalate], or any phosphate salt that can form a calcium phosphate precipitate (Schmidt, 1974; Wollum & Miller, 1980; Kingsley & Bohlool, 1981; Robert & Schmidt, 1983; Demezas & Bottomley, 1986). The following procedure has worked satisfactorily in my research laboratory with soils covering a wide range of textures and organic matter contents.

6–2.3.1 Procedure

1. Weigh out quadruplicate 10-g quantities of soil and add 95 mL of 0.1 M $(NH_4)_2HPO_4$, pH 8.0. Add two drops of antifoam B (Sigma), and five drops of Tween 20 (Sigma).
2. Shake vigorously for 15 min.
3. Add 1.3 g of an 8:5 mixture of $CaCl_2:MgCO_3$ and shake for a further 2 min.
4. Allow the soil to flocculate for 20 to 30 min and then carefully remove a 30 to 40 mL portion of the cleared supernatant from above the flocculated soil in each bottle.
5. Filter 5 to 20-mL portions through blackened 25-mm-diam. polycarbonate membranes under -0.05 MPa suction pressure as described above.
6. Treat the sample with the appropriate stain (usually a fluorescent antibody, see section 6–2.2).

6–2.3.2 Comments

If the researcher is careful, 30 to 40 mL can be recovered from one flocculated sample, subdivided into 5- to 20-mL aliquots, and several different serotypes can be enumerated with appropriate FAs. Occasionally, we have encountered problems with inorganic precipitates developing in the supernatant above the flocculating soil and impeding sample filtration. To counter this problem, remove the supernatant and acidify it to pH 4 by adding DL-lactic acid (85% syrup) to a final concentration of 0.1% (vol/vol).

With this procedure, a bacterial population existing at 10^4 per gram can be detected, giving one cell per field of view (1000X) if 10 mL of the supernatant is filtered. Different soils may require different quantities of the flocculant, different ratios of $CaCl_2:MgCO_3$ in the flocculant, and different concentrations of ammonium phosphate to optimize colloidal precipitation. For some Oregon soils, 0.05 M phosphate is superior to 0.1 M, and the optimal quantity of $CaCl_2$ can range between 0.8 to 1.3 g. Preliminary experiments should be carried out to determine optimum values for each soil.

6–2.4 Separation of Bacteria from Soil by Density Gradient Centrifugation

Several papers have been published in which bacteria were separated from soil by density-gradient centrifugation techniques (Bakken & Olsen, 1983; Bakken, 1985; Wollum & Miller, 1980; Macdonald, 1986a). These methods vary in the compound used to generate the gradient and the degree of sophistication with which the gradient is produced. The variant of Bakken (1985) is described below.

6–2.4.1 Procedure

1. Suspend 10-g samples of soil in 95-mL of deionized water; homogenize in a Waring blender for three 1-min intervals with interspersed cooling periods in an ice-bath.
2. Centrifuge the soil homogenate for 15 min in a swing-out rotor at a low centrifugal force (630–1000 × g).
3. Remove the supernatant from above the soil pellet and concentrate the bacterial cells in this supernatant by centrifuging at 10 000 × g for 20 min.
4. Vigorously stir an aqueous suspension of Percoll (1.2 g mL^{-1}, Pharmacia Fine Chemicals, Piscataway, NJ) and reduce its pH to 7.0 with 1 M HCl.
5. Develop a nonlinear gradient (1.0–1.14 g mL^{-1}) by suspending 30-mL of the Percoll suspension in a 40-mL capacity polycarbonate centrifuge tube and centrifuge for 40 min at 27 000 × g.
6. Load 2-mL quantities of the bacterial suspension from step 3 on the gradient and centrifuge at 10 000 × g for 1 h.

7. Load one or two replicate gradients with density marker beads (Pharmacia) instead of with the microbial suspension.
8. Recover 3-mL portions of the gradient after centrifugation is complete.
9. To separate microorganisms from the Percoll particles, dilute each 3-mL portion with 57 mL of filtered deionized water (0.2 μm, pore size).
10. Centrifuge at 10 000 × g for 20 min. Decant off the supernatant and resuspend the bacteria-containing pellet in 1 mL of dionized water.
11. Prepare microbial suspension for appropriate staining and microscopic observation as described in section 6–2.1.
12. The specific gravity of the zones in the gradient in which microorganisms are concentrated can be determined by comparing with the zones in which specific density marker beads had equilibrated.

6–2.4.2 Comments

Between 26 and 70% of the total bacterial cells in three different soils were recovered within the upper region of the gradient corresponding to a specific gravity of 1.0 to 1.14 g per mL. The majority of the remaining cells were in the pellet below the gradient. Since Percoll particles interfere with both filtering and microscopic observation of the bacterial cells, the latter are separated from the Percoll by diluting the suspension 1:20 in distilled water and pelleting the bacteria by centrifugation for 20 min at 10 000 × g.

6–2.4.3 Variations of the Density Gradient Procedure

Macdonald (1986a) concluded that linear gradients of Percoll were superior to nonlinear gradients for separating soil microorganisms from soil colloids. After centrifugation, microorganisms were found to be distributed uniformly across the whole gradient. Subsequently, Percoll particles were separated from the microorganisms by passing the appropriate fractions over a column of Sepharose beads (Sephacryl S1000).

6–2.5 Determining the Efficiency of Recovering Bacteria from Soil for Microscopic Enumeration

The standard procedure is to add to soil a known number of cells of an undefined mixture of soil bacterial types and then to determine what proportion of them can be recovered with a specific extraction and enumeration procedure. If the researcher is interested in determining the efficiency of recovery of a specific microorganism, the experiment can be modified if the microbe of interest possesses either a specific antibiotic, nutritional, or reporter gene marker, or if a strain-specific fluorescent antibody is available. The efficiency of recovery is calculated as follows:

$$\frac{\text{Number in spiked sample} - \text{Number in unspiked sample}}{\text{Number of cells in the spike}} \times 100$$

6–2.5.1 Preparation and Addition of the Spike

A mixed bacterial population is grown by inoculating 25-mL portions of a dilute yeast extract-peptone broth (0.4 g L^{-1} of each) with an inoculum recovered from soil by the extraction procedure described above. Cells are harvested by centrifugation at 10 000 rpm for 10 to 15 min, resuspended in deionized water to a density of approximately 5×10^8 cells mL^{-1} and fixed in formalin (2% vol/vol, final concentration). One-milliliter portions of the fixed cell suspension are mixed into 3-g quantities of slightly moist soil (−0.5 to −1.5 MPa), and allowed to infiltrate into the soil by incubating overnight in a refrigerator. Bacterial populations are recovered and enumerated from the spiked and unspiked samples using the AO or FA procedures described above.

6–2.5.2 Comments

Efficiency values ranging from 15 to 60% have been determined in our laboratory on different soils. Broth-grown cells will not penetrate into the soil matrix like the indigenous populations and therefore the efficiency of recovery may be overestimated. An alternate approach is to determine the relative proportions of the native soil population that can be recovered in consecutive extractions of the same soil sample.

6–2.5.3 Recovery of Bacteria Using Consecutive Extractions

Soil bacteria are extracted essentially as described above in section 6–2.1 except that 250-mL capacity centrifuge bottles replace the milk dilution bottles. A 5-mL sample of supernatant is recovered from above the settling soil particles and added to 95 mL of filtered deionized water. The soil suspension remaining in the centrifuge bottle is gently centrifuged at $1900 \times g$ for 5 min and the rest of the supernatant is decanted and discarded. Another 27-mL volume of filter-sterilized saline is added to the soil pellet and the extraction procedure is repeated. We have routinely carried out a total of four extractions. Staining with FA is carried out as described earlier (section 6–2.2). In the case of total bacteria, 200-fold soil dilutions are further diluted to 500-fold prior to staining and enumerating with acridine orange as described in section 6–2.1.

6–3 DETERMINING THE PROPORTION OF VIABLE SOIL BACTERIA USING A CELL ELONGATION ASSAY

6–3.1 Principle

The principle behind this method lies in the action of antibiotic inhibitors of DNA gyrase. These compounds can prevent DNA replication and cell division while allowing other cellular functions to continue (Kogure et al., 1979). In the presence of substrates and the inhibitor, viable cells in soil populations can undergo significant elongation beyond their normal size (Bottomley & Maggard, 1990). By determining the number of elongated cells, an estimate can be made of the proportion of the soil bacterial population that is viable. The rate of appearance of elongated cells also provides preliminary information about the competence of the soil community to respond to specific nutrients under particular soil conditions.

6–3.2 Procedure

The separation of bacteria from soil is carried out as described in section 6–2.1. The extent of soil dilution is influenced by the density of the organisms to be evaluated. Although 500-fold dilution can be adequate for assessing viability of the total bacteria population, this should be reduced to 100- to 200-fold if specific bacterial serotypes existing at 10^4 to 10^6 per gram are to be evaluated by immunofluorescence microscopy.

1. Add 15 mL of the 10-fold soil dilution to 245 mL of filter-sterilized (0.2-μm pore size) deionized water. Filter through 8.0- and 3.0-μm pore size membranes, as described in section 6–2.1.2.
2. To the filtered soil solution add 30 mL of a mineral salts solution of the following composition (grams per liter): $MgSO_4$, (2); $CaCl_2$, (0.8); K_2HPO_4, (5); pH 6.5.
3. Add 10 mL of a filter-sterilized solution of yeast extract (12 g L^{-1}) to give a final concentration of 0.4-g yeast extract L^{-1}.
4. Dispense portions (29.8 mL) into large test tubes (20 × 2.5 cm) or 250-mL flasks, which can be incubated and aerated in a temperature-controlled water bath.
5. To each sample add a 0.2-mL portion of a stock solution of nalidixic acid (1.5 mg mL^{-1} in 0.01 M NaOH) to provide a final concentration of 10 μg mL^{-1}. At 4 to 6-h intervals over a period of 24 h, sacrifice replicate tubes by adding formalin (2%, vol/vol, final concentration) and immediately place on ice, or in a refrigerator until processed for microscopy.
6. Include control samples of the diluted soil suspensions to which only the substrate or nalidixic acid is added.
7. Prepare samples for microscopy by staining with acridine orange (section 6–2.1) or with FA (section 6–2.2).
8. Enumerate both total and elongated bacterial cells with the aid of a calibrated whipple disc inserted into the eyepiece of the microscope. Count 25 to 30 fields of view at 1000X. Although the definition of an elongated cell is somewhat arbitrary, I consider a cell of ≥ 4 μm in length to be elongated.

6–3.3 Comments

In studies from my laboratory to assess the viability within serotypes of soil populations of *Rhizobium leguminosarum* bv. *trifolii,* several observations have been made. Substrate quality and concentration can influence the proportion of elongated cells. Concentrations of yeast extract < 100 mg per liter can limit the rate of appearance of elongated cells and their final number. Filter sterilization of substrates is recommended since autoclaved yeast extract significantly delayed the elongation of some soil rhizobial serotypes. Regardless of sterilization procedure, some substrates can delay cell elongation and produce a significantly lower total percentage of elongated cells than yeast extract (e.g., brain heart infusion and tryptone). The concentration of nalidixic acid may need to be increased or decreased in some situations either because it cannot prevent cell proliferation in the population, or because it is too toxic and prevents cell elongation. In either case, the number of viable cells is underestimated. Concentrations of nalidixic acid ranging from 10 to 30 mg L^{-1} are satisfactory for indigenous populations of *R. leguminosarum*. We have found cell elongation in soil rhizobial populations to be highly sensitive to other DNA gyrase inhibitors such as ciprofloxacin and norfloxacin. For example, in some soil samples, concentrations ≤ 1.0 mg L^{-1} were sufficient to prevent proliferation of soil rhizobia and caused cell elongation. In other soil samples, recovered on different occasions from the same site, the same concentrations of the latter inhibitors were toxic to the cells and prevented elongation from occurring. Researchers should carry out preliminary experiments to evaluate choice and concentration of substrate and gyrase inhibitors.

6–4 DETERMINING THE PROPORTION OF VIABLE SOIL BACTERIA BY FOLLOWING THE REDUCTION OF TETRAZOLIUM DYES TO FORMAZAN

6–4.1 Principle

The principle behind this method lies with the use of tetrazolium salts as electron acceptors from metabolism resulting in the formation of insoluble formazan salts that are deposited inside the cell. The following procedure was described by Norton and Firestone (1991).

6–4.2 Procedure

1. Disperse soil samples equivalent to 1 g of oven-dry weight in 20 mL of 0.05 *M* Tris-HCl, pH 7.5. Blend at low speed for 1 min.
2. To a scintillation vial, add a 2-mL portion of the homogenized soil suspension along with 1 mL of an INT solution (2 mg mL^{-1} 2-(*p*-iodophenyl)-3-(*p*-nitrophenyl)-5-phenyltetrazolium chloride, in 0.2 *M* Tris-HCl, pH 7.5), and 1 mL of a solution containing NADH and NADPH (4 mg per mL of each in 0.2 *M* Tris-HCl, pH 7.5).

3. Incubate the samples in the dark at 25 °C for 4 h.
4. After 4 h of incubation fix the samples with 1 mL of 20% formalin and store overnight at 4 °C to maximize formazan crystal deposition.
5. Disperse the samples by immersing vials in a Branson ultrasonic cleaner for 2 min at "low power" (50-W output).
6. Add an agar solution to provide a final concentration of 0.01% (wt/vol).
7. Dispense a 0.01-mL portion of suspension onto a 1-cm^2 area marked on an acid-washed slide. Air dry and heat fix the smear.
8. Stain the specimens with fluorescein isothiocyanate (FITC) as follows.

6–4.3 Staining Procedure (from Schmidt & Paul, 1982)

1. Dissolve 1 mg of FITC in a pre-filtered solution consisting of 0.25 mL of 0.5 M carbonate-bicarbonate buffer (pH 9.6), 1.1 mL of 0.01 M phosphate buffer (pH 7.2), and 1.1 mL of 0.15 M NaCl. The solution must be stored in the dark at 4 °C where it is usable for up to 6 h.
2. Stain smears for 3 min with the FITC solution.
3. Destain with 0.5 M carbonate-bicarbonate buffer for 10 min, and then with 5% (wt/vol) sodium pyrophosphate for 2 min. Rinse the slide with deionized water and allow to air dry.
4. Mount the smears in carbonate-buffered glycerol (5 mL of 1 M Na$_2$CO$_3$ in 45 mL of glycerol).
5. Visualize the cells by epifluorescence microscopy with a fluorescein filter set. Determine the proportion of cells that contain formazan deposits by viewing the specimen with phase contrast microscopy.

6–4.4 Comments

It can take some time, patience, and experience to achieve the correct intensity of light and contrast to see the formazan deposits. It can be helpful to vary the intensity of the transmitted light, and experiment with different colored filters over the light source to generate the contrast for definitive identification of formazan deposits within bacterial cells. Few attempts have been made to evaluate the proportion of soil bacterial populations that are viable and respiratorily competent. Nonetheless, three different experimental approaches, two of which are described above (Norton & Firestone, 1991; Bottomley & Maggard, 1990), and an autoradiographic approach (Ramsay, 1984) have given similar ranges of values for the proportions of soil bacterial populations that are respiratorily competent (41–57%), (64–75%), and (58–71%), respectively.

6–5 MICROSCOPIC DETERMINATION OF THE MYCELIAL LENGTH OF SOIL FUNGI

Since 1970, soil scientists have attempted to use light microscopy to measure the fungal contribution to soil microbial biomass. While efficiency of recovery has been the primary concern when dealing with soil bacteria, an additional problem is encountered with soil fungi. A balance must be achieved between homogenizing soil sufficient to release mycelia and yet insufficient to cause excessive fragmentation of the mycelia.

6–5.1 Agar Film Technique

Lodge and Ingham (1991) compared different agar film techniques for estimating mycelial length of fungi in soil and litter. The authors emphasized three issues.

1. Hand shaking was superior to blending for separating mycelia from soil; blending, however, was superior for litter samples.
2. Emphasis was placed upon the importance of carrying out preliminary experiments to determine the combination of lowest soil dilution and lowest magnification of viewing that result in the greatest estimate of mycelial length.
3. The importance of preliminary experiments to determine the magnitude of the different sources of error in the measurement of mycelial length was emphasized. For example, determinations should be made of the following sources of variation: (i) among estimates of mycelial length on different agar films from the same soil suspension, (ii) among estimates made on agar films prepared from different soil suspensions from the same soil sample, and (iii) among estimates made on agar films prepared from different soil samples. Once these values are ascertained then the number of soil samples, agar films per sample and measurements per agar film can be determined for each experimental situation. The following section describes the agar film technique of Lodge and Ingham (1991).

6–5.1.1 Procedure

1. Place 1 g of soil in 9 mL of phosphate buffer, 0.6 M, pH 7.6.
2. Shake by hand for 5 min.
3. Transfer 1 mL of soil suspension to 9 mL of buffer to generate a 100-fold dilution.
4. Transfer 1-mL portions of the soil dilutions (1:10 and 1:100) to test tubes.
5. To each sample add 1 mL of fluorescein diacetate (FDA, 20-mg per liter in 0.6 M phosphate buffer, pH 7.2) and incubate for 3 min.
6. Add 1 mL of 5% (wt/vol) liquid agar (60 °C) to each tube and mix the contents thoroughly.

7. Add a drop of the agar suspension to coverslip well slides. The well is constructed by taping two no. 1 coverslips (0.13- to 0.15-mm thick) to a microscope slide at a distance of 1-cm apart.
8. Immediately place a third coverslip over the agar film to create a film of known thickness. The amount of agar suspension should be sufficient to fill the coverslip well.
9. Examine the agar films at magnifications of 160, 450, and 1000X with phase contrast microscopy to estimate total mycelial length. Twenty fields of view should be examined along each of three transects across each of the agar films. The dilution that results in superior viewing and counting of hyphae should be used for further estimations with that particular soil.
10. Examine the agar films using epifluorescence microscopy to determine the proportion of the mycelial length that is active in the hydrolysis of FDA.

6–5.1.2 Comments

Lodge and Ingham (1991) showed that, in general, mycelial length is underestimated in soil samples diluted > 100-fold and examined at high magnification (X1000). However, particle interference in some soils made observations of low dilutions a nonpractical option. The coefficient of variation (CV) for estimates of mycelial length in agar suspensions prepared from the same soil or litter sample was excellent (8%), and rarely as high as 26%. The CVs among samples collected on the same date ranged between 19 to 30% for litter and 28 to 67% for soil.

6–5.2 Membrane Filtration Method for Determining the Mycelial Length

West (1988) used both membrane filtration and agar-film methods, in combination with several different stains (fluorescein isothiocyanate, acridine orange, phenolic aniline blue, and the polysaccharide binding agent, calcofluor M2R), for estimating mycelial length in soil. He concluded that a combination of membrane filtration and calcofluor produced the greatest estimate of mycelial length. The method is described below.

6–5.2.1 Procedure

1. Suspend soil equivalent to 1 g of oven-dry weight in 50 mL of a dilute Ringers solution (g L^{-1}): NaCl (2.25); KCl (0.11); $CaCl_2$ (0.12); $NaHCO_3$ (0.05).
2. Homogenize in a blender for 1 to 4 min.
3. Dilute the suspension another 10-fold with Ringers solution to a final value of 1:500.
4. Filter a 1-mL portion onto a "blackened" 1.0-μm pore size polycarbonate membrane (see section 6–2.1.3).

5. Stain the membrane for 24 h at room temperature with an aqueous solution of calcofluor white, M_2R (2.3 mg per mL) fluorescent brightener type A (Polysciences, Inc., Warrington, PA).
6. Destain the membrane with distilled water.
7. Mount the membrane on a drop of distilled water. Apply a no. 1 coverslip.
8. View the specimen at 1000X magnification with incident light using the UV filter set described earlier for DAPI staining (see section 6–2.1.9).

6–5.2.2 Comments

Although the greatest value for mycelial length was obtained with the fluorescent brightener, discrimination between cytoplasm-containing and "ghost" mycelia cannot be made. Other methods are required to determine the proportion of mycelial length that is metabolically competent. Two methods are described below.

6–6 DETERMINING THE PROPORTION OF METABOLICALLY ACTIVE FUNGAL MYCELIA BY FOLLOWING THE HYDROLYSIS OF FLUORESCEIN DIACETATE

Active cytoplasm within mycelia contains the enzyme esterase. The latter cleaves a nonfluorescent ester, fluorescein diacetate (FDA), into fluorescein that can be detected by epifluorescence microscopy. The method of Stamatiadis et al. (1990) is described below.

6–6.1 Procedure

1. Suspend 10 g of field moist soil in 95 mL of 60-mM phosphate buffer adjusted to the pH of the soil. Shake at 225 rev min^{-1} for 15 min.
2. Dilute 1 mL of the soil suspension into 4 mL of phosphate buffer to produce a 50-fold soil dilution.
3. Incubate 1 mL of the 50-fold soil dilution with 1 mL of filter-sterilized fluoroscein diacetate (FDA, 20 mg L^{-1}) for 3 min at room temperature.
4. Add 1 mL of a water agar solution (1.5% wt/vol in phosphate buffer, pH 7.6).
5. Transfer 0.1 mL of the suspension to a microscope slide containing a cavity of known volume.
6. As soon as possible, examine the slides under incident light microscopy at 400 to 1000X to determine the proportion of FDA-hydrolyzing mycelia.
7. Determine the total mycelial length by examining the specimen with phase contrast optics.

6–6.2 Comments

Several researchers have commented on the fact that the fluorescence of the fluorescein liberated in the mycelia decays within a few hours of storage at room temperature. This issue is not as serious a problem as originally thought, provided that the specimens are mounted in agar containing a buffer of $\geq pH$ 7.5. Although some researchers have reported that the proportion of FDA-hydrolyzing mycelia is low, and that the enzyme activity is restricted to growing mycelial tips, this is not always the case. E. Ingham (personal communication) has observed strong seasonal influences with mycelial activity ranging from virtually zero in summer to 100% of mycelial length being FDA-active in spring time.

6–7 DETERMINING THE PROPORTION OF METABOLICALLY ACTIVE FUNGAL MYCELIA BY FOLLOWING THE REDUCTION OF TETRAZOLIUM DYES TO FORMAZAN

Norton and Firestone (1991) reported success in using INT reduction to determine the proportion of respiratorily active mycelia within fungal populations recovered from microcosm soil that had been in close proximity to roots of ponderosa pine (*Pinus ponderosa* Dougl.). The method is described below.

6–7.1 Procedure

1. Prepare 20-fold soil dilutions, incubate with a solution of INT and NADH-NADPH, and fix with formalin as described in section 6–4.2.
2. Stain a 1-mL portion of soil suspension overnight at 4 °C with 0.5 mL of trypan blue (0.5% wt/vol aqueous solution).
3. Add 98.5 mL of 60-mM phosphate buffer (pH 7.6) to produce 0.1 mg soil per mL.
4. Filter 15 mL (1.5 mg of soil) onto a 25-mm (0.2-μm pore size) "blackened" polycarbonate membrane (see section 6–2.1.3).
5. Dip an acid-washed slide into a warm solution consisting of 5% (wt/vol) gelatin and 0.05% (wt/vol) $KCr(SO_4)_2 \cdot 12H_2O$, and place upon an ice-cold metal tray.
6. Place one-half of the filter (microbe-side down) on the gelatin layer.
7. Dry the slide and filter combination overnight over silica gel.
8. Equilibrate the slides for 2 to 3 h in an atmosphere of 70% relative humidity and carefully peel off the filter.
9. Spray the gelatin film with a fine mist of a solution of 2% (wt/vol) gelatin in 0.05% (wt/vol) $KCr(SO_4)_2$.
10. Place a cover slip over the gelatin film and observe microscopically at 500X with phase contrast microscopy.

6–7.2 Comments

With the method described above, a large portion (40–50%) of the fungal mycelia was found to be respiratorily active. Since the samples were recovered from the rhizosphere zone of tree seedlings growing under microcosm conditions, it is of interest to speculate about what percentages of the active mycelia were ectomycorrhizae or soil saprophytes.

6–8 DETERMINING THE WEIGHT OF SOIL BIOMASS FROM MICROSCOPIC ESTIMATES OF BIOVOLUME

Although difficult and time consuming, microscopic methods have been used to determine the magnitude of soil biomass. To achieve this objective, the researcher needs to be able to confidently measure length and width of cells, and to know a dry weight/cellular volume conversion factor. In the case of the conversion factor, there is controversy over whether the water content of soil microorganisms is influenced by soil water potential.

6–8.1 Influence of Cellular Water Content on a Dry Weight: Biovolume Conversion Factor

1. If the volume of a bacterial cell equals 1 μm^3, and the cell has a wet density of 1.1×10^{-12} g μm^{-3}, then the wet weight of that cell equals 1.1×10^{-12} g. If the cell is 80% by weight water and the specific gravity of water is 1.0, then the cell contains 0.88×10^{-12} g of water. By subtraction, the solid content of the cell weighs 0.22×10^{-12} g, and the dry weight/biovolume ratio is 0.22×10^{-12} g μm^{-3}, or 0.22 g cm^{-3} (10^{12} $\mu m^3 = 1$ cm^3).

2. If the water content of the same bacterial cell is only 70%, then the cell contains 0.77×10^{-12} g of water. The solid constituents weigh 0.33×10^{-12} g, and the dry weight/biovolume conversion is 0.33×10^{-12} g μm^{-3}, or 0.33 g cm^{-3}.

6–8.2 Comments

Studies by Bakken and Olsen (1983) and Van Veen and Paul (1979) have shown that the percent solids in laboratory-grown soil bacteria is greater than the traditionally quoted value of 20%. Bakken and Olsen (1983) determined that the dry weight/biovolume conversion factors for soil bacteria and soil fungi grown in the absence of water-stress were 0.33 g cm^{-3}, and 0.23 g cm^{-3}, respectively. Van Veen and Paul (1979) determined dry weight/biovolume conversion factors for soil fungi growing with and without water stress to be 0.34 and 0.2 g cm^{-3}, respectively.

6–8.3 Calculation of Bacterial Biomass (W_B)

Bacterial biomass ($\mu g\ g^{-1}$ oven-dry soil) is given by the following equation:

$$NV\ B^b\ 10^6 = W_B$$

N = Number of bacteria g^{-1} of oven-dry soil.
V = Average volume of bacterial cell (μm^3).
B^b = Biomass/biovolume conversion factor; 0.22 to 0.33×10^{-12} g μm^{-3}
10^6 = Factor required to convert g to μg.

6–8.4 Calculation of Fungal Biomass (W_F)

Fungal biomass ($\mu g\ g^{-1}$ oven-dry soil) is given by the following equation:

$$L\pi\ r^2\ B^f\ 10^6 = W_F$$

L = Mycelial length in $\mu m\ g^{-1}$ oven-dry soil.
r = Average radius of mycelia (μm).
B^f = Biomass/biovolume conversion factor (0.2–0.33×10^{-12} g μm^{-3}).
10^6 = Factor required to convert g to μg.

6–8.5 Comments

Obviously, the choice of the biovolume/biomass conversion factor influences the magnitude of biomass estimated from biovolume determinations. The efficiency of recovering microorganisms from soil will also impact biomass computations. In addition, determining the width and length of bacterial cells is not an easy task with stains that react only with intracellular contents such as nucleic acids. Stains, such as fluorescent antibodies, that react with specific components of cell walls or "brighteners" such as calcofluor that react with cell wall polysaccharides are more suitable for this task.

6–9 MICROSCOPIC ENUMERATION OF BACTERIA WITH FLUORESCENTLY LABELED OLIGONUCLEOTIDES DIRECTED AT SPECIFIC REGIONS OF 16S RIBOSOMAL RNA

In recent years, researchers have developed methods to label oligonucleotides with the same fluorescent compounds traditionally used for antibody labeling. Oligonucleotide sequences can be synthesized that are complementary to specific regions within the 16S rRNA molecule. Specificity of hybridization can be achieved at levels ranging from kingdom through species (Giovannoni et al., 1988). While other chapters in this book are devoted specifically to the development and use of oligonucleotide probes, I will focus generically on their microscopic use on wholecells. Although this method is in the early stages of development for use in soil, it will be only a matter of time before it becomes commonplace. Two procedures are described below.

6–9.1 Procedure (from Amann et al., 1990)

6–9.1.1 Preparation of Cell Smears

1. Soak Teflon-coated glass slides (six hybridization wells per slide) for 1 h in ethanolic KOH (10% wt/vol KOH in 95% vol/vol ethanol). Rinse in deionized water.
2. After air drying, dip the slides into a hot (70 °C) solution of 0.1% (wt/vol) gelatin in a 0.01% (wt/vol) aqueous solution of $KCr(SO_4)_2$.
3. Fix 1 mL mid-log phase cells (10^9 cells per mL) in 3 mL of formalin (4% vol/vol) in 0.2 M sodium phosphate buffer, pH 7.2. Incubate at room temperature for at least 3 h and store at 4 °C.
4. Mix the fixed cell suspension with 0.4 mL of 1% (vol/vol) Nonidet P-40 (Sigma Chem. Co., St. Louis, MO) and pellet the cells by low speed centrifugation.
5. Resuspend the cells in 1 mL of Nonidet-P40 solution.
6. Add portions (3 µL ca. 10^6 cells) to each well, air dry, and dehydrate the cells with consecutive 3 min exposures to solutions of 50, 80, and 100% (vol/vol) ethanol, respectively.
7. Store at room temperature.

6–9.1.2 Whole-cell Hybridization Procedure

1. To each well on the slide add a 9-µL portion of a hybridization mixture containing (0.9 M NaCl, 50-mM Na phosphate pH 7.0, 5-mM EDTA, 0.1% (SDS), 0.5-mg of poly(A) mL^{-1}, Ficoll, 2 mg mL^{-1}; polyvinyl pyrolidone, 2 mg mL^{-1}; bovine serum albumin, 2 mg mL^{-1}.
2. After 30 min of incubation at the hybridization temperature (37 °C), add 50-ng of the appropriate fluorescein or rhodamine conjugated oligonucleotide and continue the incubation for 2 to 5 h.
3. Flush away the hybridization mixture with a warm (37 °C) wash solution (0.9 M NaCl, 50 mM Na phosphate pH 7.0, 0.1% SDS).
4. Immerse the slide in the wash solution (37 °C) for 15 min.
5. Rinse the slide in distilled water and air dry.
6. Mount the slide in distilled water and view with epifluorescent microscopy at 1000X under oil immersion with a filter set appropriate for the specific fluorochrome conjugated to the oligonucleotide probe (rhodamine or fluorescein).

7. If photography is to be attempted, the researcher should be aware that long exposure times may be required for natural samples (≥ 30 s) because of the low level of RNA in cells of low activity.

6–9.2 Procedure (from Tsien et al., 1990)

6–9.2.1 Preparation of Smears

1. Prepare air-dried smears (30 µL of an early log-phase culture, 10^6 cells) on slides predipped into a gelatin-chromate solution.
2. Fix smears in a solution of formalin (3.7% vol/vol) in 90% (vol/vol) methanol for 10 min, rinse briefly in distilled water, and treat with 50-mM Na borohydride for 30 min in the dark with occasional agitation.
3. Rinse slides with water and air dry.

6–9.2.2 Whole-cell Hybridization Procedure

1. To each smear, add 10 µL of a hybridization solution containing 1.7 ng µL^{-1} of the appropriate fluorochrome labeled oligonucleotide probe. Apply a coverslip, and incubate overnight at 37 °C in a moisture-saturated chamber.
2. Remove the coverslip by soaking the slides in ice-cold 5X SET buffer (1X SET is 0.75 M NaCl, 0.1 M Tris-HCl, [pH 7.8], 5-mM EDTA).
3. Rinse the slides three times in 0.2X SET at 37 °C (10 min each rinse).
4. Air dry the slides in the dark prior to microscopic examination.
5. To enhance fluorescence, mount the slides in phosphate-buffered glycerol containing 0.1% p-phenylenediamine.
6. View by epifluorescence microscopy using a filter set appropriate for the specific fluorochrome conjugated to the oligonucleotide.

6–9.2.3 Comments

The intensity of fluorescence is a function of the amount of RNA in the cell (DeLong et al., 1989). As a consequence, cells recovered from many soil environments will have low metabolic activity and will fluoresce poorly. Different strategies are being developed to overcome this problem. For example, two or more different oligonucleotide probes have been used to enhance the amount of fluorescence. They are targeted to different parts of the 16S rRNA molecule of the same organismal group (Amann et al., 1990). Alternatively, specialized cameras with high sensitivity to low levels of light may become a practical solution (S. Giovannoni, 1992, personal communcation).

ACKNOWLEDGMENT

This article represents Technical Paper no. 9938 of the Oregon Agricultural Experiment Station. The author acknowledges the critical review and comments upon the manuscript by Dr. Elaine Ingham and the word processing skills of Ms. C. Pelroy.

REFERENCES

Amann, R.I., Krumholz, L., and D.A. Stahl. 1990. Fluorescent-oligonucleotide probing of whole cells for determinative, phylogenetic and environmental studies in microbiology. J. Bacteriol. 172:762–770.

Bakken, L.R. 1985. Separation and purification of bacteria from soil. Appl. Environ. Microbiol. 49:1482–1487.

Bakken, L.R., and R.A. Olsen. 1983. Buoyant densities and dry-matter contents of microorganisms. Conversion of a measured biovolume into biomass. Appl. Environ. Microbiol. 45:1188–1195.

Bohlool, B.B., and E.L. Schmidt. 1968. Nonspecific staining: Its control in immunofluorescence examinaton of soil. Science 162:1012–1014.

Bottomley, P.J., and M.H. Dughri. 1989. Population size and distribution of *Rhizobium leguminosarum* bv. *trifolii* in relation to total soil bacteria and soil depth. Appl. Environ. Microbiol. 55:959–964.

Bottomley, P.J., and S.P. Maggard. 1990. Determination of viability within serotypes of a soil population of *Rhizobium leguminosarum* bv. *trifolii*. Appl. Environ. Microbiol. 56:533–540.

DeLong, E.F., G.S. Wickham, and N.R. Pace. 1989. Phylogenetic stains: Ribosomal RNA-based probes for the identification of single cells. Science 243:1360–1363.

Demezas, D.H., and P.J. Bottomley. 1986. Autecology in rhizospheres and nodulating behavior of indigenous *Rhizobium trifolii*. Appl. Environ. Microbiol. 52:1014–1019.

Foster, R., A.D. Rovira, and T. Cock. 1983. Ultrastructure of the root-soil interface. Am. Phytopathol. Soc., St. Paul.

Giovannoni, S.J., E.F. DeLong, G.J. Olsen, and N.R. Pace. 1988. Phylogenetic group-specific oligonucleotide probes for identification of single microbial cells. J. Bacteriol. 170:720–726.

Kingsley, M.T., and B.B. Bohlool. 1981. Release of *Rhizobium* spp. from tropical soils and recovery for immunofluorescence enumeration. Appl. Environ. Microbiol. 42:241–248.

Kogure, K., U. Simidu, and N. Taga. 1979. A tentative direct microscopic method for counting living marine bacteria. Can. J. Microbiol. 25:415–420.

Lodge, D.J., and E.R. Ingham. 1991. A comparison of agar film techniques for estimating fungal biovolumes in litter and soil. Agric. Ecosyst. Environ. 34:131–144.

Macdonald, R.M. 1986a. Sampling soil microfloras: Optimization of density gradient centrifugation in Percoll to separate microorganisms from soil suspensions. Soil Biol. Biochem. 18:407–410.

Macdonald, R.M. 1986b. Sampling soil microfloras: Problems in estimating concentration and activity of mixed populations of soil microorganisms. Soil Biol. Biochem. 18:411–416.

Meyer-Reil, L.-A. 1978. Autoradiography and epifluorescence microscopy combined for the determination of number and spectrum of actively metabolizing bacteria in natural waters. Appl. Environ. Microbiol. 36:506–512.

Norton, J.M., and M.K. Firestone. 1991. Metabolic status of bacteria and fungi in the rhizosphere of Ponderosa pine seedlings. Appl. Environ. Microbiol. 57:1161–1167.

Paul, J.H. 1982. Use of Hoechst dyes 33258 and 33342 for enumeration of attached and planktonic bacteria. Appl. Environ. Microbiol. 43:939–944.

Porter, K.G., and Y.S. Feig. 1980. The use of DAPI for identifying and counting aquatic microflora. Limnol. Oceanogr. 25:943–948.

Ramsay, A.J. 1984. Extraction of bacteria from soil: Efficiency of shaking or ultrasonication as indicated by direct counts and autoradiography. Soil Biol. Biochem. 16:475–481.

Robert, F.M., and E.L. Schmidt. 1983. Population changes and persistence of *Rhizobium phaseoli* in soil and rhizospheres. Appl. Environ. Microbiol. 45:550–556.

Schmidt, E.L. 1974. Quantitative autecological study of microorganisms in soil by immuno-fluorescence. Soil Sci. 118:141–149.

Schmidt, E.L., and E.A. Paul. 1982. Microscopic methods for soil microorganisms. p. 1027–1042. *In* A.L. Page et al. (ed.) Methods of soil analysis: Chemical and microbiological methods. Agron. Monogr. 9. 2nd ed. ASA and SSSA, Madison, WI.

Stamatiadis, S., J.W. Doran, and E.R. Ingham. 1990. Use of staining and inhibitors to separate fungal and bacterial activity in soil. Soil Biol. Biochem. 22:81–88.

Starkey, R.L. 1938. Some influences of the development of higher plants upon the micro-organisms in the soil. VI. Microscopic examination of the rhizosphere. Soil Sci. 45:207–249.

Tsien, H.S., B.J. Bratina, K. Tsuji, and R.S. Hanson. 1990. Use of oligonucleotide signature probes for identification of physiological groups of methylotrophic bacteria. Appl. Environ. Microbiol. 56:2858–2865.

Van Veen, J.A., and E.A. Paul. 1979. Conversion of biovolume measurements of soil organisms grown under various moisture tensions to biomass and their nutrient content. Appl. Environ. Microbiol. 37:686–692.

West, A.W. 1988. Specimen preparation, stain type, and extraction and observation procedures as factors in the estimation of soil mycelial lengths and volumes by light microscopy. Biol. Fert. Soils 7:88–94.

Wollum, A.G., and R.H. Miller. 1980. Density centrifugation methods for recovering *Rhizobium* spp. from soil for fluorescent antibody studies. Appl. Environ. Microbiol. 39:466–469.

Chapter 7

Viruses

J. SCOTT ANGLE, *University of Maryland, College Park, Maryland*

Despite the fact that viruses may be the most numerous microorganisms in soil, we know little about their population dynamics, distribution, and significance. It is especially unclear how viruses interact with other soil microorganisms. Many reports, however, have suggested possible interactive roles of viruses in soil. A significant role, which has been studied for many years, is the direct lysis of soil bacteria. Direct lysis of soil bacteria by viruses called *bacteriophages* (abbrev. phages) has been shown to reduce populations of susceptible hosts. "Fatigued alfalfa fields" in France were attributed to the phage-mediated reduction of the soil population of rhizobia (Demolon & Dunez, 1935, 1936). More recent reports have confirmed that the soil population of specific rhizobia can be reduced several log units by the addition of an appropriate virulent phage to soil (Hashem & Angle, 1988, 1990).

Phages are also intimately involved in the survival and evolution of other soil microbes. Recent evidence has demonstrated that phages in soil may be involved in the process of transduction (Saye et al., 1990; Stotzky, 1989). Small genetic elements are carried between microorganisms when packaged within a phage particle. Direct genetic transfer, as mediated by phages, may be important in the evolution of host microorganisms infected by these particles. The infection of host organisms by phages and the establishment of a lysogenic association could significantly enhance natural variation and the subsequent rate by which species adapt to changes in their environment (Reanney, 1974). The importance of transduction, as related to natural variation, however, is controversial and the subject of intense speculation (Selander & Levin, 1980).

When a large number of phages are isolated that specifically lyse a particular species of bacteria, a phage-typing system can be developed to identify bacterial isolates beyond the species level (Brown et al., 1978; Lesley, 1982). Phage typing and identification has found particular application in the identification of *Rhizobium* and *Bradyrhizobium* (Basit et al., 1991; Lesley, 1982; Staniewski, 1970; Staniewski & Kowalska, 1974).

To examine the potential significance of phages in soil, it is necessary to have an accurate and reliable assay method. Detection and enumeration of phages in soil, however, are confounded by many practical and theoretical difficulties. Virulent particles outside the host cell can potentially be extracted and counted. Viruses in soil may be either in a virulent or lysogenic state. Many phages are integrated into the genome of host organisms and therefore cannot be detected using traditional assay methods. Reanney et al. 1983 suggested that the vast majority of phages in soil were in the lysogenic state and that only a small portion of the population was in the free, virulent state. Hence, enumeration procedures may only count a small fraction of the total population in soil. Additional problems are encountered during enumeration with the extraction of phages from soil, a problem inherent in the enumeration of any group of microorganisms in soil. Phages are tightly adsorbed to soil colloids (Burge & Enkiri, 1978; Duboise et al., 1979; Sykes & Williams, 1978) and only a small fraction may be desorbed from soil (Reanney & Marsh, 1973).

Detection and enumeration of phages in soil requires that the viruses be culturable on an appropriate host. Individual phages are defined by their ability to infect specific hosts. Phages have been isolated from soil (Table 7-1) that specifically infect a wide variety of soil microorganisms. When actual numbers of phages have been enumerated in soil, populations have been reported to be extremely low. Tan and Reanney (1976) found only a few *Bacillus* phages per gram dry soil. Casida and Liu (1974) reported that phages for *Arthrobacter* were rarely isolated from soil, unless numbers of the host in soil were artificially enriched, even when soil extracts were concentrated several-fold. While it is likely that the number of phages in soil is low due to the lysogenic state of the population, it is also possible that current methods of extraction are highly inefficient, thereby underestimating the true population. This point must always be considered when the number of phages in soil is determined.

Observations in the aquatic environment indicate that phages are extremely abundant and conventional plaque assay techniques enumerate only a small fraction of the total number present (Proctor et al., 1988; Torella & Morita, 1979). Using ultracentrifugation to concentrate and "semiquantitatively" count viruses in water, Bergh et al. (1989) demonstrated that actual numbers were 10^3 to 10^7 higher than were enumerated by the standard plaque assay. Numbers in soil are also likely to be underestimated using conventional plaque assay techniques given the difficulty in extracting viruses from soil and the diverse host range associated with an equally diverse bacterial population.

From an ecological perspective, phages are the most important viruses in soil, however, exogenous additions of human viruses to soil have also received considerable attention. These viruses are typically of enteric origin. Individuals infected with clinical or subclinical enteric viruses may shed high numbers of viral particles in feces. Although sewage treatment reduces particle viability, high numbers may remain in sludge and effluent (Grabow, 1968; Irving & Smith, 1981). Land application of effluent and

Table 7–1. Examples of phages isolated from soil.

Phage host	Reference
Bradyrhizobium japonicum	Hashem et al., 1986
Rhizobium leguminosarum biovar *trifolii*	Barnet, 1972
Rhizobium meliloti	Marants, 1974
Rhizobium leguminosarum biovar *phaseoli*	Moskalenko & Pautenshi, 1969
Arthrobacter globiformis	Casida & Liu, 1974
Bacillus stearothermophilus	Reanney & Marsh, 1973
Bacillus subtilis	Romig & Brodetsky, 1961
Pseudomonas aeruginosa	Bradley, 1966
Streptomyces spp.	Williams & Lanning, 1984
Nocardia spp.	Williams et al., 1980

sludge is widely practiced throughout the USA and effectively introduces these enteric viruses into the soil. Numerous studies over the past 20 yr have examined the leaching through (Dizer et al., 1984; Gilbert et al., 1976; Lance & Gerba, 1980; Pancorbo et al., 1988) and survival (Bitton et al., 1984; Dubois et al., 1979; Hurst et al., 1980; Farrah et al., 1981) in soil of enteric viruses.

Detection of enteric viruses in soil uses techniques similar to that used for phage. Viruses are initially extracted from soil, often with the use of an eluent capable of desorbing viruses from soil colloids. Since enteric viruses are not capable of reproducing in soil (the appropriate host is not present) populations may be low, depending upon the time after addition and the potential for survival of the virus. It may, therefore, be necessary to concentrate viruses prior to the plaque assay. Filtration, flocculation, and centrifugation procedures have been used to concentrate viruses prior to analysis. A tissue culture plaque assay, appropriate for the virus being studied, is used at this stage of detection and enumeration (Hurst et al., 1991).

Two other groups of viruses, those that infect insects and plants, have also been extensively studied. According to Farrah and Bitton (1990), more than 600 plant viruses have been identified, many of which can survive for extended periods of time in soil. These authors also discuss viruses that attack insects, especially in light of their potential use as biological control agents.

7–1 GENERAL PRINCIPLES OF ANALYSIS

Numerous methods are available for studying viruses in soil. For phages, when the population in soil is suspected to be low, an initial enrichment procedure can be used. By increasing the number of host organisms in soil, a concurrent increase in phage numbers is also observed (Bradley et al., 1961; Germida & Casida, 1983).This general concept is often followed to isolate a single phage for ecological or genetic studies. When phages in soil are present in sufficient numbers to adequately count, and it is necessary to determine the actual number of phages in soil, a direct counting procedure can be used.

It should be noted that several alternative methods exist to directly examine viruses in soil, however, they have not been convincingly shown to be useful for enumeration. Electron microscopy (Bystricky et al., 1975), immunofluorescence microscopy (Hukuhara & Akami, 1987), and the enzyme-linked immunosorbent assay (ELISA) (Fuxa et al., 1985) have all been used to study viruses in soil. While never having gained wide acceptance, future refinements may make these techniques important in the study of viruses in soil. Use of gene probes also offers promise as a means to detect enteric viruses in soil (Preston et al., 1990). Since this technique does not require bacterial cells or specific mammalian tissue cultures, it avoids the restriction imposed by limited host range. Unfortunately, the use of gene probes is not yet sufficiently developed to be employed as a standard method in enumeration of viruses in soil. Use of the polymerase chain reaction (PCR) to amplify low copy number of viral DNA (or RNA) may find application in the future, especially when methods of nucleic acid extraction from soil are improved. Theoretically, the use of PCR would allow for the detection of a single viral particle.

This chapter will describe the most widely used techniques for the detection and enumeration of phage and enteric viruses in soil. When phage or enteric viral populations are present in higher numbers (>1000 PFU g^{-1} dry soil), a direct counting method can be employed. When populations are low in soil (<1000 PFU g^{-1} dry soil), enrichment or concentration methods must be first used. The techniques in the current chapter are relatively simple and do not require expensive supplies or equipment. Most microbiological laboratories are capable of routinely carrying out these procedures.

7–2 PHAGES

7–2.1 Direct Counts

7–2.1.1 Materials

1. Diluents
   a. Physiological saline diluent: Dissolve 8.5 g NaCl in 1000 mL of distilled water. Autoclave.
   b. Buffered 10% beef extract: Dissolve 10 g of beef extract powder, 1.34 g $Na_2HPO_4 \cdot 7H_2O$ and 0.12 g citric acid in 100 mL distilled water. Dissolve on a magnetic stirrer. This solution may be safely autoclaved at 121 °C for 15 min.
2. Dilution bottles.
3. Reciprocal shaker.
4. Centrifuge capable of speeds of 2000 × g.
5. Chloroform.

6. Membrane filters, 0.20-μm pore size.
7. Biological media, specifically capable of supporting host bacterium, poured into petri plates.
8. Soft-agar overlays: Prepare broth of biological media capable of supporting the growth of the host bacterium. Add agar to broth at a concentration of 0.7%. Bring to boil to melt agar. Pipet 5.0 mL of soft agar into 10-mL test tubes. Cap and autoclave. Soft agar should be maintained at 50 °C until used.
9. Dilution tubes containing 9.0 mL of diluent.
10. Incubator.

7–2.1.2 Procedures

The method discussed below was originally described by Adams (1959). Minor modifications to the original procedure have been developed to enhance phage recoverability from soil (Lanning & Williams, 1982; Williams & Lanning, 1984). Prepare dilution blanks containing 95-mL diluent. Each of the initial dilution bottles should also be amended with 0.5% chloroform (Crosse & Hingorani, 1958). Chloroform at this concentration will kill all soil microbes except for phages, which are usually not affected by chloroform. A small percentage of phages, however, are sensitive to chloroform (Billing, 1969). When chloroform-sensitive phages are enumerated, the initial soil dilution, after shaking and centrifugation, must be filtered through a 0.20-μm pore size membrane filter. If the chloroform sensitivity of the phage is not known, and if an accurate estimate of the soil population is required, a preliminary experiment may be required to establish chloroform sensitivity. The diluent may consist of physiological saline, especially if the soil contains a high indigenous phage titer or has previously been spiked with a phage or host bacterium. When low phage populations are anticipated, a beef extract diluent may enhance extraction efficiency (USEPA, 1989).

Aseptically, add 10.0 g of field moist soil to the initial dilution bottle, cap, and shake on a reciprocal shaker at 300 oscillations per minute for 30 min. Pour the contents into a sterile centrifuge tube. Centrifuge the contents of the initial dilution at $2000 \times g$ for 15 min to separate the soil from the supernatant. The supernatant from the initial dilution should then be diluted to the extinction point (generally 10^{-9} unless previously established) in tubes containing 9.0 mL of diluent. As previously discussed, if chloroform-sensitive phages are being enumerated, the supernatant may be passed through a 0.20-μm pore size membrane filter as an alternate means of sterilizing the supernatant. It should be noted, however, that a significant portion of the phages may non-specifically adsorb to the filter, thereby giving lower phage numbers than are actually present in the soil. Collard (1970) reported that phage titers were reduced approximately 90% during passage through a membrane filter. If it is suspected that recovery efficiency will be low, efficiency can be estimated by spiking soil with a known phage concentration, followed by immediate recovery.

Phage in the various supernatant dilutions are assayed using the soft agar overlay technique. The host bacterium should be grown to mid-log phase in the broth medium that best supports its growth. Soft agar tubes (5.0 mL) should be autoclaved immediately prior to use and allowed to cool to approximately 50 °C. The soft agar is then amended with 0.5-mL of broth culture of the host and 0.5 mL of the phage suspension. The contents of the tube are vortexed for 10 s and aseptically poured onto an agar plate containing approximately 20 mL of an appropriate growth media. The plate should be gently rotated to evenly distribute the soft agar over the surface of the plate. Once the soft agar has solidified, the plates are placed into an incubator and held at the optimum temperature for the host bacterium. Plates should not be inverted since the top agar is semi-solid.

Examine the plates daily for the appearance of plaques. Plaques are cleared circular zones within the turbid lawn of bacterial growth (Fig. 7–1). Some plaques may develop quite rapidly, completely lysing the host and clarifying the agar in a short period of time. It is, therefore, important to count the plaques after they have become large enough to easily detect, but have not yet completely covered the plate. In addition to counting the plaques, it is also important to note the general size and appearance of the plaques. Typical plaques are clear with a sharp margin. Plaques, however, may also occasionally appear turbid with an irregular margin.

Results should be expressed as plaque forming units (PFU) per g of dry soil. Soil moisture content can be determined on samples weighed before and after heating in an oven at 105 °C for 3 d. The number of phage in the field moist soil can then be corrected for the soil moisture content and reported on an oven-dry weight basis.

7–2.2 Enrichment Procedures

7–2.2.1 Materials

1. 100 mL of broth culture appropriate for host bacterium.
2. All materials noted under 6–2.1.

7–2.2.2 Procedures

As previously noted, the number of free-living phage in soil may often be very low (< 1000 PFU g^{-1}). Detection and potential isolation of phages in these soils by standard dilution and plating procedures would, therefore, be difficult. In such instances when it is not important to quantify phage numbers in soil, they can be enriched prior to isolation. Bradley et al. (1961) and Romig and Brodetsky (1961) demonstrated that the enrichment procedure can be used to isolate phages that are extremely rare in soil.

To enrich specific phage populations in soil, a 100-mL broth culture should be inoculated with the host bacterium and grown to mid-log phase. Ten grams of soil is then added to the broth and the mixture incubated for a sufficient period of time to allow high numbers of phage to be produced. Following incubation, the mixture should be centrifuged at 2000 × g to

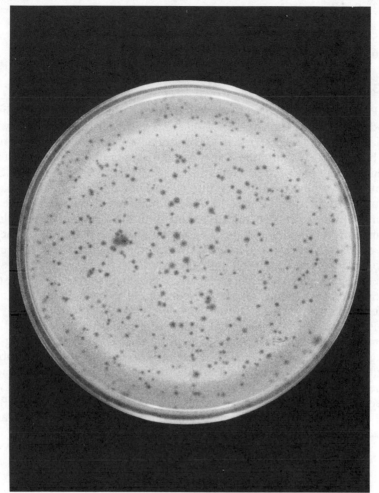

Fig. 7–1. Soft-agar overlay technique showing plaques in the lawn of the host bacterium (*Bradyrhizobium japonicum*).

pellet soil and cellular debris. The supernatant can then be filtered through a 0.20-μm membrane filter or amended with chloroform (0.5 mL per 100 mL broth). The resulting supernatant is assayed using the soft agar overlay technique as described above.

7–2.3 Purification and Storage

7–2.3.1 Materials

1. Biological media capable of supporting host bacterium poured into petri plates.
2. 9.0 mL physiological saline dilution blanks.

3. Sterile scalpel.
4. 100 mL mid-log phase broth culture of host bacterium.
5. Chloroform.
6. 0.20 μm membrane filters.

7–2.3.2 Procedures

Phages can be purified by repeated isolation of single plaques. Single, well-defined plaques from 7–2.1.2 or 7–2.2.2 are cut from the agar with a sterile scalpel and transferred to 100 mL of a log-phase broth culture of the host bacterium. The mixture is incubated for a period of time sufficient to produce a high phage titer. Generally, 12 h is adequate for this step. Since the number of phages in the mixture is not known, it is necessary that the mixture be diluted to at least 10^{-9}. One milliliter of the phage-amended culture should be placed into a 9.0-mL dilution blank and the dilution continued to extinction. Phages in each dilution are plated using the soft agar overlay technique described above. On a plate that shows discrete, individual plaques, single plaques are cut from the agar and the entire procedure repeated. The plaque isolation is usually repeated for a total of five times. At this point, phages are considered to have originated from a single isolate.

Phage numbers in the initial host suspension often exceed 10^{10} PFU per milliliter. The purified phage suspension can be stabilized for long-term storage. The suspension should be filtered through a 0.20 μm membrane filter to remove viable cells and cellular debris. Chloroform is then added to the phage suspension to a final concentration of 0.5%. Phages, when stored in chloroform-amended broth, are viable for many years. Phage titers have been shown to decline only one or two log units over a 5-yr period (Angle, 1992, unpublished data).

7–3 ENTERIC VIRUSES

7–3.1 Direct Counts

7–3.1.1 Materials

1. The same materials as described in 7–2.1 for extraction of phages from soil.
2. Disodium hydrogen phosphate (0.15 M): Add 21.3 g of Na_2HPO_4 to 1 L distilled water.
3. Hydrochloric acid (1 M): Add 83 mL concentrated HCl to 917 mL of distilled water.
4. Sodium hydroxide (1 M): Add 40 g of NaOH pellets to 1 L distilled water.
5. pH meter.
6. Appropriate cell culture.

7–3.1.2 Procedures

Prepare extracts using identical procedures to that described in section 7–2.1.2. If high titers are anticipated in soil, the extract may be assayed as per the following section using cell line cultures. Often, however, the number of enteric viruses in soil will be below the limit that can be detected with confidence. When numbers in soil are low, it is first necessary to concentrate viruses from the eluate. The following procedure describes the concentration process. This procedure has been adapted from Katzenelson et al. (1976) and Bitton et al. (1979) and uses beef extract (10%) as the extracting solution. It should be noted that the concentration procedure often results in the loss of some viable viruses. Therefore, concentration should only be employed when necessary to ensure that adequate numbers are present in the final extract.

Add 300 mL of the phage extract (see 7–2.1.2) to a 2-L beaker. To this mixture, add 700 mL of distilled water to dilute the beef extract. Beef extract at the original concentration of 10% often fails to result in appreciable flocculation, hence the necessity for this initial dilution. Place the mixture on a stirring plate and insert a combination pH electrode. Add 1.0 M of HCl until the pH is reduced to 3.5. At this pH, a visible flocculate will form. Continue to stir for an additional 30 min. Pour contents into a plastic centrifuge bottle(s) and centrifuge at 2000 × g for 15 min at 4°C. Discard supernatant and redissolve precipitate in 50 mL 0.15 M Na_2HPO_4. Pour into a 100-mL plastic centrifuge bottle and add a magnetic stirring bar. Gently stir the mixture until the precipitate has dissolved. If necessary, adjust the pH to 7.0 with the addition of 1.0 M HCl or NaOH.

Principals of viral enumeration in the concentrated supernatant are similar to that described for the phage, however, specific methods vary depending upon the individual virus being enumerated (Melnick & Wenner, 1969). Generally, monolayers of animal cells, capable of being infected by the virus, are grown on a solid support. Cells are then flooded with the concentrated viral supernatant for a period of time sufficient to allow the viruses to attach to the cells. Soft agar is then poured over the cell layer and the mixture incubated for several weeks. The presence of plaques, or several other types of distinct morphological features, indicates the presence of an infective unit.

REFERENCES

Adams, M.H. 1959. Bacteriophages. Interscience Publ., London.

Barnet, Y.M. 1972. Bactiophages of *Rhizobium trifolii* I. Morphology and host range. J. Gen. Microbiol. 15:1–15.

Basit, H.A., J.S. Angle, S. Salem, E.M. Gewaily, S.I. Kotob, and P. van Berkum. 1991. Pheontypic diversity among strains of *Bradyrhizobium japonicum* belonging to serogroup 110. Appl. Environ. Microbiol. 57:1570–1572.

Bergh, O., K.Y. Borsheim, G. Brathax, and M. Heldal. 1989. High abundance of viruses found in aquatic environments. Nature (London) 340:467–468.

Billing, E. 1969. Isolation, growth and preservation of bacteriophages. Meth. Microbiol. 38:315–329.

Bitton, G., M.J. Charles, and S.R. Farrah. 1979. Virus detection in soils: A comparison of four recovery methods. Can. J. Microbiol. 25:874–880.

Bitton, G., O.C. Pancorbo, and S.R. Farrah. 1984. Virus transport and survival after land application of sewage sludge. Appl. Environ. Microbiol. 47:905–909.

Bradley, D.E. 1966. The structure and infective process of a *Pseudomonas aeruginosa* bacteriophage containing RNA. J. Gen. Microbiol. 45:83–96.

Bradley, S.G., D.L. Anderson, and L.A. Jones. 1961. Phylogeny of actinomycetes as revealed by susceptibility to phage. Dev. Ind. Microbiol. 2:223–229.

Brown, D.R., J.G. Holt, and P.A. Pattee. 1978. Isolation and characterization of *Arthrobacter* bacteriophages and their application to phage typing of soil arthrobacters. Appl. Environ. Microbiol. 35:185–191.

Burge, W.D., and N.K. Enkiri. 1978. Adsorption kinetics of bacteriophage φ X-174 in soil. J. Environ. Qual. 7:536–541.

Bystricky, V., G. Stotzky, and M. Schiffenbauer. 1975. Electron microscopy of T_1-bacteriophage adsorbed to clay minerals: Application of the critical point drying method. Can. J. microbiol. 21:1278–1282.

Casida, L.E., and W. Liu. 1974. *Arthrobacter globiformis* and its bacteriophage in soil. Appl. Microbiol. 28:951–959.

Collard, C.A. 1970. Comparative studies of methods of isolation of actinophage starting with their natural habitat. CR Soc. Biol. 164:465–468.

Crosse, J.E., and M.K. Hingorani. 1958. A method for isolating *Pseudomonas mors-prunorum*. Nature (London) 181:60–61.

Demolon, A., and A. Dunez. 1935. Recherches sur le role du bacteriophage dans fatigue des luzernieres. Ann. Agron. 5:89–111.

Demolon, A., and A. Dunez. 1936. Nouvelles observations sur le bacteriophage et le bacteriophage et la fatigue des sols cultivées en luzerne. Ann. Agron. 6:434–454.

Dizer, H., A. Nassar, and J.M. Lopez. 1984. Penetration of different human pathogenic viruses into sand columns percolated with distilled water, groundwater, or wastewater. Appl. Environ. Microbiol. 47:409–415.

Duboise, S.M., B.E. Moore, C.A. Sorber, and B.P. Sagik. 1979. Viruses in soil systems. p. 245–285. *In* H.D. Isenberg (ed.) CRC Crit. Rev. in Microbiology. Vol. 7. CRC Press, Boca Raton, FL.

Farrah, S.R., and G. Bitton. 1990. Viruses in the soil environment. p. 529–556. *In* J. Bollag and G. Stotzky (ed.) Soil biochemistry. Vol. 6. Marcel Dekker, New York.

Farrah, S.R., G. Bitton, E. Hoffmann, O. Lanni, O.C. Pancorbo, M.C. Lutrick, and G.E. Bertrand. 1981. Survival of enteroviruses and coliform bacteria in a sludge lagoon. Appl. Environ. Microbiol. 41:459–465.

Fuxa, J.R., G.W. Warren, and C. Kawanishi. 1985. Comparison of bioassay and enzyme-linked immunosorbent assay for quantification of *Spodoptera frugiperda* nuclear polyhidrosis viruses. Soil J. Invert. Pathol. 46:133–138.

Germida, J.J., and L.E. Casida, Jr. 1983. *Ensifer adhaerens* predatory activity against other bacteria in soil, as monitored by indirect phage analysis. Appl. Environ. Microbiol. 45:1380–1388.

Gilbert, R.G., R.C. Rice, H. Bouwer, C.P. Gerba, C. Wallis, and J.L. Melnik. 1976. Wastewater renovation and reuse: Virus removal by soil filtration. Science 192:1004–1005.

Grabow, W.O.K. 1968. The virology of waste water treatment. Water Res. 2:675–701.

Hashem, F.M., and J.S. Angle. 1988. Rhizobiophage effects on *Bradyrhizobium japonicum*, nodulation and soybean growth. Soil Biol. Biochem. 20:69–73.

Hashem, F.M., and J.S. Angle. 1990. Rhizobiophage effects on nodulation, nitrogen fixation, and yield of field-grown soybeans (*Glycine max* L. Merr.). Biol. Fert. Soils 9:330–334.

Hashem, F.M., J.S. Angle, and P.A. Ristiano. 1986. Isolation and characterization of rhizobiophage specific for *Rhizobium japonicum* USDA 117. Can. J. Microbiol. 32:326–329.

Hukuhara, T., and K. Akami. 1987. Demonstration of polyhedral inclusion bodies of a nuclear polyhidrosis virus in field soil by immunofluorescence microscopy. J. Invert. Pathol. 49:130–132.

Hurst, C.J., C.P. Gerba, J.C. Lance, and R.C. Rice. 1980. Survival of enteroviruses in rapid-infiltration basins during the land application of wastewater. Appl. Environ. Microbiol. 40:192–200.

Hurst, C.J., S.A. Schaub, M.D. Sobsey, S.R. Farrah, C.P. Gerba, J.B. Rose, S.M. Goyal, E.P. Larkin, R. Sullivan, J.T. Tierney, R.T. O'Brian, R.S. Safferman, M.E. Morris, F.M. Wellings, A.L. Lewis, G. Berg, P.W. Britton, and J.A. Winter. 1991. Multilaboratory evaluation of methods for detecting enteric viruses in soil. Appl. Environ. Microbiol. 57:395–401.

Irving, L, and F. Smith. 1981. One-year survey of enteroviruses, adenoviruses, and reoviruses isolated from effluent at an activated sludge purification plant. Appl. Environ. Microbiol. 41:51–59.

Katzenelson, E., B. Fattal, and T. Hostovesky. 1976. Organic flocculation: An efficient second-step concentration method for the detection of viruses in tap water. Appl. Environ. Microbiol. 32:638–639.

Lance, J.C., and C.P. Gerba. 1980. Poliovirus movement during high rate land filtration of sewage water. J. Environ. Qual. 9:31–34.

Lanning, S., and S.T. Williams. 1982. Methods for the direct isolation and enumeration of actinophages in soil. J. Gen. Microbiol. 128:2063–2071.

Lesley, S.M. 1982. A bacteriophage typing system for Rhizobium meliloti. Can. J. Microbiol. 28:180–189.

Marants, L.A., L.N. Moskalenko, and Y.A. Rautenshtein. 1974. Some biological properties of phages of Rhizobium meliloti. Mikrobiologia 42:967–973.

Melnick, J.L., and H.A. Wenner. 1969. Enteroviruses. p. 529–602. In E.H. Lennette and N.J. Schmidt (ed.) Viral and rickettsial infections. 4th ed. Am. Public Health Assoc., New York.

Moskalenko, L.N., and I.A. Pautenshi. 1969. Lysogeny of bean nodule bacteria, Rhizobium phaseoli. Mikrobiologia 38:340–345.

Pancorbo, O.O., G. Bitton, S.R. Farrah, G.E. Gifford, and A.R. Overman. 1988. Poliovirus retention in soil columns after application of chemical and polyelectrolyte-conditioned sludges. Appl. Environ. Microbiol. 54:118–123.

Preston, D.R., G.R. Chaudhry, and S.R. Farrah. 1990. Detection and identification of poliovirus in environmental samples using nucleic acid hybridization. Can. J. Microbiol. 36:664–669.

Proctor, L.M., J.A. Fuhrman, and M.C. Ledbetter. 1988. Marine bacteriophages and bacterial mortality. Eos 69:1111–1112.

Reanney, D.C. 1974. Viruses and evolution. Int. Rev. Cytol. 37:21–52.

Reanney, D.C., P.C. Gowland, and J.H. Slater. 1983. Genetic interactions among microbial communities. p. 379–422. In J.H. Slater et al. (ed.) Microbes in their natural environments. Cambridge Univ. Press, Cambridge.

Reanney, D.C., and S.C.N. Marsh. 1973. The ecology of viruses attacking Bacillus stearothermophilus in soil. Soil Biol. Biochem. 5:399–408.

Romig, W.R., and A.M. Brodetsky. 1961. Isolation and preliminary characterization of bacteriophages of Bacillus subtilis. J. Bacteriol. 82:135–141.

Saye, D.J., D.A. Ogunseitan, G.S. Sayler, and R.V. Miller. 1990. Transduction of linked chromosomal genes between Pseudomonas aeruginosa strains during incubation in situ in a freshwater habitat. Appl. Environ. Microbiol. 56:140–145.

Selander, R.K., and B.R. Levin. 1980. Genetic diversity and structure in Escherichia coli populations. Science 210:545–547.

Staniewski, R. 1970. Typing Rhizobium by phages. Can. J. Microbiol. 16:1003–1009.

Staniewski, R., and W. Kowalska. 1974. Typing of Rhizobium meliloti mutants by means of phage. Acta Microbiol. Pol. Ser. A6:183–186.

Stotzky, G. 1989. Gene transfer among bacteria in soil. p. 165–222. In S.B. Levy and R.V. Miller (ed.) Gene transfer in the environment. McGraw-Hill Publ. Co., New York.

Sykes, I.K., and S.T. Williams. 1978. Interactions of actinophage and clays. J. Gen. Microbiol. 108:97–102.

Tan, J.S.H., and D.C. Reanney. 1976. Interactions between bacteriophages and bacteria in soil. Soil Biol. Biochem. 8:145–150.

Torella, F., and R.Y. Morita. 1979. Evidence by electron micrographs for a high incidence of bacteriophage particles in the waters of Yaquina Bay, Oregon. Ecological and taxonomic implications. Appl. Environ. Microbiol. 37:774–778.

U.S. Environmental Protection Agency. 1989. Method for recovering viruses from sludges, soils, sediments and other solids. Sept. 1989. EPA/600/4-84/013R7.

Williams, S.T., and S. Lanning. 1984. Studies in the ecology of streptomycete phage in soil. p. 473–483. *In* L. Ortiz-Ortiz et al. (ed.) Biological, biochemical and biomedical aspects of Actinomycetes. Academic Press, London.

Williams, S.T., E.M.H. Wellington, and L.S. Tipler. 1980. The taxonomic implications of the reactions of representative *Nocardia* strains to actinophage. J. Gen. Microbiol. 119:173–178.

Chapter 8

Recovery and Enumeration of Viable Bacteria

DAVID A. ZUBERER, *Texas A&M University, College Station, Texas*

One of the greatest challenges facing soil microbiologists for more than a century has been the recovery and enumeration of "total" bacterial populations in soils. The difficulties lie not in the actual methodologies (the procedures are in fact relatively simple), but in the interpretation of results that is confounded by numerous errors arising chiefly from the nature of the substratum—the soil and the microbes themselves. It is safe to say that the dilution plate counting technique for the enumeration of microorganisms is one of the oldest and most widely used techniques in microbiology. The development of the procedures dates back to the period of Robert Koch and contemporaries (about 1880) when these pioneering microbiologists were faced with the problem of developing techniques for the cultivation of bacteria on "solid" culture media. Their efforts were directed mainly at characterizing microorganisms from a wide variety of habitats at a period when the field of microbiology (bacteriology in particular) was in its infancy. One of the earliest and lasting breakthroughs in the study of bacteria was the discovery of agar (a polysaccharide derived from marine algae) as a jelling agent for culture media. A unique and microbiologically useful property of agar is that it melts at 100 °C but does not gel until cooled to about 40 °C making it suitable for incorporating live organisms in the gel. It is also resistant to decomposition by most terrestrial microorganisms making it well suited for use as a jelling agent for studies of soil microorganisms.

Shortly after Koch's demonstrations of the plate culture technique (which actually employed flat glass plates) Petri (1887; see translation in Brock, 1961) described the use of dishes with loose-fitting overlapping lids (the well-known petri plate or dish prevalent in all microbiology labs) for the long-term growth of microorganisms with minimal contamination from external sources. Thus, with these pioneering innovations, the foundation for the cultivation and enumeration of microorganisms by the so-called dilution plate method was laid.

This chapter is not intended to be an exhaustive review of all the nuances of methods for enumeration of viable bacteria from soil. Excellent reviews have been published and the reader should consult these for further details if desired (Jensen, 1962, 1968; Parkinson et al., 1971a; Wollum, 1982). The aim of this chapter is to present sufficient information regarding techniques for counting the viable population of soil bacteria through the use of plate counting techniques so that an investigator can initiate the procedures in the lab without having to consult numerous additional references. Whenever possible I have tried to include information to explain why a given manipulation is essential and attempted to point out the recognized inadequacies of the procedures so as not to leave the reader with a false impression of the validity of these counting techniques for various types of studies.

8–1 PRINCIPLES OF ENUMERATING SOIL BACTERIA

While it remains one of the most widely used techniques for the enumeration of microorganisms, the dilution plate count has been the subject of intense scrutiny and criticism almost from its inception. Chief among the criticisms is that only a small fraction (approx. 1–10%) of the number of cells observed with direct microscopic counts is recovered as "viable" bacteria in the plate count procedure. There are several reasons for such low recoveries and these will be described later.

Perhaps one of the greatest problems regarding plate counts is one of semantics. It has been common practice to refer to the numbers derived from plate counts as "total" counts of bacteria. Because plate counts yield only a small fraction (1–10%) of the true "total" bacterial population in a soil (currently, that which can be observed microscopically), any estimation of the "total" population is subject to many limitations. Since in the final analysis, a plate count reveals only those bacteria growing under the specific cultural conditions chosen by an investigator, the objection to the expression of results as "total' bacteria is entirely justified. However, such objections might be easily overcome if results of plate counts are simply expressed as numbers of "recoverable viable bacteria" obtained using a given set of parameters.

The principles of enumeration of microorganisms were established during the early period of the development of microbiology as a science. Problems in enumeration stem from two major characteristics of soil bacteria, their enormous numbers and their microscopic size. Thus, one of the first principles of enumerating soil bacteria is dilution of the population to manageable numbers. To this end, a known quantity of soil is usually dispersed in a known volume of some diluent and this first dilution is then further diluted depending on the estimated density of bacteria in the sample. Numbers of organisms are determined by inoculating appropriate culture media, in tubes or plates, with aliquots of the dilutions and, following some period of incubation, numbers of positive tubes or colonies on plates are counted. Taking the dilution factor into consideration an estimated number of bacteria per gram of sample is derived.

A second major assumption in the estimation of numbers of viable bacteria using plate count techniques is that all viable cells in a sample will produce a visible colony on or in the agar medium used for enumeration. One need only consider the loss of anaerobes from an "aerobic" plate count to realize the stringent limitations of this assumption! It is also becoming increasingly clear that many cells that are metabolically active do not for one reason or another produce colonies on solid media. Thus, it is clear that plate counts estimate only those cells that are tolerant of the medium and incubation conditions and which can exhibit "viability" in the sense of being able to divide and produce recognizable colonies. The category "viable but non-culturable" has been applied to cells that appear viable, as judged by vital staining or other procedures, but yet apparently fail to be detected in plate counting procedures.

The enumeration procedure is commonly referred to as the "dilution plate method" and usually employs either of two common versions. In the "spread-plate" technique, aliquots of dilutions are spread across an agar medium whereas in the "pour-plate" technique aliquots of dilutions are placed in sterile petri plates and molten, cooled agar is poured into the plate and mixed with the inoculum. Several critical manipulations in the dilution plate method and ramifications of variation in these will be discussed in the sections that follow.

8–1.1 Limitations of Plate Counts

It is universally accepted and well documented that viable plate counting techniques rarely account for more than a few percent of the numbers of cells as estimated by direct microscopical counts. Some limitations of the plate count had already been described by Waksman (1922). Skinner et al. (1952) discussed reasons for the large discrepancies between direct microscopic counts and plate counts. Among the reasons they listed for the inadequacies of the plate count were: clumps of bacteria giving rise to single colonies, competition on the plates causing colonies to fail to develop, i.e., mutual competition and the selectivity of the plating medium including "noncultivation of obligate anaerobes." Similar arguments have been advanced by Brock (1966) and Buck (1979) and many others. It is also obvious that one of the chief limitations of plate counts of viable soil bacteria is the failure to dislodge cells from within aggregates and the surfaces of soil particles to which they adhere rather tenaciously. This was confirmed in a recent study in which a large proportion of bacterial cells remained associated with soil particles even after a fairly rigorous, multi-step dispersion, and differential centrifugation procedure (Hopkins et al., 1991). Sedimentation of particle-bound cells during the dilution process no doubt contributes to some of the high variability often observed in plate count data. For this reason, some researchers have advocated continuous agitation of dilution tubes/bottles while the sample is withdrawn. Still others have advocated a brief period of settling before the sample is withdrawn for transfer. Again, the matter of internal consistency of procedure comes to the forefront. Samples, no matter what the treatment, must be handled consistently to avoid the situation summed up nicely by Wyant (1921): "Up to the present time each investigator has been a law unto himself in so far as the technic [sic] used in quantitative bacteriological soil analysis is concerned." Some might argue that the situation has not changed much in 70 years!

Despite past and current criticisms, it is likely that the plate count technique will remain firmly entrenched in microbiology labs throughout the world and there seems to be little wrong with this so long as users of the technique fully acquaint themselves with the shortcomings of the technique. It is, after all, a useful tool for making comparisons of the effects of various soil treatments on those bacteria that are recoverable under the conditions defined by the experimenter. One can draw an analogy between the plate count technique and the acetylene reduction assay for measuring nitrogen (N_2) fixation. Both are well suited for certain tasks within their respective applications and both have been and continue to be much criticized. The techniques themselves are not without usefulness, rather, it is the misapplication of the techniques with which one must take issue!

8–2 MATERIALS AND EQUIPMENT

8–2.1 Field

1. Shovels, core samplers, augers, or trowels.
2. Propane torch or butane lighter and alcohol if "sterilization" of field implements is deemed a necessary part of the study.
3. Sample bags: usually polyethylene. Size appropriate to needs. Zip-seal or Whirlpak bags are quite convenient.
4. Insulated chest for transport of samples.
5. Tape and marking pens.
6. Sealable containers (plastic containers or mason jars) for samples which must be placed under a controlled gas phase, e.g., anaerobic conditions.

8–2.2 Laboratory

1. Autoclave and oven.
2. Balance.
3. Dilution bottles (wide-mouthed 160-mL milk dilution bottles) and tubes (16 × 150 mm culture tubes or larger). Keep in mind how much solution will be needed. If one intends to inoculate more than nine replicate plates of a single medium, or a battery of

several media the dilution blanks must be of sufficient volume to accommodate the total volume of the aliquots to be plated, e.g., 18 mL.

4. Mechanical shaker, blender, or other appropriate device (e.g., sonicator) for mixing samples for dilution.

5. Vortex mixer. Useful for mixing solutions prepared in culture tubes.

6. Pipettes; 1-mL serological with 0.1-mL graduations; 5 or 10 mL with 1-mL graduations. Alternatively, any of the manual or motorized piston drive pipettors can greatly improve the repetitive pipetting involved in dilution and transfer of inoculum to plates. They also eliminate potential hazards of mouth pipetting!

7. Water bath.

8. Glass spreaders ("hockey sticks"): glass rod (3-mm diam.) bent at a 45-degree angle or long Pasteur pipettes sealed at the tip and bent in a similar fashion. The flat "spreading" portion should be about 40 to 50 mm long.

9. Petri plates.

10. Media: any of an extremely wide variety of general purpose or selective media appropriate to the research problem under investigation (see section 8–8).

11. Diluents (see section 8–5). Generally prepared as 99-, 95-, or 9-mL dilution blanks in 160-mL "milk dilution bottles" or 16 × 150-mm culture tubes.

8–3 COLLECTION AND PREPARATION OF SOIL SAMPLES

It is generally agreed that collection of soil samples in the field requires few precautions other than avoidance of gross cross contamination and a sound plan for obtaining samples representative of the area(s) under study (see chapters 1 and 2 in this book). Gross contamination is avoided simply by cleaning field implements between sampling units (plots etc.). If necessary, implements can be doused with alcohol and "flame-sterilized" by igniting the alcohol with a propane torch or a butane lighter. Use due caution! Probably the most widely used method of sample collection involves compositing numerous small samples from plots into larger samples that are thoroughly mixed prior to microbiological analysis. Small samples can be retrieved with trowels or core samplers and placed into plastic bags to contain the composite. The samples should be representative of the area under investigation. It is generally agreed that composites offer a better "picture" of the sampling area than a few small samples enumerated individually. However, it is incumbent upon the investigator to establish some measure of the variability within a unit. Several composite samples should be prepared for each site/plot in question. If depth is important, make certain to keep soil "horizons" separate and pay attention to the rooting pattern and depth of the resident plant community or crop.

Samples can be transported to the laboratory on ice or maintained near ambient temperature in an insulated chest. The literature is not free of contrasts on this point. Some investigators advocate the use of ice (4 °C) for transport and storage (e.g., Jensen, 1962) while others advocate maintaining samples at or near the conditions experienced at the time of collection (Casida, 1968). Cooling the samples may have deleterious effects on some bacteria. Regardless of the method chosen, it is critically important to avoid exposing samples to excessive heat and desiccation as these two properties probably cause the greatest damage to the microbial populations. Once in the laboratory, the samples are usually screened (1–2 mm mesh size) to remove large particles, root fragments etc. Avoid excessive drying during the screening process! Return samples to plastic bags, seal and maintain them at constant conditions until they can be processed for enumeration.

8–4 RELEASE OF BACTERIA FROM SOILS

The release of bacteria from soil particles and aggregates is perhaps the quintessential step in the enumeration of the soil population and it is here where most of the variation in recovery of viable bacteria can occur. The techniques for release of microorganisms include simple hand shaking (with or without glass beads), trituration with a mortar and pestle, mechanical shaking (reciprocal, rotary, and wrist-action), blending (with and without dispersants such as Calgon etc.), blending followed by gradient centrifugation, and sonication.

Typically, because of time and equipment limitations some form of shaking is used as the principal mechanism for recovery of viable cells, however, it must be realized that these methods may be among the least effective. Wollum (1982) recommended shaking the soil suspension (with glass beads in the diluent) with the bottles in a horizontal position, for 10 min on a mechanical shaker (speed and stroke not specified). Recently, Kobel-Boelke et al. (1988) reported a relatively effective shaking method for recovery of bacteria from deep aquifer sediments. They used 25 g of soil in 100 mL of sterile tap water containing 20 g of 4-mm glass beads and shook the samples for 30 min at 160 rpm. The advantages of shaking procedures are that no specialized equipment (other than the shaker) is required and each sample is prepared individually under aseptic conditions, i.e., it is not necessary to sanitize blender cups etc. between samples. Shake methods can be effective for recovery of bacteria and have compared favorably with more extensive methods in a recent study by Hopkins et al. (1991). Similar numbers of bacteria were found in the supernatants of suspensions of a sandy soil prepared by the method described by Wollum (1982) and by a more extensive blending/centrifugation technique. However, recovery of bacteria was not as good from a peat and a clay soil. Soil type, as is well known, has a major impact on the relative efficiencies of the various dispersion techniques.

Blending is a relatively effective way to disperse soil aggregates and bacteria. It is convenient to prepare the 10^{-1} dilution in the blender cup by adding a measured volume of diluent. Three 1-min blending periods separated by chilling in an ice bath usually accomplish adequate dispersion of the samples. Dispersion may be facilitated by the addition of 0.1% sodium pyrophosphate to the diluent (Jensen, 1962; Balkwill, 1990).

Sonication can also be used to effectively disperse soils for determination of microbial numbers (Stevenson, 1958; Balkwill, 1990). Stevenson (1958) reported a two- to fourfold increase in bacterial numbers following 4 min of treatment at "peak frequency." Numbers declined sharply with further treatment probably as a result of heating. On the other hand, Balkwill (1990) states that "sonication appears to have little or no tendency to kill the microbial cells (at least in surface soils), even at very high settings." Sonication is generally performed on the 10^{-1} dilution of soil followed by standard serial dilution and plating of the samples.

Regardless of which technique is used for the recovery of viable cells from soil, it is extremely important that the method be applied uniformly to all samples. Only if the conditions are faithfully duplicated can the results be of sufficient validity to justify comparisons among treatments with respect to microbial numbers.

Despite the limitations of various extraction techniques that rely on shaking of soil to disperse the bacterial population, it is likely that they will remain popular due to their simplicity and economy with regard to time and equipment. However, it falls upon the user to fully comprehend the limitations of a chosen procedure and to be conservative in the interpretation of data obtained with the procedure!

8–5 DILUENTS USED IN RECOVERY AND ENUMERATION OF SOIL BACTERIA

A wide variety of diluents has been used since the advent of counting techniques more than a century ago. It is fairly safe to say that the most widely used diluent has been sterile, distilled, or tap water. That is not to say that this is the best diluent, merely the most convenient. Because cell numbers can decline fairly rapidly in water during the dilution process (Straka & Stokes, 1957–58; Gorrill & MacNeil, 1960; King & Hurst, 1963) several solutions have been employed in an effort to prolong survival in the diluent. Chief among the alternatives are various salt solutions and buffers whose function it is to adjust the tonicity of the suspending medium to that of the cell to avoid osmotic shock. Many studies have shown the superiority of one or another of these solutions over water as the diluent. Some of the more commonly used solutions include 0.85% NaCl (physiological saline), Ringer's solution (full or quarter-strength), Winogradsky's salt solution, sterile 0.1% sodium pyrophosphate, phosphate buffered saline, phosphate buffers of various strengths and peptone-water. Soil extract has also been recommended as a diluent. Formulae for these solutions are given below:

1. **Physiological saline:** Dissolve 8.5 g of NaCl in 1000 mL of distilled water.
2. **Ringer's solution ("Krebs-Ringer")** (Umbreit et al., 1972): Dissolve 8.2 g of NaCl, 4.18 g of KCl, 3.32 g of $CaCl_2$, 1.9 g of KH_2PO_4 and 3.46 g of $MgSO_4 \cdot 7H_2O$ in 1000 mL distilled water.
3. **Winogradsky's salt solution:** (modified by Dabek-Szreniawska and Hattori, 1981): Dissolve 3.8 g of K_2HPO_4, 1.2 g of KH_2PO_4, 5.1 g of $MgSO_4 \cdot 7H_2O$, 2.5 g of NaCl and 0.05 g each of $Fe_2(SO_4)_3 \cdot nH_2O$ and $Mn_2(SO_4)_3 \cdot 4H_2O$ in 1000 mL of deionized water.
4. **0.1% Sodium pyrophosphate:** Dissolve 0.1% (wt/vol) sodium pyrophosphate in distilled water (Balkwill, 1990).
5. **Phosphate-buffered saline:** Dissolve 1.18 g of Na_2HPO_4, 0.22 g of $NaH_2PO_4 \cdot 1H_2O$ and 8.5 g of NaCl in 1000 mL distilled water. If necessary adjust pH to 7.2.
6. **Peptone-water:** Dissolve 1.0 g of peptone in 1000 mL distilled water. As this diluent contains a usable C source, it is important not to leave cells in this solution too long to avoid growth of the cells.
7. **Soil extract:** Mix 1 kg of fresh soil with 1000 to 1500 mL of distilled water in a large flask and autoclave the suspension for 30 min to 1 h. Add 0.5 g of calcium carbonate or calcium sulfate (supposed to flocculate clays), mix and filter the suspension through several layers of filter paper. Repeat filtration till a clear liquid (usually straw-colored) is obtained. Sterilize and store in the freezer or refrigerator.

Various modifications of these solutions have been prepared. For example, it is not uncommon to see quarter-strength Ringer's solution or various diluted forms of the Winogradsky salt solution used as diluents. In a recent study comparing direct microscopic counts with plate counts, Olsen and Bakken (1987) reported that distilled water as a diluent resulted in 15% fewer colonies than Calgon (0.2%), Winogradsky's salt solution and soil extract, which were all within 6% of one another. Thus, while distilled water will in all likelihood remain a widely used diluent, those choosing to use it will do so at the expense of loss of a fraction of the viable population.

Regarding the preparation of dilution blanks, it has been recommended that the diluent be sterilized prior to aseptic distribution into the tubes or bottles (Koch, 1981) to avoid loss due to evaporation during autoclaving. In practice, this difficulty can be overcome by adding sufficient diluent to the vessels prior to autoclaving such that after any loss due to evaporation the appropriate volume is attained. Jensen (1962) found that evaporative loss remained relatively constant "at or a little below 3% independent of the container used," and pointed out that this would probably vary with the autoclave used. Avoid leaving dilution blanks in the sterilizer for prolonged periods after the cycle is completed!

8–6 PREPARATION OF SERIAL DILUTIONS

8–6.1 Procedure—Ten-Fold Dilution Series

1. Weigh out a 10-g sample of moist soil for enumeration and several 10-g replicates for determination of the moisture content. Moisture content must be determined so that microbial numbers can be adjusted to a dry-weight basis.
2. Add the 10-g sample to a 95-mL dilution blank (square milk dilution bottle or suitable flask) of the appropriate diluent (see section 8–5) containing 20 to 40 3-mm glass beads. This constitutes the 1/10 or 10^{-1} dilution.
3. Shake the bottle vigorously by hand for 30 to 60 s and then place it on a shaker for 20 min. A wrist-action shaker is preferable. If a rotary shaker is used, the speed should be at least 150 to 200 rpm.
4. After shaking the 10^{-1} dilution for the appropriate time, remove it from the shaker, allow it to stand for approx. 30 s and, using a sterile 1-mL pipette, remove 1 mL from the middle region of the suspension and transfer it to a 9-mL blank to achieve the 10^{-2} dilution. Mix the contents thoroughly by vortexing and continue (always using a fresh pipette between subsequent dilutions) by repeating these steps until the desired dilution factor is obtained. As an alternative to using 9-mL blanks, 10-mL samples may be transferred to 90-mL blanks in dilution bottles. Some economy of time and glassware may be achieved by using 1-mL samples and 99-mL dilution blanks, however, accuracy may suffer with such a practice. If 100-fold dilutions are used, the intermediate 10-fold dilutions are accomplished by plating 0.1-mL samples from the dilution blanks. Hint: to avoid confusion during the preparation of dilution series it is helpful to place the "inoculated" tubes one row further back in the culture tube rack thereby minimizing the risk of transferring to the same tube twice or of skipping a tube.
5. When the dilution series is completed, distribute aliquots of the chosen dilutions (those that will likely yield 30–300 colonies) for pour- or spread-plating as described below. Some economy in glassware can be gained by using a single pipette to transfer the aliquots as long as the most dilute samples are transferred first (e.g., 10^{-7} then 10^{-6}). Alternatively, the pipette used to make the transfer during the dilution process can also be used to distribute the inoculum to plates before it is discarded after the transfer.

8–6.2 Comments

The 10-fold or decimal dilution series is probably the most common choice for enumeration of microorganisms. This is largely a matter of convenience and should not be taken to indicate that other dilutions (e.g., fourfold etc.) are not acceptable. In many soil microbiology laboratories, it is common practice to prepare the initial 1 in 10 (10^{-1}) dilution by weighing out 10 g of moist soil and adding it to 95 mL of diluent. However, this is a point where there remains considerable variation in the literature describing recent research. Many papers report using 10 g of soil in 90 or 100 mL of diluent. In the earlier volumes of this series (see Clark, 1965; Wollum, 1982) the "10 in 95 mL" method was described as standard practice in soil microbiology. Historically, early pioneers in soil microbiology used some variation in which 10 g of soil was added to 100 mL of diluent (Wyant, 1921). Quantities varied but the ratio of soil to diluent was always 1 to 10. Later, investigators began to add 10 g of soil to 90 mL of diluent to achieve the "1 in 10" dilution. The practice of using 10 g of soil in 90 mL of diluent is quite common in recent work. The origin of the practice of using 10 g in 95 mL is somewhat obscure but stems from the knowledge that soil cannot be considered a solid for dilution on a volumetric basis. Thus, 10 g of a typical mineral soil occupies approximately 5 cm^3 and when placed in 95 mL of diluent, a dilution of 10 g of soil in 100 cm^3 volume results. The important points to be made here are that regardless of the exact method of preparing the first dilution the 10-g soil sample provides a more representative sample than 1 g and regardless of whether one chooses 100, 95, or 90 mL of diluent, it is internal standardization and control that is most critical to the investigator. In other words, choose a method and stay with it throughout. Remember that any plate count method is largely a comparative method and only samples processed identically can be compared with any degree of validity! In practice, the use of 10 g of soil in 95 mL of diluent is equally convenient and is almost universally accepted among soil microbiologists. Any communications regarding the outcome of enumeration work should specify quite unambiguously how dilution series were prepared. Given the variation evident in the modern literature, it is unacceptable to merely report that "soil was serially diluted."

It is generally acknowledged that the dilution process is where the largest error in the counting methods is derived. It goes without saying that accuracy in pipetting is absolutely essential and errors arising during the dilution procedure have been well described by earlier workers (e.g., see Ingram & Eddy, 1953). Usually, dilution to 10^{-6} or 10^{-7} will be satisfactory for most soils, however, soils with high organic matter content or soils receiving organic wastes may require further dilution. Experience is the best guide in the latter cases. It is absolutely essential that a new pipette be used to transfer each sample to the next "higher" dilution (e.g., in transferring the 10^{-3} to make the 10^{-4} dilution) since it is well established that microorganisms adhere to the walls of pipettes and may accumulate for later erroneous discharge into the more dilute samples giving rise to falsely high numbers! It is also recommended procedure to fill and evacuate a pipette several times before taking the actual aliquot to be transferred to "saturate" the adsorption sites on the wall of the pipette.

The dilution of samples should be accomplished as quickly as possible to avoid death of the bacteria in the diluent. To minimize the die-off of cells in diluent a variety of solutions has been employed (see formulae in section 8–5). In practice, many investigators use sterile distilled water or tap water for convenience. A detailed review of the negative aspects of various diluents is beyond the scope of this chapter. The reader should consult the papers by Straka and Stokes (1957–58), Jensen (1962), King and Hurst (1963), and Dabek-Szreniawska and Hattori (1981) for further details concerning the advantages or disadvantages of various solutions. In general, dilute salt solutions and buffers or peptone water (0.1% peptone in water) yield greater survival of organisms during dilution (see above section). It is still appropriate to test various diluents before embarking on a project that is heavily contingent on the results of enumeration procedures. No matter which diluent is chosen, it is important that the temperature of the solution be tempered to that of the sample (usually ambient) before commencing the dilution procedure to avoid death due to thermal shock (hot or cold!) of the organisms (Gorrill & MacNeil, 1960; Strange & Dark, 1962; Busta et al., 1984).

Before moving away from the subject of dilutions, it seems appropriate to raise an issue noted by previous researchers and by almost anyone who has used the dilution plate technique to enumerate soil bacteria, namely, the nonproportionality of colony numbers and dilution factor. Ideally one would expect to see a 10-fold drop in colony numbers between successive 10-fold dilutions yet this is rarely the case. It is frequently observed that the more dilute suspensions give rise to higher colony numbers. The phenomenon was reported by Meiklejohn (1957) who speculated that the lower numbers at lower (less dilute) dilutions may have been due to mutual competition among organisms developing on the plates (i.e., more space, more colonies) or to the dilution of some inhibitory substance present in soil. Jensen (1962) attributed the phenomenon to adsorption of organisms to the inner walls of pipettes, increasing dispersion of the suspension with increasing dilution or mutual competition or other antagonistic phenomena on plates. Casida (1968) demonstrated greater colony counts by plating samples in large pie plates containing a deep layer of agar.

Observations such as these raise the question of which dilution in a series should be counted to derive the population estimate. In general, whenever possible, within an experiment the numbers should be derived from the same dilution, e.g., all counts derived from a 10^{-5} dilution. However, in reality, this is not always achievable. Perhaps under these circumstances, the choice should be to use the dilution giving the greatest number so long as it conforms to "30–300" rule (see section 8–9.1) and is obviously not a result of faulty technique. Such a practice would conform to the sentiment of Parkinson et al. (1971a) who felt that it might be "more practicable to use attainment of the highest count possible as the main standard (of the plate count; author's addition), allowing individuals to achieve this within the limitations of their own laboratories."

8–7 PLATING TECHNIQUES

8–7.1 The Pour Plate Technique

8–7.1.1 Procedure

1. Prepare a sufficient quantity of culture medium to fill the number of plates needed for counting the number of samples desired. For example, if one is going to plate four dilutions per sample with five replicate plates (15–20 mL per plate) per dilution then 300 to 400 mL of medium is required for each sample to be enumerated. Keep the medium in a water bath (42–45 °C) and be certain to mix the medium before pouring plates by gently swirling the bottle or flask so as to avoid foaming. This is necessary because the agar will settle out upon standing and failure to suspend it will result in pouring some "soupy" plates.

2. Label a set of plates for each sample. Label the bottom half of the dish and include sample ID, dilution, date, and medium if necessary.

3. Select a range of three to four dilutions that is suspected to yield plates containing between 30 and 300 colonies per plate. From each of these dilutions, after thorough mixing, transfer 1.0-mL aliquots to the labeled sterile plates. After the inocula have been distributed, pour approximately 15 to 20 mL of molten agar medium, distribute the inoculum by swirling the plate three to five times in a clockwise circular motion then three to five times counterclockwise, allow the medium to solidify then invert the plates for incubation at 25 to 28 °C (higher or lower if necessary).

4. Determine the dilution(s) which exhibit between 30 to 300 colonies (see section 8–9.1), count the colonies (use a colony counter, hand lens, or dissection microscope if necessary) and calculate the number of bacteria present. Average the numbers of colonies appearing on the replicate plates and multiply the mean colony number by the reciprocal of the dilution. For example, if an average of 150 colonies was observed on plates from the 10^{-6} dilution, the count would be expressed as 1.5×10^8 colony forming units (CFU). This figure must be adjusted to a soil dry weight basis.

Soil water content is calculated as follows:

$$\% H_2O = \frac{(\text{wet wt.} - \text{dry wt.})}{\text{oven-dry wt. of soil}} \times 100$$

Thus, if the soil that yielded the results above contained 25% moisture, the adjusted count would be 2.0×10^8 CFU per gram. This value is calculated as follows:

Colony number (mean) / Dry weight of soil

Where: Dry weight of soil = Wet weight $[1 - (\% H_2O, \text{moist soil} / 100)]$

Or: $1.5 \times 10^8 / 0.75$ g dry weight = 2.0×10^8 CFU/g dry soil

Viable counts are traditionally rounded off to two significant figures because of the inherent error (approx. 10%) in the procedures and to avoid giving a sense of false accuracy to the data.

8–7.1.2 Comments

It has been shown that the pour plate technique gives estimates of bacterial numbers that are lower than those derived with the spread plate method. One possible reason for this is the loss of viable cells due to the thermal shock caused by the molten agar. Therefore, it is critical that the temperature of the medium be as low as possible (preferably 43–45 °C) when the plates are poured to avoid killing cells due to excess heat! Keep a thermometer in a bottle of water in the bath to monitor the temperature. Avoid the "touch or feel method" of determining the proper pouring temperature for the medium.

It is generally recommended that at least five replicate plates (or more) are prepared for each dilution though in practice three replicates per dilution can provide a satisfactory compromise between accuracy and economy. Plates are incubated inverted to prevent condensation from falling to the agar surface allowing unwanted spreading of colonies due to water films. Plates for soil bacteria are generally incubated for 4 to 7 d but as long as 10 to 14 d (or even longer) to allow development and differentiation of actinomycetes. Long-term incubation requires some precautions to minimize evaporative losses of water from the medium. Keep a pan of water in the incubator or incubate the plates in sealed containers or plastic bags containing moistened paper towels. Be aware that incubation in sealed containers may lead to some undesirable changes in the gas phase such as the accumulation of CO_2 or possible depletion of O_2.

8–7.2 The Spread Plate Technique

8–7.2.1 Procedure

1. Prepare a sufficient number of plates with the appropriate culture medium. Allow the plates to "dry" (see comments below) prior to using them in the enumeration procedure.

2. Label the plates as described in section 8–7.1.1.

3. Choose the appropriate range of dilutions and transfer 0.1-mL aliquots (may be as large as 0.5–1.0 mL if suitably dried plates are used) to the agar surface in the center of the plate. Try to minimize the time between adding the inoculum and spreading it. *Remember:* since you are using 0.1-mL aliquots that you must use one dilution lower than that which is to be enumerated on the plate. For example, to prepare plates for a 10^{-4} dilution take the inoculum from the 10^{-3} dilution blank.

4. Once the inocula have been delivered to the plates, use a sterile or freshly flamed glass spreader and spread the inoculum over the agar surface by moving the spreader back and forth while rotating the plate with a

manual or motorized turntable ("lazy Susan") for approximately 15 s. Flame the spreader between successive dilutions (see also comments below). If after spreading the plates, some liquid remains visible on the agar surface, allow the plates to stand right side up until the moisture is absorbed. Afterward, invert the plates for incubation.

5. Following a suitable incubation period, count colonies as in section 8–7.1.1 and calculate the number of cells per gram (dry weight basis).

8–7.2.2 Comments

The spread plate technique differs from the pour plate technique in that aliquots from dilution tubes are spread across the surface of pre-poured, predried agar media in petri plates. The technique offers the advantage that all colonies develop on the surface of the agar such that recognition of different organisms is facilitated. It is especially useful for differentiating between colonies of actinomycetes and bacteria as the former will produce tough, leathery, often powdery colonies that adhere tenaciously to the agar surface. Surface development of colonies also facilitates subsequent isolation and characterization of the organisms. A disadvantage of the technique is that usually only 0.1 mL of inoculum is applied to the plates to prevent excessive moisture on the agar surface thus effectively limiting the use of the method to suspensions containing more than 300 cells mL^{-1} (if the rule of counting 30–300 colonies per plate is adhered to) (Clark, 1971). In an earlier study, Clark (1971) showed that the spread plate method produced counts that were 70 to 80% greater than counts of the same materials analyzed with the pour plate method. Reasons for the higher counts were unclear but it is possible that improved aeration allowed better colony development, clumps were further disaggregated during the spreading process and perhaps most likely, cells were not subject to the thermal shock attendant with immersion in molten agar media (Clark, 1971; Busta et al., 1984).

Since liquid inoculum must be absorbed by the agar medium it is recommended that plates be allowed to dry somewhat prior to inoculation so that the agar will readily absorb the fluid and minimize the chances for spreading of colonies due to excess moisture. Clark (1971) recommended drying to a point where at least 1 g of water was lost per 15 g of agar in the medium. He recommended a temperature of 50 °C to shorten drying.

Once delivered to the plate, the inoculum is spread evenly across the plate using a flamed bent glass spreader. Spreading the inoculum is facilitated if one uses a manual or motorized turntable to rotate the plate while moving the spreader back and forth across the agar surface for about 15 s. Some recommend that a new spreader be used for each plate, some recommend flaming the spreader between plates and still others advocate only changing or flaming the spreader between subsequent dilutions. Clearly, the first option is most rigorous in minimizing carryover of cells but it is unwieldy in terms of glassware. Some economy can be gained by starting with the most diluted samples and working backward to the more concentrated samples. It has been shown that no more than 1 to 4% of the inoculum is transferred with the spreader from one plate to the next (Postgate, 1969; Clark, 1971).

8–7.3 The Drop Plate Method

8–7.3.1 Procedure

1. Obtain calibrated "dropping" pipettes or use one of the piston type pipettors both of which are commercially available. The stepwise, motor driven pipettor is particularly useful for the repetitive distribution of drops to multiple plates.

2. Prepare a sufficient number of plates of the appropriate culture medium and dry them to a suitable level prior to inoculation (see section 8–7.2.2).

3. From the chosen range of dilutions, drop 0.02-mL drops about 1.5 to 2 cm from the edge of the plate from a height of 10 to 25 mm above the agar surface. To expand the area of the drop the plate may be tilted in a circular rocking motion. After inoculation and absorption of the inoculum, the plates are incubated in the inverted position. Incubation times as short as 1 to 2 d may be sufficient but the investigator may need to determine the optimum time based on experience with a given type of sample. Use of a 5x or 10x lens (dissection scope) is helpful for detecting very small colonies.

4. Counts are obtained from the drops showing the largest number of countable colonies and free of confluent growth (usually 20 or more colonies). The mean of the six counts multiplied to correct for drop volume and dilution factor produces the recoverable viable count.

8–7.3.2 Comments

The drop plate method was first described by Miles and Misra (1938) and has seen various modifications since its inception. The technique described above is that of Herbert (1990). The drop plate method is probably the least satisfactory method for counting soil bacteria because of the small area in which colonies are allowed to develop and the resulting high degree of interaction between developing colonies.

As in the spread plate technique, the agar should be suitably dried before inoculation. Herbert (1990) recommends using six dilutions and placing one drop from each dilution on each of six plates. The principal advantage of the drop plate technique is its efficient use of agar plates. The technique is probably suitable for enumeration of bacterial cell suspensions prepared as inoculum. However, because of the extreme diversity in the soil population it is highly likely that interactions between organisms have a large effect on the numbers of colonies developing in the small drop area. The technique warrants further evaluation for its utility in evaluating soil bacterial populations especially where effective selective media may be employed.

8–8 MEDIA FOR ENUMERATION OF SOIL BACTERIA

It is difficult, if not impossible, to consider an in-depth listing of the vast number of culture media that have been used in the enumeration of soil bacteria. Listed below, with their formulae, are some of the more commonly used media for soil bacteria. The reader may wish to consult any of several manuals of microbiological methods, e.g., the *Manual of Methods for General Bacteriology* (American Society for Microbiology) or various volumes of the series *Methods in Microbiology* (Academic Press, London and New York) for listings of general purpose culture media. In addition, one should consult other chapters in this volume for references to selective media for various physiological groups of soil bacteria, i.e., actinomycetes, diazotrophs, and denitrifiers etc. Useful listings of media for soil bacteria have been given by Parkinson et al. (1971b) and Wollum (1982). All media listed below are sterilized by autoclaving at 121 °C for at least 15 min.

1. **Soil Extract Agar (James, 1958):** Preparation of soil extract can be accomplished in several ways. The most common way is to autoclave (121 °C, 30 min) 1000 g of soil in 1 L of distilled water. After autoclaving the suspension is cooled and repeatedly filtered to obtain a clear liquid the color of which will vary according to the organic content of the soil. Filtration of soils rich in clay can be facilitated by the addition of about 0.5 g of $CaCO_3$ or $CaSO_4$ to flocculate the clay fraction. After filtration the volume is restored to 1000 mL by adding water. Alternatively, sufficient water may be added initially (usually 1200–1500 mL) such that the final volume of extract is near 1000 mL without additional water (James, 1958). Soil extract, thus prepared, may be sterilized "as is" and solidified with agar or it may be amended with 0.25 g of K_2HPO_4 (after James, 1958) or 1.0 g of glucose, 5.0 g of yeast extract and 0.2 g of K_2HPO_4 (Parkinson et al., 1971b; Wollum, 1982).

   Soil extract may be autoclaved repeatedly, however, it is preferable to dispense the extract in smaller volumes which are then sterilized and stored for later use. Cold extracts have been used (James, 1958; Olsen & Bakken, 1987) and whereas James (1958) indicated lower counts were obtained with a "cold" soil extract, Olsen and Bakken (1987) observed only minor differences in counts on media prepared with cold vs. autoclaved extracts. They also observed, as have others, that the source of the soil extract had little effect on bacterial counts, i.e., counts were not markedly different whether the extract was prepared from the same soil being counted or it was prepared from a different soil.

2. **Tryptic Soy Agar** (Difco manual): 15.0 g, tryptone; 5.0 g, soytone; 5.0 g, NaCl; 15 g, agar. Preformulated medium is commercially available (e.g., Difco, Chicago; BBL, Baltimore). The medium can be used in varied "strengths" (see section 8–8.4). When using di-

luted forms of the commercial medium add agar to obtain a final concentration of 15 to 20 g L^{-1}.

3. **Nutrient Agar:** 5.0 g, peptone; 3.0 g, beef extract; 15 g, agar dissolved in 1000 mL of distilled water. Glucose (5.0 g L^{-1}) is sometimes added to the medium.

4. **Peptone Yeast Extract Agar** (Goodfellow et al., 1968): 5.0 g, peptone; 1.0 g, yeast extract; 0.01 g, ferric phosphate; 15 g of agar; 1000 mL of distilled water, pH to 7.2.

5. **PTYG Medium** (Balkwill, 1990): 5.0 g, peptone; 5.0 g, tryptone; 10.0 g, yeast extract; 10.0 g, glucose; 0.6 g, $MgSO_4 \cdot 7H_2O$; 0.07 g, $CaCl_2 \cdot 2H_2O$; 15 g, agar; 1000 mL, distilled water. Balkwill (1990) recommended a 1% concentration (i.e., 1/100th-strength) of this medium for recovery of a wide variety of bacteria from both deep and shallow aquifers.

6. **M9 Minimal Medium** (Sambrook et al., 1989): To 750 mL of sterile deionized water (cooled to 50 °C or lower), add 200 mL of 5X M9 salts (see below) and make up to 1 L. Add 20 mL of an appropriate C source (20% solution sterilized by filtration). To the cooled diluted medium add 2 mL of 1 M $MgSO_4$ and 0.1 mL of 1 M $CaCL_2$. Each of these solutions should be autoclaved separately. If necessary, supplement the medium with stock solutions of appropriate amino acids, vitamins etc.

    5X M9 Salts: Dissolve the following salts to a volume of 1 L: 64.0 g, $Na_2HPO_4 \cdot 7H_2O$; 15.0 g, KH_2PO_4; 2.5 g, NaCl; and 5.0 g, NH_4Cl. Sterilize by autoclaving.

    This minimal medium may be useful in studies where a defined medium is required.

7. **Soil Solution Equivalent Medium** (Angle et al., 1991): The medium contains the following ions at the listed concentrations (millimolar): NO_3^-, 2.5; NH_4^+, 2.5; HPO_4^{-2}, 0.005; Na^+, 2.5; Ca^{2+}, 4.0; Mg^{2+}, 2.0; K^+, 0.503; Cl^-, 4.0; SO_4^{2-}, 5.0 and Fe [as ethylenediamine-di(o-hydroxyphenylacetic acid) FeEDDHA)], 0.02. The ions are added as KH_2PO_4, $CaSO_4$, $MgCl_2$ and NH_4NO_3. To adjust the pH from an initial 4.1 to 6.0 add 0.5 mM K (0.5 mL 1.0 M KOH) which supplies the appropriate amount of K and 4.0 mM Na (4.0 mL of 1.0 M NaOH). The pH of the medium is maintained at 6.0 by the addition of 10 mM 2-(N-morpholino)ethanesulfonic acid (MES) buffer. Appropriate C sources are added as needed.

    The medium is sterilized by autoclaving and the FeEDDHA is added when the medium has cooled to 50 °C because the compound is heat labile. Because of the low phosphate content of the medium it may be possible to add the C source before autoclaving thus simplifying the preparation of the medium.

    The advantage of the soil solution equivalent (SSE) medium is that it more closely simulates the soil solution that microorganisms would be exposed to in their native habitat. The solution can be prepared at different pH values to accommodate soils of differing pH and the ionic ratios can be adjusted to accommodate different soils. For example, to cultivate bacteria from a calcareous soil, the pH and Ca levels would be adjusted appropriately. The medium also has the advantage that like M9 above, it is a defined medium.

8–8.1 Addition of Inhibitors to Culture Media

Almost any culture medium can be made more selective for a particular group of microorganisms through a judicious choice of antibiotics or other antimicrobial agents such as dyes. The development of fungi in media designed for enumeration of bacteria can be suppressed by the addition of cycloheximide ($100 \, \mu g \, mL^{-1}$) which can be sterilized by autoclaving or with filter-sterilized natamycin ($21.6 \, mg \, L^{-1}$) added to cooled media (Pedersen, 1992). Gram-positive bacteria can be suppressed through the addition of penicillin ($1–10 \, \mu g \, mL^{-1}$) or vancomycin ($15 \, \mu g \, mL^{-1}$). Conversely, gram-negative bacteria may be suppressed by adding polymixin B ($5 \, \mu g \, mL^{-1}$) to the medium. Given the very wide-spread use of bacteria bearing one or more antibiotic resistance "markers" in ecological studies (see chapter 27 in this book), including releases of genetically engineered strains, a large number of antibiotics has found use in an equally large number of culture media for the isolation and enumeration of narrowly defined groups of soil bacteria. It is beyond the scope of this chapter to cover such media in detail. However, an important point that must be considered in using antibiotics etc. is their thermostability. It is imperative to determine whether a compound can withstand autoclaving without losing efficacy. Thermolabile (e.g., streptomycin) compounds must be sterilized by micropore filtration and added to cooled, molten agar media prior to pouring plates. Additionally, some antibiotics may require dissolving in special solvents before adding to media. A compatible membrane material is required for filter sterilization of such materials.

8–8.2 Some Examples of Selective Media

While it is not practical to present a long list of selective media of use in soil microbiology, a few formulas will be presented to illustrate some principles commonly employed in the design of such media.

NPCC Medium for Fluorescent Pseudomonads (Simon et al., 1973): 20.0 g Proteose peptone No. 3 (Difco); 8 mL glycerol, 1.5 g K_2SO_4, 1.5 g $MgSO_4 \cdot 7H_2O$, agar 12 g (authors specify Oxoid Ionagar No. 1), 940 mL distilled water. The pH is adjusted to 7.2. Penicillin G (75 000 units), novobiocin (45 mg), and cycloheximide (75 mg) are dissolved in 3 mL of 95% ethanol and diluted with 50 mL of sterile distilled water and added to 940 mL of the molten, cooled basal medium above. Chloramphenicol ($5–12.5 \, \mu g \, mL^{-1}$) is also added to the molten cooled medium.

The base medium is Medium B of King et al. (1954). The medium was originally formulated for the purpose of differentiating strains of *Pseudomonas* based on differential production of pigments on the medium. The original authors as well as Simon et al. (1973) were careful to point out that the type of peptone used had a significant bearing on the production of fluorescent pigments. The addition of the antibiotics and the antifungal cycloheximide make the medium relatively selective for fluorescent pseudomonads because of a high frequency of intrinsic antibiotic resistance among this group. There is sufficient glycerol and salts in this medium to exert a selective osmotic pressure as well. More than 70% of the colonies developing on this medium were reported to be fluorescent pseudomonads.

S1 Medium for Fluorescent Pseudomonads (Gould et al., 1985): 10.0 g of sucrose, 10.0 mL of glycerol, 5.0 g of Casamino acids (Difco), 1.0 g of $NaHCO_3$, 1.0 g of $MgSO_4 \cdot 7H_2O$, 2.3 g of K_2HPO_4, 1.0 g of sodium lauroyl sarcosine (SLS), 20 mg of trimethoprim (Sigma), 18 g of agar, 1000 mL of distilled water. Final pH of the medium is 7.4 to 7.6. Trimethoprim is added to the medium after it has been autoclaved and cooled.

SLS eliminates the growth of gram-positive organisms (presumably through alterations of cell surface properties) and trimethoprim is active against facultative gram-negative organisms (Gould et al., 1985). Sucrose and glycerol were reported to be essential for this medium because the combination of these compounds provides an osmotic stress that is one of the selective features of the medium. The medium was reported to be more selective than previous media and reportedly has the advantage that fluorescence can be observed on the initial isolation medium without having to resort to streaking on King B medium. The authors claim that the medium recovers greater numbers of fluorescent pseudomonads than the NPCC medium above however, each of the media have been shown to contain up to 70 to 75% fluorescent colonies. It should be pointed out that neither of the media arc completely selective and that use of these and any selective medium for that matter, requires judicious observation of culture plates during the counting process!

Selective Medium for Recovery of *Arthrobacter* from Soil (Hagedorn & Holt, 1975): 4.0 g (0.4%) of trypticase soy agar, 2.0 g (0.2%) of yeast extract, 20.0 g of NaCl, 0.1 g of Actidione (cycloheximide), 150 mg of methyl red, 15 g of agar, 1000 mL of distilled water. The medium is sterilized by autoclaving and filter-sterilized methyl red is added to the cooled medium. The medium is adjusted to the pH of the soil under study.

The authors claimed the cycloheximide (0.01%) and 2% NaCL effectively inhibited all fungi and most streptomycetes, nocardia and gram-negative bacteria while the methyl red (150 μg mL^{-1}) inhibited other gram-positive bacteria but did not affect the arthrobacters. The pH of the medium (from 5.0–8.5) apparently had no effect on the "selectivity" of the medium. Approximately 75% (55–83%) of the colonies on the selective medium were putative arthrobacters as judged by microscopical observations for rod-to-coccus morphological cycle, snapping division, pleomorphism and V-forms. In using this medium in subsequent investigations of the ecology of *Arthrobacter* in soil, the authors used the assumption that 78% of the colonies on the selective medium were arthrobacters and no further criteria were applied.

It is clear that this medium is more "semi-selective" rather than strictly selective for arthrobacters. The medium relies on the use of elevated salt concentrations, a dye and the antifungal compound cycloheximide to accomplish control of a large part of the normal flora. Still, about 25% or more of the colonies developing on this medium were not arthrobacters, thus the counting procedure using this medium is somewhat arbitrary.

8–8.3 Comments

Several points regarding selective media for the enumeration of soil bacteria are illustrated with the media described above. All of these examples rely on the use of one or more antimicrobial compounds to control major portions of the soil microflora and it is always a possibility that some members of the group under study may also be eliminated by these agents. Any formulation should be investigated rigorously to ascertain any untoward negative effects on the microbes which the medium is supposed to select. These media also rely, to some extent, on the C sources for their "selectivity." These effects extend beyond those due to microbial preferences for C sources because in some instances the C compounds also serve to impose an osmotic stress that may contribute to the selectivity of the medium. The use of various salts can have similar effects (i.e., there may be both specific ion effects and osmotic effects). The importance of the ionic balance of the medium is addressed in the "soil solution equivalent medium" described in the previous section.

Regardless of the efficacy of the many selective or semi-selective media that are available, it is clear that much more work is needed in this area to study the ecology of natural communities without having to resort to the development of antibiotic-resistant strains etc. This will, no doubt, be a difficult task, however, the development of useful media should continue to be a viable subject of research in soil microbiology.

8–8.4 "Strength" of Culture Media for Soil Bacteria

A point worthy of consideration in designing media for the isolation and enumeration of soil bacteria is the nutrient concentration in the medium. It has been recognized for some time that nutrient-poor media frequently give rise to larger numbers of bacteria than nutrient-rich media. A discussion of oligotrophy in soil is beyond the scope of this section but it is well known that many soil bacteria behave as oligotrophs (preferring very low levels of organic C sources for growth) thus they may not be recovered on nutrient-rich commercial media such as nutrient agar and tryptic soy agar prepared at their normal concentrations. For example, Hattori (1976) reported numbers of bacteria four- to sixfold greater on 1/100th-strength nutrient agar than on the undiluted medium. Maximum numbers of colonies developed on full-strength nutrient broth in "a few days" whereas it required 3 to 4 wk incubation for developing maximum numbers on diluted nutrient broth and the colonies were described as "pinhole size." He further demonstrated the sensitivity of soil isolates recovered on the diluted medium to concentrations of NaCl normally included in the medium. Olsen and Bakken (1987) also reported declining numbers of recoverable bacteria with increasing concentrations of nutrients in several media. In fact, they showed that for two of three soils studied, Winogradsky salts agar (prepared by solidifying Winogradsky's salt solution, see section 8–5) was as effective as the same solution containing added nutrients or soil extract. They found that "water agar" prepared with Difco agar (grade not specified) contained 32 and 96 mg L^{-1} of glucose and galactose, respectively, prior to hydrolysis of the medium. Hattori (1980) also reported the effect of different agars on numbers of bacteria recovered in diluted nutrient broth. He found that numbers were greater on purified agar (Difco and Oxoid; 112 and 160%, respectively, of the count on Noble agar) than on Noble agar (Difco) which was better than Bacto agar which gave 87% of the count on Noble agar. Thus, it is apparent that the composition of the agar itself can have an effect on numbers of bacteria recovered on low-nutrient media.

Another aspect of the nutrient concentration issue is whether or not the same bacteria can or will be recovered on nutrient-rich and nutrient-poor media. There is evidence that different "sets" of isolates can be obtained with the two categories of media and that a greater variety of isolates can be obtained using both a rich and a poor medium for the same samples. Balkwill et al. (1989) reported isolating about as many different isolates of bacteria from a deep aquifer sediment on a rich medium as on a low-nutrient medium. However, subsequent characterization showed that different "strains" of microbes were isolated on concentrated and dilute media and there was < 11% overlap between the two groups. Similarly, Hattori (1976) isolated bacteria on diluted nutrient broth that would not grow on the full strength medium. Olsen and Bakken (1987) indicated that a soil extract agar prepared with a cold-extracted soil extract recovered about 10 times as many "thin rods" and greater numbers of small cocci and coccobacilli than either a yeast peptone medium or a soil extract agar with added glucose. Thus it is clear that, depending on the aims of an investigation, it might be advisable to include media of high and low nutrient concentrations.

8–9 ANALYSIS AND PRESENTATION OF PLATE COUNT DATA

In-depth presentations regarding statistical concepts and calculations relevant to bacterial plate counts have been given by Eisenhart and Wilson (1943), Meynell and Meynell (1970), and Cowell and Morisetti (1969). The reader should consult these works for detailed information regarding the derivation of the statistics applicable to plate count data. A brief, helpful discussion can be found in the soil bacteriology laboratory manual of O.N. Allen (1949). However, a few points regarding the calculations and presentation of plate count data are worth mentioning.

8–9.1 The "30 to 300 Rule"

The beginning student in microbiology laboratories is often told to count dilution plates that contain between 30 to 300 colonies or to follow the so-called "30 to 300 rule" with little explanation as to the rationale of such a practice. The derivation of this practice dates back to the work of Breed and Dotherrer (1916) in which they presented a rationale for the procedure. This range of colonies encompasses a lower limit, 30, below which accuracy of the technique declines rapidly, and an upper limit of 300, above which improved accuracy increases only slowly with the counting of additional colonies. Meynell and Meynell (1970) state that 600 colonies is an optimal number to count. However, other investigators report that fatigue of the person doing the counting causes higher numbers to be subject to increasing errors in counting. Whatever the case, the practice of counting 30 to 300 colonies will likely remain a standard practice in deriving plate count data. A recent study by Tomasiewicz et al. (1980) suggests that 25 to 250 colonies is the most suitable range for counting colonies on plates.

There is a bit of conflict inherent in the 30 to 300 colony range for counting soil bacteria. As was pointed out earlier, colony numbers do not always accurately mirror the dilution factor such that more dilute samples tend to give rise to larger microbial numbers. This is often attributed to effects of competition (see section 8–6.2) among bacteria developing on plates so that fewer cells express themselves through colony development when there is more crowding at the lower dilutions. If this is so, then it appears that numbers based on dilutions showing smaller numbers of colonies within the accepted range represent the more accurate estimate of viable recoverable bacteria since they are least subject to the negative effects of competition. Whenever possible, it is most desirable to determine and compare numbers derived from a single level of dilution (e.g., 10^{-6} etc.). This obviates the confusion that may be caused by using different dilutions to compare samples.

Plate count data can be sufficiently reliable to allow gross comparisons of soils receiving different treatments etc. Thus the means of populations from different plots can be compared using a variety of statistical procedures for comparisons of means including Student's t test and analysis of variance followed by appropriate procedures for separation of multiple means. Prior to analyzing plate count data, the bacterial numbers should be subjected to a $\log_{10}$ transformation. The purpose of the transformation is to reduce the variance of the data to a level acceptable for the statistical tests. Baker (1985) presented a concise, useful discussion of various aspects of the $\log_{10}$ transformation and the presentation of plate count data for publication. She recommends performing statistical analysis on the transformed data, including calculations of confidence intervals. To avoid any confusion caused by tabulation of log values she recommends retransformation of the data to arithmetic form by taking the anti-log of the means and standard deviations of the log-transformed data. Using such a procedure, the confidence interval will not be symmetrical about the mean, rather it will be skewed, reflecting the nonadditivity of the variances in the original data (Baker, 1985).

8–10 CONCLUSION

Despite the criticisms and limitations of plate-counting methods, it is highly likely that they will remain in use in laboratories throughout the world. This is entirely appropriate because the techniques, despite their limitations, are quite useful and suitable methods to replace them have been slow to emerge, are rather laborious or do not yet exist. Problems arise mainly from overestimating the utility of the techniques not from the techniques themselves. Plate count techniques offer several advantages for certain types of microbiological studies including: convenience, simplicity of performance, relatively low costs, little specialized equipment, provision of a method for comparison of environments, and a supply of viable microorganisms for further study. Disadvantages include estimation of only a small part of the "total" population due mainly to selectivity of culture media, problems arising from clumping of cells and failure to extract cells from soil particles and aggregates, provision of a "viable" count only with relatively little information about biomass, activity etc. and questionable quantitation because of these uncertainties and other inaccuracies inherent in the methods (Buck, 1979).

It should be clear to anyone contemplating the use of plate-counting techniques that they cannot and do not provide an estimate of the "total bacterial population" in a soil. Usage of such language is strongly discouraged. Rather, it seems more desirable to adopt language that is more descriptive of the types of results which can be achieved with plate-counting techniques. That is, they provide an estimate of the "recoverable viable bacteria" in a sample under well-defined conditions imposed by the choice of culture medium, incubation conditions etc. Carefully used they serve as well as other trusted tools for microbiological analysis, carelessly used they provide little information of experimental value. They are perhaps best used in conjunction with other assessments of microbial biomass and activity where they can provide additional information regarding types and numbers of bacteria present. Used with adequate selective media they can provide a tool for the detection of relatively low numbers of indigenous bacteria intentionally released into an environment.

I hope that you, the reader, will have found the information presented here of some value, and, if nothing else, found this a useful refresher course. Any inadequacies of the coverage rest solely with the author. I regret that as of this writing, I was unable to "dig up" hard documentation as to the origin of the practice of preparing the "1 in 10" dilution using 10 g of soil in 95 mL of diluent. Perhaps one of you will know of such information. If so, I would appreciate hearing from you! For all those who indulged my phone calls (and you were many) I thank you for the insights you provided!

REFERENCES

Allen, O.N. 1949. Experiments in soil bacteriology. Burgess Publ. Co., Minneapolis.

Angle, J.S., S.P. McGrath, and R.L. Chaney. 1991. New culture medium containing ionic concentrations of nutrients similar to concentrations found in the soil solution. Appl. Environ. Microbiol. 57:3674–3676.

Baker, K.H. 1985. Logarithmic transformation of data. Am. Soc. Microbiol. News 51:443.

Balkwill, D.L. 1990. Deep-aquifer microorganisms. p. 183–211. In D.P. Labeda (ed.) Isolation of biotechnological organisms from nature. McGraw-Hill Publ. Co., New York.

Balkwill, D.L., J.K. Frederickson, and J.K. Thomas. 1989. Vertical and horizontal variations in the physiological diversity of aerobic chemoheterotrophic bacterial microflora in deep southeast coastal plain subsurface sediments. Appl. Environ. Microbiol. 55:1058–1065.

Breed, R.S., and W.D. Dotherrer. 1916. The number of colonies allowable on satisfactory agar plates. N.Y. (Geneva) Agric. Exp. Stn. Tech. Bull. 53:3–11.

Brock, T.D. 1961. Milestones in microbiology. Am. Soc. for Microbiol., Washington, DC.

Brock, T.D. 1966. Principles of microbial ecology. Prentice-Hall, Englewood Cliffs, NJ.

Buck, J.D. 1979. The plate count in aquatic microbiology. p. 19–28. In J.W. Costerton and R.R. Colwell (ed.) Native aquatic bacteria: Enumeration, activity, and ecology (Symposium). Am. Soc. for Testing and Materials, Philadelphia.

Busta, F.F., E.H. Peterson, D.M. Adams, and M.G. Johnson. 1984. Colony count methods. p. 62–83. In M.L. Speck (ed.) Compendium of methods for the microbiological examination of foods. Am. Publ. Health Assoc., Washington, DC.

Casida, L.E., Jr. 1968. Methods for the isolation and estimation of activity of soil bacteria. p. 97–122. In T.R.G. Gray and D. Parkinson (ed.) The ecology of soil bacteria. Univ. of Toronto Press, Toronto.

Clark, D.S. 1971. Studies on the surface plate method of counting bacteria. Can. J. Microbiol. 17:943–946.

Clark, F.E. 1965. Agar-plate method for total microbial count. p. 1460–1466. In C.A. Black et al. (ed.) Methods of soil analysis. Part 2. Agron. Monogr. 9. ASA, Madison, WI.

Cowell, N.D., and M.D. Morisetti. 1969. Microbiological techniques—Some statistical aspects. J. Sci. Food Agric. 20:573–579.

Dabek-Szreniawska, M., and T. Hattori. 1981. Winogradsky's salts solution as a diluting medium for plate count of oligotrophic bacteria in soil. J. Gen. Appl. Microbiol. 27:517–518.

Eisenhart, C., and P.W. Wilson. 1943. Statistical methods and control in bacteriology. Bacteriol. Rev. 7:57–137.

Goodfellow, M., I.R. Hill, and T.R.G. Gray. 1968. Bacteria in a pine forest soil. p. 500–515. In T.R.G. Gray and D. Parkinson (ed.) The ecology of soil bacteria. Liverpool Univ. Press, Liverpool.

Gorrill, R.H., and E.M. McNeil. 1960. The effect of cold diluent on the viable count of Pseudomonas pyocyanea. J. Gen. Microbiol. 22:437–442.

Gould, W.D., C. Hagedorn, T.R. Bardinelli, and R.M. Zablotowicz. 1985. New selective medium for enumeration and recovery of fluorescent pseudomonads from various habitats. Appl. Environ. Microbiol. 49:28–32.

Hagedorn, C., and J.G. Holt. 1975. Ecology of soil arthrobacters in Clarion-Webster toposequences of Iowa. Appl. Environ. Microbiol. 29:211–218.

Hattori, T. 1976. Plate count of bacteria in soil on a diluted nutrient broth as a culture medium. Rep. Inst. Agric. Res. 27:23–30.

Hattori, T. 1980. A note on the effect of different types of agar on plate count of oligotrophic bacteria in soil. J. Gen. Appl. Microbiol. 26:373–374.

Herbert, R.A. 1990. Methods for enumerating microorganisms and determining biomass in natural environments. p. 1–39. *In* R. Grigorova and J.R. Norris (ed.) Methods in microbiology. Vol. 22. Academic Press, London.

Hopkins, D.W., S.J. Macnaughton, and A.G. O'Donnell. 1991. A dispersion and differential centrifugation technique for representatively sampling microorganisms from soil. Soil Biol. Biochem. 23:217–225.

Ingram, M., and B.P. Eddy. 1953. A warning note on the use of a pipette for several dilutions on viable bacterial counts. Lab. Pract. 2:11–13.

James, N. 1958. Soil extract in soil microbiology. Can. J. Microbiol. 4:363–370.

Jensen, V. 1962. Studies on the microflora of Danish beech forest soils. Zentralbl. Bakteriol. Parasitenkd. Infektionskr. Abt. 2 116:13–32.

Jensen, V. 1968. The plate count technique. p. 158–170. *In* T.R.G. Gray and D. Parkinson (ed.) The ecology of soil bacteria, an international symposium. Univ. of Toronto Press, Toronto.

King, E.O., M.K. Ward, and D.E. Raney. 1954. Two simple media for the demonstration of pyocyanin and fluorescin. J. Lab. Clin. Med. 44:301–307.

King, W.L., and A. Hurst. 1963. A note on the survival of some bacteria in different diluents. J. Appl. Bacteriol. 26:504–506.

Kobel-Boelke, J., B. Tienken, and A. Nehrkorn. 1988. Microbial communities in the saturated groundwater environment. I: Methods of isolation and characterization of heterotrophic bacteria. Microb. Ecol. 16:17–29.

Koch, A.L. 1981. Growth measurement. p. 179–207. *In* R.N. Costilow (ed.) Manual of methods for general bacteriology. Am. Soc. Microbiol., Washington, DC.

Meiklejohn, J. 1957. Numbers of bacteria and actinomycetes in a Kenya soil. J. Soil Sci. 8:240–247.

Meynell, G.G., and E. Meynell. 1970. Quantitative aspects of microbiological experiments. p. 173–255. *In* Theory and practice in experimental bacteriology. Cambridge Univ. Press, Cambridge.

Miles, A.A., and S.S. Misra. 1938. The estimation of the bactericidal powers of blood. J. Hyg., Camb., 38:732–749.

Olsen, R.A., and L.R. Bakken. 1987. Viability of soil bacteria: Optimization of plate-counting technique and comparison between total counts and plate counts within different size groups. Microb. Ecol. 13:59–74.

Parkinson, D., T.R.G. Gray, and S.T. Williams. 1971a. Biomass measurements. p. 57–70. *In* Methods for studying the ecology of soil micro-organisms. Blackwell Sci. Publ., Oxford.

Parkinson, D., T.R.G. Gray, and S.T. Williams. 1971b. Media for isolation of soil micro-organisms. p. 105–116. *In* Methods for studying the ecology of soil micro-organisms. Blackwell Sci. Publ., Oxford.

Pedersen, J.C. 1992. Natamycin as a fungicide in agar media. Appl. Environ. Microbiol. 58:1064–1066.

Postgate, J.R. 1969. Viable counts and viability. p. 611–628. *In* J.R. Norris and D.W. Ribbons (ed.) Methods in microbiology. Vol. 1. Academic Press, New York.

Sambrook, J., E.F. Fritsch, and T. Maniatis. 1989. Molecular cloning. A laboratory manual. 2nd ed. Cold Spring Harbor Lab. Press, Cold Spring Harbor, NY.

Simon, A., A.D. Rovira, and D.C. Sands. 1973. An improved medium for isolating fluorescent pseudomonads. J. Appl. Bacteriol. 36:141–145.

Skinner, F.A., P.C.T. Jones, and J.E. Mollison. 1952. A comparison of a direct- and a plate-counting technique for the quantitative estimation of soil micro-organisms. J. Gen. Microbiol. 6:261–271.

Stevenson, I.L. 1958. The effect of sonic vibration on the bacterial plate count of soil. Plant Soil 10:1–8.

Straka, R.P., and J.L. Stokes. 1957–58. Rapid destruction of bacteria in commonly used diluents and its elimination. J. Appl. Microbiol. 5–6:21–25.

Strange, R.E., and F.A. Dark. 1962. Effect of chilling on *Aerobacter aerogenes* in aqueous suspension. J. Gen. Microbiol. 29:719–730.

Tomasiewicz, D.M., D.T. Hotchkins, G.W. Reinhold, R.B. Read, and P.A. Hartman. 1980. The most suitable number of colonies on plates for counting. J. Food Prot. 43:282–286.

Umbreit, W.W., R.H. Burris, and J.F. Stauffer. 1972. Manometric and biochemical techniques. 5th ed. Burgess Publ. Co., Minneapolis.

Waksman, S.A. 1922. Microbiological analysis of soil as an index of soil fertility: II. Methods of the study of numbers of microorganisms in the soil. Soil Sci. 14:283–298.

Wollum, A.G. 1982. Cultural methods for soil microorganisms. p. 781–802. *In* A.L. Page et al. (ed.) Methods of soil analysis. Part 2. 2nd ed. Agron. Monogr. 9. ASA and SSSA, Madison, WI.

Wyant, Z.N. 1921. A comparison of the technic recommended by various authors for quantitative bacteriological analysis of soil. Soil Sci. XI:295–303.

Chapter 9

Coliform Bacteria

RONALD F. TURCO, *Purdue University, West Lafayette, Indiana*

Domestic sewage can contain bacteria, protozoa, and virus that are pathogenic to humans. The most common disease-causing agents associated with human fecal materials are *Salmonella* spp. (causing gastroenteritis and typhoid fever), *Shigella* spp. (causing dysentery and gastroenteritis) and *Vibrio* spp. (causing gastroenteritis). The protozoan *Entamoeba histolytica* and *Giardia* spp. (causing gastroenteritis) are also encountered. Fecal material associated with livestock production can also be a source of infection for humans. Microbial quality of a given matrix is generally assessed by examining samples not for the disease-causing microorganism, but for the presence of indicators such as coliform bacteria.

The term coliform bacteria describes a group of aerobic or facultative, gram-negative, non-spore forming rods that reside in the intestinal tract of all vertebrates. The term coliform bacteria should be used to only describe bacteria fitting these criteria that ferment lactose to gas within 48 h when incubated at 35 °C. By definition coliforms are inclusive to the genera *Escherichia, Klebsiella,* and *Enterobacter* (Brock & Madigan, 1991). It should be noted however, the study of coliforms is in general, the study of *E. coli* (Thelin & Gifford, 1983). Coliforms act as indicators for the possible presence of all members of the Enterobacteriaceae family as well as other fecal contaminant organisms. Confusion over the use of the term coliform bacteria has resulted as the family Enterobacteriaceae includes genera that are lactose negative (*Shigella, Serratia, Edwardsiella, Proteus, Salmonella, Provicencia, Yersinia, Cedecea, Morganella, Hafnia*), delayed in their lactose use (*Citrobacter* and *Arizona*) or lactose positive (*Kluyvera* and *Rahnella*) but not considered coliforms. It should be apparent that the term coliform is an operational rather than a true taxonomic definition.

The presence of coliform bacteria is used as an indicator of the overall sanitary quality of soil and water environments. Use of an indicator such as coliforms, as opposed to the actual disease-causing organisms, is advantageous as the indicators generally occur at higher frequencies than the pathogens and are simpler and safer to detect. Concentrations of fecal coliforms can range from 10^6 to 10^9, 10^4 to 10^7, 10^1 to 10^5 coliforms g^{-1} feces for humans, farm animals, and rodents, respectively. Fecal coliforms can make up more than 96% of the coliforms found in human feces and up to 98% of coliforms in other warm-blooded animals (Geldreich, 1976). However, detection of coliforms in environmental samples of soil or water is complicated as members of the genus *Klebsiella* are both coliforms and common soil organisms. It should be noted that while *Klebsiella* is often thought of as only a conflicting organism in studies of environmental coliforms, 30 to 40% of humans and other animals may carry *K. pneumoniae* in their intestinal tract (Geldreich, 1976). Many authors (Doran & Linn, 1979; Thelin & Gifford, 1983) are in agreement with Geldreich (1976) who concluded that estimations of total coliforms in environmental samples show not only possible fecal contamination, but detect many organisms of limited sanitary importance. The importance of bacteria residing in soil and water as a source of bias in coliform estimation cannot be overstated. Doran et al. (1981) reported that in fields without cattle the ratio of total to fecal coliforms was 74 to 1. In contrast, grazed pasture had a total to fecal coliform ratio of 5 to 1. Other data cited by Doran et al. (1981) shows that total to fecal ratios tend to fall between 5:1 and 100:1, with the ratio expanding only when humans and animals are absent. However, others have reported that in soils receiving inputs of sewage effluent, the ratio of total to fecal coliforms can still exceed 10 to 1 (Tate & Terry, 1980). A similar finding was reported for watersheds receiving inputs of cattle fecal material (Jawson et al., 1982). This overlap in possible sources of bacteria has led to the further operational division of coliforms into total and fecal coliforms.

The differentiation between total and fecal coliforms is enabled by fermentation of lactose at an elevated incubation temperature (44.5 °C), this temperature is thought to suppress the activity and growth of coliforms from environmental sources. As pointed out by Thelin and Gifford (1983), many researchers now feel that the occurrence of fecal contamination is best estimated by the observation of lactose fermentation at a 44.5 °C incubation temperature, rather than the simpler total coliform methods. When working with soil, the determination of fecal coliforms is preferred.

Environmental regulation concerning the number of coliforms in water are focused around the maintenance of safe drinking water supplies. Sources of contamination to water supplies include runoff from feed lots and sludge-treated soils as well as discharge from septic systems. Drinking water should contain less than one coliform 100 mL^{-1}; however, soils and stream water may contain much higher levels. Our interest is in detecting the source of contamination so that water supplies are protected (Doran et al., 1981). The choice of coliform detection method should be a reflection of the experimental question being asked. While many advanced methods are available, the simple and inexpensive most probable number (MPN) approach will answer most questions (see chapter 5 by Woomer in this book). When used with a confirmation step, it provides a reliable estimation for the presence of coliforms. On the other hand, if we are interested in specific questions about the survival of these organisms in soil, more sensitive techniques such as polymerase chain reaction (PCR) can be employed (Steffan & Atlas, 1988 and chapter 34 in this book).

Temporal variations as well as distance to an input source may affect the numbers reported. Differences in the mean level of fecal coliforms depends on both time of year and soil sampling depth. Mean fecal coliform counts can range from 10 to 394 MPN g^{-1} soil for the surface of pasture soil while rarely exceeding 3 MPN g^{-1} soil at 15 cm of depth (Faust, 1982). Counts for soil sampled in autumn tend to be higher for fecal coliforms. In nonpasture soils, fecal coliforms ranged from 3 to 26 MPN g^{-1} soil with the highest counts occurring in the autumn (Faust, 1982). Faust (1976) and Faust and Goff (1977) both reported pasture lands are a significant source of fecal bacteria. While Geldreich et al. (1962) has reported the presence of fecal coliforms in virgin forest soils.

Our efforts are aimed at describing methods for the detection of coliforms in soil. An excellent discussion of methods for coliform detection in water and wastewater is provided in Standard Methods (APHA, 1989).

9–1 RECOVERY AND ENUMERATION OF FECAL COLIFORMS FROM SOIL

9–1.1 Collection of Samples—Soil

Because of the association of fecal coliforms with humans, sampling for the presence of the bacteria should be carried out using an aseptic protocol. Direct contact between the person sampling and the sampling equipment or soil materials should be avoided.

9–1.1.1 Materials

1. Sterile shovel, trowel, or scoop.
2. Sterile jars, plastic bags, and bottles.
3. Marking pen and label tape.
4. Insulated cooler chest, with ice or cold packs.
5. Plastic pails, 20 L.
6. Sterile distilled water washing solution.
7. Methanol, 50 to 70% solution with water.

9–1.1.2 The Sampling Approach

With the exception of purely exploratory sampling, a sampling plan should be devised. The sampling plan should address the collection method, rationale for the site selections, labeling, preservation, transport, and analysis of the samples. A well-documented plan will go a long way in preventing improper interpretation of the findings.

Approaches to sampling soil generally follow one of three methods: judgmental, random, or a random sampling method called systematic sampling (see chapter 1 by Wollum in this book). Selection of the type of sampling approach should always be based on the goals of the research project. Sampling plans must be realistic and reflect the situation at hand. Keith (1991) has defined the strengths and weaknesses of the three sampling systems. In brief, judgmental sampling is based on prior knowledge of a given site or visual assessment at the time of sampling. This approach requires the smallest number of samples and results in a large relative bias. Judgmental sampling is used when we address a question, for example, by collecting samples at a spill site to understand the impact of this discharge to land surface. Random sampling is generally conducted by using "on-the-site" randomization as a site selection method, is often conducted with limited pre-planning, requires a largest number of samples but has a small sample bias. The sampling sites are randomly selected by tossing a shovel or other device into the air and collecting the soil at the point it hits the ground. Systematic sampling is a type of random sampling but samples are collected at predetermined grid coordinates. This approach requires both a pre-knowledge of the site and a larger number of samples, but has a small bias. Because the grid lines are randomly applied to the site, systematic sampling plans can ignore obvious sources of pollution in providing random samples. Often combinations of systematic and judgmental sampling are used. This is typically observed when samples are collected across a field as well as at an obvious site of contamination. Several recent books and articles deal with the limitations of different sampling approaches and should be consulted in the construction of a sampling plan (Keith, 1991; James & Wells, 1990; Triegel, 1988).

Use of a field composite or taking multiple samples from across a field and bulking them together as an indicator of the so called "average population," provides only general information about the site (see chapter 2 by Parkin and Robinson in this book). Field compositing will suppress information about the spatial distribution of the resident microbial population but does provide a general estimation of contamination. The number of samples needed to properly predict the microbial population across a land surface is difficult to estimate. The sample number needed is a function of the population variance and the needed level of precision. Both Keith (1991) and Gilbert (1987) provide models to estimate sample number using different sampling approaches.

9–1.1.3 Soil Collection

Once a plan is selected, soil collection can be undertaken. Collection points in the field should be identified, numbered, and marked with flags or stakes. The population present on the surface is best sampled by use of a scoop or shovel to collect multiple samples from a fixed depth at each sampling point. This soil is mixed, and subsample (50–200 g) removed for analysis. Plastic bags can be used for the subsample. The remainder of the soil is discarded and the bucket washed with water and then rinsed with 70% methanol. If sampling at depth in the profile a push probe can be used. However, it is possible to contaminate the lower samples with materials from the surface when the probe is pushed in. A better approach is to use a large diameter probe and collect the entire core. The core is returned to the lab and sectioned by depth. The sides of the smaller cores are pared off with a sterile knife, resulting in a column of soil. The column of soil can then be used in microbial assessments while the side-wall materials can be used for physical measurements (Konopka & Turco, 1991).

Because we are addressing a specific type of bacteria the equipment should be cleaned with water and 70% methanol between all sampling points. To estimate bacterial carryover it is advisable to collect an equipment blank, a washing from the field cleaned equipment between each sampling point. The equipment blank is made by rinsing the scoop, shovel, or bucket with sterile distilled water after the cleaning procedure. This rinse water is collected, diluted, and analyzed as if it were a soil sample.

Soil samples and equipment washings should be transported in ice-cooled containers. The cooler provides a stable environment for the transport of the sample. Excessive heat, drying, or other environmental extremes should be avoided. All samples should be analyzed immediately following collection; long holding times should be avoided. If a delay in analysis is required, refrigeration (at 4 °C) is preferable to freezing. Sample storage before analysis should not exceed 1 d (with refrigeration).

9–1.2 Enumeration of Bacteria

While many procedures for the enumeration of coliform bacteria are available, we will look at the methods most suited for soil. Enumeration is traditionally carried out with multiple-tube fermentation, resulting in a MPN estimate (see chapter 5 in this book). Work with soil has generally been confined to the MPN multiple-tube fermentation method. In general, the estimates are more reliable than the membrane filter method popular in the study of water. When the membrane filter is used with soil, the soil tends to clog and obscure the filter, making estimations difficult.

9–1.2.1 Dilution of Soil Material

Soil materials recovered from the field need to be diluted before enumeration of the resident bacteria is possible. Dilution serves to separate the bacteria from the soil surface as well as reduce the numbers of bacteria to a countable level. The choice of diluents range from distilled water to complex phosphate saline solutions. These diluents need to be sterile. A major drawback to the use of distilled or tap water is that it is osmotically hypotonic. Physiological saline is isotonic to mammalian cells. The recovery of coliform bacteria, some of which may be damaged, may be lowered by either isomatic shock.

9–1.2.2 Diluents

1. Sterile-distilled water: 1000 mL of distilled water is placed into a 2-L container. The container is loosely capped and steam auto-claved at 1.05 kg cm^{-2}, 121 °C for 20 min. The containers should be cooled before removing from the autoclave. If erlenmeyer flasks are used, cotton plugs wrapped in cheese cloth serve well as the cap. Plugs should be covered with aluminum foil before auto-claving.
2. Sterile-physiological saline: dissolve 8.5 g of sodium chloride (NaCl) in 1 L of distilled water and autoclave as above.
3. Phosphate-buffered saline (with gelatin): the protein aids in stabi-lizing fragile and weakened bacterial cells. Sodium chloride 8.5 g, 0.3 g anhydrous potassium dihydrogen phosphate (KH$_2$PO$_4$), 0.6 g anhydrous, sodium hydrogen phosphate (Na$_2$HPO$_4$). Adjust pH to 6.8 using either 0.1 M sodium hydroxide (NaOH) or hydrochloric acid (HCl). Autoclave as above or add 0.1 g gelatin and then autoclave.
4. Peptone-water: dissolve 1.0 g of peptone in 1 L of distilled water, final pH should be 6.8, autoclave as above.

9–1.2.3 Supplies

1. 160 mL wide-mouth dilution bottle, 1 per sample, containing 95 mL (after autoclaving) of diluent solution.
2. 160 mL milk dilution bottles, 8 per sample, containing 90 mL (after autoclaving) of diluent solution.
3. Sterile glass or plastic disposable 10 mL graduated, serological pi-pettes and a pipette bulb. The initial transfer is facilitated with the use of a wide-bore pipette.
4. Mechanical or electronic balance. Calibrate balance according to the owners manual.
5. Flat bed mechanical shaker.

9–1.2.4 Procedure

1. For each field sample, make one 95-mL and eight 90-mL dilution blanks. The wide-mouthed dilution bottle should be included for use with the initial dilution. The caps, bottles, and diluent should be sterile.
2. Weigh out 10 g of wet soil and transfer it to the wide mouth bottle.
3. The bottle is capped and shaken by hand; check for leaks around the seal of the lid, then place the bottle horizontally on a mechani-cal shaker and shake for 10 min.
4. The bottle is removed from the shaker and 10 mL of the suspension transferred to a 90-mL blank using a sterile pipette. Wide-tip pi-pettes designed for suspension are best suited for this initial dilu-tion. This step establishes a 10^{-2} dilution.

5. This bottle is capped and shaken by hand for 1 min. The cap is removed and a second sterile 10-mL pipette is used to transfer 10 mL to a sterile 90-mL dilution blank. This process is repeated for the remaining dilution blanks. This establishes a dilution sequence ranging from 10^{-2} to 10^{-8}.

A subsample of the soil material should be oven dried for 48 h at 105 °C to determine the moisture content of the soil. This allows the experimentalist to determine the actual dilution of the approximate 1:10 initial dilution.

9–2 DETECTION AND ENUMERATION OF TOTAL COLIFORMS

9–2.1 Multiple Tube Fermentation

The total coliforms presumptive test gives an indication of the presence of all coliform (lactose-positive) bacteria. This procedure results in an estimated value based on probabilities and requires confirmatory evaluation to obtain an unequivocal answer.

9–2.1.1 Materials

1. Lauryl tryptose broth (LTB): Tryptose 20.0 g, lactose 5.0 g, K_2HPO_4 2.75 g, KH_2PO_4 2.75 g, NaCl 5.0 g, sodium lauryl sulfate 0.1 g, distilled water 1000 mL. Adjust pH to 6.8. Medium is dispensed (9 mL) into fermentation tubes containing an inverted Durham tube (6 × 50 mm) and capped. Make sure that the inserted tube is covered with media. Sterilize tubes as above. Prepare enough tubes to allow at least five tubes per sample dilution.

2. EC medium (EC): Tryptose or trypticase 20 g, lactose 5 g, bile salts mixture or bile salts no. 31.5 g, K_2HPO_4 4.0 g, KH_2PO_4 1.5 g, NaCl 5.0 g, distilled water 1000 mL. Adjust pH to 6.8 and disperse (9 mL) into fermentation tubes containing a Durham tube. Make sure the Durham tube is covered. Sterilize tubes as above. Prepare at least five tubes per dilution step.

All of these media are available from commercial sources. To maintain procedural consistency, it is recommended that commercial media be used whenever possible.

9–2.1.2 Supplies

1. Sterile 1-mL graduate pipettes, or an automatic pipette and sterile 1-mL pipetter tips.
2. Water bath with a 35 °C ± 0.5 range.
3. Test tubes with LTB media, Durham tube, and support rack.

9–2.1.3 Procedure

A total of 40 LTB-fermentation tubes, five for each dilution step are needed. The statistical validity of the procedure is enhanced when more fermentation tubes (10–15 are not uncommon) are used per dilution. Commencing at the 1×10^{-8} dilution, transfer 1 mL of diluted soil solution to a LTB-fermentation tube. This is repeated a total of at least five times at each dilution. The same pipette (pipetter tip) can be used for the transfers within a dilution. This procedure is repeated for the various dilutions within a series.

If an automatic pipetter is used, make sure the plastic tips are sterile and the device is properly calibrated. Calibration is done by following the guidelines provided with the instrument. A calibration schedule should be established for each instrument. When calibrating the pipetter, use tips that have been sterilized in a manner similar to what will be used in the experimental work. Autoclaving may expand the tip and this needs to be taken into consideration.

When the transfers are completed, the LTB tubes and rack are transferred to a water bath (35 °C ± 0.5) and incubated for approximately 24 h. The water should cover all of the media in the tube. After 24 h, each tube is picked up and gently shaken. Gas formation is indicated by bubbles in the Durham tube or bubbles suspended in the media should be recorded as positive for lactose use. Tubes showing no gas formation should be recorded as negative and returned to the rack. The negative tubes are incubated an additional 24 h and checked in the same manner.

The use of a pre-drawn spreadsheet aids in the collection of these data. Assuming an eight-fold dilution series and five fermentation tubes at each dilution, the sheet should have eight columns and five rows. Each major row should be divided in half. Tubes showing positive gas at 24 h should be scored with a plus in both halves of the box. Tubes showing no gas at 24 h should be scored with a minus in the top half of the box, and, if positive at 48 h scored in the lower half of the box. This provides a convenient method to calculate the MPN value.

9–2.2 Confirmation of Fecal Coliforms

9–2.2.1 Materials

1. Flammable or disposable transfer loop.
2. Water bath capable of 44.5 °C ± 0.5.
3. Test tubes with EC media, Durham tube, and support rack.

9–2.2.2 Procedure

Lauryl tryptose broth tubes showing gas formation at either 24 or 48 h are considered as presumptive for the presence of coliform bacteria. However, given the widespread occurrence of lactose-positive bacteria in soil, a confirmatory test is required.

While not generally reported in soils work, a completed test is a confirmation for the presence of fecal coliform bacteria. Several procedures have been proposed, but the simplest involves inoculating EC broth with an aliquot from a positive LTB tube and incubating the EC tube at 44.5 °C. Growth accompanied by gas formation after 24 h indicates a positive complete test.

The number of confirmatory EC-fermentation tubes is difficult to estimate. One EC tube is needed for each positive (showing gas production) LTB tube. A sterile metal transfer loop or disposal transfer loop is used to move a small amount of liquid media from the positive LTB tubes to the EC tube. The LTB tube should be gently shaken before the transfer. Within 30 min of inoculation the EC tubes are transferred to an incubator (44.5 ± 0.2 °C) and the results scored after 24 h.

Gas formation in the EC tube within the 24 h incubation period is a positive confirmation for fecal coliforms. Growth without gas formation, or no growth is considered a negative reaction for fecal coliforms. The data should be recorded in a manner similar to that described previously.

Direct estimation of fecal coliforms using the elevated temperature (44.5 °C) and EC tubes is not recommended. The presumptive test serves as a pre-incubation phase that allows time for the recovery of damaged or dormant cells. This incubation is especially critical when sampling from a matrix such as soil.

9–2.2.3 Calculation

Tables listing commonly used dilutions and numbers of tubes per dilution are available (Cochran, 1950; Alexander, 1982). Simple computer programs are also available (Koch, 1981; Woomer et al., 1990). The specific use and calculation of the population density is described with the table or program (see chapter 5 in this book).

9–2.2.4 Comments

The total amount of dry material used to make the media is based on a final volume of inoculant and liquid media equal to 10 mL. This ratio should be maintained when altering the amount of either. As Hass and Heller (1988) indicated as well as many others, the reliability of MPN estimates of populations is only fair. The precision of the estimation is enhanced by increasing the number of tubes tested at each dilution. For example, the use of 10 or 20 LTB tubes would improve the estimation. Alternatively, the estimation can be enhanced by narrowing the initial dilution step. In our case 10^{-1} or 10-fold dilution step was used. The estimation could be enhanced by reducing this to a two or fourfold step. Narrowing the dilution step may be warranted when some idea of population size is available, however, in general, use would be expensive and time consuming. The computer model of Woomer et al. (1990) allows an easy way to calculate MPN estimates for combinations of tube numbers and dilutions not found in published tables.

For the MPN estimation to be valid, it is assumed that the organisms are randomly distributed in the matrix and growth will be evident whenever one or more bacteria are present (Loyer & Hamilton, 1984). Hence, damaged cells, poor transfer techniques or incomplete removal of the cells from the soil surface will impact upon the reliability of the MPN estimation.

9–3 RAPID TEST FOR DETECTION OF *E. COLI* IN SOIL

9–3.1 Introduction

The dilution steps along with the presumptive and confirmatory tests for *E. coli* (fecal coliforms) will take up to 96 h. However, recent work has shown that production of the enzyme glucuronidase is limited to *E. coli* and some strains of *Salmonella* and *Shigella*. Feng and Hartman (1982) showed that hydrolysis of methylumbelliferone glucuronide (MUG) by glucuronidase will result in a fluorogenic product. Detection of the product can be made with long-wave UV light. Hence, this enzyme will provide a unique marker for the presence of *E. coli*.

9–3.2 Materials

1. Lauryl tryptose broth with methylumbelliferone glucuronide (LTB-MUG) tryptose, 20.0 g; lactose, 5.0 g; dipotassium hydrogen phosphate (K_2HPO_4), 2.75 g; potassium dihydrogen phosphate (KH_2PO_4), 2.75 g; sodium chloride, NaCl, 5.0 g; sodium lauryl sulfate, 0.1 g; distilled water, 1 L. The substrate MUG (Sigma Chemical Co., St. Louis, MO) is made up in warm water and added to the above media to achieve a final concentration of 100 µg mL^{-1}. The LTB-MUG is dispensed into test tubes and autoclaved as above.

2. Phosphate buffered saline (PBS) sodium chloride (NaCl), 8.5 g; sodium dihydrogen phosphate ($NaH_2PO_4 \cdot H_2O$), 0.39 g; disodium hydrogen phosphate, 1.0 g; distilled water, 1 L; adjust final pH to 7.2.

All of these materials are available from commercial sources. To maintain procedural consistency, it is recommended that commercial media be used whenever possible.

9–3.3 Supplies

1. 160 mL wide-mouth dilution bottle, 1 per sample, containing 95 mL (after autoclaving) of PBS buffer.
2. 160 mL milk dilution bottles, 8 per sample, containing 90 mL (after autoclaving) of PBS buffer.
3. Sterile glass or plastic disposable 10-mL graduated, serological pipettes and a pipette bulb. The initial transfer is facilitated with the use of a wide-bore pipette.

4. Mechanical or electronic balance. Calibrate balance according to the owners manual.
5. Flat bed mechanical shaker.
6. Sterile 1-mL graduate pipettes, pipette bulb, or a mechanical pipette and sterile 1-mL pipetter tips.

9–3.4 Procedure

1. Soil samples are collected as previously described (see section 9–1.1.2).
2. Soil samples are diluted as previously described except PBS is used (see section 9–1.2.4).
3. A total of 40 LB tubes described previously, five for each dilution are needed (see section 9–2.1.1).
4. Starting at 1×10^{-8} dilution, transfer 1 mL of diluted soil to the LTB-fermentation tube. This is repeated a total of five times at each dilution. The same pipette can be used at each dilution, but should be changed between dilutions.
5. Incubate at 35 ± 0.5 °C for 24 h. Check for the production of gas as before.
6. Tubes showing gas production are considered presumptive for *E. coli*.
7. Using the 1-mL pipette or mechanical pipette, transfer 0.1 mL of LTB-solution from a presumptive positive tube to the LTB-solution from a presumptive positive tube to the LTB-MUG tube. Incubate at 44.5 ± 0.2 °C for 24 h. Check the contents of the tube for fluorescence.
8. Fluorescence is determined at 366 nm long-wave UV light (6 W). (Lights are available commercially.)
9. Enumeration should follow the procedure described previously (see section 9–2.2.3).

9–3.5 Comments

Use of MUG-media in testing of environmental samples has been limited to sediments and sea water (Balebona et al., 1990) and foods (Weiss & Humber, 1988; Manafi et al., 1991). Almost any media that is used for fecal coliform detection has been modified and tested with MUG. These media include: EC medium, lactose bile, brilliant green, lactose broth, m-endo broth and others (see review by Manafi et al., 1991). In general, most authors have reported a high correlation between positive MUG reactions and other means of identifying *E. coli*. The clearest results are apparent when the tubes are checked for fluorescence within 24 h of inoculation. Many authors have indicated the importance of UV light selection. While a 366-nm UV lamp is needed, the wattage of the lamp is also important. Lamps of wattages > 6 W can cause all of the samples to look positive. Conversely, lamps of wattages < 6 W can result in a uniform negative finding.

9–4 DIRECT METHODS FOR DETECTION OF *E. COLI* IN SOIL

Methods described in this chapter rely on the enumeration of fecal contaminants following isolation and subculturing the bacteria. This is the case for traditional lactose fermentation method or the newer glucuronidase-based procedures. As indicated by Bej et al. (1990a,b) lactose fermentation procedures are remiss in that they require: maintenance of the viability of the cells between time of collection and enumeration, overlook chemical or environmentally stressed and nonculturable but viable cells, and lack of an exact specificity for the detection of *E. coli.* The use of glucuronidase-based systems gives an accurate picture for the presence of *E. coli,* but maintains the prerequisite of cell growth. Conflicting evidence has been reported concerning the incidence of D-glucuronidase-negative *E. coli* (Bej et al., 1991b). Cheng et al. (1989) reported the occurrence of up to 15% D-glucuronidase negative bacteria in a given fecal material.

The use of direct DNA extraction or extraction of bacteria (see chapter 35 in this book) and subsequent extraction of DNA, coupled to the use of polymerase chain reaction (PCR) DNA amplification (see chapter 33 in this book) offers a method for the detection of specific bacterial species without isolation or subculturing the cells. This offers the advantage of detection of viable but nonculturable organisms as well as the ability to detect dormant organisms. The PCR amplification and for detection of coliforms has been used to test aquatic (Bej et al., 1990a,b; 1991a,b) and soil systems (Josephson et al., 1991).

While a complete description of the PCR procedure is given in chapter 34 by Pepper and Pillai of this book, we will highlight a few points. The PCR amplification is a means of making many copies of a segment of DNA. The PCR amplification is in a 2^N exponential fashion where N is the number of reaction cycles. For each cycle, generally 7 to 10 min, a selected sequence in the target DNA is amplified twofold. The amplification is not dependent on cell growth, but on the simple presence of a template of DNA. Hence, the DNA in cells at low number, or DNA in cells that are viable but nonculturable can be amplified and then detected. It is theoretically possible to detect a single cell in an extracted sample of soil (Josephson et al., 1991). The amplified region of DNA is then detected using an appropriate gene probe (Bej et al., 1990a,b; 1991a,b). Limitations to use of PCR in soils center on removing the bacteria or DNA from the soil materials and the accessibility to gene probes.

ACKNOWLEDGMENT

Paper no. 13,465 of the Purdue Univ. Agric. Exp. Stn. Series.

REFERENCES

Alexander, M. 1982. Most probable number method for microbial populations. p. 815–820. *In* A.L. Page et al. (ed.) Methods of soil analysis. Part 2. Chemical and microbial properties. 2nd ed. ASA and SSSA, Madison, WI.

American Public Health Association. 1989. Standard methods for the examination of water and wastewater. 17th ed. Am. Public Health Assoc. Publ., Washington, DC.

Balebona, M.C., M.A. Morinigo, R. Cornax, J.J. Borrego, V.M. Torregrossa, and M.J. Gauthier. 1990. Modified most-probable-number technique for the specific determination of *Escherichia coli* from environmental samples using a fluorogenic method. J. Microbiol. Meth. 12:235–245.

Bej, A.K., J.L. DiCeasare, L. Haff, and R.M. Atlas. 1991a. Detection of *Escherichia coli* and *Shigella* spp. in water by using the polymerase chain reaction and gene probes for *uid*. Appl. Environ. Microbiol. 57:1013–1017.

Bej, A.K., M.H. Mahbubani, J.L. DiCeasare, and R.M. Atlas. 1991b. Polymerase chain reaction-gene probe detection of microorganisms by using filter-concentrated samples. Appl. Environ. Microbiol. 57:3529–3534.

Bej, A.K., S.C. McCarty, and R.M. Atlas. 1990a. Detection of coliform bacteria and *Escherichia coli* by multiplex polymerase chain reaction: Comparison with defined substrate and plating methods for water quality monitoring. Appl. Environ. Microbiol. 56:2429–2432.

Bej, A.K., R.J. Steffan, J. DiCesare, L. Haff, and R.M. Atlas. 1990b. Detection of coliform bacteria in water by polymerase chain reaction and gene probes. Appl. Environ. Microbiol. 56:307–314.

Brock, T.D., and M.T. Madigan. 1991. Major microbial disease. p. 556–557. *In* Biology of microorganisms. 6th ed. Prentice Hall, Englewood Cliffs, NJ.

Cheng, G.W., J. Brill, and R. Lum. 1989. Proportion of β-D-glucuronidase-negative *Escherichia coli* in human fecal samples. J. Appl. Microbiol. Biotechnol. 5:113–122.

Cochran, W.G. 1950. Estimation of bacterial densities by means of the most probable number. Biometrics 6:105–116.

Doran, J.W., and D.M. Linn. 1979. Bacteriological quality of runoff water from pastureland. Appl. Environ. Microbiol. 37:985–991.

Doran, J.W., J.S. Schepers, and N.P. Swanson. 1981. Chemical and bacteriological quality of pasture runoff. J. Soil Water Conserv. 36:166–171.

Faust, M.A. 1976. Coliform bacteria from diffuse sources as a factor in esturaine pollution. Water Res. 10:619–627.

Faust, M.A. 1982. Relationship between land-use practices and fecal bacteria in soils. J. Environ. Qual. 11:141–146.

Faust, M.A., and N.M. Goff. 1977. Basin size, water flow and land use effects on fecal coliform pollution from a rural water shed. p. 611–634. *In* D.L. Correll (ed.) Watershed research in eastern North America. Vol. 2. Chesapeake Bay Center for Environ. Studies, Smithsonian Inst., Edgewater, MD.

Feng, P.C.S., and P.A. Hartman. 1982. Fluorogenic assays for immediate confirmation of *E. coli*. Appl. Environ. Microbiol. 43:1320–1329.

Geldreich, E.E. 1976. Fecal coliforms and fecal streptococcus density relationships in waste discharges and receiving waters. p. 349–369. CRC Crit. Rev. Environ. Control. CRC Press, Boca Raton, FL.

Geldreich, E.E., R.H. Bordner, C.B. Huff, H.F. Clark, and P.W. Kabler. 1962. Type distribution of coliform bacteria in the feces of warm-blooded animals. J. Water Pollut. Control. Fed. 34:295–301.

Gilbert, R.O. 1987. Statistical methods for environmental pollution monitoring. Van Nostrand Reinhold Co., New York.

Hass, C.N., and B. Heller. 1988. Test of the validity of the poisson assumption for analysis of most-probable-number results. Appl. Environ. Microbiol. 54:2996–3002.

James, D.W., and K.L. Wells. 1990. Soil sample collection and handling: Technique based on source and degree of field variability. p. 25–45. *In* R.L. Westerman (ed.) Soil testing and plant analysis. 3rd ed. SSSA, Madison, WI.

Jawson, M.D., L.F. Elliott, K.E. Saxton, and D.H. Fortier. 1982. The effect of cattle grazing on indicator bacteria in runoff from a pacific northwest watershed. J. Environ. Qual. 11:621–627.

Josephson, K.L., S.D. Pillai, J. Way, C.P. Gerba, and I.L. Pepper. 1991. Fecal coliforms in soil detected by polymerase chain reaction and DNA-DNA hybridization. Soil Sci. Soc. Am. J. 55:1326–1332.

Keith, L.H. 1991. Environmental sampling and analysis: A practical guide. Lewis Publ., Chelsea, MI.

Koch, A.L. 1981. Growth measurements. p. 179–208. *In* P. Gerhardt (ed.) Manual of methods for general bacteriology. Am. Soc. Microbiol., Washington, DC.

Konopka, A., and R.F. Turco. 1991. Biodegradation of organic compounds in vadose zone and aquifer sediments. Appl. Environ. Microbiol. 57:2260–2268.

Loyer, M.W., and M.A. Hamilton. 1984. Interval estimation of the density of organisms using a serial dilution experiment. Biometrics 40:907–916.

Manafi, M., W. Kneifel, and S. Bascomb. 1991. Fluorogenic and chromogenic substances used in bacterial diagnostics. Microbiol. Rev. 335–348.

Steffan, R.J., and R.M. Atlas. 1988. DNA amplification to enhance detection of genetically engineered bacteria in environmental samples. Appl. Environ. Microbiol. 54:2185–2191.

Tate, R.L., and R.E. Terry. 1980. Effect of effluent on microbial activities and coliform populations of Pahokee Muck. J. Environ. Qual. 9:673–677.

Thelin, R., and G.F. Gifford. 1983. Fecal coliform release patterns from fecal materials of cattle. J. Environ. Qual. 12:57–63.

Triegel, E.K. 1988. Sampling variability in soils and solid wastes. p. 385–395. *In* L.H. Keith (ed.) Principles of environmental sampling. ACS Professional Ref. Book, Am. Chem. Soc., Washington, DC.

Weiss, L.H., and J. Humber. 1988. Evaluation of 24-hour fluorogenic assay for the enumeration of *Escherichia coli* from food. J. Food Prot. 51:766–769.

Woomer, P., J. Bennett, and R. Yost. 1990. Overcoming the inflexibility of most-probable-number procedures. Agron. J. 82:349–353.

Chapter 10

Autotrophic Nitrifying Bacteria

EDWIN L. SCHMIDT, *University of Minnesota, St. Paul, Minnesota*

L. W. BELSER, *Nelson Polytechnic Institute, Nelson, New Zealand*

Microbial development in soils is normally limited by the supply of available C and energy. Under such conditions, most of the NH_4^+ that results from the mineralization of organic N is oxidized about as rapidly as it is formed. The product of the oxidation is NO_3^-, and the process whereby NH_4^+ is oxidized to NO_3^- is referred to as *nitrification.*

Nitrification as it occurs in soils is a strictly biological process due, so far as is known, to a few genera of chemoautotrophic bacteria. These are the nitrifying bacteria, or nitrifiers. They carry out the process in two steps: one group of nitrifiers oxidizes NH_4^+ to NO_2^-; another group oxidizes the NO_2^- to NO_3^-. Nitrite, the intermediate in the process, rarely accumulates to detectable levels in soil.

The ammonia-oxidizing group of nitrifiers is presently comprised of five genera, all of which are reported to occur in soil; the nitrite-oxidizing group is made up of four genera only one of which (*Nitrobacter*) is well documented for soil occurrence (Watson et al., 1989). All are capable of strictly autotrophic growth in the laboratory on inorganic nutrients. Energy is derived from oxidation of the appropriate specific substrate, either NH_4^+ or NO_2^-, and all C requirements can be met from assimilation of CO_2. In nature it is likely that the nitrifiers incorporate some organic compounds as accessory C but remain bound to their specific energy substrate.

Nitrifying bacteria are not subject to ready examination by usual microbiological techniques. Direct plating, even on strictly inorganic media, is rarely useful for examination or isolation because organic materials introduced with the inoculum permit the more rapid growth of heterotrophic contaminants. Isolation is difficult, and most successful isolations have been preceded by extensive and careful serial enrichment procedures. Once isolated, nitrifiers are slow-growing in culture, sparse in yield, and susceptible to contamination. Thus, with few pure culture isolates available, most biochemical studies have been limited to the genera *Nitrosomonas* and *Nitrobacter* and usually to the same few strains of these genera. Ecological studies concerned with nitrifiers in soil and in other nitrifying habitats have been virtually blocked by the complexities of the habitat and the specialized biology of the nitrifiers. Two basic aspects of nitrifying populations are essentially unknown: the diversity of nitrifiers within a given population, and the nitrifying activities, *in situ*, of the individual components of a population.

The problem of diversity is particularly pertinent in relation to the NH_4^+ oxidizers. The terrestrial ammonia oxidizers include four genera other than the commonly cited *Nitrosomonas* and two of these, *Nitrosospira* and *Nitrosolobus*, are thought to be widely distributed. Coexistence of *Nitrosomonas, Nitrosospira*, and *Nitrosolobus* in the same soil was shown by Belser and Schmidt (1978a), with *Nitrosospira* as numerically abundant as *Nitrosomonas*. These data point out that the simplification of attributing NH_4^+ oxidation to *Nitrosomonas* is not only unwarranted but patently unsound since the various genera may be expected to differ in growth rate, biomass, yield, and substrate oxidizing activity. The only genus now known to participate in the oxidation of NO_2^- in soil is *Nitrobacter*. Thus, any diversity among the NO_2^- oxidizers, given current taxonomic convention, will be at the species or strain level.

The other basic, but virtually unexplored, aspect of nitrifier ecology is that of in situ activity. Activity measurements are expressed on a per cell basis and thus require numerical estimation of the nitrifying population. The usual approach to enumeration of nitrifiers is that of the most probable number (MPN) technique (see chapter 5 by Woomer in this book), which provides an indirect, statistical estimate of the number of cells involved in a particular oxidative step.

Outlined in this chapter are methods to study the nitrifying population of soils by various approaches and with various objectives in mind. The method proposed for the enumeration of nitrifiers is the classical MPN procedure despite its cumbersome incubation period and other serious limitations. But the MPN is presented for more than enumeration purposes. It serves also as the source material for procedures to examine the diversity of the NH_4^+ oxidizers indigenous to a soil and as the starting point for the non-enrichment isolation procedure recommended. The MPN procedure is also presented as the starting point for the immunofluorescence examination of nitrifying populations. If available to the investigator, highly specific fluorescent antibodies (FAs) prepared against individual nitrifier isolates can be used to examine nitrifier-positive MPN tubes to estimate the occurrence and abundance of specific strains of nitrifiers in a soil.

10–1 ENUMERATION BY MOST PROBABLE NUMBER

Refer to chapter 5 for the theory and general considerations relating to the MPN method for the estimation of microbial populations. The MPN technique has been the most widely used method to enumerate nitrifiers, but it has distinct drawbacks. As noted in chapter 5, the method has a high inherent statistical error. In addition, there is evidence that prolonged incubation is necessary to attain maximum counts (Matulewich et al., 1975; Belser & Schmidt, 1978a).

10–1.1 Ammonium Oxidizers by Most Probable Number

10–1.1.1 Materials

1. Ammonium (NH_4^+) oxidizer medium: Since this medium is free of precipitate, it is most convenient to prepare it by diluting stock solutions. The required stock solutions are given in Table 10–1. Make all of the solutions in distilled or deionized water. Adjust pH to 7.0 to 7.2 with 2.0% wt/vol K_2CO_3 prior to autoclaving. The 0.2 to 0.3 pH unit drop during autoclaving may be ignored. Add 4 mL of this medium to each culture tube, cap the tubes, and sterilize for 15 min at 121 °C.

2. Dilution bottles: Use the stock solutions listed in Table 10–1 to prepare 1 mM phosphate buffer. Add 4 mL of potassium monohydrogen phosphate (K_2HPO_4) and 1 mL of potassium dihydrogen phosphate (KH_2PO_4) stock solutions per liter of distilled or deionized water (pH 7.1–7.4). Each 20.32-cc (8-oz.) screw-capped bottle should contain 90 mL of this buffer after sterilizing for 15 min at 121 °C.

3. Modified Griess-Ilosvay reagents:
   a. Diazotizing reagent: Dissolve 0.5 g of sulfanilamide in 100 mL of 2.4 M hydrochloric acid (HCl). Store in refrigerator. Care should be used when handling this reagent because of its carcinogenic potential.
   b. Coupling reagent: Dissolve 0.3 g of N-(1-naphthyl)-ethylenediamine hydrochloride in 100 mL of 0.12 M HCl. Store solution in an amber bottle in refrigerator.

4. Nitrate spot test reagent: Dissolve 50 mg of diphenylamine in 25 mL of conc. sulfuric acid (H_2SO_4). Store in glass-stoppered dropping bottles protected from light, and prepare fresh solutions after 14 d.

5. Sterile 1% (wt/vol) potassium carbonate (K_2CO_3) in distilled or deionized water to adjust pH of ammonia-oxidation positive MPN tubes if further examination of the dilutions is desired. Nonsterile 2% (wt/vol) K_2CO_3 for initial adjustment of medium to pH 7.0 to 7.2.

10–1.1.2 Procedure

Transfer 10 g of moist soil to a blender, add 95 mL of sterile buffer, and blend for 30 or 60 s. Transfer this 10^{-1} dilution to an 8-oz bottle, shake vigorously, and transfer 10 mL to a 90-mL dilution bottle using a sterile pipette (10^{-2} dilution). Continue dilution to at least 10^{-5}. If a high population is expected, make dilutions to 10^{-7} or 10^{-8}. From the highest serial dilution prepared, transfer 1-mL aliquots to each of five (or 10; section 10–2.2) culture tubes. Inoculate an equal number of tubes from the next four lower dilutions.

Table 10–1. Stock solutions for the preparation of nitrifier media.

Chemical constituent	Concentration of stock solution	Stock solution required per liter of media	
		NH_4^+ oxidizer	NO_2^- oxidizer[†]
	g 100 mL^{-1}	—mL—	
$(NH_4)_2SO_4$	5.0	10.0	–
KNO_2	0.85	–	1.0
$CaCl_2 \cdot 2H_2O$	1.34	1.0	1.0
$MgSO_4 \cdot 7H_2O$	4.0	1.0	1.0
Bromothymol blue	0.04	5.0	–
$K_2HPO_4(0.2\ M)$	3.48	–	4.0
$KH_2PO_4(0.2\ M)$	2.72	7.5	1.0
Chelated iron		1.0	1.0
$\quad FeSO_4 \cdot 7H_2O$	0.246		
$\quad$ EDTA disodium	0.331		
Trace elements		1.0	1.0
$\quad Na_2MoO_4 \cdot 2H_2O$	0.01		
$\quad MnCl_2$	0.02		
$\quad CoCl_2 \cdot 6H_2O$	0.0002		
$\quad ZnSO_4 \cdot 7H_2O$	0.01		
$\quad CuSO_4 \cdot 5H_2O$	0.002		

† Calcium chloride and magnesium sulfate are combined, autoclaved separately, and added aseptically to sterile solution of remaining ingredients.

Incubate at 25 to 30 °C in the dark. Make initial observations after 21 d. Preliminary checks can be done visually by noting a color change in the pH indicator from blue green to yellow. This indicates active acid production (i.e., the oxidation of NH_4^+ to NO_2^-). Although this is normally sufficient evidence of growth at this stage, it may be desirable to do a spot test for NO_2^- on some tubes. Transfer a 0.1-mL aliquot aseptically to a spot plate, add one drop of diazotizing reagent and then one drop of coupling reagent. Color production (pink to red) indicates that NO_2^- is present. Compare this with an uninoculated control tube. Since the NO_2^- reagents are very acidic, the pH indicator does not interfere with this test.

Continue the incubation with weekly monitoring for a period of at least 6 wk and thereafter until there is no change in the number of positive tubes (tubes with change in color indicator) for two consecutive weeks. Final reading of positive tubes is not restricted to visual detection of color change. Check for NO_2^- by spot testing all tubes in those dilutions that have some visibly positive tubes, and the lowest dilution with no visibly positive tubes. Further check the NO_2^--negative tubes in these end dilutions for NO_3^- by diphenylamine spot test in the event that NO_2^- oxidizers also present may have converted all the NO_2^- to NO_3^-. Remember, though, that the NO_3^- spot test is not as sensitive as the test for NO_2^-.

Record the number of positive tubes in each of the appropriate end dilutions, and estimate the NH_4^+-oxidizing population using MPN Tables in chapter 5 in this book.

10–1.2 Nitrite Oxidizers by Most Probable Number

10–1.2.1 Materials

1. Stock solutions: Use those given in Table 10–1 for preparation of NO_2^- oxidizer medium. Follow the procedure of section 10–1.1.1 with the appropriate amounts of stock solution per liter of distilled or deionized water given in column 4. Add 4 mL of medium to each culture tube. No further adjustments are required after autoclaving at 121 °C for 15 min. The pH should be between 7.0 and 7.3.
2. Dilution bottles (section 10–1.2.1).
3. Modified Griess-Ilosvay reagents (section 10–2.1.1).

10–1.2.2 Procedure

Prepare dilution series and inoculations as in section 10–1.1.2. Normally, NO_2^- oxidizer counting tubes are inoculated from the same dilution series prepared for counting NH_4^+ oxidizers.

Check the tubes weekly after an initial 21 d incubation at 25 to 30 °C in the dark. Perform spot tests as described in section 10–2.1.2 to check for the disappearance of NO_2^-.

10–1.2.3 Comments

The media used are essentially those of Soriano and Walker (1968, 1973) for NH_4^+ oxidizers and of Belser (1977) for NO_2^- oxidizers. The medium of Soriano and Walker was modified here by the addition of trace elements. Both media are free of precipitates, which is a major advantage for subsequent isolation of nitrifiers, and both require less incubation time than previously recommended media (Belser & Schmidt, 1978a). More dilute versions of these media may lead to higher nitrifier counts for some soils. Donaldson and Henderson (1989) found that a 10-fold dilution of the above ammonia oxidizer medium gave higher population estimates in some, but not all, forest soil plots studied. Similarly, Woldendorp and Laanbroek (1989) reported that the above nitrite oxidizer medium modified to 0.05 mM nitrite gave considerably higher counts than at 5.0 mM nitrite for one grassland soil, but not for a second.

10–2 DIVERSITY OF NITRIFIERS

10–2.1 Introduction

Since soils have the potential to support multiple genera of NH_4^+ oxidizers (Belser & Schmidt, 1978a), it may be of interest to estimate the predominant genus or genera present. This involves mainly the microscopic examination of aliquots from positive MPN tubes. The genera *Nitrosospira*, *Nitrosolobus*, and *Nitrosomonas* can be recognized tentatively with practice. Initially this will require the availability of pure cultures as reference standards. However, it must be noted that such direct examination by ordinary optical microscopy can provide only presumptive evidence of the predominant ammonia oxidizing genera present. Conformation requires isolation, serology, or electron microscopy, or some combination thereof.

The morphological diversity seen in NH_4^+ oxidizers is not known to occur among NO_2^- oxidizers, but the question has received little attention.

10–2.2 Materials

1. Positive MPN tubes for nitrifiers (section 10–1).
2. Special equipment: Use phase contrast microscope with × 100 or × 90 oil immersion objective.
3. Reference cultures of *Nitrosomonas, Nitrosospira, Nitrosolobus,* and *Nitrobacter* (section 10–5).
4. Sodium nitrite ($NaNO_2$), 50 mM (3.45 g L^{-1} of distilled or deionized water) distributed in small screw cap tubes and sterilized at 120 °C for 15 min.

10–2.3 Procedure

Carry out microscopic examination of positive NH_4^+ oxidizer MPN tubes as soon as the pH indicator changes color. This is equivalent to about 1 µmol mL^{-1} of NO_2^- produced. This amount of oxidation should produce enough cells to be visible microscopically. Yields should be between 2 × 10^6 and 8 × 10^6 cells µmol^{-1} of NO_2^- formed (Belser & Schmidt, 1980). A population of 5 × 10^6 cells mL^{-1} would give about 1 cell per field for a typical wet mount preparation with a × 100 objective. Prepare a wet mount by placing a drop of culture on a clean microscopic slide and a cover slip on the drop. Draw off any excess solution with tissue.

When comparing unknown cultures to pure culture references, note both size and motility. *Nitrosospira* is the smallest of the three and commonly will appear to be spinning while moving in a circular path. *Nitrosolobus* is the largest, and its superficial microscopic appearance is that of several cells joined together. Electron micrographs show that these are lobular compartments of a single cell (Watson et al., 1971). These cells commonly appear to be tumbling randomly without directional motility. *Nitrosomonas* are short rods. Other rod-shaped organisms will be present as contaminants, but *Nitrosomonas* appears to be larger and more phase dark than most contaminants. Many *Nitrosomonas* are motile. Their motility is directional in nature and appears to be much different from the other genera. See Watson et al. (1989) for phase contrast micrographs of the genera.

When MPN tubes for NO_2^- oxidizers become positive, there are insufficient cells to observe microscopically. Initially, only 0.1 μmol of NO_2^- mL^{-1} was present. About 1.0 μmol mL^{-1} needs to be oxidized to yield sufficient cells; therefore, add 5.0 μmol mL^{-1} of sterile NO_2^- per tube (0.1 mL of sterile 50 mM NO_2^- per tube) after it initially becomes positive. When this is oxidized, proceed with microscopic examination.

10–2.4 Comments

Only three genera of NH_4^+ oxidizers are assumed to be common in soil; however, at least two other genera, *Nitrosococcus* (Watson et al., 1989) and *Nitrosovibrio* (Harms et al., 1976) have been isolated from soil. If one of these is suspected in an MPN tube, an isolation should be attempted. Isolation methods are given in section 10–4 below.

10–3 IMMUNOFLUORESCENCE EXAMINATION

10–3.1 Introduction

Immunofluorescence (IF) or the fluorescent antibody (FA) technique makes use of antibodies tagged with a fluorescent marker and appropriate microscopy to detect the specific microorganisms against which the antibodies were prepared. The FA has the capability to detect and enumerate specific bacteria directly in soil (Schmidt, 1974). Features that contribute to the attractiveness of FA for study of nitrifiers are its rapidity, specificity, sensitivity, and precision.

A major practical limitation to FA study of nitrifiers at the present time is the great specificity of the FA reagent. Nitrifiers isolated from the same soil have been found to exhibit considerable serological diversity (Belser & Schmidt, 1978b), with the consequence that a substantial number of FAs may be needed for a given population. The isolation of pure cultures needed for FA preparation is time consuming and so is the preparation of the FA; thus, a great deal of effort might be necessary to establish an inventory of all nitrifiers involved in a given habitat.

Detection of nitrifiers directly in soil by FA to enumerate individual strains or serotypes is difficult because of the low nitrifier densities commonly encountered in nature. However, once FAs are prepared by or become accessible to an investigator, they are particularly valuable for several aspects of study of the nitrifiers. Included here is an approach that combines FA examination with MPN to identify individual serotypes that predominate in a particular soil. Samples from the highest positive MPN dilutions are examined by IF using individual or combined FAs. Nitrification sites can be compared with respect to nitrifier serotypes or a given site can be studied for population change with respect to time.

Detailed protocols will not be presented because of the specialized nature of the reagents and instrumentation. Protocols for preparation of fluorescent antibodies are provided in chapter 38 by Zibilske in this book. Included, however, are references to original research papers that describe combined MPN-FA investigations. The FA reagents are additionally useful to characterize new isolates and detect mixed autotrophic cultures apparently otherwise (free of heterotrophs) pure.

10–3.2 Materials

1. Positive MPN tubes for nitrifiers (section 10–1).
2. Special reagents: fluorescent antibodies prepared against pure cultures of ammonia- or nitrite-oxidizers; rhodamine-gelatin counter stain; filtered saline.
3. Special equipment: fluorescence microscope; black nonfluorescent membrane filters, 0.4 or 0.45 μm, 25-mm diam., and filtration equipment; nonfluorescing mounting medium and immersion oil.

10–3.3 Procedure

General discussions of the IF technique as an approach to microbial ecology are provided by Schmidt (1974) and Bohlool and Schmidt (1980) and in chapter 28 by Wright in this book. Serotyping of nitrifier isolates by means of FA was reported by Belser and Schmidt (1978b) and by Stanley and Schmidt (1989). Belser and Schmidt (1978b) also provide methodology for partial characterization of nitrifier diversity in a single soil using FA against pure culture isolates obtained from MPN dilution tubes (section 10–4).

Methods to combine MPN analyses with FA to characterize a part of the predominant nitrifier population at a particular site are provided in several papers. Hankinson and Schmidt (1984) isolated a *Nitrosospira* from the highest positive MPN dilution tubes of an acid forest soil. The FA prepared against the isolate was used successfully to detect the isolate in MPN dilutions prepared a year later from the same site. Further estimates of population diversity and abundance of other ammonia oxidizers was made by scanning the same MPN dilution tubes with "cocktails," comprised of 3 FAs each for different serotypes of *Nitrosomonas, Nitrosospira,* and *Nitrosolobus.* Further study of the same acid forest soil site by Hankinson and Schmidt (1988) outlined procedures that led to the detection of the first acidophilic *Nitrobacter* and a neutrophilic *Nitrobacter* as members of the numerically predominant nitrite-oxidizing population. Similar coupling of MPN and FA for ammonia-oxidizer characterization is detailed for an aquatic sediment over time in the paper by Smorczewski and Schmidt (1991).

10-4 ISOLATION OF NITRIFIERS

10-4.1 Introduction

The isolation procedure recommended here is, in contrast to most isolation approaches, a nonenrichment method. Enrichment procedures as commonly used involve inoculation of soil into a nitrifying medium. Nitrifiers that grow in this medium serve as inoculum for the next serial passage in the same medium and so on until nitrification rates in the medium are greatly enhanced with respect to the initial passage. Isolations are then attempted from this nitrifier-enriched culture. During the process, selection occurs, so that those forms best adapted to that particular enrichment process are the ones isolated if the isolation is successful. It is thus possible that the form selected for and isolated was originally a minor or inactive component of the nitrifying population in the soil and merely capable of responding to the enrichment circumstances.

In the nonenrichment procedure presented, the nitrifying population encountered at the start of the analysis is diluted in accordance with the MPN protocol. Isolations are eventually attempted from the MPN dilution tubes as they become positive. Thus, the forms that were most abundant in the soil should carry through to the higher dilutions, whereas the less abundant ones will dilute out at lower dilutions. Although MPN media appear to be somewhat selective and therefore may not allow growth of all of the soil nitrifiers that occur at each dilution, the relative abundance of those that do develop can be assessed.

Positive MPN tubes selected for attempted isolation are incubated until a cell density is achieved suitable for direct microscopic count. The count includes all cells present, nitrifiers as well as contaminants. The culture is then diluted so that on the average there is < 1 cell mL^{-1} (i.e., normally a probability of 0.5 to 0.7 that a cell is present per milliliter). Thus, if there were 6×10^6 total cells mL^{-1} counted, a dilution of 10^{-7} would be made so that the probability of 1 mL containing 1 cell would be 0.6 [i.e., $(6 \times 10^6) \times (10^{-7}) = 0.6$]. One milliliter of this dilution would then be inoculated into tubes of MPN medium. Since there would be at most only one nitrifier for several milliliters of the dilution, enough tubes should be inoculated to ensure that at least 5 to 10 can be expected to contain a nitrifier. If per milliliter there were 4×10^6 nitrifiers in a total of 6×10^6 cells and a 10^{-7} dilution were used, it would require that 25 tubes be inoculated before the expected number of positive tubes would be 10 [i.e., $(4 \times 10^6) \times (10^{-7}) \times (25) = 10$]. Of these 10, two tubes, or 20%, would be expected to be contaminated [i.e., $(2 \times 10^6) \times (10^{-7}) \times (10) = 2$], since there were 2×10^6 contaminants per milliliter present. Since the nitrifiers outnumber contaminants (i.e., 4×10^6 nitrifiers vs. 2×10^6 contaminants), the expectations of getting a pure culture are good. However, in theory and in practice, pure cultures can be obtained even if the contaminants outnumber the nitrifiers. If, for example, there are

2×10^6 nitrifiers and 4×10^6 contaminants, it would require that 50 tubes be inoculated with a 10^{-7} dilution to have nitrifiers present in 10 tubes [i.e., $(2 \times 10^6) \times (10^{-7}) \times (50) = 10$]. Of these 10, four would be expected to be contaminated (40%) and six to be pure cultures [i.e., $(4 \times 10^6) \times (10^{-7}) \times (10) = 4$]. The critical factor is that the probability of a contaminant being present in 1 mL is < 1, preferably < 0.5. If the probability is 0.5, then 50% of the tubes with nitrifiers present would be expected to be pure cultures.

10–4.2 Materials

1. Positive MPN tubes (section 10–2) with at least 1 µmol mL^{-1} of substrate oxidized (section 10–2.3).
2. Special equipment: Use phase contrast microscope with $\times 100$ or $\times 90$ oil immersion objective, Petroff Hausser (P-H) counting chamber.
3. Purity check media: *Medium I:* Combine 2 g of glucose, 1 g of yeast extract, and 1 g of casein hydrolyzate in 200 mL of distilled or deionized water to provide a X10 stock solution. Sterilize for 15 min at 121 °C. Use as broth or add 1.5% (wt/vol) agar to single strength medium for plating. Autoclave single strength medium for 15 min at 121 °C. *Medium II:* Nutrient agar (half strength). Add 1.5 g of beef extract, 2.5 g of peptone and 15.0 of agar per liter distilled or deionized water. Pour plates 24 h prior to streaking.
4. Reference cultures of *Nitrosomonas, Nitrosospira, Nitrosolobus,* and *Nitrobacter* (section 10–5).
5. Dilution bottles (section 10–1.1.1).
6. Culture tubes containing either ammonium (NH_4^+) oxidizer medium or nitrite (NO_2^-) oxidizer medium. Prepare as in section 10–1.1.1, except add 9 mL tube^{-1} and NO_2^- that is 10 times more concentrated (1.0 mM) than in the NO_2^- oxidizer medium.
7. Sterile 1% (wt/vol) potassium carbonate (Na_2CO_3) in distilled or deionized water.
8. Modified Griess Ilosvay Reagents (section 10–1.1.1).

10–4.3 Procedure

The procedure is similar to that in section 10–3.3 except that a P-H counting chamber is used instead of slides. The P-H slide surface of the counting chamber is ruled into 50-µm squares. The chamber, when fitted with a cover slip, is 20 µm deep. This volume (50 by 50 by 20 µm) is 5×10^{-8} cm^3. If there were one cell per square, then there would be 2×10^7 cells mL^{-1}.

To fill the chamber, place the special reinforced cover slip on the clean, dry P-H slide. Add the suspension at the edge of the cover slip; the suspension will be drawn into the chamber by capillary action. Remove excess suspension with a tissue. Examine the slide microscopically with a $\times 100$ oil immersion objective. Approximately four small squares (these small squares are grouped in larger squares of 16 small squares, and the larger squares are bounded by double lines) will be visible per field.

Determine the average number of nitrifiers plus contaminants per small square. The more squares counted, the better the estimate will be. There are 400 small squares per slide. Multiply the average per square by 2×10^7 cells per milliliter to get the average number of nitrifiers plus contaminants per milliliter. Dilute the suspension in 1 mM phosphate buffer until the product of the dilution times the total cell density is between 0.5 and 0.7, as explained previously (section 10–4.1). Inoculate the appropriate number of tubes so that between 5 and 10 tubes would be expected to have a nitrifier present (section 10–4.1). Incubate the tubes at 25 to 30 °C in the dark.

After 3 weeks' incubation, check the tubes for production of NO_2^- (NH_4^+ oxidizers) or disappearance of NO_2^- (NO_2^- oxidizers) as described previously (section 10–1). Positive tubes should be checked microscopically for contaminants. If no contaminants are obvious by microscopic observation, transfer the culture and check its purity by cultural observations. If the culture to be checked is an NH_4^+ oxidizer, adjust its pH to 7.0 to 7.2 with sterile 1% K_2CO_3 before transferring. Transfer 1-mL aliquots to three culture tubes containing 10.0 mL of the appropriate nitrifier medium. To one of the three transfers, add 1 mL of X10 stock purity check medium (medium I), and streak or spot this same transfer onto purity check medium II. Incubate the three as before, and check daily (visually). If the purity check tube becomes turbid within 2 to 3 d, assume that the culture is contaminated. If purity check medium I shows no turbidity or very slight turbidity after 1 wk incubation and purity check medium II shows no growth on the plate after 10 d, examine the transfer microscopically. If no contaminants are visible, assume that the cultures are pure. When the other two transfers grow up (if they are NH_4^+ oxidizers) adjust the pH, and store the cultures in the refrigerator. Store nitrite oxidizers in the dark between 15 °C and room temperature. If all cultures tested are contaminated, repeat the purification procedure.

10–5 MAINTENANCE OF PURE CULTURES

10–5.1 Introduction

Nitrifiers require considerably more attention than do most bacterial cultures. In view of the difficulties and time involved in attaining pure culture isolates, the need for careful maintenance is obvious. It is helpful to reserve special glassware for the nitrifier culture collection, to use freshly sterilized pipettes for all transfers, and to give careful attention to the quality of water and reagents used in media preparation.

Procedures for the successful long-term preservation of a range of nitrifier isolates have not been reported. The experience of one of us (ELS) suggests that storage in the frozen state is promising. Cultures were deposited on cellulosic membranes and each membrane was suspended in a few milliliters of medium in a sterile plastic vial prior to freezing and storage at −40 °C. Recoveries attempted after 40 to 56 mo had the following success ratios: 10/16 *Nitrosomonas,* 6/10 *Nitrosospira,* 1/3 *Nitrosolobus,* and 1/1 *Nitrosovibrio.* Further studies with refinements in methodology and careful monitoring over time are needed before a format for frozen storage of nitrifiers can be recommended.

10–5.2 Materials

1. Culture medium for ammonium (NH_4^+) oxidizers prepared as in Table 10–1.
2. Medium for nitrite (NO_2^-) oxidizers as given in Table 10–1, however, use 20 mL (to give 2 mM NO_2^-) of potassium nitrite (KNO_2) stock solution instead of 1.0 mL as for MPN media.
3. Large screw-capped tubes (25 by 150 mm) with 25 mL of medium per tube for culture maintenance. Polypropylene screw cap closures, 24 mm, are preferred.
4. Sterile purity test media, sterile potassium carbonate (K_2CO_3), and NO_2^- test reagents as listed in section 10–4.2.

10–5.3 Procedure

Transfer cultures and check each for purity at regular intervals. Ammonia-oxidizer cultures should be transferred every 2 to 3 mo with the realization that some isolates may require still more frequent transfers. Nitrite oxidizers are generally more tolerant of neglect, but should be transferred at 4 to 5 mo intervals. Distribute media for maintenance in 25 mL quantities in the larger tubes and for purity tests in 9.0 mL quantities in 16 by 150 mm screw-capped tubes. Prepare two large tubes and one purity test tube for each culture. Autoclave, cool, and inoculate all tubes with 10% inoculum. Add 1.0 mL of sterile stock purity test medium aseptically to complete the purity test tubes, streak on purity test medium II (10–4.3) and incubate.

When the NH_4^+ oxidizer transfers have grown up, carefully readjust the cultures to the original pH indicator color with sterile K_2CO_3. Note: Add carbonate cautiously so as not to over-adjust. We have found that some ammonia-oxidizer isolates will not tolerate a pH above 7.4. Resume incubation until the medium again becomes acidic. After the second pH adjustment refrigerate the cultures. Growth of NO_2^- oxidizer transfers will appear as sediment in the tube. After visible sediment has accumulated, store the cultures in the dark between 15 °C and room temperature. For both ammonia- and nitrite-oxidizers retain the old cultures from the two preceding transfers along with current transfers in the event of contamination or other problems. In case of contamination, go back to prior transfers to see if a pure culture can be obtained. If not, purify the culture by the procedures of section 10–5.3.

10–6 NITRIFYING ACTIVITY IN SOILS

Two approaches are given to estimate the activity of the nitrifier population in soil as reflected by the process chemistry. One is essentially a static analysis, the other kinetic; each serves quite different purposes.

10–6.1 Nitrifying Potential of Soil

The procedure is a simple variant of numerous incubation-N mineralization tests used in attempts to estimate the N status of soil for soil fertility purposes. A soil is merely incubated under conditions of favorable temperature, moisture, and NH_4^+-N, and the NO_3^- formed during that incubation is measured. In unamended soil, the rate of NO_3^- formation will be limited by the rate at which NH_4^+ is formed from the mineralization of organic soil N. When amended with NH_4^+, the soil is nonlimiting with respect to nitrifiable substrate, and the nitrifying population should increase until limited by some other factor or combination of factors. The latter circumstance is taken as the "nitrifying potential" of a soil, but the interpretation of this requires that NH_4^+ unamended (control) soil samples be included in the procedure.

Nitrifying potential measured in this way is clearly limited in its application. It is useful and instructive as a means to compare soils of different properties or the effect of different management practices within the same soil. Different soils will nitrify quite differently during comparisons in the substrate-limiting condition and will respond quite differently with substrate nonlimiting. It can give a realistic idea of rates of nitrification that could be encountered in a soil and, when related to data from other soil analyses, can suggest possible relationship between the properties of a soil and prospects for the flow of N in the soil. It is not a reliable indicator of nitrifying activity because the incubation is relatively long, and normally only a single datum point is derived. Incubation conditions in the absence of plant growth and temperature and moisture fluctuations preclude anything more than a rough projection of the data to field conditions or fertilizer practice.

10–6.1.1 Materials

1. Soil freshly collected, partially dried, and screened.
2. Ammonium sulfate $[(NH_4)_2SO_4]$, 25 mg mL^{-1}.
3. Reagents for quantitative NO_3^- determination (see chapter 41 by Bundy and Meisinger in this book).

10–6.1.2 Procedure

Determine the initial NO_3^--N content of each soil to be studied (see chapter 41). To 100 g of the slightly moist soil in a beaker or tumbler, add 2 mL of $(NH_4)_2SO_4$ solution and sufficient water to bring to approximately 1/3 bar. Prepare a duplicate. With the same soil, prepare duplicate 100-g samples without NH_4^+ amendment and brought to the same moisture status with water only. Cover with foil, weigh, and record the weight of each, and incubate in the dark at 20 to 25 °C. Check weights weekly, and restore any moisture losses. Continue the incubation for 3 wk.

After incubation, determine the final NO_3^--N concentration of each sample. Express data as micrograms of NO_3^--N g^{-1} of oven-dry soil, subtract average initial from average final values. Record NO_3^--N formed during incubation by NH_4^+ amended soils as "nitrifying potential" and that by unamended soils as "control."

10–6.1.3 Comments

Conditions favoring denitrification must be avoided throughout. Freshly collected soil should be dried only enough to accommodate addition of NH_4^+ and a little water without becoming too wet. Care should be taken to maintain the natural structure of the soil during all manipulations. If the soil is structured but tends to puddle during preparation, 25 to 50% coarse sand may be mixed into the soil at the start. Soils that are high in NO_3^--N when collected (> 10 µg g^{-1}) should be leached gently in a Büchner funnel with filter paper over and under the soil. A small-diameter funnel is used so that the soil is about 5-cm deep; the funnel stem should be closed off, and water should be added gently until a head of about 5-mm results. The water is allowed to drain by gravity, and slight suction is applied at the end. The soil is dried to slightly below field moisture over a several-hour period, and NO_3^--N is determined before the experimental samples are set up.

10–6.2 Short-term Nitrifying Activity

10–6.2.1 Introduction

Activity measurements have been used to a limited extent in studying nitrification but have been used primarily in analysis of the kinetics of the process in aquatic systems. Knowles et al. (1965) using purely chemical data, estimated kinetic parameters and biomass in natural samples by comparing their data on activities in natural samples with those determined in pure culture. Similar estimates can be made for soil slurries and to indicate the maximum activity of the unenriched indigenous nitrifying populations in soils. This will not necessarily be the in situ activity since substrate is added, diffusion constraints are removed, and aeration is optimized. The method involves measuring the activity of the indigenous unenriched population over a time period that does not allow the population to increase significantly. Conditions are adjusted to maximize activity. The measured activity is then related to the activity per cell for nitrifiers in pure culture. Such estimates may then be compared with MPN counts made on the same sample. These ideas have been discussed in more detail elsewhere (Belser, 1979).

10–6.2.2 Theoretical Considerations

In pure culture, nitrifiers have been shown to oxidize their substrates according to Michaelis-Menten kinetics modified to take into account pH effects (Boon & Laudelout, 1962; Suzuki et al., 1974; Laudelout et al., 1976). There is some evidence that these kinetics are also obeyed in soils (Nishio & Furusaka, 1971; Ardakani et al., 1973). The simple Michaelis-Menten expression is

$$dS/dt = [k_o SX/(K_m + S)]$$

where $dS\ dt^{-1}$ is the rate of substrate oxidation, S is the substrate concentration, t is time, X is the nitrifier cell density, K_m is the Michaelis-Menten constant, and k_o is the maximum activity per cell. (Note: $k_o X = V_{max}$, the maximum velocity of the standard Michaelis-Menten expression.)

Under conditions where $S \gg K_m$, X can be estimated by knowing only k_o and $dS\ dt^{-1}$, the activity of the population. Under these conditions:

$$X = (dS\ dt^{-1})/k_o.$$

Under conditions where $S \gg K_m$ does not hold, X will be $> (dS\ dt^{-1})/k_o$. Thus, $(dS\ dt^{-1})/k_o$ should always be less than, or at most equal to, the indigenous population.

The oxidation rate $dS\ dt^{-1}$ can be measured easily. Only values for k_o are required for the estimate. These values are given in Table 10–2 and were obtained in pure culture (Belser & Schmidt, 1980). Values are given for three genera of NH_4^+ oxidizers, *Nitrosomonas, Nitrosospira,* and *Nitrosolobus,* and one genus of NO_2^- oxidizer, *Nitrobacter.*

10–6.2.3 Short-term Nitrification Assay

There are several methods by which short-term nitrification activities can be measured. A simple method that appears to be adequate for soils with a significant nitrifier population is recommended here. This method involves preparing soil slurries that are shaken and periodically measured for the accumulation of NO_2^-. Sodium chlorate is added to the soil slurries to block the further oxidation of NO_2^- (Belser & Mays, 1980). With chlorate present (final concentration 10 mM), the rate at which NO_2^- accumulates is equal to the rate of NH_4^+ oxidation and only NO_2^- need be measured. Analyses are made quickly and conveniently, with high sensitivity and without problems due to a high NO_3^- background.

Table 10–2. Maximum activities per cell determined during exponential growth of nitrifiers in pure culture.

Culture	Activities per cell[†]
Ammonium oxidizers[‡]	
Nitrosomonas europaea ATCC[§]	0.011
Nitrosomonas sp.[§]	0.023
Nitrosospira briensis	0.004
Nitrosolobus multiformis	0.023
Nitrite oxidizers[¶]	
Nitrobacter winogradskyi	0.012
Nitrobacter "agilis"	0.009

† Picomoles per cell per hour.
‡ Data of Belser and Schmidt (1980).
§ The ATCC strain is noticeably smaller than many other isolates of *Nitrosomonas* as typified here by *Nitrosomonas* sp. (Belser & Schmidt, 1978b, 1980).
¶ Data of Rennie and Schmidt (1977). Note: *Nitrobacter agilis* is now considered to be a strain of *N. winogradskyi*.

10–6.2.3.1 Materials

1. Sterile 0.5 mM phosphate buffer (section 10–1.1,1/2 concentration).
2. Ammonium sulfate [$(NH_4)_2SO_4$], 0.25 M (sterile).
3. Potassium chlorate ($KClO_3$) 1.0 M (sterile).
4. Merthiolate (ethylmercurithiosalicylic acid, sodium salt) 1% (wt/vol). Store at room temperature in foil-covered bottle. Replace monthly.
5. Reagents and instrumentation for quantitative NO_2^- analysis (see chapter 41 by Bundy and Meisinger).

10–6.2.3.2 Procedure. Prepare soil slurries in duplicate by adding 20 g of moist soil to individual 250-mL cotton-stoppered flasks containing 90 mL of phosphate buffer and 0.2 mL of $(NH_4)_2SO_4$ solution (total volume approximately 100 mL depending on the volume weight of the soil).

Place flasks on a rotary shaker, and add 1.0 mL of chlorate solution per flask. Let shake for several minutes then take a 5.0-mL aliquot. Add 0.05-mL merthiolate to stop the reaction. Save for nitrite analysis at the end of the assay. Periodically thereafter, collect four to five individual samples spaced at 1- to 2-h intervals, and determine nitrite concentrations when sampling is complete.

Calculate the rate at which NO_2^- accumulates per hour per gram of oven-dry soil. This gives the NH_4^+ oxidation rate. Divide this rate by the appropriate activity per NH_4^+ oxidizer in Table 10–2 to obtain a rough estimate of the NH_4^+ oxidizer population per gram of dry weight of soil. Counts can also be made on the suspension to estimate MPN counting effectiveness. Counting efficiency is estimated by dividing the MPN count by the estimated theoretical population.

10–6.2.3.3 Comments. The short-term assay is presented as an indicator of the potential activity of the nitrifying population present in a soil at the time of sampling. That population may be fully or partially active at sampling, depending on the interactions of the soil factors that regulate nitrification. The size of that population will be dictated also by the same soil factors. The assay aims to characterize the nitrifying population under conditions such that substrate and oxygen are nonlimiting and incubation time is so short as to essentially avoid increases in the standing population during the assay.

The short-term assay provides information on two facets of the nitrifying population. On the one hand the rates of ammonia oxidation obtained by short-term assay reflect the potential overall nitrification rates of a given soil. This is the case since NH_4^+ oxidation is the rate-limiting step in overall nitrification. Hence, comparisons are readily made between different soils as to nitrification potential, and, in addition, the effects of treatment, manipulation, or seasonal change on the nitrifying population of a given soil may be estimated. On the other hand, activity observed by short-term assay is clearly a function of the size of the nitrifying population. Thus, observed activity could be an indicator of population size if it were known how much of the activity could be attributed to a single cell. Such enumeration, however, requires knowledge of the composition of the nitrifying population and of the activity constants of each major component of that population. Unfortunately, neither the nature nor activities of a particular nitrifying population can be defined by current methodologies. Preliminary estimates of nitrifier numbers that have been based on short-term assay and cell activity constants suggest that actual MPN counts enumerate < 10% of the theoretical population (Sarathchandra, 1978; Belser & Mays, 1982; Berg & Rosswall, 1985).

Nitrite oxidation rates can be measured with varying concentrations of NO_2^- added to soil slurries along with nitrapyrin [2-chloro-6-(trichloromethyl)pyridine] to inhibit the oxidation of NH_4^+-N. Nitrite disappearance is followed, and both the maximum activity per gram of soil and the K_m are calculated. The K_m serves to determine the maximum activity (V_{max}), since the enzyme systems may not be saturated at the NO_2^- concentrations used.

Various modifications of nitrification potential assays have been proposed. These may supplement or supplant the procedures recommended here. Berg and Rosswall (1985) added chlorate but not ammonium to "undisturbed" soil cores and assayed for nitrite over a 12-h period. The objective was to provide an in situ nitrification rate. The authors predicted that, after further development, the method will become valuable. Killham (1987) proposed a perfusion system for pumping a buffered ammonium solution containing chlorate through a soil column. The method allows for the introduction of various inhibitors such as antibiotics and acetylene to help characterize the nitrification process. Chlorate was also used with apparent success in longer term nitrification incubation assays by Azhar et al. (1989). The equivalent of 2.13 g $NaClO_3$ kg^{-1} soil allowed continued nitrate formation for 10 d but not thereafter. Nitrite accumulated during the first 30 d, then declined. Chlorate must be used with considerable caution, however, especially in longer-term incubations. Nitrite-oxidizing bacteria are inhibited because they reduce chlorate to the toxic chlorite form. The presence of nitrite oxidizers and other soil bacteria can lead to the accumulation of chlorite and subsequent inhibition of the ammonia oxidizers as well (Hynes & Knowles, 1983).

REFERENCES

Ardakani, M.S., J.T. Rehbock, and A.D. McLaren. 1973. Oxidation of nitrite to nitrate in a soil column. Soil Sci. Soc. Am. Proc. 37:53–56.

Azhar, E.S., O. Van Cleemput, and W. Verstraete. 1989. The effect of sodium chlorate and nitrapyrin on the nitrification mediated nitrosation process in soils. Plant Soil 116:133–139.

Belser, L.W. 1977. Nitrate reduction to nitrite, a possible source of nitrite for growth of nitrite-oxidizing bacteria. Appl. Environ. Microbiol. 34:403–410.

Belser, L.W. 1979. Population ecology of nitrifying bacteria. Annu. Rev. Microbiol. 33:309–333.

Belser, L.W., and E.L. Mays. 1980. The specific inhibition of nitrite oxidation by chlorate and its use in assessing nitrification in soils and sediments. Appl. Environ. Microbiol. 39:505–510.

Belser, L.W., and E.L. Mays. 1982. Use of nitrifier activity measurements to estimate the efficiency of viable nitrifier counts in soils and sediments. Appl. Environ. Microbiol. 43:945–948.

Belser, L.W., and E.L. Schmidt. 1978a. Diversity in the ammonia-oxidizing population of a soil. Appl. Environ. Microbiol. 36:584–588.

Belser, L.W., and E.L. Schmidt. 1978b. Serological diversity within a terrestrial ammonia-oxidizing population. Appl. Environ. Microbiol. 36:489–593.

Belser, L.W., and E.L. Schmidt. 1980. Growth and oxidation of ammonia by three genera of ammonium oxidizers. FEMS Microbiol. Lett. 7:213–216.

Berg, P., and T. Rosswall. 1985. Ammonium oxidizer numbers, potential and actual oxidation rates in two Swedish arable soils. Biol. Fert. Soils 1:131–140.

Bohlool, B.B., and E.L. Schmidt. 1980. The immunofluorescence approach in microbial ecology. Adv. Microbiol. Ecol. 4:203–241.

Boon, B., and H. Laudelout. 1962. Kinetics of nitrite oxidation by Nitrobacter winogradskyi. Biochem. J. 85:440–447.

Donaldson, J.M., and G.S. Henderson. 1989. A dilute medium to determine population size of ammonium oxidizers in forest soils. Soil Sci. Soc. Am. J. 53:1608–1611.

Hankinson, T.R., and E.L. Schmidt. 1984. Examination of an acid forest soil for ammonia- and nitrite-oxidizing autotrophic bacteria. Can. J. Microbiol. 30:1125–1132.

Hankinson, T.R., and E.L. Schmidt. 1988. An acidophilic and a neutrophilic Nitrobacter strain isolated from the numerically predominant nitrite-oxidizing population of an acid forest soil. Appl. Environ. Microbiol. 54:1536–1540.

Harms, H., H.P. Koops, and H. Wehrman. 1976. An ammonia-oxidizing bacterium Nitroso-vibrio tenus nov. gen. nov. sp. Arch. Mikrobiol. 108:105–111.

Hynes, R.K., and R. Knowles. 1983. Inhibition of chemoautotrophic nitrification by sodium chlorate and sodium chlorite: a reexamination. Appl. Environ. Microbiol. 45:1178–1182.

Killham, K. 1987. A new perfusion system for the measurement and characterization of potential rates of soil nitrification. Plant Soil 97:267–272.

Knowles, G., A.L. Downing, and M.J. Barrett. 1965. Determination of kinetic constants for nitrifying bacteria in mixed culture with the aid of an electronic computer. J. Gen. Microbiol. 38:263–273.

Laudelout, H., R. Lambert, and M.L. Pham. 1976. Influence du pH et de la pression partielle d'oxygene sur la nitrification. Ann. Microbiol. (Paris) 127A:367–382.

Matulewich, V.A., P.F. Strom, and M.S. Finstein. 1975. Length of incubation for enumerating nitrifying bacteria present in various environments. Appl. Microbiol. 29:265–268.

Nishio, M., and C. Furusaka. 1971. Kinetic study of soil percolated with nitrate. Soil Sci. Plant Nutr. 17:61–67.

Rennie, R.J., and E.L. Schmidt. 1977. Autecological and kinetic analysis of competition between strains of *Nitrobacter* in soils. Ecol. Bull. 25:431–441.

Sarathchandra, S.U. 1978. Nitrification activities of some New Zealand soils and the effect of some clay types on nitrification. N.Z. J. Agric. Res. 21:615–621.

Schmidt, E.L. 1974. Quantitative autecological study of microorganisms in soils by immunofluorescence. Soil Sci. 118:141–149.

Smorczewski, W.T., and E.L. Schmidt. 1991. Numbers, activities and diversity of autotrophic ammonia oxidizing bacteria in a freshwater, eutrophic lake sediment. Can. J. Microbiol. 37:828–833.

Soriano, S., and N. Walker. 1968. Isolation of ammonia oxidizing autotrophic bacteria. J. Appl. Bacteriol. 31:493–497.

Soriano, S., and N. Walker. 1973. The nitrifying bacteria in soils from Rothamsted classical fields and elsewhere. J. Appl. Bacteriol. 36:523–529.

Stanley, P.M., and E.L. Schmidt. 1981. Serological diversity of *Nitrobacter* from soil and aquatic habitats. Appl. Environ. Microbiol. 41:1069–1071.

Suzuki, I., V. Dular, and S.C. Kwok. 1974. Ammonia or ammonium ion as substrate for oxidation by *Nitrosomonas europaea* cells and extracts. J. Bacteriol. 120:556–558.

Watson, S.W., E. Bock, H. Harms, H-P. Koops, and A.B. Hooper. 1989. Nitrifying bacteria. p. 1808–1834. *In* J.T. Staley et al. (ed.) Bergey's manual of systematic bacteriol. Vol. 3. Williams and Wilkins, Baltimore.

Watson, S.W., L.B. Graham, C.C. Remsen, and F.W. Valoi. 1971. A lobular, ammonia-oxidizing bacterium *Nitrosolobus multiformis* nov. gen. nov. sp. Arch. Mikrobiol. 76:183–203.

Woldendorp, V.W., and H.J. Laanbroek. 1989. Activity of nitrifiers in relation to nitrogen nutrition of plants in natural ecosystems. Plant Soil 115:217–228.

Chapter 11

Free-living Dinitrogen-fixing Bacteria

ROGER KNOWLES AND **WILFREDO LASERNA BARRAQUIO,**
McGill University, Macdonald Campus,
Ste. Anne de Bellevue, Québec, Canada

Biological nitrogen (N_2) fixation assumes great significance in natural and agricultural systems in view of the impending scarcity of inorganic fertilizers. Free-living N_2-fixing microorganisms are widely distributed and found in almost every ecological niche—in soils, associated with plants, in aquatic systems and sediments (Knowles, 1978). This distribution is a function of their great biochemical, taxonomic, and ecological diversity. In the first edition of this monograph, the only free-living N_2-fixing bacteria specifically considered were the azotobacters (Clark, 1965). Of the others not included, some, like the clostridia, had been fully documented N_2 fixers for many years. A few eukaryotes (e.g., *Aureobasidium*) could not be confirmed and are not now considered to be N_2 fixers. The present list of N_2-fixing microorganisms includes at least some species or strains of the aerobic or microaerophilic *Azotobacter, Azomonas, Beggiatoa, Beijerinckia, Campylobacter, Derxia, Acetobacter, Aquaspirillum, Azospirillum, Herbaspirillum, Corynebacterium (Xanthobacter), Lignobacter,* methanotrophs, *Mycobacterium (Xanthobacter), Pseudomonas,* and *Thiobacillus;* the facultatively anaerobic Enterobacteriaceae, *Alcaligenes, Bacillus, Vibrio,* and purple photosynthetic nonsulfur bacteria; the obligately anaerobic *Clostridium, Desulfovibrio, Desulfotomaculum, Propionispira,* methanogens and purple and green photosynthetic S bacteria; and finally, many cyanobacteria (blue-green algae). For a fuller discussion of this list with appropriate references, see Knowles (1978), Salkinoja-Salonen et al. (1979), and Sprent and Sprent (1990). For many of these groups of N_2-fixing microorganisms, no well-established enrichment or enumeration methods have been thoroughly tested in different laboratories. In this chapter, the emphasis is on enumeration techniques, and for this reason, some of the methods suggested here must be taken as tentative and subject to modification as may be found necessary. Most of the groups mentioned above are included except for the N_2-fixing bacilli and photosynthetic S bacteria for which simple one- or two-step enumeration procedures have not been reported.

Enrichment procedures (Veldkamp, 1970) permit isolation of specific bacteria, and rapid routine testing of N_2-fixing ability is relatively easy using the catalysis of the reduction of C_2H_2 to C_2H_4 (by the versatile nitrogenase system) as outlined in chapter 43 by Weaver and Danso in this book and in section 11–1 (Hardy et al., 1968). Many enrichment culture techniques can be easily adapted to quantitative enumeration by an extinction dilution or most probable number (MPN) method (see chapter 5 by Woomer in this book), but there are some problems. The validity of such methods depends on the ease with which cultures of the desired organism develop from small inocula (and ideally from a single cell). For some bacteria, establishment of a culture from a single cell does not occur readily. One of the best criteria for use in MPN counts is the positive C_2H_2 reduction test (Okon et al., 1977; Patriquin & Knowles, 1972; Villemin et al., 1974), but the numerous C_2H_4 analyses involved make this tedious for other than confirmatory purposes. The C_2H_2 reduction assay cannot be used to test for CH_4-supported N_2 fixers since C_2H_2 inhibits the initial oxidation of CH_4, thus depriving the methanotroph of its source of C and energy (De Bont & Mulder, 1976), as discussed later in section 11–4.

The selective culture of N_2-fixing aerobes is difficult because it is virtually impossible to exclude the development of anaerobic microsites that permit growth and activity of facultative and even obligately anaerobic bacteria. For example, Line and Loutit (1973) demonstrated that C_2H_2 was reduced to C_2H_4 by a *Clostridium* growing along with a *Pseudomonas* culture on an aerobic agar slant. The microaerophilic fixers are particularly troublesome in this respect. For this reason, MPN techniques are here described employing aerobic semisolid cultures in which subsurface plates of microaerophiles develop at appropriate O_2 concentrations within the agar.

Inconsistencies in the compositions of the media described are a reflection of the varied published sources from which the methods are derived. Fluid for serial dilutions should maintain the integrity of bacterial cells (Billson et al., 1970; Ridge, 1970), but it should not contain combined N. Thus, the use of 0.1% peptone solution, for example, is clearly undesirable. Iron may be supplied as $FeCl_3$ or $FeSO_4 \cdot 7H_2O$, or in chelated form as ferric monosodium ethylenediaminetetraacetic acid (FeEDTA) or ethylenediaminetetraacetic acid tetrasodium salt (Sequestrene, Ciba-Geigy Corp., Greensboro, NC). Other trace element supplements that are frequently necessary for successive transfer of pure cultures may or may not be necessary for the initial enrichment cultures involved in MPN methods.

11–1 THE ACETYLENE REDUCTION ASSAY

The versatile nitrogenase system reduces several low molecular weight substrates such as N_2, C_2H_2, N_2O, N_3^-, and CN^-. The sensitive methods available for the detection and measurement of C_2H_2 and its reduction product C_2H_4, permit the use of this reaction as an indirect assay of nitrogenase activity (Hardy et al., 1968). The C_2H_2 reduction (or C_2H_2–C_2H_4)

assay is about 1000 times more sensitive than the $^{15}N_2$ method, and furthermore, it is cheap and simple in its application (Hardy et al., 1968). Acetylene is introduced into the system at a concentration sufficient to saturate the nitrogenase sites (usually 0.05–0.1 atm) and completely to inhibit the reduction of air N_2. After an appropriate incubation time the amount of C_2H_4 formed is determined by gas chromatography. The assay is described in chapter 43 in this book.

The great sensitivity permits assays of pure cultures, enrichments, some MPN tubes, and active natural systems, for example, root nodules, to be completed in < 1 h (see chapter 43). However, the low activities of many other natural soil, sediment, and rhizosphere samples require longer assays, which bring a number of potential problems:

1. Acetylene is not the physiological substrate of nitrogenase, and its presence may impose N limitation on the N_2-fixing microorganisms and may cause superinduction (and therefore overestimation) of nitrogenase (David & Fay, 1977).

2. It is extremely difficult to match, in a closed system assay (C_2H_2 or $^{15}N_2$), such conditions as O_2 concentration, nutrient and moisture availability, and light intensity that exist in situ. This is particularly true in long assays where changes can occur in microbial populations (Okon et al., 1977) as well as in the other parameters just mentioned.

3. Acetylene has side effects not shown by N_2. These include the inhibition of (i) nitrogenase-dependent H_2 evolution (Hardy et al., 1968), (ii) conventional hydrogenase activity (Smith et al., 1976), (iii) uptake-hydrogenase activity (Smith et al., 1976), (iv) cell proliferation in clostridia (Brouzes & Knowles, 1971), (v) N_2 reduction by denitrifiers (Fedorova et al., 1973), (vi) oxidation of NH_4^+ to NH_2OH by *Nitrosomonas* (Hynes & Knowles, 1978) (vii) methanogenesis (Raimbault, 1975), and (viii) oxidation of CH_4 to CH_3OH by methanotrophic bacteria (De Bont & Mulder, 1976).

4. Another side effect of C_2H_2 is that it inhibits the further metabolism of C_2H_4. Thus, the actual net accumulation of "endogenous" C_2H_4 could be greater in the presence of C_2H_2 than in its absence, and this C_2H_4 would be measured as part of the C_2H_4 produced from the C_2H_2 (Witty, 1979). A control for this employing CO is described by Nohrsted (1983).

The reader is referred to chapter 43, Turner and Gibson (1980), and Knowles (1980) for a fuller discussion of these and other aspects of the C_2H_2 reduction assay.

11–2 METHODS FOR DINITROGEN FIXERS IN GENERAL

11–2.1 Principles

The great biochemical and physiological diversity of free-living N_2-fixers make the estimation of their total number at one sampling time difficult. Enumeration procedures usually require the use of specific media and incubation conditions to demonstrate the presence of at least a generic group of N_2 fixers. The development of a single isolation medium to accommodate most of the commonly occurring free-living heterotrophic N_2 fixers in soil has been most helpful. The enumeration procedure described here uses a combined C medium on which at least nine genera of N_2 fixers can be isolated (Rennie, 1981). It requires plating the sample on the combined C agar medium, incubating the plates aerobically and anaerobically, and then testing the individual colonies for acetylene-reducing activity. Classification of the acetylene-reducing isolates into major genera is achieved through a series of biochemical tests supplemented with some physiological and cultural tests (Rennie, 1980).

11–2.1.1 Prepared Materials

1. Sterile dilution blanks (see chapter 8 by Zuberer in this book).
2. Nitrogen-deficient combined C medium (Rennie, 1981):
   a. Solution A: To 900 mL of distilled water, add 0.8 g of potassium monohydrogen phosphate (K_2HPO_4), 0.2 g of potassium dihydrogen phosphate (KH_2PO_4), 0.1 g of sodium chloride (NaCl), 28 mg of ethylenediaminetetraacetic acid disodium salt Fe ($Na_2FeEDTA$), 25 mg of sodium molybdate dihydrate ($Na_2MoO_4 \cdot 2H_2O$), 100 mg of yeast extract, 5 g of mannitol, 5 g of sucrose, and 0.5 mL of sodium lactate (60%, v/v). For the agar medium, add 12 g of purified agar (Difco Noble Agar or Oxoid Ionagar)[1].
   b. Solution B: To 100 mL of distilled water, add 0.2 g of magnesium sulfate heptahydrate ($MgSO_4 \cdot 7H_2O$), and 0.06 g of calcium chloride ($CaCl_2$).
      Sterilize the solutions separately at 121 °C for 15 min, cool, and mix. Add filter-sterilized biotin (5 µg L^{-1}) and para-aminobenzoic acid (10 µg L^{-1}). Adjust the final pH to 7.0 using sterile acid or alkali.
3. Sterile glass or disposable plastic petri dishes.
4. Anaerobic jar with H_2–CO_2 GasPak kits and catalyst (Becton Dickinson and Co., BBL Microbiology Systems, Cockeysville, MD). Alternatively, a jar with gassing ports or a vacuum desiccator.
5. Supply of N_2.
6. A rotating petri dish turntable or glass hockey stick spreader.
7. 7-mL bijou vials (Sterillin Co., Ltd., Teddington, Middlesex, UK) or serum bottles (of 6- to 10-mL capacity) containing 3 mL of the liquid combined carbon medium.
8. Sterile serum stoppers (Suba seal from W. Freeman and Co., Barnsley, Yorkshire, UK) for the above vials of medium.

---

[1]Difco Noble Agar, Difco Laboratories, Detroit, MI; or Oxoid Ionagar, Oxoid, Division of Oxo Limited, London.

11–2.1.2 Procedure

Collect and prepare the sample as described in chapter 1 and prepare serial dilutions as described in chapter 8.

Dispense the melted agar medium into sterile petri dishes, and dry the agar surface by incubating at 30 to 37 °C for a few hours. Then, from the highest serial dilution prepared (10^{-4} is usually adequate), transfer an aliquot of 0.2 to 0.5 mL to each of five replicate plates. Make similar transfers from the successive lower dilutions, including the 10^{-1}. Spread the aliquot over the agar surface as it is released from the pipette by rotating the plate on a turntable, or spread it using a glass rod hockey stick sterilized with 70% alcohol and flaming. Then again dry the agar surface. Incubate the plates aerobically or anaerobically at 25 to 30 °C for 2 to 3 d. For anaerobic incubation, place the plates in the anaerobic jar (or desiccator), and generate the appropriate H_2–CO_2–N_2 atmosphere using the GasPak kit. Alternatively, flush the chamber with N_2, or evacuate and backfill it with N_2.

Record the number and cultural characteristics of colonies. Transfer individual colonies to sterile bijou vials or serum bottles each containing 3 mL of the liquid combined C medium. Seal the vials with sterile Suba seals and incubate for 24 h at 25 to 30 °C statically. If an anaerobic atmosphere is required, evacuate and backfill the vials with O_2-free N_2 or He using a manifold equipped with sterilized needles and cotton wool filters. Replace about 10% of the gas phase in the vial with C_2H_2, and continue incubation for up to 24 h. Determine the C_2H_4 produced by GC analysis of a 0.2- to 0.5-mL sample of the gas phase (see chapter 43). Record as positive those isolates that exhibit two or three times more C_2H_4 than is present in an uninoculated control.

11–2.2 Classification into Major Genera

The nine genera of N_2-fixing bacteria that can be isolated using the combined C medium include *Azotobacter, Azospirillum, Enterobacter, Klebsiella, Erwinia, Bacillus, Clostridium, Derxia,* and *Rhodospirillum* (Rennie, 1981). A computer program designed specifically for the identification of these nine genera of N_2-fixing microorganisms has been developed and tested (Rennie, 1980). It is based on interpretation of the 70 biochemical tests of the API 20E and 50E, supplemented with tests for acetylene reduction, nitrate and nitrite reduction, catalase, oxidase, motility, and growth on MacConkey's bile salt medium. Identification can be to the genus and often to species level.

11–3 METHODS FOR AZOTOBACTERACEAE

11–3.1 Principles

In general, it is not possible to use the MPN tube method for enumeration of N_2-fixing obligate aerobes because of the ease with which anaerobic and facultative N_2 fixers can show activity in O_2-deficient micro-sites (Line & Loutit, 1973). Therefore, for members of the Azotobacte-raceae, various modifications of surface plating or MPN plate procedures are used with a N-free solution solidified with purified agar or silica gel and containing an appropriate carbohydrate. Glucose is used by all species; mannitol and benzoate are used by only some species.

Of the four genera that now comprise the Azotobacteraceae (*Azoto-bacter, Azomonas, Beijerinckia,* and *Derxia*) (Tchan, 1984), reasonably good enumeration procedures exist for the first three, and two methods based on Brown et al. (1962) for *Azotobacter* and Strijdom (1966) for *Beijerinckia* are described below. Enumeration of *Derxia* in natural sam-ples is difficult, and a method is not included here.

11–3.2 Method for *Azotobacter* and *Azomonas*

11–3.2.1 Prepared Materials

1. Sterile dilution blanks (see chapter 8 in this book).
2. Medium for *Azotobacter* (Brown et al., 1962): To 1000 mL of distilled water, add 5 g of glucose, 0.2 g of magnesium sulfate heptahydrate ($MgSO_4 \cdot 7H_2O$), 0.04 of ferrous sulfate heptahy-drate ($FeSO_4 \cdot 7H_2O$), 0.005 g of sodium molybdate dihydrate ($Na_2MoO_4 \cdot 2H_2O$), 0.15 g of anhydrous calcium chloride ($CaCl_2$), and 15 g of agar. Purified agar (e.g., Difco Noble Agar or Oxoid Ionagar) is preferable, and only about 12 g of such preparations is required. Sterilize at 121 °C for 15 min. Then to the melted and cooled agar, add separately sterilized potassium monohydrogen phosphate (K_2HPO_4) solution to give a final concentration of 0.8 g L^{-1}. The final pH is 6.8 to 7.0.
3. Sterile glass or disposable plastic petri dishes.
4. Rotating petri dish turntable or glass hockey stick spreader.

11–3.2.2 Procedure

Collect and prepare the sample as described in chapter 1, and prepare serial dilutions as described in chapter 8.

Dispense the melted cooled agar medium containing the K_2HPO_4 into sterile petri dishes. Incubate the plates partially open at 30 to 37 °C under a clean air hood or in a sterile room until the surface is quite dry. Then, from the highest serial dilution prepared (10^{-4} is usually adequate), trans-fer an aliquot of 0.2 to 0.5 mL to each of five replicate plates. Make similar transfers from the successive lower dilutions, including the 10^{-1}. Spread the aliquot over the agar surface as it is released from the pipette by rotating the plate on a turntable, or spread it using a glass rod hockey stick sterilized with 70% alcohol and flaming. The agar surface should again be dried before incubation. Incubate at 25 to 30 °C for 3 to 5 d.

Many so-called oligonitrophiles (combined N scavengers) produce small colonies on plates prepared as described. Their colonies are usually < 1 mm in diameter but occasionally may be up to 3 mm diam. and clearly transparent. These generally are not N_2 fixers. *Azotobacter* (and possibly certain species of *Azomonas*) produce colonies 2 or more millimeters in diameter having a creamy mucilaginous appearance. *Azotobacter chroococcum,* probably the most commonly occurring species, can produce colonies up to 10 mm in diameter that gradually turn brown to black with age. The blackening of such colonies is intensified if sodium benzoate is used in the medium instead of glucose (Aleem, 1953).

11–3.3 Method for *Beijerinckia*

11–3.3.1 Prepared Materials

1. Sterile dilution blanks (see chapter 8 in this book).
2. Medium for *Beijerinckia* (Becking, 1961):
   a. Solution A: To 500 mL of distilled water, add 20 g of glucose, and sterilize at 121 °C for 15 min.
   b. Solution B: To 500 mL of distilled water add 1.0 g of potassium dihydrogen phosphate (KH_2PO_4), 0.5 g of magnesium sulfate heptahydrate ($MgSO_4 \cdot 7H_2O$) and 0.02 g of sodium molybdate dihydrate ($Na_2MoO_4 \cdot 2H_2O$). Adjust to pH 5.0, and sterilize at 121 °C for 15 min. Mix solutions A and B after cooling.
3. Petri dishes, each containing 1 g of finely powdered soil (Strijdom, 1966). Sterilize the soil by autoclaving (121 °C) for 1 h on each of two successive days.

11–3.3.2 Procedure

Collect and prepare the sample as described in chapter 1, and prepare serial dilutions as described in chapter 8. To each of the petri dishes containing soil, add 5 mL of the liquid medium. Then, from the highest serial dilution prepared (10^{-3} or 10^{-4}), transfer an aliquot of 1.0 mL to each of five replicate plates. Make similar transfers from the successive lower dilutions, distributing each aliquot over the surface of the medium-saturated soil. Incubate at 30 °C for up to 3 wk, adding fresh medium if necessary to keep the soil surface moist (Strijdom, 1966). Then examine each petri dish for the presence or absence of raised mucilaginous viscous growth typical of *Beijerinckia* spp. If necessary, check growth microscopically. Record all petri dishes as positive or negative for *Beijerinckia,* and calculate the MPN by reference to appropriate tables found in chapter 5.

11–4 METHOD FOR METHANOTROPHS

11–4.1 Principles

Methanotrophic bacteria (methane-using methylotrophs) of the genera *Methylomonas, Methylobacter, Methylocystis, Methylosinus,* and *Methylococcus* can fix N_2 (Oakley & Murrell, 1988) and are obligately microaerophilic when doing so (De Bont & Mulder, 1974). Acetylene inhibits the oxidation of methane but not of methanol by these organisms (De Bont & Mulder, 1976). Thus, it is not possible to test C_2H_2 reduction by enrichment cultures using an N-deficient mineral salts medium incubated under CH_4.

Methods have been described for the enumeration of methanotrophs (e.g., Megraw & Knowles, 1987) but not for N_2-fixing methanotrophs. We, therefore, suggest that the following methods be used on a trial basis only.

11–4.2 Plate Count Method for Methanotrophs

11–4.2.1 Prepared Materials

1. Sterile dilution blanks (see chapter 8 in this book).
2. Nitrogen- and C-deficient medium for methanotrophs (after De Bont & Mulder, 1974; Whittenbury et al., 1970): To 1000 mL of distilled water, add 0.5 g of potassium monohydrogen phosphate (K_2HPO_4), 0.5 g of potassium dihydrogen phosphate (KH_2PO_4), 0.2 g of magnesium sulfate heptahydrate ($MgSO_4 \cdot 7H_2O$), 0.015 g of calcium chloride ($CaCl_2$), 0.001 g of ferrous sulfate heptahydrate ($FeSO_4 \cdot 7H_2O$) or 0.004 g of Sequestrene Fe (Ciba-Geigy), 0.001 g of sodium molybdate dihydrate ($Na_2MoO_4 \cdot 2H_2O$), and 10 mL of soil extract (see chapter 8 in this book). Adjust the pH to 6.8 to 7.0. Sterilize at 121 °C for 20 min (the phosphates should be sterilized separately).
3. Sterile glass or disposable plastic petri dishes.
4. Sterile membrane filters (47 mm, 0.45 μm, e.g., Millipore HAWG), glass fiber pads, and filtration unit.
5. Anaerobic jar with gassing ports, desiccator, or plastic bags (of impermeable Saran-type material) with rubber hose connections.
6. Supplies of CH_4 and of N_2 (or a mixture of 2–4% O_2 in N_2).

11–4.2.2 Procedure

Collect and prepare the samples as described in chapter 1, and prepare serial dilutions as described in chapter 8.

Filter aliquots (5 mL) of the dilutions through membrane filters and place the membranes onto glass-fiber pads soaked with 2 mL of the N- and C-deficient medium (Megraw & Knowles, 1987). From each dilution prepare duplicate filters for incubation in the presence and absence of CH_4 as described below.

Place the plates in a chamber (anaerobic jar or desiccator) connected to cylinders of N_2 and CH_4, a vacuum pump, and manometer or vacuum gauge. Evacuate to a residual pressure of 76 to 152 mm of Hg (608–684 mm vacuum), and partially backfill the chamber to a residual air pressure of 608 (152-mm vacuum) with N_2. Introduce CH_4 to one atmosphere. Alternatively place the plates in plastic bags, and flush with a mixture of 2 to 4% O_2 and $\geq 20\%$ CH_4 in N_2 (Hill, 1973). Note that 5 to 15% CH_4 is the explosive range. Incubate a duplicate control set of membranes in the absence of CH_4. Incubate at 25 °C for 2 to 3 or more weeks, and examine for colonies appearing on the CH_4 series but not on the control series. Representative colonies may need to be checked for ability to use CH_4 (Whittenbury et al., 1970).

11–4.3 Most Probable Number Method for Methanotrophs

11–4.3.1 Prepared Materials

1. Sterile dilution blanks (see chapter 8 in this book).
2. Nitrogen- and C-deficient semisolid medium for methanotrophs, prepared exactly as in section 11–4.2.1 but adding 1.75 g of purified agar (Difco Noble Agar or Oxoid Ionagar)[1] per liter of medium. Dispense about 3 to 5 mL of the medium in each of a series of vials or serum bottles (of 6- to 10-mL capacity), plug with cotton or foam, and sterilize at 121 °C for 15 min.
3. Sterile serum stoppers for the above vials of medium.

11–4.3.2 Procedure

Collect and prepare the sample as described in chapter 1, and prepare serial dilutions as described in chapter 8.

From the highest dilution prepared, transfer a 0.1-mL aliquot to each of five vials of the semisolid medium. Then make similar transfers from the successive lower dilutions as appropriate. Do not insert serum stoppers at this stage. Incubate at 30 to 35 °C in a desiccator or other chamber filled with an atmosphere of 20% CH_4 in air until well-defined pellicle formation has occurred. Do not disturb the pellicle by shaking the tubes. Add 0.1 mL of a 3 to 5% methanol solution to give a final concentration in the medium of 0.1% methanol (De Bont & Mulder, 1974). Replace the aerobic plug with a sterile serum stopper, and replace about 10% of the gas phase with C_2H_2. Continue the incubation for a further 6 to 24 h, and determine the C_2H_4 produced by GC analysis of a 0.1- to 0.5-mL sample of the gas phase (see chapter 43 in this book). Compare with the C_2H_4 observed in uninoculated vials treated in exactly the same way. Record as positive those vials showing pellicle formation and significantly more C_2H_4 than is present in an uninoculated control.

Calculate the MPN of N_2-fixing methylotrophs by referring to the statistical tables found in chapter 5.

11–5 METHOD FOR HYDROGEN-USING DINITROGEN FIXERS

11–5.1 Principles

"All strains previously assigned to *Corynebacterium autotrophicum* for taxonomic reasons proved to be able to fix nitrogen under autotrophic conditions, and all strains isolated as N_2-fixing hydrogen bacteria so far have been found to belong to *C. autotrophicum*" (Wiegel & Schlegel, 1976). These bacteria, now classified as *Xanthobacter,* can couple the oxidation of H_2 to the fixation of N_2 and CO_2. Dinitrogen-fixing cultures are highly microaerophilic with an optimum O_2 concentration in the region of 3 to 5% (De Bont & Leijten, 1976).

The efficient enrichment and enumeration of this organism has been reported using an N-deficient liquid medium free of organic C incubated under an atmosphere of 10% O_2, 10% CO_2, 20% H_2, and 60% N_2 (De Bont & Leijten, 1976). However, some H_2-using N_2 fixers such as *Azospirillum brasilense, Azomonas agilis, Bacillus polymyxa, Beijerinckia indica,* and *Azotobacter vinelandii* do not grow chemolithotrophically (Bowien & Schlegel, 1981; Malik & Schlegel, 1981; Pedrosa et al., 1980; Wong & Maier, 1985). Except for the cyanobacteria and other phototrophic bacteria, all known aerobic H_2-using N_2 fixers are basically heterotrophs (Bowien & Schlegel, 1981).

The method described here is an MPN technique using a semisolid medium for heterotrophs incubated under aerobic conditions (Barraquio et al., 1988). The semisolid medium allows expression of O_2-sensitive nitrogenase, and hydrogenase that is often equally O_2 sensitive.

11–5.2 Prepared Materials

1. Sterile dilution blanks (see chapter 8 in this book).
2. Combined C medium described by Rennie (1981) (section 11–2.1.1) and modified by Barraquio et al. (1988) as follows: To 1000 mL of distilled water, add 2.0 g of Difco Noble agar, 0.06 g of sequestrene NaFe (13% Fe, CIBA-Geigy Corp., Greensboro, NC) to replace sodium ferrous EDTA, 5.9 mg (instead of 25 mg) of sodium molybdate dihydrate ($NaMoO_4 \cdot 2H_2O$), and 2.5 g of sodium malate. Omit the vitamins. Nickel chloride ($NiCl_2$) at a final concentration of 1 to 5 μM may be added to enhance H_2-uptake activity.

   Prepare the medium as described in section 11–2.1.1. Dispense 4 mL of the melted semisolid medium into each of a series of pre-sterilized cotton-plugged 14-mL serum bottles (Wheaton, Millville, NJ). This volume of the medium will give a depth of about 13 mm.
3. Sterile butyl rubber stoppers and aluminum crimps for the above serum bottles.

4. A supply of gaseous tritium with a specific activity of about 20 μCi (740 kBq) per cc. Appropriate license and laboratory facilities are required.

11–5.3 Procedure

Collect and prepare the sample as described in chapter 1, and prepare serial dilutions as described in chapter 8.

From the highest serial dilution prepared, transfer a 0.1-mL aliquot to each of five serum bottles of semisolid medium. Then make similar transfers from the successive lower dilutions as deemed appropriate. Do not insert butyl stoppers at this stage. Incubate statically at 30 to 35 °C for 3 d or more, and record the presence and absence of growth as determined visually. Replace the aerobic plug with a sterile butyl stopper and apply an aluminum seal. Replace about 1% (relatively low concentration so as not to inhibit greatly H_2-uptake activity) of the gas phase in the bottle with C_2H_2. Continue incubation for a further 24 h and determine the C_2H_4 produced by GC analysis of a 0.2- to 0.5-mL sample of the gas phase (see chapter 43). Record as positive those bottles that contain two or three times more C_2H_4 than is present in an uninoculated control. Use the same cultures for the tritium (H^3H) uptake assay. Inject 0.3 mL of H^3H into the bottle, and then continue incubation for another 24 h. Vortex the bottles vigorously, and then transfer 0.1-mL aliquots into 6-mL scintillation vials each containing 4-mL scintillation fluid (Ready-Solv MP, Beckman Instruments, Fullerton, CA). Determine radioactivity of the samples by liquid scintillation (Beckman LS 7500 counter) with ^3H-galactose or other tritiated compound as the standard. An alternative method of determining H_2-using activity that can be tried in laboratories not allowed to handle isotopes is to inject 1% H_2 into each serum bottle and then determine the H_2 concentration at the start and after 24 to 48 h of incubation by GC analysis (Chan et al., 1980). Record as positive those cultures that exhibit significantly more radioactivity (or less H_2 if done by the alternative method) than is present in an uninoculated control.

Calculate the MPN of H_2-using N_2 fixers by referring to the statistical tables found in chapter 5 by Woomer in this book. The MPN of H_2-using non-N_2-fixers and N_2-fixing non-H_2-utilizers can also be determined from the data.

11–6 METHOD FOR CYANOBACTERIA

11–6.1 Principles

Cyanobacteria such as *Anabaena* and *Nostoc* can compartmentalize their nitrogenase in specialized heterocysts in which a sufficiently reducing environment is maintained. They thus can fix N_2 under highly aerobic conditions (Stewart, 1973). Other cyanobacteria, e.g., the filamentous *Plectonema* and *Oscillatoria* and the unicellular *Gloeocapsa,* do not produce heterocysts and generally synthesize nitrogenase under microaerobic to anaerobic conditions (Stewart, 1973). At high light intensities (> 9.8 μmol m^{-2} s^{-1} (> 500 lx) for *Gloeocapsa*) the production of O_2 is stimulated to such an extent that N_2 fixation may be inhibited (Stewart, 1973). Although studies of soils have revealed mainly the heterocystous N_2-fixing cyanobacteria, this could be due to the inadequacy of existing methods to select for the nonheterocystous N_2 fixers.

Most studies on the distribution of the cyanobacteria use enrichment culture in C- and N-free mineral salts solutions. The enumeration method described here uses such a medium solidified with agar for a plate count and is slightly modified from that of Jurgensen and Davey (1968).

11–6.2 Prepared Materials

1. Sterile dilution blanks (see chapter 8 in this book).
2. Nitrogen- and organic C-deficient medium for cyanobacteria (after Cameron & Fuller, 1960; Wieringa, 1968): To 1000 mL of distilled water, add 0.35 g of potassium monohydrogen phosphate (K$_2$HPO$_4$), 0.2 g of magnesium sulfate heptahydrate (MgSO$_4$·7H$_2$O), 0.15 g of calcium chloride dihydrate (CaCl$_2$·2H$_2$O), 0.1 g of sodium chloride (NaCl), 0.005 g of ferrous sulfate heptahydrate (FeSO$_4$·7H$_2$O), 100 μg of sodium molybdate dihydrate (Na$_2$MoO$_4$·2H$_2$O), 5 μg of cupric sulfate pentahydrate (CuSO$_4$·5H$_2$O), 10 μg of boric acid (H$_3$BO$_3$), 70 μg of zinc sulfate heptahydrate (ZnSO$_4$·7H$_2$O), 10 μg of manganous sulfate monohydrate (MnSO$_4$·H$_2$O), and 15 g of purified agar (Difco Noble Agar or Oxoid Ionagar).[1] Autoclave at 121 °C for 15 min.
3. Sterile glass or plastic disposable petri dishes.

11–6.3 Procedure

Collect and prepare the sample as described in chapter 1, and prepare serial dilutions as described in chapter 8.

From the highest serial dilution prepared, transfer a 1.0-mL aliquot to each of five replicate petri dishes. Make similar transfers from the successive lower dilutions as necessary. Pour at least 25 mL of the melted and cooled (45 °C) agar medium into each dish, swirl, and allow to solidify. Incubate under illumination of no more than about 49 μmol m^{-2} s^{-1} (2500 lx) from cool-white fluorescent lamps, with a 16-h light period alternating with an 8-h dark period at 20 to 30 °C. Count the colonies that develop after about 6 wk of incubation. Such a count may include organisms that are unable to fix N_2 but are efficient scavengers of fixed N.

11–7 METHOD FOR PHOTOSYNTHETIC PURPLE NONSULFUR BACTERIA

11–7.1 Principles

The genera *Rhodobacter, Rhodospirillum, Rhodopseudomonas,* and *Rhodomicrobium,* all members of the Rhodospirillaceae (Imhoff & Trüper, 1989) are photoorganotrophs that carry out photoassimilation of simple organic compounds such as malate and succinate. Organic compounds or H_2 are used as photosynthetic electron donors, and under anaerobic conditions, many species fix N_2. There are no well-documented accounts of the enumeration of these N_2 fixers from natural environments.

The method that follows is modified from the work of Gibson (1975), Weaver et al. (1975), Finke and Seeley (1978), and Siefert et al. (1978) and employs an illuminated anaerobic plate count on a N-deficient malate-containing medium.

11–7.2 Prepared Materials

1. Sterile dilution blanks (chapter 8).
2. Nitrogen-deficient malate agar medium (after Gibson, 1975; Siefert et al., 1978; Weaver et al., 1975):
   a. Solution A: To 250 mL of distilled water, add 4.0 g of ethylenediaminetetraacetic acid disodium salt (EDTA disodium), 0.2 g of ferrous sulfate heptahydrate ($FeSO_4 \cdot 7H_2O$), 0.7 g of boric acid (H_3BO_3), 0.39 g of manganous sulfate monohydrate ($MnSO_4 \cdot H_2O$), 0.06 g of zinc sulfate heptahydrate ($ZnSO_4 \cdot 7H_2O$), 0.025 g of sodium molybdate dihydrate ($Na_2MoO_4 \cdot 2H_2O$), and 0.01 g of cupric nitrate trihydrate [$Cu(NO_3)_2 \cdot 3H_2O$]. Autoclave at 121 °C for 15 min.
   b. Solution B: To 1000 mL of distilled water, add 100 mg of nicotinic acid, 50 mg of thiamine hydrochloride, 10 mg of biotin, and 10 mg of *p*-aminobenzoic acid. Sterilize by membrane filtration. Different species of the Rhodospirillaceae have different growth factor requirements, but according to Van Niel (1972), these four factors should support growth of all species.
   c. Solution C: To 990 mL of distilled water, add 0.5 g of potassium dihydrogen phosphate (KH_2PO_4), 0.2 g of magnesium sulfate heptahydrate ($MgSO_4 \cdot 7H_2O$), 0.015 g of calcium chloride dihydrate ($CaCl_2 \cdot 2H_2O$), and 1.34 g of DL-malic acid (neutralized separately with sodium hydroxide [NaOH]). To solution C, add 1 mL of solution A, and adjust the pH to 6.8. Add 12 g of purified agar (Difco Noble Agar or Oxoid Ionagar),[1] heat to dissolve, and then autoclave at 121 °C for 15 min. Allow to cool to 45 °C, add 10 mL of solution B, and mix gently. This medium is then ready for pouring plates.

3. Sterile petri dishes.
4. Anaerobic jar with H_2-CO_2 GasPak kits and catalyst (Becton, Dickinson and Co., BBL Microbiology Systems, Cockeysville, MD). Alternatively, a jar with gassing ports or a vacuum desiccator.
5. If Gaspak kits are not available, supplies of H_2, CO_2, and N_2 or a premixed cylinder containing 10% H_2, 10% CO_2, and 80% N_2 is required. In these cases, a vacuum pump and gassing manifold with vacuum gauge or gauges are required.

11–7.3 Procedure

Collect and prepare the sample as described in chapter 1, and prepare serial dilutions as described in chapter 8.

From the highest dilution prepared, transfer a 1-mL aliquot to each of five petri dishes. Make similar transfers from the successive lower dilutions as necessary. Pour about 20 mL of the 45 °C agar medium, into each dish, swirl, and allow to solidify. Place the dishes in the anaerobic jar, and generate the appropriate H_2-CO_2-N_2 atmosphere using the GasPak kit. Alternatively, flush the chamber with the appropriate gas mixture, or evacuate and backfill with N_2 and introduce the appropriate partial pressures of H_2 (0.1 atm), CO_2 (0.1 atm), and N_2 (0.8 atm) using the vacuum pump and gassing manifold (e.g., section 11–4.2.2).

Incubate the plates at 25 to 30 °C under about 49 μmol m^{-2} s^{-1} (2500 lx) of fluorescent plus incandescent light for at least 14 d, and then count the purple colonies that have developed.

11–8 METHOD FOR CLOSTRIDIA

11–8.1 Principles

The obligately anaerobic spore-forming clostridia are difficult to enumerate using anaerobic plate count techniques because of gas production, splitting of the agar, and spreading of colonies. Most probable number procedures are therefore used in which a N-deficient liquid medium containing a reducing agent and redox indicator is inoculated with appropriate dilutions of the sample (Brouzes et al., 1971). The possible criteria for positive tubes are gas production, reduction of the indicator, and detection of C_2H_2-reducing activity.

Bacilli may grow under the conditions described but, in practice, positive tubes generally contain clostridia.

11–8.2 Prepared Materials

1. Sterile dilution blanks (chapter 8).
2. Medium for N_2-fixing clostridia (Brouzes et al., 1971): To 1000 mL of distilled water, add 20 g of glucose, 2 g of sodium acetate

(NaOAc), 0.5 g of calcium carbonate ($CaCO_3$), 0.01 g of sodium molybdate dihydrate ($Na_2MoO_4 \cdot 2H_2O$), 0.25 g of potassium dihydrogen phosphate (KH_2PO_4), 0.5 g of sodium thioglycollate, 0.75 g of agar, 8 mL of 0.2% aqueous phenosafranin, and 10 mL of soil extract (see chapter 8 in this book). Dispense the medium in test tubes or Pankhurst tubes (Campbell & Evans, 1969), add a small inverted Durham tube to each, cap, and sterilize at 121 °C for 15 min.
3. Sterile rubber stoppers (or serum stoppers) for the above tubes of medium.

11–8.3 Procedure

Collect and prepare the sample as described in chapter 1, and prepare serial dilutions as described in chapter 8.

Heat the tubes of medium to almost 100 °C to drive off dissolved O_2, and then cool before use. If the medium is freshly autoclaved, merely cool before use.

From the highest serial dilution prepared, transfer a 1.0-mL aliquot to each of five tubes of the liquid degassed medium. Then make similar transfers from the successive lower dilutions as desirable. Insert the sterile stoppers tightly, and incubate the tubes at 30 °C for 14 d. The headspace of the tubes may be flushed with sterile N_2 before stoppering, but this is not absolutely essential.

Record as positive those tubes in which at least a 5-mm column of gas has accumulated in the inverted tube and in which the phenosafranin is more or less decolorized. Confirmatory tests that may be applied and that usually correlate well with the above criteria are as follows: microscopic examination of cells from the bottom of a tube for presence of iodine-stainable cells containing glycogen-like reserve material or typical sporulating *Clostridium* cells, and C_2H_2 reduction assay carried out by injecting a small amount of sterile glucose into each tube and then, several hours later, injecting enough C_2H_2 to give 0.01 to 0.1 atm pC_2H_2. One day later a sample of the headspace is analyzed by gas chromatography for the presence of two or three times more C_2H_4 than is present in a similarly treated "blank" tube of uninoculated medium (see section 11–2).

Calculate the MPN of N_2-fixing clostridia by referring to the statistical tables found in section 11–2.

11–8.4 Comments

Vegetative cells of clostridia, being extremely O_2 sensitive, are killed during the preparation of the usual aerobic dilution series. The method described above, therefore, represents more likely a spore count rather than a count of total viable cells. The count may be improved, and at least some of the viable vegetative cells included, by careful heating to degas the dilution blanks and flushing the head space with sterile O_2-scrubbed N_2 before and after each transfer by pipette. Reducing agents (1.0 g of ascorbic acid and 1.0 g of sodium thioglycollate L^{-1}) may also be included in the dilution fluid. The degassed tubes of liquid medium can also be flushed with N_2 before and after inoculation. A suitable N_2-flushing device is a Pasteur pipette or stainless steel tube attached to a cotton wool or membrane filter and sterilized before attachment to a cylinder of N_2.

11–9 METHOD FOR SULFATE-REDUCING BACTERIA

11–9.1 Principles

These bacteria are very strict anaerobes that respire by reducing SO_4^{2-} to H_2S. *Desulfovibrio* spp. are highly motile, nonsporulating vibrios that are common in terrestrial and aquatic (freshwater and marine) environments. *Desulfotomaculum* spp. are spore-forming rods with a more restricted distribution, and several other genera also occur (Postgate, 1984; Widdel & Pfennig, 1984).

The SO_4^{2-} reducers do not proliferate unless the redox potential of the medium is below -100 mV, and therefore the Eh of the growth medium is adjusted to a sufficiently low value (Postgate, 1984).

Nitrogen fixation by SO_4^{2-}-reducing bacteria has been demonstrated (Riederer-Henderson & Wilson, 1970; Postgate et al., 1985) and a N-deficient medium has been described for enumeration (Patriquin & Knowles, 1972). However, the method described below probably does not provide a true count of N_2-fixing SO_4^{2-}-reducing bacteria.

11–9.2 Prepared Materials

1. Sterile dilution blanks prepared generally according to chapter 8. However, the dilution fluid should contain 1.0 g of ascorbic acid and 1.0 g of sodium thioglycollate L^{-1}, and the dilution blanks should be heated to degas, flushed with N_2, and then stoppered just before use.
2. Nitrogen-deficient lactate medium (modified from Postgate, 1984; Patriquin & Knowles, 1972; Riederer-Henderson & Wilson, 1970):
   a. Solution A: To 100 mL of distilled water, add 0.5 g of zinc sulfate heptahydrate ($ZnSO_4 \cdot 7H_2O$), 0.5 g of manganous sulfate monohydrate ($MnSO_4 \cdot H_2O$), 0.4 g of sodium molybdate dihydrate ($Na_2MoO_4 \cdot 2H_2O$), 0.4 g of ferrous sulfate heptahydrate ($FeSO_4 \cdot 7H_2O$), 0.005 g of cobaltous sulfate ($CoSO_4$), 0.005 g of boric acid (H_3BO_3), and 0.7 mg of cupric sulfate pentahydrate ($CuSO_4 \cdot 5H_2O$).
   b. Solution B: To 900 mL of distilled water, add 0.5 g of potassium dihydrogen phosphate (KH_2PO_4), 1.0 g of calcium sulfate ($CaSO_4$), 7.0 g of magnesium sulfate heptahydrate

(MgSO$_4$·7H$_2$O), 3.5 g of sodium lactate, and 0.1 g of yeast extract.

c. Solution C: To 100 mL of distilled water, add 1.0 g of ascorbic acid and 1.0 g of sodium thioglycollate. Autoclave at 121 °C for 15 min.

Add 1 mL of solution A to solution B. Adjust the pH to between 7.0 and 7.5, and autoclave. Add the previously sterilized solution C, and immediately dispense into test tubes, each of which contains a small amount of dry-sterilized steel wool. Close with sterile rubber stoppers.

11–9.3 Procedure

Collect and prepare the sample as described in chapter 1, and prepare serial dilutions as in chapter 8, taking care to flush with N$_2$ before and after each transfer.

From the highest dilution prepared, transfer a 1-mL aliquot to each of five tubes of the liquid medium. Heat the tubes to degas just before inoculation, flush the head space with N$_2$, and immediately stopper the tube. Make similar transfers from the successive lower dilutions as seems appropriate. Incubate at 25 to 30 °C for 2 to 3 wk, and then record as positive those tubes showing a dense black coloration indicative of FeS formation.

REFERENCES

Aleem, M.I.H. 1953. Counting of *Azotobacter* in soils. Plant Soil 4:248–251.

Barraquio, W.L., A. Dumont, and R. Knowles. 1988. Enumeration of free-living aerobic N$_2$-fixing H$_2$-oxidizing bacteria by using a heterotrophic semisolid medium and most-probable-number technique. Appl. Environ. Microbiol. 54:1313–1317.

Becking, J.H. 1961. Studies on nitrogen-fixing bacteria of the genus *Beijerinckia*. I. Geographical and ecological distribution in soils. Plant Soil 14:49–81.

Billson, S., K. Williams, and J.R. Postgate. 1970. A note on the effect of diluents on the determination of viable numbers of Azotobacteraceae. J. Appl. Bacteriol. 33:270–273.

Bowien, B., and H.G. Schlegel. 1981. Physiology and biochemistry of aerobic hydrogen-oxidizing bacteria. Ann. Rev. Microbiol. 35:405–452.

Brouzes, R., and R. Knowles. 1971. Inhibition of growth of *Clostridium pasteurianum* by acetylene: Implication for nitrogen fixation assay. Can. J. Microbiol. 17:1483–1489.

Brouzes, R., C.I. Mayfield, and R. Knowles. 1971. Effect of oxygen partial pressure on nitrogen fixation and acetylene reduction in a sandy loam soil amended with glucose. p. 481–484. *In* T.A. Lie and E.G. Mulder (ed.) Biological nitrogen fixation in natural and agricultural habitats. Plant Soil, Spec. Vol., Martinus Nijhoff, the Hague.

Brown, M.E., S.K. Burlingham, and R.M. Jackson. 1962. Studies on *Azotobacter* species in soil. I. Comparison of media and techniques for counting *Azotobacter* in soil. Plant Soil 17:309–319.

Cameron, R.E., and W.H. Fuller. 1960. Nitrogen fixation by some algae in Arizona soils. Soil Sci. Soc. Am. Proc. 24:353–356.

Campbell, N.E.R., and H.J. Evans. 1969. Use of Pankhurst tubes to assay acetylene reduction by facultative and anaerobic nitrogen-fixing bacteria. Can. J. Microbiol. 15:1342–1343.

Chan, Y.K., L.M. Nelson, and R. Knowles. 1980. Hydrogen metabolism of *Azospirillum brasilense* in nitrogen-free medium. Can. J. Microbiol. 26:1126–1131.

Clark, F.E. 1965. Azotobacter. p. 1493–1497. *In* C.A. Black et al. (ed.) Methods of soil analysis. Part 2. Agron. Monogr. 9. ASA, Madison, WI.

David, K.A.V., and P. Fay. 1977. Effects of long-term treatment with acetylene on nitrogen-fixing microorganisms. Appl. Environ. Microbiol. 34:640–646.

De Bont, J.A.M., and M.W.M. Leijten. 1976. Nitrogen fixation by hydrogen-utilizing bacteria. Arch. Microbiol. 107:235–240.

De Bont, J.A.M., and E.G. Mulder. 1974. Nitrogen fixation and co-oxidation of ethylene by a methane-utilizing bacterium. J. Gen. Microbiol. 83:113–121.

De Bont, J.A.M., and E.G. Mulder. 1976. Invalidity of the acetylene reduction assay in alkane-utilizing, nitrogen-fixing bacteria. Appl. Environ. Microbiol. 31:640–647.

Fedorova, R.I., E.I. Milekhina, and N.I. Il'yukhina. 1973. Evaluation of the method of gas metabolism for detecting extraterrestrial life. Identification of nitrogen-fixing microorganisms. Izv. Akad. Nauk SSSR Ser. Biol. 1973(6):797–806.

Finke, L.R., and H.W. Seeley, Jr. 1978. Nitrogen fixation (acetylene reduction) by epiphytes of fresh water macrophytes. Appl. Environ. Microbiol. 36:129–138.

Gibson, J. 1975. Uptake of C_4 dicarboxylates and pyruvate by Rhodopseudomonas sphaeroides. J. Bacteriol. 123:471–480.

Hardy, R.W.F., R.D. Holsten, E.K. Jackson, and R.C. Burns. 1968. The acetylene-ethylene assay for N_2 fixation: Laboratory and field evaluation. Plant Physiol. 43:1185–1207.

Hill, S. 1973. Method for exposing bacterial cultures on solid media to a defined gas mixture using nylon bags. Lab. Pract. 22:193.

Hynes, R.K., and R. Knowles. 1978. Inhibition by acetylene of ammonia oxidation in Nitrosomonas europaea. FEMS Microbiol. Lett. 4:319–321.

Imhoff, J.F., and H.G. Trüper. 1989. Purple nonsulfur bacteria. p. 1658–1662. In J.T. Staley et al. (ed.) Bergey's manual of systematic bacteriology. Vol. 3. Williams and Wilkins, Baltimore.

Jurgensen, M.F., and C.B. Davey. 1968. Nitrogen-fixing blue-green algae in acid forest and nursery soils. Can. J. Microbiol. 14:1179–1183.

Knowles, R. 1978. Free-living bacteria, p. 25–40. In J. Döbereiner et al. (ed.) Limitations and potentials for biological nitrogen fixation in the tropics. Plenum Press, New York.

Knowles, R. 1980. Nitrogen fixation in natural plant communities and soils. p. 557–582. In F.J. Bergersen (ed.) Methods for evaluating biological nitrogen fixation. John Wiley and Sons, New York.

Line, M.A., and M.W. Loutit. 1973. Nitrogen-fixation by mixed cultures of aerobic and anaerobic micro-organisms in an aerobic environment. J. Gen. Microbiol. 74:179–180.

Malik, K.A., and H.G. Schlegel. 1981. Chemolithotrophic growth of bacteria able to grow under N_2-fixing conditions. FEMS Microbiol. Lett. 11:63–67.

Megraw, S.R., and R. Knowles. 1987. Active methanotrophs suppress nitrification in a humisol. Biol. Fert. Soils 4:205–212.

Nohrsted, H.-Ö. 1983. Natural formation of ethylene in forest soils and methods to correct results given by acetylene-reduction assay. Soil Biol. Biochem. 15:281–286.

Oakley, C.J., and J.C. Murrell. 1988. nifH genes in the obligate methane oxidizing bacteria. FEMS Microbiol. Lett. 49:53–57.

Okon, Y., S.L. Albrecht, and R.H. Burris. 1977. Methods for growing Spirillum lipoferum and for counting it in pure culture and in association with plants. Appl. Environ. Microbiol. 33:85–88.

Patriquin, D., and R. Knowles. 1972. Nitrogen fixation in the rhizosphere of marine angiosperms. Mar. Biol. (Berlin) 16:49–58.

Pedrosa, F.O., J. Döbereiner, and M.G. Yates. 1980. Hydrogen-dependent growth and autotrophic carbon dioxide fixation in Derxia. J. Gen. Microbiol. 119:547–551.

Postgate, J.R. 1984. The sulphate-reducing bacteria. Cambridge Univ. Press, Cambridge.

Postgate, J.R., H.M. Kent, S. Hill, and T.H. Blackburn. 1985. Nitrogen fixation by Desulfovibrio gigas and other strains of Desulfovibrio. p. 225–234. In P.W. Ludden and J.E. Burris (ed.) Nitrogen fixation and CO_2 metabolism. Elsevier Sci. Publ. Co., New York.

Raimbault, M. 1975. Étude de l'influence inhibitrice de l'acétylène sur la formation biologique du méthane dans un sol de rizière. Ann. Microbiol. (Paris) 126A:247–258.

Rennie, R.J. 1980. Dinitrogen-fixing bacteria: computer-assisted identification of soil isolates. Can. J. Microbiol. 26:1275–1283.

Rennie, R.J. 1981. A single medium for the isolation of acetylene-reducing (dinitrogen-fixing) bacteria from soils. Can. J. Microbiol. 27:8–14.

Ridge, E.H. 1970. Effects of some diluents upon viable counts of Azotobacter chroococcum. J. Gen. Appl. Microbiol. 16:189–192.

Riederer-Henderson, M.-A., and P.W. Wilson. 1970. Nitrogen fixation by sulfate-reducing bacteria. J. Gen. Microbiol. 61:27–31.

Salkinoja-Salonen, M.S., E. Vaisanen, and A. Peterson. 1979. Involvement of plasmids in the bacterial degradation of lignin-derived compounds. p. 301–314. In K.N. Timmis and A. Pühler (ed.) Plasmids of medical, environmental and commercial importance. Elsevier/ North Holland Biomedical Press, Amsterdam.

Siefert, E., R.L. Irgens, and N. Pfennig. 1978. Phototrophic purple and green bacteria in a sewage treatment plant. Appl. Environ. Microbiol. 35:38–44.

Smith, L.A., S. Hill, and M.G. Yates. 1976. Inhibition by acetylene of conventional hydrogenase in nitrogen-fixing bacteria. Nature (London) 262:209–210.

Sprent, J.I., and P. Sprent. 1990. Nitrogen fixing organisms. Pure and applied aspects. Chapman and Hall, London.

Stewart, W.D.P. 1973. Nitrogen fixation by photosynthetic microorganisms. Annu. Rev. Microbiol. 27:283–316.

Strijdom, B.W. 1966. Counting Beijerinckia spp. by a modification of the dilution tube method. S. Afr. Tydskr. Landbouwet. 9:265–266.

Tchan, Y.-T. 1984. Azotobacteraceae. p. 219–234. In N.R. Krieg and J.G. Holt (ed.) Bergey's manual of systematic bacteriology. Vol. 1. Williams and Wilkins, Baltimore.

Turner, G.L., and A.H. Gibson. 1980. Measurement of nitrogen fixation by indirect means. p. 111–138. In F.J. Bergersen (ed.) Methods for evaluating biological nitrogen-fixation. John Wiley and Sons, New York.

Van Neil, C.B. 1972. Techniques for the enrichment, isolation and maintenance of the photosynthetic bacteria. p. 3–28. In A.S. San Pietro (ed.) Methods in enzymology. Vol. 23. Academic Press, New York.

Veldkamp, H. 1970. Enrichment cultures of prokaryotic organisms. p. 305–361. In J.R. Norris and D.W. Ribbons (ed.) Methods in microbiology. Vol. 3A. Academic Press, New York.

Villemin, G., J. Balandreau, and Y. Dommergues. 1974. Utilisation du test de réduction de l'acétylène pour la numération des bactéries libres fixatrices d'azote. Ann. Microbiol. 24:87–94.

Weaver, P.F., J.D. Wall, and H. Gest. 1975. Characterization of Rhodopseudomonas capsulata. Arch. Microbiol. 105:207–216.

Whittenbury, R., K.C. Phillips, and J.T. Wilkinson. 1970. Enrichment, isolation and some properties of methane-utilizing bacteria. J. Gen. Microbiol. 61:205–218.

Widdel, F., and N. Pfennig. 1984. Dissimilatory sulfate- or sulfur-reducing bacteria. p. 663–679. In N.R. Krieg and J.G. Holt (ed.) Bergey's manual of systematic bacteriology. Vol. 1. Williams and Wilkins, Baltimore.

Wiegel, J., and H.G. Schlegel. 1976. Enrichment and isolation of nitrogen-fixing hydrogen bacteria. Arch. Microbiol. 107:139–142.

Wieringa, K.T. 1968. A new method for obtaining bacteria-free cultures of blue-green algae. Antonie van Leeuwenhoek 34:54–56.

Witty, J.F. 1979. Acetylene reduction assay can overestimate nitrogen fixation in soil. Soil Biol. Biochem. 11:209–210.

Wong, T.-Y., and R.J. Maier. 1985. H_2-dependent mixotrophic growth of N_2-fixing Azotobacter vinelandii. J. Bacteriol. 63:528–533.

Chapter 12

Legume Nodule Symbionts

R. W. WEAVER, *Texas A&M University, College Station, Texas*

PETER H. GRAHAM, *University of Minnesota, St. Paul, Minnesota*

We use the word rhizobia in this chapter to define a diverse group of organisms united by a common ability to produce root or stem nodules on leguminous plants. To varying degrees, rhizobia possess the ability to fix N independently or in symbiosis with appropriate host legumes. This ability has influenced agricultural systems since the time of the Romans, and is currently of great importance in developing countries, where the price and limited availability of nitrogenous fertilizers can often preclude their use by subsistence farmers. Interest in the rhizobia is also increasing in developed countries as legumes are used more in minimum tillage and sustainable agricultural systems.

Because of their importance in agriculture, the early taxonomy of the rhizobia emphasized host nodulation (Baldwin & Fred, 1929; Fred et al., 1932), dividing these organisms into "cross-inoculation groups", isolates from which were thought to nodulate certain legume species, but not others. Such cross-inoculation groups persist as the basis for recommending inoculant cultures, but the demonstration that nodulation, host specificity, and N_2-fixation genes can be located on transmissible plasmids, has led to the recent reclassification of the rhizobia (Jordan, 1982, 1984) and to the greater use of other phenotypic and phylogenetic traits in their characterization (Graham et al., 1991). Currently, three genera and 11 species of root and stem-nodule bacteria are distinguished, viz:

- *Rhizobium leguminosarum* biovars *leguminosarum,* and *trifolii* (Jordan, 1984)
- *R. meliloti* (Dangeard, 1926)
- *R. loti* (Jarvis et al., 1982)
- *R. fredii* (Scholla & Elkan, 1984; Chen et al., 1988)
- *R. galegae* (Lindstrom, 1989)
- *R. tropici* (Martinez et al., 1991)

- *R. haukuii* (Chen et al., 1991)
- *R. etli* (Segovia et al., 1993)
- *Azorhizobium caulinodans* (Dreyfus et al., 1988)
- *Bradyrhizobium japonicum* (Jordan, 1984)
- *B. elkanii* (Kuykendall et al., 1992)

This is still an interim classification based on isolates from nodules of only a small percentage of the 19 700 species of Leguminosae. Better definition among the slow-growing rhizobia is also needed. Current species are defined principally by genotypic similarities and differences (DNA/DNA relatedness, rRNA/DNA hybridization, 16S ribosomal RNA analysis and DNA restriction length polymorphisms). Colonial and cultural characteristics and symbiotic performance with selected hosts, however, continue as the first parameters needed for the characterization of particular rhizobia.

This chapter presents methods commonly used in the study of root and stem-nodule bacteria. Carefully followed, they should enable a microbiologist, agronomist, or geneticist with little previous experience to isolate and culture rhizobia, and to prepare and use inoculant cultures. It should be emphasized, however, that the slow growth rate of these organisms and the lack of suitable selective media can complicate such studies. Guidance from an experienced individual who has worked with these organisms may be needed.

12–1 NODULE COLLECTION AND THE ISOLATION OF SYMBIONTS

12–1.1 Principles

The reasons for isolation of rhizobia from nodules may occur as part of a germplasm collection program, may be undertaken to study their diversity in specific field situations, or may be to recover particular genetic recombinants from a mixed population of cells. Date and Halliday (1987) have prepared excellent guidelines and methods for the collection of rhizobial germplasm.

Nodules often contain more than a single strain of *Rhizobium* or *Bradyrhizobium*. Double strain occupancy (Lindeman et al., 1974; Moawad & Schmidt, 1987) and nonrhizobial contaminants in the nodule are surprisingly common (Jansen van Rensberg & Strijdom, 1972; Handelsman & Brill, 1985).

The method that follows may be used for the isolation of rhizobia from the nodules of either field or greenhouse-grown plants.

12–1.2 Materials

1. Yeast mannitol agar (YMA) (section 12–2.2).
2. 95% Ethanol.

3. Sterile water.
4. 5.25% Sodium hypochlorite (full-strength commercial bleach).
5. Tweezers and scissors.
6. Petri dishes.
7. Nodules.

12–1.3 Procedure

Cut the nodules from the plant, leaving about 0.5 cm of root attached to the nodule for ease in handling. The nodules should be firm and appear in good physical condition. Avoid older nodules that have begun to senesce, as they may contain large populations of secondary organisms.

Thoroughly wash the nodules to remove all traces of soil, then immerse them in 95% ethanol for 30 s. Transfer the nodules to 5.25% sodium hypochlorite solution, and leave them immersed in this solution for 4 min. Prolonged exposure to this substance will result in complete sterilization of the rhizobia within the nodule. Remove all of the nodules from the sodium hypochlorite and place them in a petri dish containing 20 to 25 mL of sterile water. To remove all traces of the sterilant, rinse the nodules in five to six changes of sterile water. A series of petri dishes containing 20 to 25 mL of sterile water allows rapid rinsing.

Once the nodule is surface disinfected, crush it between blunt-tipped forceps, then mix the nodule contents with a drop of sterile water in the bottom of a sterile petri dish. Streak a loopful of this suspension onto YMA plates and incubate at 28 °C for 3 to 5 d for fast-growing rhizobia and 8 to 10 d for bradyrhizobia before further subculture is attempted.

Isolates on agar medium should be white to somewhat translucent, circular and raised, and may produce significant amounts of extracellular polysaccharide. Examine wet mounts of cells from selected colonies using a light- or phase-contrast microscope. Cells should be small rod-shaped organisms 1 to 3 µm in length and 0.5 to 1.0 µm in diameter. Colonies that are colored or have a distinctive aroma, and bacteria that are distinctly coccoid, produce endospores, or are grouped in chains are not likely to be rhizobia, though Norris (1958) reported a "red" *Rhizobium* from *Lotononis,* and colonial dimorphism (Sylvester-Bradley et al., 1988b) and "doughnut"-shaped colonies have been reported (Howieson et al., 1988). Select a representative colony having the characteristics of rhizobia, and subculture it onto fresh plates of YMA. If the culture appears pure, it should then be authenticated using host infection (section 12–4) and preserved.

12–1.4 Comments

Fresh nodules should be used for strain isolations wherever possible. If necessary to delay isolation for some time, nodules should be desiccated in a vial containing silica gel or anhydrous calcium chloride that is covered with a layer of nonabsorbent cotton or cheesecloth to protect the nodules from direct contact with the desiccant. Nodules may be stored in this way for several weeks before isolation is attempted, but should then be allowed to imbibe water for 1 h prior to surface disinfection. Alternatively, nodules may be frozen soon after collection. They may be left on roots or removed to conserve freezer space.

Alternate disinfectants to sodium hypochlorite include 3 to 5% hydrogen peroxide for 3 to 4 min (Vincent, 1970) or acidified mercuric chloride (0.10 g of mercuric chloride and 0.5 mL of concentrated hydrochloric acid in 100 mL of distilled water) for 3 to 4 min. When hydrogen peroxide is used, there is no need for repeated washing in sterile water. Regulations on disposal of mercuric chloride are stringent.

Strains of *Rhizobium* have generation times of 2 to 4 h on YMA medium and produce colonies that are 2 to 4 mm in diameter after 3 to 5 d incubation. Slightly slower growth rates have been reported for *A. caulinodans* (Dreyfus et al., 1988). Most bradyrhizobia are even slower growing, having a generation time of 6 to 10 h and producing colonies that do not exceed 1 mm diam. after 5 to 7 d growth. Since the growth rate of the rhizobia is slow relative to that of organisms occurring as contaminants in nodule squashes, the inexperienced worker runs the risk of selecting and propagating contaminants, thinking they are rhizobia. It is advisable to leave all isolates from nodules for at least a week before attempting subculture; even longer may be needed with some strains of *Bradyrhizobium*.

Fungal contaminants will often grow faster than the rhizobia, and overrun plates before colonies of the latter organisms are of size sufficient to identify and subculture. To limit this, it is common to include antifungal compounds in the isolation media. Congo red (10 mL of a 1:400 wt/vol aqueous solution per liter, added to the medium before autoclaving) or 0.02 g per liter of filter sterilized cycloheximide added to the medium after autoclaving may be used (Vincent, 1970). A stock solution of cycloheximide may be made by dissolving 2 g of cycloheximide in 200 mL of ethyl alcohol. *Agrobacterium* also occurs as a contaminant on YMA plates used for nodule isolations and are easily confused with *Rhizobium*. The two organisms can, however, be distinguished using the ketolactase test (Bernaerts & De Ley, 1963).

The rhizobia from many tropical pasture and tree legumes grow best under slightly acid conditions and may not be recoverable using conventional medium. Date and Halliday (1979) used a defined medium containing arabinose, pH 4.5, as the isolation medium for rhizobia from acid soils, while both Gomez de Souza et al. (1984) and Sylvester-Bradley et al. (1988a) used YMA medium acidified to pH 5.5.

While isolates are usually authenticated by their ability to nodulate specific legumes, recent papers have reported the isolation from soil of non-infective organisms with the characteristics of *Rhizobium* (Jarvis et al., 1989; Soberon-Chavez & Najera, 1989; Segovia et al., 1991). A significant percentage of such strains acquire the ability to produce nodules following the introduction of a symbiotic plasmid.

12–2 CULTIVATION OF NODULE SYMBIONTS

12–2.1 Principles

Rhizobia are aerobic to microaerophilic organisms that under most circumstances behave like typical heterotrophs. They will usually require both a source of energy and combined N for growth, and specific vitamin requirements have been shown for some species (Graham, 1963; Elkan & Kwik, 1968; Chakrabarti et al., 1981). Less typically, specific rhizobia are capable of nitrate-dependent growth under anaerobic conditions (O'Hara & Daniel, 1985; Van Berkum & Keyser, 1985), or are autotrophic (Hanus et al., 1979) and may even harvest light energy (Eaglesham et al., 1990; Ladha et al., 1990). For most purposes, however, the rhizobia can be grown on relatively simple media. These commonly contain mannitol as the energy source, and yeast or yeast extract to supply combined N, vitamins, and minerals. Glycerol (Balatti, 1982) and gluconate (Kuykendall & Elkan, 1976) have also been used to supply energy, while arabinose and galactose may be preferable energy sources where limited pH change in the medium is desired (Date & Halliday, 1979; Howieson, 1985). The majority of slow-growing rhizobia (bradyrhizobia) cannot use either sucrose or lactose as the sole C source.

Organic buffers such as MES (pKa = 6.1 at 25 °C, useful pH range 5.5–6.7) and HEPES (pKa = 7.5 at 25 °C, useful pH range 6.8–8.2) are increasingly being used to maintain medium pH at a desired level (Cole & Elkan, 1973; Howieson, 1985).

12–2.2 Materials

Media used for the cultivation of rhizobia.

1. Yeast Mannitol Agar (YMA). There are several variants of this medium. The medium described by Fred and Waksman (1928) contains (g L^{-1}): mannitol, 10; potassium hydrogen phosphate, 0.5; magnesium sulfate heptahydrate, 0.2; sodium chloride, 0.1; calcium carbonate, 0.01; yeast extract powder, 0.5; agar, 15; distilled water to 1000 mL. For yeast extract mannitol broth (YMB), simply leave out the agar. Combine all of the ingredients except the agar in one-half of the water, then adjust the pH to 7.0 with 1 N hydrochloric acid and warm the solution to 55 °C. Separately melt the agar in the remaining water by autoclaving 15 min at 0.10 MPa. Mix the two solutions and dispense in test tubes or media bottles as desired. Sterilize the medium by autoclaving as specified above.

2. Peptone Yeast Extract Agar (PYA). A rich medium that is favored for genetic studies because rhizobia tend to produce less extracellular polysaccharide than in YMA. The medium as described by Noel et al. (1984) contains (g L^{-1}): peptone hydrolysate of casein, 5.0; yeast extract

powder, 3.0; calcium chloride dihydrate, 0.147; agar, 15; and distilled water to 1000 mL. Leave out the agar for peptone yeast extract broth (PYB). The medium is prepared, the pH adjusted, and the medium autoclaved as detailed above.

3. Keyser-Munns Agar (KMA). The KMA is a defined medium used in those studies that require a precise knowledge of medium composition (Keyser & Munns, 1979). The medium contains (mg L^{-1}): mannitol, 10 000; magnesium sulfate heptahydrate, 74; calcium chloride dihydrate, 44; iron EDTA, 37; potassium chloride, 0.75; manganese chloride tetrahydrate, 0.20; zinc sulfate septahydrate, 0.12; cupric chloride dihydrate, 0.017; sodium molybdate dihydrate, 0.0048; cobalt nitrate, 0.00037; sodium glutamate, 1100; potassium dihydrogen phosphate, 68; dipotassium hydrogen phosphate trihydrate, 114; and agar, 15 000. For this medium, microelements can be prepared as a $100 \times$ stock solution, with 10 mL added per liter. Vitamins may also be necessary for some strains. Vitamins should be prepared separately (thiamine, 1 mg: calcium pantothenate, 1 mg; biotin, 0.1 mg; distilled water, 1000 mL), and filter-sterilized, using an 0.2 μm filter, before adding to already autoclaved medium at the rate of 1 mL per liter before the plates are poured. For Keyser-Munns broth (KMB) leave the agar out of KMA.

A range of additional media are used in studying specific aspects of rhizobial physiology, including melanin (Cubo et al., 1988) and siderophore (Schwyn & Nielands, 1987; Guerinot, 1991) production, and acid-pH tolerance (Howieson, 1985; Graham et al., 1982).

12–2.3 Procedure

Cultures of bradyrhizobia and rhizobia should be maintained in a manner that ensures viability, but minimizes the possibility of mutation or contamination. Frequent subculture provides an opportunity for mutation, and marked variation in the N_2-fixing ability of single colony isolates from strains is well documented (Herridge & Roughley, 1975; Weaver & Wright, 1987; Gibson et al., 1990). The ideal is to maintain cultures of those strains that are in active use, but to replace these at regular intervals from stocks that are only rarely subcultured (section 12–3). For the former, strains may be maintained for several weeks on agar slants, with tightened screw caps, held at room temperature. Most of the bacteria die during storage but sufficient numbers survive for reisolation and culturing. Lowering the storage temperature to 4 °C after adequate growth has occurred increases the time cells may be stored to several months.

For the preparation of broth cultures, the quantity of inoculant should be at least 1% of the final number of cells needed, and 5 to 10% is highly desirable. When low inoculant rates are used, a lag period of several days can occur in the growth of the culture, and culture contamination can be a major problem. One to two loopfuls of a recent (few days old) subculture should be used per 25 mL of broth. Adequate aeration must be provided during the growth of broth cultures. This is most conveniently achieved using a laboratory shaker, but when a shaker is not available, air can be passed through nonabsorbent cotton filters, and sparged through the culture. Optimum temperature for rhizobial growth is 28 °C, but room temperature is usually adequate. Extended incubation can result in the production of copious amounts of extracellular polysaccharide, and cause problems in the centrifugation of cultures. The PYB and KMB are less prone to this problem than YMB.

Rhizobium or *Bradyrhizobium* numbers in broth culture will usually exceed 10^9 cells per milliliter. Counts of total organisms can be made using the Petroff-Hausser counter (Hausser Scientific, Bluebell, PA) and the methodologies detailed by Somasegaran and Hoben (1985). Viable counts require serial dilution of the culture in sterile tap or distilled water or physiological saline, and the plating of aliquots on the surface of YMA plates.

12–3 MAINTENANCE OF CULTURES

12–3.1 Principles

Because of the relatively slow growth of rhizobia and the possibilities for contamination and mutation, many research centers work with a three-tier system of culture maintenance. In such a system, cultures in active use (see section 12–2) are backed by stock cultures that are stored at 5 °C and sealed to prevent desiccation, while the long-term survival of strains is ensured by their lyophilization (Vincent, 1970), desiccation on porcelain beads (Norris, 1963; Vincent, 1970), or storage at ultracold temperatures using a cryoprotectant (Dye, 1980; Keyser, 1987). Published data in which these methods are compared is lacking, so the actual method selected will commonly depend on the availability of facilities and personal preference. A description of the procedure for long-term cold storage using glycerol as cryoprotectant is given below.

12–3.2 Materials

1. Cultures on YMA slants.
2. 20% solution (vol/vol) of glycerol in distilled water.
3. 4-mL screw cap cryogenic storage vials.
4. Low-temperature (−15 °C) or ultra-cold (−70 °C) freezer.

12–3.3 Procedures

Prepare the 20% glycerol solution and dispense it in 2-mL amounts in cryogenic vials. Sterilize by autoclaving for 20 min at 0.10 MPa (121 °C) on the liquid cycle. Once the glycerol is cool, aseptically transfer it to slants or plates containing the culture to be preserved. Cultures of fast-growing rhizobia should be 3 to 4 d old at the time of preservation; those for slow-growing strains about 7 d old. Scrape the bacterial growth from the surface of the medium with a sterile loop, or sterile glass beads then use a sterile 1-mL pipet to transfer bacteria and glycerol back into the cryogenic vial. Vortex the mixture to ensure even dispersion of the cells. Freeze the culture slowly in a freezer at -15 °C, then transfer it to an ultracold freezer at -70 °C for long-term storage. Multiple vials should be prepared for each strain, with a new vial used each time a fresh culture of the organism is needed.

12–3.4 Comments

The concentration of glycerol used in cryoprotection varies from 10 to 50%. Apparently the concentration of glycerol is not critical, nor is it essential that the glycerol be mixed with buffer or nutrient solution. Scraping the growth from slants or plants rather than using broth cultures ensures a larger initial population of organisms in the cryoprotectant, which allows for die-off during freezing, thawing, and storage.

The temperature of storage is largely determined by available facilities. Cultures in glycerol will remain viable for more than a year at -20 °C, but with lower temperatures, survival can be extended indefinitely. Best results are obtained with cells stored over liquid N in the gaseous phase. Freezing bacteria somewhat slowly (1–2 °C per minute) is generally considered less deleterious than rapid freezing, but thawing should be as rapid as possible. Repeated freezing and thawing will result in lysis of cells. Cultures can, however, be taken from the freezer, some frozen material removed for use as inoculum, and the vial replaced in the freezer. When stored at -20 °C, glycerol solutions may not be frozen and can be used directly as an inoculum.

12–4 ENUMERATION OF NODULE SYMBIONTS IN SOIL AND INOCULANTS

12–4.1 Principles

While several selective media have been developed for use with bradyrhizobia or rhizobia (Pattison & Skinner, 1974; Barber, 1979; Habte, 1985), the enumeration of these bacteria from soil by routine dilution count procedures is often impractical. Even in selective media, the growth of bradyrhizobia and rhizobia will commonly be less than that of other soil organisms, particularly pseudomonads. However, where the rhizobial population is large relative to that of the other organisms present, for example in peat inoculants, rhizobia can be counted by serial dilution and plating. Specific serogroups of *Rhizobium* or *Bradyrhizobium* soil populations (e.g., serogroup 123 of *B. japonicum*) can also be approximated using fluorescent antibody methods (Moawad et al., 1984; McDermott & Gra-

Table 12–1. Nutrient solutions for growing plants.†

Stock solution	Chemicals	Quantity, g L^{-1}	Quantity of stock L^{-1}
Without N			
1	K_2SO_4	93	3 mL
2	Mg $SO_4 \cdot 7H_2O$	493	1 mL
3	KH_2PO_4	23	1 mL
	K_2HPO_4	145	
4	$CaCl_2$	56	1 mL
5	$CaSO_4$		1 g
6‡	$FeCl_3$	6.5	1 mL
	Na_2H_2EDTA	13	
7	H_3BO_3	0.23	1 mL
	$MnSO_4 \cdot H_2O$	0.16	
	$ZnSO_4 \cdot 7H_2O$	0.22	
	$CaSO_4 \cdot 5H_2O$	0.08	
	$Na_2MoO_4 \cdot 2H_2O$	0.025	
	$CoCl_2 \cdot 6H_2O$	0.034	
	$NiCl_2$	0.022	
With N			
8§	KNO_3	10	1 mL
	$(NH_4)_2SO_4$	133	

† The nutrient solution is a modification of Evans et al. (1972).
‡ Dissolve the two chemicals separately before combining.
§ For nutrient solution containing N substitute stock solution 8 for 1.

ham, 1989; see chapters 6 and 28 by Bottomley and Wright, respectively, in this book). McDermott and Graham (1989) reported recovery efficiencies with this procedure of from 86 to 114%.

The method most commonly used to count rhizobia in the presence of other organisms is a variant of the most probable number (MPN) technique described in chapter 5 by Woomer in this book. The material containing rhizobia to be counted is serially diluted, then aliquots of appropriate dilutions are applied to tubes or pouches containing a suitable, aseptically grown host legume. Obviously, the species of legume used will determine the range of rhizobia counted.

12–4.2 Materials

1. Disposable plastic growth pouches (Vaughn Seed Co., Downers Grove, IL).
2. Pouch holders (record holders are convenient).
3. Seed of an appropriate legume.
4. Nitrogen-free nutrient solution (see Table 12–1).
5. Plastic drinking straws.
6. 95% ethanol solution.
7. 1% sodium hypochlorite solution (20-mL household bleach diluted to 100 mL with distilled water).

Fig. 12–1. Arrangement of divided plastic pouches used in making most probable number (MPN) counts of rhizobia.

8. Sterile water.
9. Water agar for pregerminating seedlings.
10. Samples with the rhizobia to be enumerated.
11. Greenhouse or growth chamber for plant propagation.

12–4.3 Procedure

All materials must be rhizobia free, and extreme care taken to prevent the contamination of plant growth units. Pouches arriving from the manufacturer are usually rhizobia free. They may be moistened and autoclaved prior to use. Place the pouches in a record rack or support stand to hold them while seeding, inoculating, and growing seedlings (Fig. 12–1). Commercial plastic pouches are 16 by 17 cm and are provided with a paper wick (Porter et al., 1966). To economize on space and supplies, remove the paper wicks, and use a plastic heat sealer to divide the pouches into two to four compartments depending on the size of the legume host being used. Cut the paper wicks so that they again fit into the smaller compartments. The wicks should not extend to the top of the pouches as this facilitates fungal contamination and evaporative water loss. As supplied, the upper section of the wick in which the seedling is placed has small perforations for root penetration. A sterile scalpel can be used to make a somewhat larger hole through which seedling radicals can more easily pass.

To kill any bradyrhizobia or rhizobia on the surface of seed, soak the seed in 70 to 95% ethanol for 10 min and rinse with water. Soak the rinsed seed in a solution of 1% sodium hypochlorite for 10 min. Rinse the seed five to six times in sterile water to remove any residual disinfectant. Pregerminate the seeds 2 to 3 d on water agar (agar, 10.0 g L^{-1}) or on blotting paper, then transplant to the growth pouches only those seedlings that are free from contamination and show good root development. In transplanting the seeds, use sterile forceps and make sure that the radical passes through the hole made in the upper section of the wick.

Plastic drinking straws, cut so that they extend from halfway down the pouch to 1 to 2 cm above it, greatly facilitate watering the pouches, provide rigidity, and help to limit cross-contamination. Use care during the watering of plants to ensure that rhizobia are not transferred via the tip of the watering device or through splashing. Automatic pipetting machines (e.g., Brewer Model SEPCO 60501-40A-SS, BBL, Baltimore, MD) can also be used to speed watering. Add approximately 50-mL sterile nutrient solution (Table 12–1) to each pouch: use proportionately less where the pouches are compartmentalized. It is convenient to prepare stock solutions for the nutrient solution. When preparing stock solution no. 6, it is important to dissolve the two chemicals separately before combining. The $CaSO_4$ should be added last to avoid formation of precipitates. If the nutrient solution will be autoclaved, the $CaSO_4$ should not be added until after autoclaving or an insoluble precipitate will form.

Inoculate plants when 6 to 7 d old and well established. Prepare 10-fold serial dilutions of the soil whose rhizobial population is to be counted, then inoculate 1 mL of appropriate soil dilutions onto the root systems of four replicate seedlings. Positive and negative controls should be included with each batch of seedlings. The positive control, inoculated with a pure culture of the appropriate rhizobia, will confirm that growth conditions during the experiment were appropriate for nodule development; the negative control is not inoculated but if seedlings become nodulated this will indicate contamination.

Growth pouches should be kept under well-lighted conditions in a growth chamber or greenhouse, and at temperatures appropriate to the plant species being used. Vincent (1970) shows a suitable lighting system with fluorescent and incandescent tubes; high-pressure sodium lamps also provide excellent illumination. Plant husbandry and management techniques for some tropical legumes are discussed by Summerfield et al. (1977).

Grow the plants for an additional 1 to 2 wk after nodules first appear on the positive controls. Generally, 2 wk after inoculation is adequate for small-seeded legumes, and 3 wk for larger-seeded species. Plant species can vary significantly, however, in the time needed for nodulation. Record each replicate as positive (one or more nodules present) or negative (no nodules present).

The MPN counts of rhizobia in the sample are computed by statistical procedures. A software program has been developed by Woomer et al. (1990). Statistical tables for MPN counts have been included in several publications (Brockwell, 1963; Alexander, 1965; Vincent, 1970; Brockwell et al., 1975; see chapter 5 by Woomer in this book).

12–4.4 Comments

The MPN count assumes that a single rhizobial cell can multiply and form at least one nodule on the inoculated root. This assumption is not always valid (Thompson & Vincent, 1967; Boonkerd & Weaver, 1982), and populations are often underestimated. Rhizobia must be uniformly dispersed in the diluent for accurate counting; for peat inoculants, Weaver (1979) has reported that rhizobia desorb readily and become dispersed.

It is extremely important to use high-quality seed for planting in plastic growth pouches. Seed of poor quality will not perform well in growth pouches and considerable fungal contamination may occur. It is relatively easy to kill rhizobia on the surface of seed using ethanol or sodium hypochlorite. To actually surface sterilize seed is much more challenging and generally not necessary. If surface sterilization is desired, refer to methods described by Caetano-Anolles et al. (1990).

The host used in the assay can affect the results obtained in MPN counts. Thus, *Macroptilium atropurpureum* is often used to assay *Bradyrhizobium* populations in soil, but can overestimate the number of soil rhizobia infective on a particular legume because of its wide host range. Kumar Rao et al. (1982) reported 190 000 rhizobia per gram in a Kashmir soil when *Macroptilium* was used as host, but only 3270 when a pigeons pea [*Cajanus cajan* (L.) Huth] cultivar was used. *Glycine ussuriensis* is preferable to *G. max* for counting *Bradyrhizobium japonicum* because of its small seed size (Brockwell et al., 1975).

Light can affect rooting and nodulation in several legumes, including peanuts, necessitating that the lower portions of pouches or tubes be shielded.

12–5 INOCULANTS FOR FIELD EXPERIMENTATION

12–5.1 Principles

Practical investigations with rhizobia frequently culminate with greenhouse or field experiments to determine the need for inoculation, or to observe the effects of inoculation on plant growth or yield. For greenhouse or growth-chamber studies, a broth culture of the inoculant strain will usually be adequate, and may be directly inoculated onto the seed at planting. Even under these conditions, it is important that the rate of inoculation used be a realistic one. Takats (1986) and others have reported reduced nodulation at heavy inoculant dose levels. For field experiments, the large number of seeds used will often require a more sophisticated delivery system, with the inoculant prepared in advance of the planting date and required to maintain high rhizobial numbers for as long as 6 mo.

Commercial inoculants are available in most countries, and some inoculant manufacturers will custom prepare peat inoculants using specific strains. With proper storage, purchased peat inoculants will usually contain adequate numbers of rhizobia. Recommended inoculant strains for particular host legumes are shown in Table 12-2.

Factors that must be considered in preparing inoculants include suitability of the inoculant carrier, the method of inoculation to be practiced, environmental factors likely to influence rhizobial survival during shipment and application, and additional seed treatments. Peat is the preferred inoculant carrier and can usually be obtained or purchased directly from the inoculant company to ensure its suitability. In some countries, inoculant-quality peat may not be available, and a peat source or alternate inoculant carrier may need to be identified and tested. Vegetable oils, charcoal, bagasse, coir, and composted organic material have all proved suitable (Corby, 1976; Deschodt & Strijdom, 1976; Philpotts, 1976; Halliday & Graham, 1978; Pazkowski & Berryhill, 1979; Kremer & Peterson, 1983; Graham-Weiss et al., 1987; Beck, 1991), but each material must be separately tested before use. Methods used in testing such carriers have been detailed in several publications (Roughley & Vincent, 1967; Roughley, 1968) and are beyond the scope of this chapter. Because peat is a proven inoculant carrier and readily available from inoculant manufacturers in the USA, the method described in the next section uses peat as the carrier.

12-5.2 Materials

1. Peat. Peat carriers may be supplied in two grades; a very fine peat with 70 to 95% passing through a 200-mesh sieve for the preparation of seed-applied inoculants, and a more granular preparation (16–40 mesh) for application in the seed furrow (Burton, 1982).
2. Finely ground limestone.
3. Culture of *Rhizobium* or *Bradyrhizobium*.
4. Sterile polyethylene bags for packaging the inoculant.

12-5.3 Procedure

Most peats are somewhat acid, and finely ground limestone may need to be added to raise the pH toward neutrality. Limestone additions should not exceed 10% by weight of the peat. If only small quantities of peat are needed, the simplest approach may be to buy a commercial peat-based

Table 12–2. Some recommended strains of rhizobium for selected legumes.†

Legume	Strain(s)
Aeschynomene falcata	CB 2312
Arachis hypogaea	CB 756, USDA 3339, USDA 3341
A. pintoi	CB 756, CIAT 3101
Astragalus sinicus	USDA 3466
Cajanus cajan	CB 756, USDA 3384, USDA 3474
Calopogonium sp.	CB 756
Centrosema sp.	CB 1923, CIAT 1670, CIAT 3101
Cicer arietinum	CB 2855, USDA 3378, USDA 3379
Coronilla varia	CB 2012, USDA 3165, USDA 3167
Crotalaria paulina	USDA 3384
Cyamopsis tetragonolobus	CB 3035
Desmodium heterocarpon	CIAT 3418
D. heterophyllum	CB 2085, CIAT 3469
D. intortum	SEMIA 656 SEMIA 6003
Glycine max	CB 1809, USDA 110, USDA 142
	SEMIA 587, SEMIA 5018
Lens esculenta	SEMIA 344, SEMIA 360
Leucaena leucocephala	CB 81, USDA 3404, CIAT 1967
Lotononis bainesii	CB 376
Lotus corniculatus	CB 2938, SEMIA 806, SEMIA 816
L. pedunculatis	CB 2270
L. tenuis	SEMIA 805, SEMIA 806
L. uliginosus	SEMIA 821
Lupinus sp.	CB 2026
Macroptilium atropurpureum	CB 756, SEMIA 656
Medicago lupulina	USDA 1057
M. polymorphum	SEMIA 138
M. sativa	USDA 1011, USDA 1021a, USDA 1025
	SEMIA 115, SEMIA 116
Medicago sp.	CB 3061
Onobrychis sativa	CB 2000
Peuraria sp.	CB 756, CIAT 2434
Phaseolus lunatus	USDA 3259
P. vulgaris	CB 2899, USDA 2667, USDA 2669, USDA 2674,
	CIAT 166, CIAT 632, SEMIA 491
Pisum sativum	CB 1447, USDA 2370, SEMIA 335, SEMIA 374
Stylosanthes capitata	CB 2898, CIAT 870, CIAT 995, CIAT 2138
S. guyanensis	CB 756, CIAT 71
S. hamata	CB 756, CB 2126
S. humilis	CB 756
S. scabra	CB 756
Trifolium cherleri	CB 2937
T. incarnatum	CB 2937, SEMIA 208c
T. pratense	CB 1990, SEMIA 222
T. repens	CB 1990, SEMIA 235
T. semipilosum	CB 787, SEMIA 280
T. subterraneum	CB 2937, SEMIA 222
Vicia sativa	SEMIA 354, SEMIA 366
Vigna mungo	CB 1015
V. radiata	USDA 3447
V. unguiculata	USDA 3454, SEMIA 634, SEMIA 656
Zornia sp.	CIAT 71

† Recommendations are taken from the CSIRO Division of Tropical Pastures CB *Rhizobium* Strain Catalogue (1984); the USDA Beltsville Rhizobium Culture Collection Catalog (1987); the CIAT Catalogue of *Rhizobium* strains for tropical forage legumes (1986); and the Catalogue of *Rhizobium* strains of the MIRCEN centre, Porto Alegre, Brazil. Addresses for each collection are included in Takishima et al. (1989).

inoculant, and autoclave it before adding the strain of interest. Autoclaving can sometimes affect the suitability of peat as an inoculant carrier, but this is not known to be a problem with the peats used by manufacturers in the USA.

Grow the inoculant strain in YMB until mid- to late-log phase, as detailed in section 12–2.3, then thoroughly mix approximately equal parts by weight of culture broth and dry peat. Nonsterile peat will generate heat on mixing, so it should be layered to a depth of no more than 2 to 5 cm on trays, and allowed to mature for at least 24 h before packaging. For sterile peat, aseptically transfer 50 to 100 g to a sterile polyethylene bag, then add an equal weight of culture broth. Alternatively, smaller inoculant doses can be applied to already moistened and sealed bags (Somasegaran, 1985), mix the contents by gently squeezing the bag. After mixing and maturation, peats should contain at least 100 million rhizobia per gram, and be moist (0.03 MPa tension) but not saturated. The polyethylene bags used should have a wall thickness of approximately 0.0375 mm and permit some aeration (Roughley, 1976). Store the inoculant under cool conditions until used.

12–5.4 Comments

It is important to confirm that the inoculant is of high quality before beginning field experiments. The culture used to inoculate the peat should be checked for contamination, and the viable cell count determined by plate counting. Where some time has elapsed between the preparation and use of the inoculant, plate counts should also be made at planting to verify the level of inoculation.

12–6 INOCULATION OF SEED

12–6.1 Principles

The aim of inoculation is to ensure that the rhizobia applied are sufficient to ensure rapid and abundant nodulation of the host legume. The number required will vary with host and environment, but should not be < 5000 rhizobia per seed. Two basic approaches are used. In one, the inoculant is applied directly to the seed; in the other, it is applied to the soil in the vicinity of the seed. The former method is the more practical in most situations, but has limitations where unfavorable conditions (high soil temperature, acid soil pH, and competition from native soil rhizobia) necessitate the use of higher than normal inoculation rates. In such cases, granular or liquid inoculants are commonly used and applied directly to the soil.

For seed-applied rhizobia, it is important to use an adhesive to maintain close contact between rhizobia and the seed, and to reduce the rate of decline in rhizobial numbers per seed (Vincent, 1970; Materon & Weaver, 1984). A method for seed inoculation is described below.

12–6.2 Materials

1. Seed.
2. Inoculant (see section 12–4.3 for preparation).
3. Adhesive (gum arabic, methyl ethyl cellulose, or commercial product).
4. Finely ground rock phosphate or calcium carbonate.

12–6.3 Procedure

Since seed size varies from < 1 mg (*Trifolium repens* L.) to more than 1000 mg (*Stizolobium*) per seed, with the surface area per gram of the small-seeded legume significantly greater than for larger-seeded types, the amount of inoculant culture and adhesive applied per kilogram of seed will vary greatly. For small-seeded legumes, use 20 mL of adhesive solution and 10 g of peat per kilogram of seed: for larger seeded legumes use 10 mL of adhesive and 5 g of inoculant per kilogram seed. Prepare the adhesive by dissolving 40 g of gum arabic or 5 g of methyl ethyl cellulose per 100 mL of water. Some samples of gum arabic are red and strongly acid and should be avoided.

Mix the inoculant culture and adhesive, then apply the inoculant suspension to the seed, mixing until the seed is uniformly coated. Spread the inoculated seed out in a cool location to dry. While the rhizobia should stay viable for several days, it is better to only prepare as much inoculant as needed for the following day.

Where seeds are to be planted into acid soil, or face other stresses before germination, it may be advisable to pellet the seed. This protects the rhizobia from temperature and desiccation, and during germination creates a microclimate around the seed that permits nodulation under even quite acid soil conditions (Loneragan et al., 1955; Graham et al., 1974; Philpotts, 1977). Finely ground rock phosphate is usually used as the pelleting material for slow-growing rhizobia and calcium carbonate for the fast growers. Gum arabic or methyl ethyl cellulose must be used as the adhesive; most other substances do not adequately bind the pelleting material. Mix the inoculant, adhesive, and seed as before, then add 200 g of the pelleting agent per kilogram seed. Mix rapidly but smoothly for 1 to 2 min, by which time all of the seeds should be evenly coated and separated.

Care should be taken in the use of pesticides or molybdenum in association with legume inoculants. While most herbicides and insecticides will show little effect on rhizobia at commercial rates of application, fungicides such as captan are usually lethal (Curley & Burton, 1975; Graham et al., 1980), and several others will reduce nodulation on prolonged contact with rhizobia. Seed-dressing with Mo can also be toxic to rhizobia

(Graham et al., 1974; Gault & Brockwell, 1980). Where seed-dressing is required for control of root pathogens, granular or liquid inoculant preparations can be used and placed directly in the seed furrow (Weaver & Frederick, 1974; Boonkerd et al., 1978). Soil-applied inoculants may also be used for crops such as peanuts where the seed coat is fragile, and wetting the seed is likely to damage seed viability. Such inoculants are usually used at the rate of 20 mL of liquid culture or 1 g of granular inoculant per meter of row. Both can be added mechanically via a tube attached behind the furrow opener and connected to a reservoir of the inoculant, or they can be dribbled into the seed row by hand. In each case, the inoculant should lie in the furrow with the seed or 1 to 2 cm below it.

12–7 FIELD EXPERIMENTATION INVOLVING INOCULATION

12–7.1 Principles

Field studies with the rhizobia may consider whether inoculation of a particular crop species is necessary, screen promising strains or cultivars for symbiotic response, or consider the effect of environmental, management, or soil factors on some phase of the symbiosis. Because of such differences, it is not possible to provide detailed methodologies in this chapter. Many of the principles involved, however, are similar and should be discussed.

Control treatments should be included in studies seeking response to inoculation. An uninoculated treatment will show whether the soil contains indigenous rhizobia and will give an estimate of their effectiveness. It will also provide nodule samples that can be used to characterize indigenous rhizobia by serology or other means. Nitrogen fertilized treatments will show that other constraints to production have been controlled and that response to N addition occurs. This treatment may not be practical for legume-grass mixtures, where use of fertilizer N can change the relative growth of grass and legume.

Several major and minor elements are needed for N_2-fixation (Robson, 1978; O'Hara et al., 1988). Soil at the study site should be sampled, and its chemical analysis undertaken. Where fertility is not part of the study, deficiencies in any of the needed elements should be corrected by appropriate fertilization.

Plot size will vary with the plant species being studied and the planned duration and objectives of the experiment, but will commonly provide one or more small areas for subsampling during the growing season, as well as a larger area for yield determinations. For row crops, it is normal to leave at least two rows on either side of the area to be harvested. A further 50 cm to 1 m should be left at the ends of the plot and between areas that will be sampled during the growing season. This is both to limit border effects and reduce the effects of harvest damage on subsequently sampled plants.

The area harvested for yield should be 2 to 4 m^2. A randomized complete block with four to five replications is desirable. This is because several parameters measured in these studies can show high variation. One-meter strips must be left between the plots to ensure that inoculants from one plot do not contaminate others.

The parameters measured in inoculation studies will vary with the experiment and may include plant growth and N accumulation measured at several growth stages along with nodule mass, nodule distribution on the root system, and nodule occupancy. Biological yield and N content at physiological maturity is an important parameter in crop legumes where both cultivar and strain can influence N harvest index; yield and N content at each cutting are essential measures in pasture legumes. For nodule determinations, 10 plants per replicate should be sampled to limit plant-to-plant variation. Exercise care in digging the samples and cleaning soil from the roots. Nodules detach easily from the root and may be lost. In heavy-textured soils, placing the dug root/soil ball into a bucket of water and letting soak will aid in nodule recovery. Divide the root system into four to five segments (upper and lower tap root, two to three lateral root segments), and if acetylene reduction assays (see chapter 43 by Weaver and Danso in this book) are included do them by root section rather than for the plant as a whole. Following such assays, pick the nodules from each root section, weigh them, and preserve them for strain identification by using serology (see chapter 28 by Wright in this book) or other techniques. Visual ratings of nodulation may be used in breeding nurseries where several samples precludes physical removal and weighing of all plants (Rosas & Bliss, 1986), but is not recommended for smaller studies. Nodule number is more practical and usually of less value than nodule mass. It is an important parameter where environmental conditions (soil acidity, high soil temperature, and fungicide treatment of the seeds) could have affected strain survival in the inoculant and soil. McDermott and Graham (1989) found inoculant strains competitive for nodulation sites in the crown region where the inoculant was placed but of limited mobility and so unable to contribute significantly to lateral root nodulation. Crown nodules produced by the inoculant strain fixed essentially 100% of the N accumulated early in the growth cycle, but most N_2 fixation post-flowering was due to nodules on the lateral roots.

12–8 GROWTH-POUCH INFECTION ASSAYS

12–8.1 Principles

Bhuvaneswari et al. (1980, 1981) used a modification of the growth pouch procedure for the determination of MPN counts (section 12–5) to study infection. In this procedure, the position of the root tip of growth-pouch grown plants is marked at the time of inoculation, and nodulation events are scored relative to the root-tip mark (RTM).

Infection in most legumes occurs predominantly in the region in which root hairs are just beginning to differentiate. This is located some 1 to 2 cm behind the root tip, with the developing root hairs only receptive to infection for a period of 4 to 6 h. If conditions for infection are ideal and host and rhizobia are compatible, nodules will be produced in the region above the RTM made at the time of inoculation. If inoculant density is low, if the nutrient solution is too acid, or if the host and strain are limited in compatibility, the uppermost nodules will usually occur below the RTM position.

12–8.2 Materials

With slight modifications, the materials used in this procedure are the same as detailed for the MPN procedure (section 12–5.2).

12–8.3 Procedure

Moisten the wick material of the growth pouches with distilled water. Lay pouches containing wicks flat and use a roller to remove air bubbles or bulges. Group the pouches in sets of 25, wrap them in aluminum foil and sterilize normally. Avoid wrinkles by keeping the pouches flat during autoclaving.

Disinfect seeds as described previously (section 12–5.3), then pregerminate them on water agar until the radicle is a few millimeters long. For large-seeded legumes, the radicle is strong and may be more than 1-cm long when planted but for small-seeded legumes the radicle is fragile and should not be more than 2- to 3-mm long when planted. Place the growth pouches in support racks, then aseptically plant two seedlings per pouch, making sure that the root system of each protrudes through the folded section of the paper wick and is in contact with the wick below the fold. Keep the wicks moist by adding 2 to 5 mL of nutrient solution as needed.

Prepare the inoculant as previously described, adding 1 mL of a suspension containing 100 000 cells per mL directly to each root in the pouch. Mark the root tip position at the time of inoculation on the outside of the growth pack with a fine tip marker.

Grow the plants in a growth chamber or greenhouse until well nodulated, then determine for 20 to 25 replicate pouches per treatment the distance of the uppermost nodule from the RTM made at the time of inoculation. Record also the number of nodules produced above the RTM and the percentage of plants with nodules above the RTM.

12–8.3 Comments

Recent studies have shown that speed in nodulation as determined by the RTM methodology correlates closely with the competitiveness of strains determined in pot studies (Stephens & Cooper, 1988; McDermott &

Graham, 1990; Oliveira & Graham, 1990). McDermott et al. (1991) have even found that populations of *B. japonicum* serogroup 123 from Waukegan (fine-silty over sandy or sandy-skeletal, mixed, mesic Typic Hapludolls) and Webster (fine-loamy, mixed, mesic Typic Haplaquolls) soils showed the same competitive performance in growth pouches as in the field, while Chaverra and Graham (1991) have noted cultivar variation in several traits affecting speed of nodulation.

The procedure outlined above has a high coefficient of variation and is the reason for 20 to 25 replications being needed. Further, nodulation may not occur if the developing root system is not in contact with the paper wick. It is for this reason that the root system is kept moist, rather than flooded, and the growth pouches are lain flat until planted.

REFERENCES

Alexander, M. 1965. Most-probable number method for microbial populations. p. 1467–1472. *In* C.A. Black et al. (ed.) Methods of soil analysis. Part 2. Agron. Monogr. 9. ASA, Madison, WI.

Baldwin, I.L., and E.B. Fred. 1929. Nomenclature of the root-nodule bacteria of the Leguminosae. J. Bacteriol. 17:141–150.

Balatti, A.P. 1982. Culturing *Rhizobium* in large scale fermentors. p. 127–132. *In* P.H. Graham and S.C. Harris (ed.) Biological nitrogen fixation technology for tropical agriculture. CIAT, Cali, Colombia.

Barber, L.E. 1979. Use of selective agents for recovery of *Rhizobium meliloti* from soil. Soil Sci. Soc. Am. J. 43:1145–1148.

Beck, D.P. 1991. Suitability of charcoal-amended mineral soil as carrier for *Rhizobium* inoculants. Soil Biol. Biochem. 23:41–44.

Bernaerts, M.J., and J. De Ley. 1963. A biochemical test for crown gall bacteria. Nature (London) 197:406–407.

Bhuvaneswari, T.V., A.A. Bhagwat, and W.D. Bauer. 1981. Transient susceptibility of root cells in four common legumes to nodulation by rhizobia. Plant Physiol. 68:1144–1149.

Bhuvaneswari, T.V., B.G. Turgeon, and W.D. Bauer. 1980. Early events in the infection of soybean (*Glycine max* (L.) Merr) by *Rhizobium japonicum*. Plant Physiol. 66:1027–1031.

Boonkerd, N., and R.W. Weaver. 1982. Cowpea rhizobia: Comparison of plant infection and plate counts. Soil Biol. Biochem. 14:305–307.

Boonkerd, N., D.F. Weber, and D.F. Bezdicek. 1978. Influence of *Rhizobium japonicum* strains and inoculation methods on soybeans grown in rhizobia-populated soil. Agron. J. 70:547–549.

Brockwell, J. 1963. Accuracy of a plant infection technique for counting populations of *Rhizobium trifolii*. Appl. Microbiol. 11:377–383.

Brockwell, J., A. Diatloff, A. Grassia, and A.C. Robinson. 1975. Use of the wild soybean (*Glycine ussuriensis* Regel and Moack) as a test plant in dilution-nodulation frequency tests for counting *Rhizobium japonicum*. Soil Biol. Biochem. 7:305–311.

Burton, J.C. 1982. Modern concepts in legume inoculation. p. 105–114. *In* P.H. Graham and S.C. Harris (ed.) Biological nitrogen fixation technology for tropical agriculture. CIAT, Cali, Colombia.

Caetano-Anolles, G., G. Favelukes, and W.D. Bauer. 1990. Optimization of surface sterilization for legume seed. Crop Sci. 30:708–712.

Chakrabarti, S., M.S. Lee, and A.H. Gibson. 1981. Diversity in the nutritional requirements of strains of various *Rhizobium* species. Soil Biol. Biochem. 13:349–354.

Chaverra, M.H., and P.H. Graham. 1991. Cultivar variation in traits affecting the early nodulation of common bean. Crop Sci. 32:1432–1436.

Chen, W.X., G.S. Li, Y.L. Qi, E.T. Wang, H.L. Yuan, and J.L. Li. 1991. *Rhizobium huakuii* sp. nov. isolated from the roots of *Astragalus sinicus*. Int. J. Syst. Bacteriol. 41:275–280.

Chen, W.X., G.H. Yan, and J.L. Li. 1988. Numerical taxonomic study of fast-growing soybean rhizobia and a proposal that *Rhizobium fredii* be assigned to *Sinorhizobium* gen. nov. Int. J. Syst. Bacteriol. 38:392–397.

Cole, M.A., and G.H. Elkan. 1973. Transmissible resistance to penicillin G, neomycin and chloramphenicol in *Rhizobium japonicum*. Antimicrob. Agents Chemother. 4:248–253.

Corby, H.D.L. 1976. A method of making a pure culture, peat-type inoculant, using a substitute for peat. p. 169–173. *In* P.S. Nutman (ed.) Symbiotic nitrogen fixation in plants. Cambridge Univ. Press, London.

Cubo, M.T., A.M. Buendia-Claveria, J.E. Beringer, and J.E. Ruiz-Sainz. 1988. Melanin production by *Rhizobium* strains. Appl. Environ. Microbiol. 54:1812–1817.

Curley, R.L., and J.C. Burton. 1975. Compatibility of *Rhizobium japonicum* with chemical seed protectants. Agron. J. 67:807–808.

Dangeard, P.A. 1926. Reserches sur les tubercles radicaux des legumineuses. Le Botaniste (Paris) 16:1–270.

Date, R.A., and J. Halliday. 1979. Selecting *Rhizobium* for the acid, infertile soils of the tropics. Nature (London) 277:62–64.

Date, R.A., and J. Halliday. 1987. Collection, isolation, cultivation and maintenance of rhizobia. p. 1–27. *In* G.H. Elkan (ed.) Symbiotic nitrogen fixation technology. Marcel Dekker, New York.

Deschodt, C.C., and B.W. Strijdom. 1976. Suitability of a coal bentonite base as carrier of rhizobia in inoculants. Phytophylactica 8:1–6.

Dreyfus, B., J.L. Garcia, and M. Gillis. 1988. Characterization of *Azorhizobium caulinodans* gen. nov., sp. nov., a stem-nodulating nitrogen-fixing bacterium isolated from *Sesbania rostrata*. Int. J. Syst. Bacteriol. 38:89–98.

Dye, M. 1980. Functions and maintenance of a *Rhizobium* collection. p. 435–471. *In* N.S. Subba Rao (ed.) Recent advances in biological nitrogen fixation. Edward Arnold, London.

Eaglesham, A.R.J., J.M. Ellis, W.R. Evans, D.E. Fleischman, M. Hungria, and R.W.F. Hardy. 1990. The first photosynthetic N_2-fixing *Rhizobium*. Characteristic. p. 805–811. *In* P.M. Gresshoff et al. (ed.) Nitrogen fixation: Achievements and objectives. Chapman and Hall, New York.

Elkan, G.H., and I. Kwik. 1968. Nitrogen, energy and vitamin nutrition of *Rhizobium japonicum*. J. Appl. Bacteriol. 31:399–404.

Evans, H.J., B. Koch, and R. Klucas. 1972. Preparation of nitrogenase from nodules and separation into components. Methods Enzymol. 24:470–476.

Fred, E.B., I.L. Baldwin, and E. McCoy. 1932. Root nodule bacteria and leguminous plants. Univ. of Wisconsin Press, Madison.

Fred. E.B., and S.A. Waksman. 1928. Laboratory manual of general microbiology, with special reference to the microorganisms of the soil. McGraw-Hill, New York.

Gault, R.R., and J. Brockwell. 1980. Effects of the incorporation of molybdenum compounds in the seed pellet on inoculant survival, seedling nodulation and plant growth of lucerne and subterraneum clover. Aust. J. Exp. Agric. Anim. Husb. 20:63–71.

Gibson, A.H., D.H. Demezas, R.R. Gault, T.V. Bhuvaneswari, and J. Brockwell. 1990. Genetic stability in rhizobia in the field. Plant Soil 129:37–44.

Gomez de Souza, L.A., F.M.M. Magalhaes, and L.A. de Oliveira. 1984. Avaliacao do crescimento de *Rhizobium* de leguminosas florestais tropicais em diferentes meios de cultura. Pesq. Agropec. Bras. 19s/n:165–168.

Graham, P.H. 1963. Vitamin requirements of root-nodule bacteria. J. Gen. Microbiol. 30:215–218.

Graham, P.H., V.M. Morales, and R. Cavallo. 1974. Materiales excipientes y adhesivos de posible uso en inoculacion de leguminosas en Colombia. Turrialba 24:47–50.

Graham, P.H., V.M. Morales, and O. Zambrano. 1974. Seed pelleting of a legume to apply molybdenum. Turrialba 24:335–336.

Graham, P.H., G. Ocampo, L.D. Ruiz, and A. Duque. 1980. Survival of *Rhizobium phaseoli* in contact with chemical seed protectants. Agron. J. 72:625–627.

Graham, P.H., M.J. Sadowsky, H.H. Keyser, Y.M. Barnet, R.S. Bradley, J.E. Cooper, J. DeLey, B.D.W. Jarvis, E.B. Roslycky, B.W. Strijdom, and J.P.W. Young. 1991. Proposed minimal standards for the description of new genera and species of root and stem-nodulating bacteria. Int. J. Syst. Bacteriol. 41:582–587.

Graham, P.H., S.E. Viteri, F. Mackie, A.T. Vargas, and A. Palacios. 1982. Variation in acid soil tolerance among strains of *Rhizobium phaseoli*. Field Crops Res. 5:121–128.

Graham-Weiss, L., M.L. Bennett, and A.S. Paau. 1987. Production of bacterial inoculants by direct fermentation on nutrient supplemented vermiculite. Appl. Environ. Microbiol. 53:2138–2140.

Guerinot, M.L. 1991. Iron uptake and metabolism in the rhizobia/legume symbiosis. Plant Soil 130:199–209.

Habte, M. 1985. Selective medium for recovering specific populations of rhizobia introduced into tropical soils. Appl. Environ. Microbiol. 50:1553–1555.

Halliday, J., and P.H. Graham. 1978. Comparative studies of peat and coal as inoculant carriers for *Rhizobium*. Turrialba 28:348–349.

Handelsman, J., and W.J. Brill. 1985. *Erwinia herbicola* isolates from alfalfa plants may play a role in nodulation of alfalfa by *Rhizobium meliloti*. Appl. Environ. Microbiol. 49:818–821.

Hanus, F.J., R.J. Maier, and H.J. Evans. 1979. Autotrophic growth of H_2 uptake positive strains of *Rhizobium japonicum* in an atmosphere supplied with hydrogen gas. Proc. Natl. Acad. Sci. USA 76:1788–1792.

Herridge, D.F., and R.J. Roughley. 1975. Variation in colony characteristics and symbiotic effectiveness of *Rhizobium*. J. Appl. Bacteriol. 38:19–27.

Howieson, J.G. 1985. Use of an organic buffer in the selection of acid tolerant *Rhizobium meliloti*. Plant Soil 88:367–376.

Howieson, J.G., M.A. Ewing, and M.F. D'Antuono. 1988. Selection for acid tolerance in *Rhizobium meliloti*. Plant Soil 105:179–188.

Jansen van Rensburg, H., and B.W. Strijdom. 1972. Information on the mode of entry of a bacterial contaminant into nodules of some leguminous plants. Phytophylactica 4:73–78.

Jarvis, B.D.W., C.E. Pankhurst, and J.J. Patel. 1982. *Rhizobium loti,* a new species of legume root-nodule bacteria. Int. J. Syst. Bacteriol. 32:378–380.

Jarvis, B.D.W., L.J.H. Ward, and E.A. Slade. 1989. Expression by soil bacteria of nodulation genes from *Rhizobium leguminosarum*. Appl. Environ. Microbiol. 55:1426–1434.

Jordan, D.C. 1982. Transfer of *Rhizobium japonicum* Buchanan 1980 to *Bradyrhizobium* gen. nov. a genus of slow growing root-nodule bacteria from leguminous plants. Int. J. Syst. Bacteriol. 2:136–139.

Jordan, D.C. 1984. Family III. Rhizobiaceae Conn 1938. p. 234–244. *In* Bergey's manual of systematic bacteriology. Williams and Wilkins, Baltimore.

Keyser, H.H. 1987. The role of culture collections in biological nitrogen fixation. p. 413–428. *In* G.H. Elkan (ed.) Symbiotic nitrogen fixation technology. Marcel Dekker, New York.

Keyser, H.H., and D.N. Munns. 1979. Effects of calcium, manganese and aluminum on growth of rhizobia in acid media. Soil Sci. Soc. Am. J. 43:500–503.

Kremer, R.J., and H.L. Peterson. 1983. Effects of carrier and temperature on survival of *Rhizobium* spp. in legume inocula: Development of an improved type of inoculant. Appl. Environ. Microbiol. 45:1790–1794.

Kumar Rao, J.V.D.K., P.J. Dart, and M.U. Khan. 1982. Cowpea-group *Rhizobium* in soils of the semi-arid tropics. p. 291–295. *In* P.H. Graham and S.C. Harris (ed.) Biological nitrogen fixation technology for tropical agriculture. CIAT, Cali, Colombia.

Kuykendall, L.D., and G.H. Elkan. 1976. *Rhizobium japonicum* derivatives differing in nitrogen-fixing efficiency and carbohydrate utilization. Appl. Environ. Microbiol. 32:511–519.

Kuykendall, L.D., B. Saxena, T.E. Devine, and S.E. Udell. 1992. Genetic diversity in "Bradyrhizobium japonicum" Jordan 1982 and a proposal for "Bradyrhizobium elkanii" sp.-nov. Can. J. Microbiol. 38:501–505.

Ladha, J.K., R.P. Pareek, R. So, and M. Becker. 1990. Stem nodule symbiosis and its unusual properties. p. 633–640. *In* P.M. Gresshoff et al. (ed.) Nitrogen fixation: Achievements and objectives. Chapman and Hall, New York.

Lindemann, W.C., E.L. Schmidt, and G.E. Ham. 1974. Evidence for double infection within soybean nodules. Soil Sci. 118:274–279.

Lindstrom, K. 1989. *Rhizobium galegae,* a new species of legume root-nodule bacteria. Int. J. Syst. Bacteriol. 39:365–367.

Loneragan, J.F., D. Meyer, R.G. Fawcett, and A.J. Anderson. 1955. Lime pelleted clover seeds for nodulation on acid soils. J. Aust. Inst. Agric. Sci. 21:264–265.

Martinez, E., L. Segovia, F. Martins, A.A. Franco. P. Graham, and M.A. Pardo. 1991. *Rhizobium tropici:* A novel species nodulating *Phaseolus vulgaris* L. beans and *Leucaena* spp. trees. Int. J. Syst. Bacteriol. 41:417–426.

Materon, L.A., and R.W. Weaver. 1984. Survival of *Rhizobium trifolii* on toxic and non-toxic arrowleaf clover seeds. Soil Biol. Biochem. 16:533–535.

McDermott, T.R., and P.H. Graham. 1989. *Bradyrhizobium japonicum* inoculant mobility, nodule occupancy and acetylene reduction in the soybean root system. Appl. Environ. Microbiol. 55:2493–2498.

McDermott, T.R., and P.H. Graham. 1990. Competitive ability and efficiency in nodule formation of strains of *Bradyrhizobium japonicum*. Appl. Environ. Microbiol. 56:3035–3039.

McDermott, T.R., P.H. Graham, and M.L. Ferrey. 1991. Competitiveness of indigenous populations of *Bradyrhizobium japonicum* serocluster 123 as determined using a root-tip marking procedure in growth pouches. Plant Soil 135:245–250.

Moawad, H.A., W.R. Ellis, and E.L. Schmidt. 1984. Rhizosphere response as a factor in competition among three serogroups of indigenous *Rhizobium japonicum* for nodulation of field-grown soybeans. Appl. Environ. Microbiol. 47:607–612.

Moawad, H.A., and E.L. Schmidt. 1987. Occurrence and nature of mixed infections in nodules of field-grown soybeans (*Glycine max*). Biol. Fertil. Soils 5:112–114.

Noel, K.D., F. Sanchez, L. Fernandez, J. Leemans, and M.A. Cevallos. 1984. *Rhizobium phaseoli* symbiotic mutants with transposon Tn5 insertions. J. Bacteriol. 158:148–155.

Norris, D.O. 1958. A red strain of *Rhizobium* from *Lotononis bainesii* Baker. Aust. J. Agric. Res. 9:629–632.

Norris, D.O. 1963. A porcelain bead method for storing rhizobium. Emp. J. Exp. Agric. 31:255–259.

O'Hara, G.W., N. Boonkerd, and M.J. Dilworth. 1988. Mineral constraints to nitrogen fixation. Plant Soil 108:93–110.

O'Hara, G.W., and R.M. Daniel. 1985. Rhizobial denitrification: A review. Soil Biol. Biochem. 17:1–9.

Oliveira, L.A., and P.H. Graham. 1990. Speed of nodulation and competitive ability among strains of *Rhizobium leguminosarum* by *phaseoli*. Arch. Microbiol. 153:311–315.

Pattison, A.C., and F.A. Skinner. 1974. The effects of antimicrobial substances on *Rhizobium* spp. and their use in selective media. J. Appl. Bacteriol. 37:239–250.

Pazkowski, M.W., and D.L. Berryhill. 1979. Survival of *Rhizobium phaseoli* on coal based legume inoculants. Appl. Environ. Microbiol. 38:612–615.

Philpotts, H. 1976. Filter mud as a carrier of *Rhizobium* inoculants. J. Appl. Bacteriol. 41:277–281.

Philpotts, H. 1977. Effect of inoculation method on *Rhizobium* survival and plant nodulation under adverse conditions. Aust. J. Exp. Agric. Anim. Husb. 17:308–315.

Porter, F.E., I.S. Nelson, and E.K. Wold. 1966. Plastic pouches. Crops Soils 18:10.

Robson, A.D. 1978. Mineral nutrients limiting nitrogen fixation in legumes. p. 277–293. *In* C.S. Andrew and A.J. Kamprath (ed.) The mineral nutrition of legumes on tropical and subtropical soils. CSIRO, Melbourne.

Rosas, J.C., and F.A. Bliss. 1986. Host plant traits associated with estimates of nodulation and nitrogen fixation in common bean. Hortic. Sci. 21:287–289.

Roughley, R.J. 1968. Some factors influencing the growth and survival of root nodule bacteria in peat culture. J. Appl. Bacteriol. 31:259–265.

Roughley, R.J. 1976. The production of high quality peat inoculants and their contribution to legume yield. p. 125–136. *In* P.S. Nutman (ed.) Symbiotic nitrogen fixation in plants. Cambridge Univ. Press, London.

Roughley, R.J., and J.M. Vincent. 1967. Growth and survival of *Rhizobium* spp. in peat culture. J. Appl. Bacteriol. 30:362–376.

Scholla, M.H., and G.H. Elkan. 1984. *Rhizobium fredii* sp. nov., a fast-growing species that effectively nodulates soybeans. Int. J. Syst. Bacteriol. 34:484–486.

Schwyn, B., and J.B. Nielands. 1987. Universal chemical assay for the detection and determination of siderophores. Anal. Biochem. 160:47–56.

Segovia, L., D. Pinero, R. Palacios, and E. Martinez-Romero. 1991. Genetic structure of a soil population of nonsymbiotic *Rhizobium leguminosarum*. Appl. Environ. Microbiol. 57:426–433.

Segovia, L., P.W. Young, and E. Martinez. 1993. Reclassification of American *Rhizobium leguminosarum* biovar *phaseoli* type 1 strains as *Rhizobium etli* sp. nov. Int. J. System. Bacteriol. 43:374–377.

Soberon-Chavez, G., and R. Najera. 1989. Isolation from soil of *Rhizobium leguminosarum* lacking symbiotic information. Can. J. Microbiol. 35:464–468.

Somasegaran, P. 1985. Inoculant production with diluted liquid cultures of *Rhizobium* spp. and autoclaved peat: Evaluation of diluents, *Rhizobium* spp., peats, sterility requirements, storage and plant effectiveness. Appl. Environ. Microbiol. 50:398–405.

Somasegaran, P., and H.J. Hoben. 1985. Methods in legume-*Rhizobium* technology. Univ. of Hawaii NifTAL, Maui.

Stephens, P.M., and J.E. Cooper. 1988. Variation in speed of infection of "no root hair zone" of white clover and nodulating competitiveness among strains of *Rhizobium trifolii*. Soil Biol. Biochem. 20:465–476.

Summerfield, R.J., P.A. Huxley, and F.R. Minchin. 1977. Plant husbandry and management techniques for growing grain legumes under simulated tropical conditions in controlled environments. Exp. Agric. 13:81–92.

Sylvester-Bradley, R., D. Mosquera, and J.E. Mendez. 1988a. Selection of rhizobia for inoculation of forage legumes in savanna and rainforest soils of tropical America. p. 225–233. *In* D.P. Beck and L.A. Materon (ed.) Nitrogen fixation by legumes in Mediterranean agriculture. Martinus Nijhoff, Dordrecht, the Netherlands.

Sylvester-Bradley, R., P. Thornton, and P. Jones. 1988b. Colony dimorphism in *Bradyrhizobium* strains. Appl. Environ. Microbiol. 54:1033–1038.

Takats, S.T. 1986. Suppression of nodulation in soybeans by superoptimal inoculation with *Bradyrhizobium japonicum*. Physiol. Plant. 66:669–673.

Takishima, Y., J. Shimura, Y. Ugawa, and H. Shugawara. 1989. Guide to world data center on microorganisms with a list of culture collections in the world. WFCC World data center on microorganisms, Riken, Saitama, Japan.

Thompson, J.A., and J.M. Vincent. 1967. Methods of detection and estimation of rhizobia in soil. Plant Soil 26:72–84.

Van Berkum, P., and H.H. Keyser. 1985. Anaerobic growth and denitrification among different serogroups of soybean rhizobia. Appl. Environ. Microbiol. 49:772–777.

Vincent, J.M. 1970. A manual for the practical study of the root-nodule bacteria. IBP Handb. 15. Blackwell Sci. Publ., Oxford, UK.

Weaver, R.W. 1979. Adsorption of rhizobia to peat. Soil Biol. Biochem. 11:545–546.

Weaver, R.W., and L.R. Frederick. 1974. Effect of inoculum rate on competitive nodulation of *Glycine max*. L. Merrill. ii. Field studies. Agron. J. 66:233–235.

Weaver, R.W., and S.F. Wright. 1987. Variability in effectiveness of rhizobia during culture and in nodules. Appl. Environ. Microbiol. 53:2972–2974.

Woomer, P., J.B. Bennett, and R. Yost. 1990. Overcoming the inflexibility of most-probable-number procedures. Agron. J. 82:349–353.

Chapter 13

Anaerobic Bacteria and Processes

HEINRICH F. KASPAR, *Cawthron Institute, Nelson, New Zealand*

JAMES M. TIEDJE, *Michigan State University, East Lansing, Michigan*

With the exception of rice paddies, most agricultural soils can generally be termed *aerobic*. But this does not mean that aerobic soils never experience anaerobiosis. Even very dry and well-structured soils may contain anaerobic microsites, and most agricultural land experiences seasonal variation of aeration state that may range from true aerobiosis to complete anaerobiosis. At any site in a soil, anaerobic conditions are created as soon as the O_2 demand exceeds the O_2 diffusion to this site. Thus, there are two main causes of soil anaerobiosis: (i) a high rate of O_2 consumption caused by a high respiration rate that generally correlates with a relatively high organic matter content, and (ii) a low rate of gas diffusion. Anaerobiosis occurs characteristically in poorly drained or water-logged soils. It also is to be expected in soils with large aggregates or poor, massive structure, which may be enhanced through compaction caused by frequent use of heavy equipment or by grazing cattle on wet soils containing large amounts of clay.

Some benefits of microbial activities due to soil anaerobiosis have been demonstrated. Microbial degradation of DDT and dechlorination of many organochlorine compounds occur more rapidly under anaerobic conditions. However, soil anaerobiosis is generally undesirable; it retards plant growth due to production of phytotoxic compounds such as fatty acids and H_2S, lowers the pH, reduces available N through denitrification, and restricts root respiration.

Bacteria that grow anaerobically can be divided into two groups: facultative and obligate anaerobes. Facultative anaerobes can grow under both aerobic and anaerobic conditions. This chapter is not directed at this large group of microorganisms, but the methods given here can be used for the study of anaerobic metabolism of these facultative organisms. Among the obligate anaerobes of special interest to the agricultural microbiologist are those that fix N_2, decompose cellulose, reduce SO_4^{2-}, and produce CH_4. But very little is known about soil anaerobes, though they may play an important role in nutrient transformations as well as in the degradation of organic matter and xenobiotics. A review on soil anaerobes and their activities has been written by Skinner (1975).

Two more recent and prominent books on anaerobes in nature are by Zehnder (1988) and the second edition of *The Prokaryotes* (Balows et al., 1992). This chapter is an updated version of chapter 46 published in the 1982 edition of *Methods of Soil Analysis*.

13–1 PRINCIPLES

The basic limitation of anaerobic metabolism is the disposal of reducing equivalents created by the oxidation of the energy-yielding substrate. The four chemical classes of electron acceptors used give rise to four major groups of obligate anaerobes: fermenters, SO_4^{2-} reducers, proton reducers and CO_2 reducers (most NO_3^- reducers are not obligate anaerobes [see chapter 14 by Tiedje in this book]).

1. *Obligate anaerobic fermenters* are heterotrophs that dispose of the excess electrons by producing reduced fermentation products, such as H_2, alcohols and fatty acids. Thus, they behave in the same way as facultative anaerobic fermenters. A test for aerobic growth allows the separation of these two groups. Most of the obligate anaerobic fermenters in soil appear to be clostridia. Most species of this genus are strict anaerobes, though some are aerotolerant. They are spore formers and have a chemo-organotrophic metabolism. Based on their hydrolytic capacities, they can be divided into saccharolytic, proteolytic, pectinolytic, and chitinolytic clostridia. Some species commonly found in soils are disease agents, namely, *Clostridium tetani* (tetanus), *C. perfringens* (gangrene), and *C. botulinum* (botulism, intoxication rather than infectious agent). Some of the saccharolytic clostridia are able to fix N_2 (e.g., *C. pasteurianum*), and the dissimilatory reduction of NO_3^- to NH_4^+ can be carried out by many soil clostridia.

2. *Sulfate reducers* dispose of the electrons generated by the oxidation of the energy-yielding substrate by reducing S compounds, mainly SO_4^{2-}, SO_3^{2-}, $S_2O_3^{2-}$, and S^0. In recent years, much more diversity has been discovered among the sulfate-reducing bacteria (Widdel & Hansen, 1992). There are now at least 14 genera known including gram-positive and gram-negative classes. Since SO_4^{2-} reducers are responsible for the anaerobic corrosion of Fe, they are locally important in agriculture (corrosion of submerged structures containing Fe, such as drainage tiles, irrigation devices, and waste pits). Moreover, the reduction product, H_2S, is toxic to plant roots.

Some bacteria related to sulfur and sulfate-reducing bacteria also can use metals such as Fe(III) oxides and Mn(IV) oxides as terminal electron acceptors. These organisms appear to be important particularly in terrestrial iron geochemistry (Lovley, 1991).

3. *Proton reducers* oxidize fatty acids and alcohols to acetate and CO_2. The reducing equivalents released are transferred to protons, with molecular H_2 formed as the reduced product. They are chemoorganotrophs with a limited substrate range (fatty acids and alcohols) and require a low oxidation-reduction potential (Eh). Since they are very difficult to grow in media, little information is available on this group. However, proton reducers appear to be the only organisms capable of anaerobic fatty acid and alcohol decomposition in absence of inorganic electron acceptors and thus play an essential role in the complete degradation of organic matter to CO_2 and CH_4.

4. *Carbon dioxide reducers* use CO_2 as a terminal electron acceptor. The products of the CO_2 reduction are CH_4 (methanogens) or acetate (acetogens). Methane can also be produced by reductive decarboxylation of acetate. All CO_2 reducers require a low Eh for growth. They are either chemolithotrophs or chemoorganotrophs. Some strains of acetogens form spores. Acetogens and methanogens play an important role in the regulation of the pH of anaerobic environments by their production and consumption of acetate. Methanogens mediate the last and often rate-limiting step of anaerobic mineralization of organic matter.

The requirement for anaerobic conditions is the only common characteristic of all anaerobes, and yet even with this requirement, anaerobes vary in their sensitivity to oxygen. Generally, members of the group of O_2-sensitive anaerobes lack the enzymes superoxide dismutase and catalase (strict anaerobes) or only superoxide dismutase (aerotolerant anaerobes). The absence of these enzymes leads to the accumulation of the toxic compounds O_2^- and H_2O_2, respectively. A second group of anaerobes is sensitive to the high redox potential created by O_2 rather than to O_2 itself. These organisms contain essential enzymes that require a low Eh (strict anaerobes, but are not necessarily killed under aerobic conditions). Since environments—including defined growth media—are very rarely in chemical equilibrium, it is usually not possible to conclude from a measured O_2 concentration the Eh of this particular environment, and vice versa. In addition, the tolerance of many anaerobes to O_2, high Eh, or both is not known; therefore, to ensure "sufficiently anaerobic" conditions, one should remove as much O_2 as possible *and* lower the Eh by addition of reducing agents.

Generally, soils are aerobic, and anaerobic microsites are the result of a delicate equilibrium between O_2 supply and consumption. A change of the aeration conditions during sampling is likely and may lead to changes of microbial activities and population sizes. To keep these changes at a minimum, samples should be analyzed immediately after collection. If this is impossible, they are best collected and stored as intact cores in plastic bags at 4 °C. As a rule, aeration conditions should be changed as little as possible between sampling and analysis. Thus, anaerobic techniques are generally unnecessary for collection and storage of soil samples.

Fig. 13–1. Equipment for working with anaerobes. *Opposite page* →
 A—(1) Sterilizable gas filter used for filing and evacuating septum-stoppered incubation vessels; consists of Cu tubing packed with cotton, Swagelok fittings, syringe needle connector (Teflon), and disposable needle.
   (2) Anaerobic culture tube with long butyl rubber stopper and Al crimp seal (Balch & Wolfe, 1976).
   (3) Hungate culture tube that has butyl rubber septum and screw cap with hole in center to provide access to the septum.
 B—Serum bottle with long butyl rubber stopper and crimp seal. Some manufacturers produce serum bottles with a longer neck that allows a more extensive seal between stopper and bottle (shown here).
 C—Hungate gas probe apparatus for providing O_2-free gas.
   (4) Furnaces heat Cu filings to 350 °C to remove O_2 from the flowing gas.
   (5) In raised tube, the zone of reduced Cu (Cu^0) is noted by its light (orange) color.
   (6) An oxidized zone (CuO) is apparent from its dark (black) color.
   (7) The O_2-free gas is passed to a manifold that is used to distribute it to gas probes (*8*).
   (8) The gas probes consist of glass syringes filled with cotton to sterilize the gas and hypodermic needles (15.24 cm [6 inch], 8 gauge) bent to hang in tubes or flasks.
   (9) Conventional pressure cooker adapted for use as anaerobic incubator.
   (10) Inlet and outlet tapped into cover allow evacuation and refilling with O_2-free gas.

13–2 METHODS FOR REMOVAL OF OXYGEN

13–2.1 Removal of Oxygen from Gas Lines and Filter-Sterilization of Gases (Martin, 1971; Zehnder, 1976)

13–2.1.1 Materials

1. Gastight tubing, quartz glass, neoprene, Teflon, steel, Cu (refrigerator tubing) with Swagelok or Gyrolok fittings (Crawford Fittings, Niagara Falls, Canada; Hoke Inc., Cresskill, NJ).
2. Oven, thermostatically controlled, able to keep about 500 g of Cu filings at 300 to 350 °C (e.g., Fig. 13–1, C; Sargent-Welch, Skokie, IL).
3. Copper filings, obtainable as CuO wire (has to be reduced before use, e.g., VWR Scientific Inc., Columbus, Ohio).
4. Syringe needle connectors (Hamilton Co., Reno, Nev.).

13–2.1.2 Procedure

Slowly pass commercially available high-purity gas over the hot, reduced Cu. Sterilize the gas by passing it through disposable 0.22 μm filters. An alternative is to make a filter using a 10- to 20-cm long heat-resistant tube (e.g., Cu, 6.35 mm [¼ in.] o.d.) filled with cotton and provided with a Teflon syringe needle connector that allows an easy mount of sterile needles (Fig. 13–1A). The homemade filter needs to be sterilized by dry heat before use.

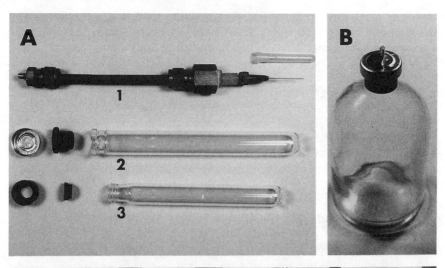

13–2.1.3 Comments

The commercially available oven can be replaced by a piece of glass-, Cu-, or steel tubing wrapped with heating tape that is controlled by a rheostat (Hungate, 1969). Upon contact with O_2, the Cu filings turn black (Fig. 13–1C), and when a significant portion has been oxidized, the column has to be regenerated to maintain efficiency. This is done by passing H_2 or a mixture of H_2 with an inert gas over the filings until they are again bright Cu colored. The reaction of H_2 with CuO is exergonic. During generation of the reduced Cu, care must be taken that the temperature does not melt the glass (for this reason quartz is recommended, though Vycor will work if the temperature is kept below 600 to 650 °C). Mixtures of H_2 and air are explosive. Before regeneration with H_2, the Cu filings must be freed of air by sparging with an inert gas. If H_2 is allowed to escape into air, it has to be diluted immediately, and sparks must be prevented by all means.

In the last decade, the expansion in work with anaerobes has led to the general commercial availability of anaerobe quality gases. Most specialists working with anaerobes prefer to pass these gases through hot Cu before use.

13–2.2 Cold Catalytic Oxygen Removal with Hydrogen

13–2.2.1 Materials

1. Container with gastight closure, able to be flushed with gas from a tank (e.g., modified pressure cooker with sealable inlet and outlet, Fig. 13–1, C).
2. Hydrogenation catalyst, usually Pd-coated asbestos (e.g., Becton, Dickinson and Co., BBL Microbiology Systems, Cockeysville, MD).
3. Nitrogen gas.
4. Gas mixture containing 7% H_2, 13% CO_2 and 80% O_2-free N_2 (section 13–3; Aranki & Freter, 1972).

13–2.2.2 Procedure

Place the cultures to be incubated anaerobically and the catalyst in the containers. After closing the jar tightly, flush with N_2 (about five times its volume). Repeat this procedure with the H_2–N_2 mixture. After the inlet and outlet are closed, the remaining O_2 is consumed by catalytic hydrogenation at normal incubation temperatures ($2H_2 + O_2 \rightarrow 2H_2O$).

13–2.2.3 Comments

This method provides good anoxic conditions in a relatively short time. The initial flushing with N_2 is done to prevent explosive mixtures of air and H_2. The system can be improved by alternate evacuation by pump or aspirator and refilling with the O_2-free gas. In this case, plates should be placed upright in the container (agar may fall off under vacuum if inverted). GasPak and Bio-Bag are convenient modifications of the cold catalyst method and are commercially available (Becton, Dickinson and Co., BBL Microbiology Systems, Cockeysville, MD; Marion Scientific Co., Kansas City, MO). They have the advantage that no flushing is necessary since the H_2 needed for O_2 consumption is chemically produced in the container. However, anaerobic conditions are obtained more slowly than in the flushed jar.

13–2.3 Other Methods

Specialists in anaerobic microbiology use the above methods, especially that in section 13–2.1, but the following classical methods are briefly mentioned because they may be the most reasonable for the occasional, simple anaerobic experiment.

Cultures of fast-growing aerobes (e.g., *Escherichia coli*) can be used to remove O_2 from closed containers (Soc. Am. Bacteriol., 1957). This method requires the least special equipment, but anaerobic conditions are established slowly. Thus, fast-growing aerobes may be able to grow in the test cultures. The method can be improved by flushing the jar at the beginning with O_2-free gas. It has been modified for use with single plates (Snieszko, 1930).

Parker (1955) used activated steel wool to remove O_2 from anaerobic culture chambers. The untreated steel wool is dipped into a detergent solution and then placed in a sealed chamber; the rusting consumes O_2. This method appears to have some advantages over the pyrogallol method in that CO_2 is not absorbed and no CO is produced.

The candle jar method also creates anaerobic conditions economically. A desiccator or other tightly closing jar is used. Add inoculated plates or open culture tubes, then add candle, light, and seal. Candle will extinguish when O_2 tension is low. This method is good for microaerophiles and those organisms that prefer high CO_2. This method is not adequate for strict anaerobes. Some CO will be produced which may inhibit some strains.

Pyrogallol has also been used to remove O_2 (Soc. Am. Bacteriol., 1957). O_2 is consumed during the alkali-catalyzed polymerization of pyrogallol (1, 2, 3-trihydroxybenzene). The alkali also removes CO_2 that can limit growth of many anaerobes. Some CO is also produced. The method is simple, effective, and appropriate for the occasional need for anaerobic conditions. Pyrogallol is toxic and contact with workers as well as cultures must be avoided.

13–3 METHODS FOR REDUCTION OF MEDIA

13–3.1 Materials

1. Reducing agents: Table 13–1 summarizes standard redox potentials at pH 7 (E_o') and stock solution preparation of reducing agents commonly used in microbiology.
2. Gassing probe (section 13–2.1, Fig. 13–1, C).

Table 13–1. Standard redox potentials at pH 7 (E_o') and stock solution preparation of reducing agents.

Reducing agent	E_o' [mV]	Preparation of stock solution	Final concentration in medium
Sodium thioglycollate	< -100	1% (wt/vol) in distilled water, autoclaved and stored under O_2-free gas	0.05%
Sodium sulfide nanohydrate	-225	1.2% (wt/vol) $Na_2S \cdot 9H_2O$ in distilled water, autoclaved and stored under O_2-free gas	0.025%
Cysteine	-340	1% (wt/vol) in distilled water, autoclaved and stored under O_2-free gas	0.025%
Titanium(III) citrate	-480	5 mL (wt/vol) aqueous solution of $TiCl_3$ added to 50 mL of 2.36% (wt/vol) aqueous solution of sodium citrate dihydrate and neutralized by addition of saturated Na_2CO_3 solution, then filter-sterilized; must be used within 1 d at concentration of 30 mL/L medium	0.5–2 mM

3. Thick-walled containers that are able to be tightly stoppered (Fig. 13–1).
4. Microbiological medium.

13–3.2 Procedure

For prereduced media, combine the heat-stable ingredients of the medium in a flask, and gently boil while sparging with O_2-free gas (e.g., $N_2 +$ 3% H_2) through the gassing probe (section 13–2.1). Continue gassing while dispensing the medium into thick-walled containers; then tightly stopper and autoclave. After slow cooling, add by syringe the heat-labile ingredients and reducing agents that have been previously filter-sterilized.

13–3.3 Comments

It is important to remove as much O_2 from the medium as possible before adding the reducing agents to minimize the toxicity that can come from use of larger quantities of reducing agent and from products produced by the reaction of reductant with O_2.

To prevent precipitation, the media must be cooled before addition of the reducing agents. Cold sterilization of the reducing agents is most conveniently done by using syringes and disposable membrane filters (0.22 μm, e.g., Millipore Corp., Bedford, MA).

Besides the above-mentioned S compounds, H_2S, dithionite, and amorphous FeS have been used as reducing agents. Amorphous FeS has been shown to remove O_2 faster than cysteine or HS^- (Brock & O'Dea, 1977). The use of S compounds may lead to false positive tubes in the enumeration of SO_4^{2-} reducers (Richard Smith, 1978, personal communication).

Titanium(III) citrate provides the lowest Eh (Zehnder & Wuhrmann, 1976). If the citrate is utilized by bacteria, it may be replaced by other ligands, such as tartrate, rhodanide, or oxalate. The Ti(III) citrate complex has a blue-violet color that is lost upon oxidation. Thus, it may also serve as a redox indicator. It is possible to use reducing agents in combinations, e.g., Na_2S, cysteine, and Ti(III) citrate (Zehnder & Wuhrmann, 1977).

Convenient thick-walled containers with secured rubber stoppers are the Hungate tube, the anaerobe culture tube, (both shown in Fig. 13–1A and obtainable from Bellco Glass, Inc., Vineland, NJ) and the serum bottle shown in Fig. 13–1B. Long, black butyl rubber stoppers provide the best seal. The common red rubber, thin-walled, sleeve-type serum bottle stoppers and silicone septa are inadequate for anaerobic work.

13–4 REDOX INDICATORS

13–4.1 Methylene Blue (Skinner, 1971)

13–4.1.1 Materials

1. Solution A: Combine 3 mL of 0.5% aqueous methylene blue solution with 100 mL of distilled water (stored in dark).
2. Solution B: Combine 0.5 g of glucose, one small crystal of thymol, and 100 mL of distilled water.
3. Solution C: Combine 1.06 g of anhydrous sodium carbonate (Na_2CO_3), 0.84 g of sodium bicarbonate ($NaHCO_3$), and 100 mL of distilled water.

13–4.1.2 Procedure

Add solution C to solution B until pH 10 is reached. (This mixture is stored cold and dark.) Mix the product with solution A to give a light blue color. Boil portions of this methylene blue indicator fluid in small beakers or test tubes until it is colorless, and then place in the containers to be monitored for anaerobiosis. Close the containers immediately, and make the atmosphere anaerobic (section 13–2).

13–4.2 Resazurin

Filter-sterilize a 0.1% aqueous stock solution of resazurin, and add to the growth medium to give a final concentration of 0.0001 to 0.0003%. Resazurin can also be autoclaved directly in the medium.

13–4.3 Phenosafranine (Bryant, 1963)

Filter sterilize a 0.1% aqueous stock solution of phenosafranine, and add to the growth medium to give a final concentration of 0.0001 to 0.0002%.

13–4.4 Comments

As mentioned in section 13–3.3, Ti(III) citrate loses its blue color upon oxidation. Its standard redox potential at pH 7 and 25 °C (E_o') is −480 mV (Zehnder & Wuhrmann, 1976). Thus, if Ti(III) citrate is used as a reducing agent and the medium stays blue-violet, highly reducing conditions are guaranteed. Phenosafranine has a E_o' of −252 mV (Jacob, 1970) and therefore changes color under less reduced conditions. Resazurin has a E_o' of −51 mV, and its colorless state may not mean sufficiently reduced conditions for growth of all obligate anaerobes. This is even more the case with methylene blue, which has a E_o' of +11 mV (equivalent to 0.05 atm O_2).

Methylene blue, resazurin, and phenosafranine gradually change color within a range of about 120 mV around their E_o' and become completely colorless 60 mV below their E_o'. The E_o of these dyes is strongly dependent on the pH of the medium. Resazurin is the most commonly used redox indicator and is the one we recommend for most soil microbiology studies. Its properties are summarized below:

$$\text{pH} < 6.5 \qquad\qquad E_o'\text{--51 mV}$$

Resazurin	→	Resorufin (oxidized)	⇌	Resorufin (reduced)
(Blue)		(Pink)		(Colorless)

In the oxidized state, it is blue above pH 6.5 and changes irreversibly to pink below this value. When mixed with reducing agent, the medium should gradually change to colorless, indicating an Eh of about −110 mV. Media that later turn pink have become oxidized and should be discarded. After extended exposure of resazurin solutions to light, they lose their property to change from colorless to pink in the presence of O_2.

A complete list of Eh indicators is given by Jacob (1970). Several of these indicators have been shown to inhibit bacterial growth. They should be used at the lowest possible concentration that the color of the medium allows. If the desired indicator is toxic, it may still be used in parallel tubes treated in an identical manner but not inoculated. A simple and meaningful alternative to chemical Eh indicators is the coculture of a strict anaerobe (Futter & Richardson, 1971).

Stock solutions of several Eh indicators are commercially available (e.g., Becton, Dickinson and Co., BBL Microbiology Systems, Cockeysville, MD).

13–5 CULTURE METHODS

13–5.1 Conventional Anaerobic Techniques

13–5.1.1 Materials

1. Anaerobic jars, preferably transparent containers with large closures and capable of being evacuated, pressurized, and sealed gas tight. Examples are desiccators, pressure cookers (Fig. 13–1C), and GasPak containers (Becton, Dickinson and Co., BBL Microbiology Systems, Cockeysville, MD).
2. Larger vessels and continuous cultures that can be sealed and made suitably anaerobic.

13–5.1.2 Procedure

Prepare the media with the reducing agent and indicator included as a liquid or solid (agar) in the same way as for culture of aerobes. Inoculate them under aerobic conditions, and place in the jar or incubator together with an indicator of anaerobiosis (section 13–4). Seal the jar or incubator, and establish anoxic conditions (section 13–2).

13–5.1.3 Comments

Conventional anaerobic culture techniques imply that the samples to be tested for strict anaerobes are temporarily exposed to aerobic conditions. Although many bacteria in soils exist in dormant forms that should be able to survive adverse conditions (Gray, 1976), the brief exposure of active anaerobic samples to air may cause a loss of viable counts. However, if they are properly done, the use of conventional techniques is usually justified by their simplicity. One of the major disadvantages of this system is that single cultures cannot be removed and examined without exposing to air all cultures in the incubator.

These methods are suitable for some fermenters, acetogens and sulfate reducers (especially sporeformers), but not for proton reducers and most methanogens.

13–5.2 Strict Anaerobic Techniques

13–5.2.1 Materials

1. Anaerobic glove box (several commercial manufacturers) and O_2-free gases.
2. Anaerobic culture tubes (preferably with septum stopper and Al crimp seal, Fig. 13–1A), roll tube spinner, streaker (Bellco Glass Inc., Vineland, NJ), continuous pipetting outfit (e.g., Becton,

Dickinson and Co., BBL Microbiology Systems, Cockeysville, MD), and O_2-free gases.

13–5.2.2 Procedures

1. Glove box (Aranki et al., 1969; Aranki & Freter, 1972): Put the soil samples into an anaerobic glove box containing an atmosphere of (usually) 7% H_2, 13% CO_2 and 80% N_2. Prepare the suspensions and dilutions of the samples with prereduced media (section 13–3.2) in the glove box. Inoculate plates by either streaking on cold, prereduced media or seeding warm agar and pouring it into dishes in analogy to conventional aerobic methods. Inoculate liquid media, e.g., most probable number (MPN) tubes (see chapter 5 in this book), and cover with foam or cotton plugs.
2. Anaerobe culture system (e.g., Holdeman & Moore, 1972a): Prepare prereduced media as described in section 13–3.2. By means of a continuous pipetting device, transfer portions of the medium to culture tubes while continuously flushing them with an O_2-free gas. Quickly close the tubes, and inoculate by syringe through the septum stopper. In the case of solid media, evenly disperse the cooling agar along the wall of the tubes by means of a spinner (roll tubes). If the agar is solidified before inoculation, surface cultures can be obtained by inoculating the agar surface with the streaker under a continuous anaerobic gas flow.

13–5.2.3 Comments

Strict anaerobic techniques allow the examination of samples without exposing them to aerobic conditions at any time. The glove box provides the most convenient handling of strict anaerobes. Media can be prepared in a conventional way, and standard techniques of bacteriology can be applied. However, the glove box method requires moderately expensive equipment and a significant amount of laboratory space. Nonetheless, the glove box has gained widespread acceptance by anaerobic microbiologists and is recommended as a basic method if extensive anaerobic work is planned.

The glove box method works on the principle that any O_2 that enters the chamber is removed when the atmosphere is circulated over the Pd catalyst by reaction with the H_2 to form water. The water vapor is removed by silica gel.

The recommended gas mixture is changed from the originally published composition of 10:90 H_2/N_2 because the new version provides $H_2 + CO_2$ for methanogens and acetogens and the CO_2 acts to reduce the chance of explosion due to H_2. Since open flames are not possible in the glove box, sterilization can be done by a resistance coil. Disposable plastic loops are convenient for culture transfer and avoid dangers of a hot resistance coil.

It is helpful to protect necks of containers by caps. It is advisable to store media and supplies in the glove box for 1 to 2 d before use. The Pd catalyst (section 13–2.2) should be regenerated once a week (160 °C for 2 h). Since it is irreversibly poisoned by H_2S, cultures producing this gas should be grown in closed bags. Also, a tray of 8- to 10-mesh activated charcoal placed under the catalyst has been shown to be effective in removing H_2S (Balch & Wolfe, 1976). Items are entered and removed through a lock that is evacuated and filled with the chamber gas. Due to the vacuum in the lock, hot media may boil out of containers unless they are tightly sealed. If thin-walled or loosely sealed containers are used, the media should be allowed to cool before transfer into the glove box. Flexible glove boxes made of plastic are easier to use and do not allow entry of significantly more O_2 than do rigid boxes. Plates to be incubated in the glove box should be placed in plastic bags to reduce desiccation.

Gas permeable materials such as plastics (e.g., syringes, pipette tips, and petri dishes) can be put in the airlock, degassed, and held overnight to allow latent O_2 to diffuse out. Recycle gases in the morning or after several hours. This procedure is also appropriate for liquids and agar plates. This procedure protects more oxygen sensitive anaerobes.

Anaerobic glove boxes are also useful for other purposes than growing cultures, e.g., incubating soil anaerobically without flooding or making manipulations or measurements in the absence of O_2. Small instruments can be placed in the glove box.

A drawback of the glove box is the fact that all cultures are exposed to the same atmosphere that must contain H_2 (e.g., proton reducers cannot be grown in an atmosphere that contains H_2, whereas methanogens and acetogens require high partial pressures of this gas for good growth). Very O_2-sensitive anaerobes (e.g., methanogens) will not grow in open media in the glove box due to the trace quantities of O_2 that diffuse through the plastic into the chamber.

The anaerobe culture system allows the cultivation of anaerobes under a variety of atmospheres, but it requires more skill than the glove box. It is the procedure recommended for the most O_2-sensitive anaerobes. Gas mixtures of 80:20 H_2/CO_2 are recommended for methanogens and 66:33 H_2/CO_2 are recommended for acetogens.

It was originally developed by Hungate (1950) and since then modified by several workers (Hungate, 1969; Bryant, 1972; Holdeman & Moore, 1972b; Macy et al., 1972, Zehnder, 1976; Balch et al., 1979; Ogg et al., 1979). For further details, refer to this literature.

13–6 ENUMERATION METHODS

13–6.1 General Comments

Since we know of no readily visible anaerobe-specific characteristic, there is no method for a direct microscopic count of anaerobic bacteria.

The great variety of physiological types among anaerobes discussed in section 13–1 makes it clear that no single method for the enumeration of anaerobes is available. Each physiological group has to be enumerated by a method based on its unique characteristics. In the following section, we provide a selection of the major methods that allow for the enumeration of the most important physiological groups of anaerobes in soils. The list does not contain all methods that have been used successfully, and more methods (especially for more specific, smaller groups such as single genera or species) are being developed. Enumerations can be made on solid media (plates, roll tubes) or by the MPN method (see chapter 5 by Woomer in this book).

It should be pointed out that numbers of bacteria give little information about their ecological importance. This is especially important with regard to soil bacteria, which can be in any physiological state between virtually complete inactivity (spores) and fast growth. To elucidate the ecological importance of bacteria, one must directly measure their *activities* under conditions as close as possible to the natural ones.

13–6.2 Heterotrophs (Molongoski & Klug, 1976)

13–6.2.1 Materials

1. Dilution fluid (grams per liter of distilled water): Combine 6.5 g of potassium monohydrogen phosphate (K_2HPO_4), 3.5 g of potassium dihydrogen phosphate (KH_2PO_4), 2.5 g of sodium carbonate (Na_2CO_3), 0.5 g of cysteine·$HCl·H_2O$, and 0.0002% resazurin. Make up to 1 L with distilled water. Adjust to pH 6.9, and autoclave.

2. Salt solution (grams per liter of distilled water): Combine 0.56 g of potassium monohydrogen phosphate (K_2HPO_4), 0.33 g of potassium dihydrogen phosphate (KH_2PO_4), 0.22 g of sodium carbonate (Na_2CO_3), 0.11 g of magnesium sulfate heptahydrate ($MgSO_4·7H_2O$), 0.22 g of ammonium sulfate [$(NH_4)_2SO_4$], 0.53 g of sodium chloride (NaCl), and 0.005 g of ferric chloride ($FeCl_3$) in distilled water to 1 L.

3. Growth medium, 1 L: Add 10 g of peptone, 10 g of yeast extract, 10 g of glucose and 15 g of agar to 900 mL of salt solution. Autoclave. Dissolve 0.6 g of calcium chloride dihydrate ($CaCl_2·2H_2O$), 0.5 g of cysteine·$HCl·H_2O$, and 1 mL of resazurin stock solution (section 13–4.2) in 50 mL of distilled water. Filter-sterilize, and add to medium. Pour plates in glove box.

13–6.2.2 Procedure

Put the soil samples into the glove box, and dilute by the serial dilution procedure using the above anaerobic dilution fluid. Prepare dilutions of 1:10, and inoculate plates with 0.1 mL of these dilutions (two plates per dilution). Incubate the inoculated plates in plastic bags in the glove box, or if necessary, remove and keep in anaerobic jars. After 14 d of incubation, examine the plates, and determine the total count per gram of dry soil.

13–6.2.3 Comments

This method includes some facultative anaerobes. To obtain the number of obligate anaerobes, one must check the colonies grown on the plates that were used for the enumeration for aerobic growth. This can be done by replica plating or streaking the colonies on the same medium and observing growth after aerobic incubation. The number of obligate anaerobes is obtained by subtraction of the aerobically growing units from the anaerobically growing units. The method may be adapted to conventional anaerobic techniques; however, one would have to accept the disadvantage pointed out in section 13–5.1.3.

13–6.3 Clostridia (Gibbs & Freame, 1965)

13–6.3.1 Materials

1. Dilution fluid, 0.1% peptone in water, autoclaved.
2. Starch solution: Dissolve 1 g of soluble starch in 200 mL of water, first making a cold slurry in a small part of the water, boiling the rest, and then stirring it into the paste.
3. Sodium sulfite (Na_2SO_3) solution, 4%.
4. Ferric citrate pentahydrate ($FeC_6H_5O_7 \cdot 5H_2O$) solution, 7%: To dissolve the ferric citrate pentahydrate, heat the solution for about 5 min.
5. Sodium hydroxide (NaOH), 10 N.
6. Growth medium (1 L): Add 10 g of peptone, 10 g of sodium acetate (NaOAc), and 5 g of yeast extract to 800 mL of water. Add this to the starch solution, and steam the mixture for 30 min. Then dissolve 1 g of glucose and 0.5 g of L-cysteine in it. Adjust the pH to 7.1 to 7.2 with 10 N sodium hydroxide (NaOH). Filter the medium through paper, fill test tubes, and then autoclave. After fast cooling, place the autoclaved tubes in the anaerobic glove box. Mix equal volumes of Na_2SO_3 and ferric citrate pentahydrate solutions. Filter-sterilize the mixture, and add 0.2 mL/10 mL of medium to the test tubes.

13–6.3.2 Procedure

Put the soil samples into the anaerobic glove box, and suspend in dilution fluid in analogy to the method for aerobes (see chapter 8 in this book). Prepare 1:10 dilutions, and inoculate tubes according to the MPN method (see chapter 5 in this book). Score the tubes turning black after incubation positive for a presumptive total clostridial count. Heat the tubes at 78 °C for 30 min, and subculture in fresh medium. Count the tubes that are still turning black after this pasteurization to determine the total clostridial count.

13–6.3.3 Comments

This method originally was developed for the enumeration of clostridia in foods, and Skinner (1971) reports its applicability for enumerating soil clostridia. By addition of 1.5% agar to the medium, the method can be modified for colony counting (deep agar or plates). However, blackening often spreads throughout the whole tube or plate, so that single colonies cannot be counted. A count of the clostridial spores can be obtained if the soil suspension is heated at 78 °C for 30 min before the dilutions.

A reservation about the method is that false positive tubes may be scored if sulfide is released from cysteine degradation. An alternative but definitive method to confirm the presence of clostridia is to microscopically examine samples from the highest positive dilutions for the presence of the characteristic clostridial spore (Caskey & Tiedje, 1979). With the reservations already made under section 13–5.1.3, a count of clostridia may be made with conventional anaerobic methods.

Readers are referred to chapter 11 by Knowles and Barraquio in this book for methods to enumerate N_2-fixing clostridia and the previous version of this chapter (chapter 46, *Methods of Soil Analysis,* Part 2, 1982) for cellulose-decomposing clostridia.

13–6.4 Sulfate Reducers (Pankhurst, 1971; also see Postgate, 1984)

13–6.4.1 Materials

1. Dilution fluid, 0.2% sodium chloride (NaCl), autoclaved.
2. Salt solution: Dissolve 0.5 g of potassium monohydrogen phosphate (K_2HPO_4), 1.0 g of ammonium chloride (NH_4Cl), 1.0 g of calcium sulfate ($CaSO_4$), 2.0 g of magnesium sulfate heptahydrate ($MgSO_4 \cdot 7H_2O$), 5.0 g of sodium lactate ($NaC_3H_5O_3$) (70% wt/wt), and 10.0 g of agar in 930 mL of tap water. Adjust to pH 8.1.
3. Ferrous ammonium sulfate [$FeSO_4(NH_4)_2SO_4 \cdot 6H_2O$] solution: Steam 1% (wt/vol) $FeSO_4(NH_4)_2SO_4 \cdot 6H_2O$ solution for 1 h on three successive days.
4. Yeast extract solution, 10% (wt/vol), autoclaved.
5. Sodium thioglycollate solution, 10% (wt/vol).
6. MPN tubes.
7. Growth medium, 1 L: Autoclave the salt solution, cool, and put into anaerobic glove box. Filter-sterilize 50 mL of ferrous ammonium sulfate [$FeSO_4(NH_4)_2SO_4 \cdot 6H_2O$] solution, 10 mL of yeast extract solution, and 10 mL of thioglycollate solution, and add to salt solution. Adjust final pH to 7.2 to 7.6. Transfer 10-mL portions of this medium to sterilized MPN tubes.

13–6.4.2 Procedure

Put soil samples into the glove box, and suspend in dilution fluid. Prepare dilutions (1:10), and inoculate tubes according to the MPN method (see chapter 5 in this book). After an incubation period of at least 4 wk (at or slightly above room temperature), count those that have turned black, and determine the number of SO_4^{2-} reducers according to the MPN tables.

13–6.4.3 Comments

The method can be modified for solid media. Although conventional anaerobic techniques (section 13–5.1) have been successfully applied, strict anaerobic techniques are recommended. Both thioglycollate and cysteine reducing agents can lead to false positives due to their decomposition yielding sulfide, but this is a less common problem with thioglycollate.

While the above more classical method has been demonstrated for sulfate reducers in soil, the improved methods of Widdel and Hansen (1992) should be considered for the fastidious autotrophic sulfate reducers which may also reside in soil.

13–6.5 Carbon Dioxide Reducers (Braun et al., 1979)

13–6.5.1 Materials

1. Cysteine-sulfide reducing agent: Dissolve 15 mL of 1 M sodium hydroxide (NaOH), 2.5 g of cysteine·HCl·H$_2$O, and 2.5 g of sodium sulfide nanohydrate (Na$_2$S·9H$_2$O) in 165 mL of water.
2. Resazurin solution (section 13–4.2).
3. Dilution fluid: Dissolve 1 g of ammonium chloride (NH$_4$Cl), 0.1 g of magnesium sulfate heptahydrate (MgSO$_4$·7H$_2$O), 0.4 g of potassium monohydrogen phosphate (K$_2$HPO$_4$), 0.4 g of potassium dihydrogen phosphate (KH$_2$PO$_4$), 2 mL of resazurin solution, and 40 mL of cysteine-sulfide reducing agent in 1 L of distilled water.
4. Mineral solution (Wolin et al., 1964): Dissolve 0.5 g of nitrilotriacetate in 10 mL of 0.5 M sodium hydroxide (NaOH), and bring to 500 mL with distilled water. Dissolve 6.2 g of magnesium sulfate heptahydrate (MgSO$_4$·7H$_2$O), 0.55 g of manganous sulfate tetrahydrate (MnSO$_4$·4H$_2$O), 1.0 g of sodium chloride (NaCl), 0.1 g of ferrous sulfate heptahydrate (FeSO$_4$·7H$_2$O), 0.17 g of cobaltous chloride hexahydrate (CoCl$_2$·6H$_2$O), 0.13 g of calcium chloride dihydrate (CaCl$_2$·2H$_2$O), 0.18 g of zinc sulfate heptahydrate (ZnSO$_4$·7H$_2$O), 0.05 g of cupric sulfate (CuSO$_4$), 0.018 g of aluminum potassium sulfate [AlK(SO$_4$)$_2$·12H$_2$O], 0.01 g of boric acid (H$_3$BO$_3$), and 0.011 g of sodium molybdate dihydrate (NaMoO$_4$·2H$_2$O) in it, and bring the volume to 1 L with distilled water.
5. Vitamin solution (Wolin et al., 1964): Dissolve 2 mg of biotin, 2 mg of folic acid, 10 mg of pyridoxine hydrochloride, 5 mg of thiamine hydrochloride, 5 mg of riboflavin, 5 mg of nicotinic acid, 5 mg of panthothenate, 0.01 mg of B$_{12}$, 5 mg of p-aminobenzoic acid, and 1 mg of thioctic acid in 1 L of distilled water.

6. Bromocresol green agar: Dissolve 1 g of agar and 0.01 g of bromocresol green in 100 mL of distilled water.
7. Flat bottles, approximately 100 mL, with stoppers.
8. Growth medium, 1 L: Dissolve 1 mL of resazurin solution, 1 g of ammonium chloride (NH_4Cl), 0.68 g of potassium dihydrogen phosphate (KH_2PO_4), 0.87 g of potassium monohydrogen phosphate (K_2HPO_4), 0.04 g of magnesium sulfate heptahydrate ($MgSO_4 \cdot 7H_2O$), 20 mL of vitamin solution, 20 mL of mineral solution, 2 g of yeast extract, 0.5 g of cysteine·HCl·H_2O, 0.25 g of sodium sulfide nanohydrate ($Na_2S \cdot 9H_2O$), and 10 g of sodium bicarbonate ($NaHCO_3$) in 950 mL of distilled water. Adjust to pH 7.8. Autoclave the medium under an 80% N_2 and 20% CO_2 atmosphere. Under a continuous stream of this gas mixture, pour 10-mL portions of the hot medium into flat bottles. Stopper the bottles, and lay them down to cool.

13–6.5.2 Procedure

Suspend the soil samples in dilution fluid under an N_2–CO_2 atmosphere. Prepare dilutions (1:10). Evenly streak portions (0.1 mL) of these dilutions on the agar under a continuous flow of 67% H_2 and 33% CO_2. Close the bottles and incubate at 37 °C. After several weeks, determine the population of CO_2 reducers by counting colonies. To distinguish between methanogens and acetogens, carefully pour 5 mL of bromocresol green agar over the colonies. Acid-producing colonies and their surrounding area turn yellow after about 1 h, whereas CH_4-producing colonies remain blue.

13–6.5.3 Comments

The distinction between methanogenic and acetogenic CO_2 reducers requires some skill. Generally, the acetogenic population is small compared with the methanogenic population, and the determination of the total CO_2-reducing population is usually sufficient. For these organisms, the anaerobe culture system is preferred (section 13–5.2).

13–7 SIMPLE METHOD TO CARRY OUT ANAEROBIC INCUBATIONS OF SOIL

13–7.1 Materials

1. Serum bottles or other containers that can be sealed gastight by a septum-type stopper (Balch et al., 1979; Fig. 13–1B).
2. Syringes that can be sealed for liquid and gas sampling (e.g., Mininert Valve, Precision Scientific Corp., Baton Rouge, LA).

3. Apparatus for removal of O_2 from gas lines (section 13–2.1).
4. Device for alternate evacuation and refilling with gas of serum bottles (aspirator or vacuum pump, three-way valve; Balch & Wolfe, 1976).
5. Gas chromatograph for O_2 measurement.

13–7.2 Procedure

Place soil samples and insoluble test materials in serum bottles. Seal these with septum stoppers, evacuate, and refill several times with an O_2-free gas (that will not affect the pH of the soil). After complete removal of O_2, which can be confirmed by gas chromatography, incubate the bottle. Liquid or gaseous additions or samplings may be made at any time through the septum. Before introducing syringe needles to the bottle, flush the syringe and needle dead volume with the O_2-free gas. Monitor anaerobic conditions by measuring the O_2 content of the headspace. Methane production is not necessarily an indicator for complete anaerobiosis, since it may occur in anaerobic microsites while most of the soil is still aerobic.

13–7.2 Comments

It is often important to study whether or how fast certain processes occur under anaerobic conditions, e.g., decomposition of pesticides or transformation of nutrients. If the process is claimed to be anaerobic, evidence of the adequacy of those conditions is necessary. Incubation of soil flooded with water but with a headspace that contains O_2 is sometimes termed *anaerobic* in the literature. Yet the condition actually established is a gradient from aerobic to anaerobic, which depends on O_2 consumption and diffusion and cannot be reliably defined. Moreover, the soil flooding approach changes the concentration of chemicals and their rate of diffusion. These are effects that may be undesirable in certain experiments. We recommend that more effort be taken to minimize O_2 exposure than simply flooding soil. The above method is a cheap and reliable way of establishing anaerobic conditions.

Entry into bottles by needle always seems to cause a trace of O_2 contamination, therefore, samples should be taken no more than necessary. Bottles can be filled with O_2-free gas to greater than 1 atm, which helps prevent entry of O_2.

ACKNOWLEDGMENTS

The authors thank James Champine for helpful suggestions on updating this chapter.

REFERENCES

Aranki, A., and R. Freter. 1972. Use of anaerobic glove boxes for the cultivation of strictly anaerobic bacteria. Am. J. Clin. Nutr. 25:1329–1334.

Aranki, A., S.A. Syed, E.B. Kenney, and R. Freter. 1969. Isolation of anaerobic bacteria from human gingiva and mouse cecum by means of a simplified glove box procedure. Appl. Microbiol. 17:568–576.

Balch, W.E., G.E. Fox, L.J. Magrum, C.R. Woese, and R.S. Wolfe. 1979. Methanogens: Reevaluation of a unique biological group. Microbiol. Rev. 43:260–296.

Balch, W.E., and R.S. Wolfe. 1976. New approach to the cultivation of methanogenic bacteria: 2-Mercaptoethane–sulfuric acid (HS-CoM)-dependent growth of *Methanobacterium ruminantium* in a pressurized atmosphere. Appl. Environ. Microbiol. 32:781–791.

Balows, A., H.G. Trüper, M. Dworkin, W. Harder, and K.H. Schleifer. 1992. The prokaryotes. 2nd ed. Vol. 1. Springer-Verlag, New York.

Braun, M., S. Schoberth, and G. Gottschalk. 1979. Enumeration of bacteria forming acetate from H_2 and CO_2 in anaerobic habitats. Arch. Microbiol. 120:201–204.

Brock, T.D., and K. O'Dea. 1977. Amorphous ferrous sulfide as a reducing agent for culture of anaerobes. Appl. Environ. Microbiol. 33:254–256.

Bryant, M.P. 1963. Symposium on digestion in ruminants: Identification of groups of anaerobic bacteria active in the rumen. J. Anim. Sci. 22:801–813.

Bryant, M.P. 1972. Commentary on the Hungate technique for culture of anaerobic bacteria. Am. J. Clin. Nutr. 25:1324–1328.

Caskey, W.H., and J.M. Tiedje. 1979. Evidence for clostridia as agents of dissimilatory reduction of nitrate to ammonia in soils. Soil Sci. Soc. Am. J. 43:931–936.

Futter, B.V., and G. Richardson. 1971. Anaerobic jars in the quantitative recovery of clostridia. p. 81–91. *In* D.A. Shapton and R.G. Board (ed.) Isolation of anaerobes. Academic Press, New York.

Gibbs, B.M., and B. Freame. 1965. Methods for the recovery of clostridia from foods. J. App.. Bacteriol. 28:95–111.

Gray, T.R.G. 1976. Survival of vegetative microbes in soil. p. 327–364. *In* T.R.G. Gray and J.R. Postage (ed.) Survival of vegetative microbes. 26th Symp. Soc. Gen. Microbiol. Cambridge Univ. Press, New York.

Holdeman, L.V., and W.E.C. Moore (ed.). 1972a. Anaerobe laboratory manual. Virginia Polytechnic Inst. and State Univ., Blacksburg.

Holdeman, L.V., and W.E.C. Moore. 1972b. Roll-tube techniques for anaerobic bacteria. Am. J. Clin. Nutr. 25:1314–1317.

Hungate, R.E. 1950. The anaerobic mesophilic cellulolytic bacteria. Bacteriol. Rev. 14:1.

Hungate, R.E. 1969. A roll tube method for cultivation of strict anaerobes. p. 117–132. *In* J.R. Norris and D.W. Ribbons (ed.) Methods in microbiology. Vol. 3B. Academic Press, New York.

Jacob, H.-E. 1970. Redox potential. p. 91–123. *In* J.R. Norris and D.W. Ribbons (ed.) Methods in microbiology. Vol. 2. Academic Press, New York.

Lovley, D.R. 1991. Dissimilatory Fe(III) and Mn(IV) reduction. Microbiol. Rev. 55:259–287.

Macy, J.M., J.E. Snellen, and R.E. Hungate. 1972. Use of syringe methods for anaerobiosis. Am. J. Clin. Nutr. 25:1318–1323.

Martin. W.J. 1972. Practical method for isolation of anaerobic bacteria in the clinical laboratory. Appl. Microbiol. 22:1168–1171.

Molongoski, J.J., and M.J. Klug. 1976. Characterization of anaerobic heterotrophic bacteria isolated from freshwater lake sediments. Appl. Environ. Microbiol. 31:83–90.

Mountfort, D.O., and H.F. Kaspar. 1986. Palladium-mediated hydrogenation of unsaturated hydrocarbons with hydrogen gas released during anaerobic cellulose degradation. Appl. Environ. Microbiol. 52:744–750.

Ogg, J.E., S.Y. Lee, and B.J. Ogg. 1979. A modified tube method for the cultivation and enumeration of anaerobic bacteria. Can. J. Microbiol. 25:987.

Pankhurst, E.S. 1971. The isolation and enumeration of sulfate-reducing bacteria. p. 223–240. *In* D.A. Shapton and R.G. Board (ed.) Isolation of anaerobes. Academic Press, New York.

Parker, C.A. 1955. Anaerobiosis with iron wool. Aust. J. Exp. Biol. Med. Sci. 33:33–38.

Postgate, J.R. 1984. The sulphate-reducing bacteria. 2nd ed. Cambridge Univ. Press, New York.

Skinner, F.A., 1971. The isolation of soil clostridia. p. 57–80. *In* D.A. Shapton and R.G. Board (ed.) Isolation of anaerobes. Academic Press, New York.

Skinner, F.A. 1975. Anaerobic bacteria and their activities in soil. p. 1–19. *In* N. Walker (ed.) Soil microbiology: A critical review. Halsted Press, New York.

Sniezko, S. 1930. The growth of anaerobic bacteria in petri dish cultures. Zentralbl. Bakteriol. Abt. 282:110–111.

Society of American Bacteriologists. 1957. Manual of microbiological methods. McGraw-Hill Book Co., New York.

Widdel, F., and T.A. Hansen. 1992. The dissimilatory sulfate- and sulfur-reducing bacteria. p. 583–624. *In* A. Balows et al. (ed.) The prokaryotes. 2nd ed. Springer-Verlag, New York.

Wolin, E.A., R.S. Wolfe, and M.J. Wolin. 1964. Viologen dye inhibition of methane formation by *Methanobacillus omelianskii*. J. Bacteriol. 87:993–998.

Zehnder, A.J.B. 1976. Oekologie der Methanbakterien, thesis Swiss Federal Inst. of Technol. no. 5716. Juris Druck and Verlag, Zuerich, Switzerland.

Zehnder, A.J.B. 1988. Environmental microbiology of anaerobes. John Wiley and Sons, New York.

Zehnder, A.J.B., and K. Wuhrmann. 1976. Titanium (III) citrate as a nontoxic oxidation-reduction buffering system for the culture of obligate anaerobes. Science 194:1165–1166.

Zehnder, A.J.B., and K. Wuhrmann. 1977. Physiology of a *Methanobacterium* strain AZ. Arch. Microbiol. 111:199–205.

Chapter 14

Denitrifiers

JAMES M. TIEDJE, *Michigan State University, East Lansing, Michigan*

Denitrification is one of the three most important fates of nitrate in the soil environment, the other two being assimilation by plants and leaching. Environmental scientists have used the term *denitrification* to describe any process in which nitrate (or nitrite) is converted to nitrogen gases. Microbiologists using this term usually restrict its meaning to a bacterial respiratory process in which electron transport phosphorylation is coupled to the sequential reduction of nitrogenous oxides. It is the organisms with this respiratory process that rapidly consume soil nitrate when oxygen is limiting; they are the primary subject of this chapter. Methods for measurement of denitrification in the field are covered in chapter 44 by Mosier and Klemedtsson in this book.

In the soil environment, there are at least six different processes that reduce nitrate (Table 14–1). While respiratory denitrification is often of the greatest quantitative significance to N budgets, awareness of the other processes is necessary because they may be important in particular situations or for other questions, such as their contribution to global warming. Table 14–1 summarizes the distinguishing characteristics of these processes, many of which can be used to determine whether the process is of potential importance.

14–1 NITRATE REDUCING PROCESSES

14–1.1 Assimilatory Nitrate Reduction

This process can be carried out by many bacteria but its significance in soil seems to be limited because it is rapidly inhibited by low concentrations (e.g., 0.1 ppm) of NH_4^+-N or organic N (Rice & Tiedje, 1989). In anaerobic environments, these compounds are typically in high concentration so this process would not be expected to coexist with respiratory denitrification. Dissimilatory processes are distinguished from assimilatory processes

Table 14–1. Processes that reduce nitrate and their distinguishing characteristics.

Process	Products	Energy conserved	Regulated by	Soil condition where expected
Assimilatory Assimilation	NH_4^+	No	NH_4^+, organic N	V. low NH_4^+
Dissimilatory Respiratory denitrification	$N_2 > N_2O > NO$	Yes	O_2	Anaerobic
Dissimilatory nitrate reduction to ammonium†	$NH_4^+ > > N_2O$	A few strains	O_2	Anaerobic
Nitrate respiration†	NO_2^-	Yes	O_2	Anaerobic
Nonrespiratory denitrification	N_2O	No	?	Aerobic
Chemodenitrification	$NO > > N_2, N_2O$	No		Acidic

† All known organisms that dissimilate nitrate to ammonium are also nitrate respirers, but most nitrate respirers accumulate nitrite.

in that the amount of N reduced in the dissimilatory process is in excess of that needed for synthesis of new biomass.

14–1.2 Respiratory Denitrification

The distinctive feature of respiratory denitrification is that nitrogen oxide reduction is coupled to electron transport phosphorylation. Therefore, there are two characteristics that define a respiratory denitrifier:

1. That nitrogen gases, principally N_2 and N_2O, are products of nitrate or nitrite reduction, and
2. That the above is coupled to a growth yield increase that is greater than if the N-oxide simply served as an electron sink.

While these are the defining criteria for a respiratory denitrifier, there are other measurable features that are usually characteristic of respiratory denitrification. These are: (i) nearly stoichiometric conversion of NO_3^- or NO_2^- to $N_2O + N_2$, (ii) a rapid rate of N gas production, and (iii) the presence of dissimilatory nitrite, nitric oxide, or nitrous oxide reductases. The defining parameters and procedures for measuring these characteristics are given in this chapter.

The denitrification pathway between NO_2^- and N_2O has been the subject of much research since the early 1980s. It now seems clear that, at least for most denitrifiers, the pathway involves the sequential conversion of $NO_3^- \rightarrow NO_2^- \rightarrow NO \rightarrow N_2O \rightarrow N_2$ with an enzyme bound nitrosyl (E–NO^+) detected during both the $NO_2^- \rightarrow NO$ (Kim & Hollocher, 1984) and $NO \rightarrow N_2O$ steps (Ye et al., 1991). Current evidence suggests that the denitrification process is organized as depicted in Fig. 14–1.

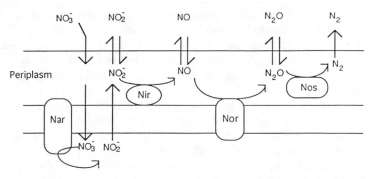

Fig. 14–1. Denitrification pathway in gram-negative bacteria and its organization relative to the cytoplasmic membrane. Nar, nitrate reductase; Nir, nitrite reductase; Nor, nitric oxide reductase; Nos, nitrous oxide reductase. (From Ye et al., 1994.)

Much work has focused on dissimilatory nitrite reductase (dNiR) since this is the first step that shunts NO_3^- to a dissimilatory fate instead of an assimilatory fate and, therefore, is a pivotal step in the biogeochemical N cycle. Two different nitrite reductases are known. One contains copper (Cu-dNiRs) and the other contains heme (heme-dNiRs). These enzymes convert NO_2^- to NO by an apparently similar mechanism, but they have completely separate evolutionary origins.

The Cu-dNiRs have been found in one-third of the respiratory denitrifiers characterized so far but are found in more divergent phylogenetic groups including *Bacillus, Rhizobium, Nitrosomonas, Rhodobacter, Thiosphaera, Achromobacter, Corynebacterium,* as well as *Alcaligenes,* and *Pseudomonas* (Hochstein & Tomlinson, 1988; Coyne et al., 1989). The Cu atoms can be easily removed from this enzyme by a copper specific chelant, diethyldithiocarbamate (DDC) (Shapleigh & Payne, 1985), making it easy to both confirm respiratory denitrification in strains with this enzyme as well as to distinguish this enzyme from the heme dNiR.

The heme diNirs contain one heme c and one heme d_1 (the active site) per monomer of this dimeric enzyme. This enzyme is found in two-thirds of the denitrifiers, including most of the *Pseudomonas* strains, but also in *Alcaligenes, Paracoccus, Thiobacillus,* and *Azospirillum.* There is no information on whether there is any ecological advantage of Cu vs. heme dNiR containing strains, or whether there is an environmentally significant difference between the two (e.g., in amounts of NO released). Nonetheless, the existence of convenient enzymatic activity assays, NiR specific antibodies and DNA probes for these structural genes makes this a convenient step to characterize respiratory denitrifiers.

Nitric oxide reductases and nitrous oxide reductases from several organisms have also been characterized. NO reductase is membrane bound and contains a cytochrome bc complex (Zumft, 1992). The lack of this enzyme in a denitrifier appears to be lethal, apparently because the NO produced accumulates to toxic levels (Braun & Zumft, 1991). Nitrous oxide reductase is a periplasmic protein and contains eight copper atoms.

Copper supplementation of medium sometimes aids growth of denitrifiers, perhaps because of the large copper requirement for this enzyme (Zumft, 1992). Nitrous oxide reductase activity is usually lost when cells are ruptured, thus it is not easy to assay cell-free extracts for this activity. There is no evidence that there is more than one type of these two enzymes in denitrifiers.

The discovery that acetylene inhibits N_2O reduction by Fedorova et al. (1973) and the simultaneous recognition by Balderston et al. (1976) and Yoshinari and Knowles (1976) of its value in denitrification assays has been the major methodological discovery that has advanced the study of denitrification. This is the basis of several methods in this chapter and chapter 44.

Several recent reviews of denitrification have been written. Those of most relevance to this chapter are by Hochstein and Tomlinson (1988); Tiedje (1988) and Zumft (1992) and a monograph by Revsbech and Sørensen (1990).

14–1.3 Dissimilatory Nitrate Reduction to Ammonium

Dissimilatory nitrate reduction to ammonium (DNRA) has recently become more widely recognized as a process distinct from assimilatory nitrate reduction (Tiedje, 1988). Both of these processes produce NH_4^+, but the former is regulated by O_2 and the latter by NH_4^+ and organic N (Table 14–1). Thus, DNRA occurs under the same conditions as respiratory denitrification. DNRA has proven not to be important in most soils. It is more significant than denitrification in environments where the available carbon/electron acceptor ratio is high, such as the bovine rumen, activated sludge digesters and some sediments (Tiedje et al., 1982). It can be stimulated in soils by excessive C addition but it still does not become the dominant NO_3^- consuming process (Caskey & Tiedje, 1979).

The major criterion that identifies DNRA is the production of NH_4^+ from NO_3^- in excess of the reduced N needed for growth. The easiest way to experimentally distinguish this process from assimilatory nitrate reduction is to measure $^{15}NH_4^+$ plus organic ^{15}N production from $^{15}NO_3^-$ in the presence of sufficient NH_4^+ (e.g., 1 mM) to repress the nitrate-assimilating pathways.

DNRA is carried out by several obligately anaerobic and facultatively anaerobic (fermentative) bacteria, including such soil inhabitants as *Clostridium, Desulfovibrio, Bacillus,* and virtually all of the Enterobacteriaceae (e.g., *Citrobacter, Enterobacter,* and *Klebsiella* are the more common of this group in soil) (Tiedje, 1988). Few population studies have been done on DNRAs in soil but those that have been done show them to be slightly more numerous than denitrifiers. Even though they may be more numerous than denitrifiers, they do not compete as well for NO_3^- probably because their V_{max}/K_m ratio for NO_3^- is much lower than it is for respiratory denitrifiers.

The NO_2^- to NH_4^+ step is not coupled to electron transport phosphorylation (ETP) in most of the studied strains, although this has now been demonstrated in at least three strains (but only one, *D. gigas,* might be expected in soil) (Tiedje, 1988). Hence, soil DNRA types probably do not benefit by enhanced growth to the extent that denitrifiers do by this reductive process. If DNRA does not benefit the soil population by ETP-linked growth, it does benefit the cells somewhat by serving as an electron sink allowing the reoxidation of NADH and by detoxifying nitrite.

14–1.4 Nitrate Respiration

Nitrate respirers possess the first step of DNRA, but accumulate NO_2^- instead of reducing it to NH_4^+. This $NO_3^- \rightarrow NO_2^-$ step is linked to ETP in nitrate respirers. The process occurs in many of the same strains that are capable of DNRA. It is not clear whether soil nitrate respirers represent a distinctly different group of organisms than DNRA types, or that particular physiological conditions can shift many typical nitrate respirers into further reduction of NO_2^- to NH_4^+. It has been shown for *Klebsiella* that C-limiting conditions result in NO_2^- accumulation while NO_3^--limiting conditions result in NH_4^+ as the principal product (Cole & Brown, 1980). If this physiological trend holds for soil nitrate respirers or DNRA types, then the condition of limiting available C found in most soils would favor NO_2^- accumulation. This product could then be removed by denitrifiers that should coexist with these NO_2^--accumulating organisms.

14–1.5 Nonrespiratory Denitrification

This is a heterogeneous collection of organisms that have the feature of reducing NO_3^- or NO_2^- to gas, but without enhanced growth. Most of these organisms produce N_2O as the only N gas and only in small quantities, e.g., 1 to 10% of the NO_3^- or NO_2^- reduced (Tiedje, 1988). A few organisms, e.g., *Propionibacterium* (Kaspar, 1982) and *Fusarium oxysporium* (Shoun et al., 1992) have shown 100% conversion of NO_2^- to N_2O, but this doesn't seem to be the typical case. Organisms capable of nonrespiratory denitrification include nitrate-assimilating bacteria, yeasts, filamentous fungi, algae, and (presumably) organisms that live in association with plants and animals since these higher organisms can emit N_2O (Bleakley & Tiedje, 1982).

These sources of N_2O are of potential significance to the atmospheric N_2O budget, since they can occur in aerobic organisms and thus exit drier soil more readily. The mechanisms of N_2O production by this group are probably varied and may result from nonspecific reactions of metalloenzymes. In the cases of *Fusarium oxysporium* the conversion of NO_2^- to N_2O is known to be due to a nitrate-nitrite inducible cytochrome P-450 monooxygenase (Shoun & Tanimoto, 1991). This is particularly interesting because it is the first eucaryote that shows substantial denitrification, although it does not enhance the organism's growth.

14–1.6 Chemodenitrification

Nonbiological production of N gases is also known to occur in soil under certain conditions (Tiedje, 1988). The most significant reaction is the acid-catalyzed destruction of nitrite, which can be significant in soils more acid than pH 5.0. The predominant product of this reaction is NO, but N_2O and N_2 have also been reported. Nitrate can also oxidize organic N-forming N_2 gas in a van Slyke type reaction. Chemodenitrification is not thought to be a major process on a global scale, but there are conditions such as acid forest soils, salting out in frozen soils, and wetting of dry soils where this reaction may be significant, particularly as a source of atmospheric NO.

14–2 KEY PHYSIOLOGICAL AND ECOLOGICAL FEATURES OF RESPIRATORY DENITRIFIERS

Respiratory denitrifiers are basically aerobes and prefer to use O_2 as their electron acceptor, but they do possess the alternative capacity to reduce nitrogen oxides when O_2 becomes limiting. Both the synthesis and activity of denitrifying enzymes is inhibited by O_2. Because denitrifying enzymes are not constitutive, their quantity in indigenous soil populations can be quite variable and probably not reflective of denitrification activity. While aerobic denitrification is known for a few organisms, there is no indication that this process occurs under aerobic conditions in soil communities. The threshold for denitrification appears to be ≤ 10 µmol O_2 (Tiedje, 1988). While oxygen is the major controller of denitrification activity, competition for C under aerobic conditions appears to be the major determinant of denitrifier population density and composition. Denitrifiers typically make up 0.1 to 5% of the total culturable population of mineral soils (Tiedje, 1988).

Most denitrifiers freshly isolated from nature have the entire pathway from NO_3^- to N_2, but many of these isolates upon culture in the laboratory will lose this capacity, particularly the $N_2O \rightarrow N_2$ step. Some isolates are known that lack the $NO_3^- \rightarrow NO_2^-$ step.

The denitrification property is widely distributed among procaryotic organisms, including organotrophs, lithotrophs, and phototrophs. The property is found in gram-positive bacteria, as well as in gram negatives, in archaebacteria, N_2 fixers, halophiles, thermophiles, sporeformers, and plant and animal pathogens. Denitrification has been reported in more than 50 genera and almost 130 species; the most recent comprehensive listing of species that denitrify was by Zumft (1992). Given this widespread distribution of the denitrifying property, it may be more instructive

Table 14–2. Theoretical energy yield and electron-accepting capacity of the steps in denitrification and dissimilatory nitrate reduction to ammonium.

Reaction[†]	No. of electrons accepted/N	Formal valence change	E_o' nitrogenous pair (mV)	$G°'$ (kJ/mol)
Denitrification[‡]				
$NO_3^- + H_2 \rightarrow NO_2^- + H_2O$	2	$+5/+3$	$+420$	-161
$NO_2^- + 1/2\ H_2 + H^+ \rightarrow NO_{(g)} + H_2O$	1	$+3/+2$	$+374$	-76.2
$2NO_{(g)} + H_2 \rightarrow N_2O_{(g)} + H_2O$	1	$+2/+1$	$+1177$	-306
$N_2O_{(g)} + H_2 \rightarrow N_{2(g)} + H_2O$	1	$+1/0$	$+1352$	-340
Total process				
$2NO_3^- + 5H_2 + 2H^+ \rightarrow N_{2(g)} + 6H_2O$	5	$+5/0$	$+749$	-1121
Dissimilatory nitrate reduction to ammonium[§]				
$NO_3^- + H_2 \rightarrow NO_2^- + H_2O$	2	$+5/+3$	$+420$	-161
$NO_2^- + 3H_2 + 2H^+ \rightarrow 2H_2O + NH_4^+$	6	$+3/-3$		-436
Total process				
$NO_3^- + 4H_2 + 2H^+ \rightarrow NH_4^+ + 3H_2O$	8	$+5/-3$		-600

† Adapted from Zumft‡ (1992) and Thauer et al.§ (1977).

to know what groups are missing representatives that denitrify (Tiedje, 1988). Three are notable: (i) obligate anaerobes, (ii) gram-positive organisms other than *Bacillus,* and (iii) the Enterobacteriaceae. Particularly striking is the apparent absence of this process in any organisms that are strong fermenters. A few weak fermenters (e.g., some strains of *Bacillus* and *Azospirillum*) also denitrify.

The theoretical energy yields and electron-accepting capacities of the respiratory denitrification and DNRA steps are shown in Table 14–2. Of particular note, is the greater electron accepting capacity of DNRA per mol of NO_3^-, which makes this an advantageous process in anaerobic environments. However, since all steps in the denitrifier respiratory pathway are coupled to phosphorylation, much of the potential energy of this pathway is realized as growth, which is not the case for DNRA organisms.

14–3 ENUMERATION OF DENITRIFIERS

14–3.1 Principle

Numerous MPN and plate methods that have been attempted for measuring denitrifier populations were reviewed in Tiedje (1982). The most reliable and convenient of these methods is to measure the denitrification process in most probable number (MPN) tubes by screening for $NO_3^- + NO_2^-$ disappearance as the presumptive test. Nitrate, NO_2^-, and N_2O have all been examined as electron acceptors, with NO_3^- found to be the most reliable. The method is based in the fact that denitrification, being a respiratory process, is faster and more competitive for NO_3^- than other NO_3^--consuming processes. Thus, loss of NO_3^- and NO_2^- under the

Table 14–3. Effect of medium on whether dissimilatory nitrate reducers to ammonium (DNRA) or denitrifiers dominate growth in MPN tubes.[†]

	Mean percent $^{15}NO_3^-$ converted to NH_4^+ and N_2O in positive tubes[‡]			
	Tryptic soy broth		Nutrient broth	
Soil dilution	$^{15}NH_4^+$	N_2O[§]	$^{15}NH_4^+$	N_2O[§]
$1:10^{-3}$	75.0	5.3	8.4	75.6
$1:10^{-5}$	62.3	5.2	11.3	69.8
$1:10^{-6}$	75.8	5.1	0.7	86.0
	Positive tubes out of 5[‡]			
$1:10^{-4}$	5	0	0	5
$1:10^{-5}$	5	0	0	4
$1:10^{-6}$	4	1	0	1

[†] Data from B.H. Bleakley. 1981. Nondenitrifying biological sources of nitrous oxide, M.S. thesis, Michigan State University.
[‡] Criteria for positive DNRA tube is $> 50\%$ of the $^{15}NO_3^-$ recovered as $^{15}NH_4^+$ and for positive denitrifier tube is $> 65\%$ of the $^{15}NO_3^-$ as N_2O.
[§] Acetylene (10%) present in headspace to block N_2O reduction.

specified conditions of this method usually corresponds to denitrification. However, there can be cases when NO_3^- and NO_2^- removal does not produce N_2 or N_2O (i.e., false positive tubes) or when NO_3^- and NO_2^- is not completely removed yet gas is produced (i.e., false negatives). Thus, a confirmatory test for denitrification is recommended in any new system under study until the reliability of the presemptive test is established.

Nutrient broth seems to be the best C source for enumeration media investigated to date. Protein-based media (no carbohydrates) support little H_2 and CO_2 production, an advantage if the confirmatory procedure is gas production in inverted tubes. Tryptic soy broth, a richer medium, favors dissimilatory NO_3^- reducers over denitrifiers (Table 14–3). Defined media result in lesser numbers. Most probable number tubes, microtiter plates, and plate counts have all been used. The MPN method is the only procedure adequately tested and practical at this time to recommend despite the inherent imprecision of this probability method.

Confirmation of the presence of denitrifiers can be either by N_2O production in the presence of C_2H_2 or by bubble formation in inverted (Durham) tubes. Bubble formation is less definitive, because one is never certain that the bubble is N_2 and not CO_2 or H_2 or that sufficient N_2 has formed to produce a bubble. In nutrient broth, however, the bubble is virtually always N_2 (Gamble et al., 1977). Organisms terminating reduction at N_2O are not counted since N_2O is water soluble and will not form a bubble. Finding at least 20% of the N from NO_3^- as N_2O in the presence of 0.1 atm C_2H_2 seems to be the best confirmatory procedure, because it avoids the above problems and generally does not fail to inhibit N_2O reduction. Organisms that dissimilate NO_3^- to NH_4^+ produce some N_2O, but in nutrient broth it is $< 20\%$ of the NO_3^- consumed.

14–3.2 Materials

1. Culture medium, 8.0 g of nutrient broth and 0.5 g of potassium nitrate (KNO_3^-) (5.0 mM) per liter. Place 10 mL of medium in 16 by 125 mm Hungate tubes (have butyl rubber septa in screw caps, Bellco Glass Inc., Vineland, NJ), and autoclave 15 min at 6.8 kg (15 lb) of pressure. For the confirmatory test, add after autoclaving, 1.0-mL acetylene (C_2H_2) aseptically by injection through a 0.22-μm filter plus syringe assembly.
2. Dilution bottles of 0.85% saline (see chapter 5 by Woomer in this book).
3. Nitrate (NO_3^-) and nitrite (NO_2^-) detection (Morgan, 1930): Dissolve 0.2 g of diphenylamine [$(C_6H_5)_2NH$] in 100 mL of conc. sulfuric acid (H_2SO_4). Store in bottle wrapped in foil, and place in refrigerator.
4. Gastight syringes (disposable tuberculin type).

14–3.3 Procedures

Collect fresh soil sample. Disperse by blending for 2 min 10 g of soil in 90 mL of water or saline plus one drop of Tween 80 (Difco Laboratories, Detroit, MI). Prepare serial dilution of suspension as described in chapter 5, usually final dilutions of 10^{-3} to 10^{-7} will suffice. Inoculate media tubes (five per dilution) with 0.1 mL of the appropriate serial dilution using 1-mL disposable syringes. Incubate at 25 to 30 °C for 14 d. At that time, withdraw 0.1 to 1.0 mL of medium by syringe and test for NO_3^- and NO_2^- by adding dropwise up to six drops of the diphenylamine reagent. A blue color indicates the presence of NO_3^- or NO_2^-; a colorless response is considered presumptive evidence of denitrification. Estimate the denitrifier populations using the MPN procedure as outlined in chapter 5.

For confirmation of denitrification, check tubes of dilutions near the apparent endpoint for accumulation of N_2O. Shake tubes to ensure equilibration of N_2O between gas and aqueous phases, then withdraw 0.5 mL of headspace by syringe and inject into a gas chromatograph equipped with a porous polymer column (e.g., Porapak Q) and a thermal conductivity (hot wire) or microthermister detector (see chapter 44 in this book). Calculate the percentage of NO_3^-–N recovered as N_2O–N. Since N_2O is water soluble, the amount in solution must be added to the headspace amount.

Total N_2O content (M) can be calculated using the Bunsen absorption coefficient (Table 14–4) according to the following equation:

$$M = C_g(V_g + V_l \cdot \alpha)$$

where M = the total amount of N_2O in the water plus gas phases, C_g = concentration of N_2O in gas phase, V_g = volume of gas phase, V_l = volume of liquid phase, and α = Bunsen absorption coefficient.

Table 14–4. Bunsen absorption coefficients in water at different temperatures for gases important in denitrification studies. Calculated from Wilhelm et al. (1977).

Temp., °C	Bunsen absorption coefficient (α)†				
	N_2O	NO	N_2	C_2H_2	O_2
5	1.06	6.46×10^{-2}	2.11×10^{-2}	1.51	4.30×10^{-2}
10	0.882	5.75×10^{-2}	1.89×10^{-2}	1.31	3.82×10^{-2}
15	0.743	5.17×10^{-2}	1.71×10^{-2}	1.16	3.43×10^{-2}
20	0.632	4.70×10^{-2}	1.57×10^{-2}	1.03	3.11×10^{-2}
25	0.544	4.31×10^{-2}	1.45×10^{-2}	0.928	2.85×10^{-2}
30	0.472	3.99×10^{-2}	1.36×10^{-2}	0.843	2.64×10^{-2}
35	0.414	3.72×10^{-2}	1.28×10^{-2}	0.771	2.46×10^{-2}

† The Bunsen absorption coefficient is the milliliters of gas at 0 °C and 760 mm of Hg (STP) that is absorbed by 1 mL of water.

14–3.4 Comments

This is the same method as proposed in the previous volume (Tiedje, 1982), but it has now been more widely evaluated (e.g., Davidson et al., 1985; Martin et al., 1988). Because of this the confirmatory test is becoming less important. Nonetheless, there can be enough false negatives (i.e., trace of NO_3^- or NO_2^- remains) to warrant recommending a confirmatory procedure until the user is confident that the presumptive test is adequate for the samples under investigation.

The above confirmatory procedure is recommended over the inverted tube method, because the identity of the gas is known and the fraction of the NO_3^- denitrified is revealed. Most MPN tubes show mixed fates of NO_3^-, with denitrifiers accounting for 50 to 90% of the NO_3^- removed under the recommended conditions (Table 14–3). With the N_2O method, any result over 20% N_2O can be correctly scored as positive for denitrification. The disadvantage of the N_2O method is the time and equipment necessary for the analysis. Lack of C_2H_2 inhibition of N_2O reduction and general C_2H_2 toxicity has not been a problem. While this confirmation method is the most quantitative and reliable, it is time consuming, and for large sample sets may not be feasible. In this case, the inverted tube method is acceptable. The procedure is described for syringes and Hungate tubes because they are essential for the C_2H_2 method. Even when C_2H_2 is not used, the syringe method of transfer requires less time.

The diphenylamine reagent for NO_3^- and NO_2^- detection is recommended over the diazotization procedure because it is a simpler one reagent test for both anions, and it is slightly less sensitive (an advantage here since less false negatives would be encountered if traces of the anions remain). It is important that the diphenylamine reagent be added dropwise since excess reagent will cause the oxidation to stop at the colorless diphenylbenzidine intermediate rather than proceeding to the blue quinoid imonium ion. A small volume of medium (≤ 1 mL) must be used since the oxidation occurs only in strong acid. The color is stable when produced by

NO_3^- but fades quickly when caused by NO_2^-; this can be used as a presumptive indicator of which anion is present. A green color has also been observed; it was correlated with lack of N_2O production and thus NO_3^- presence (Davidson et al., 1985).

The recommended NO_3^- concentration is 5.0 mM rather than the 9.9 mM of commercial NO_3^- broth, because too many false negatives have been encountered due to lack of total removal of all traces of NO_3^- and NO_2^-. Davidson et al. (1985) even found some (5%) false negatives at 5.0 mM NO_3^- at 14 d, but they lost detectable NO_3^-/NO_2^- after a slightly longer incubation. There is undoubtedly a relationship among NO_3^- concentration, length of incubation, and physiological properties of the population that may need to be optimized to ensure that traces of residual NO_3^- or NO_2^- are not causing false negatives.

Population measurements for denitrifiers should not be used as estimates for activity because denitrifying enzymes are inducible and because in situ activity is highly modified by oxygen, electron donor and NO_3^- concentrations. Population measurements are useful to understand how the population responds to environment and indexes the denitrifying capability of the population. The denitrifier enzyme assay (DEA) below should be considered as a method that better relates potential of the population to actual activity.

14–4 ENUMERATION OF DISSIMILATORY NITRATE TO AMMONIUM REDUCERS

14–4.1 Principle

These organisms occupy a similar niche as denitrifiers and carry out a similar process except that the product is NH_4^+ instead of gas. Thus, the method is similar to the respiratory denitrifier method except that conditions are used that favor organisms that convert $NO_3^- \rightarrow NH_4^+$ over denitrifiers. Tryptic soy broth is substituted for nutrient broth since this C-rich medium results in most of the $^{15}NO_3^-$ being recovered as $^{15}NH_4^+$ and not N gas (Table 14–3). Nitrate + NO_2^- removal in MPN tubes is presumptive for DNRAs. The confirmatory test is to use $^{15}NO_3^-$ and measure $^{15}NH_4^+$ produced.

14–4.2 Materials and Procedure

1. The same as for respiratory denitrifiers except that tryptic soy broth is substituted for nutrient broth.

14–4.3 Comments

This method should be considered exploratory because it has not been widely evaluated. However, if the confirmatory $^{15}NO_3^-$ method (see Table

14–3) is used there is no chance for misinterpretation of the physiological group measured.

14–5 DENITRIFIER ENZYME ACTIVITY

14–5.1 Principle

The denitrifier enzyme activity (DEA) method is a measure of the denitrifying enzyme concentration in a soil (or biomass) sample. It is based on the principle that the rate of the process is proportional to enzyme concentration when no other factors are limiting. Thus, soil is made into an anaerobic slurry to remove O_2 inhibition and diffusion limitations for substrates and products. NO_3^- is added in sufficient concentration to saturate the reductive processes. An available C source is added so electron donor is not limiting. Acetylene is added to block N_2O reduction to N_2 since N_2O can be more reliably quantified. The rate must be determined before de novo enzyme synthesis occurs. This assay is analogous to the Phase I rate originally described by Smith and Tiedje (1979) and slightly modified (Tiedje et al., 1989). Since denitrifying enzymes are inducible, enzyme content should reflect whether the soil conditions were suitable to induce the process.

14–5.2 Materials

1. 125 mL-Erlenmeyer flask with stoppers suitable for gas sampling.
2. Glucose.
3. KNO_3^-.
4. Chloramphenicol.
5. Acetylene.
6. Gastight syringes (disposable tuberculin type).
7. Oxygen-free inert gas (Ar, N_2, or He).
8. Shaker.
9. Gas chromatograph with detector suitable for N_2O quantitation.

14–5.3 Procedure

Soil (25 g) is placed in the Erlenmeyer flask containing 1 mM glucose, 1 mM KNO_3^-, and 1 g L^{-1} chloramphenicol. The flasks are capped with a gas tight stopper and made anaerobic by alternately flushing with an anaerobic gas and evacuating four times. Purified acetylene is added to the flask to achieve a final concentration of 10% (10 kPa) in the gas phase. The soil is slurried and incubated on a rotary shaker. Three replicates are recommended.

The head space gas is sampled by syringe for N_2O analysis by gas chromatography (see chapter 44 in this book). At least four measurements should be made during the incubation period to establish linearity of N_2O production. The recommended incubation period is 1 h and should not exceed 2 h unless the microbial population is much lower than typical of surface soils. The dissolved N_2O should be calculated by using the Bunsen relationship (Table 14–4). Gas samples can be stored in evacuated vials for later analysis, but controls of known concentration should be included during storage.

14–5.4 Comments

This method does not measure the natural denitrifying activity of the sample but it does reflect the environmental history of the site. This assay has been used in comparative studies to determine environmental effects on denitrification. Two new lines of evidence suggest that it may also be useful in field studies to assess activity.

DEA has been found to be highly correlated with the measured annual denitrification N loss in forest soils (Groffman & Tiedje, 1989). It is reasonable that a longer term average may be more reflective of the selection conditions the population experiences as recorded by enzyme concentration. DEA was also used to validate extrapolations of denitrification estimates from the landscape to regional scales (Groffman & Tiedje, 1992).

DEA has also been successfully used in a stochastic model along with respiration rates to predict denitrification frequency distribution and mean rates (Parkin & Robinson, 1989).

The coefficient of variation (CV) of DEA soil assays should usually be 5 to 15%. Higher variation may indicate incomplete anaerobiosis in the assay flask or a natural patchiness of denitrifiers in soil. The chloramphenicol is included to prevent protein synthesis and growth, thereby extending the linear period of N_2O production.

14–6 ISOLATION OF DENITRIFIERS

14–6.1 Principle

Isolation of denitrifiers is important when further information on denitrifier population composition, ecology, or physiology is sought. Questions about the dominate denitrifiers in particular soils can be addressed by isolating colonies that grow anaerobically on nutrient agar plus nitrate plates that were inoculated from a soil dilution series. The principle in this case is to not alter the natural denitrifier community composition and to allow each dominate denitrifier to grow and thus be registered in the analysis. Since fermentative bacteria and nitrate-respiring bacteria can also grow under these conditions, it is essential to confirm which isolates are denitrifiers (see below). In our experience, respiratory denitrifiers comprise about 15 to 30% of the colonies from soil samples obtained on this medium in this manner.

If denitrifiers with special physiological characteristics are sought (e.g., use of specific C substrates, requirement for special pH, temperature, or salt conditions) the best approach is to enrich for them under those physiological conditions.

14–6.2 Materials

1. Nutrient agar + 5 mM KNO_3 plates.
2. Chamber for anaerobic incubation.
3. Sterile dilution bottles of 0.85% saline.

14–6.3 Procedure

This procedure is taken from Gamble et al. (1977). Prepare serial dilution of soil; usually the 10^4 to 10^6 dilutions are most appropriate for fertile surface soils. Spread 1.0 mL of diluent onto pre-dried nitrate agar plates. Incubate under anaerobic conditions for 3 to 5 d. Prepare at least two dilution series from each soil and two plates for each dilution. If your goal is a more comprehensive study of the dominant denitrifying populations, pick all colonies from at least one plate with 15 to 60 colonies, and from appropriate quadrants of the other plates. Store isolates as stabs in the same agar medium. Confirm whether isolates are respiratory denitrifiers (see section 14–7).

14–6.4 Comments

When information is sought about dominant denitrifiers, it is important to avoid sampling, storage, or other potential enrichment conditions that might alter the natural population ratio among denitrifiers. Fresh samples are preferred, but if this is not possible, collection by intact soil cores, cold storage, or minimizing time elapsed from sampling all help minimize population shifts. A plating method is preferred over isolation from high-dilution, positive MPN tubes (e.g., from the enumeration protocol) because the prior growth in nutrient + nitrate broth will likely change the proportion of denitrifier types from that found in soil.

This procedure is potentially biasing because of the choice of C in the medium. The nutrient agar medium is advantageous over carbohydrate based media because it reduces the number of fermenters encountered and it reduces overgrowth by fast growers. However, it still recovers relatively fast growing gram-negative bacteria and thus may not adequately account for denitrification in the gram-positive and coryneform types.

Since many denitrifiers lose their ability to denitrify in culture (especially the $N_2O \rightarrow N_2$ step), it is important to minimize laboratory growth before confirmation of denitrification. For this reason, it is also important to immediately establish a permanent, original laboratory stock of each isolate if any further work is contemplated on these strains.

14–7 CONFIRMATION OF RESPIRATORY DENITRIFICATION

14–7.1 Principles

As noted above, two conditions must be met to describe an isolate as a respiratory denitrifier: (i) enhanced growth resulting from use of N oxides as electron acceptors; and (ii) production of nitrogen gases (NO, N_2O, or N_2) as a result of this process. There are several methods that directly test for these criteria and there are others that reliably correlate with this process and appear to also provide valid confirmation. Because there are many processes that consume nitrate, produce gases, or enhance growth (Table 14–1), it is important to confirm respiratory denitrification before claiming an organism is a true denitrifier.

14–7.1.1 Direct Confirmation by Growth and Gas Production

The following four methods are a variation of the same theme. All four are presented because each has different advantages and users may find particular ones more compatible with their facilities.

14–7.1.1.1 Growth on N_2O and Production of Bubbles. The simplest, most specific confirmation of denitrification is to test for N_2O-dependent growth *and* gas bubble (N_2) production. This method has the advantage that only respiratory denitrifiers grow using N_2O as an electron acceptor and gas product detection is simply accomplished in inverted (Durham) tubes. Gas detection is made easier because large bubbles are formed due to the insolubility of N_2 and the limited electron accepting capacity of the $N_2O \rightarrow N_2$ step which ensures that the maximum transformation occurs. Only light cell turbidity is to be expected because of the limited electron accepting capacity of N_2O. The limitation of this method is that some denitrifiers do not have the $N_2O \rightarrow N_2$ step. However, if a strain tests positive, respiratory denitrification is confirmed.

14–7.1.1.2 Growth on Nitrate and Production of Bubbles. This method is similar to the above, except that nitrate is substituted for N_2O as the electron acceptor. More care must be used with this method since nitrate respirers will also grow in this medium (but do not produce gas) and other fermenters will grow and may produce CO_2 or H_2 that can form bubbles. The use of protein-based media reduces the likelihood of finding significantly sized bubbles of these gases, however. Growth is more substantial in this medium because five electrons/N are accepted and nitrate is more soluble than N_2O. Bubbles are usually smaller than for the method in section 14–7.1.1.1. Bubble production also depends on the presence of the $N_2O \rightarrow N_2$ step as in method 14–7.1.1.1, but misses denitrifiers that lack the nitrate-reducing step (these types appear to be rare in nature).

14–7.1.1.3 Growth on Nitrate and Stoichiometric Production of N_2O in the Presence of Acetylene. This method substitutes measurement of bubbles with measurement of N_2O. N_2O is approximately 100 times more soluble in water than N_2 and hence does not form bubbles in these types of assays (Table 14–4). This method allows for a more accurate stoichiometric N balance since N_2O can be more accurately quantified than N_2. It is also a distinctive character of respiratory denitrification to find N_2O to accumulate in the presence of acetylene, but not in its absence. Nonrespiratory denitrifiers produce N_2O regardless of the presence of acetylene. Denitrifiers missing $N_2O \rightarrow N_2$ step would also behave in this later manner, but the amount of N_2O produced would be greater than for nonrespiratory denitrifiers. This method is more labor intensive and requires a gas chromatograph capable of measuring N_2O.

14–7.1.1.4 Determination of Whether Growth Yield is Proportional to N-oxide Reduction.

Of the direct methods, this method provides the most thorough characterization as a respiratory denitrifier, since growth yield is measured as a function of N reduced. This can be done by measuring total growth in the presence and absence of NO_3^-, but it is more thoroughly characterized if measured at several NO_3^- concentrations such as shown in Fig. 14–2. From these data, an estimate of growth yield can also be calculated. For most vigorous denitrifiers this value is 15 to 25 g protein mol^{-1} NO_3^- denitrified (I. Mahne and J. Tiedje, 1992, unpublished data). If N-oxide reduction is not complete, nitrate, nitrite, N_2O and N_2 can be measured so that growth yield relative to actual electron flux can be calculated. If necessary, this method can be adapted to detect respiratory growth from denitrifiers missing any of the denitrifying steps since the growth yield per step can be defined by substituting the appropriate electron acceptor in the medium.

14–7.2 Materials

(Depends on which of the above variations is chosen).

1. Nutrient broth.
2. N_2O for 14–7.1.1.1 or KNO_3 for 14–7.1.1.2, 14–7.1.1.2, 14–7.1.1.4.
3. Small test tubes for 14–7.1.1.2 or Hungate tubes with septa (Bellco Glass Inc., Vineland, NJ), for 14–7.1.1.1, 14–7.1.1.3, 14–7.1.1.4.
4. Acetylene (14–7.1.1.2).
5. Gas tight disposable syringes (14–7.1.1.3, 14–7.1.1.4).
6. Small tubes (Durham tubes) that can be inverted inside culture tubes (14–7.1.1.1, 14–7.1.1.2).

14–7.3 Procedure

Nutrient broth is the recommended medium in all above versions. In versions 14–7.1.1.1, 14–7.1.1.3, and 14–7.1.1.4, a sealed culture tube is needed to retain a specific headspace. Hungate tubes (screw cap culture tubes with septa) are the most convenient. Version 14–7.1.1.2 can either be conducted in these tubes or in an anaerobic chamber. Durham tubes are

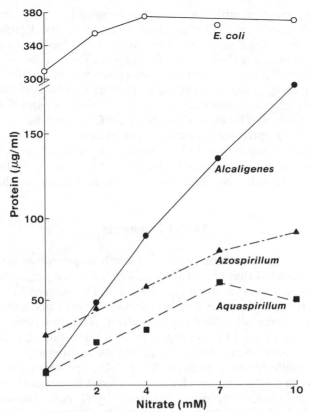

Fig. 14–2. Examples of different patterns of NO_3^--dependent growth of several denitrifiers and a nitrate-respiring, fermentive anaerobe (*E. coli*). Vigorous denitrifiers should show a response to NO_3^- concentration similar to that shown for *Alcaligenes*. Data from I. Mahne, Michigan State Univ.

small inverted tubes or vials placed in a larger culture tube (versions 14–7.1.1.1 and 14–7.1.1.2). The medium should cover the Durham tubes; when autoclaved the Durham tube will completely fill with medium. All inocula to be tested for respiratory denitrification should be inoculated in a tube with and without the electron acceptor as the control.

For version 14–7.1.1.1, the headspace is flushed with 100% N_2O, the tube is sealed and autoclaved. For version 14–7.1.1.2, the medium is made with 5 mM KNO_3; the headspace is either flushed with an anaerobic gas and sealed or capped with an aerobic culture plug and incubated in an anaerobic chamber. For version 14–7.1.1.3, the medium is amended with 5.0 mM KNO_3, placed in a sealed tube and autoclaved. After autoclaving, acetylene is injected through a sterile needle and a 0.22 μm filter assembly to achieve a 10% headspace concentration. Acetylene will block nitrous oxide reduction in almost all denitrifiers. For version 14–7.1.1.4, nutrient broth should be amended with 0, 2, 4, 7, and 10 mM KNO_3, placed in sealed tubes and autoclaved.

After inoculation with the culture to be tested, incubation should be until growth ceases or 14 d whichever comes first. Total growth can be assessed by measuring turbidity in the tubes with electron acceptor compared to the control without electron acceptor. This can be by visual estimate, optical density, or protein concentration, depending upon the precision desired. Since enhanced growth is part of the definition of respiratory denitrification, this analysis should not be overlooked. N_2 production can be estimated by bubble height in the Durham tube and compared to the bubble size produced by a well-studied denitrifier used as a positive control or by estimating the bubble volume in the Durham tube. N_2O is measured by gas chromatography (chapter 44 in this book) and nitrate and nitrite can be measured by several well-established methods (chapter 41 in this book).

14–7.4 Comments

Fig. 14–2 illustrates the different growth responses as a function of NO_3^- concentration that are encountered for different physiological types of organisms as measured by method 7.1.1.4. The pattern for typical vigorous denitrifiers is illustrated by *Alcaligenes*. Weaker denitrifiers respond as shown for *Azospirillum* and *Aquaspirillum*. *Azospirillum* is also a weak fermenter as illustrated by some growth with no nitrate, a finding that is rare for respiratory denitrifiers. *Aquaspirillum* and perhaps *Azospirillum* show some inhibition of growth at 10 mM NO_3^-, probably due to toxicity from temporary NO_2^- accumulation. The growth response of a strong fermenter and nitrate respirer is illustrated by *E. coli*. The fermentative growth is easily distinguished and respiratory nitrate reduction enhances growth but not to the extent that it does for respiratory denitrification. Nitrite can be considered as a substitute for NO_3^- in this method to avoid confusion by the growth response of nitrate respirers, but NO_2^- at concentrations of 4 to 10 mM are inhibitory to some denitrifiers (I. Mahne and J. Tiedje, 1992, unpublished data).

14–7.5 Further Characterization and Confirmation of Respiratory Denitrification

As mentioned in the introduction, there are three measurable features that are usually characteristic of respiratory denitrification. Besides confirming denitrification, these provide further information on characterizing the process for the particular strain. The first two methods and criteria are from Mahne and Tiedje (1992, unpublished data).

14–7.5.1 Conversion of 80% or More of Nitrate and Nitrite to $N_2O + N_2$

Other processes typically produce 0 to 10% but rarely more than 35% of the added N oxide as N gases. The amount of nitrogen gases produced can be measured in versions 14–7.1.1.2, 14–7.1.1.3, and 14–7.1.1.4.

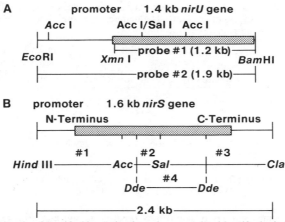

Fig. 14–3. DNA probes identified by # that have been used to identify denitrifying organisms and denitrifying genes in communities. The Cu-dNir probe (A) is from Pseudomonas G-179 and is described in Ye et al. (1993), and the heme c,d_1 dNir probe (B) is from *Pseudomonas stutzeri* JM300 and is described in Smith and Tiedje (1992). The region of the structural gene is shown by the shaded area.

14–7.5.2 A Rapid Rate of Nitrite Reduction

Denitrification is a highly evolved process coupled to respiration. Therefore, it is expected that respiratory denitrifiers should more rapidly produce N gases than is expected from other mechanisms. Mahne and Tiedje found 10 nmol N gas mg^{-1} of protein per minute to be a minimum rate for all denitrifiers tested. This rate is higher than rates found for nonrespiratory denitrifiers, and thus distinguishes respiratory denitrifiers from this group. The measurement is most easily carried out using method 14–7.1.1.3 above, but periodically analyzing the headspace for N_2O to establish the rate of gas production.

14–7.5.3 The Presence of Any of the Denitrifier Specific Enzymes

The presence of these enzymes can be assessed by reaction with antibodies specific to the denitrifier enzymes on Western blots (Coyne et al., 1989), and enzyme assays in cell-free extracts for NO_2^- or NO reduction (Ye et al., 1992). It is also now possible to use DNA probes to detect the presence of structural genes for denitrification. This method, however, does not confirm that the genes are expressed and functional. Genes for a heme nitrate reductase and copper nitrate reductase that have been used to identify the type of dNiR are shown in Fig. 14–3. Genes for nitrous oxide reductase (Viebrock & Zumft, 1988) and nitric oxide reductase (Braun & Zumft, 1992) are also now known. Other denitrification specific genes for the electron transport carriers, assembly, prosthetic group synthesis, etc. are being discovered and will likely also become available as potential

Table 14–5. Identification of bacterial strains carrying dissimilatory nitrite reductase protein (Western blot) or genes (Southern blot). The number of strains with a positive reaction out of the total of that type tested is shown.

Nitrite reductase type[†]	Western immunoblot reaction[‡]		DNA hybridization on Southern blot	
	Cu antibody	Heme antibody	Cu probe[§]	Heme probe[¶]
Cu	16/20	1/20	15/16	2/7
Heme, c,d$_1$	0/9	8/9	1/11	13/15
Nondenitrifiers				0/8

† Nitrite reductase determined by inhibition of enzyme activity with the copper chelant, diethyldithiocarbamate (DDC).
‡ Data from Coyne et al. (1989) only for strains tested by DDC method.
§ Data for 1.9-kb probe (#2) carrying *nirU*. (From Ye et al., 1993.)
¶ Data for 1.2-kb probe (#3) of *nirS*. (From Smith & Tiedje, 1992.)

Table 14–6. Taxonomic groups of denitrifiers reported in soil and arranged according to phylogenetic group and frequency of recovery.†

Frequency	Genus	Phylogenetic group or Proteobacteria subdivision	Species
Predominant	*Pseudomonas*§	Gamma rRNA group I	*fluorescens, aeruginosa, stutzeri, alcaligenes, aureofaciens, chloro-raphis, mendocina, pseudoalcali-genes*
	Alcaligenes	Beta	*faecalis, xylosoxidans, eutrophus (? Achromobacter cycloclastes)*
	Pseudomonas§	Gamma rRNA group II	*pickettii, pseudomallei*
Moderate to low	*Pseudomonas*§	Beta	*solanacearum, lemoignei*
	Bacillus	Gram positive	*azotoformans, licheniformis, stearothermophilus, cereus, halodenitrificans, polymyxa*
	Agrobacterium	Alpha	*tumefaciens, radiobacter*
	Flavobacterium‡	Bacteroides-Flavobacteria	*glenum*
Seen only under special selective conditions	*Rhizobium*	Alpha	*meliloti, trifolii, lupini, leguminosarum, fredii*
	Bradyrhizobium	Alpha	*japonicum*
	Azospirillum	Alpha	*brasilense, lipoferum*
	Nitrosomonas	Beta	*europaea*
	Nitrobacter	Beta	
	Thiobacillus	Beta	*denitrificans*
Rare	*Paracoccus*	Alpha	*denitrificans*
	Cytophaga	Gliding bacteria	*johnsonae*
	Flexibacter	Gliding bacteria	*canadensis*
	Chromobacterium	Beta	*violaceum*
	Gluconobacter	Alpha	

† Summarized from Zumft (1992) and Tiedje et al. (1989).
‡ Many reported flavobacteria have high G + C percentages and thus are probably *Pseudomonas*. Hence, the commonness of true flavobacteria as denitrifiers has probably been overestimated.
§ The taxonomy of the broad group now classified as *Pseudomonas* is undergoing revision and the second and third *Pseudomonas* groups above will likely have new and different genus names in the future.

probes. How valuable any of these probes are is highly dependent on how conserved the DNA sequences are and how specific they are for denitrifiers.

All three of the above methods can also differentiate between the heme and copper-type denitrifying nitrite reductases. Antibodies and gene probes for these two types have been used to differentiate the two classes (Table 14–5). Although not absolutely specific, the gene probes showed a higher degree of specificity than might be expected given normal divergence of sequence. The methods for immunoblotting and DNA cell blotting can be found in chapters 28 and 32 by Wright and Ogram, respectively, in this book, or the original papers. Nitrite and nitric oxide reductases can be assayed in cell-free extracts as follows. Cells are disrupted by sonication and debris removed by centrifugation at $8000 \times g$ for 20 min. The rate of nitrite disappearance or NO or N_2O appearance can be measured for the respective enzymes. The reaction mixture contains 10 mM EDTA, 40 μM phenazine methosulfate (PMS), 4 mM NADH and 50 mM HEPES (pH. 7.3). The assay is conducted under an anaerobic headspace in sealed vials. Diethyldithiocarbamate (DDC), the Cu chelant, can be added at 5 mM 20 min prior to the incubation to determine which type of nitrite reductase is present.

14–8 TAXONOMIC IDENTIFICATION

Once respiratory denitrification has been confirmed, it may be of interest to identify the organism. Respiratory denitrification is widespread among procaryotic groups, but the denitrifiers frequently encountered in soil are from limited phylogenetic groups (Table 14–6). More than 50% of the dominant soil denitrifiers seem to be from rRNA group I of the gamma group of *Pseudomonas,* and most of those within this group are fluorescent. Thus, it may be efficient to focus on this group first in confirming denitrifier identity. Consult *Bergey's Manual of Systematic Bacteriology* (Holt, 1984–1989) and *Bergey's Manual of Determinative Bacteriology* (Holt et al., 1994) for the key traits necessary to establish taxonomic identity.

ACKNOWLEDGMENT

The author's work on denitrification has been supported principally by the U.S. National Science Foundation. The author particularly acknowledges the work of Ivo Mahne on defining the criteria for respiratory denitrifiers and Bruce Bleakley for the work on the DNRA method.

REFERENCES

Balderston, W.L., B. Sherr, and W.J. Payne. 1976. Blockage by acetylene of nitrous oxide reduction in *Pseudomonas perfectomarinus.* Appl. Environ. Microbiol. 31:504–508.
Bleakley, B.H., and J.M. Tiedje. 1982. Nitrous oxide production by organisms other than nitrifiers or denitrifiers. Appl. Environ. Microbiol. 44:1342–1348.

Braun, C., and W.G. Zumft. 1991. Marker exchange of the structural genes for nitric oxide reductase blocks the denitrification pathway of *Pseudomonas stutzeri* at nitric oxide. J. Biol. Chem. 266:22785–22788.

Braun, C., and W.G. Zumft. 1992. The structural genes of the nitric oxide reductase complex from *Pseudomonas stutzeri* are part of a 30-kilobase gene cluster for denitrification. J. Bacteriol. 174:2394–2397.

Caskey, W.H., and J.M. Tiedje. 1979. Evidence for Clostridia as agents of dissimilatory reduction of nitrate to ammonium in soils. Soil Sci. Soc. Am. J. 43:931–936.

Cole, J.A., and C.M. Brown. 1980. Nitrite reduction to ammonium by fermentative bacteria: A short circuit in the biological nitrogen cycle. FEMS Microbiol. Lett. 7:65–72.

Coyne, M.S., A. Arunakumari, B.A. Averill, and J.M. Tiedje. 1989. Immunological identification and distribution of dissimilatory heme c,d_1 and nonheme copper nitrite reductases in denitrifying bacteria. Appl. Environ. Microbiol. 55:2924–2931.

Davidson, E.A., M.K. Strand, and L.F. Galloway. 1985. Evaluation of the most probable number method for enumerating denitrifying bacteria. Soil Sci. Soc. Am. J. 49:642–645.

Fedorova, R.I., E.I. Milekhina, and N.I. Ilkukhina. 1973. Possibility of using the "gas exchange" method to detect extraterrestrial life: Identification of nitrogen-fixing organisms. Izv. Akad. Nauk Arm. SSR Biol. Nauki 6:797–806.

Gamble, T.N., M.R Betlach, and J.M. Tiedje. 1977. Numerically dominant denitrifying bacteria from world soils. Appl. Environ. Microbiol. 33:926–939.

Groffman, P.M., and J.M. Tiedje. 1989. Denitrification in north temperate forest soils: relationships between denitrification and environmental factors at the landscape scale. Soil Biol. Biochem. 21:621–626.

Groffman, P.M., and J.M. Tiedje. 1992. Regional scale analysis of denitrification in north temperate forest soils. Landscape Ecol. 7:45–53.

Hochstein, L.I., and G.A. Tomlinson. 1988. The enzymes associated with denitrification. Ann. Rev. Microbiol. 42:231–261.

Holt, J.G. (ed.-in-chief). 1984–1989. Bergey's manual of systematic bacteriology, 4 volumes. Williams and Wilkins Co., Baltimore.

Holt, J.G., N.R. Krieg, P.H.A. Sneath, J.T. Staley, and S.T. Williams (ed.). 1994. Bergey's manual of determinative bacteriology. 9th ed. Williams and Wilkins Co., Baltimore.

Kaspar, H.F. 1982. Nitrite reduction to nitrous oxide by propionibacteria: Detoxication mechanism. Arch. Microbiol. 133:126–130.

Kim, C.H., and T.C. Hollocher. 1984. Catalysis of nitrosyl transfer reaction by a dissimilatory nitrite reductase (cytochrome c,d_1). J. Biol. Chem. 259:2092–2099.

Martin, K., L.L. Parsons, R.E. Murray, and M.S. Smith. 1988. Dynamics of soil denitrifier populations: Relationships between enzyme activity, most-probable-number counts, and actual N gas loss. Appl. Environ. Microbiol. 54:2711–2716.

Morgan, M.F. 1930. A sample spot-plate test for nitrate nitrogen in soil and other extracts. Science 71:343–344.

Parkin, T.B., and J.A. Robinson. 1989. Stochastic models of soil denitrification. Appl. Environ. Microbiol. 55:72–77.

Revsbech, N.P., and J. Sørensen. 1990. Denitrification in soil and sediment. Plenum Press, New York.

Rice, C.W., and J.M. Tiedje. 1989. Regulation of nitrate assimilation by ammonium in soils and in isolated soil microorganisms. Soil Biol. Biochem. 21:597–602.

Shapleigh, J.P., and W.J. Payne. 1985. Differentiation of c,d_1 cytochrome and copper nitrite reductase production in denitrifiers. FEMS Microbiol. Lett. 26:275–279.

Shoun, H., D-H. Kim, H. Uchiyama, and J. Sugiyama. 1992. Denitrification by fungi. FEMS Microbiol. Lett. 94:277–282.

Shoun, H., and T. Tanimoto. 1991. Denitrification by the fungus *Fusarium oxysporum* and involvement of cytochrome P-450 in the respiratory nitrite reduction. J. Biol. Chem. 266:11078–11082.

Smith, G.B., and J.M. Tiedje. 1992. Isolation and characterization of a nitrite reductase gene and its use as a probe for denitrifying bacteria. Appl. Environ. Microbiol. 58:376–384.

Smith, M.S., and J.M. Tiedje. 1979. Phases of denitrification following oxygen depletion in soil. Soil Biol. Biochem. 11:261–267.

Thauer, R.K., K. Jungermann, and K. Decker. 1977. Energy conservation in chemotrophic anaerobic bacteria. Microbiol. Rev. 41:100–180.

Tiedje, J.M. 1982. Denitrification p. 1011–1026. *In* A.L. Page et al. (ed.) Methods of soil analysis. Part 2. Agron. Monogr. 9. ASA, Madison, WI.

Tiedje, J.M. 1988. Ecology of denitrification and dissimilatory nitrate reduction to ammonium. p. 179–244. *In* A.J.B. Zehnder (ed.) Biology of anaerobic microorganisms. John Wiley and Sons, New York.

Tiedje, J.M., A.J. Sexstone, D.D. Myrold, and J.A. Robinson. 1982. Denitrification: Ecological niches, competition and survival. Antonie van Leeuwenhoek 48:569–583.

Tiedje, J.M., S. Simkins, and P.M. Groffman. 1989. Perspectives on measurement of denitrification in the field including recommended protocols for acetylene based methods. Plant Soil 115:261–284.

Viebrock, A., and W.G. Zumft. 1988. Molecular cloning, heterologous expression, and primary structure of the structural gene for the copper enzyme nitrous oxide reductase from denitrifying *Pseudomonas stutzeri:* J. Bacteriol. 170:4658–4668.

Wilhelm, E., R. Battino, and R.J. Wilcock. 1977. Low-pressure solubility of gases in liquid water. Chem. Rev. 77:219–262.

Ye, R.W., A. Arunakumari, B.A. Averill, and J.M. Tiedje. 1992. Mutants of *Pseudomonas fluorescens* deficient in dissimilatory nitrite reduction are also altered in nitric oxide reduction. J. Bacteriol. 174:2560–2564.

Ye, R.W., B.A. Averill, and J.M. Tiedje. 1994. Denitrification of nitrite and nitric oxide. Appl. Environ. Microbiol. 60:1053–1058.

Ye, R.W., M.R Fries, S.G. Bezborodnikov, B.A. Averill, and J.M. Tiedje. 1993. Characterization of the structural gene encoding a copper containing nitrite reductase and homology of this gene to DNA of other denitrifiers. Appl. Environ. Microbiol. 59:250–254.

Ye, R.W., I. Toro-Suarez, J.M. Tiedje, and B.A. Averill. 1991. $H_2^{18}O$ isotope exchange studies on the mechanism of reduction of nitric oxide and nitrite to nitrous oxide by denitrifying bacteria. J. Biol. Chem. 266:12848–12851.

Yoshinari, T., and R. Knowles. 1976. Acetylene inhibition of nitrous oxide reduction by denitrifying bacteria. Biochem. Biophys. Res. Commun. 69:705–710.

Zumft, W.G. 1992. The denitrifying procaryotes. *In* A. Balows et al. (ed.) The prokaryotes. Vol. I. 2nd ed. Springer-Verlag, New York.

Chapter 15

Actinomycetes

E. M. H. WELLINGTON AND **I. K. TOTH,** *University of Warwick, Coventry, UK*

Members of the actinomycete group are characterized by the production of a mycelium at some stage in their life cycle and many genera differentiate further with the formation of spores and sporangia. In most cases, the spores are for dispersal but some, for example those produced by *Micromonospora* spp., have increased resistance to adverse conditions such as heat and drought. Many species produce bioactive metabolites and in particular strains from the genera *Streptomyces* and *Micromonospora* have been used for the commercial production of antibiotics. The majority of soil-borne inoculum is nonpathogenic and saprophytic but some notable exceptions include selected species of the genera *Actinomadura, Actinomyces, Mycobacterium,* and *Nocardia.*

Actinomycetes have been isolated from a wide range of soil types collected from disparate geographical locations. In general, actinomycetes are able to survive in both wet and dry conditions, and under conditions of bulk soil pH ranging between 4 and 10. Spores of *Micromonospora* spp. have been found in sediments deposited over 100 yr ago (Cross & Attwell, 1974). The majority of soil-inhabiting actinomycetes are aerobic. However, some species of the genera *Actinomyces* and *Agromyces* are catalase negative, microaerophilic and capnophilic, requiring increased CO_2 and reduced oxygen tensions for growth. *Micromonospora* contains two obligately anaerobic species of uncertain phylogeny that have not been found in soil. While it is possible that sediment-inhabiting micromonosporas may grow microaerophilically, isolation work has focused primarily on the growth of aerobes which would be present as inactive spores in deep sediments. Aerobic genera such as *Streptomyces,* with hydrophobic spores, are less readily isolated from water-logged soils than are the aerobic micromonosporas which produce hydrophilic spores.

Thermophilic, acidophilic, and alkalophilic species of several genera have been isolated using the appropriate isolation procedures. Such strains tend to predominate in certain habitats but can also be isolated from mesophilic habitats in soil of neutral pH. Composted plant materials are a rich source of many thermophilic species (Table 15–1) and genera such as *Pseudonocardia, Saccharomonospora, Saccharopolyspora,* and *Thermomonospora* are examples of groups abundant in such defined substrates but which also occur in soil. Actinomycetes are less numerous in acid soils and certain groups such as *Micromonospora* and *Streptomyces* can be enriched by liming either in the field or as a pretreatment prior to selective isolation.

The density of actinomycete propagules can exceed $10^6 g^{-1}$ soil but this is no reflection of activity unless some attempt is made to differentiate spores from mycelium. The decomposition of plant remains is facilitated by actinomycetes since many groups synthesize cellulase and lignin-degrading enzymes (McCarthy, 1987; Crawford, 1988). Thermophilic compost-inhabiting groups have often been selected for production of such enzymes but soil mesophils from genera such as *Micromonospora* show significant cellulose degradation (McCarthy & Broda, 1984).

The major problem with the enumeration and isolation of actinomycetes is the need to counter select the wide range of faster growing eubacteria frequently present in soil. Some genera such as *Dactylosporangium, Microbispora,* and *Streptosporangium* can appear on isolation plates after 4 to 5 wk incubation. Soil fungi also grow well on media traditionally used to isolate actinomycetes and must be discouraged to avoid overgrowth of plates. For these reasons, isolation procedures need to be highly selective if they are to reveal the full diversity of actinomycete propagules present in a given soil. The extent of this diversity is indicated in Table 15–1 where more than 40 genera have been isolated using a wide range of techniques and soil samples.

The identification of these different genera is still facilitated by the use of morphology but the main groupings are now based on 16S ribosomal RNA (rRNA) sequence comparisons (Stackebrandt, 1991). The recovery of rDNA sequences directly from soil using PCR (see chapter 34 in this book) has led to the hypothesis that, even for readily cultured soil saprophytes, such as streptomycetes, current type cultures do not represent the full diversity of soil populations (Stackebrandt et al., 1991).

15–1 ENUMERATION, ENRICHMENT, AND ISOLATION

15–1.1 Principles

The dilution plate count has been used for both enumeration and isolation. Although this technique has many deficiencies it is still one of the principle methods for examination of actinomycete populations in soil. The formation of coenocytic mycelium means that only approximations of viable propagules can be made due to fragmentation. Direct methods of counting present problems due to the small size of the mycelium (0.5–1.0

Table 15–1. Actinomycete† genera found in soil.

Genus	Soil isolation procedure	Comments
Actinomadura	Nonomura & Ohara, 1971a Lavrova et al., 1972 Athalye et al., 1981	Wide distribution, higher numbers in cultivated soils
Actinomyces	Gledhill & Casida, 1969	Primary habitat human and animal tissues. *A. humiferus* found in organically rich soils
Actinoplanes, *Ampullariella*	Palleroni, 1976	Distribution world wide
Agromyces	Gledhill & Casida, 1969	Found in a wide range of soil types
Amycolata	Okazaki et al., 1983	Forest soils and rhizoplane
Amycolatopsis	Margalith & Beretta, 1961	Forest and cultivated soils
Arthrobacter	Hagedorn & Holt, 1975	Numerous and widely distributed in soil
Aureobacterium	Dias et al., 1962	Several species isolated from soil
Catellatospora	Asano & Kawamoto, 1986	Woodland soils
Cellulomonas	Stackebrandt & Keddie, 1984	Mainly isolated from soil
Corynebacterium	Von der Osten et al., 1989	Farmland soils, manure
Dactylosporangium	Shearer, 1987 Hayakawa et al., 1991b	Sediments, water, and plant litter
Frankia	Baker & O'Keefe, 1984	Usually isolated from root nodules
Geodermatophilus	Luedemann, 1968	Desert soils
Glycomyces	Labeda, 1989	Found in soils from China and USA
Gordona	Tsukamura et al., 1988	Isolated from soils and sediment
Kibdellosporangium	Shearer et al., 1986	Desert and tropical soils
Kineosporia	Parenti, 1989	Rare
Microbispora	Hayakawa et al., 1991a	Widely distributed, especially in cultivated soils
Micromonospora	Rowbotham & Cross, 1977 Nonomura & Hayakawa, 1988	Common in most soils and sediments
Microtetraspora	Nonomura & Ohara, 1971b,c	Cultivated soils
Mycobacterium	Songer, 1981 Portaels et al., 1988	Saprophytes common in a wide range of soils
Nocardia	Orchard et al., 1977	Widely distributed in soil
Nocardiodes	Prauser & Bergholz, 1974	Clay soils, savanna grassland, and plant litter
Nocardiopsis	Nonomura & Ohara, 1971a	Frequently isolated from different soil types
Oerskovia	Stackebrandt & Prauser, 1991	Organic-rich soils
Pilimelia	Makkar & Cross, 1982	In a wide range of soil types
Planobispora	Bland & Couch, 1981	Very rarely isolated
Planomonospora	Bland & Couch, 1981	Found in diverse soil types, worldwide distribution
Promicromonospora	Stackebrandt & Prauser, 1991	Compost, soils from various locations
Pseudonocardia	Goddon & Peninckx, 1984	Compost, manure, and cultivated soil
Rhodococcus	Rowbotham & Cross, 1977	Found in pastures, marine sediments, and farm land
Saccharomonospora	Lacey & Dutkiewicz, 1976	Manure, compost, and peats
Saccharopolyspora (including *Faenia*)	Lacey, 1974	Compost, stored cereals, bagasse, manure, and soil
Saccharothrix	Shearer, 1987	Common in many soil types

(continued on next page)

Table 15–1. Continued.

Genus	Soil isolation procedure	Comments
Sporichthya	Lechevalier & Lechevalier, 1989	Rare
Spirillospora	Bland & Couch, 1981	Infrequently isolated from soil
Streptomyces	El-Nakeeb & Lechevalier, 1963 Kuster & Williams, 1964	Widespread and numerous in most soils
Streptoalloteichus	Tomita et al., 1987	Rarely isolated from dry soils
Streptosporangium	Hayakawa et al., 1991b	Widely distributed often in culti-vated soils, associated with plant litter
Thermomonospora	Anderson, 1958; McCarthy & Cross, 1981	Composts, bagasse, and manure can also be found in soil

† Genera included which have been found in soil and grouped in the actinomycete branch by 16S rRNA analysis (Stackebrandt, 1991).

µm diam.). Immunofluorescence has been used to detect specific species in soil but again fragmentation and sporulation can lead to inaccuracies and poorly reproducible results. Oligofluors have also been useful in auteco-logical studies of *Streptomyces scabies* in soil (Bramwell, 1992; Hahn et al., 1992). The success of this method depends on the availability of rRNA probes and the metabolic state of the propagules; spores tend not to show significant fluorescence. Most actinomycetes have well defined life-cycles which can be examined in a qualitative way using light and scanning elec-tron microscopy (Wellington et al., 1990). The latter offers the best pos-sibilities for examination of life cycles as intact soil crumbs can be studied.

Methods can, therefore, be divided into direct and indirect tech-niques, the former requiring some form of microscopy and the latter viable culture. Enumeration methods that do not involve microscopy or viable culture include extraction and quantification of DNA and RNA (see chap-ter 35 in this book). The major consideration with nucleic acid extraction from actinomycetes in soil is the efficient lysis of spores. Studies of strep-tomycete spores have shown that bead beating of a soil suspension lyses spores enabling efficient DNA extraction (Cresswell et al., 1991).

15–1.2 Direct Methods for Observation and Enumeration

15–1.2.1 Scanning Electron Microscopy (SEM) (Wellington et al., 1990)

Soil can be sprinkled directly on to adhesive (silver dag) covering a specimen stub or in the case of a clay soil, crumbs can be sectioned with a scalpel and then placed on the adhesive. The soil is then sputter coated using gold-palladium alloy. Another approach is to insert glass or plastic cylinders into the soil and samples from different depths can be extruded or the plastic cylinder sectioned. Cores are best frozen prior to sectioning (Locci, 1969).

Mycelia are often visible at × 3000 magnification. Larger morphological structures such as sporangia (about 20 μm), sclerotia and pseudo sporangia can also be seen at this magnification. Individual spores and fragmentation patterns are observed at × 7000 to 10 000.

The extent of the mycelial colonization of soil can be estimated using the direct observation of soil particles (Clewlow et al., 1990). Place a stub in the SEM, locate the approximate position of the center of the stub and note the SEM micrometer position. Adjust the field of view to lie at one corner of a 1 cm² region centered on the stub (stubs approximately 2–5 cm diam.). Set the magnification so that the field of view is 100 μm². Alter the field of view in 1-mm steps both horizontally and vertically so that 100 views of the 1 cm² region of the stub are observed. Study each field of view carefully for signs of mycelial growth and differentiation. Each field of view is classified as no soil, soil only, low growth, medium growth and high growth. The percentage number of observations are then noted and the extent of soil colonized can be estimated.

15–1.2.2 Oligofluors for Detection by Light Microscopy

Oligonucleotide probes have the potential for in situ detection and identification of individual cells. These probes are targeted toward rRNA and can be labelled with either radioisotopes or fluorescent dyes. Such a probe/fluorescent dye complex is referred to as an oligofluor. Since rRNA content is correlated with activity of cells, probing soil populations in situ does not always result in significant hybridization signals because many intact bacterial cells are dead or of low metabolic activity. The following method has proved successful for detection of actinomycetes in soil (Hahn et al., 1992).

15–1.2.2.1 Oligonucleotide Probes. Ribosomal RNA targeted oligonucleotide probes are synthesized with a primary amino group at the 5' end (Aminolink 2; Applied Biosystems, Foster City, USA). The fluorescent dye tetramethylrhodamine isothiocyanate (Tritc, Research Organics, Cleveland, OH), is covalently linked to the amino group, and the dye-oligonucleotide conjugate (1:1) is purified from unreacted components and stored at −20 °C in double distilled water at a concentration of 50 ng μL^{-1} (Amann et al., 1990). The DNA specific dye 4′6-diamidino-2-phenylindole (DAPI) (Sigma) is stored in a 1 mg mL^{-1} solution at −20 °C.

15–1.2.2.2 In Situ Hybridization. For each sample apply 1 μL and 10-fold-dilutions of fixed cell suspensions to gelatin coated slides (0.1% gelatin, 0.01% KCr(SO$_4$)$_2$) and allow to air dry. Following dehydration in 50, 80, and 100% ethanol for 3 min, hybridize each of the preparations in 50 μL hybridization buffer (0.9 M NaCl, 0.1% SDS, 20 mM Tris, pH 7.2) and 1 μL probe (50 ng) at 45 °C for 1 h. After hybridization wash the slides in hybridization buffer for 20 min at 48 °C, rinse with distilled water, and air dry. Afterwards add 1 μL of a 1 μL mL^{-1} DAPI solution to each sample, cover with 5 μL hybridization buffer, incubate for 5 min at room temperature, rinse with distilled water, and air dry. Examine preparations with a microscope fitted for epifluorescence (with filter sets for 400 and 580 nm).

15–1.2.2.3 Cell Extraction and Fixation. Extract cells from 1 g of soil prior to cell fixation with either, (i) 10 mL 0.1% sodium pyrophosphate, pH 7.2, or (ii) 10 mL 0.1% pyrophosphate and a drop of Nonidet P-40 (Sigma), (alternatively fix cells directly in the soil sample with between 1 and 5 mL fixation buffer [4% paraformaldehyde/PBS, pH 7.2–7.4] for 3–16 h). Mix samples on a vortex mixer for 10 s and keep on ice for 2 min to allow a separation of heavy soil particles from the supernatant. Remove the supernatants and re-extract the soil pellets with either 2 mL PBS or 2 mL fixation buffer. Centrifuge the combined supernatants of the pyrophosphate extractions at 500 × g for 15 min, discard the supernatant and resuspend the pellet in 1 mL fixation buffer. After fixing for 16 h (overnight), centrifuge all samples at 8000 × g for 5 min, wash in 1 mL PBS, and resuspend the pellet in 1 mL 50% ethanol/PBS and store at −20 °C. Use 1 μL of each sample as well as 10-fold-dilutions for microscopic determination of cell recoveries after in situ hybridization.

Due to inactivity of populations in soil it may be necessary to incubate soil prior to extraction and fixation.

Incubate 1 g of soil at 30 °C for 16 h after the addition of 0.2 mL of Luria Broth (LB, Difco) medium. Alternatively, 0.2 mL of a starch or chitin-based medium (15–1.3.4, recipes (2) or (4) without agar) can be used as a selective treatment for actinomycetes. Fix and hybridize this sample, together with an unamended sample. Amendment and incubation may allow the growth of specific organisms, therefore, allowing detection.

15–1.2.2.4 Comments. Binding of oligonucleotide probes to soil does not occur, eliminating the need for counter staining and blocking techniques as used with fluorescent antibody techniques. Autofluorescence of soil organic compounds can however interfere with the detection of bacteria. Direct fixation followed by extraction with fixation buffer results in lower contamination than pyrophosphate extraction followed by fixation, the former keeping organic material in larger particles (and also minimizing loss of bacteria attached to soil particles during the extraction procedure). To achieve reliable counts detection limits will be in the range 10^5 to 10^6 cells g^{-1} soil depending on soil type and the amount of liquid needed to obtain a soil suspension. Hybridization signals of spores, however, are not observed but in a study of soil and plant colonization by Bramwell (1992), young and old streptomycete mycelium showed the same degree of fluorescence. Pure cultures of *Streptomyces* spp. show much lower permeability to oligonucleotide probes than cells grown in soil. It is important, therefore, to pretreat pure cultures of cells with lysozyme (as outlined above) but it does not seem to be necessary to treat soil grown streptomycetes in this way.

15–1.3 Methods for Isolation and Enumeration

15–1.3.1 Principles

Most isolation techniques involve a cursory extraction method coupled with a dilution of biomass to allow culturing. In many cases, actinomycete genera are present in soil at $< 10^4$ colony forming units (CFU) g^{-1}. There is a need, therefore, to develop methods that concentrate cells in a given sample and improve chances of detection. Soil biomass extraction can be either physical or chemical or a combination of the two. Mycelia will become intimately associated with soil particles and so there is an even greater need to carry out some form of dispersal to extract cells. Actinomycetes may also be present in different morphological forms within a soil sample and it is possible to differentially extract and enumerate spores and mycelium using a chemical disruption technique (Herron & Wellington, 1990).

15–1.3.2 Physical Disruption

Ringer extraction (Wellington et al., 1990): Shake 1 g of soil in 9 mL one-fourth strength Ringer solution (NaCl, 2.25 g; KCl, 0.015 g; CaCl$_2$, 0.12 g; NaHCO$_3$) on a Griffin shaker for 10 min. Serially dilute supernatant and count on selective agar.

Homogenization (Baecker & Ryan, 1987): Add 1 g soil to a 150 mL conical flask containing 99 mL sterile distilled water with Teepol (surfactant) at 10 μg mL^{-1}. Attach the flask to a high efficiency homogenizer drive, and surround the flask with crushed ice. Homogenize at high speed for 10, 20, 30, 60, and 90 s, to establish the period required to release maximum numbers of actinomycetes from a given soil. Sample at each time point, serially dilute and plate on to appropriate agar.

15–1.3.3 Chemical Disruption

Ion-exchange resin (Herron & Wellington, 1990): Transfer 100 g soil to a 500 mL centrifuge pot, add 20 g of Chelex 100 (Biorad), 100 mL of 2.5% polyethylene glycol 6000 (Sigma), 0.1% sodium deoxycholate (Sigma). Shake the pot gently with occasional fast agitation for 2 h at 4 °C on a Griffin flask shaker. Centrifuge for 30 s at 850 $\times$ g and retain pellet. This pellet contains a proportion of the mycelium, usually wrapped around fragments of organic matter and soil, a sample can be serially diluted and plated to allow enumeration and isolation. Filter the remaining supernatant through a membrane filter holder (Millipore Corp.) with no filter present, the metal support being sufficient to remove any Chelex 100 from the liquid. Add a further 100 mL PEG deoxycholate to the original pellet and re-extract as before. After the second filtration pool the supernatants and centrifuge at 2200 $\times$ g for 15 min. Allow the resulting pellet to re-suspend overnight in one-fourth strength Ringer solution before measuring the volume of the suspension and determining the viable count. This method is selective for streptomycete spores and all spores of a similar density. If the final centrifugation is increased to 17 700 × g then small mycelial fragments, rods, and cocci are pelleted and can be counted by selective plating.

15–1.3.4 Selective Media

A wide range of media have been described for the isolation and culture of soil isolated actinomycetes. The following media are the most often used in various combinations with soil pretreatments and selective inhibitors for the purposes of isolation and enumeration of specific groups of genera. In general, incubations are done at 28 to 30 °C for 7 to 20 d depending on the extent of overgrowth by fungi and bacteria.

1. Arginine glycerol salts agar (El-Nakeeb & Lechevalier, 1963): arginine, 1.0 g; glycerol, 12.5 g; K_2HPO_4, 1.0 g; $MgSO_4·7H_2O$, 0.5 g; $Fe_2(SO_4)·6H_2O$, 0.01 g; $CuSO_4·5H_2O$, 0.001 g; $ZnSO_4·7H_2O$, 0.001 g; $MnSO_4$, 0.001 g agar, 15.0 g; adjust to 1 L with distilled water and pH 6.9 to 7.1.

2. Colloidal chitin agar (Hsu & Lockwood, 1975): colloidal chitin (dry weight), 2.0 g; K_2HPO_4, 0.7 g; KH_2PO_4, 0.3 g; $MgSO_4·5H_2O$, 0.5 g; $FeSO_4·7H_2O$, 0.01 g; $ZnSO_4$, 0.001 g; $MnCl_2$, 0.001 g; agar, 20.0 g; adjust to 1 L with distilled water and pH 7.0. The colloidal chitin is made by dissolving 40 g of ground chitin in 400 mL of conc. hydrochloric acid with stirring for 30 to 40 min. Add 2 L cold water to precipitate the chitin, filter using a Buchner funnel and coarse filter paper and wash through 5 L tap water, then refilter. Repeat the washing and filtration of the chitin at least three times. Recover the filter cake, if autoclaved this paste can be stored at 4 °C, determine the dry weight by drying a sample at 100 °C.

3. Oatmeal agar: oats, 20.0 g; agar, 15.0 g; 1 L of distilled water; make up by steaming or simmering oats in 500 mL for 3 h, adjust pH to 7.2. Further additions include autoclaved soil extract (Shearer, 1987) 500 mL filtrate after shaking with 150 g soil and yeast extract, 1.0 g.

4. Starch casein (Kuster & Williams, 1964): soluble starch, 10.0 g; casein hydrolysate (Difco), 0.3 g; KNO_3, 2.0 g; NaCl, 2.0 g; K_2HPO_4, 2.0 g; $MgSO_4·7H_2O$, 0.05 g; $CaCO_3$, 0.02 g; $FeSO_4·7H_2O$, 0.01 g; agar, 20.0 g; adjust to 1 L with distilled water and pH 7.0.

5. M_3 agar (Rowbotham & Cross, 1977): KH_2PO_4, 0.466 g; Na_2HPO_4, 0.732 g; KNO_3, 0.1 g; NaCl, 0.29 g; $MgSO_4·7H_2O$, 0.1 g; $CaCO_3$, 0.02 g; Sodium propionate, 0.2 g; $FeSO_4·7H_2O$, 200 µg; $ZnSO_4·7H_2O$, 180 µg; $MnSO_4·4H_2O$, 20 µg; agar, 18.0 g; adjust to 1 L with distilled water and pH 7.0 then add cycloheximide, 50.0 mg and thiamine hydrochloride, 4.0 mg.

6. HV agar (Hayakawa & Nonomura, 1987): humic acid (in 10 mL of 0.2 M NaCl), 1.0 g; Na_2HPO_4, 0.5 g; KCl, 1.71 g; $MgSO_4·7H_2O$, 0.05 g; $FeSO_4·7H_2O$, 0.01 g; $CaCO_3$, 0.02 g; agar, 18.0 g; adjust to 1 L with distilled water and pH 7.2 than add B vitamins, thiamine hydrochloride, 0.5 mg; riboflavin, 0.5 mg; niacin, 0.5 mg; pyridoxine hydrochloride, 0.5

mg; inositol, 0.5 mg; calcium pantothenate, 0.5 mg; para-aminobenzoic acid, 0.5 mg; biotin, 0.25 mg; cycloheximide, 50.0 mg. Humic acid is extracted from 500 g of soil (A-horizon of forest soil) by suspending it in alkaline solution (5.0 g NaOH in 1 L), leave to stand for 24 h with occasional stirring. Centrifuge at 7000 rpm for 10 min to remove precipitate and acidify supernate to pH 1.0 with conc. HCl. Recover the precipitate by centrifugation at 3000 for 20 min and wash three times by centrifugation with 150 mL of water. Freeze the suspension overnight at −20 °C and, after thawing, filter and wash the granules that can then be air dried.

15–1.3.5 Selective Methods

Selective isolation procedures can be divided into two groups, firstly nonspecific methods for isolation of a range of actinomycctc groups and secondly specific procedures for certain genera. Not all genera have a specific selective isolation procedure and the efficiency of any of these techniques depends on the background population in the environmental sample and type of sample. The first consideration is for selective treatment to counter select for eubacteria and fungi by use of soil pretreatments or selective inhibitors. The second aspect is the use of specific combinations of media and pretreatments to enhance the recovery of certain groups. Evaluation of methods is imprecise due to sample variation.

15–1.3.5.1 Heat Treatments. A soil sample can be treated to reduce the numbers of nonspore forming bacteria by air-drying at room temperature, this will reduce viability of mycelial propagules and fragments. For spore-formers drying overnight at 60 °C is more selective and will give cleaner plates. Subsequent wetting to 15% wt/wt and addition of 1% wt/wt CaCO$_3$ followed by incubation for 1 to 2 d allows enrichment of actinomycetes particularly streptomycetes, micromonosporas and other sporoactinomycetes. Heat treatment of air-dried soil to 120 °C for 1 h is selective for the genera *Actinomadura, Dactylosporangium, Micromonospora, Microbispora, Microtetraspora, Saccharomonospora, Streptosporangium*, and *Thermomonospora* when used in combination with the appropriate agar (always use dried plates) and selective inhibitors given in the following section.

15–1.3.5.2 Selective Inhibitors. Few antibacterial antibiotics can be incorporated in selective media without inhibiting some component of the actinomycete population. Detergents are more effective against vegetative cells and this allows selective plating of spore extracts described in 15–1.3.3. Suspension of the sample in 0.05% wt/vol SDS at 40 °C for 20 min (Nonomura & Hayakawa, 1988) followed by dilution and plating improves recovery of actinomycete spores which germinate on isolation plates.

Antifungal antibiotics can be incorporated into all of the media 1–6 listed (15–1.3.4). Cycloheximide (50 µg mL^{-1}, Sigma) is heat stable and can be added to media before sterilization, nystatin (50 µg mL^{-1}, Sigma) is added in combination but as an aqueous suspension after autoclaving.

Fungicides such as Storite (Merck, Sharpe, and Dome) at 50 µg mL^{-1} are useful if fungal overgrowth of plates occurs (Bramwell, 1992).

15–1.3.5.3 Isolation of Streptomycetes. Fresh soil: spore extraction, (15–1.3.3) spread plate on any of the media 1 to 6 (15–1.3.4), preference is given to media 1, 2, and 4; cycloheximide and nystatin or Storite (15–1.3.5.2); penicillin G 10 µg mL^{-1}. Ringer extraction (15–1.3.2) for spore plus mycelial count, dilution plate as for spores.

Air-dried soil and soil heated to 60 °C overnight can be used and treated as above with antifungal antibiotics but antibacterials may not be needed. Plates incubated at 28 °C for 7 d will allow isolation and enumeration. Thermotolerant strains can be isolated at 40 °C and acidophils on agar adjusted to pH 4.5.

15–1.3.5.4 Isolation of Micromonospora. Heat air-dried soil to 120 °C for 1 h and dilution plate on media 1, 2, 5, or 6 (15–1.3.4). Include cycloheximide and nystatin if fungal contamination occurs and increased selectivity can be achieved by incorporation of lincomycin (20 µg mL^{-1}, Sigma) and novobiocin (20 µg mL^{-1}, Sigma) (J. Mullins & E.M.H. Wellington, 1992, unpublished data). Tunicamycin (20 µg mL^{-1}, Sigma) and nalidixic acid (20 µg mL^{-1}, Sigma) can be used as an alternative with medium 6 (Hayakawa et al., 1991a,b).

15–1.3.5.5 Isolation of *Actinomadura, Microtetraspora, Microbispora,* and *Streptosporangium*. Heat air-dried soil to 120 °C for 1 h and dilution plate on media 1, 4, or 6, all media should contain B vitamins listed for medium 6 (15–1.3.4). Gentamicin (2 µg mL^{-1}, Sigma) will aid the recovery of *Actinomadura, Microtetraspora,* and *Streptosporangium* (Shearer, 1987). Incubate plates for up to 30 d if possible.

15–1.3.5.6 Baiting Techniques for *Actinoplanes* and Related Genera. Add 1.0 g air-dried soil to a Petri dish and flood with sterile distilled water. Presterilized baits of hair or pollen are added to the water surface and the dish incubated at 20 °C for several weeks. The baits are then transferred to a transparent agar and examined at × 100 using a stereo microscope. Sporangia can be separated from the bait and rolled over the agar surface with a needle then transferred to medium 6 (15–1.3.4) and incubated at 28 °C for 7 d. Actinoplanetes can also be isolated by the use of chemotactic techniques for attraction of motile zoospores using xylose and γ-collidine (Hayakawa et al., 1991b). Dehydration and rehydration of soil can also be used to encourage release of zoospores from desiccation resistant sporangia.

15–1.3.5.7 Isolation of Nocardioforms. A range of actinomycete genera do not form spores but the substrate and aerial mycelium undergoes fragmentation. Such propagules do not always survive the drying or heating of soil, as in the case of *Nocardia* spp. Selective isolation has been achieved by the use of antibiotics, baiting with paraffin, to enrich for the many species able to degrade hydrocarbons and shaking soil with NaOH (8%).

Nocardia can be isolated by dilution plating a soil suspension (5% vol/vol) made up in Ringer solution and plating onto selective diagnostic sensitivity test agar (Oxoid CM261), pH 7.4, add sterile solution of dimethylchlortetracycline (5 µg mL^{-1}), chlortetracycline can also be included (45 µg mL^{-1}) to reduce numbers of competing bacteria (Orchard et al., 1977). Plates are incubated at 25 °C for 21 d. Rhodococci can be isolated using medium 5 (15–1.3.4) but a heating pretreatment is included; suspend soil (1:10) in Ringer solution (pH 7.0) containing gelatin (0.01%) and heat in a water bath at 55 °C for 6 min. After pretreatment mix samples by vortexing and spread 0.2 mL aliquots of neat and 10-fold dilutions onto medium 5, incubate at 30 °C for 7 d. Additional agar supplements have included *n*-alkanes, nalidixic acid, novobiocin, and potassium tellurite (Goodfellow, 1991).

15–1.4 Comments

An overview of selective procedures for the main groups of actinomycetes is provided. The efficiency of selection depends on the numbers and range of competing bacteria and other actinomycetes in the soil being sampled. In most cases where spores are produced, actinomycetes withstand a heat pretreatment. This is the most effective measure to discourage faster growing bacteria but is not always effective against fungal spores. Antibiotics added to media are frequently used to increase selectivity but often results in greatly reducing diversity of species and genera recovered. The expression of resistance may also depend on the physiological state of the propagule in the soil. Antibiotics recommended for selective isolation usually have a diagnostic value of 80- to 100% for species in the genus. Exceptions occur, such as the use of gentamicin to isolate micromonosporas (Shearer, 1987), where less than one-half of the genus have resistance. This may relate to a need for isolation of gentamicin-producing micromonosporas.

15–2 ISOLATION OF PHYSIOLOGICAL GROUPS

15–2.1 Principles

Actinomycetes are widely distributed in many environments and are generally regarded as being active saprophytes growing under well-aerated conditions in temperate and tropical soils. Some techniques have been used to recover actinomycetes outside the standard condition for growth given in 15–1.3. These include the isolation of autotrophs, microaerophils, thermophils, and acidophils. In addition, ability to degrade complex polymers, hydrocarbons, and the production of biologically active compounds has been studied. Degradative and bioactive properties are better studied by the screening of isolates than by direct detection on isolation plates. Such indicator plates tend to be inappropriate for selective isolation unless the attribute is highly selective such as growth on cresol.

15–2.2 Isolation of Autotrophs

Several actinomycetes are capable of autotrophic growth; *Amycolata autotrophica* can grow aerobically where H and CO_2 are available (Aragno & Schlegel, 1981). Thermophilic carboxydotrophic actinomycetes have been isolated from soil and compost, one of which was identified as a streptomycete and could grow on CO_2 or CO and H_2 (Bell et al., 1987). The isolation procedure used to obtain this isolate is given below.

Prepare mineral base E medium (Owens & Keddie, 1969) supplemented with 10 µg L^{-1} of *p*-aminobenzoic acid and dispense 50 mL into conical flasks sealed with Suba-seal stoppers. Introduce gases after sterilization by syringe to produce an atmosphere of 25% CO, 25% H_2, and 50% air (by volume), add 0.1 mL suspension of soil or compost and incubate statically at 50 °C. Take 0.1 mL from flasks showing growth and subculture into 20 mL of the same medium and incubate under the same conditions. Serially dilute from the flasks and plate onto mineral medium with 1.5% (wt/vol) Oxoid purified agar (Difco Bacto or Difco Noble agar). Incubate the plates for 1 wk in a stainless steel jar partly evacuated and filled with the same mixture of gases as before.

15–2.2.1 Comments

The autotrophic actinomycetes isolated by this procedure were facultative and able to grow heterotrophically on a range of media (see 15–1.3.4). Heterotrophic growth suppressed levels of calvin cycle enzymes and CO oxidoreductase in the facultative autotrophic streptomycete (Bell et al., 1987). *Amycolata autotrophica* grew better heterotrophically (Embley, 1991).

15–2.3 Isolation and Enumeration of Thermophils

Habitats such as composts and self-heated organic matter often contain large numbers of actinomycete spores. These spores are hydrophobic and more readily isolated by creating a spore cloud using a wind tunnel (Lacey, 1971) or sedimentation chamber with an Andersen sampler (Lacey & Dutkiewicz, 1976). Such methods are less effective with soil samples as spores are not so readily dislodged from the substrate although soils have been treated in this way. The spore cloud is made by placing 10 to 15 g of sample enclosed in a terylene net bag suspended by a thread to the lid of the sedimentation chamber. The bag is dropped from about 30 cm to knock off spores and a fan switched on for 5 min. The spore cloud is sampled by drawing air through an Andersen sampler (5 s with a suction of 25 L min^{-1})

which is loaded with agar plates. Half-strength nutrient or tryptone soya agars plus 0.2% casein hydrolysate (Oxoid) with cycloheximide (50 μg mL^{-1}).

The Andersen sampler contains six stages allowing separation of different sized samples; colonies of *Saccharopolyspora rectivirgula* were recovered on stages three and four (2–6 μm diam.), *Saccharomonospora viridis* on stage five (1–2 μm) and *Thermoactinomyces vulgaris* on stages five and six (< 1–2 μm) (Lacey & Dutkiewicz, 1976). Members of the latter genus form endospores and are grouped in the *Clostridium-Bacillus* branch of the 16S rRNA phylogenetic tree. *Thermomonospora* can also be recovered using the above technique although an alternative selective agar, R8, has been used to recover *Saccharomonospora viridis* (Amner et al., 1989). R8 contains MgCl$_2$·6H$_2$O, 5.1 g; yeast extract, 10.0 g; Lab m agar, 22.0 g; add after autoclaving sterile solutions of CaCl$_2$·2H$_2$O (5 *M*: 10 mL L^{-1}); L-proline (20% wt/vol; 15 mL L^{-1}); NaOH (1 *M*: 7 mL L^{-1}); cycloheximide, 50 μg L^{-1}; total volume 1 L; pH 8.0.

15–2.3.1 Comments

Thermophils can be isolated from soils by traditional dilution pour plating using the above media and those described in section 15–1.3.4. Plates prepared from spore clouds or suspensions in Ringer solution (10% wt/vol) should be incubated for about 3 d in a range of temperatures from 40 to 55 °C. With the exception of the green-spored *Saccharomonospora viridis*, many thermophilic species have a powdery yellowish-white appearance and include members from the genera *Pseudonocardia*, *Saccharomonospora*, *Saccharopolyspora*, *Saccharothrix*, *Streptomyces*, and *Thermomonospora*.

15–2.4 Isolation of Acidophilic and Alkalophilic Actinomycetes

Alkalophilic actinomycetes can be isolated from soil using a range of media adjusted to pH 10 to 12. A small number of strains classified in the genera *Amycolata*, *Saccharothrix*, and *Streptomyces* have proved capable of growth on agar adjusted to pH 11.5 and above (Mikami et al., 1985). However, members of the genus *Nocardiopsis* are more commonly associated with alkaline conditions and have been isolated from soil using media at pH 11.5 (Mikami et al., 1982; Tsujibo et al., 1988). For some species, optimal conditions for growth are above pH 10.0. Yeast extract-malt extract agar can be used for isolation of this genus containing yeast extract, 4.0 g; malt extract, 10.0 g; glucose, 4.0 g; agar, 20.0 g made up to 1 L with distilled water and adjusted to pH 11.5 with NaOH after autoclaving. Media listed in 15–1.3.4 can also be used at pH 11 to 12, in particular oatmeal and starch casein agars. A selective medium is recommended for the isolation of alkalophilic strains of *Nocardiopsis*.

15–2.4.1 Selective Agar for Alkalophilic *Nocardiopsis* Strains

Glucose asparagine agar (Nonomura & Ohara, 1971a); glucose, 1.0 g; L-asparagine, 1.0 g; K_2HPO_4, 0.3 g; $MgSO_4 \cdot 7H_2O$, 0.3 g; NaCl, 0.3 g; agar, 15.0 g; distilled water 800 mL; trace salts, 1.0 mL; autoclave this separately and adjust to pH 8.0 or higher, pour plates and allow agar to set. Then pour 4.0 mL of a sterile solution of casamino acids, 0.5 g L^{-1}, over the set agar. The trace salt solution contains $FeSO_4 \cdot 7H_2O$, 10 mg mL^{-1}; $MnSO_4 \cdot 7H_2O$, 1 mg mL^{-1}; $CuSO_4 \cdot 5H_2O$, 1.0 mg mL^{-1}; $ZnSO_4 \cdot 7H O$, 1.0 mg mL^{-1}. The antibiotics cycloheximide, nystatin (50 μg mL^{-1}); polymyxin B (4 μg mL^{-1}); penicillin (0.8 μg mL^{-1}) can be added as sterile solutions after autoclaving. Isolates can be obtained by plating soil suspensions or suspending a small amount of soil in 1 mL and spreading 100 μL aliquots onto dry agar plates, incubate at 27 °C for 14 d (Horikoshi, 1971).

15–2.5 Acidophilic Actinomycetes

Acidophilic actinomycetes can be isolated from soil using starch casein (15–1.3.4) but with the K_2HPO_4 omitted and add a sterile solution of KH_2PO_4 to 0.55% (wt/vol) and pH 4.5 (Williams et al., 1971). Streptomycetes were isolated from various soils using this medium with the pour plate technique.

15–2.6 Isolation and Enumeration of Microaerophilic Actinomycetes

Obligate anaerobes have been described within the genus *Micromonospora* but their taxonomic status is uncertain. Members of the genus *Actinomyces* have been isolated from soil by use of stoppered tubes.

15–2.6.1 Isolation of Catalase-negative *Actinomyces* Species (Gledhill & Casida, 1969)

Blend 1.0 g soil with Heart infusion broth (Difco), adjust to pH 7.8 with KOH, blend for 1 min with 1 min interval and repeat. Serially dilute suspension in Heart broth and inoculate screw-capped tubes of water agar with 10^{-2} and 10^{-6} dilutions. Incubate at 30 °C for 4 wk. Subculture onto Heart infusion agar (pH 7.8).

15–2.7 Isolation and Detection of Actinomycetes with Biodegradative Activity

Experimental data have been collated indicating that actinomycetes play a role in the degradation of lignocellulose (McCarthy, 1987). In addition, their ability to produce extracellular biodegradative enzymes has resulted in their use for production of chitinases, cellulases, xylanase, proteases, and peroxidases. Strains capable of these activities have been isolated by baiting or enrichment of soil with the target substrate, or by use of isolation media containing the substrate as sole C source. Cellulolytic and ligninolytic streptomycetes have been isolated by inoculating broth containing extracted heartwood with soil or compost (Phelan et al., 1979). Incorporation of complex substrate into selective media may also allow detection of enzyme activity by the appearance of a clearance zone in a medium containing insoluble chitin, xylan, or cellulose preparations. Clearance zones on colloidal chitin and starch casein agars (15–1.3.4) will provide evidence for production of enzymes.

15–2.7.1 Media for Detection and Isolation

1. Basal medium (Lamot & Voets, 1976): $(NH_4)_2SO_4$, 2.5 g; K_2HPO_4, 0.25 g; NaCl, 0.1 g; $MgSO_4 \cdot 7H_2O$, 0.125 g; $FeSO_4 \cdot 7H_2O$, 0.0025 g; $MnSO_4 \cdot 4H_2O$, 0.0025 g; yeast extract, 1.0 g; agar, 12.0 g; made up to 1 L and adjusted to pH 7.2

2. Substrates can be added to the basal medium as follows: cellulose, 10.0 g L^{-1} (Avicel, Sigma; ball-milled cellulose powder, ball mill Whatman CF-11 for 72 h); wood pulp, 5.0 g L^{-1}; xylan, 10.0 g L^{-1}; lignin model monomers such as Poly-blue, Poly-red, Remazol brilliant blue, at 0.2 g L^{-1}; carboxymethylcellulose (CMC-9M8F), 5.0 g L^{-1}.

To detect breakdown of substrate, examine the plates for the appearance of clearance zones that may be produced after 10 to 14 d. For CMC, flood the plates with Congo red (1% in dist. H_2O) and agitate, pour off after 15 min and flood with M NaOH and after 15 min with gentle agitation clearance zones may become evident. Plant et al. (1988) described the use of cellulose-azure for detection of cellulolytic actinomycetes; cellulose-azure (Calbiochem Chemical Co. ref. 219481) suspension (6% wt/vol) is autoclaved and added to molten basal medium to give a concentration of 2% wt/vol. Overlay 0.2 mL of this mixture over tubes of basal medium and detect activity by release of blue color into the basal layer.

3. Pectin hydrolysis can be detected using the following medium (Hankin et al., 1971); three solutions are made up and sterilized separately then combined: (i) KH_2PO_4, 4.0 g; Na_2HPO_4, 6.0 g; (ii) pectin, 5.0 g; (iii) $(NH_4)_2SO_4$, 2.0 g; yeast extract, 1.0 g; $MgSO_4 \cdot 7H_2O$, 0.2 g; $FeSO_4 \cdot 7H_2O$, 0.001 g; $CaCl_2$, 0.001 g; agar, 10.0 g. Final volume of 1 L and pH adjusted to 7.4 after autoclaving. The medium can be used for isolation and screening. To detect pectinolytic activity flood plates after 6 d with an aqueous solution of hexadecyltrimethylammonium bromide (1% wt/vol), leave to stand for 1 h and pour off. The cetrimide reveals clearance zones around colonies.

15–2.8 Detection of Antimicrobial Activity

Actinomycetes are the major group of antibiotic-producing bacteria used for commercial production of antibiotics and other biologically active compounds. There is evidence that antibiotics are produced under natural conditions (Weller & Thomashow, 1990) and that bioactive streptomycetes can be antagonistic to bacteria in soil (Turpin et al., 1992).

Attempts have been made to determine directly on isolation plates the antagonistic activity of a particular soil isolate but these have largely been unsuccessful due to problems of overlaying soil isolation plates. A broad screen for antibiotic-producing isolates can be achieved by detecting antimicrobial activity (Williams et al., 1983). Inoculate nutrient agar (Oxoid) plates (glass Petri-dishes) with a drop of a spore suspension, six actinomycete isolates can be placed on each plate. After 48 h (when all colonies are visible) incubation at 28 °C invert the plates and pour chloroform into the lids, leave for 40 min. Pour off any remaining chloroform and leave plates half open in a laminar flow cabinet for 20 min to remove any chloroform vapors. Overlay the colonies with a soft molten nutrient agar (5 mL of 70% of Oxoid nutrient agar) which has been seeded with 20 µL of log-phase culture of test bacterium. The following test bacteria were used by Williams et al. (1983) *Escherichia coli* NCIB 9132; *Pseudomonas fluorescens* NCIB 9046; *Micrococcus luteus* NCIB 196; *Bacillus subtilis* NCIB 3610; *Streptomyces murinus* ATCC 19788; *Saccharomyces cerevisiae* CBS 1171; *Candida albicans* CBS 562; *Aspergillus niger* LIV 131 (Dep. of Genetics and Microbiology, Univ. of Liverpool). Fungi were seeded into malt extract agar (Difco) and bacteria in nutrient agar (Oxoid). Incubate plates overnight at 30 °C and examine for zones of inhibition.

15–3 GROUPING AND IDENTIFICATION OF ACTINOMYCETES

15–3.1 Principles

The suprageneric grouping of the actinomycetes are now largely based on phylogenetic relationships derived from 16S rRNA sequencing (Stackebrandt, 1991). Delimitation of genera can be achieved by a combination of chemotaxonomic, morphological and physiological tests. Currently, a genus-specific rRNA probe has been described for the *Streptomyces* genus (Stackebrandt et al., 1991) but more sequence data is required to provide such probes for other groups. Group-specific rRNA probes will provide a rapid means of screening actinomycete isolates and can be used with crude rRNA or colony blots. The key to the main soil-inhabiting genera (Table 15–2) and techniques recommended below are intended as a preliminary guide to generic status. More precise and definitive identifications require extensive studies and consultation of introductions to the relevant sections in *Bergey's Manual of Systematic Bacteriology* (Goodfellow, 1989) and *The Prokaryotes* (Ensign, 1991) is recommended.

15–3.2 Cell Wall Analysis

Despite some anomalies, the cell wall chemotype (Becker et al., 1965) is still a useful and rapid technique for identifying genera and involves

Table 15–2. Selected distinguishing features of actinomycete genera.

Diagnostic amino sugar †	Morphological features	Genus
LL-DAP or 1:1 LL-DAP to *meso*-DAP	Aerial mycelium with chains of spores, stable mycelium, powdery appearance	*Streptomyces*
	No aerial mycelium, club-shaped sporangia	*Kineosporia*
LL-DAP	Fragmentation of aerial and substrate mycelium	*Nocardioides*
	Only aerial hyphae with motile propagules	*Sporichthya*
meso-DAP	a. Substrate mycelium only, fragmenting to rods and cocci:	
	Pink-orange colonies	*Rhodococcus*
	Microaerophilic	*Agromyces*
	Cream-beige colonies	*Mycobacterium*
	b. Substrate mycelium only, no fragmentation:	
	Single, heat resistant spores, orange colonies	*Micromonospora*
	Sporangia, motile spores, buff-orange colonies	*Actinoplanes*
	Finger-like sporangia, three to four motile spores	*Dactylosporangium*
	c. Powdery aerial mycelium, sporangia formed:	
	Globose or irregular sporangia	*Streptosporangium*
	Cylindrical, monosporic sporangium, motile spore	*Planomonospora*
	Spherical sporangia, many motile spores	*Spirillospora*
	d. Powdery aerial mycelium, spore chains on sporophores:	
	Short, curled chains	*Actinomadura*
	Spore chains with two spores	*Microbispora*
	Spore chains with four spores	*Microtetraspora*
	Very long spore chains	*Saccharopolyspora*
	Densely packed single spores along hypha	*Saccharomonospora*
	Single spores on branched or single sporophores or in clusters	*Thermomonospora*
	e. Aerial or substrate mycelium fragment:	
	Powdery aerial, can be extensive fragmentation	*Nocardia*
	Chains of ovoid fragments, substrate mycelium fragments	*Nocardiopsis*
		Saccharothrix
		Amycolata
		Amycolatopsis
	Irregular length fragments on aerial, budding occurs	*Pseudonocardia*

† DAP, diaminopimelic acid in whole cell hydrolysates.

determining the major amounts of LL and *meso* isomers of diaminopimelic acid (DAP) in the cell wall. A simple technique based on thin layer chromatography (Staneck & Roberts, 1974) can be used with some modifications to allow a rapid screen of many isolates. Subculture isolates onto glucose yeast extract agar; glucose, 3.0 g; yeast extract, 5.0 g; peptone, 5.0 g; agar, 15.0 g; CaCO₃, 7.5 g; make up to 1 L with dist. H₂O and adjust the pH to 6.3 before autoclaving. Inoculate a 1 in. square patch of medium and incubate at 28 °C for 10 d. Cut out the square patch of agar with growth and place in a bijou bottle containing 1 mL of 6 *M* HCl, screw cap and autoclave for 20 min at 121 °C. After cooling add 2 mL dist. H₂O and filter with

Whatman no. 1 into clean bottle. Place the bottle in a boiling water bath or oven in a fume hood and evaporate to dryness, repeat three times using 2 mL dist. H_2O to resuspend sample. Resuspend final residue in 20 μL dist. H_2O and apply 5 and 10 μL to the base of a cellulose TLC sheet (Merck no. 5552). Run for 3 h in a chromatography tank containing the following solvent system; pyridine : 6 M HCl : dist. H_2O : methanol (10:4:26:80 vol/vol). Spray with 0.2% ninhydrin in acetone and develop the plate at 100 °C for 3 min. DAP spots are olive green fading to yellow and the LL isomer moves ahead of the *meso*. A standard mixture of the isomers (DL-DAP, Sigma) can be run as reference (0.01 M solution).

15–3.3 Maintenance of Isolates

The media listed in 15–1.3.4 can all be used for subculturing oatmeal agar with yeast extract supports spore formation in many genera and if growth is poor an alternative is glucose yeast extract agar (15–3.2). Longer term storage is achieved by making spore suspensions in 10% vol/vol glycerol and freezing aliquots at −70 °C. Mycelia and fragments survive freeze-thaws but viability will decline (Wellington & Williams, 1979). Cultures can also be lyophilized.

15–3.4 Morphological Examination

Use of the key in Table 15–2 requires the morphological characteristics in addition to cell wall isomer.

Actinomycete morphology is best examined by inoculation of strains along the line of a coverslip inserted into the agar at a 45° angle so that the growth will be on the upper side of the coverslip. The coverslips should be observed dry, unstained or they can be sputter coated for SEM (see 15–1.2.1). The following features should be observed by examination of the coverslips removed from 7 and 14 d cultures grown on oatmeal agar and glucose yeast extract agar (15–3.3):

1. Substrate mycelium, this appears hyaline and ramifies over the surface of the coverslip.
2. Aerial mycelium has more refractive properties than substrate mycelium and so appears darker. It usually grows upwards at an angle to the coverslip.
3. Fragmentation of mycelium gives a zig-zag appearance with irregular lengths of rods and sometimes cocci.
4. Spores are more uniform in size (about 1 μm), often phase-bright and can have a sheath which may be ornamented and is only visible under SEM.
5. Sporangia are readily visible (about 20 μm) by light microscopy and some contain motile spores seen by addition of a drop of water.

15–3.5 Comments

Identification schemes for species exist usually in the form of diagnostic tables. Computer-assisted probabilistic schemes have been devised for *Streptomyces* (Williams et al., 1983; Langham et al., 1989; Kampfer & Kroppenstedt, 1991) and *Micromonospora* (E.M.H. Wellington and J. Mullins, 1992, unpublished data).

REFERENCES

Amann R.I., B.J. Binder, R.J. Olson, S.W. Chisholm, R. Devereux, and D.A. Stahl. 1990. Combination of 16S rRNA targeted oligonucleotide probes with flow cytometry for analyzing mixed microbial populations. Appl. Environ. Microbiol. 56:1919–1925.

Amner, W., C. Edwards, and A.J. McCarthy. 1989. Improved medium for recovery and enumeration of the farmer's lung organism, *Saccharomonospora viridis*. Appl. Environ. Microbiol. 55:2669–2674.

Anderson, A.A. 1958. A new sampler for the collection, sizing and enumeration of viable airbourne particles. J. Bacteriol. 76:471–484.

Aragno, M., and H.G. Schlegel. 1981. The hydrogen oxidizing bacteria. p. 865–893. In M.P. Starr et al. (ed.) The prokaryotes. Springer-Verlag, New York.

Asano, K., and I. Kawamoto. 1986. *Catellatospora,* a new genus of the actinomycetales. Int. J. Syst. Bacteriol. 36:512–517.

Athalye, M., J. Lacey, and M. Goodfellow. 1981. Selective isolation and enumeration of actinomycetes using Rifampicin. J. Appl. Bacteriol. 51:289–297.

Baecker, A.A.W., and K.C. Ryan. 1987. Improving the isolation of actinomycetes from soil by high-speed homogenization. S. Afr. J. Plant Soil 4:165–170.

Baker, D., and D. O'Keefe. 1984. A modified sucrose fractionation procedure for the isolation of frankiae from actinorhizal root nodules and soil samples. Plant Soil 78:23–28.

Becker, B., M.P. Lechevalier, and H.A. Lechevalier. 1965. Chemical composition of cell-wall preparations from strains of various form-genera of aerobic actinomycetes. Appl. Microbiol. 13:236–243.

Bell, J.M., C. Falconer, J. Colby, and E. Williams. 1987. CO metabolism by a thermophilic actinomycete, *Streptomyces* strain G26. J. Gen. Microbiol. 12:3445–3456.

Bland, C.E., and J.N. Couch. 1981. The family *Actinoplanaceae*. p. 2004–2010. In M.P. Starr et al. (ed.) The prokaryotes. Springer-Verlag, New York.

Bramwell, P. 1992. The characterisation and detection of plant pathogenic streptomycetes in the natural environment. PhD thesis, Univ. of Warwick.

Clewlow, L.J., N. Cresswell, and E.M.H. Wellington. 1990. A mathematical model of plasmid transfer between strains of streptomycetes in soil microcosms. Appl. Environ. Microbiol. 56:3139–3145.

Crawford, D.L. 1988. Biodegradation of agricultural and urban wastes. p. 433–439. In M. Goodfellow et al. (ed.) Actinomycetes in biotechnology. Academic Press, London.

Cresswell, N., V.A. Saunders, and E.M.H. Wellington. 1991. Detection and quantification of *Streptomyces violacolatus* plasmid DNA in soil. Lett. Appl. Microbiol. 13:193–197.

Cross, T., and R.W. Attwell. 1974. Recovery of viable thermoactinomycete endospores from deep mud cores. p. 11–20. In A.N. Barker et al. (ed.) Spore research 1973. Academic Press, London.

Dias, F.F., M.H. Bilimoria, and J.V. Bhat. 1962. *Corynebacterium barkeri* nov. spec., a pectinolytic bacterium exhibiting a biotin-folic acid inter-relationship. J. Ind. Inst. Sci. 44:59–67.

El-Nakeeb, M.A., and H.A. Lechevalier. 1963. Selective isolation of aerobic actinomycetes. Appl. Microbiol. 11:75–77.

Embley, M. 1991. The family Pseudonocardiaceae. p. 996–1027. In A. Balows et al. (ed.) The prokaryotes. 2nd ed. Springer-Verlag, New York.

Ensign, J.C. 1991. Introduction to the actinomycetes. p. 811–815. In A. Barlow et al. (ed.) The prokaryotes. 2nd ed. Springer-Verlag, New York.

Gledhill, W.E., and L.E. Casida. 1969. Predominant calalase-negative soil bacteria. II. Occurrence and characterization of *Actinomyces humiferus,* sp. nov. Appl. Microbiol. 18:114–121.

Goddon, B., and M.J. Peninckx. 1984. Identification and evolution of the cellulolytic microflora present during composting of cattle manure: the role of actinomycetes. Ann. Microbiol. 135:69–78.

Goodfellow, M. 1989. Suprageneric classification of actinomycetes. p. 2333–2339. *In* S.T. Williams et al. (ed.) Bergey's manual of systematic bacteriology. Vol. 4. Williams and Wilkins, Baltimore.

Goodfellow, M. 1991. The family Nocardiaceae. p. 1188–1213. *In* A. Balows et al. (ed.) The prokaryotes. 2nd ed. Springer-Verlag, New York.

Hagedorn, C., and J.G. Holt. 1975. Ecology of soil arthrobacters in Clarion-Webster toposequences of Iowa. Appl. Microbiol. 29:211–218.

Hahn, D., R.I. Amann, W. Ludwig, A.D.L. Akkermans, and K.H. Schleifer. 1992. Detection of micro-organisms in soil after in situ hybridization with rRNA-targeted, fluorescently labelled oligonucleotides. J. Gen. Microbiol. 138:879–887.

Hankin, L., M. Zucker, and D.C. Sands. 1971. Improved solid medium for detection and enumeration of pectolytic bacteria. Appl. Microbiol. 22:205–209.

Hayakawa, M., and H. Nonomura. 1987. Humic acid-vitamin agar, a new method for the selective isolation of soil actinomycetes. J. Ferment. Technol. 65:501–509.

Hayakawa, M., T. Sadakata, T. Kajiura, and H. Nonomura. 1991a. New methods for the highly selective isolation of *Micromonospora* and *Microbispora* from soil. J. Ferment. Bioeng. 72:320–326.

Hayakawa, M., T. Kajiura, and H. Nonomura. 1991b. New methods for the highly selective isolation of *Streptosporangium* and *Dactylosporangium* from soil. J. Ferment. Bioeng. 72:327–333.

Herron, P.R., and E.M.H. Wellington. 1990. New method for extraction of *Streptomyces* spores from soil and application to the study of lysogeny in sterile amended and non-sterile soil. Appl. Environ. Microbiol. 56:1406–1412.

Horikoshi, K. 1971. Production of alkaline enzymes by alkalophilic microorganisms. Part 1. Alkaline protease produced by Bacillus no. 221. Agric. Biol. Chem. 35:1407–1414.

Hsu, S.C., and J.L. Lockwood. 1975. Powdered chitin agar as a selective medium for enumeration of actinomycetes in water and soil. Appl. Microbiol. 29:422–426.

Kampfer, P., and R.M. Kroppenstedt. 1991. Probabilistic identification of streptomycetes using miniaturized physiological tests. J. Gen. Microbiol. 137:1893–1902.

Kuster, E., and S.T. Williams. 1964. Selection of media for isolation of streptomycetes. Nature (London) 202:928–929.

Labeda, D.P. 1989. Genus Glycomyces Labeda, Testa, Lechevalier and Lechevalier, p. 2586–2589. *In* S.T. Williams et al. (ed.) Bergey's manual of systematic bacteriology. Vol. 4. Williams and Wilkins, Baltimore.

Lacey, J. 1971. The microbiology of moist barley storage in unsealed silos. Ann. Appl. Biol. 69:187–212.

Lacey, J. 1974. Moulding of sugar-cane bagasse and its prevention. Annal. Appl. Biol. 76:63–76.

Lacey, J., and J. Dutkiewicz. 1976. Isolation of actinomycetes and fungi using a sedimentation chamber. J. Appl. Bacteriol. 41:315–319.

Lamot, E., and J.P. Voets. 1976. Cellulolytic activity of aerobic soil actinomycetes. Z. Allg. Mikrobiol. 16:345–351.

Lavrova, N.V., T.P. Preobrazhenskaya, and M.A. Sveshnikova. 1972. Isolation of soil actinomycetes on selective media with rubomycin. Antibiotiki 11:965–970.

Lechevalier, M.P., and H.E. Lechevalier. 1989. Genus Sporichthya Lechevalier, Lechevalier and Holbert 1968, 279AL. p. 2507–2508. *In* S.T. Williams et al. (ed.) Bergey's manual of systematic bacteriology. Vol. 4. Williams and Wilkins, Baltimore.

Locci, R. 1969. Preliminary examination of soil by scanning electron microscopy. Riv. Pat. Veg., S.IV. 5:167–178.

Luedemann, G. 1968. *Geodermatophilus,* a new genus of the Dermatophilaceae (Actinomycetales). J. Bacteriol. 96:1848–1858.

Makkar, N.S., and T. Cross. 1982. Actinoplanes in soil and on plant litter from freshwater habitats. J. Appl. Bacteriol. 52:209–218.

Margalith, P., and G. Beretta. 1961. Rifamycin. XI. Taxonomic study of *Streptomyces meditterranei* nov. sp. Mycopath. Mycol. Appl. 13:321–330.

McCarthy, A.J. 1987. Lignocellulose-degrading actinomycetes. FEMS Microbiol. Rev. 46:145–163.

McCarthy, A.J., and P. Broda. 1984. Screening for lignin-degrading actinomycetes and characterization of their activity against ^{14}C-lignin-labelled wheat lignocellulose. J. Gen. Microbiol. 130:2905–2913.

McCarthy, A.J., and T. Cross. 1981. A Note on a selective isolation medium for the thermophilic actinomycete Thermomonospora chromogena. J. Appl. Bacteriol. 51: 299–302.

Mikami, Y., K. Mijashita, and T. Arai. 1982. Diaminopimelic acid profiles of alkalophilic and alkali-resistant strains of actinomycetes. J. Gen. Microbiol. 128:1709–1712.

Mikami, Y., K. Mijashita, and T. Arai. 1985. Alkalophilic actinomycetes. Actinomycetes 19:176–191.

Nonomura, H., and M. Hayakawa. 1988. New methods for the selective isolation of soil actinomycetes. p. 288–293. In Y. Okami et al. (ed.) Biology of actinomycetes '88. Japan Sci. Soc. Press, Tokyo.

Nonomura, H., and Y. Ohara. 1971a. Distribution of actinomycetes in soil. XI. Some new species of the genus Actinomadura Lechevalier et al. J. Ferment. Technol. 49:904–912.

Nonomura, H., and Y. Ohara. 1971b. Distribution of actinomycetes in soil. VIII. Green-spore group of Microtetraspora, its preferential isolation and taxonomic characteristics. J. Ferment. Technol. 49:1–7.

Nonomura, H., and Y. Ohara. 1971c. Distribution of actinomycetes in soil. IX. New species of the genus Microbispora and Microtetraspora and their isolation methods. J. Ferment. Technol. 49:887–894.

Okazaki, T., N. Serizawa, R. Enokita, A. Torikata, and A. Terahara. 1983. Taxonomy of actinomycetes capable of hydroxylation of ML-236B compactin. J. Antibiot. 36:1176–1183.

Orchard, V.A., M. Goodfellow, and S.T. Williams. 1977. Selective isolation and occurrence of nocardiae in soil. Soil Biol. Biochem. 9:233–238.

Owens, J.D., and R.M. Keddie. 1969. The nitrogen nutrition of soil and herbage coryneform bacteria. J. Appl. Bacteriol. 32:338–347.

Palleroni, N.J. 1976. Chemotaxis in Actinoplanes. Archives Microbiol. 110:13–18.

Parenti, F. 1989. Genus Kineosporia Pagini and Parenti 1978, 401AL. p. 2504–2506. In S.T. Williams et al. (ed.) Bergey's manual of systematic bacteriology. Vol. 4. Williams and Wilkins, Baltimore.

Phelan, M.B., D.L. Crawford, and A.L. Pometto. 1979. Isolation of lignocellulose-decomposing actinomycetes and degradation of specifically ^{14}C-labelled lignocellulose by six selected Streptomyces strains. Can. J. Microbiol. 25:1270–1276.

Plant, E.J. R.W. Attwell, and C.A. Smith. 1988. A semi-micro quantitative assay for cellulolytic activity in microorganisms. J. Microbiol. Methods 7:259–263.

Portaels, F., A. de Muynck, and M.P. Sylla. 1988. Selective isolation of mycobacterium from soil: A statistical analysis approach. J. Gen. Microbiol. 134:849–855.

Prauser, H., and M. Bergholz. 1974. Taxonomy of actinomycetes and screening for antibiotic substances. Postepy Hig. Med. Dosw. 28:441–457.

Rowbotham, T.J., and T. Cross. 1977. Ecology of Rhodococcus coprophilus and associated actinomycetes in fresh water and agricultural habitats. J. Gen. Microbiol. 100:231–240.

Shearer, M. 1987. Methods for the isolation of non-streptomycete actinomycetes. J. Indust. Microbiol. Suppl. 2:91–97.

Shearer, M.C., P.M. Colman, R.M. Ferrin, L.J. Nesbit, and C.J. Nash. 1986. New genus of the actinomycetales: Kibdellosporangium aridum gen. nov., sp. nov. Int. J. Syst. Bacteriol. 36:47–54.

Songer, J.G. 1981. Methods for selective isolation of mycobacteria from the environment. Can. J. Microbiol. 27:1–7.

Stackebrandt, E. 1991. Unifying phylogeny and phenotypic diversity. p. 19–47. In A. Barlow et al. (ed.) The prokaryotes. 2nd ed. Springer-Verlag, New York.

Stackebrandt, E., and R.M. Keddie. 1984. Cellulomonas. p. 1325–1329. In P.H.A. Sneath et al. (ed.) Bergey's manual of systematic bacteriology. Vol. 2. Williams and Wilkins, Baltimore.

Stackebrandt, E., and H. Prauser. 1991. The family Cellulomonadaceae. p. 1323–1345. In A. Barlow et al. (ed.) The prokaryotes. 2nd ed. Springer-Verlag, New York.

Stackebrandt, E., D. Witt, C. Kemmerling, R. Kroppenstedt, and W. Liesack. 1991. Designation of streptomycete 16S and 23S rRNA-based target regions for oligonucleotide probes. Appl. Environ. Microbiol. 57:1468–1477.

Staneck, J.L., and G.D. Roberts. 1974. Simplified approach to identification of aerobic actinomycetes by thin layer chromatography. Appl. Microbiol. 28:226–231.

Tomita, K., Y. Nakakita, Y. Hoshino, K. Numata, and K. Kawaguchi. 1987. New genus of the *Actinomycetales: Streptoalloteichus hindustanicus* gen. nov., nom. rev.; sp. nov., nom. rev. Int. J. Syst. Bacteriol. 37:211–213.

Tsujibo, H., T. Sato, M. Inui, H. Yamamoto, and Y. Inamori. 1988. Intracellular accumulation of phenazine antibiotics produced by an alkalophilic actinomycete. I. Taxonomy, isolation and identification by colormetric method. J. Gen. Microbiol. 23:249–260.

Tsukamura, M., K. Hikosaka, K. Nishimura, and S. Hara. 1988. Severe progressive subcutaneous abscesses and necrotizing tenosynovitis caused by *Rhodococcus aurantiacus*. J. Clin. Microbiol. 26:201–205.

Turpin, P.E., V.K. Dhir, K.A. Maycroft, C. Rowlands, and E.M.H. Wellington. 1992. The effect of *Streptomyces* species on the survival of Salmonella in soil. FEMS Microbiol. Ecol. 101:271–280.

Von der Osten, C.H., C. Gioannetti, and A.J. Sinskey. 1989. Design of a defined medium for growth of *Corynebacterium glutamicum* in which citrate facilitates iron uptake. Biotechnol. Lett. 11:11–16.

Weller, D.M., and L.S. Thomashow. 1990. Antibiotics: Evidence for their production and sites where they are produced. 703–711. *In* New directions in biological control. Alan R. Liss, New York.

Wellington, E.M.H., and S.T. Williams. 1979. Preservation of actinomycete inoculum in frozen glycerol. Microbiol. Lett. 6:151–157.

Wellington, E.M.H., N. Cresswell, and V.A. Saunders. 1990. Growth and survival of streptomycete inoculants and the extent of plasmid transfer in sterile and non-sterile soil. Appl. Env. Microbiol. 56:1413–1419.

Williams, S.T., F.L. Davies, C.I. Mayfield, and M.R. Khan. 1971. Studies on the ecology of actinomycetes in soil. II. The pH requirements of streptomycetes from two acid soils. Soil Biol. Biochem. 4:215–225.

Williams, S.T., M. Goodfellow, E.M.H. Wellington, J.C. Vickers, G. Alderson, P.H.A. Sneath, M.J. Sackin, and A.M. Mortimor. 1983. A probability matrix for identification of some streptomycetes. J. Gen. Microbiol. 129:1815–1830.

Chapter 16

Frankia and the Actinorhizal Symbiosis

DAVID D. MYROLD, *Oregon State University, Corvallis, Oregon*

Frankia are sporulating actinomycetes capable of fixing N_2 and forming symbiotic nodules on dicotyledonous plants. The symbiosis was termed *actinorhizal* by analogy to the mycorrhizal association (Tjepkema & Torrey, 1979). This chapter reviews some of the fundamental characteristics of *Frankia* and actinorhizal symbioses and presents methodology germane to studying *Frankia*. Several review articles (Lechevalier et al., 1982; Baker & Seling, 1984; Tjepkema et al., 1986; Baker, 1988; Benson & Silvester, 1993), the recent book edited by Schwintzer and Tjepkema (1990), and a bibliography compiled by Baker (1993) provide additional background information.

16–1 CHARACTERISTICS OF *FRANKIA*

Frankia are N_2-fixing, sporulating, gram-positive, filamentous bacteria of the Actinomycetales. They have a type III cell wall and type I phospholipid content (Lechevalier et al., 1982), contain the sugar 2-*O*-methyl-D-mannose (Mort et al., 1983), and have a high G + C content of 68 to 72% (An et al., 1983). *Frankia* can be differentiated from other actinomycete genera on the basis of morphology (formation of sporangia in submerged liquid culture and formation of vesicles in N-free media), DNA homology, and 16S rRNA sequence (Lechevalier & Lechevalier, 1990).

Frankia differentiate into three cell types: vegetative hyphae, sporangia, and vesicles (Fig. 16–1). These different cell types can be produced in pure culture, in planta, and presumably in soil. Hyphae (0.5–2 μm diam.) are the actively growing cell type; they branch to form a mycelial mat and may differentiate to form vesicles and sporangia. Sporangia (10 × 30–40 μm), which contain multiple ovoid spores, commonly develop as pure cultures enter stationary phase. They also form inside nodules of some, but not all, actinorhizal plants (Schwintzer, 1990). In pure culture, vesicles are

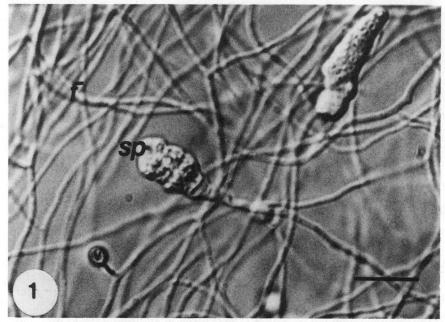

Fig. 16–1. Microscopic view of *Frankia* in pure culture, showing vegetative hyphae (f), sporangia (sp), and vesicles (v). The figure was taken with Nomarski interference contrast optics at 1,150× and the bar represents 10 µm. This photograph courtesy of J.G. Torrey and comes from the paper of Mansour et al. (1990).

spherical (4 µm), thick-walled structures borne on short stalks from the hyphae, whereas in planta their shape may vary and can be determined by the host. The thick wall is a multilaminate structure that functions as a barrier to O_2, thereby protecting the nitrogenase enzyme (Silvester et al., 1990). Vesicles are the site of N_2 fixation in free-living *Frankia* and most actinorhizae. Like spores, vesicles can give rise to vegetative hyphae (Schultz & Benson, 1989).

Benson and Schultz (1990) highlighted some of the difficulties of studying *Frankia* in pure culture. These difficulties are primarily related to its pleiomorphic growth form and are magnified in ecological studies of *Frankia*. First, *Frankia* cannot be directly isolated from soil and counted on plates, because they are slow growing and poor competitors for readily available C sources. There are no selective media available for *Frankia*. Even isolation from nodules can be problematic (see section 16–1.2). Second, because of its pleiomorphic, filamentous nature, an increase in biomass may not necessarily be manifested as an increase in numbers (e.g., viable units (VU) by plate counts or infective units (IU) by plant bioassay). Similarly, spore maturation, release, and germination may produce a large increase in IU, even though total biomass is little changed.

16–1.1 Attributes of Actinorhizae

Actinorhizal plants are found worldwide (Table 16–1). Except for *Datisca*, which has herbaceous shoots, all are woody shrubs or trees; all are perennial dicots. Most are temperate plants, with the exception of the Casuarinaceae and some members of the Myricaceae, which are native to tropical regions. Species of *Alnus* or *Elaeagnus* are, however, found in montane areas of the tropics. Many species have been introduced throughout the world; members of the Casuarinaceae and *Elaeagnus,* in particular (Baker & Schwintzer, 1990).

Nitrogen fixation by actinorhizal symbioses is substantial, ranging from 1 to > 300 kg N ha^{-1} yr^{-1} (Dawson, 1990; Hibbs & Cromack, 1990; Wheeler & Miller, 1990), which is comparable to the N input in legumes. In addition to their importance as N_2-fixing symbioses, actinorhizal plants have been used for production of wood, fiber, food, chemicals, for revegetation, as wildlife browse, and as ornamental plants (Benoit & Berry, 1990; Diem & Dommergues, 1990; Hibbs & Cromack, 1990; Wheeler & Miller, 1990).

Although there are many intriguing attributes of actinorhizae, probably the most relevant with respect to ecology include observations regarding host promiscuity and isolation of noninfective strains. It is quite common for a *Frankia* strain isolated from one host to form nodules (often effective ones) on plants from different genera or families. Based on cross-inoculation studies (see section 16–4.1), three distinct host specificity groups can be identified: (i) strains infective on *Alnus,* (ii) strains infective on *Casuarina,* and (iii) strains infective on *Elaeagnus* (Baker, 1987b; unpublished data). But these three groups may not be mutually exclusive, as strains that are capable of infecting both *Elaeagnus* and *Casuarina* have been isolated (Normand & Lalonde, 1986; Racette & Torrey, 1989) and Dobritsa et al. (1990) report a strain that can infect both *Alnus* and *Hippophae,* a member of the Elaeagnaceae. Conversely, a single host may be nodulated by *Frankia* strains isolated from other host genera or families. Using this definition, Torrey (1990) concluded that *Elaeagnus angustifolia, Gymnostoma papuanum, Myrica cerifera,* and *M. gale* are all promiscuous hosts.

Perhaps a greater frustration than cross-inoculation groups is that it has not been possible to isolate *Frankia* from most of the actinorhizal genera. In addition, some strains that have been isolated, and show morphology characteristic of *Frankia* in pure culture, will not reinfect the host plant and hence can only be considered presumptive *Frankia* strains until Koch's postulate is fulfilled. Isolates from the Rhamnaceae, e.g. *Ceanothus,* and Rosaceae, e.g. *Purshia,* are notable examples (Torrey, 1990). According to Baker (1988), *Frankia* strains able to reinfect the host have been isolated from less than one-half of all actinorhizal genera. Another example of noninfective strains are the isolates from spore-positive (sp$^+$) nodules (see section 16–4.2.1). Schwintzer (1990) reports that sp$^+$ nodules have been described in nine actinorhizal genera, with *Alnus* and *Myrica*

Table 16–1. Distribution of native actinorhizal families and genera. (Adapted from Baker [1988] and Baker and Schwintzer [1990] and updated based on Cruz-Cisneros and Valdes [1991].)

Continent	Family	Genus
North America	Betulaceae	*Alnus*
	Coriariaceae	*Coriaria*
	Datiscaceae	*Datisca*
	Elaeagnaceae	*Elaeagnus*
		Shepherdia
	Myricaceae	*Myrica*
		Comptonia
	Rhamnaceae	*Ceanothus*
	Rosaceae	*Cercocarpus*
		Chamaebatia
		Cowania
		Dryas
		Purshia
South America	Betulaceae	*Alnus*
	Coriariaceae	*Coriaria*
	Myricaceae	*Myrica*
	Rhamnaceae	*Adolphia*
		Colletia
		Discaria
		Kentrothamnus
		Retanilla
		Talguenea
		Trevoa
Africa	Myricaceae	*Myrica*
Eurasia	Betulaceae	*Alnus*
	Coriariaceae	*Coriaria*
	Datiscaceae	*Datisca*
	Elaeagnaceae	*Elaeagnus*
		Hippophae
	Myricaceae	*Myrica*
	Rosaceae	*Dryas*
Australia and Oceania	Casuarinaceae	*Allocasuarina*
		Casuarina
		Ceuthostoma
		Gymnostoma
	Coriariaceae	*Coriaria*
	Myricaceae	*Myrica*
	Rhamnaceae	*Discaria*

being the most studied. But thus far, all isolates from sp$^+$ nodules fail to induce a similar phenotype when inoculated onto the host plants (Torrey, 1987).

16–2 ISOLATION, CULTURING, AND MAINTENANCE OF *FRANKIA* STRAINS

There have been several recent reviews describing how to isolate and grow *Frankia* strains (Stowers, 1987; Diem & Dommergues, 1988; Baker, 1989; Lechevalier & Lechevalier, 1990). This section will focus upon some of the more common methods. Refer to the other reviews for additional details.

16–2.1 Isolation of *Frankia* Strains

Successful isolation of the first *Frankia* strain took more than 100 yr. The major difficulty does not appear to be any particular nutritional requirement but simply that *Frankia* is slow growing compared to most soil microorganisms. Consequently, effective isolation procedures rely primarily on enriching for *Frankia* relative to contaminants. Most often actinorhizal root nodules are used as inoculum, but a few *Frankia* strains have been isolated directly from soil (Baker & O'Keefe, 1984).

16–2.1.1 Sample Collection

Actinorhizal root nodules may be collected directly from the field or from plants grown under controlled conditions. Field nodules have the advantage of containing strains that contribute to N_2 fixation under natural conditions. Laboratory grown nodules may be more metabolically active, less-contaminated, and obtained more conveniently.

In the field, locating root nodules on some plant species is relatively easy (e.g., *Alnus,* which is generally well nodulated with many nodules near the soil surface) but more difficult with other hosts (e.g., root nodules of *Ceanothus* and *Purshia* tend to be less numerous, more widely scattered, and often found at greater depths). It is often best to begin at the base of the stem and work outward along a root. Once found, carefully excavate the root nodules and collect them along with a small portion of the attached root system. Nodules should be light in color because this is an indicator of actively growing, young tissue. Store nodules in plastic bags containing soil or a moistened paper towel to maintain a favorable water potential. It is best to keep nodules on ice to limit tissue deterioration and slow the growth of contaminating microorganisms.

Soil samples can be collected in much the same manner, however, it is advisable to take the precaution of sterilizing collecting tools between samples. The potential for cross-contamination is greater with soils, which generally have low *Frankia* populations. Soil samples collected as inoculum should be kept cool until used to limit changes in population size.

Laboratory grown nodules are produced by growing the desired host plant directly in soil collected from the field or by growing a plant in sterile potting medium and inoculating it with soil or nodule homogenate.

Samples of nodules or soil may be stored at < 4 °C for a few days until used. For periods of more than a week, it is best to freeze (-20 °C) the samples to minimize tissue deterioration and growth of contaminants. Baker (1989) states that *Frankia* within frozen nodules retain viability for at least 5 yr. It may also be possible to store desiccated nodules over silica-gel at room temperature, but less is known about long-term viability of *Frankia* when this approach is used (Lechevalier & Lechevalier, 1990).

Successful nodulation of drought-adapted Rhamnaceae has been obtained using air-dried soil samples that had been stored at room temperature for 18 mo (Tortosa & Cusata, 1991), but it is not known if this method of preservation is universally applicable.

Regardless of the source from which a *Frankia* isolate is obtained, it is important to keep a good record about the source of inoculum. Information requirements for registering *Frankia* strains (see section 16–2.3) include: date, location, host plant, location of nodule/soil, size and condition of nodule, and general site soil, topography, and vegetation description.

16–2.1.2 Isolation Procedures

Isolation of *Frankia* is basically an enrichment procedure. All methods that use nodules as the source of inoculum require a surface sterilization step. Although serial dilutions of crushed nodules have produced isolates from *Casuarina* (e.g., Diem et al., 1982), most isolations have required an additional enrichment step, such as microdissection, filtration, or density gradient centrifugation.

16–2.1.2.1 Nodule Sterilization. Clean nodules of most soil and organic debris by washing in water. Separate nodule lobes and further clean by shaking in a dilute detergent solution (Lechevalier & Lechevalier, 1990). Surface sterilization of nodules has been done with several reagents. The most common are: commercial bleach (NaOCl), either full strength (5.25%) or diluted 3 parts to 10; 30% H_2O_2; or 1 to 3% OsO_4. Although all of these sterilizing agents are potentially harmful, OsO_4 is highly toxic and should be used only in a chemical fume hood and all human contact should be avoided. Surface sterilize nodules by shaking with one of the sterilizing agents for 10 to 30 min (except OsO_4, which is more powerful and needs only 1–2 min). Sterilization efficiency can be improved by including a surfactant (e.g., 0.1% Tween 80, Triton X-100, etc.) in the sterilizing solution. Following sterilization, rinse the nodule lobes several times with sterile distilled water or buffer. Lechevalier and Lechevalier (1990) recommend that surface-sterile tissue be allowed to incubate overnight at room temperature and that the sterilization procedure be repeated to eliminate spore-forming contaminants. Surface-sterile nodule lobes can be stored frozen (-20 °C).

16–2.1.2.2 Microdissection. Callaham et al. (1978) used this procedure to make the first successful *Frankia* isolation. Basically, microdissection involves removing infected tissue from near the apex of a sterilized nodule lobe. The dissected pieces (1–3 mm diam.) are placed either in tubes of liquid medium or onto agar medium. Numerous variations of this method have arisen and some of the important modifications or recommendations are:

1. OsO_4 is the most reliable sterilizing agent (Baker, 1989).
2. First incubate dissected pieces in a fairly rich medium (e.g., yeast extract/dextrose broth) for 7 to 10 d to check for fast-growing con-

taminants and then transfer uncontaminated nodule sections to a *Frankia* isolation medium (Racette & Torrey, 1989).

3. When using liquid medium, use several replicate tubes with small volumes of medium (Baker, 1989).

4. A soft agar (0.5%) overlay may enhance success with solid medium (Diem & Dommergues, 1983).

5. Potential inhibition by plant phenolics can be minimized by adding activated C (Diem & Dommergues, 1988) or polyvinylpyrrolidone (PVP) to the medium (Lechevalier & Lechevalier, 1990).

16–2.1.2.3 Differential Filtration. Benson (1982) isolated vesicle clusters from root nodule debris, plant phenolics, and contaminating microorganisms using differential filtration. Surface sterilized nodules are crushed in sterile buffer to make an aqueous suspension and passed through an apparatus containing two nylon mesh screens (Fig. 16–2). The first screen is large enough to permit *Frankia* vesicle clusters to pass, but small enough to trap most of the plant debris. *Frankia* vesicle clusters are retained on the second screen, which allows smaller bacteria and plant phenolics to pass through. The buffer may contain PVP to complex with plant phenolics. After trapping *Frankia* vesicle clusters on the 20-μm screen, the vesicle clusters are further cleaned by rinsing with sterile buffer. The cleaned vesicle clusters are then inoculated onto agar plates or into liquid medium.

16–2.1.2.4 Density Fractionation. Separation of *Frankia* from contaminating materials has been achieved by taking advantage of density differences between *Frankia,* plant cells, contaminating microorganisms, and soil particles (Baker et al., 1979). The current modification of this method (Baker & O'Keefe, 1984) uses a discontinuous, or step, gradient of sucrose at three concentrations (1.0, 1.6, and 2.5 M). The gradient is constructed by layering the 2.5, 1.6, and 1.0 M sucrose solutions and nodule or soil suspension, from bottom to top, in 2:2:3:3 volume proportions. Baker (1989) notes that the concentrations of the sucrose solutions are critical. The suspension of soil or crushed nodules is prepared in 0.7% phenol, which seems to inhibit bacteria other than *Frankia*. The gradient is spun in a swinging-bucket rotor until equilibrium is reached, about 1 h at 600 g. *Frankia* are concentrated at the interface between the 1.6 and 2.5 M layers, although Baker (1989) suggests also sampling the interface between the 1.0 and 1.6 M layers when isolating *Frankia* from soil. Cells at the interfaces are harvested by puncturing the bottom of the tube and collecting the appropriate fractions, which can then be inoculated into isolation medium. A few additional points about this method are:

1. When isolating *Frankia* from nodule suspensions, a single step gradient of 1.6 M sucrose may be sufficient and the pellet at the bottom of the tube is used as an inoculum (Quispel & Burggraaf, 1981).

2. When isolating *Frankia* from soil, it is necessary to use an N-free selective minimal medium (Baker, 1989).

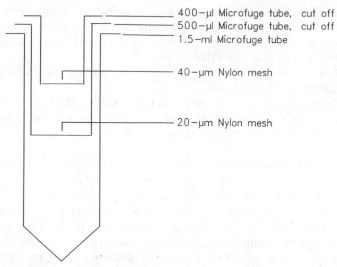

400−μl Microfuge tube, cut off
500−μl Microfuge tube, cut off
1.5−ml Microfuge tube

40−μm Nylon mesh

20−μm Nylon mesh

Fig. 16–2. Nested sieve apparatus for collecting *Frankia* vesicle clusters. Filtration is achieved by spinning in a microcentrifuge. This device is a modification of the one described by Benson (1982), who used 50 μm nylon mesh for the first screen and used larger, disposable syringes to construct his device. When using apparatus, filtration may be accomplished by pressure or vacuum as well as centrifugation.

16–2.1.2.5 Incubation and Subculturing. Incubate tubes of liquid media or agar plates in the dark at 25 to 35 °C. Because growth of *Frankia* is slow, plates should be protected from desiccation by sealing with Parafilm or enclosing them in plastic bags. Outgrowth of *Frankia* may be as rapid as 7 to 10 d, but this is uncommon; 1 to 4 mo is more usual and some colonies may take a year to grow (Lechevalier & Lechevalier, 1990). Once colonies are observed, they should be carefully examined with a dissecting microscope for the characteristic subsurface sporangia or presence of vesicles. Selected colonies can be aseptically withdrawn from liquid culture or excised from plates for subculturing.

Colonies selected for subculturing need to be thoroughly homogenized. This is most easily done by grinding the colony with sterile buffer or medium in a small, sterile Ten Broeck or Potter-Elvehjem tissue grinder. Some workers use repeated passage through a 21-gauge needle as an alternative means of fragmenting the hyphae. Homogenization increases the number of actively growing hyphal tips and also helps to release spores from sporangia. The homogenized suspension is then inoculated into liquid medium or onto agar plates.

16–2.1.2.6 Evaluation and Recommendations. There have been few systematic evaluations of isolation techniques or growth media. Burggraaf (1984) evaluated all three methods and several media and found that none was consistently more effective than another. Similarly, Rosbrook et al. (1989) found no differences in isolation success between microdissection and differential filtration methods or among several media when isolating *Frankia* from *Casuarina* root nodules. Therefore, ease of use or availability of resources are the main discriminating factors. A few points worth considering, however, are:

1. Microdissection is probably the least selective method. Passage through sucrose gradients may select for *Frankia* that are more tolerant of high osmotic stress. Differential filtration, by minimizing the effect of plant phenolics, may select for *Frankia* that are less tolerant of these compounds (Lechevalier & Lechevalier, 1990).
2. There is some evidence for a root-derived factor (dipterocarpol from alcoholic extracts of *Alnus glutinosa* roots) that may stimulate *Frankia* growth and enhance isolation success (Quispel et al., 1989).
3. *Frankia* strains isolated on one medium sometimes cannot be subcultured on that same medium, perhaps because of exhaustion of some critical factor or selective adaptation to life in vitro (Lechevalier & Lechevalier, 1990).
4. It should be remembered that more than one *Frankia* strain may be isolated from a single nodule (Benson & Hanna, 1983), consequently isolates should be checked to ensure that only one strain is present (see section 16–2.2.2.2).

To help ensure success in isolation, it is generally recommended that more than one isolation method and several isolation media be used (Baker, 1989; Lechevalier & Lechevalier, 1990). In the end, patience and perseverance appear to be primary hallmarks of successful *Frankia* isolation.

16–2.2 Cultivation and Maintenance

Growing *Frankia* in pure culture is not much different than growing other bacteria. Therefore, the following sections are not comprehensive, but rather highlight a few features particular to *Frankia*.

16–2.2.1 Common Media

More than 20 different media are described in the reviews by Baker (1989) and Lechevalier and Lechevalier (1990), and many more recipes can be found in the literature. Recipes for some of the more commonly used media are given in Tables 16–2 and 16–3. These are grouped as either isolation or culturing media, however, they are often interchangeable. Some isolation media are N free and take advantage of *Frankia*'s ability to fix N_2 in the free-living state.

Although some of the media presented in Tables 16–2 and 16–3 are quite complicated, most *Frankia* grow well on a minimal medium

Table 16–2. Common media used for isolating *Frankia*. All amounts given in mg L^{-1}.

Ingredient	*Frankia* medium†	Bennett's agar‡	S Medium§	Qmod¶
Macronutrients				
$CaCl_2 \cdot 2H_2O$			100	
$CaCO_3$	50			100
Ferric citrate			10	10
FeNaEDTA#	160			
KCl				200
KH_2PO_4	2 000			
K_2HPO_4	3 000		500	300
$MgSO_4 \cdot 7H_2O$	200			
NaCl	300		200	200
NaH_2PO_4				200
Micronutrients				
$CoSO_4 \cdot 7H_2O$			0.01	0.01
$CuSO_4 \cdot 5H_2O$	0.04		0.1	0.1
H_3BO_3	1.5		1.5	1.5
$MnSO_4 \cdot H_2O$	4.5		0.8	0.8
$Na_2MoO_4 \cdot 2H_2O$	0.25			
$(NH_4)_6Mo_7O_{24} \cdot 4H_2O$			0.2	0.2
$ZnSO_4 \cdot 7H_2O$	1.5		0.6	0.6
Carbon sources				
Beef extract		1 000		
Casamino acids	3 000	2 000		
Casein hydrolysate (NZAmine)		10 000	4 000	
Glucose††			10 000	10 000
Lecithin				5
Na pyruvate	3 000			
Peptone				5 000
Yeast extract		1 000		500
Vitamins				
Nicotinic acid	0.5			
Pyridoxine HCl	0.5			
Thiamine HCl	0.1			
pH	6.9	7.3	6.8	6.8–7.2

† Filter sterilize Na pyruvate separately; add $CaCO_3$ only to liquid media; add 10 g of agar L^{-1} for solid media. (After Benson, 1982.)

‡ Add 10 g of agar L^{-1} for solid media; may be used at one-half to one-sixteenth strength. (After Higgins et al., 1967.)

§ Add 10 g of agar L^{-1} for solid media; addition of 0.2% Tween 80 gives a media known as S+tw. (After Lechevalier et al., 1983.)

¶ Lecithin and $CaCO_3$ added after pH adjustment; lecithin supplement prepared by dissolving 0.5 g of L-α-lecithin (Sigma P-5638) in 50 mL of absolute ethanol, then add 50 mL of distilled water; add 15 g of agar L^{-1} for solid media. (After Lalonde and Calvert, 1979.)

FeNaEDTA can be made as a stock solution by dissolving 0.745 g Na_2EDTA in 100 mL of water, adding 0.556 g $FeSO_4 \cdot 7H_2O$, and autoclave to dissolve.

†† Autoclave separately to prevent carmelization.

Table 16–3. Common media used for culturing *Frankia*. All amounts given in mg L^{-1}.

Ingredient	BAP†	DPM‡	YCz§	L/2¶
Macronutrients				
$CaCl_2 \cdot 2H_2O$	10.3	10		10
FeNaEDTA#	13	23.4		
$FeSO_4 \cdot 7H_2O$			10	50
KCl			500	
KH_2PO_4	953	1 000		270
K_2HPO_4	592		1 000	
$MgCl_2 \cdot 6H_2O$				10
$MgSO_4 \cdot 7H_2O$	49.3	100	500	
NaCl				900
Na_2HPO_4				900
$NaNO_3$			2 000	
NH_4Cl	267			225
Micronutrients				
$CoCl_2$		0.025		
$CoSO_4 \cdot 7H_2O$	0.001			
$CuSO_4 \cdot 5H_2O$	0.08	0.08		3
H_3BO_3	2.86	2.86		
$MnCl_2 \cdot 4H_2O$	1.81	1.81		
$Na_2MoO_4 \cdot 2H_2O$	0.025	0.025		
$(NH_4)_6MoO_{24} \cdot 4H_2O$				4
$ZnSO_4 \cdot 7H_2O$	0.22	0.22		
Carbon sources				
Bovine serum albumin (V)				10
Glycerol				90
Na propionate	480	1,200		
Sucrose			30 000	
Tween 80				2 000
Yeast extract			2 000	
Vitamins				
Biotin	0.0225			
Ca pantothenate	0.01			
Cyanocobalamin (B_{12})				2
Folic acid	0.01			
Nicotinic acid	0.05			
Pyridoxine HCl	0.05			
Riboflavin	0.01			
Thiamine HCl	0.01			4.5
pH	6.7	6.8	6.6	7.4

† Combine Mg, Ca, Fe, micronutrients, vitamins, and N solutions and adjust pH to 6.7 using dilute KOH, autoclave. Separately sterilize phosphate buffer (pH 6.7) and C source solutions. If desired, sterile MOPS buffer (final concentration of 10.0 m*M*, pH 6.7) can also be added. Alternatively, Na pyruvate (10.0 m*M* or 1100 mg l^{-1}) can be used as a C source, in which case, the pH of all solutions should be 6.3. Modified from Fontaine et al. (1986). Omission of NH_4Cl and inclusion of MOPS buffer gives a media known as B, which is often used for inducing N_2 fixation.

‡ After Baker and O'Keefe (1984). Addition of NH_4Cl (428 mg L^{-1}) gives a media known as DPMN. Substitution of Na pyruvate (2400 mg L^{-1}) for the Na propionate gives a media known as DPM/2PF.

§ After Higgins and Lechevalier (1969).

¶ After Lechevalier et al. (1982).

FeNaEDTA can be made as a stock solution by dissolving 0.745 g of Na_2EDTA in 100 mL of water, adding 0.556 g of $FeSO_4 \cdot 7H_2O$, and autoclave to dissolve.

containing a buffered salt solution and a single C source. Most *Frankia* strains grow best on the organic acids pyruvate, propionate, and succinate, which are interchangeable in most media (Baker, 1989). Of these organic acids, propionate is often the preferred C source. Carbohydrates are generally not good substrates, (e.g., glucose may not be used by some *Frankia* strains). Some *Frankia* grow best on media that include a lipid supplement.

16–2.2.2 Culturing

Frankia grow best in liquid culture between 25 to 35 °C. They are most often grown in batch culture, although *Frankia* can be grown in chemostats. Liquid cultures may be incubated statically, shaken (which may cause hyphae to wind around each other and form clumps), or aerated. New strains often grow slower than ones that have been in culture for some time (Lechevalier & Lechevalier, 1990) and strains can be adapted to grow more rapidly by frequent transfers. Changes in growth rate, growth habit, and other culture characteristics, such as pigmentation, are not unusual with *Frankia* strains. Cultures should be homogenized when transferring them to break apart hyphae and promote more rapid growth. To maintain active cultures, transfers should be done every 2 to 6 mo.

16–2.2.2.1 Decontamination. Because of their slow-growing nature, *Frankia* cultures sometimes become contaminated. Decontamination can be achieved in several ways. Lechevalier and Lechevalier (1990) suggest the use of cycloheximide ($500\ \mu g\ mL^{-1}$) or nystatin ($10\ \mu g\ mL^{-1}$) for curing fungal contaminants and nalidixic acid ($10\ \mu g\ mL^{-1}$) for eliminating bacterial contamination. Sometimes bacterial contamination can be eliminated by physically separating filamentous *Frankia* from single-celled bacteria by filtration and washing. The efficacy of either decontamination approach is enhanced by reculturing on a N-free medium and streaking this enrichment on N-free agar to reisolate *Frankia*.

16–2.2.2.2 Production of Clonal Strains. Genetic homogeneity of *Frankia* isolates is usually assumed, but this may not be the case. Most *Frankia* isolates arise from inoculating media with a suspension of fragmented hyphae and spores. The production of clonal *Frankia* strains has been limited by the low percentage of spores that germinate on solid media. One reason for low germination rates is the recent observation that agar may inhibit germination (J.T. Leonard, 1991, personal communication). Consequently, use of an alternative gelling agent may give better results. The success of spore germination also varies among *Frankia* strains, e.g., *Frankia* isolates from *Casuarina* had germination rates ranging between 30 and 75% (Tzean & Torrey, 1989; Prin et al., 1991b), which was higher than those of *Frankia* strains isolated from other hosts. Prin et al. (1991b) found that, despite differences in colony morphology and pigmentation, there were no differences in enzyme electrophoretic patterns among 22 single-spore isolates. It remains to be seen if similar success will be had with spores of *Frankia* strains isolated from other hosts and perhaps whether such clonal material is needed.

16–2.2.3 Long-Term Preservation

Frankia can be stored over long periods of time by lyophilization, freezing in glycerol (35% [vol/vol] at −15 °C), or storing in complex media (Fontaine et al., 1986). Lyophilization appears to be the best method (Fontaine et al., 1986), although Lechevalier and Lechevalier (1990) recommend that spent culture media rather than skim milk be used. Freezing cells in glycerol was almost as effective as lyophilization but storage in complex media at room temperature was inferior to both (Fontaine et al., 1986).

16–2.3 Strain Registry

Registry numbers consist of a three letter plus a 4 to 10 number code, which is described in detail by Lechevalier (1983) and in amended forms by Lechevalier (1986) and Baker (1987a). The three letters are chosen to designate the laboratory that isolated the strain (e.g., HFP for Harvard Forest, Petersham, MA; ULQ for Laval University, Quebec, Canada; etc.) and the numbers describe the strain itself. The first four digits designate the source of the isolate, either soil or species of plant for root nodules; the last one to six digits are optional and are used by the scientist isolating the strain to provide a unique identifying number. Baker (1989) describes strain registration in more detail. When submitting a strain for registry, a form similar to that shown in Fig. 16–3 should be used. These forms should be sent to the curator of the registry: D. D. Baker; Panlabs, Inc.; 11804 North Creek Parkway; South Bothell, WA 98011-8805.

Frankia strains isolated by North American scientists may be sent to the curator of the USDA actinomycete culture collection for accession: D.P. Labeda, USDA-ARS-MWA, National Center for Agricultural Utilization Research, 1815 N. University Street, Peoria, IL 61604. Some strains of *Frankia* are also present in the IFO collection in Japan and the American Type Culture Collection (Lechevalier & Lechevalier, 1990). Most *Frankia* workers are also willing to supply isolates to interested researchers.

16–2.4 Taxonomy

The taxonomy of *Frankia* is being reevaluated. Because of the uncertainty regarding species recognition and the confusion arising from trivial strain designations (Lechevalier, 1983), the initial classification proposed by Becking (1970) has been replaced with a system of registry numbers (see section 16–2.3). More recently, Lalonde et al. (1988) suggested that enough is now known for species to be recognized within *Frankia* and proposed two species, *F. alni* and *F. elaeagni,* on the basis of differences in

TO: The Curator of the *Frankia* Catalog

I wish to register the following strain and to supply descriptive information about its isolation and characterization *in vitro*.

1. Strain designation: |___|___|___|___|
                          | lab | | host plant | | strain number |

NOTE: Please follow the rules for designating a *Frankia* strain
      as given by M.P. Lechevalier, Can. J. Bot. (1983)
      61:2964–2697 and by D. Baker, Actinomycetes (1987)
      20:85–88.

1a. Parent strain (if any): _____

2. Synonyms: _____

3. Primary literature citation describing this strain: _____

4. Other citations: _____

5. Source host plant for this strain: _____ native _____ exotic

6. Soil type (if soil was used as sourced of inoculum or this
    strain was isolated directly from soil): _____

7. Collection site:
    Latitude: _____ Longitude: _____

8. Date of collection: _____ Climatic season: _____

9. This organism was passed through one or more host plants
    before isolation: yes ___, no ___; if yes, please describe pas-
    sages: _____

10. Isolation was achieved directly from: fresh nodules ___, frozen
    nodules ___, dried nodules ___, fresh soil ___, frozen soil ___, dried
    soil ___

11. The nodules used for isolation were: sp$^+$ ___, sp$^-$ ___,
    unknown ___

12. This strain fixes N$_2$ *in vitro*: yes ___, no ___, untested ___

13a. This strain infects the following host plants: |___|, |___|, |___|, |___|, |___|, |___|, |___|,

13b. This strain has been tested on, but does not infect the fol-
     lowing host plants: |___|, |___|, |___|, |___|, |___|, |___|,
     |___|, |___|,

14a. This strain effectively fixes N$_2$ within nodules of the follow-
     ing host plants: |___|, |___|, |___|, |___|, |___|, |___|,
     |___|,

14c. This strain fixes little or no N$_2$ within nodules of the follow-
     ing host plants: |___|, |___|, |___|, |___|, |___|,
     |___|

15. The "whole-cell sugar pattern" of this strain is: ___, untest-
    ed ___

16. This strain is available to other scientists: yes ___, no ___

17. If the answer to question 16 is yes, please list names of
    other persons with whom the strain has been deposited:

18a. My name and address are: _____

18b. My phone number is: _____

18c. My fax number is: _____

18d. My e-mail address is: _____

19. Additional information about this strain that you feel is important
    to be recorded is the *Frankia* catalog may be appended to this
    document, e.g., protein gel group, serogroup, physiological
    group (A, B), 16S rRNA sequence, etc.

Fig. 16–3. Sample data form for registering a *Frankia* strain. (Adapted from Baker, 1989.)

serology, isozyme type, and soluble protein patterns. Using these same criteria, *F. alni* strains were further classified into subspecies: *pommerii,* which are sp⁻, and *vandijkii,* which are sp⁺. The two proposed *Frankia* species fit within the larger framework of nine genomic species defined by DNA:DNA hybridization (Fernandez et al., 1989), however, the subspecies do not. Currently, the use of species names has not been widely used.

16–3 QUANTIFICATION AND DIFFERENTIATION OF *FRANKIA* STRAINS

Many of the methods for quantifying *Frankia* and for differentiating among *Frankia* strains are similar to those used for other microorganisms (see chapters 25 through 32 in this book). Therefore, this section will summarize how these generic methods have been used with *Frankia* and only give details about methods that are more unique to *Frankia*.

16–3.1 Quantification

16–3.1.1 Total Biomass

Standard destructive measurements, such as dry weight, total organic C, and protein have been used to measure *Frankia* biomass. Of these methods, protein content has been judged to be most reliable and sensitive (Burggraaf & Shipton, 1982; Nittayajarn & Baker, 1989). Nondestructive methods include homogenization followed by direct microscopy or turbidity measurements, however, the most convenient and widely used approach is to determine packed cell volume (pcv). Packed cell volume is determined by centrifuging homogenized cultures at a set speed for a set period of time, e.g., microcentrifuge tubes spun at 14 000 g for 1 min. Protein content and pcv are linearly related and pcv is much quicker and easier.

16–3.1.2 Active Biomass

Active *Frankia* can be measured using plate counts, which give VU, or a plant bioassay (see section 16–5.2), which gives IU. Burggraaf and Shipton (1983) used these methods on sonicated cultures and found them to correlate quite poorly with direct counts or total protein. They generally observed VU to be 10 to 100 times higher than IU. Similar results have been found in my laboratory (Myrold et al., 1990), with VU being 30-times greater than IU. The large range in the ratio of VU:IU probably reflects differences in *Frankia* strains, culturing conditions, and homogenization techniques. Thus, calibration is required when using VU or IU measurements. Other standard methods for determining active biomass, such as ATP (see chapter 35 by Holben in this book) or INT reduction (see chapter 6 by Bottomley in this book; Prin et al., 1990) can also be used.

16–3.2 Strain Differentiation

Frankia strains were initially characterized by the host from which they were isolated. Later, hyphal morphology (type 1, sinuate; type 2, smooth) or physiology (group A, growth on many C sources; group B, use Tweens and certain organic acids) were used (Benson, 1982; Lechevalier et al., 1983). Some of the more discriminating methods that are used currently are described next. In many cases, the detailed methodology for these procedures is covered elsewhere in this book, thus, only information specific to *Frankia* is given here. Some methods provide only enough information to allow strain discrimination, whereas other methods provide information about the genetic relatedness among strains.

16–3.2.1 Serology

In contrast to rhizobia, the antigenic characteristics of *Frankia* are not particularly diverse (Simonet et al., 1990b). Three broad serotypes corresponding to the *Alnus, Elaeagnus,* and *Casuarina* host specificity groups have been defined based on heterologous reactions using polyclonal antisera (Baker et al., 1981; Lechevalier et al., 1982). Greater specificity usually has not been obtained, although Parson et al. (1985) were able to distinguish a spontaneous variant based on heterologous interactions with polyclonal antiserum in Ouchterlony double-diffusion assays. Other assay procedures, such as fluorescent labeling, ELISA, or use of monoclonal antibodies do not provide any greater degree of strain specificity (D.D. Baker, 1992, personal communication). Wright (chapter 28 in this book) describes polyclonal and monoclonal antibody production and various detection methods.

16–3.2.2 Cell Wall Chemistry

Cell wall sugars, such as 2-O-methyl-D-mannose (Mort et al., 1983), fatty acids, and phospholipids (Lechevalier, 1984) have been recognized as useful for defining *Frankia* as a genus. Fatty acid analysis can also discriminate between recognized *Frankia* spp. and subspecies and may be useful for finer separations (Simon et al., 1989). More recent research has shown the presence of several unique lipids, such as bacteriohopanetetrol (Berry et al., 1991), and some unidentified lipids that are more enriched in vesicles than in hyphae (Tunlid et al., 1989). It remains to be seen if these will be useful markers for *Frankia* strains.

16–3.2.3 Protein Patterns

The patterns resulting from the electrophoretic separation of soluble or whole-cell proteins are useful for subdividing and clustering bacterial strains (see chapter 29 by Sen in this book). One-dimensional sodium dodecyl sulfate-polyacrylamide gel electrophoresis (SDS-PAGE) was first used for *Frankia* by Benson and Hanna (1983), who were able to separate

43 isolates from one *Alnus* stand into six groups, with one group containing 35 isolates. Subsequently, Gardes and Lalonde (1987) used one-dimensional SDS-PAGE to separate 35 *Frankia* isolates into two distinct groups, which corresponded with *F. alni* and *F. elaeagni*.

16–3.2.4 Multi-Locus Allozyme Electrophoresis

A method somewhat analogous to SDS-PAGE of proteins involves the electrophoretic separation of allelic variants of chromosomally encoded enzymes (see chapter 26 by Eardly in this book). Puppo et al. (1985) first applied this method to *Frankia* and were able to achieve different levels of strain specificity with different enzymes tested. For example, strains isolated from *Casuarina* all had the same pattern for superoxide dismutase, but this pattern was different for strains isolated from other actinorhizal hosts; in contrast, there was considerable heterogeneity among *Casuarina* strains when other enzymes were used. More detailed work has shown that isozyme patterns can differentiate between *F. alni* and *F. elaeagni,* and between the two subspecies of *F. alni* (Gardes et al., 1987). Isozyme patterns have also been used to compare the diversity of strains isolated from a single *Alnus rubra* (Faure-Raynaud et al., 1990a), from individual nodules of *Alnus glutinosa* (Faure-Raynaud et al., 1991), and from single-spore isolates of a *Casuarina* strain (Prin et al., 1991b).

16–3.2.5 Plasmid Profiles

The presence of particular plasmids have been used to distinguish among bacterial strains within a species (see chapter 30 by Pepper in this book). Plasmid DNA in *Frankia* was first described by Dobritsa (1982). A more complete survey of 39 *Frankia* strains showed that only four strains had plasmids (Normand et al., 1983). Subsequently, these researchers found 11 additional plasmid-bearing *Frankia* strains in surveys of more than 200 *Frankia* strains (Simonet et al., 1984, 1985; Normand & Lalonde, 1986). The number of plasmids per strain varies from one to six, their sizes range from 8 to 60 kb (with the exception of one large 190-kb plasmid) and the numbers of plasmid copies varies from a few to more than 100 (Simonet et al., 1990a). The 190-kb plasmid carries some of the *nif* genes (Simonet et al., 1986), however, most *Frankia* plasmids are cryptic. Based on RFLP (restriction fragment length polymorphism) analysis, some strains harbor a similar plasmid (Simonet et al., 1985). To date, plasmids have been found in *Frankia* isolated from: *Alnus, Comptonia, Elaeagnus, Myrica,* and *Shepherdia* species (Mullin & An, 1990).

Mullin and An (1990) point out that the method used for DNA extraction may affect the recovery of plasmid DNA. For example, Bloom et al. (1989) used a DNA extraction protocol that enriches for plasmid DNA and found plasmids in 10 of 16 *Frankia* isolated from *Myrica pensylvanica.* It is not known whether this relatively high number of plasmid-containing strains is because of the DNA extraction protocol used or if *Frankia* from this actinorhizal host tend to have plasmids.

16–3.2.6 DNA:DNA Hybridization

Mullin and An (1990) state that DNA:DNA hybridization is the best measure of the depth of genetic diversity. In other words, it works well for detecting relatively large differences among isolates. An et al. (1985a) found DNA:DNA homology to range from 10 to 97% when DNA from 19 *Frankia* strains isolated from 12 actinorhizal host species were tested. Within *Alnus* strains, however, the range of hybridization values was much narrower, 67 to 94%. Using 43 *Frankia* strains isolated from *Alnus, Casuarina,* and *Elaeagnus,* Fernandez et al. (1989) defined three genomic species (> 70% homology) within the *Alnus* isolates, five genomic species within the *Elaeagnus* isolates, and one genomic species for the *Casuarina* isolates. Subsequent work with 27 *Frankia* strains from the *Alnus* and *Elaeagnus* host specificity groups defined five genomic species within the *Alnus* isolates and four genomic species within the *Elaeagnus* isolates (Akimov & Dobritsa, 1992). Large variation in DNA:DNA homology (12 to 99%) has been found for *Frankia* isolated from *Myrica pensylvanica* (Bloom et al., 1989), perhaps because it is a promiscuous host. Nevertheless, this large diversity within *Frankia* is undoubtedly an underestimate of the true diversity, because only isolated strains from just a few of the actinorhizal hosts have been used for DNA:DNA hybridization.

16–3.2.7 Restriction Fragment Length Polymorphism Patterns

A more discriminating test of diversity that uses DNA is the use of RFLP patterns (see chapter 31 by Sadowsky in this book). The earliest work with *Frankia* compared ethidium bromide-stained gels of total chromosomal DNA digested with restriction enzymes (An et al., 1985b). Bloom et al. (1989) applied this method to 16 *Frankia* isolates from *Myrica pensylvanica* and found nine distinct RFLP groups, which illustrates the sensitivity of this method. Application of restriction enzymes that cut at low frequency allowed more than 100 *Frankia* isolates to be grouped into 15 clusters (Beyazova & Lechevalier, 1992). Researchers also have used specific gene probes (e.g., *nifHDK* or 16S rDNA) to hybridize Southern transfers of restricted DNA (Normand et al., 1988; Nazaret et al., 1989; Nittayajarn et al., 1990; Jamann et al., 1992). This approach is very selective and has been found to give results consistent with the proposed speciation of *Frankia* (Lalonde et al., 1988; Nazaret et al., 1989) or previously described host specificity groups (Nittayajarn et al., 1990). A new modification of this approach combines the polymerase chain reaction (PCR) with subsequent restriction enzyme digestion to generate RFLP patterns (Maggia et al., 1992; Jamann et al., 1993).

16–3.2.8 DNA Sequencing and Strain Specific Probes

The most recent advance in differentiating *Frankia* strains has involved sequencing specific genes and synthesizing strain-specific gene probes (see chapter 32 by Ogram in this book). To date, two genes have been used. Based on the DNA sequence of the *nifH* gene, Simonet et al. (1990b) were able to construct 15-bp oligonucleotides that were capable of distinguishing between two *Frankia* strains isolated from *Alnus*. Hahn et al. (1989b) sequenced the portion of the *rrn* gene, which codes for rRNA's, and used this information to construct oligonucleotide probes that could differentiate between an effective and an ineffective strain (Hahn et al., 1989a). Recently, additional *rrn* sequences have been determined that allow other strains of *Frankia* to be differentiated with oligonucleotide probes (Harry et al., 1991; Nazaret et al., 1991; Bosco et al., 1992; Normand et al., 1992). Greater use will undoubtedly be made of strain-specific probes designed on the basis of sequence information. It should be noted, however, that other approaches, such as using randomly cloned pieces of *Frankia* DNA as probes, may be equally useful in differentiating *Frankia* strains.

16–4 CHARACTERIZATION OF *FRANKIA* IN SYMBIOSIS

Frankia strains can be characterized using actinorhizal host plants. Plants can be used to assess host specificity and N_2-fixing potential, or to measure differential characteristics of *Frankia* strains that are peculiar to actinorhizae, e.g., sp$^+$ vs. sp$^-$ nodules. In addition, actinorhizal hosts can be used as *Frankia* 'traps' and their nodules as enrichment sites for testing *Frankia* DNA and RNA with gene probes.

16–4.1 Host Specificity

Torrey (1990) provides an excellent guide to testing for cross-inoculation groups. Although the basic procedure is straightforward, skill is needed in raising test plants and taking necessary precautions to prevent contamination. Testing for host specificity involves inoculation of one or more actinorhizal host plants with one or more sources of *Frankia* inoculum. It is important that both positive (inoculating with a *Frankia* strain known to cause nodulation) and negative (not inoculated) controls be done. Plants are checked periodically for nodulation and sometimes the nodules are assessed for sporulation, effectiveness (nitrogenase activity), or other physiological characteristics. Patience is often required before consistent results are obtained.

16–4.1.1 Inoculum Preparation

The source of inoculum may be pure culture isolates of *Frankia*, crushed nodule suspensions, or soil. Pure cultures are probably the easiest to work with, however, standardization is needed because of the different morphologic forms of *Frankia*. Torrey (1990) suggests growing *Frankia* cultures to stationary phase in a medium that promotes spore formation. The *Frankia* should be harvested, washed, homogenized, and centrifuged to determine pcv (see section 16–3.1.1.2). About 1 to 10 μL pcv per seedling is adequate for nodulation. Experience has shown that spore suspensions are much more infective than hyphal suspensions on a pcv basis and retain a degree of viability following desiccation (Burleigh & Torrey, 1990).

Preparation of crushed nodule suspensions is similar to that described for isolation of *Frankia* (see section 16–2.1). Young, healthy nodules should be collected, washed, surface sterilized (usually H_2O_2 or NaOCl), homogenized, and the homogenate diluted in water or 1% (wt/vol) NaCl. Torrey (1990) suggests 1 g of nodule per 20 mL of diluent. Roots of the test plants may be dipped in the crushed nodule suspension or the suspension can be applied to the roots. About 0.25 to 2.5 mg crushed nodule is generally sufficient for nodulation. Soil can be applied directly or as a suspension at a rate of 1 to 5 g soil per seedling (Benoit & Berry, 1990).

16–4.1.2 Plant Propagation and Assay Systems

The details of propagating and growing all of the different types of actinorhizal plants is beyond the scope of this chapter; each has different requirements. An overview is given in Table 16–4, but the review by Benoit and Berry (1990), and the references therein, should be consulted for more complete information on growing actinorhizal plants from seed or through vegetative propagation. Techniques for the production of clonal plant material using tissue culture methods are described by Seguin and Lalonde (1990). Commercial sources of seed are available for some actinorhizal plants, but often collection is required.

Many actinorhizal hosts grow well in water culture, although some may require a solid matrix, such as sand. What follows is a general description of a water culture system. Torrey (1990) should be referred to for additional details.

Healthy seedlings with their first true leaves are inoculated and transferred to nutrient solution containers. Containers should not allow light to enter to prevent algal growth. Inoculated seedlings are started in low-N (< 10 mg-N L^{-1}) nutrient solution and transferred in a few weeks to N-free nutrient solution. Torrey (1990) points out that use of a proper nutrient solution is critical and recommends using a modified Hoagland's solution (Table 16–5). Nutrient solution is replenished as needed and renewed periodically (e.g., biweekly). Aeration of the solution is often desirable. Conditions suitable for growth of plants (adequate temperature and light) should be maintained (i.e., a growth chamber or greenhouse). Plants are checked periodically for nodulation. Usually at least 14 d is required, and sometimes several months, before nodulation occurs (Torrey, 1990).

16–4.2 Nodule Metabolism (Nitrogen fixation)

Nodule respiration, nitrogenase activity, and hydrogenase activity have been assayed to determine the efficiency of N_2 fixation (Huss-Danell,

Table 16–4. Guidelines for propagation of actinorhizal plants. (Adapted from Benoit and Berry, 1990.)

Genus	Seed treatment	Germination conditions†	Vegetative propagation‡
Alnus	Overnight soak in aerated water	20–30 °C, 16-h daylight, 7–12 d	Stem cuttings, 5000 ppm IBA or powdered rooting aid, 3–6 wk in mist, sand, or water culture
Casuarina	None	20–30 °C, 16-h daylight, 11–40 d	Softwood cuttings, 50 ppm IBA, 5 wk in mist or water culture
Ceanothus	Hot water soak	25 °C, 12–30 d	Softwood cuttings, 5000 ppm IBA or powdered rooting aid, 4–10 wk
Cercocarpus	Aerated water soak for 24 h, stratify 5 wk	15–20 °C, 1–2 wk	Semihardwood cuttings, 5000 ppm IBA, 6 wk in mist with bottom heat
Comptonia	Scarify plus 500 ppm GA$_3$	15–20 °C, 8 wk	Softwood cuttings, stem layering, root pieces covered by rooting medium
Cowania	Aerated water soak for 24 h, stratify 5 wk	15–20 °C, 1–2 wk	
Discaria			Semihardwood cuttings
Dryas			Semihardwood cuttings, layering horizontal stems
Elaeagnus		30 °C d, 20 °C night	Hardwood cuttings, layering, root cuttings
Hippophae	Stratify 90 d	30 °C d, 20 °C night, 40 d	Hardwood cuttings, layering, root cuttings
Myrica	Stratify 14–90 d, 500 ppm GA$_3$ 24 h	23–25 °C, 16 h daylight, 11–150 d	Semihardwood cuttings with 5000 ppm IBA, layering
Purshia	Stratify 5 wk	15–20 °C, 14 d	Semihardwood cuttings, 5000 ppm IBA or powdered rooting aid, 8–10 wk on shade bench
Shepherdia	Stratify 60–90 d, 0.5–1 h H$_2$SO$_4$ treatment for hard seeds	20–30 °C, 21–60 d	Semihardwood cuttings, 8000 ppm IBA dip

† GA$_3$, gibberellic acid.
‡ IBA, indole-3-butyric acid.

1990). With the one exception of the "local source" of *Frankia* from Sweden (Sellstedt & Huss-Danell, 1984; Sellstedt, 1989), hydrogenase activity is universally present in *Frankia* and actinorhizal nodules. Consequently, the focus of this section will be measurement of nitrogenase activity.

Ecological measurements of N$_2$ fixation have been made to: discriminate between effective and ineffective nodules, test strains for their N$_2$-fixing ability, compare the effectiveness of different *Frankia* strain-plant combinations, and determine the amount of N$_2$ fixed in the field. In actinorhizal symbioses, N$_2$ fixation has been assessed by mass balance, [15]N

Table 16–5. Recipes for nutrient solutions commonly used for growing actinorhizal plant seedlings.

Solution	Ingredient	Stock solution, g L^{-1}	Final solution, mL L^{-1}
1/4 Hoagland's†	*Macronutrient solutions*		
	Ca(H$_2$PO$_4$)$_2$·H$_2$O	12.6	1.25
	CaSO$_4$·2H$_2$O	1.72	50
	K$_2$SO$_4$	87	1.25
	MgSO$_4$·7H$_2$O	246	0.50
	Chelated iron		1.25
	FeSO$_4$·7H$_2$O	5.56	
	Na$_2$EDTA	7.45	
	Micronutrients		0.25
	CoCl$_2$·6H$_2$O	0.025	
	CuSO$_4$·5H$_2$O	0.080	
	H$_3$BO$_3$	2.86	
	MnCl$_2$·4H$_2$O	1.81	
	Na$_2$MoO$_4$·2H$_2$O	0.025	
	ZnSO$_4$·7H$_2$O	0.220	
	Nitrogen		0.179
	NH$_4$NO$_3$	80.05	
1/10 Evan's‡	*Solution 1*		1
	K$_2$SO$_4$	27.88	
	MgSO$_4$·7H$_2$O	49.28	
	Solution 2		1
	KH$_2$PO$_4$	2.50	
	K$_2$HPO$_4$	14.5	
	Solution 3		1
	CaCl$_2$·2H$_2$O	7.35	
	CaSO$_4$·2H$_2$O	10.33	
	Solution 4		1
	FeCL$_3$·6H$_2$O	0.483	
	Na$_2$EDTA	0.628	
	Solution 5		1
	CoSO$_4$·7H$_2$O	0.0141	
	CuSO$_4$·5H$_2$O	0.0079	
	H$_3$BO$_3$	0.143	
	MnSO$_4$·H$_2$O	0.0773	
	Na$_2$MoO$_4$·2H$_2$O	0.0050	
	ZnSO$_4$·7H$_2$O	0.0220	
	Solution 6		10
	NH$_4$NO$_3$	2.86	

† Modified from Hoagland and Arnon (1950). Because of its low solubility, it is most convenient to add CaSO$_4$·2H$_2$O directly to the final solution volume in the desired amount, e.g., 0.86 g for 1 L of 1/4 Hoagland's solution. Omit NH$_4$NO$_3$ for N-free solution. When adjusted to pH 6.0 (two to three drops of 0.1 N NaOH for 1 L of 1/4 Hoagland's solution), I have found this solution to work well in the MPN bioassay. Others have successfully used similar Hoagland's resolutions at one-fourth to full strength and pH values from 5.4 to 6.8 (Quispel, 1958; van Dijk, 1984; van Dijk & Sluimer-Stolk, 1990).

‡ This is Huss-Danell's (1978) modification of Evan's et al. (1972) solution. Solution 3 does not completely dissolve, so it must be thoroughly mixed before use. Omit NH$_4$NO$_3$ for N-free solution. The pH of 1/10 Evan's solution is about 7.0, however, it is weakly buffered. I have found 1/10 Evan's solution to be comparable to 1/4 Hoagland's solution in the MPN bioassay, however, it has been successfully used by others at one-half strength (Smolander & Sundman, 1987).

methods, and the acetylene reduction assay. Details about these methods can be found in chapter 43 or, with special reference to actinorhizal plants, in Winship and Tjepkema (1990). In addition, there have been several recent papers devoted to measuring N_2 fixation of actinorhizal plants in the field (e.g., Beaupied et al., 1990; Sougoufara et al., 1990). One important point that should not be forgotten, is that the activity of root nodules as measured on intact plants is likely to be different from that measured in excised root nodules.

Some *Frankia* strains are infective but ineffective, i.e., they are capable of forming nodules but do not fix N_2. Ineffective nodules are typically small in size, contain relatively few *Frankia,* and do not normally contain vesicles (Mian et al., 1976; Hahn et al., 1988; Berry & Sunell, 1990). Ineffective nodules often occur because of intergeneric or interspecific incompatibilities (Weber et al., 1987; van Dijk et al., 1988), however, some *Frankia* isolates form ineffective nodules even when inoculated on the same host species from which they were isolated (Hahn et al., 1988). Van Dijk and Sluimer-Stolk (1990) report that some ineffective *Frankia* have no nitrogenase activity in pure culture and their DNA does not hybridize with a *nifHDK* probe. Hahn et al. (1989a) found significant differences in the 16S rDNA sequence of an ineffective compared to an effective *Frankia* strain. In some cases, effectiveness can be lost by mutation (Faure-Raynaud et al., 1990b).

The N_2-fixing effectiveness of host-*Frankia* combinations has been studied primarily with *Casuarina* and *Alnus* spp. Over threefold differences in growth and N_2 fixation have often been found for various *Casuarina-Frankia* combinations (Reddell & Bowen, 1985; Fleming et al. 1988; Reddell et al., 1988; Sanginga et al., 1990), although Rosbrook and Bowen (1987) found no significant *Frankia* strain effect. Differences in host provenance may be greater determinants of growth and N_2-fixing effectiveness than variation among *Frankia* strains (Sougoufara et al., 1992). Similar differences in growth and N_2 fixation have been found for *Alnus-Frankia* combinations (Normand & Lalonde, 1982; Wheeler et al., 1986; Hooker & Wheeler, 1987; Weber et al., 1989). In the *Alnus-Frankia* symbiosis, greater N_2-fixing effectiveness has been reported for sp[-] compared to sp[+] symbioses (Normand & Lalonde, 1982; Wheeler et al., 1986), although this relationship is not always observed (Kurdali et al., 1990). Despite the large variation in N_2-fixing capacity of various host-*Frankia* combinations, no clearly superior host genotypes or *Frankia* strains have been recognized and any positive effect of inoculation with a superior *Frankia* strain may soon be overshadowed in the field by nodulation by indigenous *Frankia* (Hooker & Wheeler, 1987).

16–4.3 Nodule Morphology (Sporulation)

Actinorhizal nodules come in a variety of shapes and sizes: some have discrete branched lobes, some are compact, and nodules of *Casuarina* and *Myrica* can form nodule roots (Berry & Sunell, 1990). Recently, aerial nodules have been discovered on *Casuarina cunninghamiana* (Prin et al., 1991a). Nodule morphology is probably under host plant control but may reflect environmental conditions. The major feature of nodule morphology of interest to microbial ecologists, however, is sporulation.

Although virtually all *Frankia* can be induced to form sporangia in pure culture, this is not necessarily the case in planta. Van Dijk and Merkus (1976) observed that *Frankia* within actinorhizal root nodules either form many sporangia (sp$^+$) or no (or at least very few) sporangia (sp$^-$). Smolander and Sundman (1987) proposed an "intermediate" class for nodules containing few sporangia and reserving sp$^-$ only for nodules devoid of sporangia, however, Schwintzer (1990) suggests that "intermediate" nodules are probably better classified as sp$^-$.

The spore type of actinorhizal nodules is of interest for at least three reasons: (i) It is doubtful if any *Frankia* have been isolated from sp$^+$ nodules (see section 16–1.2), (ii) Interesting and complex ecological relationships have been observed between sp$^-$ and sp$^+$ strains of *Alnus* sp. and *Myrica gale* (Weber, 1986; Holman & Schwintzer, 1987; Kashanski & Schwintzer, 1987; Smolander & Sundman, 1987), and (iii) There is some evidence that sp$^-$ and sp$^+$ nodules differ in both absolute and relative nitrogenase efficiency (Wheeler et al., 1986).

Nodule spore type is determined microscopically. Either hand or microtome sections of nodules are prepared. Hand sections are easily prepared with a razor blade. A description of microtome sectioning is given by Smolander and Sundman (1987). Sections should be prepared from the base of peripheral nodule lobes, because older tissues are more likely to contain sporangia (Schwintzer, 1990). Several sections from each lobe should be examined. Once cut, sections are mounted in dilute Fabil reagent (van Dijk, 1978) or lactophenol-cotton blue (Torrey, 1987) and examined at 400× magnification. Schwintzer (1990) states that free-hand sections from sp$^+$ nodules often have a cloud of sporangia and free-spores around the edge of the nodule tissue (Fig. 16–4). When sporangia are not abundant, enough sections must be examined so that at least 50 infected cells are examined (Schwintzer, 1990).

16–4.4 Nodule DNA/RNA Assays

Immunological methods, such as the fluorescent antibody technique that is used widely with rhizobia, are not specific enough for differentiating *Frankia* strains in nodules. But with the development of molecular techniques, it is now possible to determine nodule occupancy. To date, three reports in the literature have used gene probes to detect specific *Frankia* strains in nodules.

Simonet et al. (1988) used a radiolabeled *Frankia* plasmid cloned into pBR322 to probe blots of nodule DNA obtained from *Alnus* stands. This allowed them to distinguish *Alnus* stands that had the plasmid-containing *Frankia* strain from those that did not. But there are problems associated with using plasmids as strain markers. Plasmids may be lost or transferred

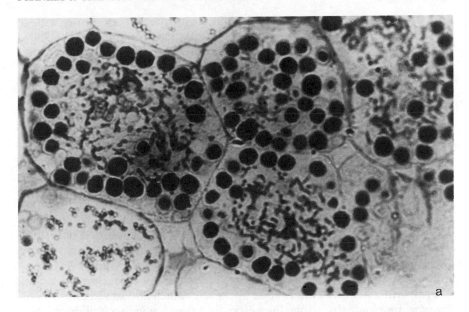

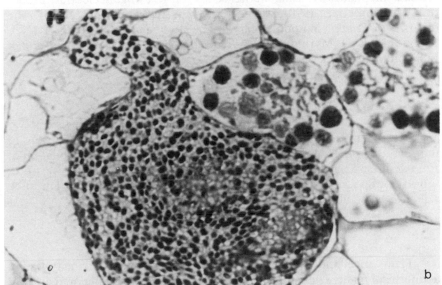

Fig. 16–4. Guide for distinguishing sp$^+$ vs. sp$^-$ nodules. Photographs of 2-μm thick cross-sections cut with a glass knife from Epon-resin embedded nodule lobes of *Alnus glutinosa*. Sections were stained with Toluidine blue and photographed under 1500× magnification. Photographs provided by C. van Dijk, N.O.I., the Netherlands.

a. Part of the cortex of a sp$^-$ nodule lobe. Cortex cells filled with clusters of thin hyphae with terminal vesicles, which is typical of effective *Alnus*-type *Frankia* infections.

b. In addition to vesicles, sp$^+$ nodules contain compact masses of *Frankia* spores that have developed from one or more sporangia, either intra- or inter cellular. In this photograph of the cortex of a sp$^+$ nodule lobe, a spore mass the size of several cortical cells is situated between cortex cells filled with vesicle clusters.

and it is possible for the same plasmid to be present in strains with a different chromosomal background or vice versa (Mullin & An, 1990). Hahn et al. (1990b) used oligonucleotide probes based on sequence differences in the variable region of 16S rRNA to detect a particular *Frankia* strain in *Alnus glutinosa* nodules. RNA was extracted from nodules and *Frankia* was detected in nodules as small as 1 mg. Simonet et al. (1990b) used oligonucleotide probes based on sequence differences in the *nifH* gene to detect and differentiate between two *Frankia* strains in *Alnus glutinosa* and *A. incana* nodules. DNA was extracted from individual nodule lobes and the *nifH* gene was specifically amplified with the PCR by using primers to the highly conserved region of the *nifH* gene. The amplified DNA was hybridized with the oligonucleotide probes to differentiate between the strains. Simonet et al. (1990b) reported that the PCR amplification step was necessary, because insufficient target DNA was obtained by extraction alone, however, D.D. Baker (1992, unpublished data) has been able to differentiate among *Frankia* strains using RFLP analysis of DNA extracted from nodules using the CTAB protocol (Murray & Thomson, 1980). These methods all require the extraction of nucleic acids and filter hybridization. An alternative approach, which allows visualization of *Frankia* in planta by using fluorescently tagged oligonucleotide probes (e.g., Amann et al., 1990), was recently reported (Prin et al., 1993). The diversity of *Frankia* in nodules can also be studied using PCR-RFLP (N.J. Ritchie and D.D. Myrold, 1993, unpublished data) or DNA sequencing of PCR products (Nick et al., 1993).

16–5 QUANTIFICATION IN SOIL

Many of the methods used to measure *Frankia* in soil are extensions of those used in pure culture or in planta. But following a small population of *Frankia* in soil against a large background of other microorganisms presents some special problems. Methods for detecting *Frankia* in soil must be selective or be able to specifically enrich for *Frankia*.

16–5.1 Plants as a Bioassay System

Plants meet the criteria given above. Within host specificity limits, actinorhizal plants select for *Frankia* when they become nodulated and the nodules are an enrichment of *Frankia*.

16–5.1.1 Use of Trap Plants

The simplest use of plants is to serve as trap plants. Actinorhizal plants can be planted in soil and then checked for nodulation. Such an approach is only qualitative, although it may be possible to make it quantitative if a dilution series of the soil with a sterile potting medium is made. In that case, an MPN estimate of *Frankia* numbers might be made.

16–5.1.2 Plant Bioassay and *Frankia* Infective Units

Virtually all of the ecological information about *Frankia* populations in soil has been collected by using some variation of a plant bioassay system. The statistical basis for most probable number (MPN) counts is given in chapter 3. After a brief summary of the autecology of *Frankia* in soil, the MPN methodology as it has been used with *Frankia* will be described.

Quantitative estimates of *Frankia* populations in soil are based on IUs as measured with a plant bioassay and to date have been devoted to *Alnus*-compatible *Frankia*. Surveys of soils from a variety of forest sites in Finland (Smolander & Sundman, 1987; van Dijk et al., 1988; Smolander, 1990), Sweden (Myrold & Huss-Danell, 1994), and the Pacific Northwest of the USA (A.B. Hilger & D.D. Myrold, 1992, unpublished data) have shown *Frankia* populations to range from 0 to 4600 IU g^{-1} soil. *Frankia* populations vary according to plant species present and soil conditions. It is particularly interesting that higher numbers have been found associated with non-actinorhizal plants (e.g., birch) than with actinorhizal plants (Smolander & Sundman, 1987; van Dijk et al., 1988; Smolander, 1990). Soil pH is positively correlated with *Frankia* populations (Smolander & Sundman, 1987). Laboratory experiments that have investigated soil and rhizosphere effects on *Frankia* populations have largely confirmed field observations. When *Frankia* were added to limed (pH 6.0) and unlimed soil (pH 4.2), survival was greater in limed soil (Smolander et al., 1988; Smolander & Sarsa, 1990; A.B. Hilger & D.D. Myrold, 1992, unpublished data). Rhizosphere effects, however, have given mixed results with some studies showing *Frankia* populations enhanced in the rhizospheres of some plants, such as birch (Smolander & Sarsa, 1990) and others showing no effect of several plant species, including birch (A.B. Hilger and D.D. Myrold, 1992, unpublished data). A survey of sand dunes in the Netherlands, which took advantage of being able to distinguish effective from ineffective *Frankia* strains, found populations of ineffective *Frankia* to be present at much higher populations than effective *Frankia* (van Dijk & Sluimer-Stolk, 1990).

16–5.1.2.1 Nodulation Capacity. The most widely used plant bioassay was developed by van Dijk (1984). The procedure involves germinating alder seedlings and growing them for 6 wk (until they have three to four leaves) in one-fourth-strength complete Hoagland's solution (Table 16–5). Nutrient solution is changed weekly. Seedlings are transferred to 250-mL culture jars containing one-half-strength N-free Hoagland's solution (eight seedlings per jar) and inoculated 2 d later. Serial dilutions of inoculum are made up in one-half-strength N-free Hoagland's solution and mixed into the culture jars. Plants are grown in a growth chamber. After 5 d, nutrient solution is replaced and thereafter replaced weekly. Nodulation is assessed 3 wk following inoculation (longer periods did not result in any more nodules forming).

Nodulation capacity (analogous to IU) was defined by van Dijk (1984) to be the number of nodules formed per unit of inoculum. It is equivalent to the slope of the line when number of nodules per jar is plotted against amount of inoculum per jar. This empirical relationship is linear up to the point where root biomass begins to limit nodulation. van Dijk (1984) chose to calculate nodulation capacity by summing the number of nodules in all jars of all dilutions in the linear portion of the curve (r) and dividing r by the fraction of the total amount of inoculum that was added to these jars (p). The resulting number is divided by the original inoculum quantity to get the nodulation capacity in terms of nodules per amount of inoculum. An estimate of the variance of the nodulation capacity is given by multiplying $(1-p)$ by $(r+1)$ and dividing by p^2. No derivation of these formulae is provided by van Dijk (1984) and it seems that doing a linear regression should provide similar estimates of the mean and variance of nodulation capacity using more widely recognized statistical techniques.

16–5.1.2.2 Most Probable Number Bioassay. Another approach for estimating infective units is based on MPN statistics. Instead of using several seedlings per container and counting total numbers of nodules, a single seedling per container is used and nodulation is scored in a plus/minus fashion. These data are analyzed using MPN tables or computer programs that calculate the appropriate MPN statistics (see chapter 5 by Woomer in this book).

A convenient experimental system for *Frankia* MPN tests was described by Hilger et al. (1991) and is shown in Fig. 16–5. Germinated seedlings of the one-leaf stage are placed in modified centrifuge tubes containing one-fourth strength N-free Hoagland's solution plus 5 mg NH_4NO_3-N L^{-1}, and inoculated with a serial dilution of soil. After 1 wk, and at 2 wk intervals thereafter, the solution is replaced with one-fourth strength N-free Hoagland's solution. Six to 8 wk following inoculation, seedlings are checked for nodule formation. Negative (not inoculated) and positive (inoculated with an effective *Frankia* strain) controls should always be included. This system works well with *Alnus rubra* and *A. incana*.

16–5.1.2.3 Comparison of Methods. The nodulation capacity assay is more widely used for measuring *Frankia* IUs, but the MPN assay may be based on a better established statistical foundation and is quite space efficient. A five-dilution, five tubes per dilution MPN assay occupies only about 0.08 m^{-2}, which may enhance the number of true replicates that can be done in a given bioassay. Both of the plant bioassay methods work well, however, and give comparable estimates of *Frankia* IUs (A.B. Hilger and D.D. Myrold, 1992, unpublished data; C. van Dijk, 1992, personal communication; Huss-Danell & Myrold, 1994).

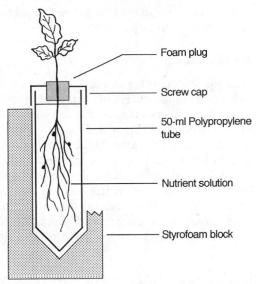

Fig. 16–5. Schematic diagram of an MPN seedling bioassay system for determining *Frankia* IU.

16–5.2 Fluorescent Antibodies

Several *Frankia* research groups have tried to use serological methods for detecting *Frankia* (see section 16–3.2.1), but no one has yet been able to produce antibodies of sufficient specificity for identifying a particular *Frankia* strain. These less specific probes might be useful for studies of *Frankia* in soil at the genus level, however, difficulties will be faced because of low soil populations.

16–5.3 DNA/RNA Probes

Perhaps the most promising approach for detecting and quantifying *Frankia* in soil will involve the use of molecular probes to detect *Frankia* DNA or RNA extracted from soil, possibly following DNA amplification by the PCR (see chapters 32, 34, and 35).

16–5.3.1 *Frankia*-Specific Probes and Polymerase Chain Reaction Primers

Frankia-specific probes and primers are based on the 16S rRNA se quences (Hahn et al., 1989a; Nazaret et al., 1991). Additional sequences are becoming available (Harry et al., 1991) and checks of the GenBank, or other sequence databases will likely turn up others. Using such information, several primers have been developed that are specific to gram-positive organisms (Table 16–6) and can be used as probes or primers in conjunction with the *Frankia*-specific primers.

16–5.3.2 DNA/RNA Extraction

There are numerous protocols for extracting DNA and RNA from soil (see chapter 35 in this book), however, only the direct extraction methods are likely to be useful for a filamentous bacterium like *Frankia*. Another important consideration when working with *Frankia* is their gram-negative cell wall structure, which can be more difficult to lyse than that of the gram-negative bacteria often used as test organisms in DNA extraction protocols. Hilger and Myrold (1991) have developed such a method to extract *Frankia* DNA from soil. The critical steps of this method remove humic materials that could interfere with subsequent probing or amplification techniques. A protocol for extracting *Frankia* RNA from soil has been published (Hahn et al., 1990a) which allows the detection of about 10^4 cells g^{-1} soil.

16–5.3.3 Polymerase Chain Reaction

Applications of PCR to *Frankia* have primarily used DNA extracted from *Frankia* isolates (Simonet et al., 1991; Bosco et al., 1992; Normand et al., 1992) or actinorhizal nodules (Simonet et al., 1990b; Nazaret et al., 1991). The PCR can be used to amplify DNA extracted from soil (Myrold et al., 1990). By using the MPN approach of dilution to extinction, the number of *Frankia* genomes can be quantified (Hilger & Myrold, 1992; Picard et al., 1992). The PCR-MPN has been used with the plant bioassay MPN to study *Frankia* populations in soil (Myrold & Huss-Danell, 1994).

16–6 CONCLUSION

Research on *Frankia,* especially its ecology in soil, is coming of age. Although there will still be methodological hurdles to overcome, many techniques are now available to allow long-standing questions to be answered. The use of molecular tools will greatly enhance these efforts, but proven methods using host plants will still play a vital role. One of the interesting areas that is just beginning to be explored is the diversity of *Frankia* in natural populations.

ACKNOWLEDGMENTS

I express appreciation to C-Y. Li, who helped me begin my research with *Frankia,* and Dwight Baker who has often given me valuable advice. Arlene Hilger and Kendall Martin have been stalwarts in my lab since we began studying *Frankia* and have been the source of much inspiration. I

Table 6–6. PCR primers that can be used to specifically amplify *Frankia* DNA. Primer pairs FR183/FR1401', FR485/FR1009', FR183/FR1009', and FR1479/FR1933' have been used successfully in my laboratory. The references cited should be consulted for other compatible primer pairs. Other possible primer pairs should be assessed for similar melting temperatures and potential to form primer dimers prior to use.

Label†	Specificity‡	Numbering§	Sequence¶	Reference
FGPS6	Eubacteria	6–25	GGA GAG TTa GAT CtT GGC TC	Bosco et al., 1992
FGPS59	*Frankia*	59–76	AAG TCG AGC GGG GAG CTT	Picard et al., 1992
FR183#	*Frankia*	183–204	CTG GTG GTG TGG AAA GAT TTA T	Modified from Hahn et al., 1989b
FGPS305'	*Frankia*	281–305	CCA GTG TGG CCG GTC GCC CTC TCA G	Picard et al., 1992
FR485	Eubacteria	485–502	CAG CAG CCG CGG TAA TAC	Hilger and Myrold, 1992, unpublished data
FGPS849††	Eubacteria	849–868	GCC TGG GGA GTA CGG CCG CA	Simonet et al., 1991
FGPS958	Actinomycetes	958–976	CTT GAC ATG CAG GGA AAT C	Simonet et al., 1991
FGPS989ac	*Frankia*, *Alnus* and *Casuarina* groups	989–1004	GGG GTC CGT AAG GGT C	Bosco et al., 1992
FGPS989e	*Frankia*, *Elaeagnus* group	989–1004	GGG GTC CtT AgG GGc t	Bosco et al., 1992
FR1009'	*Frankia*	990–1009	TGC AGG ACC CTT ACG GA(C/t) CC	Modified from Hahn et al., 1989b
SSU1000'	*Frankia*	1005–1025	AGC CAT GCA CCA CCT GTG CAG	L. Simon, 1992, personal communication
FGPS1093'	Actinomycetes	1074–1093	GCA ACA TAG GAC GAG GGT TG	Simonet et al., 1991
FGPS1176‡‡	Eubacteria	1116–1176	GGG GCA TGA TGA CTT GAC GTC	Simonet et al., 1991
FR1373'	Eubacteria	1356–1373	ACG GGC GGT GTG TAC AAG	Hilger and Myrold, 1992, unpublished data
FR1401'	Gram positive	1377–1401	TTC GGG TGT TAC CGA CTT TCG TGA C	Hilger and Myrold, 1992, unpublished data
FR1479	*Frankia* (putative)	1479–1497	GTA CCG GAA GGT GCG GCT G	Martin and Myrold, 1992, unpublished data
FGPS1493	*Frankia*§§	1493–1512	CGG TGG ATC ACC TCC TTT CT	Simonet et al., 1991
FR1933'	*Frankia* (putative)	1911–1933	TAA CTT GGC CAC AAA GAT GCT CG	Martin and Myrold, 1992, unpublished data
FGPL1948'	Eubacteria	1948–1973	ATC GGC TCg aGG TGC CAA GGg ATC CA	Bosco et al., 1992
FGPL2054'	*Frankia*§§	2038–2055	CCG GGT TTC CCC ATT CGG	Simonet et al., 1991

† No suffix designates a forward primer, homologous to the coding strand; a ' designates a reverse primer, complementary to the coding strand.
‡ Primer specificity determined either from empirical tests with other bacterial species and strains, or by comparison with *rrn* sequence information.
§ Numbering based on sequence for DNA of *Frankia* CeD isolated from *Casuarina equisetifolia* (Normand et al., 1992). This sequence is accessible from the GenBank databank, accession number M55343.
¶ Lower case bases designate disagreement with the *rrn* sequence of *Frankia* CeD (Normand et al., 1992). Bases in parentheses designate an equal mixture of the two bases at this position.
# Also known as SSU200 (L. Simon, 1992, personal communication).
†† Also known as SSU901 (Normand et al., 1992).
‡‡ Also known as FGPS1146' (Nazaret et al., 1991) and SSU1190' (Normand et al., 1992).
§§ When FGPS1493 and FGPL2054' are used together, *Geodermatophilus* DNA is amplified, however, the resulting product is different than the 561-bp product obtained with *Frankia* DNA (Simonet et al., 1991).

thank Kerstin Huss-Danell for providing an excellent sabbatical environment while I wrote this paper. While on sabbatical, I was supported by National Science Foundation grant INT-9025112 and a grant from the Swedish Council for Forestry and Agricultural Research to K. Huss-Danell. Lastly, thanks must go to the collective group of *Frankia* researchers throughout the world, who are always ready to help and advise. This article is Technical Paper No. 10,266 of the Oregon Agricultural Experiment Station.

REFERENCES

Akimov, V.N., and S.V. Dobritsa. 1992. Grouping of *Frankia* strains on the basis of DNA relatedness. System. Appl. Microbiol. 15:372–379.

Amann, R.I., L. Krumholz, and D.A. Stahl. 1990. Fluorescent-oligonucleotide probing of whole cells for determinative, phylogenetic, and environmental studies in microbiology. J. Bacteriol. 172:762–770.

An, C.S., J.W. Wills, W.S. Riggsby, and B.C. Mullin. 1983. Deoxyribonucleic acid base composition of 12 *Frankia* isolates. Can. J. Bot. 61:2859–2862.

An, C.S., W.S. Riggsby, and B.C. Mullin. 1985a. Relationships of *Frankia* isolates based on deoxyribonucleic acid homology studies. Int. J. Syst. Bacteriol. 35:140–146.

An, C.S., W.S. Riggsby, and B.C. Mullin. 1985b. Restriction pattern analysis of genomic DNA of *Frankia* isolates. Plant Soil 87:43–48.

Baker, D. 1987a. Modification of catalog numbering system for *Frankia* strains and revision of strain numbers for DDB *Frankia* collection. Actinomycetes 20:85–88.

Baker, D. 1987b. Relationships among pure cultured strains of *Frankia* based on host specificity. Physiol. Plant. 70:245–248.

Baker, D. 1988. Opportunities for autecological studies of *Frankia,* a symbiotic actinomycete. p. 271–276. *In* Y. Okami et al. (ed.) Biology of actinomycetes. Japan Sci. Soc. Press, Tokyo, Japan.

Baker, D. 1989. Methods for the isolation, culture, and characterization of the *Frankiaceae:* Soil actinomycetes and symbionts of actinorhizal plants. p. 213–236. *In* D.P. Labeda (ed.) Isolation of biotechnological organisms from nature. McGraw-Hill, New York.

Baker, D.D. 1993. A bibliography on *Frankia* and actinorhizal plants, March 1993 edition. Panlabs, Inc., South Bothell, WA.

Baker, D., and D. O'Keefe. 1984. A modified sucrose fractionation procedure for isolation of frankiae from actinorhizal root nodules and soil samples. Plant Soil 78:23–28.

Baker, D., W.L. Pengelly, and J.G. Torrey. 1981. Immunochemical analysis of relationships among isolated frankiae (Actinomycetales). Int. J. Syst. Bacteriol. 31:148–151.

Baker, D.D., and C.R Schwintzer. 1990. Introduction. p. 1–13. *In* C.R. Schwintzer and J.D. Tjepkema (ed.) The biology of *Frankia* and actinorhizal plants. Academic Press, San Diego.

Baker, D., and E. Seling. 1984. *Frankia:* New light on an actinomycete symbiont. p. 563–574. *In* L. Ortiz-Ortiz et al. (ed.) Biological, biochemical and biomedical aspects of actinomycetes. Academic Press, New York.

Baker, D., J.G. Torrey, and G.H. Kidd. 1979. Isolation by sucrose-density fractionation and cultivation in vitro of actinomycetes from nitrogen-fixing root nodules. Nature (London) 281:76–78.

Beaupied, H., A. Moiroud, A.M. Domenach, F. Kurdali, and R. Lensi. 1990. Ratio of fixed and assimilated nitrogen in a black alder (*Alnus glutinosa*) stand. Can. J. For. Res. 20:1116–1119.

Becking, J.H. 1970. *Frankiaceae* fam. nov. (Actinomycetales) with one new combination and six new species of the genus *Frankia* Brunchorst 1886. Int. J. Syst. Bacteriol. 20:201–220.

Benoit, L.F., and A.M. Berry. 1990. Methods for production and use of actinorhizal plants in forestry, low-maintenance landscapes, and revegetation. p. 281–297. *In* C.R. Schwintzer and J.D. Tjepkema (ed.) The biology of *Frankia* and Actinorhizal plants. Academic Press, San Diego.

Benson, D.R. 1982. Isolation of *Frankia* strains from alder actinorhizal nodules. Appl. Environ. Microbiol. 44:461–465.

Benson, D.R., and D.G. Hanna. 1983. *Frankia* diversity in an alder stand as estimated by sodium dodecyl sulfate-polyacrylamide gel electrophoresis of whole-cell proteins. Can. J. Bot. 61:2919–2923.

Benson, D.R., and N.A. Schultz. 1990. Physiology and biochemistry of *Frankia* in culture. p. 107–127. *In* C.R Schwintzer and J.D. Tjepkema (ed.) The biology of *Frankia* and actinorhizal plants. Academic Press, San Diego.

Benson, D.R., and W.B. Silvester. 1993. Biology of *Frankia* strains, actinomycete symbionts of actinorhizal plants. Microbiol. Rev. 57:293–319.

Berry, A.M., R.A. Moreau, and A.D. Jones. 1991. Bacteriohopanetetrol: Abundant lipid in *Frankia* cells and in nitrogen-fixing nodule tissue. Plant Physiol. 95:111–115.

Berry, A.M., and L.A. Sunell. 1990. The infection process and nodule development. p. 61–81. *In* C.R. Schwintzer and J.D. Tjepkema (ed.) The biology of *Frankia* and actinorhizal plants. Academic Press, San Diego.

Beyazova, M., and M.P. Lechevalier. 1992. Low-frequency restriction fragment analysis of *Frankia* strains (*Actinomycetales*). Int. J. System. Bacteriol. 42:422–433.

Bloom, R.A., B.C. Mullin, and R.L. Tate, III. 1989. DNA restriction patterns and solution hybridization studies of *Frankia* isolates from *Myrica pensylvanica*. Appl. Environ. Microbiol. 55:2155–2160.

Bosco, M., M.P. Fernandez, P. Simonet, R. Materassi, and P. Normand. 1992. Evidence that some *Frankia* sp. strains are able to cross boundaries between *Alnus* and *Elaeagnus* host specificity groups. Appl. Environ. Microbiol. 58:1569–1576.

Burggraaf, A.J.P. 1984. Isolation, cultivation and characterization of *Frankia* strains from actinorhizal root nodules. Ph.D. thesis. State University, Leiden, the Netherlands.

Burggraaf, A.J.P., and W.A. Shipton. 1982. Estimates of *Frankia* growth under various pH and temperature regimes. Plant Soil 69:135–147.

Burggraaf, A.J.P., and W.A. Shipton. 1983. Studies on the growth of *Frankia* isolates in relation to infectivity and nitrogen fixation (acetylene reduction). Can. J. Bot. 61:2774–2782.

Burleigh, S., and J.G. Torrey. 1990. Effectiveness of different *Frankia* cell types as inocula for the actinorhizal plant *Casuarina*. Appl. Environ. Microbiol. 56:2565–2567.

Callaham, D., P. Del Tredici, and J.G. Torrey. 1978. Isolation and cultivation *in vitro* of the actinomycete causing root nodulation in *Comptonia*. Science 199:899–902.

Cruz-Cisneros, R., and M. Valdes. 1991. Actinorhizal nodules on *Aldolphis infesta* (H.B.K.) Meissner (Rhamnaceae). Nitrogen Fix. Tree Res. Rep. 9:87–89.

Dawson, J.O. 1990. Interactions among actinorhizal and associated plant species. p. 299–316. *In* C.R. Schwintzer and J.D. Tjepkema (ed.) The biology of *Frankia* and actinorhizal plants. Academic Press, San Diego.

Diem, H.G., and Y.R Dommergues. 1983. The isolation of *Frankia* from nodules of *Casuarina*. Can. J. Bot. 61:2822–2825.

Diem, H.G., and Y.R. Dommergues. 1988. Isolation, characterization and cultivation of *Frankia*. p. 227–254. *In* N.S. Subba Rao (ed.) Biological nitrogen fixation, recent developments. Gordon and Breach Sci. Publ., New York.

Diem, H.G., and Y.R. Dommergues. 1990. Current and potential uses and management of Casuarinaceae in the tropics and subtropics. p. 317–342. *In* C.R. Schwintzer and J.D. Tjepkema (ed.) The biology of *Frankia* and actinorhizal plants. Academic Press, San Diego.

Diem, H.G., Gauthier, D., and Y.R. Dommergues. 1982. Isolation of *Frankia* from nodules of *Casuarina equisetifolia*. Can. J. Microbiol. 28:526–530.

Dobritsa, S.V. 1982. Extrachromosomal circular DNAs in endosymbiont vesicles from *Alnus glutinosa* nodules. FEMS Microbiol. Lett. 15:87–91.

Dobritsa, S.V., S.N. Novik, and O.S. Stupar. 1990. Infectivity and host specificity of strains of *Frankia*. Mikrobiologiya 59:314–320.

Evans, H.J., B. Koch, and R. Klucas. 1972. Preparation of nitrogenase from nodules and separation into components. Meth. Enzymol. 24:470–476.

Faure-Raynaud, M., M.A. Bonnefoy-Poirier, and A. Moiroud. 1990a. Diversity of *Frankia* stains [sic] isolated from actinorhizae of a single *Alnus rubra* cultivated in nursery. Symbiosis 8:147–160.

Faure-Raynaud, M., C. Daniere, A. Moiroud, and A. Capellano. 1990b. Preliminary characterization of an ineffective *Frankia* derived from a spontaneously neomycin-resistant strain. Plant Soil 129:165–172.

Faure-Raynaud, M., C. Daniere, A. Moiroud, and A. Capellano. 1991. Diversity of *Frankia* strains isolated from single nodules of *Alnus glutinosa*. Plant Soil 132:207–211.

Fernandez, M.P., H. Meugnier, P.A.D. Grimont, and R. Bardin. 1989. Deoxyribonucleic acid relatedness among members of the genus *Frankia*. Int. J. Syst. Bacteriol. 39:424–429.

Fleming, F.I., E.R. Williams, and J.W. Trunbull. 1988. Growth and nodulation and nitrogen fixation of provenances of *Casuarina cunninghamiana* inoculated with a range of *Frankia* sources. Austral. J. Bot. 36:171–181.

Fontaine, M.S., P.H. Young, and J.G. Torrey. 1986. Effects of long term preservation of *Frankia* strains on infectivity, effectivity and *in vitro* nitrogenase activity. Appl. Environ. Microbiol. 51:694–698.

Gardes, M., J. Bousquet, and M. Lalonde. 1987. Isozyme variation among 40 *Frankia* strains. Appl. Environ. Microbiol. 53:1596–1603.

Gardes, M., and M. Lalonde. 1987. Identification and subgrouping of *Frankia* strains using sodium dodecyl sulfate polyacrylamide gel electrophoresis. Physiol. Plant. 70:2237–2244.

Hahn, D., M. Dorsch, E. Stackebrandt, and A.D.L. Akkermans. 1989a. Synthetic nucleotide probes for identification of *Frankia* strains. Plant Soil 118:211–219.

Hahn, D., R. Kester, M.J.C. Starrenburg, and A.D.L. Akkermans. 1990a. Extraction of ribosomal RNA from soil for detection of *Frankia* with oligonucleotide probes. Arch. Microbiol. 154:329–335.

Hahn, D., M.P. Lechevalier, A. Fischer, and E. Stackebrandt. 1989b. Evidence for a close phylogenetic relationship between members of the genera *Frankia*, *Geodermatophilus*, and *"Blastococcus"* and emendation of the family Frankiaceae. Syst. Appl. Microbiol. 11:236–242.

Hahn, D., M.J.C. Starrenburg, and A.D.L. Akkermans. 1988. Variable compatibility of cloned *Alnus glutinosa* ecotypes against ineffective *Frankia* strains. Plant Soil 107:233–243.

Hahn, D., M.J.C. Starrenburg, and A.D.L. Akkermans. 1990b. Oligonucleotide probes against rRNA as a tool to study *Frankia* strains in root nodules. Appl. Environ. Microbiol. 56:1324–1346.

Harry, D.E., D.C. Yang, and J.O. Dawson. 1991. Nucleotide sequence and diversity in 16S ribosomal RNA from *Frankia*. Plant Soil 131:143–146.

Hibbs, D.E., and K. Cromack, Jr. 1990. Actinorhizal plants in Pacific Northwest forests. p. 343–363. *In* C.R. Schwintzer and J.D. Tjepkema (ed.) The biology of *Frankia* and actinorhizal plants. Academic Press, San Diego.

Higgins, M.L., and M.P. Lechevalier. 1969. Poorly lytic bacteriophage from *Dactylosporangium thailandensis* (Actinomycetales). J. Virol. 3:210–216.

Higgins, M.L., M.P. Lechevalier, and H.A. Lechevalier. 1967. Flagellated actinomycetes. J. Bacteriol. 94:1446–1451.

Hilger, A.B., and D.D. Myrold. 1991. Method for extraction of *Frankia* DNA from soil. Agric. Ecosys. Environ. 34:107–113.

Hilger, A.B., and D.D. Myrold. 1992. Quantitation of soil *Frankia* by bioassay and gene probe methods: response to host and non-host rhizospheres and liming. Acta Œcol. 13:505–506

Hilger, A.B., Y. Tanaka, and D.D. Myrold. 1991. Inoculation of fumigated nursery soil increases nodulation and yield of bare-root red alder (*Alnus rubra* Bong.). New Forests 5:35–42.

Hoagland, R.D., and D.I. Arnon. 1950. The water culture for growing plants without soil. California Agric. Exp. Stn. Circ. 347.

Holman, R.M., and C.R. Schwintzer. 1987. Distribution of spore-positive and spore-negative nodules of *Alnus incana* spp. *rugosa* in Maine, USA. Plant Soil 104:103–111.

Hooker, J.E., and C.T. Wheeler. 1987. The effectivity of *Frankia* for nodulation and nitrogen fixation in *Alnus rubra* and *A. glutinosa*. Physiol. Plant. 70:333–341.

Huss-Danell, K. 1978. Nitrogenase activity measurements in intact *Alnus incana*. Physiol. Plant. 43:372–376.

Huss-Danell, K. 1990. The physiology of actinorhizal nodules. p. 129–156. *In* C.R. Schwintzer and J.D. Tjepkema (ed.) The biology of *Frankia* and actinorhizal plants. Academic Press, San Diego.

Huss-Danell, D., and D.D. Myrold. 1994. Intrageneric variation in nodulation of *Alnus:* Consequences for quantifying *Frankia* infective units in soil. Soil Biol. Biochem. 26:525–531.

Jamann, S., M.P. Fernandez, and A. Moiroud. 1992. Genetic diversity of Elaeagnaceae-infective *Frankia* strains isolated from various soils. Acta Œcol. 13:395–406.

Jamann, S., M.P. Fernandez, and P. Normand. 1993. Typing method for N_2-fixing bacteria based on PCR-RFLP—application to the characterization of *Frankia* strains. Mol. Ecol. 2:17–26.

Kashanski, C.R., and C.R. Schwintzer. 1987. Distribution of spore-positive and spore-negative nodules of *Myrica gale* in Maine, USA. Plant Soil 104:113–120.

Kurdali, F., F. Rinaudo, A. Moiroud, and A.M. Domenach. 1990. Competition for nodulation and $^{15}N_2$-fixation between a Sp$^+$ and a Sp$^-$ *Frankia* strain in *Alnus incana*. Soil Biol. Biochem. 22:57–64.

Lalonde, M., and H.E. Calvert. 1979. Production of *Frankia* hyphae and spores as an infective inoculant for *Alnus* species. p. 95–110. *In* J.C. Gordon et al. (ed.) Symbiotic nitrogen fixation in the management of temperate forests. For. Res. Lab., Oregon State Univ., Corvallis.

Lalonde, M., L. Simon, J. Bousquet, and A. Seguin. 1988. Advances in the taxonomy of *Frankia:* Recognition of species *alni* and *elaeagni* and novel subspecies *pommerii* and *vandijkii*. p. 671–680. *In* H. Bothe et al. (ed.) Nitrogen fixation: Hundred years after. Fischer, Stuttgart, Germany.

Lechevalier, M.P. 1983. Cataloging *Frankia* strains. Can. J. Bot. 61:2964–2967.

Lechevalier, M.P. 1984. The taxonomy of the genus *Frankia*. Plant Soil 78:1–6.

Lechevalier, M.P. 1986. Catalog of *Frankia* strains. Actinomycetes 19:131–162.

Lechevalier, M.P., D. Baker, and F. Horriere. 1983. Physiology, chemistry, serology, and infectivity of two *Frankia* isolates from *Alnus incana* subsp. *rugosa*. Can. J. Bot. 61:2826–2833.

Lechevalier, M.P., F. Horriere, and H.A. Lechevalier. 1982. The biology of *Frankia* and related organisms. Dev. Ind. Microbiol. 23:51–60.

Lechevalier, M.P., and H.A. Lechevalier. 1990. Systematics, isolation, and culture of *Frankia*. p. 35–60. *In* C.R. Schwintzer and J.D. Tjepkema (ed.) The biology of *Frankia* and actinorhizal plants. Academic Press, San Diego.

Maggia, L., S. Nazaret, and P. Simonet. 1992. Molecular characterization of *Frankia* isolates from *Casuarina equisetifolia* root nodules harvested in West Africa (Senegal and Gambia). Acta Œcol. 13:453–462.

Mansour, S.R., A. Dewedar, and J.G. Torrey. 1990. Isolation, culture, and behavior of *Frankia* strain HFPCgI4 from root nodules of *Casuarina glauca*. Bot. Gaz. 151:490–496.

Mian, S., G. Bong, and C. Rodriquez-Barrueco. 1976. Effective and ineffective root-nodules in *Myrica faya*. proc. R. Soc. London B194:285–293.

Mort, A., P. Normand, and M. Lalonde. 1983. 2-*O*-methyl-D-mannose, a key sugar in the taxonomy of *Frankia*. Can. J. Microbiol. 29:993–1002.

Mullin, B.C., and C.S. An. 1990. The molecular genetics of *Frankia*. p. 195–214. *In* C.R. Schwintzer and J.D. Tjepkema (ed.) The biology of *Frankia* and actinorhizal plants. Academic Press, San Diego.

Murray, M.G., and W.F. Thomson. 1980. Rapid isolation of high molecular weight plant DNA. Nucleic Acids Res. 8:4321–4324.

Myrold, D.D., and K. Huss-Danell. 1994. Population dynamics of *Alnus*-infective *Frankia* in a forest soil with and without host trees. Soil Biol. Biochem. 26:533–540.

Myrold, D.D., A.B. Hilger, and S.H. Strauss. 1990. Detecting *Frankia* in soil using PCR. p. 429. *In* P.M. Gresshoff et al. (ed.) Nitrogen fixation: Achievements and objectives. Chapman and Hall, New York.

Nazaret, S., B. Cournoyer, P. Normand, and P. Simonet. 1991. Phylogenetic relationships among *Frankia* genomic species determined by use of amplified 16S rDNA sequences. J. Bacteriol. 173:4072–4078.

Nazaret, S., P. Simonet, P. Normand, and R. Bardin. 1989. Genetic diversity among *Frankia* isolated from *Casuarina* nodules. Plant Soil 118:241–247.

Nick, G., E. Paget, P. Simonet, A. Moiroud, and P. Normand. 1992. The nodular endophytes of *Coriaria* spp. form a distinct lineage within the genus *Frankia*. Mol. Ecol. 1:175–182.

Nittayajarn, A., and D.D. Baker. 1989. Methods for the quantification of *Frankia* cell biomass. Plant Soil 118:199–204.

Nittayajarn, A., B.C. Mullin, and D.D. Baker. 1990. Screening of symbiotic frankiae for host specificity by restriction fragment length polymorphism analysis. Appl. Environ. Microbiol. 56:1172–1174.

Normand, P., B. Cournoyer, P. Simonet, and S. Nazaret. 1992. Analysis of a ribosomal RNA operon in the actinomycete *Frankia.* Gene 111:119–124.

Normand, P., and M. Lalonde. 1982. Evaluation of *Frankia* strains isolated from provenances of two *Alnus* species. Can. J. Microbiol. 28:1133–1142.

Normand, P., and M. Lalonde. 1986. The genetics of actinorhizal *Frankia:* A review. Plant Soil 90:429–453.

Normand, P., P. Simonet, and R. Bardin. 1988. Conservation of *nif* sequences in *Frankia.* Mol. Gen. Genet. 213:238–246.

Normand, P., P. Simonet, J.L. Butour, C. Rosenberg, A. Moiroud, and M. Lalonde. 1983. Plasmids in *Frankia* sp. J. Bacteriol. 155:32–35.

Parson, W.L., L.R. Robertson, and C.V. Carpenter. 1985. Characterization and infectivity of a spontaneous variant isolated from *Frankia* sp. WEY 0131391. Plant Soil 87:31–42.

Picard, C., C. Ponsonnet, E. Paget, X. Nesme, and P. Simonet. 1992. Detection and enumeration of bacteria in soil by direct DNA extraction and polymerase chain reaction. Appl. Environ. Microbiol. 58:2717–2722.

Prin, Y., E. Duhoux, H.G. Diem, Y. Roederer, and Y.R. Dommergues. 1991a. Aerial nodules in *Casuarina cunninghamiana.* Appl. Environ. Microbiol. 57:871–874.

Prin, Y., L. Maggia, B. Picard, H.G. Diem, and Ph. Goullet. 1991b. Electrophoretic comparison of enzymes from 22 single-spore cultures obtained from *Frankia* strain ORS 140102. FEMS Microbiol. Lett. 77:223–228.

Prin, Y., F. Mallein-Gerin, and P. Simonet. 1993. Identification and localization of *Frankia* strains in *Alnus* nodules by *in situ* hybridization of *nif*H mRNA with strain-specific oligonucleotide probes. J. Exp. Bot. 44:815–820.

Prin, Y., M. Neyra, and H.G. Diem. 1990. Estimation of *Frankia* growth using Bradford protein and INT reduction activity estimates: Application to inoculum standardization. FEMS Microbiol. Lett. 69:91–96.

Puppo, A., L. Dimitrijevic, H.G. Diem, and Y.R. Dommergues. 1985. Homogeneity of superoxide dismutase patterns in *Frankia* strains from Casuarinaceae. FEMS Microbiol. Lett. 30:43–46.

Quispel, A. 1958. Symbiotic nitrogen fixation in non-leguminous plants. IV. The influence of some environmental conditions on different phases of the nodulation process in *Alnus glutinosa.* Acta Bot. Nerrl. 7:191–204.

Quispel, A., and A.J.P. Burggraaf. 1981. *Frankia,* the diazotrophic endophyte from actinorhizas. p. 229. *In* A.H. Gibson and W.E. Newton (ed.) Current perspectives in nitrogen fixation. Austral. Acad. Sci., Canberra, Australia.

Quispel, A., A.B. Svedsen, J. Schripsema, W.J. Bass, C. Erkelens, and J. Lugtenburg. 1989. Identification of dipterocarpol as isolation factor for the induction of primary isolation of *Frankia* from root nodules of *Alnus glutinosa* (L.) Gaertner. Mol. Plant-Microbe Interact. 2:107–112.

Racette, S., and J.G. Torrey. 1989. The isolation and infectivity of a *Frankia* strain from *Gymnostoma papuanum* (Casuarinaceae). Plant Soil 118:165–170.

Reddell, P., and G.D. Bowen. 1985. *Frankia* source affects growth, nodulation and nitrogen fixation in *Casuarina* species. New Phytol. 100:115–122.

Reddell, P., P.A. Rosbrook, G.D. Bowen, and D. Gwaze. 1988. Growth responses in *Casuarina cunninghamiana* plantings to inoculation with *Frankia.* Plant Soil 108:79–86.

Rosbrook, P.A., and G.D. Bowen. 1987. The ability of three *Frankia* isolates to nodulate and fix nitrogen with four species of *Casuarina.* Physiol. Plant. 70:373–377.

Rosbrook, P.A., A.J.P. Burggraaf, and P. Reddell. 1989. A comparison of two methods and different media for isolating *Frankia* from Casuarina root nodules. Plant Soil 120:187–193.

Sanginga, N., G.D. Bowen, and S.K.A. Danso. 1990. Genetic variability in symbiotic nitrogen fixation within and between provenances of two *Casuarina* species using the ^{15}N-labeling methods. Soil Biol. Biochem. 22:539–547.

Schultz, N.A., and D.R. Benson. 1989. Developmental potential of *Frankia* vesicles. J. Bacteriol. 171:6873–6877.

Schwintzer, C.R. 1990. Spore-positive and spore-negative nodules. p. 177–193. *In* C.R. Schwintzer and J.D. Tjepkema (ed.) The biology of *Frankia* and actinorhizal plants. Academic Press, San Diego.

Schwintzer, C.R., and J.D. Tjepkema. 1990. The biology of *Frankia* and actinorhizal plants. Academic Press, San Diego.

Seguin, A., and M. Lalonde. 1990. Micropropagation, tissue culture, and genetic transformation of actinorhizal plants and *Betula.* p. 215–238. *In* C.R. Schwintzer and J.D. Tjepkema (ed.) The biology of *Frankia* and actinorhizal plants. Academic Press, San Diego.

Sellstedt, A. 1989. Occurrence and activity of hydrogenase in symbiotic *Frankia* from field-collected *Alnus incana.* Physiol. Plant. 75:304–308.

Sellstedt, A., and K. Huss-Danell. 1984. Growth, nitrogen fixation and relative efficiency of nitrogenase in *Alnus incana* grown in different cultivation systems. Plant Soil 78:147–158.

Silvester, W.B., S.L. Harris, and J.D. Tjepkema. 1990. Oxygen regulation and hemoglobin. p. 157–176. *In* C.R. Schwintzer and J.D. Tjepkema (ed.) The biology of *Frankia* and actinorhizal plants. Academic Press, San Diego.

Simon, L., S. Jabaji-Hare, J. Bousquet, and M. Lalonde. 1989. Confirmation of *Frankia* species using cellular fatty acids analysis. System. Appl. Microbiol. 11:229–235.

Simonet, P., A. Capellano, E. Navarro, R. Bardin, and A. Moiroud. 1984. An improved method for lysis of *Frankia* with achromopeptidase allows detection of new plasmids. Can. J. Microbiol. 30:1292–1295.

Simonet, P., M.-C. Grosjean, A.K. Misra, S. Nazaret, B. Cournoyer, and P. Normand. 1991. *Frankia* genus-specific characterization by polymerase chain reaction. Appl. Environ. Microbiol. 57:3278–3286.

Simonet, P., J. Haurat, P. Normand, R. Bardin, and A. Moiroud. 1986. Localization of *nif* genes on a large plasmid in *Frankia* sp. strain ULQ 0132105009. Mol. Gen. Genet. 204:492–495.

Simonet, P., N.T. Le, E.T. du Cros, and R. Bardin. 1988. Identification of *Frankia* strains by direct DNA hybridization of crushed nodules. Appl. Environ. Microbiol. 54:2500–2503.

Simonet, P., P. Normand, A.M. Hirsch, and A.D.L. Akkermans. 1990a. The genetics of the *Frankia*-actinorhizal symbiosis. p. 77–109. *In* P.M. Gresshoff (ed.) The molecular biology of symbiotic nitrogen fixation. CRC Press, Boca Raton, FL.

Simonet, P., P. Normand, A. Moiroud, and R. Bardin. 1990b. Identification of *Frankia* strains in nodules by hybridization of polymerase chain reaction products with strain-specific oligonucleotide probes. Arch. Microbiol. 153:235–240.

Simonet, P., P. Normand, A. Moiroud, and M. Lalonde. 1984. Restriction enzyme digestion patterns of *Frankia* plasmids. Plant Soil 87:49–60.

Smolander, A. 1990. *Frankia* populations in soils under different tree species—with special emphasis on soils under *Betula pendula.* Plant Soil 121:1–10.

Smolander, A., and M-L. Sarsa. 1990. *Frankia* strains of soil under *Betula pendula:* Behaviour in soil and in pure culture. Plant Soil 122:129–136.

Smolander, A., and V. Sundman. 1987. *Frankia* in acid soils of forests devoid of actinorhizal plants. Physiol. Plant. 70:297–303.

Smolander, A., C. van Dijk, and V. Sundman. 1988. Survival of *Frankia* strains introduced into soil. Plant Soil 106:65–72.

Sougoufara, B., S.K.A. Danso, H.G. Diem, and Y.R. Dommergues. 1990. Estimating N_2 fixation and N derived from soil by *Casuarina equisetifolia* using labelled ^{15}N fertilizer: Some problems and solutions. Soil Biol. Biochem. 22:695–701.

Sougoufara, B., L. Maggia, E. Duhoux, and Y.R. Dommergues. 1992. Nodulation and N_2 fixation in nine *Casuarina* clone-*Frankia* strain combinations. Acta Œcol. 13:497–503.

Stowers, M.D. 1987. Collection, isolation, cultivation and maintenance of *Frankia.* p. 29–53. *In* G.H. Elkan (ed.) Symbiotic nitrogen fixation technology. Marcel Dekker, New York.

Tjepkema, J.D., and J.G. Torrey. 1979. Symbiotic nitrogen fixation in actinomycete-nodulated plants. Preface. Bot. Gaz. (Chicago) Suppl. 140:i–ii.

Tjepkema, J.D., C.R. Schwintzer, and D.R. Benson. 1986. Physiology of actinorhizal nodules. Annu. Rev. Plant Physiol. 37:209–232.

Torrey, J.G. 1987. Endophyte sporulation in root nodules of actinorhizal plants. Physiol. Plant. 86:581–583.

Torrey, J.G. 1990. Cross-inoculation groups within *Frankia* and host-endosymbiont associations. p. 83–106. *In* C.R. Schwintzer and J.D. Tjepkema (ed.) The biology of *Frankia* and actinorhizal plants. Academic Press, San Diego.

Tortosa, R.D., and M. Cusata. 1991. Effective nodulation of rhamnaceous actinorhizal plants induced by air dry soils. Plant Soil 131:229–233.

Tunlid, A., N.A. Schultz, D.R. Benson, D.B. Steele, and D.C. White. 1989. Differences in fatty acid composition between vegetative cells and N_2-fixing vesicles of *Frankia* sp. strain CpI1. Proc. Natl. Acad. Sci. USA 86:3399–3403.

Tzean, S.S., and J.G. Torrey. 1989. Spore germination and the life cycle of *Frankia in vitro*. Can. J. Microbiol. 35:801–806.

van Dijk, C. 1978. Spore formation and endophyte diversity in root nodules of *Alnus glutinosa* (L.) Vill. New Phytol. 81:601–615.

van Dijk, C. 1984. Ecological aspects of spore formation in the *Frankia-Alnus* symbiosis. Ph.D. thesis. State University, Leiden, the Netherlands.

van Dijk, C., and E. Merkus. 1976. A microscopical study of the development of a spore-like stage in the life cycle of the root-nodule endophyte of *Alnus glutinosa* (L.) Gaertn. New Phytol. 77:73–91.

van Dijk, C., A. Sluimer, and A. Weber. 1988. Host range differentiation of spore-positive and spore-negative strain types of *Frankia* in stands of *Alnus glutinosa* and *Alnus incana* in Finland. Physiol. Plant. 72:349–358.

van Dijk, C., and A. Sluimer-Stolk. 1990. An ineffective strain type of *Frankia* in the soil of natural stands of *Alnus glutinosa* (L.) Gaertner. Plant Soil 127:107–121.

Weber, A. 1986. Distribution of spore-positive and spore-negative nodules in stands of *Alnus glutinosa* and *Alnus incana* in Finland. Plant Soil 96:205–213.

Weber, A., E.L. Nurmiaho-Lassila, and V. Sundman. 1987. Features of intrageneric *Alnus-Frankia* specificity. Physiol. Plant. 70:289–296.

Weber, A., M-L. Sarsa, and V. Sundman. 1989. *Frankia-Alnus incana* symbiosis: Effect of endophyte on nitrogen fixation and biomass production. Plant Soil 120:291–297.

Wheeler, C.T., J.E. Hooker, A. Crowe, and A.M.M. Berrie. 1986. The improvement and utilization in forestry of nitrogen fixation by actinorhizal plants with special reference to *Alnus* in Scotland. Plant Soil 90:393–406.

Wheeler, C.T., and I.M. Miller. 1990. Current and potential uses of actinorhizal plants in Europe. p. 365–389. *In* C.R. Schwintzer and J.D. Tjepkema (ed.) The biology of *Frankia* and actinorhizal plants. Academic Press, San Diego.

Winship, L.J., and J.D. Tjepkema. 1990. Techniques for measuring nitrogenase activity in *Frankia* and actinorhizal plants. p. 263–280. *In* C.R. Schwintzer and J.D. Tjepkema (ed.) The biology of *Frankia* and actinorhizal plants. Academic Press, San Diego.

Chapter 17

Filamentous Fungi

DENNIS PARKINSON, *The University of Calgary,*
Calgary, Alberta, Canada

Fungi play important roles in several soil processes. These range from organic matter decomposition and nutrient cycling to interactions with plant roots (symbiotic or parasitic). A wide diversity of fungi have been shown to be present in soils and they represent a substantial portion of the total soil microbial biomass.

Investigators of soil fungi encounter considerable problems in qualitative and quantitative studies because the fungi exist in soil in a variety of morphological and physiological states (e.g., active, dormant and even dead hyphae, various types of spores, and resting structures such as sclerotia). In many studies, it is important to assess the active hyphal fungi in soil or organic matter samples; but, despite some progress, this is still a difficult task.

In the past decade, several reviews on methods for studying soil fungi have appeared (e.g., Parkinson, 1982; Kendrick & Parkinson, 1990; Frankland et al., 1990; Seifert, 1990; Parkinson & Coleman, 1991). These emphasize the problems stated above.

The present chapter deals particularly with methods for studying communities of filamentous fungi in soil and organic matter samples. Topics such as sampling (chapter 1 in this book), statistical analyses (chapter 2 in this book), direct microscopic methods (chapter 6 in this book) and mycorrhizal fungi (chapter 18 in this book) are dealt with in other chapters of the present volume as are immunological methods (chapter 28 in this book) which are so valuable for studies on individual species.

It must be emphasized here that, prior to embarking on a detailed study of soil fungi (as for any group of soil organisms), there should be a clear definition of the aims of the study. Subsequently, methods can be chosen that allow fulfillment of these aims. The details of the various methods provided in this chapter should be regarded as examples and not as hard-and-fast protocols. All methods will probably need modification for use with different soils.

17–1 QUALITATIVE STUDIES: ISOLATION METHODS

17–1.1 Choice of Appropriate Isolating Media

For studies on the community structure of soil fungi in the vast majority of cases, it is necessary to isolate these organisms onto nutrient media. Therefore, the choice of appropriate media for these studies is of prime importance. At one time, the aim in synecological studies of soil fungi was to develop a nonselective medium, i.e., one which would allow the isolation of any fungus present in the soil. Therefore, media such as plain water agar were used as the primary isolation medium, from which isolates were transferred to nutrient agar media for subsequent identification. This hope for the development of nonselective media has given way to the more realistic approach of using a medium (or small group of media) which allows the isolation of the maximum number of fungal taxa from the soil under study. Probably the use of a wide range of selective media would be the most efficient approach to a complete study of soil fungal community structure, but this is rarely (if ever) feasible because of the required personnel and time demands.

The most appropriate isolation medium may vary depending on the soil type being studied (and the prior experiences of the individual investigator). Therefore, it is dangerous to recommend dogmatically any particular medium. Before carrying out a large study on the mycota of a particular soil, a proper comparative study of a range of media should be carried out to determine the medium that allows isolation of the maximum number of fungal taxa.

Numerous nutrient media have been used in studies of filamentous fungi in soil. Among the most frequently used media are Czapek-Dox agar with or without small amounts of yeast extract; malt extract agar with or without materials such as peptone, dextrose, and yeast extract; and, dextrose-peptone agar with or without growth retardants (see later). Media incorporating soil extract have not been used as frequently for isolating soil fungi as for isolations of soil bacteria.

Recipes for a wide range of media have been well documented elsewhere (Parkinson et al., 1971; Johnson & Curl, 1972; Smith & Onions, 1983; Gams et al., 1987; Seifert, 1990). Recipes for some commonly used types of media are provided here:

1. Czapek-Dox medium—30.0 g of sucrose, 3.0 g of sodium nitrate ($NaNO_3$), 1.0 g of potassium monohydrogen phosphate (K_2HPO_4), 0.5 g of magnesium sulfate heptahydrate ($MgSO_4 \cdot 7H_2O$), a trace of ferrous sulfate ($FeSO_4$), 15.0 g of agar, 1000 mL of distilled water, and 0.5 to 1.0 g of yeast extract (if desired): Dissolve inorganic constituents separately; add $FeSO_4$ last. Add sucrose just before sterilization.

2. Malt extract agar: Variants of this medium range from 20.0 g of malt extract, 15.0 g of agar, and 1000 mL of distilled water to 20.0 g of malt extract, 20.0 g of dextrose, 1.0 g of peptone, 15.0 g of agar, and 1000 mL of distilled water. The amount of malt extract can be decreased if desired, and small amounts of yeast extract can be added $(0.5–1.0 \text{ g.L}^{-1})$.

3. Dextrose-peptone agar:
   a. Dissolve 10.0 g of dextrose, 5.0 g of peptone, 1.0 g of potassium dihydrogen phosphate (KH_2PO_4), 0.5 g of magnesium sulfate heptahydrate ($MgSO_4 \cdot 7H_2O$) and 20.0 g of agar in 1000 mL of distilled water, and heat. Add 3.3 mL of 1% rose bengal. Autoclave the medium, cool to about 50 °C, and add 30.0 mg of streptomycin (Martin, 1950).
   b. Dissolve 5.0 g of dextrose, 1.0 g of peptone, 2.0 g of yeast extract, 1.0 g of ammonium nitrate (NH_4NO_3), 1.0 g of potassium monohydrogen phosphate (K_2HPO_4), 0.5 g of magnesium sulfate heptahydrate ($MgSO_4 \cdot 7H_2O$), a trace of ferric chloride hexahydrate ($FeCl_3 \cdot 6H_2O$), 5.0 g of oxgall (dehydrated fresh bile), 1.0 g of sodium propionate ($NaC_3H_5O_2$), and 20.0 g of agar in 1000 mL of distilled water, and heat. Autoclave the medium, and after cooling to about 50 °C, add 30.0 mg of aureomycin and 30.0 mg of streptomycin (Papavizas & Davey, 1959).

If soil extracts are required for attempts to make media of low selectivity, the usual procedure in their preparation is to autoclave 1 kg of soil with 1 L water (20 min at either 6.8 or 9.1 kg of pressure). The material is then filtered, and the filtrate is made up to 1 L. If the resultant liquid is still cloudy, a little $CaSO_4$ is added, the liquid is allowed to stand, and then filtration is repeated.

The soil extract can be solidified with agar (1.5%) and used with or without the addition of other nutrients, such as 0.02% K_2HPO_4 or 0.1% glucose.

In soils containing large numbers of bacteria, it is frequently necessary to amend the chosen isolation medium with antibacterial substances in an attempt to reduce competition from the soil bacteria and thus increase the likelihood of isolating slow growing fungi of low competitive ability.

Acidification of the medium (pH 3.5–5.0) was frequently used, and is still used in some investigations. Acids should be added to the medium after sterilization and include 0.5% malic acid, 0.1% lactic acid, or 0.1 to 1.0 M HCl or H_2SO_4. If acidification of media is applied the investigator should ensure that this practice is not having a selective effect on the isolation of fungi.

Bacterial growth retardants such as crystal violet (10 mg L^{-1}), potassium tellurite (100 mg L^{-1}), oxgall (0.5–1.5%), rose bengal (30–700 mg L^{-1}), sodium deoxycholate and sodium propionate have been used on isolation media (Littman, 1947; Martin, 1950; Bakerspigel & Miller, 1953; Papavizas & Davey, 1959; Watling, 1971; see review by Seifert, 1990).

Fungal growth retardants have been used to restrict rapidly growing isolates and prevent slower growing fungi from being overgrown. Orthophenyl phenol, rose bengal, oxgall, dichloran (2-6-dichloro-4-nitroaniline), and benomyl (1-butycarbonyl-2-benzimadazole carbonic acid methyl ester) are examples of such retardants and these have been discussed in detail by Seifert (1990). Suggested concentrations of these compounds in nutrient media are given below.

The use of antibiotics for the suppression of bacterial development during isolations of soil fungi is now common and has tended to supplant the use of the agents mentioned above. Examples of antibiotic additions to some standard media have been given in the recipes outlined above. Streptomycin and aureomycin are commonly used antibiotics at concentration ranges of 100 to 200 and 25 to 100 mg L^{-1}, respectively. These antibiotics should be added to the nutrient agar after it has been autoclaved and cooled (prior to pouring).

A wide range of media have been developed for the selective isolation of either specific physiological groups of soil fungi (e.g., cellulolytic spp., chitinolytic spp. etc.) or specific taxa (e.g., basidiomycetes, *Fusarium* spp., *Phytophthora* spp. etc.). New media for specific studies are being developed regularly, and examples of a range of selective media were given by Johnson and Curl (1972). The most recent, detailed survey has been given by Seifert (1990).

Basidiomycetes are usually isolated from soil with relative infrequency. This is disturbing, particularly when studies are being made on forest litter soil systems where production of fruit bodies of a range of species of basidiomycetes is observed. This problem may center on the nutritional requirements of these fungi or their slow growth rates and inability to compete with faster growing more aggressive species or both. Attempts to isolate basidiomycetes from soil usually involve the addition of chemical retardants to the growth of faster growing fungi (see earlier). Orthophenyl phenol, pentachloronitrobenzene, dichloran, phenol and benomyl (to maximum of 5 mg L^{-1}) are examples of such retardants. Worrall (1991), in a comparison of several media for the selective isolation of wood-decay hymenomycetes, found that malt extract agar with benomyl (2 mg L^{-1}) and dichloran (2 mg L^{-1}) and with 100 mg L^{-1} streptomycin added after autoclaving was the most suitable medium.

Special media have been developed for isolating and culturing mycorrhizal fungi (Molina & Palmer, 1982). Three media commonly used for these purposes are:

1. Hagem's agar (Modess, 1941): 5.0 g glucose, 5.0 g malt extract, 0.5 g potassium dihydrogen phosphate (KH_2PO_4), 0.5 g magnesium sulfate heptahydrate ($MgSO_4 \cdot 7H_2O$), 0.5 g ammonium chloride (NH_4Cl), 0.5 mL of a 1% ferric citrate ($FeC_6H_5O_7 \cdot 5H_2O$) solution, 15.0 g agar, 1000 mL distilled water.

2. Melin-Norkrans medium (Norkrans, 1949): 2.5 g sucrose, 25 µg thiamine hydrochloride, 0.5 g potassium dihydrogen phosphate (KH_2PO_4), 0.15 g magnesium sulfate heptahydrate ($MgSO_4\cdot7H_2O$), 0.25 g ammonium monohydrogen phosphate [$(NH_4)_2HPO_4$], 1.2 mL of a 1% solution of ferric chloride ($FeCl_3$), 15.0 g agar, 1000 mL distilled water.
3. Modified Melin-Norkrans medium (Marx, 1969): 10 g sucrose, 3.0 g malt extract, 100 µg thiamine hydrochloride, 0.5 g magnesium sulfate heptahydrate ($MgSO_4\cdot7H_2O$), 0.25 g ammonium monohydrogen phosphate [$(NH_4)_2HPO_4$], 0.05 g calcium chloride ($CaCl_2$), 0.025 g sodium chloride (NaCl), 1.2 mL of a 1% solution of ferric chloride ($FeCl_3$), 15.0 g agar, 1000 mL distilled water.

Modified Melin-Norkrans medium has been further modified by various workers, e.g., Mexal and Reid (1973) substituted glucose for sucrose and 48 ppm Sequestrene 330 (Ciba-Geigy Corp., Greensboro, NC) for $FeCl_3$ solution. In an attempt to isolate only basidiomycete mycorrhizal fungi, Danielson (1982) added benomyl (10 mg L^{-1}) to Modified Melin-Norkrans medium.

Following primary isolation of fungi from soil, individual colonies are subcultured to provide pure cultures for identification. A wide range of media is available for the maintenance of pure cultures, many of which are "natural" media, e.g., potato extract, potato dextrose, potato-carrot, V8 juice (Campbell Soup Co., Camden, NJ) cherry extract and malt extract. For identification of specific taxa, it is important to use the media recommended by the monographs being used for identification. Details on culture storage, maintenance, and observations have been given by Gams et al., 1987; Smith, 1988; Kendrick & Parkinson, 1990, and Seifert, 1990.

17–1.2 General Soil Studies

17–1.2.1 Soil Dilution Plate Method

Until the 1950s, this method was used almost exclusively for studies on soil fungal communities. For reasons which are given later, the use of other methods has increased in the past three decades, however the soil dilution plate method continues to be used in some studies.

Ever since the work of Brierley et al. (1928), every facet of this method has been subjected to close scrutiny, e.g., initial weight of the soil sample, volume of initial suspension, type of suspending fluid, type and time of agitation to produce a homogenous soil suspension, type of diluent, type of nutrient medium, and method of plating (i.e., incorporation into the agar medium or surface plating). Wollum (1982) provided clear details on the use of this method for isolating various groups of soil microorganisms from non-rhizosphere and rhizosphere soil. It has been suggested (Parkinson et al., 1970) that if this method is used, individual workers should vary the details of the method to suit the soil type and so forth with which they are working or the types of organisms being studied. Therefore, the postulation of a standard procedure for this method would provide one of only dubious value.

One example of a methodological procedure used for a forest soil is as follows:

1. 10 g soil are dispensed in 90 mL diluent (0.2% solution of dextrin in soil extract) using mechanical shaking (250 strokes min^{-1} for 15 min).
2. From the initial suspension a dilution series is made in the conventional way using 1-mL blowout pipettes of known accuracy and 0.2% dextrin soil extract as the diluent.
3. Aliquots (0.1 mL) of chosen dilutions (e.g., 10^{-5} and 10^{-6}) are dispensed onto a chosen nutrient agar (section 17–1.1) solidified in petri dishes and spread plates are prepared.
4. The plates are incubated at the temperature required by the investigator (for studies of mesophilic fungi this is normally 22–25 °C) for at least 14 d.
5. Colonies developing on the plates are subcultured for subsequent identification.

Before embarking on a large-scale use of this method, it is important to determine the exact details of this method that are optimal for use with the soil under study.

If this method is used for soil fungal community studies, it should be remembered that:

1. It has been shown (Warcup, 1957) that the large majority of fungal colonies developing on soil dilution plates originate from spores or other propagules and not from hyphal fragments.
2. In view of the foregoing comment the method can be used for assessment of fungal spore content of soil samples but gives little information on fungi present as hyphae in those samples.
3. When this method is used, it is usually impossible to ascertain from which microhabitats in the soil the fungi originated.

17–1.2.2 Soil Plate Methods

This method, developed by Warcup (1950) is essentially a simple variation of the soil dilution plate method in which small amounts of soil are dispersed in known volumes of an appropriate nutrient agar medium.

1. A small amount of soil (0.005–0.015 g) is taken from the large soil sample, using a sterile needle with a flattened tip, and is placed in the base of a petri dish.
2. The small soil sample is thoroughly broken up (using sterile needles) and dispersed over the base of the dish. Addition of a drop of sterile water can enhance this dispersion.

3. Molten but cooled (45 °C) agar medium (8–10 mL) is added to the dish, which is then rotated gently to disperse the soil particles in the agar medium.
4. The plates prepared in this way are then incubated (usually 22–25 °C) and fungi developing from the soil particles are isolated into pure culture for identification.

Like the soil dilution plate method, this method is selective for fungi present as spores in soil samples. However, because of the simplicity and speed of execution of this method, replicate soil plates from each of several soil samples can be prepared quickly, it is valuable for preliminary examination of soil mycofloras prior to using other isolation methods.

In attempts to isolate fungi present in soil as actively growing or dormant hyphae three approaches have been developed: Immersion methods, direct hyphal isolation, and washing methods.

17–1.2.3 Immersion Methods

In an attempt to achieve direct isolation of fungi present in soil as actively growing hyphae, Chesters (1940) developed the immersion tube method. Subsequently, this method has been modified in a variety of ways (e.g., Thornton, 1952; Mueller & Durrell, 1957; Parkinson, 1957; Wood & Wilcoxson, 1960; Anderson & Huber, 1965; Luttrell, 1967).

In these methods, the isolating agar medium (frequently water agar or a chosen nutrient agar) is placed at the desired position(s) in the soil profile under study and is left in situ for a suitable period (which is determined by preliminary tests but is usually 5–7 d). The isolating medium is usually separated from the soil by an air gap, so that any fungi entering that medium must have grown actively across the air gap. After the appropriate time in the soil, the pieces of immersion apparatus are collected from the field, and in the laboratory, small pieces of the isolation medium are plated onto an appropriate nutrient agar medium. Fungi developing on these plates are assumed to have been present in the soil in an active hyphal state.

Probably the simplest form of this method and the one requiring least special equipment is that described by Mueller and Durrell (1957).

17–1.2.3.1 Procedure
1. Immersion tubes are made from autoclavable plastic centrifuge tubes in the walls of which holes (0.25 cm diam.) have been bored at desired positions. The tube is wrapped with plastic tape.
2. The tubes are filled, to within 4 cm of the top, with the chosen nutrient medium and then plugged and autoclaved.
3. The tubes are taken to the field study site, the plastic tape round each tube is pierced with a large sterile needle at positions corresponding to the holes bored in the centrifuge tube.

4. The tubes are embedded in the soil and left in situ for a predetermined number of days.
5. Following removal from the soil, the tubes are taken to the laboratory where a core of agar is taken from each tube. Each agar core is cut into small pieces each of which is plated onto the chosen nutrient medium.
6. The isolation plates are incubated and fungi developing on the plates are subcultured for subsequent identification.

The defects of this type of method appear to be:

1. Placing tubes, plates etc. into soil can cause germination of spores in the soil around the immersed materials, so that fungi isolated may be derived from these spores and not from hyphae previously active in the soil.
2. There may be interspecific competition for entry into the immersion apparatus and for subsequent colonization of the medium in the apparatus. In such cases, only fungi with high competitive abilities may be isolated.
3. Small soil invertebrates may carry fungal spores into the nutrient medium in the immersion apparatus.

Notwithstanding a small number of excellent studies (e.g., Thornton, 1956; Sewell, 1959), these immersion methods have been used too infrequently to allow full evaluation of their value in comparison with other, more frequently used, methods.

17–1.2.4 Direct Hyphal Isolation

Warcup (1955) described a method for picking hyphae from small soil samples and plating these hyphae onto nutrient medium:

1. Small soil samples or soil crumbs are saturated with sterile water and then broken with a fine jet of sterile water.
2. Heavier soil particles are allowed to sediment, and the fine particles are decanted off.
3. This procedure is repeated several times until only the heavier particles remain.
4. The remaining particles are spread in a film of sterile water and examined under a dissecting microscope.
5. Any fungal hyphae observed are picked out using sterile needles or very fine forceps, and plated onto nutrient agar.
6. It is frequently necessary to attempt to remove adherent bacteria, other propagules and organic debris from the hyphae prior to plating.
7. When plating, the position of each hyphal inoculum is marked on the base of the plate. The plates are incubated and the plated fragments are observed at least daily under the microscope to ensure that any fungal growth recorded has originated from the plated hyphae and not from adherent propagules etc.

This method is useful for restricted, specific studies; however, because of the time required to deal with each small soil sample, it would appear to be impractical for investigations that entail regular, replicate isolations from large numbers of soil samples. When it is applied, unless great care is taken, the method tends to be selective for hyphae of large diameter, for hyphae that do not fragment easily, and for dark-pigmented hyphae. It tends to be selective against fine hyaline hyphae and for hyphae that are closely associated with organic fragments.

Obviously, considerable skill is required for the efficient application of this method.

17–1.2.5 Soil Washing

In an attempt to achieve a simpler method that requires less skill and experience for isolation of fungi present as hyphae in soil samples, various methods for soil washing have been developed. These methods range from the simple washing-decanting method (Watson, 1960), where soil samples are held in 500-mL Erlenmeyer flasks and are washed with numerous changes of sterile water, to the use of automated, multisample washing machines (Hering, 1966; Bissett & Widden, 1972). All these methods attempt to achieve the removal of as many fungal spores as possible from a wide range of soil microhabitats (i.e., soil particles of different sizes and organic fragments) that can then be plated separately.

In most soil washing methods the following features are evident:

1. Soil samples are placed in separate sterile boxes, each of which contains a number of sieves of graded size (the sieve sizes being determined by the soil type being studied and by the aims of the investigation).
2. The soil samples are vigorously washed (about 2 min. wash^{-1}) with numerous changes of sterile water. The washing action is achieved either by passing sterile air through the system or by physical shaking of the apparatus. The number of washings required to achieve efficient removal of fungal spores from the soil samples being studied must be determined by preliminary tests, i.e., by plating aliquots of washing water after each washing onto replicate nutrient agar plates and counting fungal colonies developing on these plates (see Harley and Waid, 1955 for details).
3. Following thorough washing, soil from the sieves is removed onto sterile filter paper to remove excess water.
4. Individual particles of known size and nature (inorganic or organic) are then plated onto nutrient agar medium. The number of particles per plate depends on the soil type and degree of fungal development (usually one to four particles per plate).
5. Fungi developing on the plates, following incubation, are subcultured for subsequent identification.

The efficiency of soil washing methods will be related to the physical nature of the soil under study. Williams et al. (1965) studied this matter and demonstrated that sandy soil could be washed more efficiently than soils with high clay (or humus colloid) content. In all cases, it is unlikely that all spores are removed from soil samples by washing; however, a high proportion are removed.

For extensive studies on the mycoflora of soil where replicate soil samples taken at each of numerous sample times are to be studied, the construction of an automatic, multibox, washing apparatus is thoroughly worthwhile. However, in small (or very restricted) studies, the use of the simple washing-decantation method is probably appropriate. Whatever method is used, information can be gained on fungi present as hyphae (by plating washed particles) and fungi present as spores (by plating washing water).

17–1.3 Special Substrates

17–1.3.1 Roots

17–1.3.1.1 Root Washing. While various washing methods have been used since the 1930s the Harley and Waid (1955) method and its variations are the most frequently used for general studies of root surface fungi:

1. Roots taken from soil are thoroughly washed in tap water to remove macroscopic debris.
2. The root system is cut into 1 to 2 cm. pieces, the choice of the region(s) of the root system cut in this way depends on the aims of the study.
3. The 1 to 2 cm pieces are placed in 25-mL screw-capped vials and thoroughly washed in several changes of sterile water.
4. The exact number of washings required is determined by preliminary tests as for soil washing but is usually between 15 and 40 washings.
5. Each washing is usually of 2 min duration and is vigorous—being effected manually, mechanically, or in an automated apparatus.
6. After the appropriate number of washings the root pieces are transferred to petri dishes containing sterile filter paper to remove excess water from the roots.
7. The washed, surface-dried roots are cut into small segments. Two-millimeter segments have been used frequently, but smaller segments < 1 mm allow isolation of larger number of species.
8. The small segments are plated onto nutrient medium (one to four segments per plate, depending on degree of fungal colonization of the roots).
9. The plates are incubated and fungi developing from the root segments are isolated into pure culture.

17–1.3.1.2 Surface Disinfestation. This has been used commonly by plant pathologists for isolating microorganisms from internal plant tissues. When isolating fungi from roots the following general procedure can be followed:

1. Pieces of root are placed in a suitable surface disinfestant of standard concentration for standard time.
2. Then the roots must be thoroughly washed with sterile water to remove all traces of the disinfestant.
3. Excess water is removed from the roots using sterile filter paper, the roots are cut into small pieces (1–2 mm) which are plated onto nutrient agar.

Before using this method, each step must be tested to ensure that appropriate concentrations of disinfestant, time of application, and amount of washing necessary for removal of disinfectant from the roots are determined.

Commonly used sterilants include NaOCl solution, 0.1% (wt/vol) HgCl$_2$ and Nance's solution (0.1% (wt/vol) HgCl$_2$ with 0.5% Teepol[1] in distilled water). Times of surface sterilization vary greatly depending on the type of tissue being studied but are usually between 10 s and 2 min.

Zak and Bryan (1963) described a surface disinfestation method for isolating ectomycorrhizae fungi from plant roots:

1. Root samples from the field are transported to the laboratory in such a way as to prevent root drying and the temperature of the roots rising above that of the field soil.
2. In the laboratory, the roots are washed under tap water to remove as much adhering soil as possible.
3. 1.5 to 3.0 cm lengths of roots bearing mycorrhizae are cut off and thoroughly washed in detergent (sometimes ultrasonic cleansing may be needed to remove fine debris). For detergent washing 15 to 20 pieces of root are placed in a screw-capped plastic vial (4 by 2 cm) that has been perforated (top, bottom, and sides). The vial is placed in a jar of detergent solution and thoroughly shaken.
4. The vial containing root pieces is removed from the detergent and the roots are washed in tap water to remove all detergent.
5. The washed roots are then placed in 1% NaOCl solution and shaken for 10 min. (in the same manner as the detergent washing).
6. The disinfectant is removed by thorough washing with sterile distilled water.
7. Root pieces are removed aseptically from the perforated vial and placed in sterile dishes (excess water can be removed from the roots using sterile filter paper). Mycorrhizae are cut from the roots and plated onto an appropriate agar medium (see section 17–1.1).

Danielson (1991) surface sterilized root segments by dipping them in 95% ethanol and soaking in 30% H$_2$O$_2$ before rinsing in two changes of ice-cold sterile water for 1 h. Ice-cold water was used because it stops the action of H$_2$O$_2$ sooner than would water at room temperature.

---

[1] British Drug Houses, Ville St. Laurent, Montreal; or substitute sodium dedocyl sulfate.

17–1.3.1.3 Root Fragmentation. Various types of root fragmentation with or without preliminary surface sterilization have been used to attempt to isolate fungi from the inner tissues of roots. The simplest method (War-cup, 1960) involves placing small pieces of thoroughly washed root (washed as described in section 17–1.3.1.1) in drops of sterile water in petri dishes (1 drop root piece^{-1} dish^{-1}) and then dissecting the roots into as many small fragments as possible using sterile needles. The chosen nutrient agar medium is poured into each dish, after which the dishes are carefully rotated to ensure dispersal of the root fragments in the nutrient medium.

More violent disruption of washed roots can be achieved either by using a mortar and pestle or by using some type of blender (Stover & Waite, 1953; Clarke & Parkinson, 1960). Following disruption, small amounts of macerate are plated as described above.

In such fragmentation methods, it is frequently difficult to discern from which root tissues the developing fungal colonies have arisen. The possibility exists that in the maceration process, substances inhibitory to the growth of fungi may be released from the plant tissues.

If more exact information is required on the location of fungi within roots, root dissection is necessary. Waid (1957) described a method in which roots previously washed (described above) are placed in sterile water and, using a sterile scalpel etc., are dissected (working under a binocular dissecting microscope). Vascular tissue may be separated from the cortex, and these fractions can be plated separately onto the chosen nutrient agar medium.

17–1.3.2 Other Organic Matter

17–1.3.2.1 Direct Observation Followed by Isolation. Frankland et al. (1990) pointed out the value of maintaining samples (macroscopic) of different types of organic material (e.g., leaves, stems, pieces, or sections of wood) in damp chambers and recording the various fungi that sporulate or the material under these conditions. Careful use of this simple method can frequently allow isolation of species that are not obtained by use of the methods described in the previous sections. Seifert (1990) described a method that includes the following steps:

1. The base of a petri dish is lined with filter paper and with paper towelling or pulp around the inner rim. All this material is moistened with sterile water.
2. The organic material is placed in the dish which is then covered with a loosely fitting lid.
3. Dishes prepared in this way are incubated (room temperature) and examined regularly, using a dissecting microscope, for fungal sporulation that may begin as early as 2 d after initiation of damp chambers.
4. Spores of the fungi on the organic matter can be removed, using an agar-tipped needle, to nutrient agar plates.

5. If insect development on the organic matter is likely to become a problem, a solid insecticide (e.g., Vapona strips) can be placed in the damp chamber.

17–1.3.2.2 Direct Observation of Ectomycorrhizal Roots. The importance of direct observation of ectomycorrhizal roots in attempting identification of the fungal symbionts was emphasized by Zak (1973). Recent publications (e.g., Agerer, 1986; Ingleby et al., 1990; Danielson, 1991) have given details on methods for observing roots and the characters of the mycorrhizas that allow symbiont identification.

Both macroscopic and microscopic characters of mycorrhizas are important in identification of symbionts. Macroscopic characters include: color, size, form associated hyphae, hyphal strands, rhizomorphs, and sclerotia. Microscopic characters, seen in squash preparations or transverse sections include: hyphal size, presence of cystidia, hyphal ornamentation, crystals or other hyphal exudates, and hyphal wall pigmentation.

17–1.3.2.3 Serial Washing, Surface Sterilization, Fragmentation, and Dissection. All these methods have been described, above, for the isolation of fungi from roots. They can be used equally well for studies of fungi associated with other plant parts and organic materials in various stages of decomposition. For large pieces of organic matter a procedure that involves coarse blending followed by serial washing (predetermined number of washings), air drying and plating fragments (one to two per plate) onto the chosen agar medium has been used frequently.

17–1.4 Selective Methods

All the methods described above are used for synecological studies of soil fungi, i.e., where an attempt is made to isolate as many species as possible from a soil sample. However, it is desirable frequently to study the distribution of specific taxa or of specific physiological/substrate groups of fungi in soil. Therefore, methods selective for different taxa etc. have been developed.

To itemize a list of such methods is impossible in the present restricted chapter on the filamentous fungi. Scrutiny of journals such as *Phytopathology* over the past two decades would show that a considerable number of media have been developed for isolating specific soil-borne pathogens. Johnson and Curl (1972) provided a general account of methods for isolating these organisms.

Parkinson et al. (1971) provided data on selective media useful for studying different physiological/substrate groups of soil microorganisms. More recent information on these groups of soil fungi has been provided by Frankland et al. (1990) and Seifert (1990).

17-2 QUANTITATIVE METHODS

17-2.1 Introduction

For several decades, the soil dilution plate method was used exclusively in attempts to quantify soil fungal communities. However, the data obtained using this method, expressed as numbers of fungi per unit weight of soil (an unreasonable expression, given the hyphal growth form of the vast majority of the soil fungi), in reality give an indication of the number of spores present in the soil sample (see comments in section 17-1).

Other isolation methods may indicate the degree of fungal colonization of different substrates in soil and may allow comparisons of different soils using calculations of frequency of occurrence of fungi or other mathematical analyses (see Frankland et al., 1990). However, these approaches give no indications of fungal biomass.

It appears that three general types of methods are available for measuring fungal biomass: (i) direct observation methods, (ii) chemical methods, and (iii) physiological methods. Other chapters in this volume deal in detail with these three approaches (see chapters 6 and 35 in this book). Therefore, only general comments on these methods are given here.

17-2.2 Direct Observation Methods

In the early studies of soil fungi, when there was some debate on whether or not fungi were active in soil, direct observation of soil smears and of slides which had been buried in soil for appropriate time periods demonstrated fungal growth in soil. However, attempts to convert such observations into mass of fungal hyphae in soil samples were rare. Nicholas and Parkinson (1967) discussed the relative merits of several direct observation methods. The methods discussed were modified impression slides, soil sections and soil-agar films and the conclusion was that, while not ideal, the soil-agar film method was the least prone to experimental and observational errors.

With the development of superior methods for preparing soil sections, application of fluorochromes and the use of fluorescence microscopy (Altemüller & van Vliet-Lanoe, 1990), the use of soil sectioning has been shown to be valuable for studies on the spatial distribution of soil microorganisms (Postma & Altemüller, 1990). It may well prove valuable for obtaining data on microbial biomass associated with specific soil microhabitats.

17-2.2.1 Soil-Agar Film Method

The soil-agar film method (Jones & Mollison, 1948 as modified by Thomas et al., 1965) has been described for use with soil samples (Parkinson, 1982) and for litter (Frankland, 1978; Parkinson, 1982; Frankland et al., 1990). Lodge and Ingham (1991) have compared various modifications of the soil-agar film method for estimating fungal biovolumes in litter and soil, and report a *coverslip well technique* to be least variable and most preferable for making agar films. The recommendations and comments made by Frankland et al. (1990) and by Lodge and Ingham (1991) for preparation and observation of agar films can be summarized as follows:

1. The duration and intensity of methods used for dispersal of hyphae in diluent are critical factors. These must be determined in preliminary experiments.
2. For soil samples, shaking or sonication are much preferable to maceration (which breaks hyphae into small fragments requiring high magnification for detection). Litter samples require maceration, the times for which must be predetermined.
3. It is necessary to determine the optimal dilution of the soil or litter samples, i.e., the dilution giving the maximum estimate of hyphal length.
4. Different investigators recommended different ranges of magnification for measuring hyphal lengths. Bååth and Söderström (1980) found that increasing the magnification from 800× to 1250× doubled the hyphal lengths measured. Lodge and Ingham (1991) recommended using the lowest magnification under which hyphae can be readily identified, which will depend on the predominant size of hyphae (particularly diameter) and the substrate relationships of the hyphae.
5. The experimental design is important. Frankland et al. (1978) and Lodge and Ingham (1991) have discussed the importance of properly replicated observations of several replicate films (these numbers of replicates being predetermined.) Frankland et al. (1990) observed that optimum replication is usually achieved by increasing the number of field samples of soil or litter relative to the number of soil-agar films made and microscopic fields observed.
6. Conversions of hyphal length measurements to biomass require data on the proportions of hyphae falling into various diameter classes, their density, and water content. It is unsatisfactory to use values for these parameters that are taken from the older literature (e.g., Parkinson et al., 1971). Of the factors mentioned above hyphal diameter is most important (see Bååth & Söderström, 1979). Measurements should be made with care, and should be made on fresh agar films to avoid errors due to shrinkage (Jenkinson & Ladd, 1981). Moisture content and density of hyphae vary with species, age of hyphae and growth conditions, and measurements of these values should be made using mycelium collected from the field (Lodge, 1987) or using hyphae grown under carefully designed, low nutrient conditions (Frankland et al., 1990).
7. Estimates of biomass in living hyphae can be made by assuming that the presence of cell contents (observed using phase contrast microscopy) indicates live hyphae (Frankland, 1975; Frankland et

al., 1990). However, various staining methods (fluorescent stains) are used extensively (see below).

17–2.2.2 Membrane Filter Method

Hansen et al. (1974) developed a membrane filter method which is more rapid than the soil-agar film method for measuring hyphal lengths in soil samples (which can be converted to biomass as for the soil-agar films). Variations on this method have been described by Paul and Johnson (1977) and Sundman and Sivelä (1978). More recently, Elmholt and Kjøller (1987) described a general procedure for studying hyphal biomass in cultivated field soils. Details on membrane filter methods for studying general soil fungi and mycorrhizal fungi are given elsewhere (chapters 6 and 18 in this book).

Bååth and Söderström (1980) compared the soil-agar film and membrane filter methods for making measurements of hyphal length in several soils. They found that the former method gave higher hyphal length values, but in studies comparing different soils the membrane filter method was preferred because of the ease and rapidity of its execution. West (1988) found significantly greater hyphal lengths on membrane filters of grassland soil suspensions treated with fluorescent brightener than on soil-agar films stained with phenolic aniline blue and viewed with phase contrast microscopy. Bardgett (1991) found the membrane filter method particularly useful in a comparative study of different grassland soils.

Vital staining, using fluorescein diacetate (FDA), of membrane filtered soil suspensions, has been used to assess the lengths of live hyphae in soil samples (Söderström, 1977). Using this method, FDA-stained hyphae are frequently found to comprise only a small percentage of the total hyphae. Söderström (1979) suggested that the method of preparing soil suspensions (vigorous homogenization) could cause cytoplasm leakage from some hyphal fragments and thereby underestimate the length of viable hyphae. However, Ingham and Klein (1984) used a gentle extraction procedure and obtained significant correlations between lengths of FDA-stained hyphae and metabolic activity. Gray (1990), in a detailed account of the FDA-staining method, reiterates the problem of the effect of drastic homogenization destroying the integrity of some hyphae.

In the preparation of soil suspensions for membrane filtration and calculation of fungal biomass from hyphal length measurements, many of the comments made regarding soil-agar film preparation are equally relevant. Morgan et al. (1991) described an automated image analysis method for use in the membrane filter method, in which Calcafluor M2R and FDA were used to stain total and viable hyphae, respectively. The results showed excellent correlation with those obtained using manual microscopic hyphal length measurements. Using this method "operator differences" and "fatigue-induced bias" are eliminated.

17–2.3 Chemical Methods

The quantitative extraction from soil of chemical constituents of fungal cells, and the subsequent use of such data to estimate total fungal biomass is an attractive possibility. In the choice of such a cell constituent several criteria must be taken into account (Jenkinson & Ladd, 1981):

1. It should be present in live cells at concentrations that fall within a known range.
2. It should be absent from dead cells and other non-living materials in soil.
3. An accurate method must be available for assaying the compound.

With respect to studies on filamentous fungi, estimations of chitin and ergosterol have received most attention (see Frankland et al., 1990). Chitin determinations involve measuring the quantity of glucosamine released by hydrolysis of chitin (Ride & Drysdale, 1972). Unfortunately, chitin does not fulfill the first two criteria listed above. For general studies of mixed soil fungal communities where substantial arthropod communities are also present, it is difficult to obtain valid conversion factor (glucosamine content to fungal biomass) and, at present, such conversion is only appropriate in studies of single species of fungi (Herbert, 1990). In some soils, most of the glucosamine measured is present in dead organic matter (West et al., 1987). On the other hand, the use of high pressure liquid chromatography (HPLC) has enhanced the speed and sensitivity of chitin analyses (Zelles, 1988).

Ergosterol determinations are considered to be more specific indicators of fungal biomass than are glucosamine measurements (Newell et al., 1988). Various analytical methods are available for determining ergosterol; however, it is difficult to convert estimates of ergosterol to fungal biomass because of considerable interspecific and age variation in ergosterol content of hyphae (Newell et al., 1987). Nevertheless it has been used to monitor changes in fungal biomass in soil (West et al., 1987) and in litter (J. Lodge, personal communication).

Despite the attraction of chemical methods, it appears that, for studies of soil fungal or plant litter communities, direct observation methods are superior for fungal biomass determinations.

17–2.4 Physiological Methods

Anderson and Domsch (1978) used glucose stimulated respiration to determine total microbial biomass in soil samples. This method, now called *substrate-induced respiration* (SIR) is discussed elsewhere (chapter 36 in this book) as is the method for estimating active soil microbial biomass by mathematical analysis of respiration curves (Van de Werf & Verstraete, 1987).

The SIR method becomes important for determining fungal biomass in soil when it is coupled with the addition of selective inhibitors (e.g., actidione [cycloheximide] and streptomycin, protein synthesis inhibitors for fungi and bacteria, respectively) applied at predetermined optimal concentrations (Anderson & Domsch, 1973, 1975). Differences in respiration rates of antibiotic treated and untreated glucose-amended soil samples allow calculations of the relative proportions of fungi and bacteria in the total microbial biomass (Anderson & Domsch, 1975). Data obtained using this method generally show fungi to be the dominant component of the total microbial biomass, with bacterial to fungal biomass ratios ranging from 40:60 to 10:90.

For each soil type being studied it is essential to determine carefully the required concentrations of each inhibitor. The criterion for these determinations is that the sum of the inhibition effects of actidone and streptomycin when applied individually should equal the effect when these inhibitors are applied as a mixture (Anderson & Domsch, 1973). Following determination of these inhibitor concentrations, experiments with short incubation periods (6–8 h) should be used to eliminate problems with population shifts and inhibitor degradation (Anderson & Domsch, 1975).

Anderson and Domsch (1973, 1975) applied actidone and streptomycin in powder form to glucose (powder) amended soil samples. West (1986) recommended the addition of glucose and inhibitors in solution, to make them more readily available to the whole soil microbial community, particularly in soil where moisture is limiting. West (1986) also recommended that the soil + glucose and inhibitors be incubated for 3.5 h prior to the assays of inhibition. Beare et al. (1991) preincubated plant litter samples in solutions of each of the inhibitors individually or the mixture of inhibitors for 12 h at 4 °C prior to the addition of the optimal glucose concentration after the samples had been brought to room temperature (about 22 °C).

It is unclear, as yet, what proportion of the total soil microbial biomass responds quickly to glucose. But the use of the selective inhibition method can be used only with reference to the glucose-responsive component of the microbial biomass, i.e., that component which is undergoing protein synthesis immediately following glucose addition (Van de Werf & Verstraete, 1987; Wardle & Parkinson, 1990). From the data provided by Anderson and Domsch (1975), West (1986) and Wardle and Parkinson (1990) it can be seen that the method applies only to a fraction of the microbial biomass—the inhibition caused by the mixture of actidione and streptomycin is typically < 60% of glucose stimulated respiration. Therefore, two assumptions are necessary when this method is used: (i) that the bacterial to fungal biomass ratio is the same in both inhibitor sensitive and insensitive components of the total microbial biomass (West, 1986); and, (ii) that bacterial and fungal components of the total biomass respire at the same rate, per unit biomass, following addition of glucose.

Domsch et al. (1979) discussed the applicability of this method for use with a range of soils. They considered it to be objective, but commented on the need for more work to elucidate its sensitivity. This comment still holds good, particularly for studies on highly organic soils and decomposing plant litter. West (1986) found that, while the modified method was not applicable to all soils, it could be used as complementary to studies using direct observation methods.

REFERENCES

Agerer, R. 1986. Studies on ectomycorrhizae. II. Introducing remarks on characterization and identification. Mycotaxon 26:473–492.

Altemüller, H-J., and B. van Vliet-Lanoe. 1990. Soil thin section fluorescence microscopy. p. 565–578. In L.A. Douglas (ed.) Soil micromorphology. Elsevier Sci. Publ. B.V., Amsterdam.

Anderson, J.P.E., and K.H. Domsch. 1973. Quantification of bacterial and fungal contributions to soil respiration. Arch. Mikrobiol. 93:113–127.

Anderson, J.P.E., and K.H. Domsch. 1975. Measurements of bacterial and fungal contributions to respiration of selected agricultural and forest soils. Can. J. Microbiol. 21:314–322.

Anderson, J.P.E., and K.H. Domsch. 1978. A physiological method for the quantitative measurement of microbial biomass in soils. Soil Biol. Biochem. 10:215–221.

Bååth, E., and B. Söderström. 1979. The significance of hyphal diameter in calculations of fungal biovolume. Oikos 33:11–14.

Bååth, E., and B. Söderström. 1980. Comparison of the agar-film and membrane-filter methods for estimation of hyphal lengths in soil, with particular reference to the effect of magnification. Soil Biol. Biochem. 12:385–387.

Bakerspigel, A., and J.J. Miller. 1953. Comparison of oxgall, crystal violet, streptomycin, and penicillin as bacterial growth inhibitors in platings of soil fungi. Soil Sci. 76:123–126.

Bardgett, R.D. 1991. The use of the membrane filter technique for comparative measurements of hyphal lengths in different grassland sites. Agric. Ecosystems Environ. 34:115–119.

Beare, M.H., C.L. Neely, D.C. Coleman, and W.L. Hargrove. 1991. Characterization of a substrate-induced respiration method of measuring fungal, bacterial and total microbial biomass on plant residues. Agric. Ecosystems Environ. 34:65–73.

Bissett, J., and P. Widden. 1972. An automatic, multi-chamber soil-washing apparatus for removing fungal spores from soil. Can. J. Microbiol. 18:1399–1409.

Brierley, W.B., S.T. Jewson, and M. Brierley. 1928. The quantitative study of soil fungi. Int. Congr. Soil Sci. Trans. 1st (Washington, DC) 3:48–71.

Chesters, C.G.C. 1940. A method for isolating soil fungi. Trans. Br. Mycol. Soc. 24:352–355.

Clarke, J.H., and D. Parkinson. 1960. A comparison of three methods for the assessment of fungal colonization of seedling roots of leek and broad bean. Nature (London) 188:166–167.

Danielson, R.M. 1982. Taxonomic affinities and criteria for identification of the common ectendomycorrhizal symbiont of pines. Can. J. Bot. 60:7–18.

Danielson, R.M. 1991. Temporal changes and effects of amendments on the occurrence of sheathing (ecto-) mycorrhizas of conifers in oil sands tailings and coal spoil. Agric. Ecosystems Environ. 35:261–281.

Domsch, K.H., Th. Beck, J.P.E. Anderson, B. Söderström, D. Parkinson, and G. Trolldenier. 1979. A comparison of methods for soil microbial population and biomass studies. Z. Pflanzenernaehr. Bodenkd. 142:520–533.

Elmholt, S., and A. Kjøller. 1987. Measurement of the length of fungal hyphae by the membrane filter technique as a method for comparing fungal occurrence in cultivated soils. Soil Biol. Biochem. 19:679–682.

Frankland, J.C. 1975. Estimation of live fungal biomass. Soil Biol. Biochem. 7:339–340.

Frankland, J.C., J. Dighton, and L. Boddy. 1990. Methods for studying fungi in soil and forest litter. p. 343–404. In R. Grigorova and J.R. Norris (ed.) Methods in microbiology. Vol. 22. Techniques in microbial ecology. Academic Press, New York.

Frankland, J.C., D.K. Lindley, and M.J. Swift. 1978. A comparison of two methods for the estimation of mycelial biomass in leaf litter. Soil Biol. Biochem. 10:323–333.

Gams, W., H.A. van der Aa, A.J. van der Plaats-Niternik, R.A. Samson, and J.A. Stalpers. 1987. CBS course of mycology. 3rd ed. Centraalbureau voor Schimmelcultures. Baarn, the Netherlands.

Gray, T.R.G. 1990. Methods for studying the microbial ecology of soil. p. 309–342. In R. Grigorova and J.R. Norris (ed.) Methods in microbiology. Vol. 22. Techniques in microbial ecology. Academic Press, New York.

Hansen, J.F., T.F. Thingstad, and J. Goksøyr. 1974. Evaluations of hyphal lengths and fungal biomass in soil by a membrane filter technique. Oikos 25:102–107.

Harley, J.L., and J.W. Waid. 1955. A method for studying active mycelia on living roots and other surfaces in the soil. Trans. Br. Mycol. Soc. 38:104–118.

Herbert, R.A. 1990. Methods for enumerating microorganisms and determining biomass in natural environments. p. 1–39. In R. Grigorova and J.R. Norris (ed.) Methods in microbiology. Vol. 22. Techniques in microbial ecology. Academic Press, New York.

Hering, T.F. 1966. An automatic soil-washing apparatus for fungal isolation. Plant Soil 25:195–200.

Ingham, E.R., and D.A. Klein. 1984. Soil fungi: relationships between hyphal activity and staining with fluorescein diacetate. Soil Biol. Biochem. 16:273–278.

Ingleby, K., P.A. Mason, F.T. Last, and L.V. Fleming. 1990. Identification of ectomycorrhizas. ITE Res. Publ. No. 5. HMSO, London.

Jenkinson, D.S., and J.N. Ladd. 1981. Microbial biomass in soil. p. 415–471. In E.A. Paul and J.N. Ladd (ed.) Soil biochemistry. Marcel Dekker, New York.

Johnson, L.F., and E.A. Curl. 1972. Methods for research on the ecology of soil-borne plant pathogens. Burgress Publ. Co., Minneapolis.

Jones, P.C.T., and J.E. Mollison. 1948. A technique for the quantitative estimation of soil micro-organisms. J. Gen. Microbiol. 2:54–69.

Kendrick, W.B., and D. Parkinson. 1990. Soil fungi. p. 49–68. In D.L. Dindal (ed.) Soil biology guide. John Wiley and Sons, New York.

Littman, M.L. 1947. A culture medium for the primary isolation of fungi. Science 106:109–111.

Lodge, D.J. 1987. Nutrient concentrations, percentage moisture and density of field collected fungal mycelia. Soil Biol. Biochem. 19:727–734.

Lodge, D.J., and E.R. Ingham. 1991. A comparison of agar film techniques for estimating fungal biovolumes in litter and soil. Agric. Ecosystems Environ. 34:131–144.

Luttrell, E.S. 1967. A strip bait for studying the growth of fungi in soil and aerial habitats. Phytopathology 57:1266–1267.

Martin, J.P. 1950. Use of acid, rose bengal and streptomycin in the plate method for estimating soil fungi. Soil Sci. 69:215–232.

Marx, D.H. 1969. The influence of ectotrophic mycorrhizal fungi on resistance of pine roots to pathogenic infections. I. Antagonism of mycorrhizal fungi to root pathogenic fungi and bacteria. Phytopathology 59:153–163.

Mexal, J., and C.P.P. Reid. 1973. The growth of selected mycorrhizal fungi in response to induced water stress. Can. J. Bot. 51:1579–1588.

Modess, O. 1941. Zur Kenntnis der Mykorrhizabildner von Kiefer und Fichte. Symb. Bot. Usal. 5:1–146.

Molina, R., and J.G. Palmer. 1982. Isolation, maintenance, and pure culture manipulation of ectomycorrhizal fungi. p. 115–112. In N.C. Shenck (ed.) Methods and principles of mycorrhizal research. The American Phytopathol. Soc., St. Paul.

Morgan, P., C.J. Cooper, N.S. Battersby, S.A. Lee, S.T. Lewis, T.M. Machin, S.C. Graham, and R.J. Watkinson. 1991. Automated image analysis method to determine fungal biomass in soils and on solid matrices. Soil Biol. Biochem. 23:609–616.

Mueller, K.E., and L.W. Durrell. 1957. Sampling tubes for soil fungi. Phytopathology 47:243.

Newell, S.Y., T.L. Arsuffi, and R.D. Fallon. 1988. Fundamental procedures for determining ergosterol content of decaying plant material by liquid chromatography. Appl. Environ. Microbiol. 54:1876–1879.

Newell, S.Y., J.D. Miller, and R.D. Fallon. 1987. Ergosterol content of salt marsh fungi: effect of growth conditions and mycelial age. Mycologia 79:688–695.

Nicholas, D.J., and D. Parkinson. 1967. A comparison of methods for assessing the amount of fungal mycelium in soil samples. Pedobiologia 7:23–41.

Norkrans, B. 1949. Some mycorrhiza-forming Tricholoma species. Sven. Bot. Tidskr. 43:485–490.

Papavizas, G.C., and C.B. Davey. 1959. Evaluation of various media and antimicrobial agents for isolation of soil fungi. Soil Sci. 88:112–117.

Parkinson, D. 1957. New methods for the qualitative and quantitative study of fungi in the rhizosphere. Pedologie 7 (no. Spec.):146–154.

Parkinson, D., T.R.G. Gray, J. Holding, and H.M. Nagel-de-Boois. 1970. Heterotrophic microflora. p. 34–50. In J. Phillipson (ed.) IBP Handbook 18. Blackwells Sci. Publ. Ltd., Oxford.

Parkinson, D., T.R.G. Gray, and S.T. Williams. 1971. Methods for studying the ecology of soil micro-organisms. IBP Handbook 19. Blackwells Sci. Publ. Ltd., Oxford.

Parkinson, D. 1982. Filamentous fungi. p. 949–968. In A.L. Page et al. (ed.) Methods of soil analysis. Part 2. Chemical and microbiological properties. 2nd ed. ASA, and SSSA Madison, WI.

Parkinson, D., and D.C. Coleman. 1991. Microbial communities, activity and biomass. Agric. Ecosystems Environ. 34:3–33.

Paul, E.A., and R.L. Johnson. 1977. Microscopic counting and adenosine 5′-triphosphate measurement in determining microbial growth in soils. Appl. Environ. Microbiol. 34:263–269.

Postma, J., and H.-J. Altemüller. 1990. Bacteria in thin soil sections stained with the fluorescent brightener calcafluor white M2R. Soil Biol. Biochem. 22:89–96.

Ride, J.P., and R.B. Drysdale. 1972. A rapid method for the chemical estimation of filamentous fungi in plant tissue. Physiol. Pathol. 2:7–15.

Seifert, K.A. 1990. Isolation of filamentous fungi. p. 21–51. In D.P. Labeda (ed.) Isolation of biotechnological organisms from nature. McGraw Hill Publ. Co., New York.

Sewell, G.W.F. 1959. Studies of fungi in a Calluna-heathland soil. II. By the complementary use of several isolation methods. Trans. Br. Mycol. Soc. 42:354–369.

Smith, D., and A.H.S. Onions. 1983. The preservation and maintenance of living fungi. Commonwealth Mycol. Inst., Kew, England.

Smith, D. 1988. Culture and preservation. p. 75–99. In D.L. Hawksworth and B.E. Kirsop (ed.) Living resources for biotechnology. Cambridge Univ. Press, Cambridge, England.

Söderström, B.E. 1977. Vital staining of fungi in pure cultures and in soil with fluorescein diacetate. Soil Biol. Biochem. 9:59–63.

Söderström, B. 1979. Some problems in assessing the fluorescein diacetate—active fungal biomass in the soil. Soil Biol. Biochem. 11:147–148.

Stover, R.H., and B.H. Waite. 1953. An improved method of isolating Fusarium spp. from plant tissue. Phytopathology 43:700–701.

Sundman, V., and S. Sivelä. 1978. A comment on the membrane filter technique for estimation of length of fungal hyphae in soil. Soil Biol. Biochem. 10:399–401.

Thomas, A., D.P. Nicholas, and D. Parkinson. 1965. Modifications of the agar film technique for assaying lengths of mycelium in soil. Nature (London) 205:105.

Thornton, R.H. 1952. The screened immersion plate. A method for isolating soil microorganisms. Research (London) 5:190–191.

Thornton, R.H. 1956. Fungi occurring in mixed oakwood and heath soil profiles. Trans. Br. Mycol. Soc. 39:485–494.

Van de Werf, H., and W. Verstraete. 1987. Estimation of active soil microbial biomass by mathematical analysis of respiration curves: relation to conventional estimation of total biomass. Soil Biol. Biochem. 19:267–271.

Waid, J.S. 1957. Distribution of fungi within the decomposing tissues of ryegrass roots. Trans. Br. Mycol. Soc. 40:391–406.

Warcup, J.H. 1950. The soil plate method for isolation of fungi from soil. Nature (London) 166:117–118.

Warcup, J.H. 1955. Isolation of fungi from hyphae present in soil. Nature (London) 175:953–954.

Warcup, J.H. 1957. Studies on the occurrence and activity of fungi in a wheatfield soil. Trans. Br. Mycol. Soc. 40:237–262.

Warcup, J.H. 1960. Methods for isolation and estimation of activity of fungi in soil. p. 3–31. In D. Parkinson and J.S. Waid (ed.) The ecology of soil fungi. Univ. of Liverpool Press, Liverpool.

Wardle, D.A., and D. Parkinson. 1990. Response of the soil microbial biomass to glucose, and selective inhibitors, across a soil moisture gradient. Soil. Biol. Biochem. 22:825–834.

Watling, R. 1971. Basidiomycetes, Homobasidiomycetes. p. 216–236. *In* C. Booth (ed.) Methods in microbiology. Vol. 4. Academic Press, New York.

Watson, R.D. 1960. Soil washing improves the value of the soil dilution and plate count method of estimating populations of soil fungi. Phytopathology 50:792–794.

West, A.W. 1986. Improvement of the selective respiratory inhibition technique to measure eukaryote : prokaryote ratios in soils. J. Microbiol. Methods. 5:125–138.

West, A.W., W.D. Grant, and G.P. Sparling. 1987. Use of ergosterol, diaminopimelic acid and glucosamine content of soils to monitor changes in microbial populations. Soil Biol. Biochem. 19:607–612.

West, A.W. 1988. Specimen preparation, stain type, and extraction and observation procedures as factors in the estimation of soil mycelial lengths and volumes by light microscopy. Biol. Fert. Soils 7:88–94.

Williams, S.T., D. Parkinson, and N.A. Burges. 1965. An examination of the soil washing technique by its application to several soils. Plant Soil 22:167–186.

Wollum, A.G., II. 1982. Cultural methods for soil microorganisms. p. 781–802. *In* A.L. Page et al. (ed.) Methods of soil analysis. Part 2. Chemical and microbiological properties. 2nd ed. ASA, Madison, WI.

Wood, F.A., and R.D. Wilcoxson. 1960. Another screened immersion plate for isolating fungi from soil. Plant Dis. Rep. 44:594.

Worrall, J.J. 1991. Media selective for the isolation of Hymenomycetes. Mycologia 83:296–302.

Zak, B., and W.C. Bryan. 1963. Isolation of fungal symbionts from pine mycorrhizae. For. Sci. 9:270–278.

Zak, B. 1973. Classification of ectomycorrhizae. p. 43–78. *In* G.C. Marks and T.T. Kozlowski (ed.) Ectomycorrhizae, their ecology and physiology. Academic Press, New York.

Zelles, L. 1988. The simultaneous determination of muramic acid and glucosamine in soil by high-performance liquid chromatography with precolumn fluorescence derivatization. Biol. Fert. Soils 6:125–130.

Chapter 18

Vesicular-Arbuscular Mycorrhizal Fungi

DAVID M. SYLVIA, *University of Florida, Gainesville, Florida*

Mycorrhizae are symbiotic associations between beneficial soil fungi and plant roots. They have an important role in increasing plant uptake of P and other poorly mobile nutrients (O'Keefe & Sylvia, 1991). The vesicular-arbuscular mycorrhizal (VAM) fungi have a very broad host range—members of more than 90% of all vascular plant families are colonized by them. Due to the obligate nature of these organisms, however, much is still unknown about their biology in natural and managed ecosystems.

An early impetus for mycorrhizal research was the prospect for practical utilization of these fungi in forestry and agriculture. This research served mostly to demonstrate the complex biology of the symbiosis, and many efforts at utilization have been frustrated (Jeffries & Dodd, 1991). Today, there is increased research interest in understanding the basic physiology and ecology of mycorrhizae in vivo, in vitro, and in situ. Only through such efforts can we hope to reliably manage mycorrhiza in forestry and agriculture.

The purpose of this chapter is to summarize basic methods to: (i) quantify propagules of VAM fungi, (ii) select effective VAM fungi from soil, and (iii) propagate VAM inoculum and apply it in the nursery and field. A wide range of methods are currently used successfully by various research groups—each method needs to be assessed for the particular application required. For further detail and explanation, the reader is referred to several excellent reviews of the methods presented here (Schenck, 1982; Hayman, 1984; Powell & Bagyaraj, 1984; Kendrick & Berch, 1985; Jeffries, 1987; Jeffries & Dodd, 1991; Schenck & Perez, 1990a; Jarstfer & Sylvia, 1992a; Varma et al., 1992). Methods for manipulating ectomycorrhizal fungi are not discussed here.

18–1 QUANTIFICATION OF VESICULAR-ARBUSCULAR MYCORRHIZAL PROPAGULES IN SOIL

18–1.1 Introduction

Even though VAM fungi are among the most common fungi in soil, they are often overlooked because they do not grow on standard soil-dilution plating media. The spores of VAM fungi are larger than those of most other fungi (ranging from 10–1000 µm in diam.) and can easily be observed with a dissecting microscope. However, spore counts often underestimate numbers of VAM fungi since colonized roots and hyphae can also serve as propagules. For this reason, various assays have been used to obtain an estimate of total propagule number.

18–1.2 Most Probable Number Assay

The most probable number (MPN) assay was developed as a method for estimating the density of organisms in a liquid culture (Cochran, 1950). It was first used to estimate the propagule density of VAM fungi in soil by Porter (1979). Values for a MPN assay can be obtained from published tables (Halvorson & Ziegler, 1933; Fisher & Yates, 1963; de Man, 1975; Alexander, 1982); however, these tables restrict experimental design, thereby reducing the accuracy that can be obtained. A better approach is to program the equation into a computer and directly solve for the MPN value based on optimal experimental design—increased replication and decreased dilution rates will greatly enhance the accuracy of the result (chapter 5 by Woomer in this book). The general equation for calculating MPN values is presented in chapter 5 of this volume. Numerous factors affect the outcome of an MPN assay (Wilson & Trinick, 1982; Morton, 1985; Adelman & Morton, 1986; Graham & Fardelmann, 1986; An et al., 1990; O'Donnell et al., 1992); therefore, the values obtained should be considered relative rather than absolute. Nonetheless, this assay has been a useful tool for estimating propagule numbers in field soil, pot cultures, and various forms of inocula.

18–1.2.1 Important Considerations for Implementing a Most Probable Number Assay

1. Dilution factor—Preliminary studies should be conducted so that the lowest possible dilutions are used to bracket actual numbers found in the soil.
2. Sample processing—Samples should be kept cool and processed as soon as possible after collection. The sample soil needs to be relatively dry and root pieces > 2 mm in diam. removed from the sample to allow thorough mixing with the diluent soil. Obviously, these treatments will affect propagule numbers and viability, so all samples must be treated in a similar manner.

3. Diluent soil—The soil preferably should be the same as the original sample and should be pasteurized rather than sterilized. Controls with no sample added should be set up with the pasteurized soil to ensure that all VAM propagules have been eliminated.
4. Host plant—The host must be highly susceptible to VAM colonization, produce a rapidly growing, fibrous root system, and be readily cleared for observation of colonization. *Zea mays* L. is a good choice.
5. Length of assay—Plants need to be grown long enough so that roots fully exploit the soil in each container. It is better to err on the conservative side and grow plants until they are pot bound. Roots with well-developed mycorrhizae are also more easily evaluated. A typical assay may run for 6 to 8 wk.
6. Confirming negative colonization—The entire root system must be examined to confirm a negative reading.

18–1.3 Infectivity Assays

A more straight-forward approach for comparing VAM populations among soils is an infectivity assay. The draw back is that actual propagule numbers are not estimated. Plants are grown under standard conditions in soil collected from pot cultures or a field, and root colonization is estimated after 3 to 6 wk (Moorman & Reeves, 1979; Reeves et al., 1979; Schwab & Reeves, 1981; Koide & Mooney, 1987; Abbott & Robson, 1991). The amount of colonization is assumed to be proportional to the total number of VAM propagules in the soil. The length of the assay is critical. If plants grow for too short a time, the full potential for colonization is not realized; however, plants grown for too long a time may become uniformly colonized despite differences in VAM populations—preliminary studies are needed to select the proper harvest time for a given plant, soil combination.

Recently, an infection-unit method has been proposed to quantify mycorrhizal propagules (Franson & Bethlenfalvay, 1989). This study indicates that a count of infection units is a more reliable measure of the number of viable propagules than are other methods. However, this method is only applicable in short-term experiments because infection units are discernable only during the initial stages of colonization.

18–2 QUANTIFICATION OF VESICULAR-ARBUSCULAR MYCORRHIZAL COLONIZATION IN ROOTS

18–2.1 Visualizing Vesicular-Arbuscular Mycorrhizal Fungi in Roots

The VAM fungi do not cause obvious morphological changes of roots; however, they produce arbuscules and, in many cases, vesicles in roots. To observe VAM structures within the root, it is necessary to clear cortical cells of cytoplasm and phenolic compounds, and then to differentially stain the fungal tissue. Phillips and Hayman (1970) published the oft-cited method to visualize VAM fungi in roots using trypan blue in lactophenol, but the use of phenol is now discouraged (Koske & Gemma, 1989). The clearing agent for nonpigmented roots is generally KOH, but H_2O_2 (Phillips & Hayman, 1970) or NaOCl (Bevege, 1968; Graham et al., 1992) may be used for pigmented roots. Alternatives to trypan blue for staining are chlorazol black E (Brundrett et al., 1984) and acid fuchsin (Kormanik & McGraw, 1982). For nonpigmented roots, it is also possible to observe colonization nondestructively by inducing autofluorescence (Ames et al., 1982).

18–2.1.1 Procedure for Clearing and Staining Roots

1. Place root samples (approximately 0.5 g) in perforated plastic holders (e.g., OmniSette tissue cassettes, Fisher Scientific, Pittsburgh, PA) and store in cold water until they are processed.
2. Place enough 1.8 M KOH into a beaker (without samples) to allow samples to be covered and heat to 80 °C in a fume hood. Goggles, gloves, and vinyl apron should be worn for protection.
3. Place samples in the heated KOH for the desired time—15 min for tender roots such as onion, 30 min for other roots. If samples are still pigmented after the initial treatment, rinse with at least three changes of water and then place them in a beaker with 30% (wt/wt) H_2O_2 at 50 °C or 3% (wt/vol) NaOCl, acidified with several drops of 5 M HCl. Transfer roots to water as soon as samples are bleached white or become transparent. Times can vary from several seconds to several minutes. Check roots frequently to avoid destruction of the cortex.
4. Rinse with tap water using five changes of water.
5. Cover the samples with tap water and add 5 mL of conc. HCl for each 200 mL of water, stir, and drain. Repeat once.
6. Dispense enough trypan blue stain into a beaker (without samples) to cover samples and heat to 80 °C. To prepare the stain, add in order to a flask while stirring: 800 mL of glycerine, 800 mL of lactic acid, 800 mL of distilled water, and finally 1.2 g of trypan blue.
7. Place samples in the stain for 30 min, cool, and drain the stain into the large flask for reuse. Use a funnel and screen to remove debris. After several uses, additional trypan blue may be added to extend the use of the stain.
8. Rinse samples with one change of tap water to destain. Additional destaining in water may be necessary for some roots.
9. Store samples in a plastic bag in a refrigerator.

18–2.2 Estimation of Colonized Root Length

Various methods have been used to estimate root colonization by VAM fungi (Kormanik & McGraw, 1982; Hayman, 1984). The gridline-intersect method has the advantage of providing an estimate of both the proportion of colonized root and total root length (Giovannetti & Mosse, 1980). This is important because some treatments affect root and fungal growth differently. For example, when P is applied total root length may increase more rapidly than colonized root length and thus the proportion of colonized root will decrease even though the actual length of colonized root is increasing.

18–2.2.1 Gridline-Intersect Method

1. Spread a cleared and stained root sample evenly on a scribed plastic petri dish. A grid of squares is scribed on the *bottom* of the dish prior to use. The lines should be arranged so that the edge of the container does not coincide with a gridline and the two partial squares at each edge together make up a complete square. If the lines are 1.27-cm (0.5 in.) apart, then the total number of intersections is equal to total root length in centimeters (see no. 4 below).

2. Scan the gridlines under a dissecting microscope and record the total number of root intersections with the grid as well as the number of intersects with colonized roots.

3. Verify any questionable colonization with a compound microscope. To do this, cut out a small portion of the root with a scalpel, place it in water on a microscope slide, and look for VAM structures at 100 to 400×. Remember that the stains *are not specific* for VAM fungi—other fungi colonizing the root will also stain so it is important to verify the presence of arbuscules or vesicles in the root with a compound microscope.

4. Calculate total and colonized root lengths (R) using the following equation (Newman, 1966):

$$R = \pi A n / 2H$$

where A is the total area in which roots are distributed, n is the number of intersections between roots and the scribed lines, and H is the total length of the scribed lines. According to Marsh (1971), when the interline distance is 14/11 of the measuring unit (14/11 cm = 0.5 in.) then $\pi A = 2H$ and total or colonized root length in centimeter equals the number of total or colonized root intersections with the gridlines.

20–2.2.2 Modification of the Gridline-Intersect Method

McGonigle et al. (1990) argued that the gridline-intersect method is somewhat subjective because arbuscules may be difficult to distinguish with a dissecting microscope. They proposed use of a magnified-intersect method where roots are observed at 200× and arbuscules are quantified separately from vesicles and hyphae. Another limitation of the gridline-intersect method is that the intensity of colonization at each location is not estimated. To obtain an estimate of intensity a morphometric technique

(Toth & Toth, 1982) can be used where a grid of dots is placed over an image of squashed roots and colonized cortical cells are counted.

18–2.3 Chemical Determinations

Chitin determinations have been used to estimate fungal biomass in roots under controlled experimental conditions (Hepper, 1977; Bethlenfalvay et al., 1981). However, its utility in native soil is limited because chitin is ubiquitous in nature, where it is found in the cell walls of many fungi and the exoskeletons of insects, and certain soils exhibit physical and chemical properties which prevent the technique from working properly (Jarstfer & Miller, 1985). Recently, assays for ergosterol have been used to quantify ectomycorrhizal fungal biomass (Salmanowicz et al., 1989; Johnson & McGill, 1990; Martin et al., 1990) and this method should also be useful for VAM fungi. In addition, J.H. Graham (personal communication) has found that fatty acid analysis is applicable to the quantification of VAM fungal colonization.

18–2.3.1 Chitin Determinations

1. Oven-dried and ground roots (10 mg) are mixed with 4 mL of conc. KOH (120 g KOH/100 mL H_2O) and autoclaved for 1 h in 15 mL screw-cap centrifuge tubes to degrade chitin to chitosan. For an internal control, root material from noncolonized plants should be assayed to correct for hexosamines in plant tissue and organisms other than VAM fungi. In addition, a standard curve using purified chitin should be included to correct for the incomplete hydrolysis of chitin.

2. Add and mix with a vortex mixer, 8 mL of 75% (vol/vol) ice-cold ethanol (75%) to the 4 mL of autoclaved sample. Then layer 0.9 mL of a celite suspension (diatomaceous silica, suspended in 75% ethanol, 1 g per 20 mL) over the first ethanol wash.

3. The samples are centrifuged at approx. 5000 × g at 4 °C for 10 min. The pellets are then resuspended and washed sequentially in 8 mL each of ice-cold, 40% ethanol, 0.01 M HCL, and distilled water.

4. The pellet is suspended in 1.5 mL of water and assayed colorimetrically as follows (Ride & Drysdale, 1971). Equal volumes (1.5 mL) of the washed chitosan suspension, 5% (wt/vol) $NaNO_2$, and 5% (wt/vol) $KHSO_4$ are transferred to a centrifuge tube, shaken for 15 min, and centrifuged at 1500 × g for 2 min at 2 °C. Two, 1.5-mL subsamples are removed from the tube and 0.5 mL of 12.5% (wt/vol) ammonium sulfamate is added to each, shaken for 5 min, and then 0.5 mL of 0.5% (wt/vol) 3-methyl-2-benzothiazolone hydrochloride (prepared daily) is added. This mixture is heated in a boiling water bath for 3 min, cooled, and 0.5 mL of 0.5% (wt/vol) $FeCl_3$ (100 mL contained 0.83 g $FeCl_3 \cdot 6H_2O$, stored at 4 °C and discarded after 3 d) is added. After 30 min, the absorbance is read at 650 nm.

18–3 QUANTIFICATION OF VESICULAR-ARBUSCULAR MYCORRHIZAL EXTERNAL HYPHAE

18–3.1 Introduction

Even though the hyphae that grow into the soil matrix from the root are the functional organs for nutrient uptake and translocation, few researchers have obtained quantitative data on their growth and distribution. This is largely due to the technical difficulties in obtaining reliable data— there is no completely satisfactory method to quantify external hyphae of VAM fungi in soil. Three major problems have yet to be overcome: (i) there is no reliable method to distinguish VAM fungal hyphae from the myriad of other fungal hyphae in the soil, (ii) assessment of the viability and activity of hyphae is problematic, and (iii) meaningful quantification is time consuming. Nonetheless, clarification of the growth dynamics of external hyphae is essential to further understanding of their function in soil. Below are brief descriptions of some useful methods for quantification of external hyphae. Further detailed methods have been reviewed (Sylvia, 1992).

18–3.2 Indirect Methods for Total Hyphae

18–3.2.1 Colonization of "Receiver" Plants

Several researchers have used colonization of a "receiver" plant, separated from an inoculated "donor" plant by root-free soil, to estimate the rate of hyphal growth in soil (Warner & Mosse, 1983; Schüepp et al., 1987; Miller et al., 1989; Camel et al., 1991). The essence of this method is that a volume of soil is maintained free of roots by using nonbiodegradable fabric with a pore size (50–100 μm) that excludes roots, but which allows free passage of fungal hyphae. A tripartite system is constructed where one compartment contains the inoculated donor plant, the middle compartment contains root-free soil, and the third compartment contains the receiver plant. With such a system, the distance that hyphae grow through the soil can be determined; however, it will underestimate the rate of hyphal growth since colonization is not instantaneous upon contact of a hypha with a root.

18–3.2.2 Soil Aggregation

Graham et al. (1982) estimated relative hyphal development by weighing plant roots with adhering soil. They reasoned that the weight of the root ball would be proportional to the amount of hyphae that bound the unit together. Using this method, however, it is not possible to compare hyphal development among different soils or plant species since many factors affect the quantity of soil in the root ball.

18–3.2.2.1 Method

1. The soil mass is air dried and then roots are shaken vigorously to remove air-dry soil that is not firmly attached to the root system.
2. The soil that remains attached is washed from the root system into a beaker of water and after it settles, the water is decanted and the soil residue dried and weighed.
3. The roots are blotted dry and their masses determined. The amount of soil adhering to roots is expressed as the amount of dry soil attached in milligrams per gram of root.

18–3.2.3 Chitin Determination

Chitin determinations have also been used to estimate hyphal biomass in soil; however, the same limitations discussed for this method under root colonization apply here. For this method, dried soil samples are mixed with conc. KOH and autoclaved for 1 h to degrade chitin to chitosan. Subsamples of the soil-KOH suspension are transferred to centrifuge tubes and assayed for chitin as described in section 18–2.3.1. By subtracting the chitin content of control soil without VAM fungi from soil containing VAM fungi, the biomass of VAM fungi can be estimated.

18–3.3 Direct Methods for Total Hyphae

18–3.3.1 Filtration-Gridline Method

Attempts have been made to estimate hyphal development directly by extracting hyphae from soil and quantifying lengths by a modified gridline-intersect method (Abbott et al., 1984; Miller et al., 1987; Sylvia, 1988). The major problem with the filtration-gridline method is that hyphae of nonmycorrhizal fungi often are not distinguishable from those of VAM fungi. Hyphal diameter has been used to distinguish VAM fungi from other fungi in the soil. Abbott and Robson (1985) found that most hyphae in VAM fungal-inoculated pot cultures had diameters between 1 and 5 μm. They concluded that it was not possible to distinguish hyphae of VAM fungi from nonmycorrhizal fungi by morphological or staining criteria. One approach has been to subtract the length of hyphae in control treatments from VAM-inoculated treatments; however, this assumes that no interactions occur among the hyphae in the soil.

18–3.3.1.1. Method

1. Soil cores of known volume are collected at random, thoroughly mixed, subsamples (e.g., 10 g) removed, suspended in water (e.g., 500 mL), and passed through a sieve with 250-μm diam. pores.
2. The filtrate is blended for 15 s and a portion (e.g., 25 mL) passed through a membrane with pores < 5 μm diam. The exact quantity of soil, water, and suspension should be determined empirically so

that a thin layer of soil is deposited on the filter membrane without excessive clogging.

3. The membrane is briefly flooded with a trypan blue solution (0.5 g of trypan blue, 500 mL of deionized water, 170 mL of lactic acid, and 330 mL of glycerin) and rinsed with deionized water.

4. The membrane can be cut to fit on a microscope slide and observed at 300× through an eyepiece whipple disc that has a 10 by 10 lined grid. The total length of hyphae can be estimated by the gridline-intersect method (see section 18–2.2.1). Due to the spatial variability of hyphae on the membranes, it is advisable to quantify 20 to 60 fields per membrane.

18–3.3.2 Immunofluorescence Assay

Serological techniques have the potential for specific detection of mycorrhizal hyphae in soil. Kough and Linderman (1986) used an immunofluorescence assay (IFA) to detect hyphae of VAM fungi in soil and provided detailed protocols for preparation of antigens and fluorescein labelling; however, cross reactivity has limited their usefulness. The development of hybridoma technology has opened the potential for production of large quantities of monoclonal antibodies specific for a single epitope (Halk & De Boer, 1985). Wright et al. (1987) were able to obtain highly specific monoclonal antibodies for spores and hyphae of VAM fungi. Procedures for the use of monoclonal antibodies with VAM fungi are presented in chapter 28 by Wright of this book.

18–3.4 Detection of Active Hyphae

Stains that differentiate actively metabolizing cells from inactive cells have been used to estimate the "viability" of VAM hyphae. Sylvia (1988) used a solution containing iodonitrotetrazolium (INT) and NADH to locate active hyphae of two *Glomus* spp.; INT is reduced by the electron-transport system of living cells and results in formation of a red color. Schubert et al. (1987) used fluorescein diacetate (FDA) to identify functioning hyphae; FDA is hydrolyzed within living cells, releasing fluorescein which can be detected with UV illumination.

18–3.4.1 Details of the Iodonitrotetrazolium Method

1. A membrane with extracted hyphae (see section 18–3.3.1.1) is flooded with a solution containing INT and incubated for 8 h at 28 °C. The INT solution consists of equal parts of INT (1 mg mL^{-1}), NADH (3 mg mL^{-1}), and 0.2 M Tris buffer (pH 7.4).

2. The membrane is counter-stained with trypan blue, rinsed with deionized water, and cut to fit on a microscope slide.

3. The proportions of active (hyphal contents stained reddish) and inactive hyphae are determined by the gridline-intersect method.

18–4 RECOVERY OF VESICULAR-ARBUSCULAR MYCORRHIZAL FUNGAL SPORES

Hayman (1984) and Schenck and Perez (1990a) have reviewed several methods for extracting spores of VAM fungi from soil. Most of these techniques work best in sandy soils, and less well in clay or organic soils.

18–4.1 Wet Sieving and Decanting/Density Gradient Centrifugation

1. Place a 50 to 100 g sample into a 2-L container and add 1 L of water. Vigorously mix the suspension to free spores from soil. For clay soils, samples may be placed in a sodium hexametaphosphate solution instead of water (Porter et al., 1987).
2. For fungal species that form spores in roots (e.g., *Glomus intraradix* and *G. clarum*), blend the soil-root sample for 1 min in 300 mL of water to free spores from roots.
3. Let the suspension settle for 15 to 30 s (times vary depending on soil texture) and decant the supernatant through standard sieves. Sieves should be selected so as to capture the spores of interest. Use a 425-μm sieve over a 45-μm sieve for unknown field samples. Examine the contents of the top sieve for sporocarps that may be up to 1 mm in diameter. For clay soils, it is advisable to repeat the decanting and sieving procedures on the settled soil.
4. Roots can be collected from the larger-mesh sieves for evaluation of internal colonization (see section 18–2.1.1).
5. Transfer sievings to 50-mL centrifuge tubes with a steady stream of water from a wash bottle and balance opposing tubes.
6. Centrifuge at 1200 to 1300 $\times$ g in a swinging-bucket rotor for 3 min, allowing the centrifuge to stop without braking. Remove the supernatant carefully to avoid disturbing the pellet. Also remove the organic debris that adheres to the side of the tube.
7. Suspend soil particles in chilled, 1.17 M sucrose, mix contents with spatula and centrifuge immediately at 1200 to 1300 $\times$ g for 1.5 min, applying the brake to stop the centrifuge.
8. Pour the supernatant through a small-mesh sieve, rinse spores held on the sieve carefully with tap water and wash spores into a plastic Petri dish scribed with parallel lines spaced 0.5-cm apart. Spores may be counted by scanning the dish under a dissecting microscope.

18–5 IDENTIFICATION OF VESICULAR-ARBUSCULAR MYCORRHIZAL FUNGI

The identification of VAM fungi can be a frustrating process. Unless taxonomy is the major objective, it is recommended that the spores be keyed initially only to genus. Isolates of special interest can be classified to species at a later date. A description of the taxonomy of the six genera known to form VAM is present in Fig. 18–1. Careful notes should be taken on the size, color, surface characteristics, and wall morphology of the spore types recovered. Initiate pot cultures with each spore type (see section 18–7.1.5.1), keeping detailed records on the origin and subsequent pot culture history of the isolate. It is imperative that each isolate be given a unique code and that this code be used in all reports about the isolate.

Currently, the taxonomy of VAM fungi is based on a limited number of morphological characteristics of the spore. Detailed steps for identification of spores have been presented elsewhere (Morton, 1988; Schenck & Perez, 1990b). Especially useful features of the manual prepared by Schenck and Perez (1994) are a worksheet for recording spore characteristics useful in taxonomy, a key to genera, and a species guide or quick key to rapidly reduce the number of possible species. A portion of the worksheet is presented in Fig. 18–2.

With experience, some fungi can be distinguished on the basis of their morphology within roots. Abbott (1982) was able to differential 10 species of fungi by their characteristic morphology in roots; however, the differences among genera were always greater than those among species within a genera. Although this method requires careful initial observations from pot cultures, it offers promise for distinguishing fungi in plant roots.

Maintenance of good germplasm should be an essential part of any VAM-research program. The International Culture Collection of VAM Fungi (INVAM) maintains a collection of VAM isolates. Samples of VAM fungi may be submitted to INVAM for verification of classification and possible inclusion in their collection. For additional information contact Dr. J. Morton, Division of Plant and Soil Sciences, West Virginia University, Morgantown, WV 26506-6057, USA.

Biochemical and molecular methods promise to make VAM taxonomy a more precise science. Protein profiles (Rosendahl et al., 1989), serology (Wright et al., 1987), gene amplification (Simon et al., 1992), and fatty-acid analysis (J.H. Graham, 1992, personal communication) all hold great potential, but none have been sufficiently developed to be included in this chapter.

18–6 ASSESSMENT OF GROWTH RESPONSE AND SELECTION OF EFFECTIVE ISOLATES

18–6.1 Phosphorus-Response Curves and Mycorrhizal Dependency

Much of the interest in mycorrhiza has focused upon their ability to enhance P uptake and subsequently the growth of plants in soils of low to moderate P availability. The quantitative assessment of plant response to VAM fungi is affected by all the factors involved in the symbiosis, including the genotypes of both the plant and fungus, and the soil environment.

Order Glomales

Soil-borne fungi characterized by transient dichotomously-branched arbuscules in cortical cells of plant roots after establishing an obligate mutualistic symbiosis with many plant species. Spores appear to be obligately asexual, forming within or outside roots. Most species diversity is manifested in spore size, color, and microscopic features of subcellular spore walls distinct in structural properties and histochemical properties (usually in Melzer's reagent).

Suborder Glomineae

Arbuscular fungi forming intraradical vesicles in mycorrhizal roots; "chlamydospores" borne terminally, intercalarily, and laterally from one or more subtending hyphae. Members are referred to as "vesicular-arbuscular mycorrhizal (VAM) fungi."

Family Glomaceae

"Chlamydospores" (i) are produced singly, in aggregates, in an unorganized hyphal matrix, or in a highly ordered hyphal matrix, with the structural wall continuous with a wall of the subtending hypha; (ii) show morphological diversification mostly in the number and types of walls formed outside the structural wall (e.g., evanescent, expanding, mucliagenous, and unit walls); (iii) form flexible inner walls so far limited to those which are membranous, usually one or two in number, and which rarely stain positive in Melzer's reagent; (iv) seal off contents from that of the subtending hypha by different mechanisms, such as an amorphous plug, a septum, an inner membranous wall, or thickening of the structural wall; and (v) germinate usually by emergence of the germ tube through the subtending hypha.

Genus *Glomus*

Have all familiar characters except that spores are not formed in a highly organized matrix originating from a columnar base.

Family Acaulosporaceae

"Chlamydospores" (i) are born laterally from or within the neck of a sporiferous saccule that is formed terminally on a fertile hypha; usually singly but occasionally in aggregates; (ii) have an outer wall which is continuous with the subtending hyphal wall and which sloughs with age; (iii) are sessile following extraction from soil because of sloughing of attached hyphae, (iv) have a smooth to highly ornamented structural wall that does not originate from the wall of the subtending hypha; (v) form at least one flexible inner wall, but usually two or more; (vi) show most diversification in number and types of inner walls (e.g. the semi-rigid unit wall, beaded or smooth membranous wall, and amorphous wall), with the innermost wall types often producing a dextrinoid to dark red-purple reaction in Melzer's reagent; (vii) seal off contents from that of the subtending saccule neck by a plug indistinguishable from the structural wall; and (viii) germinate between a semi-rigid unit wall and the innermost pair of inner flexible walls when such walls are synthesized.

Genus *Acaulospora*

Spores borne laterally on the neck of the sporiferous saccule in a continuous transition series from those with a glomus-like hyphal attachment to those borne on a pedicel and finally to those borne on a short collar.

Genus Entrophospora

Spores are formed within the neck of the sporiferous saccule. Spore ontogenesis and spore wall diversity mirror that in *Acaulospora*.

Suborder Gigasporineae

Arbuscular fungi which also form extraradical auxillary cells singly or in clusters; globose to subglobose "azygospores" often exceeding 200 μm in diameter, forming on a bulbous sporogenous cell. Members of the group do not form vesicles and hence are "arbuscular mycorrhizal fungi."

Fig. 18–1. Continued on next page.

Genus *Gigaspora*

Spores do not differentiate any flexible innerwalls; germ tubes arise from a warty germinal wall which rarely separates from the laminated wall; auxiliary cells usually echinulate.

Genus *Scutellospora*

Spores (i) always differentiate two or more flexible inner walls which often form in adherent pairs, (ii) show most diversification in the number and types of inner walls (e.g., membranous walls of different thicknesses, coriaceous wall, amorphous wall), with the innermost flexible wall often producing a dextrinoid to dark red-purple reaction in Melzer's reagent, (iii) germinate via germ tubes arising from a persistent germination shield of variable shape and margin which always forms on the innermost pair of flexible inner walls. Auxiliary cells smooth to knobby.

Fig. 18–1. Description of the major taxa forming vesicular-arbuscular and arbuscular mycorrhizal fungi. Adapted from Morton (1988). Used with permission.

Since growth response is directly related to the P content of the plant, P-response curves should be constructed for each plant-soil combination (Abbott & Robson, 1984). Mycorrhizal and nonmycorrhizal plants are grown over a range of applied-P levels and shoot biomass is plotted against P level. The benefits of constructing these curves are: (i) ensuring that evaluations of mycorrhizal responsiveness are at soil-P levels that are limiting the growth of nonmycorrhizal plants, (ii) estimates of the amount of nutrient required for the same yield of mycorrhizal and nonmycorrhizal plants can be made, and (iii) claims for non-P effects of mycorrhizae on plant growth can be tested on plants with the same percentage of maximum growth and nutrient content.

The mycorrhizal dependency or responsiveness of a plant is usually assessed in terms of the ratio of the shoot dry mass of mycorrhizal (M) vs. nonmycorrhizal (NM) plants in a P-deficient soil as follows (Plenchette et al., 1983):

$$100 \times ((M - NM) / NM)$$

While this method for determining mycorrhizal dependency is useful, it can give misleading information since the method does not take into account differences among plant species in their responsiveness to P. When only one P level is used in the assessment of mycorrhizal dependency, there is the possibility that one plant species will respond to VAM fungi and the other will not. However, at a different P level the second plant may be quite responsive.

Controls for VAM-plant growth experiments are critical since other soil microorganisms may affect plant growth. Abbott and Robson (1984) discussed various strategies for obtaining suitable controls. Ames et al. (1987) reported that similar functional groups of microorganisms could be established if a VAM-inoculum wash is applied to both mycorrhizal and nonmycorrhizal treatments. This can be accomplished by making a

I. Many of these observations can be made with a dissecting microscope but measurements should be made with a compound microscope.
  A. Spore color: In water _____. In mountant _____. Mountant used: _____.
  B. Spore diameter: For globose spores: Range ____ μm   Mean ____μm.
    For irregularly-shaped spores: Length; Range ____ μm Mean ____ μm.
                                          Width; Range ____ μm Mean ____ μm.
  C. Composite spore wall thickness (determined on intact spores if possible: _____ μm.
  D. Attachment present? Yes __ No. __. If no, go to E. Sporiferous saccule present? Yes __ No __.
  E. Spore contents: Globular ____; Reticulate ____; Granular ____; Other: ____.
  F. Spore with mantle or other surface hyphae? Yes __ No __. If no, go to G. Width of hyphae __ μm; Color of hyphae _____; Hyphae sinuous? Yes __ No __
  G. Spores formed within the root? Yes __ No __.
  H. Auxiliary cells present? Yes __ No __. If no, go to I.
    If yes, Knobby __; Digitate __; Coralloid __; Echinulate __; Spiny __; Pigmented __ if yes, color: _____.
  I. Sporocarp present? Yes __ No __. If no, go to J.
    Sporocarp diameter _____; Peridium present? Yes __ No __ - if yes, indicate color _____.
  J. Determine the genus of your specimen. (see Fig. 18-1); Genus ____. Go to 1, 2, or 3, then II.
    1) For *Gigaspora* or *Scutellispora*:
      a. Bulbous sporogenous cell dimensions: Width __ μm Length __ μm.
      b. Subtending hyphae septate? Yes __ No __
      c. Surface ornamentation on spore present? Yes __ No __. Description _____.
      d. Germination shield present? Yes __ No __.
    2) For *Acaulospora* or *Entrophospora*:
      a. Sporiferous saccule present? Yes __ No __. If no, go to II.
      b. Sporiferous saccule collapsed? Yes __ No __. If yes, go to II.
      c. Sporiferous saccule dimensions (diameter or length / width) ____ μm.
      d. Description of sporiferous saccule contents (color; content appearance, e.g., granular; reticulate, globular; texture, e.g., smooth, rough, flaky): _____.
      e. Hyphal length between spore and sporiferous saccule ____ μm.
      f. Hyphal diam. at spore attachment ____ μm.
      g. Pore diameter on spore at point of attachment ____ μm.
      h. Hyphal attachment scar present? Yes __ No __. If yes, indicate the number present ____.
    3) For *Glomus* or *Sclerocystis*:
      a. Pore occluded? Yes __ No __.   b. Pore diameter ____ μm.
      c. Presence of a septum at the pore Yes __ No __. Protruding septum? Yes __ No __.
      d. Hyphal width adjacent to spore wall ____ μm. e. Number of attachments per spore _____.
      f. Outer wall of hypha contiguous with outer wall of spore? Yes __ No __.
      g. Type of attachment: (Check all that apply) straight __; recurved __; funnel-shaped __; branched __; septate __; constricted __; swollen __; other: _____.
II. Make these observations on broken spores with a compound microscope.
  A. Number of wall groups ____. B. Width of each wall group: A = __ B = __ C = __ D = __ E = __.
  C. Number of walls within each group: A = __ B = __ C = __ D = __ E = __.

Fig. 18-2. Continued on next page.

D. Type of wall(s) within each group: Wall Group: A __; B __; C __; D __; E __.
   A = Amorphous; C = Coriaceous; E = Evanescent (ephemeral),
   G = Germinal; L = Laminate; M = Membranous; P = Hyphal peridium;
   U = Unit; X = Expanding.
Wall reaction to Meltzer's reagent. Indicate positive (+) or negative (−) for
   each wall group. If positive, also indicate wall color and wall type ef-
   fected.
If other reagents or stains are used, record the spore wall reaction as for
   Meltzer's reagent.

Fig. 18–2. Worksheet for collecting important characters for the identification of vesicular-arbuscular mycorrhizal fungi. Adapted from Schenck and Perez (1994). Used with permission.

suspension of the inoculum in water, passing it through a fine sieve (< 10 μm) to exclude VAM propagules, and then applying equal amounts of the sieved suspension to all treatments.

18–6.2 Screening for Effective Isolates

Various isolates of VAM fungi differ in their ability to colonize host plants and promote plant growth; therefore, mycorrhizal dependency can vary widely with fungal genotype. Before embarking on an inoculation program, indigenous and exotic isolates of VAM fungi should be screened for their ability to promote the growth of the target host plant in the soil where the host will grow. For expediency, screening trials are often conducted in the greenhouse; however,

> reliable field data which show consistent growth improvements as a result of inoculation of plants subject to *routine commercial propagative practices* will do far more to convince growers of the usefulness of mycorrhizae than a plethora of carefully controlled trials where conditions have been optimized for mycorrhizal performance (Jeffries, 1987).

Often, the most effective isolates are those that colonize the plant most rapidly (Abbott & Robson, 1982; Abbott & Robson, 1984), but not necessarily those that have the greatest amount of colonization at harvest (Hung et al., 1990). For this reason, sequential harvests and assessments of colonization should be conducted. Alternatively, a nondestructive (leaf-P status) assessment of effectiveness has been proposed (Aziz & Habte, 1987).

When conducting screening trials, care must be taken to ensure that inoculum density is not a limiting factor. Ideally, inoculum potential of all isolates should be equalized (Daniels et al., 1981). However, the time required to conduct these assays is sufficient for significant changes in propagule density to occur (Daft et al., 1987; O'Donnell et al., 1992). Furthermore, when inoculum densities vary widely among isolates, standardization of the inoculum is not practical and effective isolates do not always produce large quantities of inoculum in culture (Sylvia & Burks,

1988). Nonetheless, MPN or similar assays should be conducted on inoculum used in screening trials since the results are useful for interpretation of the results of VAM-inoculation studies.

18–7 PRODUCTION AND USE OF VESICULAR-ARBUSCULAR MYCORRHIZAL INOCULA

Jarstfer and Sylvia (1992a) recently reviewed methods for the production and use of VAM inocula and this section is largely adapted from that work.

18–7.1 Soil-Based Pot Cultures

The culture of VAM fungi on plants in disinfested soil, using spores, roots or infested soil as inocula, has been the most frequently used technique for increasing propagule numbers (Menge, 1983; Menge, 1984; Schenck & Perez, 1990a).

18–7.1.1 Selection of Host Plant

Many host plants have been used under a variety of conditions (Thompson, 1986; Sreenivasa & Bagyaraj, 1988; Liyanage, 1989). Examples of plants that have been used successfully are alfalfa, maize, onion, sudan grass, and wheat. Generally, the host selected should become well colonized (> 50% of the root length), produce root mass quickly, and be able to tolerate the high-light conditions required for the fungus to reproduce rapidly. Hosts that can be propagated from seed are preferable to cuttings since they can be more easily disinfested. Most seeds may be disinfested with 10% household bleach (0.525% NaOCl) for 5 to 15 min followed with five washes of water (Tuite, 1969). Washing with water may also remove fungicides and other agrichemicals which may adversely affect VAM fungi (Tommerup & Briggs, 1981).

18–7.1.2 Soil Disinfestation

All components of the culture system should be disinfested prior to initiation of a pot culture. The method of soil disinfestation is especially important—the objective is to kill existing VAM fungi, pathogenic organisms, and weed seeds while preserving a portion of the nonpathogenic microbial community. The use of aerated steam or a commercial soil pasteurizer will eliminate VAM fungi from soils (Menge, 1983; Sylvia & Schenck, 1984). We commonly pasteurize soil by heating it to 85 °C for two, 8-h periods with 48 h between treatments. Fumigation with a methyl-bromide/chloropicrin mixture or other soil-applied biocides is effective, but requires several days for the chemical to diffuse out of the soil. Ionizing radiation (0.8–1.0 Mrad) has also been used (Jakobsen & Anderson, 1982; Jensen, 1983; Thompson, 1990). Autoclaving should be avoided because it can result in the release of toxic inorganic and organic compounds (Wolf et al., 1989).

18–7.1.3 Light, Moisture, and Temperature

Good light quality and high irradiance are necessary for maximum inoculum production (Ferguson & Menge, 1982a; Furlan & Fortin, 1977). Where natural light conditions are poor (PPFD < 500 μmol m^{-2} s^{-1}), high-intensity lamps should be used. Soil moisture affects VAM sporulation, with nonsaturated and nonstressed water conditions providing maximum sporulation (Nelsen & Safir, 1982). Excessive moisture may encourage problems with hyperparasites in the culture (Paulitz & Menge, 1986; Daniels & Menge, 1980). The best strategy is to apply water regularly to well-drained soil. Temperature is also important for pot cultures. Sporulation is positively correlated with temperature from 15 to near 30 °C for many VAM fungi (Schenck & Schroder, 1974; Schenck & Smith, 1982); however, at higher temperatures sporulation may decrease. For maximum sporulation, the temperature optimum for each isolate should be determined.

18–7.1.4 Fertilizers, Pesticides, and Pot Size

Responses to P and N fertilization are also strain dependent (Sylvia & Schenck, 1983; Johnson et al., 1984; Thompson, 1986; Liyanage, 1989; Douds & Schenck, 1990a) and are affected by the relative amounts of N and P supplied (Sylvia & Neal, 1990). The usual approach is to supply the plants with adequate nutrition, except for P, the concentration of which should be limiting for growth of nonmycorrhizal plants. For routine maintenance of pot cultures a P-free fertilizer such as Peter's (Peter Fertilizer Products, Allentown, PA) 25-0-25 can be applied; however, to determine the optimal level of P fertility, P-response curves should be developed (Abbott & Robson, 1984). The optimal level of P has been shown to differ for some species of VAM fungi (Thomson et al., 1986).

Pesticides can affect VAM colonization and sporulation. Fungicides should be carefully selected for use in pot cultures (Nemec, 1980; Menge, 1982; Jabaji-Hare & Kendrick, 1987; Dodd & Jeffries, 1989). Ants and other crawling insects also cause cross contamination of pot cultures. Insecticides and nematicides may be used effectively in VAM pot cultures (Sreenivasa & Bagyaraj, 1989).

Pot size should match the potential volume of the root system within practical space constraints. Ferguson and Menge (1982b) assessed pot volumes ranging from 750 to 15 000 cm and reported that larger containers resulted in higher spore concentrations. However, when using large volumes, any contamination will result in greater loss. In the greenhouse, pot cultures should be isolated from contaminated soil, splashing water, and crawling insects. In addition, specific isolates of VAM fungi should be kept well separated from each other.

18–7.1.5 General Procedures

To initiate pot cultures, place a layer of inoculum 1 to 2 cm below the seed or cutting. Inoculum may consist of spores, colonized roots or infested soil. Infested soil is often used to initially obtain isolates from the field; however, these "mixed-species" cultures should rapidly progress to "single-species" cultures initiated from 20 to 100 healthy, uniform spores. For critical taxonomic studies, "single-spore" cultures should be produced (Fang et al., 1983; Schenck & Perez, 1990a).

18–7.1.5.1 Method for Single-Spore Culture

1. Plastic tubes (such as available from Stuewe and Sons, Inc., Corvallis, OR) are plugged with polyester fiber batting (available at fabric stores).
2. Place moist, pasteurized soil—similar to that from which the spores were originally isolated—into the tubes to within 3 cm of the top.
3. Place a moist filter paper disk on the soil surface and transfer a single spore to it. Choose clean, bright, and nonparasitized spores. Spores can be transferred from water in petri dishes with micropipettes.
4. Add 1 to 2 cm of additional soil over the spore and plant seed of a suitable host on the surface. Place tubes in a well-lighted growth chamber or greenhouse and sample for colonization and sporulation after 12 to 16 wk.
5. Another method is to place a single spore directly on the surface of seedling root (approximately 2 cm below the crown) prior to transplanting it into a growth tube. The placement of the spore should be confirmed under a dissecting microscope before the growth medium is carefully poured around the root system.
6. The probability of success with one spore is relatively low so it is important that several tubes are established for each isolate.

18–7.2 Soil-less Media

Culturing VAM fungi in soil-less media avoids the detrimental organisms in nonsterile soil and allows control over many of the physical and chemical characteristics of the growth medium. Soil-less media are more uniform in composition, weigh less, and provide aeration better than do soil media. Growth media that have been used include bark, peat, perlite, and vermiculite (Biermann & Linderman, 1983); sand (Ojala & Jarrell, 1980), calcined montmorillonite clay (Plenchette et al., 1982), and expanded clay aggregates (Dehne & Backhaus, 1986).

Moist soil-less media do not buffer P concentration so care must be taken to avoid high solution levels of P in the root zone. Frequent addition of dilute, soluble nutrient solutions (Waterer & Coltman, 1988a; Liyanage, 1989; Douds & Schenck, 1990a), mixing of time-release fertilizer (Waterer & Coltman, 1988b) or the use of less-available forms of P (Thompson, 1986) should provide excellent cultures of VAM fungi in soil-less media.

18–7.3 Nutrient Flow and Aeroponic Systems

18–7.3.1 Introduction

At least seven species of VAM fungi have been grown in various nutrient-solution systems on hosts representing at least 21 genera (Jarstfer & Sylvia, 1992a). The primary benefit of these systems is that colonized roots and spores are produced free of any substrate, permitting more efficient production and distribution of inocula. Usually plants are inoculated with VAM fungi and grown in sand or vermiculite before they are transferred into a culture system. The plants are grown for a period of 4 to 6 wk under conditions conducive for colonization, after which they are washed and nondestructively checked for colonization (Ames et al., 1982); however, it is also possible to inoculate plants directly in the culture system (Hung et al., 1991). The P concentrations which have been reported to support VAM growth in solution cultures range from < 1 to 24 μmol.

A system that applies a fine nutrient mist to roots of intact plants (aeroponic culture) produces excellent inoculum (Sylvia & Hubbell, 1986; Hung & Sylvia, 1988). Greater concentrations of spores have been produced in aeroponic cultures when compared to soil-based pot cultures of the same age. Because the colonized-root inoculum produced in this system is free of any substrate, it can be sheared resulting in high propagule numbers per gram of root (Sylvia & Jarstfer, 1992).

18–7.3.2 Method for Producing Sheared-Root Inocula from Aeroponic Cultures

This method is modified from Jarstfer and Sylvia (1992b).

1. To inoculate the plants, distribute a spore suspension over 2 L of clean vermiculite in a disinfested pot and cover with more vermiculite. Either plant disinfested seed of the culture plant or push unrooted stem cuttings (also disinfected) into the vermiculite so the bud is level with the vermiculite surface. Sweet potato [*Ipomoea batatas* (L.) Lam] has worked well for this purpose.
2. After the culture plants have grown for at least 4 wk, remove them from the vermiculite and check for VAM colonization. Displace the plants from their pots onto a disinfected surface and separate individual plants carefully to minimize root injury. Wash the roots in several changes of water to remove as much vermiculite as possible.
3. The roots of intact plants can be nondestructively examined with epifluorescence microscopy for intracellular granular fluores-

cence,cence, indicating arbuscular formation (Ames et al., 1982). If colonization is satisfactory, the plants are transferred into an aeroponic chamber by inserting roots through holes in the chamber cover. Only colonized plants should be placed in the aeroponic chamber.

4. The aeroponic chamber should be of dimensions suitable to the desired quantity of inoculum or the space available. For sanitation reasons, an acrylic-lined chamber is desirable, but acrylic enamel paint over finished fiberglass will also work. The chamber should have a lid with 2.5-cm-diam. holes spaced 15 cm apart in a regular pattern to allow adequate distribution of the nutrient mist.

5. Two systems of misting may be used to deliver the nutrient solution to the roots of the culture plants. The first uses an atomizing disk similar to those used in cold-water humidifiers (Zobel et al., 1976). This system is limited to relatively small chambers with capacities from 60 to 80 L of nutrient solution. The second system uses a centrifugal pump to spray the nutrient solution at the upper portion of the root mass allowing the solution to flow down the roots. Multiple micro-irrigation nozzles may be used to cover the entire root system regardless of the number of plants or depth of the chamber. The pump should run intermittently to deliver a spray for 7 s every min.

6. The nutrients for aeroponic culture are dilute as the roots are in constant contact with them. Deionized or distilled water should be used to make the nutrient solutions to avoid problems with pH or toxic levels of plant micronutrients. For convenience, nutrient solutions are made by diluting stock solutions (Table 18–1).

7. When initiating a culture or routinely changing the nutrient solution, first fill the chamber with the desired amount of water then

Table 18–1. Nutrient solution for aeroponic culture of VAM fungi, based on a modified Hoagland's solution (Hoagland & Arnon, 1950).

Stock solution	Quantity of stock solution in final nutrient solution
	mL L^{-1}
0.01 M KH$_2$PO$_4$	0.3
1.00 M KNO$_3$	1.5
1.00 M Ca(NO$_3$)$_2$4H$_2$O	1.5
1.00 M MgSO$_4$7H$_2$O†	0.3
0.10 M NaCl	0.3
0.10 M NaFeEDTA	0.45
Micronutrient stock	1.0
46.20 mM H$_3$BO$_3$	
9.53 mM MnCl$_2$·4H$_2$O	
0.765 mM ZnSO$_4$·7H$_2$O	
0.32 mM CuSO$_4$·5H$_2$O	
0.066 mM Na$_2$MoO$_4$·2H$_2$O	

† Keep this solution in a separate container.

add the mixture of concentrated nutrients measured from the stock solutions. Mix well and check the pH after at least 15 min. The pH should be adjusted to 6.5 for most culture plants.

8. On a biweekly schedule, the roots of the culture plants should be cut back to 2 cm above the highest nutrient solution level and the root debris removed from the chamber, the inside walls scrubbed to remove any algae, and the nutrient solution drained quickly from the chamber and replaced with new solution.

9. To produce sheared-root inoculum, well-colonized roots harvested from aeroponic cultures (grown for 10–12 wk) are mixed 1 to 10 (wt/vol) with water and sheared with a food processor for 40 s (e.g., "Little Pro," Cuisinart, Inc., Norwich, CN). The resulting root fragments, vesicles, and free spores are collected over a fine sieve (45 μm). The sheared-root material may be mixed directly with growing medium or may be added to hydrogels such as Natrosol to make a flowable inoculum (Hung et al., 1991).

10. Spores may also be collected from fresh roots or from roots stored air dry by washing over standard sieves. For clean preparations, spores may be separated from root fragments using density-gradient centrifugation (see section 18–4.1).

18–7.4 Storage of Inoculum

Spores of VAM fungi are commonly stored at 4 °C in dried pot-culture soil (Ferguson & Woodhead, 1982); however, many VAM species do not survive this treatment. Cryopreservation of spores at −60 to −70 °C appears to be much more reliable (Douds & Schenck, 1990b). Cultures of VAM fungi should be dried slowly with the host plants and frozen in situ.

The viability of colonized-root inoculum from aeroponic culture declines with storage. Root inoculum, air-dried prior to storage at 4 °C, retains a greater density of VAM fungal propagules than roots stored moist. However, once roots are dried they cannot be sheared. For short-term storage of active roots, continue to maintain the aeroponic culture and remove plants as needed to make inoculum. When plants are harvested, excess moisture should be removed from the roots to prevent development of anaerobic conditions. Roots may be stored for short times in this moist state.

18–7.5 Application of Vesicular-Arbuscular Mycorrhizal Inocula

18–7.5.1 Introduction

Any method by which viable propagules of VAM fungi can be delivered efficiently to the rhizosphere (or potential rhizosphere) should produce the desired colonization. The composition of an inoculum should mesh economically with the method of application. In addition, to reduce the risk of spreading pathogens, the host plant used for inoculum production should not be related to the crop plant to be inoculated (Menge, 1983). For sanitation reasons, logistics, and cost, soil-less cultures or inoculum from hydroponic or aeroponic systems should be used.

18–7.5.2 General Methods

The most common application method is to place the inoculum below the seed or seedling prior to planting (Jackson et al., 1972; Hayman et al., 1981; Hayman, 1987). Hall (1979, 1980) successfully used infested soil pellets to inoculate plants in the field. To produce pellets, moistened soil-clay-sand inoculum is sandwiched between aluminum foil, cut into squares, dried, and seed is glued to the pellets with gum arabic (Hall & Kelson, 1981). Seed coating (Hattingh & Gerdemann, 1975) and somatic embryo encasement (Strullu et al., 1989) with inoculum are also methods worthy of further investigation. Sheared-roots from aeroponic cultures can be suspended in hydrogel for application as a flowable inoculum (Sylvia & Jarstfer, 1992).

18–8 MONOXENIC CULTURES FOR BASIC RESEARCH

The growth of VAM fungi in pure culture in the absence of a host has not been achieved. However, it is possible to colonize roots of intact plants or root-organ cultures to achieve monoxenic cultures that are useful for basic research on the symbiosis (Hepper, 1984). More recently, Ri T-DNA transformed roots have been used to obtain colonized root cultures (Mugnier & Mosse, 1987; Bécard & Fortin, 1988). It should be noted, however, that colonization rates in these systems are slow and only limited amounts of colonized roots have been produced.

Transformed-root culture offers the most efficient method to grow colonized roots as no plant growth regulators are required for sustained growth. Bécard and Fortin (1988) provide procedures for initiating transformed root cultures. A critical step is to obtain aseptically germinated spores of the VAM fungus or, alternatively, colonized root pieces (Williams, 1990). Hepper (1984) reviewed procedures for disinfesting and germinating spores. The most effective methods use chlorine containing compound such as hypochlorite, a surfactant, and antibacterial agents.

REFERENCES

Abbott, L.K. 1982. Comparative anatomy of vesicular-arbuscular mycorrhizas formed on subterranean clover. Aust. J. Bot. 30:485–499.

Abbott, L.K., and A.D. Robson. 1982. The role of vesicular-arbuscular mycorrhizal fungi in agriculture and the selection of fungi for inoculation. Aust. J. Agric. Res. 33:389–408.

Abbott, L.K., and A.D. Robson. 1984. The effect of mycorrhizae on plant growth. p. 113–130. In C.Ll. Powell and D.J. Bagyaraj (ed.) VA mycorrhiza. CRC Press, Boca Raton, FL.

Abbott, L.K., and A.D. Robson. 1985. Formation of external hyphae in soil by four species of vesicular-arbuscular mycorrhizal fungi. New Phytol. 99:245–255.

Abbott, L.K., and A.D. Robson. 1991. Factors influencing the occurrence of vesicular-arbuscular mycorrhizas. Agric. Ecosystem Environ. 35:121–150.

Abbott, L.K., A.D. Robson, and G. De Boer. 1984. The effect of phosphorus on the formation of hyphae in soil by the vesicular-arbuscular mycorrhizal fungus, *Glomus fasciculatum*. New Phytol. 97:437–446.

Adelman, M.J., and J.B. Morton. 1986. Infectivity of vesicular-arbuscular mycorrhizal fungi: Influence of host-soil diluent combinations on MPN estimates and percentage colonization. Soil. Biol. Biochem. 18:77–83.

Alexander, M. 1982. Most probable number method for microbial populations. p. 815–810. *In* A.L. Page et al. (ed.) Methods of soil analysis. Part 2. 2nd ed. Agron. Monogr. 9. ASA and SSSA, Madison, WI.

Ames, R.N., E.R. Ingham, and C.P.P. Reid. 1982. Ultraviolet-induced autofluorescence of arbuscular mycorrhizal root infections: an alternative to clearing and staining methods for assessing infections. Can. J. Microbiol. 28:351–355.

Ames, R.N., K.L. Mihara, and G.J. Bethlenfalvay. 1987. The establishment of microorganisms in vesicular-arbuscular mycorrhizal and control systems. Biol. Fert. Soils 3:217–223.

An, Z.-Q., J.W. Hendrix, D.E. Hershman, and G.T. Henson. 1990. Evaluation of the "Most probable number" (MPN) and wet-sieving methods for determining soil-borne populations of endogonaceous mycorrhizal fungi. Mycologia 82:576–581.

Aziz, T., and M. Habte. 1987. Determining vesicular mycorrhizal effectiveness by monitoring P status of leaf disks. Can. J. Microbiol. 33:1097–1101.

Bécard, G., and J.A. Fortin. 1988. Early events of vesicular-arbuscular mycorrhiza formation on Ri T-DNA transformed roots. New Phytol. 108:211–218.

Bethlenfalvay, G.J., R.S. Pacovsky, and M.S. Brown. 1981. Measurement of mycorrhizal infection in soybeans. Soil Sci. Soc. Am. J. 45:871–874.

Bevege, D.I. 1968. A rapid technique for clearing tannins and staining intact roots for detection of mycorrhizas caused by *Endogone* spp., and some records of infection in Australian plants. Trans. Br. Mycol. Soc. 51:808–810.

Biermann, B.J., and R.G. Linderman. 1983. Effect of container plant growth medium and fertilizer phosphorus on establishment and host growth response to vesicular-arbuscular mycorrhizae. J. Am. Soc. Hortic. Sci. 108:962–971.

Brundrett, M.C., Y. Piche, and R.L. Peterson. 1984. A new method for observing the morphology of vesicular-arbuscular mycorrhizae. Can. J. Bot. 62:2128–2134.

Camel, S.B., M.G. Reyes-Solis, R. Ferrera-Cerrato, R.L. Franson, M.S. Brown, and G.J. Bethlenfalvay. 1991. Growth of vesicular-arbuscular mycorrhizal mycelium through bulk soil. Soil Sci. Soc. Am. J. 55:389–393.

Cochran, W.G. 1950. Estimation of bacterial densities by means of the "most probable number." Biometrics 6:105–116.

Daft, M.J., D. Spencer, and G.E. Thomas. 1987. Infectivity of vesicular-arbuscular mycorrhizal inocula after storage under various environmental conditions. Trans. Br. Mycol. Soc. 88:21–27.

Daniels, B.A., P.M. McCool, and J.A. Menge. 1981. Comparative inoculum potential of spores of six vesicular-arbuscular mycorrhizal fungi. New Phytol. 89:385–392.

Daniels, B.A., and J.A. Menge. 1980. Hyperparasitism of vesicular-arbuscular mycorrhizal fungi. Phytopathology 70:584–588.

de Man, J.C. 1975. The probability of most probable numbers. Eur. J. Appl. Microbiol. 1:67–78.

Dehne, H.W., and G.F. Backhaus. 1986. The use of vesicular-arbuscular mycorrhizal fungi in plant production. I. Inoculum production. Z. Pflanzenkr. Pflanzenschutz 93:415–424.

Dodd, J.C., and P. Jeffries. 1989. Effect of fungicides on three vesicular-arbuscular mycorrhizal fungi associated with winter wheat (*Triticum aestivum* L.) Biol. Fert. Soils 7:120–128.

Douds, D.D., and N.C. Schenck. 1990a. Increased sporulation of vesicular-arbuscular mycorrhizal fungi by manipulation of nutrient regimes. Appl. Environ. Microbiol. 56:413–418.

Douds, D.D., and N.C. Schenck. 1990b. Cryopreservation of spores of vesicular-arbuscular mycorrhizal fungi. New Phytol. 115:667–674.

Fang, Y.C., A.C. McGraw, H. Modjo, and J.W. Hendrix. 1983. A procedure for isolation of single-spore cultures of certain endomycorrhizal fungi. New Phytol. 95:107–114.

Ferguson, J.J., and J.A. Menge. 1982a. The influence of light intensity and artificially extended photoperiod upon infection and sporulation of *Glomus fasciculatum* on sudan grass and on root exudation of sudan grass. New Phytol. 92:183–191.

Ferguson, J.J., and J.A. Menge. 1982b. Factors that affect production of endomycorrhizal inoculum. Proc. Fla. State Hortic. Soc. 95:39–39.

Ferguson, J.J., and S.H. Woodhead. 1982. Increase and maintenance of vesicular-arbuscular mycorrhizal fungi. p. 47–54. *In* N.C. Schenck (ed.) Methods and principles of mycorrhizal research. Am. Phytopathol. Soc., St. Paul.

Fisher, R.A., and F. Yates. 1963. Statistical tables for biological, agricultural and medical research. Oliver and Boyd, Edinburgh.

Franson, R.L., and G.J. Bethlenfalvay. 1989. Infection unit method of vesicular-arbuscular mycorrhizal propagule determination. Soil Sci. Soc. Am. J 53:754–756.

Furlan, V., and J.A. Fortin. 1977. Effects of light intensity on the formation of vesicular-arbuscular endomycorrhizas on *Allium cepa* by *Gigaspora calospora*. New Phytol. 70:335–340.

Giovannetti, M., and B. Mosse. 1980. An evaluation of techniques for measuring vesicular arbuscular mycorrhizal infections in roots. New Phytol. 84:489–500.

Graham, J.H., D.M. Eissenstat, and E.L. Drouillard. 1992. On the relationship between a plants mycorrhizal dependency and rate of vesicular-arbuscular mycorrhizal colonization. Funct. Ecol. 5:773–779.

Graham, J.H., and D. Fardelmann. 1986. Inoculation of citrus with root fragments containing chlamydospores of the mycorrhizal fungus *Glomus intraradices*. Can. J. Bot. 64:1739–1744.

Graham, J.H., R.G. Linderman, and J.A. Menge. 1982. Development of external hyphae by different isolates of mycorrhizal *Glomus* spp. in relation to root colonization and growth of Troyer citrange. New Phytol. 91:183–189.

Halk, E.L., and S.H. De Boer. 1985. Monoclonal antibodies in plant-disease research. Annu. Rev. Phytopathol. 23:321–350.

Hall, I.R. 1979. Soil pellets to introduce vesicular-arbuscular mycorrhizal fungi into soil. Soil Biol. Biochem. 11:85–86.

Hall, I.R. 1980. Growth of *Lotus pedunculatus* Cav. in an eroded soil containing soil pellets infested with endomycorrhizal fungi. N. Z. J. Agric. Res. 23:103–105.

Hall, I.R., and A. Kelson. 1981. An improved technique for the production of endomycorrhizal infested soil pellets. N. Z. J. Agric. Res. 24:221–222.

Halvorson, H.O., and N.R. Ziegler. 1933. Applications of statistics to problems in bacteriology. I. A means of determining bacterial populations by the dilution method. J. Bacteriol. 25:101–121.

Hattingh, M.J., and J.W. Gerdemann. 1975. Inoculation of Brazilian sour orange seed with an endomycorrhizal fungus. Phytopathology 65:1013–1016.

Hayman, D.S. 1984. Methods for evaluating and manipulating vesicular-arbuscular mycorrhiza. p. 95–117. *In* J.M. Lynch and J.M. Grainger (ed.) Microbiological methods for environmental microbiology. Academic Press, London.

Hayman, D.S. 1987. VA mycorrhizas in field crop systems. p. 171–192. *In* G. Safir (ed.) Ecophysiology of VA mycorrhizal plants. CRC Press, Boca Raton, FL.

Hayman, D.S., E.J. Morris, and R.J. Page. 1981. Methods for inoculating field crops with mycorrhizal fungi. Ann. Appl. Biol. 99:247–253.

Hepper, C.M. 1977. A colorimetric method for estimating vesicular-arbuscular mycorrhizal infection in roots. Soil Biol. Biochem. 9:15–18.

Hepper, C.M. 1984. Isolation and culture of VA mycorrhizal (VAM) fungi. p. 95–112. *In* C.L. Powell and D.J. Bagyaraj (ed.) VA mycorrhiza. CRC Press, Boca Raton, FL.

Hoagland, D.R., and D.I. Arnon. 1950. The water-culture method for growing plants without soil. California Agric. Exp. Stn. Cir. 347 (revised edition), Berkeley.

Hung, L.L., D.M. O'Keefe, and D.M. Sylvia. 1991. Use of a hydrogel as a sticking agent and carrier of vesicular-arbuscular mycorrhizal fungi. Mycol. Res. 95:427–429.

Hung, L.L., and D.M. Sylvia. 1988. Production of vesicular-arbuscular mycorrhizal fungus inoculum in aeroponic culture. Appl. Environ. Microbiol. 54:353–357.

Hung, L.L., D.M. Sylvia, and D.M. O'Keefe. 1990. Isolate selection and phosphorus interaction of vesicular-arbuscular mycorrhizal fungi in biomass crops. Soil Sci. Soc. Am. J. 54:762–768.

Jabaji-Hare, S.H., and W.B. Kendrick. 1987. Response of an endomycorrhizal fungus in *Allium porrum* L. to different concentration of systemic fungicides, metalaxyl and fosetyl-Al. Soil Biol. Biochem. 19:95–99.

Jackson, N.E., R.E. Franklin, and R.H. Miller. 1972. Effects of vesicular-arbuscular mycorrhizae on growth and phosphorus content of three agronomic crops. Soil Sci. Soc. Am. Proc. 36:64–67.

Jakobsen, I., and A.J. Anderson. 1982. Vesicular-arbuscular mycorrhiza and growth in barley: effects of irradiation and heating of soil. Soil Biol. Biochem. 14:171–178.

Jarstfer, A.G., and R.M. Miller. 1985. Progress in the development of a chitin assay technique for measuring extraradical soilborne mycelium of V-A mycorrhizal fungi. p. 410. *In* R. Molina (ed.) Proc. 6th North American Conf. Mycorrhizae. Forest Res. Lab, Oregon State Univ., Corvallis.

Jarstfer, A.G., and D.M. Sylvia. 1992a. Inoculum production and inoculation strategies for vesicular-arbuscular mycorrhizal fungi. p. 349–377. *In* B. Metting (ed.) Soil microbial technologies: Applications in agriculture, forestry and environmental management. Marcel Dekker, New York.

Jarstfer, A.G., and D.M. Sylvia. 1992b. The production and use of aeroponically grown inocula of VAM fungi in the nursery. Florida Sea Grant Ext. Bull. 22, Gainesville.

Jeffries, P. 1987. Use of mycorrhizae in agriculture. CRC Crit. Rev. Biotechnol. 5:319–357.

Jeffries, P., and J.C. Dodd. 1991. The use of mycorrhizal inoculants in forestry and agriculture. p. 155–185. *In* D.K. Arora et al. (ed.) Handbook of applied mycology. Vol. I: Soil and plants. Marcel Dekker, New York.

Jensen, A. 1983. The effect of indigenous vesicular-arbuscular mycorrhizal fungi on nutrient uptake and growth of barley in two Danish soils. Plant Soil 70:155–163.

Johnson, B.N., and W.B. McGill. 1990. Comparison of ergosterol and chitin as quantitative estimates of mycorrhizal infection and *Pinus contorta* seedling response to inoculation. Can. J. For. Res. 20:1125–1131.

Johnson, C.R., W.M. Jarrell, and J.A. Menge. 1984. Influence of ammonium-nitrate ratio and solution pH on mycorrhizal infection, growth and nutrient composition of *Chrysanthemum morifolium* var circus. Plant Soil 77:151–157.

Kendrick, B., and S. Berch. 1985. Mycorrhizae: Applications in agriculture and forestry. p. 109–152. *In* C.W. Robinson (ed.) Comprehensive biotechnology. Vol. 4. Pergamon Press, Oxford.

Koide, R.T., and H.A. Mooney. 1987. Spatial variation in inoculum potential of vesicular-arbuscular mycorrhizal fungi caused by formation of gopher mounds. New Phytol. 107:173–182.

Kormanik, P.P., and A.-C. McGraw. 1982. Quantification of vesicular-arbuscular mycorrhizae in plant roots. p. 37–45. *In* N.C. Schenck (ed.) Methods and principles of mycorrhizal research. Am. Phytopathol. Soc., St. Paul.

Koske, R.E., and J.N. Gemma. 1989. A modified procedure for staining roots to detect VA mycorrhizas. Mycol. Res. 92:486–505.

Kough, J.L., and R.G. Linderman. 1986. Monitoring extra-matrical hyphae of a vesicular-arbuscular mycorrhizal fungus with an immunofluorescence assay and the soil aggregation technique. Soil Biol. Biochem. 18:309–313.

Liyanage, H.D. 1989. Effects of phosphorus nutrition and host species on root colonization and sporulation by vesicular-arbuscular (VA) mycorrhizal fungi in sand-vermiculite medium. M.S. thesis, Univ. of Florida, Gainesville.

Marsh, B. 1971. Measurement of length in random arrangements of lines. J. Appl. Ecol. 8:265–267.

Martin, F., C. Delaruelle, and J.L. Hilbert. 1990. An improved ergosterol assay to estimate fungal biomass in ectomycorrhizas. Mycol. Res. 94:1059–1064.

McGonigle, T.P., M.H. Miller, D.G. Evans, G.S. Fairchild, and J.A. Swan. 1990. A new method which gives an objective measure of colonization of roots by vesicular-arbuscular mycorrhizal fungi. New Phytol. 115:495–501.

Menge, J.A. 1982. Effect of soil fumigants and fungicides on vesicular-arbuscular fungi. Phytopathology 72:1125–1132.

Menge, J.A. 1983. Utilization of vesicular-arbuscular mycorrhizal fungi in agriculture. Can. J. Bot. 61:1015–1024.

Menge, J.A. 1984. Inoculum production. p. 187–203. *In* C.L. Powell and D.J. Bagyaraj (ed.) VA mycorrhiza. CRC Press, Boca Raton, FL.

Miller, D.D., M. Bodmer, and H. Schüepp. 1989. Spread of endomycorrhizal colonization and effects on growth of apple seedlings. New Phytol. 111:51–59.

Miller, R.M., A.G. Jarstfer, and J.K. Pillai. 1987. Biomass allocation in an *Agropyron smithii-Glomus* symbiosis. Am. J. Bot. 74:114–122.

Moorman, T., and F.B. Reeves. 1979. The role of endomycorrhizae in revegetation practices in the semi-arid West. II. A bioassay to determine the effect of land disturbance on endomycorrhizal populations. Am. J. Bot. 66:14–18.

Morton, J.B. 1985. Underestimation of most probable numbers of vesicular-arbuscular mycorrhizae endophytes because of non-staining mycorrhizae. Soil Biol. Biochem. 17:383–384.

Morton, J.B. 1988. Taxonomy of VA mycorrhizal fungi: Classification, nomenclature, and identification. Mycotaxon 32:267–324.

Morton, J.B., and G.L. Benny. 1990. Revised classification of arbuscular mycorrhizal fungi (Zygomycetes): A new order, Glomales, two new suborders, Glomineae and Gigasporineae, and two new families, Acaulosporaceae and Gigasporaceae, with an emendation of Glomaceae. Mycotaxon 37:471–491.

Mugnier, J., and B. Mosse. 1987. Vesicular-arbuscular mycorrhizal infection in transformed root-inducing T-DNA roots grown axenically. Phytopathology 77:1045–1050.

Nelsen, C.E., and G.R. Safir. 1982. Increased drought tolerance of mycorrhizal onion plants caused by improved phosphorus nutrition. Planta 154:407–413.

Nemec, S. 1980. Effects of eleven fungicides on endomycorrhizal development in sour oranges. Can. J. Bot. 58:522–527.

Newman, E.I. 1966. A method for estimating the total length of root in a sample. J. Appl. Ecol. 3:139–145.

O'Donnell, J.J., D.M. Sylvia, W.D. Pitman, and J.E. Rechcigl. 1992. Inoculation of *Vigna parkeri* with mycorrhizal fungi in an acid Florida spodosol. Trop. Grassl. 26:120–129.

O'Keefe, D.M., and D.M. Sylvia. 1991. Mechanisms of the vesicular-arbuscular mycorrhizal plant-growth response. p. 35–54. *In* D.K. Arora et al. (ed.) Handbook of applied mycology. Marcel Dekker, New York.

Ojala, J.C., and W.M. Jarrell. 1980. Hydroponic sand culture systems for mycorrhizal research. Plant Soil 57:297–303.

Paulitz, T.C., and J.A. Menge. 1986. The effect of a mycoparasite on the mycorrhizal fungus, *Glomus deserticola*. Phytopathology 76:351–354.

Phillips, J.M., and D.S. Hayman. 1970. Improved procedures for clearing roots and staining parasitic and vesicular-arbuscular mycorrhizal fungi for rapid assessment of infection. Trans. Br. Mycol. Soc. 55:158–161.

Plenchette, C., J.A. Fortin, and V. Furlan. 1983. Growth response of several plant species to mycorrhizae in a soil of moderate P-fertility. I. Mycorrhizal dependency under field conditions. Plant Soil 70:199–209.

Plenchette, C., V. Furlan, and J.A. Fortin. 1982. Effects of different endomycorrhizal fungi on five host plants grown on calcined montmorillonite clay. J. Am. Soc. Hortic. Sci. 107:535–538.

Porter, W.M. 1979. The 'most probable number' method for enumerating infective propagules of vesicular-arbuscular mycorrhizal fungi in soil. Aust. J. Soil Res. 17:515–519.

Porter, W.M., A.D. Robson, and L.K. Abbott. 1987. Factors controlling the distribution of VAM fungi in relation to soil pH. J. Appl. Ecol. 24:663–672.

Powell, C. Ll., and D.J. Bagyaraj (ed.) 1984. VA mycorrhiza. CRC Press, Boca Raton, FL.

Reeves, F.B., D. Wagner, T. Moorman, and J. Kiel. 1979. Role of endomycorrhizae in revegetation practices in the semi-arid west. 1. Comparison of incidence of mycorrhizae in severely disturbed versus natural environments. Am. J. Bot. 66:6–13.

Ride, J.P., and R.B. Drysdale. 1971. A chemical method for estimating *Fusarium oxysporum* f. *lycopersici* in infected tomato plants. Physiol. Plant Pathol. 1:409–420.

Rosendahl, S., R. Sen. C.M. Hepper, and C. Azcon-Aguilar. 1989. Quantification of three vesicular-arbuscular mycorrhizal fungi (*Glomus* spp.) in roots of leek (*Allium porrum*) on the basis of activity of diagnostic enzymes after polyacrylamide gel electrophoresis. Soil Biol. Biochem. 21:519–522.

Salmanowicz, B., J.-E. Nylund, and H. Wallander. 1989. High performance liquid chromatography-assay of ergosterol: A technique to estimate fungal biomass in roots with ectomycorrhiza. Agric. Ecosystem Environ. 28:437–440.

Schenck, N.C. (ed.). 1982. Methods and principles of mycorrhizal research. The Am. Phytopathol. Soc., St. Paul.

Schenck, N.C., and Y. Perez. 1990. Isolation and culture of VA mycorrhizal fungi. p. 237–258. In D.P. Labeda (ed.) Isolation of biotechnological organisms from nature. McGraw Hill Publ. Co., New York.

Schenck, N.C., and Y. Perez. 1994. Identification procedures. Chapter 4. In Methods manual for VA mycorrhizae Synergistic Publ., Gainesville, FL. (In press.)

Schenck, N.C., and V.N. Schroder. 1974. Temperature response of Endogone mycorrhiza on soybean roots. Mycologia 66:600–605.

Schenck, N.C., and G.S. Smith. 1982. Responses of six species of vesicular-arbuscular mycorrhizal fungi and their effects on soybean at four soil temperatures. New Phytol. 92:193–201.

Schubert, A., C. Marzachi, M. Mazzitelli, M.C. Cravero, and P. Bonfante-Fasolo. 1987. Development of total and viable extraradical mycelium in the vesicular-arbuscular mycorrhizal fungus Glomus clarum Nicol. & Schenck. New Phytol. 107:183–190.

Schüepp, H., D.D. Miller, and M. Bodmer. 1987. A new technique for monitoring hyphal growth of vesicular-arbuscular mycorrhizal fungi through soil. Trans. Br. Mycol. Soc. 89:429–435.

Schwab, S.M., and F.B. Reeves. 1981. The role of endomycorrhizae in revegetation practices in the semi-arid west. III. Vertical distribution of vesicular-arbuscular mycorrhizal inoculum potential. Am. J. Bot. 68:1293–1297.

Simon, L., M. Lalonde, and T.D. Bruns. 1992. Specific amplification of 18S fungal ribosomal genes from vesicular-arbuscular endomycorrhizal fungi colonizing roots. Appl. Environ. Microbiol. 58:291–295.

Sreenivasa, M.N., and D.J. Bagyaraj. 1988. Selection of a suitable host for mass multiplication of Glomus fasciculatum. Plant Soil 109:125–127.

Sreenivasa, M.N., and D.J. Bagyaraj. 1989. Use of pesticides for mass production of vesicular-arbuscular mycorrhizal inoculum. Plant Soil 119:127–132.

Strullu, D.G., C. Romand, P. Callac, E. Teoule, and Y. Demarly. 1989. Mycorrhizal synthesis in vitro between Glomus spp. and artificial seeds of alfalfa. New Phytol. 113:545–548.

Sylvia, D.M. 1988. Activity of external hyphae of vesicular-arbuscular mycorrhizal fungi. Soil Biol. Biochem. 20:39–43.

Sylvia, D.M. 1992. Quantification of external hyphae of vesicular-arbuscular mycorrhizal fungi. p. 53–65. In A.K. Varma et al. (ed.) Methods in microbiology: Experiments with mycorrhizae. Academic Press, New York.

Sylvia, D.M., and J.N. Burks. 1988. Selection of a vesicular-arbuscular mycorrhizal fungus for practical inoculation of Uniola paniculata. Mycologia 80:565–568.

Sylvia, D.M., and D.H. Hubbell. 1986. Growth and sporulation of vesicular-arbuscular mycorrhizal fungi in aeroponic and membrane systems. Symbiosis 1:259–267.

Sylvia, D.M., and A.G. Jarstfer. 1992. Sheared-root inocula of vesicular-arbuscular mycorrhizal fungi. Appl. Environ. Microbiol. 58:229–232.

Sylvia, D.M., and L.H. Neal. 1990. Nitrogen affects the phosphorus response of VA mycorrhiza. New Phytol. 115:303–310.

Sylvia, D.M., and N.C. Schenck. 1983. Application of superphosphate to mycorrhizal plants stimulates sporulation of phosphorus-tolerant vesicular-arbuscular mycorrhizal fungi. New Phytol. 95:655–661.

Sylvia, D.M., and N.C. Schenck. 1984. Aerated-steam treatment to eliminate VA mycorrhizal fungi from soil. Soil Biol. Biochem. 16:675–676.

Thompson, J.P. 1986. Soilless cultures of vesicular-arbuscular mycorrhizae of cereals: Effects of nutrient concentration and nitrogen source. Can. J. Bot. 64:2282–2294.

Thompson, J.P. 1990. Soil sterilization methods to show VA-mycorrhizae aid P and Zn nutrition of wheat in vertisols. Soil Biol. Biochem. 22:229–240.

Thomson, B.D., A.D. Robson, and L.K. Abbott. 1986. Effects of phosphorus on the formation of mycorrhizas by Gigaspora calospora and Glomus fasciculatum in relation to root carbohydrates. New Phytol. 103:751–765.

Tommerup, I.C., and G.G. Briggs. 1981. Influence of agricultural chemicals on germination of vesicular-arbuscular endophyte spores. Trans. Br. Mycol. Soc. 76:326–328.

Toth, R., and D. Toth. 1982. Quantifying vesicular-arbuscular mycorrhizae using a morphometric technique. Mycologia 74:182–187.

Tuite, J. 1969. Plant pathological methods, fungi and bacteria. Burgess Publ. Co., Minneapolis, MN.

Varma, A.K., J.R. Norris, and D.J. Read (ed.). 1992. Methods in microbiology: Experiments with mycorrhizae. Vol. 24. Academic Press Limited, London.

Warner, A., and B. Mosse. 1983. Spread of vesicular-arbuscular mycorrhizal fungi between separate root systems. Trans. Br. Mycol. Soc. 80:353–354.

Waterer, D.R., and R.R. Coltman. 1988a. Phosphorus concentration and application interval influence growth and mycorrhizal infection of tomato and onion transplants. J. Am. Soc. Hortic. Sci. 5:704–708.

Waterer, D.R., and R.R. Coltman. 1988b. Effects of controlled-release phosphorus and inoculum density on the growth and mycorrhizal infection of pepper and leek transplants. HortScience 23:620–622.

Williams, P.G. 1990. Disinfecting vesicular-arbuscular mycorrhizas. Mycol. Res. 94:995–997.

Wilson, J.M., and M.J. Trinick. 1982. Factors affecting the estimation of numbers of infective propagules of VAMF by the most probable number method. Aust. J. Soil Res. 21:73–81.

Wolf, D.C., T.H. Dao, H.D. Scott, and T.L. Lavy. 1989. Influence of sterilization methods on selected soil microbiological, physical, and chemical properties. J. Environ. Qual. 18:39–44.

Wright, S.F., J.B. Morton, and J.E. Sworobuk. 1987. Identification of a vesicular-arbuscular mycorrhizal fungus by using monoclonal antibodies in an enzyme-linked immunosorbent assay. Appl. Environ. Microbiol. 53:2222–2225.

Zobel, R.W., P. Del Tredici, and J.G. Torrey. 1976. Method for growing plants aeroponically. Plant Physiol. 57:344–346.

Chapter 19

Isolation of Microorganisms Producing Antibiotics

JEFFRY J. FUHRMANN, *University of Delaware, Newark, Delaware*

The soil has long been recognized as a reservoir for microorganisms capable of producing a variety of antimicrobial substances. The pioneering work of Selman A. Waksman unequivocally established soil actinomycetes as prominent producers of antibiotics (Waksman, 1967). Voluminous research has revealed that many other microbial groups produce compounds having a variety of negative effects on genetically similar or dissimilar microorganisms. The types of substances represented are diverse, and include "classical" antibiotics having low molecular weights, proteinaceous bacteriocins, iron-sequestering siderophores, lytic enzymes, and volatile compounds. Although some of these compounds are of interest for their potential therapeutic value, essentially all warrant study for their possible ecological roles in soil-plant systems. With respect to the latter, particular attention has been directed toward the modifying influences of antagonistic microorganisms on phytopathogens (Fravel, 1988; Leong, 1986; Weller, 1988) and members of the Rhizobiaceae (Barnet et al., 1988; Parker et al., 1977).

Given the diversity of microorganisms and compounds represented, the development of standard assays for the detection of antibiosis is a difficult task. Numerous types of assays have been described in the literature, each accompanied by variations developed to meet unique experimental objectives and conditions. The methods described in this chapter convey the basic principles and procedures appropriate to each general approach, and draw attention to modifications having widespread application. Emphasis will be placed on procedures applicable to bacteria (including actinomycetes) and fungi. The reader is encouraged to consult the literature for additional information related to the particular organisms and experimental systems under consideration.

19–1 GENERAL PRINCIPLES

19–1.1 Preface

It is generally conceded that in vitro studies for antibiosis do not reliably predict microbial interactions in natural soil environments (Fravel, 1988; Weller, 1988; Whipps, 1987). It is, therefore, recommended that the rationale for conducting such studies be clearly established prior to their initiation. Circumstances commonly cited for initiating attempts to isolate antibiotic-producing microorganisms include pre-existing studies establishing the presence of antagonists within defined taxonomic or physiological groups of microorganisms, evidence for high frequencies of antagonistic organisms within specific niches or soil types (e.g., suppressive soils), and coincidental observations suggesting antagonistic interactions are operative within a particular experimental system.

19–1.2 Sampling Strategies

No definitive guidelines can be given for obtaining environmental samples containing a high proportion of antagonistic microorganisms. This is largely due to a persistent lack of knowledge regarding the functional distribution of such organisms in the soil environment (Labeda & Schearer, 1990; Williams & Vickers, 1986). However, there are instances in which research objectives logically suggest certain sampling strategies. Such a situation exists when one wishes to identify microorganisms for the biocontrol of certain phytopathogens or which are particularly suited to a given ecological niche. In the former case, successes have been obtained by isolating from disease suppressive soils known to contain the pathogen (Weller et al., 1988), or from the surface of naturally occurring or introduced fungal tissues such as sclerotia (Utkhede, 1984). Similarly, in searching for antagonistic rhizobacteria suited to colonizing a particular plant species, it is logical to obtain isolations from roots of that species taken from various geographical locations (Lambert et al., 1987). In many instances, however, the successful approach has simply been to sample from diverse soils or vegetative types. Potent antagonists have also been isolated from several unusual habitats, including soil insects (Attafuah & Bradbury, 1989).

19–1.3 Selection of Isolation Method

Microorganisms to be examined for production of antimicrobial substances are commonly referred to as *antagonists* or *producers;* these terms will be applied to both potential and proven antagonists. Conversely, those microorganisms being examined for sensitivity to antagonists are referred to as *test organisms* or *indicator organisms*.

Antagonists can be detected by either (i) first randomly isolating microorganisms into pure culture and subsequently testing each for activity against indicator organisms in a systematic manner (sections 19–4 and 19–5), or (ii) using a screening technique designed to detect putative antagonists prior to isolation and subsequent retesting (section 19–6). The choice between these two approaches in dictated primarily by the experimental objectives and the cultural characteristics of the organisms involved (Johnson & Curl, 1972).

Use of previously isolated organisms is the more flexible approach and necessary when the culture media or conditions appropriate for isolating the potential antagonists do not permit proper growth of the test organism(s). Strategies and procedures for the isolation of specific types of microorganisms are described elsewhere in chapter 8 of this volume.

Screening techniques have the potential for relatively high isolation efficiencies for antagonists. This benefit is best realized in focused studies, such as when only one test organism is under consideration or where there is preexisting information regarding the optimal conditions for production of specific antimicrobial compounds of interest (e.g., siderophores under iron-limited conditions). Instances where this strategy has been successful are described below (section 19–6).

19–1.4 Selection of Indicator Organisms

In many cases, the choice of indicator organism(s) to be used will be dictated by the objectives of the particular study at hand. In the remaining instances, however, some thought must be given to the selection of organisms to be used as indicators. Microorganisms differ in their susceptibilities to various antibiotic substances, and this can considerably influence the experimental results (Fravel, 1988). Assays should be conducted with more than one indicator organism, or with a known, highly susceptible genotype, if the maximum number of putative antagonists are desired. Alternatively, use of a resistant indicator genotype will enhance identification of strong antagonists and possibly reduce the number of false positives detected.

19–1.5 Assay Standardization

An assay must be carefully standardized if it is to provide reproducible and unambiguous results (Piddock, 1990). In addition to the environmental conditions used to incubate the assay cultures, it is important to control the methods employed to prepare media and inocula. Many studies have shown that small variations in media composition can influence microbial interactions (Fravel, 1988). Agar plates should be of uniform depth and age, particularly if quantitative measurements of inhibition will be made (Vidaver et al., 1972). Inocula should be cultured and processed under defined conditions to ensure that they are of a reasonably similar physiological state when used. Standardization of cell numbers is especially important, because inhibition of the test organism is generally inversely related to its initial population density (Piddock, 1990). Additionally, the growth stage of the putative producer should be carefully monitored because production of antibiotic substances generally peaks following logarithmic growth (Demain et al., 1983; Katz & Demain, 1977), although exceptions to this general rule are well documented (Katz & Demain, 1977; Tagg et al., 1976).

19–2 MICROBIOLOGICAL MEDIA

19–2.1 General Comments

Media composition strongly affects antibiosis by modifying microbial growth, differentially altering antibiotic synthesis and inactivation, and affecting the physiological sensitivity of microorganisms to particular antibiotics (Tagg et al., 1976). This situation is made more complex by the fact that one is often trying to detect antibiotics of unknown identity and biosynthetic requirements (Fravel, 1988).

Given the above uncertainties, it is strongly recommended that more than one medium be used in assaying for antibiosis. Use of several media may also help identify artifacts caused by such factors as growth-induced media acidification or nutrient depletion (Tagg et al., 1976). Described below are media that have commonly been employed in antibiosis assays, and which are recommended in cases where guidance regarding media composition is unavailable. However, the reader is encouraged to consult the literature for additional media that may be better suited to particular applications. Additionally, new media formulations may be appropriately used in certain situations, although it is recommended that a common standard medium also be included for comparison. Media components that have been identified as enhancing antibiosis include glycerol (Axelrood et al., 1988) and other soluble C sources (Gross & Vidaver, 1978), insoluble C sources (e.g., carboxymethyl cellulose) (Brewer et al., 1987), plant extracts (Fuhrmann & Wollum, 1989; Weinhold & Bowman, 1968), and various N sources (Kanner et al., 1978) including amino acids and peptones (Brewer et al., 1987). Glucose concentration in media has been shown to differentially regulate production of antibiotic substances produced by *Pseudomonas fluorescens* (James & Gutterson, 1986).

19–2.2 Common Media for Antibiosis Assays

Unless otherwise noted, all media constituents are combined and sterilized by autoclaving at 121 °C for 15 min. Agar is omitted for preparing liquid media. Dehydrated preparations of many of these media are available commercially and often used for antibiosis assays. Other media of a more specialized nature are described separately in subsequent sections. See Dhingra and Sinclair (1985) and Johnson and Curl (1972) for additional media.

1. **Corn meal agar:** Used for the culture of fungi and co-culture of bacteria and fungi. Place 40 g of corn meal in 1000 mL of distilled water and keep at 58 °C for 1 h. Clarify by filtration or centrifugation. Add 15 g agar.
2. **Czapek Dox agar:** Used for the culture of fungi and the co-culture of fungi and bacteria. Combine 3 g sodium nitrate ($NaNO_3$), 1 g potassium monohydrogen phosphate (K_2HPO_4), 0.5 g potassium chloride (KCl), 0.5 g magnesium sulfate heptahydrate ($MgSO_4 \cdot 7H_2O$), and 1000 mL water. Once all components are dissolved, add 0.1 g ferrous sulfate ($FeSO_4$) and 15 g agar. Add 30 g sucrose just prior to sterilization.
3. **Glucose nutrient agar:** Used for the culture of bacteria and co-culture of bacteria and fungi. Combine 5 g peptone, 5 g glucose, 3 g beef extract, 1 g yeast extract, and 15 g agar with 1000 mL distilled water.
4. **Glycerol arginine agar** (Porter et al., 1960): Used for the culture of actinomycetes. Combine 20 g glycerol, 2.5 g L-arginine, 1 g sodium chloride (NaCl), 0.1 g calcium carbonate ($CaCO_3$), 0.1 ferrous sulfate heptahydrate ($FeSO_4 \cdot 7H_2O$), 0.1 g magnesium sulfate heptahydrate ($MgSO_4 \cdot 7H_2O$), 20 g agar, and 1000 mL distilled water. Adjust pH to 7.0.
5. **King's B agar** (King et al., 1954): An iron-deficient medium used for the culture of bacteria and the co-culture of bacteria and fungi. Commonly used for the detection of bacterial siderophores (see section 19–7.3). Combine 20 g proteose peptone No. 3 (Difco), 10 g glycerol, 2.5 g potassium monohydrogen phosphate (K_2HPO_4), 6 g magnesium sulfate heptahydrate ($MgSO_4 \cdot 7H_2O$), 15 g agar, and 1000 mL water. Adjust pH to 7.2. Some investigators reduce the inorganic salts to 1.5 g each (e.g., Misaghi et al., 1982).
6. **Malt extract agar:** Used for the culture of fungi and co-culture of bacteria and fungi. Dissolve 20 g of malt extract in 1000 mL of distilled water with heating, adjust pH to 6.5 (optional), and add 20 g agar.
7. **Potato dextrose agar:** Used for the culture of fungi and co-culture of bacteria and fungi. Boil 200 g of peeled and sliced potatoes in 500 mL distilled water for 1 h. Clarify by filtration or centrifugation. To the filtrate, add 20 g glucose, distilled water to make 1000 mL, and 15 g agar.
8. **Soybean-Casein Digest Agar:** Used for the culture of bacteria and the co-culture of bacteria and fungi. Also known as Tryptic Soy Agar (Difco) and Trypticase Soy Agar (BBL). Combine 17 g pancreatic digest of casein, 3 g papaic digest of soy meal, 5 g of sodium chloride (NaCl), 2.5 g potassium monohydrogen phosphate (K_2HPO_4), 2.5 g glucose, 15 g agar, and 1000 mL distilled water.
9. **Yeast extract-dextrose-carbonate agar** (Wilson et al., 1967): Used for the culture of bacteria (especially for bacteriocin production).

Combine 10 g yeast extract, 15 g agar, 20 g $CaCO_3$ powder, and 800 mL water. Autoclave and add 200 mL of an autoclaved solution containing 100 g glucose L^{-1}. Cool the mixture to 50 °C, stir to disperse the $CaCO_3$.

10. **Yeast extract-mannitol agar** (Weaver & Frederick, 1982): Used for the culture of rhizobia. Combine 10 g mannitol, 0.5 g potassium monohydrogen phosphate (K_2HPO_4), 0.5 g yeast extract, 0.2 g magnesium sulfate heptahydrate ($MgSO_4 \cdot 7H_2O$), 0.01 g calcium carbonate ($CaCO_3$) (optional for *Bradyrhizobium*), 15 g agar, and 1000 mL water. Adjust pH to 7.0.

19–3 PREPARATION OF INOCULA

19–3.1 General Comments

Inocula should ideally be cultured in the same medium to be used for the antibiosis assay, although this is probably not necessary for routine tests. For critical studies or when the inocula will be tested with several different media, it is recommended that broth cultures be washed by centrifugation and resuspended in sterile water or other diluent of choice prior to standardization and use.

19–3.2 Bacterial Inocula

Inocula of non-filamentous bacteria are generally most efficiently and reproducibly prepared in broth culture. Cultures are grown to late-logarithmic or early stationary growth phase, which for most bacteria will produce a cell density of 10^9 mL^{-1} (absorbance at 600 nm ≈ 0.5 [1 cm light path]). Depending upon the intended application, these suspensions may be used directly but are generally diluted 10-fold with sterile distilled water or other diluent. If growth differs greatly among isolates, cell concentrations can be standardized using either direct microscopic counts or spectrophotometric methods.

Actinomycetes or other bacteria producing particulate or filamentous growth in broth culture may require specialized procedures. If cultured in broth media, the bacteria can generally be dispersed by vigorous agitation or treatment in a blender for approximately 30 s using low speed. Alternatively, conidia can be collected from sporulating lawns on agar plates by adding sterile water and dispersing the spores with a sterile rubber spatula. Remaining large clumps may be removed by sieving through sterile cheesecloth or screens. Suspensions should be standardized as described for non filamentous bacteria. Actinomycetes can also be transferred to test plates as disks cut from agar lawns by using a sterile cork hole borer (generally 5–10 mm diam.) (Patel, 1974).

19–3.3 Fungal Inocula

Depending on the assay procedure to be used, fungal inocula are applied either as propagule suspensions (spores or mycelial fragments) or as agar disks. Propagule suspensions are prepared using either the broth culture or agar lawn procedure described for actinomycetes. A final suspension density of 10^6 to 10^7 colony forming units (CFU) mL^{-1} is suitable for most assays. Agar disks are prepared by centrally inoculating an agar plate with a fungus and subsequently cutting 5- to 10-mm diam. disks from the periphery of the developing colony (when approximately 6–8 cm diam.) with a sterile cork hole borer.

19–4 DUAL CULTURE DETECTION METHODS

19–4.1 Introduction

The unifying characteristic of these techniques is that the indicator organism(s) and potential antagonist(s) are cultured simultaneously, at least for part of the duration of the assay, on an agar medium. Furthermore, it is assumed that pure cultures of the microorganisms of interest have been obtained and inocula have been prepared as described above.

19–4.2 Microbial Lawn Technique

19–4.2.1 Principles

This method is suited to the study of all combinations of bacteria and fungi as indicator organisms and potential antagonists. Inoculum of the indicator organism is evenly distributed on or within an agar medium contained in a petri dish, such that a uniformly dense lawn of the organism develops in the absence of antagonism. Potential antagonists are spotted onto localized areas of the plate surface so as to produce large colonies upon subsequent growth (hence, the "giant colony" technique of Robison, 1945). Antagonism is apparent as a zone of reduced or absent growth immediately surrounding the antagonistic colony.

The primary method described below involves spotting or streaking the potential antagonist(s) on the agar medium and then uniformly spraying a suspension of the test organism on the agar surface, generally following a suitable incubation period (e.g., Axelrood et al., 1988; Misaghi et al., 1982).

19–4.2.2 Materials

1. Petri dishes (90 mm diam.) containing 15 mL of agar medium suitable for the growth of all organisms under consideration (larger volumes may obscure inhibition zones; the agar surface should be dry).

2. Inocula of indicator and potential antagonists.
3. Apparatus for spraying indicator inoculum (glass units for spraying chromatography reagents are well suited, autoclavable, but relatively expensive; surface-sterilized plastic squeeze- or pump-type sprayers are also suitable for many applications; see Stansly (1947) to construct a specialized sprayer).
4. Sterile micropipet or inoculation loop capable of reproducibly delivering small volumes ($\leq$15 μL).

19–4.2.3 Procedures

Inocula of potential antagonists are applied to the surface of duplicate agar plates as localized spots or steaks. Generally, each application consists of about 10 μL of a suspension containing a total of approximately 10^6 bacterial or 10^4 fungal propagules. Fungi to be tested as potential antagonists are sometimes applied as agar disks and are commonly limited to one per plate, whereas one to several bacterial cultures are often tested as antagonists on a single plate.

Once applied to the agar plate, the potential antagonists are usually incubated for a period to permit colony development and antibiotic production prior to being oversprayed with the indicator organism. The optimum incubation time depends on the growth rates of the microorganisms involved. Fast- and slow-growing antagonists are generally incubated for 24 to 96 h and 7 to 14 d, respectively. It may be desirable to spray the plates immediately after applying the potential antagonists in cases where their growth is particularly rapid or the test organism develops slowly. Bacterial and fungal densities in spray suspensions are typically about 10^8 and 10^6 CFU mL^{-1}, respectively. Heavier applications of the test organism should be avoided as these may partially obscure zones of inhibition; in some cases, reduced cell densities of the indicator organism may enhance antibiotic detection (Piddock, 1990).

Following a suitable incubation period, the plates are examined for colonies surrounded by inhibition zones. Each inhibition zone should be rated for the presence of microbial growth (i.e., clear or turbid). The presence of any resistant clones within the inhibition zone should also be noted. Measurements of inhibition zones are made from the edge of the antagonistic colony to the point of normal growth of the indicator organism.

19–4.2.4 Variations

In situations where aerosol production must be minimized or strict control of microbial numbers is needed, measured amounts of indicator inocula may be incorporated directly into the agar medium. Described below are the more common modifications of this technique.

1. The indicator organism is added to 15 mL of molten (about 45 °C) sterile agar to a density of approximately 10^6 (bacteria) or 10^4

(fungi) per milliliter. The agar is poured into a petri dish, allowed to solidify, and potential antagonists are applied as described above (e.g., Angle et al., 1981; Trinick & Parker, 1982).

2. Petri plates containing 10 mL of noninoculated agar medium are prepared in the usual manner. The solidified base agar is then overlaid with 5 mL of molten (about 45 °C) sterile agar containing indicator inocula at a density of approximately 10^6 bacteria or 10^4 fungi per milliliter. Potential antagonists are applied as described above. Although requiring one more operation than the first variation, this modification often produces more distinct inhibition zones due to the thinner indicator layer employed (e.g., Patel, 1974; Schwinghamer, 1971).

3. In cases where the potential antagonist displays an extremely rapid growth rate relative to the indicator organism, it may be helpful to use a deferred method in which growth of the antagonist is prevented beyond what is necessary for colony development. This can be accomplished by using a modification of the chloroform fumigation technique described for the detection of bacteriocins (section 19–7.2) (e.g., Fuhrmann & Wollum, 1989). Briefly, agar plates are prepared and spotted with potential antagonists as described for the primary procedure. Following a suitable incubation period, the colonies are killed by exposure to chloroform vapor and then overlaid with 5 mL of a agar medium seeded with indicator inocula. Inhibition zones are described and measured following an additional incubation.

19–4.3 Fungal Disk Technique

19–4.3.1 Principles

This technique exploits the ability of filamentous fungi to spread and completely colonize an agar plate in the absence of antagonists. It is commonly used for assessing bacterial inhibition of fungi as well as interactions between fungal genotypes. Antagonism is observed as absent, reduced, or abnormal (e.g., matted or prostrate) fungal growth in the vicinity of the co-cultured organism.

19–4.3.2 Materials

1. Petri dishes (90 mm diam.) containing 15 mL of agar medium suitable for the growth of all organisms under consideration.
2. Inocula of indicator organisms and potential antagonists (bacteria as suspensions; fungi as agar disks).
3. Sterile micropipet or inoculation loop capable of reproducibly delivering small volumes (≤ 15 µL).

19–4.3.3 Procedures

For studies of bacterial inhibition of fungi, aliquots of the potential antagonists containing approximately 10^6 CFU are uniformly spotted near the periphery of the plate (generally four or fewer spots per plate). The spotted organisms are incubated according to the guidelines described for the Microbial Lawn Technique. A fungal disk is then placed in the center of the plate and incubation is continued. Inhibition zones are described and measured once the fungal colony is sufficiently developed for observations to be made (e.g., Howell & Stipanovic, 1980; Thomashow & Weller, 1988).

Inhibition among fungi is examined by simultaneously placing disks of two or more genotypes on separate areas of an agar plate. The distances separating the disks should be standardized; 5 cm is a commonly used value. The plates are then incubated and observed periodically for interactions among the developing colonies (e.g., Manandhar et al., 1987).

19–4.4 Cross Streak Technique

19–4.4.1 Principles

This technique is generally limited to the study of antagonistic interactions among bacteria (including actinomycetes). Potential antagonists are streaked in one direction on an agar medium. The test organism is subsequently streaked perpendicularly to the antagonist. An incubation period generally separates the two operations. Antagonism is apparent as absent or reduced growth of the indicator organism near the antagonist after a final incubation period.

When compared with preceding methods, the cross streak technique is labor intensive and possibly more subject to variations in inoculation application rates. However, this method has the compensating advantage of maximum control over the relative timing of inoculation between antagonists and test organisms. Additionally, it permits more than one indicator organism to be tested with a potential antagonist on a single culture plate.

19–4.4.2 Materials

1. Petri dishes (90 mm diam.) containing 15 mL of agar medium suitable for the growth of all organisms under consideration (the agar surface should be dry).
2. Inocula of indicator organisms and potential antagonists (both as cell suspensions).
3. Inoculation loops capable of delivering reproducible volumes of inocula.

19–4.4.3 Procedures

A suspension of the potential antagonist (about 10^8 CFU mL^{-1}) is streaked in one direction on a agar medium either along one side of the plate or so as to divide the plate into two halves. The applied organism is incubated as described for the Microbial Lawn Technique. A cell suspension of the test organism is streaked perpendicularly to the potential antagonist beginning at a standard distance from the former streak. Streaking away from the suspected producer ensures that an apparently antagonistic interaction is not an artifact resulting from a thinning of the test inoculum near the antagonist during the streaking process. Inhibition zones are described and measured following a suitable incubation period (e.g., Smith & Miller, 1974; Trinick & Parker, 1982).

19–5 CULTURE FILTRATE METHODS

19–5.1 Introduction

These methods differ from the previous techniques in that culture of the potential antagonist and test organism are separated in time and space. The potential antagonist is cultured in a broth medium, and a cell-free filtrate of the culture supernatant is prepared. The filtrate is subsequently tested for its ability to inhibit the growth of the test organism on either liquid or agar media. Although cumbersome when compared with the Dual Culture Detection Techniques (section 19–4), these methods have the advantage of minimizing the possibility of artifacts arising from factors such as media acidification and nutrient depletion.

19–5.2 Preparation of Culture Filtrates

19–5.2.1 Materials

1. Broth media prepared in erlenmeyer flasks (media additives inhibitory to the test organism must be avoided).
2. Pure cultures of potential antagonists.
3. Centrifuge.
4. Sterile membrane filter apparatus (0.2 or 0.45 μm).
5. Rotary evaporator or freeze dryer (optional).

19–5.2.2 Procedures

In the simplest case, potential antagonists are cultured in one or more broth media, generally well into the stationary phase of growth. This is because many antibiotic compounds are secondary growth products that are excreted into the medium only after the culture enters stationary growth. Most of the cells are removed by centrifugation and the acidity of supernatant is checked and adjusted to the desired level (typically neutrality). The supernatant is then sterilized by passage through a membrane filter and tested for inhibitory activity.

In some instances, it may be desirable to concentrate the supernatant obtained by centrifugation prior to sterilization and testing. A rotary evaporator operated at moderate temperatures (40–60 °C) is commonly used for this purpose, although freeze drying is better suited for use with thermolabile compounds. The supernatant may be either simply reduced in volume by some constant amount or brought to dryness and redissolved in a solvent of choice. Various extraction procedures have also been employed but will not be discussed here (e.g., Cubeta et al., 1985; Dunlop et al., 1989; Howell & Stipanovic, 1980; James & Gutterson, 1986; Thomashow & Weller, 1988).

19–5.3 Paper Disk Technique

19–5.3.1 Principles

This method is conducted as a modified Microbial Lawn or Fungal Disk Technique in which sterile paper disks, moistened with culture filtrates, substitute for colonies of potential antagonists. It is suited to the study of inhibitions between all combinations of bacteria and fungi.

19–5.3.2 Materials

1. Petri dishes (90 mm diam.) containing 15 mL of agar medium suitable for the growth of the test organism (the agar surface should be dry).
2. Inocula of test organism(s).
3. Sterile culture filtrates of potential antagonists prepared as described above.
4. Sterile, dry paper disks (1-cm diam. filter paper disks are commonly used; blank paper disks used for antibiotic sensitivity tests are also available; acceptable disks may be cut from filter paper using a paper punch).

19–5.3.3 Procedures

The paper disks are moistened with the culture filtrates either by dipping the dry disks in the filtrate and then draining or by applying a standard volume to each disk with a sterile micropipet. Using sterile forceps, the impregnated disks are arranged on the noninoculated agar surface. Alternatively, agar plates seeded with the indicator organism may be employed (section 19–4.2.4). Control disks moistened with fresh/concentrated medium should also be included. Nonseeded plates are subsequently sprayed with the test organism or a fungal disk is placed on the plate. It is generally not necessary to delay application of the indicator organism. Inhibition zones are described and measured following incubation (e.g., Angle et al., 1981; Anusuya & Sullia, 1984; Robison, 1945).

19–5.4 Agar Well Technique

19–5.4.1 Principles

This method is analogous to the Paper Disk Technique except that the culture filtrates are applied directly to wells cut in the agar medium with a sterile cork hole borer.

19–5.4.2 Materials

1. Petri dishes (90 mm diam.) containing a standard amount (15–25 mL) of agar medium suitable for the growth of the test organism.
2. Inocula of test organism(s).
3. Sterile culture filtrates of potential antagonists prepared as described above.
4. Sterile micropipet.
5. Cork hole borer (5–10 mm diam.).
6. Molten agar media (optional).

19–5.4.3 Procedures

Indicator lawns are initiated by seeding the agar medium or spraying with the test organism. Wells are then formed in the agar medium by removing disks cut with a flamed cork hole borer. Some investigators seal the bottom of the wells with a drop of molten agar medium (e.g., Barefoot & Klaenhammer, 1983). Measured volumes of culture filtrates or appropriate control solutions are pipeted into the agar wells. If fungal disks are to be used, these are then placed on the plate. Inhibition zones around the wells are described and measured in the usual manner (e.g., Holland & Parker, 1966; Howell & Stipanovic, 1983).

19–5.5 Radial Growth Technique

19–5.5.1 Principles

This method is used for assessing inhibition of filamentous fungi. Culture filtrates are added to the agar media prior to pouring assay plates. The radial growth of a fungus on medium amended with the culture filtrate is compared with that for the same fungus on unamended plates. Any inhibition of growth on the amended plates is assumed to result from antifungal substances released into the original broth culture by the antagonist.

19–5.5.2 Materials

1. Sterile petri dishes.
2. Sterile molten agar medium (concentration of the medium should be increased to compensate for any significant dilution caused by subsequent addition of culture filtrate).
3. Sterile culture filtrates.
4. Inocula of the indicator organism (fungal disks).

19–5.5.3 Procedures

Culture filtrates are incorporated into fresh molten agar medium prior to pouring into petri dishes. If nonconcentrated filtrates are used, these are commonly mixed in a 1:1 ratio with fresh agar medium prepared at double the normal concentration. Alternatively, fresh medium and filtrates can be mixed together in a range of ratios. In all cases, the concentration of fresh medium in the final mixture should be standardized. Small volumes of concentrated filtrates can be added without adjusting the concentration of the fresh medium. Control plates are amended with medium that has not been used to culture potential antagonists. Once the agar has gelled, a fungal disk is placed centrally on the plate and the culture is examined on a daily basis. More frequent observations may be necessary when working with fungi having high growth rates. Measurements of fungal colony size are made once the control colony has grown to approximately 6 cm in diameter (e.g., Dunlop et al., 1989; Fravel et al., 1987).

19–5.6 Biomass Technique

19–5.6.1 Principles

This method is conceptually similar to the Radial Growth Technique but is conducted using liquid rather than agar media. Growth in filtrate-amended and nonamended broth is assessed by either a direct or indirect estimate of biomass production. It is adaptable to the study of both bacterial and fungal inhibition.

19–5.6.2 Materials

1. Flasks containing sterile liquid medium suited to the growth of the test organism (concentration of the medium should be increased to compensate for any significant dilution caused by subsequent addition of culture filtrate).
2. Sterile culture filtrates.
3. Inocula of the indicator organism (cell suspensions or fungal disks).
4. Büchner filter apparatus or centrifuge (for fungi).
5. Spectrophotometer/turbidimeter or centrifuge (for bacteria).
6. Microbalance and oven (65 °C) (for direct biomass measurements).

19–5.6.3 Procedures

Culture filtrates are combined aseptically with fresh broth medium in a manner similar to that described for the Radial Growth Technique. Amended and non-amended broth are inoculated with the test organism using a standard volume of cell suspension or a standard number of fungal disks. Bacterial cell numbers in the initial inoculum should be limited to a total of approximately 10^5 CFU (see 19–5.6.4).

For direct biomass measurements, the cultures are incubated for a period sufficient to allow for substantial growth of the control organism at its maximum rate of biomass accumulation. Prolonged incubation should be avoided as it may mask differences among treatments. Fungal mycelia are recovered either by filtration or centrifugation. Bacteria are recovered by centrifugation. The recovered biomass is dried to constant mass at 65 °C and weighed. Inhibition is reflected in reduced biomass in the treatments amended with culture filtrates relative to the nonamended control (e.g., Elad & Chet, 1987). The reader is directed to Gerhardt (1981) for additional discussion concerning the direct measurement of bacterial biomass produced in pure cultures.

Bacterial biomass can also be monitored indirectly by turbidimetry. Absorbance at a given wavelength (generally 540–640 nm) is monitored with a spectrophotometer. Specially constructed flasks equipped with cuvets (e.g., Nephlo flasks manufactured by BellCo Glass, Inc., Vineland, NJ) simplify the taking of periodic measurements. See Koch (1981) for detailed information regarding determination of microbial biomass by turbidimetry.

19–5.6.4 Comments

Particular attention to inoculation rates is important when using bacteria as indicator organisms in liquid culture because of the possible presence of spontaneous antibiotic-resistant mutants in the inoculum population. Growth of such mutants can lead to an underestimation of antibiotic production by the potential antagonists (Piddock, 1990).

19–6 SCREENING METHODS

19–6.1 Introduction

Screening procedures are essentially modifications of the Microbial Lawn Technique that are designed to permit detection of putative antagonists prior to their isolation into pure culture, thereby potentially yielding high isolation efficiencies. They can often be performed quantitatively so as to permit an estimation of antagonistic populations. These techniques are primarily limited by the feasibility of developing a system that is sufficiently selective and yet suited to the growth of all organisms of interest. This requirement is particularly limiting in cases where conditions suitable for the selective culture of the desired group of antagonists are detrimental to growth of the test organism. For example, use of an antifungal agent such as cycloheximide for the selective culture of bacteria producing antifungal substances would be inappropriate unless a means can be found to minimize the effect of cycloheximide on the test fungi.

Strategies for limiting problems of media incompatibility include the use of nutritionally selective base agar (as opposed to one containing inhibitory compounds) in combination with an overlay agar medium specifically suited to the growth of the test organism (Freeman & Tims, 1955; Herr, 1959). Some investigators have successfully used reduced concentrations of inhibitory compounds in base agar by providing a third agar layer that is used to separate the base from the indicator layer (Pugashetti et al., 1982). Treatment of soil samples prior to dilution plating to select for certain microbial groups may also have utility (Broadbent et al., 1971; Panthier et al., 1979). In many cases, however, the only recourse may be to limit an experiment to the study of physiologically similar organisms or to use only previously isolated organisms.

19–6.2 Single Agar Layer Technique

19–6.2.1 Principles

This approach is inflexible when compared with the Multiple Agar Layer Technique described later. Its primary limitation is that the medium must be suited to the growth of both the antagonist(s) and indicator organism(s) of interest while preventing excessive growth of other undesirable microorganisms present in the environmental sample. However, provided this requirement is not prohibitive, the Single Agar Layer Technique is both simple and direct. Instances where this method have been used successfully are for the isolation of bacteria antagonistic to *Streptomyces scabies* (Weinhold & Bowman, 1968) and of fluorescent *Pseudomonas* spp. antagonistic to *Erwinia* (Burr et al., 1978).

19–6.2.2 Materials

1. Petri dishes (90 mm diam.) containing 15 mL of a suitable agar medium (larger volumes may observe inhibition zones; the agar surface should be dry).
2. Inocula of indicator organism (cell suspension).
3. Apparatus for spraying indicator inoculum (see section 19–4.2.2).
4. Pipets, dilution blanks, and angled glass rod for preparing spread plates.

19–6.2.3 Procedures

Dilutions of environmental samples containing antagonists are plated in the manner used for spread plates. Once colony development is apparent, the plates are sprayed with a suspension of the indicator organism and incubated as described for the Microbial Lawn Technique. Colonies surrounded by zones of inhibition are isolated into pure culture and subsequently retested for antagonistic activity using one of the previously described pure culture methods (sections 19–4 or 19–5) (Burr et al., 1978).

19–6.2.4 Variations

Cases may arise in which the agar medium used supports only minimal growth of the indicator organism but is not actively inhibitory. In such instances, it may be helpful to prepare the spray suspension in a broth medium that encourages growth of the indicator organism and, thereby, increases the visibility of inhibition zones.

Indicator inocula may be incorporated into the agar medium prior to pouring, thereby avoiding the spraying operation (e.g., Weinhold & Bowman, 1968). However, success of this approach depends upon the antagonists displaying high growth rates relative to the indicator organism.

Problems with spreading colonies may be minimized by using the pour plate method in which the diluted samples are incorporated directly into the agar medium. The indicator organism is then applied to the surface following an appropriate incubation period. The disadvantage of this approach is the increased difficulty of extracting and isolating antagonists from within the agar into pure culture.

19–6.3 Multiple Agar Layer Techniques

19–6.3.1 Principles

Separate layers of agar are used to afford a suitable growth environment for the antagonistic and indicator organisms, respectively. In some cases, additional layers are employed to modulate the diffusion of inhibitory media additives or microbially produced antagonistic compounds from the base agar into the indicator layer. This approach was first described by Kelner (1948) and has since been modified many times. The original method employed up to four agar layers, but two (e.g., Freeman & Tims, 1955) and three (e.g., Herr, 1959; Panthier et al., 1979) layers have also been used successfully. In general, the procedure appears better suited to the isolation of antagonistic bacteria than fungi (Pugashetti et al., 1982). The primary method described below is modified from that of Panthier et al. (1979).

19–6.3.2 Materials

1. Sterile, molten (40–45 °C) agar medium suitable for the growth of the antagonists of interest.
2. Sterile, molten water agar (20 g agar L^{-1}).
3. Dilutions of environmental samples.
4. Pipets.
5. Sterile, molten agar medium suitable for the growth of the indicator organisms.
6. Inocula of indicator organisms (cell suspension).
7. Apparatus for spraying indicator inocula (see section 19–4.2.2).

19–6.3.3 Procedures

Sample dilutions are quantitatively incorporated into the medium chosen for growth of the antagonists. The mixture is distributed into petri dishes (10 mL plate^{-1}) and allowed to solidify (base layer). The base layer is immediately covered with 5 mL of water agar (intermediate layer) which serves to confine colony development. After a suitable incubation period, 10 mL of the indicator medium is applied to the plate (indicator layer), cooled, and sprayed with a suspension of the indicator organism. After further incubation, any inhibition zones are described and, if desired, counted. Antagonistic organisms are recovered from the base layer, purified, and retested under pure culture conditions (section 19–4 or 19–5).

19–6.3.4 Variations

Sample dilutions can be incorporated into the intermediate layer or can be applied as a completely separate layer situated between the base and intermediate layer (Kelner, 1948). Similarly, indicator inocula may be mixed with the indicator layer prior to pouring or simply spread over the indicator layer, thereby avoiding the spraying operation.

19–6.3.5 Comments

The original method of Panthier et al. (1979) was for the isolation of actinomycetes antagonistic to various *Rhizobium* and *Bradyrhizobium* spp. For this objective, the sample dilutions contained 7 g phenol L^{-1} to reduce the growth of undesirable bacteria. Similarly, the base layer contained cycloheximide (200 mg L^{-1}) to inhibit fungal growth. The reader is referred to Pugashetti et al. (1982) and Kelner (1948) for other strategies for using selective agents with this technique.

19–7 METHODS FOR SELECTED CLASSES OF COMPOUNDS

19–7.1 Introduction

It is beyond the scope of this chapter to provide detailed experimental procedures for studying the entire range of known antimicrobial substances. Nevertheless, certain classes of compounds have received special treatment by investigators or have elicited development of unique and reasonably standardized protocols for their study. These compounds generally exhibit unique physical properties or modes of action, and this may, in part, explain their emphasis in the literature.

19–7.2 Bacteriocins

19–7.2.1 Principles

Bacteriocins may be defined as nonreplicating, proteinaceous biocides whose antagonistic activity is limited to genetically similar organisms (Vidaver, 1976). There is, however, uncertainty regarding how strictly this definition should be interpreted (Hardy, 1982; Vidaver, 1983). The term is usually restricted to compounds produced by prokaryotic organisms, although the term has also been applied to substances produced by protozoa and yeasts (Vidaver, 1983). Genetically similar organisms are generally defined as those belonging to different species of the same genus or similar genera within a family.

As is the case with other antimicrobial substances, bacteriocin production is strongly affected by media composition and other environmental conditions. Assays are typically conducted using agar media as these have generally been shown to give more reproducible results and greater sensitivity than liquid media. Although some organisms produce bacteriocins freely, it is sometimes necessary to induce production by treating cultures with ultraviolet radiation or chemical agents. The double agar layer method described below is modified from that of Vidaver et al. (1972).

19–7.2.2 Materials

1. Glass petri dishes containing 25 mL of agar medium suitable for the growth of the potential producers (partially dried plates [e.g., 28 °C for 4 d] have been reported to enhance clarity of inhibition zones).
2. Inocula of producer and indicator organisms as cell suspensions (generally several microorganisms genetically similar and dissimilar to the producers are used as indicator organisms).
3. Sterile micropipet or multipoint replicator.
4. Chloroform.
5. Sterile, molten (45 °C) soft agar medium (agar concentration reduced to 7.5 g L^{-1}).

19–7.2.3 Procedures

One to several potential producers are spotted onto the agar medium by micropipet (10 µL) or multipoint inoculator. Following an incubation period to allow for colony development, the spotted organisms are killed by exposure to chloroform vapor (hence the need for glass petri dishes). This is conveniently done by inverting the petri dishes, transferring 3 mL of chloroform to the dish lid, replacing the inverted bottom, and allowing the chloroform to slowly evaporate in a fume hood (1–2 h). Any residual chloroform vapor is allowed to dissipate for one to several hours (overnight). Some investigators prefer to remove the colonies with a glass microscope slide prior to treatment with chloroform (Tapia-Hernàndez et al., 1990). It should be noted that some bacteriocins are inactivated by chloroform (Tagg et al., 1976). The chloroform-treated base agar is then overlaid with 3 to 5 mL of the molten soft agar inoculated with the indicator organisms to a density of approximately 10^6 cells mL^{-1}. Inhibition zones are described and measured following an additional incubation period.

Induction of bacteriocin production can be attempted by using either or both of the following procedures (Mayr-Harting et al., 1972):

1. **Ultraviolet radiation**—Exponentially growing broth cultures are irradiated (254 nm) in a 2-mm layer in a dish for periods ranging from 30 to 300 s at a distance of 30 to 50 cm. The proper conditions are determined empirically. The irradiated culture is incubated to late logarithmic or early stationary growth phase in the dark and is then used to prepare inocula in the usual manner.

2. **Chemical treatment**—Exponentially growing cultures are treated with mitomycin C to a final concentration ranging from 0.1 to 1.0 $\mu g\ mL^{-1}$; occasionally concentrations up to 10 $\mu g\ mL^{-1}$ are used. The optimal concentration is determined empirically. The treated culture is incubated to late logarithmic or early stationary growth phase, centrifuged, and the cells are then used to prepare inocula.

19–7.2.4 Supplemental Procedures

Once putative bacteriocin producers have been identified, several supplemental tests are often performed to verify and characterize the suspected bacteriocins. Two common assays are summarized below. Not discussed here are assays for biocidal activity and for sensitivity to proteolytic enzymes, catalase, chloroform, heat, and acidity. The reader is referred to Mayr-Harting et al. (1972), Gross and Vidaver (1978), and Tapia-Hernàndez et al. (1990) for further information and alternative protocols.

1. **Assay for Phage**—To verify that inhibition did not result from phage activity, agar disks are removed from inhibition zones, homogenized, serially diluted, and aliquots are spotted onto agar lawns of the indicator organism (see chapter 7 by Angle in this book). Phage produce decreasing numbers of discrete plaques with increasing dilution, whereas bacteriocins produce a single inhibition zone that gradually decreases in intensity with increasing dilution. It should be realized that a positive assay for phage does not rule out simultaneous bacteriocin production.

2. **Assay for Dializability**—Most bacteriocins are large molecules due to their proteinaceous nature, generally having molecular weights > 12 000 (Hardy, 1982). Estimates of the minimum molecular weight of inhibitory substances can be obtained by culturing the putative producer on an autoclaved dialysis membrane placed on an agar medium. Membranes having molecular weight exclusions ranging from 6000 to 12 000 are commonly employed. After a suit-

able incubation period, the membrane and associated colony are aseptically removed from the plate, and the base layer is overlaid with seeded soft agar in the normal manner. Development of inhibition zones indicates that the associated antimicrobial substance is smaller than the corresponding exclusion size of the membrane.

19–7.3 Siderophores

19–7.3.1 Principles

Siderophores are iron (III) chelating compounds produced by many microorganisms under Fe-limited growth conditions that function to scavenge Fe needed for growth. Siderophores produced by one organism may reduce Fe availability to other microorganisms that either do not produce siderophores or compatible siderophore receptors or which produce siderophores with relatively weak chelating abilities (Leong, 1986). Therefore, antagonism results from competition for an essential nutrient rather than from an actively harmful compound.

It is essential that Fe contamination of media and glassware be minimized so as to avoid repression of siderophore production. Media which are inherently deficient in Fe are most commonly used, although procedures for removing contaminating Fe from media are available (Waring & Werkman, 1942). For bacteria, and particularly for the fluorescent *Pseudomonas* spp., King's B medium has become the standard for most applications. Glassware should be scrupulously cleaned and treated with 6 M hydrochloric acid (HCl) to remove traces of Fe. Water used for making media and rinsing glassware should be deionized or distilled at minimum, preferably double glass-distilled.

Many of the previously discussed detection and screening procedures may be used to identify microorganisms that exhibit siderophore-mediated antagonism. The procedure described below is essentially a specialized version of the Microbial Lawn Technique. Screening procedures for siderophore production by fluorescent *Pseudomonas* spp. have also been described (e.g., Burr et al., 1978).

19–7.3.2 Materials

1. Petri dishes containing 15 mL of King's B agar or other Fe deficient medium.
2. Inocula of potential antagonists and indicator organisms (cell suspensions).
3. Sterile micropipet.
4. Spray apparatus for applying indicator organisms.

19–7.3.3 Procedures

The assay is conducted as described for the Microbial Lawn Technique. Antagonistic microorganisms should be retested using both iron deficient and iron sufficient media, the latter amended to 50 μm of Fe as $FeCl_3$. Putative siderophore producers are identified as exhibiting antagonistic activity in the Fe deficient medium only.

19–7.3.4 Comments

Fluorescent *Pseudomonas* spp. often cause strong siderophore-mediated antagonisms and are characterized by the production of diffusible yellow-green pigments (siderophores) that exhibit a blue-green fluorescence under longwave (365 nm) ultraviolet radiation. This fluorescence is eliminated by the addition of ferric Fe to growth media containing the siderophores (Misaghi et al., 1982). Many other microorganisms produce siderophores which may or may not fluoresce. Additional confirmation of siderophore production may be obtained using the colorimetric assay of Schwyn and Neilands (1987).

Loss of inhibitory activity on iron sufficient media is not considered proof that an antagonism is siderophore-induced. Gill and Warren (1988) characterized a fluorescent pseudomonad that produced an antagonistic substance in iron limited culture that did not function in Fe assimilation. A fluorescent siderophore was co-produced in Fe deficient media, whereas neither compound was produced in media amended with Fe.

19–7.4 Mycolytic Enzymes

19–7.4.1 Principles

Microorganisms may antagonize fungi by producing enzymes that lyse fungal cell walls. One procedure for assessing mycolytic activity, and the one described here, is to apply dilutions of environmental samples to mycelial lawns and subsequently examine the mycelia for zones of lysis (Carter & Lockwood, 1957). Alternatively, previously isolated microorganisms are grown in media amended with compounds which induce mycolytic enzymes, and the resulting culture supernatants are tested for specific lytic enzymes by using appropriate colorimetric assays (e.g., Elad et al., 1982).

19–7.4.2 Materials

1. Petri dishes containing 15 mL of peptone agar (L^{-1}: 5 g peptone, 20 g agar) seeded throughout with propagules (about 10^5 mL^{-1}) of the test fungus.
2. Diluted environmental samples.
3. Spray apparatus for applying sample dilutions.

19–7.4.3 Procedures

The seeded agar plates are incubated until a uniform mycelial lawn is obtained. The plates are then sprayed with the sample dilutions and incubated for an additional 2 to 21 d. Lytic colonies will be surrounded by clear zones free of fungal hyphae. The associated organisms are isolated into pure culture and retested for lytic activity.

19–7.4.4 Comments

Carter and Lockwood (1957) identified a number of factors that affect the results of this screening procedure. Seeding of the agar with fungal propagules by incorporation was found superior to surface application. Prostrate mycelial growth was necessary for detection of lysis, and this depended on the nature of the test fungus and medium. *Glomerella cingulata* (Ston.) proved to be superior to the other indicator fungi tested, although positive results were obtained for most species examined. In general, media which supported sparse fungal growth, such as peptone agar, gave the greatest isolation efficiencies.

19–7.5 Volatile Compounds

19–7.5.1 Principles

Volatile compounds have been examined primarily for antagonistic effects towards fungi (Dennis & Webster, 1971; Keel et al., 1989). Certainly one factor contributing to this emphasis is the radial growth habit of fungal mycelia that can be easily monitored under conditions appropriate to the study of volatile compounds. Antagonism is observed as a reduced rate of radial enlargement of a fungal colony in the presence of the inhibitory compound(s) when compared to negative control cultures. Recent research has focused on the role of cyanide as a volatile inhibitor of microbial growth (e.g., Ahl et al., 1986; Keel et al., 1989). Described below are both a general procedure and one designed specifically for the detection of cyanide production.

19–7.5.2 Materials (General) (Dennis & Webster, 1971)

1. Separate petri dishes containing agar media suitable for growth of the antagonists and test organisms, respectively.
2. Inocula of the antagonists (cell suspensions or agar disks).
3. Inocula of the test fungus (agar disks).

19–7.5.3 Procedures (General)

The antagonists are applied individually to plates of the proper agar medium. Inocula should be applied so as to produce abundant growth. Depending on the relative growth rates of the organisms being tested, the subsequent operation is either performed immediately or following an appropriate incubation period. The petri dish lids are replaced by an inverted dish bottom containing an agar disk of the test fungus centrally located on a suitable agar medium. To provide controls, the bottom containing the test fungus is placed on a non-inoculated plate of the antagonist medium.

The two dish bottoms are sealed together with parafilm or similar material. Radial growth of the test fungus is monitored and all treatments are measured once the control colonies are approximately 6 cm in diameter. If preferred, this and the following procedure may be conducted by using divided petri dishes rather than by joining two undivided dishes.

19–7.5.4 Materials (Cyanide) (Bakker & Schippers, 1987)

1. Petri dishes containing an agar medium suitable for the antagonists that has been amended with 4.4 g glycine L^{-1} (glycine has been shown to stimulate cyanide production; soybean-casein digest agar and King's B agar have both been used successfully as basal media).
2. Inocula of the potential cyanogenic antagonists.
3. Filter paper.
4. Cyanide detection solution (5 g picric acid and 20 g of sodium carbonate [$NaCO_3$] dissolved in 1000 mL distilled water).

19–7.5.5 Procedures (Cyanide)

Single antagonists are streaked onto the glycine-containing medium and the plates are inverted. A piece of filter paper impregnated with the cyanide detection solution is placed in the inverted lid of the petri dish, and the plate is sealed with parafilm or similar material. Upon incubation, cyanogenic organisms will cause the filter paper to change color from yellow to orange-brown. Organisms rated positive for cyanide production are subsequently tested for antagonism toward test fungi according to the general procedure described above (19–7.5.3); the glycine-amended medium should be retained as the antagonist medium. Cyanide production can be verified using procedures described by Bakker and Schippers (1987).

19–8 CONCLUDING COMMENTS

Efforts to isolate microorganisms capable of producing antimicrobial compounds is clearly hampered by a lack of information regarding their spatial distribution and ecological relevance in soil environments (Labeda & Schearer, 1990; Williams & Vickers, 1986). This deficiency is reflected in the inability of in vitro assays for antagonism to reliably predict microbial interactions in natural soil environments (Fravel, 1988; Weller, 1988; Whipps, 1987). It is particularly remarkable that, despite the rich history of research concerning antibiotic production by soil microorganisms, there are only intermittent reports giving direct evidence for production of antibiotics in situ (Thomashow et al., 1990). Nevertheless, these infrequent reports, combined with the common observation that many soil isolates produce antimicrobial compounds in vitro, will undoubtedly sustain interest in this area of research. The increasing efforts to develop rational strategies for using biological agents to enhance plant productivity will certainly intensify this interest.

REFERENCES

Ahl, P., C. Voisard, and G. Défago. 1986. Iron bound-siderophores, cyanic acid, and antibiotics involved in suppression of *Thielaviopsis basicola* by a *Pseudomonas fluorescens* strain. J. Phytopathol. 116:121–134.

Angle, J.S., B.K. Pugashetti, and G.H. Wagner. 1981. Fungal effects on *Rhizobium japonicum*-soybean symbiosis. Agron. J 73:301–306.

Anusuya, D., and S.B. Sullia. 1984. The antibiotic effect of culture filtrates of some soil fungi on rhizobial growth in cultures. Plant Soil 77:387–390.

Attafuah, A., and J.F. Bradbury. 1989. *Pseudomonas antimicrobica*, a new species strongly antagonistic to plant pathogens. J. Appl. Bacteriol. 67:567–573.

Axelrood, P.E., M. Rella, and M.N. Schroth. 1988. Role of antibiosis in competition of *Erwinia* strains in potato infection courts. Appl. Environ. Microbiol. 54:1222–1229.

Bakker, A.W., and B. Schippers. 1987. Microbial cyanide production in the rhizosphere in relation to potato yield reduction and *Pseudomonas* spp-mediated plant growth-stimulation. Soil Biol. Biochem. 19:451–457.

Barefoot, S.F., and T.R. Klaenhammer. 1983. Detection and activity of lactacin B, a bacteriocin produced by *Lactobacillus acidophilus*. Appl. Environ. Microbiol. 45:1808–1815.

Barnet, Y.M., M.J. Trinick, R.A. Date, and R.J. Roughley. 1988. Ecology of the root-nodule bacteria. p. 1–22. *In* W.G. Murrell and I.R. Kennedy (ed.) Microbiology in action. John Wiley and Sons, New York.

Brewer, D., F.G. Mason, and A. Taylor. 1987. The production of alamethicins by *Trichoderma* spp. Can. J. Microbiol. 33:619–625.

Broadbent, P., K.F. Baker, and Y. Waterworth. 1971. Bacteria and actinomycetes antagonistic to fungal root pathogens in Australian soils. Aust. J. Biol. Sci. 24:925–944.

Burr, T.J., M.N. Schroth, and T. Suslow. 1978. Increased potato yields by treatment of seedpieccs with specific strains of *Pseudomonas fluorescens* and *P. putida*. Phytopathology 68:1377–1383.

Carter, H.P., and J.L. Lockwood. 1957. Methods for estimating numbers of soil microorganisms lytic to fungi. Phytopathology 47:151–154.

Cubeta, M.A., G.L. Harman, and J.B. Sinclair. 1985. Interaction between *Bacillus subtilis* and fungi associated with soybean seeds. Plant Dis. 69:506–509.

Demain, A.L., Y. Aharonowitz, and J.-F. Martín. 1983. Metabolic control of secondary biosynthetic pathways. p. 49–72. *In* L.C. Vining (ed.) Biochemistry and genetic regulation of commercially important antibiotics. Addison-Wesley Publ. Co., London.

Dennis, C., and J. Webster. 1971. Antagonistic properties of species-groups of *Trichoderma*. II. Production of volatile antibiotics. Trans. Br. Mycol. Soc. 57:41–48.

Dhingra, O.D., and J.B. Sinclair. 1985. Basic plant pathology methods. CRC Press, Boca Raton, FL.

Dunlop, R.W., A. Simon, K. Sivasithamparam, and E.L. Ghisalberti. 1989. An antibiotic from *Trichderma koningii* active against soilborne plant pathogens. J. Natl. Prod. 52:67–74.

Elad, Y., and I. Chet. 1987. Possible role of competition for nutrients in biocontrol of Pythium damping-off by bacteria. Phytopathology 77:190–195.

Elad, Y., I. Chet, and Y. Henis. 1982. Degradation of plant pathogenic fungi by *Trichoderma harzianum*. Can. J. Microbiol. 28:719–725.

Fravel, D.R. 1988. Role of antibiosis in the biocontrol of plant diseases. Annu. Rev. Phytopathol. 26:75–91.

Fravel, D.R., K.K. Kim, and G.C. Papavizas. 1987. Viability of microsclerotia of *Verticillium dahliae* reduced by a metabolite produced by *Talaromyces flavus*. Phytopathology 77:616–619.

Freeman, T.E., and E.C. Tims. 1955. Antibiosis in relation to pink root of shallots. Phytopathology 45:440–442.

Fuhrmann, J., and A.G. Wollum, II. 1989. *In vitro* growth responses of *Bradyrhizobium japonicum* to soybean rhizosphere bacteria. Soil Biol. Biochem. 21:131–135.

Gerhardt, P. 1981. Diluents and biomass measurement. p. 504–507. *In* P. Gerhardt (ed.) Manual of methods for general bacteriology. Am. Soc. Microbiol., Washington, DC.

Gill, P.R., Jr., and G.J. Warren. 1988. An iron-antagonized fungistatic agent that is not required for iron assimilation from a fluorescent rhizosphere pseudomonad. J. Bacteriol. 170:163–170.

Gross, D.C., and A.K. Vidaver. 1978. Bacteriocin-like substances produced by *Rhizobium japonicum* and other slow-growing rhizobia. Appl. Environ. Microbiol. 36:936–943.

Hardy, K.G. 1982. Bacteriocins. p. 368–378. *In* R.G. Burns and J.H. Slater (ed.) Experimental microbial ecology. Blackwell Sci. Publ., Oxford.

Herr, L.J. 1959. A method of assaying soils for numbers of actinomycetes antagonistic to fungal pathogens. Phytopathology 49:270–273.

Holland, A.A., and C.A. Parker. 1966. Studies on microbial antagonism in the establishment of clover pasture. II. The effect of saprophytic fungi upon *Rhizobium trifolii* and the growth of subterranean clover. Plant Soil 25:329–340.

Howell, C.R., and R.D. Stipanovic. 1980. Suppression of *Pythium ultimum*-induced damping-off of cotton seedlings by *Pseudomonas fluorescens* and its antibiotic, pyoluteorin. Phytopathology 70:712–715.

Howell, C.R., and R.D. Stipanovic. 1983. Gliovirin, a new antibiotic from *Gliocladium virens,* and its role in the biological control of *Pythium ultimum.* Can. J. Microbiol. 29:321–324.

James, D.W., Jr., and N.I. Gutterson. 1986. Multiple antibiotics produced by *Pseudomonas fluorescens* HV37a and their differential regulation by glucose. Appl. Environ. Microbiol. 52:1183–1189.

Johnson, L.F., and E.A. Curl. 1972. Methods for research on the ecology of soil-borne plant pathogens. Burgess Publ. Co., Minneapolis.

Kanner, D., N.N. Gerber, and R. Bartha. 1978. Pattern of phenazine pigment production by a strain of *Pseudomonas aeruginosa.* J. Bacteriol. 134:690–692.

Katz, E., and A.L. Demain. 1977. The peptide antibiotics of *Bacillus:* chemistry, biogenesis, and possible functions. Bacteriol. Rev. 41:449–474.

Keel, C., C. Voisard, C.H. Berling, G. Kahr, and G. Défago. 1989. Iron sufficiency, a prerequisite for the suppression of tobacco black root rot by *Pseudomonas fluorescens* strain CHAO under gnotobiotic conditions. Phytopathology 79:584–589.

Kelner, A. 1948. A method for investigating large microbial populations for antibiotic activity. J. Bacteriol. 56:157–162.

King, E.O., M.K. Ward, and D.E. Raney. 1954. Two simple media for the demonstration of pyocyanin and fluorescin. J. Lab. Clin. Med. 44:301–307.

Koch, A.L. 1981. Growth measurement. p. 179–207. *In* P. Gerhardt (ed.) Manual of methods for general bacteriology. Am. Soc. Microbiol., Washington, DC.

Labeda, D.P., and M.C. Shearer. 1990. Isolation of actinomycetes for biotechnological applications. p. 1–19. *In* D.P. Labeda (ed.) Isolation of biotechnological organisms from nature. McGraw-Hill Publ. Co., New York.

Lambert, B., F. Leyns, L. van Rooyen, F. Gossele, Y. Papon, and J. Swings. 1987. Rhizobacteria of maize and their antifungal activities. Appl. Environ. Microbiol. 53:1866–1871.

Leong, J. 1986. Siderophores: Their biochemistry and possible role in the biocontrol of plant pathogens. Ann. Rev. Phytopathol. 24:187–209.

Manandhar, J.B., P.N. Thapliyal, K.J. Cavanaugh, and J.B. Sinclair. 1987. Interaction between pathogenic and saprobic fungi isolated from soybean roots and seeds. Mycopathology 98:69–75.

Mayr-Harting, A., A.J. Hedges, and R.C.W. Berkeley. 1972. Methods for studying bacteriocins. Meth. Microbiol. 7A:315–422.

Misaghi, I.J., L.J. Stowell, R.G. Grogan, and L.C. Spearman. 1982. Fungistatic activity of water-soluble fluorescent pigments of fluorescent pseudomonads. Phytopathology 72:33–36.

Panthier, J.J., H.G. Diem, and Y. Dommergues. 1979. Rapid method to enumerate and isolate soil actinomycetes antagonistic toward rhizobia. Soil Biol. Biochem. 11:443–445.

Parker, C.A., M.J. Trinick, and D.L. Chatel. 1977. Rhizobia as soil and rhizosphere inhabitants. p. 311–352. *In* R.W.F. Hardy and A.H. Gibson (ed.) A treatise on dinitrogen fixation. Section IV: Agronomy and ecology. John Wiley and Sons, New York.

Patel, J.J. 1974. Antagonism of actinomycetes against rhizobia. Plant Soil 41:395–402.

Piddock, L.J.V. 1990. Techniques used for the determination of antimicrobial resistance and sensitivity in bacteria. J. Appl. Bacteriol. 68:307–318.

Porter, J.N., J.J. Wilhen, and H.D. Tresner. 1960. Method for preferential isolation of actinomycetes from soil. Appl. Microbiol. 8:174–178.

Pugashetti, B.K., J.S. Angle, and G.H. Wagner. 1982. Soil microorganisms antagonistic toward *Rhizobium japonicum.* Soil Biol. Biochem. 14:45–49.

Robison, R.S. 1945. The antagonistic action of the by-products of several soil microorganisms on the activities of the legume bacteria. Soil Sci. Soc. Am. Proc. 10:206–210.

Schwinghamer, E.A. 1971. Antagonism between strains of *Rhizobium trifolii* in culture. Soil Biol. Biochem. 3:355–363.

Schwyn, B., and J.B. Neilands. 1987. Universal chemical assay for the detection and determination of siderophores. Anal. Biochem. 160:47–56.

Smith, R.S., and R.H. Miller. 1974. Interactions between *Rhizobium japonicum* and soybean rhizosphere bacteria. Agron. J. 66:564–567.

Stansly, P.G. 1947. A bacterial spray apparatus useful in searching for antibiotic-producing microorganisms. J. Bacteriol. 54:443–445.

Tagg, J.R., A.S. Dajani, and L.W. Wannamaker. 1976. Bacteriocins of gram-positive bacteria. Bacteriol. Rev. 40:722–756.

Tapia-Hernàndez, A., M.A. Mascarùa-Esparza, and J. Caballero-Mellado. 1990. Production of bacteriocins and siderophore-like activity by *Azospirillum brasilense*. Microbios 64:73–83.

Thomashow, L.S., and D.M. Weller. 1988. Role of a phenazine antibiotic from *Pseudomonas fluorescens* in biological control of *Gaeumannomyces graminis* var. *tritici*. J. Bacteriol. 170:3499–3508.

Thomashow, L.S., D.M. Weller, R.F. Bonsall, and L.S. Pierson, III. 1990. Production of the antibiotic phenazine-1-carboxylic acid by fluorescent *Pseudomonas* species in the rhizosphere of wheat. Appl. Environ. Microbiol. 56:908–912.

Trinink, M.J., and C.A. Parker. 1982. Self-inhibition of rhizobial strains and the influence of cultural conditions on microbial interactions. Soil Biol. Biochem. 14:79–86.

Utkhede, R.S. 1984. Antagonism of isolates of *Bacillus subtilis* to *Phytophthora cactorum*. Can. J. Bot. 62:1032–1035.

Vidaver, A.K. 1976. Prospects for control of phytopathogenic bacteria by bacteriophages and bacteriocins. Annu. Rev. Phytopathol. 14:451–465.

Vidaver, A.K. 1983. Bacteriocins: The Lure and the reality. Plant Dis. 67:471–475.

Vidaver, A.K., M.L. Mathys, M.E. Thomas, and M.L. Schuster. 1972. Bacteriocins of the phytopathogens *Pseudomonas syringae*, *P. glycinea*, and *P. phaseolicola*. Can. J. Microbiol. 18:705–713.

Waksman, S.A. 1967. The actinomycetes. Ronald Press Co., New York.

Waring, W.S., and C.H. Werkman. 1942. Growth of bacteria in an iron-free medium. Arch. Biochem. 1:303–310.

Weaver, R.W., and L.R Frederick. 1982. *Rhizobium*. p. 1043–1070. *In* A.L. Page et al. (ed.) Methods of soil analysis. Part 2. 2nd ed. Agron. Monogr. 9. ASA and SSSA, Madison, WI.

Weinhold, A.R., and T. Bowman. 1968. Selective inhibition of the potato scab pathogen by antagonistic bacteria and substrate influence on antibiotic production. Plant Soil 28:12–24.

Weller, D.M., W.J. Howie, and R.J. Cook. 1988. Relationship between in vitro inhibition of *Gaeumannomyces graminis* var. *tritici* and suppression of take-all of wheat by fluorescent pseudomonads. Phytopathology 78:1094–1100.

Weller, D.M. 1988. Biological control of soilborne plant pathogens in the rhizosphere with bacteria. Ann. Rev. Phytopathol. 26:379–407.

Whipps, J.M. 1987. Effect of media on growth and interactions between a range of soil-borne glasshouse pathogens and antagonistic fungi. New Phytol. 107:127–142.

Williams, S.T., and J.C. Vickers. 1986. The ecology of antibiotic production. Microb. Ecol. 12:43–52.

Wilson, E.E., F.M. Zeitoun, and D.L. Frederickson. 1967. Bacterial phloen canker, a new disease of Persian walnut trees. Phytopathology 57:618–621.

Chapter 20

Microbiological Procedures
for Biodegradation Research

DENNIS D. FOCHT, *University of California, Riverside, California*

The geochemical cycling of organic wastes through the metabolism of microorganisms was considered as fait accompli during the first 75 yr that the discipline of microbiology had existed. Although the phenomenal growth in the production of synthetic organic chemicals following the end of World War II should have raised questions about the limitation of microbial metabolic diversity, soil was still considered as the great cleansing agent. Gale (1952) best summarized the concept of microbial infallibility by suggesting that there existed in nature a microorganism capable of metabolizing any conceivable compound that the organic chemist might choose to synthesize. The opposing doctrine of molecular recalcitrance was advanced by Alexander (1965), who noted that many synthetic chemicals—particularly the chlorinated hydrocarbon insecticides—were persistent in soil. He questioned the validity of expecting that microbial enzymes, which were a product of 1 to 2 billion years of evolution, should attack substrates that were introduced recently into the environment.

To keep the argument in perspective, it should be noted that there is nothing mystical about xenobiotic or synthetic chemicals ("xenos" der. Greek "foreign"). Some are biodegradable (e.g., organophosphate insecticides) and some are very recalcitrant (e.g., chlorinated aromatic hydrocarbons). Similarly, the same can be noted for natural organics (e.g., amino acids and carbohydrates vs. lignin and humic polymers).

As biodegradation research has come into its own as a discipline over the last 25 yr, it is apparent that most recalcitrant compounds—particularly the chlorinated aromatic hydrocarbons (CAHs)—are neither used as growth substrates nor mineralized by a single microbial species. Rather, they are metabolized by the action of two or more physiologically different microorganisms. In most of these cases, the organism that carries out the initial oxidation is unable to use the recalcitrant compound as a growth substrate, and must rely on an exogenous substrate to sustain its activity and survival. The phenomenon whereby an organism can transform a compound, but not use it as a growth substrate has been termed cometabolism (Horvath, 1972). Perhaps the term *fortuitous metabolism* (Dagley, 1984) better explains the underlying phenomenon: i.e., enzymes, which have resulted from the evolutionary selection of a microorganism to a natural substrate (e.g., benzene), fortuitously catalyze the metabolism of a structurally similar xenobiotic substrate (e.g., chlorobenzene) because of low enzyme specificity. However, at some point in the catabolic pathway, a particular enzyme with high substrate specificity will not attack the xenobiotic product, which then accumulates in pure culture.

Thus, methods designed to select for a single organism able to use the target compound as a growth substrate will fail with recalcitrant compounds. This failure, however, cannot be taken as an indication that the target compound is inherently recalcitrant to microbial attack because the xenobiotic metabolite may be metabolized by another microorganism, having enzymes with different substrate specificity. The enrichment culture procedure, which is the molecular backbone of biodegradation research, has been successful when its limitations regarding growth are taken into consideration. The use of structurally similar growth-promoting analogs has greatly enhanced the use of the enrichment culture procedure in understanding how recalcitrant compounds are metabolized in culture and in soil. This process, referred to as analog enrichment (Horvath, 1972; Brunner et al., 1985), has proven useful in the isolation of bacteria capable of cometabolizing many recalcitrant compounds, particularly DDT, PCBs, and other chlorinated aromatic hydrocarbons that are not used as growth substrates by any single microorganism.

Cyclohexane is an excellent example of a compound that is biodegradable in soil, yet does not select for isolates able to use it as a growth substrate (Perry, 1984). This enigma was solved in independent studies by Beam and Perry (1973, 1974) and deKlerk and van der Linden (1974), who showed that two species were involved in complete catabolism of cyclohexane. Bacteria isolated from enrichment culture on n-alkanes, were able to fortuitously oxidize cyclohexane to cyclohexanol by a monooxygenase, yet were unable to further metabolize the product. Another strain, obtained from enrichment culture on cyclohexanol, was able to complete the process. These elegant studies show why it would not be possible to isolate either organism by selective enrichment with cyclohexane because the first member, the cometabolizer, gets insufficient energy from the substrate and is, therefore, unable to grow and provide sufficient levels of cyclohexanol for enrichment of the commensal. A similar situation exists for the cometabolic-commensal mineralization of PCBs in soil, in which the first member obtains no energy for its growth (Focht & Brunner, 1985).

The intent of this chapter is to focus on the isolation, growth, and physiology of the microorganisms involved in biodegradation. Soil biodegradation studies will not be covered here since the methods and procedures (e.g., respiration, CO_2 evolution, extraction and chromatographic analysis of substrates and metabolites) can be found in other chapters of this book.

20–1 THE ENRICHMENT CULTURE

20–1.1 Principles

The enrichment culture, also referred to as selective or elective culture, is based on the Darwinian concept of natural selection, namely, that the organism best able to exploit the niche (i.e., use the substrate for growth and energy) under all other environmental constraints (temperature, pH, O_2, and essential nutrients) will be the one that is selected. It is not the intent of this procedure to use inorganic nutrient conditions similar to those in soil (Angle et al., 1991) or in any other environment because these conditions are suboptimal for microbial growth. The objective of the enrichment culture procedure is generally to provide conditions for the rapid isolation of the desired catabolic phenotype. This procedure will be focused only on aerobic selection as this is the oldest, fastest, and most commonly used condition for selection of xenobiotic-degraders. Moreover, the isolation and significance of anaerobic bacteria that degrade xenobiotic compounds is covered in chapter 13 by Tiedje in this book.

Methods for selection and isolation will be confined to mesophilic conditions, although the procedures herein could be readily adapted to thermophilic or psychrophilic conditions for selection from environments likely to have a different microbial flora, such as high-temperature composts or antarctic brine pools. The pH at which enrichment cultures are established may influence the predominance of one microbial group over another: for example, the procedures described herein will invariably lead to the selection of bacteria over fungi. Whether this is due exclusively to a pH effect or the greater metabolic diversity of bacteria is not clear, although the latter would appear to be the case. No eucaryotic organism capable of using an aromatic hydrocarbon as a growth substrate has ever been reported since noted by Gibson and Subramanian (1984).

A major limitation of the enrichment culture is that it will work only for those organisms that are prototrophic: i.e., those that are able to synthesize all amino acids, nucleotides, vitamins, and cofactors from the inorganic N, P, and S sources and carbonaceous substrate. A defined medium generally selects against fungi and gram-positive bacteria, which are more fastidious. These limitations can be overcome to a small degree by including trace quantities of yeast extract (YE) or some other vitamin and amino acid source. Growth supplements, however, should not be used in quantities large enough to support growth as a C source. Generally, a YE concentration of 50 mg L^{-1} will support a population density of pseudomonad bacteria of 10^8 cells mL^{-1}, which will effect faint turbidity. Thus, a concentration of 25 mg L^{-1} of YE is acceptable as a cofactor supplement. High concentrations of 500 mg L^{-1} YE have led to erroneous conclusions being made about the utilization of chlorinated aromatic hydrocarbons as "sole carbon sources" (Vandenbergh et al., 1981). Use of high concentration of YE in the enrichment process merely selects for organisms that are tolerant to the xenobiotic chemical.

20–1.2 Materials

1. Platform shaker.
2. Erlenmeyer flasks (125 or 250 mL), screw-cap needed for volatile substrates.
3. Pipets.
4. Eppendorf tubes, 1 mL.
5. Wire.
6. Aluminum foil or cotton.
7. Mineral salts medium (below).

Stock solutions	Additions, mL	Final concentration, mM
K_2HPO_4, 1 M	10	10
NaH_2PO_4, 1 M	3	3
$(NH_4)_2SO_4$, 1 M	10	10
$MgSO_4$, 1M	1	1
$Ca(NO_3)_2$, 1 M	0.1	0.1
$Fe(NO_3)_3$, 1 M	0.01	0.01
Trace minerals (below)	1	
$\quad MnSO_4$		0.001
$\quad ZnSO_4$		0.001
$\quad CuSO_4$		0.001
$\quad NiSO_4$		0.0001
$\quad CoSO_4$		0.0001
$\quad Na_2MoO_4$		0.0001

Final pH = 7.25. The second column represents the amount of each stock solution to be added to a final volume of 1 L, and the third column represents the final concentration. Add about 0.9 L of distilled water before adding any of the solutions above, or precipitates will form, and then fill to volume. Stock solutions are given without water of hydration to make it easier to use any hydrated form that is available in the laboratory.

The trace mineral solution is made up with all the compounds listed. The addition of 1.0 mL to the media above will give a final concentration of each compound in column 3.

For multichlorinated compounds or high concentrations of chlorinated compounds, it may be necessary to double the concentrations of K_2HPO_4 and NaH_2PO_4 to buffer the acidity from the production of HCl. Do not be concerned about the $MgHPO_4$ precipitate that is formed with double-strength phosphate additions after autoclaving as it will disappear upon cooling. This salt is unusual in having a lower solubility in hot water than in cold water.

20–1.3 Procedure

There is generally no need for aseptic technique, sterile media, or sterile glassware during the enrichment process unless the source of the

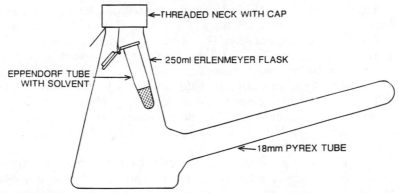

Fig. 20–1. Nephelometry flask used for growing pure cultures and measuring optical density without removing sample. Volatile substrates such as benzene are placed in a 1.0 ml Eppendorf tube that is suspended above with a wire fastened over the neck of the flask. If the Eppendorf tube is positioned in a manner that might cause spillage of solvent when reading the optical density of the tube, the spectrophotometer should be raised at the appropriate end to prevent this occurrence. Drawn to 0.4 scale.

inoculum is important. Aerial dispersal of xenobiotic-degrading bacteria is not likely to be as important in laboratories not previously engaged in biodegradation research, as in those where many isolates have been obtained. An example of the latter point was the isolation of a 2,6-dinitro-phenol-degrading bacterium, *Alcaligenes eutrophus* JMP 134 (Ecker et al., 1992), which probably originated from contaminated glassware according to H.-J. Knackmuss (personal communication). The original isolate, obtained by enrichment on 2,4-dichlorophenoxyacetic acid (Pemberton et al., 1979), was found to grow on 2,6-dinitrophenol, and was indistinguishable from the one isolated many years later. Thus, the use of sterile glassware is critical when evaluating the efficacy of inoculation on biodegradation in soil that may lack an indigenous population able to metabolize the xeno-biotic chemical. Ample precedences for lack of biodegradation in soil, in lieu of inoculation, exist with 2,4,5-trichlorophenoxyacetic acid (Chatter-jee et al., 1982), pentachlorophenol (Edgehill & Finn, 1983; Crawford & Mohn, 1985), 3-chlorobenzoate (Pertsova et al., 1984; Focht & Shelton, 1987) and parathion (Barles et al., 1979). Use of nonsterile glassware could, therefore, give false positive results from the uninoculated control.

Add 100 mL of mineral salts solution to a 250-mL Erlenmeyer flask and add the substrate. A concentration range of 100 to 500 mg L^{-1} is sufficient for nonvolatile substrates. Toxic substrates such as phenol should not exceed 300 mg L^{-1}. Volatile compounds such as toluene and benzene should be added to the Eppendorf tube suspended at the top of the flask (Fig. 20–1). Less volatile compounds such as naphthalene, biphenyl, and diphenylmethane can be added directly to the medium or to the Eppendorf tube. Acidic substrates (e.g., benzoic acids) should be neutralized with an equivalent amount of NaOH added to the media, or preferably made up previously as aqueous stock solutions of the Na or K neutralized salts since some of the acids dissolve slowly. The source and amount of inoculum is a matter of choice. Generally a 1/100 mass basis is a good working approximation to minimize soil from obscuring growth, as noted by an increase in turbidity. All flasks can be covered with aluminum foil or cotton, except those containing volatile substrates, and placed on a platform shaker at a temperature not exceeding 30 °C. (The commonly used temperature of 37 °C in clinical microbiology is too high for the growth of most gram-negative soil bacteria.)

When growth is apparent, transfer 1 mL of the enrichment culture to fresh media containing the same composition as above and incubate. When this subculture shows evidence of growth by an increase in turbidity, it is time to isolate pure cultures by aseptic technique.

20–2 ISOLATION OF PURE CULTURES

20–2.1 Principles

Questions may arise as to the necessity of making subcultures rather than directly plating or streaking out samples onto agar media containing the sole C source. If there is an abundant population of microorganisms present in the sample, then it is not necessary to enrich, and pure cultures can be isolated directly. However, in many cases, the numbers of organisms present may be too low for a direct isolation, and consequently they must be enriched to permit the population to increase to a density sufficient for isolation. Also, the presence of exogenous substrates in the sample may effect growth on agar plates without any metabolism of the desired substrates. Finally, for those very odd, yet common, cases requiring 7 to 13 mo for successful enrichment and isolation of chlorobenzene-utilizers (Reineke & Knackmuss, 1984; DeBont et al., 1986; Shraa et al., 1986; Spain & Nishino, 1987; Haigler et al., 1988) or dichlorobenzoate-utilizers (Hernandez et al., 1991), a long adaptive phase appears to be a necessity. Whether this is due to physiological adaptation, de novo enzyme synthesis, enzyme modification, mutation, or genetic exchange is unknown.

Generally, a compound is readily biodegradable if an enrichment culture is obtained quickly and with little effort. This is certainly predictable with simple aromatic acids and hydrocarbons. However, failure to obtain enrichment cultures could either be indicative of recalcitrance (e.g., CAHs) or action by a consortium (e.g., cyclohexane). On occasions, utilization of a C source in liquid media after several transfers may be apparent, yet selection and growth of isolated colonies on agar plates cannot be demonstrated. In all such cases, these microbial communities involve a syntrophic or cross-feeding effect by which the interaction is obligatory and mutually symbiotic to both.

20–2.2 Materials

1. Mineral salts agar (1.5%) for petri plates.
2. Mineral salts agar (1.5%) for slants.
3. Desiccator or sealed container.
4. Incubator set at 28 °C.
5. Inoculating loop.
6. MS media (50 mL) containing desired substrate in 250 mL Erlenmeyer flasks, sterile.

20–2.3 Procedures

Nonvolatile substrates should be added to the agar before autoclaving. Do not attempt to autoclave volatile substrates! When it is apparent that the substrate is supporting bacterial growth, remove a sample with an inoculating loop and streak the suspension according to standard microbiological procedures onto sterile mineral salts agar (1.5%) plates containing the appropriate substrate. For volatile substrates, use mineral salts agar plates containing no substrate. After streaking these plates, there are two methods that can be used for ensuring aseptic contact of the volatile substrate with the culture. Petri plates are inverted and placed on the raised floor of a desiccator with the substrate placed in a small beaker or petri plate lid underneath the raised floor at the bottom of the desiccator. The other method is to add one drop of the liquid or a few crystals of the solid (as the case may be) to the bottom inverted lid of the petri plate, place the agar lid above, and seal the two lids together with parafilm or some other temporary but effective sealer. It is advisable to use glass, and not plastic, petri plates when isolating microorganisms on volatile organic compounds.

The plates should be incubated at 28 °C or ambient temperature for several days before the presence of single isolated colonies is noted. It is not unusual to wait 7 to 10 d before colonies of sufficient size are noted with some substrates. Generally, hydrocarbon substrates select for faster (2–3 d) and more luxuriant growth than chlorinated aromatic acids.

The final proof that the compound is used for growth as a sole C source is to aseptically transfer a single isolated colony to sterile liquid media, incubate, and observe growth by the appearance of turbidity.

20–3 MAINTENANCE OF CULTURES

20–3.1 Principles

Xenobiotic-degrading bacteria frequently lose their degradative ability by spontaneous loss of plasmids, which code for many of the degradative enzymes. The influence of substrate on genetic rearrangement of a *TOL*-like plasmid in *Pseudomonas putida* has been recently documented

(Carney & Leary, 1989). The Darwinian dogma regarding the environment solely as a selective agent that is independent of influencing mutational events has recently come under question (Hall, 1990; Cairns et al., 1988). Whether or not the spontaneous loss of degradative plasmids or their rearrangement in response to substrates is suggestive of neo-Lamarckianism is not likely to be resolved before the role of environment and mutation is settled. It is important for microbiologists to recognize that transpositional changes between plasmid and chromosomal DNA occur frequently with xenobiotic-degrading bacteria, which necessitates that the selection pressure be maintained to eliminate undesirable mutants. Consequently, continuous growth on complex media is discouraged—particularly when it is apparent that reversion from the desired phenotype is high. On the other hand, there are instances in which bacteria are "refreshed" by occasional transfer to rich media. Sylvestre and Fauteux (1982) noted that a facultative anaerobe, strain B206, lost the ability to grow on 4-chlorobiphenyl or biphenyl after a series of transfers. When propagated on nutrient agar, it regained the ability to grow again on either biphenyl or 4-chlorobiphenyl. Complex media may be useful to "refresh" problematic cultures, to check for purity by observing the presence of more than one colony type, and for quickly growing up large batches of cells, but they should be used judiciously.

20–3.2 Materials

The same materials used in the previous section.

20–3.3 Procedure

The previous section described the growth of bacteria on agar plates with volatile substrates. The same basic principle applies to maintaining stock cultures of bacteria on agar slants. A closed container (e.g., desiccator containing the substrate) can be used for incubation of slants. However, once the slants are removed, they should be sealed and placed in a refrigerator to retard metabolism and conserve the small amount of substrate remaining. Alternative substrates for benzene, toluene, ethylbenze and xylene (BTEX) are toluates (methyl benzoates), which are suitable for maintaining the selection pressure for BTEX utilization (Carney & Leary, 1989).

Cultures using less volatile substrates such as biphenyl or naphthalene can be maintained on slants as described above or on slants containing the substrate, which is added immediately after the tubes have been removed from the autoclave. This can be done most rapidly by the addition of 100 µL of an acetone solution containing 70 g L^{-1} of material prior to slanting the tubes in a hood and allowing them to cool with the cap fastened loosely to permit the loss of solvent. As this procedure violates prima facie the concept of sterilization, it is necessary to frequently ascertain purity by streaking onto plates as described above. Slants, nevertheless, are more manageable than plates for sending requests of cultures and maintaining isolates over several weeks. Stock cultures are best maintained pure by removing a loopful of culture from an isolated colony on a plate and mixing it in 0.1 mL of sterile phosphate buffer (see section 20–6.2), which is then added to a sterile vial containing 30 to 50% glycerol in water, and stored at −80 °C.

20–4 GROWTH IN LIQUID CULTURES

20–4.1 Principles

In most cases, strains that are isolated on solid media will grow in liquid media. Occasionally, however, there will be strains that grow poorly in liquid media. Solid insoluble substrates, such as biphenyl, generally give low cell densities when the substrate is added to the medium prior to autoclaving, presumably because of the small surface area that is caused when the material cools into a solid lump. (As stated before, autoclaving of these substrates is discouraged in lieu of proper ventilation designs for public health reasons.) The addition of substrate by a volatile solvent sometimes facilitates this problem. However, miscible solvents such as acetone form azeotropic mixtures, such that residual solvent may be toxic to some strains. The best growth with biphenyl is achieved by powdering the substrate with a mortar and pestle and adding it to sterilized media immediately before inoculation. This nonaseptic microbiological procedure can be justified if a sufficient inoculum density is used to give a starting concentration in excess of 10^7 cells mL^{-1} on kinetic grounds as it would be difficult to argue that the introduction of an air-borne contaminant would have a density sufficient to outgrow the inoculant. The same argument would apply regarding the likelihood that contaminants could be introduced from the nonsterile substrate. As a matter of satisfaction, an uninoculated control could be easily run to verify the source of growth.

However, it cannot be stressed too strongly that this nonaseptic method should never be used for continual subculturing or maintenance of stock cultures. Once the experiment has been completed, the cells should be discarded in accordance with standard microbiological procedures. To ensure purity, stock cultures should always be maintained axenically on sterilized slants and verified on plates.

Although utilization of a growth substrate may seem rather obvious with fast-growing strains, the same cannot be said with substrates that are toxic at moderate concentrations and must be added in small quantities. Pentachlorophenol is such a case in point since it is used specifically to inhibit microbial growth. Moreover, the maximum concentration (300 mg L^{-1}) that can be used for growth of *Flavobacterium* spp. (Crawford & Mohn, 1985) does not provide much C (27% by mass) or energy, and the organism must be grown with an auxiliary C source. Thus, growth curves are required to conclusively demonstrate that an organism uses the target compound in the presence of an auxiliary growth substrate, as opposed to merely being tolerant to it. With chlorinated substrates, production of inorganic chloride would further confirm that the substrate was mineralized at the expense of growth, particularly if the release was close to complete stoichiometry of what was added. It would not be possible to demonstrate this on solid media. This section is not intended to provide a detailed kinetic examination of growth, but rather to provide a general understanding of how fast bacteria grow on xenobiotic substrates and their efficacy of utilization.

20–4.2 Materials

1. Shaking platform.
2. Erlenmeyer side-arm flasks (250 mL; Fig. 20–1) containing 30 mL media.
3. Spectrophotometer for measuring cell density turbidimetrically (525 nm).
4. Centrifuge.
5. Media for growing culture.
6. Culture.

20–4.3 Procedure

To obtain the most reproducible results, and to reduce the length of the lag phase from inoculation to the first observed increase in turbidity, it is best to use a fresh culture that has previously been growing in the same liquid media that is under investigation. The density of the inoculum should be about 10^9 cells mL^{-1}, and the flasks should be inoculated with an amount sufficient to give slight turbidity: the absorbance or optical density (OD) should be about 0.02 to 0.04 with an 18-mm thick test tube at 525 nm at the beginning. The use of dilute cell suspensions or inoculations from slants usually result in an increased apparent lag because cell densities below this concentration cannot be detected turbidimetrically. The term *apparent lag* is used here to distinguish from a real biological lag phase, which constitutes an actual failure of cells to immediately divide and begin the exponential growth phase.

The time at which measurements are taken will depend on how fast the culture grows. Cultures that grow fast (3 h or less doubling time) can be completed in a day, while those that grow slowly (12 h doubling time or more) will take several days, although they can be measured conveniently twice a day. Those strains in between will require more dedication and sleepless nights. The use of nephthelometry flasks (Fig. 20–1) obviates the risk of contamination that could occur by repeated sampling; it also maintains a constant volume. To minimize light interference, the flask should be covered completely with a thick dark plastic cover that does not let light pass through. Portfolios that are distributed at scientific meetings, work well for this purpose.

The two important kinetic parameters that can be determined from growth curves are the doubling time (t_D) and the yield coefficient (Y). The former can be determined from a semi-log plot of the optical density vs. time from a straight line of those points representing the exponential (or log) phase of growth. The latter can be approximated by the difference of the maximal turbidity (stationary growth phase) and the initial turbidity. The method for the latter is applicable only if all the substrate has gone into cell growth and not into maintenance metabolism. Thus, high substrate concentrations in which sufficient substrate remains after cells have reached stationary growth are unsuitable for determining yield coefficients. These methods are intended as approximations only. For more detailed and exacting analysis of the kinetics of growth, the reader is referred elsewhere (Simkins & Alexander, 1984; Focht & Shelton, 1987).

Absorbance measurements are more useful when related to cell mass. This relationship is established by growing cells in about 1.0 L of media to obtain sufficient mass of cells for weighing (100–500 mg). A good general rule of thumb for pseudomonad bacteria is the following relationship: 10^{10} cells mL^{-1} = 1 mg dry mass mL^{-1} = OD 4.0. Cell densities this high are rarely observed with batch cultures except with rich media under conditions of high aeration.

Cells should be harvested from the late exponential growth phase to avoid the production of lytic products and slimes that may add to the weight. It is desirable to use a refrigerated centrifuge so that autolysis will be minimized. Harvest cells by centrifugation at 6000 g for 10 min. Discard the supernatant, and wash the cells by resuspending them in about 300 mL of deionized water or about the amount that a single centrifuge bucket will hold. Centrifuge, wash, and resuspend two more times to remove the salts from the medium, which would otherwise add to the weight. Resuspend the last (third) cell pellet into duplicate volumes of 50 mL and mix thoroughly. Remove 1 mL from each sample, dilute with 9 mL of water, mix thoroughly and record the absorbance. Add the duplicate cell suspensions to two tared breakers, and wash the remaining cells into the beaker with deionized water. Evaporate the water at 90 °C or below. Avoid boiling temperatures since this will cause bumping and loss from boil over and oxidation of organic mater.

When the samples are dry, weigh the beakers and take the average value. For better precision, more samples can be taken, but it is generally unnecessary to make a standard curve over several concentration ranges. Yield coefficients are usually expressed as g cells g^{-1} substrate or g cells mol^{-1} substrate. When it is desirable to do a total C balance, the total C content of the cell mass can be determined by methods mentioned elsewhere in this book. A working approximation is that 55% of the total dry mass of the cell is comprised of C.

20–5 PREPARATION OF WASHED CELL SUSPENSIONS

20–5.1 Principles

Xenobiotic substrates that are cometabolized can be studied in the presence of the growth substrate during batch culture growth. However, this method is problematical when the growth substrate and xenobiotic compound have properties that do not distinguish them by common analytical methods. Moreover, comparisons among metabolism of several xenobiotic compounds would require a separate culture. A culture study could involve several days and a degree of uncertainty regarding sampling time and frequency.

The preparation of washed, resting cells eliminates all of the problems mentioned above. Moreover, it enables the investigator to choose what substrate the organisms should be grown on and to make comparisons between them. It is particularly advantageous to know if the metabolism of a xenobiotic compound is constitutive or inducible since this would have relevance to the performance of inoculants in soil. Thus, comparisons can be made in a single experiment, not only between xenobiotic compounds, but among those cells grown on different substrates, e.g., glucose and chlorobenzoates. The use of washed cells, which are unable to grow because of N limitations, enables the investigator to concentrate the cell density, which thereby enhances the rate of metabolism and saves time. Finally, the preparation of washed cells is a prelude to the preparation of cell-free extracts for use in enzymatic studies.

20–5.2 Materials

1. 2 L flask containing 1 L of growth medium. (Depending on the quantity of cells needed, smaller flasks and less media could be used.)
2. Platform shaker.
3. Pipets.
4. Centrifuge.
5. Phosphate buffer (0.05 M, either Na or K or both, pH 7.2).
6. Spectrophotometer.

20–5.3 Procedure

The substrate concentration of the medium should not be limiting; i.e., it should equal or exceed that necessary to achieve maximal turbidity or cell density. Cells should be harvested from the late exponential growth phase. To know all these parameters generally requires that growth curves be performed previously as in section 4. Prior to harvesting, remove an aliquot for spectrophotometric analysis of the cell density. Follow the same procedure in section 4 for harvesting cells except for using phosphate buffer instead of deionized water for washing. After the third and final centrifugation, suspend the cells in buffer to 25 mL. Make a 1/100 dilution of the suspension and measure the OD. Adjust the final OD to 4.0 with phosphate buffer according to the formula below, where V_b is the volume of buffer to be added, V_s is the volume of the cell suspension (24.9 mL), OD_s is the OD of the cell suspension (let us use 10 as an example), and $OD_{4.0}$ is the desired OD (4.0).

$$V_b = (OD_s * V_s)/OD_{4.0} - V_s$$

Thus, $V_b = 37.35$ mL in the example above to give a total washed cell volume of 62.25 mL at OD 4.0.

There is nothing fixed in stone about a washed cell suspension being 1 mg dry mass mL^{-1}. Higher or lower cell densities can be used. Many pseudomonad bacteria tend to undergo autolysis at higher cell densities. This is particularly more pronounced for experiments that may be conducted for several hours. The use of lower cell densities simply means that more time may be needed to complete the experiment.

The cell suspension can then be used in any reaction mixture to assess O_2 consumption, disappearance of substrate, or analysis of metabolites. Oxygen consumption measurements are generally completed in a few minutes (see section 7), while the other experiments may be conducted over a few hours. Contamination or demise of the culture is generally not a problem over a 24-h period. If the cell suspension is not going to be used right away, it can be kept refrigerated or on ice for several hours prior to its use.

20–6 PREPARATION AND USE OF CELL-FREE EXTRACTS

20–6.1 Principles

Although the use of washed cell suspensions can give valuable information about the rates of substrate disappearance and qualitative identification of metabolites, it is sometimes desirable to obtain information about an important, single reaction-step in the catabolic pathway. A few pertinent examples are the dehalogenation of 4-chlorobenzoate to 4-hydroxybenzoate (Elsner et al., 1991), the hydrolysis of parathion to p-nitrophenol (Munnecke, 1979), the dehalogenation of halomethanes by methane monooxygenase (Tsien & Hanson, 1992), and the dioxygenation of naphthalene (Haigler & Gibson, 1990).

The best characterized enzymes in the catabolism of aromatic compounds are the catechol dioxygenases, which incorporate both atoms of O_2 into the splitting of the aromatic ring. ortho-Pyrocatechase (EC 1.13.11.1), splits catechol between the two hydroxyl groups to give cis, cis-muconate, which has $\lambda_{max} = 260$ nm. meta-Pyrocatechase (EC 1.13.11.2) splits the ring adjacent to one of the hydroxyl groups to give muconic semialdehyde, a bright yellow-product that has $\lambda_{max} = 375$ nm at neutral to alkaline pH: the color is reversibly abolished by addition of acid. Bacteria normally possess constitutive levels of the ortho fission enzyme, presumably as a result of tryptophan catabolism. Both catechol dioxygenases can be distinguished from each other by measuring product disappearance at the two different wave lengths or by inactivating ortho activity at 55 °C for 15 min and inactivating meta activity with a 3% final solution of H_2O_2. The analysis of the two catechol dioxygenases presented below are given as an illustration for the preparation and use of cell-free extracts.

20–6.2 Materials

1. Same as for growing cells.
2. Sonic probe or French pressurized cell.
3. UV-visible spectrophotometer.
4. Methanolic or N,N-Dimethyl formamide solutions of catechol (10 mg mL^{-1}).

20–6.3 Procedure

Follow the same procedure above for growing up cells except that the final suspension should be more concentrated in about 10 mL buffer since this will facilitate breakage. Cool the cell suspension in an ice bath, and disrupt the cells by sonic bursts of 15 s duration followed by 15 s of swirling the tube in the ice bath. About 8 to 12 bursts will be necessary to get good breakage for gram-negative bacteria. When the cells have been broken, as indicated by a peculiar burnt-like odor and a glue-like appearance, clarify the extracts by centrifugation at 4 °C at 40 000 g for 30 min. Remove a small aliquot, and determine the protein concentration by one of the standard methods commonly used (Colowick & Kaplan, 1957; Bradford, 1976). The protein concentration should be between 5 to 15 mg mL^{-1}.

Since gram-positive bacteria are more resistant to mechanical breakage, the addition of an equal volume of glass beads (< 10 μm) is recommended when using sonication. Alternatively, the cell suspension can be passed through a French pressurized cell at 25 MPa (20 000 lb in.$^{-2}$).

Cell-free extracts can be purified for isolation of pure enzymes or they can be used for crude enzyme assays as is common in the determination of catechol dioxygenase activity. The measurement of this activity can be done respirometrically, as described in the following section, or spectrophotometrically, as described herein. Catechol dioxygenase activity is an extremely rapid reaction, which begins immediately upon addition of substrate. Thus, everything should be in place before substrate and cell-free extract are added.

The procedures that follow are based on a curvette volume of 1.0 mL. All additions should be changed accordingly to the volume of the curvette. Dilute the CFE with 0.067 M phosphate buffer to give a concentration of 0.1 mg protein mL^{-1}. Add 1.0 mL of the diluted cell-free extract to the curvette and allow 1 to 2 min for temperature equilibration. When the temperature is constant, add 5 µL of the catechol solution, plug the end of the curvette, invert it rapidly to insure mixing, and place it immediately into the spectrophotometer, which is hooked up to a chart recorder, or some other appropriate signal device. Metapyrocatechase activity is measured at 375 nm, while orthopyrocatechase activity is measured at 260 nm.

Product formation will stop in a few minutes as noted by a plateau. This position on the chart recorder will then represent the complete stoichiometric conversion to product. Thus, the specific activity can be determined from the rate (slope of product formation) per unit of protein added. Occasionally, stoichiometry of the *meta* ring fission product may be < 100%. Instead of a plateau, there will be a peak due to hydrolysis of the ring fission product. Should this happen, it will be necessary to preincubate the CFE at 55 °C for 15 min, which will inactivate hydrolase activity, but not metapyrocatechase activity (Dagley et al., 1960).

20–7 OXYGEN CONSUMPTION

20–7.1 Principles

Respirometry studies are rapid, provide information about a catabolic pathway, and compare relative transformation rates among different xenobiotic compounds. In the case of CFE, the stoichiometry of O_2 for an oxygenase-catalyzed reaction can be confirmed precisely since there is virtually no respiratory activity because of the absence of the particulate-linked respiratory cytochrome system. Respirometry with washed cells provides rejection or presumptory acceptance of what the intermediate products might be. However, the failure to oxidize a presumptive intermediate may also be caused by its inability to be transported into the cell. A classic example of a cell being unable to induce for a transport protein is the inability of *E. coli* to metabolize and use citrate, even though it has all the components of an active TCA cycle.

Many years ago, Warburg and Gilson manometers were commonly used in physiological work. These instruments are rarely, if ever used anymore in biodegradation research. A more rapid and less cumbersome method is the measurement of O_2 uptake in solution by an O_2 probe. Depending on the extent of information needed, each analysis takes only a few minutes, in contrast to hours by manometry. There is also little cleaning that accompanies the end of a run.

20–7.2 Materials

1. Oxygen meter with probe.
2. Recirculating water bath with tubing.
3. Chart recorder or other event marker.
4. Magnetic stirring platform.

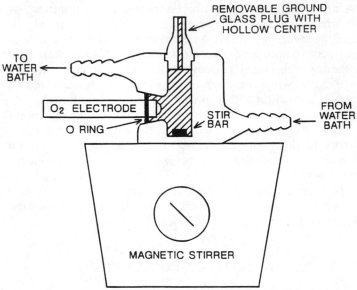

Fig. 20–2. Apparatus used for measuring consumption of dissolved O_2. Outer jacket maintains constant temperature of cell (inside) by a recirculating water bath. Drawn to half scale.

5. Magnetic stirring bar (3 by 12 mm).
6. Incubation cell and assembly (Fig. 20–2).
7. Cell suspension or cell-free extract.
8. Substrate concentrations (10–100 mg mL^{-1} in methanol or dimethyl formamide).
9. Vacuum flask and sipper.

20–7.3 Procedure

The example given below is for a system having a 2.0-mL reaction chamber. Amounts of materials can be adjusted accordingly. Higher concentrations of substrate are more suitable for washed cell suspensions where comparisons of rates are desirable. If the total stoichiometry with relation to O_2 uptake is desired, the concentration of the substrate should not exceed the oxygen demand available. For example, 1 µmol toluene (92 µg) would require 9 µmol O_2 for complete combustion to CO_2 and H_2O, yet only 0.5 µmol O_2 would be available in the entire 2.0 mL of the cell because of its low solubility (0.25 mM at 25 °C).

Cell suspensions or cell-free extracts, particularly metapyrocatechase, may lose activity with time if they are not kept on ice. However, they must first be equilibrated to the temperature of the water bath (25–30 °C) prior to adding the substrate. Since this may require several minutes between samples, it is a good idea to incubate a small batch of cells or cell-free extracts in the water bath to accommodate material for the next two runs. The first run should be with cells alone, to determine the base rate of O_2 consumption. With washed cells, this will represent primarily endogenous respiration. In the case of cell-free extracts, consumption of O_2 will be caused primarily by the electrode. Thus, it is important to take several baseline samples interspersed between the experimental samples, to obtain an accurate control level of consumption that will be subtracted from the actual experimental values. When the system is ready to go, add 10 μL of the solution substrate to the reaction chamber and record the downscale pen drift as O_2 is consumed. This rate when subtracted from the basal rate will represent the activity, which can be expressed in specific activity units per mass of cells or mass of protein.

When a run is finished, the material is removed with a sipper and the reaction chamber is rinsed several times with water. The sipper can be made from glass or stainless steel tubing that is connected to flexible tubing in line with a vacuum flask and vacuum line.

20–8 CHLORIDE DETERMINATION

20–8.1 Principles

Because chlorinated organic compounds represent the greatest contribution to soil and water pollution, there is considerable interest in the nature of halogen removal and in their catabolic pathways. There is a strong current interest in anaerobic dehalogenations as well (see chapter 13 in this book). Generation of inorganic chloride, moreover, is strong evidence for the metabolism to the compound in question. The following method is designed for pure cultures that grow in a chloride-free medium. The method is not intended for use in soil where other anions (e.g., nitrate) and ligands interfere. The method is based on a simple straightforward precipitation of inorganic chloride with $AgNO_3$ under acidic conditions that do not cause formation of insoluble phosphate or sulfate salts. The method has its drawbacks when insoluble precipitates, particularly dichlorobenzoic acids, are formed upon the addition of phosphoric acid. In some cases, the precipitate can be removed by centrifugation, while in others, it floats to the top, which necessitates removal by extraction with ethyl acetate or other suitable solvent prior to addition of silver nitrate.

20–8.2 Materials

1. 10 mM NaCl, in 100 mL of mineral salts solution.
2. 100 mL of a mineral salts solution.
3. Concentrated (15 M) H_3PO_4.
4. 1.0 M $AgNO_3$.
5. Test tubes.
6. Spectrophotometer.

20–8.3 Procedure

Prepare at least five standard concentrations to give a range of chloride from 0.2 to 3.2 mM by using the appropriate volumes of NaCl solution (0.2–3.2 mL) and the corresponding volume of mineral salts solution without NaCl (9.8–6.8 mL) to give the same volume of 10 mL in each tube. When the standards have been prepared, add 0.06 mL concentrated H_3PO_4 to all the tubes and mix thoroughly. Add 10 µL of $AgNO_3$ to one standard, mix thoroughly and measure immediately at 525 nm. Make a standard curve or calculate the regression equation. Use this for determining chloride in the culture.

From culture supernatants, add 10 mL to the spectrophotometer tube followed by 0.06 mL of concentrated H_3PO_4. If a precipitate forms, remove it by centrifugation or extraction as mentioned above. Add 10 µL $AgNO_3$ solution, and measure under the same conditions given above.

20–9 CONCLUSION

The real challenge for microbiologists is to identify and isolate the members of catabolic consortia that together, but not singly, can effect total destruction of the target molecule. This can only be done by logically envisioning a plausible catabolic pathway to establish separate enrichment cultures using the hypothetical degradation products for isolation of the desired microflora. Only when all members of the catabolic consortia can be identified, will it be possible to devise strategies for bioremediation (i.e., the biologically mediated detoxification of contaminated soil) of recalcitrant compounds. The isolation and identification of all members also makes for an attractive strategy for combining all the genes involved in the complementary catabolic pathway into a single organism. It is no longer relevant to think of biodegradation in terms of single organisms per se but rather in their contribution to the total gene pool that exists scattered in nature for the catabolism of any conceivable organic molecule.

REFERENCES

Angle, J. Scott, S.P. McGrath, and R.H. Chaney. 1991. New culture medium containing ionic concentrations of nutrients similar to concentrations found in soil solution. Appl. Environ. Microbiol. 57:3674–3676.

Alexander, M. 1965. Biodegradation: Problems of molecular recalcitrance and microbial fallability. Adv. Appl. Microbiol. 7:35–76.

Beam, H.W., and J.J. Perry. 1973. Co-metabolism as a factor in microbial degradation of cycloparaffinic hydrocarbons. Arch. Microbiol. 91:87–90.

Beam, H.W., and J.J. Perry. 1974. Microbial degradation of cycloparaffinic hydrocarbons via co-metabolism and commensalism. Arch. Microbiol. 91:87–90.

Bradford, M.M. 1976. A rapid and sensitive method for the quantitation of microgram quantities of protein utilizing the principles of protein dye-binding. Anal. Biochem. 72:248–254.

Brunner, W., F.H. Sutherland, and D.D. Focht. 1985. Enhanced biodegradation of poly-chlorinated biphenyls in soil by analog enrichment. J. Environ. Qual. 14:324–328.

Cairns, J., J. Overbaugh, and S. Miller. 1988. The origin of mutants. Nature (London) 335:142–145.

Carney, B.F., and J.V. Leary. 1989. Novel alterations in plasmid DNA associated with aromatic hydrocarbon utilization by Pseudomonas putida strain R5-3. Appl. Environ. Microbiol. 55:1523–1530.

Chatterjee, D.K., J.J. Kilbane, and A.M. Chakrabarty. 1982. Biodegradation of 2,4,5-trichlorophenoxyacetic acid in soil by a pure culture of Pseudomonas cepacia. Appl. Environ. Microbiol. 44:514–516.

Colowick, S.P., and N.O. Kaplan. 1957. Methods in enzymology III. Academic Press, New York.

Crawford, R.L., and W.W. Mohn. 1985. Microbiological removal of pentachlorophenol from soil using a Flavobacterium. Enzyme Microb. Technol. 7:617–620.

Dagley, S. 1984. Introduction. p. 1–11. In D.T. Gibson (ed.) Microbial degradation of organic compounds. Marcel Dekker, New York.

Dagley, S., W.C. Evans, and D.W. Ribbons. 1960. New pathways in the oxidative metabolism of aromatic compounds by micro-organisms. Nature (London) 188:560–566.

DeBont, J.A.M., M.J.A.W. Vorage, S. Hartmans, and W.J.J. van den Tweel. 1986. Micro-bial degradation of 1,3-dichlorobenzene. Appl. Environ. Microbiol. 52:677–680.

deKlerk, H., and A.C. van der Linden. 1974. Bacterial degradation of cyclohexane: Partici-pation of a co-oxidation reaction. Antonie van Leeeuwenhoek 40:7–15.

Ecker, S., T. Widmann, H. Lenke, O. Dickel, P. Fischer, C. Bruhn, and H.-J. Knackmuss. 1992. Catabolism of 2,6-dinitrophenol by Alcaligenes eutrophus JMP 134 and JMP 222. Arch. Microbiol. 158:149–154.

Edgehill, R.U., and R.K. Finn. 1983. Microbial treatment of soil to remove pentachlorophe-nol. Appl. Environ. Microbiol. 45:1122–1125.

Elsner, A., F. Löffler, K. Miyashita, R. Müller, and F. Lingens. 1991. Resolution of 4-chlo-robenzoate dehalogenase from Pseudomonas sp. strain CBS3 into 3 components. Appl. Environ. Microbiol. 57:324–326.

Focht, D.D., and W. Brunner. 1985. Kinetics of biphenyl and polychlorinated biphenyl metabolism in soil. Appl. Environ. Microbiol. 50:1058–1063.

Focht, D.D., and D. Shelton. 1987. Growth kinetics of Pseudomonas alcaligenes C-O relative to inoculation and 3-chlorobenzoate metabolism in soil. Appl. Environ. Microbiol. 53:1846–1849.

Gale, E.F. 1952. The chemical activities of bacteria. Academic Press, London.

Gibson, D.T., and V. Subramanian. 1984. Microbial degradation of aromatic hydrocarbons. p. 181–252. In D.T. Gibson (ed.) Microbial degradation of organic compounds. Marcel Dekker, New York.

Haigler, B.E., and D.T. Gibson. 1990. Purification and properties of NADH-ferrodoxin$_{NAP}$ Reductase, a component of naphthalene dioxygenase from Pseudomonas sp. strain NCIB 9816. J. Bacteriol. 172:457–464.

Haigler, B.E., S.F. Nishino, and J.C. Spain. 1988. Degradation of 1,2-dichlorobenzene by a Pseudomonas sp. Appl. Environ. Microbiol. 54:294–301.

Hall, B.G. 1990. Spontaneous point mutations that occur more often when advantageous than when neutral. Genetics 126:5–16.

Hernandez, B.S., F.K. Higson, R. Kondrat, and D.D. Focht. 1991. Metabolism of and inhibition by chlorobenzoates in Pseudomonas putida P111. Appl. Environ. Microbiol. 57:3361–3366.

Horvath, R.S. 1972. Microbial cometabolism and the degradation of organic compounds in nature. Bacteriol. Rev. 36:146–155.

Munnecke, D.M. 1979. Hydrolysis of organophosphate insecticides by an immobilized-en-zyme system. Biotechnol. Bioeng. 21:2247.

Pemberton, J.M., B. Corney, and R.H. Don. 1979. Evolution and spread of pesticide de-grading ability among soil micro-organisms. p. 287–299. In K.N. Timmis and P.A. Pühler (ed.), Plasmids of medical, environmental and commercial importance. Elsevier/North Holland Biomedical Press, Amsterdam.

Perry, J.J. 1984. Microbial metabolism of cyclic alkanes. p. 61–97. In R.M. Atlas (ed.) Petroleum microbiology. Macmillan, New York.

Pertsova, R.N., F. Kunc, and L.A. Golavleta. 1984. Degradation of 3-chlorobenzoate in soil by pseudomonads carrying biodegradative plasmids. Foila Microbiol. 29:242–247.

Reineke, W., and H.-J. Knackmuss. 1984. Microbial metabolism of haloaromatics: Isolation and properties of a chlorobenzene-degrading bacterium. Appl. Environ. Microbiol. 47:395–402.

Shraa, G., M.L. Boone, M.S.M. Jetten, A.R.W. van Neerven, P.J. Colberg, and A.J.B. Zehnder. 1986. Degradation of 1,4-dichlorobenzene by *Alcaligenes* sp. strain A175. Appl. Environ. Microbiol. 52:1374–1381.

Simkins, S., and M. Alexander. 1984. Models for mineralization kinetics with the variables of substrate concentration and population density. Appl. Environ. Microbiol. 47:1299–1306.

Sylvestre, M., and J. Fauteux. 1982. A new facultative anaerobe capable of growth on chlorobiphenyls. J. Gen. Appl. Microbiol. 28:61–72.

Spain, J.C., and S.F. Nishino. 1987. Degradation of 1,2-dicholorobenzene by a *Pseudomonas* sp. Appl. Environ. Microbiol. 53:1010–1019.

Tsien, H.C., and R.S. Hanson. 1992. Soluble methane monooxygenase component-B gene probe for identification of methanotrophs that rapidly degrade trichloroethylene. Appl. Environ. Microbiol. 58:953–960.

Vandenbergh, P.A., R.H. Olsen, and J.F. Colaruotolo. 1981. Isolation and genetic characterization of bacteria that degrade chloroaromatic compounds. Appl. Environ. Microbiol. 42:737–739.

Chapter 21

Algae and Cyanobacteria

F. BLAINE METTING, JR., *Battelle, Pacific Northwest Laboratories,*
Battelle Boulevard, Box 999, P7-54,
Richland, Washington

Algae are O_2-evolving photosynthetic organisms that contain plant-like chlorophylls but are distinct from higher plants in that every gametangial cell is fertile—sexual structures with differentiated tissues are not produced.[1] Thus the term algae encompasses seaweeds (macroalgae) as well as microscopic forms (microalgae). Phycology, the study of algae, has evolved historically to include the prokaryotic cyanobacteria (blue-green algae) because of their morphological, physiological, and ecological similarities to eukaryotic microalgae (Bold & Wynne, 1985). This chapter will present methods for studying microalgae and cyanobacteria that inhabit soil and rocks (endolithic habitats). General methods for studying algae in the laboratory and field can be found in a four volume series of monographs sponsored by the Phycological Society of America (Stein, 1973; Hellebust & Craigie, 1978; Gantt, 1980; Littler & Littler, 1985).

Microalgae inhabit all terrestrial surfaces. Growth rates and the complexity of community development depend primarily on available light and moisture, but are also influenced by temperature and properties of the substrate, such as pH. Numbers and biomass of microalgae vary greatly among soils, commonly ranging, respectively, between 10^9 and 10^{10} colony-forming units (CFU) per m^2 (10^3–10^6 per g), and from < 1 to 1500 kg wet weight ha^{-1} (Metting, 1993). The largest communities (in terms of biomass) are found in flooded rice (*Oryza sativa* L.) soils (Roger & Kulasooriya, 1980). Successively smaller mixed populations occur on arid and semiarid soils with minimal plant cover, temperate agricultural soils, and grasslands, forests, and other habitats where plant canopies and litter limit the flux of sunlight at the soil surface. On and in rocks, populations of

---

[1]The gametangium is the sexual organ that produces gametes. Individual cells of many single-celled soil microalgae (e.g., *Chlamydomonas*) act as gametes. In algae with multicellular gametangia, each cell is a potential gamete.

microalgae form in cracks and fissures (chasmolithic communities) and in the fabric of the rock itself (cryptoendolithic communities). Cryptoendolithic communities are remarkably similar in both hot and cold deserts throughout the world. Biomass estimates for these communities range from about 2 to 200 mg (as chlorophyll a) per m^2 (Metting, 1991).

Organic matter production by photosynthesis (CO_2 fixation), N_2 fixation, and surface consolidation are significant ecological attributes of the soil algal community. All of these activities are particularly important in semiarid and arid habitats (steppes and deserts) characterized by minimal vascular plant cover. In these situations, microalgae, together with lichens and mosses, make up what are known as cryptogamic crusts. In many situations, these crusts are the major source of biologically fixed N and can be a major factor acting to minimize erosive loss of topsoil by water runoff and wind (Metting, 1991). There is evidence that increased rates of desertification, such as in sub-Saharan Africa, can result from excessive grazing of fragile landscapes that destroys cryptogamic crusts by the trampling of livestock (Knutsen & Metting, 1991).

The most important influence that microalgae have in agriculture is biological N_2 fixation by cyanobacteria in wetland rice cultivation. Cyanobacterial N_2 fixation can provide all of the N requirements in some traditional farming systems (China, India, and Vietnam) and can contribute to the fertility of soils in which high-yielding hybrid rice varieties are grown. Estimates of 100 kg of N fixed ha^{-1} yr^{-1} or more have been reported, although values between 10 and 30 kg are more common. Both free-living species and cyanobacteria symbiotic with the water fern *Azolla* are important, although grazing by arthropods and competition from non N_2-fixing microalgae limits their overall contribution to rice production (Roger et al., 1993). On pivot-irrigated farm ground in the USA, mass-cultured green microalgae that produce copious quantities of extracellular mucilages (mostly *Chlamydomonas* spp.) have been used on a small scale as soil-conditioning agents to control erosion of sandy soils and improve infiltration of water into heavy soils (Metting et al., 1988). Because of their ecological importance and because they are physiologically similar to plants, soil microalgae and cyanobacteria have also proven to be valuable as ecological indicators for studying non-target pesticide effects (Pipe, 1992).

21–1 IDENTIFICATION OF SOIL ALGAE AND CYANOBACTERIA

21–1.1 Principles

Algae and cyanobacteria are readily recognized as distinct from other soil microorganisms by their chlorophyll a content. Using standard light microscopic methods (see chapter 6 by Bottomley in this book), cultured algae or algal cells and filaments in soil samples appear as various shades of green, blue-green, and yellow-green depending on the content of accessory pigments that can include various combinations of different chlorophylls, carotenoids, xanthophylls, and phycobilins.[2] Table 21-1 shows the major algal classes that include terrestrial microalgae and lists references to principal taxonomic treatises with descriptions of many of the species.

Identification of microscopic algae at the genus level is based on morphological properties. The major distinction among soil algae is between the prokaryotic cyanobacteria and the eukaryotic groups. It is apparent upon microscopic examination that cyanobacterial cells lack an internal organization that is homologous to the system of organelles, particularly chloroplasts and nuclei, that are readily distinguishable in eukaryotic microalgae. The distinctive coloration imparted by blue-green and red phycobilins also set the cyanobacteria apart from nearly all other soil algae. Most cyanobacteria are blue-green because of a preponderance of phycocyanins. Most unicellular cyanobacteria and many species within the common filamentous N_2-fixing genera *Anabaena* and *Nostoc* are good examples. Other cyanobacteria appear brown or black due to a large content of red phycoerythrins together with the blue-green phycocyanins. Examples include numerous filamentous N_2-fixing genera inhabiting rice soils, such as *Aulosira* and *Tolypothrix* (Roger & Kulasooriya, 1980).

Chlorophyceae (green algae) and Xanthophyceae (yellow-green algae) are not reliably differentiated from one another on the basis of color. Green algae are usually grass-green, but many soil-inhabiting species are shades of olive- or yellow-green. Mature cells and resting spores can accumulate large quantities of carotenoids and will appear various shades of orange or red. Yellow-green algae can also be grass-green or yellow-green. Two features are used to distinguish these groups, the nature of the chloroplast(s) and the presence or absence of starch. Chloroplasts in green microalgae vary widely in number (usually one), shape (cup-shaped, reticulate, axial, and variously lobed) and location (parietal, polar, and central) within the cell. In contrast, most yellow-green microalgae harbor many small, uniformly disc-shaped, parietal chloroplasts (Metting, 1981). Green algae accumulate starch as the storage product of photosynthesis; yellow-green algae store chrysolaminarin, oils, or fats (Bold & Wynne, 1985). Presence or absence of starch is readily determined by staining with a dilute solution of iodine in potassium iodide. A black staining reaction indicates the presence of starch.

Diatoms are common soil inhabitants and easily recognized by the distinct nature of the cell wall, which is known as the frustule. The frustule is made of silicon and has two halves, known as valves, that fit together like the top and bottom of a petri dish. Ornamentation of the valves is the principal criterion for taxonomic treatment at the species and subspecies levels (see section 21–2.6). Soil diatoms are usually colored various shades of brown or golden-brown.

---

[2] Algal divisions and classes are based, in part, on pigmentation together with the chemical nature of the cell wall and of the principal storage product (photosynthate).

Morphological variation among species within all classes of cyanobacteria and eukaryotic microalgae includes unicellular (single-celled), colonial, siphonous, and filamentous forms. Cell diameters range from 1 to 2 μ (e.g., the unicellular cyanobacterium *Anacystis*) to 50 μ or more (e.g., the green coccoid alga *Neospongiococcum*). Coccoid species are round unicellular forms that do not aggregate (except during cell division). Coccoid green microalgae are extremely common soil inhabitants (e.g., *Chlorococcum* and related genera). Non-coccoid unicellular species have ellipsoid, fusiform, ovoid, or cylindrical cells (Metting, 1981).

Two types of colonial algae are common in soil. In the first, cellular aggregates are indefinite in size, cell number, and relative cellular position; cell division is continuous under optimal growth conditions, and the aggregates readily fragment. Species exhibiting this morphological habit are often aggregated within a common, palmelloid matrix of extracellular polysaccharide. Examples include commonly encountered cyanobacteria (e.g., *Gloeocapsa*) and green microalgae (e.g., *Chlamydomonas* and *Palmella*). The second colonial type, known as a coenobium, has a fixed number and arrangement of cells. Sarcinoid aggregates are packets with cell numbers in multiples of four (rarely more than 16 or 32 cells). Common soil genera are the green microalgae *Chlorosarcinopsis*, *Tetracystis*, and *Fasiculochloris* (Metting, 1981).

Siphonous microalgae include all bulbous and tubular-shaped eukaryotic genera and species that are multinucleate (coenocytic). These genera are distinguished morphologically by the conspicuous lack of transverse walls or septa. The yellow-green genera *Botrydium* and *Vaucheria* are common soil algae that inhabit garden paths and bench tops in greenhouses. On soil or agar, *Botrydium* appears as a collection of bubble-like sacs that can be visible to the naked eye. *Vaucheria* is a branching filament whose oogamous sexual production of zygospores is commonly observed by microscopic examination of samples collected from soil surfaces. Other siphonous microalgae are much less common in soil.

Filamentous algae and cyanobacteria are common in soil. Filaments can be branched or unbranched and, particularly for cyanobacteria, are often enclosed together as bundles within a common gelatinous matrix. Filamentous cyanobacteria include individual species with the greatest degree of cellular differentiation within the prokaryotic kingdom. Many species have barrel-shaped vegetative cells that house photosynthetic functions, heterocysts (rounded, anaerobic N_2-fixing cells), and thick-walled resistant spores known as akinetes on a single filament whose length can be as long as a few millimeters. *Anabaena* is perhaps the best example of a genus with this morphology. *Nostoc* is closely related to *Anabaena* and is also common on soil. *Nostoc* filaments can be linear or they can be organized into variously shaped globular aggregates, the diameters of which can reach several centimeters on soil and in flooded rice fields. Many genera of filamentous green and yellow-green algae are also common soil inhabitants. Ubiquitous genera of green microalgae include *Stichococcus,* short filaments with 10 to 15 ovoid cells, and *Klebshormidium,* whose filaments can be many dozens of cells in length. *Bumilleria* and *Tribonema* are representative filamentous yellow-green algae (Metting, 1981).

Most microalgae inhabiting soil and rocks are not taxonomically distinct from planktonic forms, although certain groups are proportionately more common in terrestrial habitats and many planktonic groups have never been detected in soil. However, many coccoid, sarcinoid, and pseudo-filamentous green and yellow-green microalgae may be unique to the soil habitat. Hundreds of species in at least 185 algal and cyanobacterial genera have been identified in or isolated from soil or rocks (Metting, 1981, 1991).

21–1.2 Materials and Procedure

Use any brand of light microscope together with glass slides and other standard materials (see chapter 6 by Bottomley) for identification of soil algae and cyanobacteria. No special materials are required. Detailed (Prescott, 1978) and diagrammatic (Metting, 1981) keys are recommended to help in the preliminary identification of soil algae to the genus level. The references in Table 21–1 are recommended for confirmation of generic identity and identification of individual species.

Mount the soil sample or algal biomass collected from a culture vessel onto a glass microscope slide and examine it under the microscope beginning with scanning power. To distinguish anatomical features of eukaryotic microalgae (chloroplast number, size, and shape) or to view small cells (1–5 µ), it may be necessary to use the oil emersion with the highest power lens.

Table 21–1. Algal classes that include soil algae.

Algal class†	Common name of the group	References to taxonomic treatment of species
Chlorophyceae	Green algae	Ettl, 1983 Komárek & Fott, 1983 Starmach, 1972
Euglenophyceae	Euglenoids	Leedale, 1967
Xanthophyceae	Yellow-green algae	Ettl, 1978
Bacillariophyceae	Diatoms	Bock, 1963 Krammer & Lange-Bertalot, 1986, 1988
Cyanophyceae	Cyanobacteria (blue-green algae)	Castenholtz et al., 1989 Geitler, 1930–32 (reprinted, 1985)

† Algal classes with species that are encountered only rarely in soil include Rhodophyceae (red algae) and Eustigmatophyceae (eustigmatophytes).

21–2 DIRECT METHODS FOR ENUMERATION

21–2.1 Principles

Direct methods include techniques that permit or attempt some measurement or estimation of quantitative or qualitative (taxonomic) properties of the soil microalgal community at the time of sampling by direct examination of the soil sample. Special methods for obtaining soil samples are not necessary if the samples are subjected to microscopic examination soon after collection or if quantitative data are not the objective. For quantitative or comparative study, however, the following method is recommended for obtaining samples of the surface veneer of soil. Microalgal communities are only commonly abundant or play important roles in the upper few millimeters to centimeters of soil where light is available for phototrophic growth.

21–2.2 Materials and Procedure for Collecting Soil Veneers

A special method is required for collecting samples for studying soil algae. Quantitative data for soil algae should be reported on the basis of area. Thus, values per square centimeter should be reported. For dense populations in rice soils, values per square meter are common. It is also acceptable to report values per gram of soil, but not in lieu of reporting on the basis of area. For most soils, 1 cm^2 by 1 cm in depth is approximately equal to 3 g (air-dry) soil.

Sampling devices (samplers) for collecting soil surface veneers are not readily available from scientific supply houses. Therefore, it is necessary to have the samplers fabricated. Choose a material that is strong enough to permit the use of force in penetrating hard dry soils and that will withstand repeated autoclaving. The preferred sampler is crafted from hollow, stainless steel tubing that is cylindrical or square in cross-section, 20 cm in length, and beveled at a 45 ° angle on one end to facilitate vertical penetration of the soil. Round samplers are preferred although square ones can be used. A loose-fitting stainless steel cap is made to fit the non-beveled end of the tube. For most soils, a cross-sectional area of 1 or 3.33 cm^2 is ideal. The larger area is used for algal populations that are not readily visible. In this situation, it is common to pool three samples giving a total of 10^2 cm with which to work. A solid round (or square) dowel or other implement for use in pushing soil cores through the samplers should be fabricated to exactly fit within the sampler. For repeated collection of soil samples for studying algae, at least 20 to 30 samplers and two solid dowels should be available (Rayburn et al., 1982).

Samplers should be capped, packaged in brown paper sacks (three, four, or five per sack), and sterilized by autoclaving or baking in an oven prior to sampling. Orient the samplers in the sacks so that the capped end is grasped when opening the sack in the field and the beveled (open) ends are not exposed.

The choice of a random or systematic sampling pattern depends on the objectives of the study and limitations imposed by the site (see chapter 1 in this book). Collect samples by pushing the beveled end of a sampler into the soil to a depth of at least half the length of the tube. If necessary, use a rubber mallet to facilitate penetration of a hard soil. Place the sampler with its cap intact and containing the fresh core, beveled end first, into a clean sack, plastic bag, or other suitable container. Repeat the procedure with a new sampler and as many times as needed to collect the required number of samples.

Return the samples to the laboratory, keeping them dark and at ambient temperature. Samples may be stored on ice if it will be more than a day between sampling and analysis. Depending on the study objectives and any preliminary information or assumptions regarding the structure or density of the surface algal communities, soil veneers from the individual sample tubes can be analyzed separately (as 1- or 3.33-cm^2 samples) or they can be pooled. Alternatively, a single 1- or 3.33-cm^2 sample will often yield sufficient chlorophyll when pigmentation is used as the algal biomass index.

Process the samples at the laboratory bench and have ready the solid dowel(s), a supply of alcohol, a forceps, a supply of single-edged razor blades, and suitable borosilicate flasks, beakers, or other containers for collecting the soil samples. It is also recommended that a lined garbage can and plastic bucket or tub filled with detergent and hot water be close by for convenient disposition of used sample tubes.

Holding a sample tube in one hand, use the other hand to remove the cap and then to force the soil core through the tube by pushing with the solid dowel. When the surface of the soil core is pushed beyond the lip of the tube, use a flame-sterilized razor blade to carefully slice off a measured thickness of the veneer into a waiting flask or beaker. Continue to push the entire core through the tube into the garbage can, put the tube and cap into the wash tub, and repeat the procedure with the remaining sampling tubes.

21–2.3 Microscopy and Enumeration by Cell Counting

General methods for the use of light microscopy in the qualitative and quantitative examination of soil microalgae are the same as those described for bacteria and fungi (see chapter 6 in this book). However, because they are often much larger than heterotrophic eubacteria and are pigmented, the direct observation of microalgae and cyanobacteria in soil is less difficult. Any light microscope equipped with a high-intensity light source, a condenser, and a stage that is capable of accommodating a cell-counting chamber (haemacytometer) can be used for standard microscopic examination (Waaland *in* Gantt, 1980). A special apparatus and procedure, described below (21.2.4.2) is required for fluorescence microscopy. The limitation to optical methods is interference from soil particles, particularly when cell densities are less than about 10^4 per cm^2. Special cytological methods for cultured algae, including staining procedures, are detailed in Gantt (1980).

Two general methods for counting microalgal cells in soil or soil dilutions include the use of a standard cell-counting chamber (see chapter 6 in this book) or counting directly on glass slides that can be calibrated with the ocular and stage micrometers. Except for the larger, filamentous forms, the use of the counting chambers, with either transmission or epifluorescence microscopy, for enumeration of microalgae in soil samples is usually practical only for population densities greater than or equal to about 10^4 to 10^5 per cm^2.

21–2.4 Enumeration by Chlorophyll Autofluorescence

21–2.4.1 Principles

Chlorophyll strongly fluoresces when excited by blue and UV wavelengths. This is the basis for estimating microalgal biomass by extraction of chlorophyll from soil (21–3.3). It also permits the direct microscopic observation of algae in soil and allows distinction between living and nonliving cells (Shields, 1982). Although a transmission fluorescence microscope is suitable, epifluorescence is preferable because of its greater versatility and ease of use. For example, epifluorescence microscopy permits nearly simultaneous viewing under fluorescence and bright-field or phase-contrast images. Also, cell-counting chambers can be used with epifluorescence microscopy but, because of their thickness, cannot be used for transmission microscopy. For transmission microscopy, a special condenser illuminates the specimen so that light from the condenser does not directly enter the objective. Only light reflected, scattered, or fluoresced by the specimen enter the objective. The object thus appears to be self-luminous in a surrounding dark field.

21–2.4.2 Materials and Procedure

If a standard fluorescence microscope is not available, it is possible to convert a standard compound microscope for the purpose. Packages to upgrade a standard system for fluorescence applications are available from most microscope manufacturers. Components of a package will include a UV light source (100 W Mercury arc lamp), a dark-field condenser, and appropriate excitation and barrier filters. Some microscopes may also require modified objectives (J.R. Johansen, 1992, personal communication).

Prepare an appropriate dilution of soil, taking care to avoid transferring particles ≥ 0.1 mm into the cell-counting chamber. Focus the condenser of the light source to produce a parallel beam. Insert a blue filter between the light source and microscope. Introduce one drop of soil suspension into the counting chamber. The best magnification for epifluorescence enumeration is 100 to 400×. Count the algal cells in several fields or along several transects calculated to be statistically appropriate (see chapter 2 by Parkin and Robinson in this book). Living cells appear red against a black background, while dead cells, although sometimes difficult to distinguish, appear greenish gray. Soil particles will sometimes fluoresce yellow or green. Work in a darkened room for best results.

21–2.5 Implanted Slide Method

21–2.5.1 Principles

The procedure involves the implantation of a microscope slide vertically in the soil, leaving the top 1.5 cm of the slide projecting above the soil surface. Light attenuates downward from the exposed portion of the slide, encouraging differential microalgal growth on the surface of the slide in contact with the soil, thus providing a means by which to study the vertical distribution of the soil algal community (Pipe & Cullimore, 1980).

21–2.5.2 Materials and Procedure

Mark two glass microscope slides with small scratches at 2-mm intervals along their entire lengths and near both edges of one face on each slide. Place the two marked faces together, fastening pairs of slides together at each end with an adhesive tape. Push slide pairs into the soil until only the upper 1.5 cm is exposed. To hasten colonization of the slides or for comparative purposes, the soil may be moistened with deionized glass-distilled water or with an appropriate algal medium (section 21–5.1.3). After incubation for the desired length of time, carefully retrieve the paired slides. The best way to remove the paired slides is in a block of soil, rather than by pulling the slides from the soil. Carefully chip the soil away from the exposed surfaces of each slide and then separate each pair. Allow the slides to air dry. Remove large soil aggregates by gentle tapping against a hard surface. Spray the colonized surfaces with a molten 2% agar solution in water. Dehydrate the slides in the air stream provided by a laminar flow bench or a hair dryer. The result will be a thin covering of agar that acts as an adhesive covering the soil particles and microalgae and sealing them onto the surface of the glass.

The slides can be microscopically examined immediately (or following storage) with the addition of a few drops of water. Wrapped in aluminum foil and kept dry in the dark at room temperature, slides can be stored for up to 2 yr. Individual slides can be alternately wetted and dried through several cycles without consequence, so long as they are not kept moist for longer than a few hours at any one time. Reference to the scratch marks permits quantification of spatial distribution on the surface as a function of the original vertical distribution in soil.

21–2.6 Method for Diatom Frustules

21–2.6.1 Principles

Frustule is the term used to designate the cell wall of diatoms. Each frustule is composed of two halves, or *valves,* that fit together much like the halves of a petri dish. Diatoms are unique among algae in that their cell walls are composed of silicon. Because the taxonomy of diatoms is based on frustule ornamentation and because silicon is resistant to harsh treatment, an important aspect of the study of these microalgae is the use of chemical methods for the preparation of samples for qualitative and quantitative microscopic study. The following quantitative method has been used successfully in several studies of soil diatoms (e.g., Johansen et al., 1982, 1984).

21–2.6.2 Materials and Procedure

Prepare a composite sample of 1- or 3.33-cm^2 samples as described above (section 21–2.2). Dry the composite sample for 1 h at 105 °C in an oven. Pass the dry sample through a 1-mm mesh screen. Disperse a 1-g subsample in 20 mL of deionized glass-distilled water in a 100-mL beaker. Add 10 mL of concentrated HNO_3 and bring the mixture to a boil on a hotplate placed behind the protective sash in a forced-air fume hood. If the soil has a moderate to high organic matter content, additional oxidation is beneficial and can be achieved by the addition of about 0.5 g (or more) $K_2Cr_2O_7$ after the first 10 min of boiling in concentrated HNO_3. Be prepared to avoid spattering (bumping) as the mixture boils by repeatedly removing the beaker from the hotplate. Discontinue the heating and allow the beaker and contents to cool to room temperature when the volume has been reduced by boiling to < 20 mL.

Divide the 20-mL sample into two equal portions in 15-mL glass or plastic centrifuge tubes. Centrifuge the contents for a few seconds in a tabletop clinical centrifuge. Decant the liquid supernatants. Bring the volume back to 10 mL with water and resuspend the soil with a vortexer. Repeat the centrifugation, decanting, and resuspension six or more times. After the final rinse, bring the volume of cleaned soil to 10 mL. Repeated rinsing with deionized glass-distilled water is necessary to remove all of the nitric acid, to prevent the appearance of acid halos that otherwise interfere with microscopic examination of the details of frustule morphology. Make a 10-fold dilution of this mixture in deionized glass-distilled water for preparing slides for microscopic examination.

Semi-permanent slides for repeated study can be made in a suitable mounting medium (Hyrax and Naphrax are recommended; J.R. Johansen, 1992, personal communication). Place 0.5 mL of the diluted soil mixture on a clean 18-mm square glass coverslip. Allow to air dry overnight. The next day, add two to three drops of mounting medium on a glass slide and place it on a hotplate. The heat should be sufficient to boil the mountant within 5 to 20 s. Boil the mountant until the boiling slows but does not stop (a few seconds). Remove the slide from the heat and place the 18-mm square coverslip, diatom side down, onto the heated mountant. Place the slide back on the hotplate allowing it to boil again while tapping the coverslip lightly with a teasing needle to spread the medium to its edges. Tapping the slide several times onto the edge of the work bench by dropping it from a height of about 3 cm will help rid the mount of entrapped air bubbles.

Enumerate the diatom frustules at $1000\times$. Determine the width of the field of view and convert to millimeters. This value times 18 is the area of the slide covered by one transect. To obtain quantitative counts, start at the top or bottom of the slide at the nearest even whole number on the vernier scale of the microscope stage. Because each diatom frustule is composed of two halves (valves), count the number of valves and divide by two to obtain an estimate of the original number of frustules and, thus, diatom cells. Count the total number in nine transects, each 1-mm apart.

The number of cells per gram of dry soil is calculated by:

FC/TAM, where
F = number of frustules
C = area of coverslip (324 mm²)
T = number of transects scanned
A = area of one transect (18 mm × width of field), and
M = mass of air dry soil on coverslip (0.0025 g).

Identification of soil diatom species is performed by comparing the characteristic valve morphology with taxonomic descriptions given by Anderson and Rushforth (1976), Ashley et al. (1985), Johansen et al. (1981), and the authorities cited in Table 21–1.

21–2.7 Methods for Cyanobacteria in Rice Fields

21–2.7.1 Principles

Cyanobacteria are ubiquitous in rice fields. With minor modifications, the same direct and indirect methods described in other sections of this chapter are also used for sampling and analysis of rice fields for studying cyanobacteria and other microalgae. Modifications reflect the nature of the rice field environment and the fact that most of the important cyanobacterial species in rice fields are filamentous and often form dense mats. The methods described in this section are also suited for studying the algal component of non-agricultural wetlands, such as intermittently flooded marshes and estuaries (Roger et al., 1991).

Rice fields are usually flooded for some part or all of the cropping cycle. Thus, portions of the associated microalgal communities float and are subject to distribution patterns that are influenced by water movement, wind, and the spacing of individual rice plants. Individual filaments or portions of cyanobacterial aggregates can also move vertically in the water column and will adhere to submerged and emergent portions of the rice shoot (Roger et al., 1993).

Cyanobacteria are usually irregularly distributed in rice fields. Correlations between means and variances of different indices of cyanobacterial abundance (most probable numbers [MPNs], plate counts, wet biomass) and photodependent N_2 fixation (as acetylene reduction) in nearly all studies have shown in most cases that the variables are distributed in a lognormal pattern. This has been observed for single-locus samples collected in the same plot and for composite samples collected in replicated plots. For this reason, random sampling is discouraged. Sampling strategies should be selected only after careful examination of the conditions at the site (Roger et al., 1991).

21–2.7.2 Modified Procedure for Sample Collection, Serial Dilution, and Plating

The use of beveled sampling tubes (section 21–2.2) is adequate for non- flooded rice soils. A modification of the sampling tube is recommended for collecting samples from flooded fields. When obtaining samples, take care to minimize disturbance. To avoid muddying the waters, the person collecting the sample should stand or kneel on a dike or on a wooden board placed across the study area. In place of a beveled tube, use one with two blunt open ends. Instead of a cap at one end, close both ends with rubber stoppers. Gently push the tube down through the water column into the soil with a circular twisting motion. When the desired depth is reached: (i) plug the top, (ii) very carefully remove the tube, and (iii) plug the bottom while it is still submerged (Roger et al., 1991).

As has been mentioned, any of the direct or indirect cell counting, plating, MPN, or pigment extraction methods can be used when studying cyanobacteria in rice fields. However, because of the filamentous nature of many species and the fact that mucilagenous sheaths surrounding individual filaments or entire mats are common, it is necessary to use more forceful methods and longer periods of time when homogenizing samples prior to serial dilution for plating or inoculating MPN tubes. To facilitate disruption of filaments, Roger et al. (1991) recommend that soil samples be ground in a mortar and pestle and sieved to a particle size ≥ 0.25 mm and that composites of at least 10 samples that include the upper 5 mm of soil be stirred at 400 rpm for 15 min prior to serial dilution. For plating, agar concentrations $> 1\%$ are not recommended because they have been shown to inhibit the growth of some unicellular N_2-fixing species (Van Baalen, 1965). BG-11 medium is recommended (section 21–5.1.3.1). To discourage the growth of eukaryotic microalgae, add 20 to 50 mg cycloheximide per liter of medium. To enrich for N_2-fixing species, do not include fixed N in the medium (section 21–5.1.3).

21–2.7.3 Procedure for Visual Estimation of Coverage

For observational purposes, visual estimates often are sufficient to follow the development of algal communities during a cropping cycle when

Table 21–2. Indices of abundance for visual estimates of cyanobacteria in rice fields.

Relative abundance	Code	Index	Index percentage
No visible growth	0	0	0
Few colonies present	+	1	5
More colonies present	+ +	2	10
More colonies present	+ + +	3	15
More colonies present	+ + + +	4	20
Growth covers one-fourth of the area	1/4	5	25
Growth covers one-third of the area	1/3	6	33
Growth covers one-half of the area	1/2	7	50
Growth covers two-thirds of the area	2/3	8	67
Growth covers three-fourths of the area	3/4	9	75
Growth covers the entire area	All	10	100

it is either logistically difficult or not convenient to use more accurate methods. Roger et al. (1991) describe two methods. The square count method is convenient when algae are not abundant and when the rice plants have been transplanted in a grid pattern. The index of abundance is easier and more accurate when algal growth is denser. These techniques can also be used to estimate coverage by *Azolla* in fields where this water fern is used as a biofertilizer. Where farming practices do not include transplantation of seedlings into grid patterns, such as in the USA, where common practices include aerial seeding, it is necessary to construct a grid using spaced wooden stakes or steel rods.

21–2.7.3.1 Square Count Method. Individual rice plants in transplanted fields provide a grid that can facilitate making visual estimates of algal coverage. The square count method is used when algae are not abundant. The observer records the presence or absence (+ or −) of growth in as many squares as possible formed by individual rice plants. If desired as a preliminary measure, the percentage of squares with visible growth can be correlated to kg fresh weight ha^{-1}. Calculation of a regression equation for a representative area yields a useful tool for larger areas.

21–2.7.3.2 Index of Abundance Method. This is a semiquantitative index based on how much of the floodwater surface area is covered with algal growth. Use the subjective codes and indices shown in Table 21–2 to determine the average index percentage. This index is most accurate when standardized beforehand.

21–2.8 Procedure for Endolithic Microalgae

Except for collecting the rock specimens themselves, no special methods are used for the study of endolithic microalgae (E.I. Friedmann, 1992, personal communication). In the field, samples are chipped off of rocks and collected in Whirl-Pak plastic bags. Samples from cold deserts should be frozen for transport to the laboratory. Whole rocks, fragments, or crushed rock are used for the laboratory study of photosynthesis (O_2/CO_2 uptake/evolution) and N_2 fixation using standard methods (see chapters 38 by Zibilske and 43 by Weaver and Danso, respectively). Microalgae are isolated from rocks with standard plating techniques and are cultured in standard algal media. For extraction of chlorophyll (21–3.3.3), dimethyl sulfoxide is preferred (Bell & Sommerfeld, 1987; E.I. Friedmann, 1992, personal communication).

21–3 INDIRECT METHODS FOR ENUMERATION

21–3.1 Enumeration of Colony-forming Units on Solid Media

Standard soil dilution and plating methods (see chapter 8 by Zuberer in this book) are recommended for use in enumerating populations of soil algae as CFU on agar plates.[3] Simple modifications are used to compensate for the slower growth of microalgae and cyanobacteria relative to heterotrophic eubacteria and fungi.

Modifications of the standard plate count method for use with microalgae include:

1. The use of antibiotics to control the growth of bacteria. At lower soil dilutions (10- and 100-fold dilutions), soil particles are a source of organic compounds that support growth of heterotrophs. To discourage the growth of unwanted bacteria, supplement any algal medium with penicillin and streptomycin (21–5.1.3).
2. Pour the agar plates nearly full to reduce the likelihood of desiccation during the 1- to 3-wk (sometimes longer) incubation period. To prevent condensation from smearing the colonies, plates should be inverted and incubated on shelves under standard conditions (section 21–5.1.2). Wrapping the edges with parafilm will retard desiccation.

21–3.2 Most Probable Number Method for Enumeration

Prepare a soil dilution series as described in chapter 5. From each of six successive dilutions (10^2 through 10^6), transfer 1 mL to each of five 15-mL culture tubes containing one of the algal media described below (21–5.1.3). If the objective is the enumeration of specific groups, the appropriate medium and recommended supplements should be used (21–5.1.3). If, on the other hand, total numbers of microalgae (including cyanobacteria) are desired, soil water tubes using soil from the study site should be used (21–5.1.3.5). Incubate the tubes on culture tube racks under standard conditions (21–5.1.2). After 3 wk, record the number of

---

[3]For microalgae, colonies on agar are sometimes referred to as plant masses (Archibald & Bold, 1970).

tubes with visible algal growth. Find the MPN of microalgae in the original sample using the information provided in chapter 5 by Woomer.

21–3.3 Chlorophyll Extraction and Quantification

21–3.3.1 Principles

Extraction and spectrophotofluorometric quantification of chlorophyll is a useful index of soil algal biomass. Usefulness derives from the facts that (i) all groups of microalgae and cyanobacteria produce chlorophyll, (ii) the procedure is easy to perform and somewhat less labor and equipment intensive than the MPN or plate count methods, and (iii) data are available within hours of sampling, compared to the days or weeks required to incubate agar plates or culture tubes. The major disadvantages of the method are that (i) it fails to distinguish among microalgal species or higher taxonomic groupings (e.g., among cyanobacteria, eukaryotic microalgae, and lichen phycobionts), and (ii) as an index of biomass or metabolic activity, chlorophyll content can vary markedly among different microalgae, at various stages of growth within individual populations, and in response to environmental conditions (Metting & Barry, 1986; Sharabi & Pramer, 1973).

Two methods are described. The first is an ethanol extraction procedure developed for routine use with cultivated and irrigated silt loam and sandy loam soils (Metting & Barry, 1986). The second is a dimethyl sulfoxide (DMSO) procedure widely used for chlorophyll extraction from soils and rocks (Bell & Sommerfeld, 1987; J.R. Johansen, 1992, personal communication).

21–3.3.2 Procedure for Estimating Cell Density/Pigment Correlation in Soil

The chlorophyll method is most useful in controlled studies of the influence of experimental variables on unialgal populations on soil or when studying relatively homogeneous populations in the field. In both cases, a standard curve is prepared prior to the experiment by extracting and measuring chlorophyll from soil to which a known number of cells of a selected species with a known biomass has been added.

To prepare a curve, produce a unialgal population of the desired species in liquid culture under standard conditions (section 21–5.1.2). At the desired stage of log or stationary phase population growth, quantify the cell densities by direct microscopic enumeration (section 21–2.3). Depending on the species, culture densities can reach 10^6 per mL for filamentous cyanobacteria and as many as 10^8 or more per millimeter of unicellular eukaryotic species. Prepare a cell concentration and dilution series from the liquid culture by centrifugation and 10-fold serial dilution to yield 10 mL each of 10^3 through 10^{10} cells per mL for use in experiments in which cell densities on soil cannot be predicted. Add the 10-mL aliquots to 10 g (3 cm^2) of air-dry soil, mix thoroughly, and proceed to quantify the soil chlorophyll content as described below (section 21–3.3.3.2). The procedure is easily modified to produce a narrow range or one with more correlation points between log concentrations of cells.

21–3.3.3 Ethanol Extraction Method for Quantification of Chlorophyll in Soil

21–3.3.3.1 Materials. Have available:
1. Twice as many clean 50- or 100-mL flasks as there are samples to be processed. To half of them, add a pinch of Na_2SO_4 (approximately 1 g).
2. A Turner model 111 Filter Fluorometer or similar instrument with blanking rod and quartz cuvettes. A spectrophotometer, such as the Bausch & Lomb Spectronic 20 or 21 models, can be used in the absence of a fluorometer. However, fluorometric measurement is preferred because of its greater sensitivity, and hence usefulness, for detecting small quantities of pigment.
3. Two or more hot plates.
4. 95% ethanol, petroleum ether, and deionized glass-distilled water. Heat about 200 mL of 95% ethanol in a separate 500-mL flask and have it at hand for sample filtrates.
5. Two or more 250-mL side-arm flasks with Buchner funnels and no. 1 Whatman filter paper. Two or more 250-mL separatory funnels. Jars, jugs, or beakers for waste liquids. Canisters of 5- and 10-mL volumetric glass pipets. Two or more 25-mL graduated cylinders.

21–3.3.3.2 Procedure. Collect individual or pooled samples of the surface veneer of the soil to be studied, as outlined above (section 21–2.2). Perform the chlorophyll extraction procedure as soon as possible after collecting the samples to minimize changes in pigment concentration. There is no objective way by which to determine how much soil (how many cm^2) to use for a single extraction. In general, if algal growth is not visible to the naked eye, as much as 10 cm^2 may have to be extracted to yield sufficient chlorophyll for fluorometric quantification. If there is visible algal growth, then as little as 1 cm^2 may yield sufficient pigment. Luxuriant growth will yield an excess of chlorophyll-necessitating dilution prior to fluorometric measurement.

Perform all extraction and separation steps in a fume hood. Once familiar with the procedure, it is possible to simultaneously process a small number of flasks. However, it is recommended that the researcher first become familiar with the procedure by subjecting a single sample to the following steps.

Using the method described in section 21–2.2, add the surface veneer from the desired number of sample tubes to a clean dry flask and repeat for each desired sampling unit. Add 5 mL of cold 95% ethanol to each flask. Cover the mouth of each flask with aluminum foil. Perforate the foil with two or three small holes with a dissecting needle, safety pin, or the pointed end of a paper clip. While occasionally swirling the flask, bring the combined contents to a boil over low to moderate heat on a hotplate. To avoid bumping, remove the flask(s) from the hotplate as soon as they boil. Filter the contents through no. 1 Whatman filter paper into a 250-mL side-arm flask. Rinse the flasks with an additional 5 mL of hot 95% ethanol. Pour the leachate into a 25-mL graduated cylinder and note the volume (which should be ≤ 10 mL). Transfer the leachate to a 250-mL separatory funnel with the stopcock closed. Add an aliquot of petroleum ether equal to about twice the volume of the leachate. Gently swirl the contents for 3 to 4 min, holding the funnel at a 45° angle. Add a volume of deionized glass-distilled water equal to the total volume of leachate plus the petroleum ether, thus doubling the total liquid in the funnel. Swirl the contents for an additional 3 to 4 min. Drain the lower ethanol/water mixture and a small portion of the upper layer. Drain the remaining petroleum ether with the dissolved chlorophyll into one of the 50- or 100-mL flasks into which was placed the Na_2SO_4. Swirl the contents and fill a quartz cuvette with a representative fraction.

Excite each sample from the petroleum ether fraction at 430 nm (chlorophyll a lambda maximum) and pass to the detector mostly 669 nm fluorescence (chlorophyll a fluorescence maximum). Use the 1× slit arrangement. Dilute the sample in petroleum ether if the reading is off the scale. The density of chlorophyll in µg per cm^2 soil is equal to the fluorescence reading multiplied by 0.66 (times any appropriate dilution factors).

When using a spectrophotometer, the density of chlorophyll is given by the following relationship (Jensen in Hellebust & Craigie, 1978):

mg chlorophyll a mL^{-1} =

$$\frac{[(OD \text{ at } 666 \text{ nm}) - (OD \text{ at } 730 \text{ nm})] \times \text{mL of sample} \times 10}{890}$$

21–3.3.4 DMSO Extraction Method for Quantification of Chlorophyll in Soil

21–3.3.4.1 Materials and Procedure. The DMSO extraction procedure obviates the need for heating the samples and using separatory funnels, as described above (21–3.3.3.2). If rock samples are being extracted, it is recommended that they first be homogenized by crushing with a mortar and pestle or a rolling pin. Some benefit may also be derived from crushing and mixing soil samples.

Working in dim light to minimize degradation of the phaeophytin fraction, add about 1.5 g of rock or soil to 5 mL of DMSO in a screw-cap centrifuge tube. Mix the contents of the tube for 20 s on a vortexer and incubate for 45 to 60 min at 65 °C. Vortex the tubes a second time about half way through the incubation. Centrifuge at top speed in a clinical centrifuge for about 1 min or until the particulate matter has collected in a pellet. Record the absorbance of the supernatant solution at 665 and 750 nm. After each reading, pour the liquid contents back into the tube used

for the extraction and clean the cuvette with DMSO. Acidify the sample
with 10 μL (or two drops) of 1 N HCl. Wait 10 min and proceed to again
record absorbances at 665 and 750 nm. For each value at 665 nm, subtract
the reading at 750 nm to correct for turbidity. Calculate the concentrations
of chlorophyll a and phaeophytin from the following formulas (Lorenzen,
1967).

$$\text{mg chlorophyll } a \text{ g}^{-1} = \frac{A \times K \times (665_{t=0} - 665_{t=a}) \times V}{L}$$

$$\text{mg phaeophytin g}^{-1} = \frac{A \times K \times (R[665_{t=a}] - 665_{t=0}) \times V}{L}$$

where
A = absorption coefficient of chlorophyll a (= 11.0)
K = factor to equate reduction in absorbancy to initial chlorophyll a concentration (= 1.7, 0.7, or 2.43)
$t=0$ = absorbance after initial 45 to 60 min incubation
$t=a$ = absorbance after acidification for 10 min
L = path length of cuvette in centimeters
R = maximum ratio of $665_{t=0}$ to $665_{t=a}$ in the absence of phaeophytin
pigments

21–4 METHODS FOR ISOLATION AND PURIFICATION OF MICROALGAL CULTURES

Microalgae and cyanobacteria enriched from soil can easily be isolated
into unialgal culture and subsequently purified into axenic culture using
traditional bacteriological and traditional phycological methods developed
for planktonic species from marine and freshwater habitats (Hoshaw and
Rosowski in Stein, 1973).

Use sterile inoculating loops or flame-drawn glass Pasteur pipets to
remove individual microalgal colonies (or portions of colonies) from agar
surfaces (or liquid culture tubes) on (or in) which multiple colony types
have arisen following inoculation with aliquots from serial soil dilutions.
The colony material can then be inoculated onto the surface of a fresh agar
plate or into a culture tube containing a suitable liquid medium for subsequent purification into clonal unialgal or axenic culture.

21–4.1 Procedure for Preparing Fine Capillary Pipets

21–4.1.1 Principle

The procedure for preparing fine bored glass capillaries from individual glass Pasteur pipets is a standard phycological technique for collecting
and transferring single cells or short pieces of filaments from one tube, spot

dish, or petri dish to another. Individual capillaries are prepared one at a time as they are needed. To minimize contamination, always use pre-sterilized equipment and work at a laminar flow bench or in a clean room.

21–4.1.2 Materials and Procedure

Ahead of time, pack several borosilicate glass Pasteur pipets into a glass pipet canister (Bellco) with an aluminum closure after having first stuffed the blunt end of each pipet with a small piece of absorbent cotton. Autoclave the canisters. Dry them in an oven, and allow them to cool to room temperature before proceeding.

To make a capillary pipet, work at a laminar flow bench. Holding the blunt end of a pipet in one hand, grasp the tip of the opposite end with a pair of forceps. Hold a portion (between 3–8 to 10 cm) of the fine end of the pipet over a flame for 3 to 5 s, depending on distance from the flame until a short (1–2 cm) region begins to glow. When the glass is fluid, remove the pipet from the flame and, in one motion, pull evenly on both ends until the molten piece is reduced in thickness to the desired bore. Snap off the tip using the tweezers.[4]

To use a pipet, squeeze a 1- to 2-mL rubber bulb, attach it to the end of a pipet, and release the pressure on the bulb to create a vacuum. Touching the fine tip of the capillary to a portion of a colony on an agar surface or a cell(s) in solution is sufficient to draw the desired cells into the pipet. The cell(s) can then be ejected into a fresh culture tube or onto a fresh agar plate by compressing the rubber bulb.

21–4.2 Isolation and Purification by Repeated Washing

12–4.2.1 Principle

This traditional phycological method is used to isolate individual cells into clonal culture from mixed planktonic communities. It is also used to isolate single cells from mixed cultures in MPN tubes prepared from soil dilutions and homogenizations prepared from colonies on agar plates (Hoshaw and Rosowski *in* Stein, 1973).

21–4.2.2 Materials and Procedure

Prepare ahead of time

1. Sterile canisters of glass Pasteur pipets (21–4.1.2) and 7- × 22-mm glass spot dishes (Corning 7223) placed inside 100- × 20-mm glass petri dishes.
2. Sufficient volumes of the appropriate sterile algal medium (21–5.1), tap water, or deionized glass-distilled water.
3. Binocular dissecting microscope, forceps, and rubber bulbs.

---

[4]It is necessary to practice this technique to consistently prepare capillaries of desired thickness.

Line up a series of three (or more) dishes for a total of nine (or more) spot dishes. Add six to eight drops of the desired medium to all but one of the wells in the spot dishes. Pipet six to eight drops of the mixed algal culture to the remaining well. Perform all subsequent manipulations with the aid of a binocular dissecting microscope.

Use a flame-drawn glass pipet (section 21–4.1) to isolate individual cells or pieces of filaments from the first well and to squirt them into the adjacent well. While moving the tip of the pipet through the liquid, maintain sufficient pressure on the rubber bulb to prevent the premature uptake of liquid. Repeat until at least 12 to 15 cells have been isolated into the second well, using a fresh pipet for each cell. Repeat the procedure until as many cells as possible have been passed through as many wells as needed to ensure that at least one cell is available to inoculate fresh culture tubes (or agar surfaces) to give rise to clonal cultures.

21–4.3 Isolation and Purification by the Streak Plate Method

Prepare a suitable number of petri dishes containing the appropriate medium solidified with agar and supplemented with antibiotics to discourage bacterial contamination (section 21–5.1.3). Fill the petri plate nearly full with agar to delay desiccation during the time required for slow-growing isolates to develop adequately sized colonies from single cells. Place a small inoculum from the original spread plate or MPN tube in one corner by touching the agar surface with the inoculating loop or Pasteur pipet and streak for isolation. Incubate the plates upside down on shelves under standard conditions (section 21–5.1.2). Keeping the dishes upside down, scan the agar surface after 2 to 3 d at regular intervals at $30 \times$ power with a binocular dissecting microscope to locate individual cells or groups of cells that are distinctly separate from others. Mark the location of the cells by drawing a circle on the bottom of the closed petri dish with a marking pen. Turn the dish right side up and, using a higher power of magnification, proceed to collect the cell(s) with a sterile flame-drawn pipet. Deposit the cell(s) in a culture tube with the appropriate liquid medium. Repeat the procedure to collect as many individual cells as are reasonable. Incubate the tubes under standard conditions (section 21–5.1.2).

21–4.4 Purification by the Centrifugation Technique

Microalgae can usually be separated from adhering bacteria by repeated centrifugation and washing in sterile water with or without detergent. This method is based on the fact that microalgae settle out more readily than the much smaller bacteria.

Work with sterile equipment in a clean setting. Fill a sterile 5-, 10- or 15-mL glass tissue grinder with an appropriate aliquot from a log-phase culture of a unialgal culture. For best results with mucilagenous species, add a small drop of Tween 20 or Triton X-100 detergent. Homogenize the contents with a few short, twisting strokes, taking care not to use so much force that the contents of the reservoir are lost when plunging the barrel.

Fill a sterile 15-mL centrifuge tube with about 10 mL of the homogenate. Hold the tube on a vortexer for a few seconds. Centrifuge the tube at full speed in a tabletop clinical centrifuge (about 2000 rpm) for about 30 s. The microalgal cells will form a pellet. Pour off the supernatant. Vortex again for a few seconds. Quickly and forcefully add 10 mL of fresh sterile water to disperse the pellet. Repeat the centrifugation, decanting, vortexing, and resuspension steps. Do not add detergent after the first repetition. Place a drop from the pellet onto an agar surface after 12 repetitions (or sooner if there is a danger of losing the algal pellet altogether). Choose the appropriate algal medium (section 21–5.1.3). Supplement the medium with 0.5 g of a protein extract or 0.5 g of yeast extract to facilitate the identification of microalgal colonies that remain associated with bacteria. Streak the pellet to separate the individual cells. Incubate under standard conditions (section 21–5.1.2) and examine under a dissecting microscope, as described above (section 21–4.3).

21–4.5 Purification by the Zoospore Technique

Zoospores are the single-celled flagellated stage produced by many green and some yellow-green microalgae. A simple purification procedure depends on their motility and phototactic response (Kessler, 1984).

Inoculate a 30 mL culture tube containing about 5 mL of an appropriate sterile medium (section 21–5.1.3) with a loop full of algal mass taken from an agar surface or another liquid culture. Aseptically place a wad of sterile cotton above the level of liquid in the tube and then add enough additional sterile medium to saturate and totally immerse the cotton. Place the tubes below a source of light. If the inoculum subsequently produces zoospores, they will swim upward through the cotton barrier where they can be captured with a capillary pipet and transferred to fresh medium. Zoospores are axenic when first liberated from the parental cell wall. In traversing the cotton, many of the zoospores that were subsequently contaminated will have the few contaminating bacterial cells removed in the process.

21–5 METHODS FOR GROWTH AND STORAGE OF MICROALGAL CULTURES

21–5.1 Culture Methods and Growth Media

21–5.1.1 Principles

Most terrestrial microalgae are obligate photoautotrophs or facultative photoheterotrophs. Many are obligate photoheterotrophs requiring the provision of vitamins or other growth factors from the environment. Most facultative heterotrophs use only simple sugars or organic acids, although some are capable of growth on polysaccharides or proteins (e.g., starch and gelatin). Many are also capable of dark heterotrophic growth. A few soil microalgae are obligate heterotrophs, including some euglenoids whose saprophytic or predatory nutritional requirements make them ecologically equivalent to protozoa (Metting, 1981).

21–5.1.2 Standard Methods for Culturing Microalgae and Cyanobacteria

Microalgal cultures are maintained in liquid media or on agar slants. A wide range of conditions of irradiance, photoperiod, and temperature are employed for different experimental purposes. A dedicated room with precise temperature control, such as a fully contained, refrigerated walk-in room, is preferred. At a minimum, the room should be refrigerated and have sufficient circulation of cooled air to counteract the heat generated by the lighting fixtures. Mounting the lighting fixture ballasts outside the room is recommended to aid temperature control. Provision of central pressurized air or placement of a dedicated air compressor is recommended for aeration and agitation of liquid cultures. The following are standard culture conditions:

1. Constant air temperature of 20 or 25 °C.
2. Light provided by a bank of 40-W cool-white fluorescent bulbs of the standard type involving two lamps per ballast and providing about 3700 lux light intensity. Pairs of lamps are spaced at 10-cm intervals and placed vertically (against a wall) 30 cm behind the culture vessels or, alternatively, placed horizontally 30 cm above horizontal shelves for petri dishes or culture vessels. These spacings provide an irradiance of between 200 and 300 μE per cm^2 of photosynthetically active radiation per second at the surface of the culture vessel or petri dish, depending on the age of the bulbs.
3. Two standard photoperiods are employed about equally often. These are 12-h light/12-h dark and 16-h light/8-h dark.

Special culture methods, including continuous fermentor culture and light-temperature gradient plate culture, are given by Starr (*in* Stein, 1973). Large-scale, outdoor mass culture is described by Knutsen and Metting (1991).

21–5.1.3 Culture Media

Selected basal culture media are described below for cultivation of terrestrial (and freshwater planktonic) microalgae. Many additional media are in common use as preferred by individual researchers. The recipes for many of these are listed by Borowitzka (1988), Castenholtz et al. (1989), Starr (1978), and Stein (1973).

Any of the following media can be used as the base for an agar (solid) medium. Additionally, certain compounds and antibiotics can be added to discourage the growth of various groups of microorganisms, as follows:

1. To discourage the growth of heterotrophic bacteria in enrichment of soil samples, add 1 mL of the following filter-sterilized antibiotic stock solution per 100 mL of algal medium: 0.6 g of penicillin (1625 units mg^{-1}) and 1.0 g of streptomycin sulfate in 200 mL of deionized-glass distilled water.
2. To suppress diatom growth, add 1 to 10 mg Ge_2O L^{-1} of medium.
3. To suppress the growth of eukaryotic microalgae when enriching for cyanobacteria, add 20 to 50 mg cycloheximide L^{-1} of medium.
4. To avoid discouraging the growth of unicellular, N_2-fixing cyanobacteria, solidify the medium with no more than 1% agar.

21–5.1.3.1 BG-11 Medium for Cyanobacteria (Allen, 1968). To 1 L of deionized, glass-distilled water, add:

$NaNO_3$	1.5	g
K_2HPO_4	0.04	g
$MgSO_4 \cdot 7H_2O$	0.075	g
$CaCl_2 \cdot 2H_2O$	0.036	g
Citric acid	0.006	g
Ferric ammonium citrate	0.006	g
Na_2-EDTA	0.001	g
Na_2CO_3	0.02	g
Trace element solution	1.0	mL

Trace element solution (to make 1 L of stock solution):

H_3BO_3	2.86	g
$MnCl_2 \cdot 4H_2O$	1.81	g
$ZnSO_4 \cdot 7H_2O$	0.222	g
$Na_2MoO_4 \cdot 2H_2O$	0.39	g
$CuSO_4 \cdot 5H_2O$	0.079	g
$Co(NO_3)_2 \cdot 6H_2O$	0.494	g

Adjust to pH 7.1 after autoclaving. Some species of cyanobacteria require vitamin B_{12} which can be added after autoclaving to a concentration of 1 μg L^{-1}. Partial enrichment for N_2-fixing cyanobacteria is possible by eliminating $NaNO_3$ from the medium. It is then known as BG-11$_0$ medium. Solid BG-11 medium is prepared with 10 to 15 g of agar per liter.

21–5.1.3.2 Bold's Basal Medium for Green and Yellow-green Microalgae (Starr, 1978). This medium is also suitable for many cyanobacteria. Diatoms do not grow well in this medium (J.R. Johansen, 1992, personal communication).

Prepare the five individual macronutrient solutions (listed in a), one chelate solution (either of those in b), and three minor element stock solutions (c, d, and e). To 940 mL of deionized, glass-distilled water, add 10 mL of each of the macronutrient solutions and 1 mL of each of the EDTA and minor element solutions. Final pH is 6.6.

a. Macronutrient solutions:

$NaNO_3$	10	g 400 mL^{-1}
K_2HPO_4	3.0	g 400 mL^{-1}
KH_2PO_4	7.0	g 400 mL^{-1}
$MgSO_4 \cdot 7H_2O$	3.0	g 400 mL^{-1}
$CaCl_2 \cdot 2H_2O$	1.0	g 400 mL^{-1}

b. EDTA solution

Na_2-EDTA	50 g L^{-1}
KOH	31 g L^{-1}

c. Iron solution

$FeSO_4 \cdot 7H_2O$	4.98 g L^{-1}
conc. H_2SO_4	1 mL L^{-1}

d. Boron solution

H_3BO_3	11.42 g L^{-1}

e. Micronutrients

$MnCl_2 \cdot 4H_2O$	1.44 g L^{-1}
$ZnSO_4 \cdot 7H_2O$	8.82 g L^{-1}
MoO_3	0.71 g L^{-1}
$CuSO_4 \cdot 5H_2O$	1.57 g L^{-1}
$Co(NO_3)_2 \cdot 6H_2O$	0.49 g L^{-1}

Solid Bold's basal medium is prepared with 10 to 15 g agar L^{-1}.

21–5.1.3.3 Medium for Diatoms: Guillard's (WC) Freshwater Enrichment Basal Salt Mixture (Guillard, 1975). This all-purpose medium for freshwater algae can be modified to partially enrich for diatoms by (i) providing silicon for cell wall (frustule) formation, and (ii) providing a wide NP ratio to take advantage of the fact that many (though not all) diatoms have efficient P uptake systems compared to other microalgae. The basal medium and the vitamin mixture are both available commercially from Sigma Chemical Company, St. Louis, MO (Sigma Plant Cell Culture Catalogue). To enrich for diatoms, eliminate the $NaHCO_3$ and reduce the concentration of K_2HPO_4 10-fold (S.S. Kilham, 1992, personal communication). Prepare the following six individual macronutrient stock solutions and seven minor element solutions. Add 1 mL of each of the macronutrient and micronutrient solutions to 1000 mL of deionized, glass-distilled water. Vitamin stocks should be membrane-filtered (or autoclaved) and stored frozen. Substitution of 200 mg L^{-1} of TES or PIPES for TRIS is recommended for diatom culture as is the addition of 10 mg H_3BO_3 L^{-1} (S.S. Kilham, 1992, personal communication).

a. Macronutrient solutions

$NaNO_3$	85.01	g L^{-1}
K_2HPO_4	8.71	g L^{-1}
$MgSO_4 \cdot 7H_2O$	36.97	g L^{-1}
$CaCl_2 \cdot 2H_2O$	36.76	g L^{-1}
$NaHCO_3$	12.60	g L^{-1}
$Na_2SiO_3 \cdot 9H_2O$	28.42	g L^{-1}

e. Micronutrients

$MnCl_2 \cdot 4H_2O$	0.18 g L^{-1}
$ZnSO_4 \cdot 7H_2O$	0.022 g L^{-1}
$Na_2MoO_4 \cdot 2H_2O$	0.006 g L^{-1}
$CuSO_4 \cdot 5H_2O$	0.01 g L^{-1}
$CoCl_2 \cdot 6H_2O$	0.01 g L^{-1}
$FeCl_3 \cdot 6H_2O$	3.15 g L^{-1}
$Na_2 \cdot EDTA$	4.36 g L^{-1}

c. Vitamins

Thiamine·HCl	0.01 mg L^{-1}
Biotin	0.5 µg L^{-1}
Cyanocobalamin	0.5 µg L^{-1}

d. TRIS solution—use 2 mL L^{-1} of medium.
Tris(hydroxymethyl)-aminomethane   50 g 200 mL^{-1}

21–5.1.3.4 Soil Extract Medium (Starr, 1978). A soil extract can be used in place of the defined microelement, chelate, and organic constituents in the media described above. It is often used to improve the likelihood of enriching for microalgal species with undefined growth factor requirements that are supplied by soil. Soil extract is prepared as follows:

1. Place about 2 g of $CaCO_3$ on the bottom of a 2-L flask and add a layer of a fertile air-dried garden soil to a depth of about 3 to 5 cm.
2. Slowly add deionized glass-distilled water down the inside wall of the flask so as not to disturb the soil, until the flask is about two-thirds to three-fourths full.
3. Submit the flask to fractional sterilization (tyndallization) for 2 h in a steam chamber on two successive days. Store the flasks under refrigeration and allow the supernatant to clear for 7 d before use.

To use, add 40 mL of the soil water supernatant to 960 mL of any defined medium. The supernatant may be added before the medium is autoclaved or as a filter-sterilized supplement after autoclaving. Solid soil extract medium is prepared with 10 to 15 g agar L^{-1}.

21–5.1.3.5 Soil Water Tubes (Starr, 1978). Soil water tubes are used with the MPN method (section 21–3.2) when a quantitative index of the complete microalgal community is desired in place of enrichment for selected components. Soil water tubes are also suitable for the short-term maintenance of many microalgal species.

Soil water tubes are prepared as a simple modification of the procedure outlined immediately above (section 21–5.1.3.4). Place 1 to 2 cm of a fertile air-dried garden soil over a pinch of $CaCO_3$ in the bottom of a series of 15-mL culture tubes. Slowly add deionized glass-distilled water down the inside wall to within 4 to 5 cm of the top of the individual tubes so as not to disturb the soil. Submit racks of the tubes to fractional sterilization (tyndallization) for 2 h in a steam chamber on two successive days. Store the tubes under refrigeration, allowing the supernatants to clear for 7 d before use.

21–5.1.3.6 Organic Carbon and Energy Sources. It is often possible to isolate obligately or facultatively photoheterotrophic or heterotrophic microalgae in liquid or on agar plates where sufficient concentrations of suitable organic compounds remain following serial dilution of the soil. Microalgal cultures that grow poorly (or not at all) when inoculated into defined mineral salts media may require the provision of one or more organic compounds from the environment. For further determination of the nutritional status of these cultures, it is recommended that a subjective comparison be made between growth in one or more defined mineral media without provision of vitamins, the same media with vitamins, the same media with undefined complex organic compound mixtures (0.2% yeast extract or casamino acids), and with growth in a soil extract medium or in soil water tubes.

Microalgae and cyanobacteria grow much slower than bacteria or fungi under photoheterotrophic and heterotrophic conditions. For this reason, it is imperative to use only axenic cultures to study patterns of utilization of organic C and energy sources in the light or in the dark. Organic compounds of interest should be dissolved in deionized glass-distilled water and filter-sterilized through sterile 0.22- or 0.45-μ membrane filters. A wide range of concentrations or organic compounds are used to study heterotrophy in microalgae and cyanobacteria (Neilson, Blankley, and Lewin *in* Stein, 1973). Any defined organic compound can be tested as a potential source of organic C or energy. Standard methods for the taxonomic treatment of eukaryotic microalgae from soil include screening for utilization of 0.25 M glucose, fructose, ribose, xylose, mannose, Na-acetate, and Na-pyruvate in Bold's basal medium in the light and in the dark. Tests for the production of extracellular amylase and gelatinase are also standard for taxonomic determinations at the species level (Archibald & Bold, 1970; Metting, 1980).

21–5.2 Storage and Preservation Methods

21–5.2.1 Short-Term Storage Methods

Methods for storage and preservation of microalgal and cyanobacterial cultures are similar to those used for other microorganisms. For short-term storage (between 4 and 10 wk), cultures should be streaked onto agar slants in screw-cap culture tubes. The length of storage time possible with this method will vary depending on storage conditions and rate of desiccation of the agar medium.

The time between transfers of cultures to fresh agar slants can be lengthened by:

1. Storage at reduced levels of light. This is accomplished within the walk-in culture room by placing racks of cultures on agar slants on a separate set of shelves that are removed from the light source(s). A sheet of thin, translucent polyethylene plastic sheeting can be

draped in front of the shelves to further reduce irradiance and minimize airflow that would otherwise hasten desiccation of the agar.
2. Wrapping a piece of domestic plastic wrap or laboratory wax paper (Para-Film) tightly around the screw cap.
3. Storage at reduced temperature. Most microalgae will tolerate long periods of refrigeration (3–6 mo or more) when stored as agar slants or in a liquid medium. Soil water tubes are useful for this purpose.

Some filamentous green and yellow-green microalgae (e.g., *Zygnema* and *Vaucheria*) prefer the medium if the $CaCO_3$ is omitted. *Euglena* and other euglenoids prefer soil-water tubes to which half of a pea (*Pisum sativum* L.) cotyledon is added (Starr, 1978).

21–5.2.2 Long-Term Preservation Methods

Many microalgae and cyanobacteria can be preserved for indefinite periods by drying, freezing, or freeze-drying (lyophilization) (Holm-Hansen *in* Stein, 1973). Drying and storage at room temperature is not recommended, but it should be noted that *Nostoc* and other filamentous cyanobacteria have been revived following many decades of storage as specimens on herbarium paper in botanical collections (Metting, 1981).

Freezing with cryoprotective agents such as glycerol or dimethyl sulfide is appropriate for microalgae. For long-term preservation by freezing, a deep freeze at −25 °C is satisfactory for some species. For other cultures, cryogenic freezing at or below −40 °C is necessary. Individual cultures must be tested to determine what conditions are necessary. Species that can successfully be frozen without excessive cellular damage or death can be stored for many months.

Methods used to freeze-dry bacteria are appropriate for microalgae. Not all species, however, are equally tolerant of the procedure and each must be tested to determine suitability. In some cases, iterative modification of the standard method can successfully identify conditions specific for a given species. Species that can successfully be freeze-dried can be stored for many years. Indeed, there is evidence that cryptoendolithic communities in the dry interior valleys of Antarctica undergo cycles of natural lyophilization and may be many thousands of years old (Johnston & Vestal, 1991).

21–6 METHODS FOR ESTIMATING PHOTOSYNTHESIS

Photosynthesis (CO_2 fixation) by communities of soil algae is estimated by the same methods outlined for soil respiration described in chapter 38 in this book. The contribution of algal photosynthetic production to soil respiration is estimated by comparing CO_2 or O_2 uptake and evolution in the light and dark and assuming that any difference is due to photosynthesis. The value of estimates from in situ (field) measurements depends largely on the size of the algal community. Algal contribution is easily masked by soil respiration except in situations where a large, visible crust is present. Better estimates of potential photosynthetic contributions by algal communities are achieved by studying thin (millimeters to centimeters) surface veneers from soil cores brought into the laboratory (Shimmel & Darley, 1985).

21–7 METHODS FOR MEASURING CYANOBACTERIAL DINITROGEN FIXATION

Most cyanobacteria are capable of biological N_2 fixation. Species that form specialized cells called *heterocysts* are capable of N_2 fixation in aerobic environments. Also, many filamentous and aggregate-forming species that do not have heterocysts can fix N in aerobic environments, but only in localized microaerophilic microsites within aggregates of cells that are protected from excessive O_2 diffusion by encapsulating mucilages. Other non-heterocystous cyanobacteria with the capacity to fix N_2 can do so only under strictly anaerobic conditions (Castenholtz et al., 1989). The methods outlined in chapter 43 for direct (^{15}N) and indirect (C_2H_2 reduction) quantification of bacterial N_2 fixation should also be used for cyanobacteria.

21–8 METHODS FOR STUDYING ENDOSYMBIOTIC CYANOBACTERIA IN CYCAD ROOTS

21–8.1 Principles

Cyanobacteria form symbiotic relationships and fix N_2 in association with liverworts, fungi (lichens), *Gunnera,* the aquatic fern *Azolla,* and cycads. *Gunnera* is a genus of tropical flowering herbs mostly occurring at high elevations in which cyanobacteria inhabit glandular stem cells. As mentioned in the introduction, the *Azolla-Anabaena* symbiosis is important to flooded rice culture. Lumpkin and Plucknett (1980) have reviewed methods for studying this symbiosis.

Most cyanobacteria that form intercellular bands of filaments in coralloid roots of cycads are in the genera *Anabaena* and *Nostoc.* The association is common to the cycad genera *Zamia, Macrozamia, Encephalartos,* and *Cycas.* In some symbioses, the cyanobacteria are localized in root tissue at or near the surface where they are exposed to light and conduct photosynthesis. In others, they are excluded from light and are accordingly dependent on other mechanisms for provision of reductant and ATP from the host. Unlike nodules formed in rhizobial and *Frankia* symbioses, endosymbiotic cyanobacteria reside in normally differentiated host tissue (with the exception of *Gunnera*). In common with other bacterial symbionts, however, endosymbiotic cyanobacteria do display several structural, ultrastructural, and physiological alterations from the free-living form (Grilli-Caiola, 1975).

Methods for studying the symbiosis are fundamentally similar to those outlined in chapter 12 for legume symbionts. Although whole plant experimentation is complicated by the relatively larger size, woody nature, and much slower growth rates of cycads compared to herbaceous legumes, excision of roots or of root segments and isolation and cultivation of cyanobacteria from root nodules is relatively simple and routinely accomplished. Methods for in situ direct (N^{15}) or indirect (C_2H_2 reduction) measurement of N_2 fixation are identical to those described in chapter 43 for N_2-fixing legumes.

21–8.2 Procedure for Coralloid Roots and Root Sections

Actively growing coralloid roots or root sections are collected from natural cycad populations by digging one or a series of shallow excavations around mature plants and cutting off desired segments (Lindblad et al., 1991). Whole roots or sections can also be collected from smaller plants in the field or from a greenhouse collection by removing the entire root mass from the soil and selectively cutting desired whole roots or segments. Transport the roots to the laboratory, preferably in an inert N_2 or Ar atmosphere. Prepare a series of 1-mm-thick sections under N_2 or Ar in degassed 10 mM HEPES buffer (pH 8.0). Determine the wet weights of the sections. Analyze the sections for N_2-fixation potential using the methods outlined in chapter 43.

21–8.3 Procedure for Isolating Endosymbiotic Cyanobacteria

Wash healthy coralloid roots or root sections under running tap water. Blot the roots dry, and cut them transversely into 1-cm lengths. Surface sterilize the sections by immersion in 80% ethanol (2 min), then rinse in deionized glass-distilled water (2 min), 2.5% Na-hypochloride (2 min), and again in deionized glass-distilled water (7 min). Cut the 1-cm sections into 1-mm sections and suspend them in BG 11 medium (section 21–5.1.3.1) in a petri dish in the light. The cyanobacterial filaments will migrate into the medium from the roots (Grobbelaar et al., 1987). Individual filaments can then be isolated and purified into clonal culture by repeated washing (section 21–4.2), streaking on agar (section 21–4.3), or centrifugation (section 21–4.4).

ACKNOWLEDGMENT

The following people made important contributions to this chapter or reviewed and commented on the manuscript: Jeffrey R. Johansen, John Carroll University (diatom frustule preparation, fluorescence microscopy). Susan S. Kilham, Drexel University (diatom culture). William J. Zimmerman, University of Michigan-Dearborn (symbiotic cyanobacteria). Pierre

Roger, ORSTOM, University of Marseilles (methods for rice soils). E. Imre Friedmann, Florida State University (methods for endolithic habitats). Annette E. Pipe, Lakeland College (implanted slide method).

REFERENCES

Allen, M.M. 1968. Simple conditions for growth of unicellular blue-green algae on plates. J. Phycol. 4:1–3.

Anderson, D.C., and S.R. Rushforth. 1976. The cryptogamic flora of desert soil crusts in southern Utah, U.S.A. Nova Hedwigia 28:691–729.

Archibald, P.A., and H.C. Bold. 1970. Phycological studies XI. The genus *Chlorococcum* Meneghini. University of Texas Publ. no. 7015, Austin.

Ashley, J., S.R. Rushforth, and J.R. Johansen. 1985. Soil algae of cryptogamic crusts from the Uintah Basin, Utah, U.S.A. Great Basin Nat. 45:432–442.

Bell, R.A., and M.R. Sommerfeld. 1987. Algal biomass and primary production within a temperate zone sandstone. Am. J. Bot. 74:294–297.

Bock, W. 1963. Diatomeen extrem trockener Standorte. Nova Hedwigia 5:199–254.

Bold, H.C., and M.J. Wynne. 1985. Introduction to the algae. 2nd ed. Prentice-Hall, Englewood Cliffs, NJ.

Borowitzka, M.A. 1988. Algal growth media and sources of algal cultures. p. 456–465. *In* M.A. Borowitzka and L.J. Borowitzka (ed.) Micro-algal biotechnology. Cambridge University Press, Cambridge.

Castenholtz, R., G.L. Gherna, R. Lewin, R. Rippka, J.B. Waterbury, and B.A. Whitton, 1989. Section 19. Oxygenic photosynthetic bacteria. Group 1. Cyanobacteria. p. 1710–1799. *In* J.T. Staley (ed.) Bergey's manual of systematic bacteriology. Vol. 3. Williams and Wilkins, Baltimore.

Ettl, H. 1983. Chlorophyta I. Phytomonadina. 9 Teil. *In* H. Ettl et al. (ed.) Süßwasserflora von Mitteleuropa. Gustav Fischer Verlag, Stuttgart, Germany.

Ettl, H. 1978. Xanthophyceae. 1 Teil. *In* H. Ettl et al. (ed.) Süßwasserflora von Mitteleuropa. Gustav Fischer Verlag, Stuttgart, Germany.

Gantt, E. 1980. Handbook of phycological methods. Vol. 3. Developmental and cytological methods. Cambridge University Press, Cambridge.

Geitler, L. 1930–32. Cyanophyceae. *In* L. Rabenhortst's Kryptogammenflora. Reprinted in 1985 by Koeltz Scientific Books, Köenigstein, Germany.

Grilli-Caiola, M. 1975. A light and electron microscopic study of blue-green algae growing in the coralloid-roots of *Encephlartos altensteinii* and in culture. Phycologia 14:25–33.

Grobbelaar, N., W.E. Scott, W. Hattingh, and J. Marshall. 1987. The identification of the coralloid root endophytes of the southern African cycads and the ability of the isolates to fix dinitrogen. S. Afr. J. Bot. 53:111–118.

Guillard, R.R.L. 1975. Culture of phytoplankton for feeding marine invertebrates. p. 29–68. *In* W.C. Smith and M.H. Chanley (ed.) Culture of marine invertebrate animals. Plenum Press, New York.

Hellebust, J.A., and J.S. Craigie. 1978. Handbook of phycological methods. Vol. 2. Physiological and biochemical methods. Cambridge University Press, Cambridge.

Johansen, J.R., A. Javakul, and S.R. Rushforth. 1982. Effects of burning on the algal communities of a high desert soil near Wallsburg, Utah. J. Range Manage. 35:598–600.

Johansen, J.R., S.R. Rushforth, and J.D. Brotherson. 1981. Subaerial algae of Navajo National Monument, Arizona. Great Basin Nat. 41:433–439.

Johansen, J.R., L.L. St. Clair, B.L. Webb, and G.T. Nebeker. 1984. Recovery patterns of cryptogamic soil crusts in desert rangelands following fire disturbance. Bryologist 87:238–243.

Johnston, C.G., and J.R. Vestal. 1991. Photosynthetic carbon incorporation and turnover in Antarctic cryptoendolithic communities: are they the slowest-growing communities on earth? Appl. Environ. Microbiol. 57:2308–2311.

Kessler, J.O. 1984. A new method for concentrating and purifying swimming algae. Appl. Phycol. Forum 1:2–4.

Komárek, J., and N. Fott. 1983. Chlorophyceae (Grüenalgen) Ordung: Chlorococcales. *In* G. Huber-Pestalozzi (ed.) Das Phytoplankton des Süßwassers, Band 26, Teil 7, Heft. 1. E. Schweizerbart'sche Verlagsbuchhandlung, Stuttgart, Germany.

Knutsen, G., and F.B. Metting. 1991. Microalgal mass culture and forced development of biological crusts in arid lands. p. 487–506. *In* J. Skujins (ed.) Semiarid lands and deserts. Soil resource and reclamation. Marcel Dekker, New York.

Krammer, K., and H. Lange-Bertalot. 1986. Bacillariophyceae. 1 Teil. Naviculaceae. *In* H. Ettl. et al. (ed.) Süßwasserflora von Mitteleuropa 9. Gustav Fischer Verlag, Stuttgart, Germany.

Krammer, K., and H. Lange-Bertalot. 1988. Bacillariophyceae. 2 Teil. Bacillariaceae, Epithemiaceae, Surirellaceae. *In* H. Ettl et al. (ed.) Süßwasserflora von Mitteleuropa. Gustav Fischer Verlag, Stuttgart, Germany.

Leedale, G.F. 1967. Euglenoid flagellates. Prentice-Hall, Englewood Cliffs, NJ.

Lindblad, P., C.A. Atkins, and J.S. Pate. 1991. N$_2$-fixation by freshly isolated *Nostoc* from coralloid roots of the cycad *Macrozamia riedlei* (Fisch. ex Gaud.) Gardn. Plant Physiol. 95:753–759.

Littler, D., and M. Littler. 1985. Handbook of phycological methods. Vol. 4. Ecological field methods: Macroalgae. Cambridge University Press, Cambridge.

Lorenzen, C.F. 1967. Determination of chlorophyll and pheo-pigments: septrophotometric equations. Limnol. Oceanogr. 12:343–346.

Lumpkin, T.A., and D.L. Plucknett. 1980. *Azolla:* botany, physiology, and use as a green manure. Econ. Bot. 34:111–153.

Metting, F.B. 1980. New species of green microalgae (Chlorophycophyta) from an eastern Washington silt loam. Phycologia 19:296–306.

Metting, F.B. 1981. The systematics and ecology of soil algae. Bot. Rev. 41:195–312.

Metting, F.B. 1991. Biological surface features of semiarid lands and deserts. p. 257–293. *In* J. Skujins (ed.) Semiarid lands and deserts. Soil Resource and Reclamation. Marcel Dekker, New York.

Metting, F.B. 1993. Structure and physiological ecology of soil microbial communities. p. 3–25. *In* F.B. Metting (ed.) Soil microbial ecology. Applications in agriculture and environmental management. Marcel Dekker, New York.

Metting, F.B., and A.D. Barry. 1986. Palmelloid microalgae as soil-conditioning agents. Phase I Final Project Report 85-SBIR-8-0067. USDA, Washington, DC.

Metting, F.B., W.R. Rayburn, and P.A. Reynaud. 1988. Algae and agriculture. p. 335–370. *In* C.A. Lembi and J.R. Waaland (ed.) Algae and human affairs. Cambridge University Press, Cambridge.

Pipe, A.E. 1992. Pesticide effects on soil algae and cyanobacteria. Rev. Environ. Contamination Toxicol. 127:95–170.

Pipe, A.E., and D.R. Cullimore. 1980. An implanted slide technique for examining the effects of the herbicide diuron on soil algae. Bull. Environ. Contamination Toxicol. 24:306–312.

Prescott, G.W. 1978. How to know the freshwater algae. 3rd ed. Wm. C. Brown, Dubuque, IA.

Rayburn, W.R., R.N. Mack, and F.B. Metting. 1982. Conspicuous algal colonization of the ash from Mount St. Helens. J. Phycol. 18:537–543.

Roger, P.A., Jimenez, R., and Santiago-Ardales, S. 1991. Methods for studying blue-green algae in ricefields: distributional ecology, sampling strategies, and estimation of abundance. IRRI Res. Paper Ser., No. 150. Int. Rice Res. Inst., Los Baños, Manila, Philippines.

Roger, P.A., and S.A. Kulasooriya. 1980. Blue-green algae and rice. Int. Rice Res. Inst., Los Baños, Manila, Philippines.

Roger, P.A., W.J. Zimmerman, and T.A. Lumpkin. 1993. Microbiological management of wetland rice fields. p. 417–455. *In* F.B. Metting (ed.) Soil microbial ecology. Applications in agriculture and environmental management. Marcel Dekker, New York.

Sharabi, N.E.-D., and D. Pramer. 1973. A spectrophotometric method for studying algae in soil. Bull. Ecol. Res. Commun. NFR Sweden 17:224–227.

Shields, L.M. 1982. Algae. p. 1093–1101. *In* A.L. Page et al. (ed.) Methods of soil analysis. Part 2. Chemical and microbiological properties. 2nd ed. ASA and SSSA, Madison, WI.

Shimmel, S.M., and W.M. Darley. 1985. Productivity and density of soil algae in an agricultural system. Ecology 66:1439–1447.

Starmach, K. 1972. Chlorophyta III. Zielenice nitlowate. *In* K. Starmach and J. Sieminska (ed.) Flora Slodkowodna Polski, Tom. 10. Polska Akad. Nauk, Inst. Bot., Warsaw, Poland.

Starr, R.C. 1978. The culture collection of algae at the University of Texas at Austin. J. Phycol. 14 (supplement):47–100.

Stein, J.R. 1973. Handbook of phycological methods. Vol. 1. Culture methods and growth measurements. Cambridge University Press, Cambridge.

Tchan, Y.T. 1952. Study of soil algae. I. Fluorescence microscopy for the study of soil algae. Proc. Linn. Soc., London 77:265–269.

Van Baalen, C. 1965. Quantitative surface plating of coccoid blue-green algae. J. Phycol. 1:19–22.

Chapter 22

Nematodes

RUSSELL E. INGHAM, *Oregon State University, Corvallis, Oregon*

Nematodes arc unsegmented roundworms and one of the most ecologically diverse groups of animals on earth. They exist in nearly every habitat known from the tops of mountains to the depths of the ocean and from hot deserts to the cold of the Antarctic. Nematodes eat bacteria, fungi, algae, yeasts, diatoms, and may be predators of several small invertebrate animals, including other nematodes. In addition, they may be parasites of invertebrates, vertebrates (including humans) and all above and below-ground portions of plants. Nematodes range in length from 82 μm (marine) to 9 m (whale parasite) but most species in soil are between 0.25- and 5.5-mm long.

Nematodes may constitute as much as 90% of all multicellular animals in soil (Oostrinbrink, 1971) and often exceed several million m^{-2} (Table 21–1). While the total biomass of nematodes may often be less than that for other faunal groups (Coleman, 1976; Kusmin, 1976; MacLean, 1974) nematode metabolic activity is often higher (Kusmin, 1976; Reichle, 1977; Sohlenius, 1977).

Although nematodes are recognized as a major consumcr group in soils, the exact feeding habits of most soil nematodes arc not known. They are generally grouped into four to five trophic categories based on the nature of their food, the structure of the stoma and esophagus and method of feeding (Yeates, 1971). Plant-feeding nematodes possess stylets with a wide diversity of size and structure and are the most extensively studied group of soil nematodes because of their ability to cause plant disease and reduce crop yield. Fungal-feeding nematodes have slender stylets but are often difficult to categorize and have been included with plant feeders in many ecological studies. Bacterial-feeding nematodes are a diverse group and usually have a simple stoma in the form of a cylindrical or triangular tube, terminating in a valve-like apparatus that may bear minute teeth (Nicholas, 1975). Predatory nematodes are usually large species possessing either a large stylet or a wide cup-shaped cuticular-lined stoma armed with powerful teeth (Nicholas, 1975). Omnivores are sometimes considered as

Table 22–1. Density, biomass, and respiration estimates for nematode communities in selected ecosystems. (After Sohlenius, 1980.)

Ecosystem	Numbers	Biomass	Respiration
	millions/m^2	g fresh wt/m^2	Kcal/m^2·Y
Desert, USA	0.4	0.1	7.6
Tundra, Sweden	4.1	1.1	7.3
Beech forest, England	1.4	0.4	5.2
Tropical forest, Africa	1.7	—	—
Pine forest, Sweden	4.1	0.6	8.2
Mixed forest, Poland	7.0	0.7	33.0
Pasture, Poland	3.5	2.2	50.0
Grassland, Denmark	10.0	14.0	339.0
Potato field, Poland	5.5	0.7	14.3
Rye field, Poland	8.6	1.1	6.0

a fifth trophic category of soil nematodes. These nematodes may fit into one of the categories above but also ingest other food sources. For example, some bacterial feeders may also eat protozoa or algae and some stylet-bearing nematodes may pierce and suck algae as well as fungi or higher plants. Stages of animal-parasitic nematodes, such as hookworms, may also be found in soils but generally are not common in most soil samples.

Nematodes appear to have several important functions in soil ecosystems. Most notably, they are important pests of agricultural crops. However, even in native ecosystems, plant consumption by nematodes can be substantial, exceeding that of other herbivores (Ingham & Detling, 1984). The importance of plant consumption by nematodes on system productivity has not been adequately studied in most native ecosystems. However, nematicide studies in native North American grasslands reduced populations of plant-feeding nematodes and increased plant growth (Ingham & Detling 1990; Stanton et al., 1981). Nutrient cycling dynamics may be influenced by nematodes feeding on bacteria and fungi, regulating substrate utilization and the mineralization of nutrients (Ingham et al., 1985b). Nematodes are generally most abundant in the rhizosphere (Ingham & Coleman, 1983; Ingham et al., 1985a) and thus may interact with other rhizosphere organisms such as plant pathogens (Sikora & Carter 1987; Riedel, 1988), rhizobia (Huang, 1987), mycorrhizae (Ingham, 1988), and other nematodes (Eisenback & Griffin, 1987). Nematodes also fill an important link in the soil food web, transferring plant and microbial-based energy and nutrients to higher trophic categories in the soil food web (Hunt et al., 1987).

Thus, nematodes may be studied by soil biologists from several different disciplines. The most important methods for adequate study of soil nematodes in the field relate to: (i) taking appropriate samples, (ii) reliable extraction procedures for recovering nematodes from soil or plant tissues, and (iii) enumeration and identification of the nematodes recovered.

22–1 NEMATODE SAMPLING

The first consideration in sampling for nematodes is to design a sampling plan that is appropriate for the current objectives. For example, the sampling procedure for describing the nematode community from a particular species of plant may be quite different from that for determining if a fallow field must be treated with nematicide before a crop is planted. Available resources may also influence sampling intensity. Researchers may be able to afford to process more samples than a grower trying to diagnose the cause of plant disease. Most studies to evaluate procedures for sampling nematodes have only been concerned with estimating populations of plant-parasitic nematodes. The protocols for collecting nematode samples have been thoroughly reviewed (Barker et al., 1985a; McSorley, 1987) and are beyond the scope of this discussion. Several principles are necessary to mention, however, and many apply to sampling for other soil organisms and abiotic properties as well.

22–1.1 Nematode Distribution

Nematode species are not uniformly distributed throughout the soil but generally have a patchy distribution with areas of high densities mixed among areas of low densities (Fig. 22–1). Nematode species are often distributed differently within the same field so that areas of high density of one species may not correspond with high densities of another species (Goodell & Ferris, 1980, 1981). Nematode distribution and density may be related to soil type (e.g., high in sandy areas and lower in heavier soil), some other edaphic factor, past cropping histories or be unrelated to any apparent parameter. Inadequate sampling can result in unreliable estimates of nematode density and distribution. Inaccurate estimates can lead to expensive errors if they are used as the basis for nematode management decisions, or result in the development of errant concepts if used to describe ecological principles. The structure of the root system influences distribution of nematode populations. In annual cropping systems, the structure of preceding crops, as well as the current crop, may influence nematode distribution. Patterns of spatial distribution can be described by taking a series of soil cores and computing the mean and variance. Most distributions can be described by the negative binomial functions indicating clumped distributions. Thus, data must be transformed by normalizing functions such as $Y = \log(x + c)$ [where Y is the transformed count, x is the original count and c is a constant, usually 1] before analysis (McSorley, 1987). The density of a nematode species in soil will also affect the precision of the population estimate made from any sampling plan. If a system is to be sampled frequently, initial extensive sampling can determine the index of dispersion (k) (Southwood, 1978) associated with negative binomial distributions and assist development of an optimum sampling design for future samplings (McSorley, 1987).

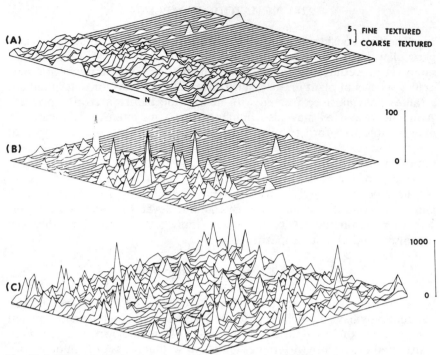

Fig. 22–1. Three-dimensional depiction of the distribution of *Helicotylenchus digonicus* (B) and *Meloidogyne arenaria* (C) in relation to soil texture (A) in an alfalfa field. (After Goodell and Ferris, 1980.)

22–1.2 Sample Collection

Nematode samples may be taken for many reasons. A bulk sample of several soil cores may be taken at one time to determine which nematode species are present. Many individual cores can be analyzed to map the population distribution in that area. Other areas may be sampled repeatedly throughout the year to determine the occurrence of temporal changes in the populations of one or several species. In all cases, however, the biology of the nematode species of interest will dictate the time, depth, and pattern of the sampling design. The primary objectives for taking nematode samples include: (i) detection and survey, (ii) diagnosis, (iii) advisory, and (iv) research (Barker & Campbell, 1981). When nematode samples are taken for general survey purposes, or, to make collections for taxonomic work, there may be little concern with the precision of the sampling design. However, if the sampling objective is for quantification of populations, quarantine, or other regulatory purposes, sampling may need to be thorough to assure detection of serious pathogens. Nematodes samples are generally taken with a cylindrical soil probe (generally 2.5 cm diam.) but shovels, trowels, or augers may be used as well. Sampling depth depends on the rooting pattern of the crop being sampled but is generally 30 to 45 cm.

A common objective for nematode sampling is to aid diagnosis of poor plant growth. If a crop is present and patches of poorly growing plants are observed, it is advisable to take separate samples from the poorly growing patches and from areas that appear healthy. This may confirm that nematodes should be implicated as a cause for the poor plant productivity if densities are high in the patches of poor growth and less in the healthy appearing areas. It is often preferable to take samples from the periphery of the patch of poor plant growth than from the center of the patch. Often plant growth in the center of these patches may be so poor that nematode densities will have declined due to inadequate root production. Higher numbers may be obtained at the periphery where populations may be rapidly increasing. When sampling for diagnosis, adequate precision is usually obtained economically by taking several samples of composited soil cores from a relatively small affected area.

Another reason for taking nematode samples is to determine if nematode populations are at a level that may damage a crop to be planted or are increasing towards damaging levels in an existing crop (advisory). If one is sampling a fallow field, or a field without visible differences in plant growth, the whole field (or subdivided portions of a large field) must be sampled. The most representative samples are obtained by taking several cores from a given area, compositing the cores into a single sample, mixing the soil thoroughly, and taking a sufficient subsample (for example 1 L or 1 kg) for analysis. The pattern in which cores are collected is dependent on the cropping system being sampled as illustrated in Fig. 22–2.

The number of samples taken from a field depends on the value of the crop and the relative nematode risk (e.g., risk to a high cash value perennial from a nematode that vectors plant viruses would be much greater than the risk to a low cash value annual crop from a nematode that reduces yield slightly). It has been estimated that 1% of crop production costs may be economical for nematode sampling. For example, if it costs $500/acre to produce a crop and $20/sample for nematode sampling and analysis, then $5/acre could be spent on sampling. Thus, each sample should represent four acres (Ferris et al., 1981a). Southwood (1978) recommended that sampling for advisory purposes should be of sufficient intensity to estimate the population with a standard error within 25% of the mean. However, the level of precision that can be attained may be governed by economic factors such as the value of the crop threatened and the relative pathogenicity of the nematode species of interest. The chances of detecting nematodes and numbers of samples needed at different population densities is discussed by Barker (1985a).

Taking nematode samples for research purposes generally requires estimates with a standard error of 10% of the mean (Southwood, 1978).

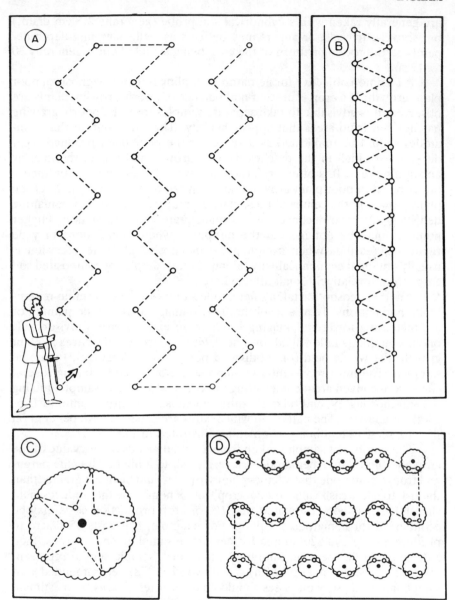

Fig. 22–2. Suggested sampling patterns for collecting soil samples for nematode assays. A) large fields or subdivisions of fields; B) for collecting soil in two center rows of four row experiments; C) single-plant plot; D) feeder-root zone of established perennials. (From Barker and Campbell, 1981.)

Most nematodes are found in the top 30 cm of soil so most soil samples can be limited to that depth. However, many species are vertically distributed with the root system and deep sampling may be necessary for deep-rooted perennials, often to depths of 0.5 to 1 m. Ingham et al. (1985a) recovered nematodes to 1.5 m in a shortgrass prairie. Extreme climates often encourage vertical migration so sampling to 30 to 45 cm is recommended for regions with hot, dry summers (Barker, 1986). However, some nematodes, such as *Meloidogyne chitwoodi,* may migrate 1.5 m upward to cause significant crop damage (Mojtahedi et al., 1991). In these situations, sampling to 90 or 120 cm may be necessary.

Most nematode analysis laboratories request that 500 to 1000 cm^3 of soil be submitted for analysis. Extraction techniques limit the amount of soil that can be processed so rarely can entire samples be extracted. Large samples should be thoroughly mixed before removing a subsample for extraction. Size of the subsample depends on the particular procedure used (see below). If soil is thoroughly mixed, only one aliquot is necessary to estimate the population in the sample. Homogeneity of mixing can be tested by extracting replicate aliquots. Care must be taken in mixing, however, to prevent nematode mortality from mechanical damage during mixing (Barker & Campbell, 1981). It is essential that samples be protected from drying out or getting too hot. Soil samples should be sealed in plastic or heavy paper bags in the field and kept in the shade or in an ice chest. Samples should be submitted for analysis as soon as possible. Many extraction and identification procedures require nematodes to be alive and poor sample handling may kill nematodes before the soil reaches the laboratory. This results in an underestimate of the population that was actually present at the time of sampling. Samples should be labeled on the outside of the bag. Labels placed next to soil often disintegrate or become illegible. Samples should be extracted as soon as possible and stored at 10 to 15 °C until extracted (Barker & Campbell, 1981).

If endoparasites are suspected, roots (or aboveground plant parts for stem, foliar, or seed-gall nematodes) should be collected, sealed in plastic bags, and sent for analysis. Plant material containing nematodes should be kept cool and moist and examined as soon as possible. Whole plants are best stored free from soil. Since shoots often decompose more quickly than roots, they should be kept in separate bags if stored for more than a day or two. Polyethylene bags are excellent storage containers.

22–1.3 Sampling Pattern

Several parameters must be considered before designing a pattern for collecting nematode samples. Goodell and Ferris (1981) examined sampling patterns that included random, stratified random and division of the field into north-south or east-west strips and evaluated each pattern with a predetermined time (cost) constraint. In this study, the optimum pattern was division of the field into north-south strips because the strips in this direction isolated a streak of fine-textured soil. Thus, if information about distribution of soil type, or other edaphic factors, is known it may be used to stratify sampling when designing a sampling pattern.

In most systems, the sampling pattern will be determined by the structure of the plant community. If the area is fallow or has a homogeneous plant community such as pasture or alfalfa, a simple random or stratified random pattern may be best. Samples are generally taken by walking in a "W" pattern and taking cores at intervals along the traversed path (Fig. 22–2). With row crops, samples should be randomly taken from within the row only and taking cores "across" different rows rather than "along" the same rows provides a more accurate estimate of the overall field mean (Noe & Barker, 1982). In systems where plants are grouped as discrete individuals, such as orchards, samples should be sampled from the drip line since it represents the area occupied by feeder roots and generally contains the highest population of nematodes (Barker, 1986). In a study of vineyards, Ferris and McKenry (1976) found that nematode densities were greatest and variation least when samples taken 30 to 45 cm from the vine and to a depth of 60 cm. When sampling for several nematode species the sampling design should be based on the most difficult species to estimate (Goodell & Ferris, 1981; McSorley & Parrado, 1982). If the same area is sampled repeatedly, selecting cores from the same location (e.g., same plants in an orchard, same distance from stem) each time will reduce the variance associated with nematode distribution (Fidler et al., 1959; Goodell, 1982).

22–1.4 Timing of Sampling Collections

Densities of most nematodes reach a maximum and minimum at different times of the year. Nematodes in temperate climates tend to reach minimum population levels in winter or early spring while, in warmer climates, populations may be at their minimum in summer and early fall. Different species cycle in different patterns, however, and may require sampling at different times (Barker, 1985a; Barker & Campbell, 1981). Timing of sampling depends on the sampling objective. Most samples are best taken at time of maximum population to provide maximum probability of detection. This provides information that can be used in management decisions for the subsequent crop. In practice, however, samples are often taken shortly before planting or in the spring when populations may be low so then management options may be implemented before populations increase. In special instances, where nematode infection may damage the quality of a belowground crop, such as potato (*Solanum tubarosum*), samples may also be take during midseason to determine if the crop may need to be harvested early to escape damage. Sufficient knowledge of the nematode species of interest must be known to optimize the sampling time. If population dynamics are studied, sampling at a minimum of 2-wk intervals are generally necessary to monitor rapid population changes that may occur.

22–1.5 Sample Size

The most appropriate method for obtaining accurate nematode sample estimates is with sufficient sample sizes. If the study objectives include precise mapping of the distribution of a nematode community or specific associations between different nematodes or nematodes and other soil organisms, many soil cores may need to be taken and processed individually. However, many sampling objectives do not require this level of precision. Sampling effort can be improved for most objectives, particularly those for advisory purposes, by increasing the number of samples taken and by increasing the number of subsamples or soil cores that comprise a single sample. Although both approaches increase the cost of sampling, it is generally more cost effective to increase the number of cores/sample than to increase the number of samples processed, provided the area represented by a single sample is not too large (Goodell, 1981). The number of cores (subsamples) taken to represent an area depends on the size of the area sampled. For advisory purposes, 10 cores are taken to represent < 1 acre, 20 to 30 cores for 1 to 5 acres and 50 to 100 cores for areas 6 to 10 acres in size. However, it is not generally recommended that one attempt to represent more than 5 acres with a single sample because of the heterogeneous distribution of nematodes (Santo et al., 1982). Large fields should be subdivided and different samples taken from the subdivisions. Stratification of sampling can minimize variance among repeated sampling from a single stratum while maximizing the differences in densities between strata (Goodell, 1982). Often fields are subdivided based on some natural stratification (Fig. 22–3). This practice allows identification of areas that need to be treated and other areas that may not need to be treated, thus limiting the amount of treatment necessary and reducing treatment costs to the grower. When the expense for more precise estimates may be permissible (e.g., high cash value crop and research studies), the relationship between sample size and level of precision may be used to determine optimum sampling effort if sufficient information about nematode distribution is available (McSorley, 1987). Estimates of the precision expected from multiple samples through subdivision or replicated sampling can be evaluated by computer simulation (Goodell & Ferris, 1981; McSorley, 1982).

22–1.6 Nematode Sampling for Ecological Studies

Few quantitative studies have been made on the precision of various sampling plans for ecological studies in native ecosystems. Most principles associated with agroecosystems are likely to apply to native systems as well but the questions addressed may be more complex. While sampling of agroecosystems may be primarily concerned with the macrodistribution of

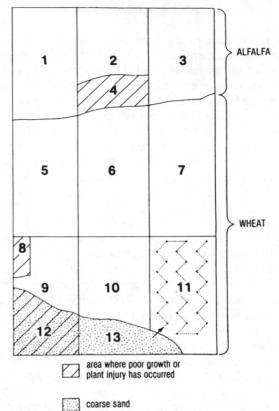

Fig. 22–3. Subdivide large fields into areas of 5 acres or less that reflect any differences in cropping history or soil type. If there are areas where there has been poor growth, treat them as separate blocks, even if they are smaller than other blocks. Assign each block a number and collect samples with the appropriate pattern from Fig. 22–2.

specific kinds of pathogenic nematodes over a large area, ecological studies may be more interested in the composition of a nematode community, its microdistribution around the root system of a single plant or within decomposing litter, or nematode distribution across ecosystems. Thus, sampling procedures may be structured differently. Since soils in native systems generally have not been tilled and can be compacted, simple soil probes may not be adequate for sampling to any significant depth and augers or Viehmeyer tubes may be necessary (Ferris et al., 1981b; Goodell, 1982). This also may apply to severely compacted agricultural fields and deeprooting crop species. If soils are rocky, sampling with any type of soil probe may be prohibited. Sampling with a trowel or small shovel may be sufficient if 5 to 10% more subsamples are included in each sample (McSorley & Parrado, 1983). Andrassy (1962) compared core sizes from 0.25 to 15 cm in diameter and determined that an effective core diameter of 1 cm was adequate for determining the density of nematodes in litter and humus horizons. When monitoring nematode community structure, composite samples should be taken from beneath the same plant species since the greater diversity of plant communities in native systems will increase the heterogeneity of the nematode community.

22–2 EXTRACTION OF NEMATODES FROM SOIL

Identification and enumeration of nematode populations requires that they must be removed from their soil or plant habitats. The most complete text for classic methods in nematology is provided by Southey (1986), but several excellent review chapters are also available (Baker, 1985b; Fortuner, 1991; Hooper, 1990; McSorley, 1987). Several procedures have been developed to extract nematodes from soil or plant tissues but no one method is ideal for all purposes. These basic procedures are often modified within individual laboratories so results from different laboratories must be compared with caution. Extraction efficiency varies with method, species, and soil type. Few studies report actual extraction efficiency but Viglierchio and Schmitt (1983) compared five methods for extraction of four species from three soil types and rarely found efficiency to exceed 50%. Similar extraction efficiencies from other studies were summarized by Barker (1985b).

The most common methods are outlined below. Each has advantages and disadvantages and is better for some nematode species than for others. The biological properties of the nematodes of interest should be known before selection of the appropriate method. However, economic, labor, and space constraints may ultimately dictate the method used. All the methods below assume that the soil has been gently sieved and mixed before obtaining a subsample for extraction.

22–2.1 Cobb Sieving and Decanting (Wet Sieving) (Cobb, 1918)

22–2.1.1 Equipment (Fig. 22–4C)

1. Two (or more) buckets or dishpans.
2. Number 10 mesh (2.00 mm), no. 50 mesh (300 µm), no. 150 mesh (106 µm) and no. 400 (38 µm) or no. 500 mesh (25 µm) screens.
3. Sink with soil trap.
4. Three 250-mL beakers.
5. Three counting dishes or bottles to store nematodes.

22–2.1.2 Procedure

1. Pour soil sample (450 g or less) into first bucket, fill bucket three-fourths full with water, break up any soil clumps with your fingers, and stir vigorously.

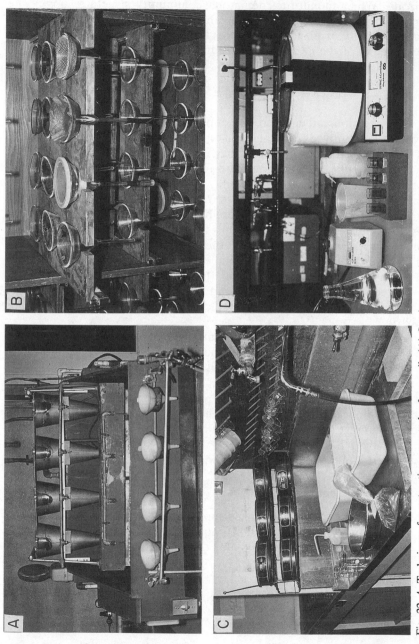

Fig. 22–4. Techniques for extracting nematodes from soil. (A) Semi-automatic elutriator for wet-sieving four soil samples simultaneously; (B) modified Baermann funnel technique; (C) Cobb sieving technique; (D) centrifugal-flotation technique.

2. Pour water through no. 10 mesh screen into second bucket. Rinse debris on no. 10 screen with a stream of water, allowing water to go into second bucket.
3. Discard debris on no. 10 screen and clean the first bucket.
4. Stir water in second bucket vigorously and pour through no. 50 screen into first bucket. Rinse debris on screen as before.
5. Backwash debris on screen into beaker labeled with screen mesh size.
6. Clean second bucket.
7. Repeat until you have three beakers of water and debris from the nos. 50, 150, and 500 screens.

Pour contents, or a portion of the contents, of each beaker into counting dishes to see which nematodes were trapped by each size screen. If there is too much water in a beaker, it can be removed by slowly decanting through a no. 500 mesh screen and backwashing the debris and nematodes back into the beaker with a small stream of water from a squirt bottle.

If the no. 500 mesh screen becomes plugged with soil, a no. 400 mesh screen may be inserted before the no. 500 mesh screen. If very little soil is trapped on the no. 500 mesh screen so that it does not plug and overflow, the procedure can be quickened by stacking the screens in order of decreasing opening size from the no. 10 on top to the no. 500 on the bottom and pouring the water through the stack of screens in one step.

Recovery can be improved by pouring the same water through the screens two or three times. Recovery is also increased if screens are held at a 45° angle when water is poured through the screen.

This method has several advantages. (i) Nematodes can be observed in a few minutes. (ii) Most species of nematodes are generally recovered in good condition. (iii) No special equipment is needed and very little space is required. The disadvantages are that it is too time consuming for large numbers of samples and that samples are usually too "dirty" with soil or organic debris to see all nematodes clearly.

Many laboratories use a shortened version of this procedure which is described below under the section on density centrifugation.

22–2.2 Modified Baermann Funnel Method

This technique was originally developed by Baermann (1917) but has been extensively modified over the years. The following procedure resembles that described by Christie and Perry (1951).

22–2.2.1 Equipment (Fig. 22–4B)

1. 12-cm internal diameter funnel to which a 7.5-cm piece of latex tubing has been attached to the stem.
2. Pinch clamp.
3. Coarse plastic or metal screen (about no. 10 mesh) shaped into the form of a basket to fit inside the funnel or a 1 cm high "rim" of

11.25-cm outside diameter, thick-walled PVC tubing with a circle of plastic window screen glued to the bottom.
4. Wet-strength tissue (e.g., laboratory tissue, toilet tissue, or milk filters) cut in a circle of sufficient diameter to be slightly larger than the basket.
5. A stand to hold the funnel so that the lower tip of the tubing is approximately 20 cm above the table.
6. Lid large enough to fit over top of funnel.

22–2.2.2 Procedure

1. Place the pinch clamp on latex tubing and fill the funnel nearly full with cool tap water. Check to see that the pinch clamp is secure and tubing does not leak. Open pinch clamp to let out air bubble in funnel stem, if necessary.
2. Line the basket with tissue/filter and place a known quantity of soil (100 g or less) on the tissue and smooth into an even layer.
3. Carefully set the basket into funnel of water and add more water to the funnel if necessary to cover soil. Pour water in between the edge of the basket and inside edge of the funnel to avoid washing soil through the tissue. Be sure pockets of air are not trapped under the basket.
4. Place lid over funnel to slow evaporation.
5. Label the funnel with the appropriate sample number.
6. Check water level daily and add water if necessary.

Nematodes will migrate through the tissue and sink through the water to collect in the stem of the funnel where they can be drained with a small volume of water into a counting dish, storage vial, centrifuge tube, etc. The length of time necessary for extraction varies with the objective, the nematode species, soil type, and amount of soil on the funnel and may vary from 1 to 7 d or more. Collecting nematodes after 3 d is usually sufficient to retrieve most nematodes that will be recovered. If nematodes are required to be alive, it is best to collect them every 24 h or less since they may die from lack of oxygen at the bottom of the funnel stem.

This method has the advantage that extraction of many samples can be initiated quickly. Nematodes are recovered in a relatively clean water sample with much less soil and debris than with Cobb sieving. However, only a small volume of soil (100 g maximum) can be processed on a single funnel and when soils with high clay contents are extracted, the water often becomes murky, making counting difficult. Nematodes must be active to be extracted. Inactive or dead individuals and certain species may not come through the filter at all. Results are not immediate since sufficient time (3 d or more) must elapse before counts can be made. While materials for this technique are relatively inexpensive, considerable space may be required to set up many funnels at a time. Some laboratories set up funnel racks on the wall in a bookshelf fashion to conserve space.

Larger volumes of soil can be processed on Baermann funnels if the soil sample is first wet sieved (see 22–2.1) and poured through a no. 500 (or no. 400) mesh screen (see initial steps for density centrifugation below). The soil and nematodes trapped by the screen are then backwashed onto the tissue filter in the funnel. Another modification of this method is called the Baermann tray or pan method (Whitehead & Hemming, 1965) where a larger basket with a larger tissue is placed inside a shallow pan filled with water. Because of the larger surface area of the basket, this method can process larger soil samples or the soil can be spread out into a thinner layer, shortening the time necessary for extraction.

22–2.3 Density (Sucrose) Centrifugation

22–2.3.1 Equipment (Fig. 22–4D)

1. Centrifuge.
2. 50 to 100 mL round bottom centrifuge tubes. Those made from polycarbonate work best.
3. 1 (or more) buckets
4. Number 20 mesh (1.00 mm) screen.
5. Number 400 mesh (38 µm) screen.
6. Number 500 mesh (25 µm) screen.
7. 2.5 M sucrose—900 g sucrose in 1 L total volume.
8. Sink with soil trap.

22–2.3.2 Procedure

While intact soil can be processed with this procedure (Caveness & Jensen, 1955), the size of the soil sample that can be extracted is severely limited. Therefore, a simplified version of Cobb sieving is generally applied first (Jenkins, 1964).

1. Rinse the soil sample (450 g or less) through a no. 20 mesh screen into a bucket, breaking up all soil clumps. Discard debris on the no. 20 screen.
2. Stir the water in the bucket vigorously and let settle for 1 min. Sandy soils may settle more rapidly and recovery is often improved with shorter settling time.
3. Slowly pour water from the bucket through a wet no. 400 mesh screen held at a 45° angle. (A no. 500 mesh screen will provide better recovery but may clog easily if the soil is heavy. Very heavy soils may require use of a no. 325 screen) Be sure screen does not plug and overflow. Tap screen gently against wall of sink to keep soil suspended if screen plugs.
4. Backwash residues remaining on the no. 400 screen through a glass funnel into a polycarbonate centrifuge tube with small stream of water from a squirt bottle. Be sure not to overfill the tube.

5. Balance tubes to the same weight by adding water. Tubes should be filled to about 1 cm from the top.
6. Centrifuge for 4 min. (Speeds used for nematode extraction depend on the radius of the centrifuge arm and are equivalent to 700 to 2900 g with 1800 g most common. Southey, 1986).
7. Carefully decant off *half* of the supernatant water without disturbing the soil plug at bottom of tube. Wipe out any organic debris sticking to sides of the tube.
8. Fill tubes with 2.5 M sucrose solution and balance tubes to equal weights by adding more sucrose from a squirt bottle. (Some labs pour off all the supernatant water and then fill with 1.25 M sucrose but this risks loss of nematodes from the tubes and requires preparation of a greater volume of sucrose. 2.5 M sucrose also has less problems with fungal growth and will not freeze if stored in a freezer.)
9. Mix thoroughly by stoppering tubes and shaking vigorously or by using a stirring rod or small electric mixer so that soil is well mixed into the sucrose solution. Be sure no soil adheres to the bottom of the tube.
10. Centrifuge for 1 min. (Longer time or higher speed may be necessary for soils high in silt or clay). Soil will settle to the bottom of the tube but nematodes will remain suspended in the sucrose.
11. Hold a no. 500 mesh screen at a 45° angle and lower one edge of the screen into a pan of water so that the level of water is above the screen but below the top of the rim. Decant sucrose solution with suspended nematodes onto the water in the screen, raise screen, and let water and sucrose drain through the screen. Rinse remaining sucrose through screen with a small stream of water and backwash nematodes into counting dish or storage vial.

This method is relatively quick (about 12 min) and most types of soil may be processed easily. Occasionally, excessive amounts of minute organic debris will be recovered with the samples making it difficult to see nematodes clearly. However, since dead nematodes are extracted as well as live nematodes, counting and interpretation of counts must be done carefully. The osmotic concentration of sucrose may have a detrimental effect on some nematodes if they are exposed to it for too long. While this technique may require less space than an extensive Baermann funnel setup, it is more labor intensive than Baermann funnel procedures and requires a centrifuge that may be expensive.

22–2.4 Elutriation (Fig. 22–4A)

Many larger nematology laboratories have constructed or purchased semiautomatic nematode extraction machines called *elutriators* (Byrd et al., 1976). In this method, the soil sample (250–500 g) is poured into one of several stainless steel funnels. Water and air are injected under pressure into the bottom of the funnel, agitating the soil sample as the funnel fills with water. When the water level reaches the top of the funnel, it over-flows, carrying fine soil particles, nematodes, and organic debris onto a trough that carries all onto a coarse screen (no. 40 mesh) which catches most of the organic matter. Water and nematodes drain through a funnel underneath the coarse screen that directs the water onto a conical surface in a sample splitter that subsamples a designated proportion of the water coming through. This subsample of water is directed onto a no. 325 or no. 400 mesh screen that catches nematodes and fine soil particles. A shaker continually agitates the no. 400 screen so that it does not plug with soil. The soil and nematodes are then backwashed and processed by sucrose centrifugation (see 22–2.3) or Baermann funnel (see 22–2.2) techniques to remove them from remaining soil and debris. Numbers are corrected for the portion of water collected by the sample splitter. Most of the equipment in the elutriator can then be inverted and cleaned automatically.

Depending on soil type and species extracted, this procedure has a relatively consistent extraction efficiency with much reduced between-operator variability. Large (250–1000 g) samples may be extracted rapidly (< 5 min) and several (usually 4) samples can be processed at one time. The self-cleaning feature saves additional time. The major disadvantages to this method are the space required and initial expense ($10,000–$20,000) needed to purchase or construct the equipment. Since only a portion of the sample is actually recovered, this method is often less efficient than some other methods and may miss nematodes present at very low densities. However, this is partially compensated by the larger soil volumes that can be extracted and that replicate subsamples can be quickly extracted. The elutriator must also be closely monitored. Soil samples with high silt or clay contents may not be sufficiently mixed and remain as a large plug in the bottom of the funnel. Screens may plug and overflow resulting in the loss of nematodes.

22–2.5 Extracting Heterodera Cysts

In areas where cyst nematodes are a problem, the cysts themselves are extracted and counted instead of, or in addition to, vermiform (worm-shaped) stages. Since these methods are not dependent on live nematodes and since cysts are resistant to drying, soil samples can be partly or fully air dried, sieved through a coarse mesh sieve and stored, if necessary. In addition, efficiency of cyst recovery is often improved by first drying the sample. A large soil sample should be collected, thoroughly mixed, and 100, 200, or 500 g subsamples extracted.

22–2.5.1 Procedure for the Flask Method (Jones, 1945)

1. Put the soil sample into a 2 L conical flask and add water to a depth of 5 cm.
2. Shake the flask with a rotary movement until the soil is thoroughly wetted and in suspension.

3. Fill the flask to the neck with water, using a powerful jet so that the suspension is well stirred and aerated. Continue to fill the flask up to the brim with a gentle trickle of water and leave to stand for several minutes until the water in the neck of the flask is clear.
4. Carefully decant the floating organic debris and cysts into a no. 20 mesh screen over a no. 60 mesh screen. Rotate the flask while pouring so that all the floating material is washed onto the sieves with minimum disturbance of the soil in the flask.
5. Wash the contents of the no. 20 mesh screen gently but thoroughly so that all cysts pass onto the no. 60 mesh screen. Discard the debris on the no. 20 mesh screen and wash the contents of the no. 60 mesh screen thoroughly. The cysts may be poured into a dish and counted immediately or rinsed onto a piece of bolting silk in a funnel. The silk can then be wrapped up and dried for examination later.

This method is simple, quick, and does not require special equipment. It is relatively efficient and recovers about 70% of cysts in soil (Jones, 1945). However, total egg populations are often underestimated because cysts completely full of eggs may not float, while empty or partially full cysts are recovered.

22–2.5.2 Procedure for Fenwick Can Method (Fenwick, 1940)

The most widely used apparatus for extracting cysts is the Fenwick can which is suitable for samples up to 300 g of soil. The can is usually made of brass and is generally 30 cm high with a sloping base and tapering towards the top. There is a drain hole 2.5 cm in diameter in the side of the can at the lowest point of the slope which is closed with a rubber stopper. Just below the rim of the can is a sloping collar with an upright rim of 6.25 cm high. The collar tapers towards the outlet which is 2.5 cm wide. A small inlet about 5 cm from the top of the can, attached to the water supply for filling the can with water is optional. A large brass funnel, 20 cm in diameter, with a stem 20 cm long, is supported above it and a no. 16 mesh screen fits into the funnel. A no. 20 mesh screen over a no. 60 mesh screen is placed below the outlet from the collar.

1. Fill the can with water and wet collecting sieves.
2. Place the soil sample on no. 20 mesh screen in the funnel and wash the soil through the screen into the can with a strong jet of water from above. Cysts and organic material (called the float) and some soil will overflow rapidly into the collar and pass down onto the collecting sieves.
3. Wash the debris and recover the cysts as described for the flask method.

Using this method, about 70% of cysts are recovered. This efficiency can be improved by elutriating the soil in the base of the can. Remove the funnel and insert a glass or metal tube, which is attached to a rubber tube from the water tap, into the can. Adjust the flow of water so that it is moderately fast but will not carry over too many soil particles. Stir the soil in the can for about 1 min. The upward current of water will carry over cysts that would otherwise be trapped by sedimentation. Wash the debris and recover cysts as described for the flask method.

22–2.5.3 Recovery and Counting of Cysts

Cysts can be recovered from the float while still moist or after drying, but to recover live eggs from cysts it is best not to allow the float to dry. Perhaps the quickest method for removing cysts from wet debris is as follows (Southey, 1986), but further processing from ethanol-glycerine (Caswell et al., 1985) may remove more of the debris.

1. Wash the debris onto a moistened filter paper in a Buchner funnel to which suction can be applied. The filter paper should be larger in diameter than the base of the funnel so that it extends 1 cm up the sides. Draw a line with a colored wax pencil along a radius of the paper before placing the paper in the funnel.
2. Add water to the funnel to a depth of about 0.5 cm and gently agitate until the debris and cysts are evenly distributed over the paper.
3. Apply suction to draw off the water.
4. Remove the filter paper from the funnel and place it momentarily on a piece of absorbent paper or cloth to remove excess water.
5. Place the filter paper on a turntable under a stereomicroscope with the radius in the field of view. Beginning at the edge of the paper, revolve the turntable and remove or count the cysts. After one revolution, move inside the paper to the next field of view and repeat the process until the whole area of the filter paper has been scanned.
6. Cysts can be separated from the debris by picking them up with an aspirator, similar to that used for collecting insects, which is connected to a suction pump and applying gentle suction.
7. Eggs and juveniles within cysts may be liberated for enumeration by crushing the cysts in a tissue homogenizer (Caswell et al., 1985).

22–3 EXTRACTION OF NEMATODES FROM PLANT MATERIAL

22–3.1 Direct Examination of Plant Material

Wash roots gently to remove as much soil as possible. Examine the tissue (shoots or roots) in water in an open petri dish under a stereomicroscope and tease tissues apart with stout needles. Nematodes released from the tissue float out and can be collected with a handling needle or fine pipet. Since nematodes often migrate from damaged tissue, it is often worthwhile to re-examine the sample after 2 or 3 h.

22–3.2 Baermann Funnel Extraction

Plant material should be chopped into small pieces and placed on a tissue filter in a typical Baermann funnel setup. If the mesh of the screen in the funnel is fine enough to hold the plant tissue and the latter is clean enough, a filter may not be necessary. It is often helpful to add a trace of wetting agent and a few milliliters of 0.15% methyl p-hydroxybenzoate solution to inhibit bacterial growth in the water within the funnel (Southey, 1986).

22–3.3 Root-Incubation Technique

1. Wash roots free from soil and cut into lengths of 5 to 10 cm; large diameter or fleshy roots can be cut longitudinally to help nematodes emerge.
2. Put the roots into containers such as screw-cap jars, closed petri dishes or sealed polyethylene bags and keep at 20 to 25 °C. The roots should be well wetted or immersed in shallow water before the containers are closed.
3. Rinse the roots and collect the water each day to examine for nematodes.
4. Add more water, seal the container and continue incubation.

This method has several applications and modifications that are discussed by Southey (1986). Most nematodes will be recovered in 4 to 7 d. Macerating the roots briefly (10 s) in a blender and placing the containers on a shaker will increase recovery. Antibiotics may be added to the water to retard microbial growth.

22–3.4 Mist Chamber Extraction (Seinhorst, 1950)

1. Chop plant material into pieces 3 to 4 mm long and place on a fine screen basket within a funnel with a long stem. The stem should set inside and near the bottom of a beaker or large test tube.
2. Place the funnel inside a mist chamber where a continuous, or periodic, fine mist of water is sprayed onto the plant material. The rate of spray must be low enough that the rate of overflow out of the beaker or test tube is not sufficient to carry nematodes along with it. A common rate is 1 gal/h under pressure of 40 lb/sq. in. At the end of the extraction period (4–7 d), pour the water from the beaker or test tube into a counting dish and examine for nematodes.

22–4 MICROSCOPIC OBSERVATION AND IDENTIFICATION
OF NEMATODES

Once nematodes have been extracted, they must be counted or identified. If the purpose of sampling was for taxonomic or regulatory objectives, only one or a few species present may be of interest. If samples were taken for making management decisions, only species expected to cause damage need to be identified and enumerated. Objectives for research in ecological studies may require more rigorous examination of the samples, however, to the extent that all juvenile and life stages of one or all species need to be differentiated and enumerated. The original objectives for taking samples must be known before microscopic examination of the samples begins.

Most routine enumeration of nematodes in samples can be performed in a shallow counting dish under a high-quality stereomicroscope illuminated from below. However the person must also have access to a compound microscope for higher magnification observations of uncertain or unknown species. Inverted compound microscopes are used by some laboratories to allow increased magnification during counting.

The most comprehensive treatment of microscopic examination of nematodes is given by Southey (1986). Shorter reviews can be found in Fortuner (1991), Hooper (1990), and Zuckerman et al. (1990). More advanced procedures that are beyond the scope of this chapter include electron microscopy (Carta, 1990; Eisenback, 1990) and molecular techniques for nematode identification (Curran, 1990; Williamson, 1990).

22–4.1 Temporary Mounts

Most observations of nematodes are superior if a living nematode is mounted under a microscope but rapid movement makes identification difficult. Nematodes can be immobilized by gently heating the slide from below with a match or an alcohol lamp. The slide must be held high enough above the flame that soot does not accumulate on the bottom of the slide. Practice is necessary to discover the appropriate amount of heat necessary to immobilize the nematode without destroying it. Nematodes can also be immobilized with fixative (see below) before preparing temporary mounts but some features may be obscured.

Most routine nematode identifications are made with temporary mounts and permanent mounts are generally reserved for archiving individuals for future reference. Nematodes can be easily observed under the microscope if mounted in a small drop of water and covered by a coverslip. Care should be taken to use sufficient water so that air bubbles do not form under the coverslip but not so much water that it flows from under the coverslip or nematodes may be carried along with it. Temporary mounts may be kept from drying out for some time by ringing the coverslip with wax from a small candle or fingernail polish. One of the most difficult techniques for beginning nematologists to develop is the transfer of selected nematodes from the counting dish to a microscope slide. A small drop of water should be placed on the slide and the nematode picked out of the counting dish with a pulp canal file or an eyelash glued to a small applicator stick. The nematode should be carefully placed in the center of the drop of water on the slide and observed under the stereomicroscope as the coverslip is placed on the slide. If the relative position of the nematode is gently marked on the coverslip with a felt-tipped pen it will be faster to locate the specimen when the slide is placed on the compound microscope.

22–4.2 Permanent Mounts

Several procedures for making permanent slides are discussed by Southey (1986). No single method is ideal for preparing permanent mounts that exactly resemble living nematodes. Discoloration or distortion generally occurs to some extent and slides often deteriorate over time but well-prepared slides will last for more than 25 yr. Advantages and disadvantages of several procedures are discussed by Santos and Abrantes (1989). The method described below (Seinhorst, 1959; Zuckerman et al., 1990) may be appropriate for most applications. Before beginning, it is extremely important that nematodes are completely free of debris.

22–4.2.1 Equipment

1. Small oven.
2. *F.A. fixative*   or   *T.A.F. fixative.*
   10 mL formalin about 40%   7 mL formalin
       formaldehyde   2 mL triethanolamine
   1 mL glacial acetic acid   91 mL water
   89 mL water
3. *Solution I*   and   *Solution II*
   20 mL 96% ethanol   95 mL 96% ethanol
   1 mL glycerine   5 mL glycerine
   79 mL Water
4. Picric acid in a saturated solution (i.e., add picric acid to water until it no longer goes into solution). Always keep picric acid covered with water as it is explosive in its dry state.
5. Small open glass dishes (see below).
6. Larger closed dishes, desiccator or large petri dishes.
7. Anhydrous glycerine.
8. Glyceel (formula as reported in Southey, 1986; glyceel is also commercially available)

Alcohol-soluble linseed oil	31.75
Nitrocellulose (ICI HX 30/50)	
(moistened with 30% wt/wt industrial	
methylated spirit)	22.67
Butyl acetate	20.41

Toluol (S-free)	20.41
Industrial methylated spirit	4.76
	100.00 g

9. Microscope slides.
10. Cover slips.

22–4.2.2 Procedure

1. Nematodes should be killed in a relaxed position for features to be readily visible. This is most easily accomplished by (a) placing nematodes in a small vial of water inside an oven at 65 °C, (b) rapidly adding nematodes in a small volume of water to hot (43 °C) double-strength fixative or (c) rapidly adding hot double-strength fixative to nematodes in a small volume of water. Nematodes may be fixed en masse and the best specimens hand picked from debris for further processing.
2. Transfer nematodes with a small amount of fixative into a small but deep glass dish such as a 37 × 25 mm Wheaton preparation dish (a glass scintillation vial cut to a height of 2 to 3 cm and firepolished also works well) and add 7 to 10 mL of solution I.
3. Addition of one to two drops of picric acid gives the nematode a yellowish tint but prevents clearing of the stylet.
4. Place the small dish open inside a larger closed dish, such as a desiccator, which contains a small vial or beaker with 95% ethanol. Put dishes inside an oven at 35 to 40 °C for 12 h. This removes most of the water from Solution I.
5. Remove dishes from the oven and remove half of the volume of Solution I with a pipet. Observe under a stereomicroscope to ensure that nematodes are not removed.
6. Add enough Solution II to almost fill the small dish, place small dish inside covered dish and return to the oven for at least 3 h.
7. Nematodes can be further processed at this point but better results will be obtained if steps 5 and 6 are repeated over a 4 to 7 d period. Remember to always view solution removal under the microscope. Often the convection currents of the evaporating alcohol will carry the nematodes up the sides of the glass dish and they may need to be rinsed down again with the pipet.
8. After several transfers, the nematodes will be in pure glycerine and can remain in this condition almost indefinitely. Dishes should remain in oven until all ethanol has evaporated and if stored before processing, returned to the oven to remove any moisture that the glycerine may have absorbed.
9. Place a small drop of anhydrous glycerine in the center of a microscope slide. (Many researchers use "Cobb" aluminum slides that hold the nematodes in glycerine "sandwiched" between two cover slips so that both sides of the slide may be viewed. (See Southey, 1986.)

10. Place a small dish of glycerine with nematodes under the stereomicroscope, pick individual specimens out with eye lash, pulp canal file etc. and place in the center of the drop of glycerine on the slide. Several nematodes may be placed on a single slide. View slide under the stereomicroscope and arrange nematodes into the center of the drop. Be sure nematodes are at the bottom of the drop and are not "floating" on top of the glycerine.

11. To prevent distortion of nematodes, particularly large nematodes, short pieces of glass wool that have been soaked in glycerine should be placed between the nematodes and the edges of the coverslip. Select pieces of the same diameter as the nematodes to be mounted.

12. Place a clean 18 to 22 mm round (preferable) or square cover slip (thickness no. 0 or 1) over the drop while observing under the microscope to ensure that nematodes do not move from the center of the drop. If glycerine flows out from under the edge of the coverslip, remove cover slip and transfer nematodes to another slide with a smaller drop of glycerine.

13. Seal the cover slip by applying glyceel around the edge with a small brush. After an hour, apply a second coat of glyceel.

14. A small diagram representing the relative locations of the final resting positions of the specimens should be included on the slide label. Unique characteristics such as different species, sexes, or life stages may be marked at the positions of the different nematodes on the label.

15. Always store slides flat. If slides are placed on their edge, nematodes will settle to the edge of the coverslip over time.

22–5 NEMATODE IDENTIFICATION

Nematodes are difficult to identify and, with the exception of well-known plant parasites, few good keys are available. In addition, classification of nematodes continues to change and different systems are accepted by different groups. A complete discussion of nematode taxonomy is beyond the scope of this chapter but a review of higher nematode taxonomy can be obtained from Freckman and Baldwin (1990) and Maggenti (1991). Although an older text and not written in key format, Goodey (1963) provides excellent descriptions of most genera found in soil and freshwater. The most recent, and probably best, comprehensive key was written by Bongers (1988) for nematodes of the Netherlands but is appropriate for other areas as well. The original publication is complete with drawings and photographs and an English translation of the key text can be obtained from the author. Since the Order Tylenchida contains most of the plant-parasitic nematodes, it has received the most attention and several texts are available. Siddiqi (1985) is a comprehensive text but Anderson and Mulvey (1979) and Mai and Lyon (1975) provide excellent pictorial keys for most plant-parasitic genera including some Dorylaimida. A videocassette for teaching identification of the most common genera of plant-parasitic nematodes is also available (see Nematology Newsletter 1992, Vol. 38:9). Although written in South Africa, Heyns (1971) provides drawings and keys to many genera that are also found in North America. Several short works include drawings and keys to many of the nematodes found in soils including both plant parasites and free-living forms (Ferris et al., 1976; Freckman & Baldwin, 1990; Smart & Nguyen, 1988). Freckman and Baldwin (1990) also includes an excellent discussion on the ecology of nematodes in soil. A bibliography of keys to terrestrial nematodes was compiled by Baird and Bernard (1983). Many of the referred works include keys to species of selected genera. Biology and identification of many insect-parasitic nematodes are found in selected chapters in Nickle (1984) and Nickle (1991).

Considerable training and experience is necessary before correct identifications can be made. In addition, few facilities exist to send nematodes to obtain or confirm identification. When nematodes are sampled in relation to plant disease objectives, species and even race identifications are often necessary. Many ecological studies, however, are often concerned with the entire nematode community and nematodes may be placed into broad taxonomic or trophic groups based on buccal and esophageal morphology. However, feeding habits are not known for most species of free-living nematodes in soil and nematodes that feed on different food items may appear morphologically similar. Nematodes that may be found in terrestrial systems belong to the Orders Tylenchida, Dorylaimida, Rhabditida, Araeolaimida, Chromadorida, Desmodorida, Enoplida, Monhysterida, and Mononchida. These orders can generally be distinguished by the structure of the buccal and esophageal regions in the anterior of the nematode. Perhaps the most distinctive feature of nematodes is the presence or absence of a stylet.

22–5.1 Nematodes with Stylets

TYLENCHIDA (Fig. 22–5A) are the most well-known nematodes in soil because they represent most species of plant-parasites. Other species feed on fungi, including mycorrhizae or are parasites of insects. They can usually be distinguished by a three-part stylet, consisting of a barb, shaft and basal knobs, and a three-part esophagus consisting of a narrow muscular procorpus, a muscular median bulb and a glandular basal region. In some genera, a narrow isthmus region separates the median bulb from the basal region and is occasionally considered a fourth part of the esophagus.

DORYLAIMIDA (Fig. 22–5B) also possess stylets but while a few well-known genera are plant parasites and may vector plant viruses, most are poorly studied. Members of the Dorylaimida may fill several trophic positions by feeding on fungi, algae, or preying on other nematodes. Many species may be omnivores. Undoubtedly, many may feed on plants without causing symptoms of disease. These nematodes are identified by a two-part

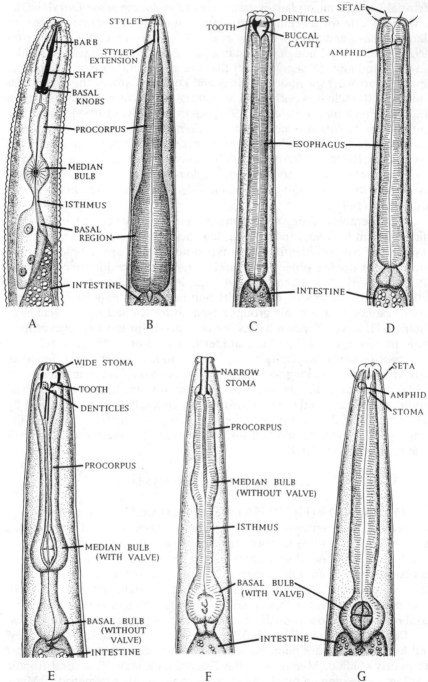

Fig. 22–5. Continued on next page.

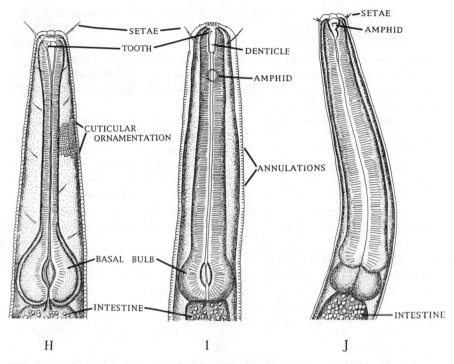

H   I   J

Fig. 22–5. Individual representatives for the nematode Orders found in soil illustrating the distinguishing features within the esophageal region. (A) Tylenchida (modified from Siddiqi), (B) Dorylaimida, (C) Mononchida, (D) Monhysterida, (E) Rhabditida with wide stoma and teeth, (F) Rhabditida with narrow, smooth stoma (modified from Mai and Lyon), (G) Araeolaimida, (H) Chromadorida (modified from Goodey), (I) Desmodorida, (J) Enoplida. (All illustrations by Kathy Merrifield.)

stylet (or spear) consisting of a stylet and a stylet extension and a two-part esophagus with a narrow procorpus and a wider basal region.

22–5.2 Nematodes without Stylets

RHABDITIDA (Fig. 22–5E, F) are the most common nematodes in soil without stylets and often dominate the entire nematode community, particularly in detrital-based food webs. Most are considered to feed on bacteria and have important roles in nutrient cycling (Ingham et al., 1985) but in fact the feeding habits of only a few species have been studied. Some species are predators of protozoa, nematodes, or other small soil invertebrates, while others are parasites or associates of insects and other invertebrates. Rhabditida morphology is very diverse. The opening inside the mouth is considered the stoma and may be wide or narrow, smooth or lined with denticles or teeth of various sizes. The walls of some stoma are heavily cuticularized while others are barely visible. The esophagus generally consists of a broad anterior portion that leads into a median bulb which may or may not contain a valve. Behind the median bulb, the esophagus often narrows into an isthmus that terminates in an expanded bulb which also may or may not contain a valve. In some species the median bulb is absent. The body never has bristles, caudal glands or spinnerets that are common in some other orders without stylets.

ARAEOLAIMIDA (Fig. 22–5G) contains only a few genera that are found in soil and all are likely to feed on bacteria. The esophagus may be relatively long and is cylindrical until it expands into a bulb just before the esophagus. The bulb may or may not have a valve. The stoma is usually narrow and tubular and is generally smooth. The head generally has four setae behind the lip region except for *Wilsonema* and *Tylocephalus* that have a cuticular expansion around the head. Circular or spiral amphids may be easy to see but some species have transverse slits that are often difficult to observe.

CHROMADORIDA (Fig. 22–5H) are primarily aquatic with a few species that occur in soil and probably feed on bacteria or algae. The stoma has a conspicuous dorsal tooth and the esophagus is cylindrical that expands into a large bulb at the base (except for *Achromadora* which has a small bulb). Setae are present around the head and the body has cuticular ornamentation. Amphids are circular in *Achromadora* but more difficult to see in other genera.

DESMODORIDA (Fig. 22–5I) rarely occur in soil. The cuticle in the head region is thickened and hyaline, forming a "helmet." Distinct annulations are visible posterior to the helmet. The esophagus is narrow anteriorly and bulboid in the posterior portion. The stoma is variable and may or may not be armed with a dorsal tooth with or without accompanying smaller denticles. Amphids are usually visible but highly variable in shape.

ENOPLIDA (Fig. 22–5J) are primarily marine with some freshwater and a few terrestrial genera. Little is known about their biology but they may feed on bacteria, diatoms, algae, or be predacious. Enoplida have a cylindrical esophagus similar to Mononchida (see below). The stoma is highly diverse and may or may not have teeth of various sizes. Setae are usually present. If amphids can be seen they are usually cup or stirrup shaped.

MONHYSTERIDA (Fig. 22–5D) are generally found in freshwater but a few genera are also occasionally found in soil. They are considered to be bacterial feeders but few have been studied. Monohysterida have a small, narrow stoma and a long cylindrical esophagus that occasionally swells to a bulb at the base. Setae (usually 4 or more) are present on or near the lip region. Circular structures (amphids) are often visible on the cuticle in the stomal region.

MONONCHIDA (Fig. 22–5C) are generally large distinctive nematodes that are major predators of other soil nematodes. Juvenile stages may subsist on bacteria but nematode prey appears to be required for the population to persist. The buccal cavity of the stoma is large and heavily cuticularized with large or small teeth. The esophagus is long and cylindrical and heavily muscularized.

REFERENCES

Anderson, R.V., and R.H. Mulvey. 1979. Plant-parasitic nematodes in Canada. Part I. An illustrated key to the genera. Agriculture Canada.

Andrassy, I. 1962. The problem of number and size of sampling unit in quantitative studies of soil nematodes. p. 65–67. In P.W. Murphy (ed.) Progress in soil zoology. Butterworth Publ., London.

Baermann, G. 1917. Eine einfache Methode zur Auffindung von Ancylostomum (Nematoden) Larven in Erdproben. Geneesk. Tijdschr. Ned. Ind. 57:131–137.

Baird, S.M., and E.C. Bernard. 1983. A bibliography of keys for the identification of plant-parasitic and free-living terrestrial nematodes. Nematol. Newsl. 29:9–18.

Barker, K.R. 1985a. Sampling nematode communities. p. 9–17. In K.R. Barker et al. (ed.) An advanced treatise on Meloidogyne. Vol. II Methodology. North Carolina State Univ. Graphics, Raleigh.

Barker, K.R. 1985b. Nematode extraction and bioassays. p. 19–35. In K.R. Barker et al. (ed.) An advanced treatise on Meloidogyne. Vol. II. Methodology. North Carolina State Univ. Graphics, Raleigh.

Barker, K.R. 1986. Determining nematode population responses to control agents. p. 283–296. In K.D. Hickey (ed.) Methods for evaluating pesticides for control of plant pathogens. Am. Phytopathol. Soc., St. Paul.

Barker, K.R., and C.L. Campbell. 1981. Sampling nematode populations. p. 451–474. In B.M. Zuckerman et al. (ed.) Plant parasitic nematodes. Vol. 3. Academic Press, New York.

Bongers, T. 1988. De nematoden van nederland. Pirola Schoorl. Natuurhist. Biblioth. KNNV nr. 46. Wageningen Agric. Univ., the Netherlands.

Byrd, D.W., Jr., K.R. Barker, H. Ferris, C.J. Nusbaum, W.E. Griffin, R.H. Small, and C.A. Stone. 1976. Two semi-automatic elutriators for extracting nematodes and certain fungi from soil. J. Nematol. 8:206–212.

Carta, L.K. 1990. Preparation of nematodes for transmission electron microscopy. p. 97–106. In W.R. Nickle (ed.) Manual of agricultural nematology. Marcel Dekker, New York.

Caswell, E.P., I.J. Thomason, and H.E. McKinney. 1985. Extraction of cysts and eggs of Heterodera schachtii from soil with an assessment of extraction efficiency. J. Nematol. 17:337–340.

Caveness, F.E., and H.J. Jensen. 1955. Modification of the centrifugal-flotation technique for the isolation and concentration of nematodes and their eggs from soil and plant tissue. Proc. Helminth. Soc. Wash. 22:87–89.

Christie, J.R., and V.G. Perry. 1951. Removing nematodes from soil. Proc. Helminth. Soc. Washington 18:106–108.

Cobb, N.A. 1918. Estimating the nema population of soil. USDA Agric. Agric. Circ. No.1.

Coleman, D.C. 1976. A review of root production processes and their influence on soil biota in terrestrial ecosystems. p. 417–434. In The 17th Symposium of The British Ecological Society, Blackwell Sci. Publ., London.

Curran, J. 1990. Application of DNA analysis to nematode taxonomy. p. 125–143. In W.R. Nickle (ed.) Manual of agricultural nematology. Marcel Dekker, New York.

Eisenback, J.D. 1990. Preparation of nematodes for scanning electron microscopy. p. 87–96. In W.R. Nickle (ed.) Manual of agricultural nematology. Marcel Dekker, New York.

Eisenback, J.D., and G.D. Griffin. 1987. Interactions with other nematodes. p. 313–320. In J.A. Veech and D.W. Dickson (ed.) Vistas on nematology. Society of Nematologists, Hyattsville, MD.

Fenwick, D.W. 1940. Methods for the recovery and counting of cysts of Heterodera schachtii from soil. J. Helminth. 18:155–172.

Ferris, H., P.B. Goodell, and M.V. McKenry. 1981. General recommendations for nematode sampling. Univ. of California Division of Agricultural Sciences Leaflet 21234, Berkeley.

Ferris, H., P.B. Goodell, and M.V. McKenry. 1981b. Sampling for nematodes. Calif. Agric. 35:13–15.

Ferris, H., and M.V. McKenry. 1976. Nematode community structure in a vineyard soil. J. Nematol. 8:131–137.

Ferris, V.R., L.M. Ferris, and J.P. Tjepkema. 1976. Genera of freshwater nematodes (Nematoda) of eastern North America. U.S. E.P.A. Aquatic Biology Section, Cincinnati, OH.

Fidler, J.H., B.M. Church, and J.F. Southey. 1959. Field sampling and laboratory examination of cereal root eelworm cysts. Plant Pathol. 8:27–34.

Fortuner, R. 1991. Field sampling and preparation of nematodes for optic microscopy. p. 75–87. In W.R. Nickle (ed.) Manual of agricultural nematology. Marcel Dekker, New York.

Freckman, D.W., and J.G. Baldwin. 1990. Nematoda. p. 155–200. In D.L. Dindal (ed.) Soil biology guide. John Wiley and Sons, New York.

Goodell, P.B. 1982. Soil sampling and processing for detection and quantification of nematode populations in ecological studies. p. 178–198. In D.W. Freckman (ed.) Nematodes in soil ecosystems. Univ. Texas Press, Austin.

Goodell, P.B., and H. Ferris. 1980. Plant-parasitic nematode distributions in an alfalfa field. J. Nematol. 12:136–141.

Goodell, P.B., and H. Ferris. 1981. Sample optimization for five plant-parasitic nematodes in an alfalfa field. J. Nematol. 13:304–313.

Goodey, T. 1963. Soil and freshwater nematodes. John Wiley and Sons, New York. (Second edition revised by J. B. Goodey.)

Heyns, J. 1971. A guide to the plant and soil nematodes of South Africa. A.A. Balkema, Cape Town, S.A.

Huang, Jeng-Sheng. 1987. Interactions of nematodes with rhizobia. p. 301–306. In J.A. Veech and D.W. Dickson (ed.). Vistas on nematology. Society of Nematologists, Hyattsville, MD.

Hooper, D.J. 1990. Extraction and processing of plant and soil nematodes. p. 45–68. In M. Luc, R.A. Sikora and J. Bridge (ed.) Plant parasitic nematodes in subtropical and tropical agriculture. C.A.B. Int., Wallingford, UK.

Hunt, H.W., D.C. Coleman, E.R. Ingham, R.E. Ingham, E.T. Elliott, J. Moore, S.L. Rose, C.P.P. Reid, and C. Morley. 1987. The detrital food web in a shortgrass prairie. Biol. Fert. Soil 3:57–68.

Ingham, R.E. 1988. Interactions between nematodes and vesicular-arbuscular mycorrhizae. Agric. Ecosyst. Environ. 24:169–182.

Ingham, R.E., R.V. Anderson, W.D. Gould, and D.C. Coleman. 1985a. Vertical distribution of nematodes in a shortgrass prairie. Pedobiologia 28:155–160.

Ingham, R.E., and D.C. Coleman. 1983. Effects of an ectoparasitic nematode on bacterial growth in gnotobiotic soil. Oikos 40:75–80.

Ingham, R.E., and J.K. Detling. 1984. Plant-herbivore interactions in a North American mixed-grass prairie. III. Soil nematode populations and root biomass on *Cynomys ludovicianus* colonies and adjacent uncolonized areas. Oecologia 63:307–313.

Ingham, R.E., and J.K. Detling. 1990. Effects of root-feeding nematodes on aboveground net primary production in a North American grassland. Plant Soil 121:279–281.

Ingham, R.E., J.A. Trofymow, E.R. Ingham, and D.C. Coleman. 1985b. Interactions of bacteria, fungi and their nematode grazers: Effects on nutrient cycling and plant growth. Ecol. Monogr. 55:119–140.

Jenkins, W.R. 1964. A rapid centrifugal-flotation technique for separating nematodes from soil. Plant Dis. Rep. 48:692.

Jones, F.G.W. 1945. Soil populations of the beet eelworm (*Heterodera schachtii* Schm.) in relation to cropping. Ann Appl. Biol. 32:351–380.

Kusmin, L.L. 1976. Free-living nematodes in tundra of the western Taimyr. Oikos 27:501–505.

MacLean, S.F., Jr. 1974. Primary production, decomposition and the activity of soil invertebrates in tundra ecosystems: A hypothesis. p. 197–206. In A.J. Holding et al. (ed.) Soil organisms and decomposition tundra. IBP Tundra Biome Steering Committee, Stockholm.

Mai, W.F., and H.H. Lyon. 1975. Pictorial key to genera of plant-parasitic nematodes. Fourth edition, revised. Comstock Publ. Assoc., Ithaca, NY.

Maggenti, A. 1991. Nemata: Higher classification. p. 147–187. *In* W.R. Nickle (ed.) Manual of agricultural nematology. Marcel Dekker, New York.

McSorley, R. 1982. Simulated sampling strategies for nematodes distributed according to a negative binomial model. J. Nematol. 14:517–522.

McSorley, R. 1987. Extraction of nematodes and sampling methods. p. 13–48. *In* R.H. Brown and B.R. Kerry (ed.) Principles and practice of nematode control in crops. Academic Press, New York.

McSorley, R., and J.L. Parrado. 1982. Estimating relative error in nematode numbers from single soil samples composed of multiple cores. J. Nematol. 14:522–529.

McSorley, R., and J.L. Parrado. 1983. Error estimates for nematode soil samples composed of cores of unequal sizes. Nematropica 13:27–36.

Mojtahedi, H., R.E. Ingham, G.S. Santo, J.N. Pinkerton, G.L. Reed, and J.H. Wilson. 1991. Seasonal migration of *Meloidogyne chitwoodi* and its role in potato production. J. Nematol. 23:162–169.

Nickle, W.R. (ed.). 1984. Plant and insect nematodes. Marcel Dekker, New York.

Nickle, W.R. (ed.). 1991. Manual of agricultural nematology. Marcel Dekker, New York.

Nicholas, W.L. 1975. The biology of free-living nematodes. Clardon Press, Oxford.

Noe, J.P., and K.R. Barker. 1982. Plant row effects on the distribution of plant-parasitic nematodes. J. Nematol. 14:460–461 (abstr.).

Oostrinbrink, M. 1971. Comparison of techniques for population estimation of soil and plant nematodes. p. 72–82. *In* J. Phillipson (ed.) Quantitative soil ecology. IBP Handbook No. 18. Blackwells, Oxford.

Reichle, D.E. 1977. The role of soil invertebrates in nutrient cycling. *In* U.L. Lohm and T. Persson (ed.) Soil organisms as components of ecosystems. Ecol. Bull. (Stockholm) 25:145–156.

Riedel, R. 1988. Interaction of plant-parasitic nematodes with soil-borne plant pathogens. p. 281–292. *In* C.A. Edwards et al. (ed.) Biological interactions in soil. Elsevier, New York.

Santo, G.S., A.P. Nyczepir, D.A. Johnson, and J.H. O'Bannon. 1982. Sampling for nematodes in soil. Washington State Univ. Agric. Res. Bull. XB 0923.

Santos, M.S.N. de A., and I.M. de O. Abrantes. Morphological characters and methods for preparing nematodes. p. 201–215. *In* R. Fortuner (ed.) Nematode identification and expert system technology. Plenum Press, New York.

Seinhorst, J.W. 1950. De betekenis van de toestand van de grond voor het optreden van aantasting door het stengelaaltje (*Ditylenchus dipsaci* (Kuhn) Filipjev). Tijdschr. PlZeikt. 56:289–348.

Seinhorst, J.W. 1959. A rapid method for the transfer of nematodes from fixative to anhydrous glycerin. Nematologica 4:67–69.

Siddiqi, M.R. 1985. Tylenchida: parasites of plants and insects. Commonwealth Agricultural Bureaux, London.

Sikora, R.A., and W.W. Carter. 1987. Nematode interactions with fungal and bacterial plant pathogens-fact or fantasy. p. 307–312. *In* J.A. Veech and D.W. Dickson (ed.) Vistas on nematology. Society of Nematologists, Hyattsville, MD.

Smart, G.C., and K.B. Nguyen. 1988. Illustrated key for the identification of common nematodes in Florida. Entomology and Nematology Dep., Inst. of Food and Agricultural Sciences, Univ. of Florida, Gainesville.

Sohlenius, B. 1977. Numbers, biomass and respiration of Nematoda, Rotatoria and Tardigrada in a 120-year-old Scots pine forest at Ivantjarnsheden, central Sweden. Swedish Coniferous Forest Proj. Tech. Rep. No. 9. Uppsala, Sweden.

Sohlenius, B. 1980. Abundance, biomass and contribution to energy flow by nematodes in terrestrial ecosystems. Oikos 34:186–194.

Southey, J.F. 1986. Laboratory methods for work with plant and soil nematodes. Her Majesty's Stationery Office, London.

Southwood, T.R.E. 1978. Ecological methods with particular reference to the study of insect populations. 2nd ed. Chapman and Hall, London.

Stanton, N.L., M. Allen, and M. Campion. 1981. The effect of the pesticide carbofuran on soil organisms and root and shoot production in shortgrass prairie. J. Appl. Ecol. 18:417–431.

Viglierchio, D.R., and R.V. Schmitt. 1983. On the methodology of nematode extraction from field samples: Comparison of methods for soil extraction. J. Nematol. 15:450–454.

Whitehead, A.G., and J.R. Hemming. 1965. A comparison of some quantitative methods of extracting small vermiform nematodes from soil. Ann. Appl. Biol. 55:25–38.

Williamson, V.M. 1990. Molecular techniques for nematode species identification. p. 107–123. *In* W.R. Nickle (ed.) Manual of agricultural nematology. Marcel Dekker, New York.

Yeates, G.W. 1971. Feeding types and feeding groups in plant and soil nematodes. Pedobiologia 8:173–179.

Zuckerman, B.M., W.F. Mai, and L.R. Krusberg. 1990. Plant nematology laboratory manual. Revised edition. Univ. of Massachusetts Agric. Exp. Stn., Amherst, MA.

Chapter 23

Protozoa

ELAINE R. INGHAM, *Oregon State University, Corvallis, Oregon*

Protozoa occur in all ecosystems of the world and several hundred species have been described (Lee et al., 1985). Protozoa are unicellular, eukaryotic, generally aerobic organisms first described by Anthonie van Leeuwenhoek (1632–1723). Taxonomically, protozoa are placed in a subkingdom in the Kingdom Protista and represent the group in which mitosis and meiosis became established (Laybourn-Parry, 1984). Eventually, photosynthetic prokaryotes, for which certain members of the nonphotosynthetic protozoa were hosts, evolved into plastids (Margulis, 1981), giving the protozoa an important evolutionary role.

This chapter summarizes factors that must be considered when enumerating protozoa, and touches on protozoan ecology and physiology as they apply to choice of enumeration method. Taxonomic considerations for identifying protozoa to genus or species are discussed and the strengths and weaknesses of various enumeration methods are described.

23–1 FIRST CONSIDERATIONS

When considering enumeration of a soil protozoa (Table 23–1), the first step is to clearly define what information is needed or desired. Will a count of the total number of protozoa be adequate? For example, a total estimate is adequate if the objective is to gain an idea of the amount of C, or N immobilized in protozoan biomass. Numbers of each group, that is, flagellates, ciliates, and amoebae, and their sizes, should be determined if the information needed is the rate at which N could be mineralized from protozoan prey groups, or if a precise measurement of biomass is needed. Depending on the species present, flagellates can be 10 to 100 times smaller than ciliates. Thus determining the number of each group can make a difference in the estimation of the amount of nutrients immobilized in protozoan biomass.

Table 23–1. Morphological characteristics of common soil protozoa.

Group† Characteristics

Flagellates–Zoomastigophorea
Diameter, 2–50 μm; biovolume 50 μm³. One or two flagella. The second flagellum usually trails posteriorly, making it appear as if the flagella arise from opposite ends of the body when in fact, both flagella arise anteriorly, as in *Bodo* or *Pleuromonas*. Flagellates sometimes display amoeboid movements and a flagellum as in *Mastigamoeba*.

Amoeba–Sarcodina
Diameter, 5–200 μm (1 by 6 mm for giant naked amoebae); biovolume, 400 μm³. Move by protoplasmic flow, either in extensions called *pseudopodia* or by whole body flow. Naked amoebae are differentiated by pseudopod type (lobose, conical, filiose, or reticular), while testate amoebae secret shells (tests) produced by the individual and constructed of soil particles bound together by secretions.

Ciliates—Ciliophora
Diameter, 10–150 μm; biovolume, 3000 μm³. Short, numerous flagella-like cilia. Differentiated by distribution of cilia, presence of undulating membranes, placement of the cytostome (mouth)

† All soil protozoa are non-photosynthetic.

Is it necessary to enumerate different types of amoebae, flagellates, or ciliates? For example, different groups of amoebae characterize different types of soil. Testate amoebae, which produce distinctive "tests" or shells, are typically found in forested soils, while presence of only naked amoebae is indicative of grassland or desert soils. Soil protozoan communities are distinct in a wide range of specific habitats, although overlap of individual species occurs between ecosystems (Stout et al., 1982). There are species with wide-spread distributions and a tremendous range of adaptability to environmental fluctuations (Corliss, 1979; Bamforth, 1985). Other species are intolerant of even small changes in specific environmental conditions (Austin et al., 1990; Foissner, 1986; Bodenheimer & Reich, 1933; Cutler & Dixon, 1927; Davis, 1981; Lousier & Bamforth, 1990; Stout, 1955, 1956, 1975, 1984). In fact, several unique protozoan communities can occur along the length of a single conifer root (Ingham et al., 1991b).

Is it important to identify type, genera, or species of protozoa? As indicators of environmental perturbation and disturbance, protozoa are perhaps unsurpassed (Foissner, 1986; also Austin et al., 1990; Domsch et al., 1983; Gupta & Germida, 1989; Lal & Saxena, 1982; Lord & Wright, 1985; Singh & Crump, 1953; Smith & Wenzel, 1947; Stout, 1956). When two or more disturbances occur at the same time, the interaction of dis-advantage-to vs. selection-for responsive species can be difficult to interpret (e.g., Ingham et al., 1986a, b). To improve the use of protozoa as indicators, however, a greater understanding of: (i) their response to disturbances beyond the normal seasonal cycle, (ii) their habitat-specificity, and (iii) their prey-preferences in specific habitats is needed. Efforts should be directed towards understanding changes in protozoan community composition in terrestrial systems.

Protozoa can be classified to genus and often to species based on morphology alone (Lee et al., 1985). While a comprehensive guide to soil ciliates, testate, and naked amoebae, and flagellates is needed (Lousier & Bamforth, 1990), protozoa can be rapidly and easily cultured. Protozoa occur in large numbers in natural ecosystems and "capture" of a representative picture of the community is not a problem.

23–1.1 Considering Soil

The next factor that needs to be examined when considering what method to use for enumeration is the type of soil. Soils range from heavy clays to pure sand. While it is not necessary to know the precise amounts of sand, silt, and clay, the general mix of soil fractions should be known, as well as the amount of organic material present. Protozoa are transparent and therefore invisible on amorphous, dark surfaces, such as clay and organic matter. Stains are often useful in highlighting protozoa in sandy soil, but organic matter and clays will hold stains, making them less useful in high clay or organic soils. Many objects that look like protozoa, but are not, occur in organic matter. Familiarity with protozoan morphology is critical when working with organic soils.

Similarly, removal of protozoa from soil by centrifugation or washing can be fairly successful when using sandy soils, but not when the soil contains clay or fine organic matter. The greater the surface area present in soil, the greater the retention of protozoa, and the greater the probability that protozoan numbers will be underestimated.

23–1.2 Environmental Selection

When enumerating protozoa, an understanding of the factors that influence protozoan survival and growth is useful. Numbers of protozoa are likely to be higher when conditions are optimal for survival or conducive to growth than when conditions are stressful to protozoa and no growth is occurring.

Numbers of flagellates and naked amoebae can equal their bacterial prey (up to several million per gram of soil). More commonly, flagellates or amoebae number in the tens of thousands per gram of soil. Although Bamforth (1976) reported 60 000 ciliates g^{-1} soil in a subtropical forest, ciliates and testate amoebae generally reproduce more slowly than either flagellates or naked amoebae and numbers can range from < 10 to as many as 1000 g^{-1} in temperate soils.

Soil temperature and moisture are important but insufficient predictors of protozoan community composition or density in soil. Equally important are the number of bacteria and the composition of the bacterial community, type and density of vegetation, amount and type of soil organic matter, depth and decomposition rate of litter, and type of soil parent material. Roots are a source of labile C and the rhizosphere has markedly higher numbers of protozoa than nonrhizosphere soil (Darbyshire & Greaves, 1967; Elliott et al., 1980; Ingham et al., 1985). Protozoan numbers consistently track bacterial numbers, but increases and decreases in protozoan numbers lag behind increases and decreases in bacterial numbers (Bryant et al., 1982; Clarholm, 1981; Elliott & Coleman, 1977). Notable peaks in both bacterial and protozoan numbers occur in the spring in semiarid grasslands and forests (Ingham et al., 1986a) and in both the fall and spring in mediterranean climates of the Pacific Northwest and California (Ingham et al., 1991a).

Certain testate amoebae are associated with calcareous soils (Bonnet, 1964) and an ecological classification has been developed based on test shapes that generally correlate with moisture regime (Bonnet, 1975). Similarly, *Pleuromonas jaculans, Acanthamoeba polyphage,* and *Crytolophosis* spp. comprised 80% or more of protozoa in a semiarid grassland soil (Frey et al., 1985) and appear to be characteristic of protozoan communities in dry climates.

As moisture increases, both population density and the number of species of protozoa present increases (Bamforth, 1981, 1984, 1985). Bamforth (1968) found that deciduous forests in Maryland and Louisiana contain a variety of ciliate species, whereas adjoining corn (*Zea mays* L.) and wheat (*Triticum aestivum* L.) fields show a paucity of species. The species present in the agricultural fields were similar to those found in desert soils. In Arizona deserts, Bamforth (1984) found the dry climate selected for a low diversity of ciliate species and a predominance of lobose (plagiostome and axial) testacean amoebae. As moisture decreased, the number of lobose species increased as compared to filiose amoebae, and amoebae with acrostomes increased as compared to those with plagiostomes (Bamforth, 1991a). Bamforth also suggested that the ratio of colpoda to polyhymenophoran species peaks within 5 to 40 d after disturbance of soils across a broad environmental gradient.

Kuikman et al. (1990) found that while protozoa encyst on drying of the soil, they respond rapidly to remoistening, and that under moderately fluctuating moisture regimes, protozoan activity results in higher plant N uptake than in soils kept continuously moist. However, when moisture stress was more frequent and severe, fewer protozoa were active and less N became available to plants than under continuously moist conditions. This suggests that optimal grazing of protozoa on bacteria, corresponding to an optimal N release rate for plant uptake, can be controlled by moisture fluctuations.

23–2 PROTOZOAN ECOLOGY

23–2.1 Ecological Roles

Soil protozoa are important because of their impact on soil processes. These roles include:

1. Mineralization of N, P, and S immobilized in bacterial and fungal biomass (Anderson et al., 1978; Bamforth, 1985; Clarholm, 1985;

Cole et al., 1978; Elliott & Coleman, 1977; Elliott et al., 1980; Gupta & Germida, 1989; Ingham et al., 1986a,b; Kuikman et al., 1990; Woods et al., 1982).

2. Enhanced nitrification rates (Griffiths, 1989).

3. Immobilization (sequestering) of C, N, P, and other nutrients in protozoan biomass (Hunt et al., 1977, 1987; Ingham et al., 1986a,b; Ingham & Horton, 1987).

4. Release of C from soil via respiration (Anderson et al., 1978; Bryant et al., 1982; Foissner, 1986; Parker et al., 1984; Kuikman et al., 1990).

5. Food source for predators such as nematodes and arthropods (Bryant et al., 1982; Ingham et al., 1985; Hunt et al., 1987).

6. Reducing bacterial numbers (Chao & Alexander, 1981; Danso et al., 1975; Habte & Alexander, 1977), control of numbers and diversity of bacteria and fungi (Darbyshire & Greaves, 1967; Ingham et al., 1985; Ingham et al., 1986a,b; Kuikman et al., 1990). Bacterial survival can be improved by adding bentonite clay that allows the formation of micro-niches in which protozoa cannot reach their bacterial prey (Heynen et al., 1988).

7. Suppression of bacterial and fungal pathogens, i.e., use as biocontrol agents (Chakraborty & Warcup, 1984; Chakraborty & Old, 1982).

8. Agents of plant disease (Dollet, 1984).

9. Indicators of disturbance, chemical impacts, presence of genetically altered bacteria, and soil degradation (Austin et al., 1990; Cairns et al., 1978; Pratt & Cairns, 1985; Foissner, 1986; Bamforth, 1991a,b).

23–2.2 Protozoan Communities

The protozoan community present in alpine litter is significantly different from the protozoan community in root-associated soil, which is different from the community present in soil from below the litter layer not-associated with the root, and all were different from protozoan communities present deeper in the soil profile (Foissner et al., 1982). For the most part, numbers of protozoa decrease as depth increases (Bamforth, 1976; Foissner & Adams, 1980).

Protozoa occupy the interstitial spaces in soil, need waterfilms at least several micrometers thick to remain active, and are usually concentrated near surfaces with high densities of bacteria. As a consequence, protozoa occur in highest numbers around roots and labile organic matter. Preferential feeding of protozoa on certain bacterial species may explain why some species are cosmopolitan, while others are limited to particular soils or ecosystems. Some ciliates, such as *Blepharisma undulans* and *Gonostomum affine* need more than one bacterial species to survive in culture (Luftenegger et al., 1985). While feeding on certain bacteria and fungi, some protozoa appear to release metabolites that depress the growth of other bacteria and fungi. As a result, soil protozoa may have a role in suppressing plant pathogens (Foissner, 1986).

Luftenegger et al. (1985) found that ciliates can be r- or K-selected, defined by response to substrate availability and growth rate. Certain soils contain predominantly K-selected species, while others contain mostly r-selected species. Bamforth (1991a) found that the ratio of r-selected colpoda to K-selected polyhymenophoran ciliates decreased as ecosystem stability increased. Petz et al. (1986) discovered a group of ciliates that feed exclusively on fungi and yeasts, the Grossglockneridae. These ciliates have a feeding tube near the oral cavity, which they use to attach to fungal hyphae, causing a distinctive swelling and a unique "hole" on the surface of the hypha.

Pratt and Cairns (1985) defined six functional trophic groups of protozoa in aquatic systems: (i) photosynthetic autotrophs, (ii) bacterivorous-detritivores which ingest their prey whole, (iii) saprotrophs that feed on dissolved organic molecules, (iv) algivores, (v) nonselective omnivores, and (vi) predators. These six groupings are somewhat inappropriate for soil as opposed to sediments, inundated soils, or wetlands. Photosynthetic protozoa do not occur in soils which seasonally dry below field capacity. Algae are rarely an important component of seasonally dry soils, although there can be blooms of cyanobacteria on the surfaces of soils in mesic systems in the spring that would allow ephemeral populations of "algivores" to develop. Many species of soil protozoa have only recently been discovered and their feeding patterns relative to their prey groups have not been determined. Given these considerations, it remains to be seen whether functional groupings based on aquatic foodwebs may provide useful information for terrestrial systems.

23–2.3 Numbers and Food Resources

Flagellates and the smaller amoebae and ciliates are usually more numerous and more widely distributed (Table 23–2), except in litter or continuously moist soil, where the larger ciliates and testate amoebae are often more abundant. Even though protozoan biomass is small as compared to bacterial biomass, about 2 g m^{-2}, (5–10 µg C g^{-1} dry soil), their ecological importance is greater than indicated by size alone.

Although some flagellates feed osmotrophically, most flagellates, naked and testate amoebae, and ciliates feed on bacteria, yeast, algae, and detrital particles (Sandon, 1927). Some flagellates, amoebae and ciliates can be maintained axenically on soluble nutrients (Lousier & Bamforth, 1990). Foissner (1986) reported that 50% of known soil ciliates feed partly or exclusively on other soil protozoa. Ciliates of the strictly soil-occurring Grossglockneridae feed almost exclusively on fungi. Although apparently rare, vampyrellid amoebae attack fungal spores while others digest the cell wall of hyphae, and consume the cytoplasm (Chakraborty & Warcup, 1984; Couteaux, 1985).

Table 23–2. Average numbers g⁻¹ dry soil of protozoa in various ecosystems.†

Ecosystem (Reference)	No. flagellates	No. amoebae	No. ciliates
Agricultural Systems			
Unplanted soil	7 000	30 000	155
Rhizosphere	33 000	100 000	875
(Darbyshire & Greaves, 1967)			
Austrian wheat	NE‡	751	1 285
(Foissner, 1986)			
Maize			
Noncultivated	340 350	25 190	11 460
Nonfertilized	833 000	1 295	2 230
+ farm manure	740 000	144 690	11 265
(in Foissner, 1986 from Detcheva's work)			
Grasslands			
Meadows	NE	878	1 027
(Foissner, 1986)			
Minnesota tallgrass	400 000	80 000	150
(Shreffler, personal communication)			
Bluegrass soil	NE	2 000	800
Grazed pasture	NE	2 000	600
(Bamforth, 1971)			
Semiarid prairie	8 000	5 000	80
(Ingham et al., 1986a)	20 000	18 000	70
Mountain meadow	28 000	24 000	138
(Ingham et al., 1989; maximum numbers)			
Alpine meadow	NE	5 000	2 000
(Foissner & Adams, 1980)			
(Berger et al., 1986)	NE	1 200	38 000
Forests			
Spruce	NE	2 800	30
Birch	NE	700	120
Pine-hemlock	NE	7 800	490
Pine	NE	1 600–6 600	40–140
Beech-maple	NE	400	160
Oak	NE	500–2 000	40–200
(Bamforth, 1971)			
Lodgepole pine	700	20	5
(Ingham et al., 1986a; spring)			
(Ingham et al., 1989; maximum numbers)	30 000	25 000	225
Mature Douglas-fir			
Mat soil (Nov)	26 532	3 018	74
Non-mat soil (Nov.)	27 903	0	0
(Cromack et al., 1988)			
Subtropical soils			
Soil	1 575 000	419 000	66 000
Rhizosphere	410 000	15 000 000	66 000
(Bamforth, 1976)			

† Only papers that give numbers for at least two groups. Papers giving numbers for single groups are numerous and not included in this table.
‡ NE = not estimated.

23–2.4 Escape from Adverse Conditions

To escape adverse conditions, protozoa form a nonmotile, inactive stage called a cyst (Lee et al., 1985). Normal seasonal fluctuations cause certain species of protozoa to encyst and become dormant, while others excyst and become active. This seasonal cycle must be understood as the background against which other disturbances select for or disadvantage certain species.

Some percentage of a protozoan population will not survive encystment and excystment (Bryant et al., 1982). The benefit of escaping inappropriate environmental conditions is offset by the loss of individuals who cannot meet the metabolic "cost" of entering and exiting dormancy. Since active organisms, that is, the trophic stages, are the ones performing the major portion of their ecosystem function, only trophic stages should be considered when determining function. Unfortunately, no widely accepted method for differentiating between trophic and dormant (encysted) stages exists (see methods section). In the following discussions, estimates of numbers and community structure include both trophs and cysts unless otherwise noted.

23–3 METHODS OF ENUMERATION

Commonly, enumeration techniques use: (i) direct observation or (ii) separation from soil (Table 23–3). An experienced investigator is needed to discriminate between protozoa and material in soil that appears similar in size and shape.

23–3.1 Direct Observation Methods

The benefits of direct observation are: (i) the only equipment needed is a dish in which to place the soil suspension, stain, sterile water, and a good phase contrast or differential contrast microscope; and (ii) no increase or decrease in numbers, size, or activity of protozoa occurs if the sample is observed immediately after collection. Rapid movement of protozoa during direct observation makes enumeration or identification difficult. Movement can be slowed by cooling, or by addition of methyl cellulose, 1% $NiSO_4$, or protamine.

In all of the following enumeration methods, phase contrast or differential interference contrast (Nomarski, Hoffman) microscopy can improve observation of many taxonomic features without the use of stains. This avoids problems, such as lysis and production of artifacts, when stains are used. Observation of live specimens can be crucial in identifying protozoan genera or species, in that the number, placement, and beating pattern of flagella and cilia on cell surfaces, or the way pseudopods are produced by amoebae, that is, a slow, steady forward motion, a periodic bleb, or a rapid

Table 23–3. Summary of methods.

All groups
23–3.1.1 Dried soil smears
23–3.1.3 Watered soil suspensions
23–3.1.4 Staining and fixation methods
23–3.2.1 Most probable number technique†
23–3.2.3 Soil suspensions on agar (not quantitative)
23–3.2.4 Density centrifugation
Flagellates
23–3.2.7 Uhlig Ice Extraction Method
Amoebae
23–3.1.2 Soil-agar films
23–3.1.5 Membrane filtration techniques
23–3.2.5 Filtration
23–3.2.6 Overlay techniques
23–3.2.10 Flotation
Ciliates
23–3.2.2 Combination watered soil suspension and MPN techniques
23–3.2.7 Uhlig Ice Extraction Method
23–3.2.8 Electric fields
Determination of Encysted Forms
23–3.2.1 Most probable number (when combined with initial treatment of soil with acid)
23–3.2.2 Combination of the watered soil suspension and MPN techniques

† Most commonly used basic method. Modifications abound.

spurt, are characteristics of taxonomic importance. These details are easily observed using interference contrast microscopy.

Scanning electron microscopy (SEM) can also increase resolution of surface characteristics of protozoa. Scanning electron microscopy is more useful than transmission EM in this capacity, because SEM visualizes surface morphological characters (Stout et al., 1982).

23–3.1.1 Dried Soil Smears (Jenkinson et al., 1976)

Only a slide and a microscope are needed for this "classic" method, but long hours of microscope time are required.

23–3.1.1.1 Procedure

1. Suspend soil in buffer (see discussion below for amounts).
2. Place an aliquot on a slide, allow the suspension to dry.
3. Observe for protozoa using 10 to 45× magnification. Phase contrast microscopy is recommended, interference contrast highly recommended, but bright field microscopy and an iris diaphragm can also be used.

23–3.1.1.2 Considerations. The amount of soil placed in the buffer—the dilution—is critical. There should be enough soil so there are numerous protozoa to observe, but not too much soil so that the protozoa are obscured by the soil particles. With each new soil, there is a certain "art" involved, based on trial and error experience with similar soil. To start with a new soil, dilute 1 g of soil (wet weight) with 1 mL of sterile water (or buffer). Observe an aliquot before the suspension dries completely. If no protozoa are observed within a half hour (time depends on one's patience), the numbers of protozoa may be too low for this method to be useful. If many protozoa are observed within a few minutes, increase the dilution until a maximum estimate is obtained. This is the optimal soil dilution for this soil at this season given the disturbances this soil has received.

As a routine approach, sterile tap water and sterile potassium phosphate buffer (pH adjusted to near that of the soil being used) should be compared to determine which gives the highest estimate of protozoa. Dilution of soil in a liquid will likely cause some lysis due to osmotic shock. However, there is no information detailing which buffers cause which protozoa to lyse, or not lyse. Until protozoologists come to a better understanding of what causes protozoa to lyse under certain conditions, and not under others, it is probably best to choose a standard buffer, adjusted to the pH of the particular soil.

23–3.1.1.3 Limitations. (i) Soil particles may obscure protozoa if the soil is not diluted appropriately, resulting in an underestimate of the number of protozoa present; (ii) Cells may shrink or lyse on drying such that volume and biomass is underestimated; (iii) The amount of soil actually observed is small, typically only 5 to 20 mg (Lousier & Parkinson, 1981); (iv) Cells may migrate to the edge of the smear during drying, altering the random distribution of cells, and invalidating enumeration.

23–3.1.2 Soil-agar Films

A modification of the Jones and Mollison (1948) method is best for observing testate forms, since nontestate forms may be lysed by the hot agar. In Jones and Mollison's original work, 0.1 mL of dilute methylene blue stain was added to 1 mL of the first dilution. Any general stain can be used, although researchers should be cognizant that these stains may lyse sensitive forms.

23–3.1.2.1 Procedure
1. Dilute the soil (usually 1:10 dilution).
2. Add 1 mL of 1.5% (wt/vol) agar.
3. Place an aliquot in a well of known depth (hemacytometers are not the best choice; use nail polish to adhere two no. 1 1/2 coverslips (0.15-mm thickness) about 1 cm apart on a glass slide and place the agar-soil suspension in the well between the coverslips (see Lodge & Ingham, 1991).
4. Cover with a coverslip.
5. Allow the film to solidify.
6. Count the number of protozoa observed in a known volume.

23–3.1.2.2 Considerations. Choosing the appropriate dilution is critical (see 23–3.1.1). Start with a 1:10 dilution and increase or decrease the dilution as appropriate. The usual problems are too few protozoa, soil particles too dense to see the protozoa, and the amount of time to examine the whole film, or more than one film, if numbers are extremely low.

23–3.1.2.3 Limitations. (i) Nontestate protozoa may not survive this treatment (Stout et al., 1982), (ii) the film may dry while observation is proceeding and nontestate cells may lyse, (iii) too few organisms per film.

23–3.1.3 Watered Soil Suspensions

The assumption in this method is that protozoa can be rinsed off soil particles, whereupon protozoa can be easily counted or identified, since they are no longer obscured by soil particles (Foissner, 1980, 1986).

23–3.1.3.1 Procedure
1. Suspend soil in a diluent (e.g., 2 g in 20 mL of sterile distilled water, or 0.2 g in 3 mL of water).
2. Observe 10 to 20 drops of the suspension successively (Bamforth & Bennett, 1985).

23–3.1.3.2 Considerations. This method works well with sandy soil, but with high clay or organic matter soils, there are at least two problems. First, protozoa are extremely difficult to rinse off clay particles or organic matter. Second, fine clay particles or amorphous organic matter that remain suspended in the water obscure the protozoa. Foissner (1986) determined that the efficiencies of recovery of ciliates, flagellates, naked amoebae, and testate amoebae added to various soils were 55 to 100%, 30 to 100%, < 10%, and 30 to 100%, respectively. These values are likely on the high side, because addition-recovery tests do not take into account the fact that organisms added to soil are not intimately associated with soil fractions and thus are more easily recovered than native populations.

23–3.1.3.3 Limitations. (i) Protozoa may be obscured by soil particles and their numbers underestimated, (ii) the method can become extremely time-consuming, for example, up to 8 h to observe a single soil sample (Griffiths & Ritz, 1988), (iii) protozoa are nearly the same refractive index as water, and can be easily overlooked on the surfaces of dark, amorphous soil material.

23–3.1.4 Staining and Fixation Methods

Stains are added to soil dilutions to create greater visual contrast between the organisms and soil particles (Stout et al., 1982; Lee et al., 1985; Couteaux & Palka, 1988).

23–3.1.4.1 Procedure
1. Add one drop to 1 mL (depending on intensity desired) of the following (expressed wt/vol unless otherwise noted) to 1 mL of soil suspension:

- Bouin-Hollande (6.25 g copper acetate in 250 mL distilled water; add 10 mL picric acid; filter; add 25 mL 40% (vol/vol) formalin and 2.5 mL glacial acetic acid).
- Schaudin's fixative (66 mL of a 6% mercuric chloride solution in 13 mL of 95% (vol/vol) ethanol).
- Carnoy's fixative (30 mL absolute ethanol, 10 mL glacial acetic acid).
- Fleming's fixative (30 mL of 1% chromic acid, 8 mL of 2% osmium tetroxide, 2 mL of glacial acetic acid).
- Glutaraldehyde (place sample in 2–5% (vol/vol) glutaraldehyde in appropriate buffer such as phosphate, cacodylate, or water, rinse in buffer, place in 1 to 4% osmium tetroxide), or
- Hollande's fixative (4 g picric acid, 3.5 g cupric acetate, 10 mL formalin, 5 mL glacial acetic acid, 100 mL distilled water).
- Protargol, nigrosin, and silver stains can be used to examine the distribution of cilia.
- Iron-hematoxylin, acid methyl-green, and Feulgen stains are used to stain nuclei.
- A variety of flagellar stains are described in Lee et al. (1985), although differential interference contrast microscopy will often obviate the need for staining.

Protozoa can be fixed to slides to prevent cells from being washed off during fixation steps (see Lee et al., 1985 for additional details).

2. Apply a small drop of Mayer's albumen (1 part glycerol, 1 part eggwhite) to the center of a slide or coverslip, let dry 5 min.
3. Apply a concentrated aliquot of stained and dried protozoa (suspend stained protozoa for 5 min, followed by centrifugation and decanting of supernatant through a 15–30–50–70–85% isopropanol series) to the albumen layer. Let alcohol just evaporate.
4. Gently cover dried aliquot with one to five drops of formol alcohol (3 parts 10% formalin, 1 part 95% ethanol) and let stand until albumin turns white (3–4 min).
5. Flood with 95% isopropanol, cover with coverslip, immerse in 95% isopropanol for 15 min, rehydrate through the isopropanol series.

23–3.1.4.2 Limitations. (i) Stains and fixatives can lyse protozoa for a variety of reasons, (ii) dilution is critical, and (iii) two to 3 h may be needed to examine a single soil sample (Griffiths & Ritz, 1988).

23–3.1.5 Membrane Filtration Technique

This is the basic, no-frills filtration technique, another "classic" method (Couteaux, 1967; Couteaux & Palka, 1988; Lousier & Parkinson, 1981).

23–3.1.5.1 Procedure
1. Dilute soil (see discussion above).
2. Filter an aliquot onto a membrane filter. The pore size should be no larger than 1 μm diam. or flagellates may be lost.
3. Using a microscope, observe for protozoa.

23–3.1.5.2 Considerations. Polycarbonate filters are recommended because the cells are easier to see and measure on the flat, smooth surface. Filtration is used mainly to enumerate testate amoebae, since these forms are less sensitive to collapse (Griffiths & Ritz, 1988). Foissner (1986) criticized filtration because it destroys the largest testate amoebae.

23–3.1.5.3 Limitations The pressure involved in filtration often destroys cells.

23–3.2 Indirect Enumeration

23–3.2.1 Most Probable Number

Most probable number (MPN) is the most widely used method to estimate total numbers (Singh, 1942; Darbyshire et al., 1974).

23–3.2.1.1 Procedure
1. Dilute soil in 10-fold or 2-fold steps to around 10^{-6}.
2. Place 4 to 12 replicate portions (0.5–1 mL each) from each dilution in wells containing 0.5 mL of soil extract agar (mix 100 mL distilled water with 100 g of the soil being tested, let soil settle, decant soil suspension, add 1.5% (wt/vol) agar, autoclave, and dispense into wells) as a source of food for bacteria.
3. Incubate at room temperature or normal temperature for the soil of interest (i.e., 4 °C if soil is from Antarctic, 30 °C if tropical soil) allow the protozoa to reproduce and reach high numbers.
4. After 4 to 7 d for flagellates, 7 to 10 d for naked amoebae and ciliates, or weeks for testate amoebae, depending on soil type, and temperature of incubation, mix, and remove a drop from each well.
5. Scan the drop using 20 to 45× magnification (phase contrast microscopy recommended, interference contrast microscopy highly recommended).
6. Record the presence or absence of flagellates, amoeba, and ciliates.
7. Calculate density in original soil from MPN tables (see chapter 5 by Woomer in this book). In addition, the dominant species in each well can be identified.

23–3.2.1.2 Considerations. The smaller the dilution step, the greater the precision of the estimate obtained (see chapter 5 in this book). However, more aliquots must be prepared and examined. In the author's experience, tissue culture plates with 24 wells are easiest to use. In this case, the 0.5 mL of soil extract agar can be put in each well several days in advance of adding the 0.5 mL of soil dilution. In these plates, the wells are arranged in six rows with four wells per row. One soil dilution series (10-fold dilutions to 10^{-6} dilution) can be placed in one plate, with four replicate wells per solution. Alternatively, one plate can be used for four dilutions with six replicate wells per dilution. Tissue culture plates with 96 wells—12 rows with eight replicates per dilution, can also be used, but the total volume per well is minimal and the liquid in wells can evaporate during incubation.

An inexpensive alternative to tissue culture plates is to pour agar in a petri plate and push sterile plastic rings into the agar, such that wells are formed above the agar surface. Unfortunately, the soil dilution can "escape" along the bottom of the plate and contaminate other wells.

Clarholm (1985) suggested that protozoa could be observed directly in the wells of the 96-well tissue culture plates but several researchers, including the author, have found significantly lower estimates of flagellate and ciliates using this approach as compared with removing aliquots from the wells, placing them on slides and observing them with a microscope. Seeing amoebae in the wells is not a problem, however.

23–3.2.1.3 Limitations. In addition to the inherent variability of the MPN technique (see chapter 5), several factors must also be considered.

1. Culture conditions may not be optimal. Certain species of protozoa preferentially feed on bacterial strains that may not be present, while other bacteria may be toxic to protozoa (Singh, 1945).
2. Protozoa prey upon each other and may reduce their populations below measurable levels.
3. Substrate for bacterial growth may not be present and the protozoa may starve. Bacteria and substrate may need to be added.
4. At low dilutions, soil particles may obscure individual protozoa.
5. The diluent may cause lysis or death of certain protozoa. Various diluents, such as phosphate buffers of various pH, or sterile water, should be tested with each new soil.
6. Soil dilutions can be prepared, for example, by adding 1 g of soil to 9 mL diluent, by adding 1 g of soil to 10 mL of diluent, or adding 1 g of soil to an amount of diluent that brings the total volume to 10 mL. This should be clearly stated in any publication to reduce confusion for those attempting to repeat the work.
7. Dormant protozoa excyst and become active during the incubation period. The difference between the numbers estimated from untreated soil (trophic plus encysted) and numbers estimated from an acid-treated (2% HCl; Cutler, 1923) subsample of the soil (encysted number) gives the number of trophic, or active, protozoa present in that soil. However, acid may destroy some cysts, may stimulate some to excyst that wouldn't excyst in the untreated soil, and may remove preferred bacterial food sources, resulting in the starvation of some excysted forms (Foissner, 1986).

23–3.2.2 Combination of the Watered Soil Suspension and MPN Techniques (Bamforth, 1991b)

Bamforth (personal communication) suggested that this method is appropriate only for ciliates since amoebae and flagellates are difficult to observe in the watered soil suspensions. However, higher estimates of active ciliates are usually obtained by this method than by the MPN technique alone.

23–3.2.2.1 Procedure

1. Examine, drop by drop, 0.2 g of soil mixed with 3 mL of sterile soil extract solution to obtain active numbers of ciliates (see above).
2. Prepare dilution series and perform MPN (see 23–3.2.1 above) to obtain total numbers of ciliates.
3. Subtract the number of active ciliates from the total estimate obtained from the MPN to determine an estimate of the number of encysted ciliates.

23–3.2.2.2 Considerations. In one approach, both an estimate of active (i.e., performing a physiological function) and total biomass, or total C immobilized in ciliate biomass, for example can be obtained. However, this method does not appear appropriate for obtaining the same information for amoebae or flagellates. Improvements of these methods are needed.

23–3.2.3 Soil Suspensions on Agar (Foissner, 1986)

Similar to direct observation of soil suspensions (23–3.1.2). This technique needs to be further improved to allow quantification of protozoan densities. Currently, it is probably the least expensive, least time-consuming, and reveals the greatest number of species, making it the best method for determining protozoan diversity.

23–3.2.3.1 Procedure

1. Place a 1-cm layer of soil, litter or plant debris (10–50 g fresh weight) evenly across the bottom of a petri dish.
2. Saturate (but do not flood) the soil with water.
3. Drain 5 to 20 mL of water by tilting the petri dish and pressing the soil gently with a finger.
4. Examine drained water with a microscope.
5. Incubate; repeat daily or weekly by adding more water and draining again.

Depending on incubation temperature and the species present in the sample, flagellates and colpoda, a common group of ciliates, usually appear within the first week. Other ciliates, followed by naked amoebae appear within 7 to 10 d, while testate amoebae appear after 2 to 4 wk.

23–3.2.3.2 Considerations. A slightly "less messy" version of this method was suggested by Stout et al. (1982). Place 10 g of soil on one-half of a 2% (wt/vol) agar plate and slowly add 20 mL of sterile distilled water to cover the soil with water. Observe protozoa in the water on the non-soil side of the dish without disturbing the soil. Stout observed a succession of protozoa with time and suggested that the use of distilled water enhanced excystment. Certain protozoan species grew most rapidly between 20 and 25 °C, some were inhibited above 25 °C, while others did not grow or excyst below 10 °C. Small flagellates and small amoebae were detected in a few days, while ciliates and testate amoebae appeared in 4 d to 4 wk. Addition of soil extract, nutrients, or bacteria increased the populations of some species, but inhibited others. To obtain ciliates that tolerate high organic matter soils, cultures were incubated anaerobically in a screw cap tube (Stout et al., 1982).

23–3.2.3.3 Limitations. This method is not quantitative.

23–3.2.4 Density Centrifugation (Griffiths & Ritz, 1988)

Centrifugation with high density solutions of sucrose, Percoll, and other materials have been used to separate bacteria from soil particles (Bone & Balkwill, 1986; Basel et al., 1983). Griffiths and Ritz (1988) tested a variety of dispersing agents, mixing methods, centrifugation methods, and staining procedures. The best general method is given below.

23–3.2.4.1 Procedure

1. Sieve soil (5 mm mesh).
2. Add 5 g fresh soil (i.e., not air dried) to a mixture of 50 mL distilled water and 50 mL of 50 mM Tris buffer, pH 7.5.
3. Shake for 10 min (wrist action shaker).
4. Settle for 60 s.
5. Remove a 1-mL aliquot (taken 5 cm below the meniscus).
6. Incubate with 0.1 mL 0.4% (wt/vol) aqueous iodonitrotetrazolium (INT) for 4 h at 25 °C.
7. Fix with 0.1 mL of 25% (vol/vol) glutaraldehyde.
8. Load onto a 5 mL Percoll column prepared in a 0.1 M phosphate buffer (pH 7.0) in sterile 15 mL polycarbonate centrifuge tubes.
9. Allow the soil-gradient to settle for 30 min.
10. Centrifuge (3000 × g for 2 h).
11. Decant the supernatant.
12. Stain and resuspend pellet with 1 mL of a 5 µg/mL aqueous solution of diamidinophenyl indole (DAPI).
13. Filter through a black 25-mm diam. 0.8 µm pore size membrane filter using gentle suction of −7 kPa.
14. Stain with acridine orange (1–10 mL, 33 µg mL^{-1} solution).
15. Mount filters on microscope slides and observe using epifluorescence (fluorescein filter set, 1000× magnification, oil immersion lens).

23–3.2.4.2 Considerations. Criteria for identifying protozoa are: (i) distinctly stained nucleus or nuclei, (ii) appropriate size and shape, (iii) cytoplasm present, (iv) no red fluorescence indicative of chlorophyll. Active organisms can be identified by the deposition of formazan from the respiratory reduction of INT (except recent reports suggest that immersion oil solubilizes formazan). Viable, but inactive, organisms have brightly stained nuclei, because acridine orange intercalates between nucleic acid bases. In the author's experience, the bright orange color from staining with acridine can overwhelm the formazan color, rendering it impossible to detect the active protozoa.

Shrinkage of cells may occur as a result of fixation in glutaraldehyde, use of Tris buffer, INT, DAPI, or acridine orange, or as a result of filtration. The use of dispersants are not recommended (Griffiths & Ritz, 1988) as they are toxic, and distort or reduce the volume of observed cells.

Griffiths and Ritz (1988) found that 92% of added organisms with a density of < 1.12 g cm^{-3} were recovered. Populations removed from 100 to 200 g of soil can be observed with this method since the protozoa are concentrated on a single filter. Counting is still beset with the problems of interference by soil particles, and differentiation of flagellates from amoebae or ciliates.

23–3.2.5 Filtration (Couteaux & Palka, 1988)

This method is useful for determining the numbers of ciliates, specifically *Colpoda* but has not been tested with other groups of protozoa.

23–3.2.5.1 Procedure
1. Fix 1.5 g of dry soil in 10 mL of Bouin-Hollande solution (see 23–3.1.4) to strengthen the cells such that they do not disintegrate under the force of filtration,
2. Add 20 mL of xylidine ponceau 2R for each 1 g of dry soil (Aldrich Chemical Co, Milwaukee, WI; or Sigma Chemical Co., St. Louis, MO) and let stain overnight.
3. Filter through a millipore filter (pore size of 0.8 μm).
4. Place filters on microscope slides, clear with immersion oil, cover with a coverslip and observe at 200 × magnification.

23–3.2.5.2 Considerations. Phase contrast or interference contrast microscopy is recommended to improve differentiation of flagellates from small amoebae. Use of polycarbonate filters gives a smoother surface and makes searching for the organisms easier.

In the original paper, the number of ciliates enumerated by this method was 10-fold greater than by MPN technique. The explanation given by Couteaux and Palka (1988) was that ciliates are lysed as a result of osmotic pressure changes during dilution for MPN. This method needs further work to determine its usefulness in enumerating the flagellates and amoebae.

23–3.2.6 Overlay Techniques (Stout et al., 1982)

Overlay techniques are not often used, perhaps because contamination between lines is a significant problem.

23–3.2.6.1 Procedure

1. Dilute soil, usually five to six twofold steps.
2. Spread an aliquot of each dilution in single, separate lines over the surface of any non-nutrient agar (e.g., 2% agar).
3. Prepare a bacterial suspension in agar. For example, mix a turbid suspension of bacteria that the protozoa will eat (e.g., *Enterobacter aerogenes*) in a mineral salts-phosphate buffer agar adjusted to the pH of the soil. Pour the agar over the lines previously spread on the non-nutrient agar, after the soil dilution is dry.
4. Incubate for 7 to 9 d at 30 °C.
5. Zones around the soil lines without bacterial growth indicates that bacteria have been eaten by amoebae (Singh, 1946; Menapace et al., 1975) and that, therefore, amoebae exist in that dilution.
6. Calculate densities of amoebae according to MPN tables.

23–3.2.6.2 Considerations. Higher numbers of amoebae were enumerated if the soil dilutions were spread over bacterial (e.g., *E. coli*) lawns (O'Dell, 1979). After incubation, the number of amoebae present could be counted from the number of plaques (clear zones) in the bacterial lawn. An alternate method used petri dishes with 1.5% (wt/vol) water agar inoculated with zig-zag streaks of mixed bacteria. A streak of soil suspension is drawn over these bacterial-streaks, and the plates incubated 4 to 6 d. Amoebae migrate to the bacterial streaks (Menapace et al., 1975).

23–3.2.6.3 Limitations. (i) All the limitations of MPN methods apply, (ii) not only amoebae cause clearing. Nematodes, virus, or other protozoa such as flagellates and ciliates could clear the bacteria.

23–3.2.7 Uhlig Ice Extraction Method (Uhlig, 1964)

This method has rarely been used, but with respect to obtaining certain indicator species, might be highly useful. The concept is that microfauna flee from the temperature gradient into the water and the liberated organisms are enumerated.

23–3.2.7.1 Procedure

1. Soil on a supporting screen is just immersed in water, with a block of dry ice hung directly above the soil.
2. The water is observed for flagellates and ciliates.

23–3.2.7.2 Considerations. Since the organisms move out of the soil, this method is not plagued by the problem of protozoa being obscured by soil particles and only active stages move out of the soil. However, amoebae are slow moving and probably freeze before they escape. While flagellate and ciliate numbers might be adequately estimated, amoebal populations will be underestimated. Unfortunately, there have been no studies comparing relative numbers of protozoa obtained by this method with other enumeration methods.

23–3.2.8 Electric Fields (Wagener et al., 1986)

Ciliates in soil suspensions migrate uni-directionally in applied electrical fields and can be driven from the soil into clear water by this method. This could be useful in enumerating ciliates, which are difficult to enumerate adequately by other means. This method needs further work to assess its usefulness.

23–3.2.9 Coulter Counting, Flow Cytometers (Stout et al., 1982)

Several researchers have tried to enumerate protozoa using Coulter counter and flow cytometers (Stout et al., 1982), but the technology cannot distinguish soil particles from soil organisms, as both are the same size. The real difficulty, as seen in the descriptions of the other methods, is to separate the organisms from soil particles. Once separated, counting can be performed by any number of methods. To date, however, flow cytometry has not been useful in determining numbers of any group of soil organism.

23–3.2.10 Flotation (Schonborn, 1977)

Another little-used, but potentially useful, separation method based on the observation that bubbling gas through a soil suspension causes testate amoebae to float. Connect a pipette to a compressed air outlet, place the tip of the pipette at the bottom of a 100-mL beaker containing 1 to 10 g of soil in 90 to 100 mL of a buffer appropriate to the soil pH. Bubble the soil suspension for several minutes and collect the liquid which overflows the container. Alternatively, siphon the surface of the suspension and examine this material for protozoa.

23–3.2.10.1 Considerations. Couteaux (1967) found this technique underestimated numbers of testate amoebae while Stout et al. (1982) found the tests were rendered unidentifiable. Some care must be taken with respect to air pressure while bubbling. More work is needed to standardize this method.

23–4 IDENTIFICATION

Identification of protozoan species is possible with all of the above techniques, by observing samples with interference contrast microscopy (Nomarski, Hoffman's, etc.) or by staining. Enumeration combined with identification is time-consuming and the amount of soil actually observed may be so small that the rare genera or species may not be observed. The main reason Foissner's group has discovered so many new soil ciliate species is that they use as much as 50 g of soil in their methods (Foissner, 1986). In only 2 yr of studying beech forest, grassland, and wheat soil, Foissner (1986) found up to 80 different ciliate, and 35 different testate amoebae, species. Quantitative enumeration, however, is not possible with Foissner's culturing technique and needs to be combined with another method. Flagellates and naked amoebae, two important groups of protozoa, were either sparse in his soils, or not adequately extracted. Another method to isolate flagellates and amoebae may need to be performed in conjunction with his culture technique to obtain a fully rounded picture of the protozoan community.

Levine et al. (1980) lists seven phyla of protozoa. The free-living soil protozoa include flagellates (Phylum Sarcomastigophora, Subphylum Mastigophora), naked and testate amoebae (Phylum Sarcomastigophora, Subphylum Sarcodina), and ciliates (Phylum Ciliophora). More recently, the slime molds have been shown to be similar to, and been classified with, amoebae (Lee et al., 1985). Symbiotic protozoa are often found in the digestive systems of soil animals. For example, cellulose-digesting flagellates colonize the guts of termites, sporozoan Gregarinda infect the seminal vesicles of earthworms, and ciliates are found in the digestive system of earthworms. The infective stages are defecated into soil and new infections result by contact (Stout et al., 1982). These groups are usually not found in keys with an aquatic emphasis.

A comprehensive taxonomic guide to soil protozoa is needed. For many years, Sandon's monograph (1927) served as the only systematic account for all four groups of protozoa in a single publication. Kudo's fifth edition (1966) and Jahn et al.'s second edition (1979) are useful general references, but do not list species and leave out most soil species. Smith's key on the terrestrial protozoa of Antarctic islands (1978) furnishes a useful guide, but its scope is limited by the restricted fauna of that harsh environment.

There are excellent papers on the taxonomy of individual groups of soil protozoa. For example, Foissner (1984, 1986) provides excellent coverage of Colpodida and hypotrichs, Corliss' *The Ciliated Protozoa* (1979), a modern synthesis of Kahl's monograph (1935), covers aquatic and medically important ciliates, but not soil forms. Naked and testate amoebae are covered by Geltzer et al. (1985), Ogden and Hedley (1980), and Chardez (1967), but these publications are difficult to obtain. Page's key *An Illustrated Guide to Freshwater and Soil Amoebae* (1976) and his *A New Key to Freshwater and Soil Gymnamoebae* (1990) do not list typical soil species.

Lee et al. (1985) included descriptions of all five groups of protozoa, but the strong aquatic emphasis of current protozoan taxonomy is quite apparent. Lousier and Bamforth (1990) deleted strictly aquatic species, making their key the most useful available for soil protozoology. Unfortunately, it differentiates only to genus, with illustrations of only common species. Foissner's paper (1986) is the most complete compendium of soil ciliate species. There simply is no key that covers all species of known, or even common, soil protozoa.

23–5 SUMMARY

There is no single method or combination of methods to which some researcher will not object as inadequate for enumerating protozoa. Couteaux et al. (1985) discusses the problems of nearly every current technique. Foissner (1986) strongly supports direct-counting approaches, while Stout et al. (1982) supports culture techniques. Investigators must choose those methods that best address their objectives and recognize the limitations to their approach.

For ciliates, direct counting methods, such as fixation of soil suspensions with osmic acid or glutaraldehyde, followed by staining with various indicators such as FDA (activity), acridine orange or DAPI (presence of nucleic acids) (Couteaux & Palka, 1988) will probably result in the maximum estimate of numbers. For testate amoebae, either membrane filtration or direct count techniques seem the most appropriate. For enumeration of flagellates and amoebae, MPN estimates appear best, but the samples must be observed within the first few days after culturing, as flagellate populations are often grazed to near-extinction by amoebae and ciliates. The method of Griffiths and Ritz (1988) seems promising for estimating numbers of all the major groups, if flagellates can be distinguished from amoebae.

Several recent publications review methods (Lee et al., 1985), taxonomic classification (Lousier & Bamforth, 1990; Lee et al., 1985; Levine et al., 1980) and ecology of protozoa (Bamforth, 1980; Laybourn-Parry, 1984; Lee et al., 1985; Foissner, 1986). These publications are recommended for those interested in further, in-depth reading about protozoa.

Protozoa respond quickly, and in predictable ways to a variety of factors, including soil development, the type and amount of humus or organic matter, moisture content, irrigation, fire, plant community, compaction, certain pesticide applications, and fertilization. Because of their rapid responses, their high numbers in soil, their ease of capture, and known responses to disturbance, these organisms appear to be highly useful indicators. With further research their predictive, and indicative, functions can be greatly increased.

ACKNOWLEDGMENT

Dr. Stuart Bamforth provided insight with respect to the usefulness of various taxonomic guides. Several anonymous reviewers helped to improve the manuscript in the preparatory stages. My thanks to Dr. Wilhelm Foissner for sending his excellent review of soil protozoa (Foissner, 1986) that summarizes much of the European literature unavailable in the USA.

REFERENCES

Anderson, R.V., E.T. Elliott, J.F. McClellan, D.C. Coleman, C.V. Cole, and H.W. Hunt. 1978. Trophic interactions in soils as they affect energy and nutrient dynamics. III. Biotic interactions of bacteria, amoebae and nematodes. Microb. Ecol. 4:361–371.

Austin, H.K., P.G. Hartel, and D.C. Coleman. 1990. Effect of genetically-altered *Pseudomonas solanacearum* on predatory protozoa. Soil Biol. Biochem. 22:115–117.

Bamforth, S.S. 1971. The numbers and proportions of testacea and ciliates in litters and soils. J. Protozool. 18:24–28.

Bamforth, S.S. 1968. Forest soil protozoa of the Florida parishes of Louisiana. Proc. La. Acad. Sci. 31:5–15.

Bamforth, S.S. 1976. Rhizosphere-soil comparisons in subtropical forests of southeastern Louisiana. Trans. Am. Microsc. Soc. 95:613–621.

Bamforth, S.S. 1980. Terrestrial protozoa. J. Protozool. 27:33–36.

Bamforth, S.S. 1981. Protist biogeography. J. Protozool. 28:2–9.

Bamforth, S.S. 1984. Microbial distributions in Arizona deserts and woodlands. Soil. Biol. Biochem. 16:133–137.

Bamforth, S.S. 1985. The role of protozoa in litters and soils. J. Protozool. 32:404–409.

Bamforth, S.S. 1991a. Implications of soil protozoan biodiversity. Soil biodiversity and function: Resolving global and microscopic scales. Soil Ecol. Soc. Meet. April 1991. Corvallis, OR.

Bamforth, S.S. 1991b. Enumeration of soil ciliate active forms and cysts by a direct count method. Agric. Ecosyst. Environ. 34:209–212.

Bamforth, S.S., and L.W. Bennet. 1985. Soil protozoa of two Utah cool deserts. Pedobiologia 28:423–426.

Basel, R.M., E.R. Richter, and G.J. Banwart. 1983. Monitoring microbial numbers in food by density centrifugation. Appl. Environ. Microbiol. 45:1156–1159.

Berger, H., W. Foissner, and H. Adam. 1986. Field experiments on the effects of fertilizers and lime on the soil microflora of an alpine pasture. Pedobiology 29:261–272.

Bodenheimer, F.S., and K. Reich. 1933. Studies on soil protozoa. Soil Sci. 38:259–265.

Bone, T.L., and D.L. Balkwill. 1986. Improved flotation technique of microscopy of *in situ* soil and sediment microorganisms. Appl. Environ. Microbiol. 51:462–468.

Bonnet, L. 1964. Le peuplement thecamoebiens des sols. Rev. Ecol. Biol. Sol. 1:123–408.

Bonnet, L. 1975. Types morphologiques, ecologie et evolution de la theque chez les Thecamoebiens. Protistologica 11:363–378.

Bryant, R.J., L.E. Woods, D.C. Coleman, B.C. Fairbanks, J.F. McClellan, and C.V. Cole. 1982. Interactions of bacterial and amoebal populations in soil microcosms with fluctuating moisture content. Appl. Environ. Microbiol. 43:747–752.

Cairns, A., M.E. Dutch, E.M. Guy, and J.D. Stout. 1978. Effect of irrigation with municipal water or sewage effluent on the biology of soil cores. N.Z. J. Agric. Res. 21:1–9.

Chakraborty, S., and K.M. Old. 1982. Mycophagous soil amoeba: Interactions with three plant pathogenic fungi. Soil Biol. Biochem. 14:247–255.

Chakraborty, S., and J.H. Warcup. 1984. Soil amoebae and saprophytic survival of *Gaeumannomyces graminis tritici* in a suppressive pasture soil. Soil Biol. Biochem. 15:181–185.

Chao, W.L., and M. Alexander. 1981. Interaction between protozoa and *Rhizobium* in chemically amended soil. Soil Sci. Soc. Am. J. 45:48–50.

Chardez, D. 1967. Des Protozoaires Thecamoebiens. Les Naturalistes Belges, Brussels. (Out-of-print.)

Clarholm, M. 1981. Protozoan grazing of bacteria in soil—impact and importance. Microb. Ecol. 7:343–350.

Clarholm, M. 1985. Interactions of bacteria, protozoa and plants leading to mineralization of soil nitrogen. Soil Biol. Biochem. 17:181–187.

Cole, C.V., E.T. Elliott, H.W. Hunt, and D.C. Coleman. 1978. Trophic interactions in soil as they affect energy and nutrient dynamics. V. Phosphorus transformations. Microb. Ecol. 4:381–387.

Corliss, J.O. 1979. The Ciliated protozoa. Characterization, classification and guide to the literature. 2nd ed. Pergamon Press, Oxford and Frankfurt.

Couteaux, M.-M. 1967. Une technique d'observation des thecamoebiens du sol pour l'estimation de leur densite absolue. Rev. Ecol. Biol. Sol. 4:593–596.

Couteaux, M.M. 1985. Relationships between testate amoebae and fungi in humus microcosms. Soil Biol. Biochem. 17:339–345.

Couteaux, M.M., and L. Palka. 1988. A direct counting method for soil ciliates. Soil Biol. Biochem. 20:7–10.

Cromack, K. Jr., B.L. Fichter, A.M. Moldenke, J.A. Entry, and E.R. Ingham. 1988. Interactions between soil animals and fungal mats. Agric. Ecosyst. Environ. 24:161–168.

Cutler, D.W. 1923. The action of protozoa on bacteria when inoculated into sterile soil. Ann. Appl. Biol. 10:137–141.

Cutler, D.W., and A. Dixon. 1927. The effect of soil storage and water content on the protozoan population. Ann. Appl. Biol. 14:247–254.

Danso, S.K.A., S.O. Keya, and M. Alexander. 1975. Protozoa and the decline of *Rhizobium* populations added to soil. Can. J. Microbiol. 21:884–895.

Darbyshire, J.F., and M.P. Greaves. 1967. Protozoa and bacteria in the rhizosphere of *Sinapis alba* L., *Trifolium repens* L. and *Lolium pereene* L. Can. J. Microbiol. 13:1057–1068.

Darbyshire, J.F., R.E. Wheatley, M.P. Greaves, and R.H.E. Inkson. 1974. A rapid micromethod for estimating bacterial and protozoan populations in soil. Rev. Ecol. Biol. Sol 11:465–475.

Davis, R.C. 1981. Structure and function of two antarctic terrestrial moss communities. Ecol. Monogr. 51:125–143.

Dollet, W.D. 1984. Plant diseases caused by flagellate protozoa (*Phytomonas*). Annu. Rev. Phytopathol. 22:115–132.

Domsch, K.H., G. Jagnow, and T.-H. Anderson. 1983. An ecological concept for the assessment of side-effects of agrochemicals on soil microorganisms. Residue Rev. 86:65–105.

Elliott, E.T., R.V. Anderson, D.C. Coleman, and C.V. Cole. 1980. Habitable pore space and microbial trophic interactions. Oikos 35:327–335.

Elliott, E.T., and D.C. Coleman. 1977. Soil protozoan dynamics in a shortgrass prairie. Biochemistry 9:113–118.

Foissner, W. 1984. Infraciliatur, silberliniensystem, and biometrie einiger neuer und wenig bekannter terrestrischer, limnischer und mariner Ciliaten (Protozoa: Ciliophora) aus den Klassen Kinetofragminophora, Colpodea und Polyhymenophora. Inst. fur Zoologie, Univ. of Salzburg, Austria.

Foissner, W. 1986. Soil protozoa: fundamental problems, ecological significance, adaptations, indicators of environmental quality, guide to the literature. Prog. Protist. 2:69–212.

Foissner, W., and H. Adams. 1980. Abundance, vertical distribution and species richness of soil ciliates and testacae of an alpine pasture and a ski trail at the Schlossalm near Bad Hofgastein (Austria). Zool. Anz. Jena 205:181–187.

Foissner, W., H. Franz, and H. Adam. 1982. Terrestrische protozoen als bodenindikatoren im boden einer planierten ski-piste. Pedobiology 24:45–56.

Frey, J.S., J.F. McClellan, E.R. Ingham, and D.C. Coleman. 1985. Filter-out-grazers (FOG): A filtration experiment for separating protozoan grazers in soil. Biol. Fert. Soil 1:73–79.

Geltzer, J.G., G.A. Korgonova, and D.A. Alexeiev. 1985. Pochvennye rakovinnye ameby i metody ich izucheniya. (Soil testacea amoebae and methods for their study). Moscow State University Press.

Griffiths, B.S. 1989. Enhanced nitrification in the presence of bacteriophagous protozoa. Soil Biol. Biochem. 21:1045–1051.

Griffiths, B.S., and K. Ritz. 1988. A technique to extract, enumerate and measure protozoa from mineral soils. Soil Biol. Biochem. 20:163–174.

Gupta, V.V.S.R., and J.J. Germida. 1989. Influence of bacterial-amoebal interactions on sulfur transformations in soil. Soil Biol. Biochem. 21:921–930.

Habte, M., and M. Alexander. 1977. Further evidence for the regulation of bacterial populations in soil by protozoa. Arch. Microbiol. 113:181–183.

Heynen, C.E., J.D. van Elsas, P.J. Kuikman, and J.A. Van Veen. 1988. Dynamics of *Rhizobium leguminosarum* biovar *trifolii* introduced into soil: The effect of bentonite clay on predation by protozoa. Soil Biol. Biochem. 20:483–488.

Hunt, H.W., C.V. Cole, D.A. Klein, and D.C. Coleman. 1977. A simulation model for the effect of predation on continuous culture. Microb. Ecol. 3:259–278.

Hunt, H.W., D.C. Coleman, E.R. Ingham, R.E. Ingham, E.T. Elliott, J.C. Moore, S.L. Rose, C.P.P. Reid, and C.R. Morley. 1987. The detrital foodweb in a shortgrass prairie. Biol. Fert. Soil 3:57–68.

Ingham, E.R., D.C. Coleman, and J.C. Moore. 1989. An analysis of food-web structure and function in a shortgrass prairie, a mountain meadow, and a lodgepole pine forest. Biol. Fert. Soils 8:29–37.

Ingham, E.R., and K.A. Horton. 1987. Bacterial, fungal and protozoan responses to chloroform fumigation in stored prairie soil. Soil Biol. Biochem. 19:545–550.

Ingham, E.R., H.B. Massicotte, and D.L. Luoma. 1991b. Protozoan communities around conifer roots colonized by ectomycorrhizal fungi. p. 32 (abstr). *In* Soil biodiversity and function: Resolving global and microscopic scales. Soil Ecol. Soc., Corvallis, OR.

Ingham, E.R., W.G. Thies, D.L. Luoma, A.R. Moldenke, and M.A. Castellano. 1991a. Bioresponse of non-target organisms resulting from the use of chloropicrin to control laminated root rot in a Northwest conifer forest: Part 2. Evaluation of bioresponses. p. 85–90. *In* USEPA Conference Proceedings. Pesticides in natural systems: Can their effects be monitored? USEPA Region 10, Seattle, WA.

Ingham, E.R., J.A. Trofymow, R.N. Ames, H.W. Hunt, C.R. Morley, J.C. Moore, and D.C. Coleman. 1986a. Trophic interactions and nitrogen cycling in a semiarid grassland soil. Part I. Seasonal dynamics of the soil foodweb. J. Appl. Ecol. 23:608–615.

Ingham, E.R., J.A. Trofymow, R.N. Ames, H.W. Hunt, C.R. Morley, J.C. Moore, and D.C. Coleman. 1986b. Trophic interactions and nitrogen cycling in a semiarid grassland soil. Part II. System responses to removal of different groups of soil microbes or fauna. J. Appl. Ecol. 23:615–630.

Ingham, R.E., J.A. Trofymow, E.R. Ingham, and D.C. Coleman. 1985. Interactions of bacteria, fungi, and their nematode grazers: Effects on nutrient cycling and plant growth. Ecol. Monogr. 55:119–140.

Jahn, T.L., E.C. Bovee, and F.F. Jahn. 1979. How to know the protozoa. 2nd ed. The pictured key nature series. Wm. C. Brown, Dubuque, IA.

Jenkinson, D.S., D.S. Powlson, and R.W. Wedderburn. 1976. The effects of biocidal treatments on metabolism in soil. Soil Biol. Biochem. 8:189–202.

Jones, P.C.T., and J.E. Mollison. 1948. A technique for the quantitative estimation of soil microorganisms. J. Gen. Microbiol. 2:54–69.

Kahl, A. 1935. Urtiere oder Protozoa. *In* F. Dahl (ed.) Die Tierwelt Deutschlands. G. Fischer Jena.

Kudo, R.R. 1966. Protozoology. Springer-Verlag, New York.

Kuikman, P.J., Van Elsas, J.D., A.G. Jassen, S.L.G.E. Burgers, and J.A. Van Veen. 1990. Population dynamics and activity of bacteria and protozoa in relation to their spatial distribution in soil. Soil Biol. Biochem. 22:1063–1073.

Lal, R., and D.M. Saxena. 1982. Accumulation, metabolism and effects of organochlorine insecticides in microorganisms. Microbiol. Rev. 46:95–127.

Laybourn-Parry, J. 1984. A functional biology of free-living protozoa. Univ. of California Press, Berkeley.

Lee, J.J., S.H. Hunter, and E.D. Bovee. 1985. An illustrated guide to the protozoa. Soc. of Protozoologists, Lawrence, KS.

Levine, N.D., J.O. Corliss, F.E.G. Cox, D. Deroux, J. Grain, B.M. Honigberg, G.F. Leedale, R. Loeblich II, J. Lom, D. Lynn, E.G. Merinfield, F.C. Page, G. Poljansky, V. Sprague, J. Vacra, F.G. Wallace, and J. Wieser. 1980. A new revised classification of the protozoa. J. Protozool. 27:36–58.

Lodge, D.J., and E.R. Ingham. 1991. A comparison of agar film techniques for estimating fungal biovolumes in litter and soil. Agric. Ecosyst. Environ. 34:131–144.

Lord, S., and S.J. Wright. 1985. The interactions of pesticides with free-living protozoa. J. Protozool. (abstr.). 31:44A.

Lousier, J.D., and S.S. Bamforth. 1990. Soil protozoa. p. 97–136. *In* D. Dindal (ed.) Soil biology guide. John Wiley and Sons, New York.

Lousier, J.D., and D. Parkinson. 1981. Evaluation of a membrane filter technique to count soil and litter testacea. Soil Biol. Biochem. 13:209–213.

Luftenegger, G., W. Foissner, and H. Adam. 1985. r- and K-selection in soil ciliates: A field experimental approach. Oecologia 66:574–579.

Margulis, L. 1981. Symbiosis in cell evolution. W.H. Freeman, San Francisco.

Menapace, D., D.A. Klein, J.F. McClellan, and J.V. Mayeux. 1975. A simplified overlay plaque technic for evaluating responses of small free-living amoebae in grassland soils. J. Protozool. 22:405–410.

O'Dell, W.D. 1979. Isolation, enumeration and identification of amoebae from a Nebraska lake. J. Protozool. 26:265–269.

Ogden, C.G., and R.H. Hedley. 1980. An atlas of freshwater testate amoebae. British Museum and Oxford Univ. Press.

Page, F.C. 1976. An illustrated guide to freshwater and soil amoebae freshwater biological association. The Ferry House, Ambleside, Cumbria, LA22 OLP.

Page, F.C. 1990. A new key to freshwater and soil gymnamoebae (with instructions for culture). Freshwater Biological Assoc., The Ferry House, Ambleside, Cumbria, LA22 OLP.

Parker, L.W., D.W. Freckman, Y. Steinberger, L Driggers, and W.G. Whitford. 1984. Effects of simulated rainfall and litter quantities on desert soil biota: Soil respiration, microflora, and protozoa. Pedobiology 27:185–195.

Petz, W., W. Foissner, E. Wirnsberger, W.D. Krautgartner, and H. Adam. 1986. Mycophagy, a new feeding strategy in autochthonous soil ciliates. (In Austrian.) Naturwissenschaften 73:560.

Pratt, J.R., and J. Cairns, Jr. 1985. Functional groups in the protozoa: Roles in differing ecosystems. J. Protozool. 32:415–423.

Sandon, H. 1927. The composition and distribution of the protozoan fauna of the soil. Oliver and Boyd, Edinburgh.

Schonborn, W. 1977. Production studies on protozoa. Oecologia 27:171–184.

Singh, B.N. 1942. Toxic effects of certain bacterial metabolic products on soil protozoa. Nature (London) 149:168.

Singh, B.N. 1945. The selection of bacterial food by soil amoebae and the toxic effects of bacterial pigments and other products on soil protozoa. Br. J. Exp. Pathol. 26:316–325.

Singh, B.N. 1946. A method of estimating the numbers of soil protozoa, especially amoebae, based on their differential feeding on bacteria. Ann. Appl. Microbiol. 33:112–119.

Singh, B.N., and L.M. Crump. 1953. The effect of partial sterilization by steam and formalin on the numbers of amoebae in field soil. J. Gen. Microbiol. 8:421–426.

Smith, H.G. 1978. The distribution and ecology of terrestrial protozoa of sub-antarctic and maritime antarctic islands. Br. Antarct. Surv. Sci. Rep. 95:1–104.

Smith, N.R., and M.E. Wenzel. 1947. Soil microorganisms are affected by some of the new insecticides. Proc. Soil Sci. Soc. Am. 12:227–233.

Stout, J.D. 1955. Environmental factors affecting the life history of three soil species of Colpoda (Ciliata). Trans. Roy. Soc. N.Z. 82:1165–1188.

Stout, J.D. 1956. Reaction of ciliates to environmental factors. Ecology 37:178–191.

Stout, J.D. 1975. The relationship between protozoan populations and biological activity in soils. Am. Zool. 13:193–201.

Stout, J.D. 1984. The protozoan fauna of a seasonally inundated soil under grassland. Soil Biol. Biochem. 16:121–125.

Stout, J.D., S.S. Bamforth, and J.D. Lousier. 1982. Protozoa. p. 1103–1120. In R.H. Miller (ed.) Methods of soil analysis. Part 2. Chemical and microbiological properties. 2nd ed. Agron. Monogr. 9. ASA and SSSA, Madison, WI.

Uhlig, G. 1964. Eine enfache methode zur extraktion der vagilen, mesopsammalen mikrofauna. Helgol. Wiss. Meeresunters. 11:178–185.

Wagener, S., C.K. Stumm, and G.D. Vogels. 1986. Electromigration, a tool for studies on anaerobic ciliates. FEMS Microbiol. Ecol. 38:197–203.

Woods, L.E., C.V. Cole, E.T. Elliott, R.V. Anderson, and D.C. Coleman. 1982. Nitrogen transformations in soil as affected by bacterial-microfaunal interactions. Soil Biol. Biochem. 14:93–98.

Chapter 24

Arthropods

ANDREW R. MOLDENKE, *Oregon State University, Corvallis, Oregon*

All literature reviews on the subject of soil arthropods basically come to the same generalizable conclusions: (i) soil is a relatively difficult medium from which to extract arthropods; (ii) the efficiency of any one extraction method varies between common soil types, because they differ significantly in chemical composition and microstructure; (iii) attempts to quantify *absolute* census counts or resident biomass for any volume of soil are unlikely to be profitable, due to biases in the extraction efficiency for each individual species (which varies by season, soil type, and horizon) which are prohibitively time-consuming to attempt to quantify; (iv) taxonomic identification is frequently not possible to specific precision, and even in the rare event when possible, the correct name often does not access ecological data sufficient to unambiguously assign it a defined functional role(s) in the soil being studied. If you are experienced in soil arthropod studies, you already are aware of these difficulties; if not, it's important to list them up front. Soil fauna is challenging to study, but is very rewarding when the complexity of the synergistic interactions between the fauna and the microbes are revealed.

Soil arthropods function in soil ecosystems in numerous ways: chemical transformation; structural architecture; mixing and transport. The importance of arthropods is not expressed in terms of percent of community respiration; arthropod contribution is usually $\ll 10\%$. Arthropods, however, are now recognized as the catalytic regulators of microbial activity (reviews: Crossley, 1977; Seastedt, 1984; Visser, 1985; Fitter et al., 1985; Edwards et al., 1988; Shaw et al., 1991). In many ways the beneficial impact of arthropods is correlated with their physical activity, for instance shredding litter, burrowing in coarse woody debris, aeration of the soil and transport of inocula. Many studies evaluate the effect of arthropods upon a biological or physical process, such as mineralization or leaching. However, Anderson et al. (1985) have shown that the most critical effects of arthropods may be in mediating chemical transformations in the soil that are usually not directly monitored. Arthropods affect soil microbial communities as the direct result of feeding (or being fed upon by other predaceous arthropods). These effects are likely to be density dependent, with normal population levels stimulating microbial activity and plant growth and either decreased or epidemic population counts depressing growth response (Finlay, 1985). Again, it is the interplay of the arthropods on the microbes, as monitored ultimately by plant growth, that is often of most interest.

24–1 PRINCIPLES

Basically arthropods can be extracted from a sample of soil either through physical methods or behavior modification. Physical methods are generally labor-intensive and work best on soils with low organic matter content; meso- (0.5–2.5 cm) and macro-arthropods (> 2.5 cm) are most efficiently extracted. Inducing the biota to leave the soil on their own is labor-frugal and works best on soils with high organic matter content; microarthropods are efficiently extracted as well as larger taxa. There is no a priori best solution to the question of how to extract the biota. I would strongly suggest preliminary experimentation with the substrate in question prior to implementation of a research design. The efficiency of extraction varies for each species (in unpredictable ways) depending upon the physical and chemical characteristics of the medium. Obviously, if the research requires an accurate measure of immobile stages (i.e., eggs, pupae, and cysts), then a physical means must be utilized.

Since no one claims that any method is 100% efficient, there is no means to assess *absolute* efficiency. Several workers have compared the relative efficiencies of different methods on specific organisms in specific soil types. Some researchers have released a known number of specific organisms into the soil and then recovered them. Efficiency studies are summarized by Edwards and Fletcher (1970, 1971) and repeated largely unchanged in Edwards (1990). Universal advice is to experiment with different types of extraction techniques relevant to your particular taxon of interest, and with the range of soil types in the study area.

24–2 METHODS

24–2.1 Evaluation of Biota in the Field

24–2.1.1 Destructive Sampling of Soil for Biota

There are no universally applicable techniques that yield robust estimates of soil macroarthropod populations. Large arthropods (and earthworms) are, almost by definition, not present in densities that can be adequately sampled by typical coring devices. The usual method is to designate specific areas (perhaps 1 m^2), place a large light-colored sheet (or denim) on the ground surface adjacent, and progressively use a spade to excavate the sample site. The contents of the spade can either be placed directly on the sheet and gingerly broken up by hand, or can be run through an archaeological sieve first. Field assistants search the substrate for the specific pest species, beneficial predator or worm in question.

24–2.1.1 Limitations

Prior knowledge of the range of depths that the target species inhabits is necessary; sample depth of more than a meter is required for worms under many conditions. This technique is very labor intensive, highly subject to individual bias, and efficient only for the largest most mobile taxa. It is seldom used to quantify more than one type of macroinvertebrate at a time, since search image efficiency becomes limiting.

A hand sorting variant of this method to characterize the community in forests with a deep litter layer is to sample a relatively large area (0.5–1.0 m^2), place it in a plastic bag, return to the laboratory and expose it in a shallow light-colored tub. A technician can pick through it thoroughly by eye aided by a large suspended magnifying lens; efficiency is probably directly dependent upon the activity of the species. Intermittent misting of the sample with formaldehyde (take prudent safety precautions) from an atomizer increases specimen visibility markedly, since the quiescent fauna are stimulated to activity.

24–2.1.2 Pitfall Trapping

The most widely used method for determining the relative abundances of larger arthropods is pitfall trapping (Greenslade, 1964; Southwood, 1978; Franke et al., 1988; Doube & Giller, 1990). A container (with or without preservatives) is sunk into the ground to a depth placing the lip at the surface of the soil (Fig. 24–1). In theory, epigeic species walking across the ground happen upon the traps by accident, fall in, and can't climb back out. As such, this trapping method measures relative activity of resident taxa (NOT density), draws captives from differing amounts of surrounding territory (dependent upon individual species mobilities) and is correlated with species-specific motor skills. As a passive collecting device it has no equal for simplicity and inexpensiveness; interpretation of the quantitative results is difficult (Gist & Crossley, 1973; Price & Shepard, 1980). Efficiency can be greatly increased in some habitats by erecting radially arranged barriers (1–2 m long) extending outwards from the trap that induce some of the fauna that encounter the barrier to funnel toward the trap along the wall surface.

However, the basic nature of the trap is seldom passive. Invertebrates are usually specifically attracted to the trap by the odor of the preservative or the odors emanating from the cut roots and fungal hyphae at the site of insertion. Trap catches are usually most numerous on the first several days of sampling, due to the freshness of the disturbance. All sorts of baits (e.g., meat, fruit, dung, and fungi) can be added to the traps to increase the

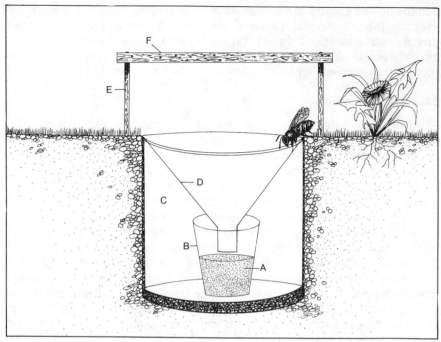

Fig. 24–1. Pitfall trap. Epigeic species will enter trap space under the rain roof (F), walk or slip down the funnel (D) and be preserved in the polypropylene or ethylene glycol (antifreeze = A) contained in the 6 to 8 oz. plastic cup (B). The entire trapping apparatus is housed in a 1 gallon plastic can (C), sunk into the ground so that its top is flush with the ground surface. The soil immediately adjacent to the gallon can (C) must be firmly compacted to prohibit subsidence and maximize accessibility to the arthropod fauna. The rain-roof (F) can be made of composition-board or aluminum sheeting—the former is easily supported by ten-penny nails (E), the latter requires reflexing the corners up over the nail heads. The trap may preserve a variety of arthropod types, including winged species that are either truly soil-associated (e.g., soil-nesting bee, as pictured) or simply attracted to scents emanating from the fluid. Traps can be run for specific portions of a day, or continuously for several weeks.

sampling range (and the degree to which they attract flying insects). Trap design can be modified to accommodate battery-powered fraction collectors to quantify periods of maximal activity. Traps can also be sunk sequentially below the soil surface to assess differential species behavior by depth (Loreau, 1987).

24–2.1.2.1 Limitations

Three major concerns in installing a pitfall grid are:

1. Minimizing damage to the resident populations of small vertebrates (mammals, reptiles, and amphibians). To this end it is critical to install a slippery funnel with a basal diameter small enough (about 2 cm, depending upon local fauna) to exclude most vertebrates.

2. Inactivation of the preservative by rain/surface flow dilution. Rain can be excluded by a square roof made of aluminum sheeting or particleboard and suspended by nails from the corners. Surface flow is more difficult to predict; avoid concave surfaces and trench around the trap to direct the water flow.

3. Vandalism by larger vertebrates (human and non-human) is unavoidable. Animals either remove traps to drink the preservative they contain or the meals they have already caught. A statistician should be consulted *prior* to installing a grid to ensure the design will be robust if occasional samples are lost. Though it is possible to estimate absolute densities from the ratio of captures within a grid versus on the periphery, the usual use of pitfalling is in estimating differing relative abundances at different sites.

24–2.2 Sampling Soil Cores (review: Kubiena, 1938; Spence, 1985)

24–2.2.1 Passive Extraction of Biota from Soil Cores

Most samples are taken by soil coring devices familiar to soil scientists. I mention the pictured device (Fig. 24–2) since we use it frequently, and it is tied to a specific extraction method mentioned below. A double-cylinder hammer-driven core sampler developed to assay bulk density (Blake & Hartge, 1986) is relatively efficient at sampling arthropods because of its: (i) relatively wide core diameter (about 7.5 cm); and (ii) design features to minimize compaction during sampling. The inner sleeve of the corer is normally aluminum, but since it is not directly driven by the hammer it can just as easily be made of PVC plastic. Each sample within the plastic collar can be covered with flexible screening, placed directly into a plastic bag, stored in a cooler, taken to the lab and subsequently placed in a high-gradient extractor without further handling.

24–2.2.2 Faunal Distribution in Soil Cores by Microscopic Methods

Soil is a diverse medium structurally and chemically. A major limitation affecting soil fauna studies is knowledge of the microenvironments inhabited by the different species. A number of techniques have been developed (but seldom used) to quantify the structure of soils on a scale relevant to arthropods (Anderson, 1978). The methods seek to accomplish three objectives: preserve the soil micro-architecture in situ, transport the sample to the laboratory unaltered, and section it for examination under the microscope. In the field, soils can either be quick frozen with water followed by liquid nitrogen (Froelich & Miles, 1986), or embedded in agar (Haarlov & Weis-Fogh, 1953) or gelatine (Anderson & Healey, 1970). Frozen soils are embedded in the laboratory; protein embedded samples are hardened in formalin prior to sectioning.

By far the most elegant, all-purpose, permanent technique is to embed the samples in epoxy (Rusek, 1985). The limitations are that the samples must be embedded in the laboratory, cured in an oven, and sectioned by

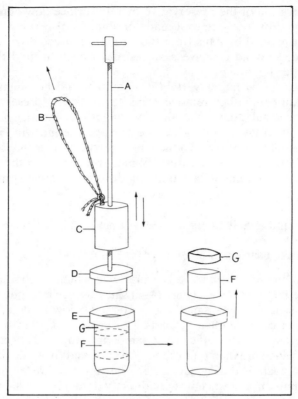

Fig. 24–2. Double-cylinder soil corer for removing soil samples with minimal compaction. The detachable core-cutter (E) is driven into the ground by the heavy weight (C) by pulling C up along the handle (A) with the rope attachment (B) and releasing it to fall onto the nested end-plate (D). When the core-cutter (E) is driven to the desired depth (bottom of wider ring flush with soil surface), the handle assembly (A–D) is removed, and the core-cutter removed from the ground. Holding the core-cutter (E) horizontally, the core itself within its sheath (rings F + G) is gently pushed out of the center and onto a plastic sheet. The lowest portion of the soil column and the uppermost portion adjacent to the nested end-plate (contained in the narrow sleeve [G]) are cut off with a pen-knife, leaving a noncompacted volumetric sample in F. A flexible screen is placed over the *top* of F and secured with a rubber band; the sample is inverted, placed in a ziplock plastic bag, and transported to the laboratory for extraction.

high-technological geological saws. (The saws must be switched from oil-lubrication to water-lubrication during the process, thus limiting access to other concurrent users of a shared facility.)

24–2.2.3 Extraction of Soil Arthropods through Behavior Modification
(reviews: Kevan, 1955; Evans et al., 1961; Murphy, 1962; Phillipson, 1970; Dunger & Fiedler, 1989; Edwards, 1990)

24–2.2.3.1 Dry Funnels. The most commonly used extraction procedure for the majority of soil arthropods involves a funnel apparatus originally proposed by Berlese, and subsequently modified by many other workers (A. Tullgren [1918], most prominently). In this method the sample is placed in a metal funnel supported on a wire mesh; an environmental stimulant is applied above the sample and a bottle with a preservative fluid catches the biota as they burrow through the sample and drop through the screening.

The kinetic stimulant usually applied is either heat combined with light (electric light bulb), or a repellant chemical (formalin, tear gas). Within extreme limits, it is widely observed that extraction efficiency increases with decreasing Wattage (25–75 W) and with decreasing sample volumes. However, since the number of funnels is usually limited, longer extraction times can result in prolonged storage of samples which may produce population artifacts. Extraction efficiency is increased by alternating heating and cooling periods. If samples resting on the mesh screening are more than 10 cm in thickness, layers of the upper already-dried soil can be removed sequentially to promote extraction from the remaining substrate.

Efficiency in very thick samples is reduced, because non-even distribution of soil components permits fauna to migrate to foci of pleasing microclimate (i.e., around a large root or coarse woody debris) within the soil column, which subsequently become surrounded by inhospitable conditions forcing the fauna to perish in situ. Large sample volumes function much better for qualitative (presence/absence) determination, than for robust quantitative estimates of population density. Such qualitative studies can be enhanced by pre-filtering the sample through a mesh-size that removes larger stones, litter and roots; vigorous agitation within a double-bottomed burlap/denim bag with a screen mesh sewn in over a bottom zippered chamber, breaks up well-decayed pieces of wood in the sample (exposing the inhabitants). This preagitation technique is used in the field when the samples are being collected, significantly reducing volumes and weights to be transported and housed in coolers.

There is no standard size or shape to the funnel apparatus itself. Mesoarthropods are generally extracted in funnels averaging 30 cm in diameter and 45 cm in depth. Microarthropods are usually extracted in funnels 5 to 12.5 cm in diameter by 7.5 to 20 cm high. Although macroarthropods are successfully extracted from funnels (if the mesh size permits), their low numbers are such that funnel extraction is an inappropriate means of determining population densities. (Macroarthropods are generally estimated through a different technique—see "pitfall trapping" above).

24–2.2.3.1.1 Mesoarthropod Extractor

If used under nonextreme conditions, metal Berlese funnels can last for several decades (even with heavy use). However, start-up costs for the necessary welding and socket wiring can exceed $150/unit (Fig. 24–3). To reduce costs of fabrication, yet maintain all-metal construction, I suggest using a standard galvanized water pail (leaving at least 3–4 cm of the rim of the base to which three 5-mm diam. dowels can be welded) and

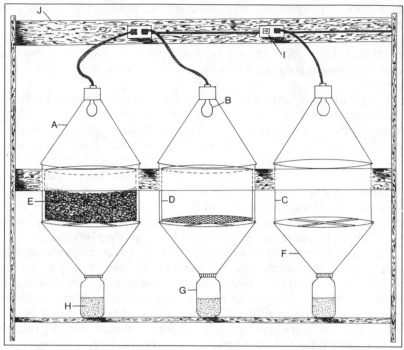

Fig. 24–3. Berlese funnel extractor. The basic structural unit consists of a metal cone (A; must be vented to release evaporated water!) with electric light bulb (B) resting upon a metal combination cylinder (C) and funnel (F). A convenient number of these metal units (i.e., 10–15) are attached to a wooden frame (J) and provided with individual electric outlets (I). The sample (E) is housed in a metal cylinder with ¼ in. mesh floor (D), supported on cross-pieces welded at the junction between C and F. The sample (E) is placed in the sample housing (D) over a dry empty collecting bottle; the throughfall soil is removed, replaced into the sample and a new collecting jar (G) with ethylene or propylene glycol as preservative (H) is placed underneath.

aluminum sheets (cut to size by shears and stapled together; used 1 by 1.6 m aluminum sheets can be purchased from the recycling department of nearly every newspaper company for a few cents each). Galvanized stove pipe with a disc of hardware mesh and a plastic funnel secured by duct tape also works efficiently. To avoid the expense of soldering (and resoldering) collecting jar tops to the base of the funnel, use a non-evaporative fixative (e.g., 50:50 antifreeze ethylene glycol/water). WARNING: DO NOT attempt to use cardboard canisters covered with aluminum foil to house the samples; it is only a matter of time before a fire occurs.

24–2.2.3.1.2 Microarthropod Extractor

Macfadyen (1953) developed a sophisticated apparatus designed to control heating and drying of the sample surface while independently controlling cooling and humidification of the sample base. There have been a number of modifications of this "high-gradient design" which significantly

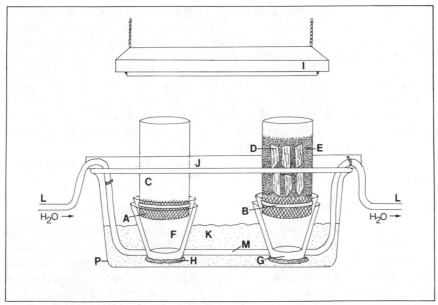

Fig. 24–4. Inexpensive high-gradient extractor design. The basic design consists of a series of samples (D + E), enclosed in sections of PVC pipe (C), held above collecting cups (F + G) which are glued (H) to the bottom of a plastic tray (P). (Gluing is not permanent; to avoid periodic regluing the cups can be affixed to ¼ in. hardware mesh with rubber washers and short bolts.) The water (K) which fills the plastic tray bathes copper tubing (M) which is connected by hoses (L), either to other such units or to a container equipped with a sump pump to which ice is periodically added. Ice water is continually circulated, cooling in turn both the copper tube (M) and the water bath (K). The heat stimulus is applied by light (I; 250 Watt heat lamp) from above (insulated by white plastic beadboard [J]), which forces arthropods in the sample (D) to travel through the soil or packing material (E), through the mesh (A; secured by rubber band B) into the collecting cup (F). The collecting cup is provided with a solution of cycloheximide (0.2 g/L) to prevent specimen decay. The specimen cup (F) is removed from its retainer cup (G) and the sample decanted with a squirt bottle into a storage vial. To prevent damage from prolonged usage and hydraulic leaks at the hose clamps, the plastic tray should have an overflow hole installed below the level of the top of the retainer cups and the pan housing the sump pump should be connected to a toilet float valve, which will automatically add water in case of a leak. (Basic idea developed in conjunction with B.L. Fichter, Dep. of Entomology, and K Cromack, Dep. of Forest Science, Oregon State University.)

reduce cost but maintain the increased levels of microarthropod extraction. The modification I employ costs about $250 for a total of 72 separate extractors (Fig. 24–4). A unit of nine extractors each can be cooled independently with simple periodic addition of ice (and an overflow valve), or several units can be hooked together in sets of 8 to 10 unit trays with ice water recycled by a 1 horsepower pump through a coil of copper tubing in an ice chest periodically refilled with block ice. Eight dish-pan units fill a laboratory/greenhouse bench and generally is all a single 20-amp circuit breaker will support.

The PVC pipe in which the sample sits is the same housing employed inside the soil corer (Fig. 24–2F); thereby sample handling is minimized. (Many soil fauna are extremely fragile taxa, and are only able to migrate through pre-existing soil pores in the extractor.) The soil core is inverted prior to extraction, to minimize the distance that has to be negotiated during the extraction process. The entire apparatus can be constructed from materials available in any large department store. Though the size of the sample units of PVC pipe may be of any dimensions (if not used in conjunction with a congruent soil corer), the pictured size fits easily into a quart zip-lock plastic sandwich bag in the field.

The possibility now exists to carefully control the gradients with thermistors and microcomputers (Andren, 1985).

24–2.2.3.1.3 Special Modifications of the Microarthropod Extractor

This apparatus has been modified successfully for more recalcitrant substrates (e.g., freshly fallen logs, roots). Two-inch thick cross-sections are cut from the experimental logs in the field with a power saw. Chiseling along the wood grain removes 2 by 2 cm (5 cm long) chunks of wood in the vicinity of borer activity. These chunks are placed into the PVC extractor, packed with moistened sterilized sawdust and extracted as above. Very large numbers of commensal arthropods and nematode worms can be discovered in this manner. It is particularly important to place antibiotics in the collecting cup when extracting from wood, since numerous dissolved sugars drip into the cup (originating largely from the fresh sawdust).

Since dry funnels require healthy invertebrates, it is crucial to minimize the length of time between field collecting and extracting. Norton and Kethley (1988) describe a portable nylon apparatus suitable for airplane travel and overnight extraction.

24–2.2.3.2 Extraction of Soil Arthropods with Wet Funnels

Hydrophilic invertebrates are not effectively separated by Berlese or high-gradient funnels, since the medium dehydrates rapidly and most forms are capable of becoming cryptobiotic. Immersing in water a sample wrapped in cheesecloth (holds soil particles in, allows egress of biota) stimulates nematodes, tardigrades, copepods, planaria and enchytraeids to locomote. If the sample is placed in a funnel over a mesh, the emerging fauna will sink to the bottom of the water column. A clamp at the bottom can be released and the fauna decanted to a plate for counting or a vial for storage (see chapter 22 by Ingham in this book). Applying heat with a light bulb to the surface may speed the process (O'Conner, 1955); the average Baermann extraction is run for 2 to 4 d.

Milne et al. (1958) reversed the process by placing the base of the soil core in a water bath and gradually raising the temperature until the insect larvae all emerged from the heated sample onto the top of the soil surface.

Simple wet-funnel extraction, like Berlese/dry funnel extraction, is not labor-intensive, it requires no expensive equipment, and it yields a wide variety of fauna. As with dry funnel extraction, efficiency is dependent upon the volume of the sample (usually 5–20 g); the greater the surface to volume ratio, the more efficient is extraction. Since extraction is dependent upon behavior induced in the invertebrate, extraction efficiency is species-specific and context-specific.

24–2.3 Extraction of Soil Arthropods by Physical Methods

24–2.3.1 Flotation in a Salt Solution

Most soil invertebrates are characterized by a specific gravity slightly greater than water (1.0), but none higher than 1.1 (Edwards, 1967). Therefore, soil immersed and agitated in a dense salt solution yields all the fauna to the surface. Unfortunately, most of the organic debris in the soil also floats to the surface, and because of its bulk it can obscure the fauna from enumeration. Efficiency can be increased by prewashing the soil sample through a series of sieves; this method is particularly effective if you are censusing for one particular organism of known size distribution. Ladell (1936) demonstrated that a sample can: (i) be washed through a series of sieves and collected in a modified beaker; (ii) the beaker is filled from the bottom with salt solution, and as it fills it is agitated by bubbling compressed gas injected along with the salt; specimens in the beaker are decanted into a collecting tube and washed; (iii) specimens (plus organic debris) are agitated with a mixture of heptane (benzene) and water, let settle, and the specimens pipetted from the organic layer, washed and preserved in 70% ethanol for analysis.

Edwards et al. (1970) adapted and mechanized the system to reduce handling time, increase efficiency of extraction and minimize operator bias. Four samples in wire mesh are simultaneously rotated and subjected to a spray of water. The resultant washed and screened sample is transferred to a container and vigorously agitated in a mixture of zinc sulfate solution (specific gravity 1.4) and 1:1 xylene + carbon tetrachloride (specific gravity 1.2); use enough of each solution to insure adequate vertical separation in the beaker. After separation of the liquids, the solvents can be bled off from a stopcock in the bottom or more liquid can be added, and then the topmost layer containing all the invertebrates decanted through a lateral vent. The mechanism works because the organic debris floats on the surface of the salt solution, whereas the specimens float on top of the organic layer. Both these methods are labor intensive.

24–2.3.2 Extraction of Microarthropods by Elutriation

This process is also driven by the difference in specific gravity of soil fauna and the soil particles. Heavy soil particles are allowed to sink in an upwards flow of water which carries the invertebrates with it to the top (Oostenbrink, 1960, 1970; Seinhurst, 1956, 1962). Historically, elutriation was used primarily for nematode extraction, nevertheless it has been adapted for arthropods (von Torne, 1962; Bieri & Delucchi, 1980).

These methods work best if the soil aggregates are broken down and the arthropods exposed to the liquid before being placed in the current; dispersing agents such as sodium citrate, sodium oxalate or Calgon are used to presoak the samples. This procedure is labor intensive, and the apparatus relatively costly to manufacture, but the major limitation is that the soil sample must be relatively small.

24–2.3.3 Extraction of Microarthropods by Centrifugation
(review: Goodey, 1957; Muller, 1962)

Though standard practice for nematode extraction, centrifugation is seldom used for arthropods though it works on the same principle. A sample is washed through a series of decreasing mesh diameter sieves. The appropriate filtrate is centrifuged in a dense solution (usually sucrose) which concentrates the fauna at the top. The fauna-containing layer is decanted, washed and preserved for analysis. Though considerably more labor intensive and time-consuming than other methods, in theory it should be excellent for sampling even non-mobile life stages such as eggs and cysts effectively.

24–2.3.4 Extraction of Microarthropods Based upon the Properties of the Cuticle

Although originally employed in the salt flotation method to increase separation of microarthropods from organic debris, the lipophilic nature of the cuticle can be used to extract arthropods directly from the soil. Aucamp & Ryke (1964) agitated aqueous soil samples in a container lined with grease. The arthropods adhered to the removable walls, which could be observed directly under the microscope.

Walter et al. (1987; Geurs et al., 1990) add heptane and then water to a sample previously fixed with ethanol and placed in a vacuum to remove air bubbles from plant debris. The sample is agitated and the invertebrates that float to the water/heptane interface, are decanted to a sieve, and preserved in alcohol. This approach is not useful in soils high in soluble organics because a tar-like layer forms at the organic/inorganic interface. As mentioned in the section on Berlese extraction, many taxa (often characteristic of more mesic soils) are not sufficiently lipophilic and probably would be overlooked by this method. The authors present evidence that for their soils, heptane extraction is significantly more efficient than high-gradient extraction.

24–3 PROCESSING THE EXTRACTED BIOTA SAMPLE

Relative to gathering samples in the field, extracting them in the lab (particularly with funnels), and identifying the material takes by far the most time. Since species-rich soils require sufficient replicate samples to adequately estimate abundances, decreasing handling time per sample is the key.

A typical fully extracted Berlese sample of 0.3 m² (to a depth of 10 cm) of soil from Pacific Northwest forests will contain 500 to 10 000 individual mites and insects embedded in a soil layer many times their volume. Taking an aliquot, diluting it, and laboriously counting all the morphotypes present severely underestimates species richness, consumes a vast amount of time, and cannot in most cases be trusted quantitatively. The following procedure is one I personally employ (Fig. 24–5) because I have found it to increase the speed of processing more than 10-fold, and allows accurate quantification of the entire arthropod fauna.

Step 1: Transfer sample obtained from under the Berlese extractor (B = arthropods + debris) to 10 to 100 mL vial (A); if previously stored in 70% ethanol, dilute the alcohol (C) to < 20%.

Step 2: Add several drops of mineral or vegetable oil (D) to vial with pipette (E).

Step 3: Cap (F) the vial and agitate to expose all the arthropods to the oil. Let set for half an hour and allow oil layer to separate on top (bringing with it nearly all the specimens out of the debris).

Step 4: With a pipette transfer oil layer (with specimens) to petri plate (4D), transfer alcohol and precipitate to another petri plate (4B). It is probably not possible (and certainly not practical) for the initial sorting and counting to identify all of the taxa to the species level. If one is interested in all the different types of taxa present in a sample, it is only necessary at this time to determine the morphospecies present in the nonorganic bottom residue which will subsequently be discarded. All the other morphospecies will be preserved in oil in a labelled bulk well.

Step 5: With pipette (or fine forceps), transfer specimens from petri plates to plastic 12 to 15 hole well-plate (H); rough sort to major taxonomic groupings at this time (e.g., beetles, flies, mites, and springtails).

Step 6: Further sort each broad taxonomic category to finer units (e.g., separate major families of beetles); employ a 96-well plate.

Step 7: Sort family groups to individual morphospecies; archive most specimens from the oil layer in mineral oil on a well-plate.

Step 8: Remove the specimens from the alcohol and precipitate and archive them in 50:50 ethanol/water, with a drop of glycerine added. These specimens (and also the spiders removed from the oil layer) will dehydrate if stored in oil. Dehydrated specimens are often not possible to identify because of the distortion that has taken place; the drop of glycerine permits insect cuticles to maintain their plasticity even if the alcohol subsequently evaporates. Soaking accidentally dehydrated specimens in warm lactic acid usually returns them to useful condition.

Step 9: Send more than a single synoptic specimen of each "morphospecies" to the consulting taxonomist. Although the specialist may conclude that your "morphospecies" is more than one taxon, with your specimens sorted and archived in this manner, it is easy to reexamine your tentative identifications in the light of the new resolution afforded by the

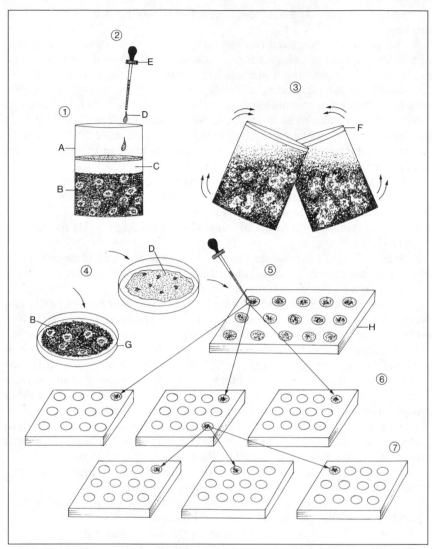

Fig. 24–5. Multi-species sample processing protocol. Step 1: Transfer sample to vial and dilute
the alcohol. Step 2: Add several drops of mineral oil to vial with pipette. Step 3: Cap the
vial and agitate; let set. Step 4: With pipette transfer oil layer (with specimens) to petri plate
(4D), and transfer alcohol (with debris) to another petri plate (4B). Step 5: With pipette (or
fine forceps), transfer specimens from petri plates to plastic 12 to 15 hole well-plate (H);
rough sort to major taxonomic groupings at this time (e.g., beetles, flies, mites, and springtails). Step 6: Further sort a broad taxonomic category to finer units (e.g., separate major
families of beetles); employ a 96-well plate. Step 7: Sort family groups to individual morphospecies. (Under certain circumstances, steps 5–7 may be replaced by transferring all the
specimens to a 50-mm plastic petri plate, withdrawing as much oil as possible with a
micropipette, and sorting with a probe/tweezers the fauna into separate clusters. Specimens
can remain "on display" for at least a decade for easy teaching or reference purposes.)

taxonomist and correct your data accordingly. A diverse assemblage of taxa can be stored in oil for over a decade with no signs of contamination or decay.

24–3.1 Comments

1. With vision unimpeded by soil particles it is easy to count the different "morphospecies" separately. Technicians with limited skill in identification affirm it is much easier to distinguish the different "morphospecies" when they are all clustered together in the drop of oil, than when they are spread out through a diversionary field-of-view.

2. Only invertebrates with hydrophilic exoskeletons will remain in the debris on the bottom of the vial. These specimens will include primarily insect larvae (usually Diptera), worms (oligochaetes and enchytraeids), molluscs (snails and slugs), isopods, larger millipedes, and occasional species in groups normally extracted with the oil (i.e., *Nanhermannia*—oribatid mite; *Zercon*—gamasid mite). So few types of fauna possess hydrophilic exoskeletons (they are usually the large-bodied species), that after processing a few samples, the researcher will develop sufficient skill to scan the debris at the bottom of the ethanol/water layer in a few seconds in search of specimens.

3. Though it is far quicker to cut your sample containers with oil at step 2, and store the majority of specimens in oil, it is also possible to cut with Epsom salt ($MgSO_4 \cdot H_2O$) and store the specimens in glycerine (except the soft-bodied forms that must be in alcohol or ethylene glycol). Glycerine is easier to remove from the synoptic specimens sent to the taxonomist.

4. The taxa most in need of authoritative identification are the non-rare ones—simply take examples of these out of alcohol-preserved samples that have never been cut with oil. Cutting with Epsom salt has the benefit of separating all the fauna from the mineral soil, but it includes a lot of organic debris in the sample as well. The same is true of centrifuging in sugar solutions. Cutting by specific gravity either with salt or sugar requires thorough washing of the specimens before preservation—the smallest amount of salt can crystallize and the smallest amount of sugar can result in bacterial growth in the sample. The organic debris (leaves, stems, and fungal hyphae) that rises with specimens in the extraction has to be removed by tweezers; even if only a little bit remains, fungal growth will be significant.

24–4 BIOTA IDENTIFICATION
(review: Behan et al., 1985; Dindal, 1990)

With probably more than 75% of the terrestrial biota associated with the soil for at least one major phase of its life cycle (Southwood, 1978), identification of soil fauna is a formidable task. In North America the most comprehensive reference text is Dindal (1990); this book is a compilation of identification keys, ecology, and bibliographic references by all of the leading specialists on different soil taxa. This is where any research group should start. Insect keys can be supplemented by the most widely used references in North America (Borer et al., 1976; Stehr, 1987, 1991), Krantz (1986) for mites, and oribatids may be supplemented with Moldenke & Fichter (1988; [copies available from the author]). Differing taxa can be identified to different levels of resolution with the Dindal book. A fundamental problem is that even with this book, and with the assistance of taxonomic specialists, identification past the level of genus is usually impossible for the most abundant taxa. Most of the species in these groups have never been given scientific names in North America. Taxonomists are a severely limiting quantity. If a research project requires identifications to be made, arrangements must be made before research is begun and the taxonomist(s) must be a participant in the design of the proposed research.

24–5 PRESERVATION AND ARCHIVING
(reviews: Martin, 1977; Steyskal et al., 1986)

24–5.1 Preservation

There are numerous fixatives and storing solutions available, each has its own specific advantages/disadvantages relative to certain taxa. Ethanol (70–80%) is by far the most frequent fixative and long-term preservative. Isopropyl (rubbing) alcohol or a 5% solution of 40% formalin are acceptable alternatives. When BOTH fixing and storing in ethanol, it is critical to: (i) decant and renew the solution prior to long-term storage (more than 1–2 wk); and (ii) add glycerine to the storage solution (several drops per 100 mL). A major expense in curating samples is the cost of vials and caps. Regardless of how they are advertised, no caps/stoppers/corks are evaporation proof. Neoprene stoppers are the most efficient (but far more costly than the vials themselves). The gelatine in the preservative protects the insect tissues, if the alcohol fully evaporates while archived.

Earthworms must be relaxed before killing and fixing. Place worms in a solution of 1 part $MgSO_4$ (saturated):3 to 4 parts of water for 1 to 2 h. Kill and fix by dipping in Bouin's solution for 2 to 20 s, blotting on paper and immersing (overnight) until stiff in FAA (90 mL 50% ethanol + 5 mL glacial acetic acid + 5 mL formalin solution). Place in 70% ethanol for permanent storage.

Bouin's and Carnoy's fixatives are widely used when the ultimate aim is sectioned microscope analysis of the samples. Bouin's fixative is made from 75 mL of picric acid solution (add 1 g picric acid crystals to 75 mL water), 25 mL of 40% formalin and 5 mL glacial acetic acid. Fix for at least 1 h, wash in water and dehydrate through an alcohol concentration series. Carnoy's fixative is made from 60 mL absolute ethanol, 10 mL of glacial acetic acid and 30 mL of chloroform ($CHCl_3$). Fix for 15 min to 2 h and wash in 95% ethanol. After fixation, specimens can be stored in 70% alcohol until embedded.

24–6 ARCHIVING

Ultimately most specimens will end up pinned, stored in alcohol or put on slides when they are examined by a taxonomist. It is VERY important to be consistent with conventional methods in labeling the specimens, since most will be incorporated into existing museum collections (see chapter 7 on preservation in Borer et al., 1976). Be sure the labels on pinned insects do not exceed 2 × 1 cm; place additional information on separate labels stacked on the pin. Be sure to use alcohol-insoluble ink on labels placed in bottles (most xeroxed labels will NOT last unless they are on high quality bond paper). Use only the right-hand side of a slide for collection information (see chapter 34 on slide making techniques in Krantz, 1986).

Keep most of your specimens in alcohol (or glycerine) for easy storage. Slide making is time-consumptive, and often crucial characters are not easy to observe—it is very helpful to have a supply of specimens in alcohol that can be placed on temporary hanging-drop slides for rapid analysis. Before examination with a dissecting stereomicroscope, specimens usually require clearing in either lactic acid (most specimens can be stored in lactic acid indefinitely); glycerol (50 mL)-water (50 mL distilled)-acetic acid (3 mL glacial); or lactophenol (50 g phenol-25 mL water + 50 mL lactic acid). The latter two clearing agents are too strong to store specimens in; rinse and store in ethanol.

Quick observation slides use a hanging-drop. The well on the slide is half-filled with lactic acid, the specimen inserted, and a square cover slip placed on top so that it projects over part of the well. The specimen caught in the meniscus can be manipulated with a teasing needle to examine all of its surfaces.

24–6.1 Assigning Morphospecies to Functional Ecological Groupings
(reviews: Kevan, 1968; Wallwork, 1970; Edwards & Lofty, 1977; Bal, 1982; Eisenbeis & Wichard, 1987)

This is no easy task—the diversity of meaningful ecological roles is large; and the information base available for most soil fauna is extremely limited (even in Europe, where more by far is known). However, two generalizations are universally affirmed: the more precise the taxonomic identification, the more likely that useful functional groupings can be accomplished; equally true, the more precise the identification the more time and effort required to achieve it (= more cost). Certain broad taxonomic groupings contain more precise functional information than others. For example, Chilopoda are all invertebrate predators and Collembola are mostly fungivores. In contrast, Coleoptera are too diverse to be labeled with a specific ecological function. The aim should NOT be to take all identifications to the same taxonomic resolution (e.g., order), but to take identifications to a level (different for each group) that is ecologically meaningful. How can you accomplish this? There are really no reference texts that are, at one and the same time, detailed enough for some groups but not overly complex and confusing for others.

We are currently developing an illustrated computer program, for use on Macintosh pc's, which starts to address this need (Moldenke et al., 1990). The intent of the design of COMTESA (*CO*mputer *T*axonomy and *E*cology of *S*oil *A*nimals) is that it is useful to all levels of users from novice to experienced soil zoologist. COMTESA can be used by the novice to differentiate basic functional groups with ease. Knowledgeable scientists and their research staff can use COMTESA to differentiate to the species or functional group levels, and can use the supplementary modules to store and retrieve information and scientific references about organisms at any level of organization (order, family, genus, . . .). Print-based keys currently available for identification of arthropods are often designed to cover wide geographic regions, such that their level of resolution is poor for the scale at which most ecological studies are conducted. The modular approach of COMTESA copes with this problem since it is conceptually divided into two parts designed to deal with the different scales of resolution. Part I distinguishes about 150 different functional/taxonomic groups and has an ecological emphasis. For instance, the xylevorous and microphytophagous species of oribatid mites are distinguished from one another while predaceous spiders are divided into component hunting guilds. This part of the key should be useful across North America with only minor alterations to meet local needs. Part II consists of modules that provide identification to the generic and species level and are specific to site/region/ecosystem. The key is driven by clicking a mouse on the proper choice of diagrammatic images. Unfortunately, it will be many years before such a system is modified to work at the species level for a widespread series of ecosystems.

24–6.2 Transforming Census Data
(review: Edwards, 1967; Phillipson, 1970; Petersen & Luxton, 1982)

Since large-bodied species are usually infrequent in samples and small-bodied species usually the most abundant, enumeration data will produce community analyses very different from analyses based on biomass or respiratory consumption. Average weights for most taxa can be obtained by weighing individuals on an electrobalance. The smallest springtails, oribatids, and prostig mites (< 5 µg each) require pooled samples of about a dozen individuals. The most appropriate biomass transformation for earthworms with digestive systems filled with soil has been addressed by Bouche (1966).

For each class of soil organisms, a specific relationship between length (easily measured with an optical micrometer) and weight exists (Reichle, 1967; Jarosik, 1989). These average conversion factors can be used for most studies, with the caveat that unusually shaped or exceptionally large species should be weighed directly.

Respiratory equivalents for dry biomass of many taxa are available as well as direct measurements of respiratory rates (Berthet, 1967) and feeding rates (O'Conner, 1963; Coleman, 1968) under controlled lab conditions. As the authors point out, these studies are limited and extrapolation to other field conditions would not be recommended (Healey, 1970; Persson & Lohm, 1977). The direct and indirect effect (regulatory effect of a fungivore's activity on its resource's metabolism) is testable in microcosms (Anderson & Ineson, 1982). Total respiratory rates of: (i) soil with endemic microflora and (ii) soil with endemic microflora PLUS differing population levels of particular fungivores, could be determined analogously to Anderson's ANREG model experimentation for N mineralization (Anderson et al., 1985). An exemplary study of individual species contribution to community respiratory rates is that of Luxton (1981).

24–7 REARING

Efficient laboratory rearing is the key to associating unidentifiable immature stages, quantifying respiration and nutrient mineralization rates, and determining the functional roles played and microhabitats selected by arthropods. There have been several techniques developed, but they have been employed in too few instances to have made an impact upon the vast number of fundamental questions remaining.

The most fundamental apparatus to maintain and observe microarthropods is a container partially filled with a mixture of Plaster of Paris and powdered charcoal. Water is periodically added to the substrate to maintain high humidity and very fine mesh is glued over a port in the cap to permit transfer of gasses. Small groups of oribatids (Evans et al., 1961; Sengbusch, 1974; Arlian & Wooley, 1970; Krantz, 1986) or gamasids can be kept in a 3 cm tall by 3 cm diam. container. Periodic feeding and especially removal of wastes is necessary to keep the culture healthy; the main cause of death in most rearing attempts is overgrowth by fungi.

Simple microcosms have been proposed by Anderson and Ineson (1982; Fig. 24–6) and refined by Taylor and Parkinson (1988), which permit measurement of both respiration and nutrient leaching. A soil sample (with or without specific arthropods) is placed in a cylinder over a nylon mesh and inserted into a slightly wider cylinder to rest on top of inert beads and a leaching port. The air-tight lid can either be provided with a rubber injection septum, a trough holding gas-absorbant chemicals, or a conductivity cell for measurement of CO_2 production. The latter reduces volumetric errors in the titration process and allows continuous monitoring of

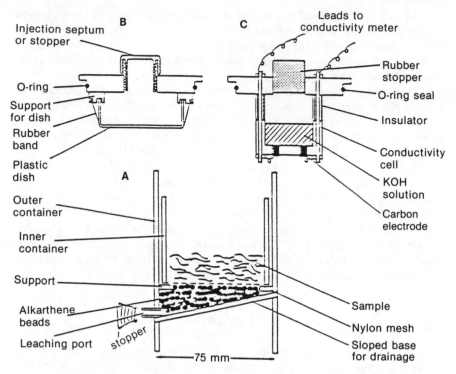

Fig. 24–6. Simple microcosm design of Anderson and Ineson (with permission from 1982), permitting measurement of both leachates and gaseous exchange from soil. (A) Base section showing inner sample container and drainage system for leachate sampling. (B) Lid to hold absorbants for gas assays. (C) Lid fitted with conductivity cell for measuring CO_2 evolution.

respiratory activity. Sensitivity must be adjusted to the rate of respiration; calibration is explained in Anderson and Ineson (1982). Leaching is achieved by flooding the substrate and drawing off the soluble nutrients through a leaching port. The leaching port may also be used to introduce inhibitors, pesticides, or nutrient amendments.

The key to unraveling the functional roles of soil arthropods in nature is to investigate their interactions with the normal component of microflora—"in order to express the ecological niche of the animals rather than their ability to adapt to artificial conditions" (Hagvar, 1988). Hagvar defaunated undisturbed forest soil samples, allowed the reestablishment of a natural microflora in the field, then investigated the interactions of mites with microflora under a set of controlled laboratory conditions.

The spatial scales in which arthropods act is poorly known. Basic elements such as how they locomote through the soil, how aggregated they

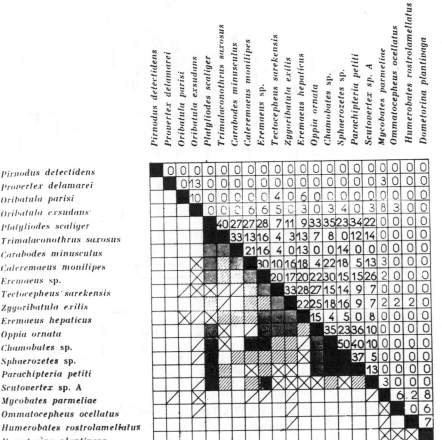

Fig. 24–7. Matrix table of oribatid mite biocoenosis (with permission from Trave, 1963). Representation to visualize the strength of species associations from samples with many replicates. In this example, *Trimalaconothrus* co-occurs with *Platyliodes* and *Sphaerozetes*, *Chamobates* and *Parachipteria* with one another more than 40% of the time; the great majority of the species co-occur in less than 10% of the sample replicates and are therefore excluded from this biocoenosis.

are and which microhabitats they select are unknown. The recent development of root boxes (analogous to rhizotrons, but somewhat less costly and easier to manipulate) should facilitate photographic documentation of soil arthropod activities (Rygiewicz et al., 1988; Unestam & Stenstrom, 1989). The design of Rygiewicz et al. (1988) is particularly promising, since it can differentiate activities within the rhizosphere associated with feeding upon roots vs. fungi.

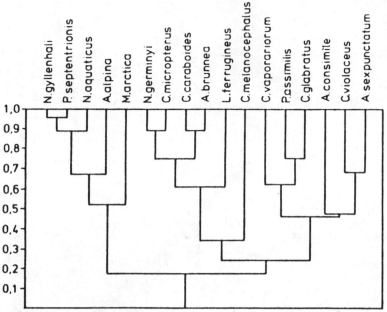

Fig. 24–8. Dendrogram depicting the similarity in habitat preference between each of 17 species of carabid ground beetles (with permission from Rafseth, 1980). Numerical index of habitat similarity on vertical axis; species *N. gyllenhali* and *P. septentrionis* are the most frequently associated.

24–8 STATISTICAL METHODS TO ANALYZE DIVERSITY
(review: Southwood, 1978; Begon et al., 1986)

Biodiversity in the soil is likely to exceed that of any other component of terrestrial ecosystems (Moldenke & Lattin, 1990). Not only is it difficult to quantify the interactions between species, but it is a practical concern simply to describe patterns of species occurrence. Moldenke (1990) tabulates the diversity likely to be found in a typical square meter of Northwest conifer forest soil. The question of whether there are basic patterns of co-occurrence, implying the existence of semi-independent microcommunities, is usually analyzed by constructing a matrix of co-occurrence frequencies (Trave, 1963; Fig. 24–7).

Degree of similarity in habitat preference can be represented as a dendrogram (Rafseth, 1980; Fig. 24–8). Similarly, the elegant feeding preference studies of Hartenstein (1962) could be transformed to matrix values reflecting feeding specialization and niche overlap (viz., Moldenke, 1975). The response of the diverse soil community to specific environmental variables can be quantified by techniques developed by phytosociologists. TWINSPAN and DECORANA are widely used examples of principal components analysis (Wauthy et al., 1989). TWINSPAN is a hierarchical analysis; Moldenke (unpublished data; McIver et al., 1991) used it to test the relative importance of a series of environmental variables in determining the composition of ground-dwelling spider communities. DECORANA is a similar technique but does not impose a hierarchical dichotomy on the results. Response of complex community composition to single or multiple environmental changes are graphed such that each point represents the full range of species within a sample and the distance between points represents the degree of similarity between any two samples.

A specific benefit to these two community representation algorithms is that they calculate particular "indicator species" for each of the discriminations they perform. This permits reducing the diversity of fauna that has to be enumerated from several hundred per square meter to perhaps a dozen that can be easily learned by a technician; subsequent analysis for the effect of a given environmental insult (i.e., acid precipitation, management practice, and herbicide application) can thus be appreciably facilitated (Eyre et al., 1990).

REFERENCES

Anderson, J.M. 1978. A method to quantify soil-microhabitat complexity and its application to a study of soil animal species diversity. Soil Biol. Biochem. 10:77–78.

Anderson, J.M., and I.N. Healey. 1970. Improvements in the gelatine embedding technique for woodland soil and litter samples. Pedobiology 10:108–120.

Anderson, J.M., S.A. Huish, P. Ineson, M.A. Leonard, and P.R. Splatt. 1985. Interactions of invertebrates, microorganisms and tree roots. p. 377–392. In A.H. Fitter et al. (ed.) Ecological interactions in the soil. Blackwell Sci. Publ., Oxford.

Anderson, J.M., and P. Ineson. 1982. A soil microcosm system and its application to measurements of respiration and nutrient leaching. Soil Biol. Biochem. 14:415–416.

Andren, D. 1985. Microcomputer-controlled extractor for soil microarthropods. Soil. Biol. Biochem. 17:27–30.

Arlian, L.G., and T.A. Wooley. 1970. Observations on the biology of Liacarus cidarus (Acari; Cryptostigmata; Liacaridae). J. Kans. Entomol. Soc. 43:297–301.

Aucamp, J.L., and P.A.J. Ryke. 1964. A preliminary report on a grease-film extraction method for soil microarthropods. Pedobiology 4:77–79.

Bal, L. 1982. Zoological ripening of soils. Centre Agric. Publ. Doc., Wageningen, the Netherlands.

Begon, M., J.L. Harper, and C.R. Townsend. 1986. Ecology: Individuals, populations and communities. Sinauer Assoc., Sunderland, MA.

Behan-Pelletier, V., S.B. Hill, A. Fjellberg, R.A. Norton, and A. Tomlin. 1985. Soil invertebrates: Major reference texts. In J. Spence (ed.) Faunal influences on soil structure. Quest. Entomol. 21:675–687.

Berthet, P. 1967. The metabolic activity of oribatid mites (Acarina) in different forest floors. p. 709–725. In K. Petrusewicz (ed.) Secondary productivity of terrestrial ecosystems. Vol. 2. Panstowowe Wydawnictwo Naukowe, Warsaw.

Bieri, M., and V. Delucchi. 1980. Eine neue konzipierte Auswaschanlage zur Gewinnung von Bodenarthropoden. Mitt. Schweiz. Entomol. Ges. 53:327–339.

Blake, G.R., and K.H. Hartge. 1986. Bulk density. p. 363–375. In A. Klute (ed.) Methods of soil analysis. Part I. 2nd ed. Agron. Monogr. 9. ASA and SSSA, Madison, WI.

Borer, D.J., D.M. Delong, and C.A. Triplehorn. 1976. An introduction to the study of insects. 4th ed. Holt, Rinehart and Winston, New York.

Bouche, M.B. 1966. Sur un nouveau procede d'obtenation de la vacuite artificielle de tube digestif des Lombricides. Rev. Ecol. Biol. Sol 3:479–482.

Coleman, D.C. 1968. Foodwebs of small arthropods in a broomsedge field, studied with radio-isotope labeled fungi. p. 203–207. In J. Phillipson (ed.) Methods of study in soil ecology. UNESCO, Paris.

Crossley, D.A., Jr., 1977. The role of terrestrial saprophagous arthropods in forest soils: current status of concepts. p. 49–56. *In* W.J. Mattson (ed.) The role of arthropods in forest ecosystems. Springer-Verlag, New York.

Dindal, D.C. 1990. Soil biology guide. John Wiley and Sons, New York.

Doube, B.M., and P.S. Giller. 1990. A comparison of two types of traps for sampling dung beetle populations (Coleoptera: Scarabaeidae). Bull. Entomol. Res. 80:259–263.

Dunger, W., and H.J. Fiedler. 1989. Methoden der bodenbiologie. VEB Gustav Fischer Verlag, Jena.

Edwards, C.A. 1967. Relationship between weight, volumes and numbers of soil animals. p. 1–10. *In* O. Graff and J.E. Satchell (ed.) Progress in soil biology. Braunschweig Verlag Fader Vieweg, Univ. of Sohn.

Edwards, C.A. 1990. The assessment of populations of soil-inhabiting invertebrates. Agric. Ecosyst. Environ. 34:145–176.

Edwards, C.A., A.E. Whiting, and G.W. Heath. 1970. A mechanical washing method for separation of invertebrates from soil. Pedobiology 10:141–148.

Edwards, C.A., and K.E. Fletcher. 1970. Assessment of terrestrial invertebrate populations. p. 57–66. *In* J. Phillipson (ed.) Proceedings of symposium on methods of study in soil ecology. UNESCO, Paris.

Edwards, C.A., and K.E. Fletcher. 1971. A comparison of extraction methods for terrestrial arthropod populations. p. 150–185. *In* J. Phillipson (ed.) IBP Handbook no. 18. Methods for the study of productivity and energy flow in soil ecosystems. Blackwell, Oxford.

Edwards, C.A., and J.R. Lofty. 1977. Biology of earthworms. Chapman and Hall, London.

Edwards, C.A., B.R. Stinner, D. Stinner, and S. Rabatin. 1988. Biological interactions in soil. Agric. Ecosyst. Environ. 24:1–380.

Eisenbeis, G., and W. Wichard. 1987. Atlas on the biology of soil arthropods. Springer-Verlag, Berlin.

Evans, G., J. Owens, G. Shields, and D. MacFarlane. 1961. The terrestrial acari of the British Isles. Adlard & Son, Bartholomew Press, Dorking, England.

Eyre, M.D., M.L. Luff, and S.P. Rushton. 1990. The ground beetle (Coleoptera: Carabidae) fauna of intensively managed agricultural grasslands in northern England and southern Scotland. Pedobiology 34:11–17.

Finlay, R.D. 1985. Interactions between soil microarthropods and endomycorrhizal associations of higher plants. p. 319–331. *In* A.H. Fitter et al. (ed.) Ecological interactions in the soil. Blackwell Sci. Publ., Oxford.

Fitter, A.H., D. Atkinson, D.J. Read, and M.B. Usher (ed.). 1985. Ecological interactions in the soil. Blackwell Sci. Publ., Oxford.

Franke, U., B. Friebe, and L. Beck. 1988. Methods for determining soil animal population densities using quadrats and pitfall traps. Pedobiology 32:253–284.

Froehlich, H.A., and D.W.R. Miles. 1986. A freezing technique for sampling skeletal, structureless forest soils. Soil Sci. Am. J. 50:1640–1642.

Geurs, M., J. Bongers, and L. Brussard. 1990. Improvements of the heptane flotation method for collecting microarthropods from silt loam soil. Agric. Ecosyst. Environ. 34:213–221.

Gist, C.S., and D.A. Crossley. 1973. A method for quantifying pitfall trapping. Environ. Entomol. 2:951–952.

Goodey, J.B. 1957. Laboratory methods for work with plant and soil nematodes. Tech. Bull. Min. Agric. no. 2. 3rd ed. HMSO, London.

Greenslade, P.J.M. 1964. Pitfall trapping as a method for studying populations of Carabidae (Coleoptera). J. Anim. Ecol. 33:301–310.

Haarlov, N., and T. Weis-Fogh. 1953. A microscopical technique for studying the undisturbed texture of soils. Oikos 4:44–57.

Hagvar, H. 1988. Decomposition studies in an easily-constructed microcosm: effects of microarthropods and varying soil pH. Pedobiology 31:293–303.

Hartenstein, R. 1962. Soil Oribatei I. Feeding specificity among soil Oribatei (Acarina). Ann. Entomol. Soc. Am. 55:202–206.

Healey, I.N. 1970. The study of production and energy flow in populations of soft-bodied micro-arthropods. p. 175–179. *In* J. Phillipson (ed.) Methods of study in soil ecology. UNESCO, Paris.

Jarosik, V. 1989. Mass versus length relationship for carabid beetles (Coleoptera; Carabidae). Pedobiology 33:87–89.

Kevan, D.K. McE. 1955. Soil zoology. Butterworths, London.

Kevan, D.K. McE. 1968. Soil animals. Witherby Ltd., London.

Krantz, G.W. 1986. A manual of Acarology. 2nd ed., amended version. Oregon State Univ. Bookstore, Corvallis.

Kubiena, A. 1938. Micropedology. College Press, Ames, IA.

Ladell, W.R.S. 1936. A new apparatus for separating insects and other arthropods from soil. Ann. Appl. Biol. 23:862–879.

Loreau, M. 1987. Vertical distribution of activity of carabid beetles in a beech forest floor. Pedobiology 30:173–178.

Luxton, M. 1981. Studies on the oribatid mites of a Danish beech wood soil VII: Energy budgets. Pedobiology 22:77–111.

Macfadyen, A. 1953. Notes on methods for the extraction of small soil arthropods. J. Anim. Ecol. 22:65–77.

Martin, J.E.H. 1977. The insects and Arachnids of Canada I: Collecting, preparing and preserving insects, mites, and spiders. Biosyst. Res. Inst., Canada Dep. Agric. Publ. no. 1643.

McIver, J.D., G.L. Parsons, and A.R. Moldenke. 1991. Litter spider succession after clearcutting in a western coniferous forest. Can. J. For. Res. 22:984–992.

Milne, A., R.E. Coggins, and R. Laughlin. 1958. The determination of numbers of leather-jackets in sample turves. J. Anim. Ecol. 27:125–145.

Moldenke, A.R. 1975. Niche specialization and species diversity along a California transect. Oecologia 21:219–242.

Moldenke, A.R. 1990. One hundred twenty thousand little legs. Wings 15:11–14.

Moldenke, A.R., and B.L. Fichter. 1988. Invertebrates of the HJ Andrews Experimental Forest, western Cascade Mountains, Oregon: IV. The oribatid mites (Acari; Cryptostig-mata). USDA-FS, Pacific Northwest Range and Exp. Stn., Rep. GTR-PNW-217. U.S. Gov. Print. Office, Washington, DC.

Moldenke, A.R., C. Shaw, and J.R. Boyle. 1990. Computer-driven image-based soil fauna taxonomy. Agric. Ecosys. Environ. 34:177–185.

Muller, G. 1962. A centrifugal-flotation extraction technique and its comparison with two funnel extractors. p. 207–211. In P.W. Murphy (ed.) Progress in soil zoology. Butter-worths, London.

Murphy, P.W. 1962. Extraction methods for soil animals. I & II. p. 75–155. In P.W. Murphy (ed.) Progress in soil zoology. Butterworths, London.

Norton, R.A., and J.B. Kethley. 1988. A collapsible full-sized Berlese-funnel system. Entomol. News 99:41–47.

O'Conner, F.B. 1955. Extraction of enchytraeid worms from a coniferous forest soil. Nature (London) 175:815–816.

O'Conner, F.B. 1963. Oxygen consumption and population metabolism of Enchytraeidae. p. 32–48. In J. Doeksen and J. van der Drift (ed.) Soil organisms. North Holland Publ. Co., Amsterdam.

Oostenbrink, M. 1960. Estimating nematode populations by some selected methods. p. 85–102. In B.N. Sasser and W.R. Jenkins (ed.) Nematology. Univ. North Carolina Press, Chapel Hill.

Oostenbrink, M. 1970. Comparison of techniques for population estimation of soil and plant nematodes. p. 249–255. In J. Phillipson (ed.) Methods of study in soil ecology. UNESCO, Paris.

Persson, T., and U. Lohm. (ed.). 1977. Energetical significance of the annelids and arthro-pods in a Swedish grassland soil. (Ecol. Bull. 23) Swedish Natural Research Council, Stockholm, Sweden.

Petersen, H., and M. Luxton. 1982. A comparative analysis of soil fauna populations and their role in decomposition processes. Oikos 39:287–388.

Phillipson, J. 1970. Methods of study in soil ecology. UNESCO, Paris.

Price, J.F., and M. Shepard. 1980. Sampling ground predators in soybean fields. p. 532–543. In M. Kogan and D.C. Herzog (ed.). Sampling methods in soybean entomology. Springer-Verlag, Berlin.

Rafseth, D. 1980. Ecological analysis of carabid communities. Biol. Conserv. 17:131–141.

Reichle, D.E. 1967. Relation of body size to food intake, oxygen consumption, and trace element metabolism in forest floor arthropods. Ecology 49:538–542.

Rusek, J. 1985. Soil microstructures—contributions on specific soil organisms. In J. Spence (ed.) Faunal influences on soil structure. Quest. Entomol. 21:497–514.

Rygiewicz, P.T., S.L. Miller, and D.M. Durall. 1988. A root mycocosm for growing ecto-mycorrhizal hyphae apart from host roots while maintaining symbiotic integrity. Plant Soil 109:282–284.

Seastedt, T.R. 1984. The role of microarthropods in decomposition and mineralization processes. Annu. Rev. Entomol. 29:25–46.

Seinhorst, J.W. 1956. The quantitative extraction of nematodes from soil. Nematologica 1:249–267.

Seinhorst, J.W. 1962. Modifications of the elutriation method for extracting nematodes from soil. Nematologica 8:117–128.

Sengbusch, H.G. 1974. Culture methods for cryptostigmatid mites (Acari: Oribatei). Proc. 4th Int. Congr. Acarol.:83–87.

Shaw, C.H., H. Lundkvist, A.R. Moldenke, and J.R. Boyle. 1991. The relationships of soil fauna to long-term forest productivity in temperate and boreal ecosystems: processes and research strategies. p. 39–77. In W.J. Dyck and C.A. Mees (ed.) Long-term field trials to assess environmental impacts of harvesting. Proc., IEA/BE T6/A6 Workshop, Florida, USA, Feb. 1990. Rep. #5. Forest Res. Inst., Rotorua, New Zealand, FRI Bull. No. 161.

Southwood, T.R.E. 1978. Ecological methods, with particular emphasis to the study of insect populations. 2nd ed. Methuen Press, London.

Spence, J. 1985. Faunal influences on soil structure. Quest. Entomol. 21:371–686.

Stehr, F.W. Vol. 1, 1987; Vol. 2, 1991. Immature insects. Kendall/Hunt Publ. Co., Dubuque, IA.

Steyskal, G.C., W.C. Murphy, and E.M. Hoover (ed.). 1986. Insects and mites: techniques for collection and preservation. USDA-ARS, Misc. Publ. 1443. U.S. Gov. Print. Office, Washington, DC.

Taylor, B., and D. Parkinson. 1988. A new microcosm approach to litter decomposition studies. Can. J. Bot. 66:1933–1939.

Trave, J. 1963. Ecologie et biologie des oribates (Acariens) saxicoles et arboricoles. Suppl. Vie et Milieu no. 14. Hermann, Paris.

Tullgren, A. 1918. Ein sehr einfacher Ausleseapparat fur terricole Tierformen. Z. Angew. Entomol. 4:149–150.

Unestam, T., and E. Stenstrom. 1989. A method for observing and manipulating roots and root-associated fungi on plants growing in nonsterile substrates. Scand. J. For. Res. 4:51–58.

Visser, S. 1985. The role of soil invertebrates in determining the composition of soil microbial communities. p. 297–318. In A.J. Fitter et al. (ed.) Ecological interactions in the soil. Blackwell Sci. Publ., Oxford.

von Torne, E. 1962. An elutriation and sieving apparatus for extracting microarthropoda from soil. p. 204–206. In P.W. Murphy (ed.) Progress in soil zoology. Butterworths, London.

Wallwork, J.A. 1970. Ecology of soil animals. McGraw Hill, New York.

Walter, D.E., J. Kethley, and J.C. Moore. 1987. A heptane flotation method for recovering microarthropods from semiarid soils with comparison to the Merchant-Crossley high-gradient extractor method and estimates of microarthropod biomass. Pedobiology 30:221–232.

Wauthy, G., V. Mundon-Izay, and M. Dufrene. 1989. Geographic ecology of soil oribatid mites in deciduous forests. Pedobiology 33:399–416.

Chapter 25

Carbon Utilization and Fatty Acid Profiles For Characterization of Bacteria

A. C. KENNEDY, *USDA-ARS, Pullman, Washington*

Understanding relationships among bacteria advances our knowledge of bacterial ecology and community structure, assists in the investigation of an unknown bacterium, and facilitates communication between laboratories concerning strains. Typically, unknown strains of bacteria are classified by comparing phenotypic or genetic characteristics to those of bacteria previously identified and representative of known taxa. The characterization of bacteria recovered from soil is time consuming, often frustrating, and sometimes difficult. With techniques such as nucleic acid or fatty acid analysis, the streamlining of diagnostic tests in microtubes or microplates, and the principles of numerical taxonomy, relationships between bacterial strains can be investigated more efficiently than ever before.

Many characteristics and behavior patterns are incorporated into identification schemes. Typically, one must first establish the major group to which it belongs; i.e., whether it is phototrophic, chemoautotrophic, chemolithotrophic, or chemoheterotrophic, aerobic or anaerobic, microaerophilic or facultative. Other characteristics of interest pertain to the chemical, nutritional, morphological, biochemical, physiological, genetic, pathogenic and antigenic properties of the bacterium (Krieg & Holt, 1984).

This chapter describes the analyses of C utilization patterns and fatty acid profiles for bacterial characterization and identification. The capacity of a strain to metabolize various C substrates can be used in its identification by matching its pattern of utilization with those of known strains (Goor et al., 1984). Fatty acid profiles also are a stable and reliable means of delineating taxa (Miller & Berger, 1985; DeBoer & Sasser, 1986). As informative as these two methodologies are, they do not stand alone as identification techniques, but are to be used with other techniques to characterize and gain further information on bacterial isolates, including those from soil.

25–1 CHARACTERIZATION OF BACTERIA

To identify a bacterium, its characteristics need to be compared with attributes found in established taxa. The most widely used classification scheme is *Bergey's Manual of Systematic Bacteriology* (Krieg & Holt, 1984), which contains many standard methodological outlines used in identification. *Bergey's Manual* is an excellent source of taxonomic relationships because it contains keys and tables that can be easily used for all species and genera of bacteria. Many contributors, all expert in their respective fields, compiled the information, which is continually being updated.

To use the information gained from C utilization and fatty acid analyses, mathematical manipulations are needed. For example, numerical taxonomy quantitatively groups strains into homogeneous groups or clusters based on likeness. The method is successful in the classification and identification of bacteria when the traits of an unknown bacterium are compared with those of known strains. Groupings or clusters are calculated objectively from similarities among phenotypic or genetic characteristics of the bacteria. These calculations can deal effectively with the abundance of information gathered and circumvents the bias that may develop from intuitive judgments or subjective determinations. There are several excellent reviews on the use of numerical taxonomy in bacterial identification (Lockhart & Liston, 1970; Colwell, 1973; Sneath & Sokal, 1973; Sneath, 1978a; Willcox et al., 1980; Colwell & Austin, 1981). Also, computerized techniques are accessible to almost everyone (Alderson, 1985; Goodfellow et al., 1985).

Identification, nevertheless, is only as good as the data collected for comparison. There are a few precautions that must be considered when using methods for characterization of bacteria. First, purity of the isolate is of utmost importance and may be difficult to achieve. For example, transfer of a single colony isolate may not be sufficient to ensure a pure culture. This especially may be the case with heavy slime producers, which may harbor contaminates, or with isolates on a selective medium concealing dormant contaminants. Second, the selection of type or reference strains is critical to successfully identify an isolate. Third, standard environmental conditions during growth of the bacterial cultures are crucial in the determination of taxonomic relationships. With these precautions in mind, analyses can be compared between laboratories and classification and identification of isolates can be achieved.

25–2 CARBON SOURCE UTILIZATION

25–2.1 Principles

Bacterial cell maintenance and growth requires energy, C, and numerous inorganic ions; however, species differ in the compounds that each can use to meet their metabolic requirements (Hugh & Leifson, 1953; McKinley et al., 1982; Krieg & Holt, 1984; Mergaert et al., 1984; Johnson et al., 1989). Carbon utilization is central to survival, growth, and competitiveness of bacteria in any community (Clark, 1965; Veldkamp et al., 1984). By virtue of the diverse array of compounds available for metabolism and requirements of specific organisms, C utilization can be used as a discriminating means to characterize and identify bacteria if those requirements are specific for that organism or a group or organisms (Fahy & Hayward, 1983; Goor et al., 1984; Mergaert et al., 1984).

Selection of the substrates is critical to the analysis and can be customized for individual investigations. Potential substrates may include sugars, amino acids, organic acids, especially those that will differentiate various taxa in question. This selection can be broad for the examination of general functional groups to specific for identification of species within closely related genera.

Chemical indicators have been used extensively to reflect growth of microorganisms on select C sources (Jacob, 1970). For example, bromothymol blue and phenol red can be used to detect metabolism of C sources, acid production, and fermentation processes. Redox dyes such as triphenyl tetrazolium chloride (TTC) can be used to detect increased respiration associated with substrate oxidation or utilization (Lederberg, 1948; Pergram, 1969; Bochner & Savageau, 1977). When the bacterium oxidizes the C substrate, NADH is formed with a resultant flow of electrons. Triphenyl tetrazolium chloride will capture these electrons to form triphenyl tetrazolium formazan, a bright red precipitate, which can be assayed spectrophotometrically.

The initial growth medium and growth conditions prior to inoculation have a profound effect on the results of these assays. Cultures used should be fresh and actively growing, preferably in the mid to late-log growth stage. The organism should be inoculated and grown at optimal conditions for that particular organism. Cells grown under starvation or stress conditions may react differently to the substrates and thus exhibit unique characteristics (Morita, 1982; Matin et al., 1989).

The use of numerical taxonomy in identification of bacterial strains is well established with a large number of taxa defined by this method (Sokal & Sneath, 1963; Colwell & Austin, 1981). Numerical taxonomy defines taxonomic groups based on the similarity or differences of strains when tested for equally weighted characteristics. Further information on the techniques and procedures in numerical taxonomy can be obtained from reviews of the subject (Sneath, 1978a,b; Colwell & Austin, 1981; Sneath et al., 1981; Alderson, 1985). Many cluster analysis computer programs also are available to assist in the calculations and identification of the unknown strains (Nie et al., 1975; Alvey et al., 1977; Wishart, 1978; Dixon & Brown, 1979; Kaufman & Massart, 1981).

Data can be expressed by three different types of numerical codes. First, a binary code can be used to represent either a positive (growth) or negative (no growth) response with 1 or 0, respectively. Alternately, a quantitative code can be used to indicate the magnitude or extent of a reaction, which would be the case of spectrophotometric values. Finally, a qualitative code can be used to indicate differences in colony morphology, gram stain, cell shape, or pigmentation.

In the computer analysis of encoded data, the relationship between pairs of isolates is assessed and used to construct clusters of related bacteria. This is accomplished by the calculation of coefficients of similarity, by which each pair of bacteria is compared to determine similarity or mismatches. Characteristics to be used in the calculations should include only those that indicate differences among the isolates tested. Redundant or correlated characteristics that do not discriminate within the set are excluded.

The two coefficients of similarity most widely used in numerical taxonomy of bacteria are the simple matching coefficient (S_{SM}) and the Jaccard coefficient (S_J). The simple matching coefficient contains the positive and negative matches between isolates and thus, may overestimate similarity or homogeneity between isolates. The Jaccard coefficient considers only the positive matches. The two equations and the other coefficients that also may be used are discussed in much greater detail elsewhere (Sneath & Sokal, 1973; Austin & Colwell, 1977; Goodfellow et al., 1985). From these calculations, the isolates can be grouped and dendograms or branched diagrams constructed (Alderson, 1985; Sneath, 1978a).

25–2.2 Materials

1. 96-well microtiter plates, 0.3 mL working volume, flat or round well bottoms, with non-reversible lids, sterile.
2. 200 mL bottles or flasks, autoclavable.
3. Filter membrane (0.2 μm) sterilization unit.
4. Multichannel, repetitive dispensing pipet.
5. Dispensing reservoir, sterile.
6. Multipoint inoculating stamp, designed to fit 96-well plate.
7. Petri plates or microtubes with appropriate medium.
8. Calcium sulfate ($CaSO_4$), 7.0 mM solution.
9. Triphenyl Tetrazolium Chloride (TTC), 15 mM solution.
10. Characterization Medium (CM).

	g/L
Ammonium phosphate monobasic ($NH_4H_2PO_4$)	1.25
Potassium chloride (KCl)	0.25
Magnesium sulfate ($MgSO_4 \cdot 7H_2O$)	0.25

Adjust to pH 7.0 with 1.0 N NaOH before autoclaving.
11. Carbon substrate (CS), 10% solution.

25–2.3 Procedure

25–2.3.1 Microtiter Plate Preparation

Dispense 100 mL of the Characterization Medium (CM; modified from Medium A; Ayers et al., 1919) into 200 mL bottles and autoclave.

Table 25–1. Possible C and energy sources for bacterial isolates.

Carbohydrates, sugars and sugar alcohols
Arabinose, cellulose, dextrin, ducitol, fructose, galactose, glucose, glycerol, glycogen, inulin, lactose, maltose, mannitol, adonitol, melibiose, raffinose, rhamnose, ribose, sorbitol, starch, trehalose, xylose

Amino acids
Alanine, arginine, asparagine, aspartic acid, glutamic acid, glutamine, glycine, histidine, isoleucine, leucine, lysine, methionine, phenylalanine, proline, serine, threonine, tryptophan, tyrosine, valine

Organic acids
Butyric acid, glyoxylic acid, lactic acid, propionic acid, pyruvic acid

Tricarboxylic acid cycle acids
Acetic acid, aconitic acid, citric acid, fumaric acid, α-ketoglutaric acid, malic acid, oxaloacetic acid, succinic acid

Water-soluble vitamins
p-aminobenzoic acid, betine, biotin, calcium pantothenate, carnitine HCl, choline, citrate, cystamine, ferulic acid, folic acid, p-hydroxybenzoic acid, inositol, nicotinic acid, putrescine 2HCl, pyridoxal ethylacetal, pyridoxamine, pyridoxine, riboflavine, sodium riboflavine $PO_4 \cdot 2H_2O$, spermidine $PO_4 \cdot 3H_2O$, thiamine NO_3, thioctic acid, vitamin B_{12}

Purines and pyrimidines
Adenine, adenosine, adenylic acid, cytidylic acid, deoxyguanosine, guanine, guanosine H_2O, hypoxanthine, orotic acid, sodium guanylate, sodium inosinate, thymidine, thymine, uracil, uridine, xanthine, xanthosine

Table 25–2. Compounds present in root exudates (Rovira, 1965).

Sugars
Arabinose, fructose, glucose, galactose, maltose, raffinose, rhamnose, ribose, sucrose, xylose, oligosaccharides

Amino compounds
α-alanine, β-alanine, α-aminobutyric acid, asparagine, arginine, aspartic acid, cystathionine, cystine/cysteine, glutamine, glycine, homoserine, leucine/isoleucine, lysine, methionine, ornithine, phenylalanine, proline, serine, threonine, tyrosine, tryptophan

Organic acids
Acetic, butyric, citric, fumaric, glycolic, malic, malonic, oxalic, propionic, succinic, tartaric, valeric

Fatty acids and sterols
Campesterol, cholesterol, linoleic, linolenic, oleic, palmitic, sitosterol, stearic, stigmasterol

Filter sterilize (through a 0.2 μm filter) 15 mM TTC solution and add 1 mL to each bottle. Filter sterilize the C substrate solution (CS) and add 1 mL to each bottle. The C substrates to be tested can be selected from among sugars, amino acids, organic acids, tricarboxylic cycle components (Table 25–1) or rhizosphere components (Table 25–2). Other organic molecules of interest may be hydrocarbons, aromatic compounds, amino sugars, and pesticidal residues, etc. The substrates selected will depend on the experimental objectives, the type of characterization desired and the known properties of the species to be tested. The number and type of substrates can be customized for the goals of the analyses or to correspond to a database used for isolate comparison.

To construct the substrate microtiter plates, pour a small quantity of each C-source medium into sterile reservoirs. Using a multichannel repetitive dispensing pipet or similar dispensing tool, dispense 200 μL of a given C-source medium into all wells of a 96-well microtiter plate. Each plate will contain a different substrate. Dispense sterile 7.0 mM $CaSO_4$ or similar diluent containing 15 mM TTC into two plates. One plate will serve as a master plate and source of inoculum and the other as a C-free control. The microtiter plates should not be poured more than a day prior to use, as they are susceptible to dehydration. Use sterile components and maintain sterile techniques throughout the assembly procedure.

Microtubes (1.0 mL) or microcentrifuge tubes (0.5–1.0 mL) can also be used if microtiter plates are unavailable. Appropriately sized racks can be found to hold any number of tubes. Alternatively, the C utilization assay can be performed on agar plates by adding agar to the CS medium and streaking the isolates on the surface of the agar.

Cultures can be initially prepared in one of two ways. For agar-grown cells, remove desired isolates from cryostorage and place on their initial selection agar medium. Incubate at 25 °C for 24 to 48 h or until significant growth has occurred. Transfer directly from plates to wells in the master microtiter plate. For broth-grown cells, add 1 mL of the appropriate sterile broth to sterile 1.5-mL microcentrifuge tubes. Transfer cells from plates or directly from cryostorage to the corresponding broth tube. Incubate on a flat bed shaker for 24 h at 25 °C and 100 rpm. Centrifuge cells using a microcentrifuge (9000 × g), decant the supernatant, and wash three to four times with sterile 7.0 mM $CaSO_4$ or appropriate diluent. Wash the cells thoroughly to prevent contamination by foreign C sources derived from the initial culture broth. Resuspend the pellet in 1 mL of 7.0 mM $CaSO_4$. Using a multichannel dispensing pipet, place 50 μL of the washed suspensions into the wells of the microtiter plates.

25–2.3.2 Microbial Characterization

Bacteria from soil dilution plates or cultures can be characterized in prepared 96-well microtiter plates. Ninety-two isolates can be processed in each plate with four wells used as controls. Two wells per plate should be left noninoculated to ensure that the medium is not capable of abiotic reduction of the TTC resulting in false positives. Other controls should include at least one well inoculated with a bacterium capable of using the substrate and one well inoculated with a bacterium not able to use the substrate. Type and reference strains representative of the species, subspecies, and genera in question are included to verify media and to better differentiate the unknown. The second $CaSO_4$ plate containing no C source (mentioned above) should be inoculated with the same isolates to serve as a negative control. Positive reactions in this plate will indicate contamination from foreign C sources derived from the initial culture medium, plate, or inoculating stamp. Isolates that cause a positive reaction in this control well may also be producing capsular substances not removed with washing. The presence of this capsular material can be determined by microscopic examination.

Using a sterile toothpick or inoculating loop, transfer the isolates to be characterized from agar plates into wells of the master plate containing $CaSO_4$ and TTC. Arrange isolates on the master plate as desired for the test plate with each well containing a different isolate. A multipoint inoculating stamp can be purchased from commercial suppliers or made by inserting thin stainless steel pins into putty-filled wells of the microtiter plate. Alternatively, insert pins into a wood, plastic, or metal template in the pattern of the wells on a plate. To inoculate the microtiter plates, immerse the prongs of a multipoint inoculating stamp into the master plate and transfer to the corresponding wells of a microtiter plate of specific C medium. Wash the inoculator prongs with water two to three times between each C plate. Also, sterilize the inoculator by immersing the prongs in ethanol and flaming before and after each transfer. Cover the plate with a lid and incubate. If gas production or information on anaerobic utilization is preferred, add mineral oil to each well and place plates in an anaerobic chamber to exclude O_2.

Incubate the plates at optimal temperature and time. For soil isolates, the incubation period may vary from 48 h to 7 d at 25 °C. Many isolates may not show color change until after 5 d. Determine growth by a change of color of the TTC (i.e., reduction of TTC to triphenyl tetrazolium formazan producing a red precipitate) in the medium. This indicates that the isolate is capable of using the C source supplied, barring any false positives from the above-mentioned controls. Reactions can be read either by eye or on a microtiter plate reader at a wavelength of 590 nm.

There are also several commercially available rapid identification systems for conducting the analyses described above that are convenient and easy to use. Their usefulness depends on the database used in the identification. Bacteria are identified by comparison of the unknown with strains in the database. Most of the presently characterized strains are not soil inhabitants, thus the database may have limited value. Commercial tests include the API systems (Analylab Products, Plainview, NY 11803)[1] and the Entero-Set 20 system (Fisher Scientific Co, Diagnostics Div., Orangeburg, NY 10962), which use plastic strips of microtubes containing dehydrated media. A dilute suspension of the organism to be tested is added to the microtubes to reconstitute the media and biochemical tests can be read within 24 h. Biolog plates (Biolog, Inc, Hayward, CA, 94545) use a 95 substrate system in microtiter plates and tetrazolium violet to colorimetrically detect metabolism within 24 h. Minitek system (BBL Microbiology Systems, Cockeysville, MD 21030) uses substrate impregnated disks in

[1]Trade names and company names are included for the benefit of the reader and do not imply endorsement or preferential treatment of the product by the USDA.

12-well plates. The test organisms are suspended in Minitek broth, added to each well, and incubated. PathoTec Rapid I-D system (General Diagnostics Div., Warner-Lambert Co., Morris Plains, NJ 07950) uses reagents impregnated on test strips. The strips are immersed in a suspension of the test organism and read after 4 h. Micro-ID system for Enterobacteriaceae (General Diagnostics Div., Warner-Lambert Co., Morris Plains, NJ 07950) uses filter paper disks impregnated with reagents and substrates on a card with 15 test chambers. A dilute suspension of a bacterium is used to inoculate the chambers and results can be read in 4 h. The Enterotube and Oxi-Ferm system (Roche Diagnostics, Nutley, NJ 07110) and Corning r/b Enteric Differential and N/F system (Corning Medical Microbiology Products, Roslyn, NY 11576) tests characteristics using compartmentalized test tubes containing agar media. The tubes are inoculated by stabbing a needle with the test organisms to the bottom of each tube then streaking the surface of the agar.

25–2.3.3 Data Analysis

Compare the data obtained from the plates with C utilization patterns from strains with known taxonomic descriptions found in the literature (Krieg & Holt, 1984) or from databases previously constructed. Use the reference and type strains for comparison.

Cluster the isolates using numerical taxonomy methods. For analysis of the test results, code the data in a numerical form constructed either as a binary code with growth represented by 1, no growth by 0, or a quantitative code, if optical density values are collected from the microtiter plate readings. Include in the analysis only those characters that showed differences. Any redundant characteristics, those that are all + or all −, should be excluded.

The phenotypic relationship between each pair of isolates can be determined. Include the type and reference strains in the calculations to define the clusters with the reference bacteria. Compare each isolate with another for positive (+, +; a) or negative (−, −; b) matches or mismatches (+, −; c). Construct coefficients to determine the relationship between pairs of isolates. The simple matching coefficient (S_{SM}), which considers all matches, either positive or negative, is defined as:

$$S_{SM} = (a+b)/(a+b+c)$$

The second equation, the Jaccard coefficient (S_J), considers only the positive matches and is equivalent to:

$$S_J = a/(a+c)$$

The S_{SM} coefficient includes a total match between isolates and estimates overall similarity. A large number of negative matches may erroneously reflect homogeneity as determined by this coefficient. The S_J coefficient reduces the chance of a false similarity by excluding the negative matches and considering only the positive matches. From either of these coefficients, similarity matrices are generated to develop cluster analyses that arrange the isolates according to their overall similarity. From this information, it is possible to construct dendrograms or branched diagrams if desired (Alderson, 1985; Sneath, 1978a).

25–3 FATTY ACID ANALYSIS

25–3.1 Principles

The lipid content of most bacteria is low, comprising < 5% of the total cell dry weight; however, lipids are essential components of every cell. The lipid component of the cell consists of free fatty acids, hydrocarbons, fatty alcohols, or other compounds of bound fatty acids such as phospholipids, peptidolipids, glycolipids, etc. Lipids have distinct structure and form characteristic, diverse patterns that are stable within taxa (Kates, 1972; O'Leary, 1975). Since the pioneering work in the late 1920s on bacterial lipids (Anderson & Chargaff, 1929), research has centered on the use of lipids in taxonomic investigations (Shaw, 1974; Drucker, 1976; Lechevalier, 1977; Moss, 1981). Lipid composition also can be used in ecological studies of bacterial communities (White et al., 1979; Bobbie & White, 1980).

Most of the data collected on lipid composition has concentrated on the fatty acids. The majority of fatty acids in bacteria range from 9 to 20 carbons in length. The profiles of these fatty acids may be characteristic of specific genera and species and thus, are reliable for species classification (Moss et al., 1974; Miller, 1982; Miller & Berger, 1985; DeBoer & Sasser, 1986). Fatty acid determinations are similar to nucleic acid determinations in taxonomic evaluations (Graham et al., 1990) and in general support the bacterial groupings in *Bergey's Manual* (Krieg & Holt, 1984; Lechevalier, 1977). Other lipids such as phospholipids, glycolipids, hydrocarbons, and isoprenoid quinones (Goldfine, 1972; Lechevalier, 1977; Collins & Jones, 1981; Guckert & White, 1986) can also be used in bacterial identification, but will not be covered in this chapter.

Fatty acids can easily be extracted and esterified with methanol to form fatty acid methyl esters (FAMEs) by several different methods (Bligh & Dyer, 1959; White & Frerman, 1967; Moss, 1981; Sasser, 1990). These FAMEs can then be analyzed quantitatively and qualitatively by high-resolution fused-silica capillary gas chromatography. This technique is useful for a large number of samples because it is rapid and can be automated.

Although the lipid composition of a bacterium is highly conserved, fatty acid content may vary with environmental conditions and standard conditions for growth and analysis are of paramount importance. Fatty acid patterns are influenced by many factors such as growth medium, culture age, incubation temperature and environment, and technique of analysis

(Marr & Ingraham, 1962; Cullen et al., 1971; Chang & Fulco, 1973; Mc-Garrity & Armstrong, 1975; Jacques & Hunt, 1980). Standardization of growth conditions is needed in any taxonomic investigation of isolates using fatty acid patterns and is extremely critical in comparison of unknown species with published databases. The purity of isolates and the inclusion of reference or type strains in the set of isolates analyzed are critical to the success of the identification.

25–3.2 Materials

1. Inoculation loop.
2. 13 by 100 mm Teflon-capped test tubes.
3. Water bath, variable temperature, 80 to 100 °C.
4. Vortexer.
5. Shaker.
6. Pasteur pipets.
7. GC vials.
8. Database libraries.
9. Centrifuge.
10. −20 °C freezer, optional.
11. Capillary column: 25 to 50 m by 0.2 mm i.d., fused silica column
12. Gas chromatograph: programmable 170 to 270 °C at 5 °C per minute equipped with flame ionizer detector and integrator.
13. Computer: for data manipulation and comparison to database libraries.

Reagents

1. Trypticase Soy broth.
2. Trypticase Soy Agar plates.
3. 4.0 N Sodium hydroxide (NaOH) in 50% methanol.
4. 6.0 N Hydrochloric acid (HCl) in 50% methanol.
5. Hexane and methyl tert-butyl ether (MTBE) (1:1, vol/vol).
6. 0.3 N Sodium hydroxide.
7. Highly purified methyl ester standards of straight chain saturated fatty acids from 9 to 20 carbons in length and five hydroxy acids.
8. Hydrogen or helium—carrier gas.

25–3.3 Procedure

Trypticase Soy medium is generally used to grow aerobic bacteria for fatty acid analysis. Grow the isolates to be tested in Trypticase Soy broth for 24 h at 25 to 28 °C. Centrifuge cells two to three times; wash each time by resuspension in 7.0 mM CaSO$_4$. An alternative procedure is to grow the cells on Trypticase Soy Agar plates for 24 h at 25 to 28 °C, followed by scraping the cell mass with sterile distilled water into a 13 by 100 mm test tube.

The procedure outlined below (Miller, 1982; Miller & Berger, 1985) allows all steps (saponification, methylation, and extraction of methyl esters) to be carried out in single Teflon-capped test tube. Along with the unknown isolates, reference strains and blanks need to be included in each group of samples. Rubber and plastic will contaminate the system; therefore, only Teflon and glass should be used. Reagents and solvents used should be of HPLC grade purity.

Saponify the cells by adding 1.0 mL of 4 N NaOH in 50% methanol to each tube. Cap the tube with a Teflon-coated cap, vortex, and place in 100 °C water bath for 5 min. Remove tube, and vortex for 10 s and then return tube to the water bath for another 30 min to complete the saponification. Remove tubes from the water bath and allow to cool at room temperature. Add 2 mL of 6.0 N HCl in 50% methanol and vortex. Recap the tubes and heat at 80 °C for 10 min to allow formation of the methyl esters of the free fatty acids. Add 1 mL hexane/methyl tert-butyl ether (1:1 vol/vol). Shake gently for 10 min. The fatty acid methyl esters will be in the organic (top) phase. Discard the aqueous (lower) phase using a Pasteur pipet. Add 3 mL of 0.25 N NaOH to the tube. Recap the tube and shake for 5 min. Place the extract of the hexane/ether organic (top) phase in a GC vial or Teflon-lined, capped, test tube and store the samples at −20 °C until analysis.

The gas chromatograph should contain a capillary column with hydrogen or helium as the carrier. A progressive temperature program ranging from 170 to 270 °C in 5 °C increments can be used for separation of the fatty acids. Inject 200 to 300 μL of the hexane/ether phase. Identify the peaks by comparison of their retention times on the column with the retention times of the methyl ester standards.

Databases for bacteria can be purchased by which pattern recognition programs allow bacterial identification from peak information (Microbial ID, Inc., Newark, DE 19711). The larger the number of strains in the database, the greater the success of correct identification. With these programs, identification is possible to the subspecies level.

Without the benefit of a computer program, identification is possible to the species level. Calculate the ratios of stable fatty acid pairs, such as 14:16, 15 Iso:15 anteiso, and compare these values with ratios from known strains. The methods of numerical taxonomy as described in section 25–2.3 also can be used to determine the similarity of isolates based on fatty acid profiles.

REFERENCES

Alderson, G. 1985. The application and relevance of non-hierarchic methods in bacterial taxonomy. p. 227–263. *In* M. Goodfellow et al. (ed.) Computer-assisted bacterial systematics. Academic Press, London.

Alvey, N.G., D.F. Banfield, R.I. Baxter, J.C. Gower, W.J. Krazanowski, P.W. Lane, P.K. Leech, J.A. Nelder, R.W. Payne, K.M. Phelps, C.E. Rogers, G.J.S. Ross, H.R. Simpson, A.D. Todd, R.W.M. Wedderburn, and G.N. Wilkinson. 1977. GENSTAT.

A general statistical program. The Statistics Department. Rothamsted Exp. Stn., Harpenden, UK.

Anderson, R.J., and E. Chargaff. 1929. The chemistry of the lipoids of *Tuberle bacilli*. VI. Tuberculostearic acid and phthioic acid from the acetone-soluble fat. J. Biol. Chem. 85:77.

Austin, B., and R.R. Colwell. 1977. Evaluation of some coefficients for use in numerical taxonomy of microorganisms. Int. J. Syst. Bacteriol. 27:204–210.

Ayers, S.H., P. Rupp, and W.T. Johnson. 1919. A study of the alkali-forming bacteria in milk. USDA Bull. 782. U.S. Gov. Print. Office, Washington, DC.

Bligh, E.G., and W.J. Dyer. 1959. A rapid method of total lipid extraction and purification. Can. J. Biochem. Phys. 37:911–917.

Bobbie, R.J., and D.C. White. 1980. Characterization of benthic microbial community structure by high-resolution gas chromatography of fatty acid methyl esters. Appl. Environ. Microbiol. 39:1212–1222.

Bochner, B.R., and M.A. Savageau. 1977. Generalized indicator plate for genetic, metabolic, and taxonomic studies with microorganisms. Appl. Environ. Microbiol. 33:434.

Chang, N.C., and A.J. Fulco. 1973. The effects of temperature and fatty acid structure on lipid metabolism in *Bacillus licheniformis*. Biochim. Biophys. Acta. 296:287.

Clark, F.E. 1965. The concept of competition in microbial ecology. p. 339–344. *In* K.F. Baker and W.C. Synder (ed.) Ecology of soil-borne plant pathogens. Univ. of California Press, Berkeley.

Collins, M.D., and D. Jones. 1981. Distribution of isoprenoid quinone structural types in bacteria and their taxonomic implication. Microbiol. Rev. 45:316–354.

Colwell, R.R. 1973. Genetic and phenetic classification of bacteria. Adv. Appl. Microbiol. 16:137–175.

Colwell, R.R., and B. Austin. 1981. Numerical taxonomy. p. 444–449. *In* P. Gerherdt (ed.) Manual of methods for general bacteriology. Am. Soc. for Microbiol., Washington, DC.

Cullen, J., M.C. Phillips, and G.G. Shipley. 1971. The effects of temperature on the composition and physical properties of the lipids of *Pseudomonas fluorescens*. Biochem. J. 125:733.

DeBoer, S.H., and M. Sasser. 1986. Differentiation of *Erwinia carotovora* spp. *carotovora* and *E. carotovora* spp. *atrosptica* on the basis of fatty acid composition. Can. J. Microbiol. 32:796–800.

Dixon, W.J., and M.B. Brown. 1979. BMDP-79 biomedical computer programs P-series. 2nd ed. Univ. of California Press, Berkeley.

Drucker, D.B. 1976. Gas-liquid chromatographic chemotaxonomy. Meth. Microbiol. 9:51–125.

Fahy, P.C., and A.C. Hayward. 1983. Media and methods for isolation and diagnostic tests. *In* P.C. Fahy and A.C. Hayward (ed.) Plant and bacterial diseases. Academic Press, Sydney.

Goldfine, H. 1972. Comparative aspects of bacterial lipids. p. 1–51. *In* A.H. Rose and D.W. Tempest (ed.) Advances in microbial physiology. Academic Press, New York.

Goodfellow, M., D. Jones, and F.G. Priest. 1985. Computer-assisted bacterial systematics. Academic Press, London.

Goor, M., J. Mergaert, L. Verdonck, C. Ryckaert, R. Vantomme, J. Swings, K. Kersters, and J. De Ley. 1984. The use of API systems in the identification of phytopathogenic bacteria. Med. Fac. Landbouwwet. Rijksuniv. Gent. 49:499–507.

Graham, J.H., J.S. Hartung, R.E. Stall, and A.R. Chase. 1990. Pathological, restriction-fragment length polymorphism, and fatty acid profile relationships between *Xanthomonas campestris* from citrus and noncitrus hosts. Phytopathology 80:829–836.

Guckert, J.B., and D.C. White. 1986. Phospholipid, ester-linked fatty acid analysis in microbial ecology. Proc. IV ISME 455–459.

Hugh, R., and E. Leifson. 1953. The taxonomic significance of fermentative versus oxidative metabolism of carbohydrates by various Gram⁻ bacteria. J. Bacteriol. 66:24–26.

Jacob, H.W. 1970. Redox potential. p. 92–123. *In* J.R. Norris and D.W. Ribbons (ed.) Methods in microbiology. Vol. 2. Academic Press, New York.

Jacques, N.A., and A.L. Hunt. 1980. Studies on cyclopropane fatty acid synthesis: Effect of carbon source and oxygen tension on cyclopropane fatty acid synthetase activity in *Pseudomonas denitrificans*. Biochem. Biophys. Acta. 619:453–470.

Johnson, K.G., M.C. Silva, C.R. MacKenzie, H. Schneider, and J.D. Fontana. 1989. Microbial degradation of hemicellulosic materials. Appl. Biochem. Biotechnol. 20:245–258.

Kates, M. 1972. Techniques of lipidology. Isolation, analysis and identification of lipids. Elsevier, New York.

Kaufmam, L., and D.L. Massart. 1981. MASLOC user's guide. Vrije Universiteit, Brussels.

Krieg, N.R., and J.G. Holt. 1984. Bergey's manual of systematic bacteriology. Vol. 1. Williams and Wilkens, Baltimore.

Lechevalier, M.P. 1977. Lipids in bacterial taxonomy—A taxonomist's view. Crit. Rev. Microbiol. 5:109–210.

Lederberg, J. 1948. Detection of fermentative variants with tetrazolium. J. Bacteriol. 56:695.

Lockhart, W.R., and J. Liston. 1970. Methods for numerical taxonomy. Am. Soc. for Microbiol., Bethesda, MD.

Marr, A.G., and J.L. Ingraham. 1962. Effect of temperature on the composition of fatty acids in *Escherichia coli*. J. Bacteriol. 84:1260–1267.

Matin, A., E.A. Auger, P.H. Blum, and J.E. Schultz. 1989. Genetic basis of starvation survival in nondifferentiating bacteria. Annu. Rev. Microbiol. 43:293–316.

McGarrity, J.T., and J.B. Armstrong. 1975. The effect of salt on phospholipid fatty acid composition in *Escherichia coli* K-12. Biochem. Biophys. Acta. 398:258–264.

McKinley, V.L., T.W. Federle, and J.R. Vestal. 1982. Effects of hydrocarbons on the plant litter microbiota of an Arctic lake. Appl. Environ. Microbiol. 43:129–135.

Mergaert, J., L. Verdonck, K. Kersters, J. Swings, J.M. Boeufgras, and J. De Ley. 1984. Numerical taxonomy of *Erwinia* species using API systems. J. Gen. Microbiol. 130:1893–1910.

Miller, L.T. 1982. A single derivitization method for bacterial fatty acid methyl esters including hydroxy acids. J. Clin. Microbiol. 16:584–586.

Miller, L.T., and T. Berger. 1985. Bacteria identification by gas chromatography of whole cell fatty acids. Hewlett-Packard Application Note. p. 228–241. Hewlett-Packard Co., Wilmington, DE 19808.

Morita, R.Y. 1982. Starvation-survival of heterotrophs in the marine environment. Adv. Microb. Ecol. 6:171–198.

Moss, C.W. 1981. Gas-liquid chromatography as an analytical tool in microbiology. J. Chromatogr. 203:337–347.

Moss, C., M.A. Lambert, and W.H. Merwin. 1974. Comparison of rapid methods for analysis of bacterial fatty acids. Appl. Microbiol. 28:80–85.

Nie, N.H., C.H. Hull, J.G. Jenkins, K. Steinbrenner, and D.H. Brent. 1975. SPSS: Statistical package for social sciences. McGraw-Hill, New York.

O'Leary, W.M. 1975. The chemistry of microbial lipids. CRC Crit. Rev. Microbiol. 4:41.

Pergram, R.G. 1969. The microbial uses of 2,3,5-tri phenyltetrazolium chloride. J. Med. Lab. Technol. 26:175–198.

Rovira, A.D. 1965. Plant root exudates and their influence upon soil microorganisms. p. 170–186. *In* K.F. Baker and W.C. Synder (ed.) Ecology of soil-borne plant pathogens. Univ. of California, Berkeley.

Sasser, M. 1990. Identification of bacteria through fatty acid analysis. p. 199–204. *In* Z. Klement et al. (ed.) Methods in phytobacteriology. Akademiai, Kiado, Budapest.

Shaw, N. 1974. Lipid composition as a guide to the classification of bacteria. Adv. Appl. Microbiol. 17:63.

Sneath, P.H.A. 1978a. Classification of microorganisms. p. 9/1–9/31. *In* J.R. Norris and M.H. Richmond (ed.) Essays in microbiology. John Wiley, Chichester, UK.

Sneath, P.H.A. 1978b. Identification of microorganisms. p. 10/1–10/32. *In* J.R. Norris and M.H. Richmond (ed.) Essays in microbiology. John Wiley, Chichester, UK.

Sneath, P.H.A., and R.R. Sokal. 1973. Numerical taxonomy. W.H. Freeman and Co., San Francisco.

Sneath, P.H.A., M. Stevens, and M.J. Sackin. 1981. Numerical taxonomy of *Pseudomonas* based on published records of substrate utilization. Antonie van Leeuwenhoek; J. Microbiol. Serol. 47:423–448.

Sokal, R.R., and P.H.A. Sneath. 1963. Principles of numerical taxonomy. W.H. Freeman and Co., San Francisco.

Veldkamp, H., H. Van Gemerden, W. Harder, and H.J. Laanbroek. 1984. Competition among bacteria: An overview. p. 279–290. *In* M.J. Klug and C.A. Reddy (ed.) Current perspectives in microbial ecology. Am. Soc. for Microbiol., Washington, DC.

White, D.C., W.M. Davis, J.S. Nickels, J.S. King, and R.J. Bobbie. 1979. Determination of the sedimentary microbial biomass by extractable lipid phosphate. Oecologia 40:51–62.

White, D.C., and F.E. Frerman. 1967. Extraction, characterization, and cellular localization of the lipids of *Staphylococcus aureus*. J. Bacteriol. 94:1854–1867.

Willcox, W.R., S.P. LaPage, and B. Holmes. 1980. A review of numerical methods in bacterial identification. Antonie van Leewenhoek; J. Microbiol. Serol. 46:233–299.

Wishart, D. 1978. CLUSTAN user manual. 3rd ed. Univ. of Edinburgh, Edinburgh.

Chapter 26

Multilocus Enzyme Electrophoresis Methods for the Analysis of Bacterial Population Genetic Structure

B. D. EARDLY, *Penn State University Berks Campus, Reading, Pennsylvania*

The field of population biology can be divided into two disciplines: population genetics and population ecology (Hedrick, 1984). Studies in population genetics address questions on the genetic composition of populations, whereas studies in population ecology focus on environmental and biological factors influencing the numbers of organisms in populations. Several texts have been written on the ecology of soil bacteria (e.g., Atlas & Bartha, 1987; Campbell, 1984), whereas the existing literature on the genetic structure of bacterial populations in soils is limited to several studies (Demezas et al., 1991; Denny et al., 1988; Eardly et al., 1990; Engvild et al., 1990; McArthur et al., 1988; Pinero et al., 1988; Segovia et al., 1991; Young, 1985). Genetic structure is a useful, although somewhat ambiguous term, that is used to describe both the phylogenetic relatedness and the genetic diversity that exists in natural populations of organisms. Most studies examining the genetic structure of natural populations are intended to reveal basic information on the evolutionary development of a particular group or species. Oftentimes this involves studies of many subpopulations that may be taken to represent an entire taxon. Methods that are used to study the genetic structure in populations, such as multilocus enzyme electrophoresis (or MLEE), should not be confused with other methods where the primary aim is to categorize strains on the basis of a particular characteristic that may be of immediate practical significance (e.g., antibiotic resistance to track a gene in a population). An understanding of the genetic structure of a bacterial population is important for several reasons; it can provide insight into genetic relationships that exist among members of a population, and it can provide a sound framework for designing further ecological, physiological, and genetic studies of that population.

The relatively small number of published studies on the population genetics of soil bacteria can be explained by the fact that, prior to 1980, there were no rapid methods for identifying stable multilocus genotypes in soil populations. Within the past decade however, molecular genetics techniques have made it possible to efficiently identify multilocus genotypes in bacteria. Only recently have these techniques been applied to soil populations. Thus, soil microbiologists are now applying the principles of population genetics (developed through the study of eukaryotic populations) to the study of prokaryotic populations in the soil. One practical consequence of these developments is that researchers are beginning to integrate both ecological and genetic approaches in their studies of soil populations (McArthur et al., 1988). By doing so, they are broadening the scope of the questions that they can ask about these populations.

Since the chromosome is the fundamental genetic element in the bacterial cell, most studies on the genetic structure in bacterial populations have focussed on chromosomal variation. The bacterial chromosome is structured as a functional patchwork of genes and gene families (Smith et al., 1991). On a finer scale, each gene is composed of a mosaic of segments, with each segment having its own unique phylogenetic history. Because of this complicated genomic arrangement, the most informative population genetic analysis methods rely on composite assessments of genetic variation involving analyses of a large number of structural genes (in numerous strains).

Studies with *E. coli* have identified more than 30 genes coding for important metabolic enzymes that have proven to be useful in MLEE analyses. It appears that these genes are distributed more-or-less randomly around the bacterial chromosome (Whittam et al., 1983). Thus a survey of genetic variation in these genes provides a reliable index of genetic variation in a given population. Other genetic characterization methods, such as restriction fragment length polymorphism analysis (see chapter 31 by Sadowsky in this book) can also be used for this type of approach, however the amount of genetic coding information that can be surveyed with these methods is relatively small. There are also other genetic characterization methods designed to estimate genetic similarity between entire genomes (e.g., DNA/DNA hybridization and restriction fragment fingerprinting; see chapters 31 and 32 by Ogram in this book), however these methods usually rely on a single estimate of overall similarity, and the results may be influenced by other complicating factors (such as the presence or absence of large plasmids). In sum, methods other than MLEE tend to lack the unique combination of resolving power and versatility that is afforded by multilocus analyses of a large number of genes.

The MLEE method has been especially informative in studies of the population genetics and epidemiological characteristics of human pathogens (for a review, see Selander and Musser, 1990). In practical application of this method, numerous strains representing a particular species are characterized by electrophoretic analyses of up to 35 metabolic enzymes. These enzymes are typically components of important metabolic pathways (e.g., glucose 6-phosphate dehydrogenase or malate dehydrogenase) and thus are widely distributed across diverse genera. The electrophoretic variation observed among the enzymes is equated with allelic variation in the genes coding for these proteins.

Subsequent interpretation of the MLEE data is based on the premise that, in the absence of genetic recombination, identical electromorphs (enzyme proteins having identical electrophoretic migration characteristics) occur only in strains that are closely related by descent. If two strains share most of their enzyme electromorphs, then it is assumed that the strains share a recent common ancestor. If their electromorph profiles are identical (i.e., they have the same electrophoretic type or ET), then they are presumed to be clones of the same strain.

The MLEE method has distinct advantages over methods that rely on phenotypic differences among bacterial strains (e.g., morphological variation or environmental stress resistance) in that the MLEE method provides an unbiased assessment of population genetic structure. This assertion is based on studies that have shown that specific enzyme electromorphs do not appear to provide a selective advantage for the strains in which they reside (Dykhuizen & Hartl, 1983), nor do they appear to be subject to convergent evolution (Kimura, 1983).

For the most part, the basic techniques and enzyme staining systems that are described here have already been described elsewhere (Harris & Hopkinson, 1976; Selander et al., 1986). This chapter was written primarily to provide specific information on MLEE techniques that have proven to be especially useful in the study of *Rhizobium* and other soil bacteria (e.g., Denny et al., 1988).

26–1 PRINCIPLES

Assuming that all organisms evolved from a single common ancestor, it is possible to estimate genetic relatedness between individuals by comparing allelic variation among several highly conserved genes. Because the net electrostatic charge of a protein (and hence its rate of migration during electrophoresis) is determined by amino acid sequence, mobility variants (electromorphs) of an enzyme can be directly equated with allelic variation in the gene coding for the protein. Through the use of MLEE, the genetic relatedness between isolates is estimated by comparing the electrophoretic mobilities of several metabolically important, water-soluble enzymes (Selander et al., 1986).

The genes examined in MLEE analyses are usually "housekeeping" enzyme genes, or genes that occur in basic metabolic pathways. The most informative MLEE studies involve analysis of electrophoretic variation at 15 to 30 enzyme loci. By examining variation at such a large number of loci, it is possible to accurately estimate both genetic diversity and phylogenetic relationships within and among natural populations of bacteria.

Aside from their metabolic importance, the genes that are commonly used for MLEE analyses usually have three other characteristics in common; (i) they are chromosomally encoded, (ii) they are present as single copies in the genome, and (iii) their products (enzymes) are easily identified (stained) in gels following electrophoresis. Plasmid-encoded "housekeeping" enzyme genes have been encountered in some MLEE studies (e.g., Eardly et al., 1990) however, recent studies of plasmid-cured strains of *R. leguminosarum* bv. *trifolii* (unpublished data) and *R. leguminosarum* bv. *phaseoli* (Young, 1985), indicate that plasmid-encoded housekeeping enzymes are relatively rare.

The choice of enzymes for a MLEE analysis depends on the species being examined. For example, many enzymes that are routinely analyzed in MLEE studies of *Rhizobium,* show no activity in similar assays of *Bradyrhizobium* (unpublished data). Because of this inconsistency across taxa, MLEE studies of a "new" species should be preceded by preliminary experiments to determine; (i) which enzymes will provide good activity, and (ii) the optimal assay conditions for these enzymes. Experience has shown that while some buffer systems produce distinct bands for certain enzymes (desirable), other buffer systems may produce diffuse or multiple bands (undesirable) for the same enzymes.

If published MLEE information for a particular species (or a close relative) is not available, preliminary screening experiments should be conducted to select the best enzymes and buffer systems for that particular species. The variables involved in this selection process include: strain genotype, enzyme, and buffer system. For the sake of efficiency, only two or three representative strains should be examined in these preliminary analyses. During these analyses, lysates from this small set of strains can be placed in repetitive sets (e.g., five sets) across the same gel. After electrophoresis, the gel can then be sliced (with fine nylon thread) horizontally (e.g., to produce four identical slabs) and vertically (e.g., to produce five identical lateral sections). The result is a large number of identical "minigels" (the sample sectioning scheme described above would produce 20 identical minigels), that can each be stained for a different enzyme. Using this approach, one could select optimal buffer conditions for 20 different enzymes using only four normal-size gels (one for each of the buffer systems listed in Table 26-1).

In all MLEE studies, it is advisable to examine as many enzymes as physically possible. For example, if only three or four enzymes are used in an analysis of a soil population, it is possible that actual differences between closely related strains may not be revealed. Allelic variation resulting from silent substitution at the level of DNA, and electrophoretically neutral substitution at the level of the polypeptide, are not revealed by MLEE analysis. A related consideration associated with the use of a small number of loci is that statistical clustering methods may produce misleading results, as these methods require a large number of loci to provide a reasonable degree of statistical confidence. At least 10, and preferably 15 loci should be examined in MLEE analyses to avoid these pitfalls. In

Table 26–1. Buffer systems for electrophoresis of bacterial enzymes (Selander et al., 1986).

System	Electrode buffer	Gel buffer	Voltage
TC 6.7	Tris-citrate (pH 6.3): 27.0 g Tris $(C_4N_{11}NO_3)$, 18.1 g of citric acid monohydrate $(C_6H_7O_7 \cdot H_2O)$, 1.0 L water; pH adjusted with sodium hydroxide (NaOH).	Tris-citrate (pH 6.7): 0.97 g of Tris, 0.63 g of citric acid monohydrate, 1.0 L of water; pH adjusted with sodium hydroxide.	250
TC 8.0	Tris-citrate (pH 8.0): 83.0 g Tris, 33.1 g of citric acid monohydrate, 1.0 L water.	Tris-citrate (pH 8.0): electrode buffer diluted 1:29.	130
TC 8.7	Borate (pH 8.2): 18.5 g of boric acid (H_3BO_3), 2.4 g of sodium hydroxide, 1.0 L water.	Tris-citrate (pH 8.7): 9.21 g of Tris, 1.05 g of citric acid monohydrate, 1.0 L of water.	250
TM 7.4	Tris-maleate (pH 7.4): 12.1 g of Tris, 11.6 g of maleic acid $(C_4H_4O_4)$, 3.72 g of ethylenediaminetetraacetic acid (EDTA) disodium salt dihydrate $(C_{10}H_{14}N_2O_8Na_2 \cdot 2H_2O)$, 2.03 g of magnesium chloride $(MgCl_2 \cdot 6H_2O)$, 1.0 L of water; pH adjusted with 5.0 g of sodium hydroxide.	Tris-maleate (pH 7.4): electrode buffer diluted 1:9.	100

selecting specific enzymes to be examined, it is best to use those that have been used widely in previous studies. This approach will not only facilitate comparison of data across studies, but it is also likely that the enzymes were originally chosen because they were readily stained and scored.

In addition to providing information on genetic relatedness and genetic diversity in natural bacterial populations, MLEE data are useful in assessing the importance of previous chromosomal recombination in a population. These analyses are based on the premise that, when blocks of genes recombine repeatedly in a population, alleles at different enzyme loci eventually become randomized. If allelic combinations are found in a random assortment in a population, then the loci are said to exist in a state of linkage equilibrium. This type of randomization would not occur if members of the population were strictly clonal. When the distribution of alleles deviates strongly from a random assortment (i.e., blocks of certain allele combinations dominate entire populations), then the enzyme loci are said to exist in a state of linkage *dis*equilibrium, implying a clonal population structure.

When polymorphic enzymes (those having multiple electromorphs) are present in a population of bacteria (as is the case in most natural populations), recombination can generate an enormous number of multilocus genotypic combinations (Selander & Levin, 1980). However, recent population studies of human pathogens (Selander & Musser, 1990) and soil bacteria (Young, 1985) have shown that only relatively few multilocus genotypes (ETs) are normally encountered. The simplest explanation for this observation is that recombination is extremely rare in these populations. This conclusion is important since it indicates that: (i) it is possible to identify distinct chromosomal lineages in these populations, and (ii) it is possible to develop conceptual frameworks describing the genetic relationships among these lineages.

Genetic relatedness or genetic distance (D) between strains in bacterial populations may be expressed as the proportion of loci at which dissimilar alleles occur, i.e., the proportion of mismatches between ETs (electrophoretic types). From a matrix of coefficients of genetic distance (i.e., an ET similarity matrix), a cluster analysis may be used to illustrate overall genetic relatedness among strains (see Selander et al., 1986 for sample calculations). Genetic distance dendograms summarizing the results of cluster analyses are a simple way to visualize the results contained in an ET similarity matrix.

For example, if two isolates have identical enzyme electromorphs at 10 out of 20 loci (enzymes), their ETs will be connected by a branch on a dendogram at a genetic distance of 0.5 (halfway across a dendogram on a genetic distance scale ranging from 0.0–1.0). In interpreting this type of dendogram (with genetic distance on the abscissa), it is important to realize that all branches may be rotated about their horizontal axes, and that vertical lists of ETs are often arranged as a matter of convenience (e.g., type strains are usually placed first).

Genetic diversity estimates can be calculated for populations and individual loci using MLEE allele frequency data (Selander et al., 1986). In addition to providing genetic diversity estimates, this data can also be used to generate statistics apportioning genetic diversity both within and between subpopulations (Nei, 1977; Whittam et al., 1983). This type of information has proven to be useful in assessing the influence of certain environmental factors on population genetic structure (Harrison et al., 1989; McArthur et al., 1988; Pinero et al., 1988).

26–2 METHODS

26–2.1 Preparation of Enzyme Extracts

26–2.1.1 Materials

1. Early stationary phase (approximately o.d. 1.0) broth cultures (150 mL). See chapter 8 by Zuberer in this book for methods of obtaining pure cultures of soil bacteria.
2. Plastic centrifuge bottles (250 mL).
3. High speed centrifuge tubes (6.5 mL).
4. High speed centrifuge with rotors for 250 mL bottles and 6.5 mL tubes.
5. Disposable transfer pipets (2 mL).
6. Ice-cold resuspension buffer (pH 6.8): 10 mM Tris ($C_4N_{11}NO_3$); 1 mM ethylenediaminetetraacetic acid (EDTA), disodium salt, dihydrate ($C_{10}H_{14}N_2O_8Na_2 \cdot 2H_2O$); 0.5 m$M$ nicotinamide adenine dinucleotide phosphate (NADP), sodium salt ($C_{21}H_{27}N_7O_{17}P_3Na$).

7. Conical centrifuge tubes, plastic (20 mL).
8. Sonicator with microtip and ice bath.
9. Ultracold (-70 °C) freezer and cryo-storage vials.

26–2.1.2 Procedures

Harvest the broth culture cells by centrifugation at 15 000 × g for 10 min. Resuspend the pellet in 1.5 mL of resuspension buffer and transfer the suspension to conical centrifuge tube (on ice). Lyse the cells by sonication using a 50% pulse for 15 s (on ice). Remove cell debris from the lysate by refrigerated (4 °C) centrifugation at 30 000 × g for 20 min. Transfer clear supernatant to cryo-storage vials, and store at -70 °C.

26–2.1.3 Comments

To minimize loss of enzyme activity, lysates should be maintained on ice at all times. Practical experience has shown that conical tubes tend to reduce foaming (and resulting loss of enzyme activity) during sonication. Transferable gum labels should also be used to avoid repetitive labeling and associated errors. Lysates may be repeatedly thawed and refrozen without a noticeable loss of activity (for most enzymes). Usually enzymes retain activity for several weeks (to months), however a few enzymes (e.g., aconitase in *Rhizobium*) lose activity over several days in storage, and thus should be analyzed immediately.

26–2.2 Electrophoresis

26–2.2.1 Materials

1. Horizonal gel electrophoresis apparatus (see comments).
2. Starch gel mold, heavy plastic (or glass), approximately 190 by 210 by 9 mm (must be able to tolerate 100 °C).
3. Electrophoresis ice bath pan, with thin glass plate support (approximately same length and width as gel mold).
4. Hydrolyzed potato starch (STARCHart Corp., P.O. Box 268, Smithville, TX 78957).
5. Vacuum pump.
6. Erlenmeyer flask (1 L).
7. Gel buffers (see Table 26–1).
8. Saran wrap or plastic film.
9. Whatman no. 3 filter paper, cut into lysate wicks (9 by 6 mm) and blotting sheets (approximately 10 by 10 cm).
10. Forceps.
11. Thawed lysates (on ice).
12. Amaranth tracking dye ($C_{20}H_{11}N_2Na_3O_{10}S_3$) solution, 1% (wt/vol).
13. Constant voltage power supply.

14. Electrode buffer wicks, approximately 10 by 20 cm. These can be either thin cellulose sponges (O-Cell-O Sponges, General Mills Inc., Tonawanda, NY 14150), reusable paper wipes, or layered no. 3 filter paper.
15. Microspatula or knife.
16. Plastic storage/staining boxes, 186 by 135 mm by 48 mm deep (Durphy Packaging, Huntington Valley, PA 19006).

26–2.2.2 Procedures

Starch gels should be prepared 6 to 24 h before use. To prepare a 190 by 210 by 9 mm gel, a suspension of 49 g hydrolyzed potato starch in 420 mL gel buffer (Table 26–1) is heated (with vigorous swirling) over a bunsen burner in a 1 L Erlenmeyer flask just beyond boiling (Selander et al., 1986). The clear suspension is then aspirated for 1 min (or until very large bubbles appear) and immediately poured into a gel mold. The gel is allowed to cool at room temperature for approximately 2 h. The gel should then be covered with plastic film and refrigerated (to avoid dessication).

Prior to gel loading, lysates should be partially thawed in tap water and maintained in an ice bath. To prepare the gel for loading, a vertical slit is cut the width (and depth) of the gel, approximately 2 cm from the top (or cathodal) edge of the gel. To load the gel, filter paper wicks are dipped briefly in lysate, and blotted on a sheet of filter paper. While still moist, the wicks are then inserted into the gel slit, in order, starting approximately 1 cm from the left edge of the gel (leaving approximately 1–2 mm between wicks). To facilitate band scoring, at least two sets of standard lysates should be interspersed among the other lysates, as the buffer migration front is slightly curvilinear in most starch gels. Standard lysates may be mixed to minimize the number of lanes required for standards. Finally, tracking dye wicks should be placed at the ends of the gel slit to identify the migration front.

After loading, the gel is placed between electrode buffer reservoirs, with the loading slit (nearest the top edge of the gel) closest to the cathodal reservoir (most enzymes are anions at the pH's used). Electrode buffer reservoirs are then filled with the appropriate buffer (Table 26–1) to a point 1 to 2 cm below the edge of the gel. One side of each electrode buffer wick is then placed in the buffer and the other side slightly overlapping (1–2 cm) the nearest edge of the gel. The plastic film (originally used to cover the gel) is placed over the gel and wicks to prevent dehydration. During electrophoresis, a constant voltage is applied (Table 26–1) and the gel is cooled either by placing an ice pan on the gel (preferred), or by running the gel in a cold room (which tends to shorten the lifespan of the power supply). Electrophoresis is continued until the dye front has migrated approximately two-thirds of the way toward the anodal edge of the gel. This may take from 2 to 8 h, depending on the buffer system used (e.g., Tris-citrate pH 6.7 vs. Tris maleate pH 7.4, respectively).

Table 26–2. Useful enzymes and buffer systems for *Rhizobium*.

Enzyme			Buffer system†
Name	EC no.	Symbol	
Aconitase	4.2.1.3	ACO	TC 6.7
Adenylate kinase	2.7.4.3	ADK	TC 8.7
Alcohol dehydrogenase	1.1.1.1	ADH	TC 8.0
β-Galactosidase	3.2.1.23	BGA	TC 6.7
Glucose 6-phosphate dehydrogenase	1.1.1.49	G6P	TC 8.0
Hexokinase	2.7.1.1	HEX	TC 8.7
3-Hydroxybutyrate dehydrogenase‡	1.1.1.30	HBD	TC 6.7
Hypoxanthine dehydrogenase	1.2.3.2	XDH	TC 8.0
Indophenol oxidase (superoxide dismutase)	1.15.1.1	IPO	TC 8.7
Isocitrate dehydrogenase‡	1.1.1.42	IDH	TC 8.0
Leucine aminopeptidase	3.4.1.1	LAP	TC 8.0
Leucine dehydrogenase	1.4.3.2	LED	TC 8.7
DL-leucyl-DL-alanine peptidase‡	3.4.x.x	LAL	TC 8.0
Leucyl-glycyl-glycine peptidase	3.4.x.x	LGG	TM 7.4
Lysine dehydrogenase	1.4.3.x	LYD	TC 8.0
Malate dehydrogenase‡	1.1.1.37	MDH	TC 6.7
Nucleoside phosphorylase	2.4.2.1	NSP	TM 7.4
6-Phosphogluconate dehydrogenase	1.1.1.44	6PG	TC 6.7
Phosphoglucose isomerase	5.3.1.9	PGI	TC 8.0
Phosphoglucomutase	2.7.5.1	PGM	TC 8.7
L-phenylalanyl-L-leucine peptidase	3.4.x.x	PLP	TM 7.4
Shikimate dehydrogenase	1.1.1.25	SKD	TM 7.4

† See Table 26–1 for an explanation of buffer systems.
‡ Useful enzymes and buffer systems for *Bradyrhizobium*.

After electrophoresis, the cathodal portion of the gel (including the 1–2 mm in contact with the loading wicks) and the anodal portion of the gel (below the migration front) are trimmed with a spatula and discarded. The remaining (central) portion of the gel is sliced with a nylon thread, to provide three or four identical slabs (each 1–2 mm thick). Prior to slicing, it is important to remove any air bubbles underneath the gel. The top slice is usually discarded, as the lower slices usually produce the most distinct bands. These slices are held at 4 °C in plastic boxes until they are stained.

26–2.2.3 Comments

A list of useful enzymes and compatible buffer systems (corresponding to Table 26–1) for *Rhizobium* are listed in Table 26–2. Other buffer systems may be substituted, however, results may differ depending on the buffer-enzyme combination used (Segovia et al., 1991).

In selecting an apparatus for starch gel electrophoresis, there are several important considerations. Perhaps the primary consideration is expense, followed closely by the number of gels that can be efficiently handled in a particular laboratory situation. Because of the large sample sizes that are frequently encountered in bacterial population genetics studies, and because of the large number of reruns that are usually necessary to compare the range of different ETs on the same gel, it is often desirable to run a large number of gels at the same time. To achieve this goal economically, suitable electrophoresis devices can usually be constructed from materials available at the local hardware and electrical supply stores (e.g., drawer organizer trays). If this approach is used, careful attention should be given to the electrical hazards that are associated with open buffer reservoirs. Several different designs for starch gel electrophoresis devices are available in the literature (Brewer & Sing, 1970; Harris & Hopkinson, 1976; May, 1980).

Interestingly, there are few commercial starch gel electrophoresis devices available, and those that are, are multiple-use devices that tend to be relatively high priced. Although it seems that typical submarine agarose electrophoresis devices should suffice, the gel trays in these are not designed to tolerate the high temperature of molten starch. Furthermore, the electrode buffer reservoirs in these apparatuses are designed to be filled to capacity, with the electrode buffer forming the electrical bridge between the reservoirs. In starch gel electrophoresis, the gel itself carries the electrical current. If standard submarine agarose electrophoresis devices are to be used, some modifications (heavier gel trays and partially filled buffer reservoirs) will be necessary.

26–2.3 Enzyme Staining

26–2.3.1 Materials (Selander et al., 1986)

1. cis-Aconitic acid ($C_6H_6O_6$).
2. Adenosine 5'-diphosphate or ADP ($C_{10}H_{15}N_5O_{10}P_2$).
3. Agar, Difco Bacto.
4. Adenosine 5'-triphosphate or ATP ($C_{10}H_{14}N_5O_{13}P_3Na_2$).
5. Ethanol (C_2H_5OH).
6. Fast Black K salt ($C_{14}H_{12}N_5O_4 \cdot 1/2ZnCl_4$).
7. o-Dianisidine dihydrochloride ($C_{14}H_{16}N_2O_2 \cdot 2HCl$).
8. Fructose 6-phosphate, disodium salt ($C_6H_{11}O_9PNa_2$).
9. β-(D) + Glucose ($C_6H_{12}O_6$).
10. Glucose 1-phosphate, disodium salt, hydrate ($C_6H_{11}O_9PNa_2$).
11. Glucose 6-phosphate, disodium salt, hydrate ($C_6H_{11}O_9PNa_2$).
12. Glucose 6-phosphate dehydrogenase, 10 U/mL (Sigma Chemical G 8878).
13. 0.25 M Glycine/carbonate (pH 10) buffer: 1.88 g of glycine ($C_2H_5NO_2$) and 2.65 g of sodium carbonate (Na_2CO_3) in 100 mL of water.
14. 0.10 M Glycyl/glycine (pH 7.5) buffer: 11.3 g of glycine in 1 L of water. Adjust pH with 1 M potassium hydroxide (KOH).
15. Hexokinase (Sigma Chemical H 5125).
16. DL-β-Hydroxybutyric acid, sodium salt ($C_4H_7O_3Na$).

17. Inosine ($C_{10}H_{12}N_4O_5$).
18. 1 M Isocitric acid solution: 2.94 g of DL-isocitric acid, trisodium salt ($C_6H_5O_7Na_3$) in 100 mL of water.
19. Isocitrate dehydrogenase, 50 U/mL (Sigma Chemical I 2002).
20. Isopropanol (C_3H_8O).
21. L-Leucine, crystalline ($C_6H_{13}NO_2$).
22. L-leucine-β-napthylamide hydrochloride ($C_{16}H_{20}N_{20}$·HCl).
23. DL-Leucyl-DL-alanine ($C_9H_{18}N_2O_3$).
24. Leucyl-glycyl-glycine ($C_{10}H_{19}N_3O_4$).
25. L-Lysine ($C_6H_{14}N_2O_2$), free base, crystalline.
26. 2.0 M Malic acid solution: 1 L of water, 268 g of DL-malic ($C_4H_6O_5$) acid, and 160 g of sodium hydroxide (NaOH), pH 7.0. Caution: potentially explosive reaction.
27. 0.1 M Magnesium chloride solution: 2.03 g of magnesium chloride hexahydrate ($MgCl_2$·$6H_2O$) in 100 mL of water.
28. 0.25 M Manganese chloride solution: 4.90 g of manganese chloride tetrahydrate ($MnCl_2$·$4H_2O$) in 100 mL of water.
29. 4-Methylumbelliferyl β-galactoside ($C_{16}H_{18}O_8$).
30. MTT solution: 1.25 g of dimethylthiazol tetrazolium bromide ($C_{18}H_{16}N_5SBr$) in 100 mL of water.
31. Nicotinamide adenine dinucleotide (NAD) solution: 1 g of NAD-free acid ($C_{21}H_{27}N_7O_{14}P_2$) in 100 mL of water.
32. Nicotinamide adenine dinucleotide phosphate (NADP) solution: 1 g of NADP sodium salt ($C_{21}H_{27}N_7O_{17}P_3Na$) in 100 mL of water.
33. Peroxidase (Sigma Chemical P 8125).
34. L-Phenylalanyl-L-leucine ($C_{15}H_{22}N_2O_3$).
35. Phosphate-citrate (pH 5.0) buffer: mix 2.0 mL 1.0 M phosphoric acid (H_2PO_3), 10.2 mL 2.0 M sodium hydroxide (NaOH), and 1.03 g citric acid monohydrate ($C_6H_8O_7$·H_2O) in 76.9 mL of water.
36. 6-Phosphogluconic acid, barium salt ($C_6H_{10}O_{10}P$·3/2Ba).
37. PMS solution: 1 g phenazine methosulfate ($C_{13}H_{11}N_2$·CH_3SO_4) in 100 mL of water.
38. 0.1 M Potassium phosphate (pH 5.5) buffer: 13.6 g of potassium phosphate, monobasic (KH_2PO_4) in 1 L water; adjust pH with sodium hydroxide (NaOH).
39. Shikimic acid, crystalline ($C_7H_{10}O_5$).
40. Snake venom (Sigma Chemical V 7000).
41. Sodium chloride (NaCl).
42. Sodium phosphate (pH 7.0) buffer: mix equal volumes of; 27.6 g of sodium phosphate, monobasic (NaH_2PO_4) in 1 L water; with 53.6 g sodium phosphate, dibasic (Na_2HPO_4) in 1 L water; dilute the mixture 1:25 with water.
43. 0.2 M Tris hydrochloride (pH 8.0) buffer: 24.2 g of Tris ($C_4N_{11}NO_3$) in 1 L of water, adjust pH with hydrochloric acid.
44. Xanthine oxidase, 2 U/mL (Sigma Chemical × 4875).

26–2.3.2 Procedures

1. Aconitase (ACO). Stained by mixing the following ingredients (in order) in a small flask and pouring the solution over a gel slab in a staining box; 15 mL 0.2 M Tris HCl buffer, 30.0 mg of *cis*-aconitic acid, 10 mL 0.1 M MgCl$_2$ solution, 1.0 mL NADP solution, 0.1 mL isocitrate dehydrogenase solution, 0.5 mL PMS solution, 1.0 mL of MTT solution, and 500 mg of molten agar (60 °C) dissolved in 25 mL 0.2 M Tris HCl buffer. The gel is then incubated for 30 min to 1 h at 37 °C.

2. Adenylate kinase (ADK). Stained by mixing the following ingredients (in order) in a small flask and immediately pouring the solution over a gel slab in a staining box; 25 mL 0.2 M Tris HCl buffer, 25 mg of ADP, 100 mg of D-glucose, 1 mg (approximately) of hexokinase, 1 mL of 0.1 M MgCl$_2$ solution, 1 mL of NADP solution, 1.5 mL of glucose 6-phosphate dehydrogenase solution, 0.6 mL of PMS solution, 0.6 mL of MTT solution, and 500 mg of molten agar (60 °C) dissolved in 25 mL of 0.2 M Tris HCl buffer. The gel is then incubated for 30 min to 1 h at 37 °C.

3. Alcohol dehydrogenase (ADH). Stained by mixing the following ingredients (in order) in a small flask and immediately pouring the solution over a gel slab in a staining box; 50 mL of 0.2 M Tris HCl buffer, 3 mL of ethanol, 2 mL of isopropanol, 2 mL of NAD solution, 0.5 mL of PMS solution, and 1 mL of MTT solution. The gel is then incubated for 30 min to 1 h at 37 °C.

4. Beta-galactosidase (BGA). Stained by mixing the following ingredients (in order) in a small flask and immediately pouring the solution over a filter paper covering the gel slab in a staining box (Harris & Hopkinson, 1976); 50 mL of phosphate-citrate buffer and 9 mg of methylumbelliferyl β-galactoside. Any air bubbles under the filter paper are removed, and the gel is incubated for 30 min at 37 °C. After incubation, the gel should be sprayed lightly with glycine-carbonate buffer and viewed immediately on an ultraviolet transilluminator (with eye protection).

5. Glucose 6-phosphate dehydrogenase (G6P). Stained by mixing the following ingredients (in order) in a small flask and immediately pouring the solution over a gel slab in a staining box; 50 mL of 0.2 M Tris HCl buffer, 100 mg of glucose 6-phosphate, 1 mL of 0.1 M MgCl$_2$ solution, 1 mL of NADP solution, 0.5 mL of PMS solution, and 1.0 mL of MTT solution. The gel is then incubated for 30 min to 1 h at 37 °C.

6. Hexokinase (HEX). Stained by mixing the following ingredients (in order) in a small flask and immediately pouring the solution over a gel slab in a staining box; 50 mL of glycyl-glycine buffer, 200 mg of D-glucose, 50 mg of ATP, 1 mL of NADP solution, 1 mL of glucose 6-phosphate dehydrogenase solution, 2 mL of 0.1 M MgCl$_2$ solution, 0.5 mL of PMS solution, and 1 mL of MTT solution. The gel is then incubated for 30 min to 1 h at 37 °C.

7. Hydroxybutyrate dehydrogenase (HBD). Stained by mixing the following ingredients (in order) in a small flask and immediately pouring

the solution over a gel slab in a staining box; 50 mL of 0.2 M Tris HCl, 200 mg DL-hydroxybutyrate, 200 mg of NaCl, 2 mL of NAD solution, 2 mL of 0.1 M MgCl$_2$ solution, 0.5 mL of PMS solution, and 1 mL of MTT solution. The gel is then incubated for 30 min to 1 h at 37 °C.

8. Hypoxanthine dehydrogenase (XDH). Stained by mixing the following ingredients (in order) in a small flask and immediately pouring the solution over a gel slab in a staining box; 50 mL of 0.2 M Tris HCl buffer, 100 mg of hypoxanthine, 2 mL of NAD solution, 0.5 mL of PMS solution, and 1 mL of MTT solution. The gel is then incubated for 30 min to 1 h at 37 °C.

9. Indophenol oxidase (IPO). Stained by mixing the following ingredients (in order) in a small flask and immediately pouring the solution over a gel slab in a staining box; 40 mL of 0.2 M Tris HCl buffer, 1 mL of 0.1 M MgCl$_2$ solution, 0.5 mL of PMS solution, and 1 mL of MTT solution. The gel is then incubated in the dark for 30 min at 37 °C and then exposed to bright light (e.g., standard transilluminator) for 30 min.

10. Isocitrate dehydrogenase (IDH). Stained by mixing the following ingredients (in order) in a small flask and immediately pouring the solution over a gel slab in a staining box; 50 mL of 0.2 M Tris HCl buffer, 2 mL of 0.1 M isocitric acid solution, 1 mL of NADP solution, 2 mL of 0.1 M MgCl$_2$ solution, 0.5 mL of PMS solution, and 1 mL of MTT solution. The gel is then incubated for 30 min to 1 h at 37 °C.

11. Leucine aminopeptidase (LAP). Stained by thoroughly mixing the following ingredients (in order) in a small flask for 10 min and immediately pouring the solution over a gel slab in a staining box; 50 mL of potassium phosphate buffer, 1 mL of 0.1 M MgCl$_2$ solution, 30 mg of L-leucine-β-napthylamide hydrochloride, and 30 mg Fast Black K salt. The gel is then incubated for 30 min to 1 h at 37 °C.

12. Leucine dehydrogenase (LED). Stained by mixing the following ingredients (in order) in a small flask and immediately pouring the solution over a gel slab in a staining box; 50 mL of sodium phosphate buffer, 50 mg L-Leucine, 2 mL of NAD solution, 0.5 mL of PMS solution, 1 mL of MTT solution. The gel is then incubated for 30 min to 1 h at 37 °C.

13. Leucyl-DL-alanine peptidase (LAL). Stained by gently mixing the following ingredients in a small flask for 10 min; 20 mg DL-leucyl-DL-alanine, 10 mg of peroxidase, 10 mg of O-dianisidine dihydrochloride, 10 mg of snake venom, 25 mL of Tris HCl buffer, and 0.5 mL 0.25 M MnCl$_2$ solution. After mixing, 500 mg of molten agar (60 °C) dissolved in 25 mL of 0.2 M Tris HCl buffer is added to the staining solution and the mixture is immediately poured over a gel slab in a staining box. The gel is then incubated for 30 min to 1 h at 37 °C.

14. Leucyl-glycyl-glycine peptidase (LGG). Stained by gently mixing the following ingredients in a small flask for 10 min; 20 mg of LGG, 10 mg of peroxidase, 10 mg of O-dianisidine dihydrochloride, 10 mg of snake venom, 25 mL 0.2 M Tris HCl buffer, and 0.5 mL of 0.25 M MnCl$_2$ solution. After mixing, 500 mg of molten agar (60 °C) dissolved in 25 mL

of 0.2 M Tris HCl buffer, is added to the stain solution and the mixture is immediately poured over a gel slab in a staining box. The gel is then incubated for 30 min to 1 h at 37 °C.

15. Lysine dehydrogenase (LYD). Stained by mixing the following ingredients (in order) in a small flask and immediately pouring the solution over a gel slab in a staining box; 50 mL of sodium phosphate buffer, 50 mg of L-Lysine, 2 mL of NAD solution, 0.5 mL of PMS solution, 1 mL of MTT solution. The gel is then incubated for 30 min to 1 h at 37 °C.

16. Malate dehydrogenase (MDH). Stained by mixing the following ingredients (in order) in a small flask and immediately pouring the solution over a gel slab in a staining box; 40 mL of 0.2 M Tris HCl buffer, 6 mL of 2.0 M DL-malic acid, 2 mL of NAD solution, 0.5 mL of PMS solution, and 1 mL of MTT solution. The gel is then incubated for 30 min to 1 h at 37 °C.

17. Nucleoside phosphorylase (NSP). Stained by mixing the following ingredients (in order) in a small flask and immediately pouring the solution over a gel slab in a staining box; 25 mL of sodium phosphate buffer, 20 mg of inosine, 1 mL of xanthine oxidase solution, 0.5 mL of PMS solution, 1 mL of MTT solution, and 500 mg of molten agar (60 °C) dissolved in 25 mL of sodium phosphate buffer. The gel is then incubated for 30 min to 1 h at 37 °C.

18. 6-Phosphogluconate dehydrogenase (6PG). Stained by mixing the following ingredients (in order) in a small flask for 10 min; 20 mL of 0.2 M Tris HCl, 10 mL of 0.1 M MgCl$_2$ solution, and 30 mg of 6-phosphogluconic acid. After mixing, the following ingredients are added to the flask, and the mixture is poured over a gel slab in a staining box; 0.5 mL of NADP solution, 0.3 mL of PMS solution, and 0.6 mL of MTT solution. The gel is then incubated for 30 min to 1 h at 37 °C.

19. Phosphoglucose isomerase (PGI). Stained by mixing the following ingredients (in order) in a small flask and immediately pouring the solution over a gel slab in a staining box; 25 mL of 0.2 M Tris HCl buffer, 10 mg of fructose 6-phosphate, 0.3 mL of MgCl$_2$, 0.3 mL of glucose 6-phosphate dehydrogenase solution, 0.6 mL of NADP solution, 0.5 mL of PMS solution, 1 mL of MTT solution, and 500 mg of molten agar (60 °C) dissolved in 25 mL of 0.2 M Tris HCl buffer. The gel is then incubated for 30 min to 1 h at 37 °C.

20. Phosphoglucomutase (PGM). Stained by thoroughly mixing the following ingredients (in order) in a small flask for 10 min; 25 mL of water, 5 mL of 0.2 M Tris HCl buffer, 5 mL of 0.1 M MgCl$_2$ solution, 5 mL of glucose 1-phosphate (Sigma Chemical G 1259), 5 mL of glucose 6-phosphate dehydrogenase solution, and 0.5 mL of NADP solution. After mixing, the following ingredients are added and the solution is poured immediately over a gel slab in a staining box; 0.5 mL of PMS solution and 1 mL of MTT solution. The gel is then incubated for 30 min to 1 h at 37 °C.

21. Phenylalanyl-L-leucine peptidase (PLP). Stained by gently mixing the following ingredients in a small flask for 10 min; 20 mg of L-phenyl-

alanyl-L-leucine, 10 mg of peroxidase, 10 mg of O-dianisidine dihydrochloride, 10 mg of snake venom, 25 mL of 0.2 M Tris HCl buffer, and 0.5 mL of 0.25 M MnCl$_2$ solution. After mixing, 500 mg of molten agar (60 °C) dissolved in 25 mL of 0.2 M Tris HCl buffer is added to the stain solution and the mixture is immediately poured over a gel slab in a staining box. The gel is then incubated for 30 min to 1 h at 37 °C.

22. Shikimate dehydrogenase (SKD). Stained by mixing the following ingredients (in order) in a small flask and immediately pouring the solution over a gel slab in a staining box; 50 mL of 0.2 M Tris HCl buffer, 30 mg of shikimic acid, 1 mL of NADP solution, 2 mL of 0.1 M MgCl$_2$ solution, 0.5 mL of PMS solution, and 1 mL of MTT solution. The gel is then incubated for 30 min to 1 h at 37 °C.

26–2.3.3 Comments

All of the chemical reagents described are available through commercial chemical suppliers (e.g., Sigma Chemical Co., St. Louis). Most reagents require some form of refrigerated storage, and many are potential carcinogens, therefore appropriate protective measures should be taken during use. It is recommended that several of the enzymes be stained with agar overlays (ACO, ADK, LAL, LGG, NSP, PGI, and PLP), as this technique sometimes improves the intensity of resulting bands (extended incubation times are also useful for this purpose). Where agar overlays are not used, gels should be suspended in staining solution prior to incubation, either by a microspatula, or by tapping the side of the staining box. Gels should be incubated in the dark, and may be incubated at room temperature, except where commercially prepared enzymes from other organisms are added to the stain preparations. Incubation reactions should be terminated when complete bands are just visible in all lanes, as overstaining sometimes obscures small migrational differences between bands. Acrylamide gels may be used instead of starch gels (Young, 1985), however these are usually much more difficult to slice.

26–2.4 Collecting and Analyzing Data

26–2.4.1 Materials

1. Transilluminator (optional).
2. Gel storage solution (per liter); 230 mL of methanol (CH$_3$OH), 50 mL of glacial acetic acid (CH$_3$COOH), 720 mL of water.
3. Computer program of statistics for population genetics, written especially for use with bacteriological data, are available upon request from Dr. T.S. Whittam, Dep. of Biology, 607 Mueller Lab, Penn State Univ., University Park, PA 16802.
4. MLEE data text file.
5. Personal computer (IBM or IBM clone) with math co-processor.

26–2.4.2 Procedures

After staining, gels should be rinsed briefly under tap water, and submerged in storage solution. Comparisons of enzyme mobilities of all ETs are made visually against one another on the same gel slice; several reruns are usually necessary (Selander et al., 1986). Because migrational distance differences can be slight, comparisons of measured Rf values between gels are not reliable. Faint bands can be most easily detected by the use of a transilluminator. By convention, electromorphs are numbered (scored) in order of decreasing anodal migration (i.e., the fastest migrating electromorphs receive the smallest scores). It is important to note that during subsequent data analysis, numerical electromorph values are treated qualitatively rather than quantitatively (e.g., electromorphs could be scored using letter codes, but numerical scores are more easily incorporated into data analysis programs). Null alleles should be confirmed by repeating all procedures using fresh lysates. Enzymes showing strong multiple bands should be avoided, as these additional electromorphs may represent plasmid-encoded genes (Eardly et al., 1990). Certain stains will occasionally produce weak bands on nontarget enzymes and these should be identified and avoided during scoring. Other precautions in scoring are discussed by Selander et al., (1986).

Each isolate is characterized by its combination of electromorphs over the enzymes assayed, and distinct profiles are designated as electrophoretic types (or ETs). For computer data analysis, the ETs are arranged in rows in a tabular format, with enzyme loci heading respective columns. The data, which is most conveniently stored as a word-processing file, is then transformed to a text file for computer analysis. The computer programs ETDIV and ETCLUS, written by T.S. Whittam, Dep. of Biology, Penn State University, are used for genetic diversity analysis and cluster analysis, respectively. Sample files and explanatory notes are supplied with the programs.

26–2.4.3 Comments

There are numerous examples of population genetic studies in the literature that may serve as useful references for interpreting MLEE data; e.g., Eardly et al., 1990; Harrison et al., 1989; Nei, 1975; Nei, 1977; Pinero et al., 1988; Whittam, 1989; Whittam et al., 1983; Whittam et al., 1989; Young, 1985. Publication-quality dendograms illustrating relationships between strains can be drawn with a variety of computer programs (e.g., the Macintosh McDraft program). If a large number of new strains are to be compared to a previously characterized collection, it is usually most efficient to determine the complete range of ETs in the new collection, prior to making comparisons with older collections.

REFERENCES

Atlas, R.M., and R. Bartha. 1987. Microbial ecology: Fundamentals and applications. The Benjamin/Cummings Publ. Co., Menlo Park, CA.

Brewer, G.J., and C.F. Sing. 1970. An introduction to isozyme techniques. Academic Press, New York.

Campbell, R.E. 1984. Microbial ecology. Blackwell Sci. Publ., Oxford.

Demezas, D.H., T.B. Reardon, J.M. Watson, and A.H. Gibson. 1991. Genetic diversity among *Rhizobium leguminosarum* bv. *Trifolii* strains revealed by allozyme and restriction fragment length polymorphism analyses. Appl. Environ. Microbiol. 57:3489–3495.

Denny, T.P., M.N. Gilmour, and R.K. Selander. 1988. Genetic diversity and relationships of two pathovars of *Pseudomonas syringae.* J. Gen. Microbiol. 134:1949–1960.

Dykhuizen, D.E., and D.L. Hartl. 1983. Functional effects of PGI allozymes in *Escherichia coli.* Genetics 105:1–18.

Eardly, B.D., L.A. Materon, N.H. Smith, D.A. Johnson, M.D. Rumbaugh, and R.K. Selander. 1990. Genetic structure of natural populations of the nitrogen-fixing bacterium *Rhizobium meliloti.* Appl. Environ. Microbiol. 56:187–194.

Engvild, K.C., E.S. Jensen, and L. Skøt. 1990. Parallel variation in isoenzyme and nitrogen fixation markers in a *Rhizobium* population. Plant Soil 128:283–286.

Harris, H., and D.A. Hopkinson. 1976. Handbook of enzyme electrophoresis in human genetics. North-Holland Publ. Co., Amsterdam.

Harrison, S.P., D.G. Jones, and J.P.W. Young. 1989. *Rhizobium* population genetics: Genetic variation within and between populations from diverse locations. J. Gen. Microbiol. 135:1061–1069.

Hedrick, P.W. 1984. Population biology: The evolution and ecology of populations. Jones and Bartlett Publ., Boston.

Kimura, M. 1983. The neutral theory of evolution. Cambridge University Press, Cambridge.

May, B. 1980. The salmonid genome: Evolutionary restructuring following a tetraploid event. Ph.D. diss. Pennsylvania State University, University Park.

McArthur, J.V., D.A. Kovacic, and M.S. Smith. 1988. Genetic diversity in natural populations of a soil bacterium across a landscape gradient. Proc. Natl. Acad. Sci. 85:9621–9624.

Nei, M. 1975. Molecular population genetics and evolution. North Holland Publ. Co., Amsterdam.

Nei, M. 1977. F-statistics and analysis of gene diversity in subdivided populations. Ann. Hum. Genet. 41:225–233.

Pinero, D., E. Martinez, and R.K. Selander. 1988. Genetic diversity and relationships among isolates of *Rhizobium leguminosarum* biovar *phaseoli.* Appl. Environ. Microbiol. 54:2825–2832.

Segovia, L., D. Pinero, R. Palacios, and E. Martinez-Romero. 1991. Genetic structure of a soil population of non-symbiotic *Rhizobium leguminosarum* isolates. Appl. Environ. Microbiol. 57:426–433.

Selander, R.K., D.A. Caugant, H. Ochman, J.M. Musser, M.N. Gilmour, and T.S. Whittam. 1986. Methods of multilocus enzyme electrophoresis for bacterial population genetics and systematics. Appl. Environ. Microbiol. 51:873–884.

Selander, R.K., and B.R. Levin. 1980. Genetic diversity and structure in *Escherichia coli* populations. Science 210:545–547.

Selander, R.K., and J.M. Musser. 1990. The population genetics of bacterial pathogenesis. p. 11–36. *In* B.H. Iglewski, and V.L. Clark (ed.) Molecular basis of bacterial pathogenesis. Academic Press, Orlando, FL.

Smith, J.M., C.G. Dowson, and B.G. Spratt. 1991. Localized sex in bacteria. Nature (London) 349:29–31.

Whittam, T.S. 1989. Clonal dynamics of *Escherichia coli* in its natural habitats. Antonie van Leeuwenhoek, J. Microbiol. Serol. 55:23–32.

Whittam, T.S., H. Ochman, and R.K. Selander. 1983. Geographic components of linkage disequilibrium in natural populations of *Escherichia coli.* Mol. Biol. Evol. 1:67–83.

Whittam, T.S., M.L. Wolfe, and R.A. Wilson. 1989. Genetic relationships among *Escherichia coli* isolates causing urinary tract infections in humans and animals. Epidemiol. Inf. 102:37–46.

Young, J.P.W. 1985. *Rhizobium* population genetics: enzyme polymorphism in isolates from peas, clover, beans, and lucerne grown at the same site. J. Gen. Microbiol. 131:2399–2408.

Chapter 27

Spontaneous and Intrinsic Antibiotic Resistance Markers

CHARLES HAGEDORN, *Virginia Polytechnic Institute & State University, Blacksburg, Virginia*

The selection and use of antibiotic resistant bacterial strains has emerged over the last three decades as a powerful research technique in ecological studies. Although this procedure has been most widely reported in association with (*Brady*) *Rhizobium* investigations, it has been applied to many other microbial species and strains. Generally, antibiotics (antibacterial agents and bacteriocides) are used to select for resistant strains that can be recovered and enumerated, although occasionally other types of materials (i.e., fungicides and antimetabolites) have also been employed for this purpose. For example, in 1963 a strain of *Serratia marcesens,* resistant to penicillin, aureomycin, and actidione was employed as a measure of pollutant flow in an estuary (Rippon, 1963). Antibiotic resistant strains have seen use in all phases of (*Brady*) *Rhizobium* research including survival (Barber, 1979), competitiveness (Skrdleta, 1970), nodule occupancy (Kuykendall & Weber, 1978), multiplication in soil (Chatel & Parker, 1973b), symbiotic properties (Pankhurst, 1977), and carrier and inoculation tests (Hagedorn, 1979). Resistant strains have also been employed in surface water studies (Pike et al., 1969; Wimpenny et al., 1972), biological disease control (Weller & Cook, 1983; Howell & Stipanovic, 1979), and wastewater treatment (McCoy & Hagedorn, 1979; Hagedorn, 1984).

Two procedures have emerged for obtaining antibiotic resistant strains. These are direct selection for spontaneous antibiotic resistance (SAR) and screening for intrinsic antibiotic resistance (IAR). With the widespread use of antibiotic resistant bacteria in ecological research, it is appropriate to consider both the advantages and disadvantages of developing such strains. Topics to be evaluated are (i) guidelines [based on published reports] for selection, of SAR and IAR strains, (ii) procedures for strain evaluation, and (iii) applications in field studies including recovery of strains from environmental samples.

The major advantage of using SAR and IAR strains is the capability for direct, quantitative recovery of the strains from a variety of ecosystems including water, plant tissue, and soil (Bromfield & Jones, 1980; Bushby, 1982). If it is desirable to demonstrate that the use of an amended bacterial strain has some particular function (such as N_2 fixation), then precise recovery and identification of that specific strain is essential. Antibiotic resistance (both SAR and IAR) can provide a basis for this selection and recovery as well as for identification. In some research approaches, the strain itself may be of only secondary importance compared to its performance as a biological tracer to study some phenomenon such as water movement (Hagedorn, 1979) or pollutant dispersal (Armstrong et al., 1982). Other advantages of the use of SAR and IAR strains in ecological studies are that the selection and evaluation of strains is fairly straightforward, requires no sophisticated equipment, can be performed routinely in most laboratories, and has wide applicability to many types of research (Hagedorn, 1986).

The most obvious problem that must be considered in the use of SAR mutants is the stability of the genetic marker. If the resistant strain reverts to a sensitive condition at a high rate, then detection will not be possible in a complex microbial population even though the amended organism is still present. Also, selection procedures could result in strains that possess an unknown number of multiple mutations. Some of these could result in losses of specific metabolic functions that would be difficult to detect (Hagedorn, 1979). In early studies with antibiotic resistant strains of *Rhizobium,* Schwinghamer (1967), and Schwinghamer and Dudman (1973) first described the loss of some physiological characteristics by the resistant isolates. Resistance to any one antibiotic (or antimetabolite) may induce multiple changes within a given gene or mutations at several different loci (Hollis et al., 1981). These gene alterations can each result in a change in some phenotypic characteristic, which may not be readily detectable. In addition, the possibility can never be completely eliminated that indigenous strains with identical resistance patterns to the amended strain exist in any given habitat. With IAR strains, it is especially important to evaluate the resistance patterns of indigenous strains to determine the degree of similarity to the amended strain. In many studies, mutagenic agents such as UV radiation or nitrosoquanidine have been employed prior to selecting strains. The possibility of inducing multiple mutations with these procedures is high, and considerable effort must be devoted to each isolate to ensure that major genetic damage has not occurred (Obaton, 1971). Also, it has been relatively easy in several cases to isolate naturally occurring strains with the same resistances or metabolic capabilities as the selected strains (Roy & Mishia, 1974; Bromfield & Jones, 1980). There has been considerable discussion of the potential mobility of the resistance genes within the general microbial population. This concern relates to the transfer of the resistance element to other bacterial strains, thus dispersing resistance genes into the environment (Talbot et al., 1980; Stewart & Koditschek, 1980; Fontaine & Hoadley, 1976). There is also the remote but real possibility that such strains (or genetic recipients) could enter food chains and thus contribute to antibiotic resistance in microbes inhabiting the gastrointestinal tract (Cenci et al., 1980). Certainly the decision to use SAR strains requires careful judgment and exact testing procedures designed to assure complete environmental safety and confidence with resultant data.

27–1 SELECTION OF SPONTANEOUS ANTIBIOTIC RESISTANT STRAINS

27–1.1 Principles

Methods for the selection of SAR and IAR strains have been adequately described by Bushby (1982), Hagedorn (1986), Eaglesham (1987), and Kuykendall (1987). Although these publications relate primarily to *Rhizobium*, the described procedures could be employed with most bacteria. Commonly employed methods include approaches for the selection and identification of both spontaneous and intrinsic resistant strains.

For strains formed spontaneously prior to exposure to the selective agent, an antibiotic can allow isolation of the strain from among a dense population of prototrophs (antibiotic sensitive microbes). Within bacterial populations, it has been estimated that one cell in every 10^8 cells is spontaneously resistant to any one antibiotic, whereas one cell in 10^{12} cells may be resistant to two combined antibiotics (Linton, 1983). Because of these low frequencies of occurrence, it may be necessary to prepare a centrifuged suspension to obtain a sufficiently high cell density for detecting mutants resistant to multiple antibiotics. Generally, spontaneous mutants are usually selected sequentially against each antibiotic rather than in combination. This allows larger numbers of resistant colonies to be detected, avoids simultaneous effects of combined antibiotics, and provides exact information on each resistant isolate.

The choice of antibiotics for selection is of critical importance to the success of obtaining genetic variants. Although early reports indicated that loss of desirable characteristics (such as nodulation) is associated with some classes of antibiotics, this has not been observed with aminoglycosides (Schwinghamer, 1964, 1967). A later report demonstrated that many bacterial strains react differently to common antibiotics and may not be adversely affected by selection for spontaneous resistance (Hagedorn, 1979). This implies that any antibiotic could potentially be used as a marker as long as the SAR strain was examined for potential changes in other characteristics. There are nine antibiotics that have been used to develop at least 95% of the SAR strains described in the literature (Table 27–1). Nalidixic acid and rifampicin have seen wide use as single markers, and because they are synthetic compounds, the occurrence of resistance to them is rare. Spectinomycin has been available for only a relatively short period, making it a useful selection compound since indigenous microbial

Table 27–1. Antibiotic classes and commonly used examples.

Aminoglycosides	Peptides
Gentamycin	Bacitracin
†Kanamycin	
Neomycin	Penicillins
†Streptomycin	†Ampicillin
†Spectinomycin	Cephalothin
Vancomycin	Penicillin G
Macrolides	Tetracyclines
Erythromycin	†Tetracycline
Miscellaneous	Synthetics
†Chloramphenicol	†Nalidixic Acid
†Novobiocin	†Rifampicin

† Indicates antibiotics most frequently cited for SAR strains.

populations possess virtually no spectinomycin resistance. Regardless of the type of bacterium being employed to develop a SAR strain, the use of double resistance markers includes virtually any combination of antibiotics listed in Table 27–1. Resistance concentrations for SARs are almost always between 50 μg mL^{-1} and 500 μg mL^{-1}, with 50 to 200 μg mL^{-1} being most frequently cited.

Closely related to antibiotic selection is the choice of microbial growth media. As more selective media are used (e.g., reduced nutrient content), the susceptibility of the target strain to the selective agent may increase. In most cases, it is preferable to select strains and pursue antibiotic screening on the same medium that will be used for recovery and identification in the planned ecological studies (Barber, 1979; Armstrong et al., 1982). For example, McCoy and Hagedorn (1979) used SAR strains of *Escherichia coli* in conjunction with a selective medium (Eosin Methylene Blue Agar) for both strain development and enumeration studies.

27–1.2 Materials

Specific materials will depend upon the characteristics of the parental culture that is being used to develop the SAR strain. Once the parental culture and appropriate media for cultivation are determined, the selection of the antibiotic(s) can be made. This selection will depend, in part, upon published experience with antibiotics used with the chosen parental culture. With the culture, media, and antibiotics in hand, the following procedures require only the most basic of laboratory materials and equipment.

27–1.3 Procedures

27–1.3.1 Isolation of SAR Strains

Antibiotic-resistant strains of bacteria are easily obtained by the process of direct selection, and it is usually unnecessary to employ any enrichment techniques (Linton, 1983). In direct selection of SAR strains, large numbers of pure strain are plated onto the agar surface of petri dishes containing various concentrations of an antibiotic. Different concentrations are used to determine the optimum level for selection. Ideally, an antibiotic concentration can be determined that will effectively prevent any growth by the wild-type parent strain, yet will permit the growth of resistant strains. By using two-fold increments of concentrations (e.g., 10, 20, 40, 80, 160, and 320 µg mL^{-1}), a suitable antibiotic concentration is usually found without difficulty. The direct isolation procedure for SAR strains arising "spontaneously" is not elaborate (Kuykendall, 1987).

Begin by developing a broth culture of the bacterial strain to high cell densities of $\geq 5 \times 10^8$ colony-forming unit (CFU) mL^{-1}. As inoculum for this culture and to maintain culture purity, choose a well-isolated colony from a plate where colonies have arisen from single cells (use a surfactant such as Tween-20 or Tween-40 (2% vol/vol) if necessary to dissociate clumps). Also maintain a source of the parental strain by using material from the same colony to develop an agar slant culture to be stored as soon as the culture has grown. Spread 0.1-mL aliquots of the dense culture onto the agar [with a sterile, bent (45°) glass rod] of antibiotic-amended growth medium. Invert the plates and incubate under conditions appropriate for the particular organism (Scholla & Elkan, 1984). Observe plates under good lighting conditions. Those petri plates giving confluent lawns of growth represent concentrations of antibiotic that are below the minimum inhibitory concentration (MIC) for that particular strain of organism. Look for plates that give distinct colonies arising from a clear background. A clear background represents complete inhibition of the parental strain.

Pick resistant colonies with a sterile loop and streak on another antibiotic-supplemented plate of the same medium. The same antibiotic concentration that was useful for SAR strain selection can be used for purification. This simple purification step should be sufficient for obtaining a pure culture of the antibiotic-resistant strain of interest. Useful genetically marked strains of bacteria should grow under selective conditions at about the same rate that the parent strain grows under nonselective conditions (i.e., in the absence of the antibiotic). The ability to detect even extremely rare resistant cells is what makes the technique of direct selection a highly useful procedure. In addition to the derivation of SAR strains, direct selection may also be used to obtain strains resistant to phages, and various bacteriocidal or bacteriostatic antimetabolites.

Success using the experimental protocol outlined above for obtaining SAR strains is high, but there are some problems that should be addressed. The wild-type or parental strain may already be resistant to the antibiotic that was used. If this is the case, a lawn of bacterial growth will be evident even at the highest concentration employed. Complete counterselection (no growth) of the parent strain may occur without the appearance of "spontaneous" resistant colonies. The numbers of cells in the bacterial population used may not have been high enough to obtain a relatively rare mutant (one that occurs at a frequency of less than 10^{-8}). Lastly, cells with resistance to the particular antibiotic of choice may not occur at a detectable frequency. Most of these problems can be avoided by developing an SAR that is resistant to more than one antibiotic. Many researchers now employ resistance to at least two antibiotics for developing useful SAR strains, and experience has shown that resistance to two antibiotics appears to be adequate for most purposes (Linton, 1983; Kuykendall, 1987). It is best to choose antibiotics from different classes (Table 27–1) to avoid problems with acquired cross-resistance within a single class.

27–1.3.2 Identification of SAR Strains

Antibiotic-resistant strains of bacteria are readily identified on the basis of their ability to grow under selective conditions that prohibit growth of their wild-type parent strain. However, because other strains of bacteria of the same or unrelated taxa may naturally possess that particular antibiotic resistance marker, positive identification is not necessarily assured on the basis of one resistance marker. An example of problems with positive strain identification based on a single genetic marker is as follows:

About one-third of the strains classified as *B. japonicum* strains in the Beltsville USDA collection are naturally resistant to high levels (≥ 250 μg mL^{-1}) of rifampicin, a potent inhibitor of transcription in bacteria (Kuykendall, 1987). Rifampicin-resistant strains, occurring at frequencies of 10^{-7} to 10^{-8}, are readily obtained from those *B. japonicum* strains that are naturally rifampicin sensitive. Naturally occurring rifampicin-sensitive strains may give a haze of growth even at 100 μg mL^{-1} rifampicin, but complete inhibition of the parent strain occurs at 250 or 500 μg mL^{-1}. Rifampicin-resistant strains of *B. japonicum* are not distinguishable from naturally occurring rifampicin-resistant *Bradyrhizobium* on the sole basis of rifampicin-resistance levels. However, these two groups may be distinguished on the basis of other phenotypic characteristics such as colony morphology and the ability of some strains of the latter group to produce rhizobitoxine (RT)-induced chlorotic symptoms in planta with RT-sensitive soybean genotypes.

In the identification of SAR strains of bacteria, there are some possible difficulties that extend beyond the question of how many markers are required for unambiguous strain identification. One potential problem is the effect of environmental factors such as temperature, pH, and available C source on the phenotypic expression of a genetic determinant by a bacterial strain. Another potential problem is the effect of biotic factors such as the presence of other species and genera of microorganisms (as in ecology studies) or the presence of related organisms (as in gene transfer experiments). For this reason, environmental factors and conditions need to be precisely defined and consistently repeated to obtain accurate results, and biotic factors, some of which may as yet remain undefined, may influence the results of experiments.

27–2 SELECTION OF INTRINSIC ANTIBIOTIC RESISTANT STRAINS

27–2.1 Principles

Intrinsic resistance (to low concentrations of many antibiotics) has received considerable attention because it allows identification of resistance-sensitivity patterns at low inhibitor concentrations, thereby avoiding the selection of mutants with high inhibition tolerance. Pinto et al. (1974) differentiated *R. meliloti* strains using natural resistance to low levels (2–10 μg mL^{-1}) of aminoglycoside antibiotics as a selective factor. Josey et al. (1979) screened several strains of *R. leguminosarum* against various concentrations of eight antibiotics and found unique resistance patterns for all but one strain. Reliable strain identification with reference to stock cultures could be made on the basis of these patterns when examining nodule isolates from host plants. Kramer and Peterson (1981) used this technique with both fast and slow-growing rhizobia and found that resistance-susceptibility patterns were reliable for identifying strains in field nodules. However, they reported that the rhizobia must first be isolated from the nodules to obtain a standard inoculum size and then typed on antibiotic-containing media. There are problems with native isolates possessing the same resistance patterns to strains being studied, and it may also be necessary to examine the stability of each resistance-sensitivity characteristic to ensure its consistency over time (Linton, 1983). Certainly the uniqueness of the selected pattern must be fully examined before the technique of intrinsic resistance can be employed for any strain (Eaglesham, 1987).

One potentially useful function of IAR is a tag for the determination of microbial competitive performance in pot and field studies, since differences in IAR can sometimes occur even in closely related strains. For example, cultures that were identical by serological criteria could be differentiated by IAR (Kramer & Peterson, 1981). It should be noted that the use of strains with intrinsic resistance to at least two antibiotics guards against possible complications arising from spontaneous mutations within inoculum cultures or nodules.

With model systems it is necessary to ensure that the strains of interest have different reactions to one or, preferably, two antibiotics. In a competitiveness experiment with *Medicago* spp. grown in tubes, a strain of *R. meliloti* resistant to kanamycin at 2 μg mL^{-1} was paired with each of three strains susceptible to kanamycin but resistant to streptomycin at 1.5 μg mL^{-1} (Pinto et al., 1974). The appropriate IAR strains could be recovered quantitatively from rhizosphere mixtures, even when present in low proportion, as well as from nodules. In a similar experiment, again with *R. meliloti,* four strains were chosen that had distinct profiles of intrinsic resistance to eight antibiotics, allowing examination of the competitive interactions of all two-, three-, and four-strain mixtures (Peterson et al.,

1983). In a study of Fix$^-$ mutants of bradyrhizobia, determining that Fix$^+$ nodules were due to revertants rather than to chance contaminants was achieved by confirming that their tolerance to antimicrobial agents was identical to that of the parent strain and dissimilar to those of other bradyrhizobia (Misplon & Bishop, 1983).

When IAR is used as a tracer to assess performance of a bacterial strain added to soil, it is usually advisable to make preliminary assessment of the IAR characteristics of similar indigenous strains. The occurrence of native strains with the same IAR profile as an amended strain does not necessarily disqualify the use of that strain provided that the native strains can be separated by other characteristics such as biochemical reactions or host plant interactions (Levin & Montgomery, 1974). In a study of clover nodule occupancy by four strains of *R. leguminosarum* biovar *trifolii* applied individually in a field trial, three of the four strains were distinguishable from the indigenous rhizobia by their IAR patterns (McLaughlin & Ahmad, 1984). The fourth strain had the same IAR "fingerprint" as 48% of the nodule isolates from uninoculated plots, and so yielded no useful data. Introduced bradyrhizobia in soil examined over a 12-y period showed that their IAR characteristics were only slightly less stable than their serological characteristics, which remained unchanged throughout the period (Diatloff, 1977).

The IAR has proven to be a useful tool for several aspects of microbial research. However, it would be misleading to indicate that by closely following published methods, the researcher is guaranteed reproducible IAR data. In general, reports of problems in producing dependable data seldom appear in the literature. Nevertheless, in several cases authors have described or implied batch-to-batch variability in the determination of IAR characteristics of strains (Misplon & Bishop, 1983; Sinclair & Eaglesham 1984). Even more disturbing, for reasons not yet understood, some bacterial strains are highly variable in their reaction to certain antibiotics (Josey et al., 1979; Beynon & Josey, 1980; Stein et al., 1982). Lack of rigorous control of inoculum size inherent with the pin-type multi-inoculator may be a major factor in within-batch variability, and differences in incubation conditions may contribute significantly to between-batch variability (Bromfield et al., 1982). In a study of bradyrhizobia, with care taken to reproduce the growth stage of the inoculum as well as its volume, and account taken of variable incubation conditions (only in the most expensive incubators is the temperature largely independent of ambient), reproducibility within batches was better than 99% and between batches was 96% (Sinclair & Eaglesham, 1984). However, it does appear that the heterogeneity reported for a natural population of *R. leguminosarum* biovar *phaseoli* (Beynon & Josey, 1980) was overestimated (Stein et al., 1982). Certainly the differentiation of strains within a population on the basis of low-level IAR is fraught with difficulties. It has been suggested that, in characterizing amended or inoculant bacterial strains, IAR can be useful in conjunction with other methods, but should not be used exclusively (Eaglesham, 1987).

There are three methods for the determination of IAR characteristics: (i) the agar (or disk)-diffusion method, using antibiotic-impregnated paper disks (e.g., Sensi-Discs, BBL, Cockeysville, MD; Dispenso-Discs, Difco, Detroit, MI) applied to bacterial-seeded agar; (ii) the agar-dilution method, using stock solutions of antibiotics added to sterilized agar media held at 45 to 50 °C, then used to culture the target strain after cooling; and (iii) the broth-dilution method, similar to method (ii) with liquid media (Eaglesham, 1987).

27.2.2 Materials

Materials for assessing IAR strains will include a wider spectrum of antibiotics than those used to develop SAR strains, and they can be chosen from those most successfully used in IAR tests (see previous section, 27–2.1). The choice of media for determining IAR patterns will partially depend upon the culture being used. If the desired bacterium is heterotrophic, routine media reported in the literature for IAR tests (e.g., yeast-mannitol and trypticase soy agar) should be used so that results can be compared with published reports. As with SAR strains, only basic laboratory materials are needed once the culture, growth media, and antibiotics have been identified.

27–2.3 Procedures

27–2.3.1 Agar-Diffusion Method

Although not recommended for slow-growing bacteria of clinical importance (Cenci et al., 1980), the agar-diffusion method has been used successfully with slow- as well as fast-growing rhizobia, and other soil bacteria. Commercially available antibiotic-impregnated paper disks are manufactured according to federal quality standards and are therefore reliable. To maintain potency, stocks of disks should be stored with a desiccant in a freezer, preferably at −65 °C. Supplies for day-to-day use can be transferred to a 4 °C refrigerator and, to prevent condensation, should be allowed to warm to room temperature before being opened to the atmosphere. Dispensers that deposit up to 12 disks spaced equally on a 15-cm-diam. petri plate are commercially available (Difco); otherwise, the disks are placed on the seeded agar surface with sterile forceps. Synergistic and antagonistic interactions are possible between antibiotics; with this in mind, disks should be separated sufficiently for each inhibition zone to be entire.

A range of disk potency is available for each antibiotic, and preliminary examination will determine the most appropriate level(s) of the chosen antibiotics. The susceptibility of a strain to an antibiotic is judged by the clear inhibition zone around the disk after incubation. The size of the clear zone is related to the rate of diffusion of the antibiotic into the agar medium and generation time of the target strain. There is a complex interaction between these parameters (Linton, 1983), and since both are influenced by several factors, care must be taken to ensure that the conditions used for different batches are as similar as possible. Composition of the culture medium can affect antimicrobial activity and the culture medium should be routinely reproducible with the same brands and reagent quality used each time, and sterilization procedures should be standardized. A constant volume of culture medium per petri plate is important and, since pH is particularly critical, it should be routinely checked with a surface electrode.

The number of organisms applied in the lawn, and their growth stage, can affect the results of the agar-diffusion test. Strain-to-strain variation in extracellular polysaccharide production can result in cultures of equal optical density (OD) but disparate cell densities. Moreover, as a result of strain differences in generation time, cultures of equal age often have significantly different cell densities. Where reproducibility problems are encountered, the construction of standard curves of OD vs. viable count for each strain of interest will help to ensure equal cell numbers in inocula, at the same growth stage, within batches or between batches. Also, unless generation times are known, care should be exercised in the interpretation of strain-to-strain variations in the sizes of inhibition zones. As a check on between-batch reproducibility, a few strains of known response to the antibiotics could be included each time as these can also serve as indicators of disk potency (Cole & Elkan, 1979). A modification of the agar-diffusion method involves incorporating the bacteria in molten agar and flooding it over the previously poured and solidified culture medium (i.e., "agar overlay"). This has been found to improve reproducibility in clinical work (Fontaine & Hoadley, 1976) and has been used with rhizobia (Peterson et al., 1983).

27–2.3.2 Agar-Dilution Method

As described for the agar-diffusion method, conditions and components must be standardized to ensure reproducible results. Particularly important are culture medium composition, pH and sterilization procedures, temperature of agar medium when antibiotic stock solution is added, volume of agar per petri plate, growth stage of test cultures, and size of inoculum (Eaglesham, 1987). A critical factor is antibiotic preparation. Many antibiotics are relatively unstable and must be stored strictly according to manufacturers' instructions. Whenever possible, freshly obtained materials should be used. Stock solutions should be used immediately, but if storage is necessary, they should be frozen, preferably at $-20 °C$, for no longer than 14 d. Some antibiotics are not readily soluble in water and stocks are prepared in alternative solvents (e.g., nalidixic acid in 1 M NaOH, rifampicin in methanol). Stock antibiotic solutions must be prepared with due regard to asepsis since they deteriorate if autoclaved; filter sterilization is recommended. Filter sterilization involves the use of disposable membrane filters, usually with a 0.22 μm pore size and a prefilter to prevent clogging. The filter assembly is autoclaved before use, and the membrane and pre-filter (both can be purchased already sterilized) are placed on the assembly for use. Negative pressure (e.g., a vacuum pump) is required to pull the solution through the membrane filter (Hagedorn, 1986). Stocks are added to the autoclaved agar medium held at 45 to 50 °C (the same precise temperature being used each time), thoroughly mixed, poured, and cooled. The plates should be used as soon as possible after preparation, preferably after surface drying in a laminar flow cabinet. Careful hand inoculation (Sinclair & Eaglesham, 1984) or the use of a multipin inoculator (Josey et al., 1979) allows several strains to be tested on each plate. An agar-overlay modification of the agar-dilution method involves relatively few cells suspended in the molten agar overlay, with the sensitivity to the antibiotic expressed as percent survival (Rai, 1983). A concentration gradient technique was found to give better reproducibility in the analysis of IAR of *R. leguminosarum* biovar *phaseoli* strains than the pin inoculation/standard agar-dilution method (Bromfield et al., 1982).

27–2.3.3 Broth-Dilution Method

With the broth-dilution method, a separate broth culture is required for every strain × antibiotic combination and, probably for this reason, has been little used in microbial ecology studies. Small volumes of broth (about 2 mL) are used generally, but even greater economy is possible using microdilution techniques that operate on 50-μL volumes. An advantage of the broth-dilution method is that turbidity measurements allow quantitative expression of data in contrast to the qualitative visual estimates with the agar-dilution method (but see the agar-overlay modification of Rai [1983]). Where a relatively small number of strain × antibiotic combinations are to be examined, the use of side-arm flasks permits determinations of antibiotic effects on generation times. As with the other methods, there are several factors that can reduce within- and between-batch reproducibility, and care is again required in the preparation and execution of the technique.

27–3 EVALUATION OF SAR AND IAR STRAINS

27–3.1 Principles

After appropriate SAR and IAR strains have been isolated with a selective screening procedure, each strain must be evaluated for similarity to the parent strain (for SAR isolates) and to similar indigenous strains (SAR and IAR isolates). The objective is to select an SAR strain with, if possible, at least a single difference from the parent (e.g., antibiotic resistance profiles). If multiple mutations must be accepted, the number should be minimized, and the *secondary* mutations must be in genomes with minimal impact on organismal behavior and function. Regardless of the manner in which the SAR strains are selected and the type of marker they contain, identical procedures should be employed to determine phenotypic similarity to the parent strain. It would not be practical to examine every SAR or IAR strain against all criteria, but efforts should focus on those that provide the best comparison (depending upon the intended use of the strains).

27–3.2 Materials

Materials needed for the following section on cultural tests (27–3.3.1) will depend largely upon the bacterium that is being used in determining the choice of media and reagents for the various growth and biochemical tests that may be required. For the following sections on greenhouse and field tests (27–3.3.2 and 27–3.3.3), the in-planta assays will require selection of an appropriate plant species, a method of inoculation (e.g., seed coat, foliar spray, or granular in-furrow), preparation of a carrier (e.g., peat) and inclusion of other organisms if necessary (e.g., fungal pathogens if the SAR or IAR strain is a fungicidal agent). Many of the specifics will need to be chosen based on the type of organism, the target (or host) plant species, and the use for which the SAR and IAR strain was intended.

27–3.3 Procedures

27–3.3.1 Cultural Tests

Initial tests should be performed in the laboratory because this is the simplest level of examination, and strains not successful at this stage can be discarded, thereby avoiding unnecessary inclusion in more complex tests. SAR and IAR strains should perform consistently with an adequate level of resistance over appropriate cultural conditions. This includes comparison of SAR strains to the parental type based on growth rates, generation times, growth over a range of temperatures and pH values, and utilization of C and N sources with minimal media and other pertinent biochemical tests. The SAR strain should be plated on media with and without the selective agent so that reversion frequencies and recovery efficiencies can be estimated. Reversion frequencies should be no $> 10\%$, whereas recovery efficiencies need not be high, as long as this characteristic is sufficiently stable to allow reasonable population estimates (McCoy & Hagedorn, 1979).

For genetic studies, cross resistance with other antibiotics is not desirable, although cross-resistant strains can be useful as an aid in identification. Bushby (1982) found that cross resistance allowed a greater spectrum of antibiotics to be added to selective agar to reduce contaminant growth. Accurate plate counts of field populations of an amended *B. japonicum* strain could be obtained by employing multiple antibiotics conferred by selections against streptomycin (Obaton, 1971). Hagedorn (1979) reported that, with *R. leguminosarum* biovar *trifolii*, cross-resistance patterns were generally restricted to those antibiotics with similar modes of activity. The development of cross resistance occurred in 5.7% of the possible resistance combinations. With any SAR strain, cross resistance is acceptable as long as the resistance levels to the other antibiotics are known.

Other characteristics that have been employed for characterization of SAR and IAR strains are phage resistance patterns, colony morphology or pigmentation, nodulation for rhizobial strains, and serological tests (Kowalski et al., 1974). Serological comparisons can be useful with techniques such as gel diffusion so that small differences in reactions to antisera can be detected (Diatloff, 1977; Vincent, 1970).

27–3.3.2 Growth Chamber or Greenhouse Tests

Studies involving the effects of specific environmental parameters on selected SAR and IAR strains are often valuable as predictors of behavior and survival in field trials. Such experiments are best performed in a growth chamber or greenhouse where behavior of parental or similar strains can be compared against the selected strain in every case. Parameters to be studied depend upon the type of field environment under consideration. Such evaluations have been conducted in soil and rhizosphere (Pugashetti & Wagner, 1980; Johnston & Beringer, 1975), competition with mixtures of *Rhizobium* strains (Kuykendall & Weber, 1978), nodule occupancy in legumes (Materon & Hagedorn, 1983), survival in soil (McCoy & Hagedorn, 1979), and on plant structures (Weller & Cook, 1983).

Depending upon experimental objectives, specific aspects of the intended environment such as moisture levels (Chatel & Parker, 1973a), temperature (Hardarson & Jones, 1979), or acidity-alkalinity (Munns, 1970) may be examined. With such studies comparisons may include survival of the SAR or IAR strain (McCoy & Hagedorn, 1979), nodulation (Bushby, 1982), N_2 fixation (Brockwell & Hely, 1966), competition in nodule occupancy (Levin & Montgomery, 1974), or suppression of plant pathogens (Weller & Cook, 1983; Howell & Stipanovic 1979). Results may be difficult to interpret because it usually is not possible to examine the parent strain since it carries no genetic marker and cannot be identified in the presence of natural populations. If interactions of the parent strain cannot be determined, there is little to compare a SAR strain against. The best use of these tests is to observe for failure to nodulate, failure to fix N, rapid elimination of the introduced strain, or failure to exhibit some other type of activity by the amended strain. In spite of the difficulty in designing and pursuing such studies, discovering the failure of a SAR or IAR strain at this point is still preferable to conducting unsuccessful field studies.

27–3.3.3 Field Studies

The ultimate test for any selected strain is an unaltered, unrestricted environment; that is, in situ in a native ecosystem. In environmental studies, it is necessary to keep treatments to a minimum because of the effort involved in preparation and maintenance of field sites, especially in cropped ecosystems where a commitment to field operations may be as long as 8 mo per year (Materon & Hagedorn, 1983). In such cases, amendment of the SAR or IAR strain to the soil can only be made once annually, so it is essential to use only those strains that have appeared suitable through every other examination. For example, many forage legume species (i.e., clovers and alfalfa) mature over a 6 to 8 mo period, yet inoculation, to be successful, can only be performed at planting. Correcting a failure at this point (i.e., poor nodulation) involves replanting the stand at the start of the next growing season (Hagedorn, 1986). Parent strains and genetic variants should be compared against each other in whatever evaluation is appropriate (i.e., disease suppression and crop yields). More specific comparisons can be performed if the parent can be detected by another method (i.e., serology) because persistence of both the amended strain and parent (for SAR) may be important in evaluating the success of the field experimentation.

Sample storage should be tested before field studies are initiated. This is also true for recovery media, incubation conditions, and enumeration of SAR or IAR isolates. It is also advisable to examine isolates from nonselective media (where possible) to determine if a significant portion of the amended strain's population is unable to survive the transfer from an environmental sample to selective media. If such a subpopulation is found, then sample preincubation may be needed or initial isolation on a less specific medium followed by transfer onto a more selective formulation (McCoy & Hagedorn, 1979). Finally, additional tests should be performed to determine if any obvious morphological or physiological characteristics have changed during the course of the environmental studies (Phillips, 1974; Levin & Montgomery, 1974). These comparisons should be against SAR or IAR stock cultures, and sufficient tests ought to be conducted to allow a reasonable assessment of whether any changes in the strain have occurred. These extra identification steps may be as straightforward as a series of biochemical tests or as elaborate as symbiotic efficiency on a host plant (Vincent, 1970) or protein profiles, depending upon the desired level of sensitivity for examining the recovered strain (Diatloff, 1977).

Containment of the amended strain at the experimental site is important. Containment may be difficult or at least of lesser importance to the outcome of an experiment in soil as compared to aquatic sites, in that bacteria applied to water could easily contaminate surface or shallow groundwater via water transport. Due to limitations of bacterial movement in soils because of various absorption phenomena, groundwater contamination is less likely. Careful monitoring of water flow is necessary to reduce this potential in aquatic ecosystems (Wimpenny et al., 1972; Armstrong et al., 1982). In field studies, it may be desirable to dispose of crop residues or examine such material for the presence of any of the amended bacteria. Limiting access to the experimental site and soil fumigation after completion of the research may also be appropriate if the test organism must be eliminated from the site.

Field experimental designs must also be chosen based both on the amended strains characteristics and the plant community to be studied. For example, in designing field trails involving annual crops, a randomized complete block design is generally appropriate. However, if the crop is a reseeding annual or a perennial where the sites are to be monitored for several years, it may be desirable to separate plots amended with the SAR and IAR strain from untreated plots (Hagedorn, 1979; Bushby, 1982). If not, the treatments will eventually become cross contaminated. This contamination can be confirmed by analysis for the amended strain outside of the plots to which it was applied (Mytton, 1975).

REFERENCES

Armstrong, J.L., J.J. Calomiris, and R.J. Seidler. 1982. Selection of antibiotic-resistant standard plate count bacteria during water treatment. Appl. Environ. Microbiol. 44:308–316.

Barber, L.E. 1979. Use of selective agents for recovery of *Rhizobium meliloti* from soil. Soil Sci. Soc. Am. J. 43:1145–1148.

Beynon, J.L., and D.P. Josey. 1980. Demonstration of heterogeneity in a natural population of *Rhizobium phaseoli* using variation in intrinsic antibiotic resistance. J. Gen. Microbiol. 188:437–441.

Brockwell, J., and F.W. Hely. 1966. Symbiotic characteristics of *Rhizobium meliloti:* An appraisal of the systematic treatment of nodulation and nitrogen fixation: Interactions between hosts and rhizobia of diverse origins. Aust. J. Agric. Res. 17:885–889.

Bromfield, E.S.P., and D.G. Jones. 1980. Studies on double strain occupancy of nodules and the competitive ability of *Rhizobium trifolii* on red and white clover grown in soil and agar. Ann. Appl. Biol. 94:51–59.

Bromfield, E.S.P., M. Stein, and R.P. White. 1982. Identification of *Rhizobium* strains on antibiotic-concentration gradients. Ann. Appl. Biol. 101:269–274.

Bushby, H.V.A. 1982. Direct quantitative recovery of *Rhizobium* from soil and rhizosphere. p. 59–67. *In* J.M. Vincent (ed.) Nitrogen fixation in legumes. Academic Press, New York.

Cenci, G., G. Morozzi, R. Danielle, and F. Scazzocchio. 1980. Antibiotic and metal resistance in *Escherichia coli* strains isolated from the environment and from patients. Ann. Sclavo 22:212–226

Chatel, D.L., and C.A. Parker. 1973a. Survival of field-grown *Rhizobium* over the dry summer period in Western Australia. Soil Biol. Biochem. 5:415–423.

Chatel, D.L., and C.A. Parker. 1973b. The colonization of host-root and soil by rhizobia. 1. Species and strain differences in the field. Soil Biol. Biochem. 5:425–432.

Cole, M.A., and G.H. Elkan. 1979. Multiple antibiotic resistance in *Rhizobium japonicum*. Appl. Environ. Microbiol. 37:867–871.

Diatloff, A. 1977. Ecological studies of root-nodule bacteria introduced into field environments—6. Antigenic and symbiotic stability in *Lotononis* rhizobia over a 12-year period. Soil Biol. Biochem. 9:85–88.

Eaglesham, A.R.J. 1987. The use of intrinsic antibiotic resistance for *Rhizobium* study. p. 185–204. *In* G.H. Elkan (ed.) Symbiotic nitrogen fixation technology. Marcel Dekker, New York.

Fontaine, T.D., and A.W. Hoadley III. 1976. Transferable drug resistance associated with coliforms isolated from hospital and domestic sewage. Health Lab. Sci. 13:238–242.

Hagedorn, C. 1979. Relationship of antibiotic resistance to effectiveness in *Rhizobium trifolii* populations. Soil Sci. Soc. Am. J. 43:921–925.

Hagedorn, C. 1984. Microbiological aspects of groundwater pollution due to septic tanks. p. 181–195. *In* G. Bitton, and C.P. Gerba (ed.) Groundwater pollution microbiology. John Wiley and Sons, New York.

Hagedorn, C. 1986. Role of genetic variants in autecological research. p. 61–74. *In* R.L. Tate (ed.) Microbial autecology. John Wiley and Sons, New York.

Hardarson, G., and D.G. Jones. 1979. Effect of temperature on competition amongst strains of *Rhizobium trifolii* for nodulation of two white clover varieties. Ann. Appl. Biol. 92:229–236.

Hollis, A.B., Kloos, W.E., and G.H. Elkan. 1981. Metabolite resistance and mutational effects in *Rhizobium*. J. Gen. Microbiol. 123:215–218.

Howell, C.R., and R.D. Stipanovic. 1979. Control of *Rhizoctonia solani* on cotton seedlings with *Pseudomonas fluorescens* and with an antibiotic produced by the bacterium. Phytopathology 69:480–484.

Johnston, A.W.B., and J.E. Beringer. 1975. Identification of *Rhizobium* strains in pea root nodules using genetic markers. J. Gen. Microbiol. 87:343–350.

Josey, D.P., J.L. Beynon, A.W.B. Johnston, and J.E. Beringer. 1979. Strain identification in *Rhizobium* using intrinsic antibiotic resistance. J. Appl. Bacteriol. 46:343–350.

Kowalski, M., G.E. Ham, L.R. Fredrick, and I.C. Anderson. 1974. Relationship between strains of *Rhizobium japonicum* and their bacteriophages from soil and nodules of field-grown soybeans. Soil Sci. 118:221–228.

Kramer, R.J., and H.L. Peterson. 1981. Nodulation efficiency of legume inoculation as determined by intrinsic antibiotic resistance. Appl. Environ. Microbiol. 43:636–642.

Kuykendall, L.D. 1987. Isolation and identification of genetically marked strains of nitrogen-fixing microsymbionts of soybeans. p. 205–220. *In* G.H. Elkan (ed.) Symbiotic nitrogen fixation technology. Marcel Dekker, New York.

Kuykendall, L.D., and D.F. Weber. 1978. Genetically marked *Rhizobium* identifiable as inoculum strains in nodules of soybean plants grown in fields populated with *R. japonicum*. Appl. Environ. Microbiol. 36:915–919.

Levin, R.A., and M.P. Montgomery. 1974. Symbiotic effectiveness of antibiotic-resistant mutants of *Rhizobium japonicum*. Plant Soil 41:669–676.

Linton, A.H. 1983. Theory of antibiotic inhibition zone formation, disc sensitivity methods and MIC determinations. p. 19–30. *In* A.D. Russell, and L.B. Quesnel (ed.) Antibiotics: Assessment of antimicrobial activity and resistance. Academic Press, New York.

Materon, L.A., and C. Hagedorn. 1983. Competitiveness and symbiotic effectiveness of five strains of *Rhizobium trifolii* on red clover. Soil. Sci. Soc. Am. J. 47:491–495.

McCoy, E.L., and C. Hagedorn. 1979. Quantitatively tracing bacterial transport in saturated soil systems. Water Air Soil Pollut. 11:467–479.

McLaughlin, W., and M.H. Ahmad. 1984. Intrinsic antibiotic-resistance and streptomycin uptake on cowpea *Rhizobium*. FEMS Microbiol. Lett. 26:299–306.

Misplon, J.A., and P.E. Bishop. 1983. Isolation and partial characterization of fix mutants from *Rhizobium* strain 32H1. Plant Soil 74:395–402.

Munns, D.N. 1970. Nodulation of *Medicago sativa* in solution culture. V. Calcium and pH requirements during infection. Plant Soil 32:90–94.

Mytton, L.R. 1975. Plant genotype × *Rhizobium* strain interactions in white clover. Ann. Appl. Biol. 80:103–107.

Obaton, M. 1971. Utilization de mutants spontanes resistants aux antibioti-ques pour letude ecologique du *Rhizobium*. C. R. Acad. Bulg. Sci. 272:2630–2633.

Pankhurst, C.E. 1977. Symbiotic effectiveness of antibiotic resistant mutants of fast- and slow-growing strains of *Rhizobium* nodulating *Lotus* species. Can. J. Microbiol. 23:1026–1033.

Peterson, E.A., J.C. Sirois, W.B. Berndt, and R.W. Miller. 1983. Evaluation of competitive ability of *Rhizobium meliloti* strains for nodulation in alfalfa. Can. J. Microbiol. 29:541–546.

Phillips, D.A. 1974. Factors affecting the reduction of acetylene by *Rhizobium* soybean cell associations *in vitro*. Plant Cell Physiol. 54:654–658.

Pike, E.B., A.W.D. Bufton, and D.J. Gould. 1969. The use of *Serratia indica*, and *Bacillus subtilis* and *niger* spores for tracing sewage dispersion in the sea. J. Appl. Bacteriol. 32:206–209.

Pinto, C.M., P.K. Yao, and J.M. Vincent. 1974. Nodulating competitiveness amongst strains of *Rhizobium meliloti* and *R. trifolii*. Aust. J. Agric. Res. 25:317–329.

Pugashetti, B.K., and G.H. Wagner. 1980. Survival and multiplication of *Rhizobium japonicum* strains in silt loam. Plant Soil 56:217–227.

Rai, R. 1983. The salt tolerance of *Rhizobium* strains and lentil genotypes and the effects of salinity on aspects of symbiotic N-fixation. J. Agric. Sci., Cambridge 100:81–87.

Rippon, J.E. 1963. The use of a coloured bacterium as an indicator of local water movement. Chem. Ind. (March) 16:445–502.

Roy, P., and A.K. Mishia. 1974. The relative frequencies of spontaneous and UV [ultraviolet]-induced antibiotic resistant mutations in *Rhizobium lupini*. Sci. Cult. 40:373–378.

Scholla, M.H., and G.H. Elkan. 1984. *Rhizobium fredii* sp. *nov.,* a fast growing species that effectively nodulates soybeans. Int. J. Syst. Bacteriol. 34:484–488.

Schwinghamer, E.A. 1964. Association between antibiotic resistance and ineffectiveness in mutant strains of *Rhizobium* spp. Can. J. Microbiol. 10:221–223.

Schwinghamer, E.A. 1967. Effectiveness of *Rhizobium* as modified by mutation for resistance to antibiotics. Antonie van Leeuwenhoek, J. Microbiol. Serol. 33:121–136.

Schwinghamer, E.A., and W.F. Dudman. 1973. Evaluation of spectinomycin resistance as a marker for ecological studies with *Rhizobium* spp. J. Appl. Bacteriol. 36:263–272.

Sinclair, M.J., and A.R.J. Eaglesham. 1984. Intrinsic antibiotic resistance in relation to colony morphology in three populations of West African cowpea Rhizobia. Soil Biol. Biochem. 26:247–252.

Skrdleta, V. 1970. Competition for nodule sites between two inoculum strains of *Rhizobium japonicum* as affected by delayed inoculation. Soil Biol. Biochem. 2:167–170.

Stein, M., E.S.P. Bromfield, and M. Dye. 1982. An assessment of a method based on intrinsic antibiotic resistance for identifying *Rhizobium* strains. Ann. Appl. Biol. 101:261–266.

Stewart, K.R., and L. Koditschek. 1980. Drug resistance transfer in *Escherichia coli* in New York bight sediment. USA Mar. Poll. Bull. 11:130–133.

Talbot, T.W., Jr., D.K. Yamamoto, M.W. Smith, and R.J. Seidler. 1980. Antibiotic resistance and its transfer among clinical and non-clinical *Klebsiella* strains in botanical environments. Appl. Environ. Microbiol. 39:97–104.

Vincent, J.M. 1970. A manual for the practical study of root nodule bacteria. IBP Handb. no. 15. Blackwell Publ. Ltd., Oxford.

Weller, D.M., and R.J. Cook. 1983. Suppression of take-all of wheat by seed treatments with fluorescent pseudomonads. Phytopathology 73:463–469.

Wimpenny, J.W.P., N. Cotton, and M. Stratham. 1972. Microbes as tracers of water movement. Water Res. 6:731–739.

Chapter 28

Serology and Conjugation of Antibodies

S. F. WRIGHT, *USDA-ARS Soil Microbial Systems Laboratory, Beltsville, Maryland*

Identification of microorganisms using antibodies has been a common practice for more than 30 yr. Antibody probes have been employed in environmental microbiology because of their usefulness in answering questions about organisms found in mixtures that comprise natural ecosystems. Principle uses of antibodies in soil microbiology are: (i) to identify specific microorganisms that are studied in situ and that must be identified as individual cells in a mixture of organisms, (ii) to quantify or identify colonies of bacteria in a mixed culture on agar media, (iii) for identification of a selected strain or isolate, and (iv) to identify *Rhizobium* or *Bradyrhizobium* strains in legume nodules.

The approach in this chapter is to present references for methods of production of antibodies, basic methods for concentration and storage of antibodies, a method to conjugate antibodies with fluorescine, and several variations of the enzyme-linked immunosorbent assay (ELISA).

Enzyme immunoassays have been used extensively for the past decade. Some of the uses in environmental microbiology are to quantify microorganisms (Nambiar & Anjaiah, 1985; Renwick & Jones, 1985) and to detect *Rhizobium* or *Bradyrhizobium* strains in nodules (Fuhrmann & Wollum, 1985; Kishinevsky & Jones, 1987; Wright et al., 1986), and viruses in water (Nasser & Metcalf, 1987).

Detailed information on the uses of fluorescent-tagged antibodies is given in chapter 6 by Bottomley in this book. Conjugation of antibodies and assays can be performed as described in this text, as modified for specific uses, or as described in information accompanying commercially available kits. References to kits are given in the text, and mention of a product by company name does not necessarily exclude the use of comparable products available from other sources. Antibody processing and assays involving antibodies are advancing rapidly and new products and kits are becoming available through companies specializing in immunological reagents or general biotechnology supplies.

Production and use of antibodies requires skills that can be learned by trial and error. However, it is highly recommended that investigators who choose this methodology to identify organisms seek training in laboratories where antibody assays are used routinely. Much time, effort, and expense can be saved by such training.

28–1 ANTIBODIES

28–1.1 Microbes as Antigens

An antigen is a molecule recognized as a foreign substance by a vertebrate. An antibody is a molecule generated in response to a foreign substance that acts to neutralize or damage the invading antigen molecule. Antigens may be as small as six amino acids or six monosaccharide units (Kabat, 1967). Proteins and polysaccharides from fungi, viruses, and bacteria that elicit antibody responses (immunogens) can be injected into an animal as unmodified whole cells or particles, partially purified compounds, or purified compounds. This chapter focuses primarily on the production and use of antibodies to identify bacteria, but the basic principles apply to identification of other microorganisms found in soils.

28–1.2 Antibodies

Vertebrates have an inherent ability to recognize large numbers (estimated in the millions) of antigens and then, depending on the host's ability to respond, to initiate antibody production against them. The mechanism for the initial and rapid recognition of so many different antigens was first proposed by Burnet (1959). A population of a type of lymphocyte (B cell) is continuously present, and each member of the population possesses a set recognition site for a given foreign molecule. Stimulation of a specific B cell by a complementary antigenic determinant may initiate a process resulting in clonal expansion of antibody-producing cell lines and release of large amounts of antibody into the peripheral blood.

Antibodies are serum glycoproteins that have both variable and constant amino acid sequence regions. Specific binding of an antigen occurs in identical variable regions that are at the tips of the two prongs of a molecule represented in the shape of a "Y" (Fig. 28–1). The molecule is a dimer containing four polypeptide chains—two light and two heavy chains. Antibodies are differentiated into classes based upon the heavy chains. Figure 28–1 illustrates the two most common classes of antibodies resulting from immunization with bacterial antigens—immunoglobulin G (IgG) and immunoglobulin M (IgM). These molecules can be differentiated by reactions employing antibodies specific for amino acids on the heavy chain of IgG (γ heavy chain with several subclass variations) or IgM (μ heavy chain) of a particular animal species immunoglobulin. Anti-mouse IgG, anti-rabbit IgG, or anti-mouse IgM antibodies are examples of products obtained

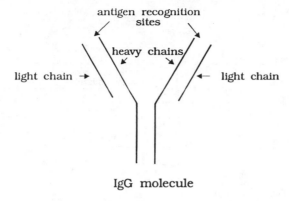

IgG molecule

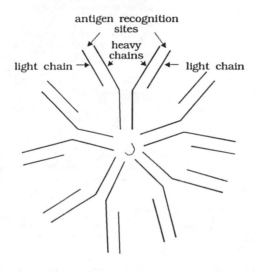

IgM molecule

Fig. 28–1. A diagrammatic representation of the structure of immunoglobulin G (IgG) and immunoglobulin M (IgM) molecules. Heavy and light chains of these molecules are labeled along with the antigen recognition sites. The IgM molecule is a pentamer of molecules similar to IgG with a total of 10 antigen recognition sites.

by injecting mouse (*Mus musculus*) or rabbit (*Oryctolagus cuniculus*) immunoglobulin in goats (*Capra hircus*). Anti-immunoglobulin antibodies are widely available from biochemical suppliers.

Immunoglobulin M antibodies are the first type to appear in blood following immunization. After exposure to the antigen by booster injections, there may be a switch to production of IgG antibodies. Early response antibodies often have lower affinity (binding strength) for antigens

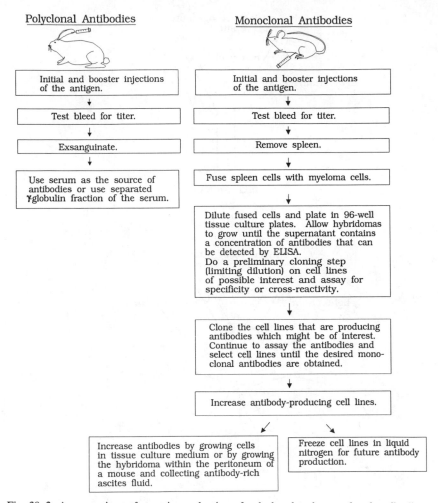

Fig. 28–2. A comparison of steps in production of polyclonal and monoclonal antibodies.

than antibodies that develop later. However, early response antibodies generally are more specific than later, higher-affinity antibodies (Marchalonis, 1982).

Two general types of antibody preparations are designated by the descriptive terms *polyclonal* and *monoclonal*. Figure 28–2 compares the production of these two types of antibodies. Polyclonal antiserum contains a mixture of antibodies to microbial antigens in serum taken from an immunized animal. Monoclonal antibodies result from the fusion of an individual antibody-secreting cell from an immunized mouse or rat and a malignant (myeloma) cell to produce an immortal hybrid cell (hybridoma) that secretes a single type of specific antibody. Figure 28–3 compares the

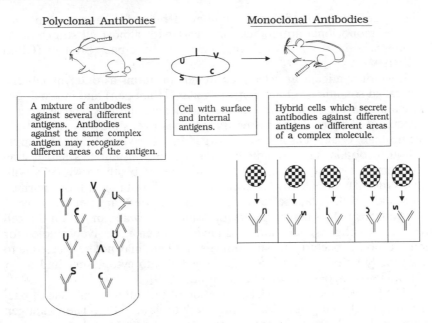

Fig. 28–3. A comparison of the products obtained from polyclonal and monoclonal antibody production. The antigen represented is a bacterium with several different antigenic sites.

products obtained from polyclonal or monoclonal antibody production procedures.

Protocols for inducing polyclonal antiserum generally lead to production of IgG antibodies. Protocols for immunizations to produce monoclonal antibodies can be adjusted for either early or late-response antibodies, and selection for IgG or IgM antibodies can be made during screening of antibody-producing cell lines.

28–1.3 Antibodies Against Microorganisms

The discussion of antibodies for identification of microorganisms will focus on gram-negative bacteria. For information on antigens of gram-positive bacteria and other microorganisms the reader is referred to Kwapinski (1969), Male et al. (1987), and Macario and Conway de Macario (1985).

Gram-negative bacteria have many antigenic components, but when whole cells are injected, type-specific or group-specific antibodies suitable for identification of such cells generally are elicited. Type-specific antigens occur among one, two, or three strains of a species. For example, there are type-specific monoclonal antibodies that react with several strains of *Rhizobium leguminosarum* bv. *trifolii* (Wright et al., 1986). Group-specific antigens are shared by a larger number of related microorganisms, for example, monoclonal antibodies that react with almost all strains of *R. leguminosarum* bv. *trifolii* or *R. meliloti* tested have been produce (Olsen et al., 1994).

Bacterial antigens are classified in general terms as outer membrane (also called somatic or O-antigen), flagellar (H antigen) and internal. Somatic antigens are lipopolysaccharide complexes, and this type of antigen is discussed in more detail below. Flagellar antigens may be more strain specific than O-antigens for some organisms (Davies, 1951). A simple method to obtain flagellar antigens for immunizing animals and testing antiserum is by inactivating somatic antigens of broth-grown cells with 0.1% (vol/vol) formaldehyde (Davies, 1951). To obtain a more purified preparation of flagella, the reader is referred to de Maagd et al., 1989.

Antigens that occur on the cytoplasmic membrane or within the cell are internal antigens. Such antigens have not been extensively studied for identification of bacteria because internal antigens may not be accessible to an antibody unless cells are first fractionated into cytoplasmic and outer membranes components (de Maagd et al., 1989).

The lipopolysaccharide (LPS) fraction of the outer membrane of bacteria is strongly antigenic (McCartney & Wardlaw, 1985). The O-antigen that caps the LPS molecule and imparts "smoothness" in colony morphology has antigenic sites of the type-specific range (Holme & Gustafsson, 1985; Carlin & Lindberg, 1985). Another LPS fraction, the lipid-A rough core portion, has antigenic sites of the group-specific range (Holme & Gustafsson, 1985; Carlin & Lindberg, 1985). The O-antigen is heat stable and, therefore, to produce antibodies against this fraction, bacterial cells are washed in phosphate-buffered saline (see 28–2.2), resuspended in the same buffer, and then heated at 100 °C for 30 min to denature many of the other antigenic determinants before being injected into the animal. Lipopolysaccharide antigens elicit strong immune responses (McCartney & Wardlaw, 1985) such that antibodies to other antigenic configurations of a bacterial cell often are not produced in amounts great enough to be detected. Also, other antibody-binding sites may be masked by O-antigen LPS (de Maagd et al., 1989) when the whole cell is used in assays to detect antibodies.

Plant viruses and beneficial or pathogenic plant fungi can be identified using antibodies (Lin et al., 1990; Culver & Sherwood, 1988; Nameth et al., 1990; Wright et al., 1987; Hardham et al., 1986; Banowetz et al., 1984). Fungal spores (Wright et al., 1987; Hardham et al., 1986) have been used as antigens. Williams and Chase (1967) describe methods using plant and bacterial viruses as antigens. Harlow and Lane (1988) provide information on purifying antigens and making weak antigens more antigenic along with practical information on other current methods in immunology.

The growth stage of bacteria and the medium used for culturing bacteria may be important in expression of an antigen or contamination with unwanted antigens. A young, motile broth culture of a *Rhizobium* strain will have flagellar antigens (Vincent, 1970). Immunodeterminants of *R. leguminosarum* bv. *trifolii* can be growth-phase-dependent (Hrabak et al., 1981). Use of complex media for production of antigen and testing isolates for reactivity with polyclonal antibodies may interfere with the reaction (Vincent, 1970).

Defining the degree of specificity of an antibody to a selected microbe is essential to use antibodies effectively for tracing microorganism in the environment. Testing for specificity should involve reacting the antibody with as large a number of similar and diverse antigens as the researcher deems necessary to be convinced of the specificity. For example, antibodies against rhizobia should be tested for reactivity against isolates from the same species and biovar and naturalized isolates from the geographic area where the test strain is to be inoculated. Spiking soils with defined numbers of the test strain and quantifying extracted cells reactive with the antibody is an approach that is sometimes useful as a test of specificity.

28–1.3.1 Polyclonal Antibodies

For routine production of polyclonal antibodies New Zealand white rabbits generally are used. The form of the antigen to be injected, route of immunization, and injection schedule are covered in many references (Campbell et al., 1974; Harlow & Lane, 1988; Kwapinski, 1972; Kishinevsky & Jones, 1987; Sutherland, 1977; Warr, 1982; Williams & Chase, 1967; Graham, 1969; Dudman, 1977; Vincent, 1970). Investigators who routinely produce antibodies against specific microorganisms are probably the best sources for the fine details of an immunization protocol. A generalized protocol for injection, test bleeding, and exsanguination is described by Warr (1982). Graham (1969) gives references for various immunization schedules using rhizobia as antigens, and Kishinevsky and Jones (1987) provide detailed methods for immunization of rabbits with *Bradyrhizobium, Rhizobium leguminosarum* bv. *trifolii*, and *R. meliloti* for antiserum used in ELISA. Quantification of bacterial antigen to be injected is generally by number of cells. For example, Kishinevsky and Jones (1987) use 0.5 to 1.0 mL of a cell suspension with 10^8 to 10^{11} cells/mL for each injection. Animals vary in the intensity of the response to an antigen, so a given protocol may not always give the expected results.

Polyclonal antibodies may contain a mixture of specific and cross-reactive antibodies against several to many organisms similar to the immunogen. Cross-reactive antibodies can be adsorbed from such antiserum (see section 28–2), and often the result is a highly specific antiserum.

28–1.3.2 Monoclonal Antibodies

Monoclonal antibody production requires more time and resources than polyclonal production. However, as indicated in Fig. 28–2, the end product can be more specific than polyclonal antiserum because of the ability to be highly selective for individual antibody-secreting cell lines. Monoclonal antibodies may not always be strain-specific (Wright et al., 1986; Wright, 1990). However, recently highly specific monoclonal antibodies to a *R. leguminosarum* bv. *trifolii* isolate and to a *R. meliloti* strain have been produced (Olsen et al., 1994).

Training for monoclonal antibody production requires, at a minimum, an intense workshop over several days, or better, a longer-term stay at a production facility. Many universities now offer monoclonal antibody production in a central laboratory as a service. Users of these services should be well-informed of the production process to describe and get a product of desired specificity.

A brief discussion of monoclonal antibody production will be presented here. For more information the reader is directed to books on production and use of monoclonal antibodies (Goding, 1986; Harlow & Lane, 1988; Hurrell, 1983; Zola, 1987).

A mouse or rat strain for which a compatible genetically modified myeloma (malignant cell for fusion with the B-cell) line exists is used. The BALB/c mouse strain is used most commonly. After immunization, the animal's spleen is removed as the source of antibody-producing cells. A suspension of splenic B lymphocytes from the mouse is fused with the appropriate myeloma tumor cells by the action of polyethylene glycol on cell membranes. The fused cells are diluted and cultured in the presence of chemicals that ensure growth of only hybrids between myeloma and splenic cells (hybridomas). Hybridomas can be seen microscopically after 4 to 5 d, and macroscopic plaques of cells are visible within 15 d post-fusion. The tissue culture supernatant contains antibodies secreted from the hybridomas and is used to test for the presence of antibodies. Screening and selection of desirable hybridoma cell lines is accomplished by testing for secreted antibody reaction with selected antigens. Selected cell line clones are obtained by limiting dilution. Such cell lines can be frozen in liquid N and regrown at a later date. Antibodies can be obtained from tissue culture fluid or mouse ascites. For the latter source, the hybridoma is injected within the peritoneum of a mouse and highly concentrated antibodies are obtained from the ascitic fluid resulting from the in vivo growth of the hybridoma.

Fusion of an antibody-producing cell with a myeloma cell is a random event and the outcome of a fusion can depend upon the skill and experience of the person performing the fusion. However, the animal must have a prefusion serum reaction against the antigen. Harlow and Lane (1988) recommend assaying test bleed serum diluted 1 in 5 or 1 in 10 buffer compared with equivalently diluted serum taken from an unimmunized animal or an animal immunized with an unrelated antigen. Obtaining a titer of preimmunization serum is also advisable. We have found high titers in some unimmunized mouse serum when tested against vesicular-arbuscular mycorrhizal spore antigens (unpublished data).

28–1.4 Procedures for Utilization of Antibodies

Investigators who are initiating the use of antibodies in enzyme-linked immunosorbent assay (ELISA) for identification of microorganisms should have the assay in place before production of antibodies is started. This may necessitate requesting the donation of an antibody and an antigen from a colleague to calibrate the ELISA against the donor's standard assay. For example, polyclonal antibody titers are often given in terms of agglutination test results (Vincent, 1970). A polyclonal antiserum against a *Rhizobium* strain should have an agglutination titer of at least 1:960 (Schmidt et al., 1968) before conjugation with a fluorochrome (see section 28–4). Likewise, a polyclonal antibody for ELISA to identify *Bradyrhizobium* or *Rhizobium* strains must have an agglutination titer of 1:1280 (Kishinevsky & Jones, 1987). ELISA results using antibodies to detect *Bradyrhizobium* peanut strains are more sensitive than agglutination by four to six orders of magnitude (Kishinevsky & Bar-Joseph, 1978). Therefore, it is expected that antiserum titer determined by ELISA also will be much higher than an agglutination titer.

28–2 ADSORPTION OF CROSS-REACTIVE ANTIBODIES FROM POLYCLONAL ANTISERUM

28–2.1 Principles

Polyclonal antiserum obtained from rabbits may be used without further purification. Because the serum is a mixture of antibodies against different antigens, it is often necessary to adsorb the antiserum with cross-reactive antigens to obtain or enhance specificity. The following is a recommendation for the ratio of serum to adsorbent, but it may be necessary to readsorb an antiserum if activity against the cross-reactive species is only partially eliminated by the suggested protocol.

28–2.2 Materials

1. Glass test tubes (10 × 75 mm).
2. Phosphate-buffered saline (PBS). Dissolve 8.0 g of sodium chloride (NaCl), 0.2 g of potassium chloride (KCl), 1.44 g of dibasic sodium phosphate (Na_2HPO_4), and 0.24 g monobasic potassium phosphate (KH_2PO_4) in 800 mL distilled or deionized water. Adjust the pH to 7.4. Bring the volume to 1 L. Dispense in convenient volumes, sterilize by autoclaving (121 °C, 15 psi, for 15 min), and store at room temperature.
3. Cells of a cross-reactive species are grown under the same conditions as the antigen, harvested by centrifugation (15 min at

10 000 × g), resuspended in PBS to a concentration of 10^{10} to 10^{11} cells/mL, heated at 100 °C for 30 min and then washed three times in PBS.
4. Antiserum diluted 1:1 and 1:2 with PBS.
5. Water bath at 37 °C.

28–2.2 Procedure

Use 1 mL of the adsorbing antigen/mL of antiserum. Mix gently, incubate for 1 h in the water bath, and then remove the serum from the adsorbing material by centrifugation (15 min at 10 000 × g). Assay the resulting adsorbed antiserum against the antigen of interest by using the assay for which the antibody will be employed and compare results with the unadsorbed antiserum. Readsorb if necessary. Larger volumes of antiserum can be adsorbed using the same ratio of antiserum/adsorbing cells.

28–3 PURIFICATION OF ANTIBODIES FROM ANTISERUM

28–3.1 Principles

Conjugation of dyes or enzymes to immunoglobulins requires that immunoglobulins be concentrated and purified. Ammonium sulfate precipitation of immunoglobulins from serum or tissue culture supernatant is one of the most commonly used methods to concentrate antibodies. Most immunoglobulins will precipitate at 35 to 40% of solution saturation with ammonium sulfate (Goding, 1986). Mouse monoclonal antibodies may require 50% saturation (Harlow & Lane, 1988). Therefore, for precipitation of immunoglobulins from rabbit serum, 35% saturation is recommended, and for mouse antibodies, 50% saturation is recommended.

Immunoglobulin G may be further purified from polyclonal antiserum for conjugation with fluorochromes or enzymes by passing the concentrated immunoglobulins through a column of diethylaminoethyl (DEAE)-cellulose. The column is packed in a low molarity buffer at pH 8.0. At pH 8.0 and lower pH, IgGs have the lowest charge of the serum proteins and often can be eluted from the column as the only protein not bound in the starting buffer. Quantification of immunoglobulin by using absorbance at 280 nm is suggested. The assumption has been made that the immunoglobulin of interest is an IgG. For concentration of an IgM monoclonal antibody, the reader is referred to other sources (Harlow & Lane, 1988; Goding, 1986).

Other methods for purification of antibodies are available including affinity chromatography using protein A or protein G. The reader is referred to texts on current methods for handling and using antibodies (Harlow & Lane, 1988; Goding, 1986; Zola, 1987).

28–3.2 Materials

1. A glass beaker to accommodate over twice the volume of antiserum to be concentrated, a stirring bar and magnetic stirrer. A convenient volume to start with is 5 to 10 mL of polyclonal antiserum. Various amounts of monoclonal antiserum might be the starting point, depending on whether tissue culture supernatant or ascites fluid is the source of the antibody. Ascites fluid should be diluted 1 part:3 parts with PBS before starting the precipitation procedure.
2. Saturated ammonium sulfate—800 g of ammonium sulfate $[(NH_4)_2SO_4]$ per liter of distilled or deionized water. Leave the solution at room temperature at least overnight. Undissolved $(NH_4)_2SO_4$ will remain at the bottom of the vessel. The pH of the solution should be adjusted to 7.5 with 2 N of sodium hydroxide (NaOH) or hydrochloric acid (HCl) on the day of use.
3. Dialysis tubing (molecular weight cut-off of 12 000–14 000). Cut the tubing to an appropriate length to accommodate about twice the volume of antibody to provide sufficient room for an increase in volume during dialysis and extra length to tie or clamp securely at both ends. Bring tubing to a low boil in distilled water and then rinse with distilled water. Soak tubing in 0.01 M ethylene diamine tetraacetate·2H$_2$O (EDTA) for 10 min and then rinse thoroughly in distilled water (5–10 times) and use immediately.
4. Phosphate-buffered saline (see section 28–3.2), diluted to half-strength and pH adjusted to 8.0. Use 3 L for each dialysis tube.
5. DEAE-cellulose (Whatman DE23, Whatman LabSales, Hillsboro, OR). Follow manufacturer's instruction for hydration, pour into a column (disposable columns in several different sizes in a trial pack are available from Pierce, Rockford, IL), and pre-equilibrate in half-strength PBS. Use approximately 2 mL of wet DEAE-cellulose for each milliliter of serum. A DEAE IgG purification kit including DEAE-cellulose and premixed application and regeneration buffers is available from Bio-Rad Chemical Division, Richmond, CA.
6. UV-visible spectrophotometer with 1 cm quartz cuvettes.

28–3.3 Procedure

Determine the volume of serum, tissue culture supernatant, or ascitic fluid. Centrifuge 30 min at 3000 × g. Transfer the supernatant to a beaker and stir the antibody solution gently while slowly adding the $(NH_4)_2SO_4$. To initially precipitate mouse antibodies from tissue culture supernate, add an equal volume of saturated $(NH_4)_2SO_4$ which brings the concentration to approximately 50%. To precipitate antibodies from rabbit serum or mouse ascites, add saturated $(NH_4)_2SO_4$ to bring the concentration to 35%. For every 65 volumes of antiserum, add 35 volumes $(NH_4)_2SO_4$, or

35 × volume of antibody/65 = volume of saturated $(NH_4)_2SO_4$. An example of the amount of saturated $(NH_4)_2SO_4$ to add to 10 mL of antiserum to obtain a 35% saturated solution is 5.4 mL. Add $(NH_4)_2SO_4$ dropwise and allow each drop to disperse before adding the next drop. Let the mixture stand for at least 1 h at 4 °C. Centrifuge at 3000 to 10 000 × g for 30 min and then carefully remove and discard the supernatant. For polyclonal antiserum or ascites, resuspend the precipitate in half-strength PBS to the original volume. For tissue culture supernatant, resuspend the pellet in 1/10 the original volume. Repeat the $(NH_4)_2SO_4$ precipitation at least two more times or until the precipitate is completely white using the new volume to calculate 35% $(NH_4)_2SO_4$. For serum or ascites, without creating bubbles, resuspend the pellet in 0.3 to 0.5 volumes of the starting volume using half-strength PBS. For monoclonal antibodies from tissue culture supernatant, resuspend the pellet in 0.1 volume of the starting volume. Transfer the antibody solution to the dialysis tubing. After removing trapped air, tie or clamp the tubing securely. Dialyze overnight in three changes of half-strength PBS. Remove the antibody solution from the tubing and centrifuge (30 min at 3000 × g) to remove any debris.

Optional further purification to obtain only IgG from polyclonal antiserum is accomplished by passing the antibody solution through the DEAE-cellulose column and eluting the IgG with half-strength PBS. The eluate is collected in approximately 2-mL fractions. Monitor eluate at 280 nm and combine all fractions containing the first protein peak. Determine the concentration of immunoglobulin by reading the absorbance at 280 nm with half-strength PBS as a blank. Divide the reading by 1.35 to obtain mg/mL of IgG.

28–4 CONJUGATION OF ANTIBODIES WITH FLUORESCEIN ISOTHYOCYANATE

28–4.1 Principles

The most commonly used method for direct detection of microorganisms is fluorescent dye-tagged antibodies. Conjugation of antibodies with fluorochromes is a technically simple procedure, but care should be taken to obtain optimal conjugation. Maximum efficiency of conjugation is achieved at pH 9.5 where a large fraction of the lysine in the globulin is more receptive to linkage with the dye (Goding, 1986). Tris, extraneous amino acids, or azide will inhibit conjugation (Goding, 1986). Isothyocyanate derivatives should be stored under desiccation because of susceptibility to hydrolysis even from moisture in the air. Almost all preparations of fluorochromes contain at least some hydrolyzed product making it necessary to add excess fluorochrome to achieve maximum conjugation. At high protein concentrations (> 10 mg/mL) conjugation is strongly favored. At lower concentrations (1–2 mg/mL) competing hydrolysis becomes significant but a suitable product can be obtained by adding excess fluorochrome. The method presented here is for conjugation with fluorescein isothyocyanate (FITC). Other dyes such as rhodamine or Texas red may be used, and methods for conjugation with these can be found in laboratory manuals (Goding, 1986; Harlow & Lane, 1988). A kit for FITC tagging of antibodies is available which contains all necessary reagents and detailed instructions (Boeringer Mannheim Corp., Indianapolis, IN).

28–4.2 Materials

1. Ammonium sulphate fractionated and DEAE-cellulose purified IgG antibodies.
2. Dialysis tubing (molecular weight cut-off 12 000–14 000).
3. Carbonate/bicarbonate buffer—8.6 g of sodium carbonate (Na_2CO_3) and 17.2 sodium bicarbonate ($NaHCO_3$) in 1 L distilled or deionized water, pH 9.5.
4. Fluorescein isothiocyanate warmed to room temperature before opening. Stock solution 10 or 1 mg/mL in dimethyl suphoxide (DMSO)—the best grade available with no water (to prevent immediate hydrolysis of the dye).
5. Sephadex G-25 columns. (Available as disposable PD-10 columns suitable for desalting a volume up to 2.5 mL from Pharmacia, Piscataway, NJ).
6. PBS—see section 28–2.2.
7. UV-visible spectrophotometer with 1 cm quartz cuvettes.

28–4.3 Procedure

Dialyze the antibody overnight in carbonate/bicarbonate buffer. Determine the concentration of IgG (protein) in mg/mL by dividing the OD_{280} by 1.35 using as a blank the carbonate/bicarbonate buffer. The amount of FITC to add depends upon the protein concentration (Goding, 1986). For concentrations > 10 mg/mL use 10 to 20 μg FITC/mg protein, and for 3 or 1 mg/mL use approximately 50 or 100 μg FITC/mg protein, respectively. Add the desired volume of FITC solution to the protein slowly with stirring. Allow the reaction to proceed at room temperature in the dark without stirring for 1 to 2 h. The reaction time depends upon the concentration of protein. At high protein concentrations 1 h may be sufficient. At lower concentrations of protein the longer time may be necessary. Up to 3 h may be desirable if the protein concentration is < 1 to 2 mg/mL.

Adjust the pH of the conjugated protein to 7.4 by slowly adding 1 N HCl. Use gel filtration (Sephadex G-25) to remove the unreacted or hydrolyzed dye from the protein according to directions supplied with the columns using PBS as the buffer. The first colored band to emerge is the conjugated protein.

The absorbance ratio of $OD_{495}:OD_{280}$ should be 0.3 to 1.0 (Goding, 1986; Harlow & Lane, 1988). If the conjugated protein does not fall within this ideal ratio, the mass of fluorochrome may need to be adjusted for future conjugations or, using the same preparation, another attempt at conjugation can be made to improve the ratio. See chapter 6 in this book for fluorescent antibody techniques.

28–5 STORAGE OF ANTIBODIES AND ANTIBODY CONJUGATES

Antibodies can be stored at 4 °C for years without significant activity loss if proteolysis is inhibited. Proteolysis may result from proteases in the antibody source or from microbial contamination. Preservatives that prevent microbial growth can be added or preparations can be filter-sterilized. Preservation by the addition of 10 mM sodium azide (NaN_3) or 0.005% merthiolate is common. Sodium azide is poisonous. It blocks the cytochrome electron transport system, and therefore, all additions of NaN_3 and handling of such amended antibodies should be done with care. Disposable plastic gloves are recommended for handling NaN_3 and all solutions containing this compound. If the concentration of antibody is < 1 mg/mL, it is also advisable to add bovine serum albumin at a final concentration of 1% (wt/vol). An additional consideration for fluorescein-conjugated antibodies is the use of lightproof containers.

Freezing and thawing can lead to irreversible aggregation. However, if freezing is the method of choice, the number of freeze-thaw cycles should be limited by freezing at −20 °C in multiple small aliquots.

28–6 IMMUNOASSAYS

28–6.1 Principles

There are a variety of immunoassays to detect and quantify antigens and antibodies. Current methods provide sensitivity much beyond that obtained by agglutination, double diffusion, and precipitation tests previously used as standard assays (Kishinevsky & Bar-Joseph, 1978). Large numbers of samples can be easily accommodated even in laboratories that only occasionally use immunoassays and by investigators who are not highly skilled immunochemists.

Soil microbiologists generally are interested in assays that detect microorganisms by reacting the antibody with an antigen on the cell surface. For this reason, excess antibody is applied, and unreacted antibody is removed by thorough washing. There are direct or indirect methods to detect antigen-antibody reactions. Direct detection is by using an enzyme or a fluorochrome conjugated to the antibody. An enzyme-linked immunosorbent assay (ELISA) employs a substrate and chromogenic hydrogen donor that results in color development proportional to the quantity of labeled antibody attached to the antigen. A fluorochrome conjugated antibody (FA) attached to an antigen is revealed by viewing the antigen-antibody complex under UV light with a fluorescence microscope. The antigen is identified by intense fluorescence, and the fluorescent color depends upon the fluorochrome used and the wavelength of UV light used to excite the fluorochrome. Use of FA techniques for direct detection of microorganisms is presented in chapter 6 of this book.

Indirect assays also use enzymes or fluorochromes to reveal the attachment of a specific antibody to an antigen. However, instead of directly detecting the antibody probe, a second antibody is added that attaches to the first antibody. The second antibody is against the animal-specific immunoglobulin type of the first antibody and is conjugated to an enzyme or fluorochrome. Second antibodies for indirect assays are widely available from chemical supply companies. Indirect assays are used commonly because the primary antibody does not have to be conjugated and can be used in a variety of different assays. Figure 28–4 shows the steps in an indirect ELISA.

Kishinevsky et al. (1984) reported on a comparison between direct and indirect ELISA of rhizobia using polyclonal antiserum. In general, good agreement was obtained between the two tests, although there were a few cases where a direct test was able to distinguish between some serologically related strains.

Enzyme immunoassays take advantage of binding properties of plastics, nitrocellulose, and other solid supports (Harlow & Lane, 1988). Proteins attach to these supports by covalent or noncovalent bonding and make it possible to manipulate the antigen-antibody complexes during the steps of the immunoassay. Ninety-six well microtiter plates of polyvinylchloride or polystyrene are available from many scientific supply houses and are used for ELISAs that use a soluble end-product. Each well is a separate incubation chamber.

Nitrocellulose is commonly used for an ELISA that employs a precipitating end-product. This type of assay is called a "dot-blot" or "immunoblot," and has been used to identify strains of soybean rhizobia (Ayanaba et al., 1986), for counting alfalfa rhizobia in nonsterile inoculants (Olsen & Rice, 1989), and to identify the antigen released from crushed spores of a mycorrhizal fungus colonizing plant roots (Wright & Morton, 1989).

Standard ELISA procedures that use soluble products for detection of horseradish peroxidase and alkaline phosphatase are given below. Enhancement of the chromogenic reaction is obtained by the use of biotinylated second antibodies that are detected by horseradish peroxidase- or alkaline phosphatase-labeled streptavidin. Streptavidin binds tightly to biotin, and the sensitivity of the ELISA reaction is enhanced because of the availability of four streptavidin-binding sites on the biotin. Also, an immunoblot procedure for identification of rhizobia in mixtures of colonies growing on agar is presented.

Results of ELISAs using a soluble chromogen are generally interpreted as positive or negative according to visual observation or by measuring absorbance. Each 96-well plate should include, at a minimum, a

1. Adsorb antigen to plate (A).

2. Wash away any antigen that is not attached (A).

3. Block unoccupied sites on the plastic with a protein such as bovine serum albumin to prevent nonspecific adsorption of antibodies. Steps 2 and 3 often are combined using the blocking agent to wash away unattached antigen (A).

4. Add polyclonal or monoclonal antibody and incubate (B).

5. Wash away unattached antibody (B).

6. Add enzyme-labeled anti-rabbit or anti-mouse antibody and incubate (C).

7. Wash away unattached second antibody (C).

8. Add the enzyme substrate along with the chromogen and incubate. The amount of color developed over a specified time can be measured on a colorimeter (D).

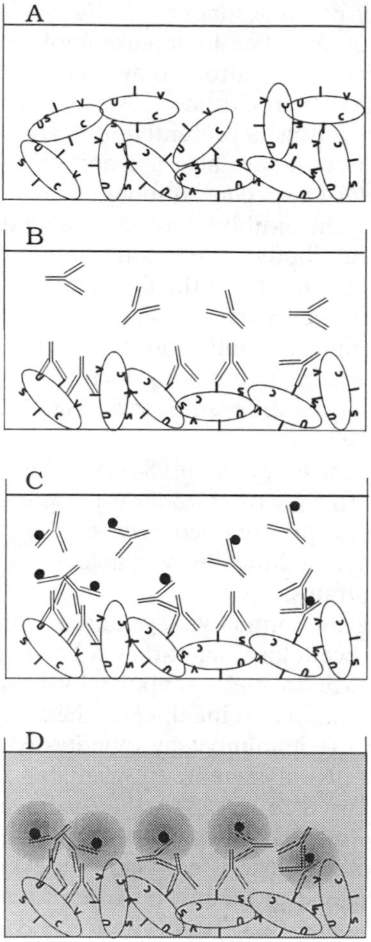

Fig. 28–4. An indirect enzyme-linked immunosorbent assay using a bacterial strain as the antigen. Diagrams illustrate reactions occurring in one well of a 96 well assay plate.

reagent control (no antigen), a positive antigen control, a negative antigen control, and a negative antibody control. Therefore, in addition to the minimum number of controls, 92 different tests can be run on one plate. Difficulty may be encountered in setting a cut-off absorbance for negative reactions. Variation in the intensity of the reaction of the positive control occurs when using the same concentration of antigen on ELISA plates prepared on different days and assayed using a standardized test (Olsen et al., 1994). It is recommended that at least two replicate samples be tested.

Day-to-day variation in results and discrimination between positive and negative values may present difficulties. Voller et al. (1980) suggest correcting samples for day-to-day variation using the following: corrected sample value = measured absorbance of sample × 1.0/absorbance of the reference positive sample. Results are reported as: (i) positive or negative according to a cutoff value determined on the basis of the highest value for a group of negative samples, or (ii) as ratios of sample values to the mean value of a group of negative samples where ratios greater than two or three times the negative value are usually considered positive. Negative values are means of absorbance readings of replicate negative antigen controls or negative antibody controls, depending on which is greater. Questionable positive results can be rerun using a higher dilution of antigen and comparing results using a nonreactive antibody to test for nonspecific binding of a primary antibody. Kishinevsky and Jones (1987) suggest using a cutoff value two (or three) times above the mean absorbance value of the negative group for ELISAs to detect rhizobia.

To avoid "cross-talk" among antigens or antibodies on a 96-well plate, care should be taken during the steps of the assay up to application of the second antibody to avoid overflow during filling and washing of wells. If different antibodies are being tested, it may be necessary to skip wells between antibody tests to avoid washover during processing.

Endogenous enzymes interfere with the use of ELISA to detect the presence of antigens in plant tissues or bacterial cells. To detect endogenous enzyme activity of a complex immunogen such as a bacterial cell or in tissue such as legume nodules, react the cells or tissue with the chromogen and enzyme substrate mixture for the ELISA and look for color development. There are several methods to pre-treat samples to inactive endogenous enzymes, and these will be presented in the discussions of principles of the various types of ELISAs presented below.

Methods using commercially available chemicals to prepare reagents for assays are given below. Also, commercial kits (Bio-Rad, Kirkegaard and Perry, and others) are available that include prepared color reagents. These may be useful for those who are initiating an assay.

The method presented below is for detection of antigens from various sources using an antibody that has been diluted to a working solution. A checkerboard titration (section 28–6.4) was performed to determine the optimum dilution for the antibody.

28–6.2 Peroxidase-Linked Immunosorbent Assay

28–6.2.1 Principles

Peroxidase-linked enzyme assays are sensitive (Tijssen, 1987) and reagents are readily available. A precaution is necessary for use of this enzyme in immunoassays involving antigens that might have an active peroxidase enzyme or are contaminated by plant material that has active peroxidase enzymes such as in legume nodules. Endogenous peroxidase in alfalfa nodules can be inactivated by drying the nodules followed by heating crushed, diluted nodule material in flowing steam for 10 min (Perry

Olsen, 1992, personal communication). Exposure to 15% (vol/vol) peroxide for 1 h inhibited activity of endogenous peroxidase in plant roots (Wright & Morton, 1988), and although there was bubble formation, the antigen was not destroyed. Peroxidase can also be blocked by incubating the specimen with 4 parts methanol to 1 part of 3% H_2O_2 for 20 min, or by using a solution of 0.1% (wt/vol) phenylhydrazine hydrochloride in PBS (Harlow & Lane, 1988).

28–6.2.2 Materials

1. 96-well polyvinyl chloride microtiter plates.
2. PBS (see section 28–2.2).
3. PBST. PBS plus 0.2 mL/L polyethylenesorbitan monolaurate (Tween-20).
4. Blocking concentrate. Bovine serum albumin, 10% (wt/vol) in PBS. The concentrate can be frozen in 10 mL aliquots. Working solution is 1:15 dilution of concentrate in PBS.
5. Antigen. Add 50 µL/well in a 96-well plate. Include positive antigen controls, blanks for reagent controls, an antigen that does not react with the antibody, and an antibody that does not react with the control antigen.
6. Antibody diluted to the appropriate dilution using 10% BSA in PBS.
7. Peroxidase-labeled antibody solution, 5 mL (amount for a 96-well plate). Make the dilution suggested by the commercial supplier in 4.5 mL of PBS plus 0.5 mL of blocking concentrate.
8. ABTS substrate solution.
   A. Stock solutions
      1. Citrate (free acid), 0.525 g.
         Distilled or deionized water, 45 mL.
         Adjust pH to 4.0 with 3 N NaOH and bring volume to 50 mL.
      2. ABTS [2,2′-azino-di-(ethyl-benzthizoline-sulfonic acid)]. Dissolve 0.015 g in 1.0 mL of deionized H_2O.
         Store at 4 °C protected from light. Stable for 4 wk. Handle ABTS carefully and use gloves. This compound may be carcinogenic.
      3. 30% H_2O_2.
   B. ABTS substrate-chromogen solution. Make up just before using. For each 96-well plate: Citrate buffer, 5 mL; ABTS solution, 100 µL; and H_2O_2, 5 µL.
9. TMB—an alternative substrate-chromogen solution.
   A. Dissolve 0.1 mg of 3′,3′,5′,5′-tetramethylbenzidine (TMB) in 0.1 mL of dimethylsulfoxide (DMSO). Add 9.9 mL of 0.1 M of $CH_3COONa\cdot3H_2O$ (sodium acetate), pH 6.0. Filter through Whatman No. 1, add H_2O_2 to a final concentration of 0.01% and use immediately.
   B. Sulfuric acid (H_2SO_4) 1.0 M.

10. Optional: ELISA plate colorimeter designed to read 96-well plates with a 450-nm filter for ABTS or a 405-nm filter for TMB. If sulfuric acid is added to TMB to stop the reaction, read at 450 nm.
11. Optional: ELISA plate washer if large numbers of assays will be performed. Various models are available commercially.

28–6.2.3 Procedure

Attach or coat antigen(s) on 96-well plates. Try different temperatures (4 °C, room temperature, 37 °C, or 65 °C) and different incubation times (2–24 h) for optimal adsorption. It may be necessary to allow the antigen to evaporate to dryness in wells. If short-term incubations are acceptable, carefully empty any moisture from the wells by inverting the plate over a sink and flicking the solution out of the wells with a wrist-snapping action. While holding the plate upside down, blot the moisture remaining on the edges of the wells on an absorbent paper towel.

Block sites on the plastic that are not occupied by components from the cells by filling each well with the working concentration of the blocking solution. This requires 200 to 300 µL/well. Incubate for 15 min and empty wells as described above. Add the primary antibody at a predetermined dilution for an absorbance reading between 0.2 and 2.0 or a strong visual signal that can readily be distinguished from any background color of the negative control (section 28–6.4). Incubate the antigen with the primary antibody for 1 h at room temperature. Empty the plate and blot on paper towels. Wash each well three times with PBST by flooding the wells, shaking the solution out, and blotting the plate on paper towels.

Add 50 µL/well of the secondary (peroxidase-linked antibody) and incubate for 1 h at room temperature. Empty the wells as described above and wash each well four times.

Add 50 µL of ABTS substrate solution and read absorbance at 405 or 414 nm or note positives after 15 min. If TMB is used as the chromogen, after the final wash in PBST, add 50 µL of the substrate to each well and incubate for 10 to 30 min at room temperature. Positives appear pale blue. At the end of the incubation period, add 50 µL of H_2SO_4 to each well which causes positives to appear bright yellow. Read absorbance at 450 nm. If a printout from a spectrophotometer reading is not available, a template of the 96-well plate can be made and positive wells marked for use as a permanent record, or simply record the positive wells by column and row designation.

28–6.3 Alkaline Phosphatase-Linked Immunosorbent Assay

28–6.3.1 Principles

Alkaline phosphatase (AP) is an alternative enzyme conjugate. The procedure is similar to that described in section 28–6.2.1 except that TBS is used instead of PBS. Tijssen (1987) warns that inorganic phosphate (P_i) from PBS or spontaneous hydrolysis of the chromogen p-nitrophenyl phosphate can interfere with enzyme activity.

Endogenous alkaline phosphatase activity of samples to be examined by ELISA may be blocked by including 0.1 mM levamisole in the substrate solution. Levamisole inhibits activity of tissue phosphatases other than the intestinal alkaline phosphatase used to prepare the labeled antibody but has been found to be ineffective in inhibiting legume nodule alkaline phosphatase (Perry Olsen, personal communication). Incubation in a pH 2.6 buffer has been found effective for inactivation of endogenous alkaline phosphatase (28–6.6). This step can be added to the routine ELISA procedure presented below if endogenous enzyme is found to be present. Endogenous AP does not appear to interfere with the use of this assay to detect rhizobia in clover legume nodules (Wright et al., 1986), but the type of sample to be used in the assay should be tested for AP activity. Inhibition of endogenous activity of bacteria on immunoblots is given in section 28–6.5. For routine solid-phase ELISA of antigen with endogenous AP, the antigen can be pre-treated before it is adsorbed to 96-well plates, or the inactivation step may be inserted before the blocking step.

28–6.3.2 Materials

1. Tris-buffered saline (TBS).
   Tris(hydroxymethyl)aminomethane, 0.01 M and NaCl, 0.25 M.
   Adjust pH to 7.4 with HCl before bringing to volume.
2. TBST. TBS plus 0.2 mL/L polyethylenesorbitan monolaurate (Tween 20).
3. Antigen and antigen attachment to 96-well plates (see sections 28–6.2.1 and 28–6.2.2).
4. Blocking concentrate made with TBS instead of PBS (see section 28–6.2.1). Working solution is a 1:15 dilution in TBS.
5. Alkaline phosphatase (intestinal mucosal enzyme)-labeled second antibody solution, 5 mL (amount for a 96-well plate). Make the dilution suggested by the commercial supplier in 4.5 mL of TBS and 0.5 mL of blocking concentrate.
6. Diethanolamine (DEA) buffer, 10%. Mix 97 mL of DEA with 1 L of 0.01% $MgCl_2$ solution, and adjust the pH to 9.8 with 1 N of HCl. The solution can be stored at room temperature in an amber bottle if NaN_3 (0.2 g/L) is added.
7. Substrate-chromogen solution. Dissolve p-nitrophenyl phosphate, 1 mg/mL in DEA buffer. Filter through a Whatman No. 1 filter before using to remove any undissolved particles.
8. Spectrophotometer designed to read 96-well plates with a 405-nm filter.
9. Optional blocking agent for endogenous alkaline phosphatase. Make up the substrate-chromogen solution in 0.1 nM of levamisole [(L[-]-2,3,5,6-Tetrahydro-6-phenylimidazo[2,1-b]thiazole].

10. Color development stop solution. Ethylenediaminetetraacetic acid (EDTA) 0.1 M.

28–6.3.3 Procedure

The procedures are the same as those described in section 28–6.2 except that the second antibody is conjugated with alkaline phosphatase. If legume nodules are to be used as the antigen, crush the nodules in TBS and allow the debris to settle for a minute before pipetting off the solution to add to the ELISA plate. Add 50 µL of the substrate-chromogen solution to each well and incubate for 10 to 30 min at room temperature. Add 50 µL of stop solution, 0.1 M of EDTA to each well and read plates at 405 nm.

28–6.4 Checkerboard Titration

Optimal concentrations for antigen(s) and primary antibody for indirect microplate ELISA can be determined by a checkerboard titration. The antigen is serially diluted and horizontal rows of wells in a 96-well plate are charged with the dilutions leaving one row with the dilution buffer only as a control. The primary antiserum is serially diluted and used in vertical columns across the plate. The indirect test is completed and absorbance readings are made. The dilutions of antigen and antibody giving an absorbance of approximately 1.0 are optimal for use when testing similar antigens for reactivity with the selected antibody.

28–6.5 Biotinylated Second Antibody for Enhanced ELISA

Biotinylated second antibodies are available commercially and come with recommendations for working dilutions for ELISA (i.e., biotin-labeled goat anti-mouse IgG to be used at a 1:10 000 working dilution). Streptavidin-peroxidase or alkaline phosphatase conjugates also are available commercially and recommended working solutions are given in product literature or supply catalogs. The ELISA is performed as described in section 28–6.2 or 28–6.3 except that there are two steps to the second-antibody incubation. Biotin conjugated antibody is added to wells and incubated for 1 h and then wells are washed three times with the buffer recommended for the enzyme conjugate. Enzyme-linked streptavidin is added to wells and incubated for 1 h followed by the final four washings with the appropriate buffer. Color development is as described above in section 28–6.2 or 28–6.3.

28–6.6 Immunoblot to Identify Colonies on a Dilution Plate

28–6.6.1 Principles

An immunoblot analysis of colonies of rhizobia growing in mixed culture on dilution plates has been developed (Olsen & Rice, 1989). This analysis takes advantage of the ability to identify an antigen immobilized on nitrocellulose by using an assay that is essentially an indirect ELISA employing a precipitating stain to reveal the enzyme.

28–6.6.2 Materials

1. Agar plates with colonies of the bacterium of interest. For *R. leguminosarum* bv. *trifolii* spread plates were incubated for 3 d.
2. Nitrocellulose (82.5 mm-diam., Bio-Rad Laboratories, Richmond, CA) or purchased in sheets and cut to size.
3. TBS pH 7.4 (see section 28–6.3.2).
4. TBS acidified to pH 2.6 with HCl.
5. Specific antibody diluted in TBS.
6. Nonfat powdered skim milk, 2.0% (wt/vol) diluted in TBS.
7. TBS plus 0.05% (vol/vol) Tween-20 (TBST).
8. Second antibody. Goat anti-rabbit immunoglobulin G conjugated with alkaline phosphatase for detection of polyclonal rabbit primary antibody or goat anti-mouse immunoglobulin G or M for a mouse monoclonal antibody.
9. Substrate-chromogen stock solutions.
   A. Dissolve 0.5 g of nitro blue tetrazolium (NBT) in 10 mL of 70% (vol/vol in deionized water) dimethylformamide.
   B. Dissolve 0.5 g of bromochloroindolyl phosphate disodium salt (BCIP) in 10 mL of 100% dimethylformamide.
   C. Alkaline phosphatase buffer: 100 mM of NaCl, 5 mM of MgCl$_2$·6H$_2$O, and 100 mM Tris (pH 9.5). All stock solutions are stable at 4 °C for at least 1 yr. Nitro blue tetrazolium and 5-bromo-4-chloro-3-indolyl phosphate are also available commercially (Bio-Rad Laboratories, Richmond, CA).
10. Color reaction stop solution, 20 mM of EDTA or rinse the membrane well with water.
11. India Ink and a disposable syringe needle.

28–6.6.3 Procedure

A nitrocellulose membrane is laid on a selected plate and allowed to wet completely and touch all colonies. If it is desired to make the membrane a template of the colony, dip the syringe needle in India ink and make marks through the membrane and into the agar at the edge in each quadrant beginning at the 1200 h position with one hole followed by two, and three holes at 1600 and 2000 h positions, respectively. Carefully peel the membrane off and immerse in deionized water to a depth of at least 5 cm in a dish. Rinse off excess unattached colonies with a stream of water until colonies are no longer visible. Air dry the membrane (colony side up) on absorbent paper and then soak the membrane for 30 min in acidified TBS to denature endogenous alkaline phosphatase or skip this step and use levamisole in the substrate-chromogen solution (section 28–6.3.2). Wash to neutrality in TBS (pH 7.4), blot lightly, and soak in skim milk for 30 min

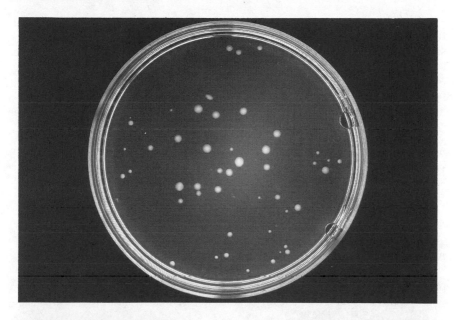

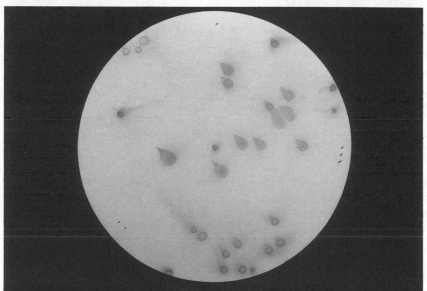

Fig. 28–5. The original agar spread plate of a mixture of strains of *Rhizobium leguminosarum* bv. *trifolii* (*top*). An immunoblot of the plate (*bottom*) which was reacted with a monoclonal antibody specific for one strain in the mixture.

at 37 °C to block nonspecific protein-binding sites. Wash the membrane three times in 250 mL of TBST and place in an appropriate dilution of the specific antibody for 2 h at room temperature. Wash three times (5 min. each) in 250 mL of TBST and transfer to the second antibody solution for

1.5 h at room temperature. Wash three times in 250 mL of TBS and transfer to freshly prepared (< 1 h old) substrate-chromogen solution consisting of 66 µL of NBT stock mixed with 10 mL of alkaline phosphatase buffer before adding 33 µL of BCIP stock. Add sufficient solution to cover the nitrocellulose and develop with agitation until the stain is suitably dark, typically 30 min. Stop the reaction with 20 mM EDTA. Positive reactions are distinct purple spots at the sites where colonies that reacted with the primary antibody are located on the original plate. Color development is complete in 60 min, and membranes can be washed and stored in the dark. A photocopy of the membrane can also be kept as a permanent record. Results represent a mirror image of the colonies on the plate. Results are shown in Fig. 28–5.

REFERENCES

Ayanaba, A., K.D. Weiland, and R.M. Zabolotowicz. 1986. Evaluation of diverse antisera, conjugates, and support media for detecting *Bradyrhizobium japonicum* by indirect enzyme-linked immunosorbent assay. Appl. Environ. Microbiol. 52:1132–1138.

Banowetz, G.M., E.J. Trione, and B.B. Krygier. 1984. Immunological comparisons of teliospores of two wheat bunt fungi, *Tilletia* species, using monoclonal antibodies and antisera. Mycologia 76:51–62.

Burnet, F.M. 1959. The clonal selection theory of acquired immunity. Vanderbilt Univ. Press, Nashville, TN.

Campbell, D.H., J.S. Garvey, N.E. Cremer, and D.H. Sussdorf. 1974. Methods in immunology. W.A. Benjamin, Reading, MA.

Carlin, N.I.A., and A.A. Lindberg. 1985. Monoclonal antibodies specific for the O-antigen of *Shigella flexneri* and *Shigella sonnei:* Immunochemical characterization and clinical usefulness. p. 137–165. *In* A.J.L. Macario and E. Conway de Macario (ed.) Monoclonal antibodies against bacteria. Vol. 1. Academic Press, New York.

Culver, J.N., and J.L. Sherwood. 1988. Detection of peanut stripe virus in peanut seed by an indirect enzyme-linked immunosorbent assay using a monoclonal antibody. Plant Dis. 72:676–679.

Davies, S.N. 1951. The serology of *Bacillus polymyxa*. J. Gen. Microbiol. 5:807–816.

de Maagd, R.A., R. de Rijk, I.H.M. Mulders, and B.J.J. Lugtenberg. 1989. Immunological characterization of *Rhizobium leguminosarum* outer membrane antigens by use of polyclonal and monoclonal antibodies. J. Bacteriol. 171:1136–1142.

Dudman, W.F. 1977. Serological methods and their application to dinitrogen fixing organisms. p. 487–508. *In* R.W.F. Hardy and A.H. Gibson (ed.) A treatise on dinitrogen fixation, Section IV. John Wiley and Sons, New York.

Fuhrmann, J., and A.G. Wollum II. 1985. Simplified enzyme-linked immunosorbent assay for routine identification of *Rhizobium japonicum* antigens. Appl. Environ. Microbiol. 49:1010–1013.

Goding, J.W. 1986. Monoclonal antibodies: Principles and practice. 2nd ed. Academic Press, New York.

Graham, P.H. 1969. Analytical serology of the Rhizobiaceae. p. 353–378. *In* J.B.G. Kwapinski (ed.) Analytical serology of microorganisms. Vol. 2. John Wiley and Sons, New York.

Hardham, A.R., E. Suzaki, and J.L. Perkin. 1986. Monoclonal antibodies to isolate-, species-, and genus-specific components on the surface of zoospores and cysts of the fungus *Phytophthora cinnamoni*. Can. J. Bot. 64:311–321.

Harlow, E., and D. Lane. 1988. Antibodies: A laboratory manual. Cold Spring Harbor, New York.

Holme, T., and B. Gustafsson. 1985. Monoclonal antibodies against group- and type-specific antigens of *Vibrio cholerae* O:1. p. 167–189. *In* A.J.L. Macario and E. Conway de Macario (ed.) Monoclonal antibodies against bacteria. Vol. 1. Academic Press, New York.

Hrabak, E.M., M.R. Urbano, and F.B. Dazzo. 1981. Growth-phase-dependent immunode-terminants of *Rhizobium trifolii* lipopolysaccharide which bind trifoliin A, a white clover lectin. J. Bacteriol. 148:697–711.

Hurrell, J.G.R. 1983. Monoclonal hybridoma antibodies: Techniques and applications. CRC Press, Boca Raton, FL.

Kabat, E.A. 1967. Structural concepts in immunology and immunochemistry. Holt, Rinehart and Wilson, New York.

Kishinevsky, B., and M. Bar-Joseph. 1978. Rhizobium strain identification in *Arachis hypogaea* nodules by enzyme-linked immunosorbent assay (ELISA). Can. J. Microbiol. 24:1537–1543.

Kishinevsky, B.D., and D.G. Jones. 1987. Enzyme-linked immunosorbent assay (ELISA) for the detection and identification of *Rhizobium* strains. p. 157–184. *In* G.H. Elkan (ed.) Symbiotic nitrogen fixation technology. Marcel Dekker, New York.

Kishinevsky, B.D., A. Maoz, D. Gurfel, and C. Nemas. 1984. A serological study of *Rhizobium* strains using direct and indirect enzyme linked immunosorbent assays (ELISA). Adv. Agric. Biotechnol. 4:346.

Kwapinski, J.B.G. 1969. Analytical serology of microorganisms. Vol. 1–2. John Wiley and Sons, New York.

Kwapinski, J.B.G. 1972. Methodology of immunochemical and immunological research. John Wiley and Sons, New York.

Lin, N.S., Y.H. Hsu, and H.T. Hsu. 1990. Immunological detection of plant viruses and a mycoplasmalike organism by direct tissue blotting on nitrocellulose membranes. Phytopathology 80:824 828.

Macario, A.J.L., and E. Conway de Macario. 1985. Monoclonal antibodies against bacteria. Vol. I. Academic Press, Orlando, FL.

Male, D., B. Champion, and A. Cooke. 1987. Advanced immunology. J.B. Lippincott Co., Philadelphia, PA.

Marchalonis, J.J. 1982. Structure of antibodies and their usefulness to non-immunologists. p. 3–20. *In* J.J. Marchalonis and G.W. Warr (ed.) Antibody as a tool. John Wiley and Sons, New York.

McCartney, A.C., and A.C. Wardlaw. 1985. Endotoxic activities of lipopolysaccharides. p. 203–238. *In* D.E.S. Stewart-Tull and M. Davies (ed.) Immunology of the bacterial cell envelope. John Wiley and Sons, New York.

Nambiar, P.T.C., and V. Anjaiah. 1985. Enumeration of rhizobia by enzyme-linked immunosorbent assay (ELISA). J. Appl. Bacteriol. 58:187–193.

Nameth, S.T., W.W. Shane, and J.C. Stier. 1990. Development of a monoclonal antibody for detection of *Leptosphaeria korrae*, the causal agent of necrotic ringspot disease of turfgrass. Phytopathology 80:1208–1211.

Nasser, A.M., and T.G. Metcalf. 1987. An A-ELISA to detect hepatitis A virus in estuarian samples. Appl. Environ. Microbiol. 53:1192–1195.

Olsen, P., S. Wright, M. Collins, and W. Rice. 1994. Patterns of reactivity between a panel of monoclonal antibodies and forage *Rhizobium* strains. Appl. Environ. Microbiol. 60:654–661.

Olsen, P.E., and W.A. Rice. 1989. *Rhizobium* strain identification and quantification in commercial inoculants by immunoblot analysis. Appl. Environ. Microbiol. 55:520–522.

Renwick, A., and D.G. Jones. 1985. A comparison of the fluorescent ELISA and antibiotic resistance identification techniques for use in ecological experiments with *Rhizobium trifolii*. J. Appl. Bacteriol. 58:199–206.

Schmidt, E.L., R.O. Bankole, and B.B. Bohlool. 1968. Fluorescent-antibody approach to study of rhizobia in soil. J. Bacteriol. 95:1987–1992.

Sutherland, I.W. 1977. Immunochemical aspects of polysaccharide antigens. p. 399–443. *In* L.E. Glynn and M.W. Stewart (ed.) Immunochemistry: An advanced textbook. John Wiley and Sons, New York.

Tijssen, P. 1987. Practice and theory of enzyme immunoassays. *In* R.H. Burdon and P.H. van Knippenberg (ed.) Laboratory techniques in biochemistry and molecular biology. Elsevier, New York.

Vincent, J.M. 1970. A manual for the practical study of root-nodule bacteria. IBP Handb. No. 15. Blackwell Scientific Publ., Oxford, Great Britain.

Voller, A., D. Bidwell, and A. Bartlett. 1980. Enzyme-linked immunosorbent assay. p. 359–371. *In* N.R. Rose and H. Friedman (ed.) Manual of clinical immunology. 2nd ed. Am. Soc. for Microbiology, Washington, DC.

Warr, G.W. 1982. Preparation of antigens and principles of immunization. p. 21–58. *In* J.J. Marchalonis and G.W. Warr (ed.) Antibody as a tool. John Wiley and Sons, New York.

Williams, C.A., and M.W. Chase. 1967. Methods in immunology and immunochemistry. Vol. I. Preparation of antigens and antibodies. Academic Press, New York.

Wright, S.F. 1990. Production and epitope analysis of monoclonal antibodies against a *Rhizobium leguminosarum* biovar *trifolii* strain. Appl. Environ. Microbiol. 56:2262–2264.

Wright, S.F., J.G. Foster, and O.L. Bennett. 1986. Production and use of monoclonal antibodies for identification of strains of *Rhizobium trifolii*. Appl. Environ. Microbiol. 52:119–123.

Wright, S.F., and J.B. Morton. 1989. Detection of VAM fungal root colonization by using a dot-immunoblot assay. Appl. Environ. Microbiol. 55:761–763.

Wright, S.F., J.B. Morton, and J.E. Sworobuk. 1987. Identification of a vesicular-arbuscular mycorrhizal fungus by using monoclonal antibodies in an enzyme-linked immunosorbent assay. Appl. Environ. Microbiol. 53:2222–2225.

Zola, H. 1987. Monoclonal antibodies: A manual of techniques. CRC Press, Boca Raton, FL.

Chapter 29

Whole-Cell Protein Profiles of Soil Bacteria by Gel Electrophoresis

DIPANKAR SEN, *Texas A&M University, College Station, Texas*

Identification of individual microbial isolates from the soil population is a challenging task. Identification at the genus and species level may be achieved by taxonomic methods using morphological, physiological, and genetical criteria, but identification at the strain level, and sometimes even at the species level, can be difficult and ambiguous. Serological methods have been applied for such purposes but their power of resolution is limited. Although monoclonal antibodies can identify a strain with greater certainty, the procedures involved are complex, expensive, and may not be suitable for routine analysis of a large number of strains (see chapter 28 by Wright in this book). Like common serological methods where polyclonal antibodies are used, plasmid profile (see chapter 30 by Pepper in this book) or multilocus enzyme analyses (see chapter 26 by Eardly in this book) may place a microbial isolate in a group but not establish its absolute identity. In many situations, no single method is available for identification of a strain with acceptable accuracy and thus more than one method may need to be used (Dughri & Bottomley, 1983). Whole-cell protein profiles of microbes can be used to complement other methods of identification. The methodology is no more complex than determining plasmid profiles or multilocus enzyme patterns and like those two methods, uses the same basic physical principle of electrophoresis.

Identification of a strain by its whole-cell protein profile depends on the premise that a particular genotype, when grown under a defined set of conditions, will express a particular group of proteins and that such a protein profile will be unique for the genotype. Whole-cell soluble proteins (Benson & Hanna, 1983; Dughri & Bottomley, 1983; Fuquay et al., 1984; Kamicker & Brill, 1986; Lochner et al., 1989; Noel & Brill, 1980) or only specific fractions of cell proteins, like cell wall proteins (Pankhurst, 1974), membrane proteins (De Maagd et al., 1988), or periplasmic proteins (Glenn et al., 1986) have been used to distinguish between strains or mutants of soil bacteria.

The degree of precision required by a researcher may dictate whether a one- or two-dimensional electrophoretic profile is developed. In the former, proteins are resolved only by their size; in the latter, proteins are separated by their charge characteristics in the first dimension followed by molecular weight resolution in the second dimension. Two-dimensional electrophoresis has been used effectively for positive identification of strains (Benson et al., 1984; Leps et al., 1980; Roberts et al., 1980) but the procedure has many more steps compared to one-dimensional electrophoresis and requires more equipment. One gel is required for each sample and comparisons of the complex patterns on different gels are relatively difficult. Optical scanners and computer-assisted analysis, however, can facilitate comparison of protein patterns.

One-dimensional sodium dodecyl sulfate polyacrylamide gel electrophoresis (SDS-PAGE) on a slab gel is the simplest method for comparing whole-cell protein profiles of bacteria. Visual comparison and photographing protein profile patterns of different samples in parallel lanes on the same slab gel is easier than on tube gels where each sample is electrophoresed separately. This method has been used extensively for identification of soil-borne bacteria. It should also be noted that in addition to strain identification, changes in whole-cell protein profiles have been used to detect synthesis of "stress" proteins and proteins encoded by bacterial plasmids (Krishnan & Pueppke, 1991; Sen et al., 1990).

In the last two decades, whole-cell protein profiles have been extensively used for identification and classification of various groups of bacteria (Kersters & De Ley, 1980). Considering the attention given to rhizobia by soil microbiologists, it is not surprising that most studies of protein profile patterns of soil bacteria have involved rhizobial species. Although the methods described are primarily used for rhizobia, they should be applicable for other microbes as well. The procedure for electrophoresis should require no modification. For various microbes, the optimal methods for cell growth, cell disruption, and preparing protein samples of sufficient concentration may have to be established during pilot experiments (Jackman, 1985).

29–1 PRINCIPLES

A charged molecule moves toward its opposite charge in an electrical field. When molecules with similar charges move through a porous matrix, the rate of movement is determined by the size and shape of the molecules and the matrical pores. Thus, macromolecules like nucleic acids and proteins are separated into zones or bands comprised of the different size fractions when a mixture of various species is subjected to electrophoresis in a porous matrix such as agarose, starch, or polyacrylamide gel. The medium also stabilizes the zones in the positions to which they have migrated. The matrix of choice should be one that does not ionize, absorb, or react with the molecules being electrophoresed, and whose porosity can be manipulated. For mixtures of proteins, the matrix of choice is polyacrylamide as it has all the required properties listed above. The pore-size range for polyacrylamide can be widely varied using different concentrations of acrylamide without making the gel too soft or too brittle to handle. Moreover, polyacrylamide gels can be prepared with concentration gradients that provide for greater resolution of the protein components.

Migration rates of native proteins depend on their net charges derived from exposed side chains and on their three-dimensional sizes due to their secondary and tertiary structures. These variables are eliminated by heat-denaturing proteins in the presence of a sulfhydryl reagent and the anionic detergent, SDS. The former breaks the disulfide bonds and the latter binds quantitatively with the polypeptide chains to produce molecular species of constant mass/charge ratio. As a consequence, the proteins are separated by SDS-PAGE solely according to their molecular weights.

To enhance focussing of the protein bands, a discontinuous gel matrix and buffers of different ionic composition and pH are employed. This is referred to as disc gel electrophoresis. Initially, the proteins are electrophoresed through a large-pore gel, called the *stacking gel*. The relative movement of ions in the stacking gel buffer and tank buffer are such that proteins are focussed without sieving into a thin compact zone at the interface of the stacking and resolving gels. The zone then enters the smaller-pore *resolving gel* where they are separated according to their sizes while remaining in focussed bands.

29–2 METHODS

29–2.1 Growth and Collection of Cells

29–2.1.1 Culture Medium and Wash Buffer

1. For *Rhizobium* and *Bradyrhizobium,* yeast-mannitol broth is a suitable culture medium (see chapter 12 by Weaver and Graham in this book). Other broth media, appropriate for the specific microorganism can also be used. Avoid any medium that stimulates excessive production of capsular materials that make it difficult to collect sufficient cells for analysis.

2. For washing harvested cells, use 10 mM Tris(hydroxy-methyl)aminoethane-HCl buffer (Tris-HCl buffer), pH 7.6.

29–2.1.2 Growth of Cells

For each microbial isolate, inoculate 30 to 50 mL of broth in a 125-mL Erlenmeyer flask. Incubate with shaking (150 rpm) at the optimum growth temperature (28 °C for rhizobia) until late exponential or stationary phase has been obtained.

29–2.1.3 Harvesting Cells

For this and all subsequent steps, keep cells and extracts below 4 °C. Centrifuge cultures at 10 000 × g for 10 min in a refrigerated centrifuge. Discard supernatant and wash the bacterial cell pellet three times by resuspending in 30 mL ice-cold 10 mM tris-HCl buffer, pH 7.6, and centrifuging each time. Add 0.5 mL of the same buffer to the final pellet and suspend the cells by vortexing. Cells may be used immediately or stored at −20 °C. A 1.5-mL microcentrifuge tube is a suitable container for storage and further processing.

29–2.2 Disruption of Cells

Cells are disrupted by repeated (10 times is generally sufficient) rapid freezing in liquid N followed by thawing in a 45 °C water bath. For this procedure and later treatment of protein samples with sample buffer, a 16-hole circular aluminum microcentrifuge tube-rack with a handle (available from USA/Scientific Plastics, Ocala, FL), that fits into a 1000-mL glass beaker, is convenient. After the cells are disrupted, protein extracts may be used immediately or the suspensions may be frozen and stored at −20 °C. An aliquot of the preparation (0.1 mL is usually sufficient) may be centrifuged and the supernatant used for determination of protein content by the method of Bradford (1976) or Lowry et al. (1951). Cells can also be disrupted using a sonicator (Roberts et al., 1980) or a French press (Sen & Weaver, 1988).

29–2.3 Preparation of Protein Extracts for SDS-PAGE

29–2.3.1 Materials

1. Sample buffer:
   0.125 M Tris-HCl buffer, pH 6.8
   4.6% (wt/vol) SDS
   20% (vol/vol) glycerol or 20% (wt/vol) sucrose
   10% (vol/vol) 2-mercaptoethanol or 80 mM Dithiothreitol
   0.01% (wt/vol) bromophenol blue

The sample buffer may be prepared in advance and kept frozen at −20 °C in suitably sized aliquots (1 or 2 mL). For each experiment, one aliquot is thawed and used. However, freshly prepared sample buffer is often recommended to avoid doubling of some bands for commercially available protein standards that are electrophoresed alongside whole-cell protein preparations as molecular weight markers.

29–2.3.2 Procedure

Centrifuge the cell preparation in a microcentrifuge at maximum speed (at least 10 000 × g) to obtain a clear supernatant containing the soluble proteins. Transfer supernatant to another microcentrifuge tube avoiding all particulate matter. Add an equal volume of sample buffer, mix gently, and heat the samples in a water bath at 95 to 98 °C for 5 min.

29–2.3.3 Comments

After harvesting and washing, cells may be suspended and disrupted in the sample buffer instead of the wash buffer. This may enhance cell disruption. However, in the presence of SDS, particularly after heating, the cells may lyse to form a gelatinous mass. It may be difficult to separate from the lysate a clear, pipettable, protein-containing solution even by centrifuging at high speeds in a microcentrifuge. The presence of the viscous mass in the protein sample distorts the protein bands. High viscosity of cell lysates is due to the release of DNA. The DNA can be degraded by the addition of DNase, however, this adds extra bands to the protein profiles. Similarly, addition of lysozyme, particularly to gram-positive bacteria, may facilitate cell disruption but also adds bands to the protein profile.

The method described above yields whole-cell soluble proteins from the bacteria and includes membrane proteins, especially when SDS is used. This preparation is used for analysis by one-dimensional PAGE described below. Protein extracts from disrupted cells (section 29–2.2) can also be used for more detailed analysis by two-dimensional PAGE by following the method of O'Farrell (1975). For specific fractions of cell proteins, like membrane proteins, periplasmic proteins, cell wall proteins etc., specialized procedures are required (De Maagd et al., 1988; Glenn et al., 1986; Krishnan & Pueppke, 1991; Pankhurst, 1974).

29–2.4 Electrophoresis

The following procedure is similar to that described by Laemmli (1970). All chemicals should be high purity electrophoresis grade and high purity water should be used. Glass-distilled water will usually suffice, although for some procedures, like silver staining, distilled, deionized water is necessary.

29–2.4.1 Materials

1. Vertical gel electrophoresis apparatus: Any commercially available (Hoefer Scientific Instruments, Bio-Rad Lab., R. Shadel Inc. etc.) or home-made apparatus with approximately 15 × 15 cm glass plates and 0.75 to 1.0 mm thick spacers and Teflon combs.
2. Power source:
   Any direct current (DC) power source capable of delivering up to 30 mA current per gel within 500 volts can be used. It is desirable for the power source to have the capability to operate in constant current and voltage modes. Having automatic cross-over, where the power source switches from constant current mode to constant voltage mode when voltage increases to a pre-set value, is an additional advantage.

Caution: While operating electrophoresis equipment attention should be paid to possible electrical hazards.

3. 2× tank buffer:

   0.05 M Tris, 0.384 M glycine, 0.2% SDS.

   SDS may be omitted from the stock solution to avoid frothing while handling. An appropriate amount of 10% SDS can be added while diluting to 1× concentration before use.

4. 4× lower (resolving) gel buffer;

   1.5 M Tris-HCl buffer, pH 8.8

5. 4× upper (stacking) gel buffer:

   0.5 M Tris-HCl buffer, pH 6.8

6. 30% acrylamide solution for resolving gel:

   29.2% acrylamide and 0.8% N,N'-methylene-bis-acrylamide (bis-acrylamide) (wt/vol) solution in water. Store in a well-stoppered, dark bottle in a refrigerator. This solution can be used for up to 1 mo.

   Caution: Acrylamide is neurotoxic. Avoid all skin contact as well as inhalation of fine crystals.

7. 30% acrylamide solution for stacking gel:

   28.8% acrylamide and 1.2% bis-acrylamide (wt/vol) solution in water. Store as no. 6.

8. 10% (wt/vol) SDS in water.

9. 10% (wt/vol) ammonium persulfate (APS) in water.

   Must be prepared fresh everyday. Solid is highly hygroscopic. Store in a well-stoppered container, preferably in a desiccator. Only small quantities (10 g) should be obtained at any one time.

10. N,N,N',N'-tetramethyl-ethylenediamine (TEMED).

    This chemical absorbs water and is oxidized. Store at room temperature in the well-stoppered original container. Only small quantities (5–10 mL) should be obtained at any one time.

11. Water-saturated isobutanol:

    Mix equal volumes (e.g., 25 mL each) of isobutanol and water in a small bottle (100 mL). Shake thoroughly and let stand. Isobutanol will float to the top. Store over water and shake and let the isobutanol separate before using each time.

12. Overlay buffer:

    0.1% SDS in 0.375 M Tris-HCl buffer, pH 8.8.

29–2.4.2 Procedure

Assembling gel mold:

1. Plates should be without scratches and scrupulously cleaned; acid washing is recommended if previous use involved the application of any grease.
2. Plates should be wiped with absolute alcohol and dried just prior to assembly.

3. All spacers etc. should be clean and tightly secured to prevent leaking of the acrylamide solution or slow leakage of air into the solution.
4. Assemble glass plates to form the gel mold following manufacturer's instructions. Better resolution of protein bands is obtained with thinner gels. However, a very thin gel may present problems during introduction of the samples and will limit the volume of the sample that can be used. A 0.75 to 1.0 mm thick gel is usually optimal.
5. Secure the glass plate sandwich(es) to the casting stand or electrophoresis box depending on the type of apparatus used.
6. Horizontally level the stand or box.

Preparation of resolving gel:

Gels should be cast at room temperature, 25 °C is the optimum temperature for polymerization of acrylamide.

1. Allow acrylamide and other solutions to come to room temperature before use to achieve proper degassing. Exposure to atmospheric or dissolved oxygen inhibits the polymerization of acrylamide. The amount of monomeric SDS-acrylamide solution used will depend on the capacity of the gel mold. Final acrylamide concentrations of 10 to 15% are usually suitable for whole-cell protein profiles. The most suitable concentration should be determined by trial.
2. Mix the required amounts of stock solutions nos. 4, 6, 8, and water in an Erlenmeyer flask for the desired concentration of acrylamide and 0.1% SDS. For example, to prepare 20 mL of 12% acrylamide, mix 5 mL of buffer (no. 4), 8 mL of acrylamide soln. (no. 6), 0.2 mL of 10% SDS (no. 8) and add water to a final volume of 20 mL.
3. Filter the mixture if it is not absolutely clear and free of particulate matter.
4. Degas the monomeric acrylamide solution under vacuum (at least 0.016 MPa) in a desiccator for 15 min preferably with stirring.
5. Remove the solution from the desiccator and add the polymerization initiators TEMED and APS to a final concentration of 0.04% for each, e.g., 8 μL of TEMED and 80 μL of 10% APS to 20 mL of acrylamide monomer solution. The amount of TEMED and APS may have to be adjusted slightly depending on the extent of possible breakdown of these reagents and can be determined by observing the polymerization time required. The quantities used should be the minimum necessary for polymerization to occur in no less than 20 min but within 1 h.
6. Proceed quickly after adding the polymerizers. Mix well by swirling the solution slowly, avoiding turbulence that introduces oxygen from the air.

7. Transfer the solution to the gel mold by pouring directly from the flask or by using a pipet. Pour in a steady stream, do not introduce air bubbles. Fill to the desired height, leaving sufficient space for the stacking gel (see below).

8. Using a syringe and a 25-gauge needle, gently overlay with a layer (2 mm) of water-saturated isobutanol by letting the isobutanol trickle down the sides of the spacers. The isobutanol layer levels the acrylamide surface and prevents contact with atmospheric oxygen. The gel can be overlaid either with water or buffer but since isobutanol is lighter than water, it is easier to overlay without disturbing the surface of the acrylamide solution.

9. Following gelation, a sharp line of separation is formed between the polyacrylamide gel and a thin layer of water above it. This line will be slightly below the top of the poured monomer solution.

10. Immediately drain off the isobutanol by inverting the gel assembly since leaving the isobutanol on top of the gel causes some dehydration of the gel. Rinse the top of the gel with distilled water and layer with the overlay buffer. The gel should be allowed to stand at least 90 min before use and may be left overnight as the process of polymerization within the gel continues for some period after visible gelation occurs. The polymerized gel should appear perfectly transparent and without any swirling patterns.

11. Any excess monomer solution left in the flask should be allowed to polymerize and thus rendered non-toxic before discarding. If necessary, add initiators (TEMED and APS) to double their concentrations in the excess solution.

Preparation of stacking gel:

The stacking gel should be prepared about 90 min before actual electrophoresis. The wells formed by the comb in the stacking gel, or at least the sample columns after loading, should be shorter than the depth of the stacking gel below the wells. Care should, therefore, be taken to leave enough space during preparation of the resolving gel. Twice the height of the teeth of the comb is used as a safe measure.

1. Prepare a 3 to 4% monomeric acrylamide mixture using the stacking gel buffer (no. 5), stacking gel acrylamide solution (no. 7), SDS (no. 8) and water, degas and handle as described above. Initiate polymerization by adding TEMED and APS to final concentrations of 0.08 and 0.04%, respectively.

2. Pour the overlay buffer off the top of the resolving gel and fill the gel mold with the stacking gel monomer.

3. Approaching from the side at an angle, introduce the well-forming comb slowly, tooth by tooth. Avoid trapping air bubbles that prevent the formation of sharply defined wells. The comb should be thoroughly cleaned before use. Minute amounts of air trapped on its surface can inhibit polymerization of acrylamide leading to the same problem. Visible gelation, easily seen around the teeth of the

comb, should occur within 20 min. Allow polymerization to proceed for 90 min.

4. Slowly pull the comb straight out of the gel, allowing air to enter the wells at the same time. Avoid distorting the rectangular shape of the wells. If the well partitions are not vertical and parallel, lanes will overlap after electrophoresis.

5. Rinse the wells gently with 1× electrophoresis tank buffer and then fill with the same buffer.

In some apparatus, the walls separating the wells may not polymerize completely due to the top of the stacking gel being partially exposed to air during polymerization. This can be avoided by adding riboflavin to a final concentration of 5 μg/mL along with TEMED and APS. Riboflavin is an initiator and an oxygen scavenger. In contrast to chemical polymerization by TEMED and APS, photochemical polymerization of acrylamide can be initiated by the addition of TEMED and riboflavin and exposing it to light with UV components (fluorescent light will suffice). The process requires a small amount of oxygen for the conversion of riboflavin to its active form and this principle is used, when necessary, to achieve complete polymerization of the stacking gel.

Although the stacking gel may appear slightly whitish compared to the perfectly transparent resolving gel, this does not affect the results. However, the gel may be made more transparent by reducing the concentration of bis-acrylamide in the stacking gel stock solution. Before discarding any excess monomer solution, allow it to polymerize as described above.

Loading the samples:

1. Depending on the model of apparatus used, attach the gel sandwich(es) to the electrophoresis box either before or after loading.

2. Load samples in the wells using 50 μL syringes with 25-gauge needles (blunt end preferred) or special gel loading pipet tips. For a large number of samples, it is easier to use the latter. Place the tip in the bottom of the well and gently layer the samples below the buffer column. The heavier sample solutions (due to presence of glycerol or sucrose) will settle to the bottom displacing the buffer upwards. Avoid cross contaminations between samples. The volume of each sample that should be loaded will depend upon the volume of the wells as well as on the protein concentration of the samples. The smallest volume (preferably no more than 30 μL) to provide 30 to 50 μg of protein should be used. When methods described above are followed, 30 μL of sample will produce satisfactory protein bands. However, if the protein content of the samples have been determined (section 29–2.2) the sample volume can be calculated to give 30 to 50 μg of protein.

3. One well may be loaded with a sample of commercially available (Bio-Rad Lab., Sigma Chemical Co., Integrated Separation Systems etc.) protein molecular weight markers. Molecular weights in the range of 14 to 66 kDa are generally useful.

Electrophoresis:

1. Fill the upper and lower buffer tanks with tank buffer.
2. Attach electrical leads to the power source, with cathode to the upper buffer tank and anode to the lower buffer tank.
3. Electrophorese at a constant current of 10 mA/gel until the samples enter the resolving gel (approximately 1 h).
4. Increase the current to 30 mA per gel and allow electrophoresis to continue until the dye front reaches the bottom of the gel (usually 3–5 h).
5. Turn off power and remove the gel sandwich(es) for fixing and staining the gels.

Electrophoresis buffer in the lower (i.e., anodic) tank may be used several times but fresh buffer must be used in the upper (i.e., cathodic) tank.

29–2.4.3 Comments

Occasionally the protein bands migrate at a faster rate in the central lanes compared to the side lanes producing what is known as a smiling effect. The lanes also tend to flare at the bottom. These distortions result from greater electrical heating in the middle of the gels compared to the sides. This may be avoided by performing the electrophoresis in a cold room (4 °C) or by circulating chilled water through a built-in cooling system in the electrophoresis apparatus. However, electrophoresing at room temperature yields satisfactory results if overheating is avoided by maintaining a low wattage. This may be achieved by switching from constant current to constant voltage if the voltage exceeds 400 volts because of an increase in the resistance of the gels.

Electrophoresis should be performed at the fastest allowable speed. Applying insufficient current will prolong electrophoresis leading to diffusion of low molecular weight bands. The current and voltage requirements may need to be determined by trial and error but reasonable limits are indicated above. Often it is best not to use the two outermost lanes for samples since they tend not to run straight and are compressed laterally due to flaring of inner lanes.

29–2.5 Visualizing Protein Profiles

29–2.5.1 Materials

1. Fixing solution:
   45% (vol/vol) methanol, 10% (vol/vol) acetic acid.
2. Staining solution:
   0.1% (wt/vol) Coomassie brilliant blue R in the fixing solution.

3. Destaining solution I:
   Same as fixing solution.
4. Destaining solution II:
   7.5% (vol/vol) methanol, 10% (vol/vol) acetic acid.
5. Destaining solution III:
   10% (vol/vol) acetic acid.

Caution: Methanol-containing solutions should be handled under the fume hood. Avoid inhalation of fumes and contact with skin.

29–2.5.2 Procedure

After completion of electrophoresis, the protein bands on the gels should be fixed immediately to prevent diffusion of bands.

1. Disassemble clamps etc. that hold the glass plates together and remove spacers.
2. Position the gel horizontally and pry open the glass plates by inserting a suitable object, such as a plastic wedge, without damaging the gel. The gel should remain adhered to one of the plates.
3. Remove the stacking gel by scraping from the glass plate. A small piece from a corner of the gel may be cut off to mark the original orientation of the gel in relation to the samples.
4. Transfer the gel to 250 mL of fixing solution in a suitable glass container (e.g., a rectangular baking dish). This may be done by inverting the glass plate with the gel over the dish and gently prying one corner of the gel off the plate.
5. Cover the dish tightly with plastic wrap and shake gently, preferably on a orbital shaker (50–60 rpm), for 1 h. The gel will shrink due to dehydration.
6. Drain off the fixing solution that may be reused as the first destaining bath. Add 250 mL of staining solution and shake for at least 1 h. Gel can be left in staining solution overnight or longer. Staining solution can be reused several times but should be filtered between uses.
7. Destain the gel by shaking in destaining solution I. Faster destaining is achieved by using small volumes (100 mL) of destaining solution and changing more frequently. Three changes of 20 min each should suffice before changing to 200 mL of destaining solution II. Shake the gel in this solution until the protein bands are clearly visible and the background is faint.
8. Transfer to 10% acetic acid. The gel will expand again and background will become clearer. Two to three changes of acetic acid should be used to remove the methanol that can destain the protein bands. The gel can be left in acetic acid indefinitely. The clarity of the bands usually improves in a few days.

29–2.6 Storing and Recording of Protein Profiles

29–2.6.1 Storage of Gels

Gels can be stored in 7% (vol/vol) acetic acid indefinitely without any appreciable loss of stain. They can be stored in covered baking or similar dishes or in sealable plastic bags. Alternatively, gels can be dried on white blotting paper or transparent cellophane sheets for storing. Several different kinds of gel driers are available commercially or gels can be sandwiched between cellophane sheets and glass plates and dried in an oven (Juang et al., 1984). Before drying, gels may be washed in 1.5% (vol/vol) glycerol for 20 min. This prevents the dried gel from smelling of acetic acid and reduces susceptibility to cracking. To avoid cracking and excessive curling, remove the gel from the drier only after it has cooled completely.

29–2.6.2 Photographic Record

Gels can be photographed in a wet or dry state. To make a quick photographic record, particularly if the gel is required for other procedures, electrophoresis duplicating paper (EDP, made by Eastman Kodak) can be used. For this procedure, the wet gel itself is used as a negative. The paper is exposed to incandescent light through the gel laid on an amber filter sheet and developed.

Wet gels can also be used for conventional photography. The gel is laid on a white-light transilluminator and photographed using color slide film (Kodak Ektachrome is suitable) for making slides or black-and-white film for making negatives for prints. For the latter, Kodak Technical Pan Film 2415 is suitable. For a Coomassie blue-stained gel, a yellow filter will improve contrast.

Gels dried on white paper may be photographed for prints or slides by reflected light (standard photographic copy stand) using the same films as described above.

29–3 GENERAL COMMENTS

The procedures described above are usually adequate for obtaining whole-cell protein profiles of most soil bacteria (Fig. 29–1). Modifications, however, may be required in the preparation of protein samples for different bacterial types. The best method for cell disruption must be determined by the investigator. For example, freezing and thawing is adequate for disruption of cells of *Rhizobium* (Dughri & Bottomley, 1983; Weaver et al., 1989) but for *Bradyrhizobium* cell disruption using a French Press yields better results (Sen & Weaver, 1988). When protein samples are prepared as described, good profiles are produced for *Rhizobium* but samples from *Bradyrhizobium* tend to produce smeared profiles. Smearing can be eliminated to a great extent if proteins in the extracts are precipitated

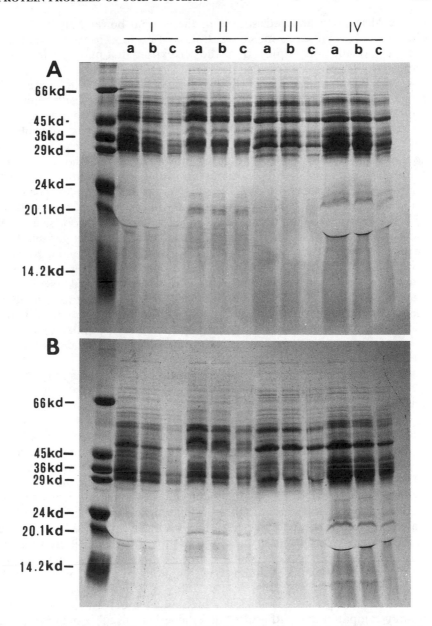

Fig. 29–1. Whole-cell protein profiles of *Rhizobium leguminosarum* bv *trifolii* obtained by sodium dodecyl sulfate-polyacrylamide gel electrophoresis. A is a 12% and B is a 7 to 17% exponential gradient gel. Both were loaded with identical material, electrophoresed together in the same box for the same length of time and stained with Coomassie brilliant blue. The leftmost lane shows molecular weight markers. Whole-cell protein preparations of strains 162Y10 (I), 162Y13 (II), 162K10 (III), and 162K13 (IV) were loaded using 45, 30, and 15 µg proteins for lanes a's, b's, and c's, respectively.

with cold acetone and redissolved in the sample buffer before electro-phoresing. This increases the concentration of proteins as well as removes other compounds, like lipids, from the cell extract that may interfere with the electrophoresis of proteins.

Protein preparations from whole cells may be expected to have a wide range of molecular weights. Sometimes a gel of uniform concentration may not give a usable profile across the total length. If the acrylamide concentration is too low (i.e., pores are large) small proteins may run out from the gel before the largest proteins enter the gel. Conversely, large proteins may not enter a resolving gel or be sufficiently separated if the acrylamide concentration is too high (i.e., pores are small). In this situation, a gel gradient with the concentration increasing from top to bottom may be more suitable (Fig. 29–1). As the proteins move through such a gel, their velocity of movement is reduced upon reaching zones where the pore sizes limit their passage. This further compacts the bands and allows larger numbers of proteins to be observed in the gel. Either linear or exponential gradient acrylamide concentrations can be prepared. The latter gives better resolution of low molecular weight proteins. Various gradient makers are commercially available or one can be fabricated. It is important to remember that when making gradient gels, the amount of polymerization initiators added must be reduced so that gelation does not start prematurely when monomers of two different concentrations are being mixed and poured into the gel mold.

Sometimes the concentration of proteins in the samples may be low giving faint bands in the stained gel. One way of overcoming such a problem is to concentrate the preparation by acetone precipitation or membrane filtration. A different approach is to use a staining method of higher sensitivity, e.g., silver staining. Silver staining can detect as little as 1/100 of the amount of protein detected by Coomassie blue staining (Merril et al., 1984). It should be noted that the affinity of these two stains for different proteins is variable. Thus, a given set of proteins may produce profiles that appear different depending on what staining procedure is used. Moreover, other macromolecules like nucleic acids, lipids etc. that may be present in a crude preparation of whole-cell proteins, will also be stained by silver and some of these will produce bands of different color. Thus, for whole-cell protein preparations, a silver-stained gel may appear overcrowded with bands, including bands of different colors, making visual analysis difficult.

In most instances, Coomassie blue-stained protein profiles may be visually compared and differences ascertained. If visual comparisons are difficult or detailed record of protein bands are required, gels can be scanned densitometrically. Peak positions and heights, indicating proteins and their relative quantities, in the traces of each sample can then be compared for qualitative and quantitative differences (Pankhurst, 1974).

Many variations of the general procedure can be found in the literature and published procedures for specific microbes should be tried when they are available. It should also be noted that any factor (e.g., growth medium composition or temperature) that affects the metabolism of a microbe may bring about changes in its protein profile (Glenn et al., 1986; Hood et al., 1988). Also, absolute reproducibility of protein profiles may be difficult to achieve due to variations in factors such as uniformity of gel concentration or staining intensity of proteins (Hood et al., 1988). One should, therefore, restrict comparisons to protein profiles on the same gel using bacterial strains grown under identical conditions. Considering the variabilities that may appear in protein profiles, caution should be exercised in drawing conclusions about identities or relatedness among strains. Whereas preliminary conclusions may be drawn about the relatedness of some microbial isolates and the number of different types in a collection, more sophisticated analysis may be required for establishing evolutionary or taxonomic relations. For the analysis of protein profiles of a large number of strains for taxonomic purposes, the need for standardized methodologies has been emphasized (Jackman, 1985) and computerized techniques for such analysis have been developed (Albritton et al., 1988; Jackman, 1985; Kersters & De Ley, 1980).

REFERENCES

Albritton, W.L., X.P. Chen, and V. Khanna. 1988. Comparison of whole-cell protein electrophoretic profiles of *Haemophilus influenzae:* Implementation of a microcomputer mainframe linked system and description of a new similarity coefficient. Can. J. Microbiol. 34:1129–1134.

Benson, D.R., S.E. Buchholz, and D.G. Hanna. 1984. Identification of *Frankia* strains by two-dimensional polyacrylamide gel electrophoresis. Appl. Environ. Microbiol. 47:489–494.

Benson, D.R., and D. Hanna. 1983. *Frankia* diversity in an alder stand as estimated by sodium dodecyl sulfate-polyacrylamide gel electrophoresis of whole-cell proteins. Can. J. Bot. 61:2919–2923.

Bradford, M.M. 1976. A rapid and sensitive method for the quantitation of microgram quantities of protein utilizing the principle of protein-dye binding. Anal. Biochem. 72:248–254.

De Maagd, R.A., C. Van Rossum, and B.J.J. Lugtenberg. 1988. Recognition of individual strains of fast-growing rhizobia by using profiles of membrane proteins and lipopolysccharadies. J. Bacteriol. 170:3782–3785.

Dughri, M.H., and P.J. Bottomley. 1983. Complementary methodologies to delineate the composition of *Rhizobium trifolii* populations in root nodules. Soil Sci. Soc. Am. J. 47:939–945.

Fuquay, J.I., P.J. Bottomley, and M.B. Jenkins. 1984. Complementary methods for differentiation of *Rhizobium meliloti* isolates. Appl. Environ. Microbiol. 47:663–669.

Glenn, A.R., R. Knuckey, and M.J. Dilworth. 1986. Periplasmic proteins of *Rhizobium:* variation with growth conditions and use in strain identification. FEMS Lett. 35:65–69.

Hood, D.W., C.S. Dow, and P.N. Green. 1988. Electrophoretic comparison of total soluble protein in the pink-pigmented facultative methylotrophs. J. Gen. Microbiol. 134:2375–2383.

Jackman, P.J.H. 1985. Bacterial taxonomy based on electrophoretic whole-cell protein patterns. p. 115–129. *In* M. Goodfellow and D.E. Minnikin (ed.) Chemical methods in bacterial systematics. Academic Press, London.

Juang, R-H. Y-D. Chang, H-Y. Sung, and J-C Su. 1984. Oven-drying method for polyacrylamide gel slab packed in cellophane sandwich. Anal. Biochem. 141:348–350.

Kamicker, B.J., and W.J. Brill. 1986. Identification of *Bradyrhizobium japonicum* nodule isolates from Wisconsin soybean farms. Appl. Environ. Microbiol. 51:487–492.

Kersters, K., and J. De Ley. 1980. Classification and identification of bacteria by electrophoresis of their proteins. p. 273–297. *In* M. Goodfellow and R.G. Board (ed.) Microbiological classification and identification. Academic Press, London.

Krishnan, H.B., and S.G. Pueppke. 1991. *nolC*, a *Rhizobium fredii* gene involved in cultivar-specific nodulation of soybean, shares homology with a heat-shock gene. Molec. Microbiol. 5:737–745.

Laemmli, U.K. 1970. Cleavage of structural proteins during the assembly of the head of bacteriophage T4. Nature (London) 227:680–685.

Leps, W.T., G.P. Roberts, and W.J. Brill. 1980. Use of two-dimensional polyacrylamide electrophoresis to demonstrate that putative *Rhizobium* cross-inoculation mutants actually are contaminants. Appl. Environ. Microbiol. 39:460–462.

Lochner, H.H., B.W. Srijdom, and I.J. Law. 1989. Unaltered nodulation competitiveness of a strain of *Bradyrhizobium* sp. (*Lotus*) after a decade in soil. Appl. Environ. Microbiol. 55:3000–3008.

Lowry, O.H., N.J. Rosebrough, A.L. Farr, and R.J. Randall. 1951. Protein measurement with Folin phenol reagent. J. Biol. Chem. 193:265–275.

Merril, C.R., D. Goldman, and M.L. van Keuren. 1984. Gel protein stains: Silver stain. p. 441–447. *In* W.B. Jakoby (ed.) Methods in enzymology. Vol. 104. Academic Press, New York.

Noel, K.D., and W.J. Brill. 1980. Diversity and dynamics of indigenous *Rhizobium japonicum* populations. Appl. Environ. Microbiol. 40:931–938.

O'Farrell, P.H. 1975. High resolution two-dimensional electrophoresis of proteins. J. Biol. Chem. 250:4007–4021.

Pankhurst, C.E. 1974. Ineffective *Rhizobium trifolli* mutants examined by immunediffusion, gel-electrophoresis and electron microscopy. J. Gen. Microbiol. 82:405–413.

Roberts, G.P., W.T. Leps, L.E. Silver, and W.J. Brill. 1980. Use of two-dimensional polyacrylamide gel electrophoresis to identify and classify *Rhizobium* strains. Appl. Environ. Microbiol. 39:414–422.

Sen, D., J.I. Baldani, and R.W. Weaver. 1990. Expression of heat induced proteins in wild type and plasmid-cured derivatives of *Rhizobium leguminosarum* bv *trifolii*. p. 583. *In* P.M. Gresshoff et al. (ed.) Nitrogen fixation: Achievements and objectives. Chapman and Hall, New York.

Sen, D., and R.W. Weaver. 1988. Protein profiles of bacteroids of *Bradyrhizobium* from peanut and cowpea. Microbios Lett. 38:61–64.

Weaver, R.W., D. Sen, J.J. Coll, C.R. Dixon, and G.R. Smith. 1989. Specificity of arrowleaf clover for rhizobia and its establishment on soil from crimson clover pastures. Soil Sci. Soc. Am. J. 53:731–734.

Chapter 30

Plasmid Profiles

IAN L. PEPPER, *University of Arizona, Tucson, Arizona*

30–1 PLASMID CHROMOSOME RELATIONSHIPS

Most of the genetic information necessary for a functional bacterial cell is contained in the bacterial chromosome. For many soil organisms this is normally a single circular chromosome consisting of about 1 to 3×10^6 nucleic acid base pairs, which code for 1000 or more different genes. Many common soil organisms have chromosomes of this size including *Pseudomonas* and *Bacillus* spp. The chromosome provides fundamental gene sequences necessary for metabolism and reproduction of the cell. In addition, bacteria may contain extrachromosomal or accessory genetic elements including plasmids (Campbell, 1981). Plasmids are additional DNA sequences that are separate from the chromosome. Normally plasmids code for genes with cellular functions that are not mandatory for cell growth and division. For example, rhizobia contain symbiotic plasmids that code for N_2 fixation. However, at times plasmids may be important for bacterial growth, such as when plasmid coded biodegradative genes are necessary for the utilization of toxic waste organic materials in soil. Plasmids are autonomous in that plasmid copy number, or number of identical plasmids per cell, are normally independent of the number of chromosome copies, and also expendable, meaning that the accessory genetic element is not essential to the growth of the organism in its normal environment (Krawiec & Riley, 1990).

Bacterial plasmids can be very small ($\approx$10 Kb) like those plasmids associated with *Streptomyces,* or, as in case of rhizobia, very large (300–1000 Kb) megaplasmids (Fig. 30–1) containing genetic information equivalent to one-fifth of the chromosome (Burkardt et al., 1987). In addition, different bacterial species or even strains of species may contain several different plasmids each with variable copy numbers. The functions of genes encoded by several plasmids are sometimes well defined. However, the function of many plasmids are not known, and these are referred to as cryptic plasmids. The relationships between plasmids and chromosomes are complex, in particular for megaplasmids, where the relative amount of

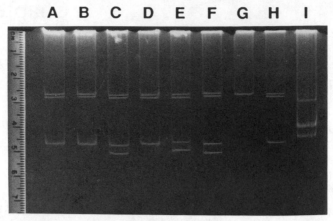

Fig. 30–1. Examples of plasmid profiles of *Rhizobium meliloti* cultures isolated from root nodules. Lane 1 is the plasmid profile of standard strain (DB1) which contains three plasmids of molecular weights 585, 250, and 165 MDa. Lanes A, B, C, and H show isolates with identical plasmid profiles. Lanes C and F also illustrate identical profiles, but these profiles are clearly different from A, B, D, and H, as is the profile in Lane G. Lane E contains a profile that is similar to those in C and F, but it is different. Strains used in Lanes E, C, and F were shown to be different by intrinsic antibiotic resistance patterns.

genetic information coded by the plasmid is of a similar order of magnitude to that of the chromosome. In addition, some plasmids can integrate into the chromosome during replication and function as part of the chromosome. During later replications, this process can be reversed, with the plasmid DNA being excised and allowed to function as a self-replicating entity within the cell. Well-known examples of this include integration and excision of the F plasmid in *Escherichia coli*. In addition, interspecific crosses between *Streptomyces* spp. occur, in which stable genetic elements in the donor species become plasmids in the recipient strains. Thus, plasmids, which technically are not vital for survival of the cell, are none the less often critical for the cell to function efficiently in its given environmental niche. This is due to the phenotypic expression of plasmid genes that confer an advantage to that cell.

30–1.1 Types of Plasmids

1. **Low-copy-number plasmids.** These are plasmids > 10 kb that normally exist with one to two copies per cell.
2. **High-copy-number plasmids.** These are normally smaller plasmids (< 10 kb) that have 10 to 100 copies per cell.
3. **Relaxed plasmids.** These are plasmids whose replication does not depend on initiation of cell replication. Therefore, these plasmids can be amplified (i.e., copy number increased) relative to cell number.

4. **Stringent plasmids.** These plasmids depend on cell replication, and plasmid replication is synchronized with replication of the bacterial chromosome. Thus, stringent plasmids have low copy numbers that cannot be amplified. Since cells growing rapidly may have three to four chromosomes, stringent plasmids can still be present with copy numbers > 1/cell.

5. **Conjugative plasmids.** These are plasmids which are self-transmissible and can be transferred from one bacterial cell to another during conjugation between the two cells. The cells can be of the same species or of different species. Conjugative plasmids are normally large and contain transfer genes known as tra genes (Slater et al., 1988).

6. **Non-conjugative plasmids.** These do not contain tra genes and are not self-transmissible. However, some plasmids can transfer to other cells by mobilization to other conjugative plasmids although not all nonconjugative plasmids are mobilizable. In this process of mobilization, transfer of nonconjugative plasmids relies on the tra genes of the conjugative plasmids.

7. **Incompatible plasmids.** Plasmids vary in their ability to co-exist within the same cell. Incompatible plasmids cannot exist together and give rise to incompatibility groups (Inc groups). Compatible plasmids belong to different Inc groups and vice versa. Currently 30 incompatibility groups have been defined for *E. coli* and 13 for *Staphylococcus aureus* (Old & Primrose, 1989).

8. **Inc P plasmids.** These plasmids are able to exist in a wide variety of bacterial species. Most plasmids exist in only a few closely related species and are known as narrow host range plasmids.

30–1.2 Functions of Plasmids

All plasmids contain gene sequences that are necessary for their own replication, and may also contain tra genes necessary for self-transmission (Nordstrom et al., 1984). Additional gene sequences code for a variety of phenotypic traits.

1. Cryptic plasmids. These are plasmids with sequences that encode for unknown phenotypic traits or sequences with no known function. These are often present as megaplasmids in rhizobia, and can only be cured or deleted with great difficulty. However, some megaplasmids in rhizobia have known functions including genes required for nodulation and N_2 fixation. Overall, most bacterial plasmids regardless of size are cryptic.

2. Resistance plasmids. Resistance or R-plasmids contain gene sequences that code for molecules that confer protection of the bacterial cell from specific deleterious substances. Specific plasmids are known that confer resistance to antibiotics, heavy metals, colicins, or bacteriophage.

3. Degradative plasmids. These plasmids are also known as catabolic plasmids and code for the breakdown of unusual metabolites. These

metabolites can be unusual C compounds of microbial origin, or xenobiotics including newly synthesized pesticides. Many soil bacterial isolates contain degradative plasmids, particularly *Pseudomonas* spp. These same plasmids are now useful in the production of genetically engineered microorganisms (GEMs) used to degrade hazardous wastes contaminating soil or water. Examples of degradative plasmids include the TOL plasmid involved in the breakdown of toluene or the pJP4 plasmid that contains gene sequences coding for the degradation of 2,4-dichlorophenoxyacetic acid.

4. Plant interactive plasmids. There are two well-known interactions involving plasmids and higher plants. The first concerns the symbiosis between the bacterium *Rhizobium* and leguminous plants. Typical *Rhizobium* spp. contain Sym (symbiotic) plasmids that code for infection and nodule formation on the plants, as well as gene sequences necessary for biological N_2 fixation. The second system involves parasitism between *Agrobacterium* spp. and higher plants. Tumor inducing (Ti) plasmids from the bacterium integrate into the host DNA, and induce tumors that result in crown gall disease. This is the only known occurrence of procaryotic DNA being integrated into eucaryotic DNA.

5. Miscellaneous plasmids. There are a variety of other kinds of plasmids with unique functions such as RNA metabolism, conjugation, or bacterial cell envelope alteration. Given that chromosomal DNA sequences can and do become integrated into plasmids during plasmid transfer, the potential functions associated with plasmids appear to be unlimited.

30–2 PLASMID PROFILE ANALYSIS

30–2.1 Introduction

The size and number of plasmids associated with a given bacterium are often unique. Therefore, if plasmids from a given soil bacterium can be obtained without shearing, gel electrophoresis allows all plasmids to be resolved into a unique plasmid profile. Following staining with ethidium bromide, the profile can be visualized and photographed using UV radiation (Hirsch et al., 1980).

Several protocols for plasmid isolation have been reported over the past decade. Most procedures used to isolate plasmid DNA depend on their circular form, and relatively small size compared to the bacterial chromosome. As the size of the plasmid increases, it becomes more difficult to separate the plasmid from the chromosome, and shearing of the plasmid becomes more prevalent. Usually bacterial cells are lysed by the addition of either alkali, EDTA, lysozyme or similar enzyme, or sodium dodecyl sulfate (SDS). Following cell lysis cellular debris and chromosomal DNA can be pelleted by centrifugation leaving the plasmid DNA in the supernatant, which is subsequently ethanol precipitated and purified. For megaplasmids, an in situ lysis in the gel procedure is used, which reduces plasmid shearing.

A comprehensive review of bacterial plasmid isolation and purification was published by Trevors (1985). Specific protocols for plasmid isolation in *Pseudomonas* (Itoh et al., 1984), and gram-positive bacteria (Anderson & McKay, 1983) are also available. Detailed here are three basic methods to obtain plasmid profiles: for megaplasmids the Eckhardt (1978) in situ lysis protocol is described. Two additional procedures are also described that rely on extraction of plasmid DNA followed by purification by gel electrophoresis or cesium chloride ultracentrifugation purification.

30–2.2 In Situ Lysis of Bacteria Containing Plasmids Greater Than 20 kb

This procedure was originally reported by Eckhardt (1978), but was modified by D. Berryhill (1985, personal communication). It is particularly useful for determination of megaplasmids in rhizobia or other gram-negative organisms. The in situ lysis method reduces shearing of megaplasmids that frequently occurs if other methods are employed.

30–2.2.1 Materials

1. Bacterial isolate on slant.
2. Yeast extract mannitol broth (YEM) (10 g of mannitol, 0.5 g of K_2HPO_4, 0.2 g of $MgSO_4\cdot7H_2O$, 0.1 g of NaCl, 0.4 g of yeast extract, 1000 mL H_2O).
3. Yeast extract broth (YE) = YEM minus mannitol.
4. Spectrophotometer.
5. Centrifuge.
6. 0.1% sarkosyl (*N*-Lauryl sarcosine) in TE buffer (50 mM of Tris, 20 mM of EDTA, pH 8.0).
7. Tris borate buffer (89 mM of Tris, 89 mM of boric acid, 2.5 mM of EDTA (disodium form), pH 8.2) in 20% Ficoll 400 000.
8. Ribonuclease—10 mg/mL in 0.4 M sodium acetate. Heat for 2 min at 98 °C before adding to lysozyme mix. Ribonuclease is 71 units/ mg.
9. Lysozyme mix (22 500 units lysozyme, 0.3 units Ribonuclease I, 0.5 mg of bromphenol blue, 0.2 g of ficoll 400 000/mL in Tris borate buffer.
10. 0.2% SDS (sodium dodecyl sulfate), 10% Ficoll 400 000 in Tris borate buffer.
11. Electrophoresis unit (vertical).
12. Agarose (0.7% agarose in Tris borate buffer).
13. Overlay mixture (0.2% SDS, 5% Ficoll 400 000 in Tris borate buffer).
14. Ethidium bromide (1 µg/mL).
15. Transilluminator (366 nm).
16. Polaroid camera with Polaroid film (52 or 57 film type).

30–1.2.2 Procedure

Each bacterial isolate is grown in 10 mL of YEM broth for 48 h at 28 °C with shaking. The cells are subsequently used as a 1% inoculant for 25 mL of YE broth without shaking at 28 °C. This step reduces polysaccharide production. After growth for 24 h, the OD is adjusted to 0.05 at 540 nm with YE broth, and 10 mL of cells centrifuged at 8000 g for 10 min at 4 °C. For rhizobia or other gram-negative organisms, the resulting pellet is resuspended in 1 mL of 0.1% sarkosyl in TE buffer and again centrifuged for 10 min at 8000 g at 4 °C. The release of DNA from gram-positive organisms is improved by resuspension of pelleted cells in 0.5 mL of ice cold acetone for 5 min (Heath et al., 1986). Cells are resedimented at 8000 g for 10 min at 4 °C, the acetone decanted and residual acetone removed by blowing with a gentle stream of air. This pellet can then be resuspended in sarkosyl in TE buffer in a similar manner to that of the rhizobia. After the tube is thoroughly drained, the resulting pellet is vigorously resuspended in 100 µL of Tris borate buffer containing 20% Ficoll 400 000 and placed on ice. A 50 µL sample of the cell suspension is transferred to a 1.5-mL microfuge tube on ice and mixed with 25 µL lysozyme mixture. A 25-µL sample of this mixture is immediately loaded into the sample well of a vertical gel (0.7% agarose) into which 50 µL of SDS had been placed. The contents of the well are covered with 100 µL of the overlay mixture and sealed with 0.7% agarose. For megaplasmids, electrophoresis is carried out at 4 °C at 5 mA for 1 h and then 40 mA for 16 h using a vertical electrophoresis unit. The vertical gel gives good results and does not shear the megaplasmids. However, horizontal gels can also give good results. Smaller plasmids require shorter duration of electrophoresis. Following electrophoresis, the gel is stained with ethidium bromide (1 µg/mL) in water for 15 min, destained in distilled water for 20 min, and photographed over 300 nm UV radiation using Polaroid film e.g., Type 52 or Type 57.

Plasmid size can be determined by comparing fragment sizes relative to migration distance along a gel. Plasmids of defined size can be used for comparison. The standard sized plasmids can be bought commercially from several biotechnology companies.

30–2.3 Extraction of Plasmids from Cells, followed by CsCl Ultracentrifugation Purification and Gel Electrophoresis

This procedure is based on that of Hirsch et al. (1980) that was developed for analysis of strains of *Rhizobium leguminosarum*. Traditionally, rhizobia have been the predominant soil bacterium to which plasmid profile analyses have been applied.

30–2.3.1 Materials

1. Bacterial isolate on slant.
2. Liquid TY medium (0.5% wt/vol Difco Bacto-Tryptone, 0.3% wt/vol Difco Bacto-Yeast Extract and 7 mM CaCl$_2$).

3. Liquid PA medium (0.4% wt/vol Difco Bacto-Peptone, 2 mM MgSO$_4$).
4. Centrifuge.
5. TE buffer (50 mM Tris HCl, 20 mM EDTA, pH 8.0).
6. Dialysis buffer (10 mM Tris HCl, 1 mM EDTA, pH 8.0).
7. Tris borate electrophoresis buffer (per liter of buffer, 10.8 g of Tris, 0.93 g of EDTA and 5.5 g of boric acid, pH 8.3).
8. Pronase (5 mg per mL in TE buffer).
9. SDS (10% sodium dodecyl sulfate wt/vol in TE buffer).
10. 3 M NaOH.
11. 2 M Tris HCl, pH 7.0.
12. 50 mL centrifuge tubes.
13. 5 M NaCl.
14. Polyethyleneglycol (PEG 6000) 50% wt/vol.
15. TE buffer containing 0.1% vol/vol diethylpyrocarbonate.
16. CsCl TE buffer (30 g of CsCl in 25 mL TE buffer).
17. CsCl.
18. Ethidium bromide (10 mg/mL).
19. Polycarbonate ultracentrifuge tubes, 50 mL.
20. Ultracentrifuge.
21. Hand held UV illumination, 366 nm.
22. 18 or 20 gauge syringe.
23. Cold 2-propanol saturated with CsCl/TE buffer at 4 °C.
24. 75% ethanol.
25. Electrophoresis unit—horizontal.
26. Agarose (0.7% in Tris borate buffer).
27. Transilluminator (366 nm).
28. Loading dye (20% wt/vol Ficoll 400, 0.125% wt/vol bromophenol blue, 50 mM EDTA).
29. Polaroid camera with Polaroid type 55 film.

30–2.3.2 Procedure

All buffers, solutions, and glassware should be autoclaved or rinsed with 75% ethanol to minimize contamination with nucleases. For small-scale crude plasmid preparations 200 mL of PA medium is inoculated with 2 mL of a stationary phase stock culture of *Rhizobium* or other organism in TY medium and shaken at 28 °C for 36 h. Bacteria at a concentration of about 10^8 CFU/mL are centrifuged at 8000 g for 10 min at 4 °C, washed in TE buffer and resuspended in 16 mL TE buffer. Two milliliters of pronase in TE buffer (pre-incubated for 1 h at 37 °C) and 2 mL SDS in TE buffer are added, and the mixture incubated at 37 °C with gentle shaking for 1 h. The clear, viscous lysate is adjusted to pH 12.4 with about 0.5 mL of 3 M NaOH, with gentle but thorough stirring using a plastic rod or pipette. After standing at room temperature for 30 min, the lysate is slowly adjusted to pH 8.5 with about 1.5 mL of 2 M Tris HCl, and transferred to a 40 mL centrifuge tube. After addition of 5 M NaCl to give a final concentration of 1 M, the contents of the tube are mixed by gentle inversion, and left on ice for 4 h or overnight. The SDS/NaCl complex is precipitated by centrifugation at 10 000 rpm for 20 min at 4 °C. The supernatant is decanted to a fresh tube, and PEG 6000 added to give a final concentration of 10%. The contents are mixed by gentle inversion and left at 4 °C overnight. The DNA is precipitated by centrifuging at 7000 rpm at 4 °C for 15 min. The supernatant is discarded, and the pellet resuspended in 0.5 mL TE buffer with diethylpyrocarbonate to inhibit nuclease activity. Samples are stored at 4 °C.

For large-scale preparations, volumes are increased 10-fold and 4.5 mL of the DNA solution is added to a solution of 30 g of CsCl in 25 mL of TE buffer, followed by 2.6 mL of ethidium bromide (10 mg/mL) solution. The remaining crude plasmid preparation is retained to test directly for the presence of plasmids on gels. The solution is mixed by gentle inversion, then centrifuged in a swinging bucket rotor for 10 min at 5000 rpm at 4 °C. The resultant skin of residual PEG 6000 is removed before transferring the solution to a 40-mL polycarbonate ultracentrifuge tube. Samples are centrifuged for 18 h at 40 000 rpm at 15 °C. The gradient is then illuminated with UV light and the lower band corresponding to plasmid DNA is removed by insertion of a 18 to 20 gauge syringe through the side of the tube. Ethidium bromide is removed from the plasmid preparation by five extractions with two volumes of cold propanol saturated with CsCl/TE buffer. This is followed by dialysis against four changes of dialysis buffer at 4 °C, using 1 L of buffer with a minimum of 4 h between changes.

30–2.3.3 Gel Electrophoresis

The samples of crude plasmid preparations can be centrifuged for 2 min at 200 rpm and 40 µL of the supernatant, or 40 µL of the CsCl purified plasmid preparation added to 10 µL of loading dye. Samples are electrophoresed in 0.7% agarose gels at 45 mA (180 V) at 4 °C for 6 h. After staining in Tris borate buffer with ethidium bromide (0.5 µg/mL), and destaining in Tris borate buffer for 20 min, plasmid bands can be visualized on a transilluminator and bands subsequently photographed using a Polaroid camera with Polaroid film.

30–2.4 Alkaline Lysis Miniprep (Birnboim & Doly, 1979)

In this protocol, bacteria are lysed via sodium dodecyl sulfate (SDS) and sodium hydroxide, which denatures chromosomal and plasmid DNA. Treatment with potassium acetate reanneals plasmid DNA, whereas the chromosomal DNA forms a precipitate with potassium and is removed by centrifugation. The plasmid DNA is collected by ethanol precipitation. This protocol is similar to that described by Ausubel et al. (1987) in *Current Protocols in Molecular Biology,* and is useful for plasmid profiles where plasmids are small (< 10 Kb). It has successfully been used on several genera of soil bacteria including *Pseudomonas* and *Alcaligenes* spp.

30–2.4.1 Materials

1. Bacterial isolate on slant.
2. LB medium (per liter, 10 g of tryptone, 5 g of yeast extract, 5 g of NaCl, 1 mL of 1 N of NaOH, 1000 mL of H_2O).
3. Microfuge plus 1.5-mL tubes.
4. Pasteur pipet.
5. Glucose/Tris/EDTA (GTE) solution (50 mM of glucose, 25 mM of Tris, 10 mM of EDTA, pH 8.0). Autoclave and store at 4 °C.
6. NaOH/SDS solution (0.2 N of NaOH, 1% wt/vol sodium dodecyl sulfate). Prepare immediately before use.
7. Potassium acetate solution (29.5 mL of glacial acetic acid, KOH pellets to pH 4.8, H_2O to 100 mL). Store at room temperature, do not autoclave.
8. 95% ethanol.
9. 70% ethanol.
10. TE buffer (10 mM of Tris, 1 mM of EDTA, pH 8.0).

30–2.4.2 Procedure

Inoculate 5 mL of sterile LB medium with a single bacterial colony. Grow to stationary phase. Spin 1.5 mL of cells 20 s in a microcentrifuge at 8000 g for 10 min at 4 °C. Remove the supernatant with a Pasteur pipet. Resuspend pellet in 100 µL of GTE solution and let sit 5 min at room temperature. Be sure cells are completely resuspended. Add 200 µL of NaOH/SDS solution, mix by tapping tube with finger, and place on ice for 5 min. Add 150 µL of potassium acetate solution and vortex briefly to mix. Place on ice for 5 min. Spin 3 min at 8000 g for 10 min at 4 °C to pellet cell debris and chromosomal DNA. Transfer supernatant to a fresh tube, mix with 0.8 mL of 95% ethanol, and let sit 2 min at room temperature to precipitate nucleic acids. Spin 1 min at room temperature to pellet plasmid DNA and RNA. Remove supernatant, wash the pellet with 1 mL of 70% ethanol and dry pellet under vacuum. Resuspend the pellet in 30 µL of TE buffer. Gel electrophoresis is conducted as described in (30–2.3.2) to allow resolution of plasmids. Time of electrophoresis is dependent on the size of the plasmids.

30–2.5 Storage of Plasmid DNA

Plasmid DNA can be stored in TE buffer at 4 °C for several weeks or preserved for several years by storing at −20 or −70 °C.

30–3 APPLICATIONS OF PLASMID PROFILE ANALYSES FOR SOIL BACTERIA

30–3.1 Plasmid Functions

Plasmids generally code for a variety of phenotypic traits, and it is often of interest to ascribe a particular phenotypic trait to a specific plasmid. If gene probes for a specific gene of interest are available, then plasmid profile analysis followed by a southern blot can identify whether a given plasmid contains that gene sequence. For example, 2,4-dichlorophenoxyacetic acid is degraded in part by plasmid pJP4, originally located in *Alcaligenes eutrophus* strain JMP134. Neilson et al. (1992) made an end-labelled probe specific to the tfdB gene contained in pJP4, which can be used to identify plasmids containing the pJP4 sequence.

Specific phenotypic traits can also be ascribed to particular plasmids by plasmid-curing studies. Here, strains of soil isolates containing multiple plasmids must be subjected to stress that results in a 99.9% die off. Examples of stress can be heat or ethidium bromide treatment. Surviving organisms are then screened for phenotypic traits. Isolates that lose a phenotypic trait of interest are then subjected to plasmid profile analysis.

30–3.2 Classification of Soil Bacteria

Plasmid profiles have been used to identify soil bacteria at the strain level. Such identification is based on the premise that a particular set of plasmids is unique to that strain with respect to size and number of plasmids. Soil bacteria such as rhizobia often contain several megaplasmids, and plasmid profiles of these bacteria have been used to identify rhizobia at the strain level (Shishido & Pepper, 1990). A typical megaplasmid profile is shown in Fig. 30–1. However, care must be taken in such analyses, because different strains may contain similar plasmid profiles. Researchers are strongly encouraged to use two methods of identification (see chapters 25, 26, 27, 28, and 29), to confirm the identity of soil isolates. Shishido and Pepper (1990) used plasmid profile analysis coupled to intrinsic antibiotic resistance patterns in the study of rhizobia soil isolates. Overall, such an approach is useful to identify dominant strains of soil bacteria of a given genus, when it is not practical to use other methods such as serology due to the unknown identity of the dominant organisms.

REFERENCES

Anderson, D.G., and L.L. McKay. 1983. Simple and rapid method for isolating large plasmid DNA from *Lactic Streptococci*. Appl. Environ. Microbiol. 46:549–552.

Ausubel, F.M., R. Brent, R.E. Kingston, D.D. Moore, J.A. Smith, S.G. Sideman, and K. Struhl (ed.) 1987. Current protocols in molecular biology. John Wiley and Sons, New York.

Birnboim, H.C., and J. Doly. 1979. A rapid alkaline extraction method for screening recombinant plasmid DNA. Nucl. Acids Res. 7:1513–1523.

Burkardt, B., D. Schillik, and A. Puhler. 1987. Physical characterization of *Rhizobium meliloti* megaplasmids. Plasmid 17:13–25.

Campbell, A. 1981. Evolutionary significance of accessory DNA elements in bacteria. Annu. Rev. Microbiol. 35:55–83.

Eckhardt, T. 1978. A rapid method for the identification of plasmid deoxyribonucleic acid in bacteria. Plasmid 1:584–588.

Heath, L.S., G.L. Sloan, and H.E. Heath. 1986. A simple and generally applicable procedure for releasing DNA from bacterial cells. Appl. Environ. Microbiol. 51:1138–1140.

Hirsch, P.R., M. VanMontagu, A.W.B. Johnston, N.J. Brewin, and J. Schell. 1980. Physical identification of bacteriocinogenic, nodulation and other plasmids in strains of *Rhizobium leguminosarum*. J. Gen. Microbiol. 120:403–412.

Itoh, Y., J.M. Watson, D. Haas, and T. Leisinger. 1984. Genetic and molecular characterization of the *Pseudomonas* plasmid pVSI. Plasmid 11:206–220.

Krawiec, S., and M. Riley. 1990. Organization of the bacterial chromosome. Microbiol. Rev. 54:502–539.

Neilson, J.W., K.L. Josephson, S.D. Pillai, and I.L. Pepper. 1992. PCR and gene probe detection of 2,4-D degrading plasmid, pJP4. Appl. Environ. Microbiol. 58:1271–1275.

Nordstrom, K., S. Molin, and J. Light. 1984. Control of replication of bacterial plasmid: Genetics, molecular biology and physiology of the plasmid R1 system. Plasmid 12:71–90.

Old, R.W., and S.B. Primrose. 1989. Plasmids as cloning vehicles for use in *E. coli*. p. 39–59. *In* R.W. Old and S.B. Primrose (ed.) Principles of gene manipulation: An introduction to genetic engineering. Blackwell Sci. Publ., Oxford.

Shishido, M., and I.L. Pepper. 1990. Identification of dominant indigenous *Rhizobium meliloti* by plasmid profiles and intrinsic antibiotic resistance. Soil Biol. Biochem. 22:11–16.

Slater, J.H., A.J. Weightman, and D. Godwin-Thomas. 1988. Plasmids. p. 44–50. *In* J.M. Lynch and J.E. Hobbie (ed.) Microorganisms in action: Concepts and applications in microbial ecology. Blackwell Sci. Publ., Oxford.

Trevors, J.T. 1985. Bacterial plasmid isolation and purification. J. Microbiol. Methods 3:259–271.

Chapter 31

DNA Fingerprinting and Restriction Fragment Length Polymorphism Analysis

M. J. SADOWSKY, *University of Minnesota, St. Paul, Minnesota*

DNA fingerprinting and restriction fragment length polymorphism (RFLP) analyses have proven extremely useful for strain identification, epidemiological studies, and the taxonomic analysis of prokaryotic and eukaryotic organisms. Both techniques require the isolation of relatively high-molecular-weight genomic or plasmid DNA, enzymatic cleavage of the isolated nucleic acids using restriction endonucleases, and electrophoretic separation of the resulting DNA fragments. The two techniques differ with respect to the means of examining the resultant restriction fragments: DNA fingerprinting uses ethidium bromide staining and visualizes all restriction fragments, whereas RFLP analysis used DNA or RNA probes that selectively bind (hybridize) to a few restriction fragments. In either case, the resulting banding patterns are generally unique to one or a few strains of a particular microbe and as such, can serve as a "fingerprint" for strain identification. Figure 31–1 shows the general scheme for DNA fingerprinting and RFLP analysis of bacterial genomic DNA. While DNA fingerprinting is relatively rapid, routine, and inexpensive to perform, RFLP analysis is more complex, expensive, and time consuming. However, RFLP analysis can show small differences between the genomic DNAs of organisms that is not evidenced by DNA fingerprinting techniques. In addition, RFLP analyses can also be useful for the construction of genetic maps and for map-based cloning in eukaryotic organisms (Young, 1990). More recently, it has been shown that DNA primers corresponding to repetitive extragenic palindromic (REP) and enterobacterial repetitive intergenic consensus (ERIC) sequences,coupled with the polymerase chain reaction (PCR) technique can be used to fingerprint the genomes of a large number of different gram-negative soil bacteria (de Bruijn, 1992; Hulton et al., 1991; Judd et al., 1993; Stern et al., 1984; Versalovic et al., 1991).

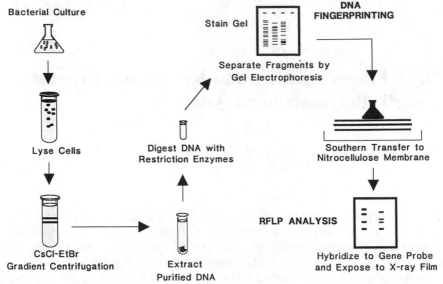

Fig. 31–1. Flow diagram of procedures required for the DNA fingerprinting and RFLP analysis of bacterial genomic DNA.

31–1 DNA FINGERPRINTING

31–1.1 Introduction

DNA fingerprinting has found wide application for a variety of organisms. While some initial studies used DNA fingerprinting of bacteria for epidemiological analyses of hospital infections (Kaper et al., 1982; Kristiansen et al., 1986; Kuijper et al., 1987; Langenberg et al., 1986; Skjoid et al., 1987; Tompkins et al., 1987), the technique has also found great application in the epidemiological and taxonomic analysis of yeast (Panchal et al., 1987; Scherer & Stevens, 1987), mycoplasmas (Chandler et al., 1982), fungi (Koch et al., 1991), viruses (Buchman et al., 1978; Christensen et al., 1987), several diverse bacterial species (Falk et al., 1984; Langenberg et al., 1986; Ramos & Harlander, 1990), and humans (Gill et al., 1987). In addition to medically important organisms, DNA fingerprinting techniques have also been used to study the taxonomic relatedness of agriculturally important microorganisms. These organisms include both bacterial and fungal pathogens (Lazo et al., 1987) and plant symbionts (Brown et al., 1989; Glynn et al., 1985; Kaijalainen & Lindstrom, 1989; Mielenz et al., 1979; Sadowsky et al., 1987; Schmidt et al., 1986).

The fingerprinting of bacterial DNA is an extremely useful technique to differentiate closely related soil microorganisms. The methodology used does not appreciably vary with the genus and species of the organism examined. However, slightly different DNA extraction/purification techniques may be required depending on the organisms under investigation.

31–1.2 DNA Isolation and Purification

The DNA fingerprinting technique requires the isolation and partial purification of relatively high-molecular-weight nucleic acids. In most instances, the source of the DNA is a pure culture of a microorganism (see chapter 8 in this book). DNA directly extracted from soils (see chapter 35 in this book) is not amenable to DNA fingerprinting techniques, since it contains genomic DNA from many microorganisms. Once isolated, the DNA must be of sufficient purity to be reproducibly cleaved by appropriate restriction endonucleases (restriction enzymes; see below). Several methods have been developed to isolate high-molecular-weight DNA. Most techniques can be adapted for either large-scale preparations or for minipreparations, depending on the quantity of DNA required. While some of these procedures are derived from modifications of the original technique of Marmur (1961) and require phenol and chloroform extractions prior to DNA precipitation, others omit these rather caustic reagents and purify DNA using cesium chloride-ethidium bromide (EtBr) equilibrium-density gradient ultracentrifugation (Sambrook et al., 1989). The CsCl-EtBr technique can be used for both small- or large-scale DNA preparations and results in the isolation of high purity DNA. While this technique is more time consuming and expensive than Marmur-type procedures, the purity of the isolated nucleic acids makes this the method of choice for DNA fingerprinting and RFLP analysis.

Recently, several rapid, small-scale DNA isolation procedures have also been developed for the fingerprinting of bacterial DNA (Heath et al., 1986; Pitcher et al., 1989; Ramos & Harlander, 1990). While some of the procedures work well with particular bacterial genera (Ramos & Harlander, 1990), others can be modified to extract DNA from widely divergent bacterial species (Heath et al., 1986; Pitcher et al., 1989). It should be noted, however, that some small-scale isolation techniques may produce DNA that is recalcitrant to restriction enzyme digestion and these procedures should be tested on a limited number of isolates before being adopted for wide-scale use.

31–1.3 Restriction Enzyme Selection and Use

Restriction enzymes are types of DNA endonucleases found in many microorganisms. Restriction enzymes have been isolated from more than 600 microorganisms (Logtenberg & Bakker, 1988; Roberts & Macelis, 1991). The type II DNAses, which recognize and cleave double-stranded DNA at specific base sequences (restriction sites), are used for DNA fingerprinting. Restriction enzymes are named for the organisms from which they are isolated. For example, *Eco*R1 was the first restriction enzyme isolated from *Escherichia coli* strain RY13 and *Hin*dIII was the third enzyme isolated from *Haemophilus influenzae* strain Rd. DNA restriction enzymes evolved as part of the host's restriction-modification system and function to cleave foreign DNA entering a cell. Methylation of a cell's

DNA makes it resistant to the degradative action of its own restriction enzymes.

Restriction enzymes that are most useful for DNA fingerprinting recognize four to six specific nucleotides, although several eight-base recognition site enzymes (e.g., *Sfi*I and *Not*I) have found recent use (Levine & Cech, 1989). Generally, an enzyme that has a shorter recognition sequence produces a larger number of fragments than one recognizing a larger sequence. Most often, the nucleic acid bases for restriction sites are arranged as palindromes (inverted repeat sequences). The number of DNA fragments produced by digesting DNA with a given restriction enzyme varies according to the mole % guanine plus cytosine (G + C) content of the DNA and the number and nature of the recognition sites. For a particular organism, however, the number of DNA fragments produced by digestion of the genome with a specific enzyme will be constant. Since bacteria have genomes with different numbers and arrangements of restriction sites (or have recognition sites that are differentially modified by methylases), bacterial strains exhibit distinct restriction enzyme fragment patterns. While the selection of specific restriction enzymes for a DNA fingerprinting application is empirically determined, inexpensive and readily available restriction enzymes, such as *Eco*RI and *Hin*dIII, are often the first choice.

31–1.4 Gel Electrophoresis

DNA fragments produced from the digestion of DNA with specific restriction endonucleases can be separated according to molecular weight by electrophoresis. Since DNA fragments all have uniform negative charge density, due to the phosphate groups in phosphodiester backbone, they migrate to the anode at a rate that is proportional to their molecular weight. Consequently, small fragments migrate faster than large fragments in an electric field. Agarose and polyacrylamide are the media of choice for separating DNA molecules. For most DNA fingerprinting applications, horizontal agarose gels are sufficient to resolve restriction fragments. Agarose, a poly galactose derivative, has high polymeric strength, low electroendosmosis (Perbal, 1984) and is nontoxic. Concentrations of agarose used in gels can be varied depending on the size of the fragments to be separated. For DNA fingerprinting, fragments are usually resolved in 0.8 to 1.0% gels. At these concentrations, however, large fragments [around 30 kilobase pairs (kb) or greater] do not appreciably enter the gel matrix under continuous electric fields.

Recently, several new electrophoresis techniques have been developed to circumvent problems associated with the electrophoresis of large DNA molecules. These techniques include pulsed-field gel electrophoresis (PFGE) (Carle & Olson, 1985; Chu et al., 1986; Schwartz & Cantor, 1984) and field-inversion gel electrophoresis (FIGE) (Carle et al., 1986; Sobral et al., 1990; Smith et al., 1988). Both techniques use agarose as the gel support matrix, but vary the electric field during the electrophoretic separation of DNA. Both the FIGE and PFGE techniques have been used to fingerprint and determine the size of bacterial and yeast genomic DNA (Bautsch, 1988; Canard & Cole, 1989; Carle & Olson, 1986; Kauc et al., 1989; Lee & Smith, 1988; Smith et al., 1986; Sobral et al., 1990.

Since polyacrylamide gels are used for the separation of DNA fragments that are < 2 kb in size (Coulson & Sulston, 1988), they are usually of limited value for DNA fingerprinting. However, polyacrylamide gels can be used to fingerprint smaller, cloned genomic DNA fragments (Coulson & Sulston, 1988).

31–1.5 Principles

The DNA fingerprinting technique is a "genotype-based" method that allows one to differentiate and identify both closely related or genetically distinct bacterial strains. The technique alleviates problems associated with phenotypic analyses (growth on specific media or immunological reaction) in that DNA expression is not required. The technique is extremely useful when applied to studies involving indigenous and exogenously added soil microorganisms, since many soil microbes are biochemically ill-defined and physiologically and morphologically nondescript. DNA fingerprinting techniques are based on the fact that every organism has a unique number and location of restriction enzyme recognition sites. Digestion of the DNA from a soil bacterium with an appropriate restriction endonuclease will produce a series of restriction fragments that are unique to that organism. When the restriction fragments are separated by electrophoresis, a banding pattern (a "DNA fingerprint") is produced that is characteristic and unique for a particular organism or strain. Similarities and differences in banding patterns may indicate the degree of genetic relatedness among organisms, with more closely related organisms having more similar banding patterns. Organisms having indistinguishable banding patterns can be regarded as being identical or near-identical. It should be noted, however, that while DNA fingerprints are stable over many generations of microbial growth (Ramos & Harlander, 1990; Scherer & Stevens, 1987), they are susceptible to changes caused by curing or rearrangement of indigenous plasmids and prophages and the loss or rearrangement of indigenous insertion-like sequences.

31–1.6 Methods

31–1.6.1 Special Apparatus

1. Horizontal gel electrophoresis apparatus. Either homemade (Sambrook et al., 1989; Perbal, 1984) or commercially available (Bio-Rad Laboratories, Richmond, CA; Gibco-BRL Life Technologies, Gaithersburg, MD; or Hoefer Scientific Instruments, San Francisco, CA) horizontal, submarine, agarose gel electrophoresis units are suitable for DNA fingerprinting analyses. Large-size gel equipment (gel bed of about 20 cm) is preferable

over small bed apparatus, since the latter better resolve restriction fragments.

2. Power supplies. Several commercially available D.C. power supplies (Bio-Rad Laboratories, Richmond, CA; Gibco-BRL Life Technologies, Inc., Gaithersburg, MD; or Hoefer Scientific Instruments, San Francisco, CA) can be used for the electrophoretic separation of DNA fragments. Regulated, D.C. power supplies in kit form (HeathKit Model IP-2717A) can also be used.

3. Ultracentrifuge, rotors, ultracentrifuge tubes, and tube sealer. An ultracentrifuge is required for the large-scale isolation and purification of genomic DNA. Several different ultracentrifuge rotors can be used, though centrifugation through large gradients (> 13.5 mL) is thought to produce more purified DNA. For this reason, it is recommended that the Beckman Type 50 Ti, Type 70.1 Ti, or Type 80 Ti fixed-angle rotors be used. Centrifugation times can be reduced, however, by using vertical or near-vertical ultracentrifuge rotors, such as the Beckman VTi 65.1 or NVT 65.

4. A water bath, capable of being set at both 37° and 50 °C, is required for restriction enzyme digestion reactions and for cooling melted agarose.

5. Refrigerated centrifuge (Sorval RC-5B or equivalent), rotor (SS-34 or equivalent) and centrifuge tubes (50 mL).

6. Microcentrifuge and tubes.

7. Photographic apparatus and film, Polaroid MP-4 camera (or equivalent) equipped with a Kodak number 23A Wratten gelatin filter. Use Polaroid Type 55 film to obtain both a positive and negative image.

8. Ultraviolet spectrophotometer and quartz cuvettes.

9. Ultraviolet transilluminator, capable of emitting long- or near-long wave (300–310 nm) UV light.

10. Micropipetors and tips.

31–1.6.2 Reagents

1. TEN buffer (50 mM Tris[hydroxymethyl]aminomethane, pH 8.0, 20 mM of disodium-ethylenediaminetetraacetic acid, and 50 mM of NaCl).

2. Lysozyme, egg white (Muramidase, EC 3.2.1.17), 5 mg/mL freshly prepared in TEN buffer.

3. Pronase solution [Protease Sigma Type XIV (Pronase E)], 5 mg/mL prepared in TEN buffer. Predigest solution by incubation for 1 h at 37 °C prior to use. Freeze unused solution.

4. Cesium chloride (99 + % pure). Finely ground in a mortar and pestle.

5. TE buffer, 10 mM of Tris[hydroxymethyl]aminomethane, pH 8.0 and 1 mM of disodium-ethylenediaminetetraacetic acid.

6. TBE electrophoresis buffer (89 mM of Tris[hydroxymethyl]aminomethane, pH 8.0, 89 mM of boric acid, and 2 mM of disodium-

ethylenediaminetetraacetic acid). Make 5X stock and dilute to 0.5X working solution.

7. Ethidium bromide solution, 10 mg/mL in distilled water. Protect from light. Caution should be exercised when handling ethidium bromide. It is a CARCINOGEN. Use gloves when handling solutions.

8. Sarkosyl solution, 20% wt/vol sodium *N*-lauroylsarcosine in distilled water.

9. Guanidium thiocyanate solution, 60 g of guanidium thiocyanate, 20 mL 0.5 *M* of disodium-ethylenediaminetetraacetic acid, pH 8.0, 20 mL of deionized water.

10. Sarkosyl solution, 10% wt/vol sodium *N*-lauroylsarcosine in distilled water.

11. Ammonium acetate solution, 7.5 *M*.

12. Isopropanol (2-propanol).

13. 70% ethanol solution.

14. Lysozyme solution, 50 mg/mL, freshly prepared in TE buffer.

15. Chloroform-isoamyl alcohol (24:1).

16. Water and salt saturated *n*-butanol solution. Saturate about 100 mL of distilled water with NaCl until a large amount of salt drops from the solution. Add 900 mL of *n*-butanol, shake briefly, and allow to separate into two phases.

17. Restriction enzyme digestion buffer (for Gibco-BRL enzyme *Eco*R1), 50 m*M* of Tris[hydroxymethyl]aminomethane, pH 8.0, 10 m*M* of MgCl$_2$, and 100 m*M* of NaCl. Make 10X stock and freeze in small aliquots.

18. Dialysis membranes (12 000–14 000 MWCO, 10 mm). Prepare nuclease- and protease-free membranes as described by Sambrook et al. (1989).

19. Agarose (ultrapure).

20. Electrophoresis sample buffer, 0.25% bromphenol blue, 0.25% xylene cyanol, and 15% Ficoll (Type 400) in distilled water. Store at room temperature.

31–1.6.3 Large-Scale Isolation and Purification of Bacterial DNA

The procedure for large-scale DNA isolation is adapted from Sadowsky et al. (1987). Grow 25-mL bacterial cultures to mid-exponential phase (1–2 d) in a growth medium that does not allow the production of appreciable amounts of polysaccharides (see chapter 1 by Wollum in this book). For many slow-growing gram-negative soil organisms, basal HM salts minimal medium (Cole & Elkan, 1973) supplemented with 1.0 g/L of yeast extract, 1.0 g/L of sodium gluconate and 1.0 g/L of arabinose suppresses polysaccharide formation. For fast-growing gram-negative organisms, TY (Beringer, 1974) or PPM media (Sadowsky & Bohlool, 1983) can be used. Centrifuge 5 to 7 mL of the bacterial culture at 10 000 × *g* for 10 min in a refrigerated centrifuge at 4 °C. Resuspend the bacterial pellet in

20 mL TEN buffer and centrifuge at 10 000 × g for 10 min at 4 °C. Resuspend the washed pellet in 8 mL of TEN buffer and add 1.0 mL of 5 mg/mL lysozyme solution. Incubate at 37 °C for 30 min. Add 1.0 mL of Pronase solution and incubate 30 min at 37 °C. Add 1.0 mL of sarkosyl solution (20% in distilled H_2O) and incubate until bacteria lyse and the solution clears (about 1 h). Add 11.0 g of CsCl powder and incubate at 37 °C for 15 min to warm tubes. Gently dissolve the CsCl in the DNA solution by inverting the capped tube several times. Add 0.6 mL of the ethidium bromide solution and mix by inversion. Add the entire DNA solution to one 13.5-mL ultracentrifuge tube using a funnel created from a 20-mL syringe equipped with an 18-gauge needle or a Pasteur pipette. Remove solution from the funnel portion of the tube with a paper towel and balance and seal tubes. Centrifuge the sealed tubes at 40 000 rpm for 40 h at 20 °C in a Beckman Type 50 Ti, Type 70.1 Ti, or Type 80 Ti fixed-angle rotor (or equivalent). If the Beckman vertical, VTi 65.1 or near-vertical, NVT ultracentrifuge rotors are used, centrifugation time can be reduced to 18 h at this speed. Visualize DNA bands in the gradient tubes with a long-wave (302, 310, or 365 nm) UV light source. Separately remove the top and bottom bands (the bottom band contains plasmid DNA and the top band linearized plasmid and chromosomal DNA) by puncturing the side of the tube with a 3-mL syringe equipped with an 18-gauge needle. Transfer the DNA solution to a 17 × 100 mm sterile, plastic, disposable tube. Add 5 mL of NaCl-saturated n-butanol solution and gently invert the tube to extract the ethidium bromide. Remove the pink upper phase and discard in an ethidium bromide waste container. Add fresh butanol solution to the tube and repeat the extraction five to six times until the lower phase is colorless (when tested against a white background). Dialyze the DNA solution overnight against 4 L of TE buffer. Transfer the dialyzed solution to a clean, sterile, plastic tube. Estimate the concentration of DNA within the tube by measuring the absorbance of a 1:20 dilution of the sample (50 µL added to 950 µL distilled water) at 260 nm. Use distilled water as the spectrophotometric blanking reference. The value directly read from the spectrophotometer is the nucleic acid concentration in mg/mL. Values should fall in the range of 0.03 to 0.2 A_{260} units. This is not the absolute DNA concentration, however, since the preparation is not free of RNA. Use immediately or freeze at −20 °C until needed. Solutions contaminated with ethidium bromide should be treated before disposal as described by Sambrook et al. (1989).

31–1.6.4 Small-Scale Isolation of Bacterial DNA ("minipreps")

The following procedure is adapted from Pitcher et al. (1989). The amount and age of bacterial cultures used in the minipreparation procedure influence how well the cells lyse. Consequently, optimal cell concentration should be empirically determined for the strains studied. A general procedure, which lyses cells of several different bacterial genera, is presented below. Grow 5-mL bacterial cultures, 1 to 2 d (to mid-exponential phase) in a medium that represses polysaccharide formation (see above). Add 0.65-mL aliquots of the culture to microcentrifuge tubes and centrifuge for 1 min at 12 000 × g. Resuspend cells in 100 μL of fresh lysozyme solution (50 mg/mL in TE buffer) and incubate at 37 °C for 30 min. It should be noted that *Staphylococcus* strains should be suspended in 100 μL of lysostaphin (50 mg/mL in 100 mM of Na-phosphate, pH 7.0) prior to incubation. Add 0.5 mL of guanidium thiocyanate solution and vortex the suspension briefly. Incubate at 37 °C for 5 to 10 min until lysis is complete (the solution clears). Cool the tubes on ice and add 0.25 mL of ice-cold 7.5 M of ammonium acetate solution. Mix by sharply inverting the tube several times and store on ice for 10 min. Add 0.5 mL of chloroform:isoamyl alcohol and briefly mix the phases by inverting the tube. Centrifuge at 12 000 × g for 10 min in the microcentrifuge. Transfer the upper aqueous phase to clean microcentrifuge tubes and add 0.54 volume cold 2-propanol. Gently invert the tubes several times and recover the precipitated DNA by centrifugation for about 30 s at 6500 × g. Wash the pelleted DNA with 70% ethanol and dry the pellets in vacuo. Dissolve the DNA pellets in 50 μL of sterile distilled water by incubating overnight at 4 °C. Use immediately or freeze at −20 °C until needed.

31–1.6.5 Restriction Endonuclease Digestion of DNA Samples

For routine DNA fingerprinting, 2.25 μg of genomic DNA is digested. This is enough DNA for about three wells. The amount of DNA solution digested varies according to the initial nucleic acid concentration. To determine the amount of DNA solution to use, divide the amount of DNA needed (2.25 μg) by the absorbance obtained at 260 nm (see 31–1.6.3 above). For sake of discussion, assume 20 μL of DNA solution was needed to obtain 2 μg of DNA. For the DNA minipreparations, however, start out by digesting 20 μL of DNA solution. The final digestion mixture volume should be three to four times the DNA solution volume (80 μL in both of our examples). Add the DNA solution to a clean sterile microcentrifuge tube and add 8 μL of 10× *Eco*R1 restriction enzyme buffer. Add 51 μL of sterile distilled water and 1 μL of restriction enzyme *Eco*R1 (1–10 units/μL). Mix by gently flicking the tube and centrifuge briefly (10 s) in a microcentrifuge. Incubate at 37 °C for 2 to 3 h in a water bath incubator. Stop the reaction by adding 8 μL of electrophoresis sample buffer and store at 4 °C until needed.

31–1.6.6 Agarose Gel Electrophoresis

The resulting DNA restriction fragments are separated into discrete bands by electrophoresis in 0.7% agarose gels. The following protocol is for one large (bed size of 20 × 25 cm) agarose gel. Add 2.1 g of agarose to 300 mL of 0.5X TBE electrophoresis buffer in a clean 1-L flask. Cover the flask with a plastic weigh boat and microwave (or heat on a heated stirplate with constant stirring) for 4 min or until the agarose is totally dissolved. Measure the volume of the solution in a clean graduated cylinder and, if necessary, bring back to 300 mL using distilled water. Cool the agarose solution in a 50 °C water bath for 15 min. Add 15 μL of the ethidium bromide solution (10 mg/mL), gently swirl to mix (avoid making bubbles) and pour the entire contents of the flask into a sealed (using Scotch tape) gel tray. Gloves should be worn at all times, since the agarose contains ethidium bromide. Immediately add an electrophoresis gel comb (20 teeth, 2.0 mm thick) to the tray and let the agarose solidify for at least 30 min at room temperature. Gently remove the gel comb by pulling the comb straight up and flood the surface of the gel with 0.5X TBE buffer. Add the gel tray to a gel box containing 0.5X TBE electrophoresis buffer. The gel box should contain sufficient buffer, such that the surface of the gel is submerged under about 4 to 5 mm of buffer. Load the wells of the gel with 29 μL of the DNA digestion-tracking dye mixture (see 31–1.6.5 above). End wells should be reserved for molecular weight standards. We routinely use 0.5 to 0.75 μg of *Eco*R1 and *Hin*dIII-digested lambda DNA in each of the end gel lanes. Electrophorese gels at 30 V, constant voltage, at room temperature overnight to obtain high resolution of restriction fragments. Smaller-sized, 11 × 14-cm, gels should be run overnight at 15 V (constant voltage). Stop the electrophoresis when the dye front has reached about 6 cm from the bottom of the gel. Restriction fragments are visualized on the 302 nm UV transilluminator. Photograph the gel using Polaroid Type 55 or Type 665 positive/negative film (try 30 s at f4.5). Be certain to use a fluorescent ruler, butted up against one side of the gel, with the 0-cm mark positioned equal with the well.

Similarities or differences in banding patterns can be seen directly from the gel or photographs. Both the presence and absence of restriction fragments on a particular portion of the gel should be considered when assessing relationships between DNAs. The similarities in banding patterns are routinely used to assign strains to DNA fingerprint groups. This is best done by "eye," since banding patterns are relatively complex. However, band presence and intensity data can be obtained using a gel densitometer and the appropriate computer software.

31–2 RFLP ANALYSES

31–2.1 Introduction

Restriction fragment length polymorphism analysis is an extremely powerful technique for strain identification, epidemiological studies, and the taxonomic analysis of prokaryotic and eukaryotic organisms (Bostein et al., 1980; Brown et al., 1989; Cook et al., 1989; Denny, 1988; Landry & Michelmore, 1987; Sadowsky et al., 1990; Scherer & Stevens, 1987; Wheatcroft & Watson, 1988). The technique uses cloned fragments of genomic (or plasmid) DNA as molecular markers to assess natural variation in an organism's DNA sequence. Since every organism has a specific number of restriction enzyme recognition sites at particular locations in the genome, digestion of an organism's DNA with a restriction enzyme results in the production of a series of DNA fragments unique to that organism. The fragments produced by DNA digestion are specific for each DNA/restriction enzyme combination and can be used as a tool to identify any given organism. Bacteria having their DNA rearranged by insertions, deletions, inversions, or base substitutions, will produce restriction fragments of different sizes. The fragment size differences are a direct result of changes in the number or location of restriction enzyme recognition sites within a given piece of DNA. These differences in fragment size are called restriction fragment length polymorphisms.

The RFLP technique differs from conventional DNA fingerprinting, in that the former technique uses cloned DNA (or RNA) as a hybridization probe to visualize the differences in fragment lengths produced by restriction enzyme digestion. While the relatively small bacterial genome can be visualized as distinct bands using the DNA fingerprinting technique, the complexity of banding patterns makes strain identification difficult, when comparing closely related organisms. The RFLP technique reduces the complexity of the banding pattern by imaging only a limited number of fragments. In addition, the RFLP technique is sensitive to small changes in an organism's genome and allows the detection of differences in genomic organization that would be indiscernible using DNA fingerprinting methodology.

The probes to use with the RFLP technique depend on the organism under study and whether the organism has any distinctive physiological properties. While chromosomally located DNA regions have generally been used for RFLP analysis, plasmid-borne sequences can be used as well. If the organisms under study lack any distinctive biological properties (such as nodulation, N_2 fixation, avirulence, or pectate lyase gene functions), randomly cloned genomic DNA can be used as a hybridization probe. However, since it is not possible to predict how much RFLP variation is present within a given group of organisms, several randomly selected gene probes should be tried. Also, to detect polymorphisms, it is frequently necessary to digest the DNA with several restriction enzymes. Strain, species, and guild-specific DNA probes can also be used to assess relationships among a particular group of organisms (Brown et al., 1989; Sadowsky et al., 1987; Sadowsky et al., 1990). Since the specificity of the hybridization reaction depends on both the probe and target sequences, care must be exercised in choosing both of these parameters. Lastly, while single copy target sequences are often detected using the RFLP technique, hybridizations to repetitive (reiterated) target DNA sequences are often useful for strain identification and characterization (Gabriel et al., 1989; Hahn & Hennecke, 1987; Rodriguez-Quinones et al., 1992).

31–2.2 Principles

The RFLP technique is a genotypic method that allows one to differentiate and identify both closely related or genetically distinct bacterial strains. The technique is useful for studies involving indigenous and exogenously added soil microorganisms, since many soil microbes are morphologically nondescript and taxonomically poorly characterized. The RFLP technique is based on the fact that every organism has a unique number and locations of restriction enzyme recognition sites. When the restriction fragments are separated by electrophoresis, transferred to membrane filters, and hybridized to specific gene probes, hybridizing fragments are visualized that are characteristic for a particular organism or strain. Similarities and differences in hybridization patterns indicate the degree of genetic relatedness among organisms. Organisms having indistinguishable hybridization patterns with several restriction enzymes are identical or closely related. Since hybridization patterns are changed by loss of indigenous plasmids and prophages or the loss or rearrangement of indigenous insertion-like sequences, the RFLP technique is the method of choice for examining the stability of an organism's genome.

31–2.3 Methods

31–2.3.1 Special Apparatus

1. Water bath shaker incubator or hybridization oven, capable of being set at 41, 46, and 65 °C is required for Southern hybridizations.
2. Refrigerated centrifuge (Sorval RC-5B or equivalent), rotor (SS-34 or equivalent), 15-mL Corex centrifuge tubes and adapters.
3. Microcentrifuge and tubes.
4. X-ray film and film holders.
5. Tuberculin syringes (1.0 mL).
6. Seal-A-Meal bags (20 × 30 cm) and sealer (Dazey Corp., Industrial Airport, Kansas or equivalent). Alternately, hybridization bags can be purchased from BRL/GIBCO, Gaithersburg, MD.
7. Plexiglass Beta Shield.
8. Nonsterile disposable gloves.
9. Filter paper blotting pads (Gibco-BRL cat. no. 1056cs) or paper towels.
10. Hybridization membranes, Nytran (S&S Inc.), nitrocellulose, or other nylon or supported media cut to 20 × 25 cm size.
11. Glass trays, Pyrex baking dishes.
12. Whatman 3 MM chromatography paper.
13. Vacuum oven, capable of being set at 80 °C. Note, a vacuum oven is not needed for nylon membranes, but must be used to prevent combustion of nitrocellulose membranes.
14. Table-top shaker.
15. Plexiglass plates 22 × 26 cm.
16. Hand-held radioactive survey meter with GM detector.

31–2.3.2 Reagents

1. 20 × SSC buffer (3 M of NaCl, 0.3 M of trisodium citrate, pH 7.0).
2. Gel depurination solution (0.25 M of HCl).
3. Gel denaturing solution (1.5 M of NaCl and 0.5 M of NaOH).
4. Gel neutralization solution (1.5 M of NaCl and 1.5 M Tris [hydroxymethyl]aminomethane, pH 7.5).
5. 100 × Denhardt's reagent [2% wt/vol polyvinylpyrrolidone (MW 40 000), 2% wt/vol Ficoll 400, and 2% wt/vol bovine serum albumin (fraction V)].
6. 50% Dextran sulfate (50% wt/vol dextran sulfate in distilled water).
7. Formamide (use Gibco-BRL 5515UA or equivalent) deionized with Biorad AG 501-X8 beads for 30 min at room temperature. Freeze at −20 °C until use.
8. Denatured, sonicated salmon sperm DNA (10 mg/mL in distilled water). Use Sigma D 9156 or equivalently prepared preparations. Boil 10 min and put on ice before use.
9. Sodium lauryl sulfate (SDS), purified grade.
10. Prehybridization solution (6X SSC, 10X Denhardt's solution, 1% SDS, and 100 µg/mL boiled salmon sperm DNA).
11. Hybridization solution (6X SSC, 1% SDS, 100 µg/mL boiled salmon sperm DNA, 50% deionized formamide, and 5% Dextran sulfate).
12. High-salt blot wash solution, 2X SSC, 0.1% SDS. Prepare 2 L.
13. Low-salt wash solution, 0.1X SSC, 0.1% SDS. Prepare 5 L.
14. Random primer, DNA labeling kit (use Amersham Corporation, Cat. no. RPN.1601Z or equivalent).
15. Hybridization Probe DNA (purified by using CsCl ultracentrifugation). Each labelling reaction requires 25 ng of linearized DNA.
16. Radioactive deoxynucleotide (use [α-^{32}P]dCTP at 3000 Ci/mmol and 10 mCi/mL).
17. Sephadex G-50 in STE buffer (10 mM Tris, pH 8.0, 1 mM disodium-ethylenediaminetetraacetic acid, and 100 mM NaCl).

31–2.3.3 Southern Transfer of DNA

Digest DNA samples with restriction enzymes and separate DNA fragments by electrophoresis as described above (sections 31–1.6.5 and 31–1.6.6). Independently digest DNA samples with three different restriction enzymes and run fragments out on three separate gels. In initial experiments, try using restriction enzymes *Eco*RI, *Hin*dIII, and *Bam*HI. The procedure that follows is a modification of that described by Southern (1975). After photographing gels (section 31–1.6.6), gently transfer gel to a Pyrex baking dish. Add 500 mL of depurination solution and gently shake for 5 min at room temperature. Discard solution and repeat washing in fresh depurination solution for 5 min. Gently rinse gel in distilled water and add 500 mL of denaturing solution. Gently shake gel for 15 min at room temperature. Discard solution and repeat washing in fresh denaturing solution for 15 min. Discard solution and add 750 mL of neutralization solution. Shake 30 min at room temperature. Soak a 52 × 22 cm sheet of Whatman 3 MM paper in 10X SSC and lay it sideways on the surface of a 26 × 22 cm plexiglass plate. Add 800 mL of 10X SSC to a Pyrex baking dish (23.5 × 32 cm) and lay the filter paper-covered plate, sideways, on the baking dish. The overhanging edges of the filter paper should be draped inside of the baking dish to wick-up the 10X SSC buffer. Gently transfer prepared gel to the soaked 3 MM filter paper. Soak a pre-cut sheet of Nytran filter membrane (20 × 25 cm) in 2X SSC. Cover the entire surface of the gel with the membrane. Express air bubbles trapped under the membrane using a gloved finger. Overlay the membrane with two, 20 × 25 cm, sheets of Whatman 3 MM filter paper and place a stack of blotting pads (or flattened paper towels) on the surface of the 3 MM filter paper. Cover the blotting pads with a 20 × 25 cm plexiglass plate and put a 500 g weight on the top of the stack. Allow the DNA to transfer to the membrane overnight. Carefully remove the blotting pads and 3 MM filter paper sheets. Mark the location of the gel lanes on the Nytran membrane using a no. 2 pencil. Peel the membrane off of the compressed gel and gently wash the filter in 2X SSC for 5 min at room temperature. Air dry the membrane at room temperature and bake at 80 °C in vacuo.

31–2.3.4 Preparation of Radioactive DNA Probes

Prepare radioactive DNA probes by the random primer labelling method. A convenient kit for this method is made by Amersham Corporation. Add 25 ng of linearized probe DNA (in 2.5 µL of dH$_2$O) with 10 µL of reaction buffer, 5 µL of primer solution, 5 µL of [^{32}P]dCTP, 24.5 µL of dH$_2$O and 2 µL enzyme solution (DNA polymerase Klenow fragment). Incubate at room temperature overnight and separate unincorporated nucleotides from the labelled DNA by passing the solution through a Sephadex G-50 column (made in a glass-wool-plugged 5-mL pipette fitted with a clamped hose on the end). Elute the column with STE buffer and collect 1-mL fractions in microcentrifuge tubes. The elution of the radioactive DNA can be followed using a hand-held radioactive survey meter. Alternatively, unincorporated nucleotides can be removed by using the spun-column procedure as described by Sambrook et al. (1989).

31–3.3.5 Hybridization of Southern Blot to Radioactive DNA Probes

Wash the baked membrane in 0.1X SSC and 0.5% SDS for 30 min at 65 °C in a plastic tray. Air dry the membrane. Place the washed filter membrane in a 20 × 30 cm seal-a-meal bag and add 25 mL of prehybridization solution. Remove all large air bubbles from the bag and seal using the bag sealer. Incubate at 41 °C overnight. Cut off one corner of the sealed bag and remove all of the prehybridization solution. Add 20 mL of hybridization solution containing all of the radioactive DNA probe to the bag. Reseal the bag and incubate overnight at 46 °C in the water bath-shaker. Decant the hybridization solution into a radioactive waste container and briefly wash the blot in 200 mL of high-salt blot washing solution. Decant the solution into a radioactive waste container. Wash the blot in 800 mL of high salt blot-washing solution for 15 min at room temperature with gentle shaking. Decant the solution and repeat the washing step. Wash the blot for 15 min in 1 L of low-salt blot washing solution at room temperature. Wash the blot for 30 min in low-salt blot washing solution at 65 °C. Repeat the above 65 °C washing a second time. Wash the blot for 5 min at room temperature in 0.1X SSC containing no SDS. Air dry the membrane and tape the blot to a small sheet of Whatman 3 MM paper. Expose the hybridized membrane to x-ray film (Kodak X-AR or equivalent) for 2 to 48 h, depending on the extent of probe hybridizing to target DNA.

31–2.3.6 Analysis of Results and Calculations

While the banding patterns obtained by hybridizing genomic DNA to specific gene probes is less complex than the patterns obtained by using DNA fingerprinting, hybridization done with several gene probes and enzymes can lead to the accumulation of a significant amount of data. It is recommended that DNAs be independently digested with several restriction enzymes (three), prior to hybridizing with DNA probes. In addition, more than one probe (located in a different portion of the genome) should be used to obtain a reliable estimate of the relatedness of the isolates.

The data collected from the large number of strain × enzyme × probe combinations need to be properly analyzed before one can obtain a clear understanding of the relatedness of organisms. Several methods are available to do this type of analysis. In the easiest cases, with one to three shared hybridizing fragments per lane, the organisms can be simply assigned to hybridization groups on the basis or absence of specific bands. Those organisms sharing more fragments at the same gel position are more related than those with less numbers of fragments in common. Differences in the number and position of hybridizing fragments (between gel lanes) indicates that the organism's genome has undergone changes (relative to some standard strain). These changes are due to substitutions of base pairs of insertions, deletions, or rearrangements of genomic (or plasmid) DNA, that such restriction enzyme recognition sites have been destroyed or created. The less the number of differences seen between the hybridization patterns of two strains, the less these strains have genetically diverged. Thus, the relatedness of bacterial strains can be obtained by comparing gel lanes for the position and fraction of conserved restriction fragments hybridizing to a specific gene probe. Upholt (1977) has developed a series of equations that can be used to estimate sequence divergence between defined DNA sequences of individuals or species. The method, however, can only be used as an estimate of sequence divergence, since the model assumes that restriction site divergence only occurs by base substitutions.

For relatively complex hybridization patterns obtained using several restriction enzyme and probe combinations, the degree of relatedness of strains may be obtained by cluster analysis. In this case, the hybridization data are recorded as the absence or presence of a particular band at a specific gel location. The data can then be analyzed using genetic distance equations (Nei, 1978), principle coordinate analysis, and cluster analysis (Sneath & Sokal, 1973) to determine the degree of relatedness between strains. A detailed discussion of the numerical taxonomic application of this statistical theory can be found in Sneath and Sokal (1973). Several PC-based computer programs can be used to perform all of the required statistical operations. These programs include: PHYLIP, a phylogenetic analysis computer program package (Dr. Joe Felsenstein, Univ. of Washington, Seattle); and NTSYS-PC (Rohlf, 1989), a commercially available, PC-based series of programs for numerical taxonomic and phylogenetic analysis (Exeter Software, Setauket, NY).

REFERENCES

Bautsch, W. 1988. Rapid physical mapping of the *Mycoplasma mobile* genome by two-dimensional field inversion gel electrophoresis techniques. Nucleic Acids Res. 16:11461–11467.

Beringer, J.E. 1974. R-factor transfer in *Rhizobium leguminosarum*. J. Gen. Microbiol. 84:188–198.

Bostein, D.R., M. Skolnick, and R.W. Davies. 1980. Construction of a genetic linkage map in man using restriction fragment length polymorphisms. Am. J. Hum. Genet. 32:314–331.

Brown, G., Z. Khan, and R. Lifshitz. 1989. Plant promoting rhizobacteria: strain identification by restriction fragment length polymorphisms. Can. J. Microbiol. 36:242–248.

Buchman, T.G., B. Roizman, G. Adams, and B.H. Stover. 1978. Restriction endonuclease fingerprinting of *Herpes Simplex* virus DNA: A novel epidemiological tool applied to a nosocomial outbreak. J. Infect. Dis. 138:488–498.

Canard, B., and S.T. Cole. 1989. Genome organization of the anaerobic pathogen *Clostridium perfringens*. Proc. Natl. Acad. Sci. (USA) 86:6676–6680.

Carle, G.F., M. Frank, and M.V. Olson. 1986. Electrophoretic separation of large DNA molecules by periodic inversion of the electric field. Science 232:65–68.

Carle, G.F., and M.V. Olson. 1985. An electrophoretic karyotype for yeast. Proc. Natl. Acad. Sci. (USA) 82:3756–3760.

Chandler, D.K.S., S. Razin, E.B. Stephens, R. Harasawa, and M.F. Barile. 1982. Genomic and phenotypic analysis of *Mycoplasma pneumoniae* strains. Infect. Immun. 38:604–609.

Christensen, L.S., K.J. Sorensen, and J.C. Lei. 1987. Restriction fragment pattern (RFP) analysis of genomes from Danish isolates of suid herpes virus 1 (Aujezsky's Disease Virus). Arch. Virol. 97:215–224.

Chu, G., D.V. Vollrath, and R.W. Davis. 1986. Separation of large DNA molecules by contour-clamped homogeneous electric fields. Science 234:1582–1585.

Cole, M.A., and G.H. Elkan. 1973. Transmissible resistance to penicillin G, neomycin, and chloramphenicol in *Rhizobium japonicum*. Antimicrob. Agents Chemother. 4:248–253.

Cook, D., E. Barlow, and L. Sequeira. 1989. Genetic diversity of *Pseudomonas solanacearum:* Detection of restriction fragment length polymorphisms with DNA probes that specify virulence and the hypersensitive response. Molec. Plant Microbe Interact. 2:113–121.

Coulson, A., and J. Sulston. 1988. Genome mapping by restriction fingerprinting. *In* K.E. Davies (ed.) Genome analysis: A practical approach. IRL Press, Oxford.

de Bruijn, F.J. 1992. Use of repetitive (repetitive extragenic palindromic and enterobacterial repetitive intergenic consensus) sequences and the polymerase chain reaction to fingerprint the genomes of *Rhizobium meliloti* isolates and other soil bacteria. Appl. Environ. Microbiol. 58:2180–2187.

Denny, T.P. 1988. Differentiation of *Pseudomonas syringae* pv. tomato from *P. syringae* with a DNA hybridization probe. Phytopathology 78:1186–1193.

Falk, E.S., B. Bjorvatn, D. Danielsson, B. Kristiansen, K. Melby, and B. Sorensen. 1984. Restriction endonuclease fingerprinting of chromosomal DNA of *Neisseria gonorrhoeae*. Acta Pathol. Microbiol. Immunol. Scand. Sect. B. 92:271–278.

Gabriel, D.W., M.T. Kingsley, J.E. Hunter, and T. Gottwald. 1989. Reinstatement of *Xanthomonas citri* (ex Haase) and *X. phaseoli* (ex Smith) to species and reclassification of all *X. campestris* pv. *citri* strains. Inst. J. Syst. Bacteriol. 39:14–22.

Gill, P., J.E. Lygo, S.J. Fowler, and D.J. Werrett. 1987. An evaluation of DNA fingerprinting for forensic purposes. Electrophoresis 8:38–44.

Glynn, P., P. Higgins, A. Squartini, and F. O'Gara. 1985. Strain identification in *Rhizobium trifolii* using DNA restriction analysis, plasmid DNA profiles, and intrinsic antibiotic resistances. FEMS Microbiol. Lett. 30:177–182.

Hahn, M., and H. Hennecke. 1987. Conservation of a symbiotic DNA region on soybean root nodule bacteria. Appl. Environ. Microbiol. 53:2253–2255.

Heath, L.S., G.L. Sloan, and H.E. Heath. 1986. A simple and generally applicable procedure for releasing DNA from bacterial cells. Appl. Environ. Microbiol. 51:1138–1140.

Hulton, C.S.J., C.F. Higgins, and P.M. Sharp. 1991. ERIC sequences: a novel family of repetitive elements in the genomes of *Escherichia coli, Salmonella typhimurium* and other enteric bacteria. Mol. Microbiol. 5:825–834.

Judd, A.K., M. Schneider, M.J. Sadowsky, and F.J. de Bruijn. 1993. Use of repetetive sequences and the polymerase chain react technique to classify genetically related *Bradyrhizobium japonicum* serocluster 123 strains. Appl. Environ. Microbiol. 59: 1702–1708.

Kaijalainen, S., and K. Lindstrom. 1989. Restriction fragment length polymorphism analysis of *Rhizobium galegae* strains. J. Bacteriol. 171:5561–5566.

Kaper, J.B., H.B. Bradford, N.C. Roberts, and S. Falkow. 1982. Molecular epidemiology of *Vibrio cholerae* in the U.S. Gulf coast. J. Clin. Microbiol. 16:129–134.

Kauc, L., M. Mitchell, and S.H. Goddgal. 1989. Size and physical map of the chromosome of *Haemophilus influenzae*. J. Bacteriol. 171:2472–2479.

Koch, E., K. Song, T.C. Osborn, and P.H. Williams. 1991. Relationship between pathogenicity and phylogeny based on restriction fragment length polymorphism in *Leptosphaeria maculans*. Mol. Plant Microbe Interact. 4:341–349.

Kristiansen, B.E., B. Sorensen, B. Bjorvatn, E.S. Falk, E. Fosse, K. Byrn, L.O. Froholm, P. Gaustad, and K. Bovre. 1986. An outbreak of Group B meningococcal disease: Tracing the causative strain of *Neisseria meningitidis* by DNA fingerprinting. J. Clin. Microbiol. 23:764–767.

Kuijper, E.J., J.H. Oudbier, W.N. Stuifbergen, A. Jansz, and H.C. Zanen. 1987. Application of whole-cell DNA restriction endonuclease profiles to the epidemiology of *Clostridium difficile*-induced diarrhea. J. Clin. Microbiol. 25:751–753.

Landry, B.S., and R.W. Michelmore. 1987. Methods and applications of restriction fragment length polymorphism analysis to plants. p. 25–44. In G. Bruening et al. (ed.) Proceedings of the Conference on Tailoring Genes for Crop Improvement. Univ. of California, Davis.

Langenberg, W., E.A.J. Rauws, A. Widjojokusumo, G.N. Tytgat, and H.C. Zanen. 1986. Identification of *Campylobacter pyloridis* isolates by restriction endonuclease DNA analysis. J. Clin. Microbiol. 24:414–417.

Lazo, G.R., R. Roffey, and D.W. Gabriel. 1987. Pathovars of *Xanthomonas campestris* are distinguishable by restriction fragment length polymorphisms. Int. J. Syst. Bacteriol. 37:214–221.

Lee, J.L., and H.O. Smith. 1988. Sizing of the *Haemophilus influenzae* Rd genome by pulsed-field agarose gel electrophoresis. J. Bacteriol. 170:4402–4405.

Levine, J.D., and C.L. Cech. 1989. Low frequency restriction enzyme in pulsed field electrophoresis. Biotechnology 7:1033–1036.

Logtenberg H., and E. Bakker. 1988. The DNA fingerprint. Endeavour New Ser. 12:28–33.

Marmur, J. 1961. A procedure for the isolation of deoxyribonucleic acid from microorganisms. J. Mol. Biol. 3:208–218.

Mielenz, J.R., L.E. Jackson, F. O'Gara, and K.T. Shanmugam. 1979. Fingerprinting bacterial chromosomal DNA with restriction endonuclease EcoR1: comparison of *Rhizobium* spp. and identification of mutants. Can. J. Microbiol. 25:803–807.

Nei, M. 1978. Estimation of average heterozygosity and genetic distance from a small number of individuals. Genetics 89:583–590.

Panchal, C.J., L. Bast, T. Dowhanick, and G.G. Stewart. 1987. A rapid, simple, and reliable method of differentiating brewing yeast strains based on DNA restriction patterns. J. Inst. Brew. 9:325–327.

Perbal, B. 1984. A practical guide to molecular cloning. John Wiley and Sons, New York.

Pitcher, D.G., N.A. Saunders, and R.J. Owen. 1989. Rapid extraction of bacterial genomic DNA with quanidium thiocyanate. Lett. Appl. Microbiol. 8:151–156.

Ramos, M.S., and S.K. Harlander. 1990. DNA fingerprinting of lactococci and streptococci used in dairy fermentations. Appl. Microbiol. Biotechnol. 34:368–374.

Roberts, R.J., and D. Macelis. 1991. Restriction enzymes and their isoschizomers. Nucleic Acids Res. 19(Suppl.):2077–2109.

Rodriguez-Quinones, F., A.K. Judd, M.J. Sadowsky, R. Liu, and P.B. Cregan. 1992. Hyperreiterated DNA regions are conserved among Bradyrhizobium japonicum serocluster 123 strains. Appl. Environ. Microbiol. 58:1878–1885.

Rohlf, F.J. 1989. NTYSYS-pc. Numerical taxonomy and multivariate analysis system. Exeter Publ. Setauket, NY.

Sadowsky, M.J., and B.B. Bohlool. 1983. Possible involvement of a megaplasmid in nodulation of soybeans by fast-growing rhizobia from China. Appl. Environ. Microbiol. 46:906–911.

Sadowsky, M.J., Cregan, P.B., and H.H. Keyser. 1990. DNA hybridization probe for use in determining restricted nodulation among Bradyrhizobium japonicum serocluster 123 field isolates. Appl. Environ. Microbiol. 56:1768–1774.

Sadowsky, M.J., R.E. Tully, P.B. Cregan, and H.H. Keyser. 1987. Genetic diversity in Bradyrhizobium japonicum serogroup 123 and its relation to genotype-specific nodulation of soybeans. Appl. Environ. Microbiol. 53:2624–2630.

Sambrook, J., E.F. Fritsch, and T. Maniatis. 1989. Molecular cloning: A laboratory manual. Cold Spring Harbor Lab. Press, Cold Spring Harbor, NY.

Scherer, S., and D.A. Stevens. 1987. Application of DNA typing methods to epidemiology and taxonomy of Candida species. J. Clin. Microbiol. 25:675–679.

Schmidt, E.L., M.J. Zidwick, and H.H. Abebe. 1986. Bradyrhizobium japonicum serocluster 123 and diversity among member isolates. Appl. Environ. Microbiol. 51:212–215.

Schwartz, D.C., and C.R. Cantor. 1984. Separation of chromosome-sized DNAs by pulsed-field gradient gel electrophoresis. Cell 37:67–75.

Skjoid, S., P.G. Quie, L.A. Fries, M. Barnham, and P.P. Cleary. 1987. DNA fingerprinting of Streptococcus zooepidemicus (Lancefield Group C) as an aid to epidemiological study. J. Infec. Dis. 155:1145–1150.

Smith, C.L., S.R. Klco, and C.R. Cantor. 1988. Pulsed-field gel electrophoresis and the technology of large DNA molecules. p. 41–72. In K.E. Davies (ed.) Genome analysis: A practical approach. IRL Press, Oxford.

Smith, C.L., P.E. Warburton, A. Gaal, and C.R. Cantor. 1986. Analysis of genome organization and rearrangement by pulsed field gradient electrophoresis. p. 45–70. In J. Setlow and A. Hollaender (ed.) Genetic engineering. Vol. 8. Academic Press, New York.

Sneath, P.H., and R.R. Sokal. 1973. Numerical taxonomy. Freeman Press, San Francisco.

Sobral, B.W., M.J. Sadowsky, and A.G. Atherly. 1990. Genome analysis of Bradyrhizobium japonicum serocluster 123 field isolates by using field inversion gel electrophoresis. Appl. Environ. Microbiol. 56:1949–1953.

Southern, E.M. 1975. Detection of specific sequences among DNA fragments separated by gel electrophoresis. J. Mol. Biol. 98:503–517.

Stern, M.J., G.F.-L. Ames, N.H. Smith, E.C. Robinson, and C.F. Higgins. 1984. Repetitive extragenic palindromic sequences: a major component of the bacterial genome. Cell 37:1015–1026.

Tompkins, L.S., N.J. Troup, T. Woods, W. Bibb, and R.M. McKinney. 1987. Molecular epidemiology of Legionella species by restriction endonuclease and alloenzyme analysis. J. Clin. Microbiol. 25:1875–1880.

Upholt, W.B. 1977. Estimation of DNA sequence divergence from comparison of restriction endonuclease digests. Nucleic Acids Res. 4:1257–1265.

Versalovic, J., T. Koeuth, and J.R. Lupski. 1991. Distribution of repetitive DNA sequences in eubacteria and application to fingerprinting of bacterial genomes. Nucleic Acids Res. 19:6823–6831.

Wheatcroft, R., and R.J. Watson. 1988. A positive strain identification method for Rhizobium meliloti. Appl. Environ. Microbiol. 54:574–576.

Young, N.D. 1990. Potential applications of map-based cloning to plant pathology. Physiol. Molec. Plant Pathol. 37:81–94.

Chapter 32

Nucleic Acid Probes

A. V. OGRAM AND **D. F. BEZDICEK,** *Washington State University, Pullman, Washington*

Much information regarding the state of a soil microbial community at a given time is contained within the nucleic acid fraction of the community. By analyzing the various components of the nucleic acid fraction, specific aspects of the structure and function of the microbial community may be determined. The technique most commonly used for this purpose is gene probe analysis, which measures the concentrations of specific genetic sequences in a nucleic acid sample. The measurement of specific gene concentrations allows inferences concerning the structure and function of microbial communities to be made that are based on the functions of the target genes.

In general, a nucleic acid probe is a single strand of either DNA or RNA that will hybridize, or bind, with its complementary sequences in the intended target nucleic acid. As might be expected, gene probes are usually the cloned versions of genes to be measured in nature. If, for example, one wishes to measure the concentration of the gene coding for the production of naphthalene dioxygenase in a soil sample, the probe of choice would be derived from a cloned naphthalene dioxygenase gene.

Gene probes may be divided into two general categories, phylogenetic and function probes. Phylogenetic probes are constructed from genetic sequences, usually fragments of genes coding for the production of ribosomal RNA (rRNA) and are highly specific to one taxonomic group. These probes may be used to measure the concentration of that phylogenetic group within the sample. Depending upon the nature of the probe, it may be specific to a phylogenetic group as broadly defined as a kingdom, or as narrowly defined as a specific strain within a particular species. Specific regions of ribosomal RNA molecules are highly conserved throughout phylogeny, and are, therefore, the best targets for phylogenetic probes. Targeting probes toward specific regions of rRNA molecules is advantageous not only because they are highly conserved, but also because a cell may contain thousands of ribosomes, thereby greatly increasing the sensitivity over targets that are present only as single copies in cells (Olsen et al., 1986). Function probes, such as those that measure genes for N_2 fixation or the degradation of a pollutant, are not specific to one taxonomic group, but rather measure the potential of the community as a whole to perform a given function. It should be noted that function probes provide measurements of the *potential* and not of the *activity* of the target gene if DNA is the target material (Ogram & Sayler, 1988). Only if mRNA is used as the target material can information regarding the actual activity of a specific function be obtained (Fleming et al., 1993). Gene probes have been used with increasing frequency in environmental and ecological research since 1980, and several reviews on this topic have recently been published (Hazen & Jimenez, 1988; Holben & Tiedje, 1988; Ogram & Sayler, 1988; Olson, 1991; Pickup, 1991; Sayler & Layton, 1990; Stotzky et al., 1990; Trevors & van Elsas, 1989).

The most commonly used gene probe techniques in environmental research are colony hybridization and the hybridization of direct DNA extracts. Colony hybridization involves dislodging the target organisms from soil and cultivating them on agar plates. Isolated colonies are lysed and transferred to a DNA hybridization membrane. Direct extraction of nucleic acids have been discussed in detail in chapter 35 by Holben in this book. Each of these techniques has its weaknesses and strengths that may be exploited depending upon the type of information desired. For most applications in community structure and function analysis, the total extraction and probing of community DNA may be preferable to colony hybridization. Colony hybridization requires that cells be cultured before analysis may begin, which may not be easily accomplished. Since many of the species in nature may not be readily culturable by standard laboratory media and growth conditions, most of the organisms in a typical analysis may be excluded from the analysis. Colony hybridization is preferred over direct DNA extractions, however, if bacterial strains possessing a particular genotype are to be isolated (Pettigrew & Sayler, 1986).

The use of nucleic acid probes in ecological and environmental analyses has several advantages over the more traditional biochemically based analyses. The primary advantage is that cultivation or incubation of samples in the laboratory may not be necessary, thereby eliminating the primary source of bias experienced in traditional analyses. While laboratory cultivation is required for colony hybridization, it is not necessary when the target nucleic acid is obtained by direct extraction. When laboratory incubations are required as part of an analysis, the longer the incubation, the greater the potential deviation from nature due to constraints placed on the sample by the artificial nature of the laboratory. This is a particularly important consideration when the concentrations of specific phylogenetic groups must be known. Current estimates place the percentage of readily culturable soil microorganisms between 0.05 to 0.1% of the total present, indicating that traditional techniques for the isolation and identification of soil microbes are highly inefficient. With gene probe technology coupled with DNA extraction from soil, the sample's nucleic acids may be extracted immediately upon arrival in the laboratory or the sample may be frozen indefinitely for processing at a later date, thereby providing an accurate picture of the sample at the time of sampling.

The greatest limitation of the application of gene probe analysis to soil microbiology is the current lack of knowledge concerning the molecular biology of soil microorganisms. The majority of work that has been done in this area has been almost exclusively limited to bacteria, with essentially no work being reported for eukaryotes. In order to apply gene probe analysis to soil microorganisms, the molecular genetics of both the gene probe and the target sequences must be known. The gene of interest must have been cloned for use as the gene probe, and the target gene must be similar enough to the gene probe for hybridization to occur. If there is more than one type of gene responsible for a given function, gene probe analysis may underestimate the total number of genes specific to the function of interest, thereby underestimating the potential for the function to occur. Conversely, if nontarget genes share enough sequence homology with the gene probe to hybridize with the probe, overestimation of the sequence will occur. As the molecular genetics of soil microorganisms becomes better known, these ambiguities will undoubtedly decrease and the number of applications of gene probe technology will increase. These considerations are primarily directed toward the analysis of indigenous microorganisms and not toward those that have been added to the soil. In general, the genetic markers used for added strains are well known, and very specific probes can be designed for their analysis.

32–1 PROBE SELECTION

Selection of the specific segment of nucleic acid to be used as a gene probe is the most crucial part of the nucleic acid hybridization and depends upon the particular application. Phylogenetic probes that are directed toward rRNA are typically synthetic oligonucleotides and designed to hybridize with specific parts of the rRNA molecule that have been identified as being conserved for the particular phylogenetic group of interest. For this reason, they are precise and little ambiguity enters into data interpretation. There is also an absolute requirement that the rRNA molecule of the target organism be sequenced, and that the sequences be made available to the potential user. Since rRNAs from relatively few organisms have been sequenced, and the design of specific oligonucleotide probes from rRNA sequences is complicated, strategies for their design are beyond the scope of this chapter (Olsen et al., 1986).

There are several considerations to be taken into account when choosing a probe that is not a ribosomally directed synthetic oligonucleotide. The primary requirement of a gene probe is that its degree of specificity for the target sequences be known. The most useful probes are usually those that are prepared from specific genes that are well characterized, so that when homology is detected between the probe and DNA isolated from a soil sample, the probability that one has detected a copy of the target sequence is quite high. Ambiguity of data interpretation often occurs when the exact nature of the probe is unknown. An example of this might be when a large, uncharacterized plasmid encoding a particular function is used as a probe. The genes of interest may comprise only a few kilobases, leaving the rest of the plasmid uncharacterized. When using such a probe, one cannot be sure whether one has detected homology with the desired genes or some other unknown genes that may be located on a completely different plasmid or the chromosome. Similar problems may arise when specific genes are cloned into a cloning vector and the entire plasmid is used as a probe. In this case, it is possible that the origin of replication or antibiotic resistance markers located on the vector may hybridize with similar genes in the target material. This problem may be overcome by first hybridizing the sample with the vector alone, thereby blocking sites homologous to the vector and not with the cloned sequences.

The safest approach is to use a small, relatively well-defined segment that has been purified of the vector and any other extraneous pieces of DNA. The user should also be aware of any homology that the probe may share with nontarget genes. An example of this might be that a probe derived from a dioxygenase gene that had been cloned from a pathway involved in the degradation of a particular pollutant may also share homology with other dioxygenase genes that are unrelated to the catabolic pathway of interest. One way to check this possibility is that if the gene of interest has been sequenced, homology with other sequenced genes may be checked through GenBank[1] (a sequence data base; Bilofsky et al., 1986) and related computer analysis packages. The use of GenBank may aid in the choice of segments of genes to be used as probes that are specific to one particular gene and share no known homology with other related known genes. Computer software packages designed for comparing different sequences from GenBank can be valuable tools for assisting in the design of appropriate probes, but it is beyond the scope of this chapter to provide a detailed review of available software. Most universities have such programs readily available, and the interested reader is encouraged to contact a computer sciences department at a convenient university regarding these programs.

There may also be cases where a probe is too specific for a given application. In the case of function probes, gene fragments used as probes may not be highly conserved across genera, and the probe may fail to detect the function in species other than the one from which it was origi-

---

[1]GenBank is a registered trademark of the U.S. Department of Health and Human Services.

nally cloned. In such cases, it may be necessary to use more than one gene coding for a specific function that have been cloned from a range of organisms possessing the function. Here again, comparisons of sequences contained in GenBank and related computer packages may aid in the selection of a gene fragment or combination of fragments to be used as probes for genes that are not highly conserved across species or genera.

The size of the probe may also be a consideration. Up to a point, the smaller the probe, the greater the accuracy with which it can be targeted toward a specific part of the target gene. Very small (10–20 bp) probes may be subject to non-specific hybridization. This reduces the level of ambiguity and the probability of hybridization with nontarget sequences. Probes that are directed toward ribosomal RNAs are typically 20 to 30 bp in length, and most function-specific probes are between 500 and 1500 bp. It should be noted that there is a certain trade-off between selectivity and sensitivity. The shorter the probe, the less sensitive is the probe simply because there are fewer sites available on a short piece of nucleic acid for a label, such as ^{32}P, than there is on a longer piece. This, however, is usually not a great problem and will be discussed further in the section on labeling.

32–2 ISOLATION AND PURIFICATION OF FRAGMENTS TO BE USED AS PROBES

32–2.1 Principles

Many DNA fragments to be used as probes have been cloned into plasmid vectors, and the researcher should isolate the probe DNA from the vector prior to labeling. This is usually accomplished by digesting the plasmid (already purified by CsCl gradient ultracentrifugation) with the appropriate restriction enzymes, separating the vector from the probe DNA (or insert) by agarose gel electrophoresis, and then isolating the fragment from the agarose. For details on the use of restriction enzymes and agarose gel electrophoresis, see chapter 31 by Sadowsky in this book.

Several methods may be used to purify DNA fragments isolated from agarose gels, with the most commonly used being extraction with phenol and chloroform. For this method, agarose having a low melting temperature ("low melt agarose") is used at a concentration of around 1%. The DNA is electrophoresed in the gel and stained with ethidium bromide, as with regular agarose gel electrophoresis. The desired band is cut out of the gel with a scalpel or razor blade, melted in a 65 °C water bath, and then extracted with phenol and chloroform. The final step is precipitation of the DNA with ethanol and resuspension in an appropriate buffer.

Another common method of isolating fragments from regular agarose is by electroelution of the DNA from the agarose and into a buffer. After electrophoresis, the desired DNA band is cut out of the gel, and placed in an electroelution device (marketed commercially by Hoeffer, San Francisco; Bio-Rad, Richmond, CA; and IBI, New Haven, CT) in which an electric current is established. The DNA will move out of the gel and into the buffer, toward the positive pole of the field. The DNA may then be recovered from the electroelution buffer by precipitation with ethanol. Electroelution is simple, but the apparatus required is expensive. A similar principle is employed using DNA-binding papers to trap the fragment during electrophoresis. With this method, the DNA is electrophoresed to the point that the insert is separated from the vector, the power turned off, and, while it is still in the electrophoresis apparatus, the gel is sliced in front of the desired band. DNA-binding paper is placed in the slice, and the electrophoresis is continued. The DNA will move out of the gel and onto the binding paper. The DNA may then be removed from the paper by elution with salt, as is specified by the vendor (an example of this is marketed by Schleicher & Schuell, Keene, NH).

A technique that is rapidly becoming popular is the use of suspensions of finely ground glass to adsorb the DNA following chemical digestion of the agarose. Examples of this system are Qiaex (Qiagen, Chatsworth, CA) and GeneClean (Bio 101, Vista, CA). These methods may be more rapid and yield higher recoveries than the other techniques mentioned. The procedures used for these techniques depend upon the specific system, and the protocols supplied by the vendor should be followed. It should be noted that recoveries of larger fragments (>20 kb) are lower than for smaller fragments with all of the above techniques (Sambrook et al., 1989).

Regardless of the method of purification from agarose, the fragments may still be contaminated with significant amounts of degradation products from the vector. This is particularly true if the desired fragment is smaller than the vector. In this case, it may be necessary to reelectrophorese the fragment in agarose and reisolate it, although even this may not result in a completely pure fragment.

32–2.2 Materials

1. Low melting temperature agarose.
2. Agarose gel electrophoresis apparatus, buffers (see chapter 31 in this book).
3. Molecular biology grade phenol equilibrated to pH 8 with 0.1 M Tris·Cl (Sambrook et al., 1989).
4. Chloroform.
5. 20 mM Tris·Cl (pH 8.0), 1 mM EDTA (pH 8.0).
6. 10 mM ammonium acetate.
7. 95% Ethanol.
8. 0.5 µg/mL ethidium bromide.
9. Water bath set at 65 °C.
10. Scalpel or razor blade.

11. Ultraviolet lamp for viewing gels.
12. Sterile 1.5 mL microfuge tubes.

32–2.3 Procedure

Note: The principles and specific procedures for the use of restriction enzymes and agarose gel electrophoresis are covered in chapter 31. This section is adapted largely from Sambrook et al. (1989).

1. Digest plasmid DNA with appropriate enzymes to separate vector from the desired insert.
2. Pour gel with 1% low melt agarose and allow to solidify at 4 °C.
3. Electrophorese DNA with appropriate buffer.
4. Stain gel with 0.5 μg/mL ethidium bromide.
5. Cut out the desired band with a scalpel or razor blade and place in sterile 1.5-mL microfuge tube.
6. Add 5 volumes of 20 mM Tris·HCl, 1 mM EDTA (pH 8.0) and heat for 5 min at 65 °C. Split into two or three tubes if necessary.
7. Allow the mixture to cool and add one volume of an equal mixture of phenol and chloroform. Mix well and centrifuge in microfuge for 5 min.
8. Remove the top (aqueous) layer, being careful not to disturb the white interface, and place in a clean tube.
9. Repeat steps 7 and 8, followed by extraction of the aqueous layer with chloroform alone.
10. Precipitate the DNA by adding 0.2 volumes of 10 M ammonium acetate and 2 volumes of ethanol. Precipitate for at least 10 min, and recover the DNA by centrifugation for 15 min. Discard the supernatant.
11. Add 500 μL 70% ethanol to the pellet, mix gently, and recentrifuge as in step 10. This ethanol wash serves to remove precipitated salts.
12. Dry the DNA pellet and resuspend in an appropriate buffer, such as TE (10 mM Tris·Cl, 1 mM EDTA, pH 8.0).

32–3 LABELING OF PROBES

32–3.1 Introduction

Once a suitable probe has been chosen for a given application, a decision must be made as to how the probe should be labeled. Several different strategies are available for labeling, and in many cases the method of choice depends upon the preference of the investigator and not upon some clear choice for a given application. The first decision that can be made is whether to use a radioactive or a nonradioactive labeling system.

Either system has advantages and disadvantages that the individual researcher should consider with regard to his/her laboratory and application. Radioactive probes are most commonly used because of their greater sensitivity, although nonradioactive labeling systems are becoming more sensitive and are increasing in popularity.

The most commonly used radioisotope in gene probe studies is ^{32}P in the form of ^{32}P-dCTP, although ^{35}S is also occasionally used. The main advantages of using a radioisotope as a label are that: (i) sensitivity is usually maximized; (ii) times required for detection are decreased; and (iii) a variety of probe generation strategies are available. The main disadvantages with ^{32}P are: (i) it is a strong beta particle emitter and is therefore a health hazard, and so appropriate precautions regarding protection and disposal are necessary; (ii) it has a half life of 14 d, which means that any probe labeled with ^{32}P has a rather short shelf life; and (iii) the high-energy beta particle emitted may fragment the probe, thereby decreasing its effective length (this is usually not a problem). Nonradioactive labels have the advantages of presenting relatively low health hazards and having indefinite shelf lives. The sensitivity of some recently developed nonradioactive labels are approaching that of radioactive labels and may therefore be preferable to using ^{32}P. Nonradioactive probes will be discussed in detail below.

A variety of labeling systems are available, and the method of choice again largely depends upon the individual researcher. Fortunately, most of these systems are commercially available as kits, which include everything the user will require except for the labeled nucleotide. For the most part, they are all easy to use and almost fool-proof. The most commonly used methods for labeling fragments larger than 500 bp (nick translation and random primer fill-in) rely on commercially available kits and are based on the incorporation of a labeled nucleotide, usually dCTP, into double-stranded DNA. These methods are relatively quick and simple, and are quite satisfactory for most applications. One disadvantage that they share is that a double-stranded DNA probe must first be denatured (the two strands separated) and added to the hybridization reaction. When this is done, there is the possibility that the two strands of the probe will hybridize to each other and not to the target molecules. To avoid this, the concentration of the probe must be kept low in the hybridization reaction, which limits the sensitivity of the system. Alternative labeling strategies have been developed that produce single stranded probes, thereby increasing sensitivity by allowing greater concentrations of the probe to be used in the hybridization reaction. These methods are also available in commercial kits (Promega, Madison, WI; Strategene, La Jolla, CA), but are somewhat more difficult to use than either nick translation or random primer fill-in because subcloning is required. These single-stranded probes also have the advantage over double-stranded probes in that they are usually made of RNA rather than DNA, and RNA/DNA hybrids are more stable than are DNA/DNA hybrids and therefore are more sensitive than DNA probes.

When high levels of sensitivity [low limits of detection] are required in the analysis, we recommend the use of single-stranded RNA probes.

Small fragments (i.e., < 50 bp) are not efficiently labeled by either nick translation or random priming, and are usually labeled by the enzymatic attachment of radioactive nucleotides to one end of the fragment. This can either be accomplished through the action of polynucleotide kinase, which adds one ^{32}P-ATP molecule to the 5' end of the target molecule, or by terminal deoxynucleotidyl transferase, which adds a "tail" of up to 20 radiolabeled nucleotides to the end of the DNA molecule. Synthetic oligonucleotides are typically end labeled by T4 polynucleotide kinase.

32–3.2 Nonradioactive Probes

Nonradioactive probes are alternatives to radiolabeling for detection of nucleic acid sequences. They usually are stable for long periods of storage and are relatively safe to handle. In practice, the nonradioactive probes are used much the same as radioactive probes. In DNA probe labeling using the biotinylated system, the ^{32}P-dCTP typically used in nick translation is replaced by biotin-dATP or biotin-dUTP (Langer et al., 1981). Other methods of DNA labeling can be used such as random priming that are outlined in protocols supplied by various manufacturers.

Biotin-labeled DNA is visualized by the application of a conjugate of streptavidin-alkaline phosphatase (Leary et al., 1983) or streptavidin-peroxidase (Syvanan et al., 1986). Color reaction is achieved by exposure to the dyes BCIP (5-bromo-4-chloro-3-indolyl phosphate) and nitroblue tetrazolium chloride (NBT), which are oxidized and reduced respectively in dephosphorylation reactions to precipitates which turn blue on the filter. The principles of hybridization stringency, or the factors controlling the degree of fidelity of hybridization between the probe and target, are essentially the same as for radioactive probes. Although modifications of these methods have been reported, especially in the choice of the tagged deoxy nucleotide, many protocols have used the alkaline phosphatase system in conjugation BCIP-NBT for color reaction.

Although the sensitivity of nonradioactive probes are relatively high for purified DNA on Southern blots, they often result in a high degree of nonspecific hybridization, known as background signal, especially in colony hybridization, dot/slot blots, and other applications for nucleic acid detection in a complex cellular matrix. Nonspecific hybridization using biotinylated DNA probes has been reduced by the use of lysozyme with or without proteinase K (Zeph & Stotzky, 1989; Zeph et al., 1991; in kit form, Bethesda Research Lab., Gaithersburg, MD). Phenol-chloroform or chloroform extraction of bacterial colonies has further improved background problems on filters containing bacterial cells (Hass & Fleming, 1988; Zeph & Stotzky, 1989).

A nonradioactive kit system other than the biotin system is Genius (trademark of Boehringer Mannheim Corp., Indianapolis, IN) which reportedly eliminates the need for lysozyme and proteinase K for removing background interference (Martin et al., 1990). This system uses a digoxigenin-labeled nucleic acid probe for hybridization to target nucleic acid. Homologous nucleic acid sequences are detected using an anti-digoxigenin antibody conjugate (Holtke et al., 1988) which permits the detection of < 0.1 pg of homologous DNA. Labeling can be conducted using a variety of methods. Detection is achieved using anti-digoxigenin antibodies conjugated to alkaline phosphatase, peroxidase, fluorescein, or rhodamine (Martin et al., 1990). Color reaction on filters can be achieved using conventional NBT and BCIP reagents. In the Genius system, an alternative chemiluminescent detection system uses Lumi-Phos 530 containing Lumigen, a 1,2-dioxetane alkaline phosphatase substrate that emits light at sites of nucleic acid hybridization. The signal can be recorded on x-ray film by following the usual protocol for radiolabeled probes. Both Lumi-Phos 530 and Lumigen are trademarks of Lumigen, Detroit, MI.

The versatility and options for nucleic acid labeling, hybridization, and detection is depicted below for the Genius System (Martin et al., 1990), but is applicable to most other protocols for nonradioactive probes as well:

Label Probe
   Random priming
   Nick translation
   Oligonucleotide tailing
   RNA probes
   Photodigoxigenin
   cDNA analysis

Hybridization
   Southern blots
   Colony hybridization
   Dot blots

Detection
   Alkaline phosphatase
   Peroxidase
   Fluorescein
   Rhodamine
   Unconjugated
   Chemiluminescence

32–3.3 Nick Translation

32–3.3.1 Principles

The most commonly used technique for labeling DNA is by nick translation. The DNA to be labeled is mixed with DNase I, DNA polymerase I, ^{32}P-dCTP, nonradioactive dGTP, dATP, dTTP, and a buffer containing Mg^{2+} (required by DNase I). The phosphate backbone of the double-stranded DNA is nicked in several places by the DNase I, and the DNA polymerase I (pol I) repairs the nick by attaching to the DNA at the site of the nick and moving in the 5' to 3' direction. As pol I moves along the DNA template, its $5' \rightarrow 3'$ exonuclease activity removes nucleotides from the nicked strand and the enzyme replaces them with those provided in the reaction mixture. This incorporates ^{32}P-dCTP into the strand, resulting in a specific activity of more than 10^8 cpm/µg. As the strand is repaired, the nick moves, or is translated, down the DNA template with the enzyme. Probes generated by nick translation usually have maximum lengths of around 500 bp due to fragmentation by DNase I, regardless of the size of the original DNA template.

After completion of the nick translation reaction, the labeled material should be separated from the unincorporated ^{32}P-dCTP by either gel exclusion chromatography (using Sephadex G-50), as described here, or by the spun column method (Sambrook et al., 1989).

Most laboratories use commercially available kits for labeling DNA by nick translation. These kits are relatively inexpensive, very convenient, and almost foolproof. Each of these kits contains all of the necessary components, including enzymes, nonlabeled nucleotides, buffer, and control DNA. The only component usually lacking from these kits is the radiolabeled nucleotide (typically ^{32}P-dCTP), which must be supplied by the user. These kits are supplied with protocols that have been optimized by the vendor for the specific system they market, and the user should follow it closely. A generalized procedure for nick translation reactions is presented so that the reader may know what is involved (Sambrook et al., 1989), however, it is recommended that researchers use one of the many commercially available kits and follow the protocol supplied by the vendor.

32–3.3.2 Materials

1. $10\times$ nick translation buffer:
   0.5 M Tris.Cl (pH 7.5)
   0.1 M MgSO$_4$
   1 mM dithiothreitol
   500 µg/mL bovine serum albumin
2. Pancreatic DNAase I (10 ng/mL in 0.15 M NaCl and 50% glycerol).
3. E. coli DNA polymerase I (5 units/µL in 50% glycerol)
4. dNTP mixture containing 20 mM dATP, 20 mM dGTP, and 20 mM dTTP.
5. Alpha labeled ^{32}P-dCTP with a specific activity of at least 3000 Ci/mmol and a concentration of 10 µCi/µL.
6. Water bath set at 16 °C.
7. 10-100 µL adjustable pipets and autoclaved pipet tips.
8. Sterile double distilled water.
9. Sterile 0.5 M EDTA (pH 8.0).
10. Autoclaved 0.5-mL microfuge tubes.

11. Microfuge.
12. Sephadex G-50 saturated and autoclaved with TEN buffer.
13. TEN buffer:
    10 mM Tris·Cl (pH 8.0)
    1 mM EDTA (pH 8.0)
    100 mM NaCl
14. Disposable 3-mL syringe.
15. Sterile glass wool.
16. 18-gauge hypodermic needle.
17. Geiger-Mueller counter.
18. Shielding appropriate for use with ^{32}P.

32–3.3.3 Procedure

1. Mix together:
    2.5 μL 10× nick translation buffer
    0.5 μg probe DNA
    1 μL dNTP mixture
    5 μL alpha labeled ^{32}P-dCTP
    2.5 μL 10 ng/mL DNAase I
    2.5 units *E. coli* DNA polymerase I
    Sterile double distilled water to a total volume of 25.5 μL
2. Mix gently by vortexing followed by a 15 to 30 s centrifugation in a microfuge.
3. Incubate for 1 h at 16 °C, after which time the reaction is stopped with the addition of 1 μL 0.5 M EDTA (pH 8).
4. Separate labeled DNA from the unincorporated nucleotides by passage through Sephadex G-50 columns. These columns are prepared as follows:
    a. Place a small amount of glass wool in the top fitting of an 18-gauge needle, and attach the needle to a 3-mL disposable syringe.
    b. Pour in enough of autoclaved Sephadex G-50 slurry (in TEN buffer) to fill the syringe to approximately the 3-mL mark.
    c. Wash the column with several milliliters of sterile TEN buffer by allowing the buffer to pass through the column.
    d. Apply the nick translation mixture to the top of the column and allow the mixture to enter the column.
    e. Begin collecting fractions of approximately 10 drops from the bottom of the column in microfuge tubes, adding more TEN buffer to the top of the column as necessary.
    f. Monitor the radioactivity in the fractions with a hand-held Geiger-Mueller counter. Labeled DNA should be eluted from the column in the void volume and the unincorporated nucleotides will follow in a second peak.
    g. Pool the fractions from the first peak (containing the labeled DNA), and safely dispose of the unincorporated nucleotides.

32–3.4 Random Primer Fill-in

Random primer fill-in is usually employed to label DNA of < 1 kb in length. This technique has several advantages over nick translation, among them being: probes are generated to a higher specific activity ($> 10^9$ cpm/μg vs. 10^8 cpm/μg); and less DNA is required (25 ng as opposed to 0.5 μg). Like nick translation, random primer labeling systems are available in convenient commercial kits, and we recommend that the researcher follow the protocol supplied by the vendor.

Labeling by random primer fill-in (or primer extension) is based on the use of several oligonucleotides (usually hexamers) of different base compositions that hybridize randomly to denatured (single stranded) DNA. The hybridized oligonucleotide is then used as a primer by the Klenow fragment of DNA polymerase I to synthesize DNA from the single stranded template. The Klenow fragment does not have the $5' \rightarrow 3'$ exonuclease activity possessed by pol I. The Klenow fragment incorporates nucleotides (including ^{32}P-dCTP) supplied in the reaction mixture as it moves down the template, thereby synthesizing a labeled strand of DNA. Once the labeling is completed, the labeled DNA may be separated from the unincorporated nucleotides by passage through Sephadex G-50, as described above in section 32–3.3.3.

32–3.5 Single Stranded Probes

For applications where sensitive probes are required, such as when low concentrations of a particular gene must be measured, single-stranded probes have some important advantages over other labeling strategies. The primary advantage is that higher concentrations of probe may be used without the risk of having the probe reanneal with itself rather than the target DNA. As with most labeling procedures, several convenient kits are available from commercial vendors and we recommend that the researcher purchase a kit and follow the supplied instructions.

Single-stranded probes are generally based on the subcloning of the fragment of interest (usually purified from an agarose gel; see above) into the polycloning region of a high copy number plasmid. The polycloning region of the plasmid is flanked by either one or two different strong promoters (such as the SP6 or the T7 bacteriophage promoters). The plasmid is then mixed in vitro with the appropriate RNA polymerase (either SP6 or T7 RNA polymerase), buffer, RNase inhibitor, nonlabeled ribonucleotides ATP, GTP, and UTP, and labeled CTP. The RNA that will be produced from this reaction will be produced from the inserted fragment and will be labeled to a high specific activity due to the incorporation of the labeled CTP. Due to the ubiquity of RNases, extreme care must be taken to either remove or inactivate all RNases in the reagents and glassware used in these procedures (Sambrook et al., 1989).

32–3.6 End Labeling of Synthetic Oligonucleotides

32–3.6.1 Principles

Synthetic oligonucleotides are generally too small to be labeled by any of the procedures listed above, and are usually labeled by attaching a radiolabeled phosphate molecule to one end of the oligomer. This process, known as end labeling, is accomplished by transferring the radiolabeled gamma phosphate group of ATP to the unphosphorylated 5′ end of the oligonucleotide through the action of the enzyme T4 polynucleotide kinase. Since only one labeled group is transferred to each oligonucleotide, the specific activity of the labeled product may be lower than that produced by other labeling schemes. This procedure has the requirement that the 5′ end of the DNA to be labeled must be unphosphorylated, which is not a problem with synthetic oligonucleotides since they are produced with unphosphorylated 5′ ends. If, however, nonsynthetic oligonucleotides are to be end-labeled, they must be dephosphorylated by calf intestinal alkaline phosphatase (CIAP). It should be noted that, unless the researcher specifies that the synthetic oligonucleotide be purified, it is frequently delivered in a solution of ammonium hydroxide and shorter oligonucleotides (so-called failure sequences) that must be removed prior to labeling. Commercial kits that contain enzymes and buffers for end labeling are available, and a general procedure taken from Sambrook et al. (1989) is presented below. Due to the small size of the labeled material, it is separated from nonincorporated oligonucleotides by precipitation with ethanol rather than by the size exclusion chromatography procedure described above in section 32–3.3.3.

32–3.6.2 Materials

1. Ethanol
2. 0.5-mL microfuge tubes.
3. Water baths set at 37° and 68 °C.
4. TE buffer (see chapter 34).
5. 10× kinase buffer:
    0.5 M Tris·Cl (pH 7.6)
    0.1 M $MgCl_2$
    50 mM dithiothreitol
    1 mM spermidine HCl
    1 mM EDTA (pH 8.0)
6. Gamma labeled ^{32}P-dATP (5000 Ci/mmol, 10 mCi/mL).
7. T4 polynucleotide kinase.
8. Sterile distilled H_2O.

32–3.6.3 Procedure

1. Mix together the following in a 0.5-mL microfuge tube:
    Synthetic oligonucleotide   1.0 μL
    10× Kinase buffer   2.0 μL

[Gamma-^{32}P]ATP	5.0 μL
Sterile distilled H$_2$O	11.4 μL
T4 polynucleotide kinase	1.0 μL (8 units)

2. Incubate at 37 °C in a water bath for 45 min.
3. Heat to 68 °C for 10 min to inactivate the polynucleotide kinase.
4. Separate the labeled oligonucleotide from nonincorporated nucleotides by precipitation with ethanol as follows:
   1. Add 40 μL of sterile distilled H$_2$O to the reaction mixture.
   2. Add 240 μL of 5 M ammonium acetate, mix, and then add 750 μL ice cold ethanol.
   3. Centrifuge at high speed in a microfuge, at 0 °C if possible. The labeled oligonucleotide should pellet out, leaving unincorporated nucleotides in the supernatant.
   4. Carefully remove the radioactive supernatant and discard appropriately.
   5. In order to remove precipitated salts, add 500 μL of 80% ethanol to the pellet, mix gently, and recentrifuge.
   6. Remove and discard the supernatant, and allow the precipitated oligonucleotide to dry by evaporation.
   7. Resuspend the dried pellet in 100 μL TE (pH 7.6).

32–4 HYBRIDIZATION

32–4.1 Introduction

Once target bacteria or DNA have been isolated from environmental samples, there are several ways in which nucleic acid probes can be used to identify specific sequences through DNA or RNA hybridization. DNA hybridization is a method of determining whether an unknown DNA sequence has genetic homology with a labeled DNA probe of known structure or function. As a general procedure for hybridization of DNA from bacterial colonies (colony hybridization), target cells containing DNA are applied to solid filters, lysed to release DNA, hybridized or annealed with a labeled DNA probe of interest, and washed to control the level of binding between the DNA probe and target DNA. The intensity of labeled probe DNA through radioactive or nonradioactive methods is assessed. The specificity of DNA probes is based upon both the specificity of probe DNA to target DNA sequences and the conditions set during hybridization and washing.

The physical-chemical properties of double-stranded DNA, and temperature and salt concentration during hybridization and washing, determine the tendency of the DNA molecule to form two single strands or double-stranded DNA. Since double-stranded DNA is held together by hydrogen bonds between bases, an increase in temperature and a decrease in salt concentration both favor single stranded DNA over double-stranded DNA. The melting point, or T_m, is defined as the temperature at which 50% of the solution DNA is single stranded. The T_m of two hybridized DNA strands can be estimated from the following equation (Bolton & McCarthy, 1962):

$$T_m = 81.5 \text{ C} + 16.6(\log_{10}Na^+) +$$

$$0.41(\%G + C) - 0.63(\%formamide) - (600/1),$$

where 1 = length of the hybrid in base pairs when the Na^+ concentration is between 0.01 to 0.4 M and the G + C content is between 30 to 75%.

The relative stability of nucleic acid hybrids decreases in the order RNA:RNA > RNA:DNA > DNA:DNA (Sambrook et al., 1989). The T_m of double-stranded DNA decreases by 1 to 1.5% with each 1% decrease in homology (Bonner et al., 1973). Washing temperatures are generally carried out at $T_m - 12$ °C.

Both salt concentration and temperature are controlled during the washing step [reassociation of the probe DNA with the target DNA] that determines "stringency" or specificity of the probe to detect homologous DNA sequences. High stringency during the washing step at a temperature of 65 to 68 °C favors closely matched hybrids, although investigators may alter stringency as required. Some investigators prefer to hybridize using 50% formamide at 42 °C, which reduces the temperature required for the same level of stringency. The same principles of hybridization apply for nucleic acids on solid filters during either Southern, colony hybridization, and dot/slot blot hybridization (Hames & Higgins, 1985).

32–4.2 Colony Hybridization and Dot/Slot Blots

32–4.2.1 Principles

Colony hybridization was first described by Grunstein and Hogness (1975) for in situ lysis and detection of bacterial DNA sequences with a DNA probe. The method involves the direct transfer of bacterial colonies from an agar plate to a nylon or nitrocellulose filter. The cells are lysed in place and fixed, followed by prehybridization, hybridization with the probe of interest, and washing. Commonly used filters include nitrocellulose and nylon (Sambrook et al., 1989). Nylon membranes are more durable and therefore can withstand repeated screening with several probes. Nylon membranes have the advantage over nitrocellulose in that they are cheaper, more durable during hybridization, and less prone to cracking upon drying (Sambrook et al., 1989). Colony hybridization has the advantage over direct extraction of DNA in that colonies can be replica plated onto other solid media and specific colonies containing the gene of interest can be identified and isolated. The method has the disadvantage in that only culturable cells harboring the genes of interest will be identified.

Dot and slot blots differ from colony hybridization in that the filters are placed in a commercial manifold system containing up to 96 wells. DNA extracts from environmental samples are pulled through the manifold by vacuum, and bound to the membrane. See chapter 35 for protocols on direct extraction of DNA from environmental samples. Through the addition of appropriate standards, some quantification of the concentration of target sequences can be achieved by comparison of probe intensity to unknown samples, as will be discussed below in section 32–5. The usual sequence of prehybridization, hybridization, and washing steps are conducted on the filter. The method has the advantage in that culturing of organisms is not required. The method has also been used for the enumeration of cultured bacterial cells in combination with the MPN technique (Fredrickson et al., 1988). Lysing of cells, however, is necessary prior to hybridization.

Appropriate controls are of the utmost importance in any hybridization. Both positive and negative controls should be used in any hybridization, and the choices for these are not always straight forward. The positive control should always be the gene of interest, either in cells for colony hybridization, or as DNA for dot and slot blots. A standard curve may be generated by applying a range of masses of the target (or probe) DNA to the hybridization filter, from which the concentration of the unknowns may be related to the standards (Hames & Higgins, 1985; Dockendorff et al., 1992), as is discussed below in section 32–5. Negative controls are typically any DNA with which the probe should have no known homology. If the probe has been purified from a vector (section 32–2), it is important to know that no vector has been co-purified with the fragment, as is sometimes the case. In order to reduce this possibility, the vector without the insert should be used. If positive hybridization results, the user will then be aware that he/she may also be detecting signals from sequences homologous to the vector that are present in the environmental sample. Another example of a negative control that may be used is a gene that codes for a function that is similar to, but not identical to, the gene of interest. This type of control allows proper determination of stringency conditions.

32–4.2.2 Colony Hybridization

32–4.2.2.1 Materials. Many protocols are available, each of which may require specific materials. The following is summarized as a general approach from Sambrook et al. (1989).

1. Nitrocellulose or alternative paper cut to fit the size of a petri dish.
2. Whatman 3MM paper.
3. 10% sodium dodecyl sulfate (SDS; optional).
4. Denaturing solution (0.5 N of NaOH, 1.5 M of NaCl).
5. Neutralizing solution (1.5 M of NaCl, 0.5 M of Tris-Cl, pH 7.4).
6. 2× standard saline citrate (SSC).
7. BLOTTO (Bovine Lacto Transfer Technique Optimizer).

32–4.2.2.2 Procedure

1. Identify filter paper with pencil and make an appropriate mark on both the agar plate and filter paper.
2. Place the filter paper on the edge of the agar plate and spread over the surface with care to avoid bubbles.
3. Cut four pieces of Whatman 3MM paper to fit into four suitable containers. Saturate each with the above four solutions.
4. With forceps, peel the filter from the agar plate and add colony side up to tray 1. Expose the filter to the SDS for 3 min.
5. Expose the filter to the denaturing, neutralization, and SSC solutions for 5 min each.
6. Lay the filters, colony side up, on a dry sheet of 3MM paper. Dry at room temperature for 30 min.
7. Sandwich the filter between two dry 3MM papers and dry at 80 °C in a vacuum oven for 1 to 2 h.
8. Hybridize with the appropriate probe as described below.

An alternative method is to puddle the above solutions onto the surface of Saran wrap and add each filter to the corresponding solutions (Sambrook et al., 1989).

32–4.2.3 Slot/Dot Blots

32–4.2.3.1 Materials. From protocols of Sambrook et al. (1989).

1. Appropriate trays, pipettes, and rotating temperature-controlled incubators as required. The investigator is referred to either of the above references for preparation of buffers and reagents. Volumes used in the following protocols can be adjusted for the number of filters.
2. Prewashing solution:
     50 mM Tris-Cl (pH 8.0)
     1 M NaCl
     1 mM EDTA
     0.1% SDS
3. Prehybridization solution:
     5× SSPE
     0.1% SDS
     5× Denhardt's solution[2]
     100 µg/mL denatured, salmon sperm DNA
     50% formamide (optional)

---

[2]50× Denhardt's solution:
5 g polyvinyl pyrrolidone
5 g bovine serum albumin
H_2O to 500 mL

32–4.2.3.2 Procedure.

Washing Step

This step removes bacterial debris and other loosely adhering fragments.

1. Float baked filter on the surface of $2 \times$ SSC until thoroughly wetted and submerge for 5 min.
2. Transfer filters to a glass dish containing 200 mL of prewash solution. Filters may be stacked on one another. Cover with Saran Wrap, transfer to a rotating shaker, and incubate for 30 min at 50 °C. Do not allow the filters to dry.
3. Gently remove adhering bacterial cells with a Kemwipe that reduces background without reducing the intensity of probe.

Prehybridization

4. Transfer filters to a glass dish containing 150 mL of prehybridization solution and incubate with gentle shaking for 1 to 2 h at 65 to 68 °C for aqueous solutions or at 42 °C for solutions containing formamide. Plastic freezing trays or heat sealable bags have been used in place of glass dishes. Filters can be stacked over one another, but make sure they are free and in contact with the solution at all times. Denhardt's solution, salmon sperm DNA, and Bovine Lacto Transfer Technique Optimizer (BLOTTO) are blocking agents to reduce nonspecific binding of probe DNA. Dextran sulfate is sometimes used to increase the rate of reassociation of nucleic acids.

Hybridization

At this point, the procedure will differ depending on whether radioactive or nonradioactive probes are used. For nonradioactive probes, refer to the manufacturer's protocol. If radioactive ^{32}P probes are used, the following is a general protocol.

1. Denature the probe by boiling for 5 min and chill quickly in crushed ice. Add the probe to the appropriate containers at a concentration of from 2×10^5 to 1×10^6 cpm of ^{32}P-labeled probe (specific activity $> 5 \times 10^7$ cpm/µg DNA) per milliliter of solution. Hybridize at the appropriate temperature (65–68 °C for most aqueous solutions) for 4 to 24 h, depending on the efficiency of reaction. Care must be taken not to contaminate the water bath with radioactive material. If sealable bags are used, they should be double wrapped.

Washing

During the final washing steps, the nonbinding probe is removed and the stringency is increased in subsequent steps to remove loosely binding DNA probe sequences.

1. After hybridization, carefully remove the solution and dispose of the bags and solution properly. Add the filters to about 300 mL of $2 \times$ SSC and 1% SDS and gently wash with gentle agitation at room

temperature for 5 min. Transfer the filters to a fresh batch of wash solution and agitate for 5 min. Repeat this process twice.

2. Wash the filters twice with gentle agitation in 300 mL of $1 \times$ SSC and 0.1% SDS at 65 °C.

3. Expose the labeled membrane to x-ray film as described below (section 32–5).

32–5 DETECTION SYSTEMS

32–5.1 Principles

Once hybridization has been completed, the extent of hybridization may be quantified by several different methods when radioactive probes are used. The most commonly used technique is by autoradiography, or the exposure of x-ray film by the radiolabeled membrane. In the case of colony hybridization, the positive signals on the x-ray film are used to identify the positive colonies that contain the genes of interest on the original plates (Pettigrew & Sayler, 1986). In dot and slot blots, the relative intensities of the standards are compared visually with that of the unknown samples, and a quantitative determination is estimated. More precise quantification may be obtained by using a scanning densitometer, although the cost of these instruments may be prohibitive. When ^{32}P is used, the sensitivity of autoradiography may be increased by the addition of intensifier screens to the film cassette. Intensifier screens are placed on the opposite side of the film from the blot, in such a manner as to allow the beta particles emitted from the blots to pass through the film and strike the intensifying screens. As a result of being struck by beta particles, the screen will emit photons, thereby increasing the amount of exposure per dot. The use of intensifier screens results in an increase in sensitivity of approximately fivefold (Sambrook et al., 1989).

Pre-flashing x-ray film can increase the sensitivity of detection, decrease the background, and increase the correlation between radioactivity present on the filter and the intensity of the image on the film (Sambrook et al., 1989). Pre-flashing is accomplished with a stroboscope or a photographic flash. Blue-light from the flash should be reduced with an orange filter (comparable to Wratten 22A), and light from the flash diffused by covering either the flash or the film with Whatman No. 1 filter paper. The distance from the flash to the film should be at lease 50 cm, and should be determined empirically by flashing film in total darkness over a range of distances. The distance should be optimized so that the difference between pre-flashed and non-preflashed film absorbance at 545 nm is 0.15 absorbance units. Optical densities should be measured in a spectrophotometer, using non-flashed film as the blank.

It is also possible to obtain quantitative information from slot and dot blots by scintillation counting. This is usually accomplished by first exposing an autoradiogram to locate the position of the dots, and then cutting the dots out of the membrane. These dots are then mixed with an appropriate amount of scintillation cocktail and counted in a scintillation counter with appropriate settings for ^{32}P. Standard dots are counted along with the samples, and a standard curve is drawn by plotting scintillation counts against mass of the set of standards. If the level of radioactivity is high enough, Cerenkov radiation may be taken advantage of to avoid using scintillation cocktail. Cerenkov counting uses water as the solvent instead of scintillation cocktail, and the resulting emissions are monitored in the same energy range as ^3H rather than ^{32}P. A counting efficiency of 10 to 30% is experienced with Cerenkov counting relative to scintillation counting. While this method is not as efficient as scintillation counting, it does allow the user to avoid scintillation cocktail, which is expensive, toxic, and must be disposed of as hazardous waste. Due to its short half-life (14 d), ^{32}P in water may be stored until it has decayed to an acceptable limit and then disposed of as nonradioactive aqueous waste.

32–5.1.2 Materials

1. Saran wrap.
2. X-ray film, such as Kodak XAR-2.
3. X-ray film cassette with intensifying screen.
4. Blot from either colony hybridization or slot/dot blot.
5. Unwanted piece of x-ray film.
6. Scotch tape.
7. Darkroom.
8. Developing and fixing liquids for developing x-ray film, or an automatic developer.
9. Scintillation vials.
10. Scintillation counter.
11. Scissors.
12. Disposable gloves.
13. Photographic flash unit, orange filter.

32–5.1.3 Procedures

1. Wrap blot in Saran Wrap and affix to used piece of x-ray film with Scotch tape.
2. Place the used film with the blot in the film cassette with intensifier screens in place. Note: Intensifier screens are only used with ^{32}P-labeled probes. The blot may then be labeled with radioactive ink so that the film will be oriented with respect to the blot. Radioactive ink may be made by making up a dilution of ^{32}P-dCTP in water, and then dipping the end of a marking pen in the dilution.
3. Pre-flash film.
4. In the dark, place the unexposed film between the blot and intensifier screen. Seal the cassette so that it is light tight, and store at −70 to −90 °C.

5. The amount of time required for exposure varies with the amount of radioactivity on the blot, and may range from a matter of hours to 2 to 3 wk. The length of exposure for a specific autoradiogram may be largely a matter of trial and error.

6. Upon completion of the exposure time, remove the cassette from the freezer and allow it to come to room temperature.

7. Develop the film either by automatic developer or developing and fixing liquids.

8. The film may then be read by densitometry or by visual inspection (as described in 32–5.1), or it may be used as a guide to cut out the dots for scintillation counting.

9. If either scintillation or Cerenkov counting is to be used, cut out the radioactive dots with a pair of scissors and place the dots in labeled scintillation vials.

10. If Cerenkov counting is to be used, add approximately 10 mL of H_2O to the vials and shake. Count in a scintillation counter on the ^3H channel. If scintillation counting is to be used, add approximately 10 mL of scintillation cocktail, shake, and count on the ^{32}P channel.

11. Construct a standard curve from the standards by plotting cpm vs. mass of standards. From this, the mass of probe homologous DNA in the samples may be determined.

REFERENCES

Bilofsky, H.S., C. Burks, J.W. Fickett, W.B. Goad, F.I. Lewitter, W.P. Rindone, C.D. Swindell, and C.-S. Tung. 1986. The GenBank genetic sequence databank. Nucleic Acids Res. 14:1–4.

Bolton, E.T., and B.J. McCarthy. 1962. A general method for the isolation of RNA complementary to DNA. Proc. Natl. Acad. Sci. 48:1390–1397.

Bonner, T.I., D.J. Brenner, B.R. Neufeld, and R.J. Britten. 1973. Reduction in the rate of DNA reassociation by sequence diversion. J. Mol. Biol. 81:123.

Dockendorff, C.T., A. Breen, O. Ogunseitan, J. Packard, and G.S. Sayler. 1992. Practical considerations of nucleic acid hybridization and reassociation in environmental analysis. p. 393–420. In M.A. Levin et al. (ed.) Microbial ecology: Principles, methods, and applications. McGraw-Hill, New York.

Fleming, J.T., J. Sanseverino, and G.S. Sayler. 1993. Quantitative relationship between naphthalene catabolic gene frequency and expression in predicting PAH degradation in soils at town gas manufacturing sites. Environ. Sci. Technol. 27:1068–1074.

Fredrickson, J.K., D.F. Bezdicek, F.J. Brockman, and S.W. Lee. 1988. Enumeration of Tn-5 mutant bacteria in soil using a most-probable-number-DNA hybridization procedure and antibiotic resistance. Appl. Environ. Microbiol. 54:446–453.

Grunstein, M., and D.S. Hogness. 1975. Colony hybridization: A method for the isolation of cloned DNAs that contain a specific gene. Proc. Natl. Acad. Sci. 72:3961–3965.

Haas, M.J., and D.J. Fleming. 1988. A simplified lysis method allowing the use of biotinylated probes in colony hybridization. Anal. Biochem. 168:239–246.

Hames, B.D., and S.J. Higgins (ed.). 1985. Nucleic acid hybridisation: A practical approach. IRL Press, Washington, DC.

Hazen, T.C., and L. Jimenez. 1988. Enumeration and identification of bacteria from environmental samples using nucleic acid probes. Microbiol. Sci. 5:340–343.

Holben, W.E., and J.M. Tiedje. 1988. Applications of nucleic acid hybridization in microbial ecology. Ecology 69:561–568.

Holtke, H.J., R. Siebl, J. Burg, K. Muhlegger, R. Mattes, and C. Kessler. 1988. Non-radioactive High Sensitivity DNA labeling and detection system (digoxigenin:anti-digoxigenin based ELISA principle). Fresenius Z. Anal. Chem. 330:337–346.

Langer, P.R., A.A. Waldrop, and D.C. Ward. 1981. Enzymatic synthesis of biotin-labeled polynucleotides: Novel nucleic acid affinity probes. Proc. Natl. Acad. Sci. USA 78:6633–6637.

Leary, J.J., D.J. Brigati, and D.C. Ward. 1983. Rapid and sensitive colorimetric method for visualizing biotin-labeled DNA probes hybridized to DNA or RNA immobilized on nitrocellulose: bioblots. Proc. Natl. Acad. Sci. USA 80:4045–4049.

Martin, R., C. Hoover, S. Grimme, C. Grogan, J. Holtke, and C. Kessler. 1990. A highly sensitive, non-radioactive DNA labeling and detection system. BioTechniques 9:762–778.

Ogram, A.V., and G.S. Sayler. 1988. The use of gene probes in the rapid analysis of natural microbial communities. J. Ind. Microbiol. 3:281–292.

Olsen, G.J., D.J. Lane, S.J. Giovannoni, N.R. Pace, and D.A. Stahl. 1986. Microbial ecology and evolution: A ribosomal RNA approach. Annu. Rev. Microbiol. 40:337–365.

Olson, B.H. 1991. Tracking and using genes in the environment. Environ. Sci. Technol. 25:604–611.

Pettigrew, C.A., and G.S. Sayler. 1986. The use of DNA:DNA colony hybridization in the rapid isolation of 4-chlorobiphenyl degradative bacterial phenotypes. J. Microbiol. Methods 5:205–213.

Pickup, R.W. 1991. Development of molecular methods for the detection of specific bacteria in the environment. J. Gen. Microbiol. 137:1009–1019.

Sambrook, J., E.F. Fritsch, and T. Maniatis. 1989. Molecular cloning: A laboratory manual. N. Ford et al. (ed.) 2nd ed. Cold Spring Harbor Lab., Cold Spring Harbor, NY.

Sayler, G.S., and A.C. Layton. 1990. Environmental application of nucleic acid hybridization. Annu. Rev. Microbiol. 44:625–648.

Stotzky, G., M.A. Devanas, and L.R. Zeph. 1990. Methods for studying bacterial gene transfer in soil by conjugation and transduction. Adv. Appl. Microbiol. 35:57–169.

Syvanan, A.C., M. Laaksonen, and H. Soderlund. 1986. Fast quantification of nucleic acid hybrids by affinity-based hybrid collection. Nucleic Acids Res. 14:5037–5048.

Trevors, J.T., and J.D. van Elsas. 1989. A review of selected methods in environmental microbial genetics. Can. J. Microbiol. 35:895–902.

Zeph, L.R., X. Lin, and G. Stotzky. 1991. Comparison of three non-radioactive and a radioactive probe for the detection of target DNA by DNA hybridization. Curr. Microbiol. 22:79–84.

Zeph, L.R., and G. Stotzky. 1989. Use of biotinylated DNA probe to detect bacteria transduced by bacteriophage P1 in soil. Appl. Environ. Microbiol. 55:661–665.

Chapter 33

Marking Soil Bacteria with *lacZY*

T. E. STALEY, *USDA, ARS, NAA, Beckley, West Virginia*

D. J. DRAHOS, *SBP Technologies, Inc., Stone Mountain, Georgia*

Acquiring ecological information from laboratory and field studies of bacteria applied to nonsterile soil requires that methods be available to follow, unambiguously, the fate of the inoculum strain in a background of often otherwise indistinguishable, indigenous microflora. The major emphases for the development of these methods have been the rather recent realization that selected indigenous strains, or genetically engineered microorganisms (GEMs), with special metabolic capabilities may be effectively applied as bioremediation agents for environmental pollution problems or have important agronomic attributes. Dinitrogen-fixing bacteria such as rhizobia, bradyrhizobia, and azospirilla are being intensely studied as at least partial replacements for commercial N fertilizer, while agrobacteria, bacilli, and pseudomonads are receiving ever more attention as biocontrol agents in place of pesticides. More recently, pseudomonads have been employed to reduce frost damage to crops (Ice⁻ strains) or simply to improve crop growth by colonizing the roots (plant growth-promoting rhizobacteria or PGPR strains). Success in any of these areas could have a significant impact on reducing our dependence on agricultural chemicals. Public concern about the ultimate fate of these released bacteria, or their engineered genes, has further increased the impetus to develop effective means for their tracking.

Pseudomonads, because of their extremely versatile metabolic capabilities, ubiquitousness, and rhizospheric competence, to say nothing of their role as phytopathogens, have become increasingly important as model microorganisms for tracking studies. Strains from this heterogenous group of rod-shaped, Gram-negative, respiratory, chemoorganotrophic bacteria have traditionally been tracked in the environment, as have strains from many other genera, by selection and use of mutants resistant to antibiotics, especially rifampicin (Rf) or nalidixic acid (Nx).[1] Because background populations of either Rf or Nx resistant bacteria often exceed 10^3 cfu g^{-1} soil, attempts have been made to use mutants resistant to both antibiotics. Unfortunately, such mutants have often been shown to have lost their pathogenicity, rhizospheric competence, or both (Godwin & Slater, 1979; Lindstrom et al., 1990; Stotzky & Babich, 1984). Mutants having resistance to an antibiotic by virtue of expression of detoxifying or membrane transport-inhibiting proteins, e.g., for kanamycin (Km), streptomycin (Sm), and tetracycline (Tc), are often debilitated (Chopra et al., 1981; Levy, 1984; Moyed et al., 1983). As with all procedures based on antibiotic selection, resistance of indigenous strains to these antibiotics, as well as to Rf and Nx, can also seriously compromise the results. For these reasons, and because of the continuing questions surrounding the stability of antibiotic-resistance markers, particularly if they are plasmid-mediated, alternative methods have constantly been pursued. Such methods, in general, may be categorized as either serological (chapter 28 in this book) or molecular/genetic. The latter can likewise be divided into direct detection methods, whereby specific nucleic acid probes hybridize to their homologous, foreign (or novel) target sequences (chapter 32), and indirect detection methods, whereby a foreign (or novel) gene product, usually an enzyme, is detected. Both of these methods have been made possible by recent advances in molecular biology that have enabled the insertion of genes into cells of strains from many genera of soil bacteria, actinomycetes, and fungi and their detection by nucleic acid (DNA and RNA) hybridization procedures.

A detailed description of all the indirect (metabolic) molecular genetic methods currently available to mark soil bacteria is beyond the scope of this chapter. Because the *lacZY*-marking system has been more extensively investigated and accepted for soil microbiological investigations than other metabolic methods, and because it is the only one that is selectable in that only bacteria containing these genes will grow (express their novel products), theoretically, on selection medium upon plating a mixed inoculum, this chapter is exclusively devoted to its description. This is not to imply that other methods, for example, *ina* (ice-nucleating activity), *lux* (luciferase), *uidA* (β-glucuronidase), and *xylE* (catechol-2,3-dioxygenase), may not be more appropriate for some applications (Drahos, 1991a). For reasons previously mentioned, this chapter pertains to the use of the *lacZY*-marking system only for pseudomonads. However, the full extent to which the IncQ-based, *lacZY* marking system described can be applied to various other bacterial strains is unknown. To date, its use has been documented in the literature only for fluorescent pseudomonads, primarily because of their importance as potential biocontrol agents. However, it is likely that application of this system will soon extend to other appropriate, Gram-negative strains from genera related to *E. coli* such as *Agrobacter-*

---

[1]Resistance (r) or sensitivity (s) is denoted by superscript letters after the antibiotic abbreviation; superscript numerals refer to levels of antibiotics in μg mL^{-1}.

ium, Azospirillum, Rhizobium, Serratia, and *Xanthomonas.* This system has, for example, already been used to mark three alfalfa-stimulating *Serratia* spp., *S. fonticola, S. liquefaciens,* and *S. proteomaculans,* obtained from Allelix-Crop Technologies (E.G. Lawrence, 1992, personal communication).

Marking bacteria, as used in the context of this chapter, necessarily involves a change in the phenotype and genotype of the parental (recipient or target) strain. This change should be nondebilitating, permanent, and easily detectable if it is to gain widespread acceptance by researchers in soil microbiology. The most recent version of the *lacZY* marking system, the IncQ-based, monocomponent transposition system developed by Monsanto scientists (a nonantibiotic, gene cassette insertion into the chromosome of pseudomonads resulting in a highly selectable and easily scorable phenotype) allows such a change. It is the purpose of this chapter to describe the features of this system, and earlier versions, that allow the production of *lacZY*-marked strains, the requirements of the parental strain for marking, and the methodology for marking. Methods for confirming the insertion of *lacZY* and the identity of the marked strain with its parental strain are also presented. Associated topics that are discussed include the effects of *lacZY* insertion on the marked (exconjugant) strain, marker stability, and recovery from nonsterile soil. As with all marking systems, the *lacZY* system has both attributes and deficiencies, which are also discussed.

Because bacterial strains marked by this IncQ-based system are recipients of a foreign gene, they are therefore considered GEMs. As a consequence, their use and transport in the USA must be done in accordance with all institutional, local, state, and federal regulations pertaining to such organisms. Unless parental strains are pathogenic, there should be little difficulty in obtaining approval by the USDA, APHIS (Form 2000, July 89) for shipment of the parental or *lacZY*-marked strain across state lines. Laboratory and greenhouse usage of *lacZY*-marked strains will require approval by the investigator's institutional biosafety committee. This approval primarily depends on the availability of a BL-1 containment facility as described in the Federal Register, May 1986, vol. 51, no. 88, p. 16958–16985. Most modern microbiology laboratories can attain this level of containment with few or no modifications. Approval for field releases of these GEMs must be approved on a case-by-case basis by the EPA.

33–1 PRINCIPLES

Escherichia coli is able to use lactose by virtue of a stretch of DNA in its chromosome that codes for two enzymes, lactose permease and β-galactosidase, which allow for transport of the disaccharide across the cell membrane and hydrolysis to glucose and galactose, respectively. These two structural genes (along with *lacA*), and their control elements, constitute the lactose (*lac*) operon of about 6000 bp of DNA. Soil bacteria, in general, rarely use lactose (Okamura et al., 1983), which is not surprising considering that this substrate is not found in soil or plants. Fluorescent pseudomonads, with the possible exception of one isolate from milk (Marin & Marshall, 1983), have never been shown to efficiently metabolize lactose, i.e., possess β-galactosidase. In a collection of more than 530 fluorescent pseudomonads, most of which were isolated from the rhizosphere of soybean (*Glycine max* L.) grown in the Midwest, none were able to be grown on minimal lactose medium (Drahos et al., 1986). In the "Analytical Profile Index" compiled by Analytab Products, Inc. (Analytab Products, 1982), the lack of lactose utilization by fluorescent pseudomonads was confirmed in an examination of more than 11 000 strains. As a consequence, insertion of the *E. coli* genes for lactose uptake (*lacY*) and hydrolysis (*lacZ*) into fluorescent pseudomonads from soil, assuming they would be functional (i.e., expressed as the enzymes), represents a novel, and potentially unambiguous, genetic means of marking these bacteria.

The basic genetic element of the IncQ-based system, developed by G. Barry (1988a) and described in this chapter, pMON7197 (Fig. 33–1), represents the latest construct of a series of plasmids designed for the delivery of *lacZY* to Gram-negative bacteria. This 26.1-kb plasmid is a refinement over an earlier construct (pMON7029) used in a bicomponent transposition system that was larger in size (about 56 kb), more difficult to manipulate, and lacked unique cloning sites into which other possible monitoring or control elements, or cryptic genes, might be inserted (Barry, 1986). Of particular importance in pMON7197 is the insertion of *lacZY* between the disabled right and left arms of the transposon, Tn*7*, which allows for the single and permanent insertion of the Tn*7-lac*7117 element into the chromosome of the target strain. In addition, pMON7197 carries all the genes necessary for the transposition event (*tnsA-E*), and resistance to gentamicin (Gm), a characteristic that is used to confirm the absence of the plasmid in the *lacZY*-marked strain.

Methods for marking fluorescent pseudomonads that were developed prior to the above mentioned Tn7-based systems depended on transformation of the target strain by broad host-range plasmids carrying *lacZY*. Hemming and Drahos (1984) found a derivative, pKT230, of the broad host-range vector, RFS1010 (IncQ), to be stable in strains of *Pseudomonas aeruginosa, P. fluorescens, P. putida,* and *P. syringae* when grown in culture media (i.e., in vitro). Subsequently, Drahos et al. (1986) reported the same in vitro results for *P. fluorescens* 701E1 transformed with derivatives of pKT230 (pMON5002 and pMON5003, both at 20.3 kb), but in 6 to 8 wk after inoculation of soybean roots grown in nonsterile soil, a detectable decline in plasmid maintenance occurred. Kawalek and Schaad (1985), using pMON5003, also found it to be stable in transformed *Xanthomonas campestris* pv *campestris* (Xcc) after 25 generations in vitro. A recent report by Scheffer et al. (1989) indicated that *P. fluorescens* isolates obtained from elm tree tissue 1 yr after inoculation had also lost their *lacZY*-carrying plasmid.

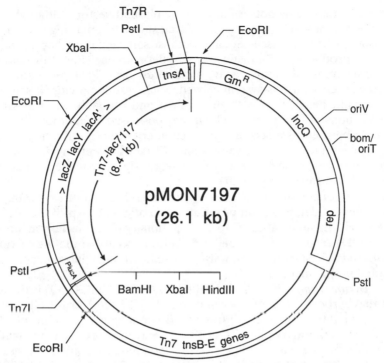

Fig. 33–1. Donor plasmid, pMON7197, used in the Tn7-based, *lacZY*-marking system (modification of Barry, 1988a).

To increase the stability of the *lacZY* marker (both in terms of improving its maintenance in the host cell and increasing its difficulty of transfer to other soil bacteria), eliminate known antibiotic resistance genes, and reduce the possibility of debilitation by the presence of functional genes on broad host-range plasmids, the transposon (Tn7)-based systems were developed. It is the monocomponent transposition system, based on the single, unstable, IncQ plasmid, pMON7197, that will be described in detail in the following sections.

33–2 MATERIALS

33–2.1 Cells

1. *Recipient (target) strain*—It should be stated from the outset that not all fluorescent pseudomonads may be suitable candidates for marking with this *lacZY* system. Besides its inability, or low efficiency of metabolism by some unknown pathway, to use lactose, the recipient strain should also be sensitive to Gm. This is important since it is this characteristic that is negatively selected during the course of purification of the exconjugants

to ensure that *lacZY* is not residing on the plasmid vector. Although fluorescent pseudomonads are particularly well suited for marking with *lacZY* for reasons previously discussed, and because selection is further facilitated due to the production of a distinct, visual fluorescence (Flu$^+$) by colonies growing on minimal lactose agar, nonfluorescing (Flu$^-$) pseudomonads may be marked with this system as well. Such pseudomonads, if sensitive to Gm, could be marked with this system and probably be successfully tracked in nonsterile soil, especially if they possessed an additional marker that could be used as a secondary selection criterion. Antibiotic resistant markers to ampicillin (Ap), tetracycline (Tc), or streptomycin (Sm), to which many soil isolates are intrinsically resistant, may be particularly useful in this regard (chapter 27).

Pseudomonads should be grown in 100-mL shake cultures of King's B (KB) medium, including an appropriate antibiotic (e.g., Ap) if the strain is resistant to it. Incubate at 25 °C to early stationary phase, usually an optical density (OD) of 0.6 to 0.8 at 580 nm, before use in the mating mixture. Luria-Bertani (LB) medium may also be used, although fluorescence is not promoted in this medium.

2. *Donor strain—E. coli* MM294 (Tn7::*lacZY*, Gm^r, IncQ) is the host strain that harbors the donor plasmid, pMON7197 (26.1 kb). Cells should be grown in 100-mL shake cultures of LB containing 15 µg Gm mL^{-1} at 37 °C to early stationary phase (OD 580 nm $\simeq$ 1.0) before use in the mating mixture. Cultures are available at no cost upon request from the Monsanto Company, Plant Protection and Improvement, 700 Chesterfield Village Parkway, St. Louis, MO 63198.

3. *Helper strain—E. coli* HB101 (Tra$^+$, Mob$^+$, Km^r, and IncP) is the host strain that harbors the helper plasmid, pRK2013 (48 kb), necessary for the mobilization of non-self-transmissible plasmids. Cells should be grown in 100-mL shake cultures of LB medium containing 50 µg Km mL^{-1} at 37 °C to early stationary phase (OD 580 nm $\simeq$ 1.0) before use in the mating mixture. Cultures are available for a fee from the American Type Culture Collection, 12301 Parklawn Drive, Rockville, MD 20852 (ATCC Number 37159).[2]

The compatibility of the recipient, donor, and helper strains should be determined before marking with this system by cross-streaking on LB agar plates, since pseudomonad strains may produce compounds inhibitory to the *E. coli* strains. In addition, the recipient pseudomonad should be tested for its ability to grow on the minimal medium described in the next section, with glucose in place of lactose.

---

[2]Mention of a trademark, proprietary product, or vendor does not constitute a guarantee or warranty of the product by the USDA, and does not imply its approval to the exclusion of other products or vendors that may also be suitable.

33–2.2 Media

1. *King's B medium (KB)*—The medium contains 20 g of proteose peptone, 10 mL of glycerol, 1.5 g of $MgSO_4 \cdot 7H_2O$, 1.5 g of K_2HPO_4, and 990 mL of distilled water. The pH will be 7.4, without adjustment, before autoclaving.

2. *Luria-Bertani medium (LB)*—The medium contains 10 g of tryptone, 5 g of yeast extract, 10 g of NaCl, and 1 L of distilled water. Adjust the pH to 7.0 with 1 *N* NaOH before autoclaving.

3. *Minimal Lactose X-gal medium (MLX)*—Autoclave 848 mL of distilled water containing 1.5% agar, cool to 55 °C, then add the following sterilized stock solutions: 100 mL of basal salts solution (10X), 2 mL of $MgSO_4 \cdot 7H_2O$ (1 *M*), 0.1 mL of $CaCl_2$ (1 *M*), 50 mL of lactose solution (20%), 1.0 mL of *N,N*-dimethylformamide (DMF) containing 5-bromo-4-chloro-3-indolyl-β-D-galactopyranoside (X-gal; 20 mg mL^{-1}). As X-gal is photosensitive, this medium should be stored in the dark and used within 1 wk.

The 10X basal salts stock solution contains 60 g of Na_2HPO_4, 30 g of KH_2PO_4, 5 g of NaCl, and 10 g of NH_4Cl L^{-1} of distilled water. The pH should be adjusted to 7.4 before autoclaving. Store at room temperature. The $MgSO_4$ and $CaCl_2$ stock solutions should be autoclaved separately and stored at room temperature. The lactose stock solution (20%) is made by dissolving 20 g of lactose in 100 mL of distilled water by warming to 40 to 50 °C. Filter sterilize through a 0.22 μm membrane and store at 5 °C. The DMF / X-gal stock solution is prepared by dissolving 20 mg X-gal in 1 mL of DMF. Add the entire volume, unsterilized, to the medium to give a final X-gal concentration of 0.002%.

Reagent grade chemicals should be used for this medium. It is particularly important to use high quality lactose, agar, and X-gal (see recommended sources under the "Chemicals and Reagents" section that follows).

33–2.3 Chemicals and Reagents

1. *Lactose* (D-lactose monohydrate)—Obtainable from Sigma Chemical Co., St. Louis, MO 63178 (No. L-1768).
2. *Agar* (Bacto)—Manufactured by Difco Laboratories, Detroit, MI 48232 (No. 1040-01).
3. *X-gal* (5-bromo-4-chloro-3-indolyl-β-D-galactopyranoside)—Obtainable from 5 Prime →3 Prime, Inc., West Chester, PA 19380 (#5306-554533); Boehringer-Mannheim, Indianapolis, IN 46250 (#703 729); or Sigma Chemical Co., St. Louis, MO 63178 (B9146). As this chemical is relatively expensive, and the prices vary considerably, several suppliers should be contacted.

4. *M9 basal salts solution (1X)* — To 1 L of distilled water add 6.0 g of Na_2HPO_4, 3.0 g of KH_2PO_4, 0.5 g of NaCl, and 1.0 g of NH_4Cl. Adjust the pH to 7.4 and autoclave.

5. *Phosphate salt solution (PSS)* — To 1 L of distilled water add 7.2 g of NaCl, 2.8 g of Na_2HPO_4, and 0.43 g of KH_2PO_4. The pH should be 7.2, without adjustment, before autoclaving.

33–3 PROCEDURE

The *lacZY* system described below is, in essence, a triparental mating of three bacterial components — donor, helper, and recipient (target) cells. The donor strain, *E. coli* MM294, harbors the "suicide" IncQ plasmid, pMON7197, which carries *lacZY* and a truncated *lacA* gene. It is this strain that, after receiving transfer (Tra$^+$) and mobilization (Mob$^+$) functions from the *E. coli* HB101 strain harboring the helper, IncP plasmid, pRK2013, actually transfers pMON7197 to the fluorescent pseudomonad strain by conjugation (mating). Upon mating, transposition of the Tn7-*lac*7117 element into the chromosome of the recipient occurs, thereby rendering it capable of efficient lactose utilization. Selection on minimal lactose agar, containing a low concentration of a chromogenic lactose analog (e.g., X-gal), allows for recovery of pseudomonad exconjugants (recipient cells successfully receiving the *lacZY* genes) with the Lac$^+$ phenotype by visually identifying blue colonies. Donor, helper, or recipient strains are unable to grow on minimal lactose agar (some cryptic growth may occur, however).

Additional portions of this section describe how the presence of the *lacZY* insert in the chromosome is confirmed, both functionally and genetically, and the procedures for ensuring the identity of chosen exconjugant strains with the parental strain.

33–3.1 Mating and Selection

Mating is done by preparing a concentrated mixture of donor, helper, and recipient cells, affording them the opportunity to conjugate on complex medium, then increasing the probability of transposition by plating on selection medium containing Gm (Fig. 33–2). Serial dispersions and grow-out steps on minimal lactose X-gal agar (MLX), without Gm (but including an antibiotic if the recipient is resistant to it), are then done with the exconjugant strains to ensure purity of the "type isolates".

Harvest broth cultures (100 mL) by centrifugation (17 000 × *g*, 10 min, room temperature), wash twice with 10-mL PSS, and adjust the cell concentration to OD 580 nm ≃ 0.1. Pellet 30 mL of the adjusted, washed cell suspensions as above and, to the *E. coli* suspensions only, add 0.2 mL LB broth. Thoroughly resuspend the *E. coli* preparations by repeated

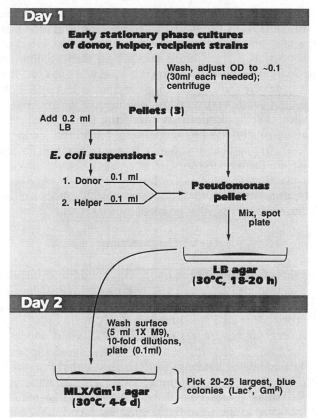

Fig. 33–2. Procedure for triparental mating for producing *lacZY* exconjugants.

pipetting, then transfer 0.1 mL of each *E. coli* suspension to the tube containing the pseudomonad pellet. Mix this mating mixture thoroughly by repeated pipetting for 20 to 30 s. Add about 0.05 mL (one drop) of the mating mixture to the center of an LB agar plate, then gently rotate and tilt the plate until the mixture covers an area of 3 to 4 cm diam. Incubate the plate at 30 °C for 18 to 24 h.

The following day, wash the surface of the mating plate with 5 mL of sterile, M9 basal salts solution (1X), prepare several 10-fold dilutions in the same solution, and spread plate 0.1 mL of each dilution on MLX agar containing 15 µg Gm mL^{-1}. Incubate the plates at 30 °C for 4 to 10 d (depending on the recipient strain used), after which smooth, circular (1–3 mm diam), slightly convex, entire-edged, and mucoidish (at least for most fluorescent pseudomonads) colonies with a sky to moderately dark blue color should appear. With sterile toothpicks, pick 20 to 25 of the largest, blue colonies (presumably those exconjugants that are most efficient in lactose uptake and utilization) and separately smear in columns all exconjugants on an MLX/Gm[15] "pick" plate.

At this point, it is still possible that the exconjugant isolates are contaminated with cells from the mating mixture, and that *lacZY* is still resident on the vector plasmid. Once visible growth has occurred on the pick plate, a small portion is transferred to 3 mL of sterile saline, thoroughly dispersed, and streaked on MLX agar, this time without Gm. Incubate the plates at 30 °C until typical colonies appear, then select well-isolated colonies for two more rounds of dispersion in saline and isolation on MLX agar alone. The color of the colonies may vary from light to dark blue, and be evenly or concentrically colored, depending on the recipient strain used [see Table 33–1 for typical colonial characteristics for *P. putida* R-20 (*lacZY*), an isolate from bean roots, grown on MLX/Cy[100] (cycloheximide) agar]. Exconjugant type isolates containing the transposed *lacZY* genes should now be pure and devoid of the Gm[r] plasmid. Stock cultures can be maintained, indefinitely, in 50% glycerol at −80 °C.

33–3.2 Confirmation

Functional confirmation of the Lac[+], Gm[s], and Flu[+] phenotype of the exconjugant type isolates is done by streaking on MLX, MLX/Gm[15], and KB agar plates, respectively. (Pseudomonas Agar F, manufactured by Difco Laboratories, may be substituted for KB agar, as it is equally efficient at promoting fluorescence.) Blue colonies should appear on MLX, indicative of the Lac[+] phenotype, while no growth should appear on MLX/Gm[15] if the plasmid vector has been lost. To ensure that the isolates' ability to synthesize its fluorescent compounds has not been affected by the transposition event, streak plates on KB agar should also be made since this medium is conducive to their synthesis. Fluorescence is also quite good on MLX agar as indicated in Table 33–1, although it takes longer to appear. In addition, sidearm culture flasks, containing minimal lactose broth, can be conveniently used to measure the growth rate (generations h^{-1}) by following the OD at 580 nm of inoculated cultures. Only the exconjugant strains should grow in this medium. It should be noted that the growth rates may vary depending on the efficiency of lactose uptake and metabolism, or other factors. If only one type isolate is needed, the best choice will likely be the one with the fastest growth rate.

Genetic confirmation of the transposition event, i.e., the physical demonstration of the insertion of *lacZY* into the recipient's chromosome, is done by DNA hybridization analysis (see chapter 32). In brief, DNA probes specific to *lac* fragments of pMON7197 are hybridized to *Eco*RI-restricted, genomic DNA from the exconjugant type isolates. Suitable fragments for probes include either one or both portions (about 4 and 6 kb) of pMON7197 between the *Eco*RI restriction sites that contain all of the lactose genes, plus some additional, shorter fragments (Fig. 33–1). Upon gel electrophoresis of the exconjugant *Eco*RI-restricted DNA and hybridization with the probe, two bands of greater molecular weight than either probe should appear if proper insertion has occurred. The explanation for

Table 33–1. Colonial characteristics of *lacZY*-marked *Pseudomonas putida* R-20 grown on MLX/Cy at 30 °C.

Characteristic†	Plate day				
	2	3	4	5	6
Diameter, mm	0.3–0.5	0.5–1	1–1.5	1.5–2	2–3
Elevation	convex	convex	flattened (slightly)	flattened (dimpled)	flattened
Color	colorless/ whitish	light blue	sky blue center, light blue edge	blue ring, whitish center	concentric blue rings
Fluorescence‡	−	+	+	±	±

† Always circular form, entire margin, mucoidish.
‡ UV light (366 nm)—blue/green color.

this is that the probability of insertion of the Tn*7* transposon into an *Eco*RI site is extremely rare. Because pMON7197 is a large (26.1 kb), low copy-number plasmid, use of the smaller (11.2 kb), higher copy-number plasmid, pMON7117 (also available from the Monsanto Company), for probe preparation will improve the yield of probe fragments. In this case, double digestion of the plasmid with the restriction enzymes, *Bgl*II and *Bam*HI, will produce three fragments, the larger of which (7 kb) will contain most of the *lacZY* genes.

32–3.3 Identification

In the unlikely case that a Lac$^+$ phenotype contaminant has been selected as an exconjugant type isolate, several tests can be performed to confirm its identity as a derivative of the parental strain. Only brief descriptions of these tests will be given, as they are amply described in easily obtainable texts, manuals, and manufacturers' literature, and in other chapters of this book.

Perhaps the easiest test is the use of the micromethod "NFT" kit for the identification of Gram-negative, nonfermentative bacteria (Analytab Products, Inc., Plainview, NY 11803). In this biochemical test, cell suspensions of both the parental and exconjugant type isolate are prepared and used to inoculate a plastic strip containing eight media cupules that detect various metabolic pathways (or enzymes) and 12 media cupules that contain various carbohydrates as sole C sources. Within 18 h, readings can be made and compared. The only differences detected should be in the cupule containing *p*-nitrophenyl-β-galacto-pyranoside (PNPG), another analog of lactose, which should turn yellow in the cupule inoculated with the exconjugant type isolate, compared to a colorless cupule for the parental strain. In addition, if the parental strain has not been assigned a binomial name, information from this test can be coded and used to classify it to at least the species level in their *Identification Codebook*.

Another easy test to confirm the identity of the parental and exconjugant type isolates is to screen for presumptive antibiotic sensitivities by use of the Dispens-O-Disc kit (Difco Laboratories). Broth cultures are spread on LB, and zones of growth inhibition are compared for 11 antibiotics. No differences in the diameters of the zones of inhibition should be found.

Additional identity confirmation can be done by performing DNA fingerprint analyses. Total genomic DNA is extracted from both the parental and exconjugant type isolates, digested with $EcoRI$, and the fragments separated by gel electrophoresis (Drahos & Barry, 1992). No differences in the fragment patterns (even though the 8.4 kb Tn7-lac7117 element is resident in the exconjugant strain) should be found, as the effect of the $lacZY$ insert is too small to be detected by this technique.

Numerous other tests of identity can be performed such as membrane fatty acid analysis, carbohydrate utilization range, two-dimensional protein profiling, RFLP analysis, etc. Many of these methods are described in detail in other chapters of this book.

33–4 EXCONJUGANT-TYPE ISOLATE VIGOR

As previously stated, one of the reasons for the development of the transposon-based delivery systems was to produce marked strains of similar vigor (integrity) as the parental strain, i.e., reduce or even eliminate any debilitation due to the introduction of a foreign gene. No evidence is available that indicates that debilitation occurs when transformed pseudomonad, exconjugant type isolates are obtained with plasmid-mediated Tn7-$lacZY$. However, the numerous reports of debilitation vis-à-vis recombinant plasmids carrying other markers, especially for antibiotics, suggest that this is a valid reason for developing the transposon-based systems. Since transposition is, by definition, an insertional mutation, the possibility always exists that the exconjugant type isolate has suffered some loss of vigor. Although injury of some sort can never be entirely ruled out, both theory and practice suggest that, in all likelihood, none will occur when fluorescent pseudomonads are marked with $lacZY$ via the Tn7-based systems.

Tn7, unlike Tn5 that has been used to deliver $B.$ $thuringiensis$ (Bti) endotoxin genes to pseudomonads (Obukowicz et al., 1986a, 1986b), normally inserts in only one site per chromosome in most bacterial species (Barth et al., 1976; Lichtenstein & Brenner, 1982; McKown et al., 1988). Barry (1986) determined that, in four exconjugant type isolates of $P.$ $fluorescens$ 701E1, all the Tn7-$lacZY$ insertion points on the chromosomes were identical, suggesting that, indeed, only one site is affected. An additional implication from these results is that excision and reinsertion did not occur that might possibly result in a debilitating mutation in some other part of the chromosome. Although it has not been documented in the literature, it is likely that this situation pertains to most pseudomonads.

It should also be noted that the procedure described in the previous section, itself, reduces the possibility of obtaining a debilitated exconjugant-type isolate. Any insertional mutation resulting in auxotrophy would be lost, or selected against due to reduced growth, under the selection conditions of the procedure.

On the practical side, at least three reports have indicated that no adverse effects have occurred in fluorescent pseudomonads marked with the Tn7-based systems. Barry (1986) found that 12 type isolates from two matings expressed similar levels of β-galactosidase, and grew in minimal glucose broth at rates similar to each other and the parental strain, *P. fluorescens* 701E1. Other nutritional requirements for one marked strain, 701L5, were found to be identical to those for the parental strain as well. Kluepfel et al. (1991), similarly found no adverse mutagenic effects on a root-colonizing, fluorescent pseudomonad, *P. aureofaciens* Ps3732RN marked with the IncQ-based, monocomponent delivery system. Barry (1988b) states that in all type isolates examined so far, none have been shown to be slower growing in vitro or poorer root colonizers than the parental strain.

33–5 MARKER STABILITY

Only brief mention has been made of the stability of the *lacZY* marker, and that only in reference to strains marked by transforming plasmids. These plasmids, generally carrying antibiotic resistance genes, are often lost in the absence of the selecting agents, such as would be encountered in the soil. As the IncQ-based procedure described in this chapter results in a chromosomal insertion, the *lacZY* genes should not be lost in the absence of lactose. In repeated studies of *P. fluorescens* 701L5 (*lacZY*), Barry (1986) found that the genes were, indeed, not lost in vitro by spontaneous deletion. In the same paper, he also directly demonstrated their high stability in the chromosome of the same bacterium that had been marked by the bicomponent system. Tn7 can normally transpose to many sites on the IncP plasmid, RP4. In a mating experiment involving this plasmid, none of the exconjugants had acquired the *lac* genes from the chromosome of *P. fluorescens* 701L5. It should also be noted that the absence of Tn7 *tns* genes in exconjugant type isolates obtained by the monocomponent system, the products from which are needed for Tn7-*lacZY* excision, further ensures the stability of *lacZY* genes. Also, no loss in stability was found in fluorescent pseudomonads marked by the bicomponent system that were reisolated from soil in greenhouse studies (Barry, 1988a). These results strongly suggest that, once inserted by the Tn7-based systems, the *lacZY* genes are stable in the recipient strains.

33–6 RECOVERY FROM NONSTERILE SOIL

Once the lacZY-marked inoculum is applied to nonsterile soil, it must be recovered from soil or root washes. Typically, dilutions are made in a buffer solution that is neither detrimental or stimulatory to the inoculum strain. Appropriate dilutions (usually 10-fold) to use will depend on the type of sample and the length of time after inoculation.

For experiments whose objective is simply to determine the rate of survival in nonsterile soil alone, i.e., without actively growing roots, ecologically realistic levels of 10^4 to 10^5 cfu g^{-1} can be used as the inoculum. Spiral plating (0.5 mL plate^{-1}), the most efficient procedure for these types of experiments, or spread plating (0.1 mL plate^{-1}) of the dilutions on MLX/Cy (cycloheximide at 100 µg mL^{-1}) will usually yield blue colonies of typical morphology in 4 to 5 d at 30 °C. As die-off in such soils is usually quite rapid, triplicate samples assayed at 0, 1, 2, 4, and 7 d will normally yield data sufficient for kinetic calculations.

For experiments whose objective is to determine rhizospheric competence or follow root colonization, some modifications of the above procedure are necessary. The reason for this is that the indigenous microflora, either as cfu g^{-1} soil or root, increases dramatically during root growth. Even though efficient root-colonizing, fluorescent pseudomonads have been shown to reach maximum levels of 10^6 cfu g^{-1} root after 2 to 3 d of growth on winter wheat (*Triticum aestivum* L.) (Drahos et al., 1988, 1991b), the ever-increasing population of indigenous microflora that is able to grow on MLX/Cy can, at other times, obliterate the visualization of lacZY-marked colonies. To overcome this problem, spread plating only should be used and the additional scorable, phenotypic marker, Flu$^+$ (fluorescence-producing), employed. Alternatively, if the lacZY-marked strain is nonfluorescent (Flu$^-$, or if background problems are still encountered with the fluorescent strain), an antibiotic (e.g., Ap at 50–100 µg mL^{-1}) to which the inoculum strain is intrinsically resistant can be added to the MLX/Cy agar. Use of these modifications on quintuplicate samples at biweekly intervals, or weekly during the early phase of root colonization experiments, will usually detect any statistically significant differences.

Several other considerations should be mentioned in regard to successfully using the lacZY tracking system for the latter type of experiments. As implied above, intrinsic antibiotic sensitivity patterns and levels of the tentative parental strains should be determined. A selection should be made, if possible, for the ones that are resistant to Ap or, possibly even better, Km or Nx (Kelch & Lee, 1978). If more than one intrinsic antibiotic marker is found, the one selected should be the one most effective in reducing the soil's indigenous background population as determined on MLX/Cy agar. Also, when scoring plates, it is often helpful to identify the Flu$^+$ colonies first, as this phenotype is easily discernible, even before typical, blue colonies appear on MLX/Cy (Table 33–1). Scoring can be done by examining the petri plate through the lid (looking down on the colonies) in a hooded box equipped with a long wavelength (366 nm) UV light. The box accommodating the Model UVL-21 Blak-Ray Lamp (Ultra-Violet Products, Inc., San Gabriel, CA 91778) is a good choice for monitoring colony fluorescence. Scoring the colonies, followed by confirmation using the blue/green color and colony morphology criteria, will produce unambiguous counts of the *lacZY*-marked strain. Additionally, the plates should be read on at least two occasions, usually after 3 and 4 d of incubation at 30 °C. As a final suggestion, positive controls, prepared by spiking soil or root washes with the *lacZY*-marked strain, should be prepared. These are particularly helpful as comparisons with the test plates, and ensure the proper functioning of the selection agar.

33–7 ATTRIBUTES AND DEFICIENCIES

The IncQ-based system described in this chapter offers several distinct advantages for marking a wide range of soil bacteria. The inherent broad host-range of the IncQ replicon allows plasmid survival in Gram-negative strains from more than 32 genera. The system uses a gentamicin resistance (Gm^r) factor, which is found only rarely in soil isolates, and therefore nearly always provides for efficient selection during the marking process. The documented preservation of metabolic vigor and rhizospheric competitiveness (Barry, 1988b; Drahos et al., 1988; Drahos et al., 1991b; Kluepfel et al., 1991) in at least one apparently very "typical" fluorescent pseudomonad, should further encourage use of this system. The high stability of the marker is also an asset. From an efficiency and practicality standpoint, this system is especially useful for large-scale assessments because many samples can be assayed by simple plating, and easily scored by visual inspection. The high sensitivity of this method, being at least 10 times greater than other culture methods, also bodes well for this system. With the use of KB + X-gal/Cy agar containing Nx and Rf, Cook et al. (1991) were able to detect 10^2 cfu g^{-1} root in field studies on the biocontrol of wheat take-all by *P. fluorescens* 2-79 (*lacZY*). The same level of detection in soil of *P. aureofaciens* Ps3732RN (*lacZY*) has been reported by Kluepfel et al. (1991) using MLX/Cy, Rf agar in their intensely monitored field release study. The relative ease with which a given target strain can be marked with *lacZY*, particularly a fluorescent pseudomonad, and the modest investment in terms of training, equipment, and supplies needed, make the process within reach of most microbiology laboratories.

One aspect of this IncQ-based system that has yet to be exploited is its potential as a multiple, and mutually verifiable, tracking system. Prior to mating, other metabolic marker genes (e.g., *uidA* or *xylE*) could be inserted into pMON7197 at its unique, multiple-cloning site (Fig. 33–1) and cointegrated into the recipient. Using appropriate, although nonselective, media for these markers would provide direct comparisons of various marker systems requiring culturing, something that has yet to be done. Further, the enumeration of *lacZY*-marked cells by culturing on MLX/Cy could be confirmed by various direct detection methods applied to a representative set of isolates scored as *lacZY*-marked colonies. Because the DNA of *lacZY*-marked cells contain unique ends of the disarmed Tn7 transposon, sequences extremely rare or nonexistent in soil bacteria, labeled probes for these ends could be prepared and used for detection of homologous sequences by DNA hybridization techniques (Kluepfel et al., 1991). Similar direct confirmation could be done with fluorescent antibody probes prepared against the *lacY* protein product, lactose permease, produced by the *lacZY*-marked strain.

Perhaps the main deficiency of this system is that, although it has not been investigated, it is probably not applicable to Gram-positive bacteria, several genera from which efficient and beneficial root-colonizing isolates have been reported. Another limitation of the system is the requirement for an additional, scorable phenotype (e.g., Flu$^+$) or second marker (e.g., an intrinsic antibiotic resistance), when high sensitivity is required for tracking in soil or on roots containing a large, indigenous bacterial population. One potential problem with this system, and in fact with all marking systems that require culturing for recovery of the inoculum strain, is the necessity for regrowth. Although it has not yet been demonstrated for pseudomonads, it is possible that prolonged exposure of the *lacZY*-marked strain to impoverishment in the soil, or to other soil edaphic or biotic factors, could lead to an underestimation of its "viable" population, particularly when plating is done on a defined medium such as MLX. Because specific probes can be generated by the polymerase chain reaction (PCR) (see chapter 34) from primers that span a portion of *lacZY* and one arm of Tn7 (e.g., *lacY* through Tn7R), direct detection in soils of such "viable but nonculturable" cells may be possible. However, the use of this technology for the accurate and precise enumeration of *lacZY*-marked cells in soils is impractical, if not impossible, at this time.

33–8 COMMENTS

Marking soil bacteria with this system and their recovery on MLX agar defines the basic *lacZY* tracking system. As previously indicated, this system has been used primarily in studies concerned with survival, rhizospheric competence, root colonization, and dispersal in the environment. It is likely that other important questions in soil microbial ecology can be addressed with this system as well. Future applications of *lacZY*-marked strains may well include their use as tracers for delivery agents of genes for antibiotic substances via root-colonizing bacteria, for toxins via endophytic bacteria, or as dually marked cointegrates for direct enumeration methods dependent on DNA hybridization and fluorescent antibodies. One important use that has only been mentioned in passing is its application to gene transfer studies. Kluepfel et al. (1991) used a labeled probe for Tn7 termini to demonstrate the absence of this marker in more than 10 000 bacterial isolates, grown on MLX/Cy agar, from field soil that had been inoculated 8 to 10 wk earlier with a *lacZY*-marked, fluorescent pseudomonad, thus indicating the lack of gene transfer to many other soil bacterial genera and species. As understanding and familiarity of this basic system, and its possible modifications increase, it is likely that other applications will be reported in the near future.

Although it is not without its limitations and deficiencies, the *lacZY* system would appear to be, at the present time, among the most reliable and sensitive methods for marking fluorescent pseudomonads. Once the appropriate strains are available, relatively routine microbiological and molecular biology procedures can be employed for the mating, selection, and confirmation of marked strains. The potential usefulness of such strains that are stably marked, nondebilitated, and easily scored would seem well worth the relatively small investments to any microbiologist interested in soil microbial ecology or the release of GEMs.

REFERENCES

Analytab Products, Ayerst Laboratories, Inc. 1982. Analytical profile index, enterobacteria-ceae and other gram-negative bacteria. Plainview, New York.

Barry, G.F. 1986. Permanent insertion of foreign genes into the chromosomes of soil bacteria. Bio/Technology 4:446–449.

Barry, G.F. 1988a. A broad-host-range shuttle system for gene insertion into the chromosomes of Gram-negative bacteria. Gene 71:75–84.

Barry, G.F. 1988b. Construction of reporter strains. p. 211–219. *In* M. Sussman et al. (ed.) The release of genetically-engineered micro-organisms. Academic Press, London.

Barth, P.T., N. Datta, R.W. Hedges, and N.J. Grinter. 1976. Transposition of a deoxyribo-nucleic acid sequence encoding trimethoprim and streptomycin resistance from R483 to other replicons. J. Bacteriol. 125:800–810.

Chopra, I., S.W. Shales, J.M. Ward, and L.J. Wallace. 1981. Reduced expression of Tn*10*-mediated tetracycline resistance in *Escherichia coli* containing more than one copy of the transposon. J. Gen. Microbiol. 126:45–54.

Cook, R.J., D.M. Weller, D.J. Drahos, P.A. Kovacevich, B.C. Hemming, G. Barnes, and E.L. Peirson. 1991. Establishment, monitoring, and termination of field tests with genetically altered bacteria applied to wheat for biological control of take-all. p. 177–199. *In* D.R. MacKenzie and C. Henry (ed.) Biological monitoring of genetically engineered plants and microbes. Agric. Res. Inst., Bethesda.

Drahos, D.J. 1991a. Current practices for monitoring genetically engineered microbes in the environment. AgBiotech News Info. 3:39–48.

Drahos, D.J. 1991b. Field testing of genetically engineered microorganisms. Biotechnol. Adv. 9:157–171.

Drahos, D.J., and G.F. Barry. 1992. Assessment of genetic stability. p. 161–181. *In* M.A. Levin et al. (ed.) Microbial ecology—Principles, methods, and applications. McGraw-Hill, New York.

Drahos, D.J., G.F. Barry, B.C. Hemming, E.J. Brandt, H.D. Skipper, E.L. Kline, D.A. Kluepfel, T.A. Hughes, and D.T. Gooden. 1988. Pre-release testing procedures: US field test of a *lacZY*-engineered soil bacterium. p. 181–191. *In* M. Sussman et al. (ed.) Release of genetically-engineered microorganisms. Academic Press, London.

Drahos, D.J., B.C. Hemming, and S. McPherson. 1986. Tracking recombinant organisms in the environment: β-galactosidase as a selectable nonantibiotic marker for fluorescent pseudomonads. Bio/Technology 4:439–443.

Godwin, D., and J.H. Slater. 1979. The influence of the growth environment on the stability of a drug resistance plasmid in *Escherichia coli* K-12. J. Gen. Microbiol. 111:201–210.

Hemming, B.C., and D.J. Drahos. 1984. β-Galactosidase, a selectable nonantibiotic marker for fluorescent pseudomonads. J. Cell Biochem. Suppl. 8B:252 (Abstract).

Kawalek, M., and N.W. Schaad. 1985. β-Galactosidase as a metabolic marker in *Xanthomonas campestris* pv. *campestris*. Phytopathology 75:1326 (Abstract).

Kelch, W.J., and J.S. Lee. 1978. Antibiotic resistance patterns of Gram-negative bacteria isolated from environmental sources. Appl. Environ. Microbiol. 36:450–456.

Kluepfel, D.A., E.L. Kline, H.D. Skipper, T.A. Hughes, D.T. Gooden, D.J. Drahos, G.F. Barry, B.C. Hemming, and E.J. Brandt. 1991. The release and tracking of genetically engineered bacteria in the environment. Phytopathology 81:348–352.

Levy, S.B. 1984. Resistance to tetracyclines. p. 191–240. In L.E. Bryan (ed.) Antimicrobial drug resistance. Academic Press, Orlando.

Lichtenstein, C., and S. Brenner. 1982. Unique insertion site of Tn7 in the E. coli chromosome. Nature (London) 297:601–603.

Lindstrom, K., P. Lipsanen, and S. Kaijalainen. 1990. Stability of markers used for identification of two Rhizobium galegae inoculant strains after five years in the field. Appl. Environ. Microbiol. 56:444–450.

Marin, A., and R.T. Marshall. 1983. Characterization of glycosidases produced by Pseudomonas fluorescens 26. J. Food Prod. 46:676–680.

McKown, R.L., K.A. Orle, T. Chen, and N.L. Craig. 1988. Sequences of Escherichia coli att Tn7, a specific site of transposon Tn7 insertion. J. Bacteriol. 170:352–358.

Moyed, H.S., T.T. Nguyen, and K.P. Bertrand. 1983. Multicopy Tn10 tet plasmids confer sensitivity to induction of tet gene expression. J. Bacteriol. 155:549–556.

Obukowicz, M.G., F.J. Perlak, K. Kusano-Kretzmer, E.J. Mayer, S.L. Bolten, and L.S. Watrud. 1986a. Tn5-mediated integration of the delta-endotoxin gene from Bacillus thuringiensis into the chromosome of root-colonizing pseudomonads. J. Bacteriol. 168:982–989.

Obukowicz, M.G., F.J. Perlak, K. Kusano-Kretzmer, E.J. Mayer, and L.S. Watrud. 1986b. Integration of the delta-endotoxin gene of Bacillus thuringiensis into the chromosome of root-colonizing pseudomonads using Tn5. Gene 45:327–331.

Okamura, Y., M. Shoda, and U. Shigezo. 1983. Activity and synthesis of β-galactosidase in various lactose utilizing bacteria. Agric. Biol. Chem. 47:1133–1134.

Scheffer, R.J., D.M. Elgersma, L.A. de Weger, and G.A. Strobel. 1989. Pseudomonas for biological control of Dutch elm disease. I. Labeling, detection and identification of Pseudomonas isolates injected into elms; comparison of various methods. Neth. J. Plant Pathol. 95:281–292.

Stotzky, G., and H. Babich. 1984. Fate of genetically-engineered microbes in natural environments. Recombinant DNA Technol. Bull. 7:163–188.

Chapter 34

Detection of Specific DNA Sequences in Environmental Samples Via Polymerase Chain Reaction

IAN L. PEPPER, *University of Arizona, Tucson, Arizona*

SURESH D. PILLAI, *Texas A&M University Research Center, El Paso, Texas*

Polymerase Chain Reaction (PCR) amplification of DNA and RNA (Saiki et al., 1985; Mullis & Faloona, 1987) has become a key protocol in many biological laboratories. Polymerase chain reaction is an enzymatic reaction catalyzed by the DNA polymerizing enzyme *Taq* polymerase. This enzyme is unique in that it is purified from a thermophilic bacterium *Thermus aquaticus,* and thus is heat stable and can withstand temperatures up to 98 °C. DNA polymerases are present in all bacterial cells, and are involved in DNA synthesis during replication and cell division.

PCR involves amplification of DNA through a repetitive process or cycles with each cycle exponentially increasing the copies of a specific target DNA sequence. The amplification of specific sequences is obtained by restricting the synthesis of new copies within specifically positioned DNA segments called *primers.* The primers are synthesized and their annealing to the denatured DNA is obtained by making use of complementary base pairing of the nitrogenous bases (adenine, guanine, cytosine, and thymine) of DNA. A typical cycle involves denaturing the double-stranded DNA into single strands, annealing short oligonucleotide primers to the now single strands and extending the primer sequences using the DNA polymerase to complete the synthesis of strands complimentary to the original single strands. This cycling is repeated to obtain an exponential increase in the copies of the specific target region of the original DNA strand. The uniqueness of PCR amplification is its specificity and sensitivity. The specificity of PCR is due to its innate ability to selectively amplify one DNA sequence in the midst of a myriad of other DNA sequences. The PCR sensitivity is based on the cycling process, which results in geometric amplification of DNA with the potential for a 10^6-fold increase of amplified DNA in a period of approximately 2 h.

The significance of PCR amplification is primarily related to its ability to exponentially increase specific DNA or RNA sequences within a relatively short period, and also the wide variety of its applications. Polymerase chain reaction amplification has found extensive applications in eukaryotic biology (Glukhov et al., 1990; Li et al., 1988; Welsh & McClelland, 1990). Only recently, however, has PCR amplification of DNA been used in environmental microbiology (Steffan & Atlas, 1988; Bej et al., 1990; Hill et al., 1991; Pillai et al., 1991; Josephson et al., 1991; Neilson et al., 1992). The use of DNA amplification as a means to study population diversity or specific bacterial groups based on specific gene sequences in the environment is attractive, since it can circumvent the need to use cultural methods for screening purposes. Since a significant percentage of the indigenous microbial population has been found to be viable but nonculturable on agar-plating media (Roszak & Colwell, 1987), cultural methods may not be of practical use. The inability to formulate culture media suitable for isolating specific bacterial groups has historically plagued soil microbiologists. The advantage of using techniques like PCR and gene-specific probes that are independent of culturability, is the ability to verify the presence or absence of specific gene sequences among the numerous sequences present in a soil environment. Thus, the use of PCR amplification has the potential to monitor the survival and persistence of specific soil bacterial groups from a wide variety of genera, based on the sequences that are amplified.

Polymerase chain reaction amplifications also have the ability to detect transcription of genes (Tsai et al., 1991; Mahbubani et al., 1991a; Wang et al., 1989) thereby providing the means to monitor in situ enzyme or protein synthesis in the environment. Thus, PCR amplification can be used to answer many diverse questions relevant to soil microbiology and its application must be guided by the aim of the researcher.

The DNA polymerase employed in the amplification process is the thermostable *Taq* DNA polymerase. Due to its thermostability, the enzyme can withstand repeated heating to 94 °C without being denatured, and hence repeated additions of the enzyme are not needed. Both *Taq* polymerase, the recombinant form, *AmpliTaq,* as well as other commercially available thermostable polymerases are well suited for the PCR process. *AmpliTaq* DNA polymerase is a thermostable 94 kDa DNA polymerase encoded by a modified form of the *T. aquaticus* DNA polymerase gene (Innis et al., 1988) cloned in an *E. coli* host. The enzyme has no detectable, associated exo- or endonuclease activity. It has a 5'-3' polymerase activity but no 3'-5' exonuclease activity (Innis et al., 1988). The different applications in which PCR is currently being employed are due to the enzyme's thermal profile and fidelity (Rose, 1990). Its optimum activity is in the same range at which stringent annealing of primers occurs (55–75 °C).

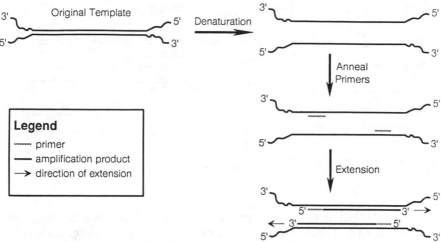

Fig. 34–1. The basic process of PCR amplification.

34–1 THEORY

PCR allows amplification of specific DNA in vitro. The principle of the methodology involves the repetitive enzymatic synthesis of DNA, using two oligonucleotide primers that hybridize to opposite strands of DNA that flank the target DNA of interest. During each cycle, the number of copies of template DNA is theoretically doubled. In practice, 25 cycles of amplification results in an approximate million-fold increase in the number of DNA copies. The primers are often unique 17 to 30 bases long oligonucleotides, carefully chosen to flank and allow amplification of the target DNA of interest. There are three steps in a PCR amplification cycle: (i) template denaturation; (ii) primer annealing, and (iii) primer extension (Fig. 34–1). All three steps occur at different but defined temperatures and time intervals. Generally, these repeating cycles are performed in an automated self-contained temperature cycler. Multiple water baths could also be used instead of the temperature cyclers. A temperature cycler, however, allows the precise temperature control required for the initiation of each step. It also provides for substantial reduction in labor.

Template denaturation occurs at a temperature greater than the melting temperature of the DNA (e.g., 94 °C). Denaturation separates template DNA into single strands allowing subsequent primer annealing. Primer annealing occurs at a lower temperature that is typically 50 to 70 °C. The higher the temperature of annealing, the more specific the annealing is, and the extent of annealing of mismatched primer to template is reduced. However, as primer annealing temperature increases, there is an associated decrease in the sensitivity of detection of the DNA sequence being amplified.

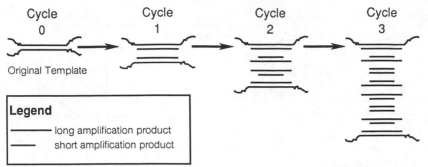

Fig. 34–2. Successive cycles of PCR showing short and long product development.

The final step of PCR is primer extension. Extension involves the synthesis of the DNA strand complimentary to the template. Extension proceeds from the 3′ end of each primer and results in a double-stranded copy of target DNA from each original single strand. Thus, a double-stranded DNA copy of the original target sequence results at the end of the first cycle. Primer extension typically occurs at 72 °C and is catalyzed by the *Taq* DNA polymerase.

During the first cycle of PCR, primers anneal to the template sequence following denaturation of target DNA into two single strands. For all cycles, two long strands are produced from the two original template strands, since there is nothing to limit extension of the strands other than perhaps the time of extension. However, all synthesized long strands (other than initial template) subsequently produce short strands of a size defined by the 5′ end of the primers. In addition, all short strands in turn result in additional short strands limited in size by the 5′ ends of the primers (Fig. 34–2). Thus, after n cycles, the number of long strands has increased linearly by $2 \times n$, whereas the number of short strands has increased exponentially by $2^n - 2n$. After several cycles, the number of short strands increases significantly, relative to long strands. For example, after 10 cycles theoretically there are 20 long strands compared to 1004 short strands. Thus, the long strands are diluted out of the amplification process resulting in a final amplification product of size defined by the 5′ ends of the primers.

34–2 PRIMER DESIGN AND AMPLIFICATION PROTOCOL

34–2.1 Design of Primers

The choice of primer sequences is critical for successful amplification of a specific DNA sequence. There are several guidelines or criteria that can aid in the design of primers to achieve successful amplification of the desired sequence. Optimal amplification can be achieved by optimizing the reaction conditions and components.

The choice of primers must be guided by the aim of the researcher. If specific detection of a target DNA at the species or strain level is required, then sequences unique to the particular species or strain are used for designing the primers. For example, the *lamB* gene codes for an outer membrane protein in *E. coli* and primers designed from this gene sequence will detect *E. coli*. However, in some cases, if for example a bioassay for environmental stress is required, then primer sequences should be selected from target DNA common to all species within a particular genus. The presence or absence of these sequences can then be interpreted as an index of a particular stress factor. Primers designed from common *nod* genes could detect all species of *Rhizobium*. Primers specific to the *giardin* gene are used to detect Giardia cysts (Mahbubani et al., 1991b). There are also cases where the investigator would prefer amplifying DNA sequences of all organisms present, and in such instances, random oligonucleotide primer sequences can be employed. For ultra pure water analyses, universal primers designed from conserved 16S rDNA sequences may be used.

In general:

1. Primers are DNA oligonucleotides, and are typically 17 to 30 bases long. They are complementary to sequences flanking the 5' ends of the complementary strands of a potential target sequence.

2. The size or length of the sequence that can be amplified can range form 100 bp to a few kilobases. (Longer sequences have been amplified using specialized amplification conditions.) It has been observed that for shorter sequences (100–500 bp), PCR amplification is easier to optimize than for longer sequences. Extremely short sequences are, however, difficult to resolve on agarose gels.

3. Each primer must be uniquely different from its complementary partner to prevent what are known as "primer dimers" (see below).

4. If possible, the primers should flank a sequence with a known restriction site for purposes of validation after amplification.

5. Primers should be selected with a GC content $> 40\%$ and preferably 50 to 60%. The melting temperature (T_m) of the primers should be near 72 °C, which is the optimum for *Taq* polymerase activity, and thus is the extension temperature. The T_m of an oligonucleotide is based on length and GC%, and calculated using the relationship $T_m = 69.3 + 0.41(GC\%) - 650/\text{oligo length}$. Higher GC contents help in allowing higher primer annealing temperatures to achieve specific annealing of primers to the template.

6. Internal secondary structure within the primers should be avoided. The presence of secondary structures within oligonucleotides can be verified, using commercially available software.

7. It is critical that primers do not contain any complementary sequences with each other. Complementary sequences must be less than two to three bases long especially at the 3' end. This may otherwise promote primer dimer artifacts during amplification. These artifacts occur when the primers anneal to each other at the

3' ends (due to complementary sequences) resulting in staggered 3' OH ends which then acts as the site for *Taq* polymerase activity. Since primers are quantitatively in excess in an amplification reaction, primers can anneal with one another, and primer dimer amplifications can predominate and result in reduced or total failure of amplification of the desired sequence. Primer dimers are short duplex products that vary in length from that of a single primer, to a fragment approaching the sum of the two primers.
8. It is advisable to use oligonucleotide primers of the highest chemical purity.

34–2.2 Special Apparatus and Reagents

Special Apparatus
1. DNA temperature cycler.

Reagents
1. Deoxynucleotide pool (dATP, dGTP, dCTP, and dTTP) 200 μ*M* of each. The nucleotides are available commercially in a lyophilized form. They are resuspended in sterile HPLC quality water.
2. 1 *M* potassium chloride
3. 1 *M* Tris-HCl (pH 8.3)
4. 15 to 100 m*M* solutions of $MgCl_2$
5. Primers (0.1 μg/μL)
6. Ampli*Taq* DNA polymerase (0.5 μL of 5 units/μL)

34–2.3 Procedures

A standard amplification protocol initially tests the efficacy of a newly designed and synthesized oligonucleotide primer pair to achieve significant amplification of the desired target sequence. This would be the starting set of conditions that can later be optimized to enhance amplification. Commercial PCR amplification kits are available (e.g., Perkin Elmer Cetus Corp., Norwalk, CT).

1. It is strongly advisable not to initially attempt amplification of sequences directly from environmental samples. Optimization of amplification succeeds when conditions are determined for pure cultures using either $CsCl_2$ or phenol-extracted DNA samples. Amplifications from environmental samples are usually successful after initial optimization trials in pure culture have been conducted.

2. It is necessary to initially quantify the target DNA for the optimization trials. Very often, it is phenol extracted and ethanol precipitated during genomic or plasmid DNA preparations. It could also be target DNA in crude cell lysates. In a standard reaction, target DNA concentration in the range of ng to pg quantities, or 10^7 colony-forming units (CFU) are used.

3. Since PCR amplification is achieved by DNA polymerization, a deoxynucleotide pool (dATP, dGTP, dCTP, and dTTP) is necessary. 200 μM of each nucleotide is provided for amplification.

4. Magnesium ion concentration is critical for enzyme efficiency and fidelity. An optimum exists between the 1 to 10 mM range. In the standard protocol, Mg^{2+} concentration is initially set at 1.5 mM.

5. The PCR Reaction Buffer (10X) contains 500 mM potassium chloride, 100 mM Tris-HCl (pH 8.3, 25 °C) and 15 mM $MgCl_2$. The buffer is available commercially or can be prepared in-house using HPLC quality water and autoclaved solutions. The Mg^{2+} may need optimization based on the primer pair being used and the target being amplified.

6. The primer concentration may be standardized in μg or in molar concentrations. Typically a standard protocol contains 0.1 to 1 μM of both primers, for example 0.5 to 2.0 μL of 0.1 μg/μL primer stocks.

7. Typical PCR amplifications that are conducted in 100 μL reactions, contain 2.0 to 2.5 Units of *Taq* or Ampli*Taq* DNA polymerase (Cetus Corp.). Thermostable DNA polymerases are currently available from different commercial vendors. The enzyme is added in 0.5 μL aliquots, or to achieve accuracy, master mixes are employed when multiple reactions are being set up.

Given below is a "standard" 100 μL PCR reaction that can routinely be employed.

	μL
H_2O	61.5–66.5
10X Reaction buffer	10
dNTPs (1.25 mM)	16
Primer #1 (0.1 μg/μL)	1.0
Primer #2 (0.1 μg/μL)	1.0
Template	5–10
Enzyme	0.5
Total volume	100

The various components of the reaction mixture are added to a 0.5-mL polypropylene tube. The quality of the tubes used in the reaction is important since uniform and efficient heat transfer through the tube is critical for the cycling conditions. The components are added to the tube sequentially using either a positive displacement pipette or PCR dedicated pipettes. To avoid nonspecific amplification products due to mispriming, the reaction is set up using ice baths. When a large number of reaction tubes are being set up, it is advisable to use so called *master mixes,* e.g., if 10 amplifications are being set up, 10-fold amounts of all reagents (except the template) are mixed together in a single tube in bulk, and then individually aliquoted into separate PCR reaction tubes. The template is finally added to the tubes. The use of master mixes reduces errors when dispensing extremely small amounts of the reagents, e.g., *Taq* polymerase to the individual tubes.

After the addition of the various reagents, a thin overlay of mineral oil is placed above the reaction mix. This prevents boiling and condensation of the reagents onto the sides and top of the tubes. The caps on the tubes are then sealed and the tubes placed in the temperature cycling instrument.

34–2.3.1 Polymerase Chain Reaction Thermal Cycling Profiles

A typical standard PCR amplification consists of repeated cycling of temperatures to achieve (i) separation of template double-stranded DNA, (ii) primers annealing to the single stranded template DNA, and (iii) extension from the 3′ ends of primers to achieve the synthesis of a copy of each of the single strands. In a typical protocol, a starting point for the temperature cycling during denaturation is 94 °C for 1.5 min, primer annealing at 37 °C for 1 min and primer extension step at 72 °C for 1 min. This is subsequently repeated for 25 cycles with the annealing temperature at 55 or 60 °C. At the end of the 25 cycles, all unextended target sequences are extended for 7 to 10 min at 72 °C. At the end of the amplification cycles, the enzyme reaction is stopped by incubation at 4 °C.

34–3 OPTIMIZATION OF AMPLIFICATION

Optimization of PCR amplification of a target sequence using a primer pair is often successfully achieved when purified target sequences are initially employed. Since PCR is a relatively complicated and incompletely understood biochemical reaction, kinetic reactions among the individual components can determine the outcome and quality of the reaction (Saiki, 1989). It is thus essential that during optimization trials, each of the parameters such as annealing temperatures, temperature holding times, Mg^{2+} concentration, primer amounts etc., be individually tested and varied one parameter at a time. This will help ensure results that are predictable and reproducible.

34–3.1 Amplification Cycles

If a "standard" 25-cycle amplification does not result in detectable amplification products, just increasing the number of cycles alone is unlikely to result in product formation. Very often, increasing the number of cycles beyond the 25 will only result in a small increase in the quantity of a product already formed. Product formation is not indefinitely proportional to the number of amplification cycles. Beyond 20 to 25 cycles, a "plateau" effect is seen due to factors such as depletion of nucleotides or primers, or enzyme inactivation. The success of an amplification is usually determined within the initial 10 cycles, beyond which, only an increase in the product already formed is seen. However, if the conditions have been optimized for a given primer pair, and the primers are being tested on an environmental sample, modified protocols such as "double PCR" or "booster PCR" or even increasing the number of cycles may result in a successful amplification.

34–3.2 Primer Concentration

Increasing or decreasing primer amounts in a sequential order have been shown to result in a significant improvement in amplification. It should be noted that often the reaction fails due to inherent defects in the chosen primer sequence and structure, rather than due to primer amounts. Optimal primer amounts are generally close to the standard amount. This is because in the standard reaction mix, the primers are in molar excess over normal template concentration ranges. Optimizing primer amounts have also been reported to be responsible for achieving specificity of amplification (Way et al., 1992).

34–3.3 Mg^{2+} Concentration

Taq DNA polymerase is a Mg^{2+}-dependent enzyme, and since template DNA, primer oligomers and dNTPs can sequester Mg^{2+} ions, the concentration of free Mg^{2+} in the reaction is critical. Specific PCR amplifications require specific concentrations of Mg ions. The optimal Mg^{2+} concentration is often within the 1 to 10 mM range when using the GeneAmp PCR buffer (GeneAmp Kit, Perkin-Elmer, Norwalk, CT) (Williams, 1989). Amplification trials using Mg^{2+} concentrations over the 1.5 to 4 mM range will help in determining the optimum concentration for the particular primer pair and template. Products may fail to be formed due to lack of free Mg^{2+} while excess Mg^{2+} may result in nonspecific product formation. Amplification of sequences from environmental samples may also require further adjustment of Mg concentrations, since samples such as soil or sludge tend to chelate free Mg^{2+}.

34–3.4 Temperature Cycling Parameters

The polymerase-mediated reaction is based on sequential heating and cooling that results in an initial separation of the target DNA into two single strands, followed by the annealing of the primers onto specific regions of the template strands and further extension of these primers. Success or failure of a specific amplification, therefore, depends on the efficiency of strand separation, primer annealing, and primer extension steps.

34–3.4.1 Strand Separation

Amplification can succeed only if the double-stranded DNA template can be separated out as two single strands to facilitate annealing of the primers. This denaturation step is usually at 94 °C for 1 min. However, templates that are long and GC rich need higher temperatures and longer denaturation times.

Denaturation conditions of 94 °C for at least 2 min are required for templates obtained from community DNA preparations (Pepper & Pillai, 1992, unpublished data). Extended extension times are also needed for environmental samples. Optimizing the length of denaturation usually provides significant results. Exposure to denaturation conditions should be limited as much as possible to preserve polymerase activity for the entire 25 cycles.

34–3.4.2 Primer Annealing

This is an important step in the overall amplification reaction. The annealing temperature determines the success of the amplification in terms of specificity and sensitivity of the amplification. Specificity of the reaction can be achieved by increasing the annealing temperature. The degree of flexibility in adjusting the annealing temperature depends on the percentage GC richness of the template. Optimizing the annealing temperature is usually done in increments of 2 °C, from a starting temperature of 40 to 42 °C. The annealing time is usually 1 min and can be adjusted if needed, after adjusting the temperature.

34–3.4.3 Primer Extension

A good starting point for optimizing this step is 72 °C for 1 min. Long target sequences (> 1 kb) requires longer extension times. *Taq* polymerase has a broad optimum range of 70 to 80 °C. When annealing is at 60 °C or greater, the annealing and extension steps can be combined together to what is termed as a two step PCR (since only denaturation and annealing/extension temperatures are used). Such an amplification strategy has been found to successfully achieve specificity of amplification (Neilson et al., 1992). This strategy also shortens the amplification protocol.

34–3.4.4 Ramp Times

The time taken to reach one temperature setting to another is termed the *ramp time*. Most commercial vendors of temperature-cycling instruments guarantee rapid and adjustable ramp times and that the displayed temperature is the temperature within the PCR reaction tube. Most often, the default settings for the ramp times on these instruments are sufficient to obtain the necessary strand separation, annealing and extension conditions. Shortening the ramp times is most often done to shorten the overall amplification process rather than as an optimizing strategy. It should be noted that the ramp times should be long enough to ensure that the template and primers are subjected to the necessary temperature conditions.

34–3.5 Nucleotide Concentrations

Since deoxynucleotides tend to bind Mg^{2+}, the concentration of dNTPs determine the availability of free Mg^{2+} in the PCR reaction. Thus, any increase in the total dNTP concentration may necessitate a corresponding increase in the Mg^{2+} concentration.

34–3.6 *Taq* Polymerase

Native *Taq* polymerase or the recombinant Ampli*Taq* have been successfully used for a variety of PCR applications. Two to 2.5 units of *Taq* polymerase are sufficient for most applications. The enzyme can be added individually to each reaction tube or a "master mix" can be prepared for several reactions, thereby avoiding errors in dispensing accurate enzyme quantities and also increasing the speed of setting up the reaction tubes.

Often, in soil samples, it is necessary to lyse cells by heating at 98 °C directly in the reaction buffer along with the other components of the reaction (Pillai et al., 1991). In such instances, the enzyme is not added initially (since this temperature can denature the enzyme), but rather added after boiling and cooling, just before the tubes are placed in the temperature cycler. The *Taq* polymerase concentration in the standard reaction mix needs optimizing especially when amplifying sequences in the presence of possible interfering substances such as soil colloids and plant tissues.

Ordering supplies such as PCR reaction tubes, HPLC grade water, nucleotides, and other PCR reagents from the same vender each time helps to maintain reproducibility of amplification. This is important since the quality of the PCR reaction tubes in terms of heat conductivity and the thermostability of the reagents are extremely critical for the temperature-based DNA amplification reaction.

34–4 IDENTIFICATION OF AMPLIFIED PRODUCTS

It is necessary that the products of an amplification reaction be identified and confirmed to correspond to the predicted products. There are three recognized methods to identify the PCR reaction products: (i) gel electrophoresis using either an agarose or polyacrylamide gels (see chapter 29 in this book), (ii) detection using specific DNA probes (see chapter 32 in this book) or (iii) HPLC separations.

34–5 QUALITY CONTROL

The extreme sensitivity of PCR allows amplification of even a few molecules of a specific DNA sequence. Unfortunately, this sensitivity is unforgiving and can result in amplification of DNA from contaminating sources rather than from the sample itself. Therefore, laboratories using PCR on a routine basis must establish a quality control program that is stringent enough to avoid false positives, and yet not so stringent as to produce false negatives.

False positives arise from contamination but can be reduced by the following:

1. Pre- and post-PCR stations. To reduce carry over of amplified DNA, samples should be prepared at a pre-PCR station in a separate room or far away from where amplified products are analyzed. A different post-PCR station is used to analyze products via electrophoresis or gene probes. The single most important step in avoiding contamination is to completely separate PCR preparation from anything to do with the amplified PCR product.
2. If possible dedicate sets of unique supplies and pipetting devices for each set of primers.
3. Autoclave buffer solutions and use HPLC grade water. Note that primers, dNTPs, and *Taq* DNA polymerase cannot be autoclaved.
4. Aliquot reagents to minimize the number of repeated samplings from given reagents. Label aliquots so that if contamination occurs, it can be traced.
5. Use disposable gloves and change frequently, particularly between pre- and post-PCR stations.
6. Spin down tubes prior to opening to remove sample from tube caps and walls.
7. Use positive displacement pipettes to avoid contamination via aerosols.
8. Premix reagents before dividing into aliquots, i.e., use master mixes. Pipette reagents for a no DNA negative control last.
9. Add all reagents before sample DNA, i.e., add DNA last, and cap each tube before proceeding to the next sample.
10. A method called *hot start* can also significantly reduce preamplification mispriming. This method involves setting up the reaction at approximately 80 °C without one critical component such as the *Taq* polymerase. The reaction is heated to 94 to 95 °C to denature the template and the reaction cooled to around 80 °C. In doing so, the chance for the primers to anneal to nonspecific regions on the template is reduced and further, there will be no extension due to the missing *Taq* polymerase.
11. Use the minimum number of PCR cycles possible for a given sample.

Contamination can be checked by the use of negative controls that are required with every assay. Two types of negative control are needed including a negative control that contains no added DNA as template. The second negative control should contain a DNA from an organism closely related to the correct target organism, but which does not contain the target DNA sequence. For example, whole cells of *E. coli* can be used as a negative control for primers that amplify *Salmonella* DNA.

Nonspecific amplification products occur when primers anneal to sites other than target sites. These products will be of a different size than the target amplification product, and thus can be differentiated from the correct amplification product on an agarose gel stained with ethidium bromide.

False negatives can also be problematic, and occur when amplification of the correct target sequence fails for any reason. For this reason, positive controls can be used to validate the primers and amplification conditions. However, positive controls must be used judiciously, since they can contribute to contamination and subsequent false positives.

34-6 SPECIFICITY OF AMPLIFICATION

The degree of specificity of primers used in PCR can intentionally be varied. If highly specific primers are required to amplify specific DNA, then unique DNA sequences must be chosen as the target for amplification, e.g., a primer pair that allows only amplification of *Salmonella* spp. specific DNA, but not *E. coli* specific DNA. Since these two groups of organisms are closely related, the design of these primers is critical to distinguish between the species of the two genera. However, if the detection of all or most bacterial DNAs in an environmental sample is required (e.g., in a bioassay) then primers termed *universal* primers are designed that allow amplification of a conserved DNA sequence present in all bacteria.

The degree of specificity can be varied by primer design, and also by changing the annealing temperature. In general, as the annealing temperature is raised, the number of base pair mismatches allowable for hybridization decreases, increasing the specificity of amplification. Increasing the annealing temperature from 50 to 55 °C will often decrease nonspecific amplification. However, along with increased specificity, there is an associated decrease in sensitivity. The maximum allowable annealing temperature is generally 10 °C less than the melting temperature which depends on GC%.

Primers can sometimes result in several different amplification products as well as an amplification product specific to the target DNA. The amplification product from the correct target sequence can be identified by appropriate size markers of standard DNA. The correct amplification product can also be identified by the use of end-labelled gene probes specific to an internal region of the amplified product. Nonspecific amplification products can arise in many ways. These include products that are due to the primers annealing to incorrect regions of the template, incomplete amplification or truncated product formation. Such nonspecific products generally arise at low-annealing temperature conditions, or when the extension time is not long enough.

34–7 SENSITIVITY OF AMPLIFICATION

Sensitivity of amplification is important when a given DNA sequence is in an environmental sample, but in low concentration or copy number. In this instance, the amplification protocol must be sensitive enough to detect the target DNA sequence and not result in a false negative. The level of sensitivity must be defined so that a negative result can be quantified. For example, current EPA drinking water standards necessitate < 1 *Salmonella* per 100 mL be detectable, therefore the use of PCR to detect *Salmonella* must somehow be capable of meeting these standards. Sensitivity can be increased by: (i) optimization of amplification, (ii) increasing the number of cycles, or (iii) concentration of the amount of target DNA in an environmental sample, for example, by filtration of a large volume of water followed by elution of bacterial cells into a smaller final volume.

Sensitivity can be evaluated in terms of whole cell lysates or pure genomic DNA preparations. In terms of whole cells, target bacteria are grown to late log phase in broth medium, serially diluted, and plated on agar to determine the number of CFUs. Concurrently, 500 µL of each dilution is added to a PCR reaction tube and centrifuged for 10 min at 14 000 rpm. All but 10 µL of the supernatant is discarded and the resulting pellet resuspended in 89.5 µL of reaction mix lacking enzyme. Cells are boiled for 10 min and 0.5 µL of *Taq* polymerase added after tubes are cooled to room temperature. PCR amplification is then conducted as described previously. Sensitivity can also be determined in terms of pure genomic DNA by using 10-fold dilutions of stock DNA preparations as a template for PCR. DNA concentration in the stock preparations are quantified using A_{260} absorbance values.

Sensitivity in terms of whole cells can be misleading since the total number of copies of a target sequence is always greater than the number of CFUs. Broth cultures inevitably contain dead or lysed cells with target sequences and each viable cell is likely to contain multiple copies of each genome (Krawiec & Riley, 1990). Therefore, equating sensitivity with CFUs tends to overestimate the actual sensitivity of the method. Sensitivity also depends on the method used to detect the amplified product. Using ethidium bromide staining of DNA PCR can often detect 10^3 to 10^4 CFUs, whereas the use of ^{32}P-labelled gene probes usually increases the sensitivity by two orders of magnitude. Using PCR in soil, a sensitivity of one cell per gram of soil has been reported (Steffan & Atlas, 1988; Pillai et al., 1991).

In terms of pure DNA, 100 ag (10^{-18}g) of a 179-bp fragment has been amplified (Josephson et al., 1991). Assuming a total genome of 4×10^6 bp which is equivalent to approximately 9 fg of DNA, the target amplification product of 179 bp is equivalent to $(179/4 \times 10^6) \times 9 \times 10^{-15}$ g or 0.4 ag. Therefore, one copy of the target DNA represents 0.4 ag and the sensitivity of detection in this study was approximately 250 copies (100 ag).

Overall, the issue of sensitivity is complex and must be evaluated for each individual set of primers. Such an evaluation is critical if PCR amplifications are to be used for diagnostic purposes.

34–8 APPLICATIONS IN ENVIRONMENTAL MICROBIOLOGY

34–8.1 Detection of Specific DNA Target Sequences in Soil Samples

Two general approaches are available that allow PCR amplification of DNA derived from soil samples. The first involves methods for the direct isolation and purification from soil samples of microbial community DNA. Bacterial cells are lysed directly in the soil matrix, followed by isolation and purification of the DNA. Protocols for direct isolation of DNA are described in chapter 35 in this book. Once a pure sample of DNA is obtained, PCR amplification is carried out as described previously.

The second approach involves extraction of all bacterial cells from a soil sample followed by separation of cells from soil colloids and PCR amplification. Two extraction methodologies are detailed below.

34–8.1.1 Extraction of Cells from Soil Using $CaCl_2$ and Sucrose Density Centrifugation Procedure (Pillai et al., 1991)

Special Apparatus
1. Sorvall SS-34 rotor.
2. Beckman J-6B centrifuge.
3. DNA temperature cycler.

Reagents
1. $CaCl_2$ (1%) sterile solution prepared using anhydrous $CaCl_2$ pellets.
2. Sucrose solution (1.33 g/mL). Solution prepared by dissolving 855 g of sucrose in 450 mL of water. Solution sterilized by autoclaving at 121 °C for 5 min.
3. PCR amplification kit.

34–8.1.1.1 Procedures
Extraction of Cells:
1. Nine milliliters of 1% $CaCl_2$ is added to 1 g of soil subsamples containing variable number of cells. The soil-$CaCl_2$ slurry is vigorously vortexed for 1 min and then allowed to settle for 1 h.
2. The upper fraction is concentrated to 5 mL by centrifugation (7650 g, 5 min) and pipetted into a 50-mL polypropylene centrifuge tube (Becton Dickinson, NJ) containing 10 mL of (1.33 g/mL) sucrose.
3. The soil-cell sucrose slurry is vigorously vortexed for 1 min. Fifteen milliliters of (1.33 g/mL) sucrose is carefully layered underneath the slurry.
4. This biphasic gradient is centrifuged at 750 g for 10 min.
5. The clear dilute sucrose upper fraction (~12 mL) that contains the bacterial cells is pipetted into a 40-mL polyallomer centrifuge tube, diluted with water, and centrifuged at 12 000 g for 20 min at 10 °C in a SS-34 rotor.

6. The bacterial cells along with soil colloids that had a similar density to the cells appear as a pellet. This pellet is directly used as the template for the PCR amplifications.

Double PCR Amplification

1. The pellet obtained after the extraction process contains bacterial cells, soil colloids, and perhaps free-naked DNA adsorbed to soil colloids.
2. The pellet is resuspended in 30 μL of H_2O and added to the PCR reaction mix containing the reaction buffer, nucleotides (Perkin Elmer Cetus, Norwalk, CT) and primers.
3. The reaction mix is heated to 98 °C in the DNA thermal cycler (Perkin Elmer Cetus, Norwalk, CT) for 10 min to lyse the cells, and subsequently cooled to room temperature during which primers anneal to their target sequences.
4. *AmpliTaq* DNA polymerase (Perkin Elmer Cetus) is added and 25 cycles of PCR performed using the DNA thermal cycler.
5. At the end of 25 cycles, 10 μL of the amplified product is added to a fresh reaction mix and further amplified for 25 cycles, thus performing a "double" PCR protocol.
6. The amplified products are detected using agarose gel electrophoresis or specific gene probes.

34–8.1.2 Extraction of Bacterial Cells Using Blending and Differential Centrifugation (Steffan & Atlas, 1988)

Special Apparatus
1. Waring blender.
2. GSA rotor.
3. Ultracentrifuge.

34–8.1.2.1 Reagents
1. 0.1 M Sodium phosphate buffer (pH 4.5).
2. 20% sodium dodecylsulfate (SDS).
3. 0.1% sodium metahexaphosphate.
4. 0.1% sodium pyrophosphate.
5. Chrombach buffer (0.33 M Tris hydrochloride, 0.001 M EDTA, and pH 8.0).
6. Lysozyme (5 mg/mL).
7. Solid ammonium acetate (2.5 M).
8. Cesium chloride (10 mg/mL).
9. TE buffer (10 mM Tris hydrochloride, 1 mM EDTA, and pH 8.0).

34–8.1.2.2 Procedure
1. One hundred grams of sediment is suspended in 300 mL of 0.1 M Sodium phosphate buffer (pH 4.5) and homogenized in a Waring blender. Samples are blended at medium speed three times for 1 min each, with 1 min cooling on ice between each blending cycle.

2. Two milliliters of 20% SDS is added to each sample, and the samples blended for an additional 5 s. The samples are placed on ice, the foaming allowed to settle for 5 min, and the samples transferred to 250-mL centrifuge bottles.

3. The bottles are shaken by hand for 1 min and centrifuged in a GSA rotor (Sorvall RS-5 centrifuge) for 10 min at 1000 g and 10 °C. The supernatants from each individual replicate are pooled in an Erlenmeyer flask and maintained on ice until further centrifugations. The pellet can be washed again with the phosphate buffer as before but without further addition of SDS. The supernatants are maintained on ice between steps.

4. The pooled supernatant is centrifuged at 10 000 g at 10 °C for 30 min to collect the bacterial cells. The pellet is suspended in 200 mL of 0.1% sodium meta hexaphosphate—0.1% sodium pyrophosphate at 5 °C.

5. The samples are shaken by hand for 1 min and centrifuged at 10 000 g for 30 min at 10 °C. These washings help to reduce organic particulates.

6. The cell pellet is finally washed in Chrombach buffer and centrifuged as above. The final pellet is transferred to a 50-mL centrifuge tube using Chrombach buffer and the volume adjusted to 25 mL.

7. The cell pellet is vortexed vigorously. Lysozyme (final concentration 5 mg/mL) is added and incubated for 2 h at 37 °C. The suspension is heated to 60 °C, and SDS is added to a final concentration of 1% and incubated for 10 min.

8. The suspension is cooled on ice for 2 h and centrifuged at 12 000 g for 20 min in a SS-34 rotor at 5 °C. The supernatant is transferred to a fresh tube and the pellet washed with Chrombach buffer and centrifuged as above. The supernatants are pooled.

9. The DNA in the cell lysate is purified using solid ammonium acetate at a final concentration of 2.5 M, and immediately centrifuged at 12 000 g for 5 °C for 30 min.

10. The DNA in the supernatant is precipitated by 2.5 volumes of ice-cold 95% ethanol and incubated at −70 °C for 1 h. The precipitated DNA is cesium chloride purified as described in chapter 31 in this book).

34–8.2 Detection of Marker Enzyme Synthesis for Bioassay Studies

Bioassay studies, to monitor the effects of manmade toxic chemicals and natural toxins on microbial survival and metabolism can also be done using PCR amplification. To perform such studies, a specific marker enzyme that is expressed or repressed based on the levels of the toxic agent is initially identified. The expression or the lack of expression is detected using a RNA-PCR amplification strategy. Since the *Taq* DNA polymerase will not amplify the expressed mRNA, the viral enzyme reverse transcriptase is initially employed to effect cDNA synthesis from RNA using either random antisense primers or 3' end specific antisense primers. This cDNA species is PCR amplified using the specific primers to detect expression of the mRNA species. Such mRNA can also be directly extracted from a soil sample (Tsai et al., 1991) and detected using the RNA-PCR protocol.

34–8.3 Evolutionary and Biodiversity Studies

The evolutionary or phylogenetic relatedness of bacterial genera and species within an environmental sample can be determined using ribosomal RNA (rRNA) sequence information and PCR amplifications. RNA-PCR amplification directly using rRNA species is relatively sensitive since there are multiple copies of rRNA. PCR amplifications can also be achieved by designing primers that flank the rRNA sequence on the DNA (rDNA). The ribosomes are cellular structures that are directly involved in protein synthesis. They are made up of two subunits in both prokaryotic and eukaryotic organisms. In prokaryotes, the ribosome is made up of a 50s and 30s subunit. The 30s subunit consists of the 16S rRNA associated with about 21 proteins. The sequence variation in the 16S rRNA can be used to genetically differentiate bacterial species or genera (Weisburg et al., 1991). There are regions of variability in the 16S rRNA among the same species. Using primers that flank these coding regions on the DNA (rDNA), or using random hexamers and reverse transcriptase for cDNA synthesis from rRNA, PCR amplification of defined regions from the different species can be achieved. Sequence variation among bacterial species or genera can be detected by either PCR product length, or by sequencing the PCR product.

34–8.4 Horizontal Gene Transfer and Genomic Rearrangements

In natural environments such as soils, horizontal gene transfer can occur by a variety of mechanisms such as transformation, conjugation, and transduction (Trevors et al., 1987). Conjugative transfer can be mediated through *tra* genes present on F-factor plasmids or through mobilization (Davis et al., 1980). Such gene transfers can be detected in an environmental sample if primers are designed to flank portions of the *tra* gene along with the sequence that may have been transferred. Detecting such an event needs careful thought to the various scenarios that might occur during a transfer event such as orientation of the transferred sequence, primers, and host sequences.

A genomic rearrangement such as a translocation in a culture obtained from an environment can also be detected using PCR amplifications. To achieve the normal exponential increase in amplification, both primers should direct synthesis in a 3' direction down both complementary strands in each amplification cycle. If only one primer is used, a linear increase in

the target sequence would be expected. A predicted rearrangement can be detected by using two separate primers, each flanking the rearrangement. If the rearrangement had indeed occurred, there would be an exponential increase in the product since both the primers would direct synthesis in the 3′ direction. If no rearrangement has occurred, there would only be a linear increase. Using agarose gel electrophoresis, the band of the predicted size would confirm the rearrangement.

REFERENCES

Bej, A.K., R.J. Steffan, J. Dicesare, L. Haff, and R.M. Atlas. 1990. Detection of coliform bacteria in water by polymerase chain reaction and gene probes. Appl. Environ. Microbiol. 56:307–314.

Davis, B.D., R. Dulbeco, H.N. Eisen, and H.S. Ginsberg. 1980. Microbiology. 3rd ed. Harper and Row Publ., Philadelphia.

Glukhov, A.I., S.A. Gronrdeev, S.V. Vinogradov, V.I. Kiselev, V.M. Kramarov, O.I. Kisseler, and E.S. Severin. 1990. Amplification of DNA sequences of Epstein-Barr and human immuno deficiency viruses using DNA-polymerase from *Thermus thermophilous*. Molec. Cell Probes 4:6:435–443.

Hill, W.E., S.P. Keasler, M.W. Trucksess, P. Feng, C.A. Keysner, and K.A. Lampell. 1991. Polymerase chain reaction identification of *Vibrio vulnificus* in artificially contaminated oysters. Appl. Environ. Microbiol. 57:707–711.

Innis, M.A., K.B. Myambo, D.H. Gelfand, and M.D. Brow. 1988. DNA sequencing with *Thermus aquaticus* DNA polymerase and direct sequencing of polymerase chain reaction amplified DNA. Proc. Natl. Acad. Sci. USA 85:9436–9440.

Josephson, K.L., S.D. Pillai, J. Way, C.P. Gerba, and I.L. Pepper. 1991. Detection of fecal coliforms in soil by PCR and DNA:DNA hybridizations. Soil Sci. Soc. Am. J. 55:1326–1332.

Krawiec, S., and M. Riley. 1990. Organization of the bacterial chromosome. Microbiol. Rev. 54:502–539.

Li, H.H., U.B. Gyllenstein, X.F. Cui, R.F. Saiki, H.A. Erlich, and N. Arnheim. 1988. Amplification and analysis of DNA sequences in single human sperm and diploid cells. Nature (London) 335:414–417.

Mahbubani, M.H., A.K. Bej, R.D. Miller, and R.M. Atlas. 1991a. Detection of bacterial mRNA using polymerase chain reaction. Biotechniques 10:48–49.

Mahbubani, M.H., A.K. Bej, M. Pealin, F.W. Schaeffer, III, W. Jakubowski, and R.M. Atlas. 1991b. Detection of *Giardia* cysts using the polymerase chain reaction and distinguishing live from dead cysts. Appl. Environ. Microbiol. 57:3456–3461.

Mullis, K.B., and F.A. Faloona. 1987. Specific synthesis of DNA in vitro via a polymerase catalyzed chain reaction. Meth. Enzymol. 255:335–350.

Neilson, J.W., K.L. Josephson, S.D. Pillai, and I.L. Pepper. 1992. PCR and gene probe detection of the 2,4-dichlorophenoxyacetic acid degradation plasmid, pJP4. Appl. Environ. Microbiol. 58:1271–1275.

Pillai, S.D., K.L. Josephson, R.L. Bailey, C.P. Gerba, and I.L. Pepper. 1991. A rapid method for processing soil samples for PCR amplification of specific gene sequences. Appl. Environ. Microbiol. 57:2283–2286.

Rose, S.D. 1990. RNA PCR. An application kit. Am. Biol. Lab. October 31–33.

Roszak, D.B., and R.R. Colwell. 1987. Survival strategies of bacteria in the natural environment. Microbiol. Rev. 51:365–379.

Saiki, R.K. 1989. The design and optimization of the PCR. p. 7–16. *In* H.A. Erlich (ed.) PCR technology principles and applications for DNA amplification.

Saiki, R.K., S. Scarf, F.A. Faloona, K.B. Mullis, G.T. Hoen, H.A. Erlich, and N. Arnheim. 1985. Enzymatic amplification of beta-globin genomic sequences and restriction site analysis for diagnosis of sickle cell anemia. Science 230:1350–1354.

Steffan, R.J., and R.M. Atlas. 1988. DNA amplification to enhance detection of genetically engineered bacteria in environmental samples. Appl. Environ. Microbiol. 54:2185–2191.

Trevors, J.T., T. Barkay, and A.W. Bourquin. 1987. Gene transfer among bacteria in soil and aquatic environments: A review. Can. J. Microbiol. 33:191–198.

Tsai, Y.C., M.J. Park, and B.H. Olson. 1991. Rapid method for direct extraction of mRNA from seeded soils. Appl. Environ. Microbiol. 57:765–768.

Wang, A.M., M.V. Doyle, and D.F. Mark. 1989. Quantitation of mRNA by the polymerase chain reaction. Proc. Natl. Acad. Sci. 86:9717–9721.

Way, J.S., K.L. Josephson, S.D. Pillai, C.P. Gerba, and I.L. Pepper. 1993. Specific detection of *Salmonella* spp. by multiplex polymerase chain reaction. Appl. Environ. Microbiol. 59:1473–1479.

Welsh, J., and M. McClelland. 1990. Fingerprinting genomes using PCR with arbitrary primers. Nucleic Acid Res. 18:7213–7218.

Weisburg, W.G., S.M. Barns, D.A. Pelletier, and D.J. Lane. 1991. 16S ribosomal DNA amplification for phylogenetic study. J. Bacteriol. 173:697–703.

Williams, J.F. 1989. Optimization strategies for the polymerase chain reaction. Biotechniques 7:762–779.

Chapter 35

Isolation and Purification of Bacterial DNA from Soil

WILLIAM E. HOLBEN, *Michigan State University*

Recently, new methods for monitoring specific bacterial populations[1] in environmental samples have become available. Several of these methods employ the techniques of molecular biology to distinguish, enumerate, and monitor individual bacterial populations within a microbial community by the detection of DNA sequences specific to those populations using appropriate molecular probes. DNA-based detection of bacterial populations helps to overcome a major limitation of microbial ecology and soil microbiology; the difficult task of specifically monitoring an individual population of microbes in the environment, and in the presence of the entire microbial community. Such capabilities are essential to understanding the complex interactions between the environment, other microorganisms, and the population(s) of interest. Prior to the development of these methods, microbial ecologists generally monitored specific microbial populations using methods that included developing mutant derivatives that could be recovered on selective media (e.g., spontaneous antibiotic resistance), or polyclonal or monoclonal antibodies raised against individual populations that were conjugated to a fluorescent dye to facilitate detection by direct microscopic analysis. DNA-based detection of microbial populations thus represents a new tool to expand the capabilities of investigators to detect and quantify microorganisms in environmental samples.

Among the advantages of DNA-based microbial monitoring methods are that a particular DNA sequence is detected directly; thus gene expression is not required. Marker genes and selectable phenotypes are also not required. This is potentially important to microbial ecologists in that it obviates the need to demonstrate that genetic manipulations of organisms required for detection by alternate strategies have not compromised the

---

[1]For the purposes of this chapter, an individual population of bacteria is defined as a group of organisms having identical, or nearly identical, genotype (genetic makeup).

---

competitiveness of the organism. Thus, DNA-based detection methodologies can be used for monitoring either genetically engineered or wildtype indigenous populations. In fact, little information about the genetic make-up of the population of interest is required since highly specific probes for detection can be generated by simply subcloning random pieces of DNA from the organism of interest and screening for specificity (Salyers et al., 1983). Alternatively, the probe sequence can be based on regions of the rRNA gene(s) that are specific to the population of interest but can readily be identified based on adjacent, conserved regions (Barry et al., 1990). Another advantage of DNA-based detection is that bacterial growth is not required for detection, thus allowing nonculturable and nonviable populations to be detected. The importance of this aspect is clear when one considers that typically only about 0.1 to 1% of the bacteria present in a soil sample can readily be cultured under laboratory conditions (Fægri et al., 1977). Having no requirement for culturing also simplifies quantitative and comparative analyses since there is no subsequent increase in the population(s) of interest compared to the other microbial populations in the community and thus relative proportions between populations are maintained.

As with any other method for microbial detection, there are limitations to these molecular approaches. Limitations do not preclude the use of this or any other detection strategy, they simply need to be recognized and understood. This will help to avoid ambiguities and potential overinterpretation of data or overestimation of detection capabilities. For example, when using probes for specific functions (e.g., a certain catabolic activity), lack of hybridization signal can only be construed as lack of homology to the sequence used as probe, not absence of the activity since isofunctional enzymes can be encoded by non-homologous DNA sequences.

One potentially limiting aspect of DNA-based detection strategies is that protocols for DNA isolation, purification, and subsequent detection and enumeration are relatively sophisticated compared to simpler detection strategies such as culturing on selective media. Relatively few samples (usually 6–24) can be simultaneously processed. This might limit some ecological studies, particularly where a large number of variables are to be studied or where statistical analyses are required. The low-end sensitivity of DNA-based detection strategies is relatively poor. Without incorporating additional protocols that amplify either the target sequence (e.g., Steffan & Atlas, 1988; Neilson et al., 1992) or the probe signal, DNA probes can detect about 10^4 copies of target sequence per gram of soil (Holben et al., 1988). Since there are generally about 10^9 bacteria in a gram of surface soil, this represents the detection of populations that constitute about 1/100 000 of the total microbial community. This level of detection might be a limitation for some analyses (e.g., risk-assessment for engineered organisms), but suffices for many other kinds of investigations where one is interested in more numerically dominant populations. Recent improvements in DNA isolation protocols and subsequent analyses have made these procedures simpler and more generally useful than as originally published.

Two alternate strategies for the isolation of total bacterial community DNA from soil samples will be presented; the first is based on the fractionation of bacteria from soil prior to lysis while the second involves direct lysis of bacteria in the presence of the soil matrix. A comparison of these methodologies is made, and recommendations for the selection of either protocol, depending on sample characteristics and the experimental question being addressed, are given. Other chapters of this volume describe in detail some of the analyses and potential applications possible with bacterial community DNA. For an overview of this emerging DNA-based microbial detection technology, the reader is referred to Holben and Tiedje (1988), Sayler and Layton (1990), and Knight et al. (1992).

35-1 GENERAL CONSIDERATIONS

There are several considerations that will affect the recovery of total bacterial community DNA from the soil environment, regardless of which of the two DNA recovery methods are employed.

35-1.1 Biomass

The bacterial biomass present in the sample has an impact on the quantity of DNA recovered due to some practical limitations of the methodologies employed. Theoretically, one could recover bacterial DNA from each bacterium present if every important reaction and handling step operated at 100% efficiency. Then, the starting sample size could be adjusted to obtain the desired yield of DNA. This, of course, is not the case. For example, with the direct lysis protocol described here, the starting sample size has been scaled down to 10 g (compared to the original protocol of Ogram et al. [1987], where 100 g samples were used) to increase the number of samples that can be processed simultaneously. Using the direct lysis protocol outlined below, 6 to 10 μg of bacterial community DNA per gram of soil can routinely be obtained for a surface soil (A horizon) that contains about 10^9 bacteria per gram, for a yield of about 60 to 100 μg per sample. However, the number of bacteria per gram of soil varies with soil type and depth. At a depth of 1 m, where microbial populations typically range from 10^6 to 10^7 bacteria per gram, one would need to process on the order of 1 kg of soil to obtain a yield of 100 μg of bacterial community DNA. Samples of this size are not readily processed by the method outlined below. The procedures, as given, are intended for use with samples such as surface soils, sediments, or sludges with total bacterial counts in the range of 10^8 to 10^{10} bacteria per gram of material. Others have developed modifications of the direct lysis protocol to recover useable amounts of DNA from samples with low biomass such as aquifer material (S. Thiem,

1992, personal communication; Smith & Tiedje, 1992; A. Ogram, 1994, personal communication).

35–1.2 Organic/Humic Content of Soils

The amount of organic/humic matter in soil has a dramatic effect on the quality (purity) of the DNA obtained, particularly when the direct lysis method is employed. The humic materials present in soil have a similar molecular weight and net charge to DNA and, thus, are readily copurified. The bacterial fractionation procedure which first separates bacteria from the bulk soil prior to cell lysis reduces, but does not eliminate, this problem. Humic contaminants interfere with subsequent enzymatic digestions of DNA (Ogram et al., 1987; Holben et al., 1988; Steffan et al., 1988), and potentially other enzymatic reactions such as DNA polymerase and ligase. Humic contaminants also confound precise quantitation of the recovered DNA because they exhibit substantial absorbance of light at 260 nm, the measure of which is generally used to quantitate DNA. It is thus preferable to use soils of moderate organic/humic content for experiments involving the isolation of total bacterial community DNA. Although there may be a correspondingly lower biomass associated with such soils, it is usually sufficient to recover usable amounts of DNA from surface soils. For cases in which high organic content soils are used, methods for more precise quantitation of DNA in the presence of humic contaminants are available. One such method, which is relatively simple and requires no sophisticated equipment, is outlined later in this chapter.

35–1.3 Clay Content of Soils

Adsorption isotherms indicate that relatively large amounts of DNA can bind to pure clays, soils, and sediments (Greaves & Wilson, 1969; Ogram et al., 1988). The binding of DNA to clays in soil can have a profound impact on the amount of bacterial community DNA recovered. For example, in comparing DNA recovered from two soils with similar organic content (2.3 and 2.7%, respectively), but different clay content (8.1 and 48%, respectively), it was found that the yield of DNA from the high clay soil was only about 15 to 25% of that from the low clay soil despite both soils having similar bacterial viable counts (unpublished observation). The mechanism of binding of DNA to clay is not well understood. Studies with flavomononucleotide binding to smectite implicate the Fe^{3+} groups of clay and the phosphate groups of the mononucleotide (Mortland et al., 1984). However, attempts to block DNA binding to clays in soil by competition with excess phosphate, or altering the pH or ionic environment, have been largely unsuccessful (1992, unpublished data). Thus, no method for effectively blocking the DNA-binding sites of clays in soil, or for removing adsorbed DNA from clays (and thus enhancing DNA recovery) has yet been described. As was the case with high organic matter soils, it is best, if possible, to avoid using high clay soils in experiments where bacterial community DNA is to be recovered.

35-2 BACTERIAL FRACTIONATION APPROACH FOR RECOVERY OF BACTERIAL COMMUNITY DNA

35-2.1 Principles

The bacterial fractionation method involves the separation of bacterial cells from the bulk of the soil prior to cell lysis and recovery of bacterial community DNA. Briefly, soil particles, debris, fungal cells, and bacterial cells are brought into suspension by homogenization in the presence of buffer. Soil particles, fungal cells, and other debris are then removed by a low-speed centrifugation step that leaves the unattached bacterial cells in suspension. High-speed centrifugation is then performed to recover the bacterial cells. This combined low-speed, high-speed centrifugation method for the recovery of bacterial cells from soils was pioneered by Fægri et al. (1977) and is termed *differential centrifugation*. Generally, multiple rounds of homogenization and differential centrifugation are performed on the same soil or sediment sample to enhance the recovery of bacteria.

In the protocol described here, the bacterial fraction is lysed using a protocol that combines the salient features of lysis protocols for various groups of bacteria, including the removal of humic contaminants using polyvinylpolypyrrolidone (PVPP), digestion by lysozyme and pronase, and incubation at high temperature (Holben et al., 1988). Following cell lysis, bacterial DNA is recovered and purified by cesium chloride-ethidium bromide equilibrium density centrifugation and precipitation in ethanol. As described here, the method includes several modifications of the previously published protocol (Holben et al., 1988) including the deletion and combination of steps resulting in a shortened protocol that gives comparable yields and purity of DNA. The yield of DNA from 50 g of agricultural surface soil with 2.3% organic matter and 8.1% clay content is in the range of 50 to 100 μg, corresponding to 1 to 2 μg per gram of soil. This protocol allows bacterial community DNA to be purified from six 50-g soil samples in 3 to 4 d. With additional effort, it is possible to process up to 12 samples simultaneously. The bacterial fractionation and cell lysis portion of the protocol (i.e., "Day 1" of the procedure) is labor intensive but the subsequent purification of DNA is largely "hands-off" time.

35-2.1.1 Advantages

An advantage of the bacterial fractionation method is that DNA recovered from soils, especially high organic content soils, tends to be less contaminated with humic materials than is DNA recovered by direct lysis since the bacteria are removed from the bulk of the soil prior to lysis. If the analysis of the bacterial community DNA isolated from a particular soil requires subsequent digestion with restriction endonucleases, and the organic content of the soil is high, isolation of DNA by bacterial fractionation may be required. The DNA recovered by this method has an average size of 50 kb (Holben et al., 1988) compared to DNA recovered by direct lysis which ranges from 30 to 40 kb in size (unpublished observation). This might be a consideration if the recovered DNA were to be used in experiments attempting to clone DNA fragments from the bacterial community DNA but is not as important when the DNA is to be used in hybridization experiments. The bacterial fractionation procedure recovers DNA only from bacteria (not fungal, protozoan, or free DNA) since the bacteria are separated from the soil prior to lysis and DNA recovery. Initially, this appeared to be an important feature of this protocol (Holben et al., 1988; Steffan et al., 1988), but more recently it appears that the direct lysis protocol also recovers primarily bacterial DNA (see below).

Originally (Holben et al., 1988), it was thought that DNA recovered by the bacterial fractionation method was representative of the entire bacterial community based on indirect evidence (Bakken, 1985). However, more recent data from this laboratory indicate that more recently grown cells are preferentially recovered from the soil environment (Holben et al., manuscript in preparation). Bacterial fractionation may preferentially recover rapidly growing bacterial cells because they do not adhere as tightly to soil particles. This can be used to an advantage if the bacterial population of interest is rapidly growing; providing a fractionation of actively growing cells from those less active, thereby increasing the sensitivity of detection of the desired population.

Another advantage of the bacterial fractionation method is that it yields viable cells from the soil bacterial community in concentrated form (Holben et al., 1988) which can be used in other types of experiments requiring live bacteria. The bacterial fractionation protocol, as described here, recovers about 33% of the bacterial cells from sandy loam soil after three rounds of homogenization and differential centrifugation (Holben et al., 1988). Additional rounds of homogenization and centrifugation will yield a diminishing return of bacterial cells for the effort involved, but may be necessary to recover more tightly adhered bacterial cells.

35–2.1.2 Disadvantages

Perhaps the main disadvantage of this procedure is that it requires a significant amount of "hands-on" activity and time to process samples through the point of cell lysis (typically 6–8 h for six samples). Although there are no particularly sophisticated procedures or specialized equipment involved, care must be taken in handling samples to minimize DNA loss and maximize reproducibility of yield. Due to the amount of handling involved, and the limitations of centrifuge rotor configuration and other logistical considerations, it is usually practical for a single person to process six soil samples simultaneously for bacterial community DNA, although up to 12 samples per day can be accommodated. The bacterial fractionation protocol yields five to six times less DNA per gram of soil than does the direct lysis procedure. This can be an important consideration, particularly when the samples have low bacterial biomass.

As mentioned above, this procedure appears to preferentially recover DNA from rapidly growing and less tightly adhered bacteria. This phenomenon may confound attempts to quantify bacterial populations unless hybridization data are compared to independent measurements using alternate methodologies (Holben et al., 1992). This selectivity would be a disadvantage for experiments in which the objective is to obtain DNA representing the entire bacterial community, for example, to assess the diversity of organisms present in the community that contain a certain gene or other target sequence.

35–3 DIRECT LYSIS APPROACH FOR THE RECOVERY OF TOTAL BACTERIAL COMMUNITY DNA

35–3.1 Principles

In the direct lysis method, bacterial cells are lysed directly in the presence of the soil matrix. High temperature, high concentrations of detergent, and mechanical disruption using minute glass beads are employed for cell lysis using this method. Direct lysis of bacterial cells in soil for the recovery of total bacterial DNA was pioneered by Ogram et al. (1987) and has since been modified by several groups for simplification, increased sample throughput, or specific applications (Steffan et al., 1988; Hilger & Myrold, 1991; Tsai & Olson, 1991; Smith & Tiedje, 1992; this chapter). As originally described, one or two 100-g soil samples could be processed for DNA recovery over the course of 3 to 4 d and several DNA concentration/ precipitation steps were involved. In the protocol described here, the released DNA is subsequently isolated and purified by cesium chloride-ethidium bromide equilibrium density centrifugation and precipitation in ethanol. Improvements include that the time required to obtain purified DNA is 2 to 3 d (mostly "hands-off" time), the protocol is much less labor intensive, and greater numbers of samples (12–24) can be processed simultaneously.

35–3.1.1 Advantages

A major advantage of the direct lysis approach is that it is less labor intensive and faster than the bacterial fractionation method. The direct lysis protocol allows one person to readily process 8 to 12 samples simultaneously, and up to 24 samples with extra effort. The time required from the starting point of sieved soil samples to the initiation of cesium chloride gradient centrifugation is only 2 to 3 h. The direct lysis procedure results in higher yields of DNA, typically 60 to 100 μg per 10 g soil sample (6–10 μg per gram of soil). This is particularly important in samples having low biomass such as aquifer materials where the ability to recover sufficient amounts of DNA can be the limiting factor to success. Bacterial community

DNA recovered by direct lysis seems to better represent the bacterial community than DNA recovered by the bacterial fractionation procedure (Holben et al., manuscript in preparation). In fact, it appears that the direct lysis protocol outlined here approaches quantitative recovery of bacterial DNA since there are about 10^9 bacteria per gram of soil, each having a genome of about 9×10^{-15} g of DNA, resulting in a calculated quantitative yield of 9 μg DNA per gram of soil. It is also significant that essentially no intact bacterial cells can be found by microscopy in a soil sample following the lysis protocol. This is an important feature for experiments involving populations that are tightly adhered to particles, not rapidly growing, or when DNA representative of the entire bacterial community is of interest.

35–3.1.2 Disadvantages

Bacterial community DNA isolated by the direct lysis procedure is generally more contaminated with humic material than is DNA isolated from the same soil by the bacterial fractionation method. However, DNA purified by direct lysis from soils having moderate levels of humic acids is generally of sufficient purity for most subsequent manipulations involving digestion with restriction enzymes, denaturation, or hybridization. If particularly fastidious reactions, such as DNA amplification using the polymerase chain reaction (PCR) are to be performed, or if DNA isolated using this method is refractory to restriction digestion or other manipulations, further purification (e.g., by additional rounds of purification on cesium chloride-ethidium bromide gradients), or isolation of DNA by bacterial fractionation may be required. DNA isolated by direct lysis tends to be of lower molecular weight (i.e., the randomly sheared DNA fragments are, on the average, smaller) than DNA isolated by bacterial fractionation. Presumably, this reflects the more harsh conditions (i.e., higher temperatures and ionic detergents) and mechanical shearing of DNA imposed by this procedure. The average DNA fragment size is still more than 25 kb and thus is suitable for most procedures involving size-fractionation of restriction enzyme-digested DNA. It is worthwhile to assess the size range of DNA obtained by this method since excessive shearing will yield smaller DNA and less satisfactory results in hybridization analyses, or if attempting to clone DNA from the microbial community. Size range can be readily determined by agarose gel electrophoresis with DNA fragments of known size included as standards.

As mentioned above and elsewhere (Ogram et al., 1987; Holben et al., 1988; Steffan et al., 1988), it was thought that, in addition to bacterial DNA, the direct lysis procedure might recover DNA from fungi, protozoa, and cell-free DNA. Recent evidence suggests that these other potential sources of DNA in soil do not contribute significantly to the DNA obtained. Other investigators have made concerted attempts to obtain fungal DNA from soil samples with little success using the direct lysis method and similar approaches (D. Harris, 1992, personal communication). It seems

Table 35–1. Considerations and recommendations for selection of the appropriate protocol for isolating bacterial community DNA from soil.

Consideration	Example	Protocol	Reason(s)
Low biomass	Aquifer material or nutrient poor soil	DL†	Better DNA recovery and low humic matter
High purity of DNA	For PCR reactions	BF‡	Bacteria removed from soil before lysis
Large number (#) of samples	Multi-parameter or high replication	DL	Faster, larger sample #'s possible
High organic/humic content of samples	Rich soils, forest litter	BF	Bacteria removed from soil before lysis
Rapidly growing or loosely adhered orgs§	Addition of specific source of C	BF	Enriches for these orgs; > sensitivity
Tightly bound orgs	EPS¶ producers or trait of organism	DL	Lyses adhered cells
DNA best represents community	Assess diversity or community-level analyses	DL	More representative

† DL = direct lysis protocol.
‡ BF = bacterial fractionation protocol.
§ orgs = organisms.
¶ EPS = exopolysaccharide.

likely that the rigid fungal cell wall and the prevalence of "ghost" fungal hyphae containing no DNA account for the difficulty in isolating DNA from fungi. The population levels of protozoa in typical agricultural surface soils are about 10^5 per gram compared to 10^9 to 10^{10} bacteria. Thus, even accounting for the larger genome size of protozoa, they could not make a significant contribution to the total amount of DNA obtained. Free (extracellular) DNA in the agricultural soils used in this laboratory, if present, does not appear to be extracted by the direct lysis protocol. This is evidenced by the fact that initial extraction of the soil for DNA prior to lysis does not recover DNA; nor does it reduce the DNA yield obtained by direct lysis compared to soil not previously extracted. In sediment samples, the initial extraction of DNA prior to cell lysis recovered about 1 μg of DNA per gram of sediment compared to the recovery of 26 μg of DNA per gram of sediment following cell lysis (Ogram et al., 1987).

In summary, the direct lysis procedure is simpler, faster, and gives higher yields of DNA that is probably more representative of the total bacterial community present in the soil sample than DNA obtained by the bacterial fractionation method. On the other hand, DNA obtained by bacterial fractionation is of higher purity, larger molecular weight, and enriched for rapidly growing (and thus, the most active) populations. The method chosen for use thus depends on the nature of the experimental question, the requirements of the subsequent analyses to be performed with the bacterial community DNA, and the characteristics of the environmental sample. Some recommendations for the appropriate bacterial community DNA recovery method based on experimental goals or soil characteristics are given in Table 35–1.

35–4 BACTERIAL FRACTIONATION PROTOCOL

35–4.1 Materials

Note: The materials list is based on the simultaneous processing of six 50-g soil samples from the stage of sieved soil to the final cesium chloride gradient purification step. The list is based on the items used in this laboratory and does not constitute a commercial endorsement of any supplies or equipment. Reasonable substitutions for particular types of centrifuge rotors, tubes, and other materials and equipment that maintain, or reasonably approximate, the specified conditions are appropriate.

1. Six standard Waring blenders with 1.2 L (40 fl. oz.) glass jars.
2. Ice/water bath for each of the blender jars.
3. Twelve 250-mL centrifuge bottles (for use with a Sorvall GSA rotor).
4. Sorvall superspeed centrifuge (e.g., model RC5B) with an SS34 (8-place) or SA600 (12-place) rotor and a GSA rotor.
5. Six small paint brushes (to facilitate resuspending bacterial pellets).
6. Twelve 50-mL polycarbonate Oak Ridge tubes.
7. Vortex mixer.
8. Water bath at 37 °C.
9. Water bath at 65 °C.
10. Refractometer (to measure the initial density of the cesium chloride gradient for DNA purification. If a refractometer is not available, density can be determined by measuring the specific gravity of the solution).
11. Sorvall ultracentrifuge with TV865B rotor or equivalent.
12. Twelve ultracentrifuge tubes for the TV865B rotor.

35–4.2 Reagents

Note: The recipes presented here are for reagents required from the stage of sieved soil to the point of the final equilibrium gradient centrifugation step. After this point, the steps are common to both DNA isolation protocols and can be found in the section titled: "Fractionation of DNA Gradients, Final Purification and Quantitation of Bacterial Community DNA."

1. 10 × Winogradsky's Salt Solution (10 × WS)
   Provides an isotonic environment for bacterial cells during homogenization and differential centrifugation steps.
   for 2 L:

- Dissolve 5.0 g of K_2HPO_4 in 800 mL of distilled H_2O.
- Dissolve the following in (a separate) 800 mL of distilled H_2O:

$MgSO_4 \cdot 7\ H_2O$	5.0 g
NaCl	2.5 g
$Fe_2(SO_4)_3 \cdot H_2O$	50 mg
$MnSO_4 \cdot 4\ H_2O$	50 mg

- Combine the above, then adjust to pH 6.0 with concentrated HCl.
- Bring the final volume to 2 L with distilled H_2O.
- Before use, dilute 1:10 with distilled H_2O and then autoclave.

2. Homogenization solution[2]

   This solution provides an isotonic environment for bacterial cells and contains ascorbic acid as a reducing agent to prevent further oxidation (which results in polymerization) of humic compounds in the soil.

   Contains:

   1 × Winogradsky's salt solution

   0.2 M sodium ascorbate (added as powder to achieve 0.2 M)

   Note: If the bacterial fractionation protocol is being used to recover viable cells, it is recommended that Winogradsky's salt solution alone be used as the homogenization solution since sodium ascorbate appears to reduce the viability of cells during the fractionation process.

3. Acid-washed Polyvinylpolypyrrolidone (PVPP)

   This insoluble polymer complexes with humic acids in the homogenization stage removing them from the aqueous phase. The acid treatment of PVPP constitutes a pretreatment which optimizes this interaction.

   - Prepare 4 L of 3 M HCl. (Most stock concentrated HCl solutions are 12.1 N so *slowly* add 992.0 mL of concentrated HCl to 3008 mL of distilled H_2O in a 4-L beaker).
   - Slowly add 300 g of PVPP with stirring, cover beaker, and stir overnight.
   - Filter suspension through Miracloth or several layers of cheesecloth (use a large Buchner funnel and a 4-L vacuum flask).
   - Resuspend the PVPP in 4 L of distilled H_2O, mix for 1 h and again filter through Miracloth or cheesecloth.
   - Resuspend the PVPP in 4 L of 20 mM potassium phosphate buffer (pH 7.4) and mix for 1 to 2 h. Check the pH of the PVPP suspension with pH paper. The desired pH is 7.0.
   - Repeat filtrations of the suspension and washes in 20 mM phosphate buffer until the PVPP suspension has a pH of 7.0.
   - Following the final filtration, spread the PVPP on lab paper and let air dry overnight.

---

[2]This should be prepared just prior to use as the reducing power of the sodium ascorbate lessens with time.

4. TE ($T_{33}E_1$)

   Protects DNA by providing a buffered environment with EDTA present to chelate divalent cations that are required for the activity of nucleases which might be present.

   Contains:
     33 mM Tris, pH 8.0
     1 mM EDTA, pH 8.0

   for 2 L:
   - Combine the following in 1 L of distilled H_2O:
       66 mL of 1 M Tris, pH 8.0
       4 mL of 0.5 M EDTA, pH 8.0
   - bring volume to 2 L, then autoclave.

     Note: The disodium form of EDTA is preferred. At high concentrations EDTA will not go into solution until it approaches the appropriate pH.

5. 5 M sodium chloride (NaCl)

   Sodium chloride is used during the cell lysis stage as a pretreatment for cells with exopolysaccharide capsules to facilitate access of lysozyme to the cell wall for more efficient lysis.

   for 500 mL:
   - Add 146.1 g of NaCl to 400 mL distilled H_2O and dissolve with stirring.
   - Bring the volume to 500 mL, then autoclave.

6. 20% Sarkosyl

   This detergent disrupts the membranes of bacterial cells facilitating the release of DNA into solution during the cell lysis stage.

   for 100 mL:
   - Add 20 g of *n*-laurylsarcosine to 50 mL of distilled H_2O, mix (slight heating will help the sarkosyl go into solution).
   - Bring the volume to 100 mL, then autoclave.

7. Tris/sucrose/EDTA

   This solution provides an appropriate environment for bacterial cells that is buffered both for pH and for osmotic potential and inactivates endogenous nucleases (which degrade DNA) by chelating divalent cations that are required for their activity. The EDTA also serves to disrupt the outer membrane of gram-negative organisms allowing lysozyme freer access to the cell wall.

   Contains:
     50 mM Tris, pH 8.0
     0.75 M sucrose
     10 mM EDTA, pH 8.0

   for 250 mL:
   - Combine the following in 200 mL of distilled H_2O:
       12.5 mL of 1 M Tris, pH 8.0
       64.2 g of sucrose
       5.0 mL of 0.5 M EDTA, pH 8.0
   - Bring volume to 250 mL with distilled H_2O, then autoclave.

8. Lysozyme solution (40 mg/mL)
   Lysozyme enzymatically attacks the cell wall of bacteria allowing rupture of the cell membrane by detergent and the release of DNA into solution.
   for 5 mL:
   - Dissolve 200 mg of lysozyme (grade 1 from chicken egg white [Sigma #L6876]) in 5.0 mL of TE. Prepare on same day and store on ice until use.

9. Pronase E (10 mg/mL)
   Pronase E comprises a mixture of protein degrading enzymes that facilitate the rupture of bacterial cells in the lysis stage.
   for 5 mL:
   - Dissolve 50 mg of pronase (type XXV from *Streptomyces griseus* [Sigma #P6911]) in 5 mL of TE. Preincubate for 30 min at 37 °C prior to use to allow the proteases to inactivate any contaminating nucleases in the mixture.

10. Ethidium bromide (10 mg/mL)
    This DNA-binding dye intercalates into DNA molecules and imposes changes in the buoyant density of DNA causing it to form a discrete band in the cesium chloride equilibrium density gradients that can be fractionated and further purified.
    for 100 mL:
    - Dissolve 1 g of ethidium bromide in 100 mL of TE. Overnight mixing with a magnetic stirrer may be required.
      Note: ethidium bromide is a potent mutagen and should be handled with care.

11. Cesium chloride balance solution (R_f = 1.3885)
    This solution is used to bring cesium chloride gradient tubes to final volume and to balance the tubes prior to ultracentrifugation.
    for approximately 300 mL:
    - Add 250 g of finely ground cesium chloride (CsCl) to 250 mL of distilled H_2O (sterile) and mix by inversion until the CsCl is dissolved.
    - Add 12.5 mL of 10 mg/mL ethidium bromide and mix.
    - Check the refractive index (R_f) and adjust as necessary to achieve an R_f value of 1.3885 (this corresponds to a density of 1.58) by adding CsCl to increase the refractive index (density) or H_2O to decrease the refractive index.
      Note: As mentioned above, refractive index is a measure of the density of the solution. In lieu of using a refractometer the investigator can determine the specific gravity of the solution and adjust the density of the solution to 1.58.

35–4.3 Procedure

Note: The following protocol describes the steps involved from the stage of sieved soil to the point of the final equilibrium gradient centrifugation step. After this point, the steps are common to both DNA isolation protocols and can be found in the section titled: "Fractionation of DNA Gradients, Final Purification and Quantitation of Bacterial Community DNA."

1. Combine each 50-g soil sample with 200 mL of homogenization solution and 15 g of acid-washed PVPP in a blender jar.

2. Homogenize for three 1-min intervals with 1 min cooling in an ice/water bath between homogenizations.

3. Pour the homogenate into a 250-mL centrifuge bottle and pellet soil, fungi, and other debris by centrifugation in a Sorvall GSA rotor at 2500 rpm (640 × g) for 15 min at 4 °C.

4. Carefully pour the supernatant into a clean 250-mL centrifuge bottle and collect the bacterial fraction by centrifugation in a Sorvall GSA rotor at 12 000 rpm (14 740 × g) for 20 min at 4 °C.

5. Add 200 mL of homogenization buffer to the soil pellet and repeat the homogenization and differential centrifugation steps two more times (i.e., repeat steps 2–4 combining the bacterial pellets in step 4).

6. Wash the cells by carefully resuspending the cell pellet in 200 mL of TE using a small, clean paint brush. Collect the bacteria by centrifugation in a Sorvall GSA rotor at 12 000 rpm (14 740 × g) for 20 min at 4 °C.

7. Gently resuspend the cell pellet in 20.0 mL of TE (again using a paint brush), transfer the cell suspension to a 50-mL Oak Ridge tube, then add 5.0 mL of 5 M NaCl and 125 μL of 20% Sarkosyl and incubate at room temperature for 10 min.

8. Collect the cells by centrifugation in a Sorvall SS34 rotor at 12 000 rpm (11 220 × g) for 20 min at 4 °C.

9. Gently resuspend the cell pellet in 3.5 mL of Tris/sucrose/EDTA with a paint brush.

10. Add 0.5 mL of lysozyme solution, mix by vortexing then incubate at 37 °C for 30 min without shaking.

11. Add 0.5 mL of pronase E, mix by vortexing, then incubate at 37 °C for 30 min without shaking.

12. Transfer to a 65 °C water bath for 10 min, then add 250 μL of 20% Sarkosyl and incubate at 65 °C for 40 min.

13. Transfer to ice and let stand for at least 30 min.

14. Clear the lysate of cellular debris by centrifugation in a Sorvall SS34 rotor at 18 000 rpm (25 260 × g) for 1 h at 4 °C.

15. Carefully transfer the supernatant to a clean Oak Ridge tube and add 9.0 mL of sterile distilled H_2O, 12.7 g of finely ground cesium chloride and 1.5 mL of ethidium bromide (10 mg/mL). Mix by gentle inversion until the cesium chloride is dissolved and adjust the refractive index to 1.3865 to 1.3885 (these values correspond to a density range of 1.55–1.58) by adding cesium chloride (to increase the value) or distilled H_2O (to decrease the value).

16. Transfer the mixture to an ultracentrifuge tube, fill the remaining volume and balance the tubes using cesium chloride balance solution, then seal the tubes and band the DNA by ultracentrifugation in a Sorvall TV865B rotor at 52 000 rpm (255 800 × g) for 9 to 16 h at 18 °C.

17. Fractionate the DNA band with a 5-mL syringe and 18 gauge needle (this procedure is detailed below in the section entitled "Fractionation of DNA Bands from Cesium Chloride Gradients"), transfer the solution to a clean ultracentrifuge tube, fill the remainder of the tube with cesium chloride balance solution and repeat the ultracentrifugation step. This second round of ultracentrifugation results in a substantial increase in purity of the DNA obtained by diluting the contaminants (which are dispersed throughout the gradient) without diluting the DNA (which forms a discrete band in a small volume).

18. Fractionate the DNA band and process through isopropanol extraction, desalting and concentrating DNA by ethanol precipitation, and DNA quantitation as described in the appropriate sections below.

35–5 DIRECT LYSIS PROTOCOL

35–5.1 Materials

Note: The materials list is based on the simultaneous processing of eight 10-g soil samples from the stage of sieved soil to the point of the final equilibrium gradient centrifugation step. Larger numbers of samples can be accommodated by either staging the ultracentrifuge runs, or by scaling the protocol to allow the use of other rotors such as the Sorvall T1270 rotor that accommodates 12 12.5-mL samples, or the Sorvall TFT45.6 rotor that accommodates 40 6.0-mL samples. This list is based on the items used in this laboratory and does not constitute a commercial endorsement of any supplies or equipment. Reasonable substitutions for particular types of centrifuge rotors, tubes and other materials and equipment that maintain, or reasonably approximate, the specified conditions are appropriate.

1. Twenty-four 50-mL polycarbonate Oak Ridge tubes.
2. Vortex mixer.
3. Water bath at 70 °C.
4. Glass beads: two sizes are used; 0.7 to 1.0 mm (Sigma #G9393) and 0.2 to 0.3 mm (Sigma #G9143).
5. Reciprocal platform shaker (Eberbach Corp., Ann Arbor, MI).
6. Sorvall superspeed centrifuge (e.g., model RC5B) with an SS34 (8-place) rotor or an SA600 (12-place) rotor.

7. Refractometer (to measure the initial density of the cesium chloride gradient for DNA purification. If a refractometer is not available, density can be determined by measuring the specific gravity of the solution).
8. Sorvall ultracentrifuge with TV865B rotor.
9. Sixteen ultracentrifuge tubes.

35–5.2 Reagents

Note: The recipes presented here are for reagents required from the stage of sieved soil to the point of the final equilibrium gradient centrifugation step. After this point, the steps are common to both DNA isolation protocols and can be found in the section titled: "Fractionation of DNA Gradients, Final Purification and Quantitation of Bacterial Community DNA."

1. Sodium phosphate buffer (1 mM at pH 7.0).
   Buffers the soil suspension during the lysis of bacterial cells.
   • This solution can be made by combining 2.1 mL of 0.2 M NaH_2PO_4 and 3.3 mL of 0.2 M Na_2HPO_4 in a total of 1 L of distilled H_2O.
2. Cesium chloride balance solution (R_f = 1.3870)
   This solution is used to bring cesium chloride gradient tubes to final volume and balance the tubes prior to ultracentrifugation.
   for approximately 300 mL:
   • Add 250 g of finely ground cesium chloride (CsCl) to 250 mL of distilled H_2O (sterile) and mix by inversion until the CsCl is dissolved.
   • Add 12.5 mL of 10 mg/mL ethidium bromide.
   • Check the refractive index and adjust as necessary to achieve an R_f of 1.3870 (this corresponds to a density of 1.56) by adding CsCl to increase the refractive index (density) or H_2O to decrease the refractive index.
     Note: Refractive index is a measure of the density of the solution. In lieu of using a refractometer, the investigator can determine the specific gravity of the solution and adjust the density of the solution to 1.56.

35–5.3 Procedure

Note: This protocol describes the steps involved from the stage of sieved soil to the point of the final equilibrium gradient centrifugation step. After this point the steps are common to both DNA isolation protocols and can be found in the section titled: "Fractionation of DNA Gradients, Final Purification and Quantitation of Bacterial Community DNA."

1. Add 20 mL of $NaPO_4$ (1.0 mM, pH 7.0) and 0.25 g of sodium dodecyl sulfate (SDS) to each 10-g soil sample in a 50-mL Oak

Ridge tube, mix by vortexing until thoroughly suspended, then incubate for 30 min at 70 °C mixing every 5 min.

2. Add 5 g of large glass beads (0.7–1.0 mm) and 5 g of small glass beads (0.2–0.3 mm) and shake for 30 min by placing horizontally on a reciprocal platform shaker at high speed (about 100 oscillations/min) at room temperature.

3. Pellet soil and cell debris by centrifugation in a Sorvall SS34 rotor at 10 000 rpm (7796 $\times$ g) for 10 min at 10 °C.

4. Transfer the supernatant to a clean Oak Ridge tube and incubate on ice for 15 to 30 min to precipitate the SDS. Clear the lysate by centrifugation in a Sorvall SS34 rotor at 10 000 rpm (7796 $\times$ g) for 10 min at 10 °C, then carefully transfer the cleared lysate to a clean Oak Ridge tube.

5. Adjust the volume of the lysate to 15.5 mL with distilled H_2O, then add 14.5 g of finely ground cesium chloride. Mix by gentle inversion until the cesium chloride is totally dissolved, then let stand at room temperature for 10 to 15 min to precipitate proteins. Clear the lysate by centrifugation at 5000 rpm (1949 $\times$ g) for 10 min at 10 °C. The precipitated proteins will form a floating layer that may appear "foamy"; this layer should be discarded.

6. Transfer the mixture to an ultracentrifuge tube containing 0.65 mL of ethidium bromide (10 mg/mL) and mix by gentle inversion. Fill the remainder of the tube with cesium chloride balance solution (R_f = 1.3870), then balance the tubes, seal and band DNA by centrifugation in a Sorvall TV865B rotor at 52 000 rpm (255 800 $\times$ g) at 18 °C for 9 to 16 h.

7. Fractionate the DNA band with a 5-mL syringe and 18 gauge needle (this procedure is detailed below in the section entitled "Fractionation of DNA Bands from Cesium Chloride Gradients"), transfer the solution to a clean ultracentrifuge tube, fill the remainder of the tube with cesium chloride balance solution and repeat the ultracentrifugation step. This second round of ultracentrifugation results in a substantial increase in purity of the DNA obtained by diluting the contaminants (which are dispersed throughout the gradient) without diluting the DNA (which forms a discrete band in a small volume).

8. Fractionate the DNA band and process through isopropanol extraction, desalting and concentrating DNA by ethanol precipitation, and DNA quantitation as described in the appropriate sections below.

35–6 FRACTIONATION OF DNA GRADIENTS, FINAL PURIFICATION AND QUANTITATION OF BACTERIAL COMMUNITY DNA

Note: At this stage, the total bacterial community DNA has been extracted from soil and is in aqueous solution. Most of the subsequent protocols and analyses are routine molecular biology protocols with some notable differences described elsewhere in this volume. For further information on routine techniques of molecular biology, or as an additional resource, the reader is referred to the many molecular biology and cloning laboratory manuals that are available (e.g., Ausubel et al., 1990; Sambrook et al., 1989).

35–6.1 Fractionation of DNA Bands from Cesium Chloride Gradients

35–6.1.1 Materials

1. Syringes (5 mL) fitted with 18-gauge needles
2. Hand-held UV light source (e.g., Blak-ray model B-100A, VWR Scientific Co.) with UV safety glasses or goggles for eye protection

35–6.1.2 Procedure

1. Stop the ultracentrifuge and carefully remove the tubes from the rotor. Avoid shaking the tubes as this will perturb the gradients. It is best to let the rotor coast to a stop from 3000 rpm to zero rather than using the brake which may cause some loss of resolution of the DNA band in the gradient.
2. Extract DNA bands in 1 to 2 mL volume under UV illumination using a 5-mL syringe and 18-gauge needle as follows:
   • Poke a hole at the top of the ultracentrifuge tube to allow air in.
   • Insert the needle and syringe just below the visible DNA band with the needle orifice pointed up and extract the DNA band by slowly withdrawing the plunger of the syringe.
   Note: Wear gloves and UV goggles or glasses to protect yourself from the ethidium bromide and UV irradiation.
3. Proceed to isopropanol extraction of ethidium bromide as described below or prepare for second banding as described above.

35–6.2 Isopropanol Extraction of Ethidium Bromide from DNA

35–6.2.1 Materials

1. Six milliliter Falcon tubes (capped polypropylene tubes).
2. Thirty milliliter Corex tubes (glass centrifuge tubes).
3. Sorvall superspeed centrifuge with an SS34 (8-place) or an SA600 (12-place) rotor.
4. Capacity for drying under vacuum (e.g., lyopholizer).
   Note: air drying can be substituted for drying under vacuum.

35–6.2.2 Reagents

1. Isopropanol saturated with 5 M sodium chloride.
   This is used to remove ethidium bromide from DNA following the ultracentrifugation steps.

for approximately 1 L:
- Prepare 1 L of 5 M sodium chloride (NaCl) in a 2 L bottle and autoclave. • After the 5 M NaCl cools, add 1 L of isopropanol to the bottle, mix thoroughly, and let sit until the organic and aqueous phases separate. There will be some volume loss as the water is mixed with the alcohol.
- Add additional isopropanol as necessary until there is some NaCl precipitate present after the phases separate.

35–6.2.3 Procedure

1. Fractionate the DNA band (in 1–2 mL volume) as described in the previous section and transfer to a 6.0-mL Falcon tube.
2. Add an equal volume of isopropanol saturated with 5 M NaCl.
3. Mix by gentle inversion and let sit until the phases separate.
4. Pipette off the top layer (isopropanol) using a Pasteur pipet and discard in an appropriate fashion (the isopropanol will be pink in color and contains ethidium bromide).
5. Repeat steps 2 to 4 until all pink color is gone and then once more (usually a total of five extractions). Proceed to desalting and concentration of DNA as described below.

35–6.3 Desalting and Concentration of DNA

35–6.3.1 Materials

1. Pipetman or equivalent pipettors for small volumes.
2. Eppendorf or equivalent microfuge and 1.5-mL microfuge tubes.

35–6.3.2 Reagents

1. 3 M Sodium Acetate (NaOAc), pH 5.2.
   Monovalent cations must be present when precipitating DNA with alcohol to provide for quantitative recovery of DNA.
   for 250 mL:
   - Dissolve 61.52 g of anhydrous NaOAc in 150 mL of distilled H_2O.
   - Adjust the pH to 5.2 with glacial acetic acid.
   - Adjust the volume to 250 mL with distilled H_2O, then autoclave.

35–6.3.3 Procedure

1. Following the removal of ethidium bromide, transfer the DNA solution to a labelled Corex tube, add two volumes of sterile distilled H_2O, then add 2 volumes of cold (-20 °C) 100% ethanol. Cover with parafilm and mix thoroughly by inversion or vortexing.

e.g., DNA volume   1.5 mL
    2 volumes distilled H_2O 3.0 mL
    2 volumes (6 × original DNA volume) ethanol 9.0 mL
Incubate overnight at −20 °C.

    Note: This first precipitation should not be incubated more than 24 h or the cesium chloride may crystallize out complicating further purification.

2. Pellet the DNA by centrifugation in a Sorvall SS34 (or SA600) rotor at 7500 rpm (4385 × g) at 4 °C for 1 h. Position the tube so that the label is toward the outside of the rotor as a reference for the location of the DNA pellet which may not be readily visualized at this stage.

3. Discard (pour off) supernatant being careful not to disturb the pellet. Invert the tube on a paper towel and drain dry (5 min).

4. Complete drying under vacuum (complete air drying may be substituted for vacuum drying).

5. Add 400 µL of sterile distilled H_2O, mix by vortexing to dissolve the DNA (it may be helpful to use a Pipetman P1000 to aid in resuspending the DNA pellet by pipetting the 400 µL of distilled H_2O up and down along sides of tube).

6. Transfer the DNA solution to a labelled 1.5-mL Eppendorf tube.

7. Collect the remaining liquid in the Corex tube by brief centrifugation and transfer to the Eppendorf tube.

8. Add 40 µL of 3 M sodium acetate (pH 5.2) and 880 µL of cold (−20 °C) ethanol to the tube, mix thoroughly by vortexing and incubate at −20 °C for at least 1 h (DNA may be stored for extended periods in 70% ethanol in the presence of monovalent cations).

9. Collect the DNA by centrifugation in the microfuge for 15 to 30 min at 4 °C.

10. Remove the supernatant with a Pasteur pipette and discard, wash the DNA pellet once with cold (−20 °C) 70% ethanol by gentle inversion, centrifuge briefly (2 min), remove the supernatant with a Pasteur pipette and discard, then dry under vacuum. Resuspend in a small volume (usually 100 µL). Proceed to quantitation of DNA as described below.

35–6.4 Spectrophotometric Quantitation of DNA

35–6.4.1 Materials

1. Spectrophotometer with UV capabilities and quartz cuvette(s).

35–6.4.2 Procedure

The concentration of DNA in the sample can be measured by monitoring the absorbance of a dilute solution of the sample at 260 and 280 nm as follows:

1. Dilute 5 μL of the DNA sample with 995 μL of distilled H_2O.
2. Measure the absorbance of this solution at 260 and 280 nm.
3. Calculate the concentration of DNA in the sample based on the value of 1.0 A_{260} unit $= 50$ μg/mL of DNA, and taking into account the 1:200 dilution factor of the sample.
4. Calculate the A_{260}/A_{280} ratio. This ratio indicates the degree of contamination of the DNA with humic (phenolic) compounds and proteins since these molecules exhibit strong absorbance at 280 nm. Pure DNA has a A_{260}/A_{280} ratio of 2.0, with a value of 1.7 to 2.0 indicating relatively pure DNA. DNA solutions can be stored at 4 °C but, preferably, at −20 °C.

Note: As mentioned previously, the precise quantitation of bacterial DNA isolated from the soil environment can be problematic due to the copurification of humic contaminants that also absorb light in the UV range. If there is a distinct brownish tinge to the DNA solution or if the A_{260}/A_{280} ratio is low, this indicates that there is significant contamination of the DNA with compounds from the soil. If such contaminants are present and precise quantitation is desirable, alternate methods of quantitation must be employed. Perhaps the simplest way to precisely quantitate DNA in the presence of humic contaminants is to measure the UV fluorescence of the DNA in the presence of ethidium bromide. This is readily accomplished using agarose gel electrophoresis by the protocol described in the next section.

35–6.5 Quantitation of DNA by Ultraviolet Fluorescence in the Presence of Ethidium Bromide

Note: For this analysis a relatively high agarose concentration of 1.2% is used. This allows the contaminants to migrate significantly far into the gel while the randomly sheared (but generally large) DNA fragments migrate as essentially a single band near the origin.

35–6.5.1 Materials

1. Submarine agarose gel apparatus. *There are a wide variety of these available commercially. It is appropriate to obtain one that will also be useful for subsequent analyses of the bacterial community DNA. A gel with minimum dimensions of 15 × 15 cm is recommended to allow sufficient resolution of differently sized DNA fragments and longer gels (e.g., 15 × 25 cm) might be more useful since more than one set of wells can be cast into a single gel as is desirable for this analysis.*
2. Power supply for agarose gel electrophoresis.
3. Microwave oven or heated stirring plate.
4. Access to a camera set-up suitable for photographing gels under UV illumination such as a transilluminator/Polaroid camera set-up which is basic equipment in most molecular biology laboratories.

5. Access to equipment for densitometric analysis of the gel photograph is desirable, as it will allow more precise quantitation, but not essential.

35–6.5.2 Reagents

1. Tris-acetate-EDTA (TAE) buffer.
   This buffer has appropriate characteristics of pH, buffering capacity, and ionic strength that will result in clearly resolved DNA bands in an agarose gel.
   Contains:
   40 mM Tris-acetate
   1 mM EDTA
   for 1 L of 20 × stock:
   • Combine the following in 800 mL of distilled H_2O:
       96.8 g of Tris-base
       22.84 mL of glacial acetic acid
       40 mL 0.5 M EDTA (pH 8.0)
   Adjust to pH 8.0, if necessary, using glacial acetic acid, bring to 1 L final volume with distilled H_2O.
2. 5 × sample loading dye (for agarose gels).
   This dye allows tracking of the progress of electrophoresis by monitoring the dye front and also makes the DNA samples sufficiently dense to sink to the bottom of the wells in the submarine gel.
   Contains:
   100 mM EDTA
   50% glycerol
   0.15% bromophenol blue
   0.15% xylene cyanole
   for 10 mL:
   • Combine the following:
       2 mL of 0.5 M EDTA, pH 8.0
       5.0 mL of 100% glycerol
       0.75 mL of 2% bromophenol blue
       0.75 mL of 2% xylene cyanole
       1.5 mL of distilled H_2O

35–6.5.3 Procedure

1. Combine an appropriate volume of 1 × TAE buffer with an appropriate amount of agarose in an Ehrlenmeyer flask (these values are determined based on the gel dimensions and the desired percentage of agarose; 1.2% for this analysis).
2. Swirl the flask to evenly distribute the agarose.
3. Heat the solution until the agarose is completely dissolved; undissolved agarose will appear as flecks in an otherwise clear solution. If using a microwave, heat at high power for 2 min or until

the mixture bubbles. Remove the flask from oven (before it boils over), carefully swirl again, and reheat until all of the agarose goes into solution.

4. Place the flask containing the molten agarose in a 55 to 65 °C water bath or on the benchtop to cool. The gel should be poured when the temperature of the solution is 55 to 65 °C (almost too hot to hold).

5. Prepare the gel apparatus for casting the gel while the agarose is cooling. There are several different types of gel boxes and these preparations will depend on the particular one being used.

6. Just prior to pouring the gel, add ethidium bromide to a final concentration of 0.5 µg/mL to the dissolved agarose and swirl to mix (for agarose gels it is convenient to have a 1 mg/mL stock solution of ethidium bromide).

7. Pour the agarose solution into the gel-casting tray and adjust the well-forming comb(s) to keep the wells properly aligned. Allow the agarose to cool and solidify (about 20–30 min) prior to use.

8. To prepare gel for running:
   • Fill the electrophoresis tank (apparatus) with buffer solution (1 × TAE containing 0.5 µg/mL ethidium bromide) and place the gel (still in the casting tray) on the tank platform. The buffer must cover the gel by 1 to 2 mm.
   • Carefully (to avoid breaking the walls of the wells) remove the comb.

9. To prepare samples for loading:
   • In a microfuge tube bring 1 µg of total bacterial community DNA solution (as determined spectrophotometrically) to a total volume of 20 µL with distilled H_2O. Add 5.0 µL of 5 × sample dye to the sample and mix.
   • A set of DNA standards should be prepared by serial dilution of a known amount of ultrapure DNA (an appropriate source would be to purchase undigested phage lambda DNA commercially). The dilution range should be from 2.0 to 0.1 µg in 20 µL volume since the concentration of humic-contaminated DNA is generally overestimated. Add 5.0 µL of 5 × sample dye to each sample and mix.

10. Load the samples and standards into the gel wells using the Pipetman pipettor. Stick the tip below the surface of the buffer but above the well bottom, and dispense the sample slowly. The sample will sink through the buffer and settle in the well. It is recommended that the DNA samples and standards be run near each other in the same gel for more precise quantitation. An appropriate way to accomplish this is to cast two sets of wells into the gel; the upper one (nearer the origin) for the DNA standard dilution series and the lower for the DNA samples of unknown concentration.

11. After the gel has been loaded, gently place the cover on the apparatus and hook up the power leads. DNA is negatively charged and will migrate toward the positive (red lead and jack in power supply) electrode. Adjust the power to 50 volts (constant voltage). Run the gel until the leading dye front (bromophenol blue) has migrated two-thirds the length of the gel or two-thirds of the way to the second set of wells.

12. Photograph the gel under UV illumination either from a transilluminator or by even distribution of UV illumination from a handheld source.

13. Determine the concentration of the DNA in the original sample by comparison to the known standards using a densitometer or by visual examination. The DNA will fluoresce bright orange and be found relatively near the origin of the gel. The humic contaminants, if significant amounts are present, will fluoresce blue-green and should migrate well ahead of the DNA in the agarose gel.

ACKNOWLEDGMENTS

The author's work has been supported by Environmental Protection Agency Grant No. CR-814575-01-0 and National Science Foundation Science and Technology Grant No. DIR-8809640. Although the research described in this article has been funded in part by the EPA, it has not been subjected to the Agency's review and therefore does not necessarily reflect the views of the agency, and no official endorsement should be inferred.

The author gratefully acknowledges the technical support provided by Bernard M. Schroeter and Robert A. Laymon.

REFERENCES

Ausubel, F.M., R. Brent, R.E. Kingston, D.D. Moore, J.G. Seidman, J.A. Smith, and K. Struhl. 1990. Current protocols in molecular biology. Vol. 2. Greene Publ. Assoc. and Wiley-Interscience, New York.

Bakken, L.R. 1985. Separation and purification of bacteria from soil. Appl. Environ. Microbiol. 49:1482–1487.

Barry, T., R. Powell, and F. Gannon. 1990. A general method to generate DNA probes for microorganisms. Bio/Technology 8:233–236.

Fægri, A., V.L. Torsvik, and J. Goksöyr. 1977. Bacterial and fungal activities in soil: Separation of bacteria and fungi by a rapid fractionated centrifugation technique. Soil Biol. Biochem. 9:105–112.

Greaves, M.P., and M.J. Wilson. 1969. The adsorption of nucleic acids by montmorillonite. Soil Biol. Biochem. 1:317–323.

Hilger, A.B., and D.D. Myrold. 1991. Method for extraction of Frankia DNA from soil. p. 107–113. In D.A. Crossley, Jr. et al. (ed.) Modern techniques in soil ecology. Elsevier Sci. Publ. Co., New York.

Holben, W.E., B.M. Schroeter, V.G.M. Calabrese, R.H. Olsen, J.K. Kukor, V.O. Biederbeck, A.E. Smith, and J.M. Tiedje. 1992. Gene probe analysis of soil microbial populations selected by amendment with 2,4-dichlorophenoxyacetic acid Appl. Environ. Microbial. 58:3941–3948.

Holben, W.E. J.K. Jansson, B.K. Chelm, and J.M. Tiedje. 1988. DNA probe method for the detection of specific microorganisms in the soil bacterial community. Appl. Environ. Microbiol. 54:703–711.

Holben, W.E., and J.M. Tiedje. 1988. Applications of nucleic acid hybridization in microbial ecology. Ecology 69:561–568.

Knight, I.T., W.E. Holben, J.M. Tiedje, and R.R. Colwell. 1992. Nucleic acid hybridization techniques for detection, identification, and enumeration of microorganisms in the environment. p. 65–91. In M.A. Levin et al. (ed.) Microbial ecology: Principles, methods, and applications. McGraw-Hill, New York.

Mortland, M.M., J.G. Lawless, H. Hartman, and R. Frankel. 1984. Smectite interactions with flavomononucleotide. Clays Clay Miner. 32:279–282.

Neilson, J.W., K.L. Josephson, S.D. Pillai, and I.L. Pepper. 1992. Polymerase chain reaction and gene probe detection of the 2,4-dichlorophenoxyacetic acid degradation plasmid, pJP4. Appl. Environ. Microbiol. 58:1271–1275.

Ogram, A., G.S. Sayler, and T. Barkay. 1987. The extraction and purification of microbial DNA from sediments. J. Microbiol. Methods 7:57–66.

Ogram, A., G.S. Sayler, D. Gustin, and R.J. Lewis. 1988. DNA adsorption to soils and sediments. Environ. Sci. Technol. 22:982–984.

Salyers, A.A., S.P. Lynn, and J.F. Gardner. 1983. Use of randomly cloned DNA fragments for identification of Bacteroides thetaiotaomicron. J. Bacteriol. 154:287–293.

Sambrook, J., E.F. Fritsch, and T. Maniatis. 1989. Molecular cloning, a laboratory manual. 2nd ed. Cold Spring Harbor Lab. Press, Cold Spring Harbor, New York.

Sayler, G.S., and A.C. Layton. 1990. Environmental application of nucleic acid hybridization. p. 625–648. In L.N. Ornston et al. (ed.) Annual review of microbiology. Annual Reviews, Palo Alto, CA.

Smith, G.B., and J.M. Tiedje. 1992. Isolation and characterization of a nitrite reductase gene and its use as a probe for denitrifying bacteria. Appl. Environ. Microbiol. 58:376–384.

Steffan, R.J., and R.M. Atlas. 1988. DNA amplification to enhance detection of genetically engineered bacteria in environmental samples. Appl. Environ. Microbiol. 54:2185–2191.

Steffan, R.J., J. Goksøyr, A.K. Bej, and R.M. Atlas. 1988. Recovery of DNA from soils and sediments. Appl. Environ. Microbiol. 54:2908–2915.

Tsai, Y-L, and B.H. Olson. 1991. Rapid method for direct extraction of DNA from soil and sediments. Appl. Environ. Microbiol. 57:1070–1074.

Chapter 36

Microbial Biomass

W. R. HORWATH AND **E. A. PAUL,** *Michigan State University, East Lansing, Michigan*

The soil microbial biomass is an important component of the soil organic matter that regulates the transformation and storage of nutrients. It is a labile component of the soil organic fraction containing 1 to 3% of the total soil C and up to 5% of the total soil N (Smith & Paul, 1990). Microbially mediated processes affect ecosystem functions associated with nutrient cycling, soil fertility, global C change, and soil organic matter turnover. The size and activity of the soil microbial biomass must be assessed to fully understand nutrient fluxes in managed and natural ecosystems.

Soil microbial biomass estimations are useful in investigations that compare temporal nutrient fluctuations along natural and perturbed gradients. The effects of tillage, crop rotations, and soil type on organic C and nutrient turnover can be assessed by following nutrient pools and activity associated with the soil microbial biomass. Microbial biomass has been shown to be a sensitive indicator of differences in sustainable cropping systems (Anderson & Domsch, 1989). The toxicity of pollutants and the degradation of organic compounds (pesticides and industrial chemicals) can be monitored by following changes in the soil microbial biomass.

This chapter describes methods that estimate the size of the soil microbial biomass and associated nutrient pools and metabolites. The methods include the chloroform fumigation incubation method (CFI), chloroform fumigation extraction method (CFE), substrate induced respiration (SIR), and adenosine triphosphate (ATP) analysis. The CFI and CFE methods are also useful to recover tracers (e.g., ^{14}C and ^{15}N) from the microbial biomass. ATP and SIR cannot be used to measure tracer incorporation into the biomass. All the methods are sensitive to minor differences in technique and must be standardized for specific soil types and operating conditions.

36–1 SOIL SAMPLING, PREPARATION, AND STORAGE

Representative soil samples that can be treated statistically will have the most significance in soil microbial biomass determinations. A knowledge of the site should be used to separate areas uncharacteristic or unrepresentative of the general landscape (see chapter 1 by Wollum in this book). Examples are low, poorly drained areas; these can be sampled separately. A minimum of four (preferably more) replicates each made from at least two separate composite samples are required to reduce the error to about 10% of the measurement (see chapter 2 by Parkin and Robinson in this book). The use of tracers to study nutrient cycling and flux rates in conjunction with biomass measurements is best accomplished in microcosms that can be completely sampled, mixed, and subsampled.

The soil is sampled by coring or removing a known dimensional quantity from the soil profile (see chapter 1). The soil should be removed from direct sunlight or placed in an ice chest. The soil can be stored overnight at 15 °C when microbial biomass determinations are to be done the following day. Soil can be stored at 4 °C for periods of a week, but the possibility of changes occurring during this storage period must be considered. Freezing of soil samples is not recommended due to the adverse biocidal effects on the soil microbial biomass. If samples must be frozen, they should be pre-incubated for 7 to 10 d before soil microbial biomass determinations are done. The drying of soil samples should be strictly avoided.

Soil samples are prepared by sieving through a 4 to 6 mm screen. This size mesh has been determined not to affect the soil microbial biomass size or activity (Jenkinson & Powlson, 1980; Ross et al., 1985). When soils are too moist for sieving they have to be dried to an adequate moisture content. The soil moisture content is determined after the sieving process. With the exception of ATP determinations, the described microbial biomass determinations do not work well for water-saturated samples.

36–2 PHYSIOLOGICAL METHODS

36–2.1 Chloroform Fumigation Incubation Method

The effect of fumigants on soil metabolism was established early during this century (Jenkinson, 1966). The respiration rate of a fumigated soil is initially less than an unfumigated sample, but as time proceeds, the respiration rate of the fumigated soil exceeds that of the unfumigated sample and eventually subsides to a lower level. The temporary flush of CO_2 from the fumigated soil is primarily due to the decomposition of microbial components from lysed microorganisms (Jenkinson, 1966). In addition, an increase in the NH_4^+ pool occurs as a result of the mineralization of nitrogenous substrates from the lysed microorganisms. The increase in CO_2 evolution and extractable NH_4^+ from fumigated samples has been used to estimate the size of the soil biomass (Jenkinson & Powlson, 1976a,b; Jenkinson, 1976; Anderson & Domsch, 1978a; Voroney & Paul, 1984).

36–2.1.1 Soil Samples

The amount of soil used will depend on its respiration rate or requirements for recovery of added tracers. Generally, 20 to 50 g (dry weight equivalent) of soil is placed in an appropriate sized container, this should allow for the addition of an extractant. Soil samples must be analyzed for initial inorganic N. Fumigated and control samples are analyzed for mineralized C and N. Analytical duplication of each soil sample is preferred. The samples to be fumigated are weighed into glass beakers and the beakers are marked with a chloroform-insensitive marker (e.g., pencil lead). The remaining treatments can be weighed into any suitable container, preferably with a closure to aid C or N extraction.

36–2.1.2 Fumigation of Soil Samples

Because of the carcinogenic-volatile properties of chloroform, all work must be done in an adequate fume hood. A beaker containing 50 mL of ethanol-free chloroform (Jenkinson & Powlson, 1976b) and antibumping granules is placed together with the soil samples into a vacuum desiccator. The desiccator is lined with moist paper towels to prevent the desiccation of soil samples during the fumigation. Commercially available ethanol-free chloroform preserved with heptachlor epoxide has been used to obtain similar results to that of purified $CHCl_3$ (Voroney et al., 1991). The desiccator is evacuated until the chloroform boils vigorously. This is repeated three times, letting air pass back into the desiccator to facilitate the distribution of the chloroform throughout the soil. The desiccator is then evacuated a fourth time until the chloroform boils vigorously for 2 min, the valve on the desiccator is closed, and the desiccator is placed in the dark at 25 °C for 18 to 24 h. Unfumigated samples are also kept in the dark in a desiccator or mason jars at 25 °C while the fumigation proceeds. Following this period, the chloroform and paper towels are removed, under the fume hood, and the desiccator evacuated 3 min for eight times letting air pass into the desiccator after each evacuation to remove residual chloroform. Never determine residual chloroform by sense of smell. Make sure the vacuum pump is periodically maintained to ensure proper operating condition.

Following the removal of chloroform, the fumigated soil samples are placed in mason jars (Fig. 36–1). Fumigated soil can be inoculated before the incubation by adding and thoroughly mixing 0.2 g of unfumigated soil to 50 g of fumigated soil (Jenkinson & Powlson, 1976b). Inoculation is often nonessential in soils with pH > 5 and high microbial populations since the fumigation procedure does not kill the entire population of the soil microbial biomass (Vance et al., 1987b). Inoculation of subsurface soils is often necessary. Soil samples are adjusted to an optimum soil moisture

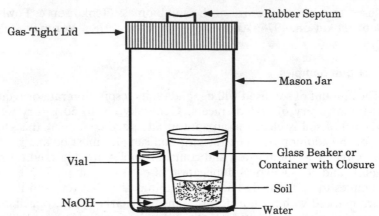

Fig. 36–1. The diagram depicts a soil sample enclosed in a mason jar with a respirometer and water. The respirometer is not used when sampling mineralized CO_2 with the gas chromatography (GC) method. When using the GC method a septum must be inserted into the mason jar lid to facilitate the sampling of the jar's headspace for CO_2 gas analysis.

content (55% of water-holding capacity). Soils prone to denitrification can be adjusted to lower water contents to reduce the gaseous loss of N (Jenkinson, 1988). Approximately 1.0 mL of water is added to the bottom of each mason jar to prevent soil desiccation (Fig. 36–1). The soils are then incubated in closed, gas tight mason jars under standard conditions (at 25 °C in the dark) for a period of 10 d.

36–2.1.3 Carbon Dioxide Mineralization Determination

A vial containing 1.0 mL of 2.0 M of NaOH is placed into each mason jar, exercising care not to partially neutralize the alkali by breathing into it. The volume or strength of the alkali can be adjusted to accommodate varying respiration rates of soils or recovery of tracers. Blanks consisting of jars without soil must be similarly maintained during the incubation period.

After the incubation period, the vials (respirometers) are titrated to determine the total C respired from the microbial biomass. An amount of $BaCl_2$ equivalent to the initial quantity of NaOH is added to each respirometer. The contents of the respirometer are then titrated to pH 7 or to a phenolphthalein endpoint using 0.1 M of HCl. The amount of CO_2-C evolved during the incubation is calculated from the volume of acid needed to attain pH 7 from the blank minus that required for the samples (1.0 mL of 2.0 M of NaOH can consume 12 mg of CO_2-C). When CO_2 levels are low, as is the case in subsurface soils, titration should be done using a double endpoint titration for bicarbonate (Jenkinson & Powlson, 1976b).

The double endpoint titration requires that the contents of the respirometer first be titrated to a pH of 9 to 10 with 1.0 M of HCl using thymolphthalein indicator solution (Sigma Chemical Co., St. Louis, MO). Add 50 μL of 0.1% carbonic anhydrase (Sigma Chemical Co., St. Louis,

MO) and titrate to pH 8.3 with 0.05 M of HCl using a pH electrode. Titrate the respirometer contents to pH 3.7 with 0.05 M of HCl and record the volume of acid. The amount of CO_2 evolved during the incubation period is calculated from the volume of acid needed to decrease the pH of the respirometer from 8.3 to 3.7 subtracted from the blank respirometer (1.0 mL of 0.05 M of HCl being equivalent to 0.6 mg of CO_2-C in NaOH solution).

Alternatively, the CO_2 accumulated in the headspace of the mason jar may be measured by gas chromatography (GC) or with an infrared gas analyzer. The GC method gives a rapid and accurate measurement of CO_2 and can be used in acidic soils (see chapter 38 by Zibiliske in this book). However, this technique is prone to error in neutral and alkaline soils (Martens, 1987) as accumulation of carbonate species in the soil solution can lead to lowered CO_2 determinations. When working with neutral and alkaline soils the CO_2 absorption method described above should be used.

36–2.1.4 Nitrogen Mineralization Determination

Fumigated, unfumigated, and samples for initial inorganic N are extracted with 1 or 2 M of KCl at a ratio of 5:1 (extractant/soil). Clay soils require a ratio of 10:1. The soil and extractant are shaken on a reciprocal shaker at 180 strokes per minute for 0.5 h. The filtered extract is then analyzed for NO_3^- and NH_4^+ (see chapter 41 by Bundy and Meisinger in this book) to determine the flush of N from the microbial biomass (Quikchem Systems, 1987). The extract can be distilled or diffused for the analysis of ^{15}N in the biomass (Brooks et al., 1989; (see chapters 40 and 42 by Hauck et al. and Hart et al., respectively in this book).

36–2.1.5 Calculation of Biomass Carbon

The amount of CO_2-C respired from fumigated and unfumigated samples is used to calculate soil microbial biomass C. Soil microbial biomass C, calculated using a control, is shown by the following equation:

$$Biomass\ C = (F_c - UF_c)/K_c$$

where

$$F_c = CO_2\ flush\ from\ the\ fumigated\ sample$$

$$UF_c = CO_2\ produced\ by\ the\ control.$$

The value of K_c is defined as the fraction of biomass C mineralized to CO_2. Biomass C can also be calculated without the subtraction of a control as shown in the following equation (Voroney & Paul, 1984).

$$Biomass\ C = F_c/K_c$$

This method of biomass C calculation is useful in soil with high basal respiration rates. The value of K_c is of considerable importance since it relates the size of the biomass to the fumigated flush of CO_2. The value of K_c can be obtained by adding a known quantity of ^{14}C-labeled microorganisms to soil and determining the proportion of the added ^{14}C that is mineralized (Anderson & Domsch, 1978a).

The culture of indigenous labeled (^{14}C and ^{15}N) microorganisms is necessary to determine realistic K values for the accurate determination of microbial biomass in soils. Chapters 7 to 9 describe techniques and provide references for isolating microorganisms from soil. The liquid medium described by Anderson and Domsch (1978a) and modified by Wardle and Parkinson (1990) is used to culture labeled organisms. One liter of medium contains 10 g of U-^{14}C-labeled D-glucose (specific activity 370 Bq mg^{-1} of glucose), 1 g of NH_4NO_3, 1 g of KH_2PO_4, 0.5 g of $MgSO_4$, 50 mg of $CaCl_2$ and 20 mg of $FeCl_3$. In addition, yeast extract (DIFCO) is added to the media, 1 g for fungi and 3 g for bacteria. Labeled NH_4NO_3 (4–10% atom % excess) may be used to enrich the microbial cultures with [15]N. The medium is autoclaved prior to the addition of labeled glucose. Bacteria and fungi are cultured in 50 mL of media in a 125-mL Erlenmeyer flask, on a culture shaker at 22 °C. Bacteria are harvested in late logarithmic phase and actinomycetes and fungi in the late linear phase. Radioactive cells are harvested by centrifugation (2000 g, 4 °C) for 20 min and washed free of adhering media with purified water by repeated centrifugation and resuspension. Fungi can be harvested by filtration and rinsed. The pellets are resuspended in water or dried and ground (0.5 mm) for introduction into soil samples.

An alternative technique to determine K factors is to label soil organisms in situ. Carbon-14-glucose (specific activity of 925 Bq mg^{-1} of C) and [15]NH_4^+ (Atom % excess of 4–10%) are added directly to the soil (Voroney & Paul, 1984). The substrates can be added in dry form using a talc carrier (Anderson & Domsch, 1978b), misted and mixed into the soil or added in solution to bring the soil moisture to 55% of water-holding capacity. The C/N ratio of the added substrates should be approximately 10:1. The amount of ^{14}C-glucose and inorganic ^{15}N added to soil will depend on the amount of glucose and N that can be immobilized by the soil biomass in a 24-h period without a major change in the microbial biomass size. The short incubation time ensures minimal new biomass production. Standard glucose (e.g., Sigma Diagnostic Kit no. 510-A) and inorganic N assays can be done on 2 M of KCl extracts of soil to determine the glucose and N amendment.

It is assumed that added organisms grown in vitro give results applicable to the native soil population. Common values for K_c, subtracting a control, range from 0.45 (Jenkinson & Ladd, 1981) to 0.41 (Anderson & Domsch, 1978a). Voroney and Paul (1984) labeled the soil biomass in situ using ^{14}C-glucose and ^{15}N-NH_4^+ and developed a K_c of 0.41 without subtracting a control. It is difficult for every researcher to develop a K_c for each soil and literature values are often used. The value of K_c used and whether a control was subtracted should be reported when reporting biomass size. We always recommend reporting of the data for fumigated and control soils iregardless of the method of calculation. This allows other workers to reinterpret the data if they wish.

The use of an appropriate control has been problematic since the conception of the CFI method. Not all of the CO_2 evolved from a fumigated soil is derived from lysed microorganisms. The basal respiration of the fumigated and control samples can also vary depending on the moisture content and amount of root fragments contained in the soil. It is, therefore, difficult to ascertain the amount of the control to subtract from the fumigated sample, especially when basal respiration is high.

Preincubation of soil samples for 7 to 10 d prior to biomass determination has been recommended to eliminate interference from sieving, wetting, and root fragments (Sparling et al., 1985). However, if time zero soil microbial biomass determinations are desired or the experiment involves tracers, preincubation of soil is not an option. Voroney and Paul (1984) suggest an alternative method excluding the use of a control since they found that in situ labeling of the soil biomass produced a K_c of 0.41 without subtracting a control.

In soils of low microbial activity the control is low. Soils with low biomass but high background activity result in negative biomass calculations when a control is subtracted. In many soils, when a control is not subtracted the biomass estimates can be too high. Jenkinson and Powlson (1976b) suggest subtracting the CO_2 produced from a 10 to 20 d incubation to reduce the amount of the control subtracted from the fumigated sample in soils described above. The possibility of subtracting a partial control based on the size of the CO_2 flush from the 0 to 10 d control using internal standards also exists. This can be calculated by adding labeled substrate to both fumigated and unfumigated samples and calculating the difference of substrate utilization between the treatments (Voroney & Paul, 1984).

36–2.1.6 Calculation of Biomass Nitrogen

Microbial biomass N (B_n) is calculated similarly to biomass C as shown:

$$B_n = (F_n - UF_n)/K_n$$

where

$$F_n = \text{The flush of } NH_4^+ \text{ due to fumigation}$$

$$UF_n = \text{The } NH_4^+ \text{ mineralized during 0 to 10 d from a control.}$$

The value of K_n is the proportion of microbial N mineralized to NH_4^+ during the 10-d incubation period. The value of B_n also has been calculated without the use of a control. The equation is:

$$B_n = F_n/K_n$$

Values of B_n calculated by either of the above equations are generally in agreement if the background NH_4^+ levels of soils taken from the field are low (Voroney & Paul, 1984).

The establishment of K_n is difficult since the N content of the microbial biomass is variable depending on substrate availability, fungal/bacterial ratio, litter C/N ratio or environmental conditions (moisture and temperature). The immobilization of mineralized N after fumigation can also complicate the calculation of K_n. Voroney and Paul (1984) related the flush of C and N from the fumigation response and developed a floating K_n based on the expression:

$$K_n = -0.014*(C_f/N_f) + 0.39$$

The inclusion of C_f/N_f ratio accounts for the reimmobilization of N during the N flush.

The values of K_n developed by adding ^{15}N-labeled microorganisms to soil range from 0.54 to 0.62 and are reviewed by Jenkinson (1988). Shen et al. (1984), obtained a value 0.68 by determining the immobilization of ^{15}N-NH_4^+ in fumigated and unfumigated soil. Jenkinson (1988) suggests a weighted mean for K_n of 0.54, for samples with C_f/N_f ratio of < 6.7. The value of K_n should be included with all reported microbial biomass N data.

36–2.2 Substrate-Induced Respiration Method

The substrate-induced respiration (SIR) method was introduced by Anderson and Domsch (1978b) to rapidly estimate the amount of C held in living, non-resting microorganisms in soil samples. The initial respiratory response to glucose as an added C and energy source was taken as an index of existing soil microorganisms before new synthesis of microorganisms occurred.

For incubations at 22 °C, a substrate-induced maximal respiration rate of 1 mL of CO_2 h^{-1} corresponds to about 40 mg of microbial C. The SIR method has been correlated against microbial biovolume measurements (West & Sparling, 1986; Beare et al., 1990) and ATP determinations (West & Sparling, 1986; Kieft & Rosacker, 1991). The SIR method as applied to replicate soil samples should be done according to the following steps.

36–2.2.1 Optimal Glucose Amendment

It is essential to standardize each soil relative to the respiratory response to glucose. The lowest glucose concentration that will give the maximum initial respiratory response (mL of CO_2 h^{-1}) is determined as measured by CO_2 evolution. The glucose amendment will range from 5 to 400 μM g^{-1} soil solution. Glucose concentrations should be reported in relation to the total soil solution as well as per gram dry weight of soil to avoid misinterpreting glucose concentrations as a result of varying soil water contents.

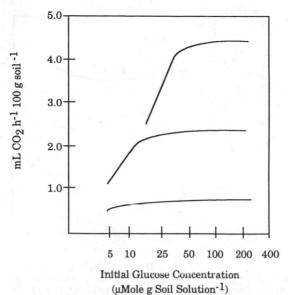

Fig. 36–2. The figure exemplifies the respiratory response from different levels of glucose amendment. The asymptote of each curve represents the minimum concentration of glucose that produces the maximal respiration rate.

Anderson and Domsch (1978b) suggest adding the glucose in dry form (0.5 g of talc plus glucose ground to a fine powder with mortar and pestle) to facilitate thorough mixing of the amendment. Our laboratory routinely adds amendments in liquid form to bring slightly dried soil to 55% of water-holding capacity with good results. West and Sparling (1986) suggest using a water-soil slurry (2:1) to minimize substrate dispersion problems and water limitation during the incubation period.

The soil is placed in a gas-tight container suitable for CO_2 headspace analysis. For soils low in organic matter, as much as 100 g of dry weight is necessary, whereas for organic layers of forest soils 10 g may be sufficient. A concentration series of glucose amendments are added to replicate soil samples (see above) to determine the lowest glucose level that yields maximum respiratory response. Figure 36–2 exemplifies a series of responses obtainable from varied soil samples (Anderson & Domsch, 1978b). When the respiratory response approaches an asymptote, the corresponding glucose level is defined as the minimum concentration of glucose invoking maximal respiratory response. The CO_2 is analyzed using a CO_2 analyzer, gas chromatography, or infrared gas analyzer to measure the amount of CO_2 respired and is expressed as mL CO_2 h^{-1} g^{-1} of dry weight soil.

The minimum concentration of glucose giving maximal respiratory response is added to replicate subsamples of soil. Enough replicate subsamples to reduce measurement error and express statistical variation are analyzed. The response of soil to the amendment is variable and, therefore,

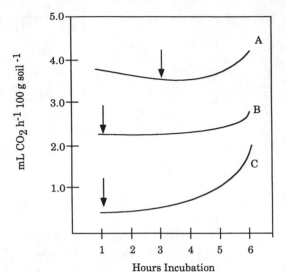

Fig. 36–3. Curves A–C represent a series of respiratory responses obtained from the predetermined glucose amendment. The arrows indicate CO_2 (mL CO_2) values used to calculate
microbial biomass C. Measurements are taken before CO_2 increases which indicates the
synthesis of new microbial biomass C.

hourly measurements are done to encompass increases, decreases or lags in
CO_2 efflux that is then followed by a normal increasing rate. The minimum
hourly rate of CO_2 production is recorded to determine biomass size (Fig.
36–3). Curves A–C of Fig. 36–3 represent possible respiratory responses as
a result of the predetermined glucose amendment. The CO_2 mineralization
rates show negative, zero, and positive slopes and associated rate maximum for each response. A positive tendency or increase in CO_2 production
is interpreted as new biomass synthesis indicating that measurements must
precede this event (Anderson & Domsch, 1975).

36–2.2.2 Calculation of Biomass Carbon

Anderson and Domsch (1978b) using 12 soils ranging from 0.778 to
39.2% C correlated the SIR method to CFI method to estimate total
microbial biomass C and developed the expression:

$$x = 40.04y + 0.37 \ (r^2 = 0.96)$$

where

$$x = \text{total microbial biomass C}$$

$$y = \text{maximum initial rate of } CO_2 \text{ respiration}$$

$$(\text{mL of } CO_2 \ g^{-1} \text{ dry weight soil}).$$

The above relationship is valid only for SIR incubations done at 22 °C.

The SIR method correlates well to other biomass methods and involves a short analysis period (1–3 h). The method has been modified through additions of mineral salts, nutrient broth, and yeast extract to produce maximal respiratory response from nutrient imbalanced soil or long-term incubations where nutrient status may change (Sparling, 1981; Smith et al., 1985). The SIR method has also been used to evaluate pesticide damage to the soil microorganisms (Anderson, 1981). Additional modifications include measuring bacterial and fungal contributions to soil and litter metabolism using selective inhibitors (Anderson & Domsch, 1973; Beare et al., 1990). A discussion of these modifications can be found in chapter 9 by Turco in this book. Under standard assay conditions (22 °C and 55% of soil water-holding capacity), this method accurately estimates biomass C by probing the respiration response of in situ soil microbial populations.

36–3 CHEMICAL METHODS

The extraction of unique compounds representative of the microbial community is an attractive method for determining the size of the microbial biomass. The extraction of ATP, nucleic acids, muramic acid, chitin, and other biomass components have been reviewed (Jenkinson & Ladd, 1981; Nannipieri et al., 1990). Methods for determining the lipid P and ergosterol content of microbial biomass are reviewed by Grant and West (1986). The determination of microbial C and N by direct extraction and the determination of ATP are covered in detail here.

36–3.1 Chloroform Fumigation Extraction Method

Microbial constituents released by fumigation and extracted directly can be used to determine the size of the soil biomass. The CFE is correlated to biomass C and N as determined by CFI (Brookes et al., 1985; Vance et al., 1987a; Gallardo & Schlesinger, 1990). This method has several potential advantages over CFI including:

1. No NH_4^+ immobilization or denitrification activity.
2. Low interference from nonmicrobial labile C and N substrates that can be used during the incubation.
3. Shorter analysis time.

However, ^{14}C specific activity of the microbial biomass C from the CFE does not correlate well with that obtained by CFI method in soil labeled with ^{14}C-labeled plant residues (Horwath, 1992, unpublished data). We have found that when extracting soils incubated (up to 1 yr) with ^{14}C substrates, the CFE specific activity is approximately one-half that of CFI indicting that different C pools are being sampled. Recent investigations (Merckx & Martin, 1987; Badalucco et al., 1990) suggest that additional anthrone-reactive and ninhydrin-reactive C of nonbiomass origin is released from fumigated soil. Even though the CFE and CFI methods release similar amounts of microbial C, the prospect of extracting dissimilar soil organic C pools with these two methods is disturbing and requires further study.

Soils held under chloroform fumigation retain protease activity, but lose dehydrogenase and C and N immobilization activity (Amato & Ladd, 1988). As a result, soluble C, organic N, and NH_4^+ levels increase until extracellular enzyme activity ceases or substrate becomes limiting. In some soils, a chloroform treatment of 1 d releases all the potentially extractable microbial products, while other soils require an exposure of up to 5 d (Brookes et al., 1985; Davidson et al., 1989).

36–3.1.1 Fumigation and Extraction

Soil samples are prepared and weighed as outlined in the sample preparation section 36–2.1.1, 10 to 20 g of dry weight, are weighed in triplicate into containers suitable for fumigation and extraction. The fumigation of soil samples is done according to CFI outlined in section 36–2.1.2, but soil fumigation is extended to attain maximum levels of soluble C and N. Fumigation periods of 1 d can be done for rapid soil biomass determinations (Brookes et al., 1985; Voroney et al., 1991). However, since soils vary in microbial activity, we recommend a 5-d fumigation as a standard time for the analytical determination of the soil biomass unless analysis of the particular soil has shown that the longer fumigation time is not necessary.

The fumigated and unfumigated soil are extracted with 0.5 M of K_2SO_4 at a ratio of 5:1 (weight of extractant to dry soil weight). The soil and extractant are usually shaken on a reciprocal shaker at 180 strokes per minute for 1.0 h. Dispersion such as in a Waring blender has been found necessary for well-aggregated soils in the determination of microbial plate counts and should be considered for CFE. After shaking, the soil suspension is filtered and the filtrate collected. The filtrate is stored at 4 °C (maximum of 1 wk) or frozen until analyzed. A blank filtrate, extractant alone, is run for each batch of samples analyzed to determine background levels of C and N in both the filter paper and extractant.

36–3.1.2 Determination of Biomass Carbon

Soluble organic C is determined on both the fumigated and unfumigated soil extracts. The soluble C is best analyzed on any suitable commercial soluble C analyzer. When such devices are not available, wet combustion techniques can be employed. Jenkinson and Powlson (1976a) describe a dichromate digestion in which an aliquot of soil extract is added to a mixture of potassium dichromate, sulfuric acid, phosphoric acid and mercury, and boiled under refluxing conditions for 30 min. The excess dichromate is titrated with ferrous ammonium sulfate using ferroin as an indicator.

Our laboratory routinely combines ^{14}C measurements with those of microbial biomass. We use a persulfate digest adapted from McDowell et al. (1987). The persulfate digest will be discussed in detail because of its ease and reduced safety requirements as well as its ability to capture the liberated $^{14}CO_2$. The method includes the following:

- 10 to 15 mL of soil extract
- 1 g of $K_2O_8S_3$ (persulfate)
- 1.0 mL of 0.025 M of H_2SO_4
- 1.0 mL of 0.1 M of NaOH placed in 15 × 45 mm vial

The digestion is done in 25 × 200 mm culture tubes equipped with a Poly Seal cap (Fisher Scientific, Fair Lawn, NJ) (Fig. 36–4). The persulfate, sulfuric acid, and filtrate (10–15 mL) are placed in the culture tube, the alkali trap is inserted, and the tube promptly capped producing a pressure-tight seal. The alkali is suspended above the digestion mixture by putting a restriction in the glass or by placing the alkali on a glass rod support. The samples are heated in a digestion block at 120 °C for 2 h. The digested samples are removed and allowed to stand overnight to complete the trapping of the liberated CO_2 into the alkali. A brown precipitate of oxidized Fe may form in the fumigated samples but does not affect the analysis. The base traps are then titrated according to section 36–2.1.3 to determine total sample C (1.0 mL of 0.1 M of NaOH can consume 600 μg CO_2-C). The base traps may be subsampled or the entire contents of the titrated traps used to determine ^{14}C activity. The blank filtrate is digested to determine background C for the control and fumigated sample. A set of glucose standards in 0.5 M of K_2SO_4 should be assayed with each sample run to verify the results of the assay.

36–3.1.3 Determination of Biomass Nitrogen

Total N in the extract (NO_3^-, NH_4^+ and organic N) is determined by Kjeldahl digestion (see chapters 40 and 42 in this book) (Brookes et al., 1985) or by ninhydrin-reactive N analysis (Amato & Ladd, 1988). Amato and Ladd (1988) used 2 M of KCl to extract the soil and determine organic N by the ninhydrin method, but this method has been modified using K_2SO_4 (Joergensen & Brookes, 1990) to allow for the analysis of both C and N from the same sample. When measuring ^{15}N-labeled biomass, the Kjeldahl procedure is used to convert organic N to NH_4^+, and distillation or diffusion is used prior to ^{15}N analysis (see chapters 40 and 42 in this book). Problems with NO_3-N if present in the extract are discussed elsewhere (chapter 42).

36–3.1.4 Calculation of Biomass Carbon

The amount of soluble C in the fumigated and unfumigated soil extract are used to determine biomass C expressed as:

$$Biomass\ C = (C_f - C_{uf})/K_{ec}$$

where

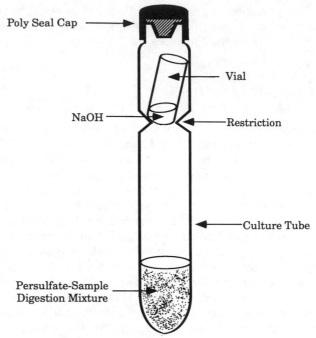

Fig. 36–4. The diagram illustrates the components of the persulfate digestion. The sample (extract of soil), digestion chemicals, and CO_2 trap are placed into the modified culture tube and sealed with a Poly Seal cap. Care must be exercised to avoid heating the cap excessively during the digestion that may distort the seal.

$$C_f = C \text{ in the fumigated extract}$$

$$C_{uf} = C \text{ in the unfumigated extract.}$$

The value of K_{ec} is the proportion of the microbial C that is extracted from the soil. The extraction of labile microbial C rendered soluble by fumigation may partition differentially between soil organic matter, clay, and extractant for different soils. Therefore, the value of K_{ec} will depend on the physical and chemical properties of the soil. Voroney et al. (1991) suggests a K_{ec} of 0.35 as a general value for microbial C-extraction efficiency.

36–3.1.5 Calculation of Biomass Nitrogen

Biomass N is calculated as the flush of N from a fumigated soil less that extracted from an unfumigated soil:

$$\text{Biomass N} = (N_f - N_{uf})/K_{en}$$

where

$$N_f = \text{total N from the fumigated soil extract}$$

$$N_{uf} = \text{total N from the unfumigated extract}$$

The value of K_{en} is the efficiency of extraction of organic microbial N and inorganic N from soil. The efficiency of extracting organic N has the same theoretical limitations associated with soluble organic C. The variable K_n equation (Voroney & Paul, 1984), discussed in section 36–2.1.6, has been used with some success to correlate to CFI (Davidson et al., 1989). Brookes et al. (1985) suggest an extraction efficiency of 0.68 across several soils for a 5-d fumigation period. A realistic value of K_{en} would have to be developed for each soil to determine exact values for biomass N. As in the case of CFI, the value of K_{en} should be included with all data sets to facilitate comparison with other microbial biomass values.

36–3.2 ATP Determinations

Much effort has been expended over the last 20 yr searching for efficient extractants of ATP from soils and sediments. Extractants such as H_2SO_4, DMSO, butanol, Tris buffer, $NaHCO_3$-$CHCl_3$, $NaHCO_3$, and $HClO_4$ have been tested (Lee et al., 1971; Conklin & MacGregor, 1972; Paul & Johnson, 1977). The ideal extractant should disrupt microbial cells rapidly, stabilize ATP by deactivating synthesizing and degradative enzymatic processes and quantitatively remove ATP from the soil matrix (Nannipieri ct al., 1990).

Multi-component extractants hold promise for the quantitative extraction of cellular ATP from soil (Jenkinson & Oades, 1979; Webster et al., 1984; Martens, 1985). Jenkinson and Oades (1979) developed an extractant consisting of trichloracetic acid, Na_2HPO_4, and paraquat (TCAPP). A phosphoric acid mixture (PA) developed by Webster et al. (1984) was found to be more efficient than 12 other methods, including TCAPP. However, this work was done on oven-dried soil, and interpreting the results is difficult. Ciardi and Nannipieri (1990) found that the PA mixture recovered ATP 1.7 to 3 times more efficiently from two soils under different agronomic management than either the TCAPP or a $NaHCO_3$-$CHCl_3$ phosphate-adenosine mixture (Martens, 1985). Arnebrant and Bååth (1991) found no difference in ATP-extractant efficiencies in forest humus with TCAPP, PA or an extractant (500 mM of H_2SO_4 and 250 mM of Na_2HPO_4) proposed by Eiland (1983). In many cases, the light output of the luciferase reaction is influenced by the extracting agent and buffers and careful use of controls and standardized conditions are necessary so that extraction efficiencies of extractants can be thoroughly scrutinized.

Paraquat used in the TCAPP can be difficult to prepare and is listed as a hazardous substance at many research institutes. The PA method has been said to be an equally efficient extractant as TCAPP and has been employed on a variety of soils and sediments (Gregorich et al., 1990; Kieft & Rosacker, 1991; Arnebrant & Bååth, 1991). A variety of commercial extractants applicable to soil are also available (Fallon & Obrigawitch,

1990) but will not be discussed. We will describe the PA extractant in detail, but emphasize that no extraction methods have been thoroughly scrutinized over a broad range of soils.

36–3.2.1 ATP Extraction Method

The components of the PA mixture (Webster et al., 1984) are designed to quench metabolic processes by destabilizing cell compartmentalization and solubilizing ATP. The H_3PO_4 extracts ATP, inactivates ATPases and saturates phosphate group binding sites. EDTA chelates metal ions and prevents inhibition of the luciferin-luciferase reaction. The adenosine molecules saturate ATP-binding sites. Urea quenches metabolic reactions by denaturing enzymes that catalyze the metabolism of nucleotides. Dimethylsulfoxide (DMSO) and detergents such as polyoxeyethylene 10 laryl ether remove and lyse cells from soil surfaces. The procedure as adapted from Webster et al. (1984) modified by Vaden et al. (1987) and recently updated (F.R. Leach, 1992, personal communication) is outlined below.

36–3.2.1.1 Reagent Preparation. Dissolve 0.5 g of polyoxeyethylene 10 laryl ether (Sigma, St. Louis, MO) in 34 mL of warm water at 45 °C and maintain at 35 °C.

10 N of phosphoric acid, 228 mL (reagent grade) L^{-1} of water.

10 M of Urea (Fisher Scientific, Fair Lawn, NJ), 600 g L^{-1} of water, warm to dissolve, use at room temperature.

Dimethylsulfoxide (DMSO) (J.T. Baker Inc., Phillipsburg, NJ).

Adenosine (Sigma Chemical Co., St. Louis, MO) 2.5 g 500 mL^{-1} of water, warm to dissolve, use at room temperature.

1 M of EDTA (Fisher Scientific, Fair Lawn, NJ), 45.22 g 100 mL^{-1} of water.

Water used to make the above solutions should be purified, filtered through a 0.22-μm filter and autoclaved.

Combine the above in the following order:

34 mL of warm polyoxeyethylene 10 laryl ether
20 mL of 10 N phosphoric acid
20 mL 10 M of Urea
20 mL of DMSO
4 mL of adenosine solution
2 mL of 1 M of EDTA

36–3.2.1.2 Extraction Procedure. Soil samples (1–2 g wet wt.) are combined with 10 mL of the extractant in a sterile centrifuge tube and sonicated for 1 min (see Webster et al., 1984) or shaken vigorously for 30 min. The sample is then centrifuged (30 000 × g, 20 min, 20 °C). The supernatant is then diluted with 0.1 Tricine buffer (pH 11.2) yielding a desired pH of between 7 and 8.

36–3.2.1.3 ATP Measurement. ATP is measured in a reaction vessel containing 50 μL of sample; 100 μL of Analytical Luminescence Laboratory's fire fly luciferase (Firelight); and 50 μL of Tricine buffer pH 7.8 containing 25 mM of Tricine, 5 mM of MgSO$_4$, 1 mM of EDTA and 1 mM of dithiothreitol. Light production from the luciferin-luciferase reaction vessel is best measured on a ATP Photometer (Integrating Photometer). Since many models of photometers exist, it is left to the reader to develop a protocol for individual models. The luciferin-luciferase light reaction can also be analyzed in a Liquid Scintillation Spectrometer with gain set at 100%, narrow window setting and photomultiplier tubes switched out of coincidence (Jenkinson & Oades, 1979).

Variations of this procedure have been reported and applied to various soil and sediment samples and it is up to the reader to decide what modifications will be incorporated into their protocols (Ciardi & Nannipieri, 1990; Kieft & Rosacker, 1991; Arnebrant & Bååth, 1991). Internal standardization of this procedure can be done using either added *E. coli* cells (10[8] cells, see Webster et al., 1984) or by amending the extractant or soil with a known quantity of ATP (5 μM). Additionally, the determination of adenylate energy charge can be done to measure metabolic energy stored in cells. Since this is not a measure of microbial biomass size, readers are referred to Vaden et al. (1987) for the analytical procedure and Nannipieri et al. (1990) for a review of this topic.

36–3.2.2 Calculation of Biomass

The relationship between C metabolized and biomass synthesized lead to the assumption that catabolic and anabolic reactions are synchronized (Tempest & Neijssel, 1987). However, studies have indicated that growth energetics and yield, especially ATP formation, are not coupled when growing conditions are adverse (Karl, 1980). Soil microorganisms are generally believed to be substrate limited, which complicates the relationship between ATP and cell biomass. Data obtained from ATP assays are difficult to relate to total microbial biomass determinations (biovolume or C).

Biomass C/ATP ratios ranging from 171 (Tate & Jenkinson, 1982) to 400 to 500 (Sparling, 1981; Martens, 1985), with many values in between, have been reported for a variety of soils. Long-term soil incubation experiments have suggested that little change in ATP concentrations occurs as the biomass is slowly starved of fresh substrate input (Brookes et al., 1987; Joergensen et al., 1990). However, results from amendment experiments show changing levels of ATP associated with substrate and time (Paul & Johnson, 1977; Nannipieri et al., 1978; Martens, 1985; Rosacker & Kieft, 1990). The discrepancy between reported results makes it difficult to compare data across a wide range of soils. It would be wise to supplement ATP determinations with one or more other biomass methods so that biomass data from different soils can be better compared.

36–4 COMPARISON OF METHODS

We have described several alternative and complementary methods since all are subject to different interpretations and require careful standardization for specific soil types. CFI, being one of the first methods developed, has been used as a baseline for correlations, but CFE is gaining acceptance because of its greater simplicity and lack of problems with the interpretation of a control. Both the CFI and CFE assay components of microbial biomass necessary to interpret nutrient-cycling processes, soil organic matter dynamics, cultural practices, and inputs associated with agronomic and natural systems.

The SIR method describes the soil microbial biomass by assaying the respiration dynamics of organisms as a result of added substrate and is useful to interpret temporal biomass changes and activity as affected by management and anthropogenic inputs. The respiratory quotient or CO_2 evolved by soil relative to the biomass as determined by SIR is particularly useful in cross comparing sustainable agriculture management techniques (Anderson & Domsch, 1989). The ATP method has been criticized because of the wide variance between ATP content and biomass C; it does, however, hold promise for studying microsites such as aggregates, rhizosphere, and deep sediment samples, since only a small amount of sample is required. It also is rapid and could be particularly useful for intra-site comparisons where a large number of replicates or treatments are involved.

We recommend careful use of internal standards and cross-referencing to other methods described in this chapter. If this is done, meaningful comparative values of microbial biomass can be determined by the method of choice. It is difficult to maintain soils at a known microbial biomass. If this were possible, the distribution of standard soil samples could eliminate problems in interpretation between laboratories.

The correlation of data from other methods to the results from CFI builds in the limits involved in this technique. Direct microscopy although slow and susceptible to differences in which person does the counting should be more often used to standardize techniques. This technique also gives estimates of sizes of organisms and fungal/bacterial ratios. Meaningful comparisons within sites on one study are fairly easy to achieve with any of the methods. Cross-site and cross-investigator comparisons, however, are also important. These should involve a thorough reporting of measurements made for $K_{c(ec)}$, $K_{n(en)}$, and control extractions. There is, unfortunately, too much information in the literature where internal standardization and cross comparison of techniques for a specific soil have not been carried out. We hope that results in the future will provide more meaningful cross-site data.

REFERENCES

Amato, M., and J.N. Ladd. 1988. Assay for microbial biomass based on ninhydrin-reactive nitrogen in extracts of fumigated soil. Soil Biol. Biochem. 20:107–114.

Anderson, J.P.E. 1981. Soil moisture and the rates of biodegradation of diallate and triallate. Soil Biol. Biochem. 13:155–161.

Anderson, J.P.E., and K.H. Domsch. 1973. Quantification of bacterial and fungal contributions to soil respiration. Arch. Mikrobiol. 93:113–127.

Anderson, J.P.E., and K.H. Domsch. 1975. Measurement of bacterial and fungal contributions to soil respiration of selected agricultural and forest soils. Can. J. Microbiol. 21:314–322.

Anderson, J.P.E., and K.H. Domsch. 1978a. Mineralization of bacteria and fungi in chloroform fumigated soils. Soil Biol. Biochem. 10:207–213.

Anderson, J.P.E., and K.H. Domsch. 1978b. A physiological method for the quantitative measurement of microbial biomass in soils. Soil Biol. Biochem. 10:215–221.

Anderson, J.P.E., and K.H. Domsch. 1989. Ratios of microbial biomass carbon to total organic carbon in arable soils. Soil Biol. Biochem. 21:471–479.

Arnebrant, K., and E. Bååth. 1991. Measurements of ATP in forest humus. Soil Biol. Biochem. 23:501–506.

Badalucco, L., P. Nannipieri, and S. Grego. 1990. Microbial biomass and anthrone-reactive carbon in soils with different organic matter contents. Soil Biol. Biochem. 22:899–904.

Beare, M.H., C.L. Neely, D.C. Coleman, and W.L. Hargrove. 1990. A substrate-induced respiration (SIR) method for measurement of fungal and bacterial biomass on plant residues. Soil Biol. Biochem. 22:585–594.

Brookes, P.C., A.D. Newcombe, and D.S. Jenkinson. 1987. Adenylate energy charge measurements in soil. Soil Biol. Biochem. 19:211–217.

Brookes, P.C., A. Landman, G. Pruden, and D.S. Jenkinson. 1985. Chloroform fumigation and the release of soil nitrogen: A rapid direct extraction method to measure microbial biomass nitrogen in soil. Soil Biol. Biochem. 17:837–842.

Brooks, P.D., J.M. Stark, B.B. McInteer, and T. Preston. 1989. Diffusion method to prepare soil extracts for automated nitrogen-15 analysis. Soil Sci. Soc. Am. J. 53:1707–1711.

Ciardi, C., and P. Nannipieri. 1990. A comparison of methods for measuring ATP in soil. Soil Biol. Biochem. 22:725–727.

Conklin, A.R., and A.N. MacGregor. 1972. Soil adenosine triphosphate: extraction, recovery and half-life. Bull. Environ. Contam. Toxicol. 7:296–300.

Davidson, E.A., R.W. Eckert, S.C. Hart, and M.K. Firestone. 1989. Direct extraction of microbial biomass nitrogen from forest and grassland soils of California. Soil Biol. Biochem. 21:773–778.

Eiland, F. 1983. A simple method for the quantitative determination of ATP in soil. Soil Biol. Biochem. 15:665–670.

Fallon, R.D., and J.B. Obrigawitch. 1990. ATP extraction from soils with commercial extractants. Soil Biol. Biochem. 22:1007–1008.

Gallardo, A., and W.H. Schlesinger. 1990. Estimating microbial biomass nitrogen using the fumigation-incubation and fumigation extraction methods in a warm-temperate forest soil. Soil Biol. Biochem. 22:927–932.

Grant, W.D., and A.W. West. 1986. Measurement of ergosterol, diaminopimelic acid and glucosamine in soil: Evaluation as indicators of microbial biomass. J. Microbiol. Methods 6:47–53.

Gregorich, E.G., G. Wren, R.P. Voroney, and R.G. Kachanoski. 1990. Calibration of a rapid direct chloroform extraction method for measuring soil microbial biomass C. Soil Biol. Biochem. 22:1009–1011.

Jenkinson, D.S. 1966. Studies on the decomposition of plant material in soil. II. Partial sterilization of soil and the soil biomass. J. Soil Sci. 28:280–302.

Jenkinson, D.S. 1976. The effect of biocidal treatments on the metabolism in soil-IV. The decomposition of fumigated organisms in soil. Soil Biol. Biochem. 8:203–208.

Jenkinson, D.S. 1988. The determination of microbial biomass carbon and nitrogen in soil. p. 368–386. In J.R. Wilson (ed.) Advances in nitrogen cycling in agricultural ecosystems. C.A.B. International, Wallingford.

Jenkinson, D.S., and J.N. Ladd. 1981. Microbial biomass in soil: Measurement and turnover. p. 415–471. *In* E.A. Paul and J.N. Ladd (ed.) Soil biochemistry. Vol. 5. Marcel Dekker, New York.

Jenkinson, D.S., and J.M. Oades. 1979. A method for measuring adenosine triphosphate in soil. Soil Biol. Biochem. 11:193–199.

Jenkinson, D.S., and D.S. Powlson. 1976a. The effects of biocidal treatments on metabolism in soil-I. Fumigation with chloroform. Soil Biol. Biochem. 8:167–177.

Jenkinson, D.S., and D.S. Powlson. 1976b. The effects of biocidal treatments on metabolism in soil-V. A method for measuring soil biomass. Soil Biol. Biochem. 8:209–213.

Jenkinson, D.S., and D.S. Powlson. 1980. Measurements of microbial biomass in intact cores and in sieved soil. Soil Biol. Biochem. 12:579–581.

Joergensen, R.G., and P.C. Brookes. 1990. Ninhydrin-reactive nitrogen measurements of microbial biomass in 0.5 M K_2SO_4 soil extracts. Soil Biol. Biochem. 22:1129–1136.

Joergensen, R.G., P.C. Brookes, and D.S. Jenkinson. 1990. Survival of the soil biomass at elevated temperatures. Soil Biol. Biochem. 22:1129–1136.

Lee, C.C., R.F. Harris, J.D.H. Williams, D.E. Armstrong, and J.K. Syers. 1971. Adenosine triphosphate in lake sediments-I. Soil Sci. Soc. Am. Proc. 35:82–86.

Karl, D.M. 1980. Cellular nucleotide measurements and applications in microbial ecology. Microbiol. Rev. 44:789–796.

Kieft, T.L., and L.L. Rosacker. 1991. Application of respiration- and adenylate-based soil microbiological assays to deep subsurface terrestrial sediments. Soil Biol. Biochem. 23:563–568.

Martens, R. 1985. Estimation of the adenylate energy charge in unamended and amended agricultural soils. Soil Biol. Biochem. 17:765–772.

Martens, R. 1987. Estimation of microbial biomass in soil by the respiration method: Importance of soil pH and flushing methods of the respired CO_2. Soil Biol. Biochem. 19:77–81.

McDowell, W.H., J.J. Cole, and C.T. Driscoll. 1987. Simplified version of the ampoule-persulfate method for the determination of dissolved organic carbon. Can. J. Fish. Aquat. Sci. 44:214–218.

Merckx, R., and J.K. Martin. 1987. Extraction of microbial biomass components from rhizosphere soils. Soil Biol. Biochem. 19:371–376.

Nannipieri, P., S. Grego, and B. Ceccanti. 1990. Ecological significance of the biological activity of soil. p. 293–355. *In* J.M. Bollag and G. Stotzky (ed.) Soil biochemistry. Vol. 6. Marcel Dekker, New York.

Nannipieri, P., R.L. Johnson, and E.A. Paul. 1978. Criteria for the measurement of microbial growth and activity in soil. Soil Biol. Biochem. 10:223–229.

Paul, E.A., and R.L. Johnson. 1977. Microscopic counting and adenosine 5'-triphosphate measurement in determining microbial growth in soils. Appl. Environ. Microbiol. 34:263–269.

Quikchem Systems. 1987. Quikchem method no. 10-107-06-2-A and 12-107-04-1-A. Quikchem Systems, Division of Lachet Chemicals, Mequon, WI.

Rosacker, L.L., and T.L. Kieft. 1990. Biomass and adenylate energy charge of a grassland soil during drying. Soil Biol. Biochem. 22:1121–1127.

Ross, D.J., T.W. Speir, K.R. Tate, and V.A. Orchard. 1985. Effects of sieving on estimations of microbial biomass, carbon and nitrogen mineralization in soil under pasture. Aust. J. Soil Res. 23:319–324.

Shen, S.M., G. Pruden, and D.S. Jenkinson. 1984. Mineralization and immobilization of nitrogen in fumigated soil and the measurement of microbial biomass nitrogen. Soil Biol. Biochem. 16:437–444.

Smith, J.L., B.L. McNeal, and H.H. Cheng. 1985. Estimation of soil microbial biomass: an analysis of the respiratory response of soils. Soil Biol. Biochem. 17:11–16.

Smith, J.L., and E.A. Paul. 1990. The significance of soil microbial biomass estimations. p. 357–396. *In* J. Bollag and G. Stotzky (ed.) Soil biochemistry. Vol. 6. Marcel Dekker, New York.

Sparling, G.P. 1981. Microcalorimetry and other methods to assess biomass activity in soil. Soil Biol. Biochem. 13:93–98.

Sparling, G.P., A.W. West, and K.N. Whale. 1985. Interference from plant roots in the estimation of soil microbial ATP, C, N and P. Soil Biol. Biochem. 17:275–278.

Tate, K.R., and D.S. Jenkinson. 1982. Adenosine triphosphate measurement in soil: An improved method. Soil Biol. Biochem. 14:331–336.

Tempest, D.W., and O.M. Neijssel. 1987. Growth yield and energy distribution. p. 797–806. *In* F.C. Neidhardt et al. (ed.) *Escherichia coli* and *Salmonella typhimurium:* Cellular and molecular biology. Am. Soc. Microbiol., Washington, DC.

Vaden, V.R., J. Webster, G.J. Hampton, M.S. Hall, and F.R. Leach. 1987. Comparison of methods for extraction of ATP from soil. J. Microbiol. Methods 7:211–217.

Vance, E.D., P.C. Brookes, and D.S. Jenkinson. 1987a. An extraction method for measuring soil microbial biomass C. Soil Biol. Biochem. 19:703–707.

Vance, E.D., P.C. Brookes, and D.S. Jenkinson. 1987b. Microbial biomass measurement in forest soils: Determination of K_c values and tests of the hypothesis to explain the failure of the chloroform fumigation-incubation method in acid soils. Soil Biol. Biochem. 19:689–696.

Voroney, R.P., and E.A. Paul. 1984. Determination of k_c and k_n *in situ* for calibration of the chloroform fumigation incubation method. Soil Biol. Biochem. 16:9–14.

Voroney, R.P., J.P. Winter, and E.G. Gregorich. 1991. Microbe/plant soil interactions. p. 77–99. *In* D.C. Coleman and B. Fry (ed.) Carbon isotope techniques. Academic Press, New York.

Wardle, D.A., and D. Parkinson. 1990. Determination of bacterial and fungal fumigation K_c factors across a soil moisture gradient. Soil Biol. Biochem. 22:811–816.

Webster, J.J. G.J. Hampton, and F.R. Leach. 1984. ATP in soil: A new extractant and extraction procedure. Soil Biol. Biochem. 16:335–342.

West, A.W., and G.P. Sparling. 1986. Correlation between four methods to estimate total microbial biomass in stored, air-dried and glucose amended soils. Soil Biol. Biochem. 18:569–576.

Chapter 37

Soil Enzymes

M. A. TABATABAI, *Iowa State University, Ames, Iowa*

Nutrient cycling in soils involves biochemical, chemical, and physiochemical reactions, with the biochemical processes being mediated by microorganisms, plant roots, and soil animals. It is well known that all biochemical reactions are catalyzed by enzymes, which are proteins with catalytic properties owing to their power of specific activation. Enzymes are catalysts, that is, they are substances that without undergoing permanent alteration cause chemical reactions to proceed at faster rates. In addition, they are specific for the types of chemical reactions in which they participate. Enzyme specificity is often dictated by the nature of the groups attached to the susceptible bonds. For example, both α-chymotrypsin and trypsin are proteolytic enzymes capable of hydrolyzing certain peptide bonds in protein. α-Chymotrypsin will hydrolyze peptide bonds in which the carbonyl group of that bond is supplied by L-tryosin, L-phenylalanine, or L-tryptophan. Trypsin will hydrolyze peptide bonds in which the carbonyl group of the peptide bond is supplied by L-arginine or L-lysine. Peptide bonds containing D-amino acids are not hydrolyzed. As another example of the specificity of enzymes, maltase hydrolyzes maltose to glucose, whereas cellubiase hydrolyzes cellubiose to glucose but not vice versa. Differences between the two substrates seem slight in that maltose is an α-glucoside and cellubiose is a β-glucoside.

Enzymes are specific activators in that they combine with their substrates in such a stereospecific fashion that they cause changes in the electronic configuration around certain susceptible bonds. These bonds then are more easily changed. Physiochemical measurements indicate that enzyme-catalyzed reactions have lower activation energy than the uncatalyzed counterparts and, therefore, have faster rates of reactions (Browman & Tabatabai, 1978; Tabatabai & Singh, 1979).

Enzymes are denatured by elevated temperature and extreme pH. Their physiochemical state and their influence on chemical reactions are markedly dependent on pH, ionic strength, temperature, and the presence or absence of inhibitors or activators.

About a century ago, Woods (as cited by Skujins, 1978) presented the first report on oxidizing enzymes, especially peroxidase, in soils at the annual meeting of the American Association for the Advancement of Science in Columbus, OH. Although the progress in soil enzymology was extremely slow until 1950, exponential progress has been made in this field within the past four decades. The history of abiontic soil enzyme research has been elegantly prepared by Skujins (1978). Therefore, soil may be looked on as a biological entity, that is, a living tissue (Quastel, 1946) with complex biochemical reactions. The enzyme activity of soils results from the activity of accumulated enzymes and from those in proliferating microorganisms. Accumulated enzymes in soils are regarded as enzymes present and active in a soil in which no microbial proliferation occurs (Kiss et al., 1975). Sources of enzymes in soils are primarily the microbial biomass, although they can also originate from plant and animal residues. Enzyme activities in soils are derived from free enzymes, such as exoenzymes released from living cells, endoenzymes released from disintegrating cells, and enzymes bound to cell constituents (enzymes present in disintegrating cells, in cell fragments, and in viable but nonproliferating cells). Proliferating microorganisms produce enzymes that are released to the soil, or that remain within the multiplying cells. Free enzymes in soils are adsorbed on organic and mineral constituents or complexed with humic substances, or both. The amount of free enzymes in the soil solution is minute compared with that in the adsorbed state. Microbial cells or cell fragments also may exist in an adsorbed state or in suspension. Therefore, soil as a system of humus and minerals contains both immobilized enzymes, stabilized by a three-dimensional network of macromolecules, and occluded microbial cells. Each of the organic and mineral fractions in soil has a special influence on enzyme activity (McLaren, 1975). Schematic diagrams showing the components of the enzyme activity in soils are presented by Skujins (1976) and Kiss et al. (1975), and the various theories and hypotheses proposed to explain the protective influence of soil constituents on enzyme activity are summarized by Tabatabai and Fu (1992).

Anyone who is concerned with studies of enzymes is concerned with the general area of catalysis, and anyone who is concerned with catalysis is most certainly concerned with the velocities of chemical reactions (chemical kinetics). Therefore, if one is to understand the kinetics of an enzyme reaction, it is fundamental to understand the meaning of the results obtained for a soil sample. Details of enzyme kinetics are not discussed in this chapter but may be found in any general biochemistry textbook. Kinetics of enzyme reactions in heterogeneous systems is discussed by McLaren and Packer (1970).

At the simplest level, when investigating enzyme reaction in soil, one must develop a so-called assay for an enzyme. This involves adding a known amount of soil to a solution containing a known concentration of substrate species and determining the rate at which the substrate is converted to a product under carefully controlled conditions of temperature, pH, and ionic strength. Before an adequate assay can be designed, however, one must know (i) the reaction catalyzed in stoichiometric detail; (ii) the species, besides the substrate, that must be present (e.g., Mg^{2+} is required for the reaction catalyzed by pyrophosphatase); (iii) the kinetic dependence of the reaction on such required species; (iv) the optimum conditions of temperature, pH, and ionic strength; and (v) a suitable means for monitoring substrate disappearance or product appearance.

One of the most difficult problems currently facing soil biologists and biochemists is the separation of the extracellular enzymatic activities from those associated with the living organisms. Soil handling has a marked effect on the results obtained in enzyme studies. Ideally, the chemical and physical properties should not be altered before enzyme assay. Although antiseptic agents have been used to retard or minimize microbial growth during the assay, especially in assays involving long incubation times, it is essential to evaluate the effect of the antiseptic agent on the enzyme-catalyzed reaction rate before its use in enzyme assay. Studies by Frankenberger and Johanson (1986) on the use of plasmolytic agents and antiseptics in soil enzyme assays showed that toluene, ethanol, and Triton X-100 effectively prevented microbial proliferation during the assay of soil amidase. Each of these agents had a different effect on the activities of several enzymes that they studied. Toluene is a plasmolytic agent as well as an antiseptic and has little effect or only slightly inhibits purified preparation of acid and alkaline phosphatases, α-glucosidase, and invertase. Toluene severely inhibits catalase and dehydrogenase activities, but enhances the activities of arylsulfatase and urease (Tabatabai & Bremner, 1970a, 1972a; Frankenberger, & Johanson, 1986). The increase of soil arylsulfatase and urease activities in the presence of toluene suggests that, because of its plasmolytic character, toluene affects the intercellular enzyme activities of the measured activities in soils. Methods based on irradiation of soil samples with high-energy radiation sterilization and use of antibiotics have also been evaluated, but such methods, like antiseptic methods, alter the enzyme-reaction rates (Tabatabai & Bremner, 1970a).

Although there are reports on extraction of enzymes from soils (for review see Tabatabai & Fu, 1992), no soil enzyme has been purified to the extent of those extracted from microorganisms, plant, and animal tissues. The extracted activities are usually associated with carbohydrate-enzyme complexes and often difficult to purify. Therefore, most of the studies on enzyme activities in soils have been based on measuring the rates of enzyme-catalyzed reactions in soil samples. A large number of enzyme-catalyzed reactions in soils have been studied, but only a few assay methods have been thoroughly evaluated to be considered standard assays. Therefore, this chapter does not deal with all the enzymes investigated in soils but does cover those assay methods that have been thoroughly evaluated. The enzymes covered are those important in C, N, P, and S cycling in soils. Additional review articles include Skujins (1967, 1976, 1978), Ladd (1978), Roberge (1978), Tabatabai and Fu (1992), and Dick and Tabatabai (1992).

37–1 PRINCIPLES

37–1.1 Mechanism of Enzyme Action

The two most remarkable properties of enzymes are their specificity and catalytic efficiency, and it is in these properties that enzymes differ most strikingly from simple catalysts. When it is possible to compare the enzymatic rates with nonenzymatic counterparts, one finds that enzymes enhance the reaction by several orders of magnitude (Segel, 1975).

Most theories and mathematical analyses of enzyme reactions are based on the concept that an enzyme acts by forming a complex or compound with the substrate. Presumably the complex of enzyme and substrate is unstable and proceeds through a step or number of steps of rearrangement to form the product plus the original enzyme. This theory of enzyme action was proposed by Michaelis and Menten (1913) and may be expressed by the following equation:

$$S + E \underset{k_2}{\overset{k_1}{\rightleftharpoons}} ES \overset{k_3}{\rightarrow} E + P \qquad [1]$$

where S is the substrate, E is the enzyme, ES is the intermediate enzyme-substrate complex, P is the product of the reaction, and k_1, k_2, and k_3 are the respective reaction velocity constants or rate constants of the three processes.

The reaction of any enzyme is generally observed by measuring the rate of the chemical reaction being catalyzed by the enzyme. Equations describing the kinetics of various types of chemical reactions may be found in physical chemistry textbooks, and rate equations for enzyme reactions may be found in Reiner (1959), Alberty (1956), and Segel (1975). Equations describing the kinetics of enzyme reactions in heterogeneous systems are described by McLaren and Packer (1970).

37–2 FACTORS AFFECTING RATES OF ENZYME REACTIONS

37–2.1 Concentration of Enzyme and Substrate

Most enzyme-catalyzed reactions may be described as either a zero-order reaction (reaction where the rate is constant and independent of substrate concentration) or first-order reaction (reaction where the rate at any time is proportional to the existing substrate concentration). The zero-order reaction expressed in a differential equation takes the form of $dx/dt = k$, upon integration $k = x/t$. The differential equation for the first-order reaction is $dx/dt = k(a - x)$, and upon integration is

$$k = \frac{1}{t} \ln \frac{a}{a-x} = \frac{2.303}{t} \log \frac{a}{a-x}. \qquad [2]$$

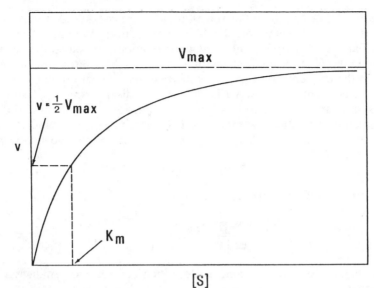

Fig. 37–1. The rectangular hyperbola describing the effect of substrate concentration [S] on reaction velocity (v).

In the first-order reaction, a is the initial substrate concentration. In both the zero-order and first-order equation, x is the concentration of converted substrate, t is the unit of time (usually in minutes), and k is the proportionality constant. Irrespective of the reaction order, the rate of the reaction is proportional to the enzyme concentration. Therefore, the value of the proportionality constant k may be considered to consist of a combination of enzyme concentration and a constant. But since the concentration of the enzyme does not change during the reaction, it may be considered as part of the proportionality constant.

When the enzyme concentration is held constant and the substrate concentration is allowed to vary over a wide range, the velocity of the enzyme-catalyzed reaction may be described by a rectangular hyperbola (Fig. 37–1). At low substrate concentrations, the velocity of the reaction follows a first-order reaction equation. As higher concentrations of substrate are reached, a maximum velocity is obtained independent of further amounts of substrate, and the zero-order reaction equation applies. The magnitude of the maximum velocity is proportional to the concentration of the enzyme; the higher the enzyme concentration, the higher the maximum velocity. A mathematical derivation of a rate equation for the effect of substrate concentration on the velocity of enzyme-catalyzed reaction was first introduced by Michaelis and Menten (1913). Derivatives of their equation are given by Neilands and Stumpf (1964) and White et al. (1968). From the Michaelis-Menten theory of enzyme action expressed in Eq. [1], it follows that the reaction rate (v) is, at any moment, equal to k_3 ES. When $S \rightarrow \infty$, the reaction rate approaches a maximum (V_{max}) equal to k_3 E_t, where $E_t = E + ES$. If initial reaction rates are measured, the substrate concentration will remain practically constant during the time of measurement. Further, if $S \gg E_t$, only a negligible amount of substrate can accumulate in the form of the intermediate complex (ES). From this it follows that: (i) the concentration of free substrate may be set equal to the total substrate concentration, and (ii) the rate of disappearance of substrate equals the rate of formation of product (steady state); when the rate of formation and disappearance of ES are equal, i.e., when $d[ES]/dt$ equals $-d[ES]/dt$, the following equation then describes the steady state:

$$k_1([E] - [ES])[S] = k_2[ES] + k_3[ES]. \qquad [3]$$

Equation [3] may be rearranged to give

$$\frac{[S]([E]-[ES])}{[ES]} = \frac{k_2-k_3}{k_1} = K_m \qquad [4]$$

where the symbols are the same as described before, and [] indicates concentration.

The term containing the three velocity constants is usually called K_m, the Michaelis constant. Numerically, K_m is equal to the substrate concentration (expressed in moles per liter) at half-maximum velocity rate ($V_{max} = 2v$). When $k_2 \gg k_3$, K_m may be set equal to the dissociation constant (k_2/k_1) of the enzyme-substrate complex, and $1/K_m$ then becomes the "affinity" constant. If one assumes that such conditions exist in soil systems, the affinity constants of several enzyme substrates have been reported (e.g., $S_2O_3^{2-}$ and CN^- for the rhodanese reaction (Tabatabai & Singh, 1979).

The Michaelis-Menten equation

$$v = V_{max}[S]/(K_m + [S]) \qquad [5]$$

satisfactorily describes the rectangular hyperbola curve (Fig. 37–1), and it agrees with the theory that enzymes act by forming an unstable complex of enzyme and substrate. Equation [5] can be written in three linear transformations:

$$\frac{1}{v} = \frac{1}{V_{max}} + \frac{K_m}{V_{max}}\frac{1}{[S]}$$

Lineweaver-Burk transformation   [6]

$$\frac{[S]}{v} = \frac{K_m}{V_{max}} + \frac{1}{V_{max}} \cdot [S]$$

Hanes-Wolf transformation   [7]

$$v = V_{max} - K_m \cdot \frac{v}{[S]}$$

Eadie-Hofstee transformation. [8]

Plots of the variables of such relationships normally give straight lines. The values of the slopes and intercepts are commonly used for determination of the constants from a set of experimental data. Once the K_m and V_{max} are known for a particular enzyme reaction under a given set of conditions, the reaction velocity, v, can be calculated for any substrate concentration. The Michaelis constant is by far the most fundamental constant in enzyme chemistry. It has the dimensions of concentration (i.e., moles per liter), and it is a constant for the enzyme only under rigidly specified conditions. The K_m value is useful in estimating the substrate concentration necessary to give a maximum velocity. The form of the Michaelis-Menten equation is such that approximately 10 and 90% of V_{max} is achieved at substrate concentration corresponding to $K_m \times 10^{-1}$ and $K_m \times 10$, respectively (Neilands & Stumpf, 1964).

Application of the Hanes-Wolf transformation (Eq. [7]) of the Michaelis-Menten equation in studies of the K_m values of soil phosphatases and arylsulfatase has been reported by Tabatabai and Bremner (1971). They showed that the apparent K_m values of these enzymes are markedly affected by shaking the soil-substrate mixture during incubation. A similar conclusion was reached by Tabatabai (1973) in determination of the apparent K_m value of urease in soils and soil fractions. Since then, studies on the Michaelis constants of these enzymes in soils have been reported by several investigators (Cervelli et al., 1973; Brams & McLaren, 1974; Thornton & McLaren, 1975; Pettit et al., 1976, 1977; Eivazi & Tabatabai, 1977). By changing the buffer and pH used by Tabatabai and Bremner (1969) in assay of phosphatase activity in soils, Cervelli et al. (1973) showed that the substrate, p-nitrophenyl phosphate, is sorbed by soils in the presence of 0.5 M NaOAc buffer (pH 4.7). They used adsorption parameters derived from the Freundlich isotherm to determine concentration of the free substrate and calculate the K_m value of acid phosphatase. Irving and Cosgrove (1976) examined the graphical techniques used in calculation of the K_m values of acid phosphatase and concluded that the linear transformation of Eadie-Hofstee (Eq. [8]) should be used. Presumably this graph shows the greatest deviation from the classical Michaelis-Menten equation, but published work shows that the linear transformations are equally applicable for estimation of the apparent K_m values of enzymes in soils (Pettit et al., 1977; Browman & Tabatabai, 1978; Dick & Tabatabai, 1978a; Tabatabai & Singh, 1979). Each transformation gives different weight to errors in the variables (Dowd & Riggs, 1965), and this is reflected in the variation of estimated K_m and V_{max} values derived for any soil enzyme by using the different plots.

The routine assay of a particular enzyme is usually carried out under conditions where the enzyme is saturated with the substrate. Under such conditions, slight changes in the substrate concentration change the measured reaction rate only slightly. Also, the formation of the product is occurring at the maximum rate, which usually permits greatest ease in its determination.

The maximum velocity can be used for determination of enzyme concentration if the molecular weight of the enzyme is known. The determination of molecular weight of soil enzymes is not possible at present, however, because of the difficulties associated with extraction and purification of enzyme proteins in soils (Tabatabai & Fu, 1992). In the soil system, an assay of enzyme activity is usually employed and expressed in units such as micrograms or micromoles of product appeared or substrate disappeared per amount of enzyme taken (e.g., per amount of dry soil or unit of organic C) per unit of time. Theoretically, measurement of either the amount of product released or the decrease in concentration of the substrate is sufficient in enzyme assays, but in practice, measurement of the product released is usually much more precise because the value is based on the difference between zero, or close to zero, and finite concentration. To measure the decrease in substrate concentration after incubation time, this decrease must be large enough to be determined accurately, and at the same time, the difference must not be so large to cause a significant change in the concentration of the substrate. Assays based on measurement of the decrease in the concentration of the substrate usually are less precise than those based on measurement of product appearance.

Soil treatments and methods of preparation have a marked effect on the results obtained in enzyme assays. The effects of air-drying, storage conditions, storage temperature, and grinding have been documented (Tabatabai & Bremner, 1970b; Speir & Ross, 1975; Zantua & Bremner, 1975b; Speir, 1977). An increase in activity can usually be obtained by subdividing the soil into separates and selecting the finer soil fraction for enzyme assay (A.D. Haig, 1955. Some characteristics of esterase- and urease-like activity in the soil. Ph.D. diss., Univ. of California, Davis.) Preparation of a homogenous soil sample is important, because soil enzymes are unevenly distributed in the soil fractions (Speir, 1977). These and other factors (e.g., shaking the soil-substrate mixture) affect enzyme activities and should be considered in studies involving kinetics of enzyme reactions in soils.

37–2.2 Temperature

The activity of any chemical reaction increases with temperature, approximately doubling for every increase of 10 °C. The rate of enzyme-catalyzed reaction increases as the temperature increases until some high temperature is reached at which the rate begins to decrease because of enzyme inactivation. Enzyme-catalyzed reactions, however, are less sensitive to temperature changes than their uncatalyzed counterparts, whereas the uncatalyzed reaction rate may double with every 10 °C elevation of temperature, the enzyme-catalyzed reaction rate will increase by a factor of < 2. The temperature dependence of the rate constant can be described by the Arrhenius equation:

$$k = A \exp\left(- E_a/ RT\right) \qquad [9]$$

where A is the preexponential factor, E_a is the energy of activation, R is the gas constants, and T is the temperature in degrees Kelvin. The Arrhenius equation can be expressed in log form

$$\log k = \left(- E_a/2.303\ RT\right) + \log A \qquad [10]$$

where $\log A$ and E_a can be determined from the intercept and slope, respectively, of a linear plot of the dependence of $\log k$ on $1/T$ at lower temperatures, generally between 10 to 50 °C.

In studies involving calculation of the activation energies and temperature coefficients (Q_{10}) of enzyme-catalyzed reactions, it is assumed that the incubation temperature has no effect on the stability (concentration) of the enzyme. Experiments to test the validity of this assumption must be performed at the temperature range desired. The validity of this assumption for soil amidase and pyrophosphatase and plant amidase and urease activities has been reported (Tabatabai & Dick, 1979; Frankenberger & Tabatabai, 1981a, 1982). Information on the stability of the enzyme at various temperatures is essential, because studies of the effect of temperature on enzyme activities in soils have shown that most soil enzymes are inactivated (denatured) at temperatures between 60 and 70 °C (Halstead, 1964; Tabatabai & Bremner, 1970a; Tabatabai & Singh, 1976; Browman & Tabatabai, 1978; Tabatabai & Dick, 1979; Frankenberger & Tabatabai, 1980a, 1981a). Usually the stability of enzymes increases as the temperature decreases below that required for the optimum reaction rate. Although air-drying of field-moist soil may decrease enzyme activities (e.g., pyrophosphatase), it seems not to affect the temperature of enzyme inactivation (Tabatabai & Dick, 1979).

37–2.3 pH

Temperature affects enzyme activity in various ways; the number of factors controlling the effect of pH on an enzyme-catalyzed reaction is even greater. Changes in H^+ concentration influence enzymes, substrates, and cofactors by altering their ionization and solubility. Variations in such properties as ionization or solubility influence the rate of catalyzed reactions. Enzymes, being proteinaceous, exhibit marked changes in ionization from fluctuation in pH. Characteristically, each enzyme has a pH value at which the rate is optimal, and at each side of this optimum the rate is lower; thus, the catalytic action of the enzymes operates in a somewhat restricted pH range. Enzymes are usually most stable in the vicinity of the optimum pH, and they are irreversibly denatured with extremes in acidity or alkalinity. It is essential to assay enzyme-catalyzed reactions at, or close to, the optimum pH of activity. In addition to controlling the ionization of enzymes and substrates, the buffer used maintains the optimum pH for the duration of the reaction; this is especially important in enzyme assays involving substrates or products with acidic or alkaline properties (e.g., formation of NH_4^+ from hydrolysis of amides, urea, L-asparagine, and L-glutamine in soils). Also, since ionic species of the buffer have a marked effect on enzyme activity, it is essential that one buffer system be used for determining the pH profile. The universal buffer has a range from pH 3 to 12 and is suitable for this type of study (Eivazi & Tabatabai, 1977; Juma & Tabatabai, 1977).

Studies of the effects of pH on the stability of enzymes are important because extreme changes in pH may irreversibly inactivate the enzymes that play an essential role in nutrient cycling and humus formation. To test the extreme changes in pH, the pH stability of urease, acid phosphatase, alkaline phosphatase, and phosphodiesterase activities of soils was investigated by Frankenberger and Johanson (1982a) by first incubating a soil sample at pH values ranging from 1 to 13 for 24 h and then measuring the activity at its optimum pH under standardized conditions. They reported that, in general, the decline in enzyme activity in a pH-profile near the optimum pH range was due to a reversible reaction that involved ionization or deionization of acidic or basic groups in the active center of the enzyme protein. Irreversible inactivation of the enzyme was particularly evident at the lower and higher ranges of acidic and alkaline conditions. Results reported showed that the pH stability of soil enzymes was highly dependent on the enzyme and soil being assayed. Frankenberger and Johanson (1982a) attributed the variation among enzymes and soils to the diversity of vegetation, microorganisms, and soil fauna as sources contributing to the enzyme activity and protective sites that allowed entrapment of the enzyme within colloidal humus and organic-mineral complexes.

37–2.4 Cofactors, Inhibitors, and Ionic Environment

Many enzymes are not catalytically active except when in combination with a nonprotein component. Generally, these nonprotein cofactors are heat-stable, dialyzable substances of low molecular weight. Cofactors are often described as activators, coenzymes, and prosthetic groups, and it is often difficult to distinguish between the meanings assigned to these terms. Prosthetic groups are usually defined as substances that are bound firmly to the enzyme. Coenzymes are usually classified as organic substances that are freely dissociable from the enzymes. Most coenzymes and prosthetic groups are changed during the sequence of the catalyzed reaction and need to be regenerated before they can again participate. Some enzymes are activated by inorganic ions furnished by mineral salts. Usually these inorganic ions are not changed during the catalyzed process but are required before the enzyme can carry out the reaction.

Pyrophosphatase is well known for its requirement of inorganic cations for its activity. Dick and Tabatabai (1983) showed that the inorganic pyrophosphatase activity in three soils decreased when exchangeable and soluble metals were removed by leaching with 1 M NH$_4$OAc (pH 8). The effect of added metal ions at various concentrations in the leached soils showed that, at certain concentrations, Ba^{2+}, Ca^{2+}, Co^{2+}, Mg^{2+}, Mn^{2+}, Ni^{2+}, and Zn^{2+} promoted, K$^+$ and Na$^+$ had no effect, and Fe^{2+} and Cr^{2+} decreased pyrophosphatase activity. At high concentrations (>50 mM), Co^{2+}, Mn^{2+}, Ni^{2+}, and Zn^{2+} inhibited pyrophosphatase activity in two soils. The concentration of metal ion needed for optimum activity of pyrophosphatase varied among the soils. The efficiency of the metal ions at optimum concentrations (average percentage increase for three soils in parentheses) in promoting pyrophosphatase was Ca^{2+} (47) $>$ Mg^{2+} (42) $>$ Ba^{2+} $=$ Co^{2+} (29) $>$ Ni^{2+} (27) $>$ Zn^{2+} (20) $>$ Mn^{2+} (16). Soil pyrophosphatase in the presence of 200 mM CaCl$_2$ or MgCl$_2$ was protected against inactivation by heat (90 °C for up to 30 min).

A large variety of chemical reagents are used to inhibit enzyme reactions. Valuable information might be provided about the essential components of complex enzyme systems by using inhibitors. Some inhibitors are able to react with active groups on the enzyme. Inhibition of enzyme activity by the inhibitors p-chloromercuribenzoate Hg^{2+}, and Ag$^+$ might suggest that SH groups in the protein molecule are essential for activity. Other reagents might react with activators or coenzymes to inhibit enzyme activity. For example, chelating agents that combine with certain metals might illustrate the necessity of a metal in a catalyzed reaction. Furthermore, many reagents inhibit enzyme reactions because they have molecular features similar to the substrate. Arsenate resembles phosphate and is able to block many of the reactions in which phosphate takes part. Competitive inhibition of soil arylsulfatase by phosphate, arsenate, tungstate, and molybdate has been demonstrated (Al-Khafaji & Tabatabai, 1979).

Salts may exert marked influences on enzyme activity. Alberty (1956) has pointed out that changes in electrolyte concentration may cause changes in the activity coefficients of the reactants and the activated complex. Also, the degree of ionization of acidic and basic groups in the protein molecule may be changed by the presence of salts. All these effects may result in either accelerating or slowing down the catalytic activity.

37–2.5 Enzyme Inhibition

The subject of enzyme inhibition requires special consideration, because soils receive a variety of organic and inorganic chemicals. Some of the compounds, such as fertilizers, pesticides, and municipal and industrial wastes, are added as parts of soil and crop management. Other compounds, such as salts of trace elements, are added to soils as impurities in lime and fertilizers and as industrial pollutants. Also, inhibition of soil enzymes for controlling the release of fertilizer has received attention since the 1970s. Most of this research, however, has been concerned with use of enzyme inhibitors for controlling urea hydrolysis in soils. Unfortunately, most of the compounds tested for inhibition of urease in soils are general enzyme inhibitors, affecting not just urease activity but a wide range of soil enzymes. Generally, little work has been done on the mechanisms of inhibition of soil enzymes by various inhibitors.

An inhibitor may produce an irreversible or reversible decrease in the enzyme rate constant. An irreversible inhibitor forms a stable compound, often by the formation of a covalent bond with a particular amino acid residue in the enzyme structure, which is essential for the catalytic activity. The phenomenon of reversible inhibition is characterized by an equilibrium between the enzyme and inhibitor defined by an equilibrium constant (K_i), which is a measure of the affinity of the enzyme for the inhibitor.

Although kinetics of enzyme inhibition is complex and varies with the enzyme, substrate, and inhibitor, three main groups of reversible inhibition can be distinguished by their characteristic effects on plots of $1/v$ vs. $1/[S]$ (Lineweaver-Burk plot). The kinetics of most enzyme reactions in the presence of varying concentrations of inhibitors are such that the linear double reciprocal plots of $1/v$ vs. $1/[S]$ show similar lines as in the absence of inhibitor except that (i) the slope, (ii) the intercept, or (iii) both are altered. The first case is called competitive, the second uncompetitive, and the third noncompetitive inhibition. The kinetic diagnostic for uncompetitive inhibition (case ii) is that in the double reciprocal plot the slopes remain constant (i.e., the lines are parallel) but the intercepts vary. This form of inhibition is rare with single substrates where one possible but not highly likely mechanism entails combination of the inhibitor exclusively with the ES complex. This type of mechanism, however, is more common in more complex cases. The other two cases are more common with single substrates and are likely mechanisms of inhibition of most soil enzymes.

Competitive inhibition depends on the lack of absolute specificity of the chemical reactivity on the active site. The enzyme active site combines more or less loosely with the inhibitor, which is possibly structurally related to the substrate, thus preventing the attachment of the substrate to the enzyme-binding site. The inhibitor and substrate, therefore, compete for the same active site forming enzyme-substrate (ES) and enzyme-inhibitor (EI) complexes, respectively; ESI complexes are not produced. The extent of inhibition depends on the relative concentrations of substrate and inhibitor in the system. This type of inhibition is indicated by the Lineweaver-Burk plot (Eq. [6]); a competitive inhibitor increases the slope of the line without changing the intercept (Fig. 37–2). Therefore, the apparent K_m value is higher in the presence than in the absence of inhibitor, but the V_{max} value remains unchanged. In this type of inhibition, the velocity of the reaction depends on the inhibitor concentration, the substrate concentration, and the relative affinities of the inhibitor and substrate for the enzyme.

Noncompetitive inhibition cannot be reversed by a simple expedient of raising the substrate concentration. In this type of inhibition, the substrate combines equally well with the enzyme or its ES complex. In this

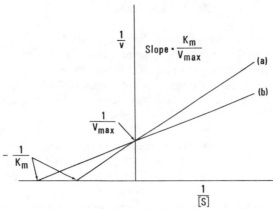

Fig. 37–2. Lineweaver-Burk plots of the Michaelis-Menten equation describing competitive inhibition; (a) in the presence of inhibitor and (b) in the absence of inhibitor.

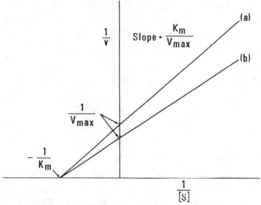

Fig. 37–3. Lineweaver-Burk plots of the Michaelis-Menten equation describing noncompetitive inhibition; (a) in the presence of inhibitor and (b) in the absence of inhibitor.

case, three complexes are formed: ES, EI, and ESI complexes, and of these only the ES complex breaks down to products. This type of inhibition is not completely overcome by increasing the substrate concentration. Lineweaver-Burk (Eq. [6]) plots yield lines showing increases in both the intercept and slope of the line in the presence than in the absence of inhibitor (Fig. 37–3). Therefore, the apparent K_m value is not changed, but the V_{max} is decreased. The amount of inhibition will accordingly depend on the inhibitor concentration and the affinity of the inhibitor to the enzyme, and it will not be influenced by the substrate concentration. Noncompetitive inhibitors apparently combine with an enzyme at a point other than the attachment of the substrate. They are then able to exert an effect on the active site even though they are situated some distance away. Enzyme inhibition by metal ions often shows noncompetitive kinetics.

37–2.6 Enzymes in Soils

Soil is a living system where all biochemical activities proceed through enzymatic processes. Enzymes accumulated in soils are present as free enzymes, such as exoenzymes released from living cells, endoenzymes released from disintegrating cells, and enzymes bound to cell constituents such as disintegrating cells, in cell fragments, and in viable but nonproliferating cells (Kiss et al., 1975). Most of the enzymes added to soils by decaying microbial tissues and plant and animal residues are likely degraded by soil proteases, and what remains is incorporated with the humus. Soil, therefore, can be viewed as a system of humus and minerals containing both immobilized enzymes and occluded microbial cells (McLaren, 1975).

Many of the fundamental questions concerning the origin, location, and persistence of soil enzymes remain unanswered. Although most organic materials are metabolized rapidly by microorganisms, both in vitro and in vivo, many enzyme-proteins persist in soils as active moieties for a long time (Skujins & McLaren, 1968). For example, experiments have shown that addition of urease to soil resulted in temporary increases in urease activity, suggesting that the added urease is either inactivated or destroyed by soil protease (Conrad, 1940c; Stojanovic, 1959; Roberge, 1970; Burns et al., 1972; Zantua & Bremner, 1976). The existence of mechanisms to protect enzymes in soils is supported by work showing that soil enzyme activity is independent of the microbial population (e.g., Paulson & Kurtz, 1969; Ramirez-Martinez & McLaren, 1966b). Also, it has been shown that urease activity could be detected in soils stored for decades and that this activity is better correlated with organic C than with microbial numbers (Skujins & McLaren, 1968, 1969). It has been suggested that native soil urease resides in organic colloidal particles with pores large enough for water, urea, NH_3, and CO_2 to pass freely, but the pores are small enough to exclude such enzyme as pronase (Burns et al., 1972); the added but not the native urease is destroyed by pronase. A model describing this system and the possible mechanisms involved in binding enzymes by soil organic matter are discussed by Paul and McLaren (1975). Recent work by Lai and Tabatabai (1992) showed that the kinetic and thermodynamic parameters of jackbean urease were significantly altered by adsorption on clay or soil mineral constituents.

Among the many factors that may affect enzyme activities in soils, cropping history, soil amendments, and some environmental factors have a substantial influence. For example, Stojanovic (1959) found marked seasonal variation in urease activity in Mississippi soils. Similarly, Cooper (1972) observed seasonal variation in arylsulfatase activity in Nigerian soils. He reported that arylsulfatase activity decreased during dry periods, increased after rainfall, and reached a maximum toward the end of the rainy season. The cycle was repeated during the dry seasons. Waterlogging of soils produces changes in E_h and pH, and consequently in the activity of many enzymes (Pulford & Tabatabai, 1988). Other studies on the extracellular enzyme activities in ecosystems have shown that vegetation, agricultural chemicals, and industrial pollutants have a marked influence on soil enzymes. Generally, the effects observed differ markedly and depend on many factors, including the soil type, dose of the chemical, and conditions of the study. Because of the importance of urea as a fertilizer in world agriculture and the potential of soil urease to cause volatilization of NH_3 produced from urea hydrolysis in soils, several workers have investigated various enzyme inhibitors for agricultural application. A detailed discussion on the inhibitors tested on urease activity in soils and enzyme interactions with various agricultural and industrial chemicals, including pesticides, is presented by Kiss et al. (1975) and Bremner and Mulvaney (1978).

Although the patterns of distribution of several enzymes in soil profiles have been studied by many workers, the relative factors that affect the activity of enzymes through the profile have not been clearly established. Like other biochemical reactions in soils, however, enzyme activities are associated with organic matter distribution profile and generally decrease with depth. This has been demonstrated for arylsulfatase (Tabatabai & Bremner, 1970b), urease (Tabatabai, 1977), inorganic pyrophosphatase (Tabatabai & Dick, 1979), amidase (Frankenberger & Tabatabai, 1981a), L-asparaginase (Frankenberger & Tabatabai, 1991b), and L-glutaminase (Frankenberger & Tabatabai, 1991d).

Several methods have been proposed for soil sterilization or inhibition of microbial proliferation that allow assay of enzyme activities (Frankenberger & Johanson, 1986). An ideal sterilization agent for extracellular enzyme detection in soil would be one that would completely inhibit all microbial activities, but not lyse the cells and not affect the extracellular enzymes. Unfortunately, however, no such universal agent is yet available.

Toluene has been the most widely used microbial inhibitor, but its usefulness is limited to the assay procedures that involve a few hours of incubation. Assay procedures involving long times of incubation with or without toluene should be avoided, because the risk involved from microbial proliferation increases as the time of incubation increases. It is believed that in assay procedures involving short incubation times, toluene inhibits the synthesis of enzymes by living cells and prevents assimilation of the reaction products. Toluene has also shown to be a plasmolytic agent in certain groups of microorganisms, in which it apparently induces the release of intracellular enzymes (Skujins, 1967). A critical examination of the effect of toluene on soil microorganisms has been made by Beck and Poschenrieder (1963). They showed that the inhibitory effect and concentration of toluene needed to suppress microbial activity are strikingly dependent on the pretreatment and moisture content of a particular soil. To suppress microbial proliferation in an air-dried, naturally moist, or dried and remoistened soil, at least 20% toluene is necessary. In a soil suspension, however, 5 to 10% toluene is sufficient. Gram-positive bacteria and

Streptomyces are considerably more resistant to toluene treatment than are gram-negative bacteria. In addition to the effect of toluene, the effect of dimethyl sulfoxide, ethanol, and Triton X-100 on soil enzyme activities was evaluated by Frankenberger and Johanson (1986). They reported that toluene, a plasmolytic agent as well as an antiseptic, had little effect or only slightly inhibited purified preparations of acid and alkaline phosphatases, α-glucosidase, and invertase, but severely inhibited catalase and dehydrogenase activities. The soil enzymes, arylsulfatase and urease, were enhanced (1.30- to 1.34-fold) in the presence of toluene, suggesting that its plasmolytic character was affecting the intercellular enzyme contribution of the measured activities in soils.

Irradiation of soil with high-energy radiation is another method used for soil sterilization. Dunn et al. (1948) introduced the electron beam for heatless sterilization of soil. McLaren et al. (1957) showed that soil can be sterilized by an electron beam of sufficient energy and intensity. They found that 2×10^6 roentgen-equivalent-physical (rep) dose was necessary for a 1-g sample; urease activity was retained in the sterilized soil. The effects of ionizing radiation on soil constituents have been reviewed by Skujins (1967), McLaren (1969), and Cawse (1975). Generally, microorganisms and enzymes are killed or inactivated as an exponential function of dose of radiation, but the dose required depends on the soil type, soil moisture, and genus of the organism.

37–3 ASSAY OF ENZYMES IN SOILS

37–3.1 Amidohydrolases (L-Asparaginase, L-Glutaminase, Amidase, and Urease)

37–3.1.1 Introduction

Several amidohydrolases are present in soils. All are involved in hydrolysis of native and added organic N to soils. Among these, L-asparaginase, L-glutaminase, amidase, and urease are the most important.

The enzyme L-asparaginase (L-asparagine amidohydrolase EC 3.5.1.1) plays an important role in N mineralization of soils. The chemical nature of N in soils is such that a large proportion (15–25%) of the total soil N is often released as NH_4^+ by acid hydrolysis (6 N of HCl). Some evidence suggests that a portion of the released NH_4^+ comes from the hydrolysis of amide (asparagine and glutamine) residues in soil organic matter (Sowden, 1958). Bremner (1955) reported that after acid hydrolysis of humic preparations, 7.3 to 12.6% of the total N was in the form of amide-N. Moreover, Sowden (1958) reported that a percentage of the NH_4^+ released during acid hydrolysis was equal to or nearly equal to the sum of aspartic acid-N plus glutamic acid-N derived from asparagine and glutamine.

L-Asparaginase activity was first detected in soils by Drobni'k (1956). This enzyme catalyzes the hydrolysis of L-asparagine, producing L-aspartic acid and NH_3 as shown below:

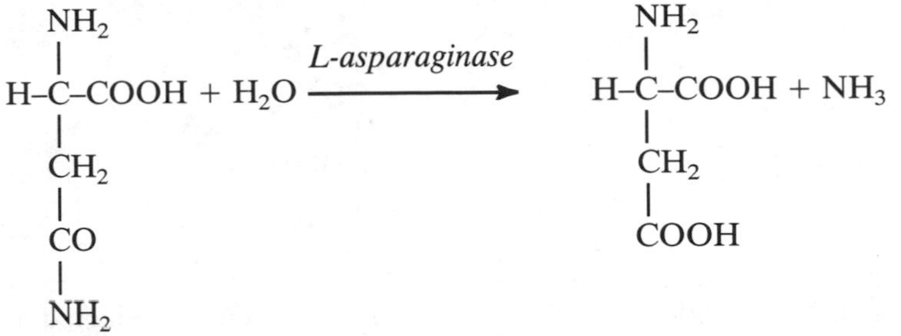

$$
\begin{array}{c}
NH_2 \\
| \\
H\text{–}C\text{–}COOH + H_2O \\
| \\
CH_2 \\
| \\
CO \\
| \\
NH_2
\end{array}
\xrightarrow{\ \ L\text{-}asparaginase\ \ }
\begin{array}{c}
NH_2 \\
| \\
H\text{–}C\text{–}COOH + NH_3 \\
| \\
CH_2 \\
| \\
COOH
\end{array}
\qquad [11]
$$

L-Asparaginase isolated from guinea pig serum specifically acts only on the L-isomer form of asparagine and not the D-isomer (Tower et al., 1963). However, the stereospecificity of L-asparaginase derived from *Alcaligenes eutrophus* is such that both D-asparagine and L-glutamine have been reported as substrates (Allison et al., 1971). This enzyme is widely distributed in nature. It has been detected in both plants and microorganisms (Wriston, 1971).

Asparaginases have been shown to vary widely in different strains of microorganisms. Campbell et al. (1967) found two asparaginases in *Escherichia coli* B and designated them EC-1 and EC-2. These two enzymes differ in solubility, chromatographic behavior, antilymphoma activity, and optimum pH. Also, *E. coli* K-12 contains two asparaginases but only one appears in cells grown under anaerobic conditions (Cedar & Schwartz, 1967). Cells of *Pseudomonas fluorescens* contain two inducible isoenzymes of asparaginase (Eremenko et al., 1975), one that hydrolyzes only L-asparagine (asparaginase A) and one that hydrolyzes L-asparagine, L-glutamine, and D-asparagine (asparaginase AG).

Soils have been tested for L-asparaginase activity by Beck and Poschenrieder (1963), simply by adding L-asparagine to soils and monitoring the NH_4^+ released, but a standard assay procedure was not developed using systematic studies of factors affecting the release of NH_4^+. The proposed method by Frankenberger and Tabatabai (1991a) involves determination of the NH_4^+-N released by L-asparaginase activity when soil is incubated with buffered (0.1 M of THAM, pH 10) L-asparagine solution and toluene at 37 °C.

L-Glutaminase is among the amidohydrolases that play an important role in supplying N to plants. This hydrolase is specific and acts on C–N bonds other than peptide bonds in linear amides.

L-Glutaminase (L-glutamine amidohydrolase, EC 3.5.1.2) activity in soils was first detected by Galstyan and Saakyan (1973). The reaction catalyzed by this enzyme involves the hydrolysis of L-glutamine yielding L-glutamic acid and NH_3 as shown:

$$\begin{array}{c}
\text{COOH} \\
| \\
\text{NH}_2\text{-C-H} + \text{H}_2\text{O} \\
| \\
\text{CH}_2 \\
| \\
\text{CH}_2 \\
| \\
\text{CO} \\
| \\
\text{NH}_2
\end{array}
\xrightarrow{\text{L-glutaminase}}
\begin{array}{c}
\text{COOH} \\
| \\
\text{NH}_2\text{-C-H} + \text{NH}_3 \\
| \\
\text{CH}_2 \\
| \\
\text{CH}_2 \\
| \\
\text{COOH}
\end{array}
\qquad [12]$$

L-Glutaminase is widely distributed in nature. It has been detected in several animals (Sayre & Roberts, 1958), plants (Bidwell, 1974), and microorganisms (Imada et al., 1973). Microorganisms that have shown to contain L-glutaminase activity include bacteria, yeasts, and fungi.

Plants and microorganisms are probable sources of L-glutaminase activity in soils. However, the main source is believed to be microbial in nature. Among the bacteria, very high levels of L-glutaminase activity have been reported in Achromobacteraceae soil isolates (Roberts et al., 1972). Fungal species that are known to produce L-glutaminase include *Tilachlidium humicola, Verticillium malthousei,* and *Penicillium urticae* (Imada et al., 1973). The proposed method by Frankenberger and Tabatabai (1991c) involves determination of the NH_4^+-N released by L-glutaminase activity when soil is incubated with buffered (0.1 M of THAM, pH 10) L-glutamine solution and toluene at 37 °C.

Amidase (acylamide amidohydrolase, EC 3.5.1.4) is the enzyme that catalyzes the hydrolysis of amides and procedures NH_3 and the corresponding carboxylic acid:

$$R \cdot CONH_2 + H_2O \rightarrow NH_3 + R \cdot COOH. \qquad [13]$$

Amidase acts on C–N bonds other than peptide bonds in linear amides. It is specific for aliphatic and aryl amides cannot act as substrates (Kelly & Clarke, 1962; Florkin & Stotz, 1964). This enzyme is widely distributed in nature. It has been detected in animals and microorganisms (Bray et al., 1949; Clarke, 1970). Amidase is present in leaves of corn (*Zea mays* L.), sorghum (*Sorghum bicolor* L. Moench), alfalfa (*Medicago sativa* L.), and soybean (*Glycine max* L.) (Frankenberger & Tabatabai, 1982). Microorganisms shown to possess amidase activity include bacteria (Clarke, 1970; Frankenberger & Tabatabai, 1985), yeast (Joshi & Handler, 1962), and fungi (Hynes, 1970, 1975). The substrates of this enzyme are sources of N for plants (Cantarella & Tabatabai, 1983). The method proposed involves determination of the NH_4-N released by amidase activity when soil is incubated with buffered (0.1 M of THAM, pH 8.5) amide solution and toluene at 37 °C (Frankenberger & Tabatabai, 1980a,b; 1981a,b).

Urease (urea amidohydolase, EC 3.5.1.5) is the enzyme that catalyzes the hydrolysis of urea to CO_2 and NH_3:

$$NH_2CONH_2 + H_2O \rightarrow CO_2 + 2NH_3. \qquad [14]$$

It acts on C–N bonds other than peptide bonds in linear amides and thus belongs to a group of enzymes that include glutaminase and amidase. Since two C–N bonds are broken in hydrolysis of urea by urease, it is evident that the stoichiometric relation in the equation is the result of component reactions. Several studies have been conducted to determine the mechanism of urease action, and the work by Gorin (1959) and Blakeley et al. (1969) has provided convincing evidence that carbamate is the intermediate in a two-step reaction. This reaction is summarized by Reithel (1971) as follows:

$$
\begin{array}{ccc}
\displaystyle O=C\!\!\begin{array}{l}{}^{\diagup NH_2}\\[-2pt]{}_{\diagdown NH_2}\end{array} + HOH \rightarrow &
\left[O=C\!\!\begin{array}{l}{}^{\diagup OH}\\[-2pt]{}_{\diagdown NH_2}\end{array} + NH_3 \rightleftharpoons O=C\!\!\begin{array}{l}{}^{\diagup ONH_4^{+}}\\[-2pt]{}_{\diagdown NH_2}\end{array} \right] & [15]
\end{array}
$$

$$+ HOH \rightarrow H_2CO_3 + 2NH_3. \qquad [12]$$

Evidence derived from kinetic data suggests that urease forms a carbamoyl complex:

$$
\begin{array}{c}
H_2N\!-\!C\!-\!enz\\
\parallel\\
O
\end{array}
$$

as one of the ES complexes, and presumably water is the acceptor in a carbamoyl transfer reaction. Therefore, carbamate is the obligatory substrate for the second step in the overall reaction (Reithel, 1971). Since the proposed mechanism is based on kinetic studies, direct evidence for this mechanism is still needed.

Urease is very widely distributed in nature. It has been detected in microorganisms, plants, and animals. Its presence in soils was first reported by Rotini (1935). Studies by Conrad (1940a,b,c; 1942a,b; 1943) provided the basic information about this enzyme in soil systems. Several reasons could account for the large amount of literature on urease. These include:

1. Urease was the first enzyme protein to be crystallized in 1926 by Sumner (1951).
2. Its reaction products (CO_2 and NH_3) are relatively easy to determine.
3. It can be purified from various sources.
4. It can give a model for a simple enzyme-catalyzed reaction.
5. It is available commercially at reasonable prices.
6. Urea is an important N fertilizer.

Besides the several reviews available on urease in the biochemistry literature (Sumner, 1951; Varner, 1960; Reithel, 1971), there are numerous articles covering some aspects of soil urease (Skujins, 1967, 1976; Kuprevich & Shcherbakova, 1971; Sequi, 1974; Kiss et al., 1975; Bremner & Mulvaney, 1978).

Soil urease activity apparently is optimum at tris(hydroxymethyl) aminomethane (THAM, Fisher Scientific Co., Chicago, or TRIS) buffer pH of 8.5 to 9.0, which is about 2 pH units greater than that for optimal urease in solution (Tabatabai & Bremner, 1972a); the K_m values of soil urease for urea range from 1.1 to 3.4 mM (Tabatabai, 1973).

A variety of methods has been used for assay of urease activity in soils (Tabatabai & Bremner, 1972a; Bremner & Mulvaney, 1978). Most of these involve determination of the NH_4^+ released on incubation of toluene-treated soil with buffered urea solution, but assays have also been performed by estimation of the CO_2 released or urea hydrolyzed on incubation (Conrad, 1940c; Porter, 1965; Skujins & McLaren, 1969; Douglas & Bremner, 1971). Several of the methods adopted have not involved use of a buffer to control pH (Porter, 1965; Douglas & Bremner, 1971) or addition of toluene to inhibit microbial proliferation (McLaren et al., 1957; Porter, 1965; Paulson & Kurtz, 1969).

The method proposed by Tabatabai and Bremner (1972a) involving estimation of the NH_4^+ released has been thoroughly evaluated. This method involves use of toluene, and it has been shown to be applicable to soils that fix NH_4^+ and is described here.

The other method described involves estimation of the rate of urea hydrolysis in soils by determination of the urea remaining after incubation of soil with a urea solution. The difference between the amount of urea added and that recovered after incubation for a specific time is taken as an estimate of urease activity. Urea may be estimated by a colorimetric method involving the reaction of urea with p-dimethyl-aminobenzaldehyde (Porter, 1965) or diacetlymonoxime (DAM) and thiosemicarbazide (TSC) (Douglas & Bremner, 1970a). Colorimetric determination of urea by DAM and TSC, however, is more sensitive than by p-dimethylaminobenzaldehyde. These methods do not give the actual urease activity in soils, because neither involves use of a buffer to control the pH of the incubation mixture (for the problems associated with such methods, see Burns, 1978), however, they are useful for estimation of the degree of urea hydrolysis in soils. The method based on determination of urea with DAM and TSC is described.

37–3.1.2 Principles

The methods, designed for assay of L-asparaginase, L-glutaminase, amidase, or urease under optimum conditions, are based on determination of the NH_4^+ released when soil is incubated with THAM buffer, L-asparagine, L-glutamine, formamide, or urea solution, and toluene at 37 °C for 2 h. The NH_4^+ released is determined by treatment of the incubated soil sample with 2 M of KCl containing Ag_2SO_4 or uranyl acetate (to stop enzyme activity) and steam distillation of an aliquot of the resulting soil suspension with MgO for 4 min. In these methods the enzymes are assayed at optimum buffer pH and substrate concentration.

The second method described for urease assay is designed for estimation of the rate of urea hydrolysis in soils. It involves colorimetric determination of the urea remaining after incubation at 37 °C for 5 h of soil with urea solution. The amount of urea hydrolyzed g^{-1} of soil 5 h^{-1} is estimated from the difference between the initial amount of urea added and that recovered after incubation.

37–3.1.3 Assay Methods

37–3.1.3.1 L-Asparaginase (Frankenberger & Tabatabai, 1991a,b)

37–3.1.3.1.1 Special Apparatus

1. Volumetric flasks, 50 mL.
2. Incubator, ordinary incubator, or temperature-controlled water bath.
3. Steam distillation apparatus (see chapter 40 in this book).

37–3.1.3.1.2 Reagents

1. Toluene, Fisher certified reagent (Fisher Scientific Co., Chicago).
2. Tris(hydroxymethyl)aminomethane (THAM) buffer 0.1 M, pH 10: Dissolve 12.2 g of THAM (Fisher certified reagent, Fisher Scientific Co., Chicago) in about 700 mL of water, titrate the pH of the solution to 10 by addition of approximately 0.1 M NaOH, and dilute with water to 1 L.
3. L-Asparagine solution, 0.5 M: Dissolve 1.65 g of L-asparagine (Sigma Chemical Co., St. Louis, MO) in about 20 mL of tris(hydroxymethyl)aminomethane (THAM) buffer, and dilute the solution to 25 mL with THAM buffer. Mix the contents while running hot tap water over the flasks.
4. Potassium chloride (2.5 M)-silver sulfate (100 ppm) (KCl-Ag_2SO_4) solution: Dissolve 100 mg of reagent-grade Ag_2SO_4 in about 700 mL of water, dissolve 188 g of reagent-grade KCl in this solution, and dilute the solution to 1 L with water.
5. Reagents for determination of NH_4^+-N (MgO, H_3BO_3-indicator solution, 0.0025 M H_2SO_4): Prepare as described in chapter 40.

37–3.1.3.1.3 Procedure

Place 5 g of soil (< 2 mm) in a 50-mL volumetric flask, add 0.2 mL of toluene and 9 mL of THAM buffer, swirl the flask for a few seconds to mix the contents, add 1 mL of 0.5 M L-asparagine solution, and swirl the flask again for a few seconds. Then stopper the flask, and place it in an incubator at 37 °C. After 2 h, remove the stopper, add approximately 35 mL of

KCl-Ag$_2$SO$_4$ solution, swirl the flask for a few seconds, and allow the flask to stand until the contents have cooled to room temperature (about 5 min). Make the contents to 50 mL by addition of KCl-Ag$_2$SO$_4$ solution, stopper the flask, and invert it several times to mix the contents. To determine NH$_4^+$-N in the resulting soil suspension, pipette a 20-mL aliquot of the suspension into a 100-mL distillation flask, and determine the NH$_4^+$-N released by steam distillation of this aliquot with 0.2 g of MgO for 4 min (chapter 40).

Controls should be performed in each series of analyses to allow for NH$_4^+$-N not derived from L-asparagine through L-asparaginase activity. To perform controls, follow the procedure described for assay of L-asparaginase activity, but make the addition of 1 mL of 0.5 M L-asparagine solution after the addition of 35 mL of KCl-Ag$_2$SO$_4$ solution.

37–3.1.3.2 L-Glutaminase (Frankenberger & Tabatabai, 1991c,d)

37–3.1.3.2.1 Special Apparatus

As described in section 37–3.1.3.1.1.

37–3.1.3.2.2 Reagents

1. Toluene, Fisher certified reagent (Fisher Scientific Co., Chicago).
2. THAM buffer (0.1 M, pH 10) and KCl-Ag$_2$SO$_4$ mixture: prepare as described in section 37–3.1.3.1.2.
3. L-Glutamine solution, 0.5 M: Dissolve 1.82 g of L-glutamine (Sigma Chemical Co., St. Louis, MO) in about 20 mL of tris(hydroxymethyl)aminomethane (THAM) buffer (0.1 M, pH 10) and dilute the solution to 25 mL with THAM buffer. Mix the contents while running hot tap water over the flask.
4. Reagents from determination of NH$_4^+$-N (MgO, H$_3$BO$_3$-indicator solution, 0.0025 M H$_2$SO$_4$): Prepare as described in chapter 40.

37–3.1.3.2.3 Procedure

Follow the procedure described for L-asparaginase activity but use 1 mL of 0.5 M L-glutamine solution instead of 1 mL of 0.5 M L-asparagine solution.

Controls should be performed in each series of analyses to allow for NH$_4^+$ not derived from L-glutamine through L-glutaminase activity. To perform controls, follow the procedure described for performing controls in assay of L-asparaginase activity but use *steam sterilized* soil and 1 mL of 0.5 M L-glutamine instead of 1 mL of 0.5 M L-asparagine.

37–3.1.3.3 Amidase (Frankenberger & Tabatabai, 1980a,b)

37–3.1.3.3.1 Special Apparatus

As described in section 37–3.1.3.1.1.

37–3.1.3.3.2 Reagents

1. Toluene, Fisher certified reagent (Fisher Scientific Co., Chicago).
2. THAM buffer (0.1 *M*, pH 8.5): Prepare as described in section 37–3.1.3.1.2.
3. Formamide solution (0.50 *M*): Add 2.0 mL of formamide (Aldrich certified) into a 100-mL volumetric flask. Make up volume by adding THAM buffer, and mix the contents. Store the solution in a refrigerator.
4. Potassium chloride (2.5 *m*)-uranyl acetate (0.005 *M*) solution: Dissolve 2.12 g of reagent-grade $UO_2(C_2H_3O_2)$ $2H_2O$ in about 700 mL of water, dissolve 188 g of reagent-grade KCl in this solution, and dilute the solution to 1 L with water.
5. Reagents for determination of NH_4^+-N (MgO, H_3BO_3-indicator solution, 0.0025 *M* H_2SO_4): Prepare as described in chapter 40.

37–3.1.3.3.3 Procedure

Follow the procedure described in section 37–3.1.3.1.3 for assay of L-asparaginase activity but use 1 mL of 0.5 *M* formamide solution instead of 1 mL of 0.5 *M* L-asparagine solution.

Controls should be performed in each series of analyses to allow for NH_4^+-N not derived from formamide through amidase activity. To perform controls, follow the procedure described in section 37–3.1.3.1.3 for performing controls in assay of L-asparaginase activity but use 1 mL of 0.5 *M* formamide instead of 1 mL of 0.5 *M* L-asparagine.

37–3.1.3.4 Urease (Tabatabai & Bremner, 1972a)

37–3.1.3.4.1 Special Apparatus

As described in section 37–3.1.3.1.1.

37–3.1.3.4.2 Reagent

1. Toluene, Fisher certified reagent (Fisher Scientific Co., Chicago).
2. THAM buffer (0.05 *M*, pH 9.0) and KCl-Ag_2SO_4 mixture: Prepare as described in section 37–3.1.3.1.2, but use 0.2 *M* H_2SO_4 instead of NaOH to titrate the THAM buffer.
3. Urea solution, 0.2 *M:* Dissolve 1.2 g of urea (Fisher certified reagent, Fisher Scientific Co., Chicago) in about 80 mL of THAM buffer, and dilute the solution to 100 mL with THAM buffer. Store the solution in a refrigerator.
4. Potassium chloride (2.5 *M*)-silver sulfate (100 ppm) (KCl-Ag_2SO_4) solution: Prepare as described in section 37–3.1.3.1.2.
5. Reagents for determination of NH_4^+-N (MgO, H_3BO_3-indicator solution, 0.0025 *M* H_2SO_4): Prepare as described in chapter 40.

37–3.1.3.4.3 Procedure

Follow the procedure described in section 37–3.1.3.1.3 for assay of L-asparaginase activity but use 1 mL of 0.2 M area solution instead of 1 mL of 0.5 M L-asparagine solution.

Controls should be performed in each series of analyses to allow for NH_4^+-N not derived from urea through urease activity. To perform controls, follow the procedure described in section 37–3.1.3.1.3 for performing controls in assay of L-asparaginase activity but use 1 mL of 0.2 M urea solution instead of 0.5 M of L-asparagine solution.

37–3.1.3.5 Urease by Determination of Urea Remaining[1]

37–3.1.3.5.1 Special Apparatus

1. Volumetric flasks, 50 mL with glass stoppers.
2. Incubator, ordinary incubator, or temperature-controlled water bath.
3. Water bath (boiling water).
4. Suction funnels (polyethylene) and filtering funnel stand. (Soil-moisture Equipment Co., Santa Barbara, CA).
5. Klett-Summerson photoelectric colorimeter or spectrophotometer.

37–3.1.3.5.2 Reagents

1. Urea (substrate) solution: Dissolve 2.0 g of urea (Fisher certified reagent, Fisher Scientific Co., Chicago) in about 700 mL of water, and adjust the volume to 1 L with water. This solution contains 2 mg of urea/mL. Store this solution in a refrigerator.
2. Phenylmercuric acetate (PMA) solution: Dissolve 50 mg of PMA (Eastman Organic Chemicals, Rochester, NY) in 1 L of water.
3. Potassium chloride-phenylmercuric acetate (2 M KCl-PMA) solution: Dissolve 1500 g of reagent-grade KCl in 9 L of water, and add 1 L of PMA solution.
4. Diacetylmonoxime (DAM) solution: Dissolve 2.5 g of DAM (Fisher Scientific Co., Chicago) in 100 mL of water.
5. Thiosemicarbazide (TSC) solution: Dissolve 0.25 g of TSC (Eastman Organic Chemicals, Rochester, NY) in 100 mL of water.
6. Acid reagent: Add 300 mL of 85% phosphoric acid (H_3PO_4) and 10 mL of concentrated sulfuric acid (H_2SO_4) to 100 mL of water, and make up the volume to 500 mL with water.
7. Color reagent: Prepare this reagent immediately before use by adding 25 mL of DAM solution and 10 mL of TSC solution to 500 mL of acid reagent.
8. Standard urea stock solution for standard curve: Dissolve 0.500 g of urea in about 1500 mL of 2 M potassium chloride-phenylmercuric

---

[1]Modified from Douglas and Bremner (1970a,b) and Zantua and Bremner (1975a).

acetate (KCl-PMA) solution, and dilute to 2 L with this same solution. If pure, dry urea is used, this solution will contain 250 μg of urea/mL. Store this solution in a refrigerator.

37–3.1.3.5.3 Procedure

Place 5 g of soil (< 2 mm, on oven-dry basis) in a 65-mL (2-oz.) glass bottle, and treat it with 5 mL of urea solution (10 mg of urea). Stopper the bottle, and incubate at 37 °C. After 5 h, remove the stopper, and add 50 mL of 2 M KCl-PMA solution (reagent 3). Stopper the bottle again, and shake it for 1 h. Filter the soil suspension, under suction, through Whatman no. 42 filter paper by using a suction funnel and filtering funnel stand.

To determine the urea remaining, pipette an aliquot (1–2 mL) of the extract containing up to 200 μg of urea into a 50-mL volumetric flask, make the volume to 10 mL with 2 M KCl-PMA solution, and add 30 mL of the color reagent. Swirl the flask for a few seconds to mix the contents, and place it in a bath of boiling water. After 30 min, remove the flask from the water bath, cool it immediately in running water for about 15 min (this can be accomplished by placing the flask in a deep tray containing cold water, 12–20 °C). Then make the volume to 50 mL with water, and mix thoroughly. Measure the intensity of the red color produced with a Klett-Summerson photoelectric colorimeter fitted with a green (no. 54) filter.

Calculate the urea content of the extract analyzed by reference to a calibration graph plotted from the results obtained with standards containing 0, 25, 50, 100, 150, and 200 μg of urea. To prepare this graph, dilute 10 mL of the standard urea stock solution to 100 mL with 2 M KCl-PMA solution in a volumetric flask, and mix thoroughly. Then pipette 0, 1, 2, 4, 6, and 8 mL of aliquots of this working standard solution into 50-mL volumetric flasks, adjust volumes to 10 mL by adding 2 M KCl-PMA solution, and proceed as described for urea analysis of the soil extract.

To calculate the amount of urea hydrolyzed in soil during incubation, divide the amount of total urea recovered by 5 (micrograms of urea recovered per gram of soil) and subtract this value from 2000 (micrograms of urea initially added per gram of soil).

37–3.1.4 Comments

In the assay methods based on determination of the NH_4^+ released, the buffer described should be used, because other buffers (e.g., phosphate, acetate, and universal buffers) give lower results for L-asparaginase, L-glutaminase, amidase, and urease activities in soils. Of the buffers tested by Tabatabai and Bremner (1972a), the buffer described prevent NH_4^+ fixation by soils (this has been confirmed by quantitative recovery of NH_4^+ added to Montana vermiculite). The THAM buffer should be prepared by using H_2SO_4, because work by Wall and Laidler (1953) showed that buffers prepared by treatment of THAM solutions with HCl have an activation effect on hydrolysis of urea by jackbean (*Canavalia ensiformis* L.) urease.

The KCl-Ag$_2$SO$_4$ solution must be prepared by the addition of KCl to Ag$_2$SO$_4$ solution as specified (Ag$_2$SO$_4$ will not dissolve in KCl solution), and the soil suspension analyzed for NH$_4^+$ must be mixed thoroughly immediately before sampling. The KCl-Ag$_2$SO$_4$ treatment used terminates the enzyme activity, and no additional NH$_4^+$ release is expected if the soil suspension is allowed to stand for 2 h before NH$_4^+$ analysis. If the soil suspension cannot be analyzed within 4 h after the addition of the KCl-Ag$_2$SO$_4$ reagent, the flask should be stored in a refrigerator.

The KCl-UO$_2$(C$_2$H$_3$O$_2$)$_2$·2H$_2$O reagent should be stirred continuously before use; if this solution is allowed to stand without stirring, KCl will precipitate slowly out of the solution with time. The purpose of adding the KCl-uranyl acetate solution to the soil sample after incubation in assay of amidase activity is to inactivate amidase and allow quantitative determination of the NH$_4^+$ released (Jakoby & Fredericks, 1964; Tabatabai & Bremner, 1972a). Potassium chloride-Ag$_2$SO$_4$ can be substituted for KCl-uranyl acetate provided the NH$_4^+$ release is determined immediately, because KCl-Ag$_2$SO$_4$ does not completely inactivate amidase activity in soils. In addition to formamide, acetamide and propionamide may be used as substrates, but with lower specificity. Assay of amidase in the absence of toluene indicated that its substrates may induce the synthesis of this enzyme by soil microorganisms (Frankenberger & Tabatabai, 1980a).

For a valid assay of enzyme activity, it is necessary to ensure that the enzyme substrate concentration is not a limiting factor in the assay procedure. The substrate concentration recommended (0.05 M L-asparagine, L-glutamine, and formamide, and 0.02 M of urea) have been shown to be satisfactory for assay of amidohydrolase activities in a variety of soils. The D-isomers of asparagine and glutamine are also hydrolyzed in soils, but at about 16 and 7% of the L-isomers, respectively, at saturating concentration of the substrates (Frankenberger & Tabatabai, 1991a,c). The sensitivity of the assay procedure that involves determination of the NH$_4^+$ released is such that precise results can be obtained with most soils even if the 2-h incubation time recommended is reduced to 1 h. No hydrolysis of L-asparagine, formamide, or urea is found when autoclaved soils are incubated with solutions of the substrates at 37 °C for 2 h under the conditions of the assay procedures described. L-Glutamine is hydrolyzed slightly during incubation, and appropriate controls should be included to account for the NH$_4^+$-N produced through chemical hydrolysis. The results obtained by these methods are not affected when the amount of toluene is increased from 0.2 to 2.0 mL.

The steps involved in the colorimetric method described for urea analysis in the second method should be strictly adhered to, but any colorimeter or spectrophotometer that permits color intensity measurement at 500 to 550 nm can be used for the procedure described. The maximum absorption of the color produced from the reaction of urea, DAM, and TSC as described is 527 nm (Douglas & Bremner, 1970a).

Numerous colorimetric procedures for analysis of urea with DAM have been developed (Yashphe, 1973), but most of these methods are actually variations of the method developed by Fearon (1939). The main disadvantages of most methods are: (i) lack of sensitivity, (ii) lack of linearity at low urea concentrations, (iii) lengthiness of procedure, (iv) low precision, and (v) instability of some of the reagents or the chromogen compound. The method described suffers from some of these disadvantages, especially from the instability of the chromogen compound and low reproducibility of the results, including the standard curve. Thus, attempts have been made to automate the development of color (Searle & Speir, 1976; Douglas et al., 1978).

An oven adjusted to 120 °C can be used instead of the water bath, which reduces the time required for maximum color development to 30 min. It is important that the flask used for color development be cooled immediately after it is removed from the water bath, because some loss of color occurs if the flask is not cooled rapidly as specified (Douglas & Bremner, 1970a). The color developed in the urea analysis described is photosensitive, but this is not a problem if the color intensity measurements are performed shortly (1 h or less) after color development. Color intensity measurement can be postponed for several hours if the solutions of the chromogen complex are stored in the dark.

Calibration curves prepared from urea standards as described show a linear relationship between urea concentration and color intensity but differ slightly from day to day. It is recommended, therefore, that urea standards be included in each series of analyses.

Nitrite interferes with the colorimetric method used for determination of urea if the concentration of NO_2^--N in the extract analyzed is more than five times the concentration of urea-N. Such concentrations of NO_2^--N are not expected in soils or in urea-treated soils, but NO_2^- interference can be eliminated by treating the aliquot of extract taken for urea analysis with 1 mL of 2% (wt/vol) sulfamic acid solution and by allowing the treated aliquot to stand for 5 min before the addition of the color reagent. The color reagent is unstable and should be prepared immediately before use. The other reagents used are stable for several weeks if stored in a refrigerator.

37–3.2 Phosphatases

37–3.2.1 Introduction

The general names *phosphatases* has been used to describe a broad group of enzymes that catalyze the hydrolysis of both esters and anhydrides of H_3PO_4 (Schmidt & Laskowski, 1961). The commission on enzymes of the International Union of Biochemistry has classified all these enzymes into five major groups (Florkin & Stotz, 1964). These include the phosphoric monoester hydrolases (EC 3.1.3), phosphoric diester hydrolases (EC 3.1.4), triphosphoric monoester hydrolases (EC 3.1.5), enzymes acting on phosphoryl-containing anhydrides (EC 3.6.1), and enzymes acting on P–N bonds (EC 3.9), such as the phosphoamidase (EC 3.9.1.1).

The phosphomonoesterases, acid phosphatase (orthophosphoric monoester phosphohydrolase, EC 3.1.3.2) and alkaline phosphatase (orthophosphoric monoester phosphohydrolase, EC 3.1.3.1) have been studied extensively. The enzymes are classified acid and alkaline phosphatases because they show optimum activities in acid and alkaline ranges, respectively. Because of the importance of these enzymes in soil organic P mineralization and plant nutrition, considerable literature has accumulated on phosphomonoesterases in soils (Speir & Ross, 1978). Most of the literature, however, is related to acid phosphatase. Consequently this enzyme has been given a prominent place in several soil biochemistry and enzymology reviews (Cosgrove, 1967; Ramirez-Martinez, 1968; Kiss et al., 1975; Speir & Ross, 1978). The general equation of the reaction catalyzed by acid and alkaline phosphatases is

$$
\begin{array}{ccc}
\overset{\displaystyle O}{\underset{\displaystyle |}{\|}} & & \overset{\displaystyle O}{\underset{\displaystyle |}{\|}} \\
R\text{-}O\text{-}P\text{-}O^- + H_2O \rightarrow R{\cdot}OH + HO\text{-}P\text{-}O^-. & & \\
\underset{O^-}{|} & & \underset{O^-}{|}
\end{array} \qquad [16]
$$

One of the most interesting properties of acid and alkaline phosphatases is their specificities. Although most of the information available on these enzymes is related to acid phosphatase in soils, these enzymes are know to hydrolyze a variety of phosphomonoesters. The hydrolysis in soils of β-glycerophosphate, phenylphosphate, β-naphthyl phosphate, and p-nitrophenyl phosphate has been reported.

Studies by Eivazi and Tabatabai (1977) and Juma and Tabatabai (1977, 1978) showed that acid phosphatase is predominant in acid soils and that alkaline phosphatase is predominant in alkaline soils. The inverse relationship between phosphatase activity and soil pH suggests that either the rate of synthesis and release of this enzyme by soil microorganisms or the stability of this enzyme is related to soil pH. Since higher plants are devoid of alkaline phosphatase activity (Dick et al., 1983; Juma & Tabatabai, 1988a,b,c), the alkaline phosphatase activity in soils seems derived totally from microorganisms.

Michaelis constants of soil acid phosphatase have been reported in several studies (Tabatabai & Bremner, 1971; Cervelli et al., 1973; Brams & McLaren, 1974; Thornton & McLaren, 1975; Eivazi & Tabatabai, 1977; Dick & Tabatabai, 1984). The apparent K_m values of acid phosphatase in soils range from 1.3 to 4.5 mM; values of alkaline phosphatase range from 0.4 to 4.9 mM. Adsorption of phosphatases on clay minerals significantly alters their kinetic parameters (Dick & Tabatabai, 1987). Similar to other kinetic studies of soil enzymes, the K_m values obtained for phosphomonoesterases are affected by shaking the soil-substrate mixture during incubation; normally the values are more uniform among soils when shaking than when static incubation technique is employed (Tabatabai & Bremner, 1971; Eivazi & Tabatabai, 1977).

Studies have shown that all heavy metals and trace elements inhibit phosphomonoesterases in soils; the degree of inhibition is related to the soil and type and concentration of the trace elements used (Tyler, 1974, 1976a,b; Juma & Tabatabai, 1977). Kinetic studies indicate that orthophosphate is a competitive inhibitor of acid and alkaline phosphatases in soils (Juma & Tabatabai, 1978).

Several methods have been proposed for estimation of phosphomonoesterase activities of soils (Skujins, 1967). The basic difference is in the substrate used and consequently in the technique employed in measuring the product of hydrolysis of the substrate by phosphatase enzymes (Dick & Tabatabai, 1978b).

Kroll and Kramer (1955) estimated soil phosphatase activity by determining the phenol released by incubation of soil with phenyl phosphate. This substrate has been used in several investigations of soil phosphatase activity (e.g., Kramer, 1957; Kramer & Yerdei, 1959; Halstead, 1964). Skujins et al. (1962) assayed soil phosphatase activity by a procedure in which the amount of β-glycerophosphate hydrolyzed by incubation of soil with this organic phosphate was estimated by analyses for extractable total and inorganic P after incubation. This method is tedious and time consuming and has low precision. Ramirez-Martinez and McLaren (1966a) proposed a method involving fluorimetric assay of the β-naphthol released by incubation of soil with β-naphthyl phosphate, but this method is complicated by sorption of β-naphthol by soil constituents and requires that the capacity of each soil analyzed to sorb β-naphthol be determined and allowed for in calculation of the results.

Of the various methods available for assay of phosphatase activity in soils, the method developed by Tabatabai and Bremner (1969) is the most rapid, accurate, and precise and is described here. It involves colorimetric estimation of the *p*-nitrophenol released when soil is incubated with buffered sodium *p*-nitrophenyl phosphate solution and toluene. The procedure used to extract the *p*-nitrophenol released by phosphatase activity develops the stable color used to estimate this phenol and gives quantitative recovery of *p*-nitrophenol added to soils.

Among the phosphatases in soils, phosphodiesterase (orthophosphoricdiester phosphohydrolase, EC 3.1.4.1) is the least studied. Phosphodiesterase catalyzes the overall reaction of the type

$$O=\overset{\displaystyle OH}{\underset{\displaystyle OR_2}{P}}-OR_1 + H_2O \rightarrow O=\overset{\displaystyle OH}{\underset{\displaystyle OR_2}{P}}-OH + R_1 \cdot OH \qquad [17]$$

where R_1 and R_2 represent either alcohol or phenol groups or nucleosides (Privat de Garilhe, 1967).

The activity of phosphodiesterase has been detected in various plants, animals, and microorganisms (Browman & Tabatabai, 1978). This enzyme is perhaps best known for its ability to degrade nucleic acid (Razzel & Khorana, 1959), and given the demonstrated ability of soils to decompose added nucleic acids (Pearson et al., 1941; Rogers, 1942) and the natural occurrence of these compounds in soils (Anderson, 1967), the potential role of phosphodiesterase activity in the soil P cycle is obvious.

Studies on inhibition of soil phosphodiesterase have shown that at 5 mM, orthophosphate, EDTA, and citrate are inhibitors of this enzyme. Similar to inhibition of phosphomonoesterase activity, orthophosphate is a competitive inhibitor of phosphodiesterase in soils. Kinetics studies indicate that the apparent K_m values of this enzyme in soils range from 1.3 to 2.0 mM, the Q_{10} is 1.7, and the average activation energy is 37 kJ/mol (Browman & Tabatabai, 1978).

Many factors affect phosphomonoesterase and phosphodiesterase activities in soils. For discussion of these factors and information about the stability and distribution of these enzymes in soils, see Eivazi and Tabatabai (1977), Browman and Tabatabai (1978), Juma and Tabatabai (1978), and Speir and Ross (1978).

Inorganic pyrophosphatase (pyrophosphate phosphohydrolase, EC 3.6.1.1) catalyzes the hydrolysis of pyrophosphate to orthophosphate. The overall reaction is

$$\begin{array}{cc} \text{O} & \text{O} \\ \| & \| \\ {}^-\text{O--P--O--P--O}^- + \text{H}_2\text{O} \rightarrow 2\text{HPO}_4^{2-}. \\ | & | \\ \text{O}^- & \text{O}^- \end{array} \qquad [18]$$

Similar to the other phosphatases described, inorganic pyrophosphatase is widely distributed in nature. Its presence has been reported in bacteria, insects, mammalian tissues, and plants (Feder, 1973). Pyrophosphatase activity in soils has also been reported by several workers (Gilliam & Sample, 1968; Hashimoto et al., 1969; Hossner & Phillips, 1971).

The activity of this enzyme in soils has received special attention, because its substrate, pyrophosphate, is used as a fertilizer. Although pyrophosphatase activity in soils has been under investigation for many years, it was not until recently that an accurate method has become available for its assay. The information available indicates that pyrophosphatase activity in soils is optimum at a buffer pH of 8.0. Formaldehyde, fluoride, oxalate, and carbonate inhibit the activity of this enzyme in soils (Dick & Tabatabai, 1978a). Studies by Stott et al. (1985) showed that pyrophosphatase activity in soils is inhibited by many metals and that AsO_4^{3-}, BO_3^{2-}, MoO_4^{2-}, PO_4^{3-}, VO^{2+}, and WO_4^{2-} are competitive inhibitors of this enzyme in soils. At low concentrations, however, many metals are activators of soil pyrophosphatase (Dick & Tabatabai, 1983). Kinetic studies with surface soils have shown that the apparent K_m values of pyrophosphatase in soils range from 20 to 51 mM, and the activation energy values range from 32 to 43 kJ/mol (Dick & Tabatabai, 1978a).

Studies by Tabatabai and Dick (1979) showed that similar to other enzyme activities in soils, pyrophosphatase activity is concentrated in surface soils and decreases with depth; it is significantly correlated with organic C in surface soils and soil profiles. Pyrophosphatase activity was also correlated with the percentage clay and mole fraction of Mg/(Mg + Ca) in water extracts of surface soils and significantly but negatively correlated with the percentage of $CaCO_3$ equivalent in surface soils. Unlike some other soil enzymes, air-drying of field-moist soils decreases pyrophosphatase activity.

Trimetaphosphate (TMP), a cyclic polyphosphate that is not sorbed by soils, is hydrolyzed by a series of biochemical reactions to yield triphosphate, pyrophosphate, and orthophosphate; phosphates that are sorbed by soils (Busman & Tabatabai, 1984):

$$P_3O_9^{3-} + H_2O \rightarrow H_2P_3O_{10}^{3-} \qquad [19]$$

$$H_2P_3O_{10}^{3-} + H_2O \rightarrow H_2P_2O_7^{2-} + H_2PO_4^- \qquad [20]$$

$$H_2P_2O_7^{2-} + H_2O \rightarrow 2H_2PO_4^- \qquad [21]$$

The hydrolysis of TMP (reaction [19]) may be catalyzed biochemically by trimetaphosphatase (trimetaphosphate hydrolase, EC 3.6.1.2) or chemically by acidic or basic solutions, especially in the presence of certain cations (Berg & Gordon, 1960; Van Wazer, 1958). Reactions [20] and [21] are catalyzed by triphosphatase and pyrophosphatase, respectively. Among the phosphatases involved in these reactions, pyrophosphatase is the most thoroughly studied (Dick & Tabatabai, 1978a).

Trimetaphosphatase activity has been reported in yeast (Kornberg, 1956; Mattenheimer, 1956; Meyerhof et al., 1953), intestinal mucosa of mammals (Berg & Gordon, 1960; Ivey & Shaver, 1977), and in plants (Stossel et al., 1981; Pierpoint, 1957). It has been suggested that trimetaphosphatase activity occurs in all living organisms (Berg & Gordon, 1960). Results reported by Blanchar and Hossner (1969) and by Rotini and Carloni (1953) on hydrolysis of TMP suggest the presence of trimetaphosphatase activity in soils.

The yield of orthophosphate (OP) was used to indicate trimetaphosphatase activity in most studies because of the difficulty in determining the primary product of TMP hydrolysis, triphosphate (TP), in the presence of other phosphates. However, the dependence of the formation of OP from TMP on other enzymes (triphosphatase and pyrophosphatase) complicates the interpretation of the results. In addition, determination of the OP released is subject to interference by the presence of the other phosphate compounds (Dick & Tabatabai, 1977b). Until recently, the lack of an adequate method for determining TMP or TP in soils has limited the number of studies on trimetaphosphatase activity in soils.

The method described for assay of trimetaphosphatase activity in soils is based on precipitation of other phosphates in a soil extract with $BaCl_2$, and determining TMP as total P remaining in the solution (Busman & Tabatabai, 1984, 1985a,b). In addition to its hydrolysis to triphosphate in soils by the enzyme trimetaphosphatase, TMP is hydrolyzed chemically by metal ions such as Ca^{2+} and Mg^{2-}. Therefore, a method is described for assaying the activity of trimetaphosphatase and for determining the rate of nonenzymatic (chemical) hydrolysis of TMP in soils. The method involves incubating 1 g of soil with 3 mL buffered (100 mM of Tris) 25 mM of TMP at 37 °C for 5 h followed by determination of the unaltered substrate. Steam-sterilized soil is used as a control to account for the amount of TMP hydrolyzed nonenzymatically.

Trimetaphosphatase activity in soils was maximum at a buffer pH of 8 and a temperature of 40 °C. Nonenzymatic hydrolysis of TMP increased up to the highest incubation temperature (80 °C) tested. Trimetaphosphatase in three Iowa soils showed apparent K_m values ranging from 6.8 to 7.2 mM of TMP, V_{max} values from 590 to 1200 mg of TMP-hydrolyzed kg^{-1} soil 5 h^{-1}, and energy of activation values from 17 to 26 kJ mol^{-1}. The activation energy values of the nonenzymatic hydrolysis ranged from 29 to 39 kJ mol^{-1}. Enzymatic and nonenzymatic hydrolysis rates in the six soils used ranged from 231 to 928 and from 101 to 888 mg of TMP-P kg^{-1} soil 5 h^{-1}, respectively. Formaldehyde and EDTA inhibited, and toluene had no effect on trimetaphosphatase activity in soils. The addition of Ca^{2+} and Mg^{2+} increased the nonenzymatic hydrolysis rates in two soils tested with Ca^{2+} being twice as effective as Mg^{2+} (Busman & Tabatabai, 1985a).

The hydrolysis of polyphosphates in soils and by plant roots has been reported (Dick & Tabatabai, 1986a,b, 1987; Juma & Tabatabai, 1988a,b). The methods employed in these studies will not be described here, and the reader may refer to the references mentioned above.

37–3.2.2 Principles

The procedures described for assay of phosphomonoesterase activities are based on colorimetric estimation of the p-nitrophenol released by phosphatase activity when soil is incubated with buffered (pH 6.5 for acid phosphatase activity and pH 11 for alkaline phosphate activity) sodium p-nitrophenyl phosphate solution and toluene. The colorimetric procedure used for estimation of p-nitrophenol is based on the fact that alkaline solutions of this phenol have a yellow color (acid solutions of p-nitrophenol and acid and alkaline solutions of p-nitrophenyl phosphate are colorless). The $CaCl_2$-NaOH treatment described for extraction of p-nitrophenol after incubation for assay of acid and alkaline phosphatases serves (i) to stop phosphatase activity, (ii) to develop the yellow color used to estimate this phenol, and (iii) to give quantitative recovery of p-nitrophenol from soils.

The principle of assay for phosphodiesterase activity in soils is similar to that of assay of acid and alkaline phosphatases; the p-nitrophenol released is extracted and determined colorimetrically. The two procedures differ, however, in that 0.1 M THAM buffer pH 12 is used for extraction of the *p*-nitrophenol released in the assay of phosphodiesterase activity instead of the 0.5 *M* NaOH used in the assay of phosphomonoesterases. The reason for this difference is that the substrate of phosphodiesterase, *bis-p*-nitrophenyl phosphate (BPNP), is not stable in NaOH solutions; BPNP is hydrolyzed with time in the presence of NaOH. The $CaCl_2$-THAM treatment described for extraction of the *p*-nitrophenol released in the assay of phosphodiesterase serves the same purpose as that of $CaCl_2$-NaOH used for extraction of this product in the assay of phosphomonoesterases.

The assay of inorganic pyrophosphatase activity is based on determination of the orthophosphate released when soil is incubated with buffered (pH 8.0) pyrophosphate solution. Generally, there are three problems associated with the measurement of orthophosphate released by enzymatic hydrolysis of pyrophosphate during the pyrophosphatase assay: (i) the orthophosphate released may be sorbed by the soil constituents and therefore not extracted; (ii) orthophosphate may continue to be hydrolyzed from the substrate (pyrophosphate) after extraction from the soil for reasons other than the enzyme (e.g., low pH); and (iii) the presence of pyrophosphate may interfere with the measurement of orthophosphate. All these problems must be overcome in any method used to assay the pyrophosphatase activity in soils. In the method described, 0.5 *M* H_2SO_4 is used to extract orthophosphate. This reagent gives quantitative recovery of orthophosphate added to soils. The colorimetric method used for determination of the orthophosphate extracted in the presence of pyrophosphate is specific for orthophosphate. This method involves a rapid formation of heteropoly blue by the reaction of orthophosphate with molybdate ions in the presence of ascorbic acid-trichloroacetic acid reagent and complexation of the excess molybdate ions by a citrate-arsenite reagent to prevent further formation of blue color from the orthophosphate derived from hydrolysis of the substrate, pyrophosphate.

37–3.2.3 Assay Methods

37–3.2.3.1 Phosphomonoesterases (Acid and Alkaline Phosphatases)
(Tabatabai & Bremner, 1969; Eivazi & Tabatabai, 1977)

37–3.2.3.1.1 Special Apparatus

1. Incubation flasks, 50-mL Erlenmeyer flasks fitted with no. 2 stoppers.
2. Incubator, ordinary incubator, or temperature-controlled water bath.
3. Colorimeter or spectrophotometer, Klett-Summerson photoelectric colorimeter fitted with a blue (no. 42) filter or a spectrophotometer than can be adjusted to a wavelength of 400 to 420 nm.

37–3.2.3.1.2 Reagents

1. Toluene, Fisher certified reagent (Fisher Scientific Co., Chicago).
2. Modified universal buffer (MUB) stock solution: Dissolve 12.1 g of tris(hydroxymethyl)aminomethane (THAM), 11.6 g of maleic acid, 14.0 g of citric acid, and 6.3 g of boric acid (H_3BO_3) in 488 mL of 1 N sodium hydroxide (NaOH) and dilute the solution to 1 L with water. Store it in a refrigerator.
3. Modified universal buffer, pH 6.5 and 11: Place 200 mL of MUB stock solution in a 500-mL beaker containing a magnetic stirring bar, and place the beaker on a magnetic stirrer. Titrate the solution to pH 6.5 with 0.1 M hydrochloric acid (HCl), and adjust the volume to 1 L with water. Titrate another 200 mL of the MUB stock solution to pH 11 by using 0.1 M NaOH, and adjust the volume to 1 L with water.
4. *p*-Nitrophenyl phosphate solution, 0.05 M: Dissolve 0.840 g of disodium *p*-nitrophenyl phosphate tetrahydrate (Sigma 104, Sigma Chemical Co., St. Louis, MO) in about 40 mL of MUB pH 6.5 (for assay of acid phosphatase) or pH 11 (for assay of alkaline phosphatase), and dilute the solution to 50 with MUB of the same pH. Store the solution in a refrigerator.
5. Calcium chloride ($CaCl_2$) 0.5 M: Dissolve 73.5 g of $CaCl_2 \cdot 2H_2O$ in about 700 mL of water, and dilute the volume to 1 L with water.
6. Sodium hydroxide (NaOH), 0.5 M: Dissolve 20 g of NaOH in about 700 mL of water, and dilute the volume to 1 L with water.
7. Standard *p*-nitrophenol solution: Dissolve 1.0 g of *p*-nitrophenol in about 700 mL of water, and dilute the solution to 1 L with water. Store the solution in a refrigerator.

37–3.2.3.1.3 Procedure

Place 1 g of soil (< 2 mm) in a 50-mL Erlenmeyer flask, add 0.2 mL of toluene, 4 mL of MUB (pH 6.5 for assay of acid phosphatase or pH 11 for assay of alkaline phosphatase), 1 mL of *p*-nitrophenyl phosphate solution made in the same buffer, and swirl the flask for a few seconds to mix the contents. Stopper the flask, and place it in an incubator at 37 °C. After 1 h, remove the stopper, add 1 mL of 0.5 M $CaCl_2$ and 4 mL of 0.5 M NaOH, swirl the flask for a few seconds, and filter the soil suspension through a Whatman no. 2v folded filter paper. Measure the yellow color intensity of the filtrate with a Klett-Summerson photoelectric colorimeter. Calculate the *p*-nitrophenol content of the filtrate by reference to a calibration graph plotted from the results obtained with standards containing 0, 10, 20, 30, 40, and 50 μg of *p*-nitrophenol. To prepare this graph, dilute 1 mL of the standard *p*-nitrophenol solution to 100 mL in a volumetric flask and mix the solution thoroughly. Then pipette 0-, 1-, 2-, 3-, 4-, and 5-mL aliquots of this diluted standard solution into a 50-mL Erlenmeyer flasks, adjust the volume to 5 mL by addition of water, and proceed as described for *p*-nitrophenol analysis of the incubated soil sample (i.e., add 1 mL of 0.5 *M* CaCl$_2$ and 4 mL of 0.5 *M* of NaOH, mix and filter the resultant suspension). If the color intensity of the filtrate exceeds that of 50 µg of the *p*-nitrophenol standard, an aliquot of the filtrate should be diluted with water until the colorimeter reading falls within the limits of the calibration graph.

Controls should be performed with each soil analyzed to allow for color not derived from *p*-nitrophenol released by phosphatase activity. To perform controls, follow the procedure described for assay of phosphatase activity, but make the addition of 1 mL of PNP solution after the additions of 0.5 *M* CaCl$_2$ and 4 mL of 0.5 *M* NaOH (i.e., immediately before filtration of the soil suspension).

37–3.2.3.2 Phosphodiesterase (Browman & Tabatabai, 1978)

37–3.2.3.2.1 Special Apparatus

1. Erlenmeyer flasks (50 mL), stoppers (no. 2), incubator, and Klett-Summerson photoelectric colorimeter described in 37–3.2.3.1.

37–3.2.3.2.2 Reagents

1. Toluene, Fisher certified reagent (Fisher Scientific Co., Chicago).
2. Tris(hydroxymethyl)aminomethane (THAM) buffer, 0.05 *M*, pH 8.0: Dissolve 6.1 g of THAM (Fisher certified reagent, Fisher Scientific Co., Chicago) in about 800 mL of water, adjust the pH to 8.0 by titration with approximately 0.1 *M* H$_2$SO$_4$, and dilute the solution to 1 L with water.
3. *Bis-p*-nitrophenyl phosphate (BPNP) solution, 0.05 *M*: Dissolve 0.906 g of sodium, *bis-p*-nitrophenyl phosphate in about 40 mL of THAM buffer pH 8.0, and dilute the volume to 50 mL with buffer. Store the solution in a refrigerator.
4. Calcium chloride (CaCl$_2$ solution, 0.5 *M*: Prepare as described for reagent 5 in 37–3.2.3.1.2.
5. Tris(hydroxymethyl)aminomethane-sodium hydroxide (THAM-NaOH) extractant solution, 0.1 *M* THAM (Fisher certified reagent, Fisher Scientific Co., Chicago), pH 12: Dissolve 12.2 g of THAM in about 800 mL of water, adjust the pH of the solution to 12 by titration with 0.5 *M* NaOH, and dilute the volume to 1 L with water.
6. Tris(hydroxymethyl)aminomethane (THAM) diluent, 0.1 *M*, pH about 10: Dissolve 12.2 g of THAM (Fisher certified reagent, Fisher Scientific Co., Chicago) in about 800 mL of water, and adjust the volume to 1 L with water.
7. Standard *p*-nitrophenol solution: Prepare as described for reagent 7 in 37–3.2.3.1.2.

37–3.2.3.2.3 Procedure

Place 1 g of soil (< 2 mm) in a 50-mL Erlenmeyer flask, and add 0.2 mL of toluene, 4 mL of THAM buffer pH 8.0, and 1 mL of BPNP solution. Swirl the flask for a few seconds to mix the contents. Stopper the flask, and incubate it at 37 °C. After 1 h, remove the stopper, add 1 mL of 0.5 M $CaCl_2$ and 4 mL of THAM-NaOH extractant solution, swirl the flask for a few seconds, and filter the suspension through a Whatman No. 2v folded filter paper. Measure the yellow color intensity of the filtrate with a Klett-Summerson photoelectric colorimeter. Calculate the p-nitrophenol content of the filtrate by reference to a calibration graph prepared as described in section 37–3.2.3.1.3.

If the color intensity of the filtrate exceeds that of 50 μg of p-nitrophenol standard, an aliquot of the filtrate should be diluted with 0.1 M THAM pH about 10 until the colorimeter reading falls within the limits of the calibration graph.

Controls should be performed with each soil analyzed to allow for color not derived from PN released by phosphodiesterase activity. To perform controls, follow the procedure described for assay of phosphodiesterase activity, but make the addition of 1 mL of BPNP solution after the addition of 0.5 M $CaCl_2$ and 4 mL of 0.1 M THAM-NaOH (i.e., immediately before filtration of the soil suspension).

37–3.2.3.3 Inorganic Pyrophosphatase (Dick & Tabatabai, 1977b, 1978a)

37–3.2.3.3.1 Special Apparatus

1. Centrifuge tubes, 50 mL, plastic.
2. Incubator, ordinary incubator, or temperature-controlled water bath.
3. Shaker, end-to-end shaker.
4. Centrifuge, high speed, 12 000 rpm (17390 g).
5. Spectrophotometer or colorimeter.

37–3.2.3.3.2 Reagents

1. Sodium hydroxide (NaOH), 1 M.
2. Modified universal buffer stock solution: Prepared as described for reagent 2 in section 37–3.2.3.1.2.
3. Pyrophosphate solution, 50 mM: Dissolve 2.23 g of sodium pyrophosphate decahydrate ($Na_4P_2O_7 \cdot 10H_2O$) (Matheson Coleman and Bell Manufacturing Chemists, Norwood, OH) in 20 mL of MUB stock solution, titrate the solution to pH 8 with 0.1 M HCl, and dilute the volume to 100 mL with water. This reagent should be prepared daily.
4. Modified universal buffer working solution, pH 8.0: Titrate 200 mL of the MUB stock solution to pH 8.0 with 0.1 M HCl, and dilute the volume to 1 L with water.

5. Sulfuric acid (H_2SO_4), 0.5 M: Add 250 mL of concentrated H_2SO_4 to 8 L of water, and dilute the volume to 9 L with water.
6. Ascorbic acid (0.1 M)-trichloroacetic acid (0.5 M) reagent (reagent A): Dissolve 8.8 g of ascorbic acid and 41 g of trichloroacetic acid (Fisher certified reagent) in about 400 mL of water, and dilute the volume to 500 mL with water. This reagent should be prepared daily.
7. Ammonium molybdate tetrahydrate [$(NH_4)_6Mo_7O_{24}$ $4H_2O$] (0.015 M) reagent (reagent B): Dissolve 9.3 g of ammonium molybdate (J.T. Backer Chemical Co., Phillipsburg, NJ) in about 450 mL of water, and adjust the volume to 500 mL with water.
8. Sodium citrate (0.15 M)—sodium arsenite (0.3 M)—acetic acid (7.5%) reagent (reagent C): Dissolve 44.1 g of sodium citrate and 39 g of sodium arsenite in about 800 mL of water, add 75 mL of glacial acetic acid (99.9%), and adjust the volume to 1 L with water.
9. Standard phosphate stock solution: Dissolve 0.4390 g of potassium dihydrogen phosphate (KH_2PO_4) in about 700 mL of water, and dilute the volume to 1 L with water. This solution contains 100 μg of PO_4^{3-}-P/mL.

37–3.2.3.3.3 Procedure

Place 1 g of soil (< 2 mm) in a 50-mL plastic centrifuge tube, add 3 mL of 50 mM pyrophosphate solution, and swirl the tube for a few seconds to mix the contents. Stopper the tube, and incubate it at 37 °C. After 5 h, remove the stopper and immediately add 3 mL of MUB pH 8 and 25 mL of 0.5 M H_2SO_4. Stopper the tube, and shake it horizontally in a reciprocal shaker for 3 min. Centrifuge the soil suspension for 30 s at 17390 g, and immediately take an aliquot of the supernatant for PO_4^{3-}-P analysis.

To analyze for the inorganic phosphate(Pi) released by inorganic pyrophosphatase, pipette a 1-mL aliquot of the supernatant solution into a 25-mL volumetric flask containing 10 mL of reagent A, immediately add 2 mL of reagent B and 5 mL of reagent C, and adjust the volume with water (swirl the flask to mix the contents after addition of the sample aliquot and each of reagents B and C). After 15 min, measure the absorbance of the heteropoly blue color developed using a spectrophotometer adjusted to a wavelength of 700 nm.

Calculate the PO_4^{3-}-P content of the aliquot analyzed by reference to a calibration graph plotted from the results obtained with standards containing 0, 5, 10, 15, 20, and 25 μg of PO_4^{3-}-P. To prepare this graph, pipette 0-, 5-, 10-, 15-, 20-, and 25-mL aliquots of the standard PO_4^{3-}-P stock solution into 100-mL volumetric flasks, make up the volumes with water, and mix thoroughly. Then analyze 1-mL aliquots of these diluted standards as described for analysis of PO_4^{3-}-P in the aliquot of the incubated sample.

Controls should be included with each soil analyzed to allow for PO_4^{3-}-P not derived from pyrophosphate through pyrophosphatase activity. To perform controls, add 3 mL of MUB pH 8.0 to 1 g of soil, and incubate for 5 h. After incubation, add 3 mL of 50 mM of pyrophosphate solution, immediately add 25 mL of 0.5 M H$_2$SO$_4$, and then extract and analyze for PO$_4^{3-}$-P as described for the assay sample.

37–3.2.3.4 Trimetaphosphatase (Busman & Tabatabai, 1984, 1985a)

37–3.2.3.4.1 Special Apparatus

1. Centrifuge tubes, 50 mL, plastic.
2. Incubator, ordinary incubator, or temperature-controlled water bath.
3. Centrifuge, high speed, 12 000 rpm (17390 g).
4. Spectrophotometer or colorimeter.

37–3.2.3.4.2 Reagents

1. Trimetaphosphatate (TMP) solution (25 mM): Dissolve 1.53 g of Na$_3$P$_3$O$_9$ (practical grade, Sigma Chemical Co., St. Louis, MO) and 2.42 g of tris(hydroxymethyl)aminomethane (THAM) in about 100 mL of water, adjust the pH to 8.0 with 0.2 M of HCl, and make up the volume to 200 mL with water.
2. Trimetaphosphate stock solution (100 mM): Dissolve 3.06 g of Na$_3$P$_3$O$_9$ (practical grade, Sigma Chemical Co., St. Louis, MO) in about 80 mL of water, and make up the volume to 100 mL with water.
3. Trimetaphosphate working solution: Dissolve 1.21 g of THAM in about 25 mL of water, add 1, 5, 10, 25, or 50 mL of the TMP stock solution (to make 1, 5, 10, 25, or 50 mM of TMP, respectively), adjust the pH to 8.0 with 0.2 M of HCl, and make up the volume to 100 mL with water.
4. Barium chloride solution (100 mM): Dissolve 24.4 g of BaCl$_2$ in about 800 mL of water, and adjust the volume to 1 L with water.
5. Ascorbic acid (0.1 M)-trichloroacetic acid (0.5 M) reagent (reagent A), ammonium molybdate tetrahydrate reagent (reagent B), sodium citrate (0.15 M)-sodium arsenite (0.3 M)-acetic acid (7.5%) reagent (reagent C), and standard phosphate stock solution: Prepare as described in section 37–3.2.3.3.2.
6. Modified Murphy-Riley reagent: Prepare as described by Dick and Tabatabai (1977a) but with 4 M of HCl substituted for 2.5 M of H$_2$SO$_4$.

37–3.2.3.4.3 Procedure

Place 1 g of soil (< 2 mm) in a 50-mL plastic centrifuge tube, add 3 mL of 25 mM of TMP solution, swirl the tube for a few seconds to mix the contents. Stopper the tube, and incubate at 37 °C. After 5 h, remove the stopper, mix on a vortex stirrer and during mixing add 10 mL of 100 mM of BaCl$_2$ slowly to precipitate other phosphate compounds. Centrifuge the tube at 12 000 × g for 5 min, and immediately remove aliquots of the supernatant for determination of orthophosphate and TMP by placing an aliquot containing < 25 µg of orthophosphate-P into a 25-mL volumetric flask and another aliquot (1–5 mL) containing < 1 mg P into a 50-mL volumetric flask, respectively. Determine orthophosphate-P as described in section 37–3.2.3.3.3.

Acidify the aliquot removed for determination of TMP with sufficient 2 M of HCl to bring the concentration to 1 M with respect to HCl and heat the flask on a steam plate (85 °C) for 1 h to hydrolyze the TMP to orthophosphate. After cooling to room temperature, adjust the volume to 50 mL with water. Determine the orthophosphate content of this flask by analyzing an aliquot by the Murphy-Riley reagent after it was neutralized with 1 M of NaOH as described by Dick and Tabatabai (1977a).

Two types of controls should be performed. In one control, the above procedure should be followed with steam-sterilized soil. A second control should be performed with each set of assays to determine precisely the amount of TMP added to the soils. To perform this control, follow the above procedures, but without soil.

The amount of TMP-P hydrolyzed enzymatically is calculated by subtracting the amount of TMP-P remaining in the assay tube (i.e., the tube containing nonsterilized soil) from the amount remaining in the control tube (i.e., the tube containing steam-sterilized soil). The amount of TMP-P hydrolyzed nonenzymatically is calculated by subtracting the amount of TMP-P remaining in the control tube (i.e., the tube containing sterilized soil) from the amount of P in the control tube containing no soil.

37–3.2.4 Comments

The solutions of the substrates used for assay of phosphomonoesterases and phosphodiesterase are stable for several days if stored in a refrigerator. The compounds used for assay of these enzymes are artificial substrates; are not expected to be found in soils. The dry substrates should be stored in a desiccator in a freezer. The standard p-nitrophenol solution is stable for a few weeks if stored in a refrigerator.

It is necessary to add $CaCl_2$ to prevent dispersion of clay and extraction of soil organic matter during the treatment with NaOH in assay of phosphomonoesterases (dispersion of clay complicates filtration, and the dark-colored organic matter extracted with NaOH interferes with colorimetric analysis for p-nitrophenol). The procedures described give quantitative recovery of p-nitrophenol added to soils. In assays of acid and alkaline phosphatases, the control analysis is so designed that it allows for the presence of trace amounts of p-nitrophenol in some commercial samples of p-nitrophenyl phosphate and for extraction of trace amounts of colored soil material by the $CaCl_2$-NaOH treatment used for extraction of p-nitrophenol. No chemical hydrolysis of p-nitrophenyl phosphate is detected under the conditions of the assay procedure described.

Acid and alkaline phosphatase and phosphodiesterase activity values are not affected by the amount of toluene added (0.1–1.0 mL). Toluene, however, increases slightly the observed activities. Air-drying of field-moist soils results in increases in acid phosphatase activity and a decrease in alkaline phosphatase activity (Eivazi & Tabatabai, 1977).

The substrate concentrations in the incubation mixtures used for assay of phosphomonoesterases and phosphodiesterase are about 5 to 10 times greater than the K_m values of these enzymes in soils. If desired, these substrate concentrations can be changed to meet the objectives of the assay.

The assay for phosphodiesterase should not be carried out for a long time because the risk from hydrolysis of the second product (p-nitrophenyl phosphate) produced from the action of phosphodiesterase on BPNP increases as incubation time increases. This is especially true in the absence of buffer (Browman & Tabatabai, 1978). This risk is minimal, however, under the conditions of the procedure described.

When assay for pyrophosphatase activity, it is important that the extraction and analysis of the orthophosphate released from pyrophosphatase be carried out immediately, because pyrophosphate hydrolyzes slowly with time in the presence of the extractant (0.5 M H_2SO_4). Also, the steps involved in determination of the orthophosphate in the presence of pyrophosphate should be adhered to. The reagents described should be tested for orthophosphate, because the presence of orthophosphate in any of the reagents may give a high value for the controls.

For the assay of trimetaphosphatase, a concentration of 25 mM of TMP was chosen because at least 20 mM is necessary for enzyme saturation and much higher concentrations would be detrimental to the precision of the assay (i.e., with little TMP hydrolyzed a high amount would remain to be determined). Tests performed on several soils showed that TMP can be recovered quantitatively by the method described and that the enzymatic and nonenzymatic hydrolysis of TMP are proportional to time of incubation (up to 5 h) and to the amount of soil used (up to 1.5 g of soil) (Busman & Tabatabai, 1984, 1985a). Other studies showed that treatment of soils with formaldehyde, EDTA, mercuric chloride, sodium nitrate, ascorbic acid, and orthophosphate inhibited trimetaphosphatase activity in soils. The activity of this enzyme is noncompetitively inhibited by EDTA, PO_4^{3-}, and analogues of PO_4^{3-}, (MoO_4^{2-}, WO_4^{2-}, VO^{2+}, AsO_4^{3-}, and $B_4O_7^{2-}$). Treatment of soils with formaldehyde, EDTA, ascorbic acid, orthophosphate, pyrophosphate, or triphosphate decrease the nonenzymatic hydrolysis of TMP, whereas treatment of soils with K_2SO_4, NH_4Cl, $CaCl_2$, and $MgCl_2$ increases the rate of nonenzymatic hydrolysis (Busman & Tabatabai, 1985b).

37–3.3 Arylsulfatase

37–3.3.1 Introduction

Several types of sulfatases occur in nature. They have been classified according to the type of organic sulfate esters they hydrolyze, with the following main groups recognized: arylsulfatases, alkylsulfatases, steroid sulfatases, glucosulfatases, chondrosulfatases, and myrosulfatases (Fromageot, 1950; Roy, 1960). Arylsulfatase (arylsulfate sulfohydrolase, EC 3.1.6.1) is the enzyme that catalyzes the hydrolysis of an arylsulfate anion by fission of the O–S bond (Spencer, 1958). The reaction

$$R \cdot OSO_3^- + H_2O \rightarrow R \cdot OH + H^+ + SO_4^{2-} \qquad [22]$$

is irreversible, and there is no evidence that any SO_4^{2-} acceptor other than water can be used or that any metal ion is required for its catalytic function. This enzyme was first discovered in 1911 in purple snails by Derrien[2], and it has been detected in plants, animals, and microorganisms (Nicholls & Roy, 1971). Because this enzyme was the first sulfatase to be detected in nature, it has received more attention than other groups of sulfatases.

Arylsulfatase was first detected in Iowa soils by Tabatabai and Bremner (1970a). Since then it has been detected in other soils around the world and in marine sediments (Cooper, 1972; Chandramohan et al., 1974; Speir & Ross, 1975; Kowalenko & Lowe, 1975; Thornton & McLaren, 1975; Houghton & Rose, 1976). Arylsulfatase is believed to be partly responsible for S cycling in soils. The suggested role of this enzyme in S mineralization is largely derived from studies showing that between 40 and 70% (avg. 50%) of the total S in surface soils of temperate regions is reduced to H_2S by HI and is converted to inorganic SO_4^{2-} with hot alkali (Freney, 1961; Tabatabai & Bremner, 1972b; Neptune et al., 1975). These findings suggested that this fraction of S in surface soils is present in the form of ester sulfate (organic sulfates) and suggested further that arylsulfatase may play an important role in the processes whereby organic soil S is mineralized and made available for plant growth.

The procedure developed by Tabatabai and Bremner (1970a), and which has been used in all other studies in assay of arylsulfatase activity in soils, involves determination of the p-nitrophenol released when the soil sample is incubated with buffered p-nitrophenyl sulfate solution and toluene at 37 °C for 1 h. Tests have indicated that arylsulfatase activity in soils is inactivated at temperatures ranging from 60 to 70 °C. The K_m values of this soil enzyme range from 0.2 to 5.7 mM. Its determination is affected by shaking the soil-buffer-substrate mixture during incubation; usually the K_m values are lower and more uniform among soils when the shaking incubation rather than the static technique is used (Tabatabai & Bremner, 1971; Thornton & McLaren, 1975).

Studies of factors affecting arylsulfatase activity have shown that the activity of this enzyme decreases with depth in soil profiles, which is associated with a decrease in organic matter content. Also, arylsulfatase activity is significantly correlated with the organic matter content of surface soils differing markedly in chemical and physical properties (Tabatabai & Bremner, 1970b). Recent studies indicate that several trace elements inhibit arylsulfatase activity in soils and that the inhibition by MoO_4^{2-}, WO_4^{2-},

---

[2]Biochemical information. Vol. 1. 1973. Boehringer Mannheim GmbH, p. 48.

AsO_4^{3-}, and PO_4^{3-} shows competitive kinetics. At 25 μmol/g of soil, common anions of soils such as NO_2^- NO_3^-, Cl^-, and SO_4^{2-} are not inhibitors of this enzyme (Al-Khafaji & Tabatabai, 1979).

Generally, air-drying of field-moist soils results in an increase in arylsulfatase activity. A possible explanation of this increase is that rewetting of air-dry soil causes a breakdown of aggregates, thus increasing the accessibility of arylsulfatase to the substrate (Tabatabai & Bremner, 1970b). Other factors that may affect arylsulfatase activity in soils are summarized by Speir & Ross (1978).

37–3.3.2 Principles

The principles of the method described are similar to those of the assay for phosphomonoesterases (acid and alkaline phosphatases) described in section 37–3.2.2. The method is based on colorimetric determination of *p*-nitrophenol released by arylsulfatase activity when soil is incubated with buffered (pH 5.8) potassium *p*-nitrophenyl sulfate solution and toluene. Usually the soil-buffer-substrate mixture is incubated at 37 °C for 1 h. The *p*-nitrophenol released is extracted by filtration after the addition of $CaCl_2$ and NaOH reagents. The colorimetric procedure used for determination of *p*-nitrophenol depends on the fact that alkaline solutions of this phenol have a yellow color (acid solutions of *p*-nitrophenol and acid and alkaline solutions of *p*-nitrophenyl sulfate are colorless). The $CaCl_2$-NaOH treatment is employed for the same purpose as that described in section 37–3.2.2.

37–3.3.3 Assay Method (Tabatabai & Bremner, 1970a)

37–3.3.3.1 Special Apparatus
1. Erlenmeyer flasks (50 mL), stoppers (no. 2), incubator, and Klett-Summerson photoelectric colorimeter described in section 37–3.2.3.1.1.

37–3.3.3.2 Reagents
1. Toluene, Fisher certified reagent (Fisher Scientific Co., Chicago).
2. Acetate buffer, 0.5 *M,* pH 5.8: Dissolve 68 g of sodium acetate trihydrate in about 700 mL of water, add 1.70 mL of glacial acetic acid (99%), and dilute the volume to 1 L with water.
3. *p*-Nitrophenyl sulfate solution, 0.05 *M:* Dissolve 0.614 g of potassium *p*-nitrophenyl sulfate (Sigma Chemical Co., St. Louis, MO) in about 40 mL of acetate buffer, and dilute the solution to 50 mL with buffer. Store this solution in a refrigerator.
4. Calcium chloride ($CaCl_2$, 0.5 *M*), sodium hydroxide (NaOH, 0.5 *M*), and standard *p*-nitrophenol solution: Prepare as described in section 37–3.2.3.1.2.

37–3.3.3.3 Procedure. Place 1 g of soil (< 2 mm) in a 50-mL Erlenmeyer flask, add 0.25 mL of toluene, 4 mL of acetate buffer, and 1 mL of *p*-nitrophenyl sulfate solution, and swirl the flask for a few seconds to mix the contents. Stopper the flask, and place it in an incubator at 37 °C. After 1 h, remove the stopper, add 1 mL of 0.5 M CaCl$_2$ and 4 mL of 0.5 M NaOH, swirl the flask for a few seconds, and filter the soil suspension through a Whatman No. 2v folded filter paper. Measure the yellow color intensity of the filtrate with a Klett-Summerson photoelectric colorimeter fitted with a blue (no. 42) filter. Calculate the *p*-nitrophenol concentration by reference to a calibration graph prepared from standard *p*-nitrophenol as described in section 37–3.2.3.1.3. if the color intensity of the filtrate exceeds the limit of the calibration graph, dilute an aliquot of the filtrate with water until the colorimeter reading falls within the limits of the graph.

Controls should be performed with each soil analyzed to allow for color not derived from *p*-nitrophenol released by arylsulfatase activity. To perform controls, follow the procedure described for assay of arylsulfatase activity, but make the addition of 1 mL of *p*-nitrophenyl sulfate solution after the addition of 1 mL of 0.5 M CaCl$_2$ and 4 mL of 0.5 M NaOH (i.e., immediately before filtration of the soil suspension).

37–3.3.4 Comments

Although several compounds (e.g., potassium phenyl sulfate, potassium nitrocatechol sulfate, and potassium phenolphthalein sulfate) can serve as substrates for arylsulfatase activity, it is difficult to extract the phenolic compounds released by enzymatic hydrolysis of these compounds. The procedure described gives quantitative (99–100%) recovery of *p*-nitrophenol added to soils. It is necessary to add CaCl$_2$ to prevent dispersion of clay and extraction of soil organic matter during the treatment with NaOH (section 37–3.2.4).

The control allows for the presence of trace amounts of *p*-nitrophenol in some commercial samples of *p*-nitrophenyl sulfate and for extraction of trace amounts of colored soil material by the CaCl$_2$-NaOH treatment. No chemical hydrolysis of *p*-nitrophenyl sulfate is detected under the conditions of the assay procedure described.

Arylsulfatase activity values by the procedure described are lower if water is used instead of pH 5.8 acetate buffer. The difference between the values obtained by using water and buffer is related to the difference between the soil pH and optimum pH (6.2) observed for soil arylsulfatase activity (Tabatabai & Bremner, 1970a).

The substrate concentration in the incubated mixture (10 mM) recommended is much higher (5- to 10-fold) than the K_m values of arylsulfatase in most soils. If desired, this substrate concentration can be changed to meet the objective of the assay.

Choice of buffer and buffer pH in the method described is based on findings of Tabatabai and Bremner (1970a) showing that the amount of *p*-nitrophenol released by incubation of soil with buffered *p*-nitrophenyl sulfate solution is considerably higher in acetate buffer than in other buffers they tested and that maximal enzymatic hydrolysis of *p*-nitrophenyl sulfate occurs with 0.5 M NaOAc buffer having pH 6.2. However, because the buffering poise of the acetate buffer is greater at pH 5.8 than at 6.2 and because the amount of p-nitrophenol released at pH 5.8 is not markedly different from that released at pH 6.2, a buffer pH of 5.8 is recommended. Arylsulfatase activity values are not affected by the amount of toluene added (0.1–1.0 mL). Toluene, however, increases the observed activity of this enzyme in soils.

37–3.4 Rhodanese

37–3.4.1 Introduction

Rhodanese (thiosulfate-cyanide sulfurtransferase, EC 2.8.1.1) is the enzyme that catalyzes the formation of thiocyanate (SCN^-) from thiosulfate ($S_2O_3^{2-}$) and cyanide (CN^-):

$$S_2O_3^{2-} + CN^- \rightarrow SCN^- + SO_3^{2-}. \qquad [23]$$

Lang (1933) discovered rhodanese in animal tissues, and since then, it has been shown that it is widely distributed in nature. Rhodanese activity was detected in soils (Tabatabai & Singh, 1976) and its possible importance in the S cycle was demonstrated by Nor and Tabatabai (1977), who found that $S_2O_3^{2-}$ is an intermediate S compound produced during oxidation of elemental S in soils. The role of rhodanese activity in S° oxidation has been reported (Deng & Dick, 1990; Dick & Deng, 1991). Rhodanese activity is correlated with organic C in soils, and its activity is affected by various soil pretreatments and inorganic salts (Singh & Tabatabai, 1978) and waterlogging (Ray et al., 1985). Studies on kinetic parameters of the rhodanese-catalyzed reaction in soils showed that the K_m values of $S_2O_3^{2-}$ and CN^- for this enzyme are similar to those for the same enzyme isolated from other biological systems (Tabatabai & Singh, 1979). The K_m values of $S_2O_3^{2-}$ and CN^- for rhodanese activity in five soils ranged from 1.2 to 10.3 and from 2.5 to 10.2 mM, respectively. The V_{max} values ranged from 511 to 1431 nmol of SCN produced g^{-1} of soil h^{-1}. The activation energy values ranged from 21.6 to 34 kJ mol^{-1}, and the average Q_{10} for temperatures ranging from 10 to 60 °C ranged from 1.25 to 1.45 (overall avg, 1.37).

37–3.4.2 Principles

The method described is based on colorimetric determination of the SCN^- produced by rhodanese activity when soil is incubated with buffered (pH 6.0) $Na_2S_2O_3$, KCN solutions, and toluene. Incubation is carried out at 37 °C for 1 h, and the reaction is terminated by the addition of a solution containing $CaSO_4$ and formaldehyde. The SCN^- produced is extracted by filtration. The procedure used for determination of the SCN^- produced is based on the reaction of SCN^- with Fe^{3+} in acidic medium to form an Fe-SCN colored complex, which is measured spectrophotometrically at 460 nm. The color is stable for at least 1 h. The formaldehyde present in the extractant stops rhodanese activity and prevents formation of the blue color due to an $Fe-S_2O_3^-$ complex.

No SCN^- formation is expected with $Na_2S_2O_3$ and KCN when incubated without soil under the conditions of the assay procedure described. Copper ions, however, catalyze the reaction of $S_2O_3^{2-}$ and CN^- with the formation of SCN^- (Nor & Tabatabai, 1975, 1976). Therefore, autoclaved soil samples should be included in the assay for rhodanese to allow for any chemical analysis.

37–3.4.3 Assay Method (Tabatabai & Singh, 1976)

37–3.4.3.1 Special Apparatus
1. Incubation flasks, stoppers, incubators, and spectrophotometer as described in section 37–3.2.3.1.1.

37–3.4.3.2 Reagents
1. Toluene, Fisher certified reagent (Fisher Scientific Co., Chicago).
2. Tris(hydroxymethyl)aminomethane-sulfuric acid (THAM-H_2SO_4) buffer (0.05 M, pH 6.0): Dissolve 6.1 g of THAM (Fisher certified reagent, Fisher Scientific Co., Chicago) in about 600 mL of water, bring the pH of the solution to 6.0 by titration with approximately 0.2 M H_2SO_4, and dilute with water to 1 L.
3. Sodium thiosulfate ($Na_2S_2O_3$) solution (0.1 M): Dissolve 24.9 g of $Na_2S_2O_3$ and 6.1 g of tris(hydroxymethyl)aminomethane (THAM) in about 600 mL of water, bring the pH of the solution to 6.0 by titration with approximately 0.2 N sulfuric acid (H_2SO_4) and dilute with water to 1 L.
4. Potassium cyanide (KCN) solution (0.1 M): Dissolve 6.5 g of KCN and 6.1 g of tris(hydroxymethyl)aminomethane (THAM) in about 600 mL of water, bring the pH of the solution to 6.0 by titration with approximately 0.2 N sulfuric acid (H_2SO_4), and dilute with water to 1 L.
5. Formaldehyde solution (37%), Fisher certified reagent.
6. Calcium sulfate-formaldehyde solution: Dissolve 2.2 g of $CaSO_4$ $2H_2O$ in about 900 mL of water, and adjust the volume to 1 L with water. Mix 100 mL of formaldehyde with 900 mL of the calcium sulfate solution.
7. Ferric nitrate-nitric acid solution [0.25 M Fe $(NO_3)_3 \cdot 9H_2O$–3.1 M HNO_3]: Dissolve 50 g of ferric nitrate nanohydrate in 100 mL of concentrated HNO_3 (analytical reagent grade, sp. grade 1.42), add this solution to 300 mL of water, and adjust the volume to 500 mL with water.

   Standard thiocyanate stock solution (20 mM): Dissolve 1.621 g of sodium thiocyanate (NaSCN) in about 800 mL of water, and dilute the solution to 1 L. One milliliter of this solution contains 20 μmol of NaSCN.

37–3.4.3.3 Procedure. Place 4 g of soil (< 2 mm) in a 50-mL Erlenmeyer flask, add 0.5 mL of toluene, 8 mL of THAM buffer, 1 mL of 0.1 M $Na_2S_2O_3$, and 1 mL of 0.1 M KCN, and swirl the flask for a few seconds to mix the contents. Stopper the flask and place it in an incubator adjusted at 37 °C. After 1 h, remove the stopper, and add 10 mL of $CaSO_4$-formaldehyde solution. Swirl the flask for a few seconds, and filter the soil suspension through a Whatman No. 2v folded filter paper. Pipette 5 mL of the filtrate into a test tube, add 1 mL of the ferric nitrate reagent, and measure the absorbance of the reddish brown color of the Fe-SCN complex formed with a spectrophotometer adjusted to wavelength of 460 nm.

Calculate the thiocyanate content of the filtrate by reference to a calibration graph plotted from the results obtained with standards containing 0, 200, 400, 600, 800, and 1000 nmol of thiocyanate. To prepare this graph, dilute 0, 2, 4, 6, 8, and 10 mL of the standard thiocyanate stock solution to 1 L in volumetric flasks, and mix the solutions thoroughly. Then pipette 5-mL aliquots of these diluted standards into test tubes, and proceed as described for analysis of the thiocyanate in the soil filtrate. If the color intensity of the filtrate exceeds that of the highest thiocyanate standard, take a smaller aliquot of the filtrate and adjust the volume to 5 mL with water before adding 1 mL of the ferric nitrate reagent.

Controls should be performed with each soil analyzed to allow for the slight yellow color derived from soil. To perform controls, follow the procedure described for assay of rhodanese activity, but use autoclaved soil samples.

37–3.5 Dehydrogenases

37–3.5.1 Introduction

Biological oxidation of organic compounds is generally a dehydrogenation process, and there are many dehydrogenases (enzymes catalyzing dehydrogenation), which are highly specific. The overall process for dehydrogenation may be presented as follows:

$$XH_2 + A \rightarrow X + AH_2 \qquad [24]$$

where XH_2 is an organic compound (hydrogen donor) and A is a hydrogen acceptor. The dehydrogenase enzyme systems apparently fulfill a significant role in the oxidation of soil organic matter as they transfer H from substrates to acceptors. Many different specific dehydrogenase systems are involved in the dehydrogenase of soils; these systems are an integral part of the microorganisms. Therefore, the result of the assay of dehydrogenase activity would show the average activity of the active population (Skujins, 1976).

The most widely used method for determination of dehydrogenase activity is that developed by Lenhard (1956). This method involves colorimetric determination of 2,3,5-triphenyl formazan (TPF) produced by the reduction of 2,3,5-triphenyltetrazolium chloride (TTC) by soil microorganisms. This assay has received considerable attention during the past decade (Skujins, 1967; Klein et al., 1971; Ross, 1971), presumably because it is believed to provide a good index of microbial activity. But this belief seems to be based entirely on Stevenson's (1959) report of a good correlation between oxygen uptake and TTC dehydrogenase activity of some Canadian soils.

Studies by Ross (1973) indicate that oxygen uptake and dehydrogenase activity are not closely related. Both activities are influenced by the bacterial population and residual plant enzyme content and are negatively correlated with catechol (phenolic) content. Work by Skujins (1973) showed that dehydrogenase activity was highly correlated with CO_2 release, proteolytic activity, and nitrification potential. The highest activity was in the top 3 cm of an arid soil, but again there was no correlation with microbial numbers. Because the dehydrogenase activity depends on the total metabolic activity of soil microorganisms, its value in different soils containing different populations does not always reflect the total numbers of viable microorganisms isolated on a particular medium (Skujins, 1976). Dehydrogenase activity often is correlated with microbial respiration when exogenous C sources are added to soils (Frankenberger & Dick, 1983).

Several treatments may affect dehydrogenase activity in soils. Toluene and $CHCl_3$ can strongly inhibit the activity of dehydrogenases in soils but have little effect at low concentrations. Inhibition of up to 70% of the original activity may occur in soils treated with 3% chloramphenicol. Because of this inhibition effect, bacteriostatic and bacteriocidal compounds are included in the reaction mixture for assay of dehydrogenases in soils. Thus, the contribution to the observed dehydrogenase activity by proliferating microorganisms is unknown but can be anticipated to be substantial, especially in assays involving soils amended with C substrates and incubated for 24 h or more. Even in short-term assays, dehydrogenase activity may be influenced not only by enzyme concentration but also by the nature and concentration of added and endogenous C substrates and by alternate electron acceptors (Ladd, 1978). Bremner and Tabatabai (1973) have shown that added Fe_2O_3, MnO_2, SO_4^{2-}, PO_4^{3-}, and Cl^- stimulated soil dehydrogenase activity, whereas NO_3^-, NO_2^-, and Fe^{3-} seemed to inhibit this activity. The apparent inhibition could be due to some of these latter compounds acting as alternative electron acceptors.

37–3.5.2 Principles

Tetrazolium salts are representatives of a unique class of compounds. These compounds have combinations of desirable properties. They are quarternary NH_4^+ salts and, as such, possess a high degree of water solubility. The water solubility varies considerably and depends on the nature and properties of the substituted groups. The solubility of TTC is sufficiently great to allow the salt to be used in a water solution. This colorless or pale-colored tetrazolium salt possess the property of being easily transformed into intensely colored, water-insoluble, methanol-soluble formazan by reduction. The apparent redox potential of TTC is about -0.08 V, which makes this compound act as an acceptor for many dehydrogenases (Mattson et al., 1947).

The method described is based on extraction with methanol and colorimetric determination of the TPF produced from the reduction of TTC in soils. The transformation occurs through a rupture of the ring, and the general reaction is

[25]

2, 3, 5-Triphenyltetrazollium chloride

37–3.5.3 Assay Method (Casida et al., 1964)

37–3.5.3.1 Special Apparatus
1. Test tubes, 16 by 150 mm.
2. Spectrophotometer or colorimeter.

37–3.5.3.2 Reagents
1. Calcium carbonate ($CaCO_3$), reagent grade.
2. 2,3,5-Triphenyltetrazolium chloride (TTC), 3%: Dissolve 3 g of TTC (Calbiochem, Los Angeles) in about 80 mL of water, and adjust the volume to 100 mL with water.
3. Methanol, analytical reagent grade.
4. Triphenyl formazan (TPF) standard solution: Dissolve 100 mg of TPF (Calbiochem, Los Angeles) in about 80 mL of methanol, and adjust the volume to 100 mL with methanol. Mix thoroughly.

37–3.5.3.3 Procedure. Thoroughly mix 20 g of air-dried soil ($<$ 2 mm) and 0.2 g of $CaCO_3$, and place 6 g of this mixture in each of three test tubes. To each tube add 1 mL of 3% aqueous solution of TTC and 2.5 mL of distilled water. This amount of liquid should be sufficient that a small amount of free liquid appears at the surface of the soil after mixing. Mix the contents of each tube with a glass rod, and stopper the tube and incubate it at 37 °C. After 24 h, remove the stopper, add 10 of methanol, and stopper the tube and shake it for 1 min. Unstopper the tube, and filter the suspension through a glass funnel plugged with absorbent cotton, into a 100-mL volumetric flask. Wash the tube with methanol and quantitatively transfer the soil to the funnel, then add additional methanol (in 10-mL portions) to the funnel until the reddish color has disappeared from the cotton plug. Dilute the filtrate to a 100-mL volume with methanol. Measure the intensity of the reddish color by using a spectrophotometer at a wavelength of 485 nm and a 1-cm cuvette with methanol as a blank. Calculate the amount of TPF produced by reference to a calibration graph prepared from TPF standards. To prepare this graph, dilute 10 mL of TPF standard solution to 100 mL with methanol (100 μg of TPF mL^{-1}). Pipette 5-, 10-, 15-, and 20-mL aliquots of this solution into 100-mL volumetric flasks (500, 1000, 1500, and 2000 μg of TPF 100 mL^{-1}), make up the volumes with methanol, and mix thoroughly. Measure the intensity of the red color of TPF as described for the samples. Plot the absorbance readings against the amount of TPF in the 100-mL standard solutions.

37–3.5.4 Comments

In general, tetrazolium salts are photosensitive, the gross change observable with visible light being the acquisition of a yellow color. Although visible light causes yellowing, UV light of short-wave length brings about reduction of TPF in aqueous solutions (Rust, 1955). The TPF produced is also light sensitive. On long exposure to visible light, TPF undergoes a color change from red to a yellow modification. Storage of the yellow modification in the dark, however, reverses the reaction, and the color gradually returns to its original red. The phenomenon is due to a *cis-trans* isomerization.

It is important that the TPF produced be extracted quantitatively from soils. This can be accomplished by washing the incubated soil in the funnel with small increments of methanol. Care should be taken to prevent turbidity in the methanol extract, because any turbidity would lead to higher dehydrogenase activity values than those expected. Also, the procedure described should be adhered to, because any changes in pH, temperature, incubation time, substrate concentration, or amount of soil incubated will lead to changes in the results obtained by this method. Also, toluene severely inhibits the activity of this enzyme in soils (Frankenberger & Johanson, 1986).

Standard curves prepared as described are reproducible, and it is not necessary to prepare a calibration graph with each set of assay. The TPF color produces is stable if stored in the dark, and none of the color changes just described can be detected under normal laboratory conditions.

37–3.6 β–Glucosidase

37–3.6.1 Introduction

The enzymes acting on glycosyl compounds (EC 3.2), including glycoside hydrolases (EC 3.2.1), have been among the hydrolases least studied in soils. The general name *glycosidases* or *glycoside hydrolases* has been used to describe a group of enzymes that catalyze the hydrolysis of different glycosides. The general equation of the reaction is

$$\text{glycosides} + \text{H}_2\text{O} \rightarrow \text{sugar} + \text{aglycons.} \qquad [26]$$

In general, a glycoside may be defined as a mixed acetal resulting from the exchange of an alkyl or aryl group for the H atom of the hemiacetal hydroxyl group of a cyclic aldose or ketose. The aglycon (or genin) is the noncarbohydrate portion attached to the carbohydrate moiety (glycosyl residue) of the glycoside. The terms *glycoside* and *glucoside* have long been used interchangeably. To avoid ambiguity, it is now customary to designate as glycoside those substances that on acid hydrolysis liberate one or several monosaccharides and an aglycon. The Commission of Enzymes of the International Union of Biochemistry has classified all these enzymes into 39 groups (Florkin & Stotz, 1964). These include enzymes such as cellulase and amylase and some important glycosidases that catalyze the hydrolysis of disaccharides. Glycosidases have usually been named according to the type of bond that they hydrolyze. Among the glycosidases, α-glucosidase (obsolete name maltase, EC 3.2.1.20), which catalyzes the hydrolysis of α-D-glucopyranosides, and β-glucosidase (obsolete name gentiobiase or cellobiase, EC 3.2.1.21), which catalyzes the hydrolysis of β-D-gluco-pyranosides, are involved in hydrolysis of maltose and cellobiose, respectively. Other important glycosidases are α-galactosidase (obsolete name melibiase, EC 3.2.1.22) and β-galactosidase (obsolete name lactase, EC 3.2.1.23). These enzymes catalyze the hydrolysis of melibiose and lactose, respectively. β-Glucosidase activity is the most predominant of the four enzymes in soils (Eivazi & Tabatabai, 1988, 1990).

Glucosidases and galactosidases are widely distributed in nature. These enzymes have also been detected in soils (Skujins, 1967, 1976). The wide distribution of β-glucosidase in fungi (Jermyn, 1958), yeast (Barnett et al., 1956), and plants (Veibel, 1950) has made possible the study of this enzyme. β-Glucosidase is more dominant in soils than α-glucosidase and α- and β-galactosidases. The hydrolysis products of β-glucosidases are believed to be important energy sources for microorganisms in soils.

Recently, Eivazi and Tabatabai (1988) surveyed the methods available for assay of β-glucosidase activity in soils and developed a simple method for determination of the activity of this enzyme. These studies also showed that this enzyme is inactivated in soils at 70 °C and that its activity is correlated with organic C content of surface soils and of soil profiles. Kinetic studies indicated that the K_m values of this enzyme for *p*-nitrophenyl-β-D-glucoside (PNG) in surface soils range from 1.3 to 2.4 mM and that the average activation energy value of β-glucosidase in three soils was 28 kJ mol^{-1}.

37–3.6.2 Principles

The principles of the method described for assay of β-glucosidase activity are similar to those of assay of phosphodiesterase (section 37–3.2.2). The method is based on colorimetric determination of the *p*-nitrophenol released by β-glucosidase when soil is incubated with buffered (pH 6.0) PNG solution and toluene. The incubation is carried out at 37 °C for 1 h. The *p*-nitrophenol released is extracted by filtration after addition of

0.5 M CaCl$_2$ and 0.1 M of THAM buffer pH 12. It is important to treat the incubated soil sample with THAM buffer pH 12 instead of the 0.5 M of NaOH used for extraction of p-nitrophenol in assay of phosphomonoesterases (section 37–3.2.2) and arylsulfatase (section 37–3.3.2), because the substrate of β-glucosidase, PNG is hydrolyzed with time in the presence of excess NaOH. The CaCl$_2$-THAM treatment described for extraction of the p-nitrophenol released in the assay of β-glucosidase activity serves the same purpose as that of CaCl$_2$-NaOH used for extraction of this enzymatic product in the assay of acid and alkaline phosphatases and arylsulfatase. The reaction involved is as follows:

[27]

the substrate is p-nitrophenyl-β-D-glucoside.

37–3.6.3 Assay Method (Eivazi & Tabatabai, 1988)

37–3.6.3.1 Special Apparatus
1. Incubation flasks, stoppers, incubator, and colorimeter or spectrophotometer as described in section 37–3.2.3.1.1.

37–3.6.3.2 Reagents
1. Toluene, MUB pH 6.0, calcium chloride (CaCl$_2$) (0.5 M), and standard p-nitrophenol solution (described in section 37–3.2.3.1.2).
2. p-Nitrophenyl-β-D-glucoside (PNG) solution, 0.05 M: Dissolve 0.654 g of PNG (Sigma Chemical Co., St. Louis, MO) in about 40 mL of MUB pH 6.0, and adjust the volume to 50 mL with the same buffer. Store the solution in a refrigerator.
3. Tris(hydroxymethyl)aminomethane (THAM) buffers (0.1 M), pH 12 and ~10: Prepared as described in section 37–3.2.3.2.2.

37–3.6.3.3 Procedure. Place 1 g of soil (<2 mm) in a 50-mL Erlenmeyer flask, add 0.25 mL of toluene, 4 mL of MUB pH 6.0, 1 mL of PNG solution, and swirl the flask for a few seconds to mix the contents. Stopper the flask, and place it in an incubator at 37 °C. After 1 h, remove the stopper, add 1 mL of 0.5 M CaCl$_2$ and 4 mL of 0.1 M THAM buffer pH

12, swirl the flask for a few seconds, and filter the soil suspension through a Whatman no. 2v folded filter paper. Measure the intensity of the yellow color of the filtrate with a Klett-Summerson photoelectric colorimeter, and calculate the amount of p-nitrophenol released as described in section 37–3.2.3.1.3.

If the color intensity of the filtrate exceeds that of the highest p-nitrophenol standard solution, an aliquot of the filtrate should be diluted with 0.1 M THAM pH ~10 until colorimeter reading falls within the limits of the calibration graph.

37–3.6.4 Comments

Choice of buffer and buffer pH in the method described is based on the finding by Eivazi and Tabatabai (1988) that the optimum pH of this enzyme in soils is at MUB pH 6. The activities of α-glucosidase and α- and β-galactosidase can be assayed by the procedure described using the corresponding substrates.

The PGN solution is stable for several days if stored in a refrigerator. It is necessary to add $CaCl_2$ to prevent dispersion of clay and extraction of soil organic matter during treatment with THAM buffer pH 12 used for extraction of the p-nitrophenol released.

Air-drying of field-moist soils results in marked increases in the activities of α- and β-glucosidases and α- and β-galactosidases in soils. The effect of trace elements and inorganic salts on the activities of these enzymes vary considerably among the elements, enzymes, and soils (Eivazi & Tabatabai, 1990).

37–3.7 Other Enzymes

The activities of several other enzymes have been detected in soils and methods have been developed for their assays. These include methods for measuring the activities of L-histidine ammonia-lyase (Frankenberger & Johanson, 1982b,c), invertase (Frankenberger & Johanson, 1983a,b), and nitrate reductase (Abdelmagid & Tabatabai, 1987; Fu & Tabatabai, 1989). Other references for assay of enzyme activities in soils may be found in the book edited by Burns (1978) and other references cited in this chapter.

REFERENCES

Abdelmagid, H.M., and M.A. Tabatabai. 1987. Nitrate reductase activity of soils. Soil Biol. Biochem. 19:421–427.

Alberty, R.A. 1956. Enzyme kinetics. Adv. Enzymol. Relat. Subj. Biochem. 17:1–64.

Al-Khafaji, A.A., and M.A. Tabatabai. 1979. Effects of trace elements on arylsulfatase activity in soils. Soil Sci. 127:129–133.

Allison, J.P., W.J. Mandy, and G.B. Kitto. 1971. The substrate specificity of L-asparaginase from *Alcaligenes eutrophus*. FEBS Lett. 14:107–108.

Anderson, G. 1967. Nucleic acids, derivatives, and organic phosphates. p. 67–90. *In* A.D. McLaren and G.H. Peterson (ed.) Soil biochemistry. Vol. 1. Marcel Dekker, New York.

Barnett, J.A., M. Ingram, and T. Swain. 1956. The use of β-glucosidase in classifying yeast. J. Gen. Microbiol. 15:529–555.

Beck, T., and H. Poschenrieder. 1963. Experiments on the effect of toluene on the soil microflora. Plant Soil 18:346–357.

Berg, G.G., and L.H. Gordon. 1960. Presence of trimetaphosphatase in the intestinal mucosa and properties of the enzyme. J. Histochem. Cytochem. 8:85–91.

Bidwell, R.G.S. 1974. Plant physiology. p. 173–206. MacMillan, New York.

Blakeley, R.L., J.A. Hinds, H.E. Kunze, E.C. Webb, and B. Zerner. 1969. Jack bean urease (EC 3.5.1.5). Determination of a carbamoyl-transfer reaction and inhibition by hydroxamic acids. Biochemistry 8:1991–2000.

Blanchar, R.W., and L.R. Hossner. 1969. Hydrolysis and sorption reactions of orthophosphate, pyrophosphate, tripolyphosphate, and trimetaphosphate anions added to an Elliot soil. Soil Sci. Soc. Am. Proc. 33:141–144.

Brams, W.H., and A.D. McLaren. 1974. Phosphatase reactions in a column of soil. Soil Biol. Biochem. 6:183–189.

Bray, H.G., S.P. James, I.M. Raffan, B.E. Ryman, and W.V. Thorpe. 1949. The fate of certain organic acids and amindes in the rabbit. Biochem. J. 44:618–625.

Bremner, J.M. 1955. Studies on soil humic acids: I. The chemical nature of humic nitrogen. J. Agric. Sci. 46:247–256.

Bremner, J.M., and R.L. Mulvaney. 1978. Urease activity in soils. p. 149–196. In R.G. Burns (ed.) Soil enzymes. Academic Press, New York.

Bremner, J.M., and M.A. Tabatabai. 1973. Effect of some inorganic substances on TTC assay of dehydrogenase activity in soils. Soil Biol. Biochem. 5:385–386.

Browman, M.G., and M.A. Tabatabai. 1978. Phosphodiesterase activity of soils. Soil Sci. Soc. Am. J. 42:284–290.

Burns, R.G. 1978. Enzyme activity in soil: Some theoretical and practical considerations. p. 295–340. In R.G. Burns (ed.) Soil enzymes. Academic Press, New York.

Burns, R.G., A.H. Pukite, and A.D. McLaren. 1972. Concerning the location and persistence of soil urease. Soil Sci. Soc. Am. Proc. 36:308–311.

Busman, L.M., and M.A. Tabatabai. 1984. Determination of trimetaphosphate added to soils. Commun. Soil Sci. Plant Anal. 15:1257–1268.

Busman, L.M., and M.A. Tabatabai. 1985a. Hydrolysis of trimetaphosphate in soils. Soil Sci. Soc. Am. J. 49:630–636.

Busman, L.M., and M.A. Tabatabai. 1985b. Factors affecting enzymic and nonenzymic hydrolysis of trimetaphosphate in soils. Soil Sci. 140:421–428.

Campbell, H.A., L.T. Mashburn, E.A. Boyce, and L.J. Old. 1967. Two L-asparaginases from Escherichia coli B: Their separation, purification, and antitumor activity. Biochemistry 6:721–730.

Cantarella H., and M.A. Tabatabai. 1983. Amides as sources of nitrogen for plants. Soil Sci. Soc. Am. J. 47:599–603.

Casida, L.E., Jr., D.A. Klein, and T. Santoro. 1964. Soil dehydrogenase activity. Soil Sci. 98:371–376.

Cawse, P.A. 1975. Microbiology and biochemistry of irradiated soils. p. 213–267. In E.A. Paul and A.D. McLaren (ed.) Soil biochemistry. Vol. 3. Marcel Dekker, New York.

Cedar, H., and J.H. Schwartz. 1967. Localization of the two L-asparaginases in anaerobically grown Escherichia coli. J. Biol. Chem. 242:3753–3755.

Cervelli, S., P. Nannipieri, B. Ceccanti, and P. Sequi. 1973. Michaelis constant of soil acid phosphatase. Soil Biol. Biochem. 5:841–845.

Chandramohan, D., K. Devendran, and R. Natarajan. 1974. Arylsulfatase activity in marine sediments. Mar. Biol. 27:89–92.

Clarke, P.H. 1970. The aliphatic amidases of Pseudomonas aeruginosa. Adv. Microb. Physiol. 4:179–222.

Conrad, J.P. 1940a. Catalytic activity causing the hydrolysis of urea in soils as influenced by several agronomic factors. Soil Sci. Soc. Am. Proc. 5:238–241.

Conrad, J.P. 1940b. Hydrolysis of urea in soils by thermolabile catalysis. Soil Sci. 49:253–263.

Conrad, J.P. 1940c. The nature of the catalyst causing the hydrolysis of urea in soils. Soil Sci. 50:119–134.

Conrad, J.P. 1942a. The occurrence and origin of ureaselike activities in soils. Soil Sci. 54:367–380.

Conrad, J.P. 1942b. Enzymatic vs. microbial concepts of urea hydrolysis in soils. Agron. J. 34:1102–1113.

Conrad, J.P. 1943. Some effects of developing alkalinities and other factors upon ureaselike activities in soils. Soil Sci. Soc. Am. Proc. 8:171–174.

Cooper, P.J.M. 1972. Aryl sulphatase activity in Northern Nigerian soils. Soil Biol. Biochem. 4:333–337.

Cosgrove, D.J. 1967. Metabolism of organic phosphate in soil. p. 216–228. In A.D. McLaren and G.H. Peterson (ed.) Soil biochemistry. Vol. 1. Marcel Dekker, New York.

Deng, S., and R.P. Dick. 1990. Sulfur oxidation and rhodanese activity in soils. Soil Sci. 150:552–560.

Dick, R.P., and S. Deng. 1991. Multivariate factor analysis of sulfur oxidation and rhodanese activity in soils. Biogeochemistry 12:87–101.

Dick, R.P., and M.A. Tabatabai. 1986a. Hydrolysis of polyphosphates by corn roots. Plant Soil 94:247–256.

Dick, R.P., and M.A. Tabatabai. 1986b. Hydrolysis of polyphosphates in soils. Soil Sci. 142:132–140.

Dick, R.P., and M.A. Tabatabai. 1987. Factors affecting hydrolysis of polyphosphates in soils. Soil Sci. 143:97–104.

Dick, W.A., N.G. Juma, and M.A. Tabatabai. 1983. Effects of soils on acid phosphatase and inorganic pyrophosphatase of corn roots. Soil Sci. 136:19–25.

Dick, W.A., and M.A. Tabatabai. 1977a. An alkaline oxidation method for determination of total phosphorus in soils. Soil Sci. Soc. Am. J. 41:511–514.

Dick, W.A., and M.A. Tabatabai. 1977b. Determination of orthophosphate in aqueous solutions containing labile organic and inorganic phosphorus compounds. J. Environ. Qual. 6:82–85.

Dick, W.A., and M.A. Tabatabai. 1978a. Inorganic pyrophosphatase activity in soils. Soil Biol. Biochem. 10:59–65.

Dick, W.A., and M.A. Tabatabai. 1978b. Hydrolysis of organic and inorganic phosphorus compounds added to soils. Geoderma 21:175–182.

Dick, W.A., and M.A. Tabatabai. 1983. Activation of soil pyrophosphate by metal ions. Soil Biol. Biochem. 15:359–363.

Dick. W.A., and M.A. Tabatabai. 1984. Kinetic parameters of phosphatases in soils and organic waste materials. Soil Sci. 137:7–15.

Dick, W.A., and M.A. Tabatabai. 1987. Kinetics and activities of phosphates-clay complexes. Soil Sci. 143:5–15.

Dick, W.A., and M.A. Tabatabai. 1992. Significance and potential uses of soil enzymes. p. 95–127. In F.B. Metting, Jr. (ed.) Soil microbial ecology: Application in agriculture and environmental management. Marcel Dekker, New York.

Douglas, L.A., and J.M. Bremner. 1970a. Extraction and colorimetric determination of urea in soils. Soil Sci. Soc. Am. Proc. 34:859–862.

Douglas, L.A., and J.M. Bremner. 1970b. Colorimetric determination of microgram quantities of urea. Anal. Lett. 3:79–87.

Douglas, L.A., and J.M. Bremner. 1971. A rapid method of evaluating different compounds as inhibitors of urease activity in soils. Soil Biol. Biochem. 3:309–315.

Douglas, L.A., H. Sochtig, and W. Flaig. 1978. Colorimetric determination of urea in soil extracts using an automated system. Soil Sci. Soc. Am. J. 42:291–292.

Dowd, J.E., and D.S. Riggs. 1965. A comparison of estimates of Michaelis-Menten kinetic constants from various linear transformations. J. Biol. Chem. 240:863–869.

Drobni'K, J. 1956. Degradation of asparagine by the soil enzyme complex. Cesk. Mikrobiol. 1L47.

Dunn, C.G., W.L. Campbell, H. Fram, and A. Hutchins. 1948. Biological and photochemical effect of high energy, electrostatically produced roentgen rays and cathode rays. J. Appl. Phys. 19:605–616.

Eivazi, F., and M.A. Tabatabai. 1977. Phosphatases in soils. Soil Biol. Biochem. 9:167–172.

Eivazi, F., and M.A. Tabatabai. 1988. Glucosidases and galactosidases in soils. Soil Biol. Biochem. 20:601–606.

Eivazi, F., and M.A. Tabatabai. 1990. Factors affecting glucosidase and galactosidase activities in soils. Soil Biol. Biochem. 22:891–897.

Eremenko, V.V., A.V. Zhukov, and A.Y. Nikolaev. 1975. Asparaginase and glutaminase activity of Pseudomonas fluorescens during continuous culturing. Microbiology 44: 550–555.

Fearon, W.R. 1939. The carbamido diacetyl reaction: A test for citruline. Biochem. J. 33:902–907.

Feder, J. 1973. The phosphatases. p. 475–508. In E.J. Griffith et al. (ed.) Environmental phosphorus handbook. John Wiley and Sons, New York.

Florkin, M., and E.H. Stotz. 1964. p. 126–134. In Comprehensive biochemistry. Vol. 13. Elsevier North-Holland, New York.

Frankenberger, W.T., Jr., and W.A. Dick. 1983. Relationships between enzyme activities and microbial growth and activity indices in soil. Soil Sci. Soc. Am. J. 47:945–951.

Frankenberger, W.T., Jr., and J.B. Johanson. 1982a. Effect of pH on enzyme stability in soils. Soil Biol. Biochem. 14:433–437.

Frankenberger, W.T., Jr., and J.B. Johanson. 1982b. L-Histidine ammonia-lyase activity in soil. Soil Sci. Soc. Am. J. 46:943–948.

Frankenberger, W.T., Jr., and J.B. Johanson. 1982c. Distribution of L-histidine ammonia-lyase activity in soils. Soil Sc. 136:347–353.

Frankenberger, W.T., Jr., and J.B. Johanson. 1983a. Method of measuring invertase activity in soils. Plant Soil 74:301–311.

Frankenberger, W.T., Jr., and J.B. Johanson. 1983b. Factor affecting invertase activity in soils. Plant Soil 74:313–323.

Frankenberger, W.T., Jr., and J.B. Johanson. 1986. Use of plasmolytic agents and antiseptics in soil enzyme assays. Soil Biol. Biochem. 18.209–213.

Frankenberger, W.T., Jr., and M.A. Tabatabai. 1980a. Amidase activity in soils: I. Methods of assay. Soil Sci. Soc. Am. J. 44:282–287.

Frankenberger, W.T., Jr., and M.A. Tabatabai. 1980b. Amidase activity in soils: II. Kinetic parameters. Soil Sci. Soc. Am. J. 44:532–536.

Frankenberger, W.T. Jr., and M.A. Tabatabai. 1981a. Amidase activity in soils: III. Stability and distribution. Soil Sci. Soc. Am. J. 45:333–338.

Frankenberger, W.T., Jr., and M.A. Tabatabai. 1981b. Amidase activity in soils: IV. Effects of trace elements and pesticides. Soil Sci. Soc. Am. J. 45:1120–1124.

Frankenberger, W.T., Jr., and M.A. Tabatabai. 1982. Amidase and urease activities in plants. Plant Soil 64:153–166.

Frankenberger, W.T., Jr., and M.A. Tabatabai. 1985. Characteristics of an amidase isolated from a soil bacterium. Soil Biol. Biochem. 17:303–308.

Frankenberger, W.T., Jr., and M.A. Tabatabai. 1991a. L-Asparaginase activity of soils. Biol. Fert. Soils 11:6–12.

Frankenberger, W.T., Jr., and M.A. Tabatabai. 1991b. Factors affecting L-asparaginase activity in soils. Biol. Fert. Soils 11:1–5.

Frankenberger, W.T., Jr., and M.A. Tabatabai. 1991c. L-Glutaminase activity of soils. Soil Biol. Biochem. 23:869–874.

Frankenberger, W.T., Jr., and M.A. Tabatabai. 1991d. Factors affecting L-glutaminase activity in soils. Soil Biol. Biochem. 23:875–879.

Freney, J.R. 1961. Some observations on the nature of organic sulphur compounds in soils. Aust. J. Agric. Res. 12:424–432.

Fromageot, C. 1950. Sulfatases. p. 517–526. In J.B. Sumner and K. Myrbäck (ed.) The enzymes. Vol. 1. Part 1. Academic Press, New York.

Fu, M.H., and M.A. Tabatabai. 1989. Nitrate reductase activity in soils: Effect of trace elements. Soil Biol. Biochem. 21:943–946.

Galstyan, A.S., and E.G. Saakyan. 1973. Determination of soil glutaminase activity. Dokl. Akad. Nauk SSSR 209:1201–1202.

Gilliam, J.W., and E.C. Sample. 1968. Hydrolysis of pyrophosphate in soils: pH and biological effects. Soil Sci. 106:352–357.

Gorin, G. 1959. On the mechanisms of urease action. Biochim. Biophys. Acta 34:268–270.

Halstead, R.L. 1964. Phosphatase activity of soil as influenced by lime and other treatments. Can. J. Soil Sci. 44:137–144.

Hashimoto, I., J.D. Hughes, and O.D. Philen, Jr. 1969. Reaction of triammonium pyrophosphate with soils and soil minerals. Soil Sci. Soc. Am. Proc. 33:401–405.

Hossner, L.R., and D.P. Phillips. 1971. Pyrophosphate hydrolysis in flooded soil. Soil Sci. Soc. Am. Proc. 35:379–383.

Houghton, C., and F.A. Rose. 1976. Liberation of sulphate from sulphate esters by soils. Appl. Environ. Microbiol. 31:969–976.

Hynes, M.J. 1970. Induction and repression of amidase enzymes in *Aspergillus nidulans.* J. Bacteriol. 103:482–487.

Hynes, M.J. 1975. Amide utilization in *Aspergillus nidulans:* Evidence for a third amidase enzyme. J. Gen. Microbiol. 91:99–109.

Imada, A., S. Igarasi, K. Nakahama, and M. Isono. 1973. Asparaginase and glutaminase activities of microorganisms. J. Gen. Microbiol. 76:85–80.

Irving, G.C., and D.J. Cosgrove. 1976. The kinetics of soil acid phosphatase. Soil Biol. Biochem. 8:335–340.

Ivey, F.J., and K. Shaver. 1977. Enzymic hydrolysis of polyphosphate in the gastrointestinal tract. J. Agric. Food Chem. 25:128–130.

Jakoby, W.B., and J. Fredericks. 1964. Reactions catalyzed by amidases. Acetamidase. J. Biol. Chem. 239:1978–1982.

Jermyn, M.A. 1958. Fungal cellulases. Aust. J. Biol. Sci. 11:114–126.

Joshi, J.G., and P. Handler. 1962. Purification and properties of nicotinamidase from Torula cremoris. J. Biol. Chem. 237:929–935.

Juma, N.G., and M.A. Tabatabai. 1977. Effects of trace elements on phosphatase activity in soils. Soil Sci. Soc. Am. J. 41:343–346.

Juma, N.G., and M.A. Tabatabai. 1978. Distribution of phosphomonoesterases in soils. Soil Sci. 126:101–108.

Juma, N.G., and M.A. Tabatabai. 1988a. Hydrolysis of organic phosphates by corn and soybean roots. Plant Soil 107:31–38.

Juma, N.G., and M.A. Tabatabai. 1988b. Phosphatase activity in corn and soybean roots: Conditions for assay and effects of metals. Plant Soil 107:39–47.

Juma, N.G., and M.A. Tabatabai. 1988c. Comparison of kinetic and thermodynamic parameters of phosphomonoesterases of soils and of corn and soybean roots. Soil Biol. Biochem. 20:533–539.

Kelly, M., and P.H. Clarke. 1962. An inducible amidase produced by a strain of *Pseudomonas aeruginosa.* J. Gen. Microbiol. 27:305–316.

Kiss, S., M. Dracan-Bularda, and D. Radulescu. 1975. Biological significance of enzymes accumulated in soil. Adv. Agron. 27:25–87.

Klein, D.A., T.C. Loh, and R.L. Goulding. 1971. A rapid procedure to evaluate the dehydrogenase activity of soils low in organic matter. Soil Biol. Biochem. 3:385–387.

Kornberg, S.R. 1956. Tripolyphosphate and trimetaphosphate in yeast extracts. J. Biol. Chem. 218:23–31.

Kowalenko, C.G., and L.E. Lowe. 1975. Mineralization of sulfur from four soils and its relationships to soil carbon, nitrogen, and phosphorus. Can. J. Soil Sci. 55:9–14.

Kramer, M. 1957. Phosphatase-Enzym-Aktivität als Anzeiger des biologisch nutzbaren phosphors im Boden. Naturwissenschaften 44:13.

Kramer, M., and G. Yerdei. 1959. Application of the method of phosphatase activity determination in agricultural chemistry. Sov. Soil Sci. (English translation) 9:1100–1103.

Kroll, L., and M. Kramer. 1955. Der Einfluss der Tonmineralien auf die Enzym-Aktivität der Bodenphosphatase. Naturwissenschaften 42:157–158.

Kuprevich, V.F., and T.A. Shcherbakova. 1971. Comparative enzymatic activity in diverse types of soil. p. 167–201. *In* A.D. McLaren and J.J. Skujins (ed.) Soil biochemistry. Vol. 2. Marcel Dekker, New York.

Ladd, J.N. 1978. Origin and range of enzymes in soil. p. 51–96. *In* R.G. Burns (ed.) Soil enzymes. Academic Press, New York.

Lai, C.M., and M.A. Tabatabai. 1992. Kinetic parameters of immobilized urease. Soil Biol. Biochem. 24:225–228.

Lang, K. 1933. Die rhodanbildung im tierkörper. Biochem. Z. 259:243–256.

Lenhard, G. 1956. Die Dehydrogenaseaktivität des Bodens als Mass für die Mikroorganismentätigkeit im Boden. Z. Pflanzenernaehr. Dueng. Bodenkd. 73:1–11.

Mattenheimer, H. 1956. Die Sybstratspezifitat "anorganischer" Polyund Metaphosphatasen: III. Papier-chromatographische Unter suchungen beim enzymatischen Abbau von anorganischen Polyund Metaphosphaten. Hoppe-Seyler; s. Z. physiol. Chem. 303:125–138.

Mattson, A.M., C.O. Jensen, and R.A. Dutcher. 1947. Triphenyl-tetrazolium chloride as a dye for vital tissue. Science 106:294–295.

McLaren, A.D. 1969. Radiation as a technique in soil biology and biochemistry. Soil Biol. Biochem. 1:63–73.

McLaren, A.D. 1975. Soil as a system of humus and clay immobilized enzymes. Chem. Scr. 8:97–99.

McLaren, A.D., and E. Packer. 1970. Some aspects of enzyme reactions in heterogeneous systems. Adv. Enzymol. Relat. Subj. Biochem. 33:245–308.

McLaren, A.D., L. Reshetko, and W. Huber. 1957. Sterilization of soil by irradiation with an electron beam, and some observations on soil enzyme activity. Soil Sci. 83:497–502.

Meyerhof, O., R. Shatas, and A. Kaplan. 1953. Heat of hydrolysis of trimetaphosphate. Biochim. Biophys. Acta 12:121–127.

Michaelis, L., and M.L. Menten. 1913. Die Kinetik der Invertinwirtung. Biochem. Z. 49:333–369.

Neilands, J.B., and P.K. Stumpf. 1964. Outlines of enzyme chemistry. 2nd ed. John Wiley and Sons, New York. p. 57–159.

Neptune, A.M.L., M.A. Tabatabai, and J.J. Hanway. 1975. Sulfur fractions and carbon-nitrogen-sulfur-phosphorus relationships in some Brazilian and Iowa soil. Soil Sci. Soc. Am. Proc. 39:51–55.

Nicholls, R.G., and A.R. Roy. 1971. Arylsulfatases. p. 21–41. *In* P.D. Boyer (ed.) The enzymes. Vol. 5. 3rd ed. Academic Press, New York.

Nor, Y.M., and M.A. Tabatabai. 1975 Colorimetric determination of microgram quantities of thiosulfate and tetrathionate. Anal. Lett. 8:537–547.

Nor, Y.M., and M.A. Tabatabai. 1976. Extraction and colorimetric determination of thiosulfate and tetrathionate in soils. Soil Sci. 122:171–178.

Nor, Y.M., and M.A. Tabatabai. 1977. Oxidation of elemental sulfur in soils. Soil Sci. Soc. Am. 41:736–741.

Paul, E.A., and A.D. McLaren. 1975. Biochemistry of the soil subsystem. p. 1–36. *In* E.A. Paul and A.D. McLaren (ed.) Soil biochemistry. Vol. 3. Marcel Dekker, New York.

Paulson, K.N., and L.T. Kurtz. 1969. Locus of urease activity in soil. Soil Sci. Soc. Am. Proc. 33:897–901.

Pearson, R.W., A.G. Norman, and C. Ho. 1941. The mineralization of the organic phosphorus of various compounds in soils. Soil Sci. Soc. Am. Proc. 6:168–175.

Pettit, N.M., L.J. Gregory, R.B. Freedman, and R.G. Burns. 1977. Differential stabilities of soil enzymes: Assay and properties of phosphatase and arylsulphatase. Biochim. Biophys. Acta 485:357–366.

Pettit, N.M., A.R.J. Smith, R.B. Freedman, and R.G. Burns. 1976. Soil urease: Activity, stability and kinetic properties. Soil Biol. Biochem. 8:479–484.

Pierpoint, W.S. 1957. The phosphatase and metaphosphatase activities of pea extracts. Biochem. J. 65:67–76.

Porter, L.K. 1965. Enzymes. p. 1536–1549. *In* C.A. Black et al. (ed.) Methods of soil analysis. Part 2. Agron. Monogr. 9. ASA, Madison, WI.

Privat de Garilhe, M. 1967. Enzymes in nucleic acid research. Holden-Day, San Francisco. p. 259–278.

Pulford, I.D., and M.A. Tabatabai. 1988. Effect of waterlogging on enzyme activities in soils. Soil Biol. Biochem. 20:215–219.

Quastel, J.H. 1946. Soil metabolism. The Royal Institute of Chemistry of Great Britain and Ireland, London.

Ramirez-Martinez, J.R. 1968. Organic phosphorus mineralization and phosphatase activity in soils. Folia Microbiol., Praha 13:161–174.

Ramirez-Martinez, J.R., and A.D. McLaren. 1966a. Determination of phosphatase activity by a fluorimetric technique. Enzymologia 30:243–253.

Ramirez-Martinez, J.R., and A.D. McLaren. 1966b. Some factors influencing the determination of phosphatase activity in native soils and in soils sterilized by irradiation. Enzymologia 31:23–38.

Ray, R.C., N. Behera, and n. Sethunathan. 1985. Rhodanese activity of flooded and non-flooded soils. Soil Biol. Biochem. 17:159–162.

Razzell, W.E., and H.G. Korana. 1959. Studies on polynucleotides: III. Enzymatic degradation. Substrate specificity and properties of snake venom phosphodiesterase. J. Biol. Chem. 234:2105–2113.

Reiner, J.M. 1959. Behavior of enzyme systems. Burgess Publ. Co., Minneapolis.

Reithel, F.J. 1971. Ureases. p. 1–21. *In* P.D. Boyer (ed.) The enzymes. Vol. 4. Academic Press, New York.

Roberge, M.R. 1970. Behavior of urease added to unsterilized, steam-sterilized and gamma radiation-sterilized black spruce humus. Can. J. Microbiol. 16:865–870.

Roberge, M.R. 1978. Methodology of soil enzyme measurement and extraction. p. 341–370. *In* R.G. Burns (ed.) Soil enzymes. Academic Press, New York.

Roberts, J., J.S. Holcenber, and W.C. Dolowy. 1972. Isolation, crystallization, and properties of Achromobacteraceae, glutaminase-asparaginase with antitumor activity. J. Biol. Chem. 247:84–90.

Rogers, H.T. 1942. Dephosphorylation of organic phosphorus compounds by soil catalysts. Soil Sci. 54:439–446.

Ross, D.J. 1971. Some factors influencing the estimation of dehydrogenase activities of some soils under pasture. Soil Biol. Biochem. 3:97–110.

Ross, D.J. 1973. Some enzyme and respiratory activities of tropical soils from New Hebrides. Soil Biol. Biochem. 5:559–567.

Rotini, O.T. 1935. La transformazione enzimatica dell'-urea nell terreno. Ann Labor. Ric. Ferm. Spallanzani 3:143–154.

Rotini, O.T., and L. Carloni. 1953. La transformazione die metafosfati in ortofasfati promossa dal terreno Agrario. Ann. Sper. Argrar. 7:1789–1799.

Roy, A.B. 1960. The synthesis and hydrolysis of sulfate esters. Adv. Enzymol. 22:205–235.

Rust, J.B. 1955. Chemical properties of the tetrazolium salts. Ann. N.Y. Acad. Sci. 17:379–384.

Sayre, F.W., and E. Roberts. 1958. Preparation and some properties of a phosphate-activated glutaminase from kidneys. J. Biol. Chem. 233:1128–1134.

Schmidt, G., and M. Laskowski, Sr. 1961. Phosphate ester cleavage (Survey). p. 3–35. *In* P.D. Boyer et al. (ed.) The enzymes. 2nd ed. Academic Press, New York.

Searle, P.L., and W.T Speir. 1976. An automated colorimetric method for the determination of urease activity in soil and plant material. Commun. Soil Sci. Plant Anal. 7:365–374.

Segel, I.H. 1975. Enzyme kinetics. John Wiley and Sons, New York.

Sequi, P. 1974. Enzymes in soil. Ital. Agric. 111:91–109.

Singh, B.B., and M.A. Tabatabai. 1978. Factors affecting rhodanese activity in soils. Soil Sci. 125:337–342.

Skujins, J. 1967. Enzymes in soil. p. 371–414. *In* A.D. McLaren and G.H. Peterson (ed.) Soil biochemistry. Vol. 1. Marcel Dekker, New York.

Skujins, J. 1973. Dehydrogenase: An indicator of biological activities in arid soils. Bull. Ecol. Res. Commun. NFR 17:235–241.

Skujins, J. 1976. Extracellular enzymes in soil. CRC Crit. Rev. Microbiol. 4:383–421.

Skujins, J. 1978. History of abiontic soil enzyme research. p. 1–49. *In* R.G. Burns (ed.) Soil enzymes. Academic Press, New York.

Skujins, J.J., L. Braal, and A.D. McLaren. 1962. Characterization of phosphatase in a terrestrial soil sterilized with an electron beam. Enzymologia 25:125–133.

Skujins, J.J., and A.D. McLaren. 1968. Persistence of enzymatic activities in stored and geologically preserved soils. Enzymologia 34:213–225.

Skujins, J.J., and A.D. McLaren. 1969. Assay of urease activity using ^{14}C-urea in stored, geologically preserved and an irradiated soils. Soil Biol. Biochem. 1:89–99.

Sowden, F.J. 1958. The forms of nitrogen in the organic matter of different horizons of soil profiles. Can. J. Soil Sci. 38:147–154.

Speir, T.W. 1977. Studies on a climosequence of soils in tussock grasslands 10. Distribution of urease, phosphatase and sulphatase activities in soil fractions. N. Z. J. Sci. 20:151–157.

Speir, T.W., and D.J. Ross. 1975. Effects of storage on the activities of protease, urease, phosphatase and sulphatase in three soils under pasture. N. Z. J. Sci. 18:231–237.

Speir, T.W., and D.J. Ross. 1978. Soil phosphatase and sulphatase. p. 197–250. *In* R.G. Burns (ed.) Soil enzymes. Academic Press, New York.

Spencer, B. 1958. Studies on sulphatases: 20. Enzymic cleavage of arylhydrogen sulphates in the presence of $H_2 {}^{18}O$. Biochem. J. 69:155–159.

Stevenson, I.L. 1959. Dehydrogenase activity in soils. Can. J. Microbiol. 5:229–235.

Stojanovic, B.J. 1959. Hydrolysis of urea in soil as affected by season and by added urease. Soil Sci. 88:251–255.

Stossel, P., G. Lazarovits, and E.W.B. Ward. 1981. Cytochemical staining and in vitro activity of acid trimetaphosphatase in etiolated soybean hypocotyls. Can. J. Bot. 59:1501–1508.

Stott, D.E., W.A. Dick, and M.A. Tabatabai. 1985. Inhibition of pyrophosphatase activity in soils by trace elements. Soil Sci. 139:112–117.

Sumner, J.B. 1951. Urease, p. 873–892. In J.B. Sumner and K. Myrbäck (ed.) The enzymes. Vol. 1. Part 2. Academic Press, New York.

Tabatabai, M.A. 1973. Michaelis constants of urease in soils and soil fractions. Soil Sci. Soc. Am. Proc. 37:707–710.

Tabatabai, M.A. 1977. Effects of trace elements on urease activity in soils. Soil Biol. Biochem. 9:9–13.

Tabatabai, M.A., and J.M. Bremner. 1969. Use of p-nitrophenyl phosphate for assay of soil phosphatase activity. Soil Biol. Biochem. 1:301–307.

Tabatabai, M.A., and J.M. Bremner. 1970a. Arylsulfatase activity of soils. Soil Sci. Soc. Am. Proc. 34:225–229.

Tabatabai, M.A., and J.M. Bremner. 1970b. Factors affecting soil arylsulfatase activity. Soil Sci. Soc. Am. Proc. 34:427–429.

Tabatabai, M.A., and J.M. Bremner. 1971. Michaelis constants of soil enzymes. Soil Biol. Biochem. 3:317–323.

Tabatabai, M.A., and J.M. Bremner. 1972a. Assay of urease activity in soils. Soil Biol. Biochem. 4:479–487.

Tabatabai, M.A., and J.M. Bremner. 1972b. Forms of sulfur, and carbon, nitrogen, and sulfur relationships in Iowa soils. Soil Sci. 114:380–386.

Tabatabai, M.A., and W.A. Dick. 1979. Distribution and stability of pyrophosphatase in soils. Soil Biol. Biochem. 11:655–659.

Tabatabai, M.A., and M.H. Fu. 1992. Extraction of enzymes from soils. p. 197–227. In G. Stotzky and J-M. Bollag (ed.) Soil biochemistry. Vol. 7. Marcel Dekker, New York.

Tabatabai, M.A., and B.B. Singh. 1976. Rhodanese activity of soils. Soil Sci. Soc. Am. J. 40:381–385.

Tabatabai, M.A., and B.B. Singh. 1979. Kinetic parameters of the rhodanese reaction in soils. Soil Biol. Biochem. 11:9–12.

Thornton, J.I., and A.D. McLaren. 1975. Enzymatic characterization of soil evidence. J. Forensic Sci. 20:674–692.

Tower, D.B., E.L. Peters, and W.C. Curtis. 1963. Guinea pig serum L-asparaginase. J. Biol. Chem. 238:983–993.

Tyler, G. 1974. Heavy metal pollution and soil enzymatic activity. Plant Soil 41:303–311.

Tyler, G. 1976a. Influence of vanadium on soil phosphatase activity. J. Environ. Qual. 5:216–217.

Tyler, G. 1976b. Heavy metal pollution, phosphatase activity and mineralization of organic phosphorus in forest soils. Soil Biol. Biochem. 8:327–332.

Van Wazer, J.R. 1958. Phosphorus and its compounds. Vol. 1. Chemistry. Interscience Publ., New York.

Varner, J.E. 1960. Urease. p. 247–256. In P.D. Boyer et al. (ed.) The enzymes. Vol. 4. Academic Press, New York.

Veibel, S. 1950. β-Glucosidase. p. 583–620. In J.B. Sumner and K. Myrbäck (ed.) The enzymes. Vol. 1. Part 1. Academic Press, New York.

Wall, M.C., and K.J. Laidler. 1953. The molecular kinetics of the urea-urease system—IV. The reaction in an inert buffer. Arch. Biochem. Biophys. 43:299–306.

White, A., P. Handler, and E.L. Smith. 1968. Principles of biochemistry. 4th ed. McGraw-Hill Publ., New York. p. 223–246.

Wriston, J.C., Jr. 1971. L-Asparaginase. p. 101–121. In P.D. Boyer (ed.) The enzymes. Vol. 4. Academic Press, New York.

Yashphe, J. 1973. Estimation of micro amounts of urea and carbamyl derivatives. Anal. Biochem. 52:143–153.

Zantua, M.I., and J.M. Bremner. 1975a. Comparison of methods of assaying urease activity in soils. Soil Biol. Biochem. 7:291–295.

Zantua, M.I., and J.M. Bremner. 1975b. Preservation of soil samples for assay of urease activity. Soil Biol. Biochem. 7:297–299.

Zantua, M.I., and J.M. Bremner. 1976. Production and persistence of urease activity in soils. Soil Biol. Biochem. 8:369–374.

Chapter 38

Carbon Mineralization[1]

L. M. ZIBILSKE, *University of Maine, Orono, Maine*

Biological oxidation of organic C in soil occupies a key position in the global C cycle and provides the principal means by which terrestrial, fixed C is returned to the atmosphere. The term *mineralization* has been advanced by Alexander (1977) and Paul and Clark (1989) as the conversion of an element from an organic to inorganic form. Applied specifically to C, mineralization may be defined as the release of CO_2 from metabolizing organisms. This concept is comparable to *soil respiration,* defined as the sum of all soil metabolic activities that produce CO_2 (Lundegårdh, 1927), or that result in the uptake of O_2 or release of CO_2 by metabolically active soil organisms (Anderson, 1982). While oxygen uptake is a part of soil respiration, it is not accommodated by the general definition of mineralization.

Caution must be exercised when considering mineralization as an index of biodegradation. Witkamp (1969) and Dobbins and Pfaender (1988) recognized that respiration or mineralization measurements may not reflect those activities that do not result in complete oxidation of C, the uptake of metabolic intermediates, or the loss of CO_2 by means other than volatilization. Under these circumstances, respiration measurements might not accurately reflect the actual degree of substrate degradation. However, respiration continues to be the most popular of methods to gauge microbial activity and substrate decomposition in soils. Accordingly, the methods covered in this chapter are similar in several respects to those previously presented by Anderson (1982) and Stotzky (1965), which include the measurement of O_2 uptake. Since this volume includes a separate chapter on isotopic techniques (see chapter 39 in this book), these will not be covered here.

[1]Maine Agric. Exp. Stn. external publ. no. 1801.

38-1 GENERAL PRINCIPLES

The application of C mineralization techniques to soil metabolism studies usually relies on the measurement of CO_2 evolution or O_2 uptake. Quantification of mineralization processes is used to provide information about: (i) the physiological status or catabolic potential of soil microbial populations; (ii) decomposition of specific organic substrates in soil; (iii) soil biomass; and (iv) the relative contributions of microbes, fauna, plant roots, and abiotic sources (e.g., the dissolution of carbonate minerals) to total C flux from the soil. All of the foregoing, as well as standing dead plant matter (Redmann & Abouguendia, 1978), constitute sources of CO_2 in terrestrial systems. Carbon dioxide sinks are also present in soil and can influence measurements of C mineralization. The activities of chemo-lithotrophic bacteria and photosynthesis by plants may affect measurements of CO_2 flux in some experiments.

38-1.1 Field vs. Laboratory Experiments

Results from field experiments are expected to be more reflective of the natural conditions under which soil organic substrates are mineralized. The more defined environmental conditions in laboratory-based investigations usually bear less resemblance to natural conditions. This distinction is usually acceptable since laboratory-based experiments are designed to assess the physiological mechanisms of CO_2 production or O_2 uptake, to determine the decomposition kinetics for specific substrates, or to partition total soil respiration into components attributable to different microbial groups, plant roots, and soil fauna.

Control of the environment in C mineralization studies ranges from none at all, such as atmospheric gas sampling at the soil surface or some distance above the surface, to strict control of small soil samples contained in respirometer flasks in laboratory experiments. Both field and laboratory investigations are normally important to fully characterize the nature of decomposition processes in soil.

38-2 EXPERIMENTAL PRINCIPLES

38-2.1 Alkali Trapping and Detection of Carbon Dioxide

One of the most popular methods of determining CO_2 uses alkali absorption followed by titrimetric, gravimetric, or conductimetric analysis. Alkali trapping methods take advantage of the weakly acidic character of CO_2. The trapping reaction is an acid-base neutralization which produces carbonate (CO_3^{2-}) when the CO_2-laden gas contacts an alkali absorber. When NaOH is used as the absorber, the reaction is described by the equilibrium:

$$CO_2 + 2\,Na^+ + 2\,OH^- \rightleftharpoons CO_3^{2-} + 2\,Na^+ + H_2O$$

As long as the absorbing solution contains a large excess of OH^-, the equilibrium is shifted far to the right and C is retained in solution as carbonate.

Carbonate dissolved in aqueous alkali can also be determined by the conductimetric method. This approach is based on the difference in the mobilities of CO_3^{2-} and OH^- in an electric field in the trapping solution. Increasing amounts of CO_3^{2-} in the solution tend to increase the electrical resistance due to the lower mobility of the CO_3^{2-}. The increase in electrical resistance (decrease in conductivity) of the solution is related to the amount of CO_2 trapped.

The trapped CO_2 is often determined indirectly by titrating the residual OH^- with a standardized acid. Granular alkali is also used as the CO_2 absorber. Commercially prepared soda lime is a coarse-grained mixture of NaOH, $Ca(OH)_2$, and CaO (Cropper et al., 1985) that irreversibly absorbs CO_2. This material can be easier to manipulate under field conditions than liquid alkali (Edwards, 1982b), and has been found to give comparable results to aqueous alkali (Cropper et al., 1985; Minderman & Vulto, 1973). Solid and liquid alkali trapping methods are detailed in sections 38–3.1 and 38–4.1, respectively. Another method used for CO_2 absorption and quantification involves the use of disposable chromatographic tubes (Buyanovsky & Wagner, 1983). These small (approximately 10 cm in length by 7 mm in diam.), commercially assembled tubes are filled with a silica gel containing hydrazine and crystal violet as a oxidation-reduction indicator that turns blue as it becomes saturated with CO_2. Further information on the use of this technique has been detailed by Anderson (1982).

38–2.2 Detection of Carbon Dioxide in Gaseous Samples

Direct analysis of CO_2 in gaseous samples by gas-solid chromatography (GC) or by infrared gas analyzer (IRGA) is common. With IRGA, the concentration gradient of CO_2 at increasing heights above the soil surface can be measured in real-time experiments. Such "aerodynamic" or "micrometeorological" methods (de Jong et al., 1979) offer the least disturbance of the soil while determining CO_2 flux. Modifications of this technique (Baldocchi et al., 1986; Kanemasu et al., 1979) provide a means to determine CO_2 directly. Infrared gas analysis has been used to determine CO_2 in soil air diffusing through a rubber balloon (Cary & Holder, 1982), and in air sweeping across the soil surface in a wind tunnel design (Schwartzkopf, 1978). Static field systems are also adaptable to analysis by portable IRGA instruments (Naganawa et al., 1989). Unmodified commercial IRGA instruments are most often used, but some of the units are modifications of photosynthesis measuring equipment.

Gas chromatography rivals absorption methods as the most popular means for analysis of soil-derived CO_2. The application of GC to soil atmosphere measurements has been reviewed by Smith (1983). Analysis of dissolved gases by GC has been described by Robards et al. (1992). Gas chromatographic analysis of gas samples for CO_2 content is detailed in section 38–3.2.

38–2.3 Detection of Oxygen

Several variations based on manometric techniques have been developed for the study of soil respiratory gas dynamics. The methods are based on the measurement of hypobaric conditions produced in the atmosphere of a closed vessel as a result of O_2 uptake from the headspace air by respiring organisms. The principles and procedures for Warburg and Gilson respirometric methods have been well described by Umbreit et al. (1972). These principles were adapted to soil studies in several procedural and vessel designs. The manometric respirometer of Klein et al. (1972), redrawn by Anderson (1982), is an instrument suitable for the study of soil respiration that can be easily and inexpensively fabricated in the laboratory. Conventional, commercially available manometric respirometers are not designed for soil respiration studies but rather for studies on cellular suspensions. The instrument does not usually provide for large samples, although the Warburg apparatus has been used to study soil respiration in small (0.5 g) samples (Salonius, 1983).

Electrolytic respirometers are incubation devices that replace O_2 used during respiration by electrolysis. Instead of measuring the movement of the liquid level in a manometer, the rising liquid level is used to control an electrolytic reaction that volumetrically replaces O_2 used by respiring organisms. The amount of O_2 produced is calculated from the amount of current passing through the system during the electrolysis. Annis and Nicol (1975) described an electrolytic apparatus into which soil cores could be placed for respiration measurements. Various designs based on this method have been used for respiration studies (Jawson et al., 1989; Knapp et al., 1983).

The most commonly used technique for the determination of O_2 is gas chromatography. By the use of a column packing material different from that discussed for the detection of CO_2, O_2 can be detected in gas samples by the thermal conductivity detector (TCD). The fact that both gases can be determined by essentially the same methodology suggests another reason for the popularity of GC in respiration studies. Detailed procedures for the analysis of O_2 by GC are given in section 38–3.2.

Various electrodes have been used to determine the concentration of O_2 in soils, sediments, and with the newer microelectrodes, inside soil aggregates. Some of the problems of membrane-covered electrodes with response time, stability, and interferences, have been improved with the newer designs of microelectrodes (Revsbech & Jørgensen, 1986). Most of these contain both the anode and cathode in one tube and can be used to determine O_2 over small distances. Some of these electrodes are also sturdy enough to be used inside soil aggregates (Sextone et al., 1985) with the assistance of a micromanipulator. The reader is referred to Revsbech and Jørgensen (1986) for detailed information concerning the use of oxygen electrodes in soil investigations.

38–2.4 Aeration

Investigations of respiratory gas dynamics are conducted in what have been termed *static* and *dynamic* systems (Baldocchi et al., 1986; Cropper et al., 1985; de Jong et al., 1979; Sakamoto & Yoshida, 1988; Witkamp, 1969). Both systems make use of a chamber for the containment of an amount of soil (measured either by enclosed soil weight or ground surface area covered by the chamber). More information pertaining to chambers is provided in section 38–2.5. In static systems, the chamber also isolates an amount of atmosphere from the environment during incubation, preventing unrestricted dilution of chamber gases by the surrounding atmosphere. Static systems, in which the concentrations of respiratory gases within the chamber can change rapidly during incubation, are generally only appropriate for relatively short-term laboratory and field studies.

Systems in which there is displacement of gases during incubation are termed *dynamic*. Apart from the *aerodynamic* method mentioned previously, dynamic systems are characterized by air flow through a confining chamber during incubation that volumetrically replaces the air in the headspace of the container. Several systems, particularly those under computer control, periodically flush the atmosphere in static chambers following sampling (Brooks & Paul, 1987; Hendricks et al., 1987; Nadelhoffer, 1990). These are essentially modified static systems. Continuously flowing systems (Insam, 1990) are also used. The concentration of respiratory gases in the effluent gas stream is measured by one of the methods discussed previously.

38–2.5 Chambers

Several different chamber designs have been used for C mineralization research. Field chambers may be as simple as large metal cans open on one end, or as elaborate as plexiglass vessels fitted with air sampling and replacement syringes working in concert to maintain isobaric conditions in a fan-circulated chamber atmosphere (Parkinson, 1981). Inverted, double-walled funnels have also been used (Mathes & Schriefer, 1985). For laboratory incubations, canning or mason jars continue to be popular containers for soil or organic materials (Blet-Charaudeau et al., 1990; Brooks & Paul, 1987). The ease with which these can be altered for particular sampling and aeration requirements and their low cost are likely reasons for this popularity. Some containers have been designed to allow both respiration and leaching measurements on soil or soil cores (Anderson & Ineson, 1982; Hendricks et al., 1987; Loos et al., 1980; Nadelhoffer, 1990). At least one vessel, based on the design of Bartha and Pramer (1965), is available commercially.

38–2.6 Comparison of Methods

Several of the analytical and incubation methods discussed have been directly compared for efficacy under particular conditions. A review of these comparisons can underscore the strengths and weaknesses of the methods and may be useful in deciding which are appropriate for a given experiment. Comparisons between methods are complicated, however, by the fact that all measurements may not be made concurrently, sampling error is not always determined, and some methods measure different components of soil CO_2 flux (de Jong et al., 1979). Under controlled laboratory conditions, van Cleve et al. (1979) determined the maximum sensitivities to CO_2 of IRGA, GC-TCD, the Gilson respirometer, and base absorption methods to be 0.31, 3.8, 3.6, and 44 μg of CO_2, respectively. Divergence in detectability of CO_2 appears to be related to the temperature of incubation. At incubation temperatures lower than 15 °C, a dynamic IRGA system was found to yield results similar to base absorption methods (Cropper et al., 1985), but at higher temperatures greater amounts of CO_2 were determined with the IRGA. Infrared gas analysis usually detects about twice the amount of CO_2 as base absorption methods (Cropper et al., 1985; Sakamoto & Yoshida, 1988). Incubation conditions apparently affect these comparisons. When IRGA was compared to differential respirometry (Edwards, 1982a) IRGA was found to give generally lower estimates of respired CO_2. This was attributed to the removal of CO_2 in the respirometer while CO_2 was allowed to accumulate in containers for the IRGA assay. Another comparison (van Cleve et al., 1979) found IRGA and base absorption methods to yield the highest estimates of CO_2. The lowest estimate in this study was produced by GC analysis, with manometry intermediate in CO_2 detection.

Cropper et al. (1985) found no difference between 1 N KOH and soda lime in CO_2 absorption capacity in static systems. Using 0.2 N of KOH, Edwards (1982b) reported 62% less CO_2 absorption in KOH than in soda lime in field experiments, and 78% less absorption by KOH in laboratory incubations. Edwards and Sollins (1973) found static systems using a base solution absorber to underestimate CO_2 when compared to dynamic systems. Minderman and Vulto (1973) also found that soda lime CO_2 absorption exceeded that of either 0.5 or 2 N of NaOH. The difference in strength of the base solutions used, however, may explain some of these differences. Cheng and Coleman (1989) and Gupta and Singh (1977) found that alkali strength was an important factor affecting CO_2 trapping efficiency. Soda lime may be more convenient for some field studies since it is less likely to spill.

Carbon mineralization in static systems occurs under somewhat different circumstances than in dynamic systems. This is attributed chiefly to the lack of CO_2 accumulation and maintenance of O_2 levels as the headspace in dynamic systems is continually replaced. Other effects have also been demonstrated. Elevated levels of CO_2 in soil have been shown to reduce the activity of some soil insects (Swift et al., 1979). Reduced NH_4^+

oxidation rates in soil were observed when NaOH concentrations exceeded 50 mM (Kinsbursky & Saltzman, 1990). This reduction occurred when CO_2 concentration was < 0.1 mL L^{-1} due to alkali trapping. Containers of water are often included in sealed vessels to counteract the dehydrating effect of alkali solutions (Bottner, 1985; Jenkinson & Powlson, 1976; Vance & Nadkarni, 1990). In dynamic systems, the air flow is often passed through water for rehydration (after being stripped of CO_2 in alkali traps) before it enters the incubation container. Most of the problems related to gas imbalances in static systems do not appear to be as significant in dynamic systems.

38-3 FIELD METHODS

38-3.1 Carbon Dioxide Detection by Soda Lime Absorption

Absorption of respired CO_2 from a defined soil surface area is used as an index of the activities of soil organisms and plant root respiration. This estimate is made by isolating the surface of the soil for a period and monitoring the production of CO_2. Use of soda lime as the CO_2 absorbent eliminates handling of liquids and the necessary titration of residual base. The amount of CO_2 absorbed through a period is determined by the gain in soda lime dry weight between the beginning and end of an exposure period. The increase in weight is directly related to the absorption of CO_2.

38-3.1.1 Special Apparatus

1. Containers should be large enough to cover 600 cm^2 of soil surface and have a total volume of at least 15 000 cm^3. An appropriate container is a cylindrical can (28 cm diam., 25 cm height) with one open end. If exposure to sunlight is expected during incubation, the cans should be protected from heating. This may be accomplished by painting the cylinders reflective white (Seto, 1982), wrapping them in aluminum foil or, preferably, placing a flat board (about 50 cm square) on top of the cylinders during incubation in the field.

2. Soda lime vessels, one per container. Glass jars with a 5.5 cm diam. are adequate for the container dimensions mentioned above. Vessels should be made of oven-safe glass with screw-caps for air-tight closure. The surface area of the vessel should be at least 5% of the soil surface covered by the container. Jar height should be 7 to 10 cm.

3. Supports for soda lime jars, one per container. These may be fashioned from pieces (12 cm square) of galvanized wire mesh bent down on the corners to form four, 2-cm legs.

38-3.1.2 Reagent

1. Soda lime (6–12 mesh). The amount of soda lime used should be slightly more than 0.06 g cm^{-2} of soil surface covered by the container. For

example, the container that covers 600 cm^2 of soil would require 36 g of soda lime.

38–3.1.3 Procedure

1. Soda lime must be dried to constant weight before use. Weigh slightly more soda lime than needed into tared, oven-safe jars. Place in a 100 °C oven for 24 h. Replace lids, allow to cool, and weigh.

2. Choose sites to be sampled that are clear of living vegetation and coarse debris that will prevent the container from making a good seal against the soil surface. Place the wire mesh support on the ground and press the legs into the ground for stability. Open a jar of soda lime and place it on the wire mesh support. Immediately cover the jar with the container, forcing the lip with a twisting motion approximately 2 cm into the ground. To prepare blanks, open a soda lime jar briefly (to approximate the time it takes to open a jar and place a container over it), then close it again and cover with a container. Prepare two blanks for every 10 sample containers used.

3. Incubate for 24 h. Retrieve the soda lime vessels and cap tightly. Recover the blanks, open briefly, and close tightly as done before during initial placement. If additional measurements are to be made, select different sites and repeat the incubation process with fresh soda lime vessels.

4. In the laboratory, open the jars and place in a 100 °C oven. Dry for 24 h, tightly replace caps, and weigh when cool. Soda lime may be used again, but only until the accumulated weight gain equals 7% of the original dry weight.

38–3.1.4 Calculation of Results

For each mole of CO_2 absorbed by soda lime, 1 mol of water is generated. Oven drying drives off this water along with water absorbed from the air during incubation. For this reason, the actual weight gain by the soda lime is about 41% less than that calculated by stoichiometry. Therefore, the weight gain determined after oven drying must be multiplied by 1.41 to express the data in terms of CO_2. Experimental determinations of this factor were found to differ insignificantly from the theoretical value and most researchers have used either 1.4 (Edwards & Ross-Todd, 1983; Weber, 1985) or 1.41 (Carlyle & U Ba Than, 1988; Gordon et al., 1987) to adjust for the chemical production of water during CO_2 absorption. Data are often expressed as mean rates of gas accumulation; mass per unit area per unit time, i.e., g CO_2 m^{-2} h^{-1}. Carbon dioxide flux estimates can then be made by adjusting the rate by the bulk density of the soil, corrected for stone content if necessary (Edwards & Ross-Todd, 1983).

38–3.1.5 Comments

Simplicity, accuracy, and the fact that no special instruments are required for the soda lime method make it an attractive choice for field soil respiration experiments. The fact that no titration must be done to determine CO_2 absorbed by soda lime eliminates errors associated with titration. An accurate balance and drying oven are all that are required to estimate C mineralization with the soda lime method. However, liquid alkali may be substituted for soda lime and analyzed as described in 38–4.1.4.

Several variations of the method have been used, reflected in the different kinds of chambers that have been used for the containment of soil during the assay. Foil-lined (Weber, 1985) and foil-covered (Gordon et al., 1987) plastic containers, plastic containers alone (Cropper et al., 1985; Edwards, 1982b), and tin-coated steel have all been used. Most modifications adhere to the relationships set out by Edwards (1982b) for chamber size, soda lime vessel size, and amount of soda lime used. Edwards (1982b) reported the absorption capacity of soda lime to be 28% of the original dry weight, and recommended that the soda lime be discarded when it reaches a total weight gain (after drying) of 7% of the original. Drying multiple vessels of soda lime at the same time does not contribute to cross contamination since soda lime does not absorb CO_2 when heated. After oven drying, the soda lime must absorb water to be activated before it will absorb CO_2. This normally occurs after placement under the chambers in the field by the absorption of water vapor from the soil. It may be advisable (Edwards, 1982a) to include a small beaker of water with the soda lime under the container if soil moisture is low.

The mesh size of soda lime used has been found to affect CO_2 absorption. Minderman and Vulto (1973) compared 1.5–2 to 2–5 mm grain sizes and found greater CO_2 absorption with the smaller grain size. The authors also found better CO_2 absorption with soda lime when compared to liquid alkali traps.

The chambers can induce changes in the biological characteristics of the soil under the cover when they are left in place for long periods. Soil excavated from sites under permanent chambers was found to have 85% fewer roots compared to soil under frequently moved chambers (Gupta & Singh, 1981). Hence, respiration rates tend to decrease over time with permanent chamber placement. Absorption of CO_2 may affect other microbial activities, such as nitrification, as observed in closed incubation vessels (Kinsbursky & Saltzman, 1990). If other indices of biological activity are being measured while trapping CO_2, possible effects of the lack of ambient levels of CO_2 must be considered.

The process of chamber placement can affect initial C mineralization rates. Effective isolation of a defined surface area of soil depends on ensuring that gases from outside the chamber do not diffuse into the chamber during incubation. Wildung et al. (1975) found that their chamber effectively isolated the soil when it was pressed 20 cm into the soil. Most other investigators have not inserted their chambers as deeply, but insertion depth may be important to consider for certain soil conditions, such as placement into very coarse-textured soils. A method that has been used to circumvent the disturbance effect of chamber placement has been reported by Cropper et al. (1985). One week prior to measurement, plastic cylinders (open on both ends) of the same diameter as the chambers to be used, are located on the sites where the chambers will be placed. Green vegetation is cut away from the area and the cylinder placed on the soil surface. If the presence of a litter layer does not allow pressing the cylinder into the soil, a knife is used to cut through the litter around the circumference of the cylinder so that it can be pressed into the soil. Later, the cylinders are replaced with the sampling chambers, aligning the chamber lip with the trench made by the open cylinders.

If GC or IRGA facilities are available, the static chamber method as described above can be easily adapted to generate gas samples in shorter-term experiments. By fitting the top of the chamber with a rubber septum, gas samples can be taken periodically by syringe or evacuated gas collection vials for later analysis in the laboratory. If GC is used, this adaptation offers the added capability of determining O_2 in addition to CO_2. Details on sampling and analysis of gas samples by GC follow.

38–3.2 Soil Carbon Dioxide and Oxygen by Gas Chromatography

Subsurface concentrations of CO_2 and O_2 in soil gas may be effectively sampled with the use of gas wells. These generally comprise tubes placed vertically in the ground through which gas samples can be aspirated from desired depths. In most cases, simple tubes are driven into the soil to the desired depth and cleared of any obstructing soil material. Theoretically, soil gas enters these conduits from distinct soil depths and is collected through the tube with a gas-tight syringe for analysis.

Soil tubes can either be installed for short periods or left in place for indefinite periods. In either instance, consideration should be given to whether the sample taken might be contaminated by gas from other locations, including channels created along the side of the tube during installation. The size of the sample should be as small as possible to ensure that soil outside the location of interest does not contribute gas to the sample.

38–3.2.1 Special Apparatus

1. Straight stainless steel tubes, 3 or 4 mm in diameter, cut to lengths approximately 15 cm longer than the depth to which they will be inserted. Rigid plastic tubing may be substituted. The lower end of the tubes may be bored with 1 to 2 mm holes to aid diffusion of gas into the tube. However, the open end of the tube is normally sufficient for gas entry.

2. Steel rods, slightly smaller in diameter than the inner diameter of the stainless steel tubes. These are used to assist installation of the gas sampling tubes.

3. Evacuated gas collection vials; 3 cc (Vacutainers, Becton-Dickinson, Rutherford, NJ; Venoject, Terumo Medical Co., Elkton, MD, or equivalent). Small bottles, closed with a butyl rubber stopper, crimp-sealed

with an aluminum ring and evacuated may be used in place of commercially prepared tubes.

4. Temperature measurement equipment. A thermocouple wire that can be passed into the soil tube is preferred. Readings are made with an inexpensive meter.

5. Gas chromatograph, equipped with a thermal conductivity detector, strip chart recorder (single or dual channel) or digital integrator. High purity He carrier gas; two packed columns (stainless steel, approx. 2 m in length by 3.17 or 6.4 mm diam.), one of Porapak (Waters Associates, Inc., Milford, MA) Q or R (for CO_2) and the other of Molecular Sieve 5A (for O_2). Chromosorb 102 (Celite Co., Wayne, NJ) may be substituted for Porapak. A common configuration of a two column, one detector instrument is one in which eluent from one column enters one side of the detector while eluent from the other column enters the other side. In this way, the baseline signal from one side of the detector serves as the reference for the signal generated by the eluting sample from the other column that enters the other side of the detector.

6. Gas-tight syringes used to inject gas samples into the gas chromatograph; 1.0 cc, and others as appropriate to produce the standard curve; plastic syringes (20 cc) to aspirate stale air from the tubes prior to sampling; 10 cc gas-tight syringes to sample the wells and transfer gas to the evacuated tubes. Alternatively, 1 cc plastic syringes with 25 gauge needles and rubber stoppers to plunge needles into during storage. These may be used to sample the gas wells without transferring the sample to an evacuated tube for transfer to the laboratory (see 38–3.2.3, no. 2).

38–3.2.2 Reagent

1. Reference gas containing CO_2 and O_2. This is used to calibrate the GC and can be obtained from commercial sources.

38–3.2.3 Procedure

1. To establish the gas sampling wells, select the sites to be sampled and push the steel rod into the ground to the desired depth. Carefully remove the rod by pulling it straight out, without disturbing the walls of the shaft. Slide the stainless steel tube onto the rod and push the sheathed pair into the shaft, once again to the desired depth. Remove the steel rod by holding the tube and pulling out the rod, leaving the tube in place in the shaft. The upper end of the tube should protrude approximately 15 cm above the soil surface. Fit a piece of heavy-walled butyl rubber tubing (approximately 10 cm) to the top of the tube and close it tightly at the open end with a short piece of glass rod. Since rubber tubing can deteriorate in direct sunlight, it may be necessary to protect the sampling end of the well by inverting a small plastic bottle over the rubber tubing section of the well between sampling periods. The gas wells should not be left open between samplings since this would allow direct gas exchange between the surface and soil subsurface.

2. To sample the wells, remove the inverted bottle and the plug. Attach a 20-cc syringe to the tubing. There must be an air-tight seal between all parts of the system (see Fig. 38–1). Withdraw two tube volumes of air, leaving the syringe attached to the end of the rubber tubing. Pierce the rubber tubing with the needle of a 10 cc gas-tight syringe and withdraw 10 cc of gas. Remove the syringe from the tubing, and inject a sample into the vacutainer, pressurizing it slightly. Inject another sample into a second, reserve vacutainer. This sample may be used if the first is lost, becomes contaminated, or for duplicate analysis. Determine the temperature of the gas in the well. If the convective transfer of heat through the access tube can be assumed to be minimal, the estimate may be effectively made by passing a thermocouple wire into the tube and taking a measurement when the reading is stable. If 1-cc syringes are used to collect the gas sample, follow the procedure for purging the stale air from the soil tubes and take a 1 cc gas sample through the rubber tubing; immediately plunging the needle of the syringe into a rubber stopper for storage.

3. Gas samples are analyzed by injection into the sample ports of the GC and measuring the recorder response to the electrical signal produced by the detector as the sample elutes from the column. These measurements are made under standardized conditions. Carrier gas flow rate, bridge current in the detector, and operating temperatures of the columns, detector, and sample ports should be optimized for the particular instrument beforehand. This operating configuration should be established and time allowed for stabilization. Conditions typical for this analysis are: He carrier flow rate of 30 mL min^{-1} (for 3.175 mm outside diam. columns, 60 mL min^{-1} for 6.35 mm outside diam. columns); column oven temperature of 50 °C; injector port temperature of 50 °C; detector temperature of 105 °C; and the bridge current set at the minimum necessary to obtain the required sensitivity. Under these conditions, CO_2 elutes from a 1.83 m, 3.18 mm inside diam. stainless steel Chromosorb 102 (80/100) packed column 1 min after injection, following a large $N_2 + O_2$ peak that elutes at approximately 0.5 min. Oxygen and Ar appear together from the molecular sieve 5A column approximately 1.25 min after injection, just before the N_2 peak.

4. Calibrate the GC by determining the response to known concentrations of the reference gases. Using a gas-tight microliter syringe, inject increasing volumes of the reference gas and record the resultant peak heights. Standards should be determined at the beginning of an analysis session and at intervals during the analysis of the gas samples.

5. After stable operating conditions are achieved and the instrument has been calibrated, the samples may be analyzed. Using a 1-cc gas-tight syringe, withdraw 1-cc from the vacutainer. Expel 0.5 cc into the air immediately before inserting the needle into the sample port of the GC. If samples have been taken in 1-cc syringes and transported with the needles inserted into rubber stoppers, apply slight pressure to the syringe plunger while removing the needle from the stopper. Expel 0.5 cc from the syringe and inject the remaining 0.5 cc into the GC. Another 0.5 cc sample (from

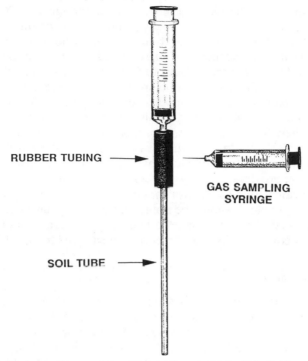

Fig. 38–1. Soil gas sampling apparatus. The stainless steel tube is forced into the soil to the desired depth and gas samples are removed through the rubber tubing with the use of a gas-tight syringe.

the same vacutainer) is injected into the sample port of the other column (molecular sieve) to determine O_2 in the sample.

38–3.2.4 Calculation of Results

Sample peaks are compared to reference peaks to determine the concentration of the particular gas in the sample. A plot of CO_2 or O_2 concentrations by the peak heights should yield a straight line. The equation of this line may be used to convert peak heights of the samples into CO_2 or O_2 concentration values. The following are real data that illustrate the preparation and use of a standard curve. A commercially prepared calibration gas that is 1% CO_2 by volume contains approximately 18.1 µg CO_2 mL^{-1} at 24 °C. When a series of samples (e.g., 25, 50, 75, 100, 200, 300, 400, and 500 µL) of this gas is injected, peak heights from 72 to 1205 mm are produced. Plotting CO_2 amounts by the corresponding peak heights yields a straight line. Regression analysis indicates that these data are well described by the linear equation: $\hat{Y} = 133 X + 25$, where Y is the peak height (in mm) and X is the amount of CO_2 (in µg) in the injected sample.

Converting injection volumes of the standard to amounts of CO_2 requires adjustment for the temperature of the standard gas at the time of analysis. Sample peak heights are substituted into the equation and corresponding values of µg of CO_2 calculated. These calculated CO_2 values must be multiplied by 2 to obtain the amount of CO_2 per cc of sample injected (0.5 cc samples were analyzed). An adjustment for the temperature of the gas at the time of sampling must be made. For example, if 0.5 cc of a soil gas sample produced an 850 mm peak, this would correspond to 6.203 µg of CO_2. Multiplied by 2, this gives approximately 12.41 µg of CO_2 mL^{-1} sample gas. If a 15 °C temperature was measured in the well when the sample was collected, the adjusted concentration of the gas in the original sample is approximately 12.80 µg CO_2 mL^{-1}. It may also be necessary to adjust the calculation by a factor that takes into account the water vapor content of the sample. It should be noted that soil CO_2 may be much higher than in this illustration. The standard curve generated should be in the range of CO_2 concentrations observed in the samples to be analyzed.

38–3.2.5 Comments

Several designs for soil gas sampling probes have been described. These range from relatively simple tubes (Castelle & Galloway, 1990; Kursar, 1989; Wagner & Buyanovsky, 1983) to more elaborate conduits and wells (Anderson, 1982; Smith, 1983). The choice of sampling apparatus will depend on the requirements of the particular experiment. Simple tubes may be appropriate if many wells are to be installed with a minimum of site disturbance; other designs, such as inverted funnels with attached tubing, must be installed by excavating the soil on the site. Some time will be required to allow the soil around these installations to return to typical conditions for the site. It is recommended that soil tubes for long-term experiments be installed at least 1 wk before the first sample is taken. This avoids the problems that may arise from disturbance caused by placement of the tubes into the soil. The number of soil gas sampling tubes required will also depend on the goals of the experiment. Natural horizontal and vertical variations in soil characteristics suggest that several replications will be appropriate.

Samples may be stored for a short period in the same syringes used to aspirate gas from the tubes. This is usually a satisfactory method of storage for a few hours, but if longer storage times are foreseen, the samples should be transferred to evacuated rubber-stoppered glass vials. Over-pressurization of the vials with sample gas is recommended to reduce the chances of introducing ambient air into the tubes during transit to the laboratory and to prevent or minimize negative pressures that develop in the tubes as samples are withdrawn for injection. The negative pressure may not be noticeable to the operator using low-volume gas syringes for sample injection (Smith, 1983). An alternative method is to over-pressurize with a tracer gas. A known quantity of internal standard gas (Robertson & Tiedje, 1985) may be introduced into the glass tubes prior to use. The amount of sample injected into the vial may be decreased proportionately. Leaks may be detected by a reduction in the tracer gas peak height on the chromatogram. Some commercial evacuated tubes may contain CO_2. It may be necessary, therefore, to flush the tubes with He and re-evacuate before use. The use of separate columns, one for CO_2 and one for O_2 detection, permits isothermal determination of both gases in one sample. Trace amounts of methane may occur in some samples. When methane is present, it will appear as a partially resolved peak on the falling slope of the N peak from the molecular sieve column and in the same position on the $N_2 + O_2$ peak from the Chromosorb 102 column. If the determination of small amounts of methane is necessary, a GC with the more sensitive flame ionization detector should be used. No rubber fittings should be used in the construction of the incubation apparatus when CH_4 or other volatile organics are to be analyzed (Marinucci & Bartha, 1979).

If the GC is equipped with temperature-programming capability, both CO_2 and O_2 can be resolved in the same column of a C black molecular sieve (e.g., Carbosieve). Highly refined and conditioned column packings are now available commercially that can sharpen sample peaks by reducing sample tailing in long (approximately 10 m) columns. Effective separation of O_2, Ar, and N_2 can be achieved isothermally in wide-bore molecular sieve-coated fused silica columns. Porous layer open tubular (PLOT) columns are alternatives to conventionally packed columns for CO_2 and O_2 analysis. These columns have the advantages of reducing analytical time, increasing sensitivity, permitting at least partial separation of O_2 and Ar, and being adaptable to existing, relatively large internal volume TC detectors.

The most commonly used gas chromatographic detector for CO_2 and O_2 determinations is the thermal conductivity detector. Other gas detectors have also been used. The [63]Ni electron capture detector (Kaspar, 1984; Kaspar & Tiedje, 1980; Myrold, 1988; Parkin et al., 1984; Rice et al., 1988; Robertson & Tiedje, 1985), which has a somewhat lower response to CO_2 (but a greater response to N_2O than the TCD) has been used for soil aeration-denitrification studies. The ultrasonic detector (Blackmer & Bremner, 1977; Keeney et al., 1985) is sensitive to permanent gases as well as to many other gases. The considerable expense of the ultrasonic detector when compared to the TCD appears to be the greatest deterrent to its widespread use for C mineralization studies.

Gas chromatography offers the advantage of detection of both CO_2 and O_2 in a short period. It is most often used for relatively short duration experiments, but computer automation has been employed in systems that repetitively sample and analyze headspace gas constituents of numerous vessels in laboratory settings (Brooks & Paul, 1987; Hendricks et al., 1987). Computerization has also been applied to IRGA techniques (Heinemeyer et al., 1989; Shelton & Parkin, 1989) as well as to conductimetric analysis (Nordgren, 1988; Nordgren et al., 1988). Automation of measurements in field experiments has been accomplished with an IRGA system (Schwartzkopf, 1978).

38–4 LABORATORY METHODS

Laboratory methods are most often used to examine the decomposition rate of particular substrates, or to determine other aspects of microbial metabolism and ecology under controlled environmental conditions. Laboratory results and field results usually differ. The difference is primarily related to the disturbance factor brought about by the manipulation of the soil that must be considered in the interpretation of data obtained during laboratory investigations. Incubation in the laboratory is popular because it offers the advantages of environmental control and a relative ease of metabolic measurement.

38–4.1 Dynamic Method for Carbon Dioxide

38–4.1.1 Principles

The kinetics of soil C mineralization may be readily observed in the laboratory with the use of incubation chambers through which CO_2-free air constantly sweeps across the soil, removing CO_2 and replenishing O_2 in the headspace above the soil. Dynamic incubation systems are particularly appropriate for longer-term experiments. Conditions in the incubation vessels of dynamic systems are often moderated such that the relative concentrations of CO_2 and O_2 are maintained at near-ambient ratios. The use of a gassing manifold (Cheng & Coleman, 1989, 1990; Martens, 1985; Stotzky, 1965; Weaver, 1974) allows the use of multiple incubation vessels. By passing the exiting air stream through a base solution, CO_2 is stripped from the air. Titrimetric determination of the residual alkali yields the amount of CO_2 in the trap. Although the method usually lacks the sensitivity of some other methods, it can be modified (van Cleve et al., 1979), if necessary, to improve detectability of CO_2.

38–4.1.2 Special Apparatus

1. Compressed air or a vacuum source; laboratory compressed air or vacuum may be used, as well as a small aquarium pump or other diaphragm pump.

2. Air distribution manifold; both a method of distributing flowing air from a single source to the individual incubation vessels and a method for regulating the pressure or vacuum. This can be made of plastic drilled to accept small pieces of rigid plastic tubing glued or otherwise affixed to the manifold for an air-tight seal.

3. Screw clamps are used along the distribution lines to provide fine control of air flow between the air source and incubation vessels. For very small gassing systems, fine metering of air flow may be required, which necessitates the use of commercially obtained metering valves.

4. Heavy walled plastic or butyl rubber tubing (impervious to CO_2; Boddy, 1983); pieces of tubing are fitted to the manifold outlets and attached to the incubation vessels.

5. Carbon dioxide scrubbing system; 25-mm test tubes with two-hole rubber stoppers to fit; 5 cm lengths of glass tubing to fit stopper holes. These tubes are arranged serially (Fig. 38–2) between the air source and incubation vessels such that air entering the vessels must first pass through the scrubbing system. In a vacuum system, the scrubber system is placed immediately after the air inlet point, before entering the incubation vessels. The empty tube serves as a backflow protector. Incoming air should be directed below the surface of the liquid in the test tubes using longer sections of glass tubing. Shorter tubing pieces are used for the exiting air above the liquid surface.

6. Carbon dioxide trapping system; 20-mm test tubes with two hole rubber stoppers to fit; these are connected in series to the outlet air flow from the incubation vessels; 5 cm lengths of glass tubing to fit one of the stopper holes. Pasteur pipettes or plastic, disposable 1-cc tuberculin syringes and needles, without plungers, fitted into the other stopper hole. Prepare the syringes by cutting at the 1-cc graduation mark such that the finger grips are removed, leaving a smooth cylinder. Adjustment of the syringe position may be necessary so that the tip of the needle is near the bottom of the test tube when the stopper is inserted (Fig. 38–3). Sections of small glass tubing may be substituted for the syringe assemblies. The outlet air line from the incubation flask is directed into the tubes through the syringe barrel and needle. The tip of the needle is just above the bottom of the tube, well below the surface of the base solution.

7. Erlenmeyer flasks, 500 mL; two-hole rubber stoppers to fit; 5 cm lengths of glass tubing to fit one of the stopper holes. Glass tubing, approximately 15 cm in length, pushed into the other stopper hole and pushed through such that the lower tip is about 0.5 cm above the surface of the soil when the stopper is fitted to the flask.

8. Burette (50 mL) clamped to a stand.

38–4.1.3 Reagents

1. Carbon dioxide-free water; prepare by boiling distilled water for approximately 5 min, removing from heat source and allowing to cool somewhat, then tightly stoppering and cooling to room temperature.

2. NaOH, 1 N; prepare with CO_2-free water. Add 40 g of NaOH pellets to 500 mL of water; swirl and cool during dissolution. When cool, dilute to 1 L and stopper tightly. The base is used to remove CO_2 from the eluent air flow. It does not require standardization.

3. NaOH, 4 N. Add 160 g of NaOH pellets to 500 mL of water; swirl and cool during dissolution. When cool, dilute to 1 L and stopper tightly. This base is used in the source air scrubber system to remove CO_2 from the air flowing into the incubator vessels.

4. Standardized hydrochloric acid (HCl, 0.5 N); add 41.5 mL of concentrated HCl (approximately 37%, specific gravity 1.19) to 500 mL of water; swirl and dilute to 1 L. Stopper tightly. Standardize the dilute acid. The standard acid is used to titrate residual base in the CO_2 traps.

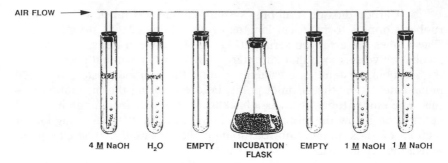

Fig. 38–2. An air flow incubation system with gas scrubbing tubes to strip the air of CO_2 before entry into the incubation flask, and CO_2 trapping tubes to strip the effluent air of respiratory CO_2 emitted by the soil.

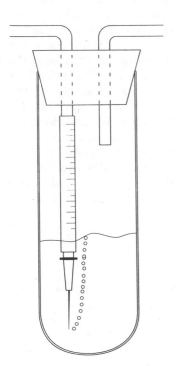

Fig. 38–3. Detail of the air flow system scrubbing tubes showing the use of a plastic syringe with attached needle to conduct the gas stream through the scrubbing liquid.

5. Concentrated sulfuric acid (H_2SO_4); to be used if necessary in the source air scrubber system to remove hydrocarbon contaminants and ammonia.

6. Barium chloride ($BaCl_2$), 3 N; add 183.2 g of $BaCl_2 \cdot 2H_2O$ to 0.5 L of CO_2-free water; stir to dissolve and stopper tightly.

7. Phenolphthalein indicator solution. Add 1 g of phenolphthalein to 100 mL of 95% (vol/vol) ethyl alcohol.

38–4.1.4 Procedure

1. Weigh soil into the flasks and tightly stopper, making certain that the longer glass tube in the stopper is just above the surface of the soil. Include one or more empty incubation flasks among the samples to serve as blanks. Attach the flask to the aeration and scrubbing systems as shown in Fig. 38–2.

2. Arrange the scrubbing train tubes in the following order from the air source: concentrated H_2SO_4, 50 mL (if necessary); 4 N of NaOH, 50 mL; CO_2-free water (if desired), 50 mL; an empty tube.

3. Set up the CO_2 trapping system with a series of one 25-mm empty tube and two, 20-mm test tubes, each of the latter containing 25 mL of 1 N of NaOH.

4. Start and regulate the air flow so that flow rates through all attached flasks are identical. The initial flow rate may be elevated to purge the flasks of ambient air before attaching the air outlet tube to the CO_2 trapping tubes. Once attached, the flow rate should be adjusted such that one flask headspace volume is exchanged about every 8 min (Clark & Gilmour, 1983). See 38–4.2.2.3–4 for discussion of vessel volume estimation methods.

5. During incubation, prepare fresh CO_2 trapping tubes and exchange them with the first tube in the series. Experimentation should be done beforehand to ensure that no more than 60% of the trapping capacity is neutralized between determinations.

6. Quantitatively transfer the contents of the trap to a 125-mL erlenmeyer flask. Wash the tube with CO_2-free water and add the washings to the flask. Add an excess of the $BaCl_2$ solution and three or four drops of the phenolphthalein indicator. Titrate with the standardized HCl.

38–4.1.5 Calculation of Results

The following formula is used to determine the amount of CO_2 retained in the base solution during the incubation period (Stotzky, 1965):
mg $CO_2 = (B-V) \cdot (NE)$ in which:

B = volume (mL) of the standard acid needed to titrate the trap solution from the empty flasks (blanks) to the endpoint,

V = volume (mL) of the standard acid needed to titrate the trap solution from the sample flasks to the endpoint,

N = normality of the acid, in milliequivalents mL^{-1},

E = the equivalent weight of C in CO_2; E = 6 if the data are to be expressed in terms of C (i.e., mg CO_2–C), or E = 22 if the data are to be expressed in terms of CO_2 (i.e., mg CO_2).

The data may be expressed on a unit basis of soil dry weight (i.e., mg CO_2–C g soil^{-1}). Data may be summed to give a cumulative amount of CO_2 evolved, or may be expressed as the mean rate of CO_2 production during the interval of incubation examined (i.e., CO_2 g soil^{-1} h^{-1}). It should be noted that for rate expressions the time (δt) is usually measured from the last determination.

38–4.1.6 Comments

Dynamic systems are somewhat more complex to use than static or intermittently aerated ones, but they offer several advantages. The advantages are associated with the constant removal of respired CO_2 and the replenishment of O_2 in the system. By constantly exchanging the atmosphere in closed vessels, a more natural situation is established. Gas exchange problems can have chemical and biological significance. Retention of CO_2 increases in soil with a pH > 6.5 (Sparling & West, 1990). The manner in which CO_2 is removed from the chamber during incubation has been shown to affect the rate at which it is evolved from neutral to high pH soils (Martens, 1987). This effect was attributed to the increased pCO_2 in the gas phase during static periods that influenced the aqueous carbonate equilibrium, causing more CO_2 to be dissolved in the aqueous phase. The periodic flushing events of these semi-static systems were not long enough to shift the equilibrium toward the gaseous phase of CO_2. Constantly aerated systems appear to favor the removal of CO_2 from the aqueous phase in neutral and high pH soils.

Accumulating CO_2 or depleting O_2 in closed incubation systems may also directly affect metabolic activity. Macfadyen (1973) observed decreased respiration as CO_2 concentrations increased. By controlling rates of aeration, it is possible to prevent conditions that may inhibit or stimulate respiration (Cheng & Coleman, 1989). Another problem that can affect results is the influence of CO_2 trapping on autotrophic processes. Ammonium oxidation was reduced 30% when NaOH traps exceeded 50 mM in a static system (Kinsbursky & Saltzman, 1990). Keeney et al. (1985) reported increasing inhibition of nitrification as CO_2 concentrations were increased from 0.3 to 100%, with no nitrification occurring in 100% CO_2. Gas exchange rates that affect the balance of CO_2 and O_2 in the atmosphere in contact with the soil may affect nitrification and other processes in incubation systems. Excessive air flow rates can intensify pressure leaks in the incubation system and cause siphoning of the base solution when the flow is interrupted during alkali trap changes (Weaver, 1974), both of which result in the loss of CO_2. Passing the CO_2-scrubbed air through water to rehydrate it should be monitored cautiously. Without adequate temperature control in the system, condensation of the water vapor may occur, increasing the water content of the soil.

Incubation in the laboratory is usually designed to examine aspects of soil metabolism, with little emphasis on reproducing the horizontal characteristics of soil and in situ CO_2 diffusion patterns. Therefore, soil can be distributed in the incubation chamber to minimize the constraints on gas exchange between the soil and atmosphere. This can be done by incubating taller vessels on their sides, which allows a given amount of soil to be spread out over a larger area (Gilmour & Gilmour, 1985).

NaOH and KOH solutions are the most commonly used base solutions to trap respired CO_2, but $Ba(OH)_2$ (Blet-Charaudeau et al., 1990; Witkamp, 1969) and BaO_2 (Fine et al., 1986) have also been infrequently employed. This may be partly explained by the potential for producing films of $BaCO_3$ on the quiescent surface of the base solution that can interfere with further absorption of CO_2 (Blet-Charaudeau et al., 1990). This problem may not be as significant in dynamic systems as it is in static systems since the base is agitated by the bubbling action of the eluting air stream. Further, the use of KOH for ^{14}C work may be compromised by the presence of naturally occurring [40]K (D.D. Focht, 1991, personal communication). Methods for radiorespirometry are covered in chapter 39.

In choosing combinations of sample size, vessel size, and the other conditions of incubation, the researcher should consider the potential consequences of those choices on the chemical and biological properties of the system that may affect C mineralization. For example, it may be necessary to cover the incubation flasks with an opaque material to preclude photosynthesis by algae and cyanobacteria.

38–4.2 Static Methods for Carbon Dioxide and Oxygen

38–4.2.1 Principles

The use of sealed chambers for incubation of soil in the laboratory provides a simple means of determining CO_2 using base absorption, or CO_2 and O_2 if gas chromatography is used for analysis. This method differs from dynamic or flow systems in that a given volume of atmosphere is entrapped above the soil in a closed, non-aerated chamber. Evolved CO_2 is either allowed to accumulate in the headspace of the container for GC analysis (Christensen, 1987; Linn & Doran, 1984; West & Sparling, 1986) or is trapped in base solutions for titrimetric or conductimetric determination (Anderson & Ineson, 1982; Chapman, 1971; Gloser & Tesařová, 1978; Nordgren et al., 1988). Incubation chambers must be opened frequently for aeration to avoid potential problems with unnaturally high CO_2 accumulations or severely depleted O_2 supply in the incubating soil. The emphasis in the following closed-chamber methods is on setup and incubation strategies; the analytical methods for CO_2 and O_2 determinations by gas chromatography and titrimetry having been covered in previous sections.

38–4.2.2 Static Incubation-Gas Chromatographic Analysis

38–4.2.2.1 Special Apparatus
1. Vessels suitable for incubation may be constructed from commercially available glass mason or preserving jars (about 1 L capacity) with

threaded rings that seal a rubber-gasketed lid tightly against the lip of the jar. Holes must be drilled through the lids so that they will accept rubber serum vial stoppers. Drill the holes somewhat smaller in diameter than that of the serum vial stopper. Remove defects in the holes by filing so that the rubber stopper is not cut as it is forced through the hole with a twisting motion. The stopper must form an airtight seal with the lid.

2. Rubber serum vial stoppers, 9 mm size; available in bulk from commercial sources. Larger stoppers generally provide longer service than smaller sizes. At least one stopper per vessel must be provided.

3. Gas chromatograph and calibration gases as described in section 38–3.2.1 and 38–3.2.2, respectively.

38–4.2.2.2 Procedure

1. Place 100 g of the soil into the jars; wipe the lip of the jar to remove any debris and close tightly with a threaded ring and lid. The jar may be laid on the side and shaken gently to distribute the soil over the larger surface area of the side.

2. After all the vessels have been prepared, take the first gas sample to establish baseline CO_2 concentrations. Incubate the jars under the desired conditions and withdraw gas samples periodically for GC analysis of respired CO_2 or O_2 uptake.

3. Gas samples are taken by syringe and directly injected into the sample port of the GC (or into the sample loop if the GC is so equipped). Gas may be mixed prior to sampling by inserting the needle of a 50-cc syringe through the septum and alternately withdrawing and re-injecting container air. Remove the mixing syringe and take a 1 cc sample with a gas-tight syringe. Purge 0.5 cc from the syringe and immediately inject the remaining 0.5 cc into the GC. Alternately, the gas sample may be injected into a vacutainer for later analysis (Neilson & Pepper, 1990). Follow procedures in 38–3.2.3 to generate the data to determine CO_2 and O_2 in the injected sample.

4. Open the vessels after sampling as necessary to aerate the soil for 15 to 30 min. Close the vessels and continue the incubation. If vessels are aerated, another sample must be taken immediately after they are closed again to mark the new baseline amounts of respiratory gases in the headspace.

38–4.2.2.3 Calculation of Results.

Follow the procedures in 38–3.2.4 for calculating the concentrations of CO_2 and O_2 in the gas sample. Gas concentrations are adjusted for headspace gas volume in the vessel and the temperature at sampling time to express the data in mass units (i.e., µg CO_2 g soil^{-1}) or on a molar basis (Ross et al., 1975). If the headspace volume is small, uncomplicated vessel volume can be estimated by the weight difference determined between an empty vessel and the same vessel filled with water (Nadelhoffer, 1990). Soil sample volume must be estimated as well. This may be done by including a replicate mass of the soil to be incubated during headspace determination. Soil sample volume may be estimated independently by calculation if the bulk density and water content are known.

Data are converted to mass units by multiplying the temperature-corrected gas concentration determined for the injected sample (0.5 mL) by 2 and then by the headspace volume of the container to obtain the total mass of CO_2 or O_2 in the headspace. This is then divided by the mass of the soil in the container to express the value on a dry soil basis. The time interval is accounted for as required to determine production rates or total CO_2 produced.

38–4.2.2.4 Comments. A manometric method (Kroeckel & Stolp, 1985) has been reported in which the vessel volume is calculated from the rise of the liquid level in a manometer in response to pressurization of the closed system by air injection. Another pressurization method for the determination of headspace volume involves the use of a pressure transducer with excitation and measurement circuitry (Myrold, 1988; Parkin et al., 1984; Rice et al., 1988). In this method, the pressure transducer is attached to stiff tubing ending in a syringe needle. This needle is inserted through a septum into the vessel. The vessel is pressurized with air by use of another syringe. The transducer produces a mV reading that stabilizes within a few seconds when the vessel is empty or within a minute when it contains soil (T.B. Parkin, 1991, personal communication). Measurement of several vessels of known volume produces a graph that can be used to estimate the unknown volume of another vessel. The circuitry required to operate a pressure transducer is found in such common laboratory equipment as multimeters and data loggers. A + 12V DC source is also required. The cost of a suitable pressure transducer is relatively small (approximately $50 U.S.). Pressure methods for determining vessel volume are also effective in determining vessel leaks.

Vessel sample size relationships may be determined beforehand such that the CO_2 level does not rise above 2% during the incubation period (Sparling, 1981). This would make the aeration procedure unnecessary. Harper and Lynch (1985) and Bowen (1990) flushed incubating vessels with sterile air when O_2 concentration fell below 10%. Both of these methods attempt to prevent the problems with microbial metabolism that can occur where significant alterations in respiratory gas concentrations arise during incubation. A ratio of 10:1 or greater, headspace volume (mL):soil weight (g), is often used for closed systems (cf. Bottner, 1985; deCatanzaro & Beauchamp, 1985; Jenkinson & Powlson, 1976; van Gestel et al., 1991).

Combining C mineralization determinations with estimations of other microbial activities is commonly desired. The following method is designed for laboratory experiments in which it is necessary to remove portions of the soil periodically during the incubation for other analyses, such as N or S mineralization or biomass determinations.

38–4.2.3 Static Incubation-Titrimetric Determination

38–4.2.3.1 Special Apparatus

1. Sealable, large-mouth glass containers about 4 L in volume make suitable incubation vessels. Mason or preserving jars, as described in section 38–4.2.2.1 (used without drilling holes into the lids), may be appropriate for smaller amounts of soil, or if only one sample is to be incubated in the jar. Larger jars allow easier access to the beakers. The lids normally provided with large jars obtained commercially may not seal properly. A few disks of aluminum foil cut to fit snugly inside the sealing side of the lid usually corrects this. If not, large rubber stoppers can be used to seal the jars.

2. Small beakers (50 mL). These will contain water to maintain the humidity within the vessel. Larger beakers (100 mL) to accommodate several, individual soil samples within the large incubation jar.

3. Burette (50 mL) clamped to a stand.

38–4.2.3.2 Reagents

1. Reagents needed are found in section 38–4.1.3; CO_2-free water, 1 N of NaOH, standardized 0.5 N of HCl, 3 N of $BaCl_2$, and the phenolphthalein indicator solution.

38–4.2.3.3 Procedure

1. Weigh 25 g replicates of the soil into 100-mL beakers. Carefully arrange the soil sample replicates and one beaker containing 50 mL of CO_2-free water in the jar.

2. Quickly pipet 1 N NaOH into a 100 mL beaker. Use 10 mL of the base for each 25 g of soil in the jar. Place this beaker in a readily accessible location in the jar. Seal the jar tightly and incubate appropriately. Set up and incubate a blank containing only water and NaOH-containing beakers.

3. Periodically, open the jar and remove the NaOH beaker for CO_2 determination. Allow the jar to remain open during the titration procedure. This is done to replenish O_2 in the jar for the next incubation period. Gently mix the contents of beaker and transfer 25 mL to a 125-mL erlenmeyer flask. Add 25 mL of CO_2-free water, an excess of 3 N $BaCl_2$, and three or four drops of phenolphthalein indicator. Titrate with 0.5 N of HCl. Titrate the NaOH from the blank in the same manner. An alternate method for titration of base solutions involves the use of carbonate dehydratase (EC 4.2.1.1) also called carbonic anhydrase (Underwood, 1961). In this method, the enzyme is added to the diluted NaOH trap and titrated sequentially to pH 10, 8.3, and 3.7. The volume of the dilute acid used between the last two points is used to calculate the amount of CO_2 trapped. The advantages of this method are the rapidity with which titration can be carried out and that the determination of the endpoint is made with a pH meter. The latter eliminates the need for subjective determination of endpoints with chromogenic indicators.

4. If other procedures are to be performed on the soil, remove one of the replicate beakers of soil. Add water, if necessary, to the beaker in the

jar. Place a fresh beaker of NaOH in the jar, reducing the amount of base used if a soil sample was removed. Seal the jar and continue the incubation. Replenish the NaOH trap in the blank and continue the incubation.

38–4.2.3.4 Calculation of Results. Results are calculated as described in section 38–4.1.5, adjusting the calculated results as necessary if soil samples were removed from the jar during incubation.

38–4.2.3.5 Comments. Placement of a small sponge (Yoda & Nishioka, 1982) or piece of filter paper (Klein et al., 1984) into liquid alkali has been suggested to improve CO_2 absorption, presumably by increasing the surface area of the absorbing liquid. The only other commonly used acid for titrating the base solutions is H_2SO_4. As sulfuric acid is added in the presence of excess Ba^{2+}, $BaSO_4$ precipitates and increases the cloudiness of the solution, which may affect the visual detection of the endpoint. Carbon dioxide trapped in base solutions may be purged by acidification of the alkali. This is useful for purging ^{14}C–CO_2 from base solutions and trapping it again in a scintillation fluor (Donnelly et al., 1990).

REFERENCES

Alexander, M. 1977. Introduction to soil microbiology. 2nd ed. John Wiley and Sons, New York.

Anderson, J.M., and P. Ineson. 1982. A soil microcosm system and its application to measurements of respiration and nutrient leaching. Soil Biol. Biochem. 14:415–416.

Anderson, J.P.E. 1982. Soil respiration. p. 831–871. In A.L. Page et al. (ed.) Methods of soil analysis. Part 2. 2nd ed. Agron. Monogr. 9. ASA and SSSA, Madison, WI.

Annis, P.C., and G.R. Nicol. 1975. Respirometry system for small biological samples. J. Appl. Ecol. 12:137–141.

Baldocchi, D.D., S.B. Verma, D.R. Matt, and D.E. Anderson. 1986. Eddy-correlation measurements of carbon dioxide efflux from the floor of a deciduous forest. J. Appl. Ecol. 23:967–975.

Bartha, R., and D. Pramer. 1965. Features of a flask and method for measuring the persistence and biological effects of pesticides in soil. Soil Sci. 100:68–70.

Blackmer, A.M., and J.M. Bremner. 1977. Gas chromatographic analysis of soil atmospheres. Soil Sci. Soc. Am. J. 41:908–912.

Blet-Charaudeau, C., J. Muller, and H. Laudelout. 1990. Kinetics of carbon dioxide evolution in relation to microbial biomass and temperature. Soil Sci. Soc. Am. J. 54:1324–1328.

Boddy, L. 1983. Carbon dioxide release from decomposing wood: Effect of water content and temperature. Soil Biol. Biochem. 15:501–510.

Bottner, P. 1985. Response of microbial biomass to alternate moist and dry conditions in a soil incubated with ^{14}C- and ^{15}N-labeled plant material. Soil Biol. Biochem. 17:329–337.

Bowen, R.M. 1990. Decomposition of wheat straw by mixed cultures of fungi isolated from arable soils. Soil Biol. Biochem. 22:401–406.

Brooks, P.D., and E.A. Paul. 1987. A new automated technique for measuring respiration in soil samples. Plant Soil 101:183–187.

Buyanovsky, G.A., and G.H. Wagner. 1983. Annual cycles of carbon dioxide level in soil air. Soil Sci. Soc. Am. J. 47:1139–1145.

Carlyle, J.C., and U Ba Than. 1988. Abiotic controls of soil respiration beneath an eighteen-year-old Pinus radiata stand in Southeastern Australia. J. Ecol. 76:654–662.

Cary, J.W., and C. Holder. 1982. A method for measuring oxygen and carbon dioxide in soil. Soil Sci. Soc. Am. J. 46:1345–1347

Castelle, A.J., and J.N. Galloway. 1990. Carbon dioxide dynamics in acid forest soils in Shenandoah National Park, Virginia. Soil Sci. Soc. Am. J. 54:252–257

Chapman, S.B. 1971. A simple conductimetric soil respirometer for field use. Oikos 22:348–353

Cheng, W., and D.C. Coleman. 1989. A simple method for measuring CO$_2$ in a continuous air-flow system: Modifications to the substrate-induced respiration technique. Soil Biol. Biochem. 21:385–388

Cheng, W., and D.C. Coleman. 1990. Effect of living roots on soil organic matter decomposition. Soil Biol. Biochem. 22:781–787

Christensen, B. 1987. Decomposability of organic matter in particle size fractions from field soils with straw incorporation. Soil Biol. Biochem. 19:429–435.

Clark, M.D., and J.T. Gilmour. 1983. The effect of temperature on decomposition at optimum and saturated soil water contents. Soil Sci. Soc. Am. J. 47:927–929.

Cropper, W.P., Jr., K.C. Ewel, and J.W. Raich. 1985. The measurement of soil CO$_2$ evolution *in situ*. Pedobiologia 28:35–40.

deCatanzaro, J.B., and E.G. Beauchamp. 1985. The effect of some carbon substrates on denitrification rates and carbon utilization in soil. Biol. Fert. Soils 1:183–187.

de Jong, E., R.E. Redmann, and E. A. Ripley. 1979. A comparison of methods to measure soil respiration. Soil Sci. 127:300–306.

Dobbins, D.C., and F.K. Pfaender. 1988. Methodology for assessing respiration and cellular incorporation of radiolabeled substrates by soil microbial communities. Microb. Ecol. 15:257–273.

Donnelly, P.K., J.A. Entry, D.L. Crawford, and K. Cromack, Jr. 1990. Cellulose and lignin degradation in forest soils: Response to moisture, temperature, and acidity. Microb. Ecol. 20:289–295.

Edwards, N.T. 1982a. A timesaving technique for measuring respiration rates in incubated soil samples. Soil Sci. Soc. Am. J. 46:1114–1116.

Edwards, N.T. 1982b. The use of soda-lime for measuring respiration rates in terrestrial systems. Pedobiologia 23:321–330.

Edwards, N.T., and B.M. Ross-Todd. 1983. Soil carbon dynamics in a mixed deciduous forest following clear-cutting with and without residue removal. Soil Sci. Soc. Am. J. 47:1014–1021.

Edwards, N.T., and P. Sollins. 1973. Continuous measurement of carbon dioxide evolution from partitioned forest floor components. Ecology 54:406–412.

Fine, P., A. Feigin, and Y. Waisel. 1986. A closed, well-oxygenated system for the determination of the emission of carbon dioxide, nitrous oxide, and ammonia. Soil Sci. Soc. Am. J. 50:1489–1493.

Gilmour, C.M., and J.T. Gilmour. 1985. Assimilation of carbon by the soil biomass. Plant Soil 86:101–112.

Gloser, J., and M. Tesařová. 1978. Litter, soil, and root respiration measurement. An improved compartmental analysis method. Pedobiologia 18:76–81.

Gordon, A.M., R.E. Schlentner, and K. van Cleve. 1987. Seasonal patterns of soil respiration and CO$_2$ evolution following harvesting in the white spruce forests of interior Alaska. Can. J. For. Res. 17:304–310.

Gupta, S.R., and J.S. Singh. 1977. Effect of alkali concentration volume and absorption area on the measurement of soil respiration in a tropical sward. Pedobiologia 17:233–239.

Gupta, S.R., and J.S. Singh. 1981. Soil respiration in a tropical grassland. Soil Biol. Biochem. 13:261–268.

Harper, S.H.T., and J.M. Lynch. 1985. Colonization and decomposition of straw by fungi. Trans. Br. Mycol. Soc. 85:655–661.

Heinemeyer, O., H. Insam, E.A. Kaiser, and G. Walenzik. 1989. Soil microbial biomass and respiration measurements: An automated technique based on infra-red analysis. Plant Soil 116:191–195.

Hendricks, C.W., E.A. Paul, and P.D. Brooks. 1987. Growth measurements of terrestrial microbial species by a continuous-flow technique. Plant Soil 101:189–195.

Insam, H. 1990. Are the soil microbial biomass and basal respiration governed by the climatic regime? Soil Biol. Biochem. 22:525–532.

Jawson, M.D., L.F. Elliott, R.I. Papendick, and G.S. Campbell. 1989. The decomposition of ^{14}C-labeled wheat straw and ^{15}N-labeled microbial material. Soil Biol. Biochem. 21:417–422.

Jenkinson, D.S., and D.S. Powlson. 1976. The effects of biocidal treatments on metabolism in soil—V. A method for measuring soil biomass. Soil Biol. Biochem. 8:209–213.

Kanemasu, E.T., M.L. Wesley, B.B. Hicks, and J.L. Heilman. 1979. Techniques of calculating energy and mass fluxes. p. 156–182. *In* B.J. Barfield and J.F. Gerber (ed.) Modifications of the aerial environment of crops. Am. Soc. Agric. Eng., St. Joseph, MI.

Kaspar, H.F. 1984. A simple method for the measurement of N_2O and CO_2 flux rates across undisturbed soil surfaces. N. Z. J. Sci. 27:243–246.

Kaspar, H.F., and J.M. Tiedje. 1980. Response of electron-capture detector to hydrogen, oxygen, nitrogen, carbon dioxide, nitric oxide and nitrous oxide. J. Chromatogr. 193:142–147.

Keeney, D.R., K.L. Sahrawat, and S.S. Adams. 1985. Carbon dioxide concentration in soil: Effects on nitrification, denitrification, and associated nitrous oxide production. Soil Biol. Biochem. 17:571–573.

Kinsbursky, R.S., and S. Saltzman. 1990. CO_2-nitrification relationships in closed soil incubation vessels. Soil Biol. Biochem. 22:571–572.

Klein, D.A., P.A. Mayeux, and S.L. Seaman. 1972. A simplified unit for evaluation of soil core respirometric activity. Plant Soil 36:177–183.

Klein, T.M., N.J. Novick, J.P. Kreitinger, and M. Alexander. 1984. Simultaneous inhibition of carbon and nitrogen mineralization in a forest soil by simulated acid precipitation. Bull. Environ. Contam. Toxicol. 32:698–703.

Knapp, E.B., L.F. Elliott, and G.S. Campbell. 1983. Microbial respiration and growth during the decomposition of wheat straw. Soil Biol. Biochem. 15:319–323.

Kroeckel, L., and H. Stolp. 1985. Influence of oxygen on denitrification and aerobic respiration in soil. Biol. Fert. Soils 1:189–193.

Kursar, T.A. 1989. Evaluation of soil respiration and soil CO_2 concentration in a lowland moist forest in Panama. Plant Soil 113:21–29.

Linn, D.M., and J.W. Doran. 1984. Effect of water-filled pore space on carbon dioxide and nitrous oxide production in tilled and nontilled soils. Soil Sci. Soc. Am. J. 48:1267–1272.

Loos, M.A., A. Kontson, and P.C. Kearney. 1980. Inexpensive soil flask for ^{14}C-pesticide studies. Soil Biol. Biochem. 12:583–585.

Lundegårdh, M. 1927. Carbon dioxide evolution of soil and crop growth. Soil Sci. 23:417–453.

Macfadyen, A. 1973. Inhibitory effects of carbon dioxide on microbial activity in soil. Pedobiologia 13:140–149.

Marinucci, A.C., and R. Bartha. 1979. Apparatus for monitoring the mineralization of volatile ^{14}C-labeled compounds. Appl. Environ. Microbiol. 38:1020–1022.

Martens, R. 1985. Limitations in the application of the fumigation technique for biomass estimations in amended soil. Soil Biol. Biochem. 17:57–63.

Martens, R. 1987. Estimation of microbial biomass in soil by the respiration method: importance of soil pH and flushing methods for the measurement of respired CO_2. Soil Biol. Biochem. 19:77–81.

Mathes, K., and T. Schricfer. 1985. Soil respiration during secondary succession: influence of temperature and moisture. Soil Biol. Biochem. 17:205–211.

Minderman, G., and J.C. Vulto. 1973. Comparison of techniques for the measurement of carbon dioxide evolution from soil. Pedobiologia 13:73–80.

Myrold, D.D. 1988. Denitrification in ryegrass and winter wheat cropping systems of western Oregon. Soil Sci. Soc. Am. J. 52:412–416.

Nadelhoffer, K.J. 1990. Microlysimeter for measuring nitrogen mineralization and microbial respiration in aerobic soil incubations. Soil Sci. Soc. Am. J. 54:411–415.

Naganawa, T., K. Kyuma, H. Yamamoto, Y. Yamamoto, H. Yokoi, and K. Tatsuyama. 1989. Measurement of soil respiration in the field: Influence of temperature, moisture level, and application of sewage sludge compost and agro-chemicals. Soil Sci. Plant Nutr. 35:509–516.

Neilson, J.W., and I.L. Pepper. 1990. Soil respiration as an index of soil aeration. Soil Sci. Soc. Am. J. 54:428–432.

Nordgren, A. 1988. Apparatus for the continuous, long-term monitoring of soil respiration rate in large numbers of samples. Soil Biol. Biochem. 20:955–957.

Nordgren, A., E. Bååth, and B. Söderström. 1988. Evaluation of soil respiration characteristics to assess heavy metal effects on soil microorganisms using glutamic acid as a substrate. Soil Biol. Biochem. 20:949–954.

Parkin, T.B., H.F. Kaspar, A.J. Sextone, and J.M. Tiedje. 1984. A gas-flow soil core method to measure field denitrification rates. Soil Biol. Biochem. 16:323–330.

Parkinson, K.J. 1981. An improved method for measuring soil respiration in the field. J. Appl. Ecol. 18:221–228.

Paul, E.A., and F.E. Clark. 1989. Soil microbiology and biochemistry. Academic Press, San Diego.

Redmann, R.E., and Z.M. Abouguendia. 1978. Partitioning of respiration from soil, litter and plants in a mixed-grassland ecosystem. Oecologia 36:69–79.

Revsbech, N.P., and B.B. Jørgensen. 1986. Microelectrodes: Their use in microbial ecology. Adv. Microb. Ecol. 9:293–352.

Rice, C.W., P.E. Sierzega, J.M. Tiedje, and L.W. Jacobs. 1988. Stimulated denitrification in the microenvironment of a biodegradable organic waste injected into soil. Soil Sci. Soc. Am. J. 52:102–108.

Robards, K., V.R. Kelly, and E. Patsalides. 1992. Determination of dissolved gases in water by gas chromatography. p. 53–86. In J.C. Giddings et al. (ed.) Advances in chromatography. Marcel Dekker, New York.

Robertson, G.P., and J.M. Tiedje. 1985. An automated technique for sampling the contents of stoppered gas-collection vials. Plant Soil 83:453–457.

Ross, D.J., B.A. McNeilly, and L.F. Molloy. 1975. Studies on a climosequence of soils in tussock grasslands 4. Respiratory activities and their relationships with temperature, moisture, and soil properties. N. Z. J. Sci. 18:377–389.

Sakamoto, K., and T. Yoshida. 1988. In situ measurement of soil respiration rate by a dynamic method. Soil Sci. Plant Nutr. 34:195–202.

Salonius, P.O. 1983. Effects of air drying on the respiration of forest soil microbial populations. Soil Biol. Biochem. 15:199–203.

Schwartzkopf, S. 1978. An open chamber technique for the measurement of carbon dioxide evolution from soils. Ecology 59:1062–1068.

Seto, M. 1982. A preliminary observation on CO_2 evolution from soil in situ measured by an air current method—An example in rainfall and plowing sequences. Jpn. J. Ecol. 32:535–538.

Sextone, A.J., N.P. Revsbech, T.B. Parkin, and J.M. Tiedje. 1985. Direct measurement of oxygen profiles and denitrification rates in soil aggregates. Soil Sci. Soc. Am. J. 49:645–651.

Shelton, D.R., and T.B. Parkin. 1989. A semiautomated instrument for measuring total and radiolabeled carbon dioxide evolution from soil. J. Environ. Qual. 18:550–554.

Smith, K.A. 1983. Gas chromatographic analysis of the soil atmosphere. p. 407–454. In K.A. Smith (ed.) Soil analysis: Instrumental techniques and related procedures. Marcel Dekker, New York.

Sparling, G.P. 1981. Microcalorimetry and other methods to assess biomass and activity in soil. Soil Biol. Biochem. 13:93–98.

Sparling, G.P., and A.W. West. 1990. A comparison of gas chromatography and differential respirometer methods to measure soil respiration and to estimate the soil microbial biomass. Pedobiologia 34:103–112.

Stotzky, G. 1965. Microbial respiration. p. 1550–1572. In C.A. Black (ed.) Methods of soil analysis. Part 2. Agron. Monogr. 9. ASA, Madison, WI.

Swift, M.J., O.W. Heal, and J.M. Anderson. 1979. Decomposition in terrestrial ecosystems. D.J. Anderson et al. (ed.) Studies in ecology. Vol. 5. Blackwell Scientific, Oxford, England.

Umbreit, W.W., R.H. Burris, and J.F. Stauffer. 1972. Manometric and biochemical techniques: A manual describing methods applicable to the study of tissue metabolism. 7th ed. Burgess Publ. Co., Minneapolis.

Underwood, A.L. 1961. Carbonic anhydrase in the titration of carbon dioxide solutions. Anal. Chem. 33:955–956.

van Cleve, K., P.I. Coyne, E. Goodwin, C. Johnson, and M. Kelley. 1979. A comparison of four methods for measuring respiration in organic material. Soil Biol. Biochem. 11:237–246.

van Gestel, M., J.N. Ladd, and M. Amato. 1991. Carbon and nitrogen mineralization from two soils of contrasting texture and microaggregate stability: Influence of sequential fumigation, drying and storage. Soil Biol. Biochem. 23:313–322.

Vance, E.D., and N.M. Nadkarni. 1990. Microbial biomass and activity in canopy organic matter and the forest floor of a tropical cloud forest. Soil Biol. Biochem. 22:677–684.

Wagner, G.H., and G.A. Buyanovsky. 1983. Use of gas sampling tubes for direct measurement of $^{14}CO_2$ in soil air. Int. J. Radiat. Isot. 34:645–648.

Weaver, R.W. 1974. A simple, inexpensive apparatus for simultaneous collection of CO_2 evolved from numerous soils. Soil Sci. Soc. Am. Proc. 38:853.

Weber, M.G. 1985. Forest soil respiration in eastern Ontario jack pine ecosystems. Can J. For. Res. 15:1069–1073.

West, A.W., and G.P. Sparling. 1986. Modifications to the substrate-induced respiration method to permit measurement of microbial biomass in soils of differing water contents. J. Microbiol. Methods 5:177–189.

Wildung, R.E., T.R. Garland, and R.L. Buschborn. 1975. The interdependent effects of soil temperature and water content on soil respiration rate and plant root decomposition in arid grasslands soils. Soil Biol. Biochem. 7:373–378.

Witkamp, M. 1969. Cycles of temperature and carbon dioxide evolution from litter and soil. Ecology 50:922–924.

Yoda, K., and M. Nishioka. 1982. Soil respiration in dry and wet seasons in a tropical dry-evergreen forest in Sakaerat, NE Thailand. Jpn. J. Ecol. 32:539–541.

Chapter 39

Isotopic Methods for the Study of Soil Organic Matter Dynamics

DUANE C. WOLF, *University of Arkansas, Fayetteville, Arkansas*

J. O. LEGG, *University of Arkansas, Fayetteville, Arkansas*

THOMAS W. BOUTTON, *Texas A & M University, College Station, Texas*

The vast majority of all organic C and N in the world's terrestrial environment is present in the form of soil organic matter, which contains approximately 1.5×10^{18}g of C (Post et al., 1982) and 0.095×10^{18}g of N (Post et al., 1985). In addition to its importance as a reservoir of C, N, and other nutrients, this pool of soil organic matter has many properties that define the structural and functional attributes of natural and agricultural ecosystems.

The benefits of soil organic matter in crop production have been recognized for centuries (Allison, 1973), although the reasons for the beneficial effects have not been clearly understood. Organic matter is highly important in relation to soil aggregation, tilth, cation exchange capacity, nutrient supply, soil water, biological activity, and other soil characteristics. Repeated additions of organic matter to soils, normally occurring from plant and animal residues, and subsequent biological, chemical, and physical processes that occur, create a highly complex and dynamic system. This complex system attracted the attention of chemists as early as the 18th century when acid and alkali extraction procedures were first used to remove organic matter from soils (Russell, 1961). In the 1940s, isotopes of C and N came into use in studies of transformations of these elements during organic matter decomposition. Paul and van Veen (1978) reviewed these early studies and proposed a model to describe the rate of organic matter decomposition. It is important to emphasize that, in the decomposition of labeled compounds, other organic compounds may be synthesized simultaneously from the mineralized C and N. The position of the labeled atom in the material undergoing mineralization can influence the results from decomposition studies.

Soil organic matter not only has fundamental importance at the ecosystem level, but also has considerable importance for global biogeochemistry due to the size of the pool and its linkages to atmospheric CO_2 via primary production and decomposition. There is currently much interest and speculation regarding whether soil organic matter is a net carbon sink or source in the global carbon cycle under present environmental conditions as well as under conditions of elevated atmospheric CO_2 concentration and altered climate (Prentice & Fung, 1990; Schlesinger, 1990). As a result of the controversial role of soil organic matter in global biogeochemistry, methodologies for quantifying fluxes of soil organic matter have become extremely important.

In this chapter, we will address isotopic methods of assessing fluxes associated with soil organic matter. There are basically four approaches available: (i) use of organic matter labeled with ^{14}C or ^{13}C; (ii) use of organic matter labeled with ^{15}N; (iii) use of natural variation in ^{13}C in organic matter; and (iv) use of ^{14}C injected into the atmosphere during nuclear weapons testing. Approaches 1 and 2 are most useful for short-term studies of 1 to 10 yr, and approaches 3 and 4 are useful for examining fluxes on time scales ranging from tens to thousands of years. Since methods involving the use of "bomb" ^{14}C have been reviewed recently (Harrison et al., 1990; Goh, 1991), we will address some of the more common methods of using ^{13}C, ^{14}C, and ^{15}N to study organic matter decomposition. The References section will provide citations for obtaining further information on pertinent isotopic methodology as well as related analytical techniques.

39–1 DECOMPOSITION OF ^{14}C-LABELED ORGANIC MATTER IN SOILS

39–1.1 Introduction

Historically, the use of ^{14}C-labeled materials to study soil organic matter decomposition has been accomplished by adding radiolabeled plant material, microorganisms, microbial products, or specific compounds to the soil and measuring the amount of $^{14}CO_2$ evolved during an aerobic incubation (Ladd & Martin, 1984; Wolf & Legg, 1984). Various methods are available to produce the radiolabeled materials and to collect the evolved $^{14}CO_2$.

The objective of this section is to present methods for the production of ^{14}C-labeled materials and define several methods available for the collection and assay of the $^{14}CO_2$ evolved during aerobic laboratory incubation studies. It is not our intent to define all available techniques, but rather to provide details of the more commonly used procedures. Using even a weak beta-emitting radioactive isotope such as ^{14}C requires particular attention to safety considerations. The use of radioisotopes generally requires a permit or license and training in radiation safety. The aspects of health and safety will not be addressed in this section, but any researchers undertaking a project involving ^{14}C should contact their local Radiation Safety Officer for details on licensing and permits.

39–1.2 Obtaining ^{14}C-Labeled Organic Materials

Obtaining ^{14}C-labeled materials for studies on organic matter decomposition is the first and possibly the most difficult step. The choice of organic material may be dictated by the objective of the given research project and the availability of specific research facilities in which to carry out labeling experiments. Products that can be labeled range from the simplest of pure compounds purchased from several commercial sources (Table 39–1) to whole plants grown in a $^{14}CO_2$ environment or microorganisms and microbial products produced in the laboratory on a wide range of labeled organic substrates. Obviously the cost, time, and facilities will have an impact on the organic material available for any given experiment.

39–1.2.1 Labeling Plant Material

One method used to produce ^{14}C-labeled plant material is to grow plants in a $^{14}CO_2$ environment in a growth chamber and harvest the plants after a suitable growth period. Another approach is to inject a ^{14}C-labeled precursor directly into the plant and let the metabolic activity of the plant incorporate the label into various biochemical fractions. Both methods require an appropriate containment facility to prevent contamination of the laboratory environment.

39–1.2.1.1 Materials
1. $^{14}CO_2$ or ^{14}C-labeled precursor.
2. Suitable growth and containment facility.

39–1.2.1.2 Procedure. Details of construction and operation of $^{14}CO_2$ growth chambers have been presented by several researchers (Andersen et al., 1961; Cheshire & Griffiths, 1989; Harris & Paul, 1991; Jenkinson, 1960; Scully et al., 1956; Smith et al., 1962; Warembourg & Kummerow, 1991).

At harvest, the plant may be separated as desired into various components such as shoots, leaves, and roots. Once the plant material is harvested, it should be lyophilized. Oven drying the plant material has the potential to release $^{14}CO_2$ into the laboratory environment and is not recommended. After freeze-drying, the material can be ground and sieved to the required size.

A second method of labeling plant material is to treat growing plants with ^{14}C-labeled precursors of lignin or cellulose biosynthesis. Haider et al. (1977) used a syringe to inject ^{14}C-labeled *p*-coumaric acid, a precursor in the synthesis of lignin, into the base of young corn (*Zea mays* L.)

Table 39–1. Partial listing of commercial sources of ^{14}C-labeled materials or liquid scintillation counting cocktails.

Company name and address	Telephone number	^{14}C-Labeled material	LSC cocktail
American Radiolabeled Chemicals, Inc. 11624 Bowling Green Dr. St. Louis, MO 63146	1-800-331-6661	X	
Amersham Corp. 2636 S. Clearbrook Dr. Arlington Heights, IL 60005	1-800-323-9750	X	
Curtin Matheson Scientific, Inc. P.O. Box 1416 Houston, TX 77251-1416	1-800-879-2670	X	X
Dupont Co., Biotechnology Systems 549 Albany St. Boston, MA 02118	1-800-551-2121	X	X
Fisher Scientific 900 Stewart Ave. Plano, TX 75074	1-800-766-7000		X
ICN Biomedicals, Inc. 3300 Hyland Ave. Costa Mesa, CA 92626	1-800-854-0530	X	X
Isolab, Inc. Drawer 4350 Akron, OH 44321	1-800-321-9632		X
Isotope Products Laboratories 1800 N. Keystone St. Burbank, CA 91504	1-818-843-7000	X	
Moravek Biochemicals, Inc. 577 Mercury Lane Brea, CA 92621	1-800-447-0100	X	X
Packard Instrument Co. 800 Research Parkway Meriden, CT 06450	1-203-238-2351	X	X
Research Products International Corp. 410 N. Business Center Dr. Mount Prospect, IL 60056	1-800-323-9814		X
Sigma Chemical Co. 3050 Spruce St. St. Louis, MO 63103	1-800-325-3010	X	X

plants. The plants were injected three times and allowed to grow an additional 3 wk before harvest. The plant tops were harvested and small molecular weight compounds extracted before the material was used for decomposition studies. Similarly, lignin components of white oak (*Quercus albus*), red maple (*Acer rubrum*), and cattail (*Typha latifola*) have been labeled by feeding plants aqueous solutions of L-[U-^{14}C]phenylalanine or [2'-^{14}C(side chain)] ferulic acid through their cut stems (Crawford, 1978; Crawford & Crawford, 1976). Cellulosic components of lignocelluloses were labeled by substituting solutions of D-[U-^{14}C]glucose (Crawford et al., 1977).

39–1.2.1.3 Comments. Safety is critical when using radioisotopes, and it is important to prevent contamination of the laboratory (Coleman & Corbin, 1991). Always wear a laboratory coat, chemical safety goggles, and disposable gloves. Work on absorbent paper and use a pipette bulb for all pipetting. Avoid being exposed to ^{14}C-labeled aerosols, dust, and volatile material. All grinding and sieving procedures should be conducted in a properly vented fume hood. It is obvious that there should be no eating, drinking, or smoking in the laboratory at any time. The growth medium may be contaminated with ^{14}C, and suitable disposal procedures must be followed.

39–1.2.2 Labeling Microbial Biomass and Microbial Products

Most studies using ^{14}C-labeled microbial biomass involve incubation of a pure culture of a given microorganism with a suitable labeled substrate such as uniformly ^{14}C-labeled glucose. The microbial biomass is harvested and lyophilized following an appropriate incubation period. The dry biomass can be ground and sieved before it is added to the soil in an incubation experiment.

39–1.2.2.1 Materials
1. ^{14}C-labeled substrate such as D-[U-^{14}C]glucose.
2. Pure culture of microorganism (see American Type Culture Collection Catalog).
3. 250-mL Erlenmeyer flasks.
4. Orbital shaker.
5. System to collect evolved $^{14}CO_2$ (see Fig. 39–1, 39–2, 38–2).
6. Freeze dryer.
7. Autoclave.

39–1.2.2.2 Procedure. Prepare a growth medium suitable for the microorganism of interest. Add an appropriate amount of ^{14}C-labeled substrate to the medium, add the inoculant, and connect the flask to the CO_2 collection system (see Fig. 39–1 and chapter 38 in this book). Place the 250-mL flask on the shaker and incubate until sufficient microbial growth has occurred. The microbial biomass can be harvested by centrifugation and then lyophilized, ground, and sieved before it is used in soil decomposition studies. Wagner and Krzywicka (1975) used $^{14}CO_2$ to label algal biomass and D-[U-^{14}C]glucose was used by Reyes and Tiedje (1973) to produce labeled *Saccharomyces cerevisiae*. Labeled whole cells have been fractionated and their components used in soil decomposition studies (Hurst & Wagner, 1969; Nakas & Klein, 1979). Measurement and turnover of microbial biomass have been evaluated (Jenkinson & Powlson, 1976; Kassim et al., 1981, 1982) and reviewed (Jenkinson & Ladd, 1981; Wagner, 1975).

Additionally, ^{14}C-labeled microbial products such as polysaccharides, polyphenols, and proteins have been isolated and used in decomposition studies. Oades and Wagner (1971) and Zunino et al. (1982) used a mineral

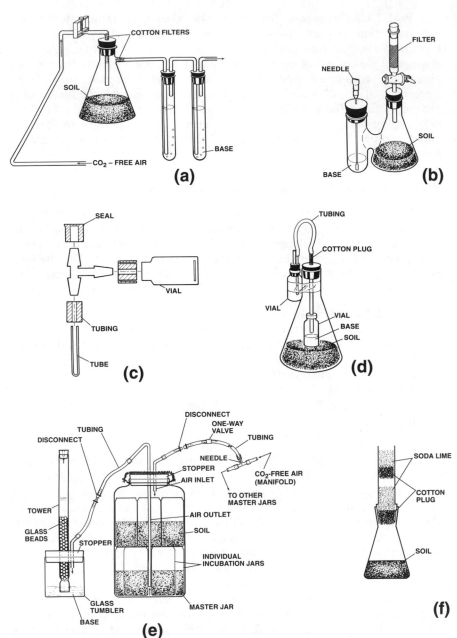

Fig. 39–1. Common units to collect $^{14}CO_2$: (a) flow-through system (Atlas & Bartha, 1972); (b) biometer flask (Bartha & Pramer, 1965); (c) tube with attached scintillation vial (Sissons, 1976); (d) static flask unit (Loos et al., 1980); (e) flow-through system (Stotzky et al., 1993); (f) static flask chamber (Anderson, 1982).

salts medium supplemented with D-[U-^{14}C]glucose to grow *Leuconostoc* spp. The polysaccharide material was collected and used in decomposition studies. Soil fungi have been grown on a mineral salts medium with D-[U-^{14}C]glucose to produce melanins used in laboratory incubations (Martin & Haider, 1979; Martin et al., 1972, 1982). The ^{14}C-labeled proteins produced by *Chlorella pyrenoidosa* and *Microcoleus* spp. also have been obtained (Verma et al., 1975).

39–1.2.2.3 Comments. The spent microbial growth medium containing ^{14}C-labeled material must be disposed of in a suitable manner. It is important to realize that a radioactive waste container with spent medium can be a source of $^{14}CO_2$ in the laboratory if microbial growth occurs.

39–1.2.3 Specific ^{14}C-Labeled Compounds

In certain studies related to organic matter decomposition, it is desirable to amend the soil with specific compounds such as an amino acid, phenol, or sugar (Haider & Martin, 1975). These compounds may be labeled in a specific position in the molecule, or they may be labeled uniformly. In most cases, the compound can be obtained from commercial sources, a few of which are listed in Table 39–1. Some companies also have a service available to synthesize special compounds.

39–1.3 Methods and Approach to Incubations of ^{14}C-Labeled Organic Materials

It is not possible to present all of the methods that have been used to determine decomposition of ^{14}C-labeled organic materials in soil. References should be consulted for additional methods or specific details of more specialized incubation systems (Anderson, 1982; Marvel et al., 1978; Stotzky, 1965). Some of the common systems are presented in Fig. 39–1. In general terms, the soil is placed in a flask and amended with the ^{14}C-labeled organic material and attached to a CO_2 collection unit (see chapter 38 in this book). For aerobic incubation studies, O_2 must not be a limiting factor. For this reason, a continuous flow-through system has been used. The evolved $^{14}CO_2$ can be trapped in a base such as KOH or NaOH and the ^{14}C activity assayed by liquid scintillation counting techniques. Several static systems have also been used and have the advantages of a simple design and conservation of space.

39–1.3.1 Materials

1. Liquid scintillation spectrometer.
2. Liquid scintillation counting (LSC) cocktail.
3. Scintillation vials.
4. Pipettes.
5. KOH or NaOH containing 10 mg/L Tropaeolin O (Aldrich Chemical Co.).

6. Soil, field moist and sieved.
7. ^{14}C-labeled organic material.
8. ^{14}CO$_2$ collection unit (see Fig. 39–1, 39–2, and 38–2).

39–1.3.2 Procedures

The amount of ^{14}C activity to add to the soil will depend upon the amount of ^{14}C expected to be found in the fraction of interest or evolved as ^{14}CO$_2$, analytical efficiency of the laboratory procedure that determines whether the activity in the fraction of interest is diluted or concentrated, and the minimal activity in the LSC cocktail to give efficient counting (Voroney et al., 1991). It is generally necessary to determine the counting efficiency for each procedure and LSC cocktail used and it may be necessary to do a preliminary or pilot study to determine the appropriate levels of activity to add. In an example given by Voroney et al. (1991), to detect a difference of 20 counts/min (cpm) at a 95% confidence level, one would have to count a sample containing 1000 cpm for 10 min at an 85% counting efficiency. In many cases, it may not be possible to obtain a level of 1000 cpm in the LSC cocktail; using longer counting times and lower confidence intervals will enable the researcher to work with samples containing activity as low as 100 cpm in the LSC cocktail.

The ^{14}C-labeled organic material should be added at a rate not to exceed 2% of the soil weight (Jenkinson, 1971). Typically, a 100-g sieved soil sample should be weighed onto nonabsorbent waxed paper or aluminum foil. If the ^{14}C-labeled organic material to be added is a dry solid, it can be weighed and added directly to the soil and carefully mixed into the soil. Because of the potential for contamination, the operation should be completed in a fume hood. If the ^{14}C-labeled material is in aqueous solution, it can be added to the soil and thoroughly mixed. Once the labeled material is added, the soil should be adjusted to the required soil water potential by adding distilled water and the soil should be carefully mixed to ensure uniform water distribution. Excessive mixing should be avoided as puddling of the soil could result that would reduce oxygen diffusion into the soil.

The moist soil containing the ^{14}C-labeled material should be transferred into a 250-mL Erlenmeyer flask and attached to the CO$_2$ collection unit. Appropriate controls (soil without organic amendment) and blanks (no soil or organic amendment in the flask) should be prepared in the same manner and attached to the CO$_2$ collection unit. It may also be appropriate to include a soil amended with unlabeled organic material.

Several CO$_2$ collection units that have been used are shown in Fig. 39–1. The unit shown in Fig. 39–2 is similar to the unit given in chapter 38, Fig. 38–2. The details of the reactions involved are given in chapter 38, and the same samples used to determine the total amount of CO$_2$ evolved can and often are used to determine the amount of ^{14}CO$_2$ evolved. One modification used for the ^{14}CO$_2$ determination is that Tropaeolin O is added to the base to give it an orange color indicating a pH > 12.7. If the base

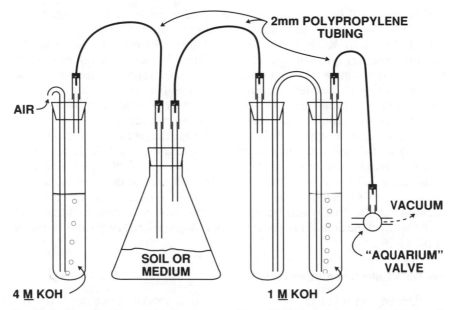

Fig. 39–2. Apparatus for studying the decomposition of ^{14}C-labeled organic materials in soil (Stevenson, 1986).

becomes partially neutralized and the pH declines to < 11, the indicator will turn yellow which indicates that the CO_2 removal efficiency has been compromised. In studies using ^{14}C-labeled materials, it is recommended that air be drawn through the system by a vacuum. Thus, if a leak develops in the system, there is little possibility of contamination of the laboratory atmosphere with [14]CO_2. A leak in the system would substantially reduce the specific activity of the collected CO_2. In experiments using ^{14}C-labeled material, it is also desirable to include a secondary or backup tube to collect any [14]CO_2 that might not be collected in the first tube. The air flow rate is adjusted to one bubble per second to provide adequate oxygen levels to maintain aerobic conditions. Faster rates may result in incomplete trapping of the CO_2. Gas exchange rates of 20 volumes per hour have been used (Wagner & Chahal, 1966), but a level of one volume exchange per hour is adequate in most cases. Relatively economical "aquarium" valves can be obtained at local pet supply stores and have proved durable and satisfactory.

At various intervals, the CO_2 collection tubes are replaced, and the amount of both total CO_2 evolved and [14]CO_2 evolved can be determined. To determine the total CO_2 evolved, see chapter 38. To determine the [14]CO_2 evolved, add an aliquot of the NaOH or KOH containing the evolved [14]CO_2 to a scintillation vial containing a suitable LSC cocktail. A partial list of commercial suppliers of LSC cocktails is given in Table 39–1. New generation LSC cocktails do not contain aromatic solvents, are non-flammable, and do not result in toxic vapors. The cocktails are biodegradable and result in fewer disposal problems. The specific volumes of base and cocktail will be determined by the load capacity of the cocktail, the activity in the sample, and the base concentration. Generally, 1 mL of 0.5 M NaOH or KOH can be counted using 10 mL of LSC, but the manufacturer can provide specific guidelines. For determining the radio-activity in the sample, the liquid scintillation spectrophotometer should be used according to the manufacturer's specifications. Discussions of scintil-lation counting procedures (L'Annunziata, 1979) and measurement of radionuclides (L'Annunziata, 1984) should be consulted.

An alternative approach is to determine the amount of activity re-maining in the soil following a given incubation period. This approach generally requires combustion of the soil to convert the ^{14}C-labeled organic material to $^{14}CO_2$ and collecting the $^{14}CO_2$ from the combustion and count-ing it. The $^{14}CO_2$ can be collected in base and the base added to a suitable cocktail, or the $^{14}CO_2$ can be collected directly in certain cocktails and counted. Specific details are provided by L'Annunziata (1979) and Voroney et al. (1991).

39–1.3.3 Calculation of Results

Depending upon the specific scintillation spectrometer used, the re-sults may be given as counts per unit time such as second (cps) or minute (cpm). The counts must be corrected for counting efficiency and back-ground and historically have been given as disintegrations per unit time (dps or dpm). The SI base unit is becquerel (Bq). One becquerel is equiv-alent to a nuclear transformation per second, or dps. Useful conversions include 1 μCi = 37 kBq = 3.7×10^4 dps = 2.22×10^6 dpm (Corbin & Swisher, 1986). Once the appropriate dilution factors are taken into ac-count, the percentage of the added ^{14}C evolved as $^{14}CO_2$ or remaining in the soil can be calculated (Eq. [1] and [2]). Modern-day scintillation counters allow the researcher to program the instrument to complete many of the calculations.

$$^{14}C \text{ in sample} = \frac{\text{Sample cps} - \text{Background cps}}{\text{Counting efficiency (expressed as a decimal fraction)}} \text{ (dilution factor)} \quad [1]$$

$$\begin{array}{l} \% \text{ added } ^{14}C \\ \text{evolved as } ^{14}CO_2 \end{array} = \frac{^{14}C \text{ evolved from sample}}{^{14}C \text{ added to soil}} \text{ (100)} \quad [2]$$

39–1.3.4 Comments

It is important to report percentage of the added ^{14}C evolved as $^{14}CO_2$ or the percentage added ^{14}C remaining in the soil rather than percentage decomposition. Historically, percentage decomposition was calculated from $^{14}CO_2$ evolution data, but the amount of $^{14}CO_2$ evolved may be substantially influenced by the amount of ^{14}C incorporated into microbial biomass or soil organic matter. The microbial products formed may be subsequently mineralized and thus influence the amount and rate of labeled substrate recovered as $^{14}CO_2$.

If, in addition to $^{14}CO_2$, volatile organics are lost from the soil and absorbed by the base used to collect $^{14}CO_2$, the amount of $^{14}CO_2$ evolved will be overestimated. Kearney and Kontson (1976) placed a polyurethane filter preceding the base trap and were able to sorb the evolved volatile products for subsequent ^{14}C determination.

Another technique for collecting $^{14}CO_2$ uses chromatographic tubes (chapter 38) that can be counted by liquid scintillation techniques. Additionally, Mayaudon (1971) presents a detailed discussion of radiorespirometry techniques in soil systems. Also, a direct soil counting technique has been used for determining ^{14}C remaining in soil treated with an herbicide (Lavy, 1975; Scott & Phillips, 1972).

39–2 ^{13}C NATURAL ABUNDANCE TECHNIQUE: BACKGROUND AND PRINCIPLES

39–2.1 Introduction

Approximately 98.89% of all C in nature is ^{12}C, and 1.11% is ^{13}C. The relative proportions of these two stable isotopes in nature vary slightly around these average values as a result of isotopic fractionation during physical, chemical, and biological processes (Boutton, 1991b). The $^{13}C/^{12}C$ ratio of organic C found in terrestrial environments is determined largely by the C isotope fractionation that occurs during photosynthesis. Plants with the C_3 photosynthetic pathway exhibit greater discrimination against ^{13}C than plants with the C_4 pathway. These natural isotopic differences between plants can be used to study the dynamics of organic matter in soil.

39–2.1.1 Stable Isotope Terminology

Because natural variation in the ratio of $^{13}C/^{12}C$ is small, stable C isotope ratios are expressed in relative terms as $\delta^{13}C_{PDB}$ values:

$$\delta^{13}C_{PDB} \ (^o/oo) = \left[\frac{R_{sample} - R_{PDB}}{R_{PDB}} \right] \times 10^3 \qquad [3]$$

where R_{sample} is the mass 45 ($^{13}C\ ^{16}O\ ^{16}O$) to mass 44 ($^{12}C\ ^{16}O\ ^{16}O$) ratio of the sample and R_{PDB} is the $^{13}C/^{12}C$ ratio of the international PDB limestone standard, which has a value of 0.0112372 (Craig, 1957). The PDB standard was a *Belemnitella americana* limestone fossil from the Cretaceous Pee Dee formation in South Carolina. Corrections are made for the presence of ^{18}O and ^{17}O in the CO_2. Thus, $\delta^{13}C_{PDB}$ is a relative index that indicates the parts per thousand (per mil, or $^o/oo$) difference between the $^{13}C/^{12}C$ ratio of the sample and that of the PDB standard. For example, a $\delta^{13}C_{PDB}$ value of -10 °/oo indicates a sample with a $^{13}C/^{12}C$ ratio 10 parts per thousand lower than the PDB standard; a $\delta^{13}C_{PDB}$ value of $+5$ °/oo indicates a sample with a $^{13}C/^{12}C$ ratio 5 parts per thousand greater than the PDB standard.

39–2.1.2 Stable Carbon Isotope Ratios of Plants and Soils

The ^{13}C natural abundance technique for studying soil organic matter dynamics uses natural differences in $\delta^{13}C_{PDB}$ values between plants with the C_3 and C_4 pathways of photosynthesis. Atmospheric CO_2 has a $\delta^{13}C_{PDB}$ value of approximately -8 °/oo (Levin et al., 1987). During photosynthesis, plants with the C_3 pathway discriminate against atmospheric $^{13}CO_2$ to a greater extent than C_4 plants (O'Leary, 1988). The C_3 plants have $\delta^{13}C_{PDB}$ values ranging from approximately -32 to -20 °/oo (mean = -27 °/oo), while C_4 plants have $\delta^{13}C_{PDB}$ values ranging from -17 to -9 °/oo (mean = -13 °/oo). Thus, C_3 and C_4 plants have distinct stable C isotope ratios and differ from each other by approximately 14 °/oo on average (Smith & Epstein, 1971). Plants with Crassulacean acid metabolism (CAM) usually have $\delta^{13}C_{PDB}$ values typical of C_4 plants; however, under certain environmental and developmental circumstances, some CAM species are able to switch to a C_3 mode of photosynthesis. These "facultative" CAM species will have $\delta^{13}C_{PDB}$ values depending upon the relative proportions of C fixed via CAM and C_3 modes. The CAM plants have $\delta^{13}C_{PDB}$ values ranging from approximately -28 to -10 °/oo, but are most commonly -20 to -10 °/oo.

Most terrestrial plant species are C_3. Most temperate zone and all forest communities are dominated by C_3 species. However, C_4 and CAM plants are significant components of many plant communities, particularly in warm, arid, or semiarid environments (Osmond et al., 1982). For example, tropical and subtropical grasslands consist almost exclusively of C_4 grasses, and CAM plants (e.g., Cactaceae, Euphorbiaceae) are important in many desert communities. In general, the proportion of C_4 species in a flora increases as latitude and altitude decrease (e.g., Teeri & Stowe, 1976; Boutton et al., 1980).

Although there are small isotopic differences between different parts of the same plant (up to 2 °/oo different from the whole plant) and between specific biochemical fractions within plants (up to 8 °/oo different from the whole plant), the C isotopic signature of the whole plant is largely preserved as dead plant tissue decomposes and enters the soil organic matter pool (Nadelhoffer & Fry, 1988; Melillo et al., 1989). Thus, soil organic matter in C_4 plant communities will have $\delta^{13}C_{PDB}$ values near -13 °/oo, while organic matter from soils in C_3 communities will be near -27 °/oo. This natural isotopic "label" in the soil organic matter enables reconstruction of the prior history of plant communities (Dzurec et al., 1985) and also permits estimation of soil organic matter dynamics in situ over relatively long periods without any type of experimental disturbance.

39–2.1.3 Natural ^{13}C and Measurement of Organic Matter Dynamics

If a community dominated by C_3 plants has been compositionally stable for a relatively long time (e.g., 500–1000 yr), then the soil organic matter in that community is in isotopic equilibrium with that C_3 vegetation; that is, they should have approximately the same δ^{13}C values. If that C_3 community (e.g., a forest) is converted to a C_4 plant community (e.g., a corn field or tropical grass pasture), then the isotopic composition of the soil organic matter will begin to shift towards that of C_4 vegetation as the C_3 component decays out of the soil and is replaced by C_4 organic matter inputs. The rate at which the original mass of C_3-derived organic matter (which is uniquely and readily identifiable by its characteristic δ^{13}C value) decays out of the system through time is a direct measure of the turnover rate of organic matter in that system. It should be noted that cultivation of the soil will accelerate organic matter turnover. Thus, changing the δ^{13}C value of the organic matter inputs (i.e., from $C_3 \rightarrow C_4$, or from $C_4 \rightarrow C_3$) is equivalent to in situ labeling of the soil organic matter (Balesdent et al., 1987). Measurements of turnover rates using the ^{13}C natural abundance technique are best suited to time periods of tens to thousands of years because: (i) the C isotopic difference between C_3 and C_4 plants is relatively small; and, (ii) the mass of existing soil organic C derived from the previous vegetation is large relative to the annual increments of organic C derived from the new vegetation (Balesdent et al., 1988). However, significant differences in δ^{13}C have been detected in upper A horizons in as little as 3 mo following a single input of C_4 litter into a C_3 plant system (Insam et al., 1991).

The use of the ^{13}C natural abundance technique obviously will be limited to situations where there has been a change from $C_3 \rightarrow C_4$ or $C_4 \rightarrow C_3$ vegetation. However, these situations are common and have been used to study organic matter dynamics where: (i) C_3 rain forest has been converted to C_4 pasture or C_4 crops; (ii) C_4 grassland has been converted to C_3 crops; (iii) C_4 savanna has been converted to C_3 woodland; and (iv) C_3 cropland has been converted to C_4 cropland (Balesdent et al., 1987, 1988, 1990; Vitorello et al., 1989; Martin et al., 1990; Skjemstad et al., 1990; Cerri et al., 1991).

39–2.2 ^{13}C Natural Abundance Technique: Methodology

39–2.2.1 Special Apparatus

1. Vacuum manifold capable of achieving 10^{-3} torr for evacuating and sealing combustion tubes.
2. Gas-oxygen torch for making quartz combustion tubes and sealing the tubes after being loaded with sample and evacuated.
3. Shade 8 or darker welding goggles to protect eyes while heating quartz with gas-oxygen torch.

4. High vacuum system capable of 10^{-3} torr for cryogenic separation and purification of CO_2 produced during combustion of organic matter in sealed quartz tubes.
5. Analytical balance readable to 0.01 mg.
6. Programmable muffle furnace.
7. Dewar flasks for holding liquid N or dry ice slush during CO_2 isolation procedure.
8. Nier-type, dual inlet, triple collector gas isotope ratio mass spectrometer for measuring ratios of isotopic species of CO_2 produced by combustion of organic matter.

39–2.2.2 Reagents

1. HCl (0.5 M) to volatilize carbonate C from soils containing pedogenic or lithogenic carbonates.
2. NaCl, NaI, $ZnBr_2$, or CsCl to produce high density (1.2–1.8 g cm^{-3}) liquid for isolation of undecomposed particulate organic debris in soils.
3. Quartz or vycor tubing (9 mm o.d. × 7 mm i.d.) for combustion of soil organic matter samples (Quartz Scientific, Fairport Harbor, OH).
4. Wire-form CuO with a low C background suitable for microanalysis (Fisher Scientific C474-500 or equivalent).
5. Reduced Cu granules, -10 to $+40$ mesh (Aldrich Chemical Co. catalog no. 31,140-5 or equivalent).
6. Quartz wool.
7. Liquid N.
8. Ethanol-dry ice slush (-78 °C).
9. Carbon isotope standards calibrated relative to the international PDB standard (available from NIST, Gaithersburg, MD, or from IAEA, Vienna, Austria).

39–2.2.3 Procedure

39–2.2.3.1 Field Sampling. The study site must consist of an area known to have been converted from a C_3-dominated to a C_4-dominated plant community (or vice versa) at a precisely known time, and there should be a remnant of the original plant community nearby to provide baseline samples. Ideally, one would like to sample sites on the same soil and in close proximity to one another that have undergone the same type of conversion at different times. Sampling such a chronosequence would provide detailed kinetics of the turnover process. To characterize the isotopic composition of the organic matter inputs, live plant tissue as well as litter should be sampled from both the baseline site with the original vegetation and the derived site with the new plant community. Plant tissue should be dried at 70 °C, ground to pass a 0.4-mm (40-mesh) screen, and set aside for isotopic analysis.

Soil samples can be obtained either by digging a pit or taking cores. Pit sampling reveals soil structure and horizonation more readily than cores, does not compress the profile, and enables more accurate sampling for bulk density. Sampling depth intervals according to horizonation enables $\delta^{13}C$ to be related to pedogenesis, and bulk density measurements at several depths in the profile will allow C dynamics to be expressed on Mg C/ha basis (1 Mg = 10^6g). Samples should be taken from several depth intervals to a depth of approximately 1 m. Samples from several pits or cores can be bulked to obtain representative $\delta^{13}C$ values of a large area or kept separate for analysis of spatial variability.

39–2.2.3.2 Preparation of Soil Samples. Soil samples should be passed through a 2-mm screen and rocks and large roots removed. Undecomposed particulate organic debris, or the "light fraction" (Stevenson & Elliott, 1989), is removed from the soil samples by flotation in high density (1.2–1.8 g cm^{-3}) inorganic solutions. Suitable inorganic chemicals for the preparation of high density liquids include NaCl, NaI, $ZnBr_2$, CsCl, and Na metatungstate. Approximately 100 g of soil is added to a 600-mL beaker, and the beaker is filled with a saturated NaCl solution (density $\cong$ 1.2 g cm^{-3}). The soil is stirred vigorously, and particulate organic debris floats to the surface. After the soil settles, the organic debris can be siphoned or strained off the surface of the liquid. By repeating this process five times, virtually all particulate organic debris is removed from the soil, and the removal can be verified by examining the soil with a dissecting microscope. The particulate organic debris, which is largely roots, can be pooled with the larger roots removed by sieving, treated with 0.5 M HCl to remove any adhering carbonate C, dried at 70 °C, ground to pass a 0.4-mm (40-mesh) screen, and set aside for isotopic analysis. The NaCl is then removed from the root-free soil by repeated washing in distilled water. Failure to remove residual salt may interfere with determination of organic C content later in the procedure.

If soils have pedogenic or lithogenic carbonates present, this C must be destroyed. Although the $\delta^{13}C_{PDB}$ value of pedogenic carbonate is related to the C_3 to C_4 composition of the plant community, both pedogenic and lithogenic carbonate are significantly enriched in ^{13}C and would seriously confound stable C isotope measurements of the soil organic matter. Approximately 100 g of soil is placed in a 600-mL beaker which is then filled with 0.5 M HCl and stirred. Soil is left in the 0.5 M HCl for 3 d, and the HCl solution is replaced daily. When carbonates have been removed, soils are washed repeatedly in distilled water to remove excess HCl. This acid pretreatment to eliminate carbonates has no effect on either the organic C content of the soil, or on the $\delta^{13}C$ value of the soil organic C (Boutton et al., unpublished data).

If turnover of only the bulk soil organic matter is to be measured, then the soils can be dried, ground to pass a 0.5-mm screen, and set aside for isotopic analysis. However, because different organic matter fractions exhibit different turnover rates, most investigators now choose to process soil samples into more defined categories such as particle-size fractions, aggregate size classes, or the classical humic fractions. The most useful information is gained when biologically significant fractions are analyzed. In some studies, particle-size fractionation has been followed by extraction of specific humic substances from the particle-size fractions. Others have described in detail the methods for the isolation of particle-size fractions (Jackson, 1969; as modified by Tiessen & Stewart, 1983), aggregate size fractions (Kemper & Rosenau, 1986), and the humic fractions (Schnitzer, 1982). Following any of these more detailed procedures, the soil organic matter fractions must be dried and ground to pass a 0.5-mm screen prior to isotopic analysis.

39–2.2.3.3 Conversion of Organic Carbon to Carbon Dioxide for Mass Spectrometry. Due to instrumental requirements, C must be converted to CO_2 for stable isotope ratio measurements by mass spectrometry. The most common and simplest method to convert organic C to CO_2 is by combustion with an excess of CuO in an evacuated, sealed quartz tube at 850 °C (Boutton, 1991a). This method does not change the C isotope composition of the original sample, produces quantitative yields of C (which permit determination of percentage C in the sample), and is rapid and relatively inexpensive.

Quartz or vycor tubing (9 mm o.d. × 7 mm i.d.) is cut to 20-cm lengths, and the tubes are sealed at one end with a gas-oxygen torch. When heating quartz or vycor to the softening point, always work under a fume hood to exhaust the resulting toxic gases and wear quartz-working goggles (Wale Apparatus, Hellertown, PA) or shade 8 or darker welding goggles to protect eyes from intense glare. The prepared tubes sealed on one end only, a porcelain crucible containing the wire-form CuO catalyst (e.g., Fisher Scientific, catalog no. C474-500), and another crucible containing quartz wool are then heated in a muffle furnace at 850 °C for 1 h to remove potential organic contaminants. Upon cooling, the CuO and quartz wool can be stored separately in clean jars and the combustion tubes stored in a desiccator until ready for use.

On a piece of weighing paper, weigh out 1.0 g of CuO catalyst. Then, tare the balance and weigh out enough sample to provide approximately 2 to 3 mg of C. If this step can be carried out on an analytical balance readable to 0.01 mg or on a microbalance, the percentage C in the sample can be determined later. The amount of soil required to provide 2 to 3 mg of C will vary mostly as a function of depth in the profile. Plant tissue, roots, and litter usually contain 40 to 50% C, and a 4- to 5-mg sample of these materials is adequate to provide 2 to 3 mg of C. Too much C can result in explosion of tubes during combustion. When the appropriate amount of sample has been weighed, mix the sample and CuO thoroughly, and use a long-stem funnel to deliver the sample/CuO mixture to the bottom of a pre-combusted quartz tube. To hold the soil in place during evacuation of the tubes later, a plug of pre-combusted quartz wool can be inserted into the combustion tube and positioned above the soil/CuO mix-

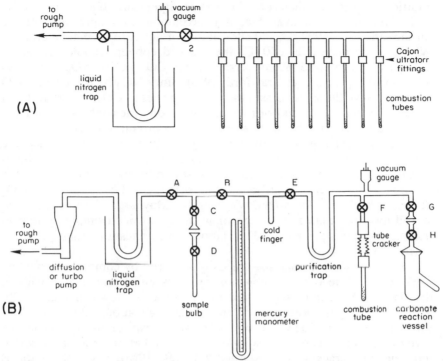

Fig. 39–3. Vacuum systems for sealing quartz combustion tubes under vacuum (A), and for isolation and volumetric measurement of CO_2 produced by combustion of organic matter (B). Circles with x's represent valves. Reproduced from Fig. 2 in Boutton (1991a), with permission from Academic Press.

ture. A quartz wool plug is not necessary for plant tissue, roots, or litter samples. Again using the long-stemmed funnel, add 0.5 g of reduced Cu granules (e.g., Aldrich Chemical Co., catalog no. 31,140-5) to the combustion tube. The long-stem funnel should be cleaned thoroughly between samples. Finally, the sample tubes must be identified clearly with an engraving tool or a high temperature marking pen capable of resisting 900 °C.

Combustion tubes loaded with samples are attached to a vacuum manifold (Fig. 39–3a) with Cajon Ultratorr O-ring fittings (Cajon Company, Macedonia, OH). Samples should be exposed to the vacuum slowly to avoid sucking the samples out of the tubes. When a vacuum of $< 10^{-2}$ torr has been achieved, the combustion tubes can be sealed with a gas-oxygen torch. Sealed tubes are then placed inside individual ceramic or inconel tubes in the muffle furnace to shield each tube from possible explosions during combustion.

The muffle furnace is heated to 900 °C and held at that temperature for 2 h. During this time, organic C is oxidized to CO_2. After 2 h, the muffle furnace is cooled to 650 °C, maintained at that temperature for two additional hours, and then cooled to room temperature. If a programmable muffle furnace is not available, it is acceptable to simply turn off the furnace after 2 h at 900 °C, and allow it to cool slowly to room temperature. While the furnace is at 650 °C, the reduced copper granules eliminate halogens, and catalyze the conversion of any CO to CO_2, NO_x to N_2, and SO_x to $CuSO_4$ (Frazer & Crawford, 1963). Since NO_2 (mass 46) and N_2O (mass 44) have the same masses as isotopic species of CO_2, it is critical that these gases be eliminated prior to mass spectrometric analysis. High resolution mass spectrometric analysis of the gases produced by combustion of several different organic compounds by this method have revealed the presence of only CO_2, H_2O, and N_2 (Boutton et al., 1983). Sealed tubes that have been combusted should not be stored for more than 5 d prior to isolation of CO_2 because carbonate forms slowly and $\delta^{13}C_{PDB}$ of the CO_2 decreases by 1 to 3 °/oo after 2 wk (Engel & Maynard, 1989). If tubes are stored for more than 5 d, the problem can be avoided simply by recombusting the tubes prior to CO_2 isolation and purification (Engel & Maynard, 1989).

Prior to isotopic analysis, the CO_2 produced by combustion must be isolated from the other combustion products and purified by cryogenic distillation. A vacuum system for this purpose is shown in Fig. 39–3b. When operating this vacuum system or handling combusted quartz tubes, safety glasses should be worn to protect against explosions, implosions, or cryogenic liquids. The combustion tube is scored at one end and inserted into the tube cracker (Des Marais & Hayes, 1976), a sample bottle is attached to the manifold, all valves are opened, and the entire vacuum system is pumped down to $< 10^{-3}$ torr. Then, a Dewar flask containing liquid N (-196 °C) is placed around the purification trap, and valve E is closed. The top of the combustion tube is broken off by flexing the tube cracker, and the gases produced during combustion are released into the vacuum system; water and CO_2 freeze into the purification trap. After 4 min, valve F is closed, valve E is opened, and all noncondensible gases (mostly N_2) are pumped away. When the vacuum is restored to 10^{-3} torr, valve B is closed, and the liquid N Dewar flask is removed from the purification trap and replaced with a Dewar containing an ethanol-dry ice slush (-78 °C). The CO_2 sublimes but water remains frozen in the trap. To measure the volume of CO_2 produced from the organic matter, the CO_2 is transferred into the manometer cold finger by cooling it with a Dewar of liquid N; 2 min should be allowed for the CO_2 to freeze into the cold finger. When the transfer is complete, valve E is closed, the liquid N Dewar is removed from the cold finger, and the CO_2 is allowed to expand into the mercury manometer calibrated previously with known volumes of CO_2. The manometer reading is noted and, together with the weight of the sample combusted, is used to calculate the percentage C of the sample. Valve A is then closed, and valves B, C, and D are opened. The CO_2 is transferred into the sample bulb by immersing it in a liquid N Dewar for 2 min. When the transfer is complete, valves C and D are closed, and the sample bulb is detached from the vacuum system and attached to the inlet system of the mass spectrometer.

An electronic capacitance manometer could be used in place of the mercury manometer shown in Fig. 39–3b. The capacitance manometer is more expensive but eliminates the need for mercury and is significantly more accurate for determining percentage C. Accuracy for electronic manometers ranges from 0.15 to 1% of reading, depending upon the model, while that for a mercury manometer ranges from approximately 2 to 5% of reading.

Organic matter can be combusted to CO_2 in Pyrex tubing at 550 °C (Sofer, 1980; Vitorello et al., 1989) with procedures identical to those outlined above. The primary advantage of this alternative is that Pyrex tubing is approximately 5 to 10% of the cost of quartz tubing. If Pyrex is used, great care should be taken to ensure good contact between sample and CuO, and combustion time should be increased to 12 h. Some reports indicate that accuracy and precision of $\delta^{13}C_{PDB}$ values obtained by combustion in Pyrex tubes at 550 °C are poorer than those obtained with quartz at 850 °C (Boutton et al., 1983; Le Feuvre & Jones, 1988; Swerhone et al., 1991). Furthermore, combustion at 550 °C may not give quantitative yields of CO_2, eliminating the possibility of determining percentage C during CO_2 isolation and purification (Boutton et al., 1983).

39–2.2.3.4 Mass Spectrometric Analysis and Isotopic Indices. Stable C isotope ratios are measured on the CO_2 generated by the above procedure with a dual-inlet, triple-collector gas isotope ratio mass spectrometer. The high precision of these instruments is due to simultaneous collection of the ion beams (masses) of interest and to repeated measurements of sample and standard gases by alternate switching during a single isotope ratio determination. The theory and methods of determining the isotopic composition of CO_2 by mass spectrometry have been reviewed (Craig, 1957; Deines, 1970; Mook & Grootes, 1973; Gonfiantini, 1981; Santrock et al., 1985). The $\delta^{13}C_{PDB}$ value is determined using Eq. [3].

Most of the error in isotopic measurements results from sample preparation. Mass spectrometer precision (1 SD), as determined by repeated analyses of the same gas sample, is often as low as 0.01 °/oo. By contrast, different preparations of aliquots of the same sample generally will have a precision (1 SD) of 0.1 °/oo for both plants and soils.

The PDB standard was derived from a limestone of marine origin with an absolute $^{13}C/^{12}C$ ratio of 0.0112372 (Craig, 1957). As the basis of the PDB scale, it has a $\delta^{13}C_{PDB}$ value of 0 °/oo. The PDB standard no longer exists, but several other primary standards were calibrated against it before the supply was exhausted, so it is still possible to express $\delta^{13}C$ values relative to PDB. Primary C isotope standards are available from the National Institute of Standards and Technology (NIST, formerly the National Bureau of Standards or NBS) or from the International Atomic Energy Agency (IAEA) in Vienna, Austria.

For some mass balance calculations, it is more appropriate to use the absolute ratio (R) or the fractional abundance (F) of ^{13}C in a sample. The absolute ratio is calculated by rearrangements of Eq. [3]:

$$R_{sample} = {}^{13}C/{}^{12}C = \left[\frac{\delta^{13}C_{sample}}{1000} + 1 \right] \times R_{PDB} \qquad [4]$$

where $R_{PDB} = 0.0112372$. The fractional abundance is the fraction of total C in a sample that is ^{13}C:

$$F = \frac{{}^{13}C}{{}^{13}C + {}^{12}C} = \frac{R_{sample}}{R_{sample} + 1} \qquad [5]$$

Additional details on these indices and their relationship with $\delta^{13}C_{PDB}$ are provided by Hayes (1983).

39–2.2.3.5 Calculating Sources of Soil Organic Matter. The relative proportions of soil C derived from C_3 and C_4 sources can be determined by simple mass balance calculations. Assuming that we have a situation where a C_4 plant community has replaced a C_3 community, then the proportion of C (p) derived from the C_4 community at some later point in time (t) can be calculated as:

$$F = (p)F_{C4} + (1\text{-}p)\ F_{C3} \qquad [6]$$

where F is the fractional abundance (see Eq. [5]) of the soil organic matter fraction of interest at time t after the transition from $C_3 \rightarrow C_4$, F_{C4} is the fractional abundance of the C_4 organic matter inputs (often an average of shoot, roots, and litter), F_{C3} is the fractional abundance of the soil organic matter fraction of interest prior to the change in vegetation, and 1-p is the proportion of C_3 plant-derived C still present in the soil at time t. Since the $\delta^{13}C_{PDB}$ scale is not linear, it is technically more correct to use fractional abundance (F) as an isotopic index in mass balance calculations such as Eq. [6]; however, over the range of $\delta^{13}C_{PDB}$ values encountered in plant-soil systems at natural abundance, $\delta^{13}C_{PDB}$ is sufficiently linear that Eq. [6] can be rewritten as:

$$\delta = (p)\ \delta_{C4} + (1\text{-}p)\ \delta_{C3} \qquad [7]$$

where $\delta^{13}C_{PDB}$ values have been substituted for the fractional abundances used in Eq. [6]. Equation [7] can be rearranged and simplified to:

$$p = \frac{\delta - \delta_{C3}}{\delta_{C4} - \delta_{C3}} \qquad [8]$$

Since the percentage C of each of the organic matter fractions was measured above as part of the sample combustion procedure outlined earlier, the relative proportions of C from C_4 (p) and C_3 (1-p) sources can be used to compute the actual masses of C from each organic matter source. If the total mass of C M (with units of mg C/g soil) is known, then the mass of C from C_4 vegetation (M_{C4}) can be calculated as:

$$M_{C4} = M \ (p) \hspace{5cm} [9]$$

Similarly, the mass of C from C_3 vegetation (M_{C3}) can be calculated as:

$$M_{C3} = M \ (1-p) \hspace{4.5cm} [10]$$

Thus, if the study site consists of plots that have been switched from C_3 to C_4 at different times or if multiple times are available from the same site, the decay of the C_3 C out of the soil system and the rate of entry of C_4 C into the soil system can be described with respect to time (Balesdent et al., 1987; Skjemstad et al., 1990; Andreux et al., 1990; Cerri et al., 1991).

If soil bulk density measurements are taken, it is possible to convert the data acquired above into C content per unit area:

$$Mg \ C/ha = (g \ C/g \ soil) \times \varrho_b \times 1 \times 10^4 \hspace{2cm} [11]$$

where ϱ_b is the bulk density of the soil layer under consideration (Mg soil/m^3), and 1 is the thickness (meters) of the soil layer under consideration. Carbon dynamics for the entire soil can be determined by solving Eq. [11] for each depth interval (e.g., 0–0.1 m, 0.1–0.2 m, etc.) and then summing the results.

Although we have assumed a situation where a C_4 plant community has replaced a C_3 community, the calculations for the reverse situation ($C_4 \rightarrow C_3$) are directly analogous to those described above.

39–2.2.3.6 Kinetics of Soil Organic Matter Turnover. Although many complex processes are responsible for soil organic matter turnover, it is generally accepted that the overall process can be described reasonably well according to first order rate kinetics which assume a single homogeneous pool of soil organic C (Jenkinson & Rayner, 1977; Paul & van Veen, 1978; Paul & Clark, 1989). When studying organic matter dynamics where a C_4 community has replaced a C_3 community, the decay of C_3 C out of the system can be approximated by the negative exponential or first order decay model:

$$A_t = A_o \ e^{-kt} \ or \ \ln \ (A_t/A_o) = -kt \hspace{2cm} [12]$$

where A_t is the mass of C_3-derived C at some time t after the $C_3 \rightarrow C_4$ switch, A_o is the mass of C_3-derived C at time 0, k is the fractional rate constant with units of time^{-1}, and t is the length of time elapsed since the

$C_3 \rightarrow C_4$ switch. The value of the fractional rate constant k is equal to the slope of the line obtained by plotting ln (A_t/A_o) against time. The half-life $(t_{1/2})$ of C_3-derived C in the system is then equal to 0.693/k, and the mean residence time or turnover time is equal to $-1/k$ (Paul & Clark, 1989). Because of the poor fit of a single compartment model to most experimental results, more complex multiexponential (multicompartmental) models would permit representation of the kinetics of labile and recalcitrant pools of soil C (e.g., Andreux et al., 1990).

39–2.2.4 Comments

The ^{13}C natural abundance technique for measuring soil organic matter dynamics is complementary to the tracer approaches using ^{14}C- or ^{15}N-enriched organic matter. Natural ^{13}C allows work to be done on large spatial scales as opposed to small plots or pots, it involves minimal perturbation to the system, and it can elucidate kinetics over relatively long periods of tens to hundreds of years. Furthermore, this technique usually operates on time scales that allow all soil organic matter pools, even those that are recalcitrant, to become "labeled." In tracer experiments with ^{14}C or ^{15}N, the duration of the experiments (usually < 5 yr) is such that the more recalcitrant fractions may not become labeled. Because this technique involves only naturally occurring stable isotopes of C, there are no hazards, regulations, or disposal problems associated with its use.

The technique also has some weaknesses that should be addressed. One of the major problems is that it is difficult to measure the exact $\delta^{13}C_{PDB}$ value of the organic matter inputs following the vegetation change (e.g., δ_{C4} in Eq. [8]). The $\delta^{13}C_{PDB}$ values of plant tissue vary slightly in response to environmental conditions and can show small differences between plant parts and between biochemical fractions. However, $\delta^{13}C_{PDB}$ values of litter samples integrated through time should serve as a reasonable estimate of the isotopic composition of the organic matter input.

Another potential problem is that, in well-drained mineral soils that have supported a stable plant community for a long time, $\delta^{13}C_{PDB}$ of the soil organic matter increases by 1 to 2 %oo from the surface to approximately 1 m in depth (Stout et al., 1981). It has been suggested that this ^{13}C-enrichment is a consequence of microbial metabolism, with microbes typically being slightly more enriched in ^{13}C than the substrate they grow on. Since organic matter increases in age with depth in the profile (Scharpenseel & Neue, 1984), the enrichment in the deeper layers may simply reflect the consequences of a longer history of microbial metabolism. While the natural abundance method assumes that the isotopic composition of the soil organic matter will equilibrate to the $\delta^{13}C_{PDB}$ value of the organic matter inputs of the new plant community, it is clear that the organic matter input alone does not determine the equilibrium $\delta^{13}C_{PDB}$ values of the soil. Further work is needed to elucidate the consequences of ^{13}C enrichment with soil depth on this technique.

39–3 DECOMPOSITION OF ^{15}N-LABELED ORGANIC MATTER IN SOILS

39–3.1 Introduction

A study of the decomposition of ^{15}N-labeled organic matter primarily involves the processes of immobilization and mineralization. Major aspects of these processes are covered in chapter 42. Indigenous soil organic N occurs in many complex forms that have varying mineralization rates. Much of the organic N is in a "passive" phase (Jansson & Persson, 1982) which is essentially inert over short periods. Incorporation of ^{15}N-labeled organic matter into soils permits an evaluation of the mineralization potential of newly incorporated organic N relative to that of the indigenous soil N, as well as determination of the possible size of the inert pool of soil N. Many studies of the nature and composition of soil organic matter have been carried out (Allison, 1973), but factors influencing the stability of the organic N constituents are still not clearly understood. The use of isotopic methods thus far has not enhanced our current knowledge of the stabilizing factors, but information has been obtained concerning the rate at which newly incorporated N becomes stabilized in forms similar to that of the indigenous soil organic N (Legg et al., 1971). Studies of the transformations of labeled organic N incorporated into soils have been carried out in field, greenhouse, and laboratory experiments. Although field experiments provide useful information from soils under natural conditions (e.g., Ladd & Amato, 1986), time requirements are large compared to greenhouse and laboratory studies. For that reason, emphasis will be placed on the latter two modes of study. Many variations of the described methods exist, and reference should be made to the literature cited for additional material.

Methods for the determination of mineral N are covered in chapters 41 and 42, organic N in chapter 40 and in section 39–4.3, and ^{15}N measurements are given in chapter 40. Additional discussions of methods are presented by Bremner and Mulvaney (1982), Keeney and Nelson (1982), Hauck (1982), and Mulvaney (1993).

39–3.2 Labeling Organic Matter with [15]N

Generally, the method selected for labeling organic matter will depend upon the type and purpose of the experiment. The methods may be categorized as (i) labeling plants by supplying growing plants with ^{15}N-labeled fertilizer or (ii) labeling soil biomass by supplying soil microorganisms with ^{15}N plus an energy source (see also chapter 40). In any case, the enrichment with ^{15}N should be sufficient to be easily detectable in the total soil N. This depends upon the amount of ^{15}N added, its enrichment, and the total N content of the soil. The minimum amount of N per sample for many mass spectrometers is about 0.5 to 1 mg. It is useful to have an excess ^{15}N percentage of 0.500 or more if the total N of the sample is so small that an addition of a measured amount of unlabeled $(NH_4)_2SO_4$ is needed to meet the minimum N requirement. Calculation of results from this dilution technique has been covered by Hauck (1982).

39–3.2.1 Growing Plants with ^{15}N-Labeled Fertilizer

39–3.2.1.1 Growing Plants in Soil to be Labeled. A greenhouse pot experiment may be set up in which ^{15}N-labeled fertilizer is added to the soil, plants are grown, and harvested plant material is chopped or ground and returned to the soil. This process can be repeated as many times as desired, provided a short incubation period is allowed between crops so that some N is mineralized to reduce N deficiency in the succeeding crop. No further additions of N need be made, and additions from seed in succeeding crops can be eliminated by growing a crop, such as oat (*Avena sativa* L.), to sufficient maturity for the seed to be harvested and used for the next crop. A typical example will be given that can be modified as needed.

39–3.2.1.2 Procedure. A bulk sample of field soil (top 15–20 cm) is obtained, sieved through a 2-mm sieve to remove extraneous plant material, and mixed well. To 2-kg soil (oven-dry basis) contained in plastic bags within the pots, add 200 mg of enriched N in solution (5–10% excess ^{15}N) and other nutrient elements as needed. These materials may be easily mixed throughout the soil in the plastic bags. The form of labeled N (nitrate, ammonium, or urea) is optional. Plant 15 oat seeds in each pot of soil, cover, and water to near field capacity. Other plants can be used, but the plant must be grown to maturity so the seeds produced can be used for subsequent planting. After germination, plants may be thinned to an equal number for each pot. Return the plant material to the soil surface. During plant growth, water should be added daily by weighing the pots and adjusting the soil to field capacity. When the seed heads are sufficiently mature for germination, plants are cut at the soil surface, seeds are collected, and the stems and leaves are cut into about 2-cm pieces on a sheet of paper or plastic. The soil is removed from the pot, broken up on a plastic sheet, and roots are cut into small pieces. Soil and plant material are replaced in the plastic bag and mixed. Water is added as required for field capacity, the tops of the bags are loosely closed to reduce evaporation and allow aeration, and incubation can proceed on the greenhouse bench for 2 to 3 wk. Seeds from the pots are kept separate, allowed to air dry during the incubation period, and then used to replant the pot from whence they came.

This procedure can be repeated as many times as desired. At definite cropping intervals, triplicate pots may be removed from the experiment and analyzed. In this case, the plant material is oven dried (60 °C), weighed, ground in a Wiley mill, and analyzed for total N and atom % ^{15}N. Plant roots may be harvested, washed, and handled in the same manner. The ^{15}N-labeled soil may be used immediately or air dried and reserved for further study.

39–3.2.1.3 Growing Plants without Soil. An initial advantage of labeling plant material with ^{15}N in the absence of soil is that the ^{15}N is not diluted by the uptake of soil N. The labeled plant material can then be added to any number of soils for decomposition studies. One disadvantage is that the ^{15}N has not had the opportunity to recycle into less readily mineralized forms as in the preceding method. Sand-culture systems generally have been used to grow the plants, although solution-culture with proper aeration should also be suitable. The procedure described is similar to that used by Ladd et al. (1981).

39–3.2.1.4 Procedure. Weigh 3-kg quantities of clean, washed sand (air-dry basis) in plastic pots without drainage holes and place in the greenhouse or growth chamber. Plant seed of the desired crop and moisten the sand. After germination and thinning of plants to a suitable number, apply a nutrient solution to maintain the sand at 10% gravimetric water content. According to Gauch (1972), the nutrient solution proven satisfactory for many types of plants contains (meq/L) Ca, 10; Mg, 4; K, 4; NO_3, 10; HPO_4, 4; and SO_4, 4, with trace elements (mg/L) Cu, 0.02; Zn, 0.05; Mn, 0.5; B, 0.5; and Fe, 3 (supplied as NaFe-EDTA). The labeled nitrate (5–10 atom % ^{15}N) may be contained in the nutrient solution or added separately at intervals during the growth period. Harvest the plants at the desired stage of maturity and wash roots to remove sand. Dry the plant material and grind to the fineness needed for the soil to be labeled. For laboratory incubation studies, the plant material should be ground to pass a 1-mm sieve. For greenhouse and field studies, where large volumes of soil are involved, coarser material may be used.

39–3.2.2 Labeling Soil Biomass

Microbial biomass forms a highly important constituent of soils in that it provides for the transformation of all organic materials entering the soil, as well as being a small but labile pool of plant nutrients. Jenkinson and Ladd (1981) estimated that about 2 to 3% of the organic C in soils they examined was present as microbial biomass.

Labeling soil biomass directly, rather than through the decomposition of ^{15}N-labeled plant materials, averts any complications that might arise from the presence of labeled plant compounds resistant to decomposition. The time required for adequate labeling is also quite short (Chichester et al., 1975; Kelley & Stevenson, 1985). A simple incubation procedure with the soil to be labeled, a readily available C source, and a ^{15}N-labeled inorganic compound results in a labeled soil within a week. The procedure described is similar to that of Kelley and Stevenson (1985) but with a higher C/N ratio.

39–3.2.2.1 Procedure. Weigh 100 g of soil (oven-dry basis) that has been air dried and passed through a 2-mm sieve, into a 250-mL Erlenmeyer flask. Wet the soil with a solution containing 1.0 g of glucose C and 10 mg of N as $(^{15}NH_4)_2SO_4$ (5 atom % ^{15}N). Incubate the soil for 1 wk at 30 °C. Remove the soil from the incubator, air dry, and store for future use. Larger soil samples may be used, or several flasks may be included in the incubation to provide enough labeled soil for the experiment.

39–3.3 Determination of Mineralization Rates

The decomposition of soil organic N is measured in terms of net mineralization rates (see chapter 42). Since the labeled organic N is not uniformly distributed throughout the organic matter of the soil, estimates of mineralized N derived from both the indigenous and labeled organic N are made. Both plant uptake studies and laboratory incubation tests have been used to determine mineralization rates. Similar results have been obtained by these two methods (e.g., Broadbent & Nakashima, 1967; Legg et al., 1971). Since plant uptake studies are more laborious and time consuming, only laboratory procedures will be outlined. It is possible to use either aerobic or anaerobic incubation to determine N mineralization (see chapter 41). In most cases, aerobic incubation has been the method of choice for the long-term, consecutive incubations and extractions that are required for organic matter decomposition studies. The aerobic system is especially useful with soil organic matter labeled with both C and N (e.g., Broadbent & Nakashima, 1974). For these reasons the aerobic system is the preferred method.

The time requirements for carrying out long-term incubations or plant uptake experiments have varied from several months to several years. Generally, it is necessary to continue tests until the relative amounts of mineralized N derived from indigenous and labeled organic sources become stabilized. The calculated "availability ratios" (Broadbent & Nakashima, 1967) vary among soils, but the reasons for such variations have not been elucidated. The described procedure essentially follows that of Broadbent and Nakashima (1967) and allows determinations of mineralization rates at regular intervals for extended periods. Size of the soil sample to be incubated may vary, depending upon the expected mineralization rate.

39–3.3.1 Procedure

Weigh out triplicate 50-g samples of labeled soil (oven-dry basis) on paper sheets, moisten, mix, and transfer to leaching tubes with fritted glass bottoms. By moistening and mixing the soil, the finer particles will not segregate out and form layers in the tubes that may impede leaching. For clayey soils that are not easily leached, it may be necessary to mix the soil with sand or expanded vermiculite to facilitate leaching. A thin layer of glass wool at the top of the soil column will reduce the dispersive action of the leaching solution on the soil particles. The leaching tubes are fitted with rubber stoppers suitable for use on small suction flasks.

Before incubation, leach the soil tubes with 80 mL of saturated $CaSO_4$ solution, applied in four equal aliquots, to remove initial mineral N. Remove excess solution with suction, and incubate the samples at 35 °C in a humid atmosphere for 2 wk. Transfer the leachate to semi-micro Kjeldahl flasks and determine the inorganic N by distillation with MgO and Devarda's alloy (see chapter 40). Alternatively, the leachate may be made up to 100-mL volume and aliquots taken for analysis. The ammonia from the distillation, after titration to determine the N content, is prepared for ^{15}N analysis in the mass spectrometer (chapter 40).

After the 2-wk incubation, leach the soil samples again in the prescribed manner and determine the inorganic N and ^{15}N content of the leachate. Any number of incubations may be carried out to obtain a pattern of the organic N decomposition that is occurring. If the amount of N mineralized in 2 wk becomes insufficient for ^{15}N analysis, the incubation period may be extended, with precautions taken for maintaining the water content of the soil.

At the conclusion of the incubation and leaching part of the experiment, remove soil from the leaching tubes and dry and prepare it for total N analysis (see chapters 40 and 41) as well as ^{15}N content. A comparison of the data for the original soil and that for the residual soil plus mineralized inorganic N will indicate whether any appreciable undetermined loss of labeled N has occurred during the prolonged incubation period.

The above procedure can be modified in many ways, such as size of soil sample and different leaching solutions. Quite often, adaptations are easily made to accommodate the laboratory equipment available without any sacrifice in precision of results.

39–3.4 Preparation of Samples for [15]N Analysis

The general procedures involved in ^{15}N analysis will not be given in this chapter. Refer to chapter 40, Hauck (1982), and Mulvaney (1993) for details. Each mass spectrometry laboratory has developed its own special apparatus for preparing and analyzing ^{15}N samples; therefore, it is necessary for anyone unfamiliar with a particular mass spectrometry laboratory to determine the most appropriate form of the research samples.

39–3.5 Calculations

The "availability ratio" concept developed by Broadbent and Nakashima (1967) has been found useful in both plant uptake and mineralization studies to determine the relative mineralization rates of indigenous and ^{15}N-labeled organic N (e.g., Chichester et al., 1975; Legg et al., 1971). For extracts of mineral N after incubation, the equation is as follows:

$$\text{Availability ratio} = \frac{\text{Labeled N (extract) / Total N (extract)}}{\text{Labeled N (soil) / Total N (soil)}} \qquad [13]$$

If the ^{15}N-labeled soil organic N has the same availability to microorganisms as the indigenous organic N, the availability ratio will be one. On the other hand, if the labeled N is more susceptible to mineralization than the indigenous N, the availability ratio will be greater than one. A ratio of less than one is conceivable, but not likely.

The component values in the equation are those at the beginning of each incubation; therefore, in a succession of incubations, the values for labeled and total N in the soil must be corrected to account for the N mineralized and extracted in the previous incubation. This correction assumes that no other losses of any consequence have occurred. An analysis of soils after the experimental period will indicate whether such losses actually occurred.

When the organic N of soils is first labeled, the initial availability ratios obtained either by cropping or mineralization are generally high but quickly decrease to a level much closer to unity. This decrease indicates that the more labile fraction of the organic N is soon mineralized, leaving a more resistant fraction that appears to become more stable with time. The relationship between availability ratios and the degree of stability of labeled organic N incorporated into the soil has not been completely elucidated.

Other useful calculations can be made by employing the general equation for estimating quantities involved in isotopic equilibria:

$$A = B(1-y)/y \qquad [14]$$

Fried and Dean (1952) used this equation to obtain a measure of the availability of a soil nutrient, A, in terms of a given rate of a labeled fertilizer, B, where the proportion of the nutrient derived from the fertilizer, y, could be determined in the plant. The A value for N can be determined when a soil is being labeled with ^{15}N, as in section 39–3.2.1.1, by determining the fraction of total N in the plants that was derived from the fertilizer.

Jansson (1958) used the same equation in mineralization studies of recently incorporated labeled organic N. In this case, A is the amount of soil organic N in the active phase, B is the amount of newly immobilized labeled N, and y is the proportion of mineralized N after incubation that is derived from the labeled organic N. In long-term incubation or plant uptake experiments, increasing A values are generally observed. This increase reflects a stabilization of the labeled N and an equilibration with increasing amounts of soil N in a passive form.

Understanding the dynamics of residue decomposition and soil organic N turnover can be enhanced by using reaction kinetics to describe the microbial transformations involved. Paul and Clark (1989) provide a clear explanation of the mathematical equations and the utilization of specific types of data for determining zero-order, first-order, and hyperbolic reactions. Such calculations for degradation rates require relatively short-time intervals between measurements; otherwise, microbial synthesis and soil organic matter formation may become complicating factors.

39–3.6 Comments

Studies of the decomposition of organic N in soils require several different approaches owing to the complexity of the problem, and it is not possible to cover all of them in a single chapter. Notable among these are the procedures that have been devised to extract and separate organic compounds from the soil (Schnitzer, 1982).

Some of the early work with ^{15}N described the changes taking place in the quantities of labeled N found in different extracts with time (e.g., Stewart et al., 1963). Another means of separating recently incorporated ^{15}N in soils is organomineral sedimentation fractionations (e.g., Chichester, 1970). Modeling also has become important to increased understanding of organic matter turnover and N cycle rates (e.g., Jenkinson & Rayner, 1977; Paul & van Veen, 1978; Myrold & Tiedje, 1986). Such procedures, along with the basic methods described, provide ample opportunities for even greater advances in studies of organic N decomposition.

39–4 EXTRACTION OF LABELED ORGANIC FRACTIONS IN STUDIES OF SOIL ORGANIC MATTER DYNAMICS

39–4.1 Introduction

In the preceding sections, it has been shown how the soil organic matter is labeled and how subsequent biological transformations are followed by various procedures. Much of the labeled organic matter becomes difficultly mineralizable and apparently enters stable forms similar to indigenous organic matter. Chemical extractions of soil organic matter have been used to determine the movement of labeled compounds into the various extracted fractions, as well as unknown nonhydrolyzable forms. Currently, there are no standardized procedures for separating the organic from the inorganic soil phase, and many variations in extractants, treatments, and conditions have evolved over the years. In general, the extractions involve hot mineral acids or bases, followed by separation into various fractions for analysis. Several excellent reviews cover the details of several methods for extraction and fractionation of soil organic matter (Bremner, 1965; Schnitzer, 1982; Stevenson, 1965; 1982b). As Stevenson (1965) stated:

> The great difficulty in all fractionation procedures is that the methods employed either separate out products which are not definite chemical entities, or they form artifacts which do not have the properties of the original material. Nevertheless, the various fractionation procedures have proved useful for

studying soil organic matter, and they will probably continue to be used in the future.

Quite often, different extraction procedures are used for C and N fractionations owing to differences in determinations of isotopic composition and interferences that may occur. For that reason, separate extraction procedures will be described for C and N in the following sections.

39–4.2 Extraction of Organic Matter Containing Labeled Carbon

39–4.2.1 Introduction

Numerous methods have been used to extract C-containing components found in soil organic matter. Many of the procedures have been summarized by Stevenson (1982a). The classical procedure involves alkaline extraction of the soil and precipitation of the humic acid fraction by acidification. The fulvic acid fraction is soluble in both the base and acid. Recently, substantial research has focused on determination of the fraction of organic C found in the microbial biomass that is the more biologically active and dynamic fraction of organic C found in soil. Procedures for estimating C levels present in microbial biomass are given in chapter 36.

The procedure given in the following section is a generalized scheme for the classical method of separation of humic and fulvic acid fractions found in soil organic matter. The more detailed discussions of the procedure given by Schnitzer (1982) and Stevenson (1982a) should be consulted. The procedure is easily adapted to determine the amount of ^{14}C in humic and fulvic acids as long as appropriate safety precautions are followed when working with radioactive material.

39–4.2.2 Materials

1. Soil containing ^{14}C.
2. Hydrochloric acid (HCl), 0.05 M, 2 M.
3. Sodium hydroxide (NaOH), 0.5 M.
4. N_2 source.
5. Horizontal or wrist-action shaker.
6. Centrifuge.
7. Polypropylene or polyethylene centrifuge tubes.
8. pH meter.
9. Dry combustion or wet oxidation unit.
10. Liquid scintillation spectrometer.
11. Liquid scintillation counting cocktail.
12. Scintillation vials.

39–4.2.3 Procedures

Weigh 40 g (dry weight equivalent) of labeled soil into a 250-mL centrifuge tube and add 200 mL of 0.05 M HCl. Stir the mixture with a stirring rod. The dilute acid removes any free carbonates and polyvalent cations and increases organic matter extraction efficiency. Centrifuge and discard the supernatant. This and all subsequent centrifugations are carried out at 1500 × g. Add 200 mL of distilled water, stir, centrifuge, and discard the supernatant. Add 200 mL of 0.5 M NaOH and displace the air in the centrifuge tube with N_2. Shake the centrifuge tube for 12 to 24 h at room temperature. Separate the dark-colored supernatant that contains the humic and fulvic acids from the soil by centrifugation and decant the supernatant and filter it through glass wool to remove suspended organic solids. Collect the supernatant in a 1-L beaker. For maximum organic matter removal, repeated extraction with 0.5 M NaOH is required. Generally, two to three extractions are adequate for most soils. Following the alkaline extractions, add 200 mL of distilled water to the residual soil, shake for 10 min, centrifuge, and add the rinse water to the supernatant in the beaker. Add 2 M HCl to the alkaline extract and adjust to pH 1 using a pH meter to monitor the change. The dark-colored precipitate is humic acid and the straw-colored solution is the fulvic acid. Allow the mixture to set overnight at room temperature or refrigerate. Centrifuge the pH 1 suspension to separate the fractions. Once the humic acid has been isolated, add 200 mL of distilled water, mix, centrifuge, and add the supernatant to the fulvic acid fraction. Both fractions should be freeze-dried, weighed to determine yield, and stored in a desiccator. The residual soil contains the humin fraction and it can also be freeze-dried for subsequent analysis.

The amount of total C and ^{14}C in the humic and fulvic acids can be determined by dry combustion or wet oxidation of the organic materials. To determine the amount of total C and ^{14}C in the humin fraction, the residual soil can be analyzed. The specific details for the oxidation procedures are given by Nelson and Sommers (1982). The amount of CO_2 produced from the oxidation can be determined by methods also given in chapter 38 and the details for determining [14]CO_2 levels are given in 39–1.3.2.

39–4.2.4 Comments

The humic and fulvic acid fractions will contain inorganic components or ash. Typical ash values are ≤10% for humic acid and ≥40% for fulvic acid. If the investigator is only interested in the amount of ^{14}C from the original substrate incorporated into the humic or fulvic acid fraction, it is not necessary to determine the percentage ash. However, because of the high ash contents in the extracted materials, the percentage C values will generally be much lower than values reported on a dry, ash-free basis. Procedures for purification of humic and fulvic acids have been detailed by Schnitzer (1982) and Stevenson (1982a).

Air or oven drying is not recommended for humic or fulvic acids. The best method for drying humic acid is lyophilization. Because of the large volume of liquid containing the fulvic acid, a flash evaporator is often used to concentrate the fulvic acid prior to freeze drying.

The excessive Cl$^-$ levels in the organic matter fractions can cause problems during combustion or wet oxidation of the organic materials. It is necessary to frequently change the halide trapping units in the combustion or oxidation trains.

39–4.3 Extraction of Organic Matter Containing Labeled Nitrogen

39–4.3.1 Introduction

Classical methods for the extraction and fractionation of organic N were based upon the assumption that much of the N is proteinaceous in nature, and procedures developed for characterizing various chemical groups in proteins were used. With ^{15}N-labeled soils, such methods required certain modifications to accommodate isotopic analyses (Bremner, 1965; Cheng & Kurtz, 1963). In the modified procedure, the soil hydrolysate is neutralized without prior removal of excess acid, and the different forms of N in the neutralized hydrolysate are measured as ammonium that is readily converted to N_2 for isotope-ratio analysis. The methods described in the following sections are basically a condensed version of the ones presented by Bremner (1965) and Stevenson (1982b), and these publications should be consulted for further information. The procedure is relatively simple and permits rapid estimation of total N, ammonium N, hexosamine N, amino acid N, and (serine + threonine) N in soil hydrolysates.

39–4.3.2 Acid Hydrolysis of Soils

39–4.3.2.1 Special Apparatus
1. Micro-Kjeldahl digestion unit.
2. Steam distillation apparatus.
3. Distillation flasks: 50- and 100-mL Pyrex Kjeldahl flasks with 19/38 ground glass joints and glass hooks.
4. Microburette, 5 mL, graduated at 0.01-mL intervals.

39–4.3.2.2 Reagents
1. Sulfuric acid (H_2SO_4), concentrated.
2. Hydrochloric acid (HCl), approximately 6 M: Add 513 mL of concentrated HCl (specific gravity 1.19 g/cm^3) to about 400 mL of water, cool, and dilute to volume in a 1-L volumetric flask.
3. n-Octyl alcohol.
4. Potassium sulfate-catalyst mixture: Prepare an intimate mixture of 200 g of K_2SO_4, 20 g of $CuSO_4 \cdot 5H_2O$, and 2 g of Se. Powder the reagents separately before mixing, and grind the mixture in a mortar to powder the cake that forms.
5. Sodium hydroxide (NaOH), approximately 10 M.
6. Sodium hydroxide, approximately 5 M.

7. Sodium hydroxide, approximately 0.5 M.

8. Boric acid indicator solution: Place 80 g of H_3BO_3 in a 5-L flask marked at a 4-L volume, add about 3800 mL of water, and heat and swirl the flask until the H_3BO_3 is dissolved. Cool the solution and add 80 mL of mixed indicator solution. The indicator is prepared by dissolving 0.099 g of bromocresol green and 0.066 g of methyl red in 100 mL of ethanol. To the boric acid + indicator solution, add 0.1 M NaOH cautiously until the solution assumes a reddish purple tint (pH about 5.0), and make the solution to 4 L with water. Mix thoroughly before use.

9. Sulfuric acid, 0.0025 M standard.

10. Magnesium oxide (MgO): Heat heavy MgO in a muffle furnace at 600 to 700 °C for 2 h. Cool in a desiccator containing KOH pellets and store in a tightly stoppered bottle.

11. Ninhydrin: Grind 10 g of reagent grade ninhydrin in a mortar and store in a small widemouth bottle.

12. Phosphate-borate buffer, pH 11.2: Place 100 g of sodium phosphate ($Na_3PO_4 \cdot 12H_2O$), 25 g of borax ($Na_2B_4O_7 \cdot 10H_2O$), and about 900 mL of water in a 1-L volumetric flask, and shake the flask until the phosphate and borate are dissolved. Dilute the solution to 1 L and store in a tightly stoppered bottle.

13. Citric acid: Grind 100 g of reagent grade citric acid ($C_6H_8O_7 \cdot H_2O$) in a mortar, and store in a small widemouth bottle.

14. Citrate buffer, pH 2.6: Mix 2.06 g of powdered sodium citrate dihydrate ($Na_3C_6H_5O_7 \cdot 2H_2O$) and 19.15 g of powdered citric acid in a mortar, and grind to a fine powder with a pestle. Store in a small widemouth bottle.

15. Periodic acid ($HIO_4 \cdot 2H_2O$) solution, approximately 0.2 M: Dissolve 4.6 g of $HIO_4 \cdot 2H_2O$ in 100 mL of water and store in a glass stoppered bottle.

16. Sodium metaarsenite ($NaAsO_2$) solution, approximately 1.0 M: Dissolve 13 g of powdered, reagent grade $NaAsO_2$ in 100 mL of water and store in a tightly stoppered bottle.

17. Standard (NH_4^+ + amino sugar + amino acid)-N solution: Dissolve 0.189 g of $(NH_4)_2SO_4$, 0.308 g of glucosamine·HCl, and 0.254 g of alanine in water. Dilute the solution to 2 L in a volumetric flask and mix thoroughly. If prepared from pure, dry reagents, this solution contains 20 µg of NH_4^+-N, 10 µg of amino sugar-N, and 20 µg of α-amino acid-N/mL. Store the solution for no more than 7 d in a refrigerator at 4 °C.

18. Standard (serine + threonine)-N solution: Dissolve 0.150 g of serine and 0.170 g of threonine in water. Dilute the solution to 2 L in a volumetric flask and mix. If prepared from pure, dry reagents, this solution contains 10 µg of serine-N and 10 µg of threonine-N/mL. Store the solution for no more than 7 d in a refrigerator at 4 °C.

39–4.3.3 Preparation and Sampling of Soil Hydrolysate

Place a sample of finely ground ($\leq$100 mesh) soil containing about 10 mg of N in a round-bottom flask fitted with a standard taper (24/40) ground-glass joint. Add two drops of octyl alcohol and 20 mL of 6 M HCl, and swirl the flask until the acid is thoroughly mixed with the soil. Place the flask in an electric heating mantle, connect the flask to a Liebig condenser fitted with a 24/40 ground-glass joint, and heat the soil-acid mixture so that it boils gently under reflux for 12 h.

After completion of hydrolysis, wash the reflux condenser with a small quantity of distilled water, allow the flask to cool, and remove the flask from the condenser. Filter the mixture through a Buchner funnel fitted with Whatman no. 50 filter paper, using a suction filtration apparatus that permits collection of the filtrate in a 200 mL tall-form beaker marked to indicate a volume of 60 mL. Wash the residue with 5- to 10-mL portions of distilled water until the filtrate reaches the 50-mL mark on the beaker. Immerse the lower half of the beaker in crushed ice, and neutralize to pH 6.5 $\pm$ 0.1 by cautious addition of NaOH, using a pH meter to follow the course of neutralization. Add the alkali slowly with constant stirring to ensure that the hydrolysate does not become alkaline at any stage of the neutralization process. Use 5 M NaOH to bring the pH to about 5 and complete the neutralization using 0.5 M NaOH. Transfer the neutralized hydrolysate by means of a small funnel to a 100-mL volumetric flask, and dilute to volume with the washings obtained by rinsing the beaker, electrodes and stirrer several times with small quantities of distilled water. Stopper the flask and invert several times to mix the contents.

To determine the different forms of N in the hydrolysate, after thorough mixing, usually a 5- to 10-mL sample is pipetted into a 50- or 100-mL distillation flask and the flask is connected to a steam distillation apparatus. It is necessary to use pipettes with wide tips that permit rapid delivery to avoid sampling errors. The form of N under analysis is determined from the NH_3-N liberated by steam distillation for 2 to 4 min.

39–4.3.4 Total Hydrolyzable Nitrogen

Place 5 mL of the neutralized hydrolysate in a 50-mL distillation flask, add 0.5 g of K_2SO_4-catalyst mixture and 2 mL of concentrated H_2SO_4, and heat the flask cautiously on a micro-Kjeldahl digestion unit until the water is removed and frothing ceases. Increase the heat until the mixture clears, and complete the digestion by boiling gently for 1 h.

After digestion, allow the flask to cool, and add about 10 mL of water (slowly and with shaking). Cool the flask under a cold-water tap, and place it in a beaker containing crushed ice. Add 5 mL of H_3BO_3 indicator solution to a 50-mL Erlenmeyer flask that is marked to indicate a volume of 35 mL, and place the flask under the condenser of the steam distillation apparatus so that the tip of the condenser is about 4 cm above the surface of the H_3BO_3. Connect the cooled distillation flask to the distillation apparatus, place 10 mL of 10 M NaOH in the entry funnel, and run the alkali slowly into the distillation flask. When about 0.5 mL of alkali remains in the funnel, rinse the funnel rapidly with about 5 mL of water, and allow about 2 mL to run into the distillation flask before sealing the funnel. Commence steam distillation and stop when the distillate reaches the 35-mL mark on the receiver flask (distillation time about 4 min). Rinse the condenser, and determine the NH_4^+-N in the distillate by titration with 0.0025 M H_2SO_4 from a microburette (1 mL = 70 µg NH_4^+-N). The color change at the endpoint is from green to a faint, permanent pink.

39–4.3.5 Acid-insoluble Nitrogen

This form of N is the difference between total soil N (Bremner & Mulvaney, 1982) and total hydrolyzable N (section 39–4.3.4). It can also be determined directly by acid digestion of the soil residue remaining after hydrolysis (Cheng & Kurtz, 1963).

39–4.3.6 Amino Acid-Nitrogen

Place 5 mL of the hydrolysate (section 39–4.3.3) in a 50-mL distillation flask, add 1 mL of 0.5 M NaOH, and heat the flask in boiling water until the volume of the sample is reduced to 2 to 3 mL (approximately 20 min). Allow the flask to cool, add 500 mg of citric acid and 100 mg of ninhydrin, and place the flask in a vigorously boiling water bath, so that its bulb is completely immersed in boiling water. After about 1 min, swirl the flask for a few seconds without removing it from the bath, and allow it to remain in the bath for an additional 9 min. Then cool the flask, add 10 mL of phosphate-borate buffer and 1 mL of 5 M NaOH, and connect the flask to the steam distillation apparatus. Determine the amount of NH_3-N liberated by steam distillation as in section 39–4.3.4 (distillation period about 4 min).

39–4.3.7 Ammonia-Nitrogen

Place 10 mL of the hydrolysate (section 39–4.3.3) in a 50- or 100-mL distillation flask, add 0.07 ± 0.01 g of MgO, and connect the flask to the steam distillation apparatus. Determine the amount of NH_3-N liberated by steam distillation as in section 39–4.3.4, but collect the distillate in a 50-mL Erlenmeyer flask that contains 5 mL of H_3BO_3-indicator solution and marked to indicate a volume of 20 mL. Discontinue distillation when the distillate reaches the 20-mL mark (distillation period about 2 min).

39–4.3.8 (Ammonia + Amino Sugar)-Nitrogen

Place 10 mL of the hydrolysate (section 39–4.3.3) in a 100-mL distillation flask, add 10 mL of phosphate-borate buffer, and connect the flask to the distillation apparatus. Determine the amount of NH_3-N liberated by steam distillation as described in section 39–4.3.4 (distillation period about 4 min).

39–4.3.9 Amino Sugar-Nitrogen

This form of N is taken as the difference between the amounts of N recovered in the preceding two sections.

39–4.3.10 (Serine + Threonine)-Nitrogen

Proceed as described in section 39–4.3.8, but after removal of $(NH_3 + amino sugar)$-N by steam distillation with phosphate-borate buffer, detach the flask from the distillation apparatus, and rinse the steam inlet tube with 3 to 5 mL of water. Collect the rinse water in the distillation flask, and cool the flask under a cold water tap. Add 2 mL of periodic acid solution, swirl the flask for about 30 s, add 2 mL of sodium arsenite solution, and connect the flask to the distillation apparatus. Determine the amount of NH_3-N liberated by steam distillation as described in section 39–4.3.4 (period of distillation about 4 min).

39–4.3.11 Comments

One of the main advantages of the described hydrolysis method is that the N in the different fractions is measured as NH_4^+-N, and this can be readily converted to N_2 for isotope-ratio analysis. Total N in a given sample may be insufficient for mass spectrometer analysis, and duplicate analyses may have to be combined for ^{15}N determinations. If the ^{15}N percentage is relatively high, it may be diluted with a measured amount of unlabeled N to provide sufficient total N for mass spectrometer requirements (Hauck, 1982). The use of ^{15}N-labeled soils provides a means of tracing the movement of added N into the various organic fractions and determination of the rate at which this occurs if a proper time sequence is employed. Allen et al. (1973) present some typical data that can be obtained by this means.

The recommended hydrolysis procedure can be modified in several ways, but the more extensive discussion of the procedure by Bremner (1965) and Stevenson (1982b) should be consulted before doing so. It should also be pointed out that the hydrolysis method presented here causes greater decomposition of amino sugars than more conventional methods. The correction factor for hydrolysis losses of amino sugars is about 1.4 (Bremner, 1965).

39–5 CONCLUSIONS

During the past 50 yr, the use of isotopes in the study of soil organic matter dynamics has led to a tremendous increase in knowledge of the system that would not have been possible otherwise. Numerous methods for the use of isotopes have been developed over the years with specific objectives in mind and equipment available at the time. This chapter outlines basic methods currently applicable to organic matter studies, recognizing that improvements and modifications are constantly being made.

The References section provides detailed information on techniques that have been developed.

REFERENCES

Allen, A.L., F.J. Stevenson, and L.T. Kurtz. 1973. Chemical distribution of residual fertilizer nitrogen in soil as revealed by nitrogen-15 studies. J. Environ. Qual. 2:120–124.

Allison, F.E. 1973. Soil organic matter and its role in crop production. Elsevier Scientific Publ. Co., New York.

Andersen, A., G. Nielsen, and H. Sorensen. 1961. Growth chamber for labeling plant material uniformly with radiocarbon. Physiol. Plant. 14:378–383.

Anderson, J.P.E. 1982. Soil respiration. p. 831–871. In A.L. Page et al. (ed.) Methods of soil analysis. Part 2. 2nd ed. Agron. Monogr. 9. ASA and SSSA, Madison, WI.

Andreux, F., C. Cerri, P.B. Vose, and V.A. Vitorello. 1990. Potential of stable isotope, ^{15}N and ^{13}C, methods for determining input and turnover in soils. p. 259–275. In A.F. Harrison et al. (ed.) Nutrient cycling in terrestrial ecosystems. Elsevier Applied Sci., New York.

Atlas, R.M., and R. Bartha. 1972. Degradation and mineralization of petroleum by two bacteria isolated from coastal waters. Biotechnol. Bioeng. 14:297–308.

Balesdent, J., A. Mariotti, and D. Boisgontier. 1990. Effect of tillage on soil organic carbon mineralization estimated from ^{13}C abundance in maize fields. J. Soil Sci. 41:587–596.

Balesdent, J., A. Mariotti, and B. Guillet. 1987. Natural ^{13}C abundance as a tracer for studies of soil organic matter dynamics. Soil Biol. Biochem. 19:25–30.

Balesdent, J., G.H. Wagner, and A. Mariotti. 1988. Soil organic matter turnover in long-term field experiments as revealed by carbon-13 natural abundance. Soil Sci. Soc. Am. J. 52:118–124.

Bartha, R., and D. Pramer. 1965. Features of a flask and method for measuring the persistence and biological effects of pesticides in soil. Soil Sci. 100:68–70.

Boutton, T.W. 1991a. Stable carbon isotope ratios of natural materials: I. Sample preparation and mass spectrometric analysis. p. 155–171. In D.C. Coleman and B. Fry (ed.) Carbon isotope techniques. Academic Press, New York.

Boutton, T.W. 1991b. Stable carbon isotope ratios of natural materials: II. Atmospheric, terrestrial, marine, and freshwater environments. p. 173–185. In D.C. Coleman and B. Fry (ed.) Carbon isotope techniques. Academic Press, New York.

Boutton, T.W., A.T. Harrison, and B.N. Smith. 1980. Distribution of biomass of species differing in photosynthetic pathway along an altitudinal transect in southeastern Wyoming grassland. Oecologia 45:287–298.

Boutton, T.W., W.W. Wong, D.L. Hachey, L.S. Lee, M.P. Cabrera, and P.D. Klein. 1983. Comparison of quartz and Pyrex tubes for combustion of organic samples for stable carbon isotope analysis. Anal. Chem. 55:1832–1833.

Bremner, J.M. 1965. Organic forms of nitrogen. p. 1238–1255. In C.A. Black et al. (ed.) Methods of soil analysis. Part 2. Agron. Monogr. 9. ASA, Madison, WI.

Bremner, J.M., and C.S. Mulvaney. 1982. Nitrogen-total. p. 595–624. In A.L. Page et al. (ed.) Methods of soil analysis. Part 2. 2nd ed. Agron. Monogr. 9. ASA and SSSA, Madison, WI.

Broadbent, F.E., and T. Nakashima. 1967. Reversion of fertilizer nitrogen in soils. Soil Sci. Soc. Am. Proc. 31:648–652.

Broadbent, F.E., and T. Nakashima. 1974. Mineralization of carbon and nitrogen in soil amended with carbon-13 and nitrogen-15 labeled plant material. Soil Sci. Soc. Am. Proc. 38:313–315.

Cerri, C.C., B. Volkoff, and F. Andreux. 1991. Nature and behavior of organic matter in soils under natural forest, and after deforestation, burning and cultivation, near Manaus. Forest Ecol. Manage. 38:247–257.

Cheng, H.H., and L.T. Kurtz. 1963. Chemical distribution of added nitrogen in soils. Soil Sci. Soc. Am. Proc. 27:312–316.

Cheshire, M.V., and B.S. Griffiths. 1989. The influence of earthworms and cranefly larvae on the decomposition of uniformly ^{14}C labeled plant material in soil. J. Soil Sci. 40:117–124.

Chichester, F.W. 1970. Transformation of fertilizer nitrogen in soil: II. Total and ^{15}N labeled nitrogen of soil organomineral sedimentation fractions. Plant Soil 33:437–456.

Chichester, F.W., J.O. Legg, and G. Stanford. 1975. Relative mineralization rates of indigenous and recently incorporated ^{15}N-labeled nitrogen. Soil Sci. 120:455–460.

Coleman, D.C., and F.T. Corbin. 1991. Introduction and ordinary counting as currently used. p. 3–9. In D.C. Coleman and B. Fry (ed.) Carbon isotope techniques. Academic Press, New York.

Corbin, F.T., and B.A. Swisher. 1986. Radioisotope techniques. p. 265–276. In N.D. Camper (ed.) Research methods in weed science. Southern Weed Sci. Soc., Champaign, IL.

Craig, H. 1957. Isotopic standards for carbon and oxygen and correction factors for mass spectrometric analysis of carbon dioxide. Geochim. Cosmochim. Acta 12:133–149.

Crawford, D.L. 1978. Lignocellulose decomposition by selected Streptomyces strains. Appl. Environ. Microbiol. 35:1041–1045.

Crawford, D.L., and R.L. Crawford. 1976. Microbial degradation of lignocellulose: The lignin component. Appl. Environ. Microbiol. 31:714–717.

Crawford, D.L., R.L. Crawford, and A.L. Pometto. 1977. Preparation of specifically labeled ^{14}C-(lignin)- and ^{14}C-(cellulose)-lignocelluloses and their decomposition by the microflora of soil. Appl. Environ. Microbiol. 33:1247–1251.

Deines, P. 1970. Mass spectrometer correction factors for the determination of small isotopic composition variations of carbon and oxygen. Int. J. Mass Spectrom. Ion Phys. 4:283–295.

Des Marais, D., and J.M. Hayes. 1976. Tube cracker for opening glass-sealed ampoules under vacuum. Anal. Chem. 48:1651–1652.

Dzurec, R.S., T.W. Boutton, M.M. Caldwell, and B.N. Smith. 1985. Carbon isotope ratios of soil organic matter and their use in assessing community composition changes in Curlew Valley, Utah. Oecologia 66:17–24.

Engel, M.H., and R.J. Maynard. 1989. Preparation of organic matter for stable carbon isotope analysis by sealed tube combustion: A cautionary note. Anal. Chem. 61:1996–1998.

Frazer, J.W., and R. Crawford. 1963. Modifications in the simultaneous determination of carbon, hydrogen, and nitrogen. Mikrochim. Acta 3:561–566.

Fried, M., and L.A. Dean. 1952. A concept concerning the measurement of available soil nutrients. Soil Sci. 73:263–271.

Gauch, H.G. 1972. Inorganic plant nutrition. Dowden, Hutchinson & Ross, Stroudsburg, PA.

Goh, K.M. 1991. Bomb carbon. p. 147–151. In D.C. Coleman and B. Fry (ed.) Carbon isotope techniques. Academic Press, New York.

Gonfiantini, R. 1981. The δ-notation and the mass spectrometric measurement techniques. p. 35–84. In J.R. Gat and R. Gonfiantini (ed.) Stable isotope hydrology. IAEA, Vienna, Austria.

Haider, K., and J.P. Martin. 1975. Decomposition of specifically carbon-14 labeled benzoic and cinnamic acid derivatives in soil. Soil Sci. Soc. Am. Proc. 39:657–662.

Haider, K., J.P. Martin, and E. Rietz. 1977. Decomposition in soil of ^{14}C-labeled coumaryl alcohols; free and linked into dehydropolymer and plant lignins and model humic acids. Soil Sci. Soc. Am. J. 41:556–562.

Harris, D., and E.A. Paul. 1991. Techniques for examining the carbon relationships of plant-microbial symbioses. p. 39–52. In D. C. Coleman and B. Fry (ed.) Carbon isotope techniques. Academic Press, New York.

Harrison, A.F., D.D. Harkness, and P.J. Bacon. 1990. The use of bomb-^{14}C for studying organic matter and N and P dynamics in a woodland soil. p. 246–258. In A.F. Harrison et al. (ed.) Nutrient cycling in terrestrial ecosystems. Elsevier Applied Sci., New York.

Hauck, R.D. 1982. Nitrogen-isotope ratio analysis. p. 735–779. In A.L. Page et al. (ed.) Methods of soil analysis. Part 2. 2nd ed. Agron. Monogr. 9. ASA and SSSA, Madison, WI.

Hayes, J.M. 1983. Practice and principles of isotopic measurements in organic geochemistry. p. 5.1–5.31. In W.G. Meinschein (ed.) Organic geochemistry of contemporaneous and ancient sediments. Soc. for Econ. Paleontologists and Mineralogists, Bloomington, IN.

Hurst, H.M., and G.H. Wagner. 1969. Decomposition of ^{14}C-labeled cell wall and cytoplasmic fractions from hyaline and melanic fungi. Soil Sci. Soc. Am. Proc. 33:707–711.

Insam, H., M.M. Ding, and A. Mariotti. 1991. Utilization of a ^{13}C-enriched tracer for carbon flux studies in a tropical Eucalyptus exserta forest. p. 515–519. In Stable isotopes in plant nutrition, soil fertility, and environmental studies. IAEA, Vienna, Austria.

Jackson, M.L. 1969. Soil chemical analysis—Advanced course. 2nd ed. 11th printing. Published by the author, Madison, WI.

Jansson, S.L. 1958. Tracer studies on nitrogen transformations in soil. Annu. Roy. Agric. Coll. Sweden 24:101–361.

Jansson, S.L., and J. Persson. 1982. Mineralization and immobilization of soil nitrogen. p. 229–252. In F.J. Stevenson (ed.) Nitrogen in agricultural soils. Agron. Monogr. 22. ASA, CSSA, and SSSA, Madison, WI.

Jenkinson, D.S. 1960. The production of ryegrass labeled with carbon-14. Plant Soil 13:279–290.

Jenkinson, D.S. 1971. Studies on the decomposition of ^{14}C-labeled organic matter in soil. Soil Sci. 111:64–70.

Jenkinson, D.S., and J.N. Ladd. 1981. Microbial biomass in soil: Measurement and turnover. p. 415–471. In E.A. Paul and J.N. Ladd (ed.) Soil biochemistry. Vol. 5. Marcel Dekker, New York.

Jenkinson, D.S., and D.S. Powlson. 1976. The effects of biocidal treatments on metabolism in soil. V: A method for measuring biomass. Soil Biol. Biochem. 8:209–213.

Jenkinson, D.S., and J.H. Rayner. 1977. The turnover of soil organic matter in some Rothamsted classical experiments. Soil Sci. 123:298–305.

Kassim, G., J.P. Martin, and K. Haider. 1981. Incorporation of a wide variety of organic substrate carbons into soil biomass as estimated by the fumigation procedure. Soil Sci. Soc. Am. J. 45:1106–1112.

Kassim, G., D.E. Stott, J.P. Martin, and K. Haider. 1982. Stabilization and incorporation into biomass of phenolic and benzenoid carbons during biodegradation in soil. Soil Sci. Soc. Am. J. 46:305–309.

Kearney, P.C., and A. Kontson. 1976. A simple system to simultaneously measure volatilization and metabolism of pesticides from soils. J. Agric. Food Chem. 24:424–426.

Keeney, D.R., and D.W. Nelson. 1982. Nitrogen-inorganic forms. p. 643–698. In A.L. Page et al. (ed.) Methods of soil analysis. Part 2. 2nd ed. Agron. Monogr. 9. ASA and SSSA, Madison, WI.

Kelley, K.R., and F.J. Stevenson. 1985. Characterization and extractability of immobilized ^{15}N from the soil microbial biomass. Soil Biol. Biochem. 17:517–523.

Kemper, W.D., and R.C. Rosenau. 1986. Aggregate stability and size distribution. p. 425–442. In A. Klute (ed.) Methods of soil analysis. Part 1. 2nd ed. Agron. Monogr. 9. ASA and SSSA, Madison, WI.

Ladd, J.N., and M.A. Amato. 1986. The fate of nitrogen from legume and fertilizer sources in soils successively cropped with wheat under field conditions. Soil Biol. Biochem. 18:417–425.

Ladd, J.N., and J.K. Martin. 1984. Soil organic matter studies. p. 67–98. In M.F. L'Annunziata and J.O. Legg (ed.) Isotopes and radiation in agricultural sciences. Vol. 1. Academic Press, London.

Ladd, J.N., J.M. Oades, and M. Amato. 1981. Distribution and recovery of nitrogen from legume residues decomposing in soils sown to wheat in the field. Soil Biol. Biochem. 13:251–256.

L'Annunziata, M.F. 1979. Radiotracers in agricultural chemistry. Academic Press, New York.

L'Annunziata, M.F. 1984. The detection and measurement of radionuclides. p. 141–231. In M.F. L'Annunziata and J.O. Legg (ed.) Isotopes and radiation in agricultural sciences. Vol. 1. Academic Press, London.

Lavy, T.L. 1975. Effects of soil pH and moisture on the direct radioassay of herbicides in soil. Weed Sci. 23:49–52.

Le Feuvre, R.P., and R.J. Jones. 1988. Static combustion of biological samples sealed in glass tubes as a preparation for δ^{13}C determination. Analyst 113:817–823.

Legg, J.O., F.W. Chichester, G. Stanford, and W.H. DeMar. 1971. Incorporation of ^{15}N-tagged mineral nitrogen into stable forms of soil organic nitrogen. Soil Sci. Soc. Am. Proc. 35:273–276.

Levin, I., B. Kromer, D. Wagenback, and K.O. Munnich. 1987. Carbon isotope measurements of atmospheric CO_2 at a coastal station in Antarctica. Tellus 39B:89–95.

Loos, M.A., A. Kontson, and P.C. Kearney. 1980. Inexpensive soil flask for ^{14}C-pesticide degradation studies. Soil Biol. Biochem. 12:583–585.

Martin, A., A. Mariotti, J. Balesdent, P. Lavelle, and R. Vuattoux. 1990. Estimate of organic matter turnover rate in a savanna soil by ^{13}C natural abundance measurements. Soil Biol. Biochem. 22:517–523.

Martin, J.P., and K. Haider. 1979. Biodegradation of ^{14}C-labeled model and cornstalk lignins, phenols, model phenolase humic polymers, and fungal melanins as influenced by a readily available carbon source and soil. Appl. Environ. Microbiol. 38:283–289.

Martin, J.P., K. Haider, and D.C. Wolf. 1972. Synthesis of phenols and phenolic polymers by *Hendersonula toruloidea* in relation to humic acid formation. Soil Sci. Soc. Am. Proc. 36:311–315.

Martin, J.P., H. Zunino, P. Peirano, M. Caiozzi, and K. Haider. 1982. Decomposition of ^{14}C-labeled lignins, model humic acid polymers, and fungal melanins in allophanic soils. Soil Biol. Biochem. 14:289–293.

Marvel, J.T., B.B. Brightwell, J.M. Malik, M.L. Sutherland, and M.L. Rueppel. 1978. A simple apparatus and quantitative method for determining the persistence of pesticides in soil. J. Agric. Food Chem. 26:1116–1120.

Mayaudon, J. 1971. Use of radiorespirometry in soil microbiology and biochemistry. p. 202–256. *In* A.D. McLaren and J. Skujins (ed.) Soil biochemistry. Vol. 2. Marcel Dekker, New York.

Melillo, J.M., J.D. Aber, A.E. Linkins, A. Ricca, B. Fry, and K.J. Nadelhoffer. 1989. Carbon and nitrogen dynamics along the decay continuum: Plant litter to soil organic matter. p. 53–62. *In* M. Clarholm and L. Bergstrom (ed.) Ecology of arable land. Kluwer Acad. Publ., Dordrecht, Netherlands.

Mook, W.G., and P.M. Grootes. 1973. The measuring procedure and corrections for the high precision mass spectrometric analysis of isotopic abundance ratios, especially referring to carbon, oxygen, and nitrogen. Int. J. Mass Spectrom. Ion Phys. 12:273–298.

Mulvaney, R.L. 1993. Mass spectrometry. p. 11–57. *In* R. Knowles and T.H. Blackburn (ed.) Nitrogen isotope techniques. Academic Press, San Diego.

Myrold, D.D., and J.M. Tiedje. 1986. Simultaneous estimation of several nitrogen cycle rates using [15]N: Theory and application. Soil Biol. Biochem. 18:559–568.

Nadelhoffer, K.J., and B. Fry. 1988. Controls on natural nitrogen-15 and carbon-13 abundances in forest soil organic matter. Soil Sci. Am. J. 52:1633–1640.

Nakas, J.P., and D.A. Klein. 1979. Decomposition of microbial cell components in a semiarid grassland soil. Appl. Environ. Microbiol. 38:454–460.

Nelson, D.W., and L.E. Sommers. 1982. Total carbon, organic carbon, and organic matter. p. 539–580. *In* A.L. Page et al. (ed.) Methods of soil analysis. Part 2. 2nd ed. Agron. Monogr. 9. ASA and SSSA, Madison, WI.

Oades, J.M., and G.H. Wagner. 1971. Biosynthesis of sugars in soils incubated with ^{14}C-glucose and ^{14}C-dextran. Soil Sci. Soc. Am. Proc. 35:914–917.

O'Leary, M.H. 1988. Carbon isotopes in photosynthesis. BioScience 38:328–336.

Osmond, C.B., K. Winter, and H. Ziegler. 1982. Functional significance of different pathways of CO_2 fixation in photosynthesis. p. 479–547. *In* O.L. Lange et al. (ed.) Plant physiological ecology II: Water relations and carbon assimilation. Springer-Verlag, Berlin.

Paul, E.A., and F.E. Clark. 1989. Soil microbiology and biochemistry. Academic Press, New York.

Paul, E.A., and J.A. van Veen. 1978. The use of tracers to determine the dynamic nature of organic matter. p. 61–102. Trans. 11th Int. Congr. Soil Sci. Vol. 3. Edmonton, Canada.

Post, W.M., W.R. Emanuel, P.J. Zinke, and A.G. Stangenberger. 1982. Soil carbon pools and world life zones. Nature (London) 298:156–159.

Post, W.M., J. Pastor, P.J. Zinke, and A.G. Stangenberger. 1985. Global patterns of soil nitrogen storage. Nature (London) 317:613–616.

Prentice, K.C., and I.Y. Fung. 1990. The sensitivity of terrestrial carbon storage to climate change. Nature (London) 346:48–51.

Reyes, V.G., and J.M. Tiedje. 1973. Metabolism of ^{14}C uniformly labeled organic material by woodlice (Isopoda: Oniscoidea) and soil microorganisms. Soil Biol. Biochem. 5:603–611.

Russell, E.W. 1961. Soil conditions and plant growth. John Wiley and Sons, New York.

Santrock, J., S. Studley, and J. Hayes. 1985. Isotopic analyses based on the mass spectrum of carbon dioxide. Anal. Chem. 57:1444–1448.

Scharpenseel, H.W., and H.U. Neue. 1984. Use of isotopes in studying the dynamics of organic matter in soils. p. 273–310. *In* Organic matter and rice. Int. Rice Res. Inst., Manila, Philippines.

Schlesinger, W.H. 1990. Evidence from chronosequence studies for a low carbon storage potential of soils. Nature (London) 348:232–234.

Schnitzer, M. 1982. Organic matter characterization. p. 581–594. *In* A.L. Page et al. (ed.) Methods of soil analysis. Part 2. 2nd ed. Agron. Monogr. 9. ASA and SSSA, Madison, WI.

Scott, H.D., and R.E. Phillips. 1972. Diffusion of selected herbicides in soil. Soil Sci. Soc. Am. Proc. 36:714–719.

Scully, N.J., W. Chorney, G. Kostal, R. Watanabe, J. Skok, and J.W. Glattfeld. 1956. Biosynthesis in ^{14}C-labeled plants: Their use in agricultural and biological research. Vol. 12. p. 377–385. *In* Proc. Int. Conf. Peaceful Uses of Atomic Energy, Geneva, Switzerland, 8–20 Aug. 1955. United Nations, New York.

Sissons, C.H. 1976. Improved technique for accurate and convenient assay of biological reactions liberating ^{14}CO$_2$. Anal. Biochem. 70:454–462.

Skjemstad, J.O., R.P. Le Feuvre, and R.E. Prebble. 1990. Turnover of soil organic matter under pasture as determined by ^{13}C natural abundance. Aust. J. Soil Res. 28:267–276.

Smith, B.N., and S. Epstein. 1971. Two categories of ^{13}C/^{12}C ratios for higher plants. Plant Physiol. 47:380–384.

Smith, J.H., F.E. Allison, and J.F. Mullins. 1962. Design and operation of a carbon-14 biosynthesis chamber. USDA-ARS Misc. Publ. 911. U.S. Gov. Print. Office, Washington, DC.

Sofer, Z. 1980. Preparation of carbon dioxide for stable carbon isotope analysis of petroleum fractions. Anal. Chem. 52:1389–1391.

Stevenson, F.J. 1965. Gross chemical fractionation of organic matter. p. 1409–1421. *In* C.A. Black et al. (ed.) Methods of soil analysis. Part 2. Agron. Monogr. 9. ASA, Madison, WI.

Stevenson, F.J. 1982a. Humus chemistry, genesis, composition, reactions. John Wiley and Sons, New York.

Stevenson, F.J. 1982b. Nitrogen—organic forms. p. 625–641. *In* A.L. Page et al. (ed.) Methods of soil analysis. Part 2, 2nd ed. Agron. Monogr. 9. ASA and SSSA, Madison, WI.

Stevenson, F.J. 1986. Cycles of soil: Carbon, nitrogen, phosphorus, sulfur, micronutrients. John Wiley and Sons, New York.

Stevenson, F.J., and E.T. Elliott. 1989. Methodologies for assessing the quantity and quality of soil organic matter. p. 173–199. *In* D.C. Coleman et al. (ed.) Dynamics of soil organic matter in tropical ecosystems. NifTAL Project, Dep. Agronomy and Soil Science, Univ. of Hawaii, Honolulu.

Stewart, B.A., L.K. Porter, and D.D. Johnson. 1963. Immobilization and mineralization of nitrogen in several organic fractions of soil. Soil Sci. Soc. Am. Proc. 27:302–304.

Stotzky, G. 1965. Microbial respiration. p. 1550–1572. *In* C.A. Black et al. (ed.) Methods of soil analysis. Part 2. Agron. Monogr. 9. ASA, Madison, WI.

Stotzky, G., M.W. Broder, J.D. Doyle, and R.A. Jones. 1993. Selected methods for the detection and assessment of ecological effects resulting from the release of genetically engineered microorganisms to the terrestrial environment. *In* S.L. Neidleman and A.I. Laskin (ed.) Adv. Appl. Microbiol. 38:1–93.

Stout, J.D., K.M. Goh, and T.A. Rafter. 1981. Chemistry and turnover of naturally occurring resistant organic compounds in soil. p. 1–73. *In* E.A. Paul and J.N. Ladd (ed.) Soil biochemistry. Vol. 5. Marcel Dekker, New York.

Swerhone, G.D.W., K.A. Hobson, C. van Kessel, and T.W. Boutton. 1991. An economical method for the preparation of plant and animal tissue for δ^{13}C analysis. Commun. Soil Sci. Plant Anal. 22:177–190.

Teeri, J.A., and L.G. Stowe. 1976. Climatic patterns and the distribution of C$_4$ grasses in North America. Oecologia 23:1–12.

Tiessen, H., and J.W.B. Stewart. 1983. Particle size fractions and their use in studies of soil organic matter: II. Cultivation effects on organic matter composition in size fractions. Soil Sci. Soc. Am. J. 47:509–514.

Verma, L., J.P. Martin, and K. Haider. 1975. Decomposition of carbon-14-labeled proteins, peptides, and amino acids; free and complexed with humic polymers. Soil Sci. Soc. Am. Proc. 39:279–284.

Vitorello, V.A., C.C. Cerri, F. Andreux, C. Feller, and R.L. Victoria. 1989. Organic matter and natural carbon-13 distribution in forested and cultivated Oxisols. Soil Sci. Soc. Am. J. 53:773–778.

Voroney, R.P., J.P. Winter, and E.G. Gregorich. 1991. Microbe/plant/soil interactions. p. 77–99. *In* D.C. Coleman and B. Fry (ed.) Carbon isotope techniques. Academic Press, New York.

Wagner, G.H. 1975. Microbial growth and carbon turnover. p. 269–305. *In* E.A. Paul and A.D. McLaren (ed.) Soil biochemistry. Vol. 3. Marcel Dekker, New York.

Wagner, G.H., and K.S. Chahal. 1966. Decomposition of carbon-14 labeled atrazine in soil samples from Sanborn Field. Soil Sci. Soc. Am. Proc. 30:752–754.

Wagner, G.H., and A.M. Krzywicka. 1975. Decomposition of algal tissues in soil. p. 202–207. *In* G. Kilbertus et al. (ed.) Biodegradation et humification. Pierron Publ., Sarreguemines, France.

Warembourg, F.R., and J. Kummerow. 1991. Photosynthesis/translocation studies in terrestrial ecosystems. p. 11–37. *In* D.C. Coleman and B. Fry (ed.) Carbon isotope techniques. Academic Press, New York.

Wolf, D.C., and J.O. Legg. 1984. Soil microbiology. p. 99–139. *In* M.F. L'Annunziata and J.O. Legg (ed.) Isotopes and radiation in agricultural sciences. Vol. 1. Academic Press, London.

Zunino, H., F. Borie, S. Aguilera, J.P. Martin, and K. Haider. 1982. Decomposition of ^{14}C-labeled glucose, plant and microbial products and phenols in volcanic ash-derived soils of Chile. Soil Biol. Biochem. 14:37–43.

Chapter 40

Practical Considerations in the Use of Nitrogen Tracers in Agricultural and Environmental Research

R. D. HAUCK, *Tennessee Valley Authority, Muscle Shoals, Alabama*

J. J. MEISINGER, *USDA-ARS, Beltsville, Maryland*

R. L. MULVANEY, *University of Illinois, Urbana, Illinois*

Agricultural research is challenged to develop crop production systems that are productive and economically viable over time, socially and politically acceptable, and conserving of natural resources. Efficient use of N, regardless of its source, is imperative for achieving these ends. Many of the questions centered about efficient N use cannot be answered satisfactorily, if at all, without use of the stable isotopes, ^{14}N and ^{15}N, as tracers. First used in agronomic studies by Norman and Werkman (1945), N tracer methodology has been used almost routinely in recent years to study N transformation processes in plants and animals, air, soils, and waters, and to understand, evaluate, and monitor the effects of different approaches to N management in production agriculture.

Stable (nonradioactive) isotopes are used almost exclusively when tracer techniques are needed in agricultural and environmental studies. The extremely short half-lives of the four radionuclides of N, ranging from 0.0125 to 603 s, militate against their use for biological studies. Of these radionuclides, only ^{13}N has been used to a limited extent as a highly sensitive tracer in metabolic studies (mainly of biological N_2 fixation) that can be completed within about 2 h.

Use of the stable isotopes of masses 14 and 15 as tracers is based on the fact that these isotopes occur naturally in an almost constant ratio. Except for slight variations in the N isotopic composition of natural nitrogenous substances, the ratio of $^{14}N/^{15}N$ is about 272:1 (i.e., naturally occurring N contains about 0.366 at. % ^{15}N or about 3660 ppm ^{15}N). The significance of natural variations from these values will be discussed later.

Stable N tracers are materials that contain an unnaturally high or low concentration of either ^{14}N or ^{15}N. Until the 1970s, almost all N-tracer experiments were conducted with substances that were enriched with ^{15}N, i.e., had higher than natural concentrations of ^{15}N. Since that time, several agricultural studies have been conducted with ^{15}N-depleted (^{14}N-enriched) materials. Use of either type of tracer material requires that its N-isotopic composition at the time of sampling be significantly different from that of other N in the system under study. Adding ^{14}N- or ^{15}N-enriched material to a system results in an increase in the total respective ^{14}N or ^{15}N content of the system. When the tracer N mixes with other N having the same form as the tracer, both the tracer and nontracer N lose their isotopic identities (an exception is N_2 formed during denitrification [Hauck, 1982]) and the mixture will have a ratio of ^{14}N/^{15}N which is different from that of its components. The change in N-isotope ratio permits the investigator to follow the course of the tracer within the system. The extent of change in isotope ratio is used to calculate the extent to which the tracer has become part of the system. Measuring the total amount of tracer in the system may permit in some studies accurate determination of N loss. Generally, addition of a tracer-labeled material to a system (e.g., soil) enables one to determine the amount of N from the labeled material taken up by plants, the distribution of that N in plant parts or soil physical and chemical fractions, and the relative contributions of the added and resident N to N components moving from soil into air or waters.

Most N-tracer studies in agriculture involve some variation of the above technique of isotope-dilution analysis. Several important isotope dilution expressions and their application were discussed by Hauck and Bremner (1976). Although the technique is simple in concept, problems are encountered in correctly interpreting the results obtained. Such problems often result from questionable assumptions made when calculating quantities involved in gross N transformations. Hauck and Bremner (1976) listed three fundamental assumptions central to the use of N isotopes as tracers in biological systems and 12 assumptions specifically relevant to soil N-transformation studies. Mulvaney (1991) discussed the significance of three common assumptions made in N-isotope pool dilution studies: (i) tracer N is distributed uniformly throughout the system or portion of system under study; (ii) the processes under study occur at constant rates; and (iii) N leaving the labeled N pool does not return. Other selected references from the extensive literature in which N tracer data interpretation is discussed include papers by Kirkham and Bartholomew (1954, 1955), Jansson (1958, 1966), Hauck (1973, 1978), Fried (1978), Koike and Hattori (1978), Blackburn (1979), Guiraud (1984), Shen et al. (1984), Jenkinson et al. (1985), Hart et al. (1986), Myrold and Tiedje (1986), Barraclough and Smith (1987), Guiraud et al. (1989), and chapter 42 in this book.

Sample variability usually is the greatest single source of methodological error in N-tracer research. Analytical error (i.e., the determination of N-isotope ratio) generally is a negligible fraction of the cumulative error involved in sample collection, processing, and chemical analysis. Reference to most of the ^{15}N literature in which procedural errors are discussed can be found in reviews by Hauck and Bremner (1964, 1976), Bremner (1965a), Bremner et al. (1966), Martin and Ross (1968), Fiedler and Proksch (1975), Hauck (1982), Fiedler (1984), and Mulvaney (1993).

Until recently, methods for N-isotope-ratio analysis were developed largely by Rittenberg and Sprinson, and most of the analytical procedures currently used in many laboratories are modifications of methods described in their papers (Rittenberg et al., 1939; Rittenberg, 1948; Sprinson & Rittenberg, 1948, 1949). Subsequent articles providing additional details of ^{15}N methodology in agricultural studies include those by Burris and Wilson (1957), San Pietro (1957), Capindale and Tomlin (1957), Junk and Svec (1958), Smith et al. (1963), Cho and Haunold (1966), Martin and Ross (1968), Fiedler and Proksch (1975), Hauck and Bremner (1976), Edwards (1978), Bergersen (1980), Buresh et al. (1982), Chen et al. (1991), Mulvaney (1993), and in the bibliography compiled by Hauck and Bystrom (1970).

Since the 1970s, improved instrumentation for sample preparation and N-isotope-ratio analysis has permitted both automation of analytical procedures and the routine determination of isotopic composition of microgram quantities of N. Details of these recent developments are reviewed by Mulvaney (1993).

The methods outlined here for N-isotope-ratio analysis of samples obtained from studies using N tracers are similar to but far less comprehensive than those given in the treatments by Bremner (1965a) and Hauck (1982) in the first and second editions of this book. Knowledge of procedural detail as well as new developments is imperative for successful use of N tracers by investigators who perform all phases of their research. However, researchers who make use of analytical services now available may not require detailed knowledge of the analytical procedures used. The level of detail given here may be sufficient to heighten awareness of procedural problems and determine the extent to which analytical precision and accuracy are affected by all phases of the research operation. Accordingly, in this chapter less emphasis is placed on the determination of N-isotope ratio and more on the overall conduct of N-tracer research, including the preparation of tracer materials, calculation of tracer needs, and collection and preparation of sample, depending on kind of study and objectives.

40–1 PREPARING ^{15}N-LABELED MATERIALS

40–1.1 Principles

Most N-tracer materials are derived from NH_4^+ or NO_3^- enriched in ^{15}N (i.e., containing an unnaturally high concentration of ^{15}N). Materials with an unnaturally low ^{15}N concentration (i.e., ^{14}N-enriched or ^{15}N-depleted) also can be prepared. The commercial preparation of the stable N isotopes resulting in the concentration of either depends on their differences in mass and, therefore, their physical properties; ^{14}N and ^{15}N in the form of NH_3, NH_4^+, or N oxides behave differently in exchange columns or distillation columns. Until the 1960s, ^{15}N was concentrated mainly by counter-current exchange of NH_3 with NH_4^+, the ^{15}N atoms being extracted from the NH_3 vapor and concentrated in the liquid phase. Currently, ^{15}N is concentrated mainly by exchange of NO with HNO_3 or by cryogenic distillation of NO, the labeled product being collected as HNO_3 which may be subsequently reduced to NH_3. A variety of tracer materials is prepared from the labeled NH_3 or NO_3^- through chemical synthesis or biosynthesis.

The term atom % ^{15}N or at.% ^{15}N commonly is used to denote ^{15}N concentration:

$$\text{at.\% } ^{15}N = \frac{\text{no. of } ^{15}N \text{ atoms}}{\text{no. of } (^{14}N + ^{15}N) \text{ atoms}} \times 100 \qquad [1]$$

Reference is made in the following discussion to atom % ^{15}N and atom % ^{15}N excess; either term can be used for expressing ^{15}N concentration. Atom % ^{15}N excess refers to the measured atom % concentration of ^{15}N minus the natural background concentration. Users of this term assume a constant value for natural ^{15}N abundance, which is incorrect, or they determine this value for the natural material that is being used as a standard, which may not be the reference material used by others. A recommended approach is to relate all N-isotope ratios to the accepted standard value of 0.3663 at.% ^{15}N for atmospheric N_2 (Junk & Svec, 1958) and not use the term atom % ^{15}N excess (Hauck, 1982).

The term delta ^{15}N ($\delta^{15}N$) commonly is used when expressing ^{15}N concentration in the range of natural abundance; it is the ^{15}N concentration expressed as parts per thousand differences from the $^{15}N/^{14}N$ ratio in a standard, usually atmospheric N_2.

$$\delta^{15}N = \frac{(^{15}N/^{14}N)x - (^{15}N/^{14}N)\text{atm}}{(^{15}N/^{14}N)\text{atm}} \times 1000 \qquad [2]$$

where $(^{15}N/^{14}N)x$ and $(^{15}N/^{14}N)$atm are the N-isotope ratios of the sample and atmospheric N_2, respectively. The terms per mil difference and per mil enrichment also have been used. Sometimes $\delta^{15}N$ values are based on the expression

$$\delta^{15}N = \frac{(\text{at. \% } ^{15}N)x - (\text{at. \% } ^{15}N)\text{atm}}{(\text{at. \% } ^{15}N)\text{atm}} \times 1000 \qquad [3]$$

Values of $\delta^{15}N$ calculated from the expression based on the use of isotope ratios (Eq. [2]) are slightly lower than those based on the use of ^{15}N concentrations (Eq. [3]), but the differences due to method of calculation are well within the limits of analytical error and can be considered negligible.

The level of ^{15}N (or ^{14}N) enrichment needed in the labeled material is determined by the amount of isotope dilution that occurs during the experiment, and this, in turn, is determined by several factors, including the nature, objectives, and duration of the study. Usually, ^{15}N is purchased in a concentration higher than needed for most studies even when it is in the form to be used. When preparing relatively large amounts of labeled material of low enrichments (e.g., tens to hundreds of grams in the range of 1 to 5 at. % ^{15}N), highly enriched material is intimately mixed with an amount of the same material of natural N-isotopic composition such that the mixture has the desired N-isotopic composition. This procedure is especially more convenient and less time-consuming when the purchased ^{15}N is in a chemical form other than needed and must be chemically converted. For example, highly enriched NH_3 is used to synthesize a small amount of highly enriched urea that is then diluted to the needed isotopic composition and amount with unlabeled urea. To ensure homogeneity, the separate components differing in N-isotopic composition are dissolved together and recrystalized from solution. Depending on experimental objectives, the labeled material can be used as fine crystals, redissolved and used in liquid form, or compressed into particles within a specified size range through granulation or pelleting, followed, if necessary, by crushing and screening.

40–1.2 Determining Nitrogen-15 Concentration Needed

The general formula for assaying a compound in a mixture of compounds using isotope dilution calculations is:

$$X2 = [(C1/C2) - 1] \cdot X1 \cdot (M2/M1) \qquad [4]$$

where

$X1$ = the weight of the tracer compound added,
$X2$ = the weight of the unknown (unlabeled) compound,
$C2$ = the isotopic concentration (expressed as at. % excess) of the compound recovered from the mixture,
$C1$ = the isotopic concentration of the original tracer compound.

The term $M2/M1$ corrects for the change in molecular weight of the compound as its isotopic composition changes upon dilution. This calculation is unnecessary when amounts are expressed as equivalents, and, where tracer concentrations are low, the correction is negligible.

The experimental accuracy of agronomic studies may be affected by the level of ^{15}N enrichment in the applied N source. For example, where N tracer is used to determine plant uptake of applied N or the amount of applied N remaining in soil, standard errors often increase with decrease in ^{15}N concentration in the applied N. This can readily be seen from Eq. [5] for calculating the percent recovery of applied N by crop plants.

$$\% \text{ N recovered} = \frac{100 \ P(c-b)}{f(a-b)} \qquad [5]$$

where

P = total plant N (e.g., in milliequivalents),
f = fertilizer N applied (e.g., in milliequivalents), and
a, b, and c = the at. % ^{15}N concentrations of the fertilizer, soil, and plants, respectively.

When ^{15}N-enriched materials are used, the value for a always is larger than for c (plants cannot take up more fertilizer N than applied). Though not always possible, when a, b, and c are determined with the same level of precision, the absolute error is constant but the relative error in determining these values increases as the values for a and c approach b, i.e., as $(c-b)$ approaches zero. Mathematical expressions for estimating error in N uptake studies clearly show a marked increase in error as the values for a and c decrease, i.e., as the ^{15}N enrichment of the fertilizer decreases. For ^{15}N-depleted fertilizer, error increases with increase in ^{15}N enrichment, i.e., as the ^{15}N concentration in the fertilizer approaches the level of natural abundance. The significance of this error in relation to other errors can be estimated from Eq. [6] (adapted from Hauck, 1978), which is for propagation of errors for assessing the relative errors in calculating the mass of ^{15}N:

$$\frac{\Delta LN}{LN} = \sqrt{\left[\frac{\Delta DM}{DM}\right]^2 \left[\frac{\Delta TN}{TN}\right]^2 \left[\frac{\Delta C}{(c-b)}\right]^2 \left[\frac{\Delta a}{(a-b)}\right]^2 \left[\frac{(c-a)\ (\Delta b)}{(c-b)\ (a-b)}\right]^2} \qquad [6]$$

where

LN = plant labeled N (mass ^{15}N/ha),
DM = plant dry matter (mass/ha),
TN = total N concentration (g N/kg),
a, b, c = at. % ^{15}N concentrations in the ^{15}N source, control plants, ^{15}N-treated plants, respectively, and
Δ = the uncertainty in the respective parameter.

This equation shows that the relative uncertainty in the mass of labeled N will be controlled by the component with the largest relative uncertainty (largest coefficient of variation, CV). In the majority of isotope experiments enough labeled N will be added to give an enrichment/depletion of at least 0.2% in the final material of interest; therefore, the terms in Eq. [6] involving $(c-b)$ and $(a-b)$ will be very small. However, these terms can become large in samples with small enrichments, such as from studies tracing N through variation in natural abundance, or studies of labeled N entering a large and slowly reactive pool (e.g., soil organic matter). For most agronomic studies, typical CVs for plant dry matter would be 6 to

15%, for total N concentration, 1 to 3%, and for ^{15}N enrichment, 0.1 to 0.3% when ^{15}N enrichment is 0.2 at. % excess or higher. A general ranking of relative variations would be: field plots > soil/plant samples > subsamples > total N determinations > ^{15}N determinations (Bartholomew, 1964; Hauck, 1978; Broadbent & Carlton, 1980; Pruden et al., 1985; Saffigna, 1988).

Materials with a low level of ^{15}N enrichment (about 1 at. % or less) and ^{15}N-depleted materials are useful for conducting single-season N uptake studies and tracing the movement of NO_3^- derived from the labeled material. In many studies, a low-level tracer can be used for determining the amount of added material remaining in the system and, occasionally with soils of low organic matter content, for determining the amount of N applied one season taken up by plants in the succeeding season.

Some typical calculations will illustrate the basis for determining the needed level of ^{15}N enrichment. Assume that soil at the experimental site contains 0.2% N (about 4500 kg of N/ha), to which is added ^{15}N-enriched fertilizer (1.0000 at. %) at a rate equivalent to 150 kg N/ha. Assume that 25% of the applied N remains in soil after harvest of the first crop. Question: Can the applied N remaining in soil be detected with a satisfactory level of precision?

A simplified Eq. [4] is used but in the following calculation the term $X1$ is the amount of tracer N added, $X2$ is the amount of (unlabeled) soil N, $C2$ is the isotopic concentration of N in the soil sample taken after harvest (expressed as at. % excess), and $C1$ is the isotopic concentration (at. % excess) of the added tracer N. The term $M2/M1$ that corrects for the change in molecular weight of the tracer N is not used. Assume soil to have a natural ^{15}N abundance of 0.3663 at. % and analytical precision to be ± 0.002 at. % ^{15}N.

$$X2 = [(C1/C2) - 1] \cdot X1$$

$$4500 = \left[\frac{(1.0000 - 0.3663)}{C2} - 1 \right] \cdot 150 \cdot 0.25$$

$$121 \; C2 = 0.6337$$

Solving for $C2$, the at. % ^{15}N excess (^{15}N concentration in excess of natural abundance) of the total soil N is found to be 0.0052, which is barely within the limits of detection. In the above example, use of a fertilizer containing 2.0000 at. % ^{15}N results in an average ^{15}N concentration in total soil N of 0.3795 or 0.0132 at. % excess, which is well within analytical precision and sufficiently different from natural ^{15}N abundance to decrease some of the interpretive errors associated with spatial variability among soil samples. Use of ^{15}N-depleted fertilizer (e.g., containing 0.009 at. % ^{15}N) results in a value for $C2$ that is only 0.003 at. % ^{15}N different from natural abundance; therefore, this tracer material would not be suitable for studying N remaining in soil under the experimental conditions cited.

Assume as before that 150 kg of N/ha is added to soil containing 4500 kg of total N/ha, and that 25% of the added N remains immobilized in the organic matter (mostly biomass) after harvest. Assume that 75% of the immobilized N is in the labile (biodegradable) fraction and has an isotopic composition identical to that of the initially added N, and that the remaining 25% of the immobilized N is in refractory, unavailable forms and need not be considered in the calculations. Further, assume that in the second growing season 10% of the immobilized labeled N in the biomass (labile organic matter) and 2% of the total soil N originally present are mineralized and that plants take up 50% of the N made available from each source (one labeled, the other unlabeled).

Question: What is the minimum ^{15}N concentration of the N added the first season such that plants grown during the second season contain ^{15}N at a concentration no lower than 0.4663 at. %?

The general isotope dilution formula, Eq. [4], is the basis of Eq. [7] used to answer the above question.

$$X = \frac{(TN)(c-b)}{a} \qquad [7]$$

where X is the amount of labeled N in the plant, TN is the total amount of N in the plant (from all sources), a is the at. % excess in the N originally added, and b and c are the ^{15}N concentrations expressed as at. % in the original soil and plants, respectively.

Labeled N originally added, 150 kg/ha, ^{15}N concentration, a, is unknown
Soil N, 4500 kg/ha, at. % ^{15}N = 0.3663
Labeled N immobilized, first season, $150 \cdot (0.25) = 37.5$ kg/ha
Labeled N in biomass, $37.5 \cdot (0.75) = 28.13$ kg/ha
Original soil N mineralized, second season, $4500 \cdot (0.02) = 90$ kg/ha
Labeled biomass N mineralized, second season, $28.13 \cdot (0.1) = 2.813$ kg/ha
Total N in plants, second season, $(90 + 2.813) \cdot (0.5) = 46.4065$ kg/ha
Average ^{15}N concentration in plants, 0.4663 at. %.

$$a = \frac{46.4065(0.4663 - 0.3663)}{2.813\,(0.5)} = 3.2994$$

Minimum ^{15}N concentration in added N required:

$$(3.2994 + 0.3663) = 3.6657 \text{ at. \%}$$

Use of somewhat higher concentrations of ^{15}N would be advisable to compensate for differences in the estimated vs. actual dilution occurring during the study.

Expressions other than Eq. [7] can be used to calculate the needed amount of ^{15}N. Many modifications of Eq. [4] for calculating isotope dilution are found in the literature and several are summarized by Hauck and

Bremner (1976). The assumptions that were made in the above example are based on typical values for immobilization, mineralization, and plant uptake as determined from N-tracer studies. Such assumptions have been found useful in anticipating the extent of isotope dilution to estimate ^{15}N needs, depending on study objectives and experimental circumstances. The above calculations do not consider the almost negligible fraction of N released from refractory soil constituents, return of crop residue to the available N pool, between-season and second-year N inputs, or corrections for the change of atom mass of the labeled N. The general isotope dilution formula can be modified as needed to make crude estimates including such factors. More precise estimates of N mineralized from crop residues differing in C/N ratio can be made using the equations given by Vigil and Kissel (1991).

40–1.3 Fertilizers

40–1.3.1 Enrichment Level and Form

Nitrogen tracer studies in agriculture and related fields usually are conducted using labeled materials in the chemical forms available at time of purchase or in forms readily synthesized from the purchased stock. Except for ^{15}N-depleted materials, most often ^{15}N is purchased at an enrichment level higher than needed for a study, followed by dilution to the needed isotopic composition. The relatively high cost of ^{15}N has been a main reason why most agronomic N-tracer studies have been conducted with small grains (high population density) in greenhouse pots or micro-size field plots. Field experiments with maize (*Zea mays* L.) requiring larger plots (e.g., 6×10 m) commonly are conducted with lower-cost ^{15}N-depleted materials or materials of low enrichment (< 1 at. % ^{15}N). Often, the labeled N source is applied as a dilute solution to facilitate uniform application (see 40–2.3.1). However, spurious results can be obtained when the physical form of the fertilizer affects its immediate chemical and biochemical reactions in soil microsites, and these reactions, in turn, significantly affect subsequent transformations, movement, and plant uptake of N (Hauck, 1984). Unpublished studies with ^{15}N-labeled soil or fertilizers have shown significant differences in the behavior in soil between fertilizer particles vs. their dilute solution, including rate of nitrification and effect on urease activity, ammonia volatilization, and solubilization followed by mineralization of soil organic matter, as affected by fertilizer type, rate, and disposition in soil. However, these results were obtained in laboratory and greenhouse studies and the practical significance of differences in effects of solids vs. dilute solutions has not been established in the field. Little attention has been paid to the potential importance of microsite reactions as affected by the physical characteristics of the N source, neither in field studies using ^{15}N nor in nontracer studies. Because published data on this subject appear not to be available, one can only speculate on its significance. In all studies under the program described by Hauck and

Kilmer (1975), granulated ^{15}N-enriched or ^{15}N-depleted fertilizers were used to obviate unknown differences in fertilizer behavior resulting from differences in physical form of the labeled and unlabeled fertilizers used. Whenever feasible, we suggest using a physical form of a labeled material that closely matches that of its commercial equivalent used by farmers.

40–1.3.2 Conventional

Conventional fertilizers as used here means water-soluble, fast-acting, usually inorganic materials used in crop production, such as $(NH_4)_2SO_4$, NH_4NO_3, $(NH_4)_2HPO_4$, and urea.

40–1.3.2.1. Dilution to Needed ^{15}N Concentration. The general isotope dilution formula, Eq. [4], can be used to determine the amounts of unlabeled and labeled materials that are mixed to obtain the desired ^{15}N concentration. Assume that 100 g of urea with a ^{15}N concentration of 2.0000 at. % is to be made from urea containing 10.3663 at. % ^{15}N.

Equation [4] can be rewritten as follows:

$$Z = \frac{(X1 \cdot C1) + (X2 \cdot C2)}{100} \qquad [8]$$

where

$X1$ and $X2$ = the weights of labeled and unlabeled N
$C1$ and $C2$ = the ^{15}N concentrations of $X1$ and $X2$, respectively, expressed as at. %
Z = the at.% ^{15}N concentration of the mixture

For the example given, $(X1 + X2) = 100$.

$$2.0000 = \frac{X1(10.3663) + (100 - X1)(0.3663)}{100}$$

$X1 = 16.337$ (g N as labeled urea)
$X2 = 83.663$ (g N as unlabeled urea)

Correcting for the differences in mass, the corresponding values are 16.326 and 83.674. Such correction usually need not be applied unless an exact ^{15}N concentration is needed. In any case, the exact ^{15}N concentration of the final mixture should be determined through isotope-ratio analysis.

40–1.3.2.2 Product Preparation. The labeled source material and its unlabeled diluent are dissolved together to ensure uniform mixing. Dry mixing is not recommended. Following dissolution (usually in water) the diluted source material is crystallized through evaporation, and then redissolved in, or washed with methyl alcohol, recrystallized, and dried. Use of methyl alcohol is recommended because it is usually more chemically pure than ethyl alcohol.

The dry, labeled N source can be used as fine crystals, redissolved for sprinkle application, or made into particles. Granulation of materials such as $(NH_4)_2SO_4$ requires experience and apparatus not usually found in the research laboratory. However, labeled source material can be compacted into hard cylinders or cakes in a Carver press at about 10 000 psi. The cakes are then crushed and screened to the required particle size. Moistening with water sometimes is required for adequate compaction.

40–1.3.3 Slow-Release

Many chemicals have been evaluated as slow-release N fertilizers (e.g., see Hauck, 1985) but the ones that have been studied the most extensively are the ureaforms, isobutylidene diurea, oxamide, and sulfur-coated urea. Only a few of the studies have been made with ^{15}N-labeled materials, e.g., with ureaform (Brown & Volk, 1966), oxamide (Westerman et al., 1972), and oxamide and isobutylidene diurea (Rubio & Hauck, 1986). The synthesis of these materials labeled with ^{15}N is described by Hauck (1994). Preparation of relatively small (kilogram) amounts of S-coated urea for use in field studies is not recommended because the laboratory-scale preparation most probably will result in a product with significantly different dissolution rates than commercial products. Differences between commercially and laboratory synthesized products are less apparent for products such as oxamide whose slow-release characteristics are a function of their molecular structure rather than a physical coating process.

40–1.3.4 Organic Residues

40–1.3.4.1 Plant Tissues. The value of ^{15}N-labeled plant tissues as tracers should not be overlooked. Large quantities of grain and stover are produced as valuable by-products of agronomic studies using ^{15}N conducted over a wide range of fertilizer practices and crop management systems. These plant parts can be used in many types of studies (e.g., determining the distribution of applied N in different plant parts and in different N fractions) as affected by N application rate and management. The materials should be available to researchers unable to conduct their own field experiments but who would make measurements other than those made by the original investigator(s) having different research objectives. Or, plant tissues with a sufficient level of ^{15}N enrichment can be used elsewhere from the originating site to study N release and movement from different crop residues. Further details on the use of ^{15}N-labeled plant residues for field studies of mineralization/immobilization are given in chapters 39 and 42.

40–1.3.4.2 Animal Wastes. Large amounts of ^{15}N-labeled animal manure suitable for field studies of manure-N movement have been produced as by-products of cattle-feeding experiments (Faust et al., 1963). Currently, at least three groups of investigators have fed ^{15}N-enriched grain to cattle or poultry specifically to produce ^{15}N-labeled animal manures for studies in soil columns or field. How the excreta are handled, processed, and stored depends on study objectives. Important considerations include preventing NH_3 loss from NH_4^+ salts and hydrolyzed urea and protein derivatives, accurately determining the N-isotopic composition of different manure components, and separating undigested material such as plant grain and fiber from the manure components of interest. Kirchmann (1985), for example, found that the N transformations of ^{15}N-labeled poultry manure were markedly affected by manure pretreatment. Some investigators have added ^{15}N-labeled urea to urine to study NH_3 evolution from simulated urine patches. Although this approach has merit for the purpose intended, it cannot be used to study the reactions and movement of different N forms in manure. Uniform methodologies for producing, handling, and studying ^{15}N-labeled animal wastes have not been established. Suggestions for preparing and characterizing labeled manures can be found in Rauhe and Bornak (1970), Rauhe et al. (1973), and Kirchmann (1989, 1990).

40–1.3.5 Soil

Eight methods have been used to accurately determine the amount of atmospheric N_2 fixed by plants (Hauck, 1979). The most common N-tracer techniques for measuring N_2 fixation in the field involve isotope dilution. The main problem associated with all isotope dilution approaches for this purpose is that their accuracy depends on how well the nonfixing control plants simulate the N uptake patterns of the N-fixing plants under study (for discussion, see Rennie, 1986; Weaver, 1986; Phillips et al., 1986; Vose & Victoria, 1986). The use of labeled soil for field measurement of biological N_2 fixation probably requires the fewest assumptions that cannot be experimentally validated. Legg and Sloger (1975) labeled soil organic matter with ^{15}N by stimulating rapid turn-over of applied ^{15}N-enriched NH_4^+ and glucose. Using the remineralized amount of initially immobilized ^{15}N as an index of mineralizeable soil organic N, they estimated from isotope dilution calculations the amount of total N in soybeans [*Glycine max*(L.)] that was fixed from the atmosphere. The calculations assumed uniform labeling of the soil organic matter fractions or the fractions that release plant-available N during crop growth. This is not a valid assumption, but the release of ^{15}N from soil organic matter accurately reflects mineralizeable N when the ^{15}N, initially immobilized in the biomass, has equilibrated with other mineralizeable soil organic matter fractions. Such equilibration can be determined by following the ^{15}N through the main chemical fractions (e.g., hydrolyzable distillable, nonhydrolyzable nondistillable, etc.) until no further change in ^{15}N distribution is observed.

Three or more years may be required to prepare ^{15}N-labeled soil that can be used to accurately study the release of mineralizeable N. Once prepared and characterized, the labeled soil is a valuable tool that can be used to study, for example, N release as affected by a particular treatment or the relative uptake of labeled soil N and unlabeled applied N, as affected by plant species.

40-2 FIELD STUDY TECHNIQUES

40-2.1 Principles

Agricultural scientists study N cycle processes to: (i) improve production, (ii) control N losses to the environment, and (iii) refine and expand understanding of these processes. They have used two fundamentally different approaches to these objectives (i.e., through use of either labeled or nonlabeled N sources). Nonlabeled N studies focus on the total N inputs, total N outputs, and the net change in various soil N pools (organic and inorganic). Studies using labeled N focus on the *interaction* of the labeled N within the soil-plant system by tracing the course of the ^{15}N atoms throughout the system. Consider the addition of ^{15}N-labeled ammoniacal fertilizer to a soil where 10% of the ^{15}N is exchanged with native soil N from the soil biomass during the cycles of mineralization-immobilization. If only simple isotopic interchange between fertilizer N and biomass N occurs (i.e., a one-for-one substitution of ^{15}N with ^{14}N atoms), the data for total N would not be affected by the substitution because total N available for plant uptake, leaching, or involvement in other processes would not be changed. However, such simple interchange does not occur. As revealed by the tracer technique, a significant sequestering of fertilizer N into a slowly available form occurs with concomitant decreases in ^{15}N involvement in other processes (e.g., plant uptake). Thus, different information is obtained about the soil N cycle through the use of ^{15}N-labeled vs. nonlabeled fertilizer.

The above differences in labeled and nonlabeled results demonstrate the need to carefully define the research objectives of a study and determine how the use of N tracers will contribute toward meeting these objectives before beginning expensive field-scale tracer experiments. Factors to consider concerning the use of labeled vs. nonlabeled N include the added costs of tracer studies and the availability of trained personnel and analytical facilities, in addition to the research objectives. In selecting either, it is important to understand that the two approaches are not equivalent.

Basic problems encountered with use of labeled N in field studies result from (i) nonuniform distributions of N within the field and (ii) natural and experimentally induced variation in composition of subsamples. The variation from nonuniform distribution of N may be spatial (differences in horizontal and vertical distribution), temporal (differences in the rates of change among different N forms over time), and process level (as affected by qualitative and quantitative differences in N cycle processes within the area under study). Spatial variability arises from nonuniform application of labeled N and the natural heterogeneity in the occurrence and extent of N-cycle processes. For example, soil denitrification has been shown to be lognormally distributed (chapter 2) by several investigators (Folorunso & Rolston, 1984; Duxbury & McConnaughey, 1986; Parkin et al., 1987). One detailed study isolated a single microsite that accounted for 85% of the denitrifying activity associated with particulate organic matter which was < 0.1% of the sample weight (Parkin, 1987). Other investigators have reported water flux and hydraulic conductivity to be lognormally distributed (Warrick et al., 1977; Jones & Wagenet, 1984) and soil NO_3^--N content to be log-normal (Tabor et al., 1985; White et al., 1987; Cameron & Wild, 1984). Not surprisingly, therefore, the distribution of labeled N is highly heterogeneous, even for uniformly applied N tracer materials (Broadbent & Carlton, 1980; Selles et al., 1986).

Temporal variability is an inherent result of N transformations occurring with different constituents at different rates, times, and places in soil. This happens, for example, when particulate organic matter becomes a denitrification microsite while a nearby area contains an active root taking up ^{15}N. These lognormally distributed N transformations change over time; the particulate organic matter site may become a zone where N is immobilized and the root uptake site may become a decomposing root channel for macropore transport of ^{15}N out of the root zone. As a result of such events, soil samples collected at the end of a study typically contain quite different ^{15}N contents in plant roots, the soil NO_3-N pool, the soil total N pool, and other N components of the soil system. For example, Saffigna (1988) reported ^{15}N concentrations for 100 d-old wheat roots of 6.01 at. % while the ^{15}N concentrations in the soil NO_3^- and total N pools were 1.62 and 0.49 at. %, respectively. This nonuniform distribution of labeled N among different components of the system under study permits one to calculate the net extent to which the labeled N has participated in a particular N transformation or movement process. However, large spatial variations in the N and N-isotopic composition within a particular component resulting from the heterogeneous dynamic features of soil N processes can result in serious sampling errors and errors of data interpretation that can be minimized only through careful sampling. Unquestionably, the need for collecting representative soil and plant samples is greater when conducting field studies with labeled N than with nonlabeled N.

40–2.2 Managing Field Variability

Problems associated with field variability can be reduced by changing type and size of plot and borders, and improving plot sampling and sample preparation techniques.

No set of "best-experimental practices" can be recommended for all sites because the best practices may be determined largely by the dominant N cycle processes occurring at the site. For example, if N transport through surface run-off and erosion is a major pathway, then using confined microplots which alter infiltration and surface run-off should be avoided.

Clear definition of research objectives is imperative before experimental techniques are decided upon. Often, objectives may conflict, requiring that objectives be ranked in order of priority. The impact of compromise should be considered (e.g., cost may determine plot size, number of replications, level of ^{15}N enrichment, labor supply, and duration of study, among other considerations). The better the matching of techniques with objectives, the greater will be the accuracy and completeness of the information that can be obtained through use of N tracer methodology.

40–2.3 Application Techniques

Which application technique to use will depend on the objectives of the study, physical form (and for solids, the particle size) of the N source, and availability of a suitable application device (e.g., for applying anhydrous or aqua ammonia or animal manure slurries). Other considerations include manner of placement (e.g., incorporated band vs. surface broadcast), time of application, and presence or absence of vegetation. Obviously, no standard application technique can be recommended but the investigator may be guided by the suggestion and comments that follow.

40–2.3.1 Most Common Technique

A goal common to most field studies with labeled N is to apply the tracer material in a manner that best represents production practices with the understanding that it is applied as uniformly as possible to minimize variability among replicate samples. Because it is difficult to scale down farm operations to experimental plot dimensions, investigators have paid less attention to duplicating production application techniques than to striving for uniform application, the latter being especially important in studies of residual N or N balance.

Most investigators have found that applying labeled fertilizers in solution is the most satisfactory way because solutions are easy to handle under field conditions and can be dispensed in several ways from surface sprays to banded injections. Solutions also allow use of the same liquid without further processing that was prepared when ^{15}N-enriched material was diluted uniformly to the desired isotopic composition.

Critical aspects for solution applicators are: (i) accurate calibration of equipment before use (so that appropriate concentrations of labeled solution can be prepared and the correct amounts applied), (ii) preselection of spray nozzles for uniformity, and (iii) frequent monitoring of application rates during field operation.

40–2.3.2 Comments

Water-soluble ^{15}N-labeled N sources most often are applied in solution by injection or spraying. Jokela and Randall (1987) used a manual refilling syringe to inject a labeled solution 8-cm deep on a 19-cm grid over a 3.5-m^2 microplot, which simulated a broadcast incorporated application, requiring an application time of about 10 min per plot. A microplot point injector for liquid fertilizer placement in high residue systems was described by Benjamin et al. (1988). Broadbent and Carlton (1980) used a positive displacement pump system, which delivered precise volumes of N solutions to an injection shank that was capable of applying ^{15}N to a field plot with a CV of < 1%. Woodcock et al. (1982) gave a detailed description of a peristaltic pump system mounted on a portable frame for 4-m² wheat plots. Small plot constant-pressure spray equipment, commonly used for herbicide research, can also be used for broadcast surface applications.

Applying labeled fertilizer as a dilute solution may result in anomalous results under conditions where the chemistry and biochemistry occurring within the fertilizer-soil microsite soon after application affect the subsequent reactions and movements of the applied tracer N. For example, in acid soils urea-N nitrifies at a faster rate than an equivalent amount of N from $(NH_4)_2SO_4$ when applied as solid particles but not as dilute solution; alkaline-hydrolyzing fertilizer particles solubilize more organic matter at the fertilizer-soil microsite than acid-hydrolyzing ones (Hauck, 1982). As indicated earlier (section 40–1.3.1), the practical significance of such differences in the behavior of solid fertilizers as compared with dilute solutions of these chemicals has not been established, but such differences have consistently been observed in laboratory, greenhouse, and unpublished field studies and merit consideration when planning studies using tracer N.

Solid N tracer materials have been successfully broadcast on field plots by hand-spreading in most studies with ^{15}N-depleted $(NH_4)_2SO_4$ and NH_4NO_3 (Hauck & Kilmer, 1975) and slow-release fertilizers (Brown & Volk, 1966; Westerman et al., 1972), and under minimal tillage (Meisinger et al., 1985), among others. For successful application, the fertilizer should be in good physical condition [preferably −6 + 16 mesh (1.23–0.58 mm)] particles, and for large plots, the area should be traversed several times during application. Another technique is to subdivide the plot into small areas and apply separately weighed tracer material to each small area (Olson, 1980). Hand broadcast applications offer the advantages of savings in time and money (if plot numbers are small) and the ability to apply ^{15}N under suboptimal conditions or in remote areas where special equipment or apparatus may not be practical. Disadvantages include the extra cost, time, or effort needed to prepare solid materials of the desired N-isotopic composition and physical form, the extra care needed to apply the solid material uniformly, and, sometimes, a greater labor requirement. When the labeled material is in the form of fine crystals, application on windy days presents problems.

Labelled N in gaseous forms is not commonly used in field studies because of inherent handling difficulties. However, Sanchez and Blackmer (1987) described an apparatus for applying ^{15}N-labeled anhydrous ammonia (NH_3) to small field plots (4.6 m²) that uses a flexible stainless steel capillary tube to inject $^{15}NH_3$ from a cylinder held at constant temperature. This apparatus permits field research with ^{15}N-labeled NH_3 (the commercial N fertilizer used in largest amount in U.S. agriculture) and also allows study of volatile additives such as certain nitrification inhibitors. The apparatus gave CVs of 3% or less and a soil environment that was considered by the authors to be representative of that around conventional ammonia applicators (Sanchez & Blackmer, 1987). Vanden Heuvel and Harrold (1990) described an apparatus for dispensing $^{15}NH_3$ with an applicator knife pulled by an electric winch along a stationary track. The apparatus permits mechanical injection of $^{15}NH_3$ to field plots without need for a tractor, marks the injection band location, and dispenses NH_3 from the same cylinders used by Vanden Heuvel (1988) for preparing the labeled NH_3. A point injector for applying ^{15}N-labeled fluid fertilizers such as urea ammonium nitrate is described by Benjamin et al. (1988).

40–2.4 Plot Type and Size

Field research with labeled N has been conducted in plots ranging in size from < 1 to 200 m^2. The appropriate plot size will depend on the study's goals and objectives, duration of the study, and available funding and labor supply.

40–2.4.1 Confined Microplots

Confined microplots are small areas of soil (0.1–1 m^2) that are isolated from surrounding soil by metal or plastic barriers. The soil is generally kept undisturbed, although the surface plow layer may be mixed. The barriers are pressed into the soil or placed around an intact block of soil exposed by excavation. They are most frequently used for short-term ^{15}N balance work where the principle objective is to study a specific N-cycle pathway or to construct a ^{15}N budget. Simulating field-scale plant growth is usually a secondary objective. Advantages of confined microplots include: (i) precise definition of system boundaries that allows simplified soil sampling (e.g., complete removal of soil); (ii) restriction of lateral dispersion; (iii) prevention of surface run-off loss (important for ^{15}N crop residue studies); and (iv) lowering cost for ^{15}N materials. Disadvantages include: (i) possible alteration of soil drainage characteristics by confining barriers; (ii) restriction of plant root systems (depends on species); and (iii) possible alteration of surface run-off processes. Confined microplots were used frequently before 1970 (see Legg & Meisinger, 1982) with small stature crops such as small grains and forage grasses. Their use subsequently decreased, especially during the 1980s because of the desire to more closely approximate farm-scale crop growth in ^{15}N studies. Although the comparison of confined vs. unconfined microplots is rarely reported, Saffigna (1988) found similar ^{15}N recoveries with wheat in 0.5-m diam. cylinders vs. 1 m^2 unconfined microplots. He gave further details on the use of confined vs. unconfined microplots, including installation techniques and materials.

40–2.4.1.1 Suggestions. We suggest that confined microplots be used when the primary goal is to study N pathways (e.g., residue mineralization and other organic N transformations, N transport within the soil or ammonia loss) and when simulation of field crop growth is not essential. We suggest using small stature crops (when vegetative cover is needed), large microplots with shallow confinements, and soils selected to minimize possible compaction during installation.

40–2.4.2 Large Field Plots

Large plots are usually employed in long-term N tracer studies and when large-scale treatments are being studied such as tillage, irrigation, or crop rotation variables. Problems with lateral movement of ^{15}N can be minimized by using large plots (> 20 m^2) and conventional two-row borders. Researchers choosing this option usually employ ^{15}N-depleted materials to reduce isotope costs (Hauck & Kilmer, 1975; Bigeriego et al., 1979; Broadbent & Carlton, 1980; Hills et al., 1983; Kitur et al., 1984; Meisinger et al., 1985). Depleted ^{15}N materials are available in large quantities, require no dilution, and are somewhat cheaper than equivalent ^{15}N-enriched materials. They have a tracer value equivalent to a material containing 0.7 at. % ^{15}N. Depleted N sources can easily be traced into crop plants during the year of application. They have also been used to follow N into the soil NO_3^--N pool (Bigeriego et al., 1979; Broadbent & Carlton, 1980) and the soil organic N pool on soils with moderate to low organic N contents, i.e., soils with < 1.5 g N/kg (Broadbent & Carlton, 1980; Kitur et al., 1984). Kilmer et al. (1974) gave an outline of research topics for use of ^{15}N-depleted vs. ^{15}N-enriched fertilizers, with special reference to studies of N fertilizer use and water quality. Most studies with ^{15}N-depleted materials have been conducted with $(NH_4)_2SO_4$ as the carrier (available in largest amount) although some studies have used the more expensive ^{15}N-depleted NH_4NO_3.

40–2.4.2.1 Suggestions. Large Plot Techniques. Use of large field plots and ^{15}N-depleted materials is suggested for studies on soils with low total N contents (< 1 g N/kg) and where typical field crop growth is to be attained, as affected by large-scale variables (e.g., tillage) expressed over several years. Large plots would probably not be suitable on soils of high organic N content (> 2 g N/kg), for intensive studies of N transformations over short time intervals, or with studies where the ^{15}N concentration in the main N pool of interest probably would be diluted to a level < 0.1 at.% different from that of natural ^{15}N abundance (the background level).

40–2.4.3 Unconfined Microsubplots

A third plot type is comprised of unconfined, ^{15}N-treated microplots (1–10 m^2) placed within larger plots treated in the same manner but with unlabeled N. Either ^{15}N-enriched or ^{15}N-depleted materials are used on the microsubplots to estimate the percent of plant N derived from the added source. The larger surrounding field plot is used to estimate total N removals and dry matter production. Nitrogen transformation rates and pathways also can be obtained on subplots treated with materials enriched with ^{15}N (usually in the concentration range of 2–5 at. %).

Unconfined microsubplots offer the advantages of lowering isotope costs while maintaining the ability to detect ^{15}N from enriched sources in different soil N pools after considerable dilution. Use of the microsubplot technique is suggested for studies where single-season plant uptake of applied N is a primary goal, for N source, rate, and placement studies, with soils of high total (organic) N content (> 2 mg/kg), and when soil sampling and ^{15}N balance studies are of secondary interest. Labeled subplots have been used to a limited extent in long-term studies of residual N but, usually, the dispersion of ^{15}N during field preparation and planting will limit their usefulness for long-term studies, especially for studies where plot boundaries are difficult to maintain, (e.g., studies involving tillage, irrigation, or rotation variables).

40–2.4.3.1 Microsubplot Size. A common question with unconfined microplots is: what size microplot will produce valid ^{15}N data while avoiding costly investment in labeled N? Two factors influence the microplot area: (i) lateral plant root distribution and (ii) lateral movement of ^{15}N from the microplot via surface/subsurface flow in soil water or through inadvertent removal of crop residue (e.g., by wind). The optimum microplot size is determined, in part, by crop plant to be used, soil type, weather conditions, and ^{15}N placement. Several investigators have studied microplot size requirements for corn, which typically has a lateral root growth of 50 to 80 cm from the crown (Allmaras & Nelson, 1971; Follett et al., 1974). Olson (1980), from a study using ^{15}N surface-broadcast and then incorporated into a silt loam soil in Kansas, concluded that accurate data on ^{15}N uptake by corn (*Zea mays* L.) could be obtained by sampling the center row of a three-row plot (71 cm between rows) and leaving 71 cm of unsampled border at each end of the center row. He also suggested that if residual fertilizer N or N balance work was to be studied, the microplot size should be increased. In Minnesota, Jokela and Randall (1987) in a 2-year study found that reliable plant ^{15}N data could be obtained from the center row of a three-row plot (76 cm between rows) if at least 38 cm of border was left at the ends of the center row. This plot size was adequate for both soil types studied, a well-drained silt loam and a poorly drained, tiled clay loam. In Iowa, Sanchez et al. (1987) found 2 by 2 m microplots adequate for measuring 1st-yr crop recovery of spring-applied, banded N, but significant lateral movement of N that was surface-applied in the fall occurred; this movement was attributed to lateral flow during the winter when soils were saturated. The lateral movement increased in the 2nd and 3rd yr for all treatments, which was attributed to redistribution of labeled plant residues and to tillage. Lateral movement was greatest with nonincorporated treatments and with moldboard plow tillage. In Nigeria on three sandy loam soils cropped to corn, Stumpe et al. (1989) found that harvesting the two center rows of a four-row plot (75 cm row spacing) would provide valid ^{15}N uptake data if at least 50 cm of row bordered the microplot on each end. Further, soil samples taken from the center area were used to measure total ^{15}N balance (average total recoveries of 99% with a CV of 8% were obtained). In England, Powlson et al. (1986) evaluated 12-row microplots spaced 16.7 cm apart (2 m^2 plots) cropped to wheat on a silty clay loam. Finding uptake of nonlabeled N by plants in the two outer rows, they advised sampling the center six rows and leaving a 50-cm border on each end of these rows. Little or no contamination was observed in the third border row, but this row was not harvested as a precaution.

40–2.4.3.2 Suggestions. The above results clearly show that unconfined microplots can yield valid ^{15}N plant uptake data for the year of application, but border areas are essential. The size of the surrounding border will depend on crop and soil type, ^{15}N application method, and the tillage system under study. It is best to determine the optimum size for each individual experimental site. However, in the absence of detailed site-specific information, we would conservatively suggest that about one-fourth of the plot dimension should surround the harvested microplot, (i.e., the harvest area for a 2 by 2 m plot would be the center square meter).

40–2.5 Sample Collection and Preparation

Collection and preparation of representative soil, plant, and water samples are important parts of any field study using ^{15}N. As indicated previously (40–1.2) in the discussion relating to Eq. [5], the relative uncertainty of data for total ^{15}N uptake decreases as sample accuracy increases because the errors resulting from field heterogeneity are larger than those involving chemical and isotope-ratio analysis.

40–2.5.1 Plants

Field studies with ^{15}N have been conducted with a wide range of plants, including cereal, forage, oilseed, fiber, sugar, vegetable, and tree crops. Each crop will have its own unique plant sampling problems resulting from differences in size, differences in harvested product, and relative importance of quality factors. Discussion of specific sampling techniques for all of these crops would exceed the scope of this chapter. Rather, the reader is referred to the individual crop-specific literature for detailed examples of sampling techniques suitable for a given crop.

The general principles of plant sampling, drying, and grinding are discussed by Jones and Steyn (1973) and LeClerg et al. (1962). Plant sampling in ^{15}N studies usually involves sampling the total aboveground portion of the crop that is then divided into its constituent plant parts that are relatively homogeneous with respect to N content or which represent meaningful harvestable products. For example, corn is commonly divided into grain (about 1.5% N) and stover (about 0.8% N) to obtain more homogeneous analytical samples. This grouping also produces data meaningful for corn grain systems, where the stover is recycled, and for corn silage systems. Tree crops may be divided into many groups to isolate the labeled N into the most actively growing tissue and thus avoid a major dilution problem. Obtaining homogeneous samples is especially important where subsequent analytical methods use small (10–50 mg) samples. If it is not possible to divide the plant into homogeneous tissue, then it is important *not* to subsample the heterogeneous mixture until all of the sample has been dried and thoroughly ground to a small particle size (e.g., < 100 mesh). This minimizes physical segregation of plant parts and the risk of using a biased subsample.

The optimum sample size needed to estimate the dry matter production or N content of a field plot will also vary with crop and soil type and growing conditions. Sometimes only 5 to 10% of a plot may be needed to estimate a certain parameter, but in most cases, 10 to 60% of the plot is sampled. Gomez and Gomez (1984) and LeClerg et al. (1962) emphasized the need for estimating the variances among plots in an experiment, between samples within a field plot, and between analytical samples to design an efficient sampling system. In most cases, even with a small percentage of the plot sampled, the greatest variability will be among plots within an experiment (Gomez & Gomez, 1984). This variability can be managed by increasing the number of replicates. A discussion of the statistical aspects of optimizing sample numbers, replication numbers, site selection, and general field plot technique can be found in LeClerg et al. (1962) and in Gomez and Gomez (1984).

40–2.5.1.1 Recommendation. We recommend that plants be sampled by harvesting the total aboveground biomass, plus any harvested root/ tuber products, then subdivided into plant parts of similar N contents (homogeneous analytical sample) and which represent N flow paths (e.g., harvested products vs. recycled residues). The fresh samples should be immediately oven-dried, then weighed, finely ground, mixed, and subsampled. It is suggested that at an early stage of field research using ¹⁵N, the variance among field plots, among samples within plots, and among subsamples to be taken for chemical analysis be estimated to design an efficient sampling strategy that considers site-specific field variability, crop variability, and facilities at the experimental site for drying and otherwise handling plant samples.

40–2.5.1.2 Comments. During multiseason studies or studies where plants are sampled more than once, special care should be taken to ensure that the sampling or harvesting operations do not cause cross-contamination from plant material inadvertently carried from one plot to another. Such contamination can be particularly serious when harvesting and transporting across the field large amounts of material (e.g., corn stover) with a relatively high ¹⁵N enrichment.

40–2.5.2 Soils

Sampling soils for representative ^{15}N content presents a difficult challenge when the tracer, as is usual, is very heterogeneously distributed in soil. Several approaches to soil sampling have been taken by investigators using ^{15}N in field studies, to be discussed below. For general discussions of sampling soil, see Cline (1944) and Petersen and Calvin (1986).

40–2.5.2.1 Complete Excavation. Complete soil removal is undoubtedly the best method to obtain quantitative soil ^{15}N data and is used when total ^{15}N content in a microplot is to be determined with high accuracy. The early work of Carter et al. (1967), in a study where ^{15}N had been added to soil confined in steel cylinders 68 cm in diameter, clearly showed that complete excavation of soil was superior to taking soil cores. The amount of ^{15}N recovered from sets of seven 1.9 cm diam. cores (composited and thoroughly mixed before chemical analysis) ranged from 87 to 135% of the N applied (average, 113%); and from representative subsamples of the entire soil excavated from the cylinders, the recovery values ranged from 98 to 103% (average, 100%). However, the complete removal of soil is only practical with confined microplots < 1 m^2 because of the large volume of soil that must be processed (a plot 1 m^2, 30 cm deep will yield a volume of soil weighing about 500 kg). Complete excavation is usually done only to a depth of 15 cm (which usually contains most of the labeled N) or, for more complete recovery, to a 45-cm depth.

40–2.5.2.2 Subplot Excavation. For large unconfined microplots (1 m^2 or more) the soil is often sampled by excavating a subplot area. Powlson et al. (1986) took two soil cores (4.75 cm diam. and 23 cm deep) from the edges of each 1 m^2 microplot with a power driven auger. Deeper cores (23–40 cm and 60–100 cm) were taken from the spaces between microplots. Moraghan et al. (1984), while constructing a ^{15}N budget for plots treated with band-applied fertilizer, reduced sampling errors by sampling 14 blocks of soil, each 20 by 45 by 30 cm deep from within unconfined microplots. In a study of point placement of ^{15}N-labeled urea supergranules to flooded rice soil, Mohanty et al. (1989) removed two blocks of soil (40 by 30 cm by 15 cm deep) containing the supergranules.

If leaching of labeled N is of interest, intense sampling of lower soil depths is necessary (e.g., Kissel & Smith, 1978; Bigeriego et al., 1979; Khanif et al., 1984) because macropore transport may considerably increase the heterogeneous distribution of nitrate moving downward through soil.

Priebe and Blackmer (1989) cite evidence for preferential horizontal and vertical movement of solution through macropores of an Iowa soil (mesic Aquic Hapludolls) under a reduced tillage, continuous corn management system. Undisturbed soil cores (10 cm diam., 50 cm long) were taken from the field (a novel method is described). The movement of water and solute through the cores was followed over a 24-h period with

^{18}O-labeled water and ^{15}N-labeled urea. Determinations of labeled water and N compounds (urea, NO_3^-, and NH_4^+) in both effluent and soil (5-cm increments) lead Priebe and Blackmer to suggest that preferential flow of water through soil macropores may be an important factor affecting N movement.

40–2.5.2.3 Small-core Sampling. Sampling large plots for ^{15}N is usually done with conventional small diameter soil cores (< 5 cm diam.). Investigators using large plots to which ^{15}N is added usually are interested in tracing the applied N into both the nitrate and total N pools. A problem with taking small cores is that the core samples may not be intercepting in a representative manner labeled N that is heterogeneously distributed in soil. The greater the spatial variability and the fewer the number of cores taken, the greater the risk of error from nonuniform sampling. Following review of several field studies, Meisinger (1984) characterized the spatial variability of soil NO_3^- as (i) being a large, small-scale component, that is, 50 to 75% of the total variability is already present within a few square meters (Beckett & Webster, 1971; Cameron et al., 1971), and (ii) as having a large CV that usually ranges between 30 and 60% with 45% being a typical value. The spatial variability of total N is less than that of NO_3^--N, having CVs ranging between 10 and 25%, typically 15%. By estimating the approximate CV and by assuming that the final bulking of soil cores will produce a sample with a mean that is approximately normally distributed (central limit theorem holds), one can calculate the number of cores needed to estimate the plot mean with a given degree of precision. Gomez and Gomez (1984) give the appropriate formula for this calculation.

$$n = (Z)^2(CV)^2/(d)^2 \qquad [9]$$

where

n = the number of cores required
Z = standard normal deviate for the alpha level ($Z = 1.96$ for 0.05 and 1.65 for 0.10)
CV = coefficient of variation (as a decimal)
d = margin of error in the plot mean (as a decimal)

Using a CV of 45% for NO_3^--N and a desired precision of $\pm 10\%$ ($d = 0.10$) in 90% of the plots (alpha = 0.10), one can calculate that 55 cores would be required. Obviously, core sampling for NO_3^- will involve substantial labor. Alternatively, the researcher can adjust expectations and calculate, for example that 14 cores per plot would be sufficient to estimate the NO_3^--N content with an accuracy of about 20% on 90% of the plots. For total N (CV = 15%), the same 14 cores per plot should estimate the plot mean to within 8% on 95% of the plots. It is apparent that many core samples must be taken to permit accurate determination of soil ^{15}N content.

40–2.5.2.4 Soil Sample Preparation. The method of processing soil after sampling should not alter the N pools under study and should ensure that representative subsamples can be taken. Large soil masses resulting from complete plot excavation can be mechanically mixed with a concrete mixer, wet sieved, then subsampled and dried (e.g., Powlson et al., 1986; White et al., 1986). Some investigators have added water to the bulk soil to produce a slurry that can be more easily mixed and subsampled (Kissel & Smith, 1978). Using this technique, subsamples should be chemically analyzed immediately to prevent change through microbial action. Whatever the procedure, care must be taken to avoid loss of N or change of N form during sample preparation. Alternatively, the soil can be dried, pulverized, sieved, and mixed for subsampling. The best order of these operations will depend on local facilities, but in all cases thoroughly mixing the sample before subsampling is imperative. Saffigna (1988) reported that the CV for determining ^{15}N-labeled NO_3^--N and total N of greenhouse soils decreased after thorough mixing from 42% to 3% and from 10% to 2%, respectively. Thorough mixing and fine pulverizing of soil samples is especially important when using analytical instruments with small (e.g., 50 mg) sample requirements. When one considers that a 50-mg sample represents 450×10^6 mg of heterogeneous field soil from a microplot 1 m^2 by 30 cm deep, it is apparent that soil sample collection, mixing, and subsampling should be given careful attention.

40–2.5.2.5 Soil Sampling Suggestions. The appropriate soil sampling technique will depend on the goals of the study, the ^{15}N application technique, and the size of the plot. For studies that require the most accurate estimates of soil ^{15}N content (e.g., ^{15}N budgets, residual soil ^{15}N) or studies with localized ^{15}N placements (banded or point placements), we suggest either (i) the complete excavation method for plots < 1 m^2 in area, or (ii) the excavation of smaller subplots within larger microplots (1–10 m^2), with the excavated area corresponding in size to the distance between plant rows or between localized zones of application (e.g., between adjacent fertilizer bands). For studies that require less accurate soil ^{15}N estimates, have received uniform ^{15}N applications over several seasons, or were conducted on large field plots (> 20 m^2), we suggest high intensity (10–40 cores/plot) small-core sampling with more intense sampling when $^{15}NO_3^-$-N is to be determined and fewer cores taken for determining total ^{15}N.

40–2.5.3 Soil Water

No less formidable than collecting representative soil samples is the collection of representative samples of soil water. The leaching of water and solute through soil is a complex, dynamic process that no longer can be described as a simple piston flow process. Water percolation often involves a small but rapidly moving portion of water that penetrates deeply into the soil through large pores and by-passes much of the soil solution (e.g., see Priebe & Blackmer, 1989). Interconnected with this large-pore water is the small-pore water, which is also displaced downward during percolation, but at a slower rate. As water and NO_3^- percolate through soil, a portion of the solution will move to relatively deep horizons but most of the NO_3^- usually will remain at shallower depths (Boswell & Anderson, 1964; Thomas & Phillips, 1979). Sampling multipore systems is difficult because the chemical composition of the large-pore water is different from that of the small-pore water. Interpreting soil water data is also difficult because different field samplers collect different portions of the large-pore vs. small-pore water.

Soil water samples generally are obtained using a suction device, a free-drainage device, or some type of lysimeter. As with most field methods, there is no ideal method to sample soil water. Only a general overview of methods will be given here; the reader is directed to several recent articles for more detailed discussion (e.g., Rhoades & Oster, 1986; Soileau & Hauck, 1987; Litaor, 1988; Starr et al., 1991).

40–2.5.3.1 Porous Ceramic Cups. Though frequently used to collect water samples, porous ceramic cups preferentially sample the large-pore sequences immediately surrounding the cup, resulting in sampling error. If the objective is to sample all the water passing a given depth, then several samples must be collected over time; otherwise, sample bias is highly probable (Hansen & Harris, 1975; Alberts et al., 1977). Porous cups are relatively easy to install but their usefulness is limited by the large errors inherent in spatial variability which use of this sampling approach cannot overcome (Biggar & Nielsen, 1976). Broadbent and Carlton (1980) used ceramic cups in field ^{15}N studies to estimate leaching in conjunction with water flux measurements. Nitrate leaching data derived from porous cups are generally regarded as "point samples" that provide *relative* comparisons among treatments but quantitative estimates of NO_3^- leaching require a simultaneous detailed study of water flux at the site (Biggar & Nielsen, 1976; Rhoades & Oster, 1986). Linden (1977) described the installation and use of porous cup samplers. Rhoades and Oster (1986) recommended reducing some of the sampler variability by preselection for uniform permeabilities and size and by using uniform sampling intervals and suction.

40–2.5.3.2 Trough Extractors. Large ceramic extractors (3 m by 15 cm) have also been placed at the bottom of 20 to 45 cm long troughs that are then placed in soil about 1.5-m deep to collect leachate (Duke & Haise, 1973; Hergert, 1986; Montgomery et al., 1987). These devices have been placed under disturbed soil with standard trenching techniques and under undisturbed soil blocks with the aid of horizontal tunnels and inflatable bladders that force them up into contact with the soil. These extractors attempt to intercept macropore water with the trough and to sample small-pore water with the evacuated ceramic tube. Montgomery et al. (1987) compared large extractors in filled-in lysimeters with the lysimeter percolate and concluded that the extractors provided a good estimate of deep percolation and NO_3^- flux on a loamy fine sand in North Dakota. However, they also reported that similar extractors covered with disturbed soil were suitable only for estimating NO_3^- concentrations (rather than flux), the concentration values obtained comparing well with those estimated using ceramic cups. Hergert (1986) reported that the extractors adequately estimated drainage under low leaching conditions (52 mm/yr), but adjustment of the vacuum level was needed under high leaching conditions (165 mm/yr). The vacuum inside the trough should be maintained equal to that in the surrounding soil to avoid too much flow into the trough (excess vacuum) or too little flow (suction too low). Large extractors are difficult to install and require regular care and attention, but they can overcome some of the problems associated with small cup extractors by sampling a large area, collecting macropore events, and sampling small-pore sequences. To our knowledge no ^{15}N studies have been reported with these large extractors.

40–2.5.3.3 Tile Discharge. Sampling tile line discharge is another method of collecting percolate samples from soils with high water tables or restricted drainage. This procedure involves conventional tile line installation, sometimes with provision for isolating the plots with surface/subsurface flow barriers, and installation of automatic collection devices (Zwerman et al., 1972; Gast et al., 1978; Bergstrom, 1987). Hallberg et al. (1986) reviewed the literature on tile drain monitoring, and several research studies have used tile drains to assess the impact of agricultural practices on water quality (Gast et al., 1978; Baker & Johnson, 1981). Tile drain data have been shown to include macropore flow events (Richard & Steenhuis, 1988), but they more typically represent flow with long lag times (months or years depending on the hydrologic cycle) between imposition of surface treatments and tile discharge. The lag times are caused by macropore processes causing slow solute equilibrium within the unsaturated zone and by long travel times for drainage to reach the tile line within the saturated zone. Tile discharge data are best used to measure NO_3^- concentrations over a long-term study for large-plot treatments. Over the period 1973 to 1979, Nelson and Randall (1983) and Buzicky et al. (1983) studied movement of ^{15}N-depleted $(NH_4)_2SO_4$ in 36 individually drained field plots (6.1 × 9.2 m), each isolated to·a depth of 1 m with plastic sheeting. Tile drainage waters flowed to underground collection chambers where they were sampled. Estimating the mass of NO_3^- leached is difficult with tile drain data because of uncertainties in delineating exact drainage areas and uncertainties in the quantity of percolation that escapes between tile lines (Hallberg et al., 1986; Bergstrom, 1987).

40–2.5.3.4 Lysimeters. Three general types of lysimeters have been used to collect soil water: disturbed (or filled-in), undisturbed (or monolith), and pan (or ebermayer). The literature on lysimetry is extensive and the reader is referred to classic reviews, such as Khonke et al. (1940) and Allison (1955), and a recent review by Soileau and Hauck (1987) for detailed discussions on lysimeters.

Agricultural scientists have used filled-in lysimeters with caution because of the radical disturbance of percolation paths during construction. They are best suited for coarse-textured soils with little or no structural development and may be equipped with vacuum extractors or be free-draining. Montgomery et al. (1987) produced bulk densities similar to those under field conditions by saturating the soil intermittently with water entering from the bottom of the lysimeter. Filled-in lysimeters have been used with ^{15}N sources to study leaching as affected by time of fertilizer application and forage crop species (Jones et al., 1977) and to evaluate nitrification inhibitors and fertilizer placement (Walters & Malzer, 1990). Of particular interest in the work of Walters and Malzer (1990) is the fact that the values for amount of N leached were threefold greater when calculated from nontracer data obtained on the leachate containing ^{15}N than when isotope dilution calculations were used. The investigators attributed this discrepancy to ^{15}N equilibration in the soil biomass, thereby reducing the amount of labeled N available for leaching. Most investigators have noted increased NO_3^--N losses and higher percolation the first few years after lysimeter filling. Filled lysimeters are best suited to soils where macro-pore flow is thought to be insignificant and to long-term studies that allow ample time for settling and structural development.

Undisturbed monolith lysimeters are being used more frequently today to preserve macro-pore processes and more closely approach field conditions. Constructing monolith lysimeters is described in several reports, including the excavation approach (Khonke et al., 1940; Brown et al., 1985), pressing cylinders into moist soil with static weight (Tackett et al., 1965) or hydraulic pressure (Cassel et al., 1974), and a drilling technique to isolate a core and simultaneously encase it in plastic (Bergstrom, 1987). Undisturbed lysimeters include both free-draining types (Chichester & Smith, 1978; Bergstrom, 1987) and vacuum-assisted types (Cassel et al., 1974; Brown et al., 1985). Undisturbed lysimeters offer the advantages of precise definition of system boundaries, inclusion of macro-pore processes, and the ability to monitor both nitrate concentration and the mass of N leached. They suffer the disadvantages of being expensive, and labor intensive, and are quite variable even for monoliths collected near each other. Nitrogen tracers have been used in monolith lysimeters to study nitrate leaching as affected by soil type and precipitation (Owens, 1960) and no-tillage vs. conventional tillage culture (Chichester & Smith, 1978). We suggest using undisturbed monolith lysimeters when determining the mass of NO_3^--N leached is the primary research objective, macro-pore transport processes are important, and sufficient labor and time are available.

Soil solution samples are also frequently collected from beneath undisturbed blocks of soil by trenching, followed by inserting a horizontal collection device (ebermayer lysimeter). The collector usually is a free-draining glass or metal pan (e.g., Tyler & Thomas, 1977; Shaffer et al., 1979) that samples gravitational water flowing under saturated conditions. Jordan (1968) modified the free-draining pan by placing glass wool, a fiberglass screen, and moist soil in the pan to collect water under "zero-tension" conditions. Haines et al. (1982) found that at the 30-cm depth, a low tension ceramic plate collected twice as much percolate as a

Jordan-type, zero-tension lysimeter, but the NO_3^--N concentration from the ceramic plate was only about one-third of that from the Jordan collector. Barbee and Brown (1986) and Shaffer et al. (1979) also compared free-draining vs. suction cup samplers and concluded that porous cups were ineffective in well-structured soil where much of the percolation was associated with rapid leaching through large pores. The ebermayer-type lysimeters offer the advantages of better focus on rapid transport processes, easier installation, and lower labor and maintenance expenses than conventional lysimeters. Disadvantages include a high degree of variability and a lack of system boundaries that make estimation of the mass of N leached nearly impossible. The ebermayer-type lysimeter is useful when macro-pore flow and NO_3^--N concentrations are of primary interest, the soil is well structured, leaching events occur primarily under ponded conditions, and labor and time are limited.

40–2.5.3.5 Suggestions. No single method of sampling soil water can be recommended for all research applications. Several of the techniques outlined above differ in that they sample different proportions of the large-pore vs. small-pore soil water and, therefore, give different estimates of the quantity and composition of the soil water. Porous ceramic cups are best suited for *qualitative* comparisons of NO_3^--N concentrations among treatments in soils where macro-pore transport is minimal. Tile drain samples are suited to long-term studies in soils with high water tables where comparison of NO_3^--N concentrations is of main interest. If sufficient money and labor are available, some type of undisturbed lysimeter is best suited for estimating the mass of N leached and the NO_3^--N concentration. Filled lysimeters are less variable and can give useful relative comparisons among treatments for NO_3^--N leaching in nonstructured soils. Collection pans installed beneath undisturbed soil are useful for studying macro-pore processes and NO_3^--N concentrations but give variable results and their use is questionable for accurately estimating the mass of N leached. Soil physicists are still learning about macro-pore flow in soils and its impact on NO_3^- transport. This knowledge should aid in designing improved field methods for studying NO_3^- transport in soil water.

40–2.5.4 Soil Gases

Accurately collecting and determining gaseous forms of N and their exchange between soil, air, and the atmosphere above continues to be an unresolved problem, especially in field studies. The two N cycle processes of greatest interest contributing to gaseous exchange are biological dinitrogen fixation (chapter 43) and denitrification (chapter 44). Many methodologies and techniques have been developed and used with moderate success, depending on experimental objectives. The use of ^{15}N is indispensable for many of these methods. Rather than make a cursory review of this subject, we refer the reader to the review by Mosier and Heinemeyer

(1985) and the publication, *Field Measurement of Dinitrogen Fixation and Denitrification* (Hauck & Weaver, 1986), which in turn refer to other articles in this area.

40–3 PREPARING FOR AND MEASURING NITROGEN-ISOTOPE RATIO

40–3.1 Principles

Of the several methods available for determining N-isotopic composition, use of the mass spectrometer continues to be the method of choice because of its accuracy, precision, convenience, and general applicability in studies of N-cycle processes. The only practical alternative is emission spectroscopy, which has the advantage of being technically simpler (e.g., high vacuum is not required) and less costly (emission spectrometers suitable for use in N tracer research are considerably cheaper than mass spectrometers). An attractive feature offered by emission spectroscopy is that this method permits N-isotope ratio measurements to be made on microgram amounts of sample N. However, systems for automatic sample gas preparation and analysis by mass spectrometry have now been developed that permit accurate and acceptable determination of N-isotope ratio in < 10 µg of N. Other methods of N-isotope analysis, such as those using infrared spectroscopy, nuclear magnetic resonance, electronparamagnetic resonance, or microwave spectroscopy, have importance in N research but are not used in routine determination of N-isotope ratio.

Before the N-isotopic composition of a sample can be determined by mass spectrometry or emission spectroscopy, all N forms in the sample must be quantitatively converted to a gas that is simple in molecular structure and isotopic composition, of low molecular weight, readily prepared from inorganic and organic compounds, preferably unreactive with components of the isotope-analyzing system, and easily pumped from the instrument. Dinitrogen (N_2) best meets these requirements and is used for the isotope-ratio analysis of N in most solids and liquids. The traditional and probably most commonly used procedure for converting combined N to N_2 for isotope-ratio analysis involves a three-step procedure: (i) conversion of sample N to NH_4^+-N by means of acid digestion (e.g., Kjeldahl procedure) or by other processes (e.g., hydrolysis); (ii) oxidation of NH_4^+-N to N_2; and (iii) determination of the isotopic composition of the N_2. This procedure originally was developed by Rittenberg and colleagues (Rittenberg et al., 1939; Rittenberg, 1948; Sprinson & Rittenberg, 1948, 1949) for use in biochemical research but has been applied extensively to the agricultural and biological sciences. Following the development of improved instruments and their coupling with emission spectrometers and some automatic mass spectrometer systems, dry combustion techniques for direct conversion of sample N to N_2 are gaining favor. Reference to other techniques for generating N_2 for special application can be found in Rittenberg (1948), Bremner (1965a), Kennedy (1965), and in the methodology section of the bibliography by Hauck and Bystrom (1970).

Many modifications in analytical methodology have been developed to improve the speed, convenience, and accuracy of N-isotope-ratio analysis. The three-step procedure, especially, remains a complicated and time-consuming process. Possible errors encountered in both the three-step and direct combustion procedures are discussed by Hauck (1982) and Mulvaney (1993), among others. The intent here is not to give complete procedural details of the many steps leading from sample preparation to determination of N content and isotopic composition. Special procedures, such as for small sample handling and diffusion techniques, are not discussed. For in-depth discussions of preparing for and measuring N-isotope ratio, the reader is referred to reviews by Bremner (1965a), Hauck (1982), and Mulvaney (1993), and to the many references cited in these reviews.

40–3.2 Conversion of Labeled Nitrogen to Ammonium

40–3.2.1 Principles

The method to be chosen for converting labeled N to NH_4^+ depends on the forms of N in the sample under study and whether there is a need to distinguish between them. For total N, a Kjeldahl method usually is used. Steam distillation methods that result in the collection of NH_4^+-N are used for the determination of NH_4^+-N, NO_2^--N, and NO_3^--N. Because only inorganic and total N are determined on most of the samples obtained from agronomic and environmental studies using ^{15}N, procedures will be outlined for only these N forms.

40–3.2.2 Total Nitrogen

Numerous modifications of the Kjeldahl method have been described (see Bremner & Mulvaney, 1982). The basic method is a two-step process: (i) digestion with concentrated H_2SO_4 to convert organic N to NH_4^+-N, and (ii) determination of the amount of NH_4^+-N in the digest. The speed and completeness of digestion are increased by adding a salt (commonly K_2SO_4) to increase the temperature of the digesting solution and a catalyst (Cu, Hg, or Se, alone or in combination) to accelerate oxidation of organic matter. Upon addition of alkali to the digest, followed by steam distillation, NH_3 is liberated, trapped, and the NH_4^+-N content determined by titration with standard acid. Although both macro- and semimicro-methods have been used for converting ^{15}N-labeled organic N to NH_4^+-N, semimicro-methods usually are preferable because they provide an appropriate amount of N (about 1 mg) for N-isotope-ratio analysis and are less subject to error resulting from contamination by residues from preceding samples (cross-contamination). The permanganate-reduced Fe modification of the Kjeldahl digestion is recommended here because it can be used for determining total N, including NO_2^--N and NO_3^--N, in either plant or soil samples.

40–3.2.2.1 Special Apparatus. Micro (30 or 50 mL) or semimicro (100 mL) digestion flasks are used with an appropriate steam distillation apparatus (for detailed descriptions and discussions, see Bremner & Mulvaney, 1982; Hauck, 1982; Keeney & Nelson, 1982).

40–3.2.2.2 Reagents. Prepare as follows:
1. Salt-catalyst mixture. Intimately mix and grind to a powder 100 g of K_2SO_4, 10 g of $CuSO_4·5H_2O$, and 1 g of Se, using a laboratory jar mill or comparable device.
2. Concentrated sulfuric acid (H_2SO_4).
3. Dilute sulfuric acid. To 500 mL of water in a 2-L Pyrex flask, slowly add with continuous stirring 500 mL of conc. H_2SO_4.
4. Permanganate solution. Dissolve 25 g of $KMnO_4$ in 500 mL of water and store in a dark container.
5. Reduced Fe. Finely divided, to pass a 100-mesh screen.
6. Sodium hydroxide solution (about 10 N). Place 3.2 kg of NaOH into a heavy walled 10-L Pyrex bottle marked to indicate a volume of 8 L, add 4 L of CO_2-free water, and swirl contents until the NaOH is dissolved. Stopper the bottle until the solution is cool, then fill to the 8-L mark with CO_2-free water. Connect the bottle with a suitable dispensing device that prevents CO_2 absorption by the alkali solution.
7. Boric acid-indicator solution. Dissolve 80 g of H_3BO_3 in about 3000 mL of water contained in a 5-L flask marked to indicate a volume of 4 L. Heating and vigorous stirring will speed dissolution. Cool and add 80 mL of mixed indicator solution (99 mg of bromcresol green plus 66 mg methyl red in 100 mL ethyl alcohol). Add 0.1 N NaOH carefully until the solution turns reddish purple (pH about 5.0), make to the 4-L mark with water, and mix.
8. Standard acid. 0.01 N (e.g., 0.00714 N H_2SO_4).

40–3.2.2.3 Procedure. Place soil or plant sample containing about 1 mg of N in a micro- or semimicro-Kjeldahl digestion flask, add 1 mL of $KMnO_4$ solution and swirl for about 30 s. Holding the flask at an angle, very slowly pipette 2 mL of dilute H_2SO_4 down the side of the flask, swirling continuously. Cool for about 5 min. Through a dry funnel with stem reaching into the bulb of the digestion flask, add 0.50 ± 0.01 g of reduced Fe, swirl, and let the mixture stand about 15 min until the effervescence ceases. Insert a small Erlenmeyer flask or other loosely fitting glass obstruction in the flask opening and simmer (reflux) gently for 45 min, avoiding significant water loss. Cool, remove glass insert, add two glass beads, 1.1 g of salt-catalyst mixture, and 3 mL of conc. H_2SO_4. Heat gently, then vigorously, turning the flask occasionally. After about 30 min the digest forms a clear greenish yellow solution, the clearing time depending on type of sample. Continue boiling until the entire digestion time is at least 1.5 times the clearing time (some soils with refractory N forms may require digestion for as long as 5 h, although < 1 h is usual for plant tissues and 1–2 h for soils). Cool slightly. To the still warm digest, add water drop by drop down the side of the flask until violent reaction subsides, then add an additional 15 to 25 mL of water. If the digest has solidified before any water addition, heat gently until the salt cake disintegrates and redissolves, after which the digest can be diluted. Transfer the diluted digest to an appropriate steam distillation apparatus, add 20 mL of 10 N NaOH, and distill into a 50-mL Erlenmeyer flask containing 5 mL of H_3BO_3-indicator solution. Collect about 35 mL of distillate. Determine the NH_4^+-N concentration in the distillate by titration with standard acid (the color change at the end point is from green to pink).

40–3.2.2.4 Comments. Significant error in the determination of N-isotope-ratio can result if all forms of N in the sample are not quantitatively converted to NH_4^+ even though failure to include all of the N is within the experimental error of the Kjeldahl procedure (commonly ranging between 1.5 and 4%). For example, the NO_3^--N content of a sample may represent a small fraction (< 1%) of a sample's total N content, but if its N-isotopic composition is different from that of other total N components, incomplete recovery of NO_3^--N will markedly affect the accuracy by which the isotopic composition of the total N is measured. Modifications of the basic Kjeldahl method are used to obviate such difficulty, for example, use of reduced Fe-permanganate or salicylic acid pretreatment for complete recovery of NO_2^--N and NO_3^--N or longer digestion times (up to 18 h) for conversion of N in refractory compounds to NH_4^+-N. Further details and examples of the magnitude of errors associated with incomplete digestion of sample are given by Hauck (1982).

To achieve the highest level of precision and accuracy, samples must be ground to pass a sieve of 100 mesh or smaller. This ensures that tissue components of different physical and chemical characteristics are intimately mixed so that aliquot samples are virtually identical. Use of finely ground samples becomes increasingly more important as sample size decreases. Special grinding techniques are required for the preparation of samples to be analyzed for ^{15}N content by some of the automated instrument systems now available. Grinding can be accomplished conveniently by rotating glass bottles containing stainless steel rods of different diameters along with the sample to be ground (Hauck, 1982; Kelley, 1994). Soil, plant tissue, and many seeds can be ground to face-powder consistency with little effort. Cross-contamination from residues in a grinding mill is eliminated because each bottle is a separate grinding chamber.

40–3.2.3 Specific Nitrogen Forms

Simple, rapid, and reproducible methods have been developed for determining the amount and isotope ratio of different N forms in soil extracts and hydrolyzates, plant extracts, and waters including exchangeable and nonexchangeable NH_4^+, NO_2^-, NO_3^-, α-amino acids, amino sugars, hydroxyamino acids, and urea. An outline of these procedures is given in Table 40–1, with references to articles that provide more detail. See also Bremner (1965b) for detailed discussion of determining inorganic forms of N. All of the methods described in Table 40–1 involve steam distillation to liberate NH_3 and are readily applicable to many kinds of agricultural and environmental studies. Problems occur when the sample concentration of N is low requiring the use of large aliquots or multiple distillations to secure sufficient N for accurate N-isotope-ratio analysis (Hauck, 1982). These problems are less severe today because improved instrumentation permits accurate determination of isotope ratio on smaller amounts (< 50 µg) of N.

40–3.2.3.1 Ammonium, Nitrite, and Nitrate N. The recommended procedures for chemical analysis of these N forms in preparation for N-isotope-ratio analysis are those described by Keeney and Nelson (1982). They are based on the finding that NH_4^+-N can be quantitatively liberated from solution containing a small amount of MgO by steam distillation in the presence of alkali-labile organic N compounds and that NH_4^+-N + NO_2^--N + NO_3^--N can be determined accurately by the same method if finely divided Devarda's alloy is added to the solution containing these N forms immediately before MgO addition and steam distillation. Correct use of these procedures in ^{15}N research requires that each N form be completely liberated from solution and recovered in the distillate, that steam distillation rates and times be carefully controlled, and that cross-contamination is negligible. The importance of understanding the several sources of error inherent in the steam distillation procedures cannot be overstated and the reader is urged to review the procedural details and commentaries given in the articles by Hauck (1982), Keeney and Nelson (1982), and Mulvaney (1993), among others.

40–3.3 Direct Conversion of Labeled Nitrogen to Dinitrogen

The dry combustion or Dumas technique typically involved heating the sample (550–650 °C) with CuO to convert inorganic and organic N to N_2 and N oxides (mainly N_2O), which are subsequently reduced to N_2 with hot Cu. Many modifications have been developed to overcome some of the limitations of this method (e.g., incomplete conversion of NO_3^- to N_2) for use in N-isotope-ratio analysis (e.g., see Fiedler and Proksch, 1975). For example, heating to 800 to 900 °C in quartz or vycor sealed tubes greatly improves quantitative conversion of all N forms to N_2. Limitations of the dry combustion method also have been successfully resolved with the development of the automatic C/N analyzer (ANCA), with which complete conversion to N_2 is achieved through flash combustion ($\cong 1700$ °C) in the presence of a catalyst (Cr_2O_3). The combustion products are swept over Cu at 600 °C and then purified chromatographically.

Commercial systems are available for direct coupling to a mass spectrometer, permitting automatic determination of both N content and N-isotope ratio. The combustion method merits consideration because it

Table 40–1. Method for conversion of different forms of soil N to NH_4^+ and separation of the NH_4^+ for N isotope-ratio analysis.

Form of N	Method	Reference[†]
	Total N	
In soils containing no NO_3^- or NO_2^-	Kjeldahl digestion of soil and distillation of digest with alkali	Bremner & Mulvaney, 1982
In soils containing NO_3^-	Kjeldahl digestion and distillation after pretreatment of soil with salicylic acid and thiosulfate to convert NO_3^--N to amino-N	Bremner & Mulvaney, 1982
In soils containing NO_3^- or NO_2^-	Kjeldahl digestion and distillation after pretreatment of soil with acidified permanganate to oxidize NO_2^- or NO_3^- and with reduced Fe and H_2SO_4, to reduce NO_3^- to NH_4^+.	Bremner & Mulvaney, 1982
	Inorganic N	
Exchangeable NH_4^+	Steam distillation of soil extract with MgO	Keeney & Nelson, 1982
NO_3^-	Steam distillation of soil extract with MgO and Devarda alloy after destruction of NO_2^- with sulfamic acid and removal of NH_4^+ by steam distillation with MgO[‡]	Keeney & Nelson, 1982
Exchangeable NH_4^+ and NO_3^-	Steam distillation of soil extract with MgO and Devarda alloy after destruction of NO_2^- with sulfamic acid[‡]	Keeney & Nelson, 1982
NO_3^- or NO_2^-	Steam distillation of soil extract with MgO and Devarda alloy after removal of NH_4^+ by steam distillation with MgO	Keeney & Nelson, 1982
Exchangeable NH_4^+ and NO_3^- or NO_2^-	Steam distillation of soil extract with MgO and Devarda alloy	Keeney & Nelson, 1982
Nonexchangeable NH_4^+	Steam distillation of soil with KOH after removal of exchangeable NH_4^+ and labile organic N compounds by KOBr and treatment of residue with HF to decompose minerals	Keeney & Nelson, 1982
	Hydrolyzable N	
Total	Kjeldahl digestion of soil hydrolyzate and steam distillation of digest with alkali	Stevenson, 1982
NH_4^+	Steam distillation of soil hydrolyzate with MgO	Stevenson, 1982
NH_4^+ and hexosamine	Steam distillation of soil hydrolyzate with pH 11.2 buffer	Stevenson, 1982
Serine and threonine	Steam distillation of soil hydrolyzate with pH 11.2 buffer after steam distillation with same buffer to remove (NH_4^+ − hexosamine)-N and treatments with periodate to convert (serine + threonine)-N to NH_4^+-N and with arsenite to reduce excess periodate	Stevenson, 1982
Amino acid	Steam distillation of soil hydrolyzate with pH 11.2 buffer after treatments with NaOH to decompose hexosamines and remove NH_4^+ and with ninhydrin to convert α-amino-N to NH_4^+-N	Stevenson, 1982

† Includes specific chapter and page numbers in monograph.
‡ If NO_2^- is absent, the treatment with sulfamic acid is omitted.

involves few steps and may be more accurate and precise than the Kjeldahl-NaOBr method for determining total N and producing N_2 for isotope-ratio analysis. For a comparison of Kjeldahl and Dumas-type procedures for use in determining total N and ^{15}N contents of several organic N sources, see Minigawa et al. (1984). A recent example of the many modifications of the Dumas method is the use of CaO to simplify removal of CO_2 and water during the conversion of combined N to N_2 (Kendall & Grim, 1990).

40–3.4 Measuring Nitrogen-Isotope Ratio

40–3.4.1 Mass Spectrometry

A mass spectrometer separates molecular ions into a spectrum according to their masses or, more accurately, according to their mass to charge ratio (m/e). The separation for N molecules is achieved by electron bombardment of N_2 to form molecular ions (mainly positively charged) of m/e 28, 29, and 30, corresponding to $^{14}N_2$, $^{14}N^{15}N$, and $^{15}N_2$, respectively. The positive ions are drawn from the ionizing chamber into the magnetic sector by a small positive potential on a repeller and a larger negative potential across accelerating electrodes. The ions are accelerated through a magnetic (or magnetic/electrostatic) field where the combination of accelerating voltage and field magnetic strength determine the curvature of each ion path. By adjusting the acceleration voltage, ions of a given m/e can be focused on a target. The intensities of the ion beams generated corresponding to m/e 28, 29, and 30 are directly related to the isotopic composition of the N_2 being analyzed.

Sample N_2 is obtained either by direct combustion or the oxidation of NH_4^+-N with alkaline NaOBr or LiOBr. Appropriate measures are taken to ensure that the N_2 is free of impurities that could give rise to interfering molecular ions (e.g., CO or NO with m/e of 28 and 30, respectively). Procedural details and discussion of the hypobromite reaction in relation to N-isotope-ratio analysis are given by Hauck (1982) and Mulvaney (1993).

Two automated systems of N-isotope-ratio analysis by mass spectroscopy have been developed. The automated Rittenberg analysis system (ARA-MS) involves the same three-step procedure used in conventional N-isotope analysis, the main difference being that the hypobromite oxidation of NH_4^+-N and the determination of N_2 isotopic composition are done automatically. The ARA-MS system is described in detail by Mulvaney (1993). Mass spectrometers also have been coupled to automatic N and C analyzers (ANCA) that directly combust sample N to N_2. Plant, soil, or other solid samples (usually containing 20 to 150 µg of N) are encased in miniature Sn capsules for loading into an ANCA autosampler, from which individual samples are dropped into a combustion chamber. After combustion, the products (CO_2, N_2, NO_x, and H_2O) are swept into a tube containing hot Cu wire to convert NO_x to N_2, then through traps to remove CO_2 and H_2O. The N_2 is further purified by gas chromatography before an aliquot of the effluent is admitted into the mass spectrometer. One of the advantages of ANCA-MS is that both N amount and isotopic composition can be obtained during the same sample run.

40–3.4.2 Emission Spectrometry

In an emission spectrometer, N_2 molecules emit characteristic light in the UV spectrum when excited at low pressure in an electrodeless discharge tube. The light emitted by the excited molecules is resolved by a monochromator and the characteristic light signals are detected by a photomultiplier. The amplified signal peaks are recorded and related directly to the abundance of the three molecular species of N, permitting calculation of N-isotope ratio.

Emission spectroscopy is used for N-isotope-ratio analysis mainly in eastern Europe, Japan, and South Africa. As indicated earlier, one advantage of its use is that only a few micrograms of N are needed for ^{15}N analysis but with the advent of ANCA-MS systems that can determine the isotopic composition of microgram quantities of N with a high level of precision, this advantage of emission spectroscopy is less attractive. However, the sample requirement for some nonautomated emission spectrometers can be reduced with special techniques to as low as 0.2 μg which is an important advantage when ^{15}N content is to be determined on small amounts of nitrogenous material separated by biochemical techniques such as thin layer chromatography. For selected references to an abundant literature on this subject, see Hauck and Bremner (1976). A fully automated emission spectrometer system capable of determining N content and isotopic composition of 10 to 20 samples per hour (2–10 μg N per sample) is described by Therion et al. (1986).

40–4 SOURCES OF NITROGEN-15 SUPPLY AND ANALYTICAL SERVICE

At least 13 commercial sources of ^{15}N-enriched compounds were available two decades ago (Hauck & Bremner, 1976). Most of these suppliers now are no longer in business or are of questionable reliability as suppliers. Currently, recommended suppliers of ^{15}N for use in agricultural studies are Cambridge Isotope Laboratories, Isotec, Inc., and ICON Services, Inc. In recent years, the U.S. Department of Energy was the main producer and supplier of both ^{15}N-enriched and ^{15}N-depleted compounds. The labeled materials were produced at the Los Alamos Scientific Laboratory in New Mexico and could be purchased from Monsanto Research Corporation's Mound Laboratory (now EG & G Mound Applied Technologies, Miamisburg, OH). The Department of Energy (DOE) has been requested by another U.S. producer of isotopically enriched compounds to withdraw from the production and distribution of such compounds and a formal petition and request was released for public comment. Despite letters of support from ^{15}N users, DOE is phasing out its direct involvement as a supplier of ^{15}N.

A large number of inorganic and organic compounds, including N gases, labeled with ^{15}N are offered by Cambridge Isotope Laboratories (CIL), 20 Commerce Way, Woburn, MA 01801.

Merck and Co., Inc./Isotopes, 4545 Oleatha Ave., St. Louis, MO 63116 still supplies a variety of ^{15}N-labeled organic compounds but reportedly is divesting its holdings in this market. MSD ISOTOPES, the stable isotope division of Merck Frosst Canada Inc., Montreal, Quebec, has been offered for sale and its status as a supplier of ^{15}N is unclear.

Although several laboratories will perform N-isotope-ratio analyses upon special arrangement, two suppliers of ^{15}N-labeled compounds also provide analytical service particularly directed toward agricultural research. Labeled materials can be purchased from ICON Services, Inc., 19 Ox Bow Lanes, Summit, NJ 07901. This supplier through its marketing arm, U.S. Services Inc., makes available automated N analyses provided by Isotope Services, Inc. (ICI). Samples can be submitted in the form of NH_4^+ (e.g., from Kjeldahl digests), from which N_2 is generated via an automated Rittenburg procedure, or as dry plant or soil material, from which N_2 is produced via automated Dumas combustion for both total N and N-isotope-ratio determinations. Isotec, Inc., a primary producer of isotopically enriched materials, offers a wide range of ^{15}N-enriched compounds plus ^{15}N-depleted NH_4NO_3 and $(NH_4)_2SO_4$, but may not continue to offer analytical services using an automated Dumas-type procedure. The company is a subsidiary of Matheson, USA and is located at 3858 Benner Road, Miamisburg, OH 45342. In addition to these commercial ventures, analytical services also are offered by university laboratories.

Buyers of ^{15}N-labeled compounds should be aware that companies differ in their method of quoting prices. All prices should be compared on the basis of contained ^{15}N rather than on the weight of total material. Some prices quoted may include container, handling, packaging, or shipping charges while others may not. A guaranteed analysis also should be insisted upon. Wise use of ^{15}N begins with a wise purchase of the amount and enrichment level needed to achieve research objectives at minimum acceptable cost.

REFERENCES

Alberts, E.E., R.E. Burwell, and G.E. Schuman. 1977. Soil nitrate-nitrogen determined by coring and solute extraction techniques. Soil Sci. Soc. Am. J. 41:90–92.

Allison, F.E. 1955. The enigma of soil nitrogen balance sheets. Adv. Agron. 7:213–250.

Allmaras, R.R., and W.W. Nelson. 1971. Corn (*Zea mays* L.) root configuration as influenced by some row-interrow variants of tillage and straw mulch management. Soil Sci. Soc. Am. Proc. 35:974–980.

Baker, J.L., and H.P. Johnson. 1981. Nitrate-nitrogen in tile drainage as affected by fertilization. J. Environ. Qual. 10:519–522.

Barbee, G.C., and K.W. Brown. 1986. Comparison between suction and free-drainage soil solution samplers. Soil Sci. 141:149–154.

Barraclough, D., and M.J. Smith. 1987. The estimation of mineralization, immobilization, and nitrification in nitrogen-15 field experiments using computer simulation. J. Soil Sci. 38:519–530.

Bartholomew, W.V. 1964. Guides in extending the use of tracer nitrogen in soils and fertilizer research. p. 81–96. *In* Soil and fertilizer nitrogen research, a projection into the future. A symposium, Publ. no. T 64-4 SF. TVA, Wilson Dam, AL.

Beckett, P.H.T., and R. Webster. 1971. Soil variability: A review. Soils Fert. 34:1–15.

Benjamin, J.G., R.M. Cruse, A.D. Blaylock, and L.V. Vogl. 1988. A small-plot point injector for liquid fertilizer application. Soil Sci. Soc. Am. J. 52:1194–1195.

Bergersen, F.J. 1980. Measurement of nitrogen fixation by direct means. p. 65–110. *In* F.J. Bergersen (ed.) Methods for evaluating biological nitrogen fixation. John Wiley and Sons, Chichester, England.

Bergstrom, L. 1987. Nitrate leaching and drainage from annual and perennial crops in tile-drained plots and lysimeters. J. Environ. Qual. 16:11–18.

Bigeriego, M., R.D. Hauck, and R.A. Olson. 1979. Uptake, translocation, and utilization of ^{15}N-depleted fertilizer in irrigated corn. Soil Sci. Soc. Am. J. 43:528–533.

Biggar, J.W., and D.R. Nielsen. 1976. Spatial variability of the leaching characteristics of a field soil. Water Resour. Res. 12:78–84.

Blackburn, T.H. 1979. Method for measuring rates of NH_4^+ turnover in anoxic sediments, using a ^{15}N-NH_4^+ dilution technique. Appl. Environ. Microbiol. 37:760–765.

Boswell, F.C., and O.E. Anderson. 1964. Nitrogen movement in undisturbed profiles of fallowed soils. Agron. J. 56:278–281.

Bremner, J.M. 1965a. Isotope-ratio analysis of nitrogen in nitrogen-15 tracer investigations. p. 1256–1286. *In* C.A. Black et al. (ed.) Methods of soil analysis. Part 2. Agron. Monogr. 9. ASA, Madison, WI.

Bremner, J.M. 1965b. Inorganic forms of nitrogen. p. 1179–1237. *In* C.A. Black et al. (ed.) Methods of soil analysis. Part 2. Agron. Monogr. 9. ASA, Madison, WI.

Bremner, J.M., H.H. Cheng, and A.P. Edwards. 1966. Assumptions and errors in nitrogen-15 tracer research. p. 429–442. *In* Report of the FAO/IAEA Tech. Meet. (Braunschweig, Germany, 1963). Pergamon Press, Elmsford, NY.

Bremner, J.M., and C.S. Mulvaney. 1982. Nitrogen—total. p. 595–624. *In* A.L. Page et al. (ed.) Methods of soil analysis. Part 2. 2nd ed. Agron. Monogr. 9. ASA and SSSA, Madison, WI.

Broadbent, F.E., and A.B. Carlton. 1980. Methodology for field trials with nitrogen-15-depleted nitrogen. J. Environ. Qual. 9:236–242.

Brown, K.W., J.C. Thomas, and M.W. Aurelius. 1985. Collecting and testing barrel sized undisturbed soil monoliths. Soil Sci. Soc. Am. J. 49:1067–1069.

Brown, M.A., and G.M. Volk. 1966. Evaluation of ureaform fertilizer using nitrogen-15-labeled materials in sandy soils. Soil Sci. Soc. Am. Proc. 30:278–281.

Buresh, R.J., E.R. Austin, and E.T. Craswell. 1982. Analytical methods in ^{15}N research. Fert. Res. 3:37–62.

Burris, R.H., and P.W. Wilson. 1957. Methods for measurement of nitrogen fixation. p. 355–366. *In* S.P. Colowick and N.O. Kaplan (ed.) Methods in enzymology. Vol. 4. Academic Press, New York.

Buzicky, G.C., G.W. Randall, R.D. Hauck, and A.C. Caldwell. 1983. Fertilizer N losses from a tile drained Mollisol as influenced by rate and time of 15-N depleted fertilizer application. p. 213. *In* Agronomy abstracts. ASA, Madison, WI.

Cameron, D.R., M. Nyborg, J.A. Toogood, and D.H. Laverty. 1971. Accuracy of field sampling for soil tests. Can. J. Soil Sci. 51:165–175.

Cameron, K.C., and A. Wild. 1984. Potential aquifer pollution leaching following the plowing of temporary grassland. J. Environ. Qual. 13:274–278.

Capindale, J.B., and D.H. Tomlin. 1957. Mass-spectrometric assay of elementary nitrogen. Nature (London) 180:701–702.

Carter, J.N., O.L. Bennett, and R.W. Pearson. 1967. Recovery of fertilizer nitrogen under field conditions using nitrogen-15. Soil Sci. Soc. Am. Proc. 31:50–56.

Cassel, D.K., T.H. Krueger, F.W. Schroer, and E.B. Norum. 1974. Solute movement through disturbed and undisturbed soil cores. Soil Sci. Soc. Am. Proc. 38:36–40.

Chen, D., P.M. Chalk, and J.R. Freney. 1991. External-source contamination during extraction-distillation in isotope-ratio analysis of soil inorganic nitrogen. Anal. Chim. Acta 245:49–55.

Chichester, F.W., and S.J. Smith. 1978. Disposition of ^{15}N-labeled fertilizer nitrate applied during corn culture in field lysimeters. J. Environ. Qual. 7:227–233.

Cho, C.M., and E. Haunold. 1966. Some problems encountered in the preparation of nitrogen-15 gas samples and mass spectrometric work. p. 443–445. *In* Use of isotopes in soil organic matter studies. Report of the FAO/IAEA Tech. Meet. (Braunschweig, Germany. 9–14 Sept. 1963). Pergamon Press, Elmsford, NY.

Cline, M.G. 1944. Principles of soil sampling. Soil Sci. 58:275–288.

Duke, H.R., and H.R. Haise. 1973. Vacuum extractors to assess deep percolation losses and chemical constituents of soil water. Soil Sci. Soc. Am. Proc. 37:963–964.

Duxbury, J.M., and P.K. McConnaughey. 1986. Effect of fertilizer source on denitrification and nitrous oxide emissions in a maize-field. Soil Sci. Soc. Am. J. 50:644–648.

Edwards, A.P. 1978. A guide to the use of ^{14}N and ^{15}N in environmental research. Spec. Rep. 78-18. U.S. Army Cold Regions Research and Engineering Lab., Hanover, NH.

Faust, H., H. Gurtler, H. Huebner, H. Mielke, W. Rommel, M. Ulbrich, and K. Wetzel. 1963. Report on cooperative research on feeding ammonium bicarbonate to cattle. Wiss. Karl Marx Univ. Leipzig 12:711–718.

Fiedler, R. 1984. The measurement of ^{15}N. p. 233–282. *In* M.F. L'Annunziata and J.O. Legg (ed.) Isotopes and radiation in agricultural sciences. Academic Press, London.

Fiedler, R., and G. Proksch. 1975. The determination of nitrogen-15 by emission and mass spectrometry in biochemical analysis: A review. Anal. Chim. Acta 78:1–62.

Follett, R.F., R.R. Allmaras, and G.A. Reichman. 1974. Distribution of corn roots in sandy soil with a declining water table. Agron. J. 66:288–292.

Folorunso, O.A., and D.E. Rolston. 1984. Spatial variability of field-measured denitrification gas fluxes. Soil Sci. Soc. Am. J. 48:1214–1219.

Fried, M. 1978. Critique of field trials with isotopically labeled nitrogen fertilizer. p. 43–62. *In* D.R. Nielsen and J.G. MacDonald (ed.) Nitrogen in the environment. Vol. 1. Nitrogen behavior in field soil. Academic Press, New York.

Gast, R.G., W.W. Nelson, and G.W. Randall. 1978. Nitrate accumulation in soils and loss in tile drainage following nitrogen applications to continuous corn. J. Environ. Qual. 7:258–261.

Gomez, K.A., and A.A. Gomez. 1984. Statistical procedures for agricultural research. 2nd ed. John Wiley and Sons, New York.

Guiraud, G. 1984. Contribution du marquage isotopique a l'evaluation des transferts d'azote entre les compartiments organiques et mineraux dans les systemes sol-plantes. These Etat, Universite P. et M. Curie, Paris.

Guiraud, G., C. Marol, and M.C. Thibaud. 1989. Mineralization of nitrogen in the presence of a nitrification inhibitor. Soil Biol. Biochem. 21:29–34.

Haines, B.L., J.B. Waide, and R.L. Todd. 1982. Soil solution nutrient concentrations sampled with tension and zero-tension lysimeters: Report of discrepancies. Soil Sci. Soc. Am. J. 46:658–661.

Hallberg, G.R., J.L. Baker, and G.W. Randall. 1986. Utility of tile-lined effluent studies to evaluate the impact of agricultural practices on groundwater. p. 298–325. *In* Proc. of Agric. Impacts on Groundwater, Omaha, NE. 11–13 Aug. 1986. Natl. Well Water Assoc., Dublin, OH.

Hansen, E.A., and A.R. Harris. 1975. Validity of soil-water samples collected with porous ceramic cups. Soil Sci. Soc. Am. Proc. 39:528–536.

Hart, P.B.S., J.H. Rayner, and D.S. Jenkinson. 1986. Influence of pool substitution on the interpretation of fertilizer experiments with ^{15}N. J. Soil Sci. 37:389–403.

Hauck, R.D. 1973. Nitrogen tracers in nitrogen cycle studies—past use and future needs. J. Environ. Qual. 2:317–327.

Hauck, R.D. 1978. Critique of field trials with isotopically labeled nitrogen fertilizer. p. 63–77. *In* D.R. Nielsen and J.G. MacDonald (ed.) Nitrogen in the environment. Vol. 1. Nitrogen behavior in field soil. Academic Press, New York.

Hauck, R.D. 1979. Methods for studying N transformations in paddy soils: Review and comments. p. 73–93. *In* Nitrogen and rice. Int. Rice Res. Inst., Manila, Philippines.

Hauck, R.D. 1982. Nitrogen—Isotope-ratio analysis. p. 735–779. *In* A.L. Page et al. (ed.) Methods of soil analysis. Part 2. 2nd ed. Agron. Monogr. 9. ASA and SSSA, Madison, WI.

Hauck, R.D. 1984. Significance of nitrogen fertilizer microsite reactions in soil. p. 507–519. *In* R.D. Hauck et al. (ed.) Nitrogen in crop production. ASA, CSSA, and SSSA, Madison, WI.

Hauck, R.D. 1985. Slow-release and bioinhibitor-amended nitrogen fertilizers. p. 293–322. *In* O.P. Engelstad (ed.) Fertilizer technology and use. 3rd ed. SSSA, Madison, WI.

Hauck, R.D. 1994. Synthesis of ^{15}N-labeled isobutylidene diurea, oxamide, and ureaforms for use in agronomic studies. Commun. Soil Sci. Plant. Anal. 25:191–197.

Hauck, R.D., and J.M. Bremner. 1964. The methodology of N[15] research. p. 97–110. *In* Soil and fertilizer nitrogen research, a projection into the future, A symposium. Publ. no. T64-4 SF. TVA, Wilson Dam, AL.

Hauck, R.D., and J.M. Bremner. 1976. Use of tracers for soil and fertilizer nitrogen research. Adv. Agron. 28:219–266.

Hauck, R.D., and M. Bystrom. 1970. [15]N—A selected bibliography for agricultural scientists. The Iowa State Univ. Press, Ames.

Hauck, R.D., and V.J. Kilmer. 1975. Cooperative research between the Tennessee Valley Authority and land-grant universities on nitrogen fertilizer use and water quality. p. 655–660. *In* E.R. Klein and P.D. Klein (ed.) Proc. 2nd Int. Congr. Stable Isotopes, Oak Brook, IL. U.S. Energy Res. Dev. Admin., Agronne, IL.

Hauck, R.D., and R.W. Weaver (ed.) 1986. Field measurement of dinitrogen fixation and denitrification. SSSA Spec. Publ. 18. SSSA, Madison, WI.

Hergert, G.W. 1986. Nitrate leaching through sandy soil as affected by sprinkler irrigation management. J. Environ. Qual. 15:272–278.

Hills, F.J., F.E. Broadbent, and O.A. Lorenz. 1983. Fertilizer nitrogen utilization by corn, tomato, and sugarbeet. Agron. J. 75:423–426.

Jansson, S.L. 1958. Tracer studies on nitrogen transformations in soil with special attention to mineralisation-immobilisation relationships. Kgl. Lantbruks-Hoegsk. Ann. Roy. Agric. Coll., Swed. 24:101–361.

Jansson, S.L. 1966. Nitrogen transformation in soil organic matter. p. 283–296. *In* Report of the FAO/IAEA Tech. Meet., (Braunschweig, Germany. 1963). Pergamon Press, Elmsford, New York.

Jenkinson, D.S., R.H. Fox, and J.H. Rayner. 1985. Interactions between fertilizer nitrogen and soil nitrogen—the so-called 'priming' effect. J. Soil Sci. 36:425–444.

Jokela, W.E., and G.W. Randall. 1987. A nitrogen-15 microplot design for measuring plant and soil recovery of fertilizer nitrogen applied to corn. Agron. J. 79:322–325.

Jones, A.J., and R.J. Wagenet. 1984. In situ estimation of hydraulic conductivity using simplified methods. Water Resour. Res. 20:1620–1626.

Jones, J.B., and W.J.A. Steyn. 1973. Sampling, handling, and analyzing plant tissue samples. p. 249–270. *In* L.M. Walsh and J.D. Beaton (ed.) Soil testing and plant analysis. SSSA, Madison, WI.

Jones, M.B., C.C. Delwiche, and W.A. Williams. 1977. Uptake and losses of ^{15}N applied to annual grass and clover in lysimeters. Agron. J. 69:1019–1023.

Jordan, C.F. 1968. A simple, tension-free lysimeter. Soil Sci. 105:81–86.

Junk, G., and H.J. Svec. 1958. The absolute abundance of the nitrogen isotopes in the atmosphere and compressed gas from various sources. Geochim. Cosmochim. Acta 14:234–243.

Keeney, D.R., and D.W. Nelson. 1982. Nitrogen—inorganic forms. p. 643–698. *In* A.L. Page et al. (ed.) Methods of soil analysis. Part 2. 2nd ed. Agron. Monogr. 9. ASA and SSSA, Madison, WI.

Kelley, K.R. 1994. Conveyor-belt apparatus for fine grinding of soil and plant materials. Soil Sci. Soc. Am. J. 58:144–146.

Kendall, C., and E. Grim. 1990. Combustion tube method for measurement of nitrogen isotope ratios using calcium oxide for total removal of carbon dioxide and water. Anal. Chem. 62:526–529.

Kennedy, I.R. 1965. Release of nitrogen from amino acids with ninhydrin for ^{15}N analysis. Anal. Biochem. 11:105–110.

Khanif, Y.M., O. Van Cleemput, and L. Baert. 1984. Interaction between nitrogen fertilization, rainfall and groundwater pollution in sandy soil (Belgium). Water, Air, Soil Pollut. 22:447–452.

Khonke, H., F.R. Dreibelbis, and J.M. Davidson. 1940. A survey and discussion of lysimeters and a bibliography on their construction and performance. Misc. Publ. no. 372. USDA, Washington, DC.

Kilmer, V.J., R.D. Hauck, and O.P. Engelstad. 1974. Nitrogen isotopes and water quality research. p. 35–43. *In* Contribution of irrigation and drainage to world food supply. Spec. Conf. Proc., Biloxi, MS. 14–16 Aug. 1974. Am. Soc. Civil Eng., New York.

Kirchmann, H. 1985. Losses, plant uptake and utilisation of manure nitrogen during a production cycle. Acta Agric. Scand. (Suppl. 24).

Kirchmann, H. 1989. A 3-year balance study with aerobic, anaerobic and fresh ^{15}N-labeled poultry manure. p. 113–125. *In* J.A. Hansen and K. Henniksen (ed.) Nitrogen in organic wastes applied to soils. Academic Press, London.

Kirchmann, H. 1990. Nitrogen interactions and crop uptake from fresh and composted ^{15}N-labeled poultry manure. J. Soil Sci. 41:379–385.

Kirkham, D., and W.V. Bartholomew. 1954. Equations for following nutrient transformations in soil, utilizing tracer data. Soil Sci. Soc. Am. Proc. 18:33–34.

Kirkham, D., and W.V. Bartholomew. 1955. Equations for following nutrient transformations in soil, utilizing tracer data: II. Soil Sci. Soc. Am. Proc. 19:189–192.

Kissel, D.E., and S.J. Smith. 1978. Fate of fertilizer nitrate applied to coastal bermudagrass on a swelling clay soil. Soil Sci. Soc. Am. J. 42:77–80.

Kitur, B.K., M.S. Smith, R.L. Blevins, and W.W. Frye. 1984. Fate of ^{15}N-depleted ammonium nitrate applied to no-tillage and conventional tillage corn. Agron. J. 76:240–242.

Koike, I., and A. Hattori. 1978. Simultaneous determinations of nitrification and nitrate reduction in coastal sediments by ^{15}N dilution technique. Appl. Environ. Microbiol. 35:853–857.

LeClerg, E.L., W.H. Leonard, and A.G. Clark. 1962. Field plot technique. 2nd ed. Burgess Publ. Co., Minneapolis.

Legg, J.O., and J.J. Meisinger. 1982. Soil nitrogen budgets. p. 503–566. *In* F.J. Stevenson (ed.) Nitrogen in agricultural soils. Agron. Monogr. 22. ASA, CSSA, and SSSA, Madison, WI.

Legg, J.O., and C. Sloger. 1975. A tracer method for determining symbiotic nitrogen fixation in field studies. p. 661–667. *In* Proc. 2nd Int. Conf. on Stable Isotopes, Oakbrook, IL. Argonne Nat. Lab., U.S. Dep. of Energy, Washington, DC.

Linden, D.R. 1977. Design, installation and use of porous ceramic samplers for monitoring soil-water quality. U.S. Dep. Agric. Tech. Bull. 1562. U.S. Gov. Print. Office, Washington, DC.

Litaor, M.I. 1988. Review of soil solution samplers. Water Resour. Res. 24:727–733.

Martin, A.E., and P.J. Ross. 1968. Significance of errors in ^{15}N measurements in soil:plant research. Int. Congr. Soil Sci. Trans., 9th (Adelaide) 3:521–529.

Meisinger, J.J. 1984. Evaluating plant-available nitrogen in soil-crop systems. p. 391–416. *In* R.D. Hauck et al. (ed.) Nitrogen in crop production. ASA, CSSA, and SSSA, Madison, WI.

Meisinger, J.J., V.A. Bandel, G. Stanford, and J.O. Legg. 1985. Nitrogen utilization of corn under minimal tillage and moldboard plow tillage. I. Four-year results using labeled N fertilizer on an Atlantic Coastal Plain soil. Agron. J. 77:602–611.

Minagawa, M., D.A. Winter, and I.R. Kaplan. 1984. Comparison of Kjeldahl and combustion methods for measurement of nitrogen isotope ratios in organic matter. Anal. Chem. 56:1859–1861.

Mohanty, S.K., S.P. Chakravorti, and A. Bhadrachalam. 1989. Nitrogen balance studies in rice using ^{15}N-labelled urea and urea supergranules. J. Agric. Sci., Cambridge 113:119–121.

Montgomery, B.R., L. Prunty, and J.W. Bauder. 1987. Vacuum trough extractors for measuring drainage and nitrate flux through sandy soils. Soil Sci. Soc. Am. J. 51:271–276.

Moraghan, J.J., T.J. Rego, R.J. Buresh, P.L.G. Vlek, J.R. Burford, S. Singh, and K.L. Sahrawat. 1984. Labelled nitrogen fertilizer research with urea in the semi-arid tropics. Plant Soil 80:21–33.

Mosier, A.R., and O. Heinemeyer. 1985. Current methods used to estimate N_2O and N_2 emissions from field soils. p. 79–99. *In* H.L. Golterman (ed.) Denitrification in the nitrogen cycle. Plenum Press, New York.

Mulvaney, R.L. 1991. Some recent advances in the use of nitrogen-15 for research on nitrogen transformations in soil. p. 283–296. *In* Stable isotopes in plant nutrition, soil fertility and environmental studies. FAO/IAEA, Vienna, Austria.

Mulvaney, R.L. 1993. Mass spectrometry. p. 11–57. *In* R. Knowles and T.H. Blackburn (ed.) Nitrogen isotopes techniques. Academic Press, San Diego.

Myrold, D.D., and J.M. Tiedje. 1986. Simultaneous estimation of several nitrogen cycle rates using [15]N: Theory and application. Soil Biol. Biochem. 18:559–568.

Nelson, W.W., and G.W. Randall. 1983. Fate of residual nitrate-N in a tile-drained Mollisol. p. 215. *In* Agronomy abstracts. ASA, Madison, WI.

Norman, A.G., and C.H. Werkman. 1943. The use of N[15] in determining N recovery from plant materials decomposing in soil. J. Am. Soc. Agron. 35:1023–1025.

Olson, R.V. 1980. Plot size requirements for measuring residual fertilizer nitrogen and nitrogen uptake by corn. Soil Sci. Soc. Am. J. 44:428–429.

Owens, L.D. 1960. Nitrogen movement and transformations in soils as evaluated by a lysimeter study utilizing isotopic nitrogen. Soil Sci. Soc. Am. Proc. 24:372–376.

Parkin, T.B. 1987. Soil microsites as a source of denitrification variability. Soil Sci. Soc. Am. J. 51:1194–1199.

Parkin, T.B., J.L. Starr, and J.J. Meisinger. 1987. Influence of sample size on measurement of soil denitrification. Soil Sci. Soc. Am. J. 51:1492–1501.

Peterson, R.G., and L.D. Calvin. 1986. Sampling. p. 33–51. In A. Klute (ed.) Methods of soil analysis. Part 1. Physical and mineralogical methods. 2nd ed. Agron. Monogr. 9. ASA, Madison, WI.

Phillips, D.A., M.B. Jones, and K.W. Foster. 1986. Advantages of the nitrogen-15 dilution technique for field measurement of symbiotic dinitrogen fixation in legumes. p. 11–21. In R.D. Hauck and R.W. Weaver (ed.) Field measurement of dinitrogen fixation and denitrification. SSSA Spec. Publ. 18. ASA, CSSA, and SSSA, Madison, WI.

Powlson, D.S., G. Pruden, A.E. Johnston, and D.S. Jenkinson. 1986. The nitrogen cycle in the Broadbalk wheat experiment: Recovery and loss of ^{15}N-labelled fertilizer applied in spring and inputs of nitrogen from the atmosphere. J. Agric. Sci., Cambridge 107:591–609.

Priebe, D.L., and A.M. Blackmer. 1989. Preferential movement of oxygen-18-labeled water and nitrogen-15-labeled urea through macropores in a Nicollet soil. J. Environ. Qual. 18:66–72.

Pruden, G., D.S. Powlson, and D.S. Jenkinson. 1985. The measurement of ^{15}N in soil and plant material. Fert. Res. 6:205–218.

Rauhe, K., and H. Bornak. 1970. Die Wirkung von ^{15}N-markiertem Rinderkot mit verschiedenen Zusatzen im Feldversuch unter besonderer Berucksichtigung der Reproduktion der organischen Substanz im Boden. Albrecht-Thaer-Arch. 14:937–948.

Rauhe, K., E. Fichtner, F. Fichtner, E. Knappe, and W. Drauschke. 1973. Quantifizierung der Wirkung organischer und mineralischer Stickstoffdünger auf Pflanze und Boden unter besonderer Berücksichtigung ^{15}N-markierter tierischer Exkremente. Arch. Acker-, Pflanzenbau Bodenkd. 17:907–916.

Rennie, R.J. 1986. Advantages and disadvantages of nitrogen-15 isotope dilution to quantify dinitrogen fixation in field-grown legumes—A critique. p. 43–58. In R.D. Hauck and R.W. Weaver (ed.) Field measurement of dinitrogen fixation and denitrification. SSSA Spec. Publ. 18. ASA, CSSA, and SSSA, Madison, WI.

Rhoades, J.D., and J.D. Oster. 1986. Solute content. p. 985–1006. In A. Klute (ed.) Methods of soil analysis. Part 1. Physical and mineralogical methods. 2nd ed. Agron. Monogr. 9. ASA and SSSA, Madison, WI.

Richard, T.L., and T.S. Steenhuis. 1988. Tile drain sampling of preferential flow on a field scale. J. Contam. Hydrol. 3:307–325.

Rittenberg, D. 1948. The preparation of gas samples for mass spectrographic analysis. p. 31–42. In D.W. Wilson et al. (ed.) Preparation and measurement of isotopic tracers. J. W. Edwards, Ann Arbor, MI.

Rittenberg, D., A.S. Keston, F. Rosebury, and R. Schoenheimer. 1939. Studies in protein metabolism. II. The determination of nitrogen isotopes in organic compounds. J. Biol. Chem. 127:291–299.

Rubio, J.L., and R.D. Hauck. 1986. Uptake and use patterns of nitrogen from urea, oxamide, and isobutylidene diurea by rice plants. Plant Soil 94:109–123.

Saffigna, P. 1988. N-15 methodology in the field. p. 433–451. In J.R. Wilson (ed.) Advances in nitrogen cycling in agricultural ecosystems. Proc. Symp. at Brisbane, Australia. 11–15 May 1987. CAB Int., Wallingford, UK.

Sanchez, C.A., and A.M. Blackmer. 1987. A method for application of nitrogen-15-labeled anhydrous ammonia to small plots. Soil Sci. Soc. Am. J. 51:259–261.

Sanchez, C.A., A.M. Blackmer, R. Horton, and D.R. Timmons. 1987. Assessment of errors associated with plot size and lateral movement of nitrogen-15 when studying fertilizer recovery under field conditions. Soil Sci. 144:344–351.

San Pietro, A. 1957. The measurement of stable isotopes. p. 473–488. In S.P. Colowick and N.O. Kaplan (ed.) Methods in enzymology. Vol. 4. Academic Press, New York.

Selles, F., R.E. Karamanos, and R.G. Kachanoski. 1986. The spatial variability of nitrogen-15 and its relation to the variability of other soil properties. Soil Sci. Soc. Am. J. 50:105–110.

Shaffer, K.A., D.D. Fritton, and D.E. Baker. 1979. Drainage water sampling in a wet, dual-pore soil system. J. Environ. Qual. 8:241–246.

Shen, S.M., G. Pruden, and D.S. Jenkinson. 1984. Mineralization and immobilization of nitrogen in fumigated soil and the measurement of microbial biomass nitrogen. Soil Biol. Biochem. 16:437–444.

Smith, J.H., J.O. Legg, and J.N. Carter. 1963. Equipment and procedures for N-15 analysis of soil and plant material with the mass spectrometer. Soil Sci. 96:313–318.

Soileau, J.M., and R.D. Hauck. 1987. A historical review of U.S. lysimeter research with emphasis on fertilizer percolation losses. p. 208–304. In Yu-S Fox (ed.) Infiltration principles and practices. Water Resources Res. Center, Univ. of Hawaii, Honolulu.

Sprinson, D.B., and D. Rittenberg. 1948. Preparation of gas samples for mass spectrometric analysis of isotope abundance. p. 82–93. U.S. Nav. Med. Bull., March–April Suppl.

Sprinson, D.B., and D. Rittenberg. 1949. The rate of utilization of ammonia for protein synthesis. J. Biol. Chem. 180:707–714.

Starr, J.L., J.J. Meisinger, and T.B. Parkin. 1991. Experience and knowledge gained from vadose zone sampling. p. 279–290. In R.G. Nash and A. Leslie (ed.) Agrichemical residue sampling design and techniques: Soil and groundwater. Am. Chem. Soc. Symp., 23–24 Apr. 1990. Boston, MA. Am. Chem. Soc., Washington, DC.

Stevenson, F.J. 1982. Nitrogen—organic forms. p. 625–641. In A.L. Page et al. (ed.) Methods of soil analysis. Part 2. 2nd ed. Agron. Monogr. 9. ASA and SSSA, Madison, WI.

Stumpe, J.M., P.L.G. Vlek, S.K. Mughogho, and F. Ganry. 1989. Microplot size requirements for measuring balances of fertilizer nitrogen-15 applied to maize. Soil Sci. Soc. Am. J. 53:797–800.

Tabor, J.A., A.W. Warrick, D.E. Myers, and D.A. Pennington. 1985. Spatial variability of nitrate in irrigated cotton: II. Soil nitrate and correlated variables. Soil Sci. Soc. Am. J. 49:390–394.

Tackett, J.L., E. Burnett, and D.W. Fryrear. 1965. A rapid procedure for securing large, undisturbed soil cores. Soil Sci. Soc. Am. Proc. 29:218–220.

Therion, J.J., H.G.C. Human, C. Clase, R.I. Mackie, and A. Kistner. 1986. Automatic nitrogen-15 analyser for use in biological research. Analyst 111:1017–1021.

Thomas, G.W., and R.E. Phillips. 1979. Consequences of water movement in macropores. J. Environ. Qual. 8:149–152.

Tyler, D.D., and G.W. Thomas. 1977. Lysimeter measurements of nitrate and chloride losses from soil under conventional and no-tillage corn. J. Environ. Qual. 6:63–66.

Vanden Heuvel, R.M. 1988. Improved apparatus for preparing nitrogen-15 labeled anhydrous ammonia. Soil Sci. Soc. Am. J. 52:1483–1486.

Vanden Heuvel, R.M., and S. Harrold. 1990. Field apparatus for knife injection of nitrogen-15-labeled liquid anhydrous ammonia. Soil Sci. Soc. Am. J. 54:531–534.

Vigil, M.F., and D.E. Kissel. 1991. Equations for estimating the amount of nitrogen mineralized from crop residues. Soil Sci. Soc. Am. J. 55:757–761.

Vose, P.B., and R.L. Victoria. 1986. Re-examination of the limitations of nitrogen-15 isotope dilution technique for the field measurement of dinitrogen fixation. p. 23–41. In R.D. Hauck and R.W. Weaver (ed.) Field measurement of dinitrogen fixation and denitrification. SSSA Spec. Publ. 18. ASA, CSSA, and SSSA, Madison, WI.

Walters, D.T., and G.L. Malzer. 1990. Nitrogen management and nitrification inhibitor effects on nitrogen-15 urea: II. Nitrogen leaching and balance. Soil Sci. Soc. Am. J. 54:122–130.

Warrick, A.W., G.J. Mullen, and D.R. Nielsen. 1977. Scaling field measured soil hydraulic properties using a similar media concept. Water Resour. Res. 13:355–362.

Weaver, R.W. 1986. Measurement of biological dinitrogen fixation in the field. p. 1–10. In R.D. Hauck and R.W. Weaver (ed.) Field measurement of dinitrogen fixation and denitrification. SSSA Spec. Publ. 18. ASA, CSSA, and SSSA, Madison, WI.

Westerman, R.L., L.T. Kurtz, and R.D. Hauck. 1972. Recovery of ^{15}N-labeled fertilizers in field experiments. Soil Sci. Soc. Am. Proc. 36:82–86.

White, R.E., R.A. Haigh, and J.H. McDuff. 1987. Frequency distributions and spatially dependent variability of ammonium and nitrate concentrations in soil under grazed and ungrazed grassland. Fert. Res. 11:193–208.

White, P.J., I. Vallis, and P.G. Saffigna. 1986. The effect of stubble management on the availability of ^{15}N-labeled residual fertilizer nitrogen and crop stubble nitrogen in an irrigated black earth. Aust. J. Exp. Agric. 26:99–106.

Woodcock, T.M., G. Pruden, D.S. Powlson, and D.S. Jenkinson. 1982. Apparatus for applying ^{15}N-labelled fertilizer uniformly to field micro-plots. J. Agric. Eng. Res. 27:369–372.

Zwerman, P.J., T. Greweling, S.D. Klausner, and D.J. Lathwell. 1972. Nitrogen and phosphorus content of water from tile drains at two levels of management and fertilization. Soil Sci. Soc. Am. Proc. 36:134–137.

Chapter 41

Nitrogen Availability Indices

L. G. BUNDY, *University of Wisconsin, Madison, Wisconsin*

J. J. MEISINGER, *USDA-ARS, Beltsville, Maryland*

Available N in soils is that N present in forms, concentrations, and spatial position that allow utilization by plants growing in the soil. Most plant species can effectively use either NH_4 or NO_3, but a few species use NH_4 preferentially. Organic N compounds such as urea and amino acids can also be taken up by plants directly, but the contribution of these organic N compounds to the overall plant N requirement is minimal. Since NH_4 is converted to NO_3 in most soils, NO_3 is usually the predominant form of available N in the plant root zone.

Available N in soils can originate from many sources including fertilizer N additions and mineralization of organic N from soil organic matter, crop residues, and organic wastes (Keeney, 1982; Meisinger, 1984). Most soils contain 0.08 to 0.4% of N, and 97 to 99% of this occurs as organic N compounds in soil organic matter (Keeney, 1982; Dahnke & Johnson, 1990). If 1 to 3% of this organic N is mineralized annually, 8 to 120 kg of N ha^{-1} is released in a plant-available form (Bremner, 1965a).

Nitrogen availability indices are N analyses or chemical or biological tests to measure or predict the amounts of available N released from soil under a specific set of conditions.

41–1 CURRENT STATUS OF NITROGEN AVAILABILITY INDICES

41–1.1 Previous Summaries

Estimating soil N availability has been a goal of soil scientists since the early 1900s. Summaries of early research (Harmsen & Van Schreven, 1955; Allison, 1965; Bremner, 1965a) reached the general conclusions that: (i) chemical extraction methods were likely to be unsuccessful because they could not imitate the action of soil microorganisms; (ii) biological incubations under standardized conditions were most successful because they used the same microbial agents active under field conditions; (iii) crop response methods were too costly and labor intensive for practical use, but were absolutely essential for calibration of laboratory methods; and (iv) methods based on soil NO_3-N levels were of "very limited value" because the soil NO_3-N pool was too transient to be useful.

A 1973 review by Dahnke and Vasey emphasized the need to include some measure of preplant mineral N. However, they emphasized that preplant N is quite variable from year to year depending on factors such as previous crop yield, previous N fertilization, nongrowing season precipitation, and soil-leaching characteristics. Therefore, they recommended annual deep preplant soil sampling to assess N availability in subhumid climates. They also reviewed chemical and biological indexes but reached no consensus regarding a recommended method.

In the 1980s, reviews by Stanford (1982) and Keeney (1982) considered residual nitrate, short- and long-term incubations, and chemical extractants. The general conclusions of these reviews were that: (i) long-term biological mineralizations were most suitable but were not practicable, (ii) short-term mineralizations were acceptable but were affected by sample handling and pretreatment (Stanford, 1982) with NH_4-N production after 7 d of anaerobic incubation being recommended (Keeney, 1982); and (iii) mild chemical extracts were acceptable for the ranking of soils with NH_4-N hydrolyzed by 0.01 M $CaCl_2$ upon overnight (16 h) autoclaving being the recommended method (Keeney, 1982). Keeney and Stanford also endorsed the preplant soil nitrate test in climates without extensive overwinter leaching. This was supported by the fact that most western U.S. states were already using some type of preplant residual NO_3-N test in 1982.

Evaluating plant-available N from a systems perspective was discussed by Meisinger (1984), who emphasized the need to systematically include all available N pools. This review discussed underlying principles, steady-state approaches, crop N requirements, N-efficiency estimates, and methods of estimating residual NO_3-N and mineralizable N. Meisinger (1984) concluded that improving N-availability assessments should involve: use of both residual NO_3-N and N-mineralization tests, expanded use of local soil properties through soil taxonomic information, and an integration of this information for each specific site by use of computer models and local weather data.

41–1.2 Selective Review of Recent Work

Since the earlier reviews cited in section 41–1.1 were prepared, the need for reliable prediction of N availability in soils has increased due to economic and environmental incentives to use available N more efficiently in crop production and minimize losses of N from cropland to the environment. These incentives have stimulated continued research on N availability indices in several areas including: (i) development of new chemical indices; (ii) modification and improvement of aerobic incubation procedures to provide more accurate assessments of N availability and facilitate an improved understanding of the characteristics of mineralizable N in soils; and (iii) development and evaluation of field tests for N availability. A selective review of recent work in these areas is provided below.

41–1.2.1 Chemical Methods Development and Comparison with Other Indices

Chemical methods of assessing N availability are appealing because they offer a simple and rapid approach to estimation of crop N needs for use in production and a convenient assessment of relative differences among various experimental treatments in research projects. To be successful, chemical procedures must detect innate differences among soils or treatments in the quantities of N likely to mineralize during a specific period (often the crop N uptake period) and must reflect the influences of environmental conditions on the rates and amounts of N mineralized. Given these demanding specifications, none of the many chemical indices proposed have been adopted for routine use in production or research, and Fox and Piekielek (1984) and Meisinger (1984) stated a widely accepted view that any single soil index is unlikely to provide sufficiently accurate predictions of the economic optimum fertilizer N recommendations for corn (*Zea mays* L.).

Recent work with chemical methods includes development and evaluation of new techniques and extensive comparisons of various chemical indices with other methods of assessing N availability including aerobic and anaerobic incubation tests, soil inorganic N measurements, and crop N uptake in field studies. New methods proposed include determination of NH_4-N released by heating soil samples with 2 M KCl (Gianello & Bremner, 1986a; Øien & Selmer-Olsen, 1980), measurement of NH_4-N released by steam distillation of soil samples with pH 11.2 phosphate-borate buffer (Gianello & Bremner, 1988), and UV absorbance of $NaHCO_3$ soil extracts at 200 nm (Hong et al., 1990).

41–1.2.2 Undisturbed Soil Core Incubation

Nitrogen mineralized during incubation of undisturbed soil cores can provide a more reliable assessment of N availability than standard methods using disturbed soil samples (Rice et al., 1987; Cabrera & Kissel, 1988a, b). Measurements of N mineralized from disturbed soil samples often overestimate field N availability due to stimulation of mineralization by drying, crushing, and sieving the soil. The relatively large number of undisturbed soil cores needed to assess N mineralization and the process required to obtain these samples (Myers et al., 1989) probably makes this approach impractical for field-scale predictions of N mineralization (Cabrera & Kissel, 1988a). However, Cabrera and Kissel (1988a) found that estimates of N mineralization in undisturbed soil could be obtained using data from disturbed samples or measured soil characteristics. Use of the undisturbed core technique is essential for N mineralization studies where variation in the degree of soil disturbance is an integral component of the systems under evaluation such as in tillage studies, pastures, forests, and fallow soils.

41–1.2.3 Field Tests for Nitrogen Availability

A major emphasis in recent work on N availability indices (NAI) is the development and evaluation of field tests for N availability. These studies involve measurement of inorganic N (usually NO_3) in soil before planting a test crop or at a specific time during the crop growing season and establishment of the relationship between the soil inorganic N content and the amounts of N available to a test crop. Once this relationship is known, results from the soil N test are used to provide N recommendations for crops.

Preplant soil profile NO_3 tests have long been recommended as a method of assessing N availability and predicting crop N needs in subhumid areas in the western USA and Canada (Hergert, 1987; Dahnke & Johnson, 1990). Although early work (King & Whitson, 1901, 1902; Buckman, 1910; Call, 1914) and periodic studies during the 1950 to 1970 era (see Keeney, 1982; Stanford, 1982; Meisinger, 1984; Dahnke & Johnson, 1990 for reviews) showed that soil profile NO_3 provided important amounts of available N to crops in at least some growing seasons in humid climates, soil NO_3 tests were not adopted in the higher rainfall areas of the USA. Recent studies of soil profile NO_3 accumulation and retention and its effects on crop N needs (Sheppard & Bates, 1986; Bundy & Malone, 1988; Roth & Fox, 1990; Liang et al., 1991) reconfirms earlier findings that substantial amounts of NO_3 can remain in the medium-to fine-textured soils during the overwinter period and can contribute substantial amounts of available N to subsequent crops. These results suggest that preplant measurement of soil profile NO_3 can provide a useful assessment of the probable contribution of profile NO_3 to crop N needs in areas with humid climates (Bock & Kelley, 1992; Bundy et al., 1992).

Another approach to use of soil NO_3 measurements to assess N availability in the field is the pre-sidedress soil NO_3 test (PSNT) for corn proposed by Magdoff et al. (1984). This procedure involves determination of NO_3-N concentration in the 0- to 30-cm soil layer when corn plants are 15 to 30 cm tall, and is discussed in detail in section 41–2.1.2. The delayed sampling time used in the PSNT method essentially provides for an in situ incubation in that soil and climatic factors affecting N availability are allowed to interact during the period prior to soil sampling (Magdoff et al., 1990). Therefore, the PSNT has potential for predicting gains in available N due to net N mineralization from soil organic matter, crop residues, including those from previous legumes, and manures or other organic N sources.

Similarly, available N losses through leaching or denitrification occurring before soil sampling should be reflected by the PSNT. This procedure has been evaluated in several locations in humid regions of the USA and is used to develop N-fertilizer recommendations for corn in several states.

It should be noted that the PSNT differs from the preplant soil profile NO_3 test in that the PSNT provides an index of N availability that can be used with appropriate calibration data to predict crop N response or the amount of additional N required by the crop. In contrast, preplant profile NO_3 tests provide a direct measurement of available N in some portion of the crop root zone, and this available N is usually directly credited against the crop N needs.

41–2 METHODS

41–2.1 Field Methods

41–2.1.1 Residual Profile Nitrate Tests

41–2.1.1.1 Field Sampling. Field sampling techniques for residual profile nitrate tests are based on studies of spatial variability of soil nitrate in the Western and Great Plains regions of the USA and Canada (Cameron et al., 1971; Bole & Pittman, 1976; Reuss et al., 1977). Results from these and other studies indicate that a composite of about 20 cores per field will estimate the mean soil NO_3 content within 15% about 80% of the time (Meisinger, 1984; Dahnke & Johnson, 1990). The sampling intensity needed to achieve this level of accuracy and precision is not greatly affected by field size, but the sampling area should be generally uniform with respect to past management, topography, and soil characteristics.

41–2.1.1.2 Procedure
1. Composite 20 soil cores from 10- to 40-ha areas with similar soils, drainage, and past management. Sample to a minimum depth of 60 cm in 30-cm depth increments and composite samples separately from each 30-cm layer. Follow suggestions from individual states or regions on the value of sampling soil to depths > 60 cm. If deeper samples are recommended, composite soil from these depths in 30-cm increments as indicated above.
2. Thoroughly mix soil from each depth increment in the field, discard nonsoil material, and place a minimum of 0.25 L of field moist soil into an appropriately labeled sample bag.
3. Dry the soil by spreading the sample out in a thin layer (about 1 cm thick) at room temperature with moderate air flow (e.g., household fans) for 24 h. Alternatively, the soil can be dried at low temperatures (< 40 °C) in a forced-air oven.
4. Crush the dried soil to pass a 2-mm screen and thoroughly mix the sample.

41–2.1.1.3 Laboratory Analysis. Analyze the soil for NO_3-N as described in section 41–2.3.1.1 or 41–2.3.1.2.

41–2.1.1.4 Interpretation of Test Results. Most areas where research has been conducted to evaluate use of a residual profile NO_3 soil test have established the relationships between test values and crop yields. Calibration information indicating the amounts of added N needed at various test values has also been determined from crop N response experiments (Hergert, 1987; Bock & Kelley, 1992). Although profile NO_3 test results are usually used to make direct credits against crop N requirements (Keeney, 1982), Hergert (1987) noted that the algorithms or other calibration data used to make N recommendations from N test results may differ substantially from the experimental data. The interpretation of soil profile NO_3 tests to develop fertilizer recommendations for crops varies among states and regions, and local information should be consulted for specific recommendations. In general, it involves calculation of the total amount of NO_3-N in the profile depth sampled and direct crediting of this N against the estimated N need of the crop to be grown.

1. Calculate the profile NO_3-N content in kg of NO_3-N ha^{-1} from the soil NO_3-N concentration and the assumption that 1 ha (30 cm of soil) weighs 4.48×10^3 Mg. If site-specific values for soil bulk density are available, these should be used in the calculation.
2. Adjust total profile NO_3-N content for background NO_3 levels or for expected differences in use efficiency of NO_3-N found at greater soil depths.
3. Calculate the recommended application rate from the difference between the N requirement of the crop to be grown and the adjusted soil profile NO_3-N content. Adjustments for legume or manure N contributions must be made separately from those based on the soil NO_3 test results.

41–2.1.1.5 Comments. Residual profile NO_3 tests differ in principle from the PSNT discussed in the following section (41–2.1.2) in that the preplant profile tests measure NO_3-N remaining in the root zone at or before crop establishment, and this NO_3 is credited against the N needs of the subsequent crop. In contrast, the PSNT provides an index of N availability measured during the early part of the growing season, which must be related to crop needs through test calibration information. Advantages of the residual NO_3 tests are that they provide a direct measurement and accounting of available N in the plant root zone that does not depend on predicting the environmentally dependent release of available N from various sources. In addition, the fall or early spring sampling time used for the preplant test is frequently more convenient for users and allows more time for soil analysis and fertilizer applications than the PSNT. The associated disadvantage is that preplant soil NO_3 tests are not useful for assessing available N contributions from organic N sources such as soil organic matter, previous legume crops, and organic wastes. Other disadvantages of preplant soil NO_3 tests are the risk of NO_3 loss from the root zone during the interval between soil sampling and crop N uptake and the increased difficulty and expense of obtaining the soil profile samples.

41–2.1.2 Pre-Sidedress Nitrate Test

This NAI is designated as the PSNT because soil nitrate is sampled a few weeks before the normal time for sidedress N applications to corn, and to clearly distinguish it from preplant nitrate tests. The PSNT also differs from preplant tests because only the surface 30 cm is sampled, while preplant tests sample to 60 cm or more. The literature also refers to the PSNT as the late spring nitrate test, the Vermont or Iowa test, and the June or mid-June nitrate test. Magdoff et al. (1984) first suggested the corn PSNT test. They reported a good relation ($R^2 = 0.74$) between corn silage yields and soil NO_3-N concentrations in the surface 30 cm when the corn plants were 15 to 30 cm tall. A review and evaluation of the PSNT test in humid regions was recently provided by Bock and Kelley (1992).

41–2.1.2.1 Principles. The PSNT is based on a timely sampling of the spring accumulation of soil NO_3-N under natural field conditions, just before the warm-season corn crop begins its period of rapid N uptake (Magdoff, 1991; Meisinger et al., 1992b). The PSNT assumes that most of the corn fertilizer N will be applied as a sidedressing and that only small quantities of N (< 50 kg of N ha^{-1}) were applied at planting, either as starter fertilizer or uniformly applied broadcast N.

The NO_3-N content of a typical agricultural soil represents the net balance between nitrate production processes vs. nitrate loss processes. In a humid-temperate climate, a typical silt loam surface soil will have nitrate contents that are: lowest in winter (due to leaching), rise in the spring and early summer (due to commencement of mineralization), decrease during summer (due to crop uptake), and increase again in the fall (Harmsen & Van Schreven, 1955; Stevenson, 1986). Figure 41–1 summarizes recent data from Sarrantonio and Scott (1988) that exemplifies this pattern (solid symbols) for central New York conditions. Surface soil NO_3-N concentration after winter leaching (or excess irrigation) would commonly be 5 to 10 mg of NO_3-N kg^{-1} but this may increase two- to sixfold in the late spring, depending on recent additions of manure or crop residues (Magdoff et al., 1984; Fox et al., 1989; Meisinger et al., 1992a).

The PSNT has succeeded in humid regions because leaching is not an efficient nitrate removal process in fine- and medium-textured soils. That is, nitrate is not as mobile and subject to complete leaching losses as once thought, due to preferential water flow. Furthermore, preferential flow is enhanced by reduced tillage systems (especially no-tillage; e.g., Tyler & Thomas, 1977; Thomas et al., 1989) that have greatly expanded in the past 10 yr. Preferential flow is characterized by rapid and deep percolation through large-pore sequences that can by-pass spring mineralized N produced in small-pore sequences.

The PSNT has also succeeded because it has been applied to a warm-season crop (corn), which allows ample time for the spring mineralization period before the crop begins its period of rapid N uptake (see dashed symbols of Fig. 41–1). However, the close juxtaposition of the spring NO_3-N maximum and the corn N uptake period can also cause logistical

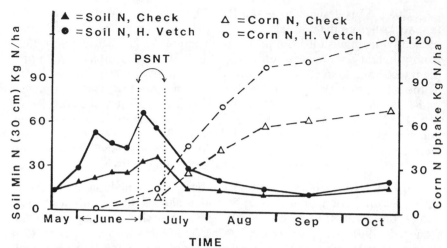

Fig. 41–1. Soil mineral N contents (solid symbols) and corn N uptake (open symbols) throughout the growing season in a central New York silt loam soil previously receiving no fertilizer
N (triangular symbols) or after a hairy vetch winter cover crop (circular symbols) (Sarrantonio & Scott, 1988).

problems. Only about 2 to 3 wk are available to measure the soil NO_3-N
and to apply the appropriate sidedress fertilizer N.

The above principles explain why the corn PSNT has been successful
in humid-temperate climates, but they also contain the basis for limitations
of this test. For example, the PSNT would be less useful in: (i) unusual
spring weather conditions that affect mineralization, such as cool or dry
springs; (ii) on highly leachable nonstructured soils, such as loamy sands;
(iii) with cool-season crops that rapidly assimilate spring mineralized N,
such as winter wheat; and (iv) in circumstances where soil N transformations perturb the NO_3-N pool, such as short-term denitrification events or
short-term immobilization events.

41–2.1.2.2 Interpretation. The PSNT must be interpreted in relation
to field crop response to determine its usefulness as a NAI. Field evaluations have been reported from 34 locations in Vermont (Magdoff et al.,
1984), nine experiments over 2 yr in Iowa (Blackmer et al., 1989), 47
treatment-year combinations in Maryland (Meisinger et al., 1992a), and
272 site-years of combined data throughout the Northeast (Magdoff et al.,
1990). These field studies have convincingly shown that the PSNT test can
successfully identify N-sufficient sites. There is a remarkable consensus
that PSNT NO_3-N concentrations of 20 to 25 mg of NO_3-N kg^{-1} or more
are associated with N sufficiency for corn.

Using the PSNT as a quantitative index of fertilizer N needs has met
with varied success. For example, Pennsylvania, Iowa, and Vermont use
PSNT values below the critical level as a direct input into fertilizer N

recommendations (Beegle et al., 1989; Jokela, 1989; Blackmer et al., 1991). Others have concluded that variability in the PSNT vs. relative yield relation limits its usefulness as a quantitative index (Fox et al., 1989; Meisinger et al., 1992a). We recommend that the quantitative adoption of the PSNT should be appraised by each individual state, because this issue involves a critical examination of: (i) current field calibration data, (ii) fertilizer N recommendation philosophies, and (iii) alternative fertilizer N-recommendation systems. Each of these elements has unique solutions that are specific to the soils, climate, and cropping systems of each state.

41–2.1.2.3 Method. The PSNT basically involves collecting a representative soil sample at a specific time and determining the NO_3-N concentration. Modifications of the preplant soil-nitrate sampling procedure have been made to accommodate the unique aspects of PSNT sampling. We have recommended an automated laboratory NO_3-N method suited to fast analysis of many samples because the PSNT requires rapid sample turn-around. A manual NO_3-N method is also described.

41–2.1.2.3.1 Field Sampling

The PSNT requires collection of a representative soil sample from the surface 30 cm when the corn is 15 to 30 cm tall. This is not a small task because soil NO_3-N variability is large, CVs routinely range between 30 and 80%, with 45% being a common value (Meisinger, 1984). The PSNT assumes that most of the N will be applied as a sidedressing with only modest N rates applied at planting (< 50 kg of N ha^{-1}) as banded starter fertilizer or as uniform broadcast N application (e.g., herbicides with UAN).

41–2.1.2.3.1.1 Procedure
1. Use the same basic soil sampling protocol as section 41–2.1.1.2, except sample only the surface 30 cm.
2. Pay special attention to identify areas that received manure, sludges, or other organic amendments within the past 3 yr and sample them separately. Carefully avoid areas with preplant fertilizer N (starter bands) by sampling between rows.
3. Dry, crush, and mix soil sample as per section 41–2.1.1.2.

41–2.1.2.3.1.2 Laboratory Analysis. Analyze the soil for NO_3-N as described in section 41–2.3.1.1 or 41–2.3.1.2.

41–2.1.2.3.1.3 Comments. The field sampling problem for NO_3-N has been summarized by Meisinger (1984), who concluded that large numbers of cores would be needed to accurately estimate NO_3-N means. Common soil sampling instructions call for 10 to 20 cores from an area selected to contain similar soils, past management, etc. With a CV of 45%, this sampling intensity would likely estimate the NO_3-N mean to ±20% on 9 out of 10 fields. Doubling the number of cores would improve the estimate to about ±14% on 9 out of 10 fields. There is a need for research studies to design efficient sampling plans for the PSNT.

Location of the samples in relation to preplant N application zones is important, due to the high N concentrations near fertilizer bands. The best way to avoid problems is to avoid the fertilizer bands, which can be accomplished with N applied as starter near the row. At this time, we do not recommend PSNT sampling if anhydrous ammonia has been applied preplant because the location of these injection zones is not known. However, Iowa researchers (Davis & Blackmer, 1990) may devise a suitable sampling scheme in the future.

The time of sample collection is critical. Sampling when the corn is 15 to 30 cm tall allows for an increase in soil NO_3-N due to spring mineralization, plus time for analysis and sidedress N applications, before rapid corn N uptake causes soil NO_3-N levels to dramatically decline (see Fig. 41-1). If sampled too early or too late, the soil NO_3-N levels will probably be biased low. The short period available for PSNT sampling places a premium on soil sampling and limits the number of sites an individual can sample. Experience in the northeastern states indicates that one consultant can service about 1000 ha (about 2500 acres) using conventional manual sampling techniques.

The 30-cm PSNT sample depth represents a compromise between sampling the entire root zone vs. conventional plow layer sampling. Rootzone sampling would greatly increase labor requirements. The 30-cm depth has been shown to produce less variability in critical NO_3-N concentrations than shallow 0- to 15-cm samples (Blackmer et al., 1989). Sampling the entire 30-cm depth is important because many soils will contain high NO_3-N levels in the surface layers due to recent mineralization of surface residues or surface broadcast applications of fertilizer N.

The objectives of soil sample drying and processing are to stop biological activity as soon as feasible and to homogenize the sample. Microbial activity can be stopped by rapid air drying, by low-temperature oven drying, or freezing; but freezing is usually only a temporary preservation measure. After drying, it is important to mix the sample well because only 10 g of soil will be analyzed out of 2000 g collected. Saffigna (1988) reported a reduction in the within sample NO_3-N CV from 25 to 5% due to simple soil mixing. One should also remember that the 10-g analytical sample ultimately represents nearly 10^{10} g of soil if a 2.5-ha area were sampled to 30 cm.

41–2.2 Laboratory Methods

Laboratory methods of assessing N availability include biological and chemical techniques (see section 41–1.1). The biological methods usually involve incubation of soil under conditions that promote N mineralization from organic sources, and measurement of the inorganic N produced. In this chapter, both short-term (ammonium production under water-logged conditions) and long-term (aerobic incubation with periodic leaching) incubation methods will be recommended. The potential use of several promising chemical methods will also be discussed.

41–2.2.1 Biological Methods

41–2.2.1.1 Theory and Potential Application of Methods. As noted in the introduction (41–1.1), biological N availability indices are based on the assumption that the same biological processes that cause release of plant-available N in the field are also responsible for production of inorganic N in the laboratory procedures. The results from biological N availability indices must be viewed as relative indications of soil N availability.

41–2.2.1.2 Ammonium Production During Waterlogged Incubation

41–2.2.1.2.1 Principles

In this method, soil N availability is estimated from NH_4-N produced during a 7-d waterlogged incubation. The procedure was initially proposed by Waring and Bremner (1964), and this technique was the biological index recommended by Keeney (1982). While limitations of biological indices stated earlier also apply to this method, it has several advantages that make it attractive where a rapid biological index is needed to provide a relative assessment of N availability. These advantages include its simplicity and ease of adaptation to laboratory routine, a short incubation time (7 d), little or no influence of sample pretreatment on test results, elimination of concerns related to optimum water content and water loss during incubation, and minimal apparatus and reagent requirements (Keeney, 1982).

Although previous reviews have cited numerous reports of satisfactory relationships between results from this method and other indices (Keeney, 1982; Stanford, 1982; Meisinger, 1984), several studies have found poor correlations between NH_4-N production under waterlogged conditions and field measurements of N availability (Fox & Piekielek, 1984; McCracken et al., 1989; Hong et al., 1990). Boone (1990) suggests that the apparent differences between N availability measured via anaerobic incubation and field data are not contradictions but instead reflect differences in N transformations measured by the two methods. Specifically, field measurements represent the net effect of N mineralization and N immobilization under aerobic conditions, while the waterlogged incubation likely measures N mineralization from aerobic soil organisms killed by the anaerobic test conditions.

41–2.2.1.2.2 Method

41–2.2.1.2.2.1 Special Apparatus. Automated analytical equipment for continuous flow analysis of NH_4-N (see section 41–2.3.2.1) or perform NH_4-N analyses using a manual method (section 41–2.3.2.2). If NH_4-N analyses are to be performed by steam distillation, use the procedures described by Keeney (1982).

41–2.2.1.2.2.2 Reagents. Potassium chloride (KCl) solution, approximately 4 *M:* Dissolve 3 kg of KCl in 10 L of water.

41–2.2.1.2.2.3 Procedure. Place 12.5 ± 0.1 mL of water in a 16 by 150-mm test tube and add 5 g of soil. Stopper the tube and place it in an incubator at 40 °C. After 7 d, add 12.5 ± 0.1 mL of 4 M KCl to the tube and shake the contents for 1 h on a mechanical shaker. Filter the contents of the tube through acid-washed Whatman no. 42 filter paper. Determine the NH_4-N concentration in the filtrate using the automated method described in section 41–2.3.2. Determine the amount of NH_4-N in the soil before incubation by the same procedure used for the incubated samples and calculate mineralizable N from the difference between the two analyses.

41–2.2.1.2.2.4 Comments. For routine use of the method, soil samples can be measured on a volume basis rather than a weight basis (Keeney, 1982). Determination of initial NH_4-N in the soil can often be omitted without greatly affecting the results, since most agricultural mineral soils contain limited amounts of exchangeable NH_4-N. Initial NH_4-N should be determined in any correlation or calibration studies with the method. Ammonium-N determinations on filtered extracts using the automated technique recommended may produce lower mineralizable N values than direct steam distillation of the incubated samples with MgO (Sahrawat & Ponnamperuma, 1978; Keeney, 1982).

41–2.2.1.3 Inorganic Nitrogen Production During Long-Term Aerobic Incubation

41–2.2.1.3.1 Principles

Use of cumulative inorganic (NH_4 + NO_3)-N production in long-term incubations to estimate potentially mineralizable N in soils was initially proposed by Stanford and Smith (1972). They suggested that the rate of N mineralization is proportional to the amount of potentially mineralizable N in the soil, that the pattern of N mineralization follows first-order kinetics, and that an N mineralization potential value (N_o) can be estimated from $\log(N_o - N_t) = \log N_o - kt/2.303$. Stanford (1982) concluded that the N_o values are a definable soil characteristic that may be useful for estimating the N-supplying capacities of soils under specific environmental conditions. Work by Stanford et al. (1973) showed that the rate constant k has a Q_{10} of 2 within the temperature range of 5 to 35 °C, and Stanford and Epstein (1974) found that soil water content near field capacity was directly correlated with N mineralization. The possibility of making adjustments for temperature and soil water content stimulated research interest in estimating N mineralization in the field based on N mineralization potentials (Smith et al., 1977).

The technique involves measurement of inorganic N produced during aerobic incubation of soil or soil amended with sand or vermiculite under near-optimum conditions of temperature, moisture, and aeration for up to 30 wk. Inorganic N is usually removed by periodic leaching of soil samples incubated in a combination filtration-incubation container. The total incubation time required varies with the objective and conditions of individual experiments, but the incubation should be continued until an adequate description of the relationship between cumulative N mineralization and time of incubation is obtained (Stanford, 1982). Although the initial work of Stanford and Smith (1972) involved a 30-wk incubation, subsequent studies (Stanford et al., 1974) indicate that adequate estimates of potentially mineralizable N can often be obtained with 8- to 10-wk incubation periods.

Initially, Stanford and Smith (1972) suggested that N mineralized during the first 7 to 14 d of incubation should be disregarded because of the strong and variable influence of sample pretreatment on the amounts of N mineralization during this period. However, subsequent work indicates that the pattern of N mineralization during the first few weeks of incubation may be critical for understanding the N availability status of soils. In general, N released during the first few weeks of incubation likely reflects the readily mineralizable or active fraction of soil organic N, while that released later probably originates from a much larger but more stable soil organic matter pool. The concepts of two separate soil organic N pools, each with its own decomposition rate, are reflected in much of the recent work using long-term incubation methods to assess soil N availability and develop appropriate models to describe the N mineralization process (Molina et al., 1983; Lindemann & Cardenas, 1984; Beauchamp et al., 1986; Deans et al., 1986; Bonde & Rosswall, 1987; Bonde & Lindberg, 1988; Ellert & Bettany, 1988; Boyle & Paul, 1989). In addition, several studies have used simultaneous measurements of N and C mineralization to describe the observed patterns of N mineralization during long-term incubations (Molina et al., 1983; Robertson et al., 1988; Houot et al., 1989).

Because of the substantial time and apparatus requirements for this method, it is usually used only when long-term N mineralization information is essential. For example, long-term incubation data provide a standardized method of assessing the potential long-term N supplying capacities of soils (Stanford, 1982). In addition to possible use for estimating N mineralization in field soils, the long-term incubation technique is useful for evaluating the effects of past management practices or experimental treatments on soil N-supplying capability. Long-term incubations are also frequently used in work to model soil N mineralization and characterize various components of the labile N pool in soils.

41–2.2.1.3.2 Method

41–2.2.1.3.2.1 Special Apparatus
1. Filter units, polystyrene, 150-mL (Falcon 7111, Becton Dickinson Co., Lincoln Park, NJ)[1], fitted with cellulose acetate filter

---

[1]Mention of product names or companies are for the benefit of the reader and do not imply endorsement or preferential treatment by ASA, the Univ. of Wisconsin-Madison, or the USDA, to the exclusion of other similar products, which may also be satisfactory.

membranes having a 0.22 μm pore size and a bubble point of 373 kPa. The pressed fiberglass prefilter media supplied with this unit are also needed.

2. Fiberglass prefilters:
   a. 47-mm diam., 0.5 μm pore size (Micron Separations, no. G15, Westboro, MA).
   b. 47-mm diam., 1.3-mm thick, with acrylic binders for additional wet strength (no. 66078, Gelman Sciences Inc., Ann Arbor, MI).
3. Fiberglass screen, 1-mm mesh size cut to fit inside the 65-mm diam. filter units (Gallagher Tent and Awning Co., Madison, WI).
4. Vacuum pump and manifold (with stopcocks for control of vacuum to filter units) capable of applying a suction of 80 kPa to each filter unit.
5. Vacuum regulator and manometer to control and monitor vacuum level.
6. Drip irrigation system (Bonde & Rosswall, 1987) for addition of leaching solution to filter units. Apparatus consists of 150-mL funnels connected with Tygon tubing to the center port of the filter unit's detachable lid. The tubing is fitted with a screw clamp to control flow rate from the funnel.
7. Filter flasks, Erlenmeyer, 250-mL, heavy walls, with sidearms.
8. Analytical equipment for determination of inorganic N in leachates (see section 41–2.3).

41–2.2.1.3.2.2 Reagents
1. Calcium chloride ($CaCl_2$) solution, approximately 0.01 *M*. Dissolve 15 g of calcium chloride dihydrate ($CaCl_2 \cdot 2H_2O$) in 10 L of water.
2. Nutrient solution, minus N (Stanford & Smith, 1972), approximately 0.002 *M* $CaSO_4 \cdot 2H_2O$; 0.002 *M* $MgSO_4$; 0.005 *M* $Ca(H_2PO_4)_2 \cdot H_2O$; and 0.0025 *M* K_2SO_4. Dissolve 3.4 g of $CaSO_4 \cdot 2H_2O$, 2.4 g of $MgSO_4$, 12.6 g of $Ca(H_2PO_4)_2 \cdot H_2O$, and 4.4 g of K_2SO_4 in 10 L of water.
3. Silica sand, acid washed, 0.42 to 0.59 mm particle diameters.

41–2.2.1.3.2.3 Procedure. Assemble the filter units by removing the threaded funnel from the filter unit base and positioning the unit's cellulose acetate membrane and O-ring gasket in the base of the filter unit. Place the Micron Separations G15 fiberglass prefilter on top of the membrane and O-ring gasket, and replace the funnel assembly. This prefilter prevents soil particles from clogging the membrane pores and avoids microbial growth on the membrane surface (Beauchamp et al., 1986). Place the pressed fiberglass filter provided with the filter unit on top of the Micron Separations prefilter to protect the prefilter and membrane from clogging. Add deionized water to the filter unit to thoroughly moisten the prefilters and membrane before adding the soil.

Uniformly mix field-moist samples of the soil to be evaluated (30 g oven-dried weight) with 30 g of acid-washed silica sand and transfer the mixture to the filter unit. Determine the water content in a subsample of the field-moist soil by oven drying (105 °C) before preparing the soil-sand mixture. Place two layers of the 1-mm mesh fiberglass screen on top of the soil-sand mixture, and place the Gelman no. 66078 fiberglass prefilter on the fiberglass screen. The screen and prefilter are permanently positioned in the filter unit to minimize soil dispersion during leaching. Attach the lid of the filter unit to the funnel, place the cotton-plugged fitting provided with the unit on the off-center port in the unit lid to allow air exchange during incubation. Seal the vacuum port in the filter unit base with the rubber cap provided.

Place the filter unit containing the soil-sand sample on a 250-mL Erlenmeyer filtering flask with its sidearm connected to a vacuum manifold adjusted to deliver a suction of 80 kPa. Attach the tube from the drip irrigation system to the center port in the filter unit lid. Add 100 mL of 0.01 M $CaCl_2$ solution to the irrigation system funnel, adjust the screw clamp on the inlet tube to a flow rate of approximately 150 mL h^{-1}, and apply suction to the filter unit. After the 0.01 M $CaCl_2$ solution has been added to the sample in the filter unit, place 25 mL of the minus-N nutrient solution into the irrigation system funnel and continue to collect the leachate. When leaching is complete (approximately 1 h), continue suction on the leaching unit for 0.5 h to provide a uniform moisture content in the soil-sand mixture for the subsequent incubation period. Disconnect the filter unit from the vacuum and irrigation systems, seal the center port in the unit lid with parafilm, and place the filter units in an incubator maintained at 35 °C and containing an open pan of water for humidity control. Measure the volume of leachate (± 1 mL) collected from each sample to allow calculation of N mineralized on a soil basis. Analyze an aliquot of the leachate for inorganic N as described in sections 41–2.3.

The procedure described in the preceding paragraph is performed immediately prior to the initial incubation and is repeated after each incubation interval. The frequency of leaching and the total duration of the incubation period depend on the objectives of the individual investigation. General information on the rates and amounts of soil N mineralization can often be obtained with four to six leachings over an 8- to 12-wk incubation period. Detailed studies may require 12 to 14 leachings and incubation times up to 40 wk.

41–2.2.1.3.2.4 Comments. The procedure described follows the basic concepts of long-term incubation and periodic leaching proposed by Stanford and Smith (1972), and employs improvements developed in more recent long-term N mineralization studies. Filter units similar to those described above have been employed in numerous long-term incubation studies (MacKay & Carefoot, 1981; Bonde & Rosswall, 1987; Bonde et al., 1988; Marion & Black, 1988; Robertson et al., 1988; Nadelhoffer, 1990). These membrane filter units provide the advantages of standardized conditions and more uniform control of soil moisture content during incubation. A filter membrane with a bubble point exceeding the suction applied for leaching provides accurate control of soil water content during incubation (MacKay & Carefoot, 1981). Increased variability in N mineralization measurements has been attributed to less precise control of soil moisture where filter membranes were not used in the incubation devices (Nadelhoffer, 1990). Although MacKay and Carefoot (1981) initially suggested overnight equilibration at an appropriate suction to achieve reproducible N mineralization values (CV = 3.9%), much shorter equilibration periods appear to be adequate. Using the 0.5-h equilibration suggested above in the author's laboratory, average CV values of < 5.0% are routinely obtained for 40-wk cumulative N mineralization values in experiments involving 20 treatments and four replications. These results suggest that extensive laboratory replication of incubated treatments may not be required for many studies. An alternative technique for controlling soil moisture contents in N mineralization studies by adjusting soil-to-sand mixing ratios has been described by Lueking and Schepers (1986).

Simultaneous measurements of CO_2 evolution are often of interest in long-term incubation N mineralization studies to obtain C mineralization information (Paustian & Schnurer, 1987; Gale & Gilmour, 1988; Robertson et al., 1988; Boyle & Paul, 1989; Houot et al., 1989; Nadelhoffer, 1990). The units described (Falcon 7111) can be accommodated inside a 1.9-L wide-mouth terephthalate jar (No. YB-06043-75, Cole-Parmer Instrument Co., Chicago, IL) along with a 50-mL Erlenmeyer flask containing 20 mL of 0.5 M NaOH to absorb CO_2 and a 70-mL vial of water to minimize soil water loss during the incubation. When the incubated soil samples are removed from the jar for periodic leaching, the alkali traps are replaced, and the sorbed CO_2 is determined using an automated colorimetric method adapted from Chaussod et al. (1986). Enclosing the filter units inside jars during incubation also provides very precise control of sample water content relative to units incubated without enclosure. In the author's laboratory, enclosed samples lost only about 2 to 3 mL of water during 4 wk of incubation based on the average volume of leachate retained at the next leaching.

Use of field-moist soil samples is recommended to avoid the enhanced N mineralization frequently observed during the first week of incubation with air-dried soils (Stanford & Smith, 1972; Beauchamp et al., 1986; Cabrera & Kissel, 1988c). Beauchamp et al. (1986) reported that, due to the initial flush of mineralized N during the first 7 d of incubation, the total amount of N mineralized in 42 d from air-dried soil was greater than from frozen or field-moist soils. The flush of N released when air-dried soils are incubated may originate from soil microbes killed by the air-drying pretreatment (Richter et al., 1982). Beauchamp et al. (1986) concluded that this N release is an experimental artifact and should not be considered part of the true soil mineralization potential. Use of air-dried samples may produce satisfactory results where a relative comparison of soil N availability is desired.

A concern with incubation procedures involving periodic leaching to remove mineralized N is that some soluble organic N may be removed during the leaching process (Smith et al., 1980; Beauchamp et al., 1986; Smith, 1987). The amounts of organic N found in leachates has varied from 37 to 70% of the total N leached (Smith et al., 1980) to 10 to 20% of the total N leached (Beauchamp et al., 1986). Beauchamp et al. (1986) found no differences in soluble organic N among soils or between air-dried or field moist sample pretreatments. Working with air-dried soils, Smith (1987) found that most of the soluble organic N was removed in the initial leaching, and that organic N represented 5% or less of the inorganic N produced during an 84-d incubation. Smith (1987) also found that the soluble organic N in leachates was not highly susceptible to mineralization. Although several authors have noted the need to include soluble organic N in estimates of total N mineralization, it appears that adequate estimates of soil N availability can be obtained from inorganic N production alone (Smith, 1987).

41-2.2.2 Chemical Extraction Methods

In view of the many chemical methods available and the low correlation of many of these techniques with field-measured N availability, no single chemical method is recommended here. The selection and use of these techniques should be determined from the objectives of the work in which the methods are used. If the goal is to obtain a rapid, relative indication of N availability among soils differing in past management, several recent techniques including digestion with 2 M KCl at 100 °C (Gianello & Bremner, 1986a), steam distillation with pH 11.2 phosphate-borate buffer (Gianello & Bremner, 1988), and UV absorbance at 200 nm of $NaHCO_3$ soil extracts (Hong et al., 1990), are likely to produce satisfactory results. These methods have advantages over previously proposed chemical indices in that they are convenient, rapid, and well correlated with N availability determined by incubation methods (Gianello & Bremner, 1986b) or field measurements (Hong et al., 1990).

If the objective is to predict soil N availability for determination of N recommendations in crop production, methods other than chemical indices should be considered because chemical indices are usually not well correlated with field-measured N availability (Hong et al., 1990). A possible exception is the procedure evaluated by Hong et al. (1990) involving determination of UV absorbance at 200 nm of $NaHCO_3$ soil extracts. This measurement reflects both NO_3 (Norman et al., 1985) and organic matter (Fox & Piekielek, 1978) in the soil extract, but the method has not been widely evaluated. In addition, Fox and Piekielek (1984) found that N mineralized in laboratory incubation methods was not well correlated with field measurements of soil N availability. This suggests that chemical indices well correlated with aerobic or anaerobic incubation methods may not provide good estimates of N availability in the field. Hong et al. (1990) showed that the best relationships between field-measured N availability and the indices evaluated in their work occurred with either direct or indirect measurements of soil NO_3 at corn planting or at the pre-sidedress growth stage. This finding and the wide-spread interest in development and use of soil NO_3 tests to predict crop N requirements (Magdoff et al., 1984; Hergert, 1987; Blackmer et al., 1989; Fox et al., 1989; Dahnke & Johnson, 1990; Magdoff et al., 1990; Magdoff, 1991) suggests that appropriately timed soil NO_3 measurements will likely provide the best assessment of field N availability.

41–2.3 Methods for Inorganic Nitrogen

41–2.3.1 Nitrate Method

Nitrate is extracted from a representative soil sample by shaking 10 g of soil with 100 mL of 2 M KCl for 1 h, followed by filtration. The basic analytical steps are: NO_3 reduction to NO_2 by passage through a copperized-cadmium column, and determination of NO_2 with a modified Griess-Illosvay procedure by diazotizing with sulfanilamide and coupling with N-(1-naphthyl)-ethylenediamine to form a purple-colored dye, which is measured by absorbance at 520 nm. The method is described in detail by Bremner (1965b) and Keeney and Nelson (1982); it has excellent sensitivity and few interferences with filtered extracts. The general working range is 0.01 to 5 mg of NO_3-N L^{-1} of soil extract (Henriksen & Selmer-Olsen, 1970; Keeney & Nelson, 1982) and a multirange manifold has been described to extend the range to 100 mg of NO_3-N L^{-1} (Jackson et al., 1975). Both an automated and a manual method of this procedure are described below. Alternative NO_3-N analysis methods may also be acceptable and are briefly discussed in the comments section below.

41–2.3.1.1 Laboratory Analysis, Automated. The following automated Cu-Cd reduction procedure has been adapted for soil analysis from EPA method no. 353.2 (USEPA, 1983) and the American Public Health Association Standard Method no. 4500-NO_3-F (APHA, 1989), which were designed for NO_3^--N analysis of water and wastewater using a Technicon autoanalyzer system (Technicon Instrument Corp., 1977b). The procedure has been successfully applied to soil extracts as reported by Henriksen and Selmer-Olsen (1970), Ananth and Moraghan (1987), Gelderman and Fixen (1988), and Tel and Heseltine (1990).

41–2.3.1.1.1 Special Apparatus

Automated analytical equipment for continuous flow instrumentation consisting of: a sampler, a manifold, and NO_3-N analytical cartridge, a proportioning pump, a flow cell colorimeter capable of measuring absorbance at 520 nm (although 510–540 nm may be used), and a recorder or other data output storage device. The detailed configuration of the apparatus depends on manufacturer's specifications for specific instruments. Consult the manufacturer's operating instructions for specifications on component operation, flow rates, tube sizes, and analysis rates.

41–2.3.1.1.2 Reagents

1. Potassium chloride, 2 *M:* Dissolve 1490 g of KCl in 8 L of distilled water and dilute to 10 L.
2. Copper sulfate solution, 0.08 *M:* Dissolve 20 g of $CuSO_4 \cdot 5H_2O$ in water and dilute to 1 L.
3. Dilute HCl, 6 *N:* Dilute 50 mL of conc. HCl to 100 mL.
4. Granulated Cd: 40 to 60 mesh.
5. Ammonium chloride-EDTA solution: Dissolve 85 g of NH_4Cl and 0.1 g of disodium ethylenediamine tetraacetate in distilled water; adjust pH to 8.5 with conc. NH_4OH; dilute to 1 L, and add 0.5 mL of the wetting agent polyoxyethylene 23 lauryl ether (Brij-35). Store in refrigerator to minimize bacterial growth.
6. Color reagent: To approximately 800 mL of distilled water, add, while stirring, 100 mL conc. phosphoric acid, 40 g of sulfanilamide, and 2 g of *N*-1-naphthylethylenediamine dihydrochloride. Stir until dissolved and dilute to 1 L. Store in brown bottle in refrigerator when not in use. This solution is stable for several months.
7. Stock nitrate solution: Dissolve 7.218 g of pure dry KNO_3 in distilled water and dilute to 1 L. This solution contains 100 mg of NO_3-N L^{-1}. Store in refrigerator.
8. Working nitrate solution: Using the stock nitrate standard solution (reagent no. 7 above), prepare working nitrate standards by pipetting 0, 1, 5, 10, 20, 30, and 50 mL into 1-L volumetric flasks and diluting to volume with soil extracting solution (2 *M* KCl, reagent no. 1 above). These working standards contain 0, 0.1, 0.5, 1, 2, 3, and 5 mg of NO_3-N mL^{-1}. Store in refrigerator.
9. Stock nitrite solution: Dissolve 0.493 g of pure dry $NaNO_2$ in distilled water and dilute to 100 mL. This solution contains 100 mg of NO_2-N mL^{-1}. Store in refrigerator at 4 °C.
10. Using stock nitrite solution (reagent no. 10 above), dilute 1.0 mL to 500 mL with 2 *M* KCl (reagent no. 1 above). This solution contains 2.0 mg of NO_2-N mL^{-1}. This solution is not stable — prepare as needed.

41–2.3.1.1.3 Procedure

1. Prepare Cd column: Place about 10 g of granulated Cd in a beaker and wash with about 50 mL 6 *N* HCl for 1 min, followed by a thorough rinse with distilled water (Cd should be silver). Add about 100 mL of 0.08 *M* $CuSO_4$ solution and swirl for about 5 min or until blue color fades, decant solution and repeat with fresh $CuSO_4$ until a brown colloidal precipitate of Cu forms. Gently rinse with distilled water to remove all precipitated Cu, repeat distilled water rinse about 10 times until all blue and light-gray color disappear from the wash (Cd should be black).

2. Set up continuous flow analysis system for $(NO_3 + NO_2)$-N as per manufacturer's instructions.

3. Extract soil by placing 10 g of dry soil in a suitable 250-mL vessel, add 100 mL of 2 M KCl, stopper, and shake vigorously. Continue shaking for 1 h on a mechanical shaker or with frequent manual shaking, filter the soil-KCl suspension through a Whatman no. 42 filter to obtain a clear extract, discard the first portion of the filtrate but collect the remainder for analysis.

4. If the soil extract pH is below 5 or above 9, adjust the pH to between 5 and 9 with either HCl or KOH.

5. Allow continuous flow analysis system to warm up according to manufacturer's instructions, allow all reagents and soil extracts to reach room temperature, obtain a stable baseline output with all reagents, supplying 2 M KCl extracting solution in place of unknowns. A new reduction column may have to be conditioned with a nitrate working standard prior to use; see manufacturer's instructions.

6. Place appropriate NO_3-N working standards in sampler (use at least 5), which will bracket the expected NO_3-N concentration in the soil extracts. Place unknown soil extracts in the sampler with periodic insertion of a mid-range NO_3-N standard for quality control.

7. Switch sample line to sampler and start analysis with due attention to output device gain to be sure all working standards are within range.

8. Calculate results by preparing an appropriate standard curve derived from plotting peak heights (or other output signal) vs. NO_3-N concentration. Compute the final soil NO_3-N concentrations (mg NO_3-N kg^{-1} soil) by converting soil extract peak heights to NO_3-N concentrations via the standard curve and multiplying by 10 (for the 10:1 dilution from soil extraction). For more precise work, soil hygroscopic moisture may be taken into account (usually 1–4%, depending on clay content) by drying a separate soil sample at 105 °C.

41–2.3.1.1.4 Comments

The precision of this method has been reported to be ±0.06 mg of NO_3-N L^{-1} (Henriksen & Selmer-Olsen, 1970), with CVs being 1 to 2% for typical soil extract NO_3-N concentrations (Skjemstad & Reeve, 1978; APHA, 1989).

This method determines $(NO_3 + NO_2)$-N in the soil extract. Separation of NO_3 from NO_2 can be accomplished by conducting the same analysis without the Cu-Cd column, which will quantify the NO_2-N directly. However, soils generally contain insignificant levels of NO_2, so the determination and interpretation of $(NO_3 + NO_2)$-N data are equivalent to NO_3-N for most applications.

The efficiency of the Cu-Cd column can be determined by passing the 2 mg of NO_3-N L^{-1} standard through the column and comparing it to the 2 mg of NO_2-N L^{-1} standard, which was not passed through the column. If $< 100\%$ efficiencies are detected, the column is still likely to be usable because the quantity of NO_2-N produced will still be proportional to the NO_3-N content. After several hundred samples, the column efficiency will reach a point where Cd metal should be re-copperized; however, this procedure is only effective a few times before a new Cd must be used. The potential problem of NO_2-N reduction within the column can be evaluated by passing the NO_2-N standard through the column and comparing it to the original NO_2-N standard. Nitrite reduction can be controlled by increasing the flow rate.

The NH_4Cl reagent has EDTA included to prevent interferences from high concentrations of Fe, Zn, and other metals. Most soil extracts, however, will not contain high enough concentrations of these metals to cause interference. Indeed, Henriksen and Selmer-Olsen (1970) found no interference from 1% solutions of several heavy metal salts. The EDTA has been included as a precaution to ensure quantitative NO_3-N analysis even in unusual situations (e.g., soils treated with sludges containing heavy metals), but several investigators have reported it to be unnecessary for routine soil analysis (Henriksen & Selmer-Olsen, 1970; Dorich & Nelson, 1984).

The 2 M KCl extract is recommended because it is the most common soil extractant for inorganic N and has proven satisfactory for a wide range of soils. Several other extracts have also been successfully employed such as: 1 M KCl (Bremner, 1965b), 0.5 M K_2SO_4 (Magdoff et al., 1984), 0.025 M $Al_2(SO_4)_3$ (Roth & Fox, 1990), and various Ca salts (Bremner, 1965b; Sims & Jackson, 1971). If a different extractant is selected, it should be evaluated with standard NO_3-N additions to representative soils to ensure quantitative extraction and freedom from interferences. Prepare the working standards with the same solution used for soil extraction. The method described is suitable for NO_3-N analysis in the 0.01 M $CaCl_2$ leachates obtained from aerobic incubation experiments (41–2.2.1.3).

The 1-h shaking time recommended can be reduced to 15 min if only NO_3 is of interest. However, 1-h shaking time must be used if exchangeable NH_4-N is to be determined. Where simultaneous extraction of large numbers of samples is desired, 2.5 g of soil and 25 mL of 2 M KCl can be shaken for 1 h in stoppered 25 by 200 mm test tubes. If the tubes are stored vertically in a refrigerator (5 °C) overnight, a sample of the extract can be decanted for analysis without filtration for most soils. This modification has performed satisfactorily in the authors' laboratory for both NO_3-N and NH_4- N analyses.

Turbid, colloidal, or highly colored extracts can lead to an accumulation of suspended material on the reduction column. This potential problem is minimized by the 10:1 dilution of the extraction step and filtration. Filtration should use a low nitrate filtering media. Two recent reports (Sparrow & Masiak, 1987; Scharf & Alley, 1988) have observed significant

NO_3-N contamination from some filtering media, which can be controlled by using quantitative grade filter paper (Whatman no. 42 or equivalent) or glass fiber filters, by pre-washing the filters, or by using centrifugation instead of filtration. The installation of a dialyzer can also control problems from colloidal material (Henriksen & Selmer-Olsen, 1970), although this reduces sensitivity.

Smith and Scott (1991) have also reviewed the background and general use of continuous-flow and discrete analysis systems in soil science. Several other NO_3-N methods suited to automation have been applied to soil extracts, such as those using hydrazine to reduce NO_3-N to NO_2-N, followed by the Griess-Illosvay reaction (Best, 1976; Markus et al., 1985). These methods can also provide satisfactory soil NO_3-N analyses, but take precautions to minimize interferences; for example, Mg can interfere with hydrazine reduction (Ananth & Moraghan, 1987).

For any NO_3-N procedure, a careful quality control program should be followed that includes: (i) periodic comparative NO_3-N analyses by independent procedures (instrumental vs. "wet chemistry" procedures), (ii) periodic NO_3-N recovery checks from known NO_3-N additions to soils, and (iii) regular "blind analysis" of standard soils containing known levels of NO_3-N.

41–2.3.1.2 Laboratory Analysis, Manual. For a simplified NO_3-N analysis, we recommend the manual Cu-Cd reduction method described in detail by Keeney and Nelson (1982) and further investigated by Dorich and Nelson (1984). The manual method is designed for use with small numbers of samples, with traditional laboratory equipment, and with uncomplicated reagents. The method is reliable, sensitive, and has few interferences.

41–2.3.1.2.1 Special Apparatus

1. Spectrometer: With 1-cm light path capable of measuring absorbance at 540 nm, although 510 to 540 nm could be used.
2. Reducing columns: 1 by 30 cm Pyrex tubes with fritted glass plates and stopcock at lower end and an upper reservoir capable of holding 75 mL. Connect a glass tube (4 mm i.d.) implanted in a two-hole no. 00 stopper to the column stopcock. A 4-mm (i.d.) glass tube in the other hole of the stopper is connected to a flow regulator and vacuum source (see Keeney and Nelson, 1982, for details).

41–2.3.1.2.2 Reagents

1. Reagents 1 through 3 are the same as section 41–2.3.1.1.2.
4. Cadmium metal: Approximately 1-mm diam. granules.
5. Ammonium chloride, 4 *M:* Dissolve 202 g of NH_4Cl in 800 mL deionized water and dilute to 1 L.
6. Ammonium chloride, 0.1 *M:* Dilute 50 mL of concentrated NH_4Cl solution (reagent no. 5 above) to 2 L with deionized water.

7. Diazotizing reagent: Dissolve 0.5 g of sulfanilamide in 100 mL of 2.4 M HCl. Store in refrigerator at 4 °C.
8. Coupling reagent: Dissolve 0.3 g of N-(1-naphthyl)-ethylenediamine dihydrochloride in 100 mL of 0.12 M HCl. Store this solution in a brown bottle in a refrigerator at 4 °C.
9. Standard NO_3-N solutions and standard NO_2-N solution are the same as reagents no. 7 through 10 of section 41–2.3.1.1.2.

41–2.3.1.2.3 Procedure

1. Prepare Cu-Cd as described in procedure 1 of section 41–2.3.1.1.3, treating enough Cd for several columns.
2. Fill Pyrex reducing column with the dilute NH_4Cl solution and pour in the Cu-Cd particles to a depth of 20 cm. Ensure that all air bubbles are removed from the column, drain off excess NH_4Cl, and wash thoroughly with dilute NH_4Cl solution (about 10 pore volumes), using a slow flow rate of about 8 mL min^{-1}. Always ensure that the Cu-Cd metal is covered with at least 1 cm of dilute NH_4Cl solution.
3. Add 1 mL of 4 M NH_4Cl solution to top reservoir and lower liquid to top of column. Add 75 mL of 0.1 M NH_4Cl solution to reservoir, attach a 100-mL volumetric flask to the stopper on the outlet, and pass the NH_4Cl through the column at a flow rate of 110 mL min^{-1} by manipulating the stopcock and vacuum flow regulator.
4. Drain off excess NH_4Cl from column until solution is just above the Cu-Cd metal. Add 1 mL of 4 M NH_4Cl solution to top of column, then pipette an aliquot of 2 M KCl soil extract containing < 20 µg of NO_3-N (usually 2–5 mL) onto the top of the column. Attach a 100-mL volumetric flask to the stopper, allow the extract to enter the Cu-Cd column by draining some NH_4Cl from the bottom, rinse the inside of the column with 2 mL of 0.1 M NH_4Cl, then add 75 mL of 0.1 M NH_4Cl to the top reservoir and begin passage through the column at a flow rate of 110 mL min^{-1}.
5. Collect the entire eluant in the 100-mL volumetric flask, containing 2 mL of diazotizing reagent, and mix. After 5 min, add 2 mL of coupling reagent, mix, bring to volume with 0.1 M NH_4Cl, and let stand for 20 min.
6. Measure absorbance at 540 nm with instrument zeroed against a reagent blank.
7. Prepare a NO_3-N standard curve by passing 4 mL of each NO_3-N working standard (reagent no. 8 of section 41–2.3.1.1.2) through the column and developing the color as described above. A 4-mL aliquot of each NO_3-N working standard should contain 0, 0.4, 2, 4, 8, 12, and 20 µg of NO_3-N and should produce a linear plot of absorbance vs. NO_3-N.
8. Calculate the soil NO_3-N concentration (mg NO_3-N kg^{-1} of soil) by converting the soil extract absorbance reading to NO_3-N via the

standard curve and making appropriate adjustments for the volume of the soil extract analyzed and the soil:2 M KCl extraction ratio used (usually a 1:10 ratio).

41–2.3.1.2.4 Comments

The comments for the automated Cu-Cd reduction method (section 41–2.3.1.1.4) should be reviewed since most are applicable to the manual method as well.

The reagents for the manual method are less complex than the automated method and are in keeping with the objective of recommending a simple NO_3-N method. The elimination of the EDTA and pH adjustment of the NH_4Cl solution should not cause metal interference problems for most soil extracts (Henriksen & Selmer-Olsen, 1970; Dorich & Nelson, 1984). However, if heavy metals are a concern, then a buffered (pH 8.5) NH_4Cl solution containing EDTA is recommended as prescribed by Gales and Booth (1975).

Careful control of the eluant flow rate is important because slow rates can cause further reduction of NO_2-N and low recoveries, while high flow rates can produce poor conversion of NO_3-N to NO_2-N. Dorich and Nelson (1984) studied effluent flow rates and column lengths and concluded that a 110 mL min^{-1} flow rate with a 20-cm column gave rapid, precise, and accurate results with 2 M KCl soil extracts. The size of the Cd granules can vary from 0.25 to 1 mm, but if granules < 1 mm are used, the column flow rates may have to be increased to avoid further reduction of NO_2-N within the column.

Particular care should be given to the preparation and use of the Cu-Cd columns. Difficulties are most likely to arise from unsatisfactory copperization of the Cd, from incomplete washing off of excess Cu from the treated Cd, or from air entering the column. If the Cu-Cd metal is inadvertently exposed to air, the column should be repacked (procedure 2 above) to completely exclude air bubbles. Further readings on the Cd reduction method for NO_3-N and the Griess-Illosvay NO_2-N reactions can be found in Bremner (1965b) and Keeney and Nelson (1982).

Other manual NO_3-N methods for small numbers of samples include steam distillation (Bremner, 1965b), micro-diffusion (Bremner, 1965b), nitrosation of salicylic acid (Cataldo et al., 1975; Vendrell & Zupancic, 1990), nitrate specific electrode (Keeney & Nelson, 1982; Gelderman & Fixen, 1988), chromotropic acid (Sims & Jackson, 1971), and direct UV absorption (Cawse, 1967). The steam distillation method has been discussed in detail by Bremner (1965b) and Keeney and Nelson (1982) and is usually the standard to which other methods are compared. Steam distillation is an excellent method and we recommend it where its requirements for special apparatus and relatively long analysis times are not serious problems.

41–2.3.2 Ammonium Method

41–2.3.2.1 Laboratory Analysis, Automated. Ammonium in soils or soil extracts is determined by reacting NH_4 with phenol and hypochlorite in alkaline solution to form an intense blue color that is measured by absorbance at 630 nm. The colorimetric procedure is widely known as the Berthelot reaction or the indophenol blue method. It has been used to determine NH_4-N in a wide range of applications. Agricultural uses of the technique include determination of NH_4-N in Kjeldahl digests (Mann, 1963) and determination of NH_4 in soils and soil extracts using manual (Kempers, 1974; Keeney & Nelson, 1982; Dorich & Nelson, 1983) and automated techniques (Rice et al., 1984; Markus et al., 1985; Gentry & Willis, 1988; Tel and Heseltine, 1990). The method has excellent sensitivity and precision and is suited to determination of NH_4 in most solutions used to extract inorganic N from soils. Potential interferences by di- and trivalent cations can be minimized by addition of EDTA (Dorich & Nelson, 1983) or potassium sodium tartrate (Technicon Instrument Corp., 1977a) during the analysis. The working concentration range for NH_4-N in sample solutions is 0.2 to 10 mg of N L^{-1}. This sample concentration range can be expanded at least 10-fold through use of a sample dilution loop or automatic dilution during the continuous flow analysis.

The procedure described here was developed for analysis of water and wastewater using a Technicon autoanalyzer (Technicon Instruments Corp., 1977a; Kopp & McKee, 1978). This procedure has performed well for analysis of 2 M KCl soil extracts and leachates from long-term soil incubation experiments in the author's laboratory. This method has been adapted for automated NH_4 analyses by other workers (Rice et al., 1984; Markus et al., 1985; Gentry & Willis, 1988).

41–2.3.2.1.1 Special Apparatus

Automated analytical equipment for continuous flow analysis of NH_4-N consisting of: a sampler, a manifold and NH_4-N analytical cartridge, a proportioning pump, a flow cell colorimeter equipped to measure absorbance at 630 nm, and a recorder or other output storage device. The detailed configuration of the apparatus depends on manufacturer's specifications for specific instruments. Consult the manufacturer's operating instructions for specifications on component operation, flow rates, tube sizes, and analysis rates.

41–2.3.2.1.2 Reagents

1. Potassium chloride (KCl), 2 M: Dissolve 1490 g of KCl in 8 L of distilled water and dilute to 10 L.

2. Alkaline phenol solution: Dissolve 200 g of NaOH in about 500 mL of distilled water. Cool the solution during preparation by placing the vessel in circulating cold water. In a fume hood, slowly add 243 g of phenol to the NaOH solution with continuous stirring and continue cooling the vessel with cold water. Add 0.5 mL of Brij-35 wetting agent and dilute the cooled solution to 1 L with distilled water. Store in a refrigerator when not in use.

3. Sodium hypochlorite solution: Commercially available household bleach containing at least 5.25% (wt/vol) available Cl is satisfactory.

4. Potassium sodium tartarate solution: Dissolve 150 g of $KNaC_4H_4O_6 \cdot 4H_2O$ in about 800 mL of distilled water, add 0.5 mL of Brij-35, and dilute to 1 L. Store in a refrigerator.

5. Stock ammonium standard solution: Dissolve 0.3821 g of dry ammonium chloride (NH_4Cl) in a 1-L volumetric flask, and dilute to volume with distilled water. This solution contains 100 mg of NH_4-N L^{-1}. Store in a refrigerator.

6. Working ammonium standard solutions: Using the stock ammonium standard solution (no. 5 above), prepare working standard solutions by pipetting 0, 2, 4, 6, 8, and 10 mL into 100-mL volumetric flasks and diluting the solutions to volume with the soil extracting solution (2 M KCl, reagent no. 1 above, or other extractant used for the samples to be analyzed). These standard solutions contain 0, 2, 4, 6, 8, and 10 mg of NH_4-N L^{-1}. Store in a refrigerator.

41–2.3.2.1.3 Procedure

1. Set up automated continuous flow analysis system for NH_4-N according to manufacturer's instructions.

2. Prepare soil extracts according to procedures described for NO_3-N analysis (section 41–2.3.1.1.3). If exchangeable NH_4-N is to be determined, the extraction must be performed by shaking soil with 2 M KCl (1:10 soil: solution ratio) for 1 h.

3. Allow the automated analysis system to warm up according to manufacturer's instructions, and allow refrigerated reagents and sample solutions to reach room temperature before use. Obtain a stable baseline output from the analysis system using all reagents (described previously) and extracting solution in place of soil extract samples.

4. Analyze working standard solutions containing 0 to 10 mg of NH_4-N L^{-1} (reagent no. 6, section 41–2.3.2.1.2), and adjust gain to obtain appropriate output readings. The output response should be linear within the range of NH_4-N concentrations used.

5. Place soil extracts to be analyzed into the instrument sampler, include a mid-range NH_4-N working standard after each 8 to 10 samples for quality control, and start the analysis. Some micropro-

cessor-controlled analytical systems provide for automatic standard analysis and recalibration if required.

6. Determine the sample NH_4-N concentration from a comparison of sample output peak heights with the standard curve obtained from working standard peak heights and NH_4-N concentrations. Calculate the soil NH_4-N concentration from the soil extract concentration and the dilution factor used in the soil extraction procedure (10:1 in the 2 M KCl extraction described above). Microprocessor-controlled instruments usually perform these calculations using system software.

41–2.3.2.1.4 Comments

The general use of continuous flow analysis systems for inorganic N determinations in soils has been reviewed by Smith and Scott (1991). The precision and accuracy of the NH_4 method described compares favorably with steam distillation and other manual methods (Rice et al., 1984; Markus et al., 1985; Adamsen et al., 1985). Typical CV values for automated methods range from 2 to 5% (Adamsen et al., 1985).

Use of salicylate as a substitute for phenol in the procedure has been proposed for both manual (Nelson, 1983; Kempers & Zweers, 1986) and automated (Adamsen et al., 1985; Gentry & Willis, 1988) versions of the indophenol ammonium method. Procedures using salicylate have equal or improved sensitivity relative to procedures using phenol and avoid the volatility and safety concerns associated with the phenol reagent.

Improved sensitivity of the procedure through use of nitroprusside $[Na_2Fe(CN)_5NO \cdot 5H_2O]$ to increase the intensity of color development was first proposed by Lubochinsky and Zalta (1954) and has been incorporated into many current procedures for NH_4 using the indophenol procedure (Dorich & Nelson, 1983; Adamsen et al., 1985; Kempers & Zweers, 1986; Gentry & Willis, 1988; Tel & Heseltine, 1990). A 10-fold increase in sensitivity of the method is usually provided by addition of nitroprusside (Keeney & Nelson, 1982).

41–2.3.2.2 Laboratory Analysis, Manual. Where NH_4 analyses are needed for a relatively small number of samples or where automated analytical equipment is not available, the steam distillation method described by Bremner (1965b) and Keeney and Nelson (1982) and the indophenol blue colorimetric procedure recommended by Keeney and Nelson (1982) provide excellent alternatives to the automated techniques described above. The manual method described here is virtually identical to the Keeney and Nelson (1982) colorimetric procedure.

41–2.3.2.2.1 Special Apparatus

1. Variable wavelength spectrometer, equipped with a 1-cm light path and capable of absorbance measurements at 636 nm.

41–2.3.2.2.2 Reagents

1. Potassium chloride (KCl), 2 *M:* Prepare as described in section 41–2.3.2.1.2.
2. Stock ammonium standard solution: Prepare as described in section 41–2.3.2.1.2. Immediately before use, dilute 4 mL of the stock NH_4 solution to 200 mL. The resulting working solution contains 2 mg of NH_4-N L^{-1}.
3. Phenol-nitroprusside reagent: Dissolve 7 g of phenol and 34 mg of sodium nitroprusside [disodium pentacyanonitrosylferrate, $Na_2Fe(CN)_5NO\cdot2H_2O$] in 80 mL of distilled water, and dilute to 100 mL. Mix well, and store in a dark-colored bottle in a refrigerator.
4. Buffered hypochlorite reagent: Dissolve 1.480 g of sodium hydroxide (NaOH) in 70 mL of distilled water, add 4.98 g of sodium monohydrogen phosphate (Na_2HPO_4) and 20 mL of sodium hypochlorite (NaOCl) solution (5–5.25% NaOCl). Use less or more hypochlorite solution if the NaOCl concentration is higher or lower than that indicated. Check the pH to ensure a value between 11.4 and 12.2. Add a small amount of additional NaOH if required to raise the pH. Dilute to a final volume of 100 mL.
5. Ethylenediaminetetraacetic acid (EDTA) reagent: Dissolve 6 g of ethylenediaminetetraacetic acid disodium salt (EDTA disodium) in 80 mL of distilled water, adjust to pH 7, mix well, and dilute to a final volume of 100 mL.

41–2.3.2.2.3 Procedure

1. Prepare filtered soil extract as described in section 41–3.3.1.1.3.
2. Pipette an aliquot (not more than 5 mL) of the filtered 2 *M* KCl extract containing between 0.5 and 12 μg of NH_4-N into a 25-mL volumetric flask. Aliquots of ≤3 mL normally contain sufficient NH_4-N for quantitation. Add 1 mL of the EDTA reagent, and mix the contents of the flask. Then allow the flask contents to stand for 1 min. Add 2 mL of the phenol-nitroprusside reagent, followed by 4 mL of the buffered hypoclorite reagent, and immediately dilute the flask to volume with distilled water and mix well. Place the flask in a water bath maintained at 40 °C, and allow it to remain 30 min. Remove the flask from the bath, cool to room temperature (about 10 min), and determine the absorbance of the colored complex at a wavelength of 636 nm against a reagent blank solution.
3. Determine the NH_4-N concentration of the sample by reference to a calibration curve plotted from the results obtained with a 25-mL standard sample containing 0, 2, 4, 6, 8, 10, and 12 μg of NH_4-N. To prepare this curve, add an appropriate amount of 2 *M* KCl solution (same volume as that used for aliquots of soil extract) to a series of 25-mL volumetric flasks. Then add 0, 1, 2, 3, 4, 5, and 6 mL of the 2 mg of NH_4-N L^{-1} solution to the series of flasks, and

measure the intensity of blue color developed with these standards by the procedure described for analysis of the extract.

41–2.3.2.2.4 Comments

Several factors affect the successful use of the indophenol blue method for measurement of NH_4 in soil extracts. Divalent cations such as Ca^{2+} and Mg^{2+} in KCl extracts of soil interfere with the described method unless EDTA is added. The maximum amounts of these cations that can be tolerated when EDTA is included is about 4 mg of Mg^{2+} or 6 mg of Ca^2_+ per 25 mL of final diluted volume (Dorich & Nelson, 1983). If a precipitate develops in samples after the addition of the buffered hypochlorite reagent, a smaller sized aliquot of soil extract should be used to reduce the amounts of divalent cations in the system, or 2 mL of the EDTA reagent should be added to provide additional capacity for complexing divalent cations. For most soils, interference from divalent cations has not been observed if 5 mL or less of KCl extract is analyzed, but the method of Kempers (1974) can be used if precipitation of divalent cations is a major problem.

A solution pH of 11.4 to 12.0 during color development is necessary for accurate and sensitive NH_4 measurements. Although the NaOH-hypochlorite reagent is buffered at the optimum pH for color development, basic or acidic extracts may influence the results obtained. All basic or acidic extracts should be neutralized before NH_4 determination to ensure satisfactory color development. The phenol-nitroprusside reagent must be added before the buffered hypochlorite reagent to obtain proper color development. The phenol-nitroprusside reagent is stable in the dark, but weekly preparation is recommended for optimum sensitivity in NH_4 analysis.

A sufficient NaOCl concentration in the buffered hypochlorite reagent is critical for accurate NH_4 measurements. The hypochlorite reagent should be prepared immediately before use to obtain optimum results, because the NaOCl concentration in this reagent decreases on standing. Commercial bleach is a satisfactory source of hypochlorite, but technical-grade NaOCl solution from laboratory supply companies may give more reproducible results.

Color development at 40 °C for 30 min, gives the most sensitive and reproducible determinations of NH_4 in KCl extracts of soils. Alternatively the color may be developed at room temperature (about 25 °C) if the time is extended to 1 h. However, higher molar absorptivity values are obtained when color is developed at 40 °C. The color produced at 25 or 40 ° is stable for at least 7 h (Dorich & Nelson, 1983).

REFERENCES

Adamsen, F.J., D.S. Bigelow, and G.R. Scott. 1985. Automated methods for ammonium, nitrate, and nitrite in 2 *M* KCl-phenylmercuric acetate extracts of soil. Commun. Soil Sci. Plant Anal. 16:883–898.

Allison, F.E. 1965. Estimating the ability of soils to supply nitrogen. Agric. Chem. 11:46–48, 139.

American Public Health Association. 1989. Nitrogen (nitrate). p. 4-137 to 4-139. In L.S. Clescern et al. (ed.) Standard methods for the examination of water and wastewater. 17th ed. Am. Public Health Assoc., Washington, DC.

Ananth, S., and J.T. Moraghan. 1987. The effect of calcium and magnesium on soil nitrate determination by automated segmented-flow methods. Soil Sci. Soc. Am. J. 51:664–667.

Beauchamp, E.G., W.D. Reynolds, D. Brasche-Villeneuve, and K. Kirby. 1986. Nitrogen mineralization kinetics with different soil pretreatments and cropping histories. Soil Sci. Soc. Am. J. 50:1478–1483.

Beegle, D., G. Roth, and R. Fox. 1989. Nitrogen soil test for corn in Pennsylvania. Pennsylvania State Univ., University Park, Agron. Facts no. 17 (revised).

Best, E.K. 1976. An automated method for determining nitrate-N in soil extracts. Queensl. J. Agric. Anim. Sci. 33:161–166.

Blackmer, A.M., T.F. Morris, D.R. Keeney, and R.D. Voss. 1991. Estimating nitrogen needs for corn by soil testing. Iowa State Univ. Publ. Pm 1381.

Blackmer, A.M., D. Pottker, M.E. Cerrato, and J. Webb. 1989. Correlations between soil nitrate concentrations in late spring and corn yields in Iowa. J. Prod. Agric. 2:103–109.

Bock, B.R., and K.R. Kelley. 1992. Predicting N fertilizer needs for corn in humid regions. TVA Bull. Y-226. Natl. Fert. and Environ. Res. Ctr., Tennessee Valley Authority, Muscle Shoals, AL.

Bole, J.B., and U.J. Pittman. 1976. Sampling southern Alberta soils for N and P soil testing. Can. J. Soil Sci. 56:531–535.

Bonde, T.A., and T. Lindberg. 1988. Nitrogen mineralization kinetics in soil during long-term aerobic laboratory incubations: A case study. J. Environ. Qual. 17:414–417.

Bonde, T.A., and T. Rosswall. 1987. Seasonal variation of potentially mineralizable nitrogen in four cropping systems. Soil Sci. Soc. Am. J. 51:1508–1514.

Bonde, T.A., J. Schnurer, and T. Rosswall. 1988. Microbial biomass as a fraction of potentially mineralizable nitrogen in soils from long-term field experiments. Soil Biol. Biochem. 20:447–452.

Boone, R.D. 1990. Soil organic matter as a potential net nitrogen sink in a fertilized cornfield, South Deerfield, Massachusetts, USA. Plant Soil 128:191–198.

Boyle, M., and E.A. Paul. 1989. Carbon and nitrogen mineralization kinetics in soil previously amended with sewage sludge. Soil Sci. Soc. Am. J. 53:99–103.

Bremner, J.M. 1965a. Nitrogen availability indexes. p. 1324–1345. In C.A. Black et al. (ed.) Methods of soil analysis. Part 2. Agron. Monogr. 9. ASA and SSSA, Madison, WI.

Bremner, J.M. 1965b. Inorganic forms of nitrogen. p. 1179–1237. In C.A. Black et al. (ed.) Methods of soil analysis. Part 2. Agron. Monogr. 9. ASA and SSSA, Madison, WI.

Buckman, H.O. 1910. Moisture and nitrate relations in dry-land agriculture. J. Am. Soc. Agron. 2:121–138.

Bundy, L.G., and E.S. Malone. 1988. Effect of residual profile nitrate on corn response to applied nitrogen. Soil Sci. Soc. Am. J. 52:1377–1383.

Bundy, L.G., M.A. Schmitt, and G.W. Randall. 1992. Predicting N fertilizer needs for corn in humid regions: Advances in the Upper Midwest. p. 73–89. In B.R. Bock and K.R. Kelley (ed.) Predicting N fertilizer needs for corn in humid regions. TVA Bull. Y-226. Natl. Fert. and Environ. Res. Ctr., Tennessee Valley Authority, Muscle Shoals, AL.

Cabrera, M.L., and D.E. Kissel. 1988a. Evaluation of a method to predict nitrogen mineralized from soil organic matter under field conditions. Soil Sci. Soc. Am. J. 52:1027–1031.

Cabrera, M.L., and D.E. Kissel. 1988b. Length of incubation time affects the parameter values of the double exponential model of nitrogen mineralization. Soil Sci. Soc. Am. J. 52:1186–1187.

Cabrera, M.L., and D.E. Kissel. 1988c. Potentially mineralizable nitrogen in disturbed and undisturbed soil samples. Soil Sci. Soc. Am. J. 52:1010–1015.

Call, L.E. 1914. The effect of different methods of preparing a seed bed for winter wheat upon yield, soil moisture, and nitrates. J. Am. Soc. Agron. 6:249–259.

Cameron, D.R., M. Nyborg, J.A. Toogood, and D.H. Laverty. 1971. Accuracy of field sampling for soil tests. Can. J. Soil Sci. 51:165–175.

Cataldo, D.A., M. Haroon, L.E. Schrader, and V.L. Young. 1975. A rapid colorimetric determination of nitrate in plant tissue by nitration of salicylic acid. Commun. Soil Sci. Plant Anal. 6:71–80.

Cawse, P.A. 1967. The determination of nitrate in soil solutions by ultraviolet spectrophotometry. Analyst (London) 92:311–315.

Chaussod, R., B. Nicolardot, and G. Catroux. 1986. Mesure en routine de la biomasse microbienne des sols par la methode de fumigation au chloroforme. Sci. Sol 25:201–211.

Dahnke, W.C., and G.V. Johnson. 1990. Testing soils for available nitrogen. p. 127–139. In R.L. Westerman (ed.) Soil testing and plant analysis. 3rd ed. SSSA, Madison, WI.

Dahnke, W.C., and E.H. Vasey. 1973. Testing soils for nitrogen. p. 97–114. In L.M. Walsh and J.D. Beaton (ed.) Soil testing and plant analysis. ASA, CSSA, and SSSA, Madison, WI.

Davis, J.G., and A.M. Blackmer. 1990. Sampling soils for late-spring nitrate in fields fertilized with anhydrous ammonia. p. 266. In Agronomy abstracts. ASA, Madison, WI.

Deans, J.R., J.A.E. Molina, and C.E. Clapp. 1986. Models for predicting potentially mineralizable nitrogen and decomposition rate constants. Soil Sci. Soc. Am. J. 50:323–326.

Dorich, R.A., and D.W. Nelson. 1983. Direct colorimetric measurement of ammonium in potassium chloride extracts of soil. Soil Sci. Soc. Am. J. 47:833–836.

Dorich, R.A., and D.W. Nelson. 1984. Evaluation of manual cadmium reduction methods for determination of nitrate in potassium chloride extracts of soils. Soil Sci. Soc. Am. J. 48:72–75.

Ellert, B.H., and J.R. Bettany. 1988. Comparison of kinetic models for describing net sulfur and nitrogen mineralization. Soil Sci. Soc. Am. J. 52:1692–1702.

Fox, R.H., and W.P. Piekielek. 1978. Field testing of several nitrogen availability indexes. Soil Sci. Soc. Am. J. 42:747–750.

Fox, R.H., and W.P. Piekielek. 1984. Relationships among anaerobically mineralized nitrogen, chemical indexes, and nitrogen availability to corn. Soil Sci. Soc. Am. J. 48:1087–1090.

Fox, R.H., G.W. Roth, K.V. Iversen, and W.P. Piekielek. 1989. Soil and tissue nitrate tests compared for predicting soil nitrogen availability to corn. Agron. J. 81:971–974.

Gale, P.M., and J.T. Gilmour. 1988. Net mineralization of carbon and nitrogen under aerobic and anaerobic conditions. Soil Sci. Soc. Am. J. 52:1006–1010.

Gales, M.E., and R.L. Booth. 1975. A copper-cadmium column for manually determining nitrate. In News of environmental research in Cincinnati. 28 Feb. 1975. USEPA, Cincinnati.

Gelderman, R.H., and P.E. Fixen. 1988. Recommended nitrate-N test. p. 10–12. In W.C. Dahnke (ed.) Recommended chemical soil test procedures for the North Central region. North Dakota Agric. Exp. Stn., Fargo, North Central Regional Publ. 221.

Gentry, C.E., and R.B. Willis. 1988. Improved method for automated determination of ammonium in soil extracts. Commun. Soil Sci. Plant Anal. 19:721–737.

Gianello, C., and J.M. Bremner. 1986a. A simple chemical method of assessing potentially available organic nitrogen in soil. Commun. Soil Sci. Plant Anal. 17:195–214.

Gianello, C., and J.M. Bremner. 1986b. Comparison of chemical methods of assessing potentially available organic nitrogen in soil. Commun. Soil Sci. Plant Anal. 17:215–236.

Gianello, C., and J.M. Bremner. 1988. A rapid steam distillation method of assessing potentially available organic nitrogen in soil. Commun. Soil Sci. Plant Anal. 19:1551–1568.

Harmsen, G.W., and D.A. Van Schreven. 1955. Mineralization of organic nitrogen in soil. Adv. Agron. 7:299–398.

Henriksen, A., and A.R. Selmer-Olsen. 1970. Automatic methods for determining nitrate and nitrite in water and soil extracts. Analyst 95:514–518.

Hergert, G.W. 1987. Status of residual nitrate-nitrogen soil tests in the United States. p. 73–88. In J.R. Brown (ed.) Soil testing: Sampling, correlation, calibration, and interpretation. ASA Spec. Publ. 21. ASA, CSSA, and SSSA, Madison, WI.

Hong, S.D., R.H. Fox, and W.P. Piekielek. 1990. Field evaluation of several chemical indexes of soil nitrogen availability. Plant Soil 123:83–88.

Houot, S., J.A.E. Molina, R. Chaussod, and C.E. Clapp. 1989. Simulation by NCSOIL of net mineralization in soils from the Deherain and 36 Parcelles fields at Grignon. Soil Sci. Soc. Am. J. 53:451–455.

Jackson, W.L., C.E. Frost, and D.M. Hildreth. 1975. Versatile multi-range analytical manifold for automated analysis of nitrate-nitrogen. Soil Sci. Soc. Am. Proc. 39:592–593.

Jokela, W. 1989. The Vermont nitrogen soil test for corn. Coop. Ext. Serv., Univ. of Vermont Publ. no. FS 133, Burlington.

Keeney, D.R. 1982. Nitrogen-availability indexes. p. 711–733. In A.L. Page et al. (ed.) Methods of soil analysis. Part 2. 2nd ed. Agron. Monogr. 9 ASA and SSSA, Madison, WI.

Keeney, D.R., and D.W. Nelson. 1982. Nitrogen inorganic forms. p. 643–698. In A.L. Page et al. (ed.) Methods of soil analysis. Part 2. 2nd ed. Agron. Monogr. 9. ASA and SSSA, Madison, WI.

Kempers, A.J. 1974. Determination of submicrogram quantities of ammonium and nitrates in soil with phenol, sodium nitroprusside and hypochlorite. Geoderma 12:201–206.

Kempers, A.J., and A. Zweers. 1986. Ammonium determination in soil extracts by the salicylate method. Commun. Soil Sci. Plant Anal. 17:715–723.

King, F.H., and A.R. Whitson. 1901. Development and distribution of nitrates and other soluble salts in cultivated soils. Wisconsin Agric. Exp. Stn. Bull. no. 85.

King, F.H., and A.R. Whitson. 1902. Development and distribution of nitrates in cultivated soils. (Second paper.) Wisconsin Agric. Exp. Stn. Bull. no. 93.

Kopp, J.F., and G.D. McKee. 1978. Methods for chemical analysis of water and wastes. Nitrogen ammonia-Method 350.1. USEPA Environ. Monitoring and Support Lab., Cincinnati.

Liang, B.C., M. Remillard, and A.F. MacKenzie. 1991. Influence of fertilizer, irrigation, and non-growing season precipitation on soil nitrate-nitrogen under corn. J. Environ. Qual. 20:123–128.

Lindemann, W.C., and M. Cardenas. 1984. Nitrogen mineralization potential and nitrogen transformations of sludge-amended soil. Soil Sci. Soc. Am. J. 48:1072–1077.

Lubochinsky, B., and J.P. Zalta. 1954. Colorimetric microdetermination of ammoniacal nitrogen. Bull. Soc. Chim. Biol. 46:1363.

Lueking, M.A., and J.S. Schepers. 1986. Achieving desired moisture conditions in potentially mineralizable nitrogen incubation studies. Soil Sci. Soc. Am. J. 50:1370–1373.

MacKay, D.C., and J.M. Carefoot. 1981. Control of water content in laboratory determination of mineralizable nitrogen in soils. Soil Sci. Soc. Am. J. 45:444–446.

Magdoff, F. 1991. Understanding the Magdoff pre-sidedress nitrate test for corn. J. Prod. Agric. 4:297–305.

Magdoff, F.R., W.E. Jokela, R.H. Fox, and G.F. Griffin. 1990. A soil test for nitrogen availability in the northeastern United States. Commun. Soil Sci. Plant Anal. 21:1103–1115.

Magdoff, F.R., D. Ross, and J. Amadon. 1984. A soil test for nitrogen availability to corn. Soil Sci. Soc. Am. J. 48:1301–1304.

Mann, L.T., Jr. 1963. Spectrophotometric determination of nitrogen in total micro-Kjeldahl digests. Anal. Chem. 35:2179–2182.

Marion, G.M., and C.H. Black. 1988. Potentially available nitrogen and phosphorus along a chaparral fire cycle chronosequence. Soil Sci. Soc. Am. J. 52:1155–1162.

Markus, K.K., J.P. Mckinnon, and A.F. Buccafuri. 1985. Automated analysis of nitrite, nitrate, and ammonium nitrogen in soils. Soil Sci. Soc. Am. J. 49:1208–1215.

McCracken, D.V., S.J. Corak, M.S. Smith, W.W. Frye, and R.L. Blevins. 1989. Residual effects of nitrogen fertilization and winter cover cropping on nitrogen availability. Soil Sci. Soc. Am. J. 53:1459–1464.

Meisinger, J.J. 1984. Evaluating plant-available nitrogen in soil-crop systems. p. 391–416. Evaluating plant-available nitrogen in soil crop systems. p. 391–416. In R.D. Hauck et al. (ed.) Nitrogen in crop production. ASA, Madison, WI.

Meisinger, J.J., V.A. Bandel, J.S. Angle, B.E. O'Keefe, and C.M. Reynolds. 1992a. Presidedress soil nitrate test evaluation in Maryland. Soil Sci. Soc. Am. J. 56:1527–1532.

Meisinger, J.J., F.R. Magdoff, and J.S. Schepers. 1992b. Predicting N fertilizer needs for corn in humid regions: Underlying principles. p. 7–27. In B.R. Bock and K.R Kelley (ed.) Predicting N fertilizer needs for corn in humid regions. TVA Bull. Y-226. Natl. Fert. and Environ. Res. Ctr., Tennessee Valley Authority, Muscle Shoals, AL.

Molina, J.A.E., C.E. Clap, M.J. Schaffer, F.W. Chichester, and W.E. Larson. 1983. NC-SOIL, A model of nitrogen and carbon transformations in soil: Description, calibration, and behavior. Soil Sci. Soc. Am. J. 47:85–91.

Myers, R.G., C.W. Swallow, and D.E. Kissel. 1989. A method to secure, leach, and incubate undisturbed soil cores. Soil Sci. Soc. Am. J. 53:467–471.

Nadelhoffer, K.J. 1990. Microlysimeter for measuring nitrogen mineralization and microbial respiration in aerobic soil incubations. Soil Sci. Soc. Am. J. 54:411–415.

Nelson, D.W. 1983. Determination of ammonium in KCl extracts of soils by salicylate method. Commun. Soil Sci. Plant Anal. 14(11):1051–1062.

Norman, R.J., J.C. Edberg, and J.W. Stucki. 1985. Determination of nitrate in soil extracts by dual-wavelength ultraviolet spectrophotometry. Soil Sci. Soc. Am. J. 49:1182–1185.

Øien, A., and A.R. Selmer-Olsen. 1980. A laboratory method for evaluation of available nitrogen in soil. Acta Agric. Scand. 30:149–156.

Paustian, K., and J. Schnurer. 1987. Fungal growth response to carbon and nitrogen limitation: Application of a model to laboratory and field data. Soil Biol. Biochem. 19:621–629.

Reuss, J.O., P.N. Soltanpour, and A.E. Ludwick. 1977. Sampling distribution of nitrates in irrigated fields. Agron. J. 69:588–592.

Rice, C.W., J.H. Grove, and M.S. Smith. 1987. Estimating soil net nitrogen mineralization as affected by tillage and soil drainage due to topographic position. Can. J. Soil Sci. 67:513–520.

Rice, C.W., M.S. Smith, and J.M. Crutchfield. 1984. Inorganic N analysis of soil extracts by automated and distillation procedures. Commun. Soil Sci. Plant Anal. 15:663–672.

Richter, J., A. Nuske, W. Habenicht, and J. Bauer. 1982. Optimized N-mineralization parameters of loess soils from incubation experiments. Plant Soil 68:379–388.

Robertson, K., J. Schnurer, M. Clarholm, T.A. Bonde, and T. Rosswall. 1988. Microbial biomass in relation to C and N mineralization during laboratory incubations. Soil Biol. Biochem. 20:281–286.

Roth, G.W., and R.H. Fox. 1990. Soil nitrate accumulations following nitrogen-fertilized corn in Pennsylvania. J. Environ. qual. 19:243–248.

Saffigna, P. 1988. N-15 methodology in the field. p. 433–451. In J.R. Wilson (ed.) Advances in nitrogen cycling in agricultural ecosystems. Proc. Symp. at Brisbane, Australia. 11–15 May 1987. CAB Int., Wallingford, UK.

Sahrawat, K.L., and F.N. Ponnamperuma. 1978. Measurement of exchangeable NH_4 in tropical rice soils. Soil Sci. Soc. Am. J. 42:282–283.

Sarrantonio, M., and T.W. Scott. 1988. Tillage effects on availability of nitrogen to corn following a winter green manure crop. Soil Sci. Soc. Am. J. 52:1661–1668.

Scharf, P., and M.M. Alley. 1988. Centrifugation: A solution to the problem posed by ammonium and nitrate contamination of filters in soil extract analysis. Soil Sci. Soc. Am. J. 52:1508–1510.

Sheppard, S.C., and T.E. Bates. 1986. Changes in nitrate concentration over winter in three southern Ontario soil profiles. Can. J. Soil Sci. 66:537–541.

Sims, J.R., and G.D. Jackson. 1971. Rapid analysis of soil nitrate with chromotropic acid. Soil Sci. Soc. Am. Proc. 35:603–606.

Skjemstad, J.O., and R. Reeve. 1978. The automatic determination of ammonia, nitrate plus nitrite, and phosphate in water in the presence of added mercury (III) chloride. J. Environ. Qual. 7:137–141.

Smith, J.L., R.R. Schnabel, B.L. McNeal, and G.S. Campbell. 1980. Potential errors in the first-order model for estimating soil nitrogen mineralization potentials. Soil Sci. Soc. Am. J. 44:996–1000.

Smith, K.A., and A. Scott. 1991. Continuous-flow, flow-injection, and discrete analysis. p. 183–227. In K.A. Smith (ed.) Soil analysis: Modern instrumental techniques. 2nd ed. Marcel Dekker, New York.

Smith, S.J. 1987. Soluble organic nitrogen losses associated with recovery of mineralized nitrogen. Soil Sci. Soc. Am. J. 51:1191–1194.

Smith, S.J., L.B. Young, and G.E. Miller. 1977. Evaluation of soil nitrogen mineralization potential under modified field conditions. Soil Sci. Soc. Am. J. 41:74–76.

Sparrow, S.D., and D.T. Masiak. 1987. Errors in the analyses for ammonium and nitrate caused by contamination from filter papers. Soil Sci. Soc. Am. J. 51:107–110.

Stanford, G. 1982. Assessment of soil nitrogen availability. p. 651–688. In F.J. Stevenson (ed.) Nitrogen in agricultural soils. Agron. Monogr. 22. ASA and SSSA, Madison, WI.

Stanford, G., J.N. Carter, and S.J. Smith. 1974. Estimates of potentially mineralizable soil nitrogen based on short-term incubations. Soil Sci. Soc. Am. Proc. 38:99–102.

Stanford, G., and E. Epstein. 1974. Nitrogen mineralization-water relations in soils. Soil Sci. Soc. Am. Proc. 38:103–107.

Stanford, G., M.H. Frere, and D.H. Schwaninger. 1973. Temperature coefficient of soil nitrogen mineralization. Soil Sci. 115:321–323.

Stanford, G., and S.J. Smith. 1972. Nitrogen mineralization potentials of soils. Soil Sci. Soc. Am. Proc. 38:99–102.

Stevenson, F.J. 1986. Cycles of soil. John Wiley and Sons, New York.

Technicon Instrument Corp. 1977a. Ammonia in water and wastewater. Industrial Method no. 98-70W/A. Tarrytown, NY.

Technicon Instrument Corp. 1977b. Nitrate and nitrite in soil extracts. Industrial Method no. 487-77A. Tarrytown, NY.

Tel, D.A., and C. Heseltine. 1990. The analysis of KCl soil extracts for nitrate, nitrite and ammonium using a TRAACS 8000 analyzer. Commun. Soil Sci. Plant Anal. 21:1681–1688.

Thomas, G.W., M.S. Smith, and R.E. Phillips. 1989. Impact of soil management practices on nitrogen leaching. p. 247–276. In R.E. Follett (ed.) Nitrogen management and ground water protection. Elsevier Sci. Publ. Co., New York.

Tyler, D.D., and G.W. Thomas. 1977. Lysimeter measurements of nitrate and chloride losses from soil under conventional and no-tillage corn. J. Environ. Qual. 6:63–66.

Vendrell, P.F., and J. Zupancic. 1990. Determination of soil nitrate by transnitration of salicylic acid. Commun. Soil Sci. Plant Anal. 21:1705–1713.

U.S. Environmental Protection Agency (USEPA). 1983. Nitrogen, nitrate-nitrite. p. 353.2-1 to 352.2-7. In Methods of chemical analysis of water and wastes. EPA-600 4-79-020. 1983 ed. ESEPA, Cincinnati.

Waring, S.A., and J.M. Bremner. 1964. Ammonium production in soil under waterlogged conditions as an index of nitrogen availability. Nature (London) 201:951–952.

Chapter 42

Nitrogen Mineralization, Immobilization, and Nitrification

STEPHEN C. HART, *Northern Arizona University, Flagstaff, Arizona*

JOHN M. STARK, *Utah State University, Logan, Utah*

ERIC A. DAVIDSON, *The Woods Hole Research Center, Woods Hole, Massachusetts*

MARY K. FIRESTONE, *University of California, Berkeley, California*

The biogeochemical cycling of N in ecosystems can be divided into an external and an internal N cycle. The external cycle includes those processes that add or remove N from ecosystems, such as: dinitrogen (N_2) fixation, dry and wet N deposition, N fertilization, N leaching, runoff erosion, denitrification, and ammonia volatilization. The internal N-cycle consists of those processes that convert N from one chemical form to another or transfer N between ecosystem pools. Processes of the internal N-cycle include: plant assimilation of N, return of N to soil in plant litterfall and root turnover, N mineralization (the conversion of organic N to inorganic N), microbial immobilization of N (the uptake of inorganic N by microorganisms), and nitrification (the production of nitrite $\{NO_2^-\}$ and nitrate $\{NO_3^-\}$ from ammonium $\{NH_4^+\}$ or organic N) (Fig. 42–1).

The significance of internal N-cycling processes can be illustrated by comparing the rates of these processes relative to external N-cycling rates. For example, Paul and Clark (1989) estimate that the sum of all output fluxes of the external N-cycle globally is about 0.25×10^{15} g-N yr^{-1}, while net N mineralization in soils is more than 14 times this amount (about 3.5×10^{15} g-N yr^{-1}). However, because net N mineralization is the difference between actual N mineralization and microbial immobilization of N, gross N mineralization rates may be over two orders of magnitude greater than all output fluxes of N combined (see below).

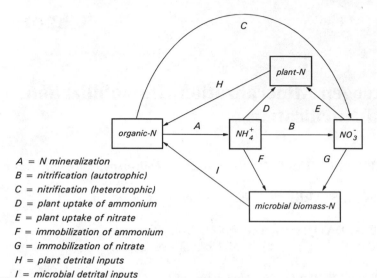

A = N mineralization
B = nitrification (autotrophic)
C = nitrification (heterotrophic)
D = plant uptake of ammonium
E = plant uptake of nitrate
F = immobilization of ammonium
G = immobilization of nitrate
H = plant detrital inputs
I = microbial detrital inputs

Fig. 42–1. Nitrogen transfers and transformations of the internal N-cycle within terrestrial ecosystems. Note that many different processes affect the size of ammonium and nitrate pools in soil. Net rates of N mineralization and nitrification are determined in the absence of plant N-uptake and detrital N-inputs (i.e., net N mineralization = {A + C} − {F + G}; net nitrification = {B + C} −G) (Davidson et al., 1992).

Understanding the factors that control the rates of internal N-cycling processes is important because of the effects these processes have on ecosystem structure and function, as well as on environmental quality. For instance, most of the N assimilated by plants is derived from inorganic-N pools, NH_4^+ and NO_3^-, and these forms of N are produced via the processes of N mineralization and nitrification. Because N is frequently the most limiting nutrient to terrestrial plants (Binkley & Hart, 1989; Paul & Clark, 1989; Vitousek & Howarth, 1991), it follows that differences in the rates of N mineralization, immobilization, and nitrification can have a profound effect on primary productivity.

Most estimates of N mineralization and nitrification have been obtained by measuring net rates that result from two or more processes occurring simultaneously. In the absence of plant assimilation, leaching, dissimilatory nitrate reduction, and denitrification, net rates of mineralization are determined from the change in the soil inorganic-N pool size over time (t).

Net N mineralization =

$$(NH_4^+\text{-}N + NO_3^-\text{-}N)_{t+1} - (NH_4^+\text{-}N + NO_3^-\text{-}N)_t.$$

A negative value indicates net immobilization. Similarly, net rates of nitrification are calculated from the net change in the soil NO_3^- pool size:

$$\text{Net nitrification} = (NO_3^-\text{-}N)_{t+1} - (NO_3^-\text{-}N)_t.$$

In addition to these pool-size based definitions, net rates of N mineralization and nitrification can also be defined in terms of the differences in actual or *gross* process rates:

Net N mineralization =

gross N mineralization − microbial immobilization of inorganic N

Net nitrification = gross nitrification − microbial immobilization of NO_3^-.

Whereas net rates can be estimated by measuring the changes in inorganic-N pool sizes, gross rates can only be estimated by techniques using two N isotopes (e.g., ^{14}N and ^{15}N).

Estimates of gross N-transformation rates in soils date back to the work by Kirkham and Bartholomew (1954, 1955), who first provided the analytical equations necessary for the calculation of gross rates. Shortly thereafter, Jansson (1958) published his now classical studies of gross N transformations in agricultural soils. Despite the importance of this treatise on N mineralization-immobilization relationships in soil, few studies have empirically measured gross rates of soil N transformations, partly because of the high cost of ^{15}N analysis. However, thousands of estimates of net N-transformation rates have been made.

Although most soil scientists and ecologists recognize the differences between net and gross rates, frequently the word "net" is dropped from the discussion and net rates are treated as if they were gross rates. With the recent advent of Automated Combustion Mass Spectrometry (see chapter 40 by Hauck et al. in this book), the cost of ^{15}N analyses has decreased substantially, so more data on gross rates in soils are likely to be forthcoming.

The purpose of this chapter is to outline both the methodologies and the concepts behind the methodologies used to assess soil internal N-cycle processes. Both laboratory and field methods will be discussed, as will those methods appropriate for determining gross and net N-transformation rates. While developing our recommendations, we have considered the quantity and quality of the information provided by each method, as well as the amount of effort and expense required to use them (Table 42–1).

42–1 MEASUREMENT OF GROSS NITROGEN-TRANSFORMATION RATES

42–1.1 General Introduction

Three ^{15}N techniques have been used to study gross N- transformation processes: (i) ^{15}N natural abundance techniques, where slight differences in isotopic enrichment of soil pools, resulting from biological discrimination, are observed over time; (ii) ^{15}N tracer techniques, where a substrate pool

Table 42–1. Summary of the relative quantity and types of information provided on internal N-cycling processes by various methods relative to the effort required to use the method.

Method	Conditions of Use	Provides information on:[†]			Effort[‡]	Information[‡]	Reference
		Mineralization	Immobilization	Nitrification			
^{15}N natural abundance	Field or laboratory	Gross	Gross	Gross	Very high	Potentially high	Herman & Rundel, 1989
^{15}N tracer	Field or laboratory	Gross or net	Gross or net	Gross or net	Very high	High	Myrold & Tiedje, 1986
Isotope-dilution:							
$^{15}NH_4^+$	Field or laboratory	Gross	Gross NH_4^+ consumption	No	High	High	Davidson et al., 1991
$^{15}NO_3^-$	Field or laboratory	No	Gross NO_3^- consumption	Gross	High	High	
Resin cores	Field	Net	Net	Net	High	Moderate	DiStefano & Gholz 1986
Buried bags	Field	Net	Net	Net	Moderate	Moderate	Davidson et al., 1990
Closed-top tubes	Field	Net	Net	Net	Moderate	Moderate	Raison et al., 1987
Aerobic incubations	Laboratory	Net	Net	Net	Moderate	Low to moderate	Hart & Binkley, 1985
Soil-slurry nitrification potential	Laboratory	No	No	Potential (V_{max})	Moderate	Moderate	Belser & Mays, 1980

† The columns for mineralization, immobilization, and nitrification indicate if these processes are measured quantitatively by the methods, and whether the estimated rate is an actual (gross), net, or potential value.

‡ The columns for effort and information provide a relative scale for comparing the time and cost of the methods with the probable value of the information obtained.

is labeled and movement of the isotope through the system is monitored over time; and (iii) ^{15}N isotope-dilution techniques, where a product pool is labeled and the rate at which production alters the isotopic enrichment of the pool is monitored.

Although it is possible to use differences in the natural abundances of ^{15}N in substrate and product pools to estimate gross rates of N mineralization and nitrification, we know of only one study (Herman & Rundel, 1989) that has used this approach. The high cost of high-precision ^{15}N analysis, uncertainties regarding the magnitude and consistency of fractionation constants, and the limited use of this method makes the application of this approach expensive and in need of more research before it can become an accepted technique.

Tracer techniques have been widely used to determine directions of mass flow and connections among ecosystem compartments (see chapter 41 by Bundy and Meisinger in this book), but they have also been used to measure gross rates of flows. For example, tracer estimates of gross nitrification are determined by adding ^{15}NH$_4^+$ to the soil and then measuring the isotopic composition of the NO$_3^-$ pool after a period of time. Based on the initial isotopic composition of the NH$_4^+$ pool and the amount of ^{15}N that ends-up in the NO$_3^-$ pool, the rate of flow of total N (^{14}N + ^{15}N) into the NO$_3^-$ pool is calculated.

There are three problems with the tracer technique for measuring gross N-process rates. One is that addition of the isotope increases the size of the substrate pool (the NH$_4^+$ pool in the above example for nitrification), which may stimulate the process of interest and result in artificially high rate estimates. A second problem occurs if there is flow of new material into the substrate pool. In the above example, this flow would be N mineralization, which indeed would be occurring concurrently with nitrification. In this case, the isotopic composition of the substrate pool will decline over the study period and the initial isotopic composition will not be representative of the final composition; thus, at the end of the time period more ambient isotope (^{14}N) will accompany each unit of added isotope (^{15}N), and rates will be underestimated. A third error occurs if there is consumption of the product pool (NO$_3^-$ in the above example) during the study period. In this case, the amount of isotope measured in the product pool will be less than the amount that actually transferred into that pool during the time period. This will also result in an underestimation of rates. Simulation models (see Myrold & Tiedje, 1986; Vitousek & Andariese, 1986; Barraclough & Smith, 1987; Bjarnason, 1988) may be used to account for changing enrichments of substrate and product pools over time; however, a stimulation of rates due to substrate addition may still occur.

The problems characteristic of ^{15}N tracer methods are largely overcome when using ^{15}N isotope-dilution techniques for estimating gross rates of N mineralization and nitrification. Because the product pool is labeled with the isotope rather than the substrate pool, rates of NH$_4^+$ and NO$_3^-$ production should not be stimulated by an increase in substrate. The equations used to calculate rates with the ^{15}N isotope-dilution method use the changing size and isotopic composition of the product N-pool resulting from simultaneous production and consumption processes (Kirkham & Bartholomew, 1954). Gross rates of NH_4^+ and NO_3^- consumption can also be calculated using the ^{15}N isotope-dilution model, but these may be stimulated by addition of the ^{15}N substrate. The following section describes ^{15}N isotope-dilution, the recommended method for estimating gross rates of N mineralization, immobilization, and nitrification in soil.

42–1.2 Nitrogen-15 Isotope-Dilution

42–1.2.1 Principles

Adding $^{15}NH_4^+$ to the soil allows estimation of gross rates of N mineralization (NH_4^+ production) and NH_4^+ consumption (microbial immobilization of N being only one of the possible consumptive processes; see below). Immediately after adding $^{15}NH_4^+$ and after a short incubation period (few days or less), the total ($^{14}N + {}^{15}N$) NH_4^+ and $^{15}NH_4^+$ are measured in KCl extracts of soil. The gross N mineralization rate is calculated from the rate of *dilution* in ^{15}N enrichment of the NH_4^+ pool as organic-^{14}N is mineralized to $^{14}NH_4^+$ by ammonifiers and from the change in the total NH_4^+ pool size. Estimation of gross nitrification and NO_3^- consumption is determined in a similar manner with $^{15}NO_3^-$ additions. Dilution of the ^{15}N enrichment of the NO_3^- pool occurs as autotrophic nitrifiers oxidize $^{14}NH_4^+$ to $^{14}NO_3^-$, or when heterotrophic nitrifiers oxidize organic-^{14}N or $^{14}NH_4^+$ to $^{14}NO_3^-$. When NH_4^+ and NO_3^- are consumed, the pool sizes may change, but consumptive processes do not alter significantly the ^{15}N enrichment of the substrate pool. Because only production of NH_4^+ and NO_3^- can significantly change the ^{15}N enrichment of the respective pools, gross rates of production of NH_4^+ and NO_3^- can be separated from the respective concurrent consumption rates.

The ^{15}N isotope-dilution model presented here was developed by Kirkham and Bartholomew (1954) for use with mixed soils and has been applied recently to intact soil cores (Davidson et al., 1991). As described previously, other methods exist for estimating gross N-transformation rates in soil using additions of ^{15}N-labeled compounds.

Three assumptions of the Kirkham and Bartholomew (1954) model are: (i) no isotopic fractionation occurs during microbial transformations of soil N, (ii) added ^{15}N may be immobilized by microorganisms, but it cannot be remineralized during the course of the incubation; and (iii) rates of microbial transformations of N are constant during the incubation period. The first assumption is not strictly true (Delwiche & Steyn, 1970), but the error due to fractionation during an incubation of a few days is small relative to the large decreases in ^{15}N enrichment of the product pool that occur from production. The second assumption may be violated if the incubation extends beyond a few days; simulation modeling of data derived from a laboratory ^{15}N-incubation experiment (Bjarnason, 1988) and field data (Bristow et al., 1987) indicate that immobilization and significant remineralization can occur within a week of adding the ^{15}N label. Violation of the third assumption results in only minor errors over relatively short-incubation periods (Kirkham & Bartholomew, 1955; Bjarnason, 1988), but significant errors could arise if soil samples experience a major disturbance prior to or during labeling, such as mixing, wetting of very dry soil, or addition of large amounts of substrate. To meet these assumptions, we propose a method that involves a short incubation (1–3 d) of minimally disturbed intact cores sampled when the soil is relatively moist and to which small amounts of ^{15}N solutions are added.

Another concern is the distribution of added ^{15}N throughout the soil sample. Because ambient inorganic-^{14}N is not uniformly distributed in soil, uniform enrichment of ^{15}N in NH_4^+ or NO_3^- pools is not possible. A sensitivity analysis has shown that a non-uniform distribution of added ^{15}N causes only minor errors in calculating gross rates, provided that the non-uniform distribution is random (Davidson et al., 1991). Only if the ^{15}N distribution is biased and coincident with non-random distribution of the process (e.g., more ^{15}N near the top of the soil core where microbial activity is greatest), does significant error arise. The ^{15}N solutions should be added to the soil as uniformly as possible to avoid such bias.

The isotope-dilution model of Kirkham and Bartholomew (1954) also yields an estimate of gross rates of consumption of NH_4^+ and NO_3^- (Table 42–1). Consumption of NH_4^+ may include microbial assimilation, NH_3 volatilization, leaching, clay fixation, and nitrification. Consumption of NO_3^- may include microbial assimilation, leaching, denitrification, and dissimilatory NO_3^- reduction. For many applications, several of these fates can be assumed to be quantitatively insignificant. Unlike gross mineralization and gross nitrification, the rates of gross consumption of NH_4^+ and NO_3^- are susceptible to overestimation by this method because of substrate stimulation. Although we present the equations for calculating gross consumption rates, we caution that they may yield overestimates of consumption rates in unamended soil.

42–1.2.2 Materials

42–1.2.2.1 Special Apparatus

1. Spinal needles: 18 gauge, with "through-hole" side-port; length can vary depending on application, but 15 cm is probably a practical upper limit (available from Popper & Sons, Inc., New Hyde Park, New York).
2. Syringes: 1-mL glass tuberculin syringes work well.
3. Cylinders for obtaining intact soil cores: diameter of ≥ 4 cm to minimize soil compaction during placement; length depends on desired sampling depth, but a core longer than 15 cm would be difficult to uniformly inject with ^{15}N; a variety of thin-walled plastic (e.g., polyvinyl chloride) and aluminum materials can be used to make the cores; it may be helpful to bevel and sharpen the

bottom edge to reduce soil compaction within the core during placement; snugly fitting caps are needed to seal the core ends; a small sledge hammer.

4. Filter rack, funnels, and Whatman no. 1 (or finer) paper filters, pre-leached with potassium chloride (KCl) solution or deionized water.

5. Stainless steel wire: cut pieces 62 mm in length.

6. Surgical gloves, tweezers, and new paper clips.

7. Whatman no. 3 paper filters: 7-mm diam. disks cut with a clean paper punch; it may also be desirable to pre-leach filter disks with KCl and deionized water (see below).

8. Four-ounce specimen containers with tight-fitting lids used in hospitals for urine samples (e.g., Fisher Scientific, Tustin, CA 92681; catalog no. 11-840A).

9. Acid-washed glass beads (4-mm diam.); acid wash: 0.5 to 1.0 M hydrochloric acid (HCl) solution; beads then rinsed with deionized water and dried.

10. Tin (Sn) 8 × 5 mm capsules (Europa Scientific, Inc., Cincinnati, OH 45212; catalog no. D1008) and disposable microtiter plates (Baxter Diagnostics, Scientific Products Division, McGraw Park, IL 60085-6787; catalog no. B1190-19) for direct combustion ^{15}N analysis; or 4-mL plastic auto-analyzer vials with snugly fitting caps (for conventional {NaOBr} ^{15}N analyses).

11. Automatic micropipette that can deliver 10 μL of solution.

12. Two measuring scoops: one that can deliver approximately 0.2 g of MgO, and one that can deliver approximately 0.4 g of Devarda's alloy.

42–1.2.2.2 Reagents

1. Nitrogen-15 labeled ammonium sulfate $(NH_4)_2SO_4$ solution: about 30 mg of N L^{-1}, highly enriched (e.g., 99%) in ^{15}N.

2. Nitrogen-15 labeled potassium nitrate (KNO_3) solution: about 30 mg of N L^{-1}, highly enriched (e.g., 99%) in ^{15}N.

3. Potassium chloride solution, approximately 2 M: dissolve 1.5 kg of reagent grade KCl in 10 L of deionized water.

4. Magnesium oxide (MgO), heavy powder.

5. Devarda's alloy, finely ground (< 250 μm).

6. Sulfuric acid (H_2SO_4) solution, approximately 5 M for analysis of trapped ^{15}N by conventional mass spectroscopy. Use potassium hydrogen sulfate (K_2HSO_4), approximately 2.5 M for analysis of trapped ^{15}N by direct combustion; make by carefully adding 7 mL of concentrated H_2SO_4 to 50-mL deionized water, add 22 g of potassium sulfate {K_2SO_4}, add more deionized water, mix until salt is dissolved, and bring to final volume of 100 mL.

7. Other reagents as required for NH_4^+ and NO_3^- analyses of KCl extracts of soil (see chapters 40 and 41 in this book).

42–1.2.3 Procedure

Drive the cylinders into the soil to the appropriate depth using a sledge hammer. Pounding on a piece of wood placed over the top of the cylinders generally works well, and reduces the chance of splitting the cylinders. Remove the core, place caps on both ends, and set it aside in the shade. Repeat this procedure until you have four cores taken adjacent to each other from within each study plot (e.g., 2×2 m).

Place the four capped cores upside-down and remove the caps from the deep end (which is now facing up). Insert a spinal needle into the center of a core until the tip reaches within 1 cm of the opposite end. Keep the obturator (metal wire) inside the needle while inserting it into the soil, and remove the obturator before attaching a syringe. Fill a syringe with 1 mL of the $(^{15}NH_4)_2SO_4$ solution and attach it to the needle. Carefully push down on the syringe plunger as you pull up on the syringe and needle, so that a column of solution is injected into the soil. If possible, twist the syringe as you pull up so that the side ports of the needle turn as it is pulled through the soil. Repeat this step until 6 mL of solution has been injected into holes distributed across the core, taking care to keep all drops of ^{15}N falling onto the soil core when the syringe is taken on and off of the needle. This procedure provides a roughly uniform distribution of the label for a 4×9 cm core (about 100 g dry-weight equivalent of soil). Wider cores may require more numerous injections, and deeper cores may require injections of greater volume to provide adequate uniformity of label distribution. If the core holds approximately 100 g of dry soil, this procedure increases the ambient NH_4^+ pool size by about 2 mg of N kg^{-1} of dry soil. Measuring single-core aliquots of ^{15}N solution into small containers before going to the field and making injections until the aliquot is consumed is an accurate and efficient way of injecting the same volume into each core.

Of the four cores from each plot, inject two with $(^{15}NH_4)_2SO_4$ solution and two with $K^{15}NO_3$ solution. Use different needles and syringes for the NH_4^+ and NO_3^- solutions. Replace the caps on the cores, turn them right-side-up, and rebury one "NH_4^+" and one "NO_3^-" core in the plot. The purpose of capping and inverting the cores is to prevent ^{15}N loss from the soil resulting from the puddling of ^{15}N solution on the bottom of the core during the injection process. Remove the soil from each of the other two injected cores and mix each sample in a plastic bag. Place a scoop from each sample into a pre-weighed specimen container of KCl (e.g., about 15 g of dry-weight equivalent soil in 75 mL of KCl). Keep the containers on ice in a cooler. These samples will be the "time-0" extract for determining the initial pool sizes of inorganic-N and their initial atom % ^{15}N enrichments, as well as the extent of ^{15}N recovery. Record the time of extraction (usually about 15 min after ^{15}N addition). Retrieve the buried cores after the desired incubation (we recommend 1–3 d to prevent significant violation of the third assumption of the isotope-dilution model; see section 42–1.2.1). Remove the soil and extract a scoop in KCl as described above.

Place these extracts in the cooler as well, and transport them back to the laboratory.

In the laboratory, re-weigh each cup of KCl + soil to determine the weight of field-moist soil extracted. Measure the gravimetric water content of the soil sample remains in the plastic bags to adjust field-moist weights to dry-weight equivalents. Filter each KCl extract through Whatman no. 1 filter (or finer for fine-textured soils), and analyze an aliquot of each filtrate for NH_4^+ and NO_3^- (see chapters 40 by Hauck et al. and 41 by Bundy and Meisinger in this book). Because most filter papers have substantial risk of NH_4^+ and NO_3^- contamination (Qasim & Flowers, 1989), they should be leached with KCl or deionized water prior to use. Large numbers of filters can be rapidly leached by placing them in a Buchner funnel, applying a low vacuum, and adding volumes of 2 M KCl and then deionized water. The filters can be used moist if they are handled with care. If rocks are present in the soil, sieve (2- or 4-mm mesh) the soil left in the cup after filtering and subtract the rock weight from the calculation of the dry-weight equivalent of soil extracted. Keep track of all soil used for analyses so that the total weight of the soil in the cylinders can be calculated.

Several methods of ^{15}N analysis are available (see chapter 40 in this book); the mass spectrometer to be used affects your choice of methods for sample preparation (see below). We briefly present the method of Brooks et al. (1989), which is suitable for ^{15}N analysis by many service laboratories.

Place two acid-washed glass beads into new plastic urine cups. Measure into the specimen cups aliquots (20–50 mL) of each KCl extract of soil from the cores that were injected with ^{15}N. Calculate the mass of NH_4^+-N or NO_3^--N in this aliquot based on previously measured concentrations. Add enough KCl extract so that 40 to 100 µg of N is contained in the specimen cup for direct combustion ^{15}N analysis, and 100 to 800 µg of N for conventional ^{15}N analysis. The trapping capacity is 350 and 1400 µg of N for the $KHSO_4^-$ and H_2SO_4-acidified filter disks, respectively. The amount of N diffused should never exceed 50 to 60% of this capacity. For many unfertilized soils, the mass of inorganic N contained in a 50-mL aliquot of KCl extract will be below the above-mentioned minimum value required for ^{15}N analysis. If so, it may be necessary to add an aliquot of NH_4^+ or NO_3^- "carrier" solution of known concentration and background ^{15}N enrichment. Carrier solution may also have to be added if the ^{15}N atom % enrichment of the solution is too high for accurate and precise analysis by mass spectrometry (see chapter 40 in this book).

Wearing clean surgical gloves, stick a wire through each filter disk and move the disk to the center of the wire. Insert a wire with a filter disk into a styrofoam sheet to hold the disk upright, and pipette a 10-µL aliquot of the acid onto the filter paper (remember to use $KHSO_4$ for direct combustion ^{15}N analysis and H_2SO_4 for conventional ^{15}N analysis; see below). For the samples that received $^{15}NH_4^+$, immediately add a scoop (about 0.2 g) of MgO to the KCl extract contained in the specimen container. Quickly (but carefully) insert the acidified filter disk-wire combination into the top of the container (pinching the container allows the wire to be wedged into the ledge on the cup so that the filter is suspended above the solution), and immediately close the container lid. Next, gently swirl the solution in the container, to avoid any splashing of solution onto the filter, but to get the MgO to disperse into a white cloud within the solution. The MgO makes the solution basic, causing NH_3 vapor to be released and subsequently captured on the acidified filter disk. Remove the disk after 6 d and dry the filter disks overnight by inserting the wires into a styrofoam sheet that is placed in a desiccator over concentrated H_2SO_4. For samples that received $^{15}NO_3^-$, do not insert a filter disk-wire combination, and do not close the lid after adding the MgO. Instead, allow the specimen container to sit open for 6 d to permit the NH_3 to escape. Then pipette acid onto the filter, place the wire and filter in the container, add a scoop (about 0.4 g) of Devarda's alloy, close the lid, swirl, and remove the disk after an additional 6 d. Dry the filter disks overnight as described previously. The Devarda's alloy reduces NO_3^- to NH_4^+, and NH_3 is captured on the acidified filter disk.

If conventional ^{15}N analysis is to be used, place the dried filter disks into an auto-analyzer vial by pushing off the filter paper disk using a clean, paper clip, and cap the vial. If direct combustion ^{15}N analysis is to be used, push the dried filter disk into the Sn capsules using a paper clip. Carefully, wrap the Sn capsule around the filter disk with clean tweezers, taking care not to touch the filter disk with the tweezers. Finally, place the Sn capsule into a microtiter plate and cover. These plates hold 96 capsules securely.

42–1.2.4 Calculations

Determination of the fraction of ^{15}N injected that was recovered (F_{15N}) from the cores extracted at time-0 is made using the following equation:

$$F_{15N} = \frac{^{15}N\ excess\ (mg\ kg^{-1})\ *\ core\ weight\ (kg)}{^{15}N\ injected\ (mg)}$$

where

^{15}N excess = {the atom % ^{15}N enrichment of the N pool enriched with ^{15}N minus the atom % ^{15}N enrichment of that pool prior to ^{15}N addition (background enrichment)}, multiplied by the N pool size (mg-N kg^{-1}), divided by 100%;

^{15}N atom % enrichment = mole fraction of ^{15}N to $^{14+15}N$ * 100%.

Gross rates of mineralization (m) and NH_4^+ consumption (c_A) are calculated from the following equations:

$$m = \frac{[NH_4^+]_0 - [NH_4^+]_t}{t} * \frac{log\ (APE_0/APE_t)}{log\ ([NH_4^+]_0/[NH_4^+]_t)}$$

$$c_A = m - \frac{[NH_4^+]_t - [NH_4^+]_0}{t}$$

where

m = gross N mineralization rate (mg of N kg^{-1} soil day^{-1}):
c_A = NH$_4^+$ consumption rate (mg of N kg^{-1} soil day^{-1});
t = time (days);
APE$_0$ = atom % ^{15}N excess of NH$_4^+$ pool at time-0
APE$_t$ = atom % ^{15}N excess of NH$_4^+$ pool at time-t
   where APE = the atom % ^{15}N enrichment of a N pool enriched with ^{15}N
minus the atom % ^{15}N enrichment of that pool prior to ^{15}N addition;
[NH$_4^+$]$_0$ = total NH$_4^+$ concentration (mg kg^{-1}) at time-0;
[NH$_4^+$]$_t$ = total NH$_4^+$ concentration (mg kg^{-1}) at time-t.

Notes:
   t = 1 to 3 d as suggested here;
   Background ^{15}N enrichments can be assumed to be 0.37 atom % ^{15}N;
for more precise work they can be measured in KCl extracts of soil taken
from unlabeled cores sampled at the same time as the time-0 cores;
   Equations are valid only for cases when m is not equal to c; Kirkham
and Bartholomew (1954) provide another equation for the condition when
m = c, which is seldom encountered;
   The atom % ^{15}NH$_4^+$ enrichment values are determined by an isotope-
ratio mass spectrometer using the diffusion filter disks from the time-0 and
time-t KCl extracts; the total NH$_4^+$ concentrations are determined by distillation or colorimetric analysis of the KCl extracts.

   To calculate gross rates of nitrification and NO$_3^-$ consumption, substitute NO$_3^-$ concentrations and atom % ^{15}N enrichments in the above
equations. The symbols "m" and "c_A" are replaced by "n" and "c_N,"
which are the gross nitrification rate and gross NO$_3^-$ consumption rate,
respectively.

42–1.2.5 Comments

   Table 42–2 shows estimates of gross N-transformation rates determined under field conditions using ^{15}N isotope-dilution in different wildland ecosystems. Note that the estimates vary over one to two orders of
magnitude among the different ecosystems. Coefficients of variation (standard deviation expressed as a proportion of the mean) for six plots per site
(with one rate estimate per plot, or n = 6) generally range from 0.15 to
0.75. Increasing the number of plots or increasing the number of cores
taken per plot is necessary if greater precision is desired.
   We do not know of any comparable in-field estimates of gross N-transformation rates in agricultural ecosystems. However, several measurements of gross rates of N mineralization, immobilization, and nitrification
have been made in the laboratory using sieved agricultural soils. These

Table 42–2. Field estimates of gross rates of N transformations in different wildland ecosystems using ^{15}N isotope-dilution.[†]

Ecosystem	Soil	NH_4^+		NO_3^-	
		Production	Consumption	Production	Consumption
		mg-N m^{-2} d^{-1}			
Oak woodland/grassland[‡]	Mineral	610–1085	820–1390	75–420	100–440
10-yr conifer forest[§]	Mineral	10–170	110–160	40–110	30–75
60-yr conifer forest[¶]	Mineral		155		
	O₂ horizon		200		
> 100-yr conifer forest[§]	Mineral	100–420	270–535	25–85	80–245
	O₂ horizon	135–250	135–310	15–105	25–100
Northern hardwood forest[††]	Mineral		290,430	80,315	135,170
Young lowland tropical forest[‡‡]	Mineral	210	35	220	130
Old lowland tropical forest[‡‡]	Mineral	240	40	235	135

[†] Gross mineralization = NH_4^+ production; gross nitrification = NO_3^- production; consumption rates may include several concurrent processes (see text).

[‡] Davidson et al., 1990; range in values reflects seasonal variation and variation resulting from the presence or absence of an oak canopy.

[§] Davidson et al., 1991; range in values reflects seasonal variation.

[¶] Schimel and Firestone (1989).

[††] Calculated from data in Zak et al., 1990; values given are for microplots with and without spring ephemerals, respectively.

[‡‡] Zou et al (1992).

laboratory estimates typically range between < 1 and 7 mg-N kg^{-1} d^{-1} (Nishio et al., 1985; Myrold & Tiedje, 1986; Schimel, 1986; Nishio & Fujimoto, 1989). Coefficients of variation for field estimates of gross N-transformation rates in agricultural ecosystems using intact soil cores are likely to be at the lower end of the range found for wildland soils. Nevertheless, we recommend at least six independent estimates of each N process per study site.

The time-0 extraction of injected cores is necessary when the fraction of ^{15}N recoverable at time-0 (F_{15N}) varies among study plots. In a soil where clay fixation of added $^{15}NH_4^+$ was significant, this fraction (F_{15N}) varied from 0.4 to 0.8 across a 50 × 50 m study area (Davidson et al., 1991). If preliminary work shows that recovery at time-0 is reliably uniform, injection and analysis of the time-0 cores should be eliminated in subsequent studies, resulting in significant savings of time and expense.

The time-0 extraction is the appropriate correction for abiotic consumptive processes of the added ^{15}N that occur immediately after addition of the label. If some of the ^{15}N added is fixed abiotically at time-0, and this is released back into the soil solution during the incubation period, the actual rate of isotopic dilution will be underestimated, resulting in an underestimate of gross N-transformation rates. Conversely, if ^{15}N displaces ^{14}N fixed in soil clays continuously over the entire incubation period rather than instantaneously after ^{15}N addition, gross N-transformation rates will be overestimated. Using sterilized soils, Davidson et al. (1991) found that abiotic fixation of added ^{15}N was completed prior to the time-0 extraction, so the time-0 extraction was the appropriate method for correcting isotope dilution estimates. However, other soil types may behave differently, so we recommend testing the dynamics of abiotic ^{15}N consumption in each soil to which the isotope-dilution method is to be applied (see Davidson et al., 1991).

In heterogeneous soils, the method outlined above may not be appropriate because time-0 and time-t cores may contain substantially different amounts of soil or inorganic N. Because the same quantity of ^{15}N would be added to each core, soil ^{15}N concentrations would be different initially, resulting in large errors in rate estimates. Davidson et al. (1991) describe a variation of the isotope-dilution method that is suitable for heterogeneous soils. In this variation, the time-0 core is used only to calculate F_{15N}; it is assumed that this same fraction of added N could be recovered in the time-t core. The atom % excess of ^{15}N at time-0 (APE_0) is calculated using the equation:

$$APE_0 = \frac{^{15}N\ injected\ (mg) * F_{15N}}{Total\text{-}N\ injected\ (mg)\ *F_{15N} + initial\ soil\text{-}N\ content\ (mg)} * 100\%$$

The initial soil-N content is determined from the product of the initial N-pool concentration and the dry-mass of soil in the time-t core. The initial N-pool concentration content is estimated from a KCl extraction of soil taken between a larger-diameter core (e.g., 8 cm) surrounding the time-t core, and the time-t core. If $^{15}NH_4^+$ is injected, the NH_4^+ concentration is measured; if $^{15}NO_3^-$ is injected, the NO_3^- concentration is measured. The rest of the procedure is identical to that described in section 42–2.2.3.

The incubation time, amount of ^{15}N injected, and the enrichment of the solution can be varied. The advantages and disadvantages of these choices are discussed by Davidson et al. (1991). Addition of 6 mL of ^{15}N solution to a soil core 4 × 9 cm causes an increase in soil water content of approximately 0.06 kg kg^{-1} depending on the bulk density of the soil. This wet-up may stimulate microbial activity and increase N transformation rates, especially if the soil is very dry. A smaller volume of a more concentrated ^{15}N solution can be injected; however, this may result in less a uniform distribution of the label. A simulation model developed by Stark (1991), based on calculated increases in NH_4^+ diffusion rates with increasing soil water content, predicted that stimulation of nitrification by solution addition was a problem in the soil modeled only at soil water potentials < −0.35 MPa. However, the effect of wetting must be expected to vary depending on the process rate being measured, the initial substrate concentration, the soil characteristics, and the initial soil water content. Currently, an attempt is being made to develop a ^{15}N isotope-dilution method using ^{15}N-labeled gases that would allow uniform labeling of a soil without any increase in the ambient soil water content (Stark & Firestone, 1991); however, the method is still in the early stages of development.

Although we have described the ^{15}N isotope-dilution method as a field method using intact soil cores, it may also be applied to intact cores within the laboratory under controlled conditions, or to mixed soil, depending on the investigators objectives. If mixed soil is to be used, we recommend adding the ^{15}N label as a fine mist. The soil is spread out in a thin layer in a large polyethylene bag and the label is sprayed evenly over the soil layer. The soil is then mixed in the bag, and the above process repeated at least one more time. Mixing the soil may serve to reduce sample heterogeneity, but may substantially increase rates of N mineralization over short incubation periods (Schimel et al., 1989).

42–2 FIELD METHODS FOR ESTIMATING NET RATES OF NITROGEN TRANSFORMATIONS

42–2.1 General Introduction

Numerous methods are available for estimating the net production of inorganic N in soil under field conditions (Table 42–1). The net accumulation of inorganic N in the absence of plant roots is thought to provide a good index of N availability to plants (see chapter 41 in this book). An accurate assessment of the N supplying power of a soil requires estimating the net production of inorganic N under field conditions because of the strong effect of site environmental factors on soil N-transformation rates (Binkley & Hart, 1989). In addition to providing estimates of plant-available N, measurements of net N mineralization and nitrification rates have also been used to assess the potential for N loss from ecosystems (Vitousek et al., 1982).

All in-field methods for assessing net rates of N mineralization and nitrification use some form of soil containment that prevents plant N-uptake. Usually intact soil cores are used in these methods because soil disturbance (sieving or mixing soils prior to incubation) has been shown to significantly increase net N production rates (Binkley & Hart, 1989). Concerns with using intact soil cores have centered around the effect of root severing and subsequent root decomposition on net N-transformation estimates. However, removing severed roots by soil sieving probably alters net soil N-transformations rates to even a greater extent. Net rates of N mineralization are calculated from the difference in NH_4^+ plus NO_3^- pool sizes between post- and pre-incubated soils. Similarly, net nitrification rates are calculated from the net change in NO_3^- pool sizes over an incubation period. Inorganic-N pool sizes are determined in concentrated salt (i.e., 2 M KCl) extracts of soil.

DiStefano and Gholz (1986) combined an ion exchange resin technique with a soil containment method to form an in-field soil incubation system that would be open to water but not ion flow, while still excluding plant uptake. The incubation design consists of an intact soil core (taken with a sharpened polyvinyl chloride [PVC] or metal tube) that is sandwiched between two ion exchange resin bags. The resin bag on top of the tube deionizes throughfall (or leachate from overlying soil horizons) entering the tube, while the bottom resin bag captures ions leaching from the core. After the incubation period, the top resin bag can be discarded (unless a measure of inorganic N entering the tube is desired), and the soil and bottom resin bag are extracted with a concentrated salt solution. The extractable NH_4^+ and NO_3^- from the soil and the bottom resin bag minus the pre-incubation soil quantities provides an estimate of the net amount of N mineralized. Net nitrification rates are determined in an analogous manner considering only changes in NO_3^-. This technique is potentially superior to other soil containment methods (see below) because it allows water content in the soil tube to fluctuate during the incubation period while preventing the loss of NH_4^+ and NO_3^- from the confined soil. Further, it also may provide a more realistic value for net N mineralization and nitrification than the other in-field incubation methods by allowing the products of mineralization and nitrification to be removed from the core. This would minimize any feedback effect of NH_4^+ and NO_3^- accumulation on N transformations, as has been observed in some soils (Raison et al., 1987; Binkley et al., 1992). The major disadvantage of the resin-core method is that it takes substantially more effort to use than the other in-field methods for measuring net N mineralization and nitrification. This increased effort may be significant for sites where a large amount of replication is needed. The resin-core method may be most efficacious in high precipitation environments, or when in-field incubations must be conducted for extended periods of time (i.e., several months to a year).

Incubations of soils in buried polyethylene bags has become a widely used in-field measure of net N mineralization and nitrification. In this "buried-bag" method, soil samples are typically taken intact with a soil-coring device and placed in polyethylene bags that are then tied shut. The plastic bags (generally 15- to 30-μm thick) are permeable to gases (O_2, CO_2, etc.) but impermeable to liquids (Gordon et al., 1987). The bags are then buried usually in the hole from which they came and incubated from 1 to 2 mo (or over winter). Repeated buried-bag incubations provide annual patterns of net N mineralization and nitrification. The method is superior to laboratory incubations primarily because of its sensitivity to on-site temperature regimes. The buried-bag method integrates on-site soil water dynamics only if the soil water content at the beginning of the incubation period is representative of soil water conditions for the entire incubation period.

The first in-field containment method for measuring net rates of N mineralization and nitrification dates back to the 1950s. Lemée (1967) developed a technique where soil samples were incubated on-site in aluminum cans that had been perforated on the sides, and were closed on the top but open on the bottom. The design allowed the soil water content within the can to be similar to that of the bulk soil, while preventing direct uptake of NH_4^+ and NO_3^- by plant roots and direct leaching of N from the incubated soil by precipitation. Adams and Attiwill (1986a,b) later modified the method by using PVC tubes with perforated sides, capped on top with inverted plastic petri dishes. One potential problem with perforated cylinders is the loss of NH_4^+ and NO_3^- ions out of the cylinder due to mass flow of water and ion diffusion. Because plant uptake of water and inorganic N is prevented within the cylinder, water potential and NH_4^+ and NO_3^- concentration gradients are likely to occur favoring movement of inorganic N ions out of the cylinder. Raison et al. (1987) used cylinders made of unperforated galvanized steel or PVC that were left uncovered or were covered with thin plastic film. They noted that all containment methods altered soil water dynamics to some degree; however, by using both open and covered systems, the effects of water content on net N mineralization and nitrification could be assessed, and rates of maximum N leaching could be estimated. The major advantage of the closed-top, solid cylinder technique over other containment methods is that it allows the assessment of net N-transformation rates in both surface soils and subsoils without causing substantial disruption of soil structure.

We recommend both the closed-top, solid cylinder and the buried-bag incubation methods for obtaining quantitative estimates of net N mineralization and nitrification rates in the field with the smallest amount of effort (Table 42–1). If subsoils are to be assessed (e.g., below 15 cm of mineral soil), we advise using the closed-top, solid cylinder method because soil structure is maintained by not requiring the removal of the soil core prior to incubation. However, if only surface soils are to be assessed (e.g., surface organic layers or 0–15 cm of mineral soil), we advise using the buried-bag method because results will be more easily compared with the vast numbers of studies that have used this method at other sites in the past.

42–2.2 Accumulation of Inorganic Nitrogen in Closed-Top Solid Cylinders

42–2.2.1 Principles

The method involves the field incubation of an intact soil sample within a thin-walled solid steel or PVC cylinder, the top of which is covered with thin polyethylene film or an inverted styrofoam cup. The length of the incubation period may be as short as a week or as long as over winter depending on objectives and changing environmental conditions. Soil subsamples taken before and after incubation are extracted with 2 M KCl and analyzed for NO_3^- and NH_4^+ by distillation or colorimetric analysis. The amount of total inorganic-N ($NO_3^- + NH_4^+$) and NO_3^- accumulated within the core after the incubation period relative to the amount present in the soil prior to incubation provides an estimate of net N mineralization and net nitrification, respectively.

42–2.2.2 Materials

42–2.2.2.1 Special Apparatus
1. Thin-walled PVC or steel cylinders (e.g., 5-cm inner diam., 24-cm length), with one sharpened end to form a cutting edge (approximately 0.5-mm thick). The length may vary depending on the soil depth of interest. A minimum of 24 cylinders are required per site (48 being preferable), or 3 per (5 × 10 m or smaller) plot.
2. Polyethylene film (< 40 μm thick) or styrofoam cups.
3. Filtration rack, funnels, and Whatman no. 1 (or finer) filter paper pre-leached with KCl or deionized water.

42–2.2.2.2 Reagents
1. Potassium chloride (KCl) solution, approximately 2 M: dissolve 1.5 kg of reagent grade KCl in 10 L of deionized water.
2. Other reagents as required for NH_4^+ and NO_3^- analyses of KCl-extracts of soil (see chapters 40 and 41 in this book).

42–2.2.3 Procedure

Twenty-four cylinders are hand driven into the soil at random locations within the site so that about 2 cm of the cylinder remains above the soil surface. These soil cylinders are then removed from the soil, the bottom 2 cm of soil is discarded (as the bottom 2 cm of soil may not be retained within the cylinder after it is removed from the ground). This gives a 0 to 20-cm sampling depth for the assay. Soils from three cores are composited in the laboratory for extraction and analysis (giving eight composite samples per site; see below). These soil samples provide the initial concentrations of NH_4^+ and NO_3^- for the incubation. The process is then repeated except this time the cylinders are left in the ground, their tops covered with thin plastic film (or inverted styrofoam cups), and allowed to incubate in the field for 1 mo. After 1 mo, the incubated cylinders are removed, and in the laboratory soils from three incubated cylinders are mixed to provide eight composite samples for laboratory extraction and analysis.

Soils are kept cool (2–5 °C) after removal from the field until analyzed in the laboratory to impede biological activity. Soil processing should commence within 2 d of collection. If this is not feasible, we recommend that the soil samples be frozen until they can be processed (see Hart & Firestone, 1989). Field-moist soils are sieved through a coarse-mesh screen (4-mm openings). The resulting fine and coarse fractions from each soil core are weighed, and the fine fractions from three soil cores are composited. A well-mixed, 10-g dry-soil equivalent subsample is extracted with 100 mL of 2 M KCl. The suspension is stirred several times over the next 24 h to ensure complete replacement of NH_4^+ by K^+ on the soil cation-exchange sites. Alternatively, the soil suspensions are shaken on a mechanical shaker for 1 h. The suspensions are filtered through filter paper (Whatman no. 1 or finer) that has been previously leached with KCl or deionized water to remove any NH_4^+ and NO_3^- contamination in the filter paper. The filtered extracts are then analyzed for NH_4^+ and NO_3^- (see chapters 40 and 41 in this book). Extracts should be kept cool (2–5 °C) until analyzed. If prolonged storage is necessary, we recommend freezing the extracts. Separate soil samples are taken from each soil (pre- and post-incubation) to determine the gravimetric soil water contents.

Net N mineralization is calculated by subtracting the initial quantity of soil inorganic-N from the post-incubation quantity of soil inorganic-N. Similarly, net nitrification is calculated as the change in NO_3^- pool size over the incubation period. Rates expressed on a per gram of oven-dry soil basis can be converted to a soil volume (area) basis by using the mean dry mass of soil contained within the cylinders to calculate soil bulk density. Repeated monthly incubations conducted over the course of a year can provide annual estimates of net N mineralization and nitrification.

42–2.2.4 Comments

Edmonds and McColl (1989) found slightly higher soil water contents in confined soils than in unconfined soils from clearcut, young, and mature *Pinus radiata* stands in South Australia. They also found, as did Raison et al. (1987), that soil temperature was marginally elevated within the metal cylinders used for soil confinement, particularly for surface soils in open areas during the summer months. However, soil temperature inside and outside the metal cylinders never varied by more than a few degrees C. Both research groups suggest that artifacts due to differences in the soil microenvironment within the soil cylinder are small, especially relative to most treatment effects.

Raison et al. (1987) reported little compaction resulted from driving the cylinders (of the dimensions give above) into the soil, even in wet, fine-textured soils. Net N mineralization and nitrification rates of a greater or smaller portion of the soil profile can be assessed using longer or shorter cylinders. Raison et al. (1987) readily sampled soils to a depth of 40 cm with this method, and noted that even deeper soil depths could be assayed. Cylinders constructed from PVC may shatter when driving them into stony or dry, clayey soils, so steel cylinders are recommended for these soil types. Use of this method for organic soil horizons may require larger diameter cylinders (approximately 30-cm diam.) to accommodate larger organic debris of varying sizes. These organic materials may require some chopping prior to KCl extraction in order for efficient subsampling (Hart & Firestone, 1991). Soil contained within the cylinders can be analyzed by depth increments if a depth profile of net rates of N mineralization and nitrification is desired (see Binkley & Hart, 1989). Because of the high spatial variability of net N mineralization and nitrification rates assessed under field conditions (coefficients of variation from 0.2–3.0), the above sampling scheme may still result in unacceptable precision for monthly incubations in some sites (Raison et al., 1987). Because variances tend to be greater for inorganic N concentrations of incubated soils, additional effort should be put into increasing the number of incubated samples if increased precision is required.

42–2.3 Accumulation of Inorganic Nitrogen in Polyethylene Bags (Buried-bag Method)

42–2.3.1 Principles

Several variations of this method exist in the literature, including the incubation of sieved soils rather than the incubation of intact soil cores described here (see Binkley & Hart, 1989). The principle of the method is similar to the closed-top, solid cylinder incubation method (section 42–2.2.1), except the soil core to be incubated is first removed from the ground and enclosed in a polyethylene bag that is then tied shut. The soil core is replaced in its original hole, and incubated for 1 to 2 mo or over the winter. As in the closed-top, solid cylinder incubation method, soil subsamples taken before and after incubation are extracted with 2 M KCl and analyzed for NO_3^- and NH_4^+ by distillation or colorimetric analysis. The amount of total inorganic N ($NO_3^- + NH_4^+$) and NO_3^- accumulated within the core after the incubation period relative to the amount present in the soil prior to incubation provides an estimate of net N mineralization and net nitrification, respectively.

42–2.3.2 Materials

42–2.3.2.1 Special Apparatus

1. Thin-walled PVC or steel cylinders (e.g., 5-cm inner diam., 12-cm length), with one sharpened end to form a cutting edge (approximately 0.5-mm thick). The length may vary depending on the soil depth of interest; however, the requirement of placing a soil core within a plastic bag gives a practical maximum in sampling depth of about 15 cm. A minimum of 24 cylinders are required per site (48 being preferable), or 3 per (5 × 10 m or smaller) plot.
2. Polyethylene bags (15–30 µg thick).
3. Filtration rack, funnels, and Whatman no. 1 (or finer) filter paper pre-leached with KCl or deionized water.

42–2.3.2.2 Reagents

1. Potassium chloride (KCl) solution, approximately 2 M: dissolve 1.5 kg of reagent grade KCl in 10 L of deionized water.
2. Other reagents as required for NH_4^+ and NO_3^- analyses of KCl-extracts of soil (see chapters 40 and 41 in this book).

42–2.3.3. Procedure

Twenty-four cylinders are hand driven into the soil at random locations within the site so that about 2-cm of the cylinder remains above the soil surface. These soil cylinders are then removed from the soil giving a 0 to 10-cm sampling depth for the assay. Soils from three cores are composited for laboratory extraction and analysis (giving eight composite samples per site). These soil samples provide the initial concentrations of NH_4^+ and NO_3^- for the incubation. The process is then repeated except this time the soil cores are enclosed in polyethylene bags that are tied shut with plastic-wire twist ties. The bagged soil cores are then reburied in their original holes and allowed to incubate in the field for 1 mo. Any surface litter removed prior to soil coring should be replaced over the core after reburial. If no surface litter was initially present, place a small amount of adjacent soil over the core to prevent any direct isolation on the bag that may result in excessive soil heating or breakdown of the plastic. After 1 mo, the bagged cylinders are removed, and soil from three of the cores are mixed to provide eight composite samples for laboratory extraction and analysis. Bagged soil cores showing visible tears in the plastic should not be analyzed. The procedures for sample processing, NH_4^+ and NO_3^- analysis, and net rate calculations are identical to those described for the closed-top, solid cylinder method (see section 42–2.2.3).

42–2.3.4 Comments

Many of the comments made for the closed-top, solid cylinder method apply to the buried-bag method as well (see section 42–2.2.4). We do not know of any studies that have directly compared estimates of net N mineralization and nitrification determined using closed-top, solid cylinder and buried-bag methods, but we assume that they would provide similar values using our recommended procedures. Differences in the results obtained using these two methods might occur in soils that have a fluctuating water table near the soil surface; however, in such a situation both of these methods may produce meaningless results.

42–3 LABORATORY METHODS FOR ESTIMATING NET NITROGEN TRANSFORMATION RATES

42–3.1 General Introduction

All of the biological laboratory incubation methods uses as indices of available N in soil involve the processes of N mineralization and immobilization, and some also involve nitrification (see Binkley & Hart, 1989). Detailed descriptions of many of these methods, particularly those commonly used for agricultural soils, are included in chapter 41. Here, we describe a version of the aerobic laboratory incubation method, which is the most frequently used assay for examining internal N-cycling processes in wildland soils.

42–3.2 Aerobic Incubation

42–3.2.1 Principles

In this method, field-moist soil is mixed, sieved, and placed in a beaker or plastic cup. The cup is usually covered with thin plastic wrap or a perforated lid to allow gas exchange but minimize water loss, and the soil is incubated for 28 d at room temperature (25 °C) in the dark. Ammonium and NO_3^- concentrations are determined by extracting soil subsamples with salt solutions (e.g., 2 M KCl) before and after incubation. Net N mineralization rates are calculated by subtracting initial inorganic-N concentrations from final inorganic-N concentrations; net nitrification rates are calculated by subtracting initial NO_3^- concentrations from final NO_3^- concentrations. The aerobic laboratory incubation method is somewhat analogous to the in-field buried-bag and closed-top, solid cylinder methods described above (section 42–2.1). All three methods measure net N-transformations rates in the absence of plant roots under constant soil water conditions. However unlike laboratory incubations, buried-bag and closed-top, solid cylinder incubations are sensitive to fluctuations in field soil temperature. Furthermore, in the laboratory incubation method described below, there is substantially greater soil disturbance prior to the incubation that also makes this approach less reliable for the estimation of field N availability.

42–3.2.2 Materials

42–3.2.2.1 Special Apparatus

1. Polyethylene film (< 40 μm thick) and 100-mL beakers or specimen containers (e.g., Fisher Scientific, Tustin, CA 92681; catalog no. 11-840A) with a small hole placed in the lid.
2. Filtration rack, funnels, and Whatman no. 1 (or finer) filter paper pre-leached with KCl and deionized water.

42–3.2.2.2 Reagents

1. Potassium chloride (KCl) solution, approximately 2 *M:* dissolve 1.5 kg of reagent grade KCl in 10 L of deionized water.
2. Other reagents as required for NH_4^+ and NO_3^- analyses of KCl-extracts of soil (see chapters 40 and 41 in this book).

42–3.2.3 Procedure

Soil samples are taken using a shovel to the depth of interest (typically 0–15 cm). Compositing several soil samples taken at random locations within a site will provide a good estimate of the mean value but will not give any information regarding within-site variability. We suggest randomly selecting a minimum of eight sampling locations (5 × 10 m or smaller plots) within a given site, and compositing three samples taken within each of these locations. This will give eight samples per site for laboratory incubation. After sampling, place the soil samples in sealed polyethylene bags in a cooler for transport back to the laboratory. Soil processing should commence as soon as possible upon returning to the laboratory, and soil should be kept in a cold room (2–5 °C) in the interim.

Field-moist soils are sieved through a coarse-mesh screen (4-mm openings). The resulting coarse fraction is discarded. For each sample, two well-mixed, 10-g dry-soil equivalent subsamples are weighed into beakers or specimen containers. One subsample is extracted immediately with 100 mL of 2 *M* KCl (see section 42–2.2.3 for details). The other sample is covered either with polyethylene film or with a cover containing a small hole to minimize water loss yet maintain gas exchange. It is recommended that the container + soil system be weighed prior to and after each week of incubation to monitor water loss. If water loss is high (> 5% relative change in water content), deionized or distilled water should be added to return the soil to its initial water content. Do not mix the soil when replacing water that is lost during the incubation because this disturbance will likely alter net N-transformation estimates. Add the water slowly with a syringe, atomizer, or eyedropper. Soil samples are incubated in the dark at room temperature in the laboratory or in an incubator maintained at 25 °C if the laboratory temperature varies by more than 2 °C. A third subsample is used to determine the gravimetric soil water content.

After the 28 d, extract the soil with 100 mL of 2 *M* KCl and filter using the same procedure as was used for the pre-incubation (initial) soil samples. Net N mineralization is calculated by subtracting the initial quantity of soil inorganic N from the post-incubation quantity of soil inorganic N. Similarly, net nitrification is calculated as the change in NO_3^- pool size over the incubation period. Rates are expressed on a per gram of oven-dry soil basis.

42–3.2.4 Comments

Net N mineralization and nitrification rates of forest floor or other organic horizon materials also may be determined using this method with a few modifications. Sampling of organic horizons may require the use of large-diameter coring devices or the excavation of larger areas to obtain representative samples. In this case, sample compositing within each sample location may not be necessary. Large organic materials are hard to subsample uniformly and may not readily fit into incubation containers. Chopping these materials in a Waring blender helps reduce the size of these materials, allowing efficient subsampling and extraction without significantly altering NH_4^+ and NO_3^- concentrations (Hart & Firestone, 1991). The low density and high water-holding capacities of organic materials also necessitates using smaller subsamples for incubation and extraction (e.g., 5-g dry-soil equivalent). Soil water content is also determined at a lower temperature (65–70 °C) than for mineral soils to reduce the loss of volatile organics.

More variations of the laboratory incubation method exist in the literature than probably any other method for assessing N availability in soil, and these variations can produce contrasting results for a given soil (see Binkley & Hart, 1989). The choice of incubation conditions (e.g., temperature, water content, sample preparation procedures, and incubation period) ultimately depends on the objectives of the investigator. Does the investigator want a maximum rate that is characteristic of the material being mineralized, or does the investigator want a rate that may reflect net N transformations in the field as closely as possible? Other considerations include the effort required to apply the method and the within-soil variability in the estimates produced. Most investigators weigh these latter two factors heavily when choosing a laboratory method; ease of use and low variability are often the two major reasons for conducting a laboratory rather than a field assay. The method described above is a compromise among these conflicting demands. Field water contents are used to be reflective of site water conditions, yet sieved soil samples are used to reduce within-soil variability and allow small and easily handled incubation containers. Ambient laboratory temperatures are suggested (25 °C) to eliminate the need for temperature-controlled chambers. One-month incubation periods are suggested so that estimates can be obtained fairly quickly, but be relatively insulated from the transient effects of soil disturbance on net rate estimates during sample processing. Using the above recommended version of the aerobic incubation technique, coefficients of variation (CV) are likely to range between 0.2 and 0.6 (Binkley & Hart, 1989).

If the investigator wants a laboratory estimate that is more reflective of field rates, we suggest modifying the above method so that the incubation temperature is reflective of mean field soil temperature over the period of interest. We also recommend incubating intact soil cores (see Lamb, 1980) to avoid the effects of soil mixing on net N-transformation rate estimates. If the investigator desires a maximum rate of net nitrification, we suggest using the shaken soil-slurry method described below (section 42–4.2). Because of the large diversity in soil organic-N substrates and the microbial populations involved in the N mineralization process, different soils will show maximal rates of net N mineralization under different sets of environmental conditions. Therefore, no method for determining maximum rates of net N mineralization can be suggested. However, frequent soil mixing during the incubation, and incubation of soil at water contents near field capacity (-0.033 MPa) and at elevated temperatures (e.g., > 25 °C) will likely result in higher rates for a given soil than the rates determined using the recommended procedures described above. Finally, concurrent measurement of CO_2 evolution (as an index of microbial activity) during the incubation may aid in interpreting the net N mineralization results obtained using the aerobic incubation method (see Binkley & Hart, 1989).

42–4 LABORATORY METHODS FOR ASSESSING NITRIFICATION

42–4.1 General Introduction

The term *nitrification potential* has been applied to nitrification rates measured using a wide variety of laboratory incubation techniques, including aerobic incubations of soil in beakers (Robertson & Vitousek, 1981), incubations involving periodic leaching of soil in microlysimeters (Robertson, 1982), incubations of soil slurries shaken in flasks (Belser & Mays, 1980), and incubations of soil in perfusion columns (Killham, 1987). All of these methods provide estimates of net nitrification rather than gross nitrification; however, each method differs in how similar net rates are to gross rates. As we have noted many times already, sample handling prior to incubation, such as sieving, air-drying, wetting-up, or even storing soil samples at low temperatures usually perturbs the samples. This perturbation, combined with incubation conditions (temperature, water content, etc.) that are different than those encountered in the field, results in rates that are usually different from actual field rates. Hence, the term nitrification potential, as commonly used in the literature, does not necessarily mean the maximum possible nitrification rate in a given soil (i.e., V_{max}); rather, this term implies that a set of operationally defined experimental conditions were used to measure the process rate.

When measurements of potential nitrification are made by the various techniques, it is usually not clear either how close net nitrification rates are to gross rates, or how close the laboratory rates are to field rates. This uncertainty results in considerable ambiguity in interpretation of results. For example, addition of high C substrates to soils during laboratory incubations typically results in lower rates of net nitrification (e.g., Johnson & Edwards, 1979; Lamb, 1980). The most common explanation is that high C availability causes immobilization of NH_4^+, and gross nitrification rates decline due to substrate limitations; however, high C availability should also increase NO_3^- immobilization. If this is the case, then a decline in net nitrification rates could be due to both reduced gross nitrification and increased NO_3^- immobilization. Unfortunately, little data are available to show the relative contribution of the two processes in controlling changes in net nitrification. When the results of potential nitrification assays are interpreted, it is critical that the possible occurrence of both processes is considered.

Of the nitrification potential assays mentioned above, the aerobic laboratory incubation is the most commonly used technique. This method has also been used as a measure of N availability to agricultural plants (chapter 41); variations in the methodology as applied to wildland soils were discussed above in section 42–3.2. A major disadvantage of the aerobic laboratory incubation to determine a nitrification potential of a soil is that the net nitrification rates produced are sensitive to changes in soil water content and NH_4^+ concentrations. This is because even at water contents near field capacity, microbial processes may be diffusion limited. Thus, increases in either soil water content or NH_4^+ concentrations will increase substrate supply, which may have significant effects on rates of nitrification or NH_4^+ and NO_3^- immobilization.

Nitrification potentials determined by the shaken soil-slurry method (Belser, 1979) may be the easiest to interpret and most reproducible of all laboratory nitrification assays. This method involves shaking a sieved soil sample in a dilute ammonium phosphate solution (approximately 1:7 soil/ solution ratio) and measuring NO_3^- accumulation over a relatively short period (usually ≤ 24 h). Nitrate immobilization by microorganisms is inhibited by high NH_4^+ concentrations (approximately 1 mM), and denitrification is inhibited because vigorous shaking continuously aerates the slurry. Because NO_3^- consumption processes are largely eliminated, net rates are equivalent to gross rates.

Although this method is often referred to as an enzyme assay, the information produced is applicable to nitrification under field conditions. Changes in nitrifier populations (size, species composition, etc.) are unlikely to occur because the assay is short term. Furthermore, substrate and moisture limitations are eliminated in this method, and thus the nitrification rate measured approximates the maximum nitrification rate (V_{max}) possible at the specific temperature of the incubation.

The nitrification potential determined from a shaken soil-slurry has also been used to estimate nitrifier population sizes (Belser, 1979). The population size may be a good integrator of nitrification rates over the-course of a year because nitrifiers depend on the nitrification process for energy; high nitrification rates are necessary to support large nitrifier populations. If nitrification rates are depressed by substrate limitations or inhibitory compounds, eventually population sizes should also decline due to lack of energy for maintenance requirements. Information on nitrifier population size can be useful in interpreting net nitrification data from aerobic laboratory incubations. For example, the presence of large nitrifier populations in soils showing little or no net nitrification might indicate that high rates of NO_3^- immobilization are occurring.

In addition to the autotrophic bacteria, heterotrophic microorganisms are known to produce NO_2^- or NO_3^- from NH_4^+ and organic compounds (Killham, 1986). Although, heterotrophs may be able to oxidize NH_4^+, they are most commonly considered as agents of oxidation of organic-N to NO_2^- or NO_3^- (Fig. 42-1). The extent to which this process occurs in the natural environment is still being debated; however it appears to be most common in acid soils (pH < 4.5); particularly those high in organic matter (Focht & Verstraete, 1977; Haynes, 1986). Heterotrophic nitrifiers include eukaroytic organisms such as fungi. Evidence demonstrating the occurrence of heterotrophic nitrification in soil samples includes: NO_2^- or NO_3^- production in the presence of inhibitors of NH_4^+ oxidation; NO_2^- or NO_3^- production that is stimulated by addition of organic-N substrates (i.e., peptone) but not stimulated by addition of NH_4^+; and failure to detect $^{15}NO_2^-$ or $^{15}NO_3^-$ after addition of $^{15}NH_4^+$ to soils undergoing active nitrification. An approach combining more than one of these lines of evidence is best for demonstrating the occurrence of heterotrophic nitrification because all contain some potential for error (Schimel et al., 1984). The nitrification potential assay described below, like most other methods for assessing nitrification in soil, integrates the rates of both autotrophic and heterotrophic NO_3^- production.

We recommend the shaken soil-slurry method for assessing nitrification potentials in soil. We feel that this method provides results that are the easiest to interpret of all the laboratory methods for assessing nitrification in soil. It is also a reasonably rapid assay that requires no highly specialized equipment or techniques, and only a moderate amount of effort (Table 42-1).

42–4.2 Shaken Soil-Slurry Method for Assessing Nitrification Potentials in Soil

42–4.2.1 Principles

This method assesses the maximum rate (V_{max}) of nitrification for a soil sample. Samples are incubated under laboratory conditions that have been optimized with respect to water content, NH_4^+, aeration, and P availability. The method can be used as an index to the size of the NH_4^+-oxidizer community in a soil sample.

42–4.2.2 Materials

42–4.2.2.1 Special Apparatus

1. Orbital shaker that holds several 250-mL Erlenmeyer flasks securely.
2. 250-mL glass Erlenmeyer flasks.
3. 15-mL centrifuge tubes or filtration rack, funnels, and Whatman no. 40 filter paper.
4. 5-mL or 10-mL automatic pipette.
5. 5-mL pipette tips with ends cut off to create 0.5-cm orifice (prevents clogging when used with soil slurries).
6. 10-mL disposable polypropylene culture tubes with caps or other suitable container to store collected liquid samples.

42–4.2.2.2. Reagents

1. Potassium monobasic phosphate (KH_2PO_4) stock solution, 0.2 M. Dissolve 27.22 g of KH_2PO_4 in 1 L of water.
2. Potassium dibasic phosphate (K_2HPO_4) stock solution, 0.2 M. Dissolve 34.84 g of K_2HPO_4 in 1 L of water.
3. Ammonium sulfate $\{(NH_4)_2SO_4\}$ stock solution, 50 mM. Dissolve 6.607 g $(NH_4)_2SO_4$ in 1 L of water.
4. Other reagents as required for NO_3^- analyses of dilute salt solutions (see chapters 40 and 41 in this book).

42–4.2.3 Procedure

Combine the following in a 1-L volumetric flask, and then bring up to volume: 1.5-mL KH_2PO_4 stock solution, 3.5-mL K_2HPO_4 stock solution, and 15-mL $(NH_4)_2SO_4$ stock solution. Adjust to pH 7.2 by adding dilute H_2SO_4 or NaOH solutions dropwise while the combined solution is stirred. This results in a solution containing 1.5 mM of NH_4^+ and 1 mM of PO_4^{3-}. One liter of this solution is enough for 10 samples.

Place 15 g of sieved (2- or 4-mm mesh), field-moist soil into a 250-mL Erlenmeyer flask (use only 10 g of material that is high in organic matter, such as soil from O horizons). Determine the gravimetric soil water content of a separate subsample. Add 100 mL of the combined solution, and cap the flask with a vented cap. A rubber stopper containing a hole will allow gas exchange but minimize evaporation. Place all flasks on an orbital shaker and shake at approximately 180 rpm for 24 h.

Sample each flask four times during the 24-h period. Two methods may be used to collect solution samples from the slurries: centrifugation or filtration. For most soil samples, centrifugation is the fastest and easiest method; however, if the soil sample has a large amount of fine organic debris that will not settle during centrifugation (such as many forest soils), filtration may be necessary.

Centrifugation: At each sampling time, remove about 10 mL of slurry from each flask with the automatic pipette fitted with the modified pipette tips. It is important to shake each flask immediately before sampling because the aliquot removed must contain the same soil/solution ratio as the rest of the slurry. Place the sample in a 15-mL centrifuge tube and spin at 8000 × g for 8 min. Draw 5 mL of clear supernatant out of the centrifuge tube with the autopipette (using regular tips) and place in a disposable polypropylene culture tube. Cap and freeze the tubes until the solutions can be analyzed for NO_3^-.

Filtration: At each sampling time, filter 10 to 15 mL of slurry (or enough to obtain sufficient filtrate for analysis) using no. 40 Whatman filter papers preleached with deionized water. Collect the filtrate in disposable polypropylene culture tubes. Cap and freeze the tubes until the solutions can be analyzed for NO_3^-.

Thaw the solutions immediately prior to analysis, and invert the tubes several times to mix. Arrange the tubes so that the entire time series of samples from a single flask are together (this will minimize the confounding rate estimates with analytical drift). Analyze the solutions for NO_3^- by distillation or colorimetic analysis (see chapters 40 and 41). For each soil sample (flask), calculate the rate of NO_3^- production (mg-N L^{-1} h^{-1}) by linear regression of solution concentration versus time. Calculate the rate per unit dry soil according to the following equation:

$$\text{Rate} \atop (\text{mg-N kg}^{-1}\text{ h}^{-1}) = \text{Rate} \atop (\text{mg-N L}^{-1}\text{ h}^{-1}) * \frac{0.1\text{ L} + \text{volume of water in field-moist soil sample}}{\text{kg oven-dry soil in flask}}$$

This equation is valid only if the soil/solution ratio in aliquots removed during sampling are the same as the soil/solution ratio in the original slurry; thus, it is critical that the slurries are not allowed to settle before sampling.

Nitrate consumption in the slurries via denitrification or microbial assimilation should be inhibited by aeration and addition of NH_4^+. However, in soils with very high C availability, some denitrification or NO_3^- immobilization may occur due to high rates of heterotrophic activity. The amount of NO_3^- consumption that is occurring can be determined by inhibiting nitrification and measuring the decline in NO_3^- concentrations. Acetylene (C_2H_2) concentrations > 10 Pa can be used to inhibit NH_4^+ oxidation, and the inhibition may last for several days (Berg et al., 1982). To check for NO_3^- consumption, weigh 15 g of field-moist soil into a 250-mL flask, add 100 mL of the solution described in the standard procedure, and recap the flask with a rubber stopper. Inject 15 mL of C_2H_2 into the flask through a rubber septum placed in the stopper or by inserting a long needle down the side of the stopper. Pump the syringe several times to mix the gas with the air in the headspace, and place it on the shaker with the other samples to be assayed. If ambient NO_3^- concentrations are suspected to be low in the soil samples, it may be necessary to add 50 to 100 µg NO_3^--N to the solution. This control flask is sampled at the same time as the other flasks; however, care must be taken that the C_2H_2 from the control flask does not reach the other samples because rates may be inhibited in these flasks as well. The control flask should be kept stoppered between sampling, and should be removed to a hood or another part of the laboratory during sampling. A decline in NO_3^- concentration over the incubation period indicates NO_3^- consumption.

The amount of NO_3^- consumption in the samples due to denitrification can be evaluated independently by adding an additional 15 mL of C_2H_2 to one of the control flasks following addition of the solution, leaving the flask stoppered continuously throughout the incubation, and measuring the rate of N_2O accumulation in the headspace. Samples of the headspace gas can be obtained with a syringe if a septum is placed in the rubber stopper used to cap the flask. Concentrations of N_2O are measured by gas chromatography with an electron capture detector (see chapter 44).

42–4.2.4 Comments

We have found that for a variety of soils the rate of NO_3^- production is linear from 2 h to more than 36 h. Therefore, the most efficient sampling scheme for estimating the slope (rate) is to concentrate sampling efforts at the beginning and end of the incubation (i.e., at approximately 2, 4, 22, and 24 h, as scheduling allows). At least 1 h is needed for the soil and solution to equilibrate. A 24-h assay provides sufficient sensitivity for most soils; shorter incubation times may be used for soils showing high nitrification activity. Incubation times longer than 24 h may permit nitrifier population growth or depletion of NH_4^+ concentrations.

Soils containing significant amounts of vermiculite or illite will fix a substantial amount of the added NH_4^+. We have seen as much as 90% of the NH_4^+ disappear within the first 2 h of incubation. When assaying soils such as these it may be necessary to increase the amount of NH_4^+ added to ensure that rates are close to V_{max}. It is a good idea to measure NH_4^+ concentrations in at least a subset of samples to verify that substrate concentrations are not limiting. If the appropriate Michaelis-Menten constant (K_m) is known, the following equation can be used to calculate how close rates are to V_{max}:

$$V/V_{max} = \frac{[NH_4^+]}{(K_m + [NH_4^+])}$$

For a variety of soils, K_m values for ammonium oxidation fall in the range of 5 to 50 μM; thus, NH_4^+ concentrations in excess of 450 μM should assure that measured rates are at least 90% of V_{max}.

Although the solution used in the assay is buffered at pH 7.2, the solution is dilute enough that the buffering capacity of most soils overwhelms the buffering capacity of the solution. The pH of the resulting slurry is usually only a few tenths of a unit different than if distilled water were added.

Sodium chlorate (1.1 g of $NaClO_3$ L^{-1}) is sometimes added to the solution to increase the sensitivity of the assay. This modification may be particularly useful for soils that have high ambient NO_3^- concentrations, or for soils that have extremely low nitrification rates. Chlorate acts as a competitive inhibitor of nitrite (NO_2^-) oxidation; thus when it is present in slurries, NO_2^- rather than NO_3^- is the end product of nitrification (Belser & Mays, 1980). Much lower rates can be detected when chlorate is added because ambient soil NO_2^- concentrations are usually very low, and because detection limits are lower for NO_2^- than for NO_3^-. Unfortunately, the effectiveness of the chlorate block declines as NO_2^- concentrations increase (Belser & Mays, 1980). For example, at NO_2^- concentrations of 0.14 mg L^{-1} (0.01 mM), chlorate was only 93% effective in blocking the conversion of NO_2^- to NO_3^-.

Storage of solution samples prior to analysis may result in conversion of NO_2^- to NO_3^- as well. We found that freezing samples resulted in a 27% decline in NO_2^- concentrations, and acidification (2 mL concentrated H_2SO_4 L^{-1} of sample solution) resulted in a 100% decline. While most of the NO_2^--N lost was converted to NO_3^-, approximately 10% was lost from solution (possibly as N-containing gases). Because of these potential problems, we recommend that chlorate be used only when incubations are short (< 0.1 mg of NO_2^--N produced L^{-1}) and samples can be analyzed immediately. If storage for short periods is absolutely necessary, the samples should be preserved with chloroform (2 mL of $CHCl_3$ L^{-1} of sample solution) and refrigerated (5 °C) rather than frozen.

Rates calculated for the control sample should be close to zero. If rates are negative, NO_3^- consumption is occurring, and the nitrification potential assay will underestimate V_{max}. The amount of C_2H_2 used in the test for NO_3^- consumption is several orders of magnitude greater than the minimum shown to inhibit autotrophic nitrification (10 Pa). We are suggesting high concentrations to ensure that sufficient C_2H_2 reaches all microsites in the sample, and so that any losses due to denitrification can be quantified directly. The control described here will not work with soils that have significant amounts of heterotrophic nitrification, because C_2H_2 does not inhibit this source of NO_3^-.

42–5 CODA

As the German physicist Werner Heisenberg's Uncertainty Principle attests, it is impossible to observe a system in its natural condition because the mode of observation alters the system's state to some degree. Methodologies for measuring both gross and net rates of soil N transformations in the field or laboratory are no exception to this rule. For instance, the exclusion of plant roots in the long-term from soil may reduce the available C supply to the mineralizers, resulting in lower gross rates estimates. In contrast, the severing of plant roots when taking soil cores might in the short-term enhance gross rates by increasing the available C supply. Mixing or sieving soil may also affect gross rates by increasing both available C supply and reducing the spatial heterogeneity of available C, NH_4^+, and NO_3^- within the soil (Schimel et al., 1989). Even the small additions of $^{15}NH_4^+$ and $^{15}NO_3^-$ required for the estimate of gross rates by isotope-dilution may increase the rate of microbial immobilization (Davidson et al., 1991). This finding also suggests that the exclusion of plant roots from samples used to estimate net rates may result in increased inorganic pools and rates of microbial N immobilization, and thus underestimate actual net rates (Davidson et al., 1992).

Because no method for assessing soil N transformations provides an unequivocal estimate, the choice of any particular method depends on the individual researcher's objectives and the amount of effort willing to be spent on the assessment. Future work is needed on all of the methods described in this chapter regarding their applicability to a wide range of soils, the degree and under what conditions the methods co-vary with each other, and the experimental and environmental factors important in controlling their estimates.

REFERENCES

Adams, M., and P. Attiwill. 1986a. Nutrient cycling and nitrogen mineralization in eucalypt forests of south-eastern Australia. I. Nutrient cycling and nitrogen turnover. Plant Soil 92:319–339.

Adams, M., and P. Attiwill. 1986b. Nutrient cycling and nitrogen mineralization in eucalypt forests of south-eastern Australia. II. Indices of nitrogen mineralization. Plant Soil 92:341–362.

Barraclough, D., and M.J. Smith. 1987. The estimation of mineralization, immobilization and nitrification in nitrogen-15 field experiments using computer simulation. J. Soil Sci. 38:519–530.

Belser, L.W. 1979. Population ecology of nitrifying bacteria. Ann. Rev. Microbiol. 33:309–333.

Belser, L.W., and E.L. Mays. 1980. Specific inhibition of nitrite oxidation by chlorate and its use in assessing nitrification in soils and sediments. Appl. Environ. Microb. 39:505–510.

Berg, P., L. Klemedtsson, and T. Rosswall. 1982. Inhibitory effect of low partial pressures of acetylene on nitrification. Soil Biol. Biochem. 14:301–303.

Binkley, D., and S.C. Hart. 1989. The components of nitrogen availability assessments in forest soils. Adv. Soil Sci. 10:57–112.

Binkley, D., P. Sollins, and R. Bell. 1992. Comparison of methods for estimating soil nitrogen transformations in adjacent conifer and alder-conifer forests. Can. J. For. Res. 22:858–863.

Bjarnason, S. 1988. Calculation of gross nitrogen immobilization and mineralization in soil. J. Soil Sci. 39:393–406.

Bristow, A.W., J.C. Ryden, and D.C. Whitehead. 1987. The fate at several time intervals of ^{15}N-labeled ammonium nitrate applied to an established grass sward. J. Soil Sci. 38:245–254.

Brooks, P.D., J.M. Stark, B.B. McInteer, and T. Preston. 1989. Diffusion method to prepare soil extracts for automated nitrogen-15 analysis. Soil Sci. Soc. Am. J. 53:1707–1711.

Davidson, E.A., S.C. Hart, and M.K. Firestone. 1992. Internal cycling of nitrate in soils of a mature coniferous forest. Ecology 73:1148–1156.

Davidson, E.A., S.C. Hart, C.A. Shanks, and M.K. Firestone. 1991. Measuring gross nitrogen mineralization, immobilization, and nitrification by ^{15}N isotopic pool dilution in intact soil cores. J. Soil Sci. 42:335–349.

Davidson, E.A., J.M. Stark, and M.K. Firestone. 1990. Microbial production and consumption of nitrate in an annual grassland. Ecology 71:1968–1975.

Delwiche, C.C., and P.L. Steyn. 1970. Nitrogen isotope fractionation in soils and microbial reactions. Environ. Sci. Technol. 4:929–935.

DiStefano, J.F., and H.L. Gholz. 1986. A proposed use of ion exchange resins to measure nitrogen mineralization and nitrification in intact soil cores. Commun. Soil Sci. Plant Anal. 17:989–998.

Edmonds, R.L., and J.G. McColl. 1989. Effects of forest management on soil nitrogen in *Pinus radiata* stands in the Australian Capital Territory. For. Ecol. Manage. 29:199–212.

Focht, D.D., and W. Verstraete. 1977. Biochemical ecology of nitrification and denitrification. p. 134–214. *In* M. Alexander (ed.) Advances in microbial ecology. Vol. 1. Plenum Press, New York.

Gordon, A., M. Tallas, and K. Van Cleve. 1987. Soil incubations in polyethylene bags: Effect of bag thickness and temperature on nitrogen transformations and CO_2 permeability. Can. J. Soil Sci. 67:65–75.

Hart, S.C., and D. Binkley. 1985. Correlations among indices of forest soil nutrient availability in fertilized and unfertilized loblolly pine plantations. Plant Soil 85:11–21.

Hart, S.C., and M.K. Firestone. 1989. Evaluation of three *in situ* soil nitrogen availability assays. Can. J. For. Res. 19:185–191.

Hart, S.C., and M.K. Firestone. 1991. Forest floor-mineral soil interactions in the internal nitrogen cycle of an old-growth forest. Biogeochemistry 12:73–97.

Haynes, R.J. 1986. Nitrification. p. 127–165. *In* R.J. Haynes (ed.) Mineral nitrogen in the plant-soil system. Academic Press, New York.

Herman, D.J., and P.W. Rundel. 1989. Nitrogen isotope fractionation in burned and unburned chaparral soils. Soil Sci. Soc. Am. J. 53:1229–1236.

Jansson, S.L. 1958. Tracer studies on nitrogen transformations in soil with special attention to mineralization-immobilization relationships. Annu. Roy. Agric. Coll. Sweden 24:101–361.

Johnson, D.W., and N.T. Edwards. 1979. The effects of stem girdling on biogeochemical cycles within a mixed deciduous forest in eastern Tennessee: II. Soil nitrogen mineralization and nitrification rates. Oecologia 40:259–271.

Killham, K. 1986. Heterotrophic nitrification. p. 117–126. *In* J.I. Prosser (ed.) Nitrification. Spec. Publ. Soc. Gen. Microbiol. Vol. 20. IRL Press, Washington, DC.

Killham, K. 1987. A new perfusion system for the measurement and characterization of potential rates of soil nitrification. Plant Soil 97:267–272.

Kirkham, D., and W.V. Bartholomew. 1954. Equations for following nutrient transformation in soil utilizing tracer data. Soil Sci. Soc. Am. Proc. 18:33–34.

Kirkham, D., and W.V. Bartholomew. 1955. Equations for following nutrient transformation in soil utilizing tracer data. 2. Soil Sci. Soc. Am. Proc. 19:189–192.

Lamb, D. 1980. Soil nitrogen mineralization in a secondary rain-forest succession. Oecologia 47:257–263.

Lemée, G. 1967. Investigations sur la minéralisation de l'azote et son évolution annuelle dans des humus forestiers *in situ*. Oecol. Plant. 2:285–324.

Myrold, D.D., and J.M. Tiedje. 1986. Simultaneous estimation of several nitrogen cycle rates using ^{15}N: theory and application. Soil Biol. Biochem. 18:559–568.

Nishio, T., and T. Fujimoto. 1989. Mineralization of soil organic matter in upland fields as determined by a $^{15}NH_4^+$ dilution technique, and absorption of nitrogen by maize. Soil Biol. Biochem. 21:661–665.

Nishio, T., T. Kanamori, and T. Fujimoto. 1985. Nitrogen transformations in an aerobic soil as determined by a $^{15}NH_4^+$ dilution technique. Soil Biol. Biochem. 17:149–154.

Paul, E.A., and F.E. Clark. 1989. Soil microbiology and biochemistry. Academic Press, New York.

Qasim, M., and T.H. Flowers. 1989. Errors in the measurement of extractable soil inorganic nitrogen caused by impurities in filter papers. Commun. Soil Sci. Plant Anal. 20:747–757.

Raison, R., M. Connell, and P. Khanna. 1987. Methodology for studying fluxes of soil mineral-N *in situ*. Soil Biol. Biochem. 19:521–530.

Robertson, G.P. 1982. Factors regulating nitrification in primary and secondary succession. Ecology 63:1561–1573.

Robertson,G.P., and P.M. Vitousek. 1981. Nitrification potentials in primary and secondary succession. Ecology 62:376–386.

Schimel, D.S. 1986. Carbon and nitrogen turnover in adjacent grassland and cropland ecosystems. Biogeochemistry 2:345–357.

Schimel, J.P., and M.K. Firestone. 1989. Nitrogen incorporation and flow through a coniferous forest soil profile. Soil Sci. Soc. Am. J. 53:779–784.

Schimel, J.P., M.K. Firestone, and K.S. Killham. 1984. Identification of heterotrophic nitrification in a Sierran forest soil. Appl. Environ. Microbiol. 48:802–806.

Schimel, J.P., L.E. Jackson, and M.K. Firestone. 1989. Spatial and temporal effects on plant-microbial competition for inorganic nitrogen in a California grassland. Soil Biol. Biochem. 21:1059–1066.

Stark, J.M. 1991. Environmental factors versus ammonium-oxidizer population characteristics as dominant controllers of nitrification in a California oak woodland-annual grassland soil. Ph.D. diss. Univ. of California, Berkeley.

Stark, J.M., and M.K. Firestone. 1991. A novel isotopic dilution technique for determining gross N-transformation rates in dry soil. p. 278. *In* Agronomy abstracts. ASA, Madison, WI.

Vitousek, P.M., and R.W. Howarth. 1991. Nitrogen limitation on land and in the sea: How can it occur? Biogeochemistry 13:87–115.

Vitousek, P.M., and S.W. Andariese. 1986. Microbial transformations of labeled nitrogen in a clear-cut pine plantation. Oecologia 68:601–605.

Vitousek, P.M., J.R. Gosz, C.C. Grier, J.M. Melillo, and W.A. Reiners. 1982. A comparative analysis of potential nitrification and nitrate mobility in forest ecosystems. Ecol. Monogr. 52:155–177.

Zak, D.R., P.M. Groffman, K.S. Pregitzer, S. Christensen, and J.M. Tiedje. 1990. The vernal dam: Plant-microbe competition for nitrogen in northern hardwood forests. Ecology 71:651–656.

Zou, X., D.W. Valentine, R.J. Stanford, Jr., and D. Binkley. 1992. Resin-core and buried-bag estimates of nitrogen transformations in Costa Rica lowland forests. Plant Soil 139:275–283.

Chapter 43

Dinitrogen Fixation

R. W. WEAVER, *Texas A&M University, College Station, Texas 77843*

SETH K. A. DANSO, *International Atomic Energy Agency,*
Vienna, Austria

The quantity of dinitrogen (N_2) fixed by diverse groups of microorganisms is of great importance to the N cycle of the biosphere. It has been estimated that approximately 83% of the N fixed annually originates from biological N_2 fixation and by contrast 14% is from manufacture of fertilizers (Burns & Hardy, 1975). Well-nodulated plants may fix in excess of 200 kg of N ha^{-1} yr^{-1} and estimates for bacteria associated with grass roots generally are in the range of a few kg ha^{-1}. Because of the extensive nature of grasslands, however, the total quantity of N_2 contributed by grasslands to the soil is equivalent to that of legumes (Burns & Hardy, 1975).

The primary limiting factor to N_2 fixation by microorganisms in nature is availability of energy to drive the N_2-fixation process. In soil systems, the major energy source is from plant photosynthates made available to microorganisms in nodule structures or, in the case of nonsymbiotic N_2 fixation, from root exudates. Without plants, N_2 fixation is minimal and of little agronomic importance. The purpose of this chapter is to describe methods that may be used to measure or estimate the quantity of N_2 fixed in systems that include plants. Measurement of N_2 fixation by pure cultures of microorganisms is described in chapter 11 by Knowles and Barraquio in this book. Methods that will be described include acetylene reduction, total N difference, isotope dilution, and $^{15}N_2$ incorporation.

43–1 ACETYLENE REDUCTION

43–1.1 Principles

Discovery that nitrogenase is able to reduce acetylene (C_2H_2) to ethylene (C_2H_4) resulted in the development of a highly sensitive method having great utility for laboratory measurements and some application for field measurement of N_2 fixation. Burris (1974, 1975) provides an excellent overview of the method's development. The method has been used extensively for the past 20 yr to estimate nitrogenase activity that has sometimes been extrapolated to amounts of N_2 fixed. A theoretical conversion factor of 3 mol of C_2H_4 produced to 1 mol of N_2 reduced was first suggested (Hardy et al., 1968). During N_2 fixation, some electrons are consumed in production of H_2 but H_2 production is restricted in the presence of C_2H_2. Thus, a theoretical conversion factor of 4 mol of C_2H_4 produced to 1 mol of N_2 reduced is more commonly used today (Boddey, 1987). Actually, conversion factors range between 1.5 to 25 and must be determined experimentally for each system (Hardy et al., 1973).

Use of C_2H_2 reduction (AR) to quantitatively measure rates of N_2 fixation has fallen into disfavor for both legume (Witty & Minchin, 1988) and grass systems (Giller, 1987). The reasons are multifold. The assay is a point in time measurement; therefore, it is not an appropriate method for measuring seasonal rates of N_2 fixation because the rate may vary with time. Problems also occur in using the method for point in time measurements. In the case of legume nodule systems, exposure of nodules to C_2H_2 may reduce O_2 permeability of the nodule resulting in reduced nitrogenase activity (Sheehy et al., 1983). Because acetylene inhibits loss of H_2 by nitrogenase, N_2 fixation would be overestimated unless the particular organism has an efficient uptake hydrogenase (Arp, 1992). The principal utility of the method was that a nodulated root could be removed from the plant and assayed in a closed system. However, removal of the plant top and disturbance of the root system are detrimental to AR activity of many legumes because of perturbation of nodule oxygen barriers (Minchin et al., 1986; Herdina & Silsbury, 1990). Under field conditions there is always the risk of losing many nodules during excavation of the root. In situ methods for legumes have been proposed (Lindstrom, 1984), but generally are not used because of labor requirements.

In grass systems, the primary problem is the vulnerability of the N_2-fixation system to changes in the gaseous environment that invariably occur during collection of soil or root samples. Additionally, the long incubation periods needed for detection of AR activity also allow time for changes in O_2 content, microbial populations, and production of C_2H_4 that occur independently of AR (Nohrstedt, 1983). Methods for estimating the magnitude of this problem are described by Nohrstedt (1983) and Witty (1979).

43–1.2 Acetylene Reduction for Nodulated Root Systems

43–1.2.1 Principles

Procedures are described separately for nodulated and non-nodulated root systems because requirements for the two systems are considerably different. For nodulated plants, measurable AR activity occurs for a single effective nodule almost immediately on exposure to C_2H_2. The closed system that is described is useful for determining if nodules are active in N_2

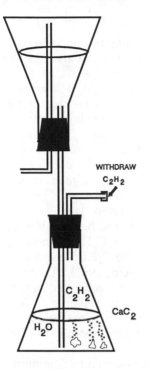

WITHDRAW
C_2H_2

C_2H_2

CaC_2

H_2O

Fig. 43–1. Apparatus for generating C_2H_2 from CaC_2. The top reservoir may be supported by a ring stand.

fixation, but it should not be considered quantitative in measuring nitrogenase activity or extrapolating to a quantity of N_2 fixed. To quantitatively measure nitrogenase activity, it is necessary to use a flow through system that will not be described in this chapter because of limited practical application (Sheehy, 1991; Gerbaud, 1990; McNeill et al., 1989; Weisz & Sinclair, 1988; Mederski & Streeter, 1977). Even with flow through systems it is necessary to experimentally determine a conversion factor if AR activity is to be converted to amount of N_2 fixed.

43–1.2.2 Materials

1. Calcium carbide (CaC_2), available from sporting goods stores.
2. Ethylene for preparation of standards.
3. Apparatus for generating C_2H_2 (Fig. 43–1).
4. Wide mouth canning jars (500 mL) with lids.
5. Rubber septa, 5.5 mm diam.
6. Plastic syringes, 50 mL.
7. Plastic syringes, 5 mL.
8. Plastic syringes, 1 mL.
9. Vacutainers, 5 mL.
10. Nodulated roots.
11. Gas chromatograph equipped with H_2 flame ionization detector.
12. Porapak Q column, 1.5 m long and 32 mm diam.
13. Nitrogen carrier gas.

43–1.2.3 Procedure

Calculate the quantity of C_2H_2 needed to provide an atmosphere of 10% C_2H_2, on a volume basis, in the vessel used for incubation of the sample. Generate the C_2H_2 from CaC_2 by adding it to the water in the apparatus (Fig. 43–1) described by Tann and Skujins (1985). The apparatus allows collection of the generated C_2H_2 in a reservoir created by the displacement of water. The apparatus may be constructed from convenient size flasks depending on the amount of C_2H_2 to be generated. The flask on the top should have a larger capacity than the flask on the bottom so that all of the water displaced may be contained. Fill the lower flask with water and assemble the apparatus. Generate C_2H_2 by dropping large crystals of CaC_2 into the lower flask and connect the upper flask. For each gram of CaC_2 added approximately 130 mL of C_2H_2 is generated. It is generally better to generate C_2H_2 from CaC_2 than using tank C_2H_2, because it is not as contaminated with other gases. If large quantities of C_2H_2 are needed, it may be most practical to use tank C_2H_2. Purity of the C_2H_2 can influence the AR activity (Tough & Crush, 1979; Hyman & Arp, 1987). Acetylene may be partially purified by passing the gas through concentrated sulfuric acid and water and by freezing out some contaminants (Hyman & Arp, 1987). Purity of the C_2H_2 will not likely be a problem for most qualitative applications, but a determination should be made experimentally by comparison with purified C_2H_2.

The vessel used for incubation should be large enough to contain the sample and provide an adequate volume of air such that O_2 will not become depleted during the incubation. Generally, a large and well-nodulated root or several smaller roots may be contained in a 500-mL canning jar and incubated for 1 h without depleting O_2 enough to interfere with the rate of AR. If nodules are removed from the root, the nodules should not occupy more than 10% of the volume of the incubation vessel to avoid O_2 depletion because of their high respiration. For extended incubations, the O_2 concentration should be monitored and supplemented as needed.

After the incubation, remove a gas sample for analysis. Usually, it is convenient to remove approximately 5 mL of gas with a needle and syringe and inject it into a 3-mL vacutainer for storage under pressure. Storage under pressure is desirable so that samples removed in a syringe will be under pressure and contamination by air minimized. The vacutainer should be prepared for receiving the gas sample. First open the container so that gases that have been released from the rubber septum during storage under vacuum will be released. Some of these gases tend to interfere with gas chromatographic analyses. Replace the septum and re-evacuate the container using a vacuum pump or a 50-mL syringe and hypodermic needle. Do not assume that the containers are fully evacuated as received from the manufacturer. It is a good practice to analyze the gases within a few days, because longer-term storage may result in absorption of some C_2H_4 into the rubber septum.

Calibrate the gas chromatograph for the range of C_2H_4 concentrations expected. One-milliliter capacity plastic syringe may be used for injections. Generally, an operating temperature of 50 °C is adequate for the column and a flow rate of 50 mL min^{-1} for a N_2 carrier gas allows for good sensitivity and separation of gases. Retention times for C_2H_4 and C_2H_2 are approximately 30 and 60 s, respectively.

43–1.3 Acetylene Reduction for Grass Systems

43–1.3.1 Principles

A major advantage of the AR method is the sensitivity of analytical methods for the detection of C_2H_4. Rates of AR that equate to < 0.5 kg of N_2 fixed ha^{-1} in 100 d can be measured when 24-h incubations are used. Thus, sensitivity of the method for detecting agronomically significant amounts of N_2 fixation is not limiting provided an incubation time of 1 d can be tolerated. Closed systems must be used to allow enough C_2H_4 to accumulate, however, in providing a closed system, the material being assayed is disturbed to some extent. The technique used for exposure of plant root systems to C_2H_2 depends on the type of plant involved and the degree of root disturbance that can be tolerated. Disturbing the root system is particularly critical because it alters the gaseous environment that strongly influences the rate of AR activity (Zuberer & Alexander, 1986) and for incubation times of hours, alterations in microbial populations may occur. Also, for such long incubation times, removal of plant tops may have a negative effect by reducing availability of plant photosynthates to roots. Leaving the plant top on unfortunately is impractical for many field investigations. The technique we have chosen to describe relies on the use of cores and worked quite well for field surveys of AR by pasture grasses and should work equally well for other plant types (Weaver et al., 1980). The results of using the technique cannot be considered quantitative for in situ rates because of disturbing the system and poor correlation with measurements based on incorporation of $^{15}N_2$ (Morris et al., 1985). A comprehensive review by Boddey (1987) has been published that covers in detail problems with measurement of AR activity of non-nodulating N_2-fixing bacteria in association with grasses.

43–1.3.2 Materials

1. Calcium carbide (see section 43–1.2.3).
2. Gas impermeable plastic bags (saran).
3. Device for taking soil cores.
4. Self-adhering insulation tape.
5. Duct tape.
6. White no. 000 rubber stoppers.
7. Large screw clamp as used for truck radiator hose.
8. Tin cans at least 8-cm diam.

43–1.3.3 Procedure

The technique depends on being able to take intact soil cores and leaving them intact. Tin cans serve as the coring device. New cans are used, and may be purchased from a canning company with one end open. Remove the closed end of the can with a can opener. The reinforced edge (rim remains after removing lid) of the can is used in forcing the sharp end of the can into the soil. Cans of 8 cm in diameter by 11 cm in height are convenient. Insert the can into the soil using a specially-constructed core driver. The core driver is a steel plate 2-cm thick that is grooved to a 1-cm depth to fit the edge of the can having the rim. Weld a 1.5-cm thick steel rod to the center of the steel plate for a cylindrical sliding weight of 3 kg that can be used to drive the can into the soil. It will be necessary to cut the tops from the plants being sampled. The apparatus for taking the cores helps in pushing the cans into the soil without collapsing the cans. Use of a hammer or a block of wood does not allow for uniform pressure on the can and the can is more likely to collapse. The can is not strong enough to be inserted into dry soil or soil containing stones. Anyway, our experience has been that dry soil does not support measurable rates of AR activity. After inserting the can, remove it from the soil with the aid of a shovel.

The soil core is placed inside a 17 by 28 cm plastic bag constructed from saran or mylar and laminated with polyethylene so that the bag is heat sealable. The saran or mylar makes the bag relatively impermeable to gases (Burris, 1974). Prepare one end of the bag for removal of gas samples. It is convenient to use approximately 2-mm thick discs cut from white no. 000 rubber stoppers. A disc is attached to the bag with a piece of gray duct tape that is at least three times the width of the disc. The disc is centered on the sticky side of the duct tape then the tape is sealed to the bag being careful not to have wrinkles that could allow gas to escape around the edges of the tape. Place the soil core into the bag, compress the bag to remove most of the air, and heat seal. Separate the upper and lower portions of the bag by tightly wrapping self-adhering insulation tape horizontally around the middle of the bag and core. Place a radiator hose clamp over the insulation tape and tighten to complete the seal. Separating the core into the two compartments facilitates mixing of gases through the soil core at the end of the incubation.

The volume of air space in the containers must be estimated so that enough C_2H_2 can be added to achieve a 10% atmosphere. The volume of a soil filled core with bag in place may be estimated by taking an extra core sample and injecting a known amount of C_2H_2 into the assembly and determining the degree of dilution by gas chromatography. The calculated value will be used to determine the amount of C_2H_2 to add to each assembly to achieve an approximate concentration of 10%. Since this concentration of C_2H_2 is saturating for nitrogenase, the actual concentration in the container is not critical. Add the calculated quantity of C_2H_2 to each container. The gas is only added to one end of the assembly and allowed to move into the core by diffusion so as to minimize disturbance of the gaseous atmosphere within the core. Incubate the assemblies under the desired temperature conditions for 12 to 24 h. At the end of the incubation, mix the gaseous contents by alternately squeezing each end. The exact volume of each container and a representative sample is needed to compute how much C_2H_4 was produced. Remove a sample for gas chromatography analyses and store the sample in a vacutainer as indicated under section 43–1.2.3. An accurate gas volume for each container is determined by the degree of C_2H_2 dilution.

43–2 NITROGEN DIFFERENCE

43–2.1 Introduction

Nitrogen balance experiments were originally used to determine the role of bacteria and plants in N_2 fixation. It may be of interest to read the historical review by Bergersen (1980b) of the development of evidence demonstrating N_2 fixation. Nitrogen increases can be readily demonstrated for inoculated plants grown on media lacking mineral N. Demonstration of N_2 fixation in soil systems is much more complex because the plant now has two sources of N; N from the soil and N_2 fixation. The contribution of N from the soil is estimated by growing plants that do not have the benefit of N_2 fixation and assuming the N_2-fixing plant obtains a similar amount of N from the soil. The amount of N in the plant not benefiting from N_2 fixation is subtracted from the plant benefiting from N_2 fixation to determine the quantity of N_2 fixed.

The method is not sensitive enough for determining the quantity of N_2 fixed over time intervals of less than a few days. Therefore, it is of little use as a point in time measurement for N_2 fixation.

A description of techniques to measure N_2 fixation under controlled conditions of the laboratory and in the field follow. The reader may also refer to descriptions of methods by Vincent (1970), Brockwell (1980), Somasegaran and Hoben (1985), and Wynne et al. (1987).

43–2.2 Nitrogen-Difference Method for Controlled Environment

43–2.2.1 Principles

Plants are grown under conditions that allow for control of mineral N inputs and prevention of contamination of N_2-fixing microorganisms. It is not practical or necessary to grow the plants axenically. The technique described is a variation of that described by Leonard (1943). Basically, plants are grown in pots provided water and nutrients by subirrigation. Subirrigation prevents N_2-fixing microorganisms from being washed into the pot during irrigation. Seeds are inoculated with the microorganism of interest and grown in a plant support medium deficient in mineral N. Plant nutrients other than N are provided by a nutrient solution. Uninoculated control plants are needed to demonstrate that contamination with nodulating microorganisms did not occur and to evaluate the combined contribution of mineral N from the plant growth medium and that from the seed. Sometimes, it is also useful to have control plants receiving mineral N to demonstrate that lack of mineral N was the main factor limiting plant growth and to compare with the symbiotic system's ability to provide N. Dinitrogen fixed is the difference in total N content between inoculated and uninoculated plants.

43–2.2.2 Materials

1. High quality legume seeds.
2. Culture of microsymbiont.
3. Plastic pots (1 L) with lids.
4. Absorbent cotton (rolls).
5. Coarse vermiculite.
6. Sodium hypochlorite (household bleach) or ethanol.
7. Chemicals for nutrient solution (see Table 12–1 in this book).

43–2.2.3 Procedure

Seed and materials used to culture plants must be free of nodulating organisms. Unless contaminated after receipt, plastic pots, absorbent cotton, vermiculite, and chemicals are free of nodulating microorganisms. It is not necessary to autoclave plant nutrient solutions made with distilled water. Used or contaminated materials, however, will need to be autoclaved. Containers may be reused by washing with a soapy solution and rinsing in a solution of household bleach (5% NaOCl) diluted 1:4 with water on a volume basis. Rinse thoroughly with water to remove the bleach.

Two 1-L plastic containers are used for each growth unit (Fig. 43–2). A lid is placed on one container and the other container sits on top of it. The lower container serves as a reservoir for water. The upper container is for the N-deficient plant growth medium. Cut a hole approximately 2 cm in diameter in the center of the lid and the bottom of the top container. Cut a second hole of approximately 6-mm diam. in the lid close to edge of the lid. Make a cotton wick by cutting a 20-cm long strip of absorbent cotton that is wide enough to snugly fit through the holes in the center of the container and lid. The wick will serve to draw nutrient solution into the rooting medium and should extend from the bottom of the nutrient solution up to approximately 6 cm into the pot containing the plants. Set the upper container on the lid of the nutrient reservoir. Snap the lid on the lower container so that it fits snugly to prevent cross-contamination of rhizobia from other pots. Fit a piece of Tygon tubing into the smaller hole of the lid for replenishment of nutrients. Place a cap over the tubing to reduce the likelihood of contamination. After assembling the two contain-

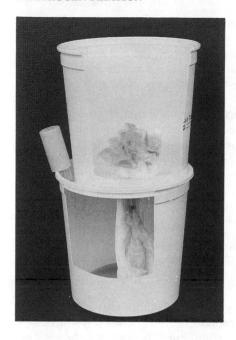

Fig. 43–2. Cutaway of modified Leonard jar used in growing legumes for evaluation of N_2 fixation.

ers, fill the top container with potting material. Vermiculite, perlite, or moderately coarse sand perform well, because they provide adequate capillaries for supply water while remaining well aerated. Hold the cotton wick upright while adding the potting materials so that the wick extends into the potting material. Fill the container to within 2 cm of the top. Prepare the N-free nutrient solution according to the composition provided in Table 12–1 in this book. Wacek and Alm (1978) provide a description of an alternate container. Moisten the potting mixture by pouring nutrient solution onto its surface. Fill the bottom reservoir to within 2 cm of the top. After this initial wetting, water and nutrients are supplied by capillary rise through the cotton wick.

To minimize experimental error, plant with high quality seed that has been thoroughly washed and treated (see chapter 12 by Weaver and Graham) so that they are free of any nodulating organisms. Plant seed carefully so they are all at equal depth and equally spaced from other seeds. Plant four to six seeds in each pot for large-seeded legumes and 20 to 30 for small-seeded legumes. After planting, inoculate the seeds by dribbling a suspension of inoculum onto the seed. Use the plant nutrient solution to dilute the inoculum to the desired concentration. Cover the seed and place a 1-cm layer of 5 to 10 mm size gravel on the top of the pot to reduce evaporation and reduce the possibility of air-borne contaminants reaching the rooting zone. After seedlings emerge, thin to one or two seedlings per pot for large-seeded legumes and 10 to 15 per pot for small-seeded plants.

Each pot should contain the same number of plants. When thinning, remove both unusually large and small plants. It is best to remove the seedlings within 7 d of emergence by pulling them from the pot and not leaving the root in the pot to contribute a source of N. Care must be exercised not to contaminate other pots because the media clinging to the roots have high populations of nodulating organisms.

Grow the plants under conditions that provide temperature and light regimes suitable for good plant growth and nodulation. Grow the plants for at least 1 mo and preferably 6 wk or longer to allow time for nodulation and accumulation of symbiotically fixed N_2. At harvest, determine plant dry weight, total plant N, and perhaps total numbers of nodules and nodule mass depending on the experimental objectives. See chapter 12 for comments on nodule storage.

43–2.3 Nitrogen Difference Method-Field

43–2.3.1 Principles

Quantitative measurement of N_2 fixation of plants grown in the field is challenging because there is no absolute method. Basically, three approaches may be taken: constructing a complete N balance for the system, subtracting the quantity of N accumulated in a plant that does not fix N_2 from the plant that fixes N_2 (N difference), and isotope dilution. The first method will not be described because it is impractical for most situations (Weaver, 1986) and the third method is described in section 43–6.

The primary assumption for the N-difference method is that the N_2-fixing plant accumulates the same amount of mineral N from the soil as the non-fixing plant. Since roots are not normally harvested the expectation is limited to plant shoots. If plant shoots are to contain the same amount of mineral N from the soil at least two conditions must hold. The two plant types must explore the same soil volume and they must be actively growing during the same time period. It is impractical to attempt to measure small contributions from N_2 fixation by this method considering soil variability and experimental error. Probably, N_2 fixation amounts of at least 20 kg ha^{-1} would be required to be statistically significant (Weaver, 1986). Generally, this method would only be useful for symbiotic N_2 fixation. Therefore, the technique described will be for symbiotic systems.

43–2.3.2 Materials

1. Dinitrogen-fixing plant.
2. Non-N_2-fixing control plant.
3. Inoculants (see chapter 12).
4. Field site.
5. Planting equipment.

43–2.3.3 Procedure

For maximum opportunity to differentiate between inoculants, select a field site that is as homogeneous as possible to reduce experimental error and is low in plant available N. If most of the N in the plant comes from mineral N, it will not be possible to distinguish between the quantity of N fixed by different treatments. To reduce the chance for contamination between plots, the site should not be prone to flooding or erosion. It may be advisable to grow a cereal crop on the site to reduce the amount of mineral N present or to incorporate some material with a wide C/N ratio to immobilize the mineral N. It is highly unlikely that differentiation between N_2-fixing ability of strains can be accomplished on a site that contains a resident population of the symbiont. It is advisable to measure the population size of the resident microorganisms capable of nodulating the plant well ahead of planting (see chapter 12).

Uninoculated control treatments should be included for each N_2-fixing cultivar used in addition to a non-fixing control plant. The uninoculated control plant may be used to determine the ability of the resident microorganisms to fix N_2 and if any inoculation response is achieved. Generally, it is not used as the non-fixing control because it is unlikely that nodules will not form somewhere on the root system by the end of the season. The non-fixing control plant should be well adapted to the soil and climate. For grain legumes a non-modulating isoline may be a good choice. Often a non-nodulating isoline of a particular cultivar will not be available but the non-nodulating plant may still be a good control plant if it has a similar maturity date to the nodulating plant. Choose non-fixing plants that have similar growth characteristics to the fixing plant. For instance, in the case of soybean (*Glycine max* L.), use a non-fixing plant that is planted in a row and is not unduly tall (shading could be a problem) relative to the soybean or has an unusually large root system.

Plot size, planting pattern, and experimental design depend on the particular experiment and field conditions. Generally, single row plots are not used because of potential border affects and increased likelihood of contamination between rows. If the soil is not physically moved between rows by cultivation, erosion or water, it is unlikely that contamination will be a problem. Generally, three or four row plots are used for grain legumes with only the middle row or rows harvested. For accurate seed yield measurements and to reduce experimental error, it is advisable to harvest 3 to 5 m of row. For measurement of N in total shoot biomass it is generally not feasible to harvest more than 1 m of row and it should be done before many of the leaves drop at physiological maturity.

In the case of forage or pasture plants, the size of the plots should be approximately 4 m^2. Plots will have to be larger on erodable soils or where water may run across plots to reduce contamination between plots. A cereal or forage grass may serve as the non-fixing control plant. Often the plots will be clipped more than once during the growing season. Thus, the non-fixing plant must be able to regrow following clipping. The central 1 m^2 area of the plot should be clipped for measurements.

For actual evaluation of biological N$_2$ fixation, the total N contained in plant biomass should be measured. Generally, it is not feasible to measure the N contained in the root system, resulting in underestimation of N$_2$ fixation. Sometimes only seed yield or dry matter production is of interest but it is not possible to infer actual amounts of N fixed in the whole plant from any of these measurements alone. Often, it is useful to make some observations on nodulation. Differences between treatments often occur during the first 3 to 4 wk of plant development. It is relatively easier to dig the root system at this time to observe nodulation than later in the season. Plants from the border rows may be sampled for nodule counts. The coefficient of variation (CV) often is quite large for measurements of nodule numbers and mass.

43–2.3.4 Comments

The selection of the non-fixing plant largely determines the accuracy of the estimation. Unfortunately, there is no certain way of selecting the correct control plant or even knowing that the correct plant was selected after the experiment. Consequently, the results of the measurements cannot be considered absolute and some argue that relative differences may not even be correct. The largest potential for error is when approximately 50% of the N in the legume comes from N$_2$ fixation and the least error occurs when < 20% or more than 80% comes from fixation. Some authors argue that more than one non-fixing plant should be used and the results averaged. The true value would still not be known, but presumably the determined value would be closer to the true value. Generally, this approach does not merit the extra labor and space required.

43–3 NITROGEN-15 ISOTOPE TECHNIQUES

43–3.1 Introduction

The basis for using isotopes as tracers or measuring biological reactions is that different isotopes of an element possess the same chemical properties even though they differ in atomic mass. Thus, one isotope may be substituted for another in a reaction and the product(s) analyzed for the substituted atom. Isotope identification is made possible through differences in radioactivity or atomic weight (see chapter 39 by Legg et al. in this book).

Two isotopes of N, ^{14}N and ^{15}N are the most widely used in agricultural research because they are stable and therefore useful for long-term studies. Their use does not involve the potential risks and health hazards commonly associated with radioisotopes. The major drawbacks, high cost of ^{15}N-labeled compounds, along with high equipment and maintenance costs, are no longer as serious as they used to be. By judicious use, the cost of ^{15}N-labeled fertilizers required for N_2 fixation studies can be less than $100 for a greenhouse experiment and, unless several treatments are being studied, approximately $1000 for a field study. The commercial analysis of samples presently ranges between $5 to $10, depending on the number of samples.

Two ^{15}N-labeling techniques are available for measuring N_2 fixation: soil labeling with ^{15}N and of more limited use, incubation in an atmosphere of $^{15}N_2$ gas. The advantages with the ^{15}N soil-labeling method include the fact that N_2 fixation can be measured directly in the field, and integrated estimates of N_2 fixed for various time periods can be assessed. The labeling of soil with ^{15}N is currently the most popular and perhaps the most reliable method for measuring N_2 fixation in field crops and has been reviewed extensively (Chalk, 1985; Danso, 1988; Hardarson et al., 1987; Hauck & Weaver, 1986). The ^{15}N soil-labeling method is further classified into the A-value, isotope dilution and ^{15}N natural abundance methods. Respectively, these methods involve: (i) the addition of a higher amount of ^{15}N-labeled fertilizer to the reference than to the fixing plant, (ii) same amount of a ^{15}N-enriched fertilizer is added to reference and fixing plants, and (iii) no labeled fertilizer is applied. Because the isotope dilution method is less controversial, simpler and more popular than the A-value method, we shall not discuss the A-value method in this chapter. Vose and Victoria (1986) have discussed it.

43–3.2 Isotope Dilution

43–3.2.1 Principle

The N in biological systems is composed predominantly of two stable isotopes, ^{15}N and ^{14}N, which constitute about 0.3663 and 99.6337%, respectively, of the N in the atmosphere and natural compounds (Rennie et al., 1978). The nearly constant ^{15}N/^{14}N ratio in nature makes it possible to use materials with artificially altered ^{15}N/^{14}N ratios to trace biological pathways and quantify N products in biological systems by examining for ^{15}N/^{14}N isotope ratio changes. A material containing more than 0.3663% ^{15}N is referred to as ^{15}N enriched or labeled, and the term atom % ^{15}N excess represents the difference between the atom % ^{15}N in the enriched material and natural abundance (0.3663%). A material containing < 0.3663 atom % ^{15}N is referred to as ^{15}N-depleted. Because ^{15}N-depleted fertilizers are less popular for N_2-fixation studies than ^{15}N-enriched materials, we will only describe the use of ^{15}N-enriched materials.

Isotope dilution of ^{15}N occurs when the higher than natural ^{15}N abundance in an enriched material becomes diluted by the lower ^{15}N/^{14}N ratio of a natural substance; the extent of ^{15}N dilution will depend upon the relative ^{15}N enrichments and amounts. Where one of the two ^{15}N compounds (x) being mixed is enriched (a_1 atom % ^{15}N excess) and the other (y) is of ^{15}N natural abundance (zero atom % ^{15}N excess), the final ^{15}N dilution or enrichment (a) can be expressed mathematically as:

$$a = [x \, (a_1)] \, / \, (x + y) \qquad\qquad [1]$$

To measure N_2 fixation by the isotope dilution method, a ^{15}N-enriched fertilizer is added to soil to increase the $^{15}N/^{14}N$ ratio of available soil N and thus create a difference between the $^{15}N/^{14}N$ ratios of soil N and atmospheric N_2. In principle, the $^{15}N/^{14}N$ ratio in a plant that totally depends on soil N reflects the integrated $^{15}N/^{14}N$ ratio of the available soil N for the period of growth. This is not the case with a N_2-fixing plant that, in addition to soil N, assimilates N_2; the higher $^{15}N/^{14}N$ ratio absorbed from the ^{15}N-labeled soil will be diluted by the lower $^{15}N/^{14}N$ of the fixed N_2. The relative difference in the $^{15}N/^{14}N$ ratio in a reference non-N_2-fixing plant and that of the fixing plant grown under identical conditions is then an indication of the capacity to fix N_2, and the following equation (McAuliffe et al., 1958; Fried & Middelboe, 1977) is used to calculate the proportion (% Ndfa) and amount (Ndfa) of N_2 fixed in the fixing crop:

$$\% \, Ndfa = \left(1 - \frac{\text{atom \% }^{15}N \text{ excess in fixing plant}}{\text{atom \% }^{15}N \text{ excess in reference plant}}\right) \times 100 \quad [2]$$

$$Ndfa = \left(1 - \frac{\text{atom \% }^{15}N \text{ excess in fixing plant}}{\text{atom \% }^{15}N \text{ excess in reference plant}}\right) \times \begin{array}{c}\text{Total N in}\\ \text{fixing plant}\end{array} \quad [3]$$

Plants absorb variable amounts of N at different stages of their growth, and the $^{15}N/^{14}N$ ratio in soil generally does not remain constant with time. Thus, the $^{15}N/^{14}N$ ratio of a soil extract at any one time or an arithmetic mean from extracts at different times is not a reliable indicator of the integrated $^{15}N/^{14}N$ ratio of plant absorbed N from a ^{15}N-labeled soil. Instead, a reference plant that totally depends on soil N is used. The accuracy of the estimates of N fixed will depend on how the value of $^{15}N/^{14}N$ assessed by the reference plant accurately reflects the $^{15}N/^{14}N$ ratio of soil-derived N in the fixing plant. It is important to ensure that the reference plant does not fix N_2, and this can be checked by examining for nodulation (if it is a nodulating species) or C_2H_4 production from C_2H_2 (see sections 43–1 and 43–2). The reference plant should absorb its N from the same N pool as the fixing plant, which may be assessed by examining for similarity in rooting pattern. The fixing and reference plants must also grow and obtain their N in a similar pattern with time, a requirement that can be verified by harvesting the different plants at various growth stages and comparing their growth and N-uptake patterns. In addition, the time of ^{15}N application, planting and harvesting should be the same for fixing and reference plants. More detailed criteria for selecting reference plants have been provided by Fried et al. (1983), Danso et al. (1986), and Peoples et al. (1989). Many different species have served as reference plants. These include non-nodulating legume isolines (currently available for grain legumes only), uninoculated legumes (provided the soil is devoid of effective homologous rhizobia and cross contamination can be avoided) and several non-legume species (e.g., grasses and many cereal crops).

One of the greatest strengths of the ^{15}N methodology has often been overlooked! It can be used to reliably rank plants for their % N derived from N_2 fixation on the basis of their relative ^{15}N enrichments (Danso et al., 1986). The lower the atom % ^{15}N excess in a cultivar or species, the greater is the contribution from N_2 fixation to the total N in the plant. A potential source of error, the reference plant is eliminated with this approach. It is important to realize, however, that this ranking approach is applicable only for plants grown under similar soil and ^{15}N-enrichment conditions and cannot be used to compare N_2-fixing plants harvested at different times.

Regarding the form of ^{15}N-labeled fertilizer to apply, there is as yet no universal recommendation or a standard ^{15}N-labeled fertilizer (Chalk, 1985). Several different ^{15}N fertilizer forms are being used. The only crucial requirement is, that the relative availability of soil N and the applied fertilizer N should be similar for both reference and fixing plants. The traditional agricultural fertilizers, K $^{15}NO_3$, $(^{15}NH_4)_2SO_4$ and $CO(^{15}NH_2)_2$ have been used.

The choice of the ^{15}N fertilizer used is frequently dictated by availability and differences in purchase price. The use of ^{15}N labeled slow-release N fertilizers is becoming increasingly popular. The incorporation of slow N-release fertilizer formulations results in more stable $^{15}N/^{14}N$ ratios in soil with time than with the traditional mineral fertilizers (Witty, 1983; Witty & Ritz, 1984). A stable $^{15}N/^{14}N$ ratio enhances the suitability of many reference plants and thus improves the accuracy of N_2-fixation estimates. Most slow-release formulations labeled with ^{15}N are not commercially available, however, thus restricting widespread usage (Danso, 1988). The alternative is to use ^{15}N immobilized within organic forms, such as by incorporating the residues of plants previously grown on ^{15}N-fertilized soil (Witty & Ritz, 1984; chapters 39 by Legg et al. and 40 by Hauck et al. in this book).

43–3.2.2 Materials

1. Seeds of N_2-fixing plant.
2. Seeds of non-N_2-fixing reference plant.
3. Inoculant containing N_2 fixers (see chapter 12 by Weaver and Graham in this book).
4. Nitrogen-labeled fertilizer, 5 to 10 atom % ^{15}N excess.
5. Watering can or knapsack sprayer.
6. Tap water for preparing solutions.

43–3.2.3 Procedure

The field site selected should be as homogeneous as possible, otherwise significant treatment differences cannot be observed easily. The N level in the soil may be low or high, depending on the objective of the study. The sensitivity of the isotope-dilution method, however, is greater at high than at low levels of N_2 fixation (Danso & Kumarasinghe, 1990), and the depressive effect of N on N_2 fixation should be considered. Because the precision of ^{15}N-derived data is generally higher than yield-dependent data (Hardarson et al., 1984), isotope plots do not need to be as large as yield plots for the same precision. Isotope plot sizes ranging from 1 to 5 m^2, consisting of four to five rows for grain crops, and for forage and pasture crops, 1 to 2 m^2 plots are satisfactory. A randomized split plot design is often convenient. The advantage with this design is that it allows contrasting treatments such as different N rates and rhizobial inoculation treatments to be isolated within separate blocks, thus minimizing effects due to erosion, and cross contamination. For trees, individual trees have frequently served as single plots. This has been dictated by cost and labor considerations as a result of the massive sizes of trees. Many multi-purpose trees are, however, small enough (especially when they are young) to allow using a few trees per plot.

To quantify N_2 fixation, a reference plant should be used. When genotypes with different maturity periods are being compared, a reference plant to match each maturity group is necessary or the $^{15}N/^{14}N$ ratio of mineral N must be constant. To avoid large errors due to soil variability, it is advisable to plant reference plants in close proximity to fixing plants, possibly on adjacent plots.

When, how, and how much of the ^{15}N label to apply to soil are influenced by the type of ^{15}N-labeling material used and the sensitivity of the instrument for analyzing ^{15}N. If a conventional mineral fertilizer is used, it can be applied to soil just prior to planting or immediately after thinning seedlings. In addition to allowing thinning to be done before ^{15}N application, post-planting ^{15}N application allows for ^{15}N application to be stopped or delayed if seeding emergence is poor or if replanting becomes necessary. Replanting should be done on all plots; otherwise serious errors will result due to the imposed difference in times of initial ^{15}N uptake among plants in the different treatments. Post-planting ^{15}N application requires greater care during application to prevent damaging plants, or to avoid serious errors from mixing of soil and rhizobia under different treatments and must be accomplished during the seedling stage. For ease of application and to ensure a fairly homogeneous distribution of the normally small amounts of ^{15}N-labeled fertilizer used, it is advisable to dissolve the appropriate amount of the ^{15}N-labeled fertilizer in water and spray the fertilizer solution onto plots (see chapter 39). It may even be advisable to prepare stock solutions from which the required aliquots are taken. The spraying may be done with a small knapsack sprayer, a normal garden watering can, or plastic squeeze bottles with perforated lids. A trial run is needed to decide on the most appropriate watering rate that will moisten the soil but not puddle it (usually between 100 to 500 mL m^{-2} depending on soil type) is required.

It is important to consider the type of instrument available when deciding how much ^{15}N to apply to soil (see chapter 39). High precision dual inlet mass spectrometers that switch back and forth from measuring a reference gas and the sample in question can measure ^{15}N abundance with a precision of 0.00004 atom %. Thus, an enrichment of 0.0001 would be adequate to indicate ^{15}N uptake. Crasswell and Eskew (1991) have recently compared several different types of commercially available instruments for precision of ^{15}N analysis. With a single inlet mass spectrometer, an enrichment of 0.004 atom % [15]N would be necessary and a similar ^{15}N enrichment would be needed with an automated N analyzer coupled to a mass spectrometer. With an emission spectrometer, an enrichment of 0.01 atom % would be needed. Approximately 1 kg of ^{15}N ha^{-1} (10 kg of N ha^{-1} of 10 atom % [15]N excess or 20 kg of N ha^{-1} of 5 atom % [15]N excess) is adequate for detection in most single-year isotope dilution experiments using most analytical systems. Only in exceptional cases (e.g., where N_2 fixation was exceedingly high or if the emission spectrometer used is of very low sensitivity) would 1.5 to 2 kg of ^{15}N ha^{-1} (equivalent to 0.15–0.2g m^{-2}) have to be applied.

For grain legumes and seasonal crops, a one-time ^{15}N application is generally satisfactory. In the case of perennials, such as pastures and N_2-fixing trees, it is recommended to split the required ^{15}N fertilizer in equal amounts and apply at quarterly, half yearly or yearly intervals (Danso et al., 1988). Where applicable, the ^{15}N additions can be made to coincide with harvests (Vallis et al., 1967). In case ^{15}N-labeled plant residues are used for soil labeling, these should preferably be chopped into small pieces (2–5 cm pieces) and rototilled into soil (chapters 38 and 39). Sometimes the labeling can be achieved in situ, by rototilling plants growing on ^{15}N-labeled soil back into the soil (Fried et al., 1983). In all cases, a greener plant residue with a C/N ratio closer to that of soil than using a matured crop is preferable. The high C/N ratios in fully matured crops, especially in cereals, could immobilize soil N and result in unusually high levels of N_2 fixation (Papastylianou & Danso, 1991).

In addition to being planted at the same time, it is necessary to harvest both fixing and reference plants at the same time. This means that for crops of different maturity groupings, matching reference plants are needed for each. Because of problems associated with quantitative recovery of roots especially in field-grown plants, most estimates of N_2 fixation are on aboveground plant parts. It is thus assumed that total N or N_2 fixed in roots constitutes a small fraction of the total N in whole plants. This assumption may not be true for all plants. Some trees have been shown to contain more N belowground than above (Sanginga et al., 1990), and the N partitioned to belowground parts of some pastures can be substantial (Heichel et al., 1981). Thus, wherever possible, roots should also be harvested. For greenhouse-grown plants and some field experiments, the freshly harvested plant sample is small enough to be handled in a single bulk for drying and grinding. With bulky samples, it is not easy to get homogeneous subsamples when entire plants are involved (chapter 40). It is advisable to split the freshly harvested plants into different plant parts for subsampling. The different parts are then dried separately in the oven at 70 °C for 48 h, and ground through a 100-mesh size sieve. One approach is to analyze separately each plant part for N and $^{15}N/^{14}N$ isotope ratios, and use the following equation (not an arithmetic mean) to calculate the overall or weighted atom % ^{15}N excess to compute N fixed for the whole plant (Eq. [4]):

Weighted atom % ^{15}N excess =

$$\frac{AE(A) \times T_N (A) + AE(B) \times T_N (B) + AE(C) \times T_N (C) \text{ etc.}}{T_N (A + B + C)} \quad [4]$$

Where AE is atom % ^{15}N excess, T_N is total N, and A,B, and C represent three harvested plant parts. This approach results in many samples being analyzed. An easier and reliable approach involves drying and grinding the individual plant parts separately as described above, but subsequently mixing to get a composite sample for N and $^{15}N/^{14}N$ analysis. From the dry weights of the individual parts before grinding, the relative dry weight ratios to use for preparing the composite mixture for N and $^{15}N/^{14}N$ ratio analysis can be calculated. It is easier to get more homogeneous mixtures from dried, ground material than from fresh material.

43–3.3 Nitrogen-15 Natural Abundance Method

43–3.3.1 Principles

The ^{15}N natural abundance method operates on the same principle as the isotope dilution method (43–3.2.1). The only difference is that with the ^{15}N natural abundance method, no ^{15}N-enriched material is added to soil. In this case, the higher $^{15}N/^{14}N$ ratio in soil than atmospheric N_2 is the consequence of natural N-cycle processes. During N turnover reactions in soil, ^{14}N is preferentially lost into the atmosphere compared to ^{15}N, creating a natural ^{15}N enrichment of soil N relative to atmospheric N_2 (Shearer & Kohl, 1986). It is important to note that the differences between the $^{15}N/^{14}N$ ratios of soil N and atmospheric N_2 are very small, being several orders of magnitude lower than where ^{15}N-enriched materials have been added. These small differences are simplified by multiplying by a thousand and expressed in δ units (Amarger et al., 1979).

$$\text{Thus, } \delta^{15}N = \left(\frac{R \text{ sample} - R \text{ standard}}{R \text{ standard}}\right) \times 1000 \quad [5]$$

$$\text{where } R = \frac{^{15}N}{^{14}N + ^{15}N} \quad [6]$$

and R standard = R_{air} = approximately 0.0003663, i.e., 1 $\delta^{15}N$ unit = 0.00037 atom % $\delta^{15}N$ excess, compared to a value of 2730 δ units contained in 1 atom % ^{15}N excess. The $\delta^{15}N$ units for most soils are between −2 and +15.

The major drawback with the ^{15}N natural abundance method is that of low sensitivity. To be able to accurately measure these small differences in ^{15}N, the analytical procedures have to be precise (Peoples et al., 1989). Care must be taken at each step of sample preparation to assure complete recovery of N. Also, the measurement requires a dual inlet, dual collector mass spectrometer equipped with a mode for switching between the gas sample and reference gas. The method is of limited use on soils with extremely low $\delta^{15}N$. Such soils are commonly found under natural vegetation (Hansen & Pate, 1987). Here, the variability in $\delta^{15}N$ may be large enough to obscure any differences between $\delta^{15}N$ of soil N and atmospheric N_2, and the ^{15}N natural abundance method should not be used. Another potential problem with the ^{15}N natural abundance method is that isotopic discrimination occurring between ^{15}N and ^{14}N within the plant can cause significant errors in estimates of N_2 fixation. This is not a serious problem with the isotope dilution method because the isotopic discrimination effect is obscured by the high ^{15}N enrichment used. A procedure devised to correct for isotopic discrimination at the ^{15}N natural abundance level involves analyzing for the $^{15}N/^{14}N$ ratio in a control fixing plant grown in a N-free medium (see section 43–3.3.3).

The ^{15}N natural abundance method, despite the above-stated problems, has provided several useful estimates of N_2 fixation. The method has particular merit for studying N_2 fixation in systems that should not be disturbed through the application of ^{15}N fertilizer (Danso et al., 1992). Where the $\delta^{15}N$ values allow, the method can provide useful information on N_2 fixation in forest trees, and, in addition to some quantitative measurements, provide useful evidence of the potential of some species with unknown N_2-fixing ability (Shearer & Kohl, 1986). The method's added value in such exploratory studies is that reference plants may not be needed; differences in the $\delta^{15}N$ values of putative N_2 can be used to rank them. The use of the ^{15}N natural abundance method does not require the purchase of ^{15}N fertilizers, and the possible suppressive effect of added ^{15}N-labeled fertilizer on N_2 fixation is eliminated.

43–3.3.2 Materials

1. Seeds of N_2-fixing plant.
2. Seeds of non-N_2-fixing reference plant.
3. Inoculant containing N_2-fixing microorganisms.
4. Complete nutrient solution minus N (see Table 43–1).
5. Plastic pots (1 L size or bigger).
6. Sieved (2 mm diam.) soil collected from 0–15 cm soil layer.
7. Hydroponic system.

43–3.3.3 Procedure

Because of the small differences in ^{15}N abundance being measured, the need to select a homogeneous site is crucial. The ^{15}N natural abundance method will work only in soils in which the variation in $\delta^{15}N$ is smaller than differences in N_2 fixed among treatments. For establishing the potential ability of species to fix N_2, measurements of the $\delta^{15}N$ in leaves or aboveground samples are sufficient (Shearer et al., 1983). A lower $\delta^{15}N$ compared to putative non-N_2 fixers growing in the same locality suggests N_2 fixation.

For measuring N_2 fixation by the ^{15}N natural abundance method, two types of reference plants are needed. One type is for assessing the $^{15}N/^{14}N$ ratio of soil N. The reference crop selection in this case is similar to that described for the isotope-dilution technique (see section 43–3.2.1), except in established natural stands, where one is compelled to rely on non-N_2-fixing plants occurring naturally within the locality. To reduce uncertainty of appropriateness of the reference plant and thus the probability of serious errors, it is advisable to use the mean of more than one of the naturally occurring non-N_2-fixing plant species rather than use only one of them as a reference. Another reference is needed for measuring the discrimination between ^{14}N and ^{15}N that occurs during N_2 fixation (Amarger et al., 1979; Shearer & Kohl, 1986). This is achieved by planting the inoculated N_2-fixing plant hydroponically or in washed sand watered regularly with N-free medium (Amarger et al., 1979) to ensure that the plant is completely dependent on N_2 fixation.

The analysis of N and $^{15}N/^{14}N$ ratios after harvest should be done either on individual plant parts or on composite plant samples (see section 43–3.2.3), but the estimates should reflect N_2 fixed in the whole plant (see section 43–3.2.3). A slightly modified isotope-dilution equation is used to calculate N_2 fixation as follows:

$$\% \text{ Ndfa} = (x - y)/(x - f) \times 100 \qquad [7]$$

where

$x = \delta^{15}N$ of non-fixing reference plant
$y = \delta^{15}N$ in the fixing plant grown in soil
$f = \delta^{15}N$ in the fixing plant grown hydroponically.

43–4 USE OF DINITROGEN-15 GAS

43–4.1 For Measuring Dinitrogen Fixation

43–4.1.1 Principles

The detection of ^{15}N in tissues of biological systems exposed to $^{15}N_2$ gas is the only direct, unequivocal method for demonstrating that N_2 fixation occurred. Uptake of ^{15}N from $^{15}N_2$ gas also allows the fate of fixed or absorbed N to be followed in various intermediates and products within plants. To measure uptake of $^{15}N_2$ gas, it is necessary to enclose the biological system in a gas-tight container with a sufficient enrichment of ^{15}N and for long enough to be able to detect ^{15}N enrichment within the biological system.

A major disadvantage of the $^{15}N_2$ gas exposure method is that it usually involves short-term assays, and higher plants, in particular, require incubation conditions quite different from those of normal field-grown crops. Thus, the major uses of $^{15}N_2$ are to demonstrate N_2 fixation in systems where activity is expected based on other indirect methods, to determine the compounds into which N is incorporated and to follow N transport in plant-microbe systems. Another use has been to determine calibration constants for C_2H_2 reduction assays.

There is no standard incubation chamber, and several types have been used depending on the N_2-fixing system and the purpose of the study. A principal requirement is that conditions within the chamber should match as closely as possible the in situ conditions. For small biological systems such as bacterial cultures or enzymes, a serum bottle or test tube with a serum stopper is adequate (Spiff & Odu, 1972). For intact plants, a desiccator or a perspex chamber (Ross et al., 1964) may be constructed. To avoid rapid changes in the composition of the atmosphere during assay of plants, Knowles (1980) has suggested a ratio of gas phase volume (mL) to sample fresh weight (g) of at least 300. Too large containers also are undesirable; more $^{15}N_2$ is required, making the study unduly expensive. In some cases, when only the root needs to be exposed to $^{15}N_2$, the pot may serve as an incubation chamber after sealing around the stem and the $^{15}N_2$ may be injected directly into the rhizosphere (Douglas & Weaver, 1986).

The $^{15}N_2$ gas may be purchased already prepared from several suppliers (e.g., Merck Sharp and Dohme, Montreal, Canada and Monsanto Research Corp., Miamisburg, OH). It is also possible to generate the $^{15}N_2$ in the laboratory. Ammonium sulphate enriched with ^{15}N is placed in an evacuated bottle, and LiOBr is injected to generate $^{15}N_2$ which is collected in a connected evacuated flask. The $^{15}N_2$ gas is then transferred to a flask containing an acidified solution to remove any traces of NH_3 before use. See Bergersen (1980a) for a detailed description of an apparatus for preparing $^{15}N_2$ from a $^{15}N_4$ salt. However, unless many studies are planned, it is probably easier and cheaper to purchase the ^{15}N gas. The ^{15}N enrichments in purchased gases have generally varied from around 40 to 90 atom % ^{15}N excess. The required enrichment is dictated by the ^{15}N measuring equipment, volume of the exposure chamber, the N_2-fixing activity of the system, duration of exposure, and other available sources of N. In most systems, a final ^{15}N enrichment in the gas phase of 10 to 50 atom % ^{15}N excess should be adequate. It is necessary to accurately know the ^{15}N enrichment of the gas phase within the incubating chamber to be able to calculate the proportion and amount of N fixed. Thus, the gas within the chamber must be sampled at the beginning and preferably also at the end (using the sampling ports) of the incubation for precise ^{15}N determination. The equation for calculating N fixed is:

$$\text{Ndfa} = \frac{\text{Atom \% } ^{15}\text{N excess in tissues of fixing system}}{\text{Atom \% } ^{15}\text{N in gas sample within chamber}} \times \begin{array}{l} \text{Total N} \\ \text{in fixing} \\ \text{system (g)} \end{array} \quad [8]$$

The length of time necessary to detect ^{15}N enrichment depends on the N_2-fixing activity of the system, the ^{15}N enrichment of the gaseous atmosphere, and the sensitivity of the analytical technique used to determine ^{15}N abundance. With active biological systems, such as the *Azolla-Anabaena* symbiosis, exposure times of a few minutes with an atmosphere enriched to 10 atom % ^{15}N excess are adequate. On the other end of the scale, for associative symbioses between grasses and heterotrophic bacteria with low levels of N_2 fixation, exposure times of several days with 50 atom % ^{15}N excess or more may be needed (Eskew et al., 1981).

The simplest N_2-fixing systems including soil samples and bacterial cultures require in addition to the $^{15}N_2$ gas, an oxygen-scavenging system. For the lower plants (e.g., blue-green algae), only a light source may be necessary. Higher plants normally have to be exposed to $^{15}N_2$ for prolonged periods to obtain valid estimates of N_2 fixation. Also, the composition of gases changes rapidly due to consumption of CO_2 for photosynthesis and the liberation of O_2. Chambers for higher plants, therefore, required supplementary set up with adequate controls for light, CO_2, O_2, gas circulation, water, and temperature control. Such an apparatus has been described by Witty and Day (1978) for laboratory investigations, but a simpler apparatus was used by Montange et al. (1981). Ideally, the N_2 partial pressure should be in the range 0.8 to 1.0 and certainly ≥ 0.4 to saturate N_2-fixing sites (Knowles, 1980). The CO_2 and O_2 concentrations should be similar to those found in nature, that is, 0.03 and 20%, respectively. Instead of sampling for CO_2 and O_2 and supplementing through sampling ports, solenoid valves that allow for the automatic analyses and addition of CO_2 and O_2 (when needed) in the circulating gas may be installed.

43–4.1.2 Materials

1. Wide mouth laboratory bottle, approximately 40-mm diam. and 60-mL capacity.
2. Serum stopper seal to fit bottle.
3. Plastic syringe fitted with valve, 50 mL.
4. Nitrogen-free nutrient solution (see Table 12–1).
5. *Azolla* plant.
6. Dinitrogen gas.
7. Mass or emission spectrometer.

43–4.1.3 Procedure

Azolla has been selected because its high rate of growth and N_2 fixation make it possible to detect ^{15}N uptake after short exposure periods of

< 1 h. Within this period rapid changes in the gaseous composition and growth conditions are unlikely, and relatively simple incubation chambers and conditions can thus be used. A laboratory bottle is only one of the many containers that can be used.

Place 1 g of a growing culture of *Azolla* together with 9 mL of nutrient solution into a 60-mL bottle and cap tightly with a serum stopper. Pierce the stopper and withdraw 20 mL of air from the chamber with a 50-ml syringe. Refill with 20 mL of $^{15}N_2$ purchased from a commercial source or generated in the laboratory (see section 43–4.1.1). It is also possible to pre-evacuate the container, flush a few times with He and refill with a pre-mixed air having the required N_2 pressure and ^{15}N enrichment.

To estimate the required ^{15}N enrichment, use the isotope-dilution equation (see Eq. [1], section 43–3.2). The "y" in the equation is the total N within *Azolla* at incubation, "x" is the increment in total N during the period of incubation, "a_1" is the required ^{15}N enrichment in the chamber and "a" the ^{15}N enrichment in *Azolla* after harvest. For a safe level of detection by both mass and emission spectrometry, aim for 0.1 atom % ^{15}N excess in *Azolla*. Assuming growth is under optimal conditions, *Azolla* can double its biomass within 3 to 5 d (avg. 4 d). Since there are 96 h in 4 d, the increase in *Azolla* biomass and N content within a 1-h incubation period would be about 1% of the initial N.

$$\text{Thus: } [0.01\,(a_1)]\,/\,(0.01 + 1) = 0.1$$

$$\text{and } a_1 = \frac{(0.1 \times 1.01)}{0.01} = 10.1 \text{ atom \% } ^{15}N \text{ excess}$$

Because the chamber was not evacuated, a higher ^{15}N enrichment will be required to compensate for dilution from ^{14}N pre-existing in the chamber. Again, the isotope-dilution equation is used to obtain the required ^{15}N enrichment. The air space available in the chamber is approximately 50 mL, 20 of which is occupied by $^{15}N_2$.

$$\text{that is, } [20\,(a_1)]\,/\,(20 + 30) = 10.1$$

$$a_1 = 10.1 \times 50\,/\,20 = 25.2\%$$

Therefore, the 20 mL of N_2 to be injected should contain a ^{15}N enrichment of approximately 25 atom % ^{15}N excess to give a final ^{15}N enrichment of about 10 atom % ^{15}N excess in the chamber. It is necessary to know accurately the ^{15}N enrichment in the chamber. For this, a sample gas has to be analyzed at the beginning and at the end of incubation.

Remove the *Azolla* plant from the container after 1 h incubation. Dry, grind, and analyze the tissues for ^{28}N, ^{29}N, and ^{30}N composition. It is essential that all three molecular species be determined since distribution of the three ionic species in the head space is not in equilibrium (Knowles, 1980).

43–4.2 Calibration of Acetylene-Reduction Method

43–4.2.1 Principles

In ecological and agricultural studies where the goal is to measure the input of N, it is necessary to convert AR values obtained to estimates of N fixed, usually expressed in kg of N ha^{-1} over a given period. Based on theoretical comparison of electron flow, Hardy et al. (1968) suggested a conversion factor of 3:1 for C_2H_2/N_2 (see section 43–1.1). There are, however, several reasons why the simple theoretical ratio is inadequate and should not be used. Acetylene is approximately 60 times more soluble in water than N_2; thus, the enzyme is much more readily saturated with C_2H_2 than with N_2. Nitrogenase also catalyzes the reduction of protons to H_2 gas when N_2 is the substrate. On the other hand, in the presence of 0.1 atmosphere C_2H_2, the reduction of protons is almost completely inhibited, and all of the electron flow is diverted to reduction of C_2H_2. Determined conversion ratios have thus varied widely (see section 43–1.1), and the use of theoretical conversion factor can result in serious errors (Rennie et al., 1978). The most reliable or preferred method for calibrating the AR method is the incorporation of ^{15}N from $^{15}N_2$ under similar conditions. In this exercise, a C_2H_2/N_2 conversion ratio will be determined for the *Azolla-Anabaena* symbiosis.

43–4.2.2 Materials

1. Materials listed in section 43–4.1.2.
2. Source of purified C_2H_2 (see sections 43–1.2 and 43–1.2.3).
3. Ethylene for preparation of standards.
4. Vacutainers, 5 mL.
5. Gas chromatograph fitted with appropriate column (see section 43–1.2.2).
6. Nitrogen carrier gas.
7. Emission or mass spectrometer.

43–4.2.3 Procedure

Incubate the *Azolla* in an atmosphere containing 10% C_2H_2 as described in section 43–1.2.3. Analyze for C_2H_2 and C_2H_4 on a gas chromatograph and harvest after 1 h. Calculate µmoles C_2H_4 produced per gram of dry weight *Azolla*.

For the $^{15}N_2$ incubation, use the same procedures described in sections 43–4.1.2 and 43–4.1.3. The following equations are used to calculate the amount and µmoles of N_2 fixed:

µg N_2 fixed =

$$\frac{\text{atom \% } ^{15}N \text{ excess in } \textit{Azolla}}{\text{atom \% } ^{15}N \text{ excess in gas phase}} \times \text{total N in } \textit{Azolla} \text{ (mg)} \qquad [9]$$

$$\mu g\ N_2\ fixed/28 = \mu moles\ N_2 \qquad [10]$$

The μmoles N_2 fixed is then compared with μmoles C_2H_4 produced to arrive at the actual conversion ratio.

REFERENCES

Amarger, N., A. Mariotti, F. Mariotti, J.C. Durr, C. Bourguignon, and B. Lagacherie. 1979. Estimate of symbiotically fixed nitrogen in field grown soybeans using variations in ^{15}N natural abundance. Plant Soil 52:269–280.

Arp, D.J. 1992. Hydrogen cycling in symbiotic bacteria. p. 432–460. *In* G. Stacey et al. (ed.) Biological nitrogen fixation. Chapman and Hall, New York.

Bergersen, F.J. 1980a. Measurement of nitrogen fixation by direct means. p. 65–110. *In* F.J. Bergersen (ed.) Methods for evaluating biological nitrogen fixation. John Wiley and Sons, New York.

Bergersen, F.J. 1980b. Methods, accidents, and design. p. 3–10. *In* F.J. Bergersen (ed.) Methods for evaluating biological nitrogen fixation. John Wiley and Sons, New York.

Boddey, R.M. 1987. Methods for quantification of nitrogen fixation associated with Gramineae. CRC Crit. Rev. Plant Sci. 6:209–266.

Brockwell, J. 1980. Experiments with crop and pasture legumes—Principles and practice. p. 417–488. *In* F.J. Bergersen (ed.) Methods for evaluating biological nitrogen fixation. John Wiley and Sons, New York.

Burns, R.C., and R.W.F. Hardy. 1975. Nitrogen fixation in bacteria and higher plants. Springer-Verlag, New York.

Burris, R.H. 1974. Methodology. p. 10–33. *In* A. Quispel (ed.) The biology of nitrogen fixation. American Elsevier Publ. Co., New York.

Burris, R.H. 1975. The acetylene reduction technique. p. 249–258B. *In* D.P. Stewart (ed.) Nitrogen fixation by free-living micro-organisms. International Biological Programme. Vol. 6. Cambridge Univ. Press, New York.

Chalk, P.M. 1985. Estimation of N_2 fixation by isotope dilution: An appraisal of techniques involving ^{15}N enrichment and their application. Soil Biol. Biochem. 17:389–410.

Crasswell, E.T., and D.L. Eskew. 1991. Nitrogen and nitrogen-15 analysis using automated mass and emission spectrometers. Soil Sci. Soc. Am. J. 55:750–756.

Danso, S.K.A. 1988. The use of ^{15}N enriched fertilizers for estimating nitrogen fixation in grain and pasture legumes. p. 345–358. *In* D.P. Beck and L.A. Materon (ed.) Nitrogen fixation by legumes in mediterranean agriculture. Martinus Nijhoff Publ., Dordrecht, Netherlands.

Danso, S.K.A., G.D. Bowen, and N. Sanginga. 1992. Biological nitrogen fixation in trees in agro-ecosystems. Plant Soil 141:177–196.

Danso, S.K.A., G. Hardarson, and F. Zapata. 1986. Assessment of dinitrogen fixation potentials of forage legumes with ^{15}N techniques. p. 26–57. *In* I. Haque et al. (ed.) Proceedings workshop on potentials of forage legumes in farming systems of sub Saharan Africa. International Livestock Centre for Africa (ILCA), Addis Ababa, Ethiopia.

Danso, S.K.A., G. Hardarson, and F. Zapata. 1988. Dinitrogen fixation estimates in alfalfa-ryegrass swards using different nitrogen-15 labeling methods. Crop. Sci. 28:106–110.

Danso, S.K.A., and K.S. Kumarasinghe. 1990. Assessment of potential sources of error in nitrogen fixation measurements by the nitrogen-15 isotope dilution technique. Plant Soil 125:87–93.

Douglas, L.A., and R.W. Weaver. 1986. Partitioning of nitrogen-15-labeled biologically fixed nitrogen and nitrogen-15-labeled nitrate in cowpea during pod development. Agron. J. 78:499–502.

Eskew, D.L., A.R.J. Eaglesham, and A.A. App. 1981. Heterotrophic [15]N_2 fixation and distribution of newly fixed nitrogen in a rice-flooded soil system. Plant Physiol. 68: 48–52.

Fried, M., S.K.A. Danso, and F. Zapata. 1983. The methodology of measurement of N_2 fixation by non-legumes as inferred from field experiments with legumes. Can. J. Microbiol. 29:1053–1062.

Fried, M., and V. Middelboe. 1977. Measurement of amount of nitrogen fixed by a legume crop. Plant Soil 47:713–715.

Gerbaud, A. 1990. Effect of acetylene on root respiration and acetylene reducing activity in nodulated soya bean. Plant Physiol. 93:1226–1229.

Giller, K.E. 1987. Use and abuse of the acetylene reduction assay for measurement of "associative" nitrogen fixation. Soil Biol. Biochem. 19:783–784.

Hansen, A.P., and J.S. Pate. 1987. Evaluation of the ^{15}N natural abundance method and xylem analysis for assessing N_2 fixation of understorey legumes in Jarrah (*Eucalyptus marginata* Donn ex Sm.) forest in S.W. Australia. J. Exp. Bot. 38:1446–1458.

Hardarson, G., S.K.A. Danso, and F. Zapata. 1987. Biological nitrogen fixation in field crops. p. 165–192. *In* B.R. Christie (ed.) Handbook of plant science in agriculture. CRC Press, Boca Raton, FL.

Hardarson, G., F. Zapata, and S.K.A. Danso. 1984. Field evaluation of symbiotic nitrogen fixation by rhizobial strains using ^{15}N methodology. Plant Soil 82:369–375.

Hardy, R.W.F., R.C. Burns, and R.D. Holsten. 1973. Applications of the acetylene-ethylene assay for measurement of nitrogen fixation. Soil Biol. Biochem. 5:47–81.

Hardy, R.W.F., R.D. Holsten, E.K. Jackson, and R.C. Burns. 1968. The acetylene-ethylene assay for N_2-fixation: Laboratory and field evaluation. Plant Physiol. 43:1185–1207.

Hauck, R.D., and R.W. Weaver. 1986. Field measurement of dinitrogen fixation and denitrification. Spec. Publ. 18. SSSA, Madison, WI.

Heichel, G.H., D.K. Barnes, and C.P. Vance. 1981. Nitrogen fixation of alfalfa in the seeding year. Crop Sci. 21:330–335.

Herdina, and J.H. Silsbury. 1990. Estimating nitrogenase activity of faba bean (*Vicia faba* L.) by acetylene reduction (AR) assay. Aust. J. Plant Physiol. 17:489–502.

Hyman, R., and D.J. Arp. 1987. Quantification and removal of some contaminating gases from acetylene used to study gas utilizing enzymes and microorganisms. Appl. Environ. Microbiol. 53:298–303.

Knowles, R. 1980. Nitrogen fixation in natural plant communities and soils. p. 557–582. *In* F.J. Bergersen (ed.) Methods for evaluating biological nitrogen fixation. John Wiley and Sons, New York.

Leonard, L.T. 1943. A simple assembly for use in the testing of cultures of rhizobia. J. Bacteriol. 45:523–525.

Lindstrom, K. 1984. Analysis of factors affecting *in situ* nitrogenase (C_2H_2) activity of *Galega orientalis, Trifolium pratense,* and *Medicago sativa* in temperate conditions. Plant Soil 79:329–341.

McAuliffe, C., D.S. Chamblee, H. Uribe-Arango, and W.W. Woodhouse, Jr. 1958. Influence of inorganic nitrogen on nitrogen fixation by legumes as revealed by ^{15}N. Agron. J. 50:334–337.

McNeill, A.M., J.E. Sheehy, and D.S.H. Drennen. 1989. The development and use of a flow-through apparatus for measuring nitrogenase activity and photosynthesis in field crops. J. Exp. Bot. 40:187–194.

Mederski, H.J., and J.G. Streeter. 1977. Continuous, automated acetylene reduction assays using intact plants. Plant Physiol. 59:1076–1081.

Minchin, F.R., J.E. Sheehy, and J.F. Witty. 1986. Further errors in the acetylene reduction assay: Effects of plant disturbance. J. Exp. Bot. 37:1581–1591.

Montange, D., F.R. Warembourg, and R. Bardin. 1981. Utilisation du $^{15}N_2$ pour estimer la fixation d'azote et sa repartition chez les legumineuses. Plant Soil 63:131–139.

Morris, D.R., D.A. Zuberer, and R.W. Weaver. 1985. Nitrogen fixation by intact grass-soil cores using $^{15}N_2$ and acetylene reduction. Soil Biol. Biochem. 17:87–91.

Nohrstedt, Hans-Organ. 1983. Natural formation of ethylene in forest soils and methods to correct results given by the acetylene-reduction assay. Soil Biol. Biochem. 15:281–286.

Papastylianou, I., and S.K.A. Danso. 1991. Nitrogen fixation and transfer in vetch and vetch-oats mixtures. Soil Biol. Biochem. 23:447–452.

Peoples, M.B., A.W. Faizah, B. Rerkasem, and D.F. Herridge. 1989. Methods for evaluating nitrogen fixation by nodulated legumes in the field. Monogr. 11. Australian Centre for Int. Agric. Res., Canberra.

Rennie, R.J., D.A. Rennie, and M. Fried. 1978. Concepts of ^{15}N usage in dinitrogen fixation studies. p. 107–133. *In* Isotopes in biological dinitrogen fixation. Proceedings of an Advisory Group Meeting, Vienna. STI/PUB/478. IAEA, Vienna.

Ross, P.J., A.E. Martin, and E.F. Henzell. 1964. A gas-tight growth chamber for investigating gaseous nitrogen changes in the soil:plant:atmosphere. Nature (London) 204: 444–447.

Sanginga, N., F. Zapata, S.K.A. Danso, and G.D. Bowen. 1990. Effect of successive cuttings on uptake and partitioning of ^{15}N among plant parts of *Leucaena leucocephala*. Biol. Fertil. Soils 9:37–42.

Shearer, G., and D.H. Kohl. 1986. N_2-fixation in field settings: Estimations based on natural ^{15}N abundance. Aust. J. Plant Physiol. 13:699–756.

Shearer, G., D.H. Kohl, R.A. Virginia, B.A. Bryan, J.L. Skeeters, E.T. Nilsen, M.R. Sharifi, and P.W. Rundel. 1983. Estimates of N_2-fixation from variation in the natural abundance of ^{15}N in Sonoran Desert Ecosystems. Oecologia 56:365–373.

Sheehy, J.E. 1991. Theory of a crop enclosure system for measuring nitrogen fixation, photosynthesis, respiration, and biological processes in the soil. Ann. Bot. 67:123–130.

Sheehy, J.E., F.R. Minchin, and J.F. Witty. 1983. Biological control of the resistance to oxygen flux in nodules. Ann. Bot. 52:565–571.

Somasegaran, P., and H.J. Hoben. 1985. Methods in legume—*Rhizobium* technology. NIFTAL Project, Univ. of Hawaii, P.O. Box 0, Paia, Maui, HI.

Spiff, E.D., and C.T.I. Odu. 1972. An assessment of non-symbiotic nitrogen fixation in some Nigerian soils by the acetylene reduction technique. Soil Biol. Biochem. 4:71–77.

Tann, C.C., and J. Skujins. 1985. Soil nitrogenase assay by ^{14}C-acetylene reduction: Comparison with the carbon monoxide inhibition method. Soil Biol. Biochem. 17:109–112.

Tough, H.J., and J.R. Crush. 1979. Effect of grade of acetylene on ethylene production by white clover (*Trifolium repens* L.) during acetylene reduction assays of nitrogen fixation. N. Z. J. Agric. Res. 22:581–583.

Vallis, I., K.P. Haydock, R.J. Ross, and E.F. Henzell. 1967. Isotopic studies on the uptake of nitrogen by pasture plants. III. The uptake of small additions of ^{15}N-labeled fertilizer by Rhodes grass and Townsville lucerne. Aust. J. Agric. Res. 18:865–877.

Vincent, J.M. 1970. A manual for the practical study of root-nodule bacteria. Int. Biol. Programme Handb. 15. Blackwell Scientific Publ., Oxford.

Vose, P.B., and R.L. Victoria. 1986. Re-examination of the limitations of nitrogen-15 isotope dilution technique for the field measurement of dinitrogen fixation. p. 23–41. *In* R.D. Hauck and R.W. Weaver (ed.) Field measurement of dinitrogen fixation and denitrification. Soil Sci. Soc. Am. Spec. Publ. 18. SSSA, Madison, WI.

Wacek, T.J., and D. Alm. 1978. Easy-to-make "Leonard Jar." Crop. Sci. 18:514–515.

Weaver, R.W. 1986. Measurement of biological dinitrogen fixation in the field. p. 1–10. *In* R.D. Hauck and R.W. Weaver (ed.) Field measurement of dinitrogen fixation and denitrification. Soil Sci. Soc. Am. Spec. Publ. 18. SSSA, Madison, WI.

Weaver, R.W., S.F. Wright, M.W. Varanka, O.E. Smith, and E.C. Holt. 1980. Dinitrogen fixation (C_2H_2) by established forage grasses in Texas. Agron. J. 72:965–968.

Weisz, P.R., and T.R. Sinclair. 1988. Soybean nodule gas permeability, nitrogen fixation and diurnal cycles in soil temperature. Plant Soil 109:227–234.

Witty, J.F., and J.M. Day. 1978. Use of ^{15}N$_2$ in evaluating asymbiotic N_2 fixation. p. 135–150. *In* Isotopes in biological dinitrogen fixation. Proceedings of an Advisory Group Meeting, Vienna. STI/PUB/478. IAEA, Vienna, Austria.

Witty, J.F. 1979. Overestimate of N_2-fixation in the rhizosphere by the acetylene reduction method. p. 137–144. *In* J.L. Harley and R.S. Russell (ed.) The soil-root interphase. Academic Press, London.

Witty, J.F. 1983. Estimating N_2-fixation in the field using ^{15}N-labeled fertilizer: Some problems and solutions. Soil Biol. Biochem. 15:631–639.

Witty, J.F., and K. Ritz. 1984. Slow-release ^{15}N fertilizer formulations to measure N_2-fixation by isotope dilution. Soil Biol. Biochem. 16:657–661.

Witty, J.F., and F.R. Minchin. 1988. Measurement of nitrogen fixation by the acetylene reduction assay; myths and mysteries. p. 331–334. *In* D.P. Beck and L.A. Materon (ed.) Nitrogen fixation by legumes in mediterranean agriculture. Martinus Nijhoff Publ., Dordrecht, the Netherlands.

Wynne, J.C., F.A. Bliss, and J.C. Rosas. 1987. Principles and practice of field designs to evaluate symbiotic nitrogen fixation. p. 371–389. *In* G.H. Elkan (ed.) Symbiotic nitrogen fixation technology. Marcel Dekker, New York.

Zuberer, D.A., and D.B. Alexander. 1986. Effects of oxygen partial pressure and combined nitrogen on N_2-fixation (C_2H_2) associated with *Zea mays* and other gramineous species. Plant Soil 90:47–58.

Chapter 44

Measuring Denitrification in the Field

A. R. MOSIER, *USDA/ARS, Fort Collins, Colorado*

LEIF KLEMEDTSSON, *IVL, Goteborg, Sweden*

Denitrification is the stepwise process by which NO_3^- is reduced to N_2:

$$NO_3^- \rightarrow NO_2^- \rightarrow NO \rightarrow N_2O \rightarrow N_2$$

The process can both be microbial and chemical, but the microbial process dominates in most soils (Broadbent & Clark, 1965). Microbial denitrification is an alternative respiration process by bacteria in the soil (Focht & Verstraete, 1977) that can use NO_3^- under O_2-limiting conditions. These bacteria prefer to use O_2 as their electron acceptor, if it is available (see chapter 14 by Tiedje in this book).

To measure denitrification in the field, the large temporal and spatial variation of the process must be considered (Folorunso & Rolston, 1984). This requires an understanding of the factors controlling the process. Soil atmosphere O_2 concentration is the main factor controlling denitrification in soils that are not water-logged (Tiedje, 1988). Oxygen status of the soil depends on the O_2 diffusion rate and the consumption rate of the gas in the soil. The diffusion rate depends upon temperature, soil structure, and water content. As the O_2 diffusion rate in water is about 10 000 times slower than in air, soil water content is a primary controlling factor. Oxygen consumption rate depends on the easily degradable organic C content of the soil and soil water and temperature, which affects soil microbial activity (Bremner & Shaw, 1958). The O_2 concentration in the bulk soil as well as in soil aggregates will have a large spatial and temporal variation, which will contribute to the variation of denitrification (Parkin, 1987).

Regulation of denitrification is furthermore complicated by the dependence of microorganisms on organic substrate, alternative N electron acceptors, for which they have to compete with other microorganisms and plants. Additionally, in soils not receiving nitrate fertilizer, denitrifiers depend on the net mineralization and the nitrification activity in the soil.

All of these can be looked upon as regulative factors that interact in a synergistic manor with the soil oxygen status creating "hot spots" where much of the denitrification activity is concentrated. These hot spots are within soil aggregates, or on decomposing organic material in the soil (Parkin, 1987; Seech & Beauchamp, 1988).

Because denitrification is predominantly confined to active sites in arable soils, it has the largest spatial and temporal variability of any of the N cycle processes (Tiedje et al., 1989). Denitrification variation imposes a large problem in accurately measuring aerial and yearly emissions. Therefore, special consideration of temporal and spatial variation is needed when measuring denitrification or N_2O-emissions in the field. Before undertaking denitrification measurements, we suggest reading Hauck and Weaver (1986), Klemedtsson et al. (1990), Mosier & Heinemeyer (1985), Parkin & Robinson (chapter 2 in this book), and Tiedje et al. (1989) as reference materials on the measurement and variation of denitrification. Because of this variability and the problems in measuring N_2 production against normal atmospheric background, quantifying denitrification is difficult and no "standard, absolute" methods exist. There is also no consensus as to a "best" method for quantifying denitrification in the field.

44–1 METHODS

This is not a comprehensive review of denitrification techniques but rather a description of techniques that we think are best suited for field measurements within the constraints of current technology. We suggest two different methods for measuring denitrification in the field, acetylene inhibition and use of ^{15}N isotopes, using either in situ flux or intact core measurements.

44–1.1 The Acetylene Inhibition Method

In the early 1970s, Federova et al. (1973) discovered that acetylene inhibited the reduction of N_2O to N_2 in the denitrification process. This discovery formed the basis for the development of the acetylene inhibition method (Balderston et al., 1976; Yoshinari & Knowles, 1976).

$$NO_3^- \rightarrow NO_2^- \rightarrow N_2O - / C_2H_2 \rightarrow N_2$$

Previously it was difficult to measure N_2 production during denitrification, because of the high background concentration of the gas in the atmosphere (78%). By inhibiting the last step in the denitrification process with acetylene, low denitrification rates can be measured in ambient atmosphere by measuring N_2O emissions, given the low natural background concentration of N_2O of about 310 ppb (Duxbury, 1986; Klemedtsson et al., 1990; Mosier & Heinemeyer, 1985; Ryden et al., 1979; Tiedje et al., 1989).

44–1.2 The Nitrogen-15 Method

An alternative to the acetylene block method is using the stable isotope [15]N. This concept was developed almost 40 yr ago but is now much more practical since the cost of isotopes has decreased and isotope ratio mass spectrometers are now more stable, more readily available, more "user friendly," and less expensive than in past decades.

Using ^{15}N-labeled fertilizer to quantify N gas production directly from denitrification permits direct measurement of N_2 and N_2O from the fertilizer applied (Rolston et al., 1976, 1978, 1982) and from the total mineral N in the soil (Hauck & Bouldin, 1961). The method is useful and relatively sensitive since the isotopic composition of N_2 in the atmosphere is 99.26% mass 28, 0.73% mass 29, and 0.0013% mass 30, small changes in mass 30 within a confined atmosphere are readily measured. Direct measurement techniques using ^{15}N, involve applying highly enriched (20–80%) ^{15}N fertilizer to a designated plot of soil and measuring N gas production following denitrification by quantifying the increase in ^{15}N-labeled gases within a closed chamber. Using the ^{15}N method for studying denitrification is constrained by the same assumptions and sources of error that apply to other uses of [15]N. These considerations are discussed in detail by Bremner (1965), Hauck and Bremner (1976), Bremner and Hauck (1982), Buresh et al. (1982), Hauck (1982) and will not be detailed herein.

44–2 EXPERIMENTAL PROTOCOLS

Both acetylene inhibition and ^{15}N can be used with many of the possible techniques for measuring denitrification. Either method may be used with static cores or closed chambers or both used simultaneously to provide a cross check. We suggest the following techniques because a minimum of field equipment is required to make measurements.

44–2.1 Static Core Protocol

In the early 1980s, different soil core techniques were developed, using the acetylene inhibition method to measure denitrification under field conditions. Two basically different soil core systems were developed. In the first, intact soil cores are incubated statically with acetylene inside a container (Aulakh et al., 1982; Robertson & Tiedje, 1987; Svensson et al., 1984; Ryden et al., 1987). In the second, confined intact soil cores are incubated such that soil air and acetylene are recirculated through the macropores of the soil (Parkin et al., 1984). Positive and negative aspects of both core systems have been detailed in a review by Tiedje et al. (1989). They conclude that both approaches estimate the denitrification process equally well if the inherent limitations of the techniques are considered. We recommend using the static core system in the field as it is less complicated and thus accommodates greater replication.

Several different types of static core systems have been used to measure denitrification rates. We suggest using encased cores: (i) to protect them from destruction during the incubation; (ii) to use the same approach on soils with different structure stability, and (iii) for ease of use in the field. According to Tiedje et al. (1989), the main problem with static cores is limited diffusion of acetylene into the soil core and the N_2O out of the core. Since encasement of the core presents a barrier to gas diffusion, the encasement should have holes or slits to facilitate gas diffusion.

Core sampling devices should be constructed so that one can determine if soil compaction has occurred during sampling. If compaction exceeds 5% the core should be discarded. If compaction is a persistent problem then it may be reduced by decreasing the length and increasing the diameter of the soil core.

Soil cores should be incubated at the in situ temperature in the field or laboratory. We recommend this because using a Q_{10} quotient to calculate changes in the denitrification rates with increasing temperature can result in underestimation. An increase in temperature increases the activity of the denitrifiers, and also promotes anaerobiosis by affecting soil respiration. If soil cores are transported to the laboratory for incubation, they should be transported on ice (Tiedje et al., 1989). It is important that the soil cores are transported without vibrations that might change soil structure.

Denitrification rates in field soils do not always follow a normal distribution in time and space (see chapter 2 by Parkin and Robinson in this book). As a general rule, if denitrification rates have a coefficient of variation (CV) above 100%, then one can assume that the distribution is skewed and that it may be lognormally distributed, at least during some periods over the year. Under these conditions one should use at least 20 replicates for each treatment to be investigated. Analysis of data can be handled as described in Parkin et al. (1988, 1990) or in Svensson et al. (1991). If the denitrification rates have a CV that is < 100%, then one can use normal statistical analysis and < 10 replicates.

44–2.1.1 Acetylene Sources

First, acetylene is a combustible gas that must be handled with care. Second, acetylene from commercial tanks contains relatively large concentrations of acetone (Hyman & Arp, 1987). Unfortunately, acetone is a good substrate for soil heterotrophic organisms and must be removed when using the acetylene in denitrification experiments. One way to remove acetone from commercial grade acetylene is to pass the gas through gas dispersion tubes in two consecutive concentrated sulfuric acid traps followed by another trap containing distilled water (Mosier et al., 1985). The acetylene stream exiting the water trap should be analyzed for acetone to ensure that no acetone is being added to the plot.

High purity acetylene can also be produced from calcium carbide (CaC_2). The acetylene formed from CaC_2 can be assumed to be clean even though small amounts of H_2, CH_4, C_2H_4, and PH_3 are produced (Hyman

& Arp, 1987). These contaminants do not appear to alter soil denitrification (Tiedje et al., 1989).

44–2.1.2 Soil Core Collection

Minimally disturbed, intact, soil cores (0–10 cm) are removed from the appropriate plots using a 1-m long, drop hammer soil sampler. The diameter of the soil core should be 3 to 8 cm. Collection is facilitated by a sampling device that contains an inner liner made from PVC or stainless steel pipe that is slightly larger than the inside diameter of the sampler tip. The liner is perforated with numerous 2-mm diam. holes made by drilling the pipe to provide ready equilibration of the air within the soil and atmosphere. The liner also serves as the core encasement during incubation. Each soil core can be quickly removed from the corer and placed in an incubation chamber. Each core is placed in a pvc or glass container that is a centimeter or less larger diameter than the outside diameter of the core liner. The headspace above the soil should be about 100 cm^3. The top of the incubation cylinder is a rubber stopper drilled to fit a gastight gas chromatographic septum through which a needle can be inserted for sampling the gas inside the incubation cylinder (Burton & Beauchamp, 1984). Intact soil cores from 10 to 20, 20 to 30 etc. can be collected in the same manner if information concerning denitrification activity deeper in the soil profile is desired.

Acetylene (free of contaminants) is added to the gas, normal air unless an oxygen-free atmosphere is desired, headspace above or around the soil core, to a final concentration of 5 to 10% vol/vol (5–10 kPa). The acetylene in the gas headspace is mixed with the gas in the larger macro pores by alternatively reducing and increasing the pressure by pumping with a large gas tight syringe (30–50 mL volume). The over pressure due to the added acetylene is vented with a needle.

Each cylinder is incubated for up to 24 h, in a hole of slightly larger dimensions, in the ground adjacent to the study area. Headspace atmosphere in the core container is sampled after 3, 12, or 24 h, after mixing the macropore gas with the head space gas with a large syringe. In soils where the nitrate supply may be quickly depleted, short time sampling intervals of 2 or 3 h are suggested. Since acetylene also blocks nitrification, soils may become nitrate deficient and underestimate denitrification. The gas samples could either be analyzed directly or stored in gas tight vials (Aulakh et al., 1991; Ryden et al., 1987).

44–2.1.3 Analysis

Air samples are analyzed for N_2O using a gas chromatograph equipped with a ^{63}Ni electron capture detector (ECD), which is operated at high temperature (300–400 °C) using a system like that described by Mosier and Mack (1980). Although detector response and linearity vary from one manufacturer to another, most systems that we know of provide satisfactory capability for N_2O. Valco Instruments pneumatically or electrically operated 10-port sampling and four-port switching valves are incorporated into the GC system. The sampling and switching valves may be controlled by either the GC microprocessor or an external data acquisition-control unit. Detector, sampling valve, switching, valve and column oven temperatures are 300 to 400, 100, 175, and 70 °C, respectively (detector temperature may vary according to manufacturer recommendations). The flow rate of carrier, backflush, and detector purge gases (95% argon + 5% methane) is 18 cm^3/min.

Gas samples are introduced into a 1 to 5 cm^3 gas sampling loop (size depends upon the sensitivity of the ECD being used) through an inlet system (Mosier & Mack, 1980). Both CO_2 and water vapor are removed from the gas samples. The two absorbent traps are prepared by packing 10-mm-diam. Millipore syringe filter holders with Ascarite and $Mg(ClO_4)_2$, and connected to the GC inlet by vacuum-tight luer fittings. The traps are used to reduce potential variability in N_2O analysis caused by different CO_2 and water vapor concentrations between samples. The traps also ensure that water vapor does not accumulate in the sample loop and eliminates any effect of high CO_2 concentrations on N_2O analysis.

To reduce analysis time and eliminate interference from water vapor (when a dryer is not used) and fluorocarbons in the sample gas mixtures, we use a 1 m by 2 mm i.d. nickel precolumn in combination with a 3 m by 2 mm i.d. analytical column. Both columns, packed with Porapak Q (80/100 mesh) are attached directly to the 10-port sampling valve (Mosier & Mack, 1980) as are the sample loop and carrier gas sources. With the valve in the starting position, the precolumn is backflushed with carrier gas while the sample loop is first evacuated and then refilled to local atmospheric pressure ±0.5 mm Hg with air from a sample syringe. Sample loop pressure is measured precisely with a Setra Systems Inc. 304B electronic manometer. When the sampling valve is rotated, the contents of the sample loop are swept through the precolumn into the analytical column that is vented to the atmosphere through the four-port switching valve located at the exit of the column. The sampling valve is time programmed to return to position A after N_2O clears the precolumn (about 2.1 min). This allows the sample loop to be evacuated and refilled and the precolumn to be backflushed while the sample N_2O continues to be separated from other atmospheric constituents on the analytical column. After O_2 is eluted from the analytical column (about 3 min), the time programmed four-port switching valve diverts the analytical column flow into the detector. Without the switching valve the large amount of O_2 in the gas sample may overload the detector to give a broad, tailing peak for O_2 and reduced analytical precision for N_2O. Oxygen also reduces the life of some EC detectors. Ten to 12 samples per hour can be run with this system, depending upon the retention characteristics of each column, at a routine precision of <3 ppb (vol/vol) for ambient concentrations of N_2O. The column system is also amenable to automation through an additional inlet valving system or direct sampling from gas collection systems (Mosier & Heinemeyer, 1985).

Table 44–1. Bunsen absorption coefficients for N_2O in water at different temperatures (Tiedje, 1982).

	Temperature °C						
	5	10	15	20	25	30	35
α†	1.06	0.88	0.74	0.63	0.54	0.47	0.41

† α is the milliliter of N_2O at 0 °C and 760 mm of Hg (STP) that is absorbed by 1 mL of water.

When N_2O concentrations are below 100 ppm the above ECD system is used. When using the acetylene inhibition method, one may need to bypass the ECD with the acetylene in the sample, after it exits the chromatographic column. Acetylene will alter both the sensitivity and stability of some ECDs.

When N_2O concentration is > 100 ppm a thermal conductivity (Ryden et al., 1978) or ultrasonic (Bremner & Blackmer, 1982) may be used rather than an ECD because of the nonlinear nature of the ECD. The same inlet and chromatography system as described above may be used or the system described by Ryden et al. (1978) is an alternative. According to Ryden et al. (1978), N_2O was separated from other gases in the sample by injection through a 0.46-mL gas sampling loop onto a column of Porapak Q (550 by 0.16 cm, i.d.) heated to 50 °C. The thermal conductivity detector used a filament current of 220 mA and helium carrier gas at 30 cm^3/min.

44–2.1.4 Calculations

The flux (Q) of N_2O per hour and core is calculated as:

$$Q = (M_s - M_1) / (T_s - T_1) \qquad [1]$$

where M_1 is the total N_2O produced in the soil core during the first gas sampling interval (T_1), generally 3 h after adding acetylene, and M_s is the total N_2O produced during the second gas sampling interval (T_s), 12 or 24 h after initiation. The M_s and M_1 are calculated according to Tiedje (1982) using the Bunsen coefficient (α), examples are listed in Table 44–1, from the equation:

$$M = C_s(V_g + V_1\,\alpha) \qquad [2]$$

where C_s is the N_2O concentration in the total gas phase, measured by gas chromatography (ng N_2O-N/mL); V_1 is volume of liquid phase; V_g = total gas volume, which is the sum of the air-filled pore space (V_{ap}) in the soil core ($V_{ap} = V_p - V_1$, where V_p is total pore space), the headspace above the core (V_h) and the free volume between the soil core and the incubation container (V_f).

The total pore space (air and water filled) V_p is calculated from the soil bulk density, (B_d), soil particle density (P_d = 2.65 g/cm^3) and the volume of the soil core (V_c) after soil water content has been determined by drying the soil at 105 °C over night:

$$V_p = (B_d/P_d) * V_c \qquad [3]$$

We recommend that N_2O emission from the soil core be expressed as ng N_2O-N d^{-1} g^{-1} dry weight of soil using the equation:

$$F = (Q * 24)/W \qquad [4]$$

where W is the dry weight of soil in each core.

44–2.2 Protocol for Closed Chamber Field Gas Flux Measurements Using Acetylene or Nitrogen-15

Gas fluxes measured with chambers estimate the soil-atmosphere exchange of gases and are not necessarily a measure of the instantaneous denitrification rate (Jury et al., 1982). With either the acetylene or the ^{15}N technique, the area of soil from which gas flux is to be measured must be defined. We suggest driving a cylinder (10–60 cm length and 15–30 cm in diameter) into the soil to establish each site. The length and diameter or shape must be suited to the field management with respect to zonal tillage, irrigation, type of crop, traffic patterns, and crop row spacing and orientation. Use of the acetylene method for long-term measurements requires establishment of multiple sites to permit rotation to new sites every 1 to 7 d, to prevent microbial adaptation to acetylene. Prolonged exposure of the soil to acetylene may lead to accelerated acetylene utilization or incomplete blockage of N_2O reduction to N_2.

44–2.2.1 Acetylene

Acetylene may be applied by either the radial diffusion of acetylene gas directly (Ryden & Dawson, 1982) or addition of 3 to 6 g of CaC_2 (Aulakh et al., 1991) to each of four 20-cm deep 2-cm diam. holes around each measurement site. Acetylene must be applied before each measurement when using cylinder gas or every 48 h when using CaC_2. Two to 3 h after applying the acetylene, a closed chamber, (Hutchinson & Mosier, 1981) is fitted and sealed over the established flux site. Use of open-base cylinders to which chamber covers can be fitted at the time of sampling prevents disturbance of soil and roots that could alter denitrification (Mosier et al., 1991). A gas sample is immediately withdrawn by syringe from the chamber, and samples are withdrawn at 15-minute intervals for 1 h.

44–2.2.1.1 Analyses. Gas samples should be returned to the laboratory and analyzed by gas chromatography as described in section 44–3.1.3. There may be occasion to analyze samples for N_2 and other fixed gases. The most direct and simple method for analyzing gas samples for oxygen and N is using a thermal conductivity or ultrasonic detector following a 2 m by 0.16 cm, i.d. stainless steel or nickel column packed with 50 to 80 mesh molecular sieve 5A. The gas sample may either be injected through a conventional GC inlet or sampling loop of 0.5 to 2 mL. The gas sampling valve and sampling loop concept provide more precise analyses than direct injection by syringe. If $> 1\%$ precision is not required for the experiment then syringe injection may be used to reduce GC system cost. To analyze atmospheric gases including O_2, N_2, and N_2O along with CO_2, CH_4, and H_2 a split column GC configuration (Beard & Guenzi, 1976) using parallel columns of Porapak Q and molecular sieve 5A, or dual Porapak Q columns using differential temperatures (Bremner & Blackmer, 1982) may be used.

44–2.2.1.2 Calculations. Nitrous oxide flux can be computed from the concentration change with time using the following equation (Hutchinson & Mosier, 1981), to correct for the reduction in soil N_2O concentration gradient with time as the gas accumulates inside the chamber:

$$N_2O \text{ flux} = \frac{V(C_1 - C_0)^2}{A\, t_1\, (2C_1 - C_2 - C_0)} \ln \frac{(C_2 - C_1)}{(C_1 - C_0)} \qquad [5]$$

where V is the volume of the chamber, A is the area of soil covered, C_0 is the initial N_2O concentration in the chamber and C_1 and C_2 are the N_2O concentrations after time t_1 and t_2 where $t_2 = 2\, t_1$.

To accurately quantify daily N_2 and N_2O emissions when flux rates are changing frequently it is necessary to make flux measurements at least every 6 h within each 24 h period (Aulakh et al., 1991) but when flux rates are small and vary with the diel variation in soil temperature, one late morning measurement should characterize the daily flux rate (Mosier, 1989).

44–2.2.2 Nitrogen-15-Modified Hauck Technique

The direct N gas measurement technique using ^{15}N was first proposed by Hauck et al. (1958). Hauck and Bouldin (1961) showed calculations for ^{15}N enrichment of NO_3^- undergoing denitrification and N_2 evolution but it was not until two decades later that the applicability of the method became evident. Siegel et al. (1982) revitalized the technique and modified the calculations. Mulvaney (1984) adapted the calculation to triple-collector mass spectrometers (MS), and Mulvaney and Boast (1986) refined the calculation equations to permit using lower ^{15}N abundance N sources. Mulvaney and Kurtz (1982) showed that N_2O evolution from soil could be measured with this technique.

The method has been used in a variety of agricultural systems ranging from upland crops to flooded rice (*Oryza sativa* L.) (Mosier et al., 1986a; Mulvaney & Vanden Heuvel, 1988; Banerjee et al., 1990; Mosier et al., 1989; Lindau et al., 1990; Buresh & DeDatta, 1990; Bronson & Mosier, 1991; Heinemeyer et al., 1988; Mohanty & Mosier, 1990; Aulakh et al., 1991).

The "Hauck Technique" involves applying highly ^{15}N-enriched fertilizer to soil (> 20 atom % ^{15}N) and using a chamber cover to isolate the atmosphere above the ^{15}N-fertilized soil for a designated time, to permit determining the rate of change of ^{15}N atoms (N gases) in the chamber atmosphere over time. The calculations (see section 44–2.2.2.3) use the fact that denitrification N-gases, principally N$_2$ under most conditions, evolving from the soil into the chamber head-space, do not randomly mix isotopically with the N-gases in the chamber (78% N$_2$ containing 0.366 atom % atom ^{15}N). Using this nonrandom ^{15}N distribution, the technique permits calculation of not only the amount of N-gas evolved ·from the added ^{15}N-enriched fertilizer but also that from natural abundance soil N constituents as well. The ^{15}N mole fraction of the NO$_3$ in the soil that serves as the N-gas source is calculated directly from N$_2$ mass spectral data (29/28 and 30/28 + 29 ratios when using a dual collector MS and 29/28 and 30/28 ratios when using a triple collector MS) (Mulvaney, 1984). It is, therefore, not necessary to disturb the soil in the plot during the course of the experiment to determine the soil ^{15}NO$_3^-$ content. The soil ^{15}NO$_3^-$ when denitrification occurred must be known to calculate the total amount of N$_2$ evolved from the site at any given time. The technique is reasonably sensitive as 5 g of N ha^{-1} d^{-1} can be detected when the soil NO$_3^-$ ^{15}N content is above 50 atom %.

The technique, as described here, requires several assumptions. The first is that the total amount of ^{28}N$_2$ inside the gas collection chamber does not change during the sample collection period. Siegel et al. (1982) noted that providing the chamber ^{28}N$_2$ is 100 times or more greater than the amount of gas produced, the calculations are correct. The equations of Mulvaney and Boast (1986) do not rely upon this assumption. The second, and major assumption is that the ^{15}N label of the NO$_3^-$ in the soil is uniform. We assume that NO$_3^-$ formed from the ^{15}N-labeled fertilizer added to the soil mixes uniformly with the unlabeled NO$_3^-$ already in the soil or with unlabeled NO$_3^-$ formed from organic N mineralization. Nonuniform mixing of the ^{15}N label may cause underestimation of the total N denitrified. It seems unlikely that complete equilibration of fertilizer N and soil N ever occurs, but studies by Mulvaney (1988) and Mosier and Schimel (1993) indicate that accurate N-gas production estimates can be made when the soil NO$_3^-$ pool is not uniformly labeled.

The depth that cylinders are driven into the soil is important when using ^{15}N techniques. It is prudent to conduct N-balance measurements along with direct N-gas flux measurements when using ^{15}N. The root zone of the vegetation should be enclosed within the cylinder so that plants outside the treated area cannot withdraw labeled N from it. However, if the whole root zone of the plant is not enclosed within the cylinder, the area around the treated site must be sampled (Mosier et al., 1986a,b). Since rice plants provide a conduit for gaseous transport of N$_2$ and N$_2$O, plants must be included inside rice field microplots (Mosier et al., 1990).

After establishing microplots, N fertilizer containing 50 to 80 atom % ^{15}N is added to the soil. Addition is made by· either injecting the appropriate amount of solution a few centimeters below the soil surface in a grid across the microplot (Schimel & Firestone, 1989); removing the top 10 cm of microplot soil and mixing it thoroughly with the ^{15}N fertilizer (Mosier et al., 1986a); or pipetting the fertilizer solution uniformly over the surface of the microplot; or applying ^{15}N directly into floodwater for rice (Lindau et al., 1990).

In upland crops, significant denitrification will not occur until the field is irrigated or sufficient rainfall is received to limit O_2 diffusion in the soil. At this time daily monitoring is required. If N_2O emissions comprise $> 5\%$ of the total $N_2 + N_2O$, then flux measurement periods of no more than 2 h can be used. If N_2O makes up $< 5\%$ of the total, as in a rice field or heavy textured soil, then overnight cover placement is appropriate (Mosier, 1989).

44–2.2.2.1 Gas Collection. Gas collection chambers (Hutchinson & Mosier, 1981) are placed over the microplots every day at midnight, 0600, 1200, and 1800 h each day for 2 h. Forty cubic centimeter gas samples are taken immediately after covering, and again after 1 and 2 h with 60 cm[3] polypropylene syringes fitted with a one-way, gas tight, plastic stopcock. Chambers are removed after 2 h. This procedure is followed for 3 to 4 d following each water event until no fertilizer-derived mineral N remains in the soil. In flooded rice, chambers are placed at 1800 h, samples taken at 0, 4, and 14 h later, and chambers removed after the 14-h sampling.

44–2.2.2.2 Analysis. One or 2 mL of gas collected from a soil chamber atmosphere is delivered to the MS by a one-way stopcock fitted syringe. To analyze the gas sample on the MS it is first necessary to construct an inlet system on the instrument so that oxygen, CO_2, and water vapor are removed from the gas sample. An inlet is simple and can be constructed for about $500. This inlet (Fig. 44–1) consists of a series of stainless steel components; a 2 to 5 mL sample loop, an oxygen scrubber, and a cold trap, each connected by a gas-tight valve to permit isolating each portion of the inlet. The inlet is arranged to permit introducing a gas sample directly from a syringe fitted with a two-way luer fitted stopcock into the sample loop. The loop can be immersed in a cryogen (liquid N or a dry ice cooled solution) to remove water vapor, CO_2 or N_2O depending upon the desired analysis. For our purpose we want to look at total N-gas evolution from the soil so will not use a cryogenic trap at this stage. From the sample loop the gas passes into the oxygen scrubber. This scrubber is made by packing a 0.5 m long by 4 mm i.d. ($\frac{1}{4}$ in. o.d.) stainless steel tube, with 10 to 15 g of Cu-coated silica O_2 trapping material (Chemical Research Suppliers, Inc., 900 Westwood Ave., Addison, IL 60101) and plugging the ends of the tube with quartz wool. This scrubber can be operated at room temperature but efficiency is much greater if an elevated temperature is used. At a temperature of 250 to 300 °C, the trap effectively removes O_2 and reduces N_2O and NO to N_2. After several hundred samples, the trap is readily regenerated by passing H_2 gas through the trap while heating at 250 to 300 °C (follow manufacturer's regeneration instructions as H_2 can be explosive).

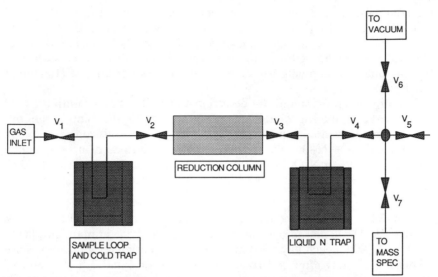

Fig. 44–1. Diagram of a simple spectrometer gas sampling inlet, (V) indicates vacuum-tight valve, sample loop is 2 to 5 cm³ volume to be used with or without a cold trap, reduction column is heated to 250 to 300 °C.

About 30 s after introducing the sample into the O_2 scrubber the sample is passed through the high-efficiency liquid N trap (a coil of tubing not just a U-tube) to remove CO_2 and H_2O. The purified sample N_2 is then passed into the MS inlet. Both sample and reference gases are introduced into the MS through the gas inlet system so that they are directly comparable. If a dual inlet MS is available, a time zero field gas sample is introduced into the MS through the sample inlet into the reference side to serve as the reference gas. If a single inlet instrument is used then samples and reference gases must be run sequentially.

44–2.2.2.3 Calculations. Rather than go through equation derivation, we will refer you to Siegel et al. (1982), Mulvaney (1984), and Mulvaney and Boast (1986). Table 44–2 summarizes the basic equations used by Mosier et al. (1986a,b, 1989) to calculate the total N gas flux from the soil. Remember that this is total $N_2 + N_2O + NO$, not just the N_2 derived from the added fertilizer.

The equations from Siegel et al. (1982), Mulvaney (1984), and Mulvaney and Boast (1986) were derived to quantify N_2 or N_2O production. If the N in N_2 and N_2O evolved during the same gas collection period are greatly different when the calculations are not theoretically valid for measuring $N_2 + N$ oxides. Mosier et al. (1986a,b) and Bronson et al. (1992), in field studies, did not find significant differences in the isotopic composition of N_2 and N_2O measured during the same gas sampling time when separate measurements were made on the N_2 and N_2O produced. Careful cross checks should be made with each experiment to ensure that large differ-

Table 44–2. Equations used to calculate N gas emissions from mass spectrometric isotope ratio measurements of gas samples.

1. $\Delta r = (^{29}N_2/^{28}N_2)$ sample $- (^{29}N_2/^{28}N_2)$ reference
2. $\Delta r' = (^{30}N_2/^{28}N_2 + {^{29}N_2})$ sample $- (^{30}N_2/^{28}N_2 + {^{29}N_2})$ reference
3. Sample = air sample from collection chamber at some time, t, after installing chamber.
4. Reference = air sample from field, i.e., normal air sample, taken from the chamber immediately after installation, $^{29}N_2/^{28}N_2$ and $^{30}N_2/^{28}N_2 + {^{29}N_2}$ are ion current ratios determined by the mass spectrometer.
5. $^{15}X_N$ = mole fraction of ^{15}N in the soil NO_3^- pool
   $= 2.015 \, (\Delta r'/\Delta r)/(1 + (2.015 \, (\Delta r'/\Delta r))$
6. d = the fraction of total N gas in the gas collection chamber attributable to denitrification.
   $= \Delta r'/(^{15}X_N)_2$
7. Total denitrification N gas evolved from the soil into the collection chamber = total N_2 in the chamber volume $\times$ d.
8. N_2 Flux $= \Delta C/A \, t$
   where
         A = soil surface area covered by chamber
         t = time that the chamber covered the soil
         ΔC = the change in concentration of $^{30}N_2$ and $^{29}N_2$ in the chamber during time t

Table 44–3. Mass spectrometer data to use in the following calculations.

	Isotope ratios			
	Sample gas 4 h Plot 1		Reference gas 0 h Plot 1	
	29/28	30/28 + 29	29/28	30/28 + 29
Rep 1	0.00735	0.0000751	0.00731	0.0000331
Rep 2	0.00735	0.0000741	0.00729	0.0000327
Rep 3	0.00735	0.0000749	0.00730	0.0000328
Mean	0.00735	0.0000751	0.00730	0.0000329

$r = 0.00735 - 0.00730 = 0.00005$
$r' = 0.0000751 - 0.0000329 = 0.0000418$
$^{15}X_N = 0.651$
$d = 0.00009863$

ences, $> 10\%$ in isotopic composition of the N in N_2 and N_2O, do not appear.

The following is an example calculation using data from gas samples collected during a 1987 rice denitrification experiment in Griffith, N.S.W., Australia, and analyzed on a VG-622 mass spectrometer (a single inlet, dual collector instrument) at the CSIRO Division of Plant Industry in Canberra. The MS was fitted with the gas sampling inlet system described above. Samples from Griffith were collected 0 and 4 h after placing chambers over rice plots that had been fertilized with 79 atom % ^{15}N-urea in the floodwater at pannicle initiation (see Mosier et al., 1989). The gas samples were injected into Vacutainers and taken to Canberra for analyses. The MS data are shown in Table 44–3 and calculations follow:

To calculate the amount of N_2 inside the chamber (7775 mL): assume that air is 78% N_2; correct volume to standard temperature and pressure (STP); 273 K and 760 mm of Hg; assume that the temperature was 20 °C (293 K) and the atmospheric pressure was 760 mm of Hg. With these assumptions calculate a total of 5652 mL of N_2 inside the chamber. The air temperature should be measured during each sampling time and used in the calculations. The atmospheric pressure for the sampling location should also be used, that is, about 640 mm of Hg in Fort Collins, CO and 760 mm of Hg in Baton Rouge, LA. To calculate the mass of this N_2, we know that 1 mol of a gas at STP occupies 22 400 mL; then 5652 mL of gas is 0.252 mol, 7.065 g.

To calculate the amount of N gas evolved from the soil into the gas chamber: N_2 evolved = N_2 in chamber × d

$$= 7.065 \text{ g} \times 0.00009863$$
$$= 697 \text{ μg N plot}^{-1} \text{ 4 h}^{-1}$$

N_2 flux = 697 μg of N/ (707 cm^2 × 4 h) = 0.247 μg N cm^{-2} h^{-1}. On a hectare basis, then 24.7 g of N h^{-1} evolved from the soil during this 4-h sampling period that began 52-h after urea fertilization.

44–2.3 Measuring Dinitrogen Emissions from Applied Nitrogen-15-Labeled Fertilizer

If the objective of a study is to determine the amount of denitrification only from the applied N fertilizer, the "Hauck" calculations do not need to be used. The analytical technique, however, remains the same as that described above. The studies of Rolston et al. (1976, 1978, 1982) show the utility of this technique where they fertilized 1-m^2 plots with 40 to 50 atom % ^{15}N KNO_3 and imposed a variety of irrigation and organic C amendments to the plots. By enclosing the air space above the plots for 1 to 4 h each day and measuring the increase in concentration within the chamber headspace of N_2O and $^{15}N_2$, they were able to show the effects of the various treatments on N loss by denitrification.

The isotopic composition of N in gas samples was determined for samples pulled through Ascarite, magnesium perchlorate, and a commercial O_2 scrubber by direct injection into the mass spectrometer. With the knowledge of the volume of the chamber headspace, the concentration of N_2 in the gas (measured by gas chromatography), the time that the plot was covered, and the atom % ^{15}N of the N in the chamber gas, the flux of N_2 from the fertilizer was calculated. The isotopic composition of the N_2 evolved from the fertilizer is just as described above for the Hauck Technique, the N_2 produced from the added nitrate does not mix isotopically with the N_2 in the air. It is, therefore, necessary to calculate atom % ^{15}N by the equation:

$$\text{atom } \% \text{ } ^{15}N = [29 + 2(30)]/[2(28) + 2(29) + 2(30)] \qquad [6]$$

This requires measuring the 29/28 isotopic ratios as well as the 30/28 ratios or $30/28 + 29$ ratios of the N_2 and has a sensitivity of about 100 g of N $ha^{-1} d^{-1}$.

44–3 PROBLEMS WITH GAS SAMPLING AND STORAGE CONTAINERS

Affordable, leak-free, noncontaminating gas sampling and storage devices are a continual concern. We generally recommend using Monoject polypropylene syringes fitted with one-way nylon stopcocks as gas-sampling devices when samples are to be analyzed within a few hours or injected into storage containers. These syringes are uniform in quality from batch to batch and are relatively inexpensive. Gases do, however, diffuse through the walls of the polypropylene syringes. Therefore, they cannot be used as storage devices. They also cannot be used to transport samples within areas where the atmospheric pressure changes. Nylon syringes leak less through their walls but still leak when changing atmosphere between the sampling and analytical sites (an elevation change of about 1200 m).

Vacutainers have been widely used as gas transport and storage devices because they are relatively inexpensive and convenient to use. Additional problems, to those discussed above, exist. First, the tubes are not completely evacuated as they come from the manufacturer, thus contain a significant amount of gas before the sample is added. The pressure in the tubes is quite constant, however. By measuring existing pressure in a large set of each lot member of evacuated vials, before and after injecting 12 mL of air, usually 75 to 85% of the contents of the Vacutainers come from the injected sample. This initial dilution does not affect the $^{15}X_N$ calculations, see Table 44–3 for definition of terms, but does reduce the calculated d value by 15 to 25%. This same dilution applies to N_2O as well. A second problem with the Vacutainers is that the N_2O content of the tubes is above that of normal air. When the tubes are filled with 12 mL of normal atmospheric air, the tube air contains > 800 ppb vol/vol N_2O rather than 310 ppb normally found in air. This is not a problem when measuring total N gas on the MS, but a correction must be made when quantifying N_2O emissions from the soil. The excess N_2O is readily correctable by collecting time zero gas samples from the field plots at each sampling time and analyzing them to determine the tube N_2O content of time zero air. These two problems are reduced by washing new Vacutainers with soap and water, drying them, then reevacuating and resealing them with silicone caulking.

Recently, we began using glass serum bottles fitted with gray butyl rubber stoppers (Wheaton) as gas transport and storage containers. We evacuate the bottles and reseal them with silicone caulking. We have found no uptake or production of N_2O from the butyl rubber stoppers.

44–4 GAS DIFFUSION PROBLEMS

Although the above techniques have been used in a variety of agricultural situations, caution is advised in their use and interpretation because of the effect of soil texture and water content on gas diffusion (Letey et al., 1980). In flooded soils and probably in wet clayey soils, the movement of the gases produced in the soil to the atmosphere above is controlled by water. Since a gas diffuses about 10 000 times more slowly in water than in air, the time required for a gas to move from its production site in the soil to the atmosphere may depend upon soil water content. In the case of flooded rice, gases may be entrapped in the soil and reach the soil surface only slowly unless the gas production zones are in close proximity to plant roots (Lindau et al., 1988). Rice plants do facilitate the movement of gases from the root zone to the atmosphere (Cicerone & Shetter, 1981) and have been shown to transport N_2 and N_2O from flooded soil to the atmosphere (Mohanty & Mosier, 1990; Mosier et al., 1990). Buresh & DeDatta (1990) have also shown, in a rice field, that $^{15}N_2$ was present in the soil of ^{15}N-fertilized plots. Although not a quantitative measure, this does show that quantification of denitrification by measuring N-gas flux may not always be complete and some measure of N entrapped in the soil needs to be made. Mosier et al. (1989) used a multiple phase equilibration (McAullife, 1971) technique to estimate N gases in a rice field soil porewater and dissolved in the floodwater.

REFERENCES

Aulakh, M.S., J.W. Doran, and A.R. Mosier. 1991. Field evaluation of four methods for measuring denitrification. Soil Sci. Soc. Am. J. 55:1332–1338.

Aulakh, M.S., D.A. Rennie, and E.A. Paul. 1982. Gaseous nitrogen losses from cropped and summer-fallow soils. Can. J. Soil Sci. 62:187–195.

Balderston, W.L., B. Sherr, and W.J. Payne. 1976. Blockage by acetylene of nitrous oxide reduction in *Pseudomonal perfectomarinus*. Appl. Environ. Microbiol. 31:504–508.

Banerjee, N.K., A.R. Mosier, K.S. Uppal, and N.N. Goswami. 1990. Use of encapsulated calcium carbide to reduce denitrification losses from urea-fertilized flooded rice. *In* Proc. Int. Denitrification Workshop, Giessen, Germany. March 1989. Mitt. Dtsch. Bodenforsch. Ges. 60:245–248.

Beard, W.E., and W.D. Guenzi. 1976. Separation of soil atmospheric gases by gas chromatography with parallel columns. Soil Sci. Soc. Am. J. 40:319–321.

Bremner, J.M. 1965. Isotope-ratio analysis of nitrogen in nitrogen-15 tracer investigations. p. 1256–1286. *In* C.A. Black (ed.) Methods of soil analysis. Agron. Monogr. 9. Part 2. ASA, Madison, WI.

Bremner, J.M., and A.M. Blackmer. 1982. Composition of soil atmospheres. p. 873–895. *In* A.L. Page et al. (ed.) Methods of soil analysis. Agron. Monogr. 9. Part 2. 2nd ed. ASA and SSSA, Madison, WI.

Bremner, J.M., and R.D. Hauck. 1982. Advances in methodology for research on nitrogen transformations. p. 479–484. *In* F.J. Stevenson (ed.) Nitrogen in agricultural soils. Agron. Monogr. 22. ASA, Madison, WI.

Bremner, J.M., and K. Shaw. 1958. Denitrification in soil. I. Methods of investigation. J. Agric. Sci. 51:22–39.

Broadbent, F.E, and F.E. Clark. 1965. Denitrification. p. 344–359. *In* W.V. Bartholomew and F.E. Clark (ed.) Agron. Monogr. 10. ASA, Madison, WI.

Bronson, K.F., and A.R. Mosier. 1991. Effect of encapsulated calcium carbide on dinitrogen, nitrous oxide, methane, and carbon dioxide emissions in flooded rice. Biol. Fertil. Soils 11:116–120.

Bronson, K.F., A. R. Mosier, and S.R. Bishnoi. 1992. Nitrous oxide emissions in irrigated corn as affected by nitrification inhibitors. Soil Sci. Soc. Am. J. 56:161–165.

Buresh, R.J., E.R. Austin, and E.T. Craswell. 1982. Analytical methods in ^{15}N research. Fert. Res. 3:37–62.

Buresh, R.J., and S.K. DeDatta. 1990. Denitrification losses from puddled rice soils in the tropics. Biol. Fertil. Soils 9:1–13.

Burton, A., and E.G. Beauchamp. 1984. Field techniques using the acetylene blockage of nitrous oxide reduction to measure denitrification. Can. J. Soil Sci. 64:555–562.

Cicerone, R.J., and J.D. Shetter. 1981. Sources of atmospheric methane: Measurements in rice paddies and a discussion. J. Geophys. Res. 86:7203–7209.

Duxbury, J.M. 1986. Advantages of the acetylene method of measuring denitrification. p. 73–91. In R.D. Hauck and R.W. Weaver (ed.) Field measurement of dinitrogen fixation and denitrification. SSSA Spec. Publ. 18. SSSA, Madison, WI.

Federova, R.I., E.I. Melekhina, and N.I. Ilyuchina. 1973. Evaluation of the method of "gas metabolism" for detecting extra terrestrial life. Identification of nitrogen-fixing organisms. Izv. Akad. Nauk SSSR. Ser. Biol. 6:791.

Focht, D.D., and M. Verstraete. 1977. Biochemical ecology of nitrification and denitrification. p. 135–214. In M. Alexander (ed.) Advances in microbial ecology. Vol. 1. Plenum Press, New York.

Folorunso, O.A., and D.E. Rolston. 1984. Spatial variability of field measured denitrification gas fluxes. Soil Sci. Soc. Am. J. 48:1214–1219.

Hauck, R.D. 1982. Nitrogen-isotope-ratio analysis. p. 735–779. In A.L. Page et al. (ed.) Methods of soil analysis. Part 2. Agron. Monogr. 9. 2nd ed. ASA and SSSA, Madison, WI.

Hauck, R.D., and R.W. Weaver. 1986. Field measurement of dinitrogen fixation and denitrification. SSSA Spec. Publ. 18. SSSA, Madison, WI.

Hauck, R.D., and D.R. Bouldin. 1961. Distribution of isotopic nitrogen in nitrogen gas during denitrification. Nature (London) 191:871–872.

Hauck, R.D., and J.M. Bremner. 1976. Use of tracers for soil and fertilizer nitrogen research. Adv. Agron. 28:219–266.

Hauck, R.D., S.W. Melsted, and P.E. Yankwich. 1958. Use of N-isotope distribution in nitrogen gas in the study of denitrification. Soil Sci. 86:287–291.

Heinemeyer, O., K. Haider, and A.R. Mosier. 1988. Phytotron studies to compare nitrogen losses from corn-planted soil by the 15-N balance or direct dinitrogen and nitrous oxide measurements. Biol. Fert. Soils 6:73–77.

Hutchinson, G.L., and A.R. Mosier. 1981. Improved soil cover method for field measurement of nitrous oxide flux. Soil Sci. Soc. Am. J. 45:311–316.

Hyman, M.R., and D.J. Arp. 1987. Quantification and removal of some contaminating gases from acetylene used to study gas-utilizing enzymes and microorganisms. Appl. Environ. Microbiol. 53:298–303.

Jury, W.A., J. Letey, and T. Collins. 1982. Analysis of chamber methods used for measuring nitrous oxide production in the field. Soil Sci. Soc. Am. J. 46:250–256.

Klemedtsson, L.K., G. Hansson, and A.R. Mosier. 1990. The use of acetylene for the quantification of N_2 and N_2O production from biological processes in soil. p. 167–180. In J. Sorensen and N.P. Revsbeck (ed.) Denitrification in soil and sediment. Plenum Press, New York.

Letey, J., W.A. Jury, A. Hadas, and N. Valoras. 1980. Gas diffusion as a factor in laboratory incubation studies on denitrification. J. Environ. Qual. 9:223–227.

Lindau, C.W., R.D. DeLaune, W.H. Patrick, Jr., and P.K. Bollich. 1990. Fertilizer effects on dinitrogen, nitrous oxide, and methane emissions from lowland rice. Soil Sci. Soc. Am. J. 54:1789–1794.

Lindau, C.W., W.H. Patrick, Jr., R.D. DeLaune, K.R. Reddy, and P.K. Bollich. 1988. Entrapment of nitrogen-15 dinitrogen during soil denitrification. Soil Sci. Soc. Am. J. 52:538–540.

McAullife, C. 1971. GC determination of solutes by multiple phase equilibration. Chem. Technol. (Jan.) 0:46–50.

Mohanty, S.K., and A.R. Mosier. 1990. Nitrification-denitrification in flooded rice soils. In 14th Int. Congr. Soil Sci. 4:326–331.

Mosier, A.R. 1989. Chamber and isotope techniques. p. 175–187. *In* M.O. Andreae and D.S. Schimel (ed.) Exchange of trace gases between terrestrial ecosystems and the atmosphere. John Wiley and Sons, Chichester.

Mosier, A.R., and D.S. Schimel. 1993. Nitrification and denitrification. p. 181–208. *In* R. Knowles and T.H. Blackburn (ed.) Nitrogen isotope techniques. Academic Press, San Diego.

Mosier, A., D. Schimel, D. Valentine, K. Bronson, and W. Parton. 1991. Methane and nitrous oxide fluxes in native, fertilized and cultivated grasslands. Nature (London) 350:330–332.

Mosier, A.R., S.K. Mohanty, A. Bhadrachalam, and S.P. Chakravorti. 1990. Evolution of dinitrogen and nitrous oxide from the soil to the atmosphere through rice plants. Biol. Fertil. Soils 9:31–36.

Mosier, A.R., S.L. Chapman, and J.R. Freney. 1989. Determination of dinitrogen emission and retention in floodwater and porewater of a lowland rice field fertilized with ^{15}N-urea. Fert. Res. 19:127–136.

Mosier, A.R., W.D. Guenzi, and E.E. Schweizer. 1986a. Determination of dinitrogen and nitrous oxide from irrigated crops in north-eastern Colorado. Soil Sci. Soc. Am. J. 50:831–833.

Mosier, A.R., W.D. Guenzi, and E.E. Schweizer. 1986b. Field denitrification estimation by nitrogen-15 and acetylene inhibition techniques. Soil Sci. Soc. Am. J. 50:831–833.

Mosier, A.R., and O. Heinemeyer. 1985. Current methods used to estimate N_2O and N_2 emissions from field soils. p. 79–99. *In* H.I. Golterman (ed.) Denitrification in the nitrogen cycle. Plenum Publ. Corp., New York.

Mosier, A.R., and L.K. Mack. 1980. Gas chromatographic system for precise, rapid analysis of N_2O. Soil Sci. Soc. Am. J. 44:1121–1123.

Mosier, A.R., F.M. Melhuish, and W.S. Meyer. 1985. Direct measurement of denitrification using acetylene blockage and infrared gas analysis in a root zone lysimeter. p. 101–115. *In* W.A. Muirhead and E. Humphreys (ed.) Root zone limitations to crop production on clay soils. Aust. Soc. Soil Sci., Riverina Branch, Griffith, Australia.

Mulvaney, R.L. 1988. Evaluation of nitrogen-15 tracer techniques for direct measurement of denitrification in soil. III. Laboratory studies. Soil Sci. Soc. Am. J. 52:1327–1332.

Mulvaney, R.L. 1984. Determination of ^{15}N-labeled dinitrogen and nitrous oxide with triple-collector mass spectrometers. Soil Sci. Soc. Am. J. 48:690–692.

Mulvaney, R.L., and C.W. Boast. 1986. Equations for determination of nitrogen-15 labeled dinitrogen and nitrous oxide by mass spectrometry. Soil Sci. Soc. Am. J. 50:360–363.

Mulvaney, R.L., and L.T. Kurtz. 1982. A new method for determination of ^{15}N-labeled nitrous oxide. Soil Sci. Soc. Am. J. 46:1178–1184.

Mulvaney, R.L., and R.M. vanden Heuvel. 1988. Evaluation of nitrogen-15 tracer techniques for direct measurement of denitrification in soil: IV. Field studies. Soil Sci. Soc. Am. J. 52:1332–1337.

Parkin, T.B. 1987. Soil microsites as a source of denitrification variability. Soil Sci. Soc. Am. Proc. 26:238–242.

Parkin, T.B., S.T. Chester, and J.A. Robinson. 1990. Calculating confidence intervals for the mean of lognormally distributed variable. Soil Sci. Soc. Am. J. 54:321–326.

Parkin, T.B., J.J. Meisinger, S.T. Chester, J.L. Starr, and J.A. Robinson. 1988. Evaluation of statistical estimation methods for lognormally distributed variables. Soil Sci. Soc. Am. J. 52:323–329.

Parkin, T.B., H.F. Kaspar, A.J. Sexstone, and J.M. Tiedje. 1984. A gas-flow soil core method to measure field denitrification rates. Soil Biol. Biochem. 16:323–330.

Robertson, G.P., and J.M. Tiedje. 1987. Nitrous oxide sources in aerobic soils. Nitrification, denitrification and other biological processes. Soil Biol. Biochem. 19:187–193.

Rolston, D.E., A.M. Sharpley, D.W. Toy, and F.E. Broadbent. 1982. Field measurement of denitrification: III. Rates during irrigation cycles. Soil Sci. Soc. Am. J. 46:289–296.

Rolston, D.E., D.L. Hoffman, and D.W. Toy. 1978. Field measurement of denitrification: I. Flux of N_2 and N_2O. Soil Sci. Soc. Am. J. 42:863–869.

Rolston, D.E., M. Fried, and D.A. Goldhamer. 1976. Denitrification measured directly from nitrogen and nitrous oxide gas fluxes. Soil Sci. Soc. Am. J. 40:256–266.

Ryden, J.C., J.H. Skinner, and D.J. Nixon. 1987. Soil core incubation system for the field measurement of denitrification using acetylene-inhibition. Soil Biol. Biochem. 19:753–757.

Ryden, J.C., and K.P. Dawson. 1982. Evaluation of the acetylene-inhibition technique for the measurement of denitrification in grassland soils. J. Sci. Food Agric. 33:1197–1206.

Ryden, J.C., L.J. Lund, and D.D. Focht. 1978. Direct in-field measurement of nitrous oxide flux from soils. Soil Sci. Soc. Am. J. 42:731–737.

Ryden, J.C., L.J. Lund, J. Letey, and D.D. Focht. 1979. Direct measurement of denitrification loss from soils. II. Development and application of field methods. Soil Sci. Soc. Am. J. 43:110–118.

Schimel, J.P., and M.K. Firestone. 1989. Nitrogen incorporation and flow through a coniferous forest soil profile. Soil Sci. Soc. Am. J. 53:779–784.

Seech, A.G., and E.G. Beauchamp. 1988. Denitrification in soil aggregates of different sizes. Soil Sci. Soc. Am. J. 52:1616–1621.

Siegel, R.S., R.D. Hauck, and L.T. Kurtz. 1982. Determination of $^{30}N_2$ and application to measurement of N_2 evolution during denitrification. Soil Sci. Soc. Am. J. 46:68–74.

Svensson, B.H., L. Klemedtsson, S. Simkins, K. Paustian, and T. Rosswall. 1991. Soil denitrification in three cropping systems characterized by differences in nitrogen and carbon supply. I. Rate-distribution frequencies, comparison between systems and seasonal N-losses. Plant Soil 138: 257–271.

Svensson, B.H., L. Klemedtsson, and T. Rosswall. 1985. Preliminary field denitrification studies of nitrate-fertilized and nitrogen fixing crops. p. 157–169. In H.L. Golterman (ed.) Denitrification and the nitrogen cycle. NATO Conf. Ser. I. Ecology Vol. 9. Plenium Press, London.

Tiedje, J.M. 1988. Ecology of denitrification and dissimilatory nitrate reduction to ammonium. p. 179–243. In A.J.B. Zehnder (ed.) Biology of anaerobic micro-organisms. John Wiley and Sons, New York.

Tiedje, J.M. 1982. Denitrification. p. 1011–1024. In A.L. Page et al. (ed.) Methods of soil analysis. Agron. Monogr. 9. Part 2. 2nd ed. ASA and SSSA, Madison, WI.

Tiedje, J.M., S. Simkins, and P.M. Groffman. 1989. Perspectives on measurement of denitrification in the field including recommended protocols for acetylene based methods. p. 217–240. In M. Clarholm and L. Bergstrom (ed.) Ecology of arable land. Kluwer Academic Publ. Dordrecht, Holland.

Yoshinari, T., and R. Knowles. 1976. Acetylene inhibition and nitrous oxide reduction by denitrifying bacteria. Biochem. Biophys. Res. Commun. 69:705–710.

Chapter 45

Sulfur Oxidation and Reduction in Soils

M. A. TABATABAI, *Iowa State University, Ames, Iowa*

Soil analysis from various parts of the world indicates that S occurs in soils mainly in organic combinations, and that most cultivated soils do not contain sufficient organic matter to supply the crops need of S (Tabatabai & Bremner, 1972a; Neptune et al., 1975; Tabatabai, 1984). In addition, the release rate of inorganic sulfate from soil organic matter is too slow to meet the crops demand of this element (Tabatabai & Bremner, 1972b; Tabatabai & Al-Khafaji, 1980). Other factors that contribute to the S need are higher yields, more intensive land use by double cropping, use of high analysis fertilizer, increasing use of irrigation practices, declining use of S as a fungicide and insecticide, and decreasing SO_2 emissions into the atmosphere because of recent environmental regulations. To meet the S requirements of crop plants, the alternative is to amend the soils with S-containing fertilizer (Walker, 1964; Hagstrom, 1984).

A wide variety of S-containing fertilizers in the form of solid and liquid are available. Among these, elemental S (S^0) offers the best solution to the problem of cost and grade because of its purity (about 100% S). However, problems of dustiness, unpleasantness, and fire hazard have limited its use as a fertilizer (Beaton & Fox, 1971). To avoid the disadvantages associated with the finely divided S^0, a product containing 90% S^0 and 10% of bentonite clay and certain proprietary additives has been developed (Sulphur Institute, 1977). Before it can be used by crops, however, S^0 has to be oxidized to sulfate. Elemental S is oxidized in soils by chemical and biochemical processes, and several factors affect these processes. Several of those factors (e.g., soil moisture and temperature, particle size, and soil texture) have been studied by several workers (for review of literature, see Weir, 1975; Konopka et al., 1984, Wainwright, 1984).

Unlike studies on forms of S and the factors affecting oxidation of S^0 in soils, not much information is available on the reduction of sulfate in soils (Konopka et al., 1984). Therefore, most of the methods available for estimation of sulfate reduction are those used in studies of transformations of S in sediments (Jørgensen, 1978a,b; Fossing & Jørgensen, 1989).

Although these methods are not widely used for estimation of sulfate reduction in soils, they will be described in this chapter.

45–1 SULFUR OXIDATION

Elemental S is oxidized in soils by biotic processes and to a much lesser degree by abiotic processes. As Wainwright (1984) indicated, a wide spectrum of microorganisms is capable of oxidizing S^0, including members of the genus *Thiobacillus,* several heterotrophs, the photosynthetic S bacteria, and the colorless, filamentous S bacteria. Of those, only the thiobacilli and heterotrophs have been shown to play an important role in S oxidation in most agricultural soils.

Early work by Guittonneau and Keilling (1932) indicated that polythionates (oxidizable S) are produced in S^0-amended soils, but because of analytical problems, no specific intermediate S compounds could be determined. In review of the literature on oxidation of S^0 by the genus *Thiobacillus,* Vishniac and Santer (1957) suggested that $S_2O_3^{2-}$, $S_4O_6^{2-}$, and SO_3^{2-} are formed as intermediate products. The first two of these compounds have been isolated from the reaction involving bacteria (London & Rittenberg, 1964; Starkey, 1956). Using ^{35}S-labeled thiosulfate, Trudinger (1959) showed that $S_4O_6^{2-}$ is the first intermediate in oxidation of thiosulfate to sulfate by a *Thiobacillus* sp. Gleen and Quastel (1953) have reported similar results for thiosulfate oxidation on perfusion through soils. Formation of such compounds during oxidation of S^0 in soils deserves investigation because work by Audus and Quastel (1947) showed that $S_2O_3^{2-}$ inhibits germination and subsequent root growth of several plants and that there is a considerable degree of selectivity of this toxic action of $S_2O_3^{2-}$ on growth of plant roots. For example, at a concentration of 250 mg/L S as $S_2O_3^{2-}$ (on solution basis) root growth of garden pea (*Pisum sativum* L.) was inhibited by 50%.

Studies by Nor and Tabatabai (1977) showed that the amount of S_2O_3-, S_2O_6-, and SO_4-S produced during incubation (30 °C) of S^0-amended soils under aerobic conditions varied with time and soil used. They showed that $S_2O_3^{2-}$ was produced within the first few days of incubation and that $S_4O_6^{2-}$ accumulated in some soils. The rate of S oxidation increased with increasing incubation temperatures (5, 15, and 30 °C) and with increasing the rate of S application (50, 100, and 200 µg of S/g soil). For 100 µg of S/g soil, the rates of oxidation of S^0 in 10 Iowa surface soils ranged from 39 to 75 µg of S/g soil after incubation at 30 °C for 70 d; rates were more rapid in alkaline soils than in acid soils. There was little change in pH of soils even when the S application rate was increased from 50 to 200 µg/g soil. The rate of S oxidation was lower in air-dried soils than in field-moist soils (Nor & Tabatabai, 1977).

The rate of S^0 oxidation is strongly affected by soil moisture, temperature, and pH; particle size; and soil type (Janzen & Bettany, 1987a,b; Nor & Tabatabai, 1977). Therefore, it is difficult to extrapolate the S^0 oxidation rate obtained from laboratory experiments to soils under field conditions. Work by Janzen and Bettany, (1987a) showed that the effect of S^0 particle size on S^0 oxidation rate can be minimized by using particle size in the range of 0.106 to 0.150 mm. The effect of soil moisture on S^0 oxidation varies among soil types with the water potential optima ranging from −0.08 to −0.27 MPa, depending on the soil type (Nevell & Wainwright, 1987; Janzen & Bettany, 1987b; Deng & Dick, 1990). The optimum temperature for S^0 oxidation under aerobic conditions has been reported to be at 30 °C (Waksman & Joffe, 1922; Li & Caldwell, 1966; Nor & Tabatabai, 1977). Deng and Dick (1990) reported optimum temperature for S^0 oxidation of 25 to 30 °C.

In most studies of S^0 oxidation in soil, the rate is usually determined by measuring the amount of SO_4^{2-} produced during a specific period after application of S^0 to soil. Therefore, the results are expressed as the proportion of S^0 oxidized during a given time. Janzen and Bettany (1987a) have identified two limitations related to the manner in which the oxidation rate is expressed. They argued that (i) oxidation rate are specific to the particle size of S^0 used. Therefore, rates determined by using one particle size are not applicable to particles of a different size. Furthermore, S^0 oxidation rates reported by different investigators using various sizes of S^0 particles cannot be compared, and (ii) current methods of defining S^0 oxidation rates are biased by confounding effects of diminishing S^0 during oxidation. Because S^0 oxidation is exclusively a function of surface area of the particle used, the amount of SO_4^{2-} produced during incubation from S^0 present, not of the total mass. In pure culture studies with thiobacilli, Laishley et al. (1983) reported that oxidative activity did not penetrate into the core of S^0 particles, perhaps because of steric hindrance. This has been demonstrated in several studies by demonstrating linear relationships between the amount of SO_4^{2-} produced and the total surface area of the applied S^0, irrespective of the amount applied (Fox et al., 1964; Janzen et al., 1982; Koehler & Roberts, 1983; Laishley et al., 1983). One problem encountered in expressing the S^0 oxidation rate as a function of the initial surface area is, however, the change in the surface area during incubation. As Janzen and Bettany (1987a) stated, the expression of S^0 oxidation rates on the basis of instantaneous rather than initial surface area allows independence of both time and particle size. The decline in surface area over time can be mathematically determined if it is assumed that the S^0 particles are relatively uniform spheres. Based on these assumptions, Janzen and Bettany (1987a) derived the following equations to calculate the proportion of S^0 oxidized and the rate constant:

$$m / m_0 = 1 - [1 - 2\, kt / gD_0]^3 \qquad [1]$$

where m / m_0 is the proportion of S^0 oxidized as a function of a rate constant (k), time (t), density (g), and initial particle diameter (D_0). Rearrangement of Eq. [1] gives:

$$k = \{1 - [1 - m/m_0]^{\frac{1}{3}}\} \cdot gD_0 / 2t \qquad [2]$$

The rate constant calculated from Eq. [2] has the desired units (g S cm^{-2} d^{-1}) and is independent of both particle diameter and time.

45–1.1 Principles

Estimation of S^0 oxidation rate in soils involves determination of SO$_4^{2-}$ produced when soil is incubated with S^0 under specific conditions. The variables that should be considered include particle size of S^0, and time and temperature of incubation. These variables must be selected so that sufficient SO$_4^{2-}$ is produced for accurate determination, SO$_4^{2-}$ produced from organic S mineralization is minimized, acidification of the soil sample is prevented, and S^0 is applied accurately (Janzen & Bettany, 1987a).

Any particle size of S^0 can be used for this purpose. In general, S^0 oxidation rate increases with decreasing particle size. If very fine (e.g., < 150 mesh, 106 μm) S^0 is used, it can be diluted with the same size glass beads to facilitate weighing (Nor & Tabatabai, 1977). More recent work showed that a relatively coarse (0.199 mm) S^0 can be used successfully for estimation of S^0 oxidation rate in soils (Janzen & Bettany, 1987a; Deng & Dick, 1990). According to Janzen and Bettany (1987a), this relatively coarse material has several advantages over other materials they considered: (i) it could be readily produced within narrow limits, (ii) its diameter could be accurately confirmed by microscopic analysis, and (iii) because of its lower specific surface area, coarse S^0 could theoretically be applied at high levels without adversely affecting oxidation rate through secondary effects, such as acidification. As a result, application of required S^0 to a small amount of soil sample is accurate and convenient.

45–1.2 Methods
(Nor & Tabatabai, 1977; Janzen & Bettany, 1987a; Deng & Dick, 1990)

45–1.2.1 Special Apparatus

1. French square bottles (8 oz, 250 mL).
2. Stoppers (no. 7).

45–1.2.2 Reagents

1. Elemental S: Sublimed S^0 (J.T. Baker Chemical Co., Phillipsburg, NJ).
2. Elemental S^0 powder, 0.119 mm: Grind prilled S^0 and collect the particles that passed through a 0.150-mm sieve, but were retained in a 0.106-mm sieve. Wet-sieve in ethanol to ensure complete removal of fine particles adhering to the desired particles. Rinse the ethanol-washed S^0 several times with deionized water and dry it overnight at 65 °C.

3. Glass beads: can be obtained from Cataphote, Inc. (P.O. Box 2369, Jackson, MS).
4. Calcium phosphate monohydrate solution [Ca(H$_2$PO$_4$)$_2$·H$_2$O], 500 mg P/L. Dissolve 2.02 g of Ca(H$_2$PO$_4$)$_2$·H$_2$O in about 700 mL of deionized water, and make to volume of 1 L with deionized water.

45–1.2.3 Procedure

Mix a 20-g sample of soil (< 2 mm) with S^0 in a 250-mL (8 oz) French square bottle to give the S^0 concentration desired. For convenience, 20 mg S^0, 0.199 mm or fine S^0 mixed with a similar size glass beads can be used. If air-dried soil is used, add 5 mL of deionized water to bring the moisture content to 30 to 50% of water-holding capacity. Stopper the bottle and incubate it for the time and temperature desired. It is recommended to incubate for 6 d at room temperature (23 °C) or at 30 °C. Remove the stopper and aerate the bottle every 2 d. After incubation, add 100 mL of 500 mg/L of P as Ca(HPO$_4$)$_2$, stopper the bottle and shake it on an end-to-end shaker for 1 h. Filter the soil-solution mixture through a Whatman No. 42 filter paper under suction. For complete removal of any particulate, filter the soil extract through a GA-8 membrane filter (Gelman Instrument Co., Ann Arbor, MI). Analyze this filtrate for S$_2$O$_3^{2-}$ and S$_4$O$_6^{2-}$ by the colorimetric methods described by Nor and Tabatabai (1975, 1976) or total inorganic S (S$_2$O$_3^{2-}$, S$_4$O$_6^{2-}$, and SO$_4^{2-}$) by the methylene blue method (Tabatabai, 1982), or for SO$_4^{2-}$ by ion chromatographic method (Dick & Tabatabai, 1979; Tabatabai, 1992). The amount of SO$_4^{2-}$ produced can be calculated from the amount of S$_2$O$_3^{2-}$ + S$_4$O$_6^{2-}$ + SO$_4^{2-}$ obtained by the methylene blue method by subtracting the values of S$_2$O$_3^{2-}$ and S$_4$O$_6^{2-}$ obtained by the colorimetric methods.

45–2 SULFATE REDUCTION

The inorganic S fraction in soils may occur as sulfate and compounds of lower oxidation state such as sulfide, polysulfide, sulfite, thiosulfate, and S^0. In well-drained, well-aerated soils, most of the inorganic S normally occurs as sulfate, and the amounts of reduced S compounds are generally < 1% (Freney, 1961). Under anaerobic conditions, particularly in tidal swamps and poorly drained or waterlogged soils, the main form of inorganic S in soils is sulfide and often S^0 (Brümmer et al., 1971a,b; Harmsen, 1954; Hart, 1959). Reduced inorganic S compounds produced in or added to soils can be oxidized under aerobic and anaerobic conditions. The main reduced S compounds added to soils are sulfide (S^{2-}), such as metal sulfides, elemental sulfur (S^0), thiosulfate (S$_2$O$_3^{2-}$), tetrathionate (S$_2$O$_4^{2-}$), and sulfite (SO$_3^{2-}$). Oxidation of these compounds in soils can be chemical (abiotic) or microbiological (biotic) or both. The chemical oxidation of aqueous sulfide by O$_2$ is relatively slow, with reported half-lives ranging from about one to several hours (Zehnder & Zinder, 1980). The studies by

Krebs (1929), however, showed that the reaction is catalyzed by traces of transition metal ions. Also, various mixtures of S^0, polysulfides, $S_2O_3^{2-}$, $S_4O_6^{2-}$, SO_3^{2-}, and SO_4^{2-} have been reported as products at near neutral pH values, indicating that a complex set of reactions is involved (Zehnder & Zinder, 1980).

Accurate determination of the reduced forms of inorganic S in soils is difficult, partly because of the ease with which they can be oxidized on exposure to air, but mainly because of the limitations of current analytical methods. No procedure has been entirely satisfactory for determination of sulfide in soils. Gilboa-Garber (1971) proposed a method for direct spectrophotometric determination of inorganic sulfide in biological materials. This method involves colorimetric determination of sulfide as methylene blue after extraction with Zn acetate. A modification of this method was successfully used by Howarth et al. (1983) for determination of sulfide in porewaters of core samples of a salt marsh. Smittenberg et al., (1951) proposed a method for determination of sulfide, sulfite, thiosulfate, polysulfide, S^0, and organically bound S in soils. It involves digestion of a soil sample with HCl for determination of the monosulfidic S compounds and digestion with Sn and HCl for determination of the total oxidizable S. In both methods the H_2S released is determined colorimetrically as methylene blue. None of these methods, however, gives accurate results; digestion of soil with HCl does not release the S from acid-insoluble metal sulfides, causing underestimation of the sulfide present, and digestion with Sn and HCl releases S from organic S, causing overestimation of the total oxidizable S (Melville et al., 1971). The nature of this S fraction, however, has not been identified. Much of the reducible S fraction in soils is extractable with 0.5 M NaOH and is distributed between the fulvic acid and humic acid fractions. Therefore, digestion of soil with Sn and HCl may prove adequate where the amounts of oxidizable S are relatively large, but this method is unsatisfactory for use with soils containing small amounts of reduced inorganic S compounds, especially with surface soils containing appreciable amounts of organic S.

Studies of pyrite formation and the measurement of sulfate reduction in salt marsh sediments by Howarth and Merkel (1984) showed that the reduction with Cr(II) of pyrite and S^0 to H_2S is more specific and sensitive than oxidation of these forms of S to sulfate. The Cr(II) reduction method has been used in studies of inorganic S compounds in modern sediments and shales and of formation of ^{35}S-labeled S^0 and pyrite in coastal marine sediments during short-term $^{35}SO_4^{2-}$ reduction measurements (Howarth & Jørgensen, 1984; Canfield et al., 1986).

The forms and concentration of S compounds in marine sediments are markedly different from those in soils. Zhabina and Volkov (1978) have developed a scheme for determination of various forms of S compounds in sea sediments. It involves systematic fractionation of S^{2-}, SO_4^{2-}, organic S, and FeS_2 in marine sediments. Wieder et al. (1985) studied the specificity and efficiency of the procedures for fractionation of total S into organic and organic constituents by analyzing a series of known standards. They reported that acid volatilization was specific for FeS. Chromium reduction recovered $> 90\%$ of the S from FeS, S^0, and FeS_2. Acetone extraction followed by Cr reduction of the filtrate was specific for S^0. Hydriodic acid reduction recovered $> 90\%$ of the S from FeS, SO_4^{2-}, and p-nitrophenyl sulfate. The Zn-HCl procedure partially recovered S from SO_4^{2-}, S^0, and FeS_2. None of the above procedures reduced L-methionine. Analysis of both moist and oven-dried peat by Wieder et al. (1985) showed that oven-drying of peat samples increased the ester-sulfate S and SO_4^{2-} fractions and decreased the estimated C-bonded S, which was calculated from the difference between the total S and HI-reducible S.

A polarographic method for determination of the reduced S species in marine porewaters has been proposed by Luther et al. (1985). They have demonstrated that, with polarographic techniques, it is possible to measure thiosulfate, sulfide, bisulfide, and polysulfide ions with a mercury electrode. They considered the polysulphide ions, S_x^{2-}, to be composed of one S atom and in the -2 oxidation state, $S(2-)$, and the remaining $(x-1)$ S atoms in the zero-valent oxidation state $S(0)$. The number of S atoms in each is measurable by this technique. Tetrathionate and other polythionates can be measured by this technique, but Luther et al. (1985) could not detect these reduced S species in the porewaters that they tested. By using this differential pulse polarographic technique, Luther et al. showed that salt marsh and subtidal porewater profiles contain significant concentrations of thiosulfate, bisulfide, and polysulfide. This technique has been successfully used in demonstrating the seasonal cycling of S and Fe in porewaters of a Delaware salt marsh (Luther & Church, 1988).

The methods developed for estimating sulfate reduction are mainly used in studies of coastal marine sediments (Fossing & Jørgensen, 1989; Jørgensen, 1978a). Even though these methods are not commonly used in studies of sulfate reduction in soils, they will be described here for potential use in studies involving sulfate reduction in flooded soils.

45–2.1 Principles

Estimation of sulfate reduction involves addition of SO_4^{2-} into a homogenized or undisturbed sediment core or soil sample and determination of the reduced inorganic S produced by reduction with Cr^{2+} in an acid solution and determination of the H_2S released as methylene blue and by counting the radioactivity of the H_2S released when $^{35}SO_4^{2-}$ is used. This Cr-reducible S (single-step method) comprises H_2S, S^0, FeS, and FeS_2. Alternately, sulfate reduction can be measured by a two-step method in which the acid volatile sulfide (H_2S + FeS) and Cr-reducible S (S^0 + FeS_2) are sequentially distilled from the sample. The fraction of $^{35}SO_4^{2-}$ reduced during incubation is calculated from the sum of ^{35}S in the acid volatile sulfide and Cr-reducible S. The single-step distillation is simpler and faster than the consecutive distillation of acid volatile S and Cr-reducible S. Comparison of the two methods by Fossing and Jørgensen (1989) showed that the single-step method resulted in higher (4–50%) sulfate reduction rates than those obtained from the sum of ^{35}S in acid volatile S and Cr-reducible S. The difference was largest when the sediment had been dried after acid volatile S but before Cr-reducible S distillation. Relative to the ^{35}S in the H_2S released by the acid volatile S distillation alone, the ^{35}S of the H_2S released from the total reduced inorganic S (Cr-reducible S) distillation resulted in 8 to 87% higher reduction rates. Studies of sulfate reduction rates in a range of marine sediments and salt marshes have shown that 10 to 15% of the reduced radiolable $^{35}SO_4^{2-}$ was recovered from Cr-reducible S (for review, see Fossing & Jørgensen, 1989), but the percentage varies widely, depending on the type of sediment and on the S chemistry. According to Fossing and Jørgensen (1989), the mechanism(s) involved in ^{35}S incorporation into S^0 and FeS_2 during short-term incubation is not clear, but ^{35}S labeled Cr-reducible S has to be taken into account in measurements of sulfate reduction rates. Because the single-step method is more accurate, convenient, and more rapid than the two-step method, the former method will be described here.

45–2.2 Methods (Fossing & Jørgensen, 1989)

45–2.2.1 Special Apparatus

1. Modified Johnson-Nishita apparatus (Tabatabai, 1982), the apparatus described by Canfield et al. (1986), or that described by Zhabina and Volkov (1978).

45–2.2.2 Reagents

1. $Zn(OAc)_2$, 5% in 1% acetic acid.
2. Nitrogen gas and ferric ammonium sulfate and p-aminodimethylanaline sulfate solutions: Prepare as described by Tabatabai (1982).
3. Concentrated HCl.
4. Ethanol (95%).
5. Chromic chloride hexahydrate ($CrCl_3\cdot6H_2O$): Prepare the highly reactive Cr^{2+} solution from the more stable Cr^{3+} by percolating 1 M $CrCl_3\cdot6H_2O$ in 0.5 M HCl through a Jones reactor with amalgamated Zn granules:

$$2\ Cr^{3+} + Zn \rightarrow 2\ Cr^{2+} + Zn^{2+}$$

An efficient reduction is varified by a color change from dark green [Cr(III)] to bright blue [Cr(II)]. For construction of the Jones reactor, see Kolthoff and Sandell, 1963, p. 569.

The Jones reductor can be constructed from a glass-column (40 cm long, 1.5 cm i.d.) with an integral sinter at the bottom and stopcocks in both ends (Fossing & Jørgensen, 1989). Wash the granular Zn (0.3–1.5 mm grain size) three times with 1 M HCl and twice with deionized water.

Amalgamate the Zn for a few minutes with a saturated solution of $HgCl_2$ (ca. 0.25 M), transfer it to the glass column, and wash it with three column volumes of 0.5 M HCl. Remove the reduced Cr solution by suction into a large polyethylene syringe in which it can be stored under reduced conditions (bright blue) for several weeks. Otherwise, the Cr(II) solution should be freshly prepared every 2 to 3 d and stored in a ground-glass stoppered bottle (Canfield et al., 1986).

An alternate and faster method of producing large volumes of Cr(II) solution is recommended by Fossing and Jørgensen (1989). In this method, simply fill a glass bottle with 1 M HCl-rinsed "mossy zinc" (Aldrich Chemicals, Milwaukee, WI) and then fill the bottle with the Cr(III) solution (1 M $CrCl_3 \cdot 6H_2O$ in 0.5 M HCl) under continuous flow of N_2. The mossy zinc does not need to be amalgamated. Chromium(III) is reduced to Cr(II) within 10 min and is kept under N_2 until it is drawn into syringes through an outlet at the bottom of the bottle. After use, the mossy Zn can be regenerated by washing with 1 M HCl.

45–2.2.3 Procedure

Sulfate reduction can be measured in undisturbed sediment core (3 cm in diameter) by the core injection technique described by Jørgensen and Fenchel (1974) and Jørgensen (1978a). In this method, a volume of 2 µL carrier-free $^{35}SO_4^{2-}$ (70 kBq) is injected at a specific depth into replicate cores from each site or station. The sediment core is incubated for 18 to 24 h at room temperature (23 °C), then it is cut into segments and transferred to 20 mL of 20% (wt/vol) of Zn acetate [$Zn(OAc)_2$] and frozen to terminate the reaction and fix the sulfides. The reduced S is then distilled as H_2S from the sediment into $Zn(OAc)_2$ traps as described below and the radioactivity of SO_4^{2-} and of the precipitated ZnS is determined. Sulfate concentration in porewater of sediments or soil samples can be determined turbidimetrically in replicated parallel core samples as described by Tabatabai (1974).

Segments from each depth interval are pooled from the replicated cores and homogenized. The homogenized sediment is centrifuged and $^{35}SO_4^{2-}$ radioactivity is measured in a subsample of the supernatant. The sediment pellet is washed twice with a salt solution (e.g., 0.1 M NaCl) to remove $^{35}SO_4^{2-}$. The washed sediment is homogenized and 1 to 2 g is transferred into a distillation flask (modified Johnson-Nishita apparatus, or the apparatus described by Canfield et al., 1986; or that described by Zhabina and Volkov, 1978) and mixed with 5 mL of distilled water and 5 mL of methanol. The distillation flask is degassed for 20 min with N_2, 16 mL of 1 M $CrCl_2$ in 0.5 M HCl and 8 mL of 12 M HCl are added, and the sediment slurry is gently boiled for 40 min. During this distillation, the total reduced inorganic S is dissolved and distilled into $Zn(OAc)_2$ sequential traps containing 10 mL 5% $Zn(OAc)_2$, buffered with 0.1% acetate and with a drop of antifoam.

More than 98% of the distilled H_2S is recovered as ZnS in the first trap. After distillation, the two traps are pooled, 5 mL is subsampled, mixed with 5 mL of scintillation fluid (Dynagel, Baker Chemical), and ^{35}S is counted. The concentration of the Cr-reducible S is determined spectrophotometrically at 670 nm in an aliquot of the trapping solution after adding 2 mL of ferric ammonium sulfate solution, 10 mL of p-aminodimethylanaline solution, and adjusting the volume to 100 mL with deionized water.

The sulfate reduction rate is calculated according to the following equation:

$$SRR = \frac{(SO_4^{2-})\, a \times 24 \times 1.06}{(A + a)\, h} \text{ nmol } SO_4^{2-} \text{ cm}^{-3}\, d^{-1}$$

where a is the total radioactivity of ZnS, A is the total radioactivity of SO_4^{2-} after incubation, h is the incubation time in hours, (SO_4^{2-}) is sulfate concentration in nmol per cm^3 sediment, and 1.06 is a correction factor for the expected isotope fractionation (Jørgensen & Fenchel, 1974; Fossing & Jørgensen, 1989).

REFERENCES

Audus, L.J., and J.H. Quastel. 1947. Selective toxic action of thiosulfate on plants. Nature (London) 60:264–265.

Beaton, J.D., and R.L. Fox. 1971. Production, marketing, and use of sulfur products. p. 335–379. In R.A. Olson (ed.) Fertilizer technology and use. SSSA, Madison, WI.

Brümmer, G., H.S. Grunwaldt, and D. Schroeder. 1971a. Contributions to the genesis and classification of Marsh soils. II. On the sulphur metabolism of muds and salt marshes. Z. Pflanzenernaehr. Dueng. Bodenkd. 128:208–220.

Brümmer, G., H.S. Grunwaldt, and D. Schroeder. 1971b. Contributions to the genesis and classification of Marsh soils: III. Contents, oxidation status, and mechanisms of bounding of sulphur in polder soils. Pflanzenernaehr. Dueng. Bodenkd. 129:92–108.

Canfield, D.E., R. Raiswell, J.T. Westrich, C.M. Reaves, and R.A. Berner. 1986. The use of chromium reduction in the analysis of reduced inorganic sulfur in sediments and shales. Chem. Geol. 54:149–155.

Deng, S., and R.P. Dick. 1990. Sulfur oxidation and rhodanese activity in soils. Soil Sci. 150:552–560.

Dick, W.A., and M.A. Tabatabai. 1979. Ion chromatographic determination of sulfate and nitrate in soils. Soil Sci. Soc. Am. J. 43:899–904.

Fossing, H., and B.B. Jørgensen. 1989. Measurement of bacterial sulfate reduction in sediments: Evaluation of a single-step chromium reduction method. Biogeochemistry 8:205–222.

Fox, R.L., H.M. Atesalp, D.H. Kampbell, and H.F. Rhoades. 1964. Factors influencing the availability of sulfur fertilizers to alfalfa and corn. Soil Sci. Soc. Am. Proc. 28:406–408.

Freney, J.R. 1961. Some observations on the nature of organic sulphur compounds in soils. Aust. J. Agric. Res. 21:424–432.

Gilboa-Garber, N. 1971. Direct spectrophotometric determination of inorganic sulfide in biological materials and in other complex mixtures. Anal. Biochem. 43:129–133.

Gleen, H., and J.H. Quastel. 1953. Sulfur metabolism in soils. Appl. Microbiol. 1:70–77.

Guittonneau, G., and J. Keilling. 1932. L' evolution et al solubilisation du soufre elementaire dans la terra arable. Ann. Agron. 2:690–725.

Hagstrom, G.R. 1984. Fertilizer sources of sulfur and their use. p. 567–581. In M.A. Tabatabai (ed.) Sulfur in agriculture. ASA, Madison, WI.

Harmsen, G.W. 1954. Observations on the formation and oxidation of pyrite in the soil. Plant Soil 5:324–348.

Hart, M.G.R. 1959. Sulphur oxidation in tidal mangrove soils of Sierra Leone. Plant Soil 11:215–236.

Howarth, R.W., A. Giblin, J. Gate, B.J. Peterson, and G.W. Luther III. 1983. Reduced sulfur compounds in the pore waters of a New England salt marsh. Environ. Biogeochem. 35:135–152.

Howarth, R.W., and B.B. Jørgensen. 1985. Formation of ^{35}S-labeled elemental sulfur and pyrite in coastal marine sediments (Limfjorden and Kysing Fjord, Denmark), during short-term $^{35}SO_4^{2-}$ reduction measurements. Geochim. Cosmochim. Acta 48:1807–1818.

Howarth, R.W., and S. Merkel. 1984. Pyrite formation and the measurement of sulfate reduction in salt marsh sediments. Limnol. Oceonogr. 9:598–608.

Janzen, H.H., and J.R. Bettany. 1987a. Measurement of sulfur oxidation in soils. Soil Sci. 143:444–452.

Janzen, H.H., and J.R. Bettany. 1987b. The effect of temperature and water potential on sulfur oxidation in soils. Soil Sci. 144:81–89.

Janzen, H.H., J.R. Bettany, and J.W.B. Stewart. 1982. Sulfur oxidation and fertilizer sources. p. 229–240. In Proc. Alberta Soil Science Workshop, Edmonton, Alta.

Jørgensen, B.B. 1978a. A comparison of methods for the quantification of bacterial sulfate reduction in coastal marine sediments: I. Measurement with radiotracer techniques. Geomicrobiol. J. 1:11–27.

Jørgensen, B.B. 1978b. A comparison of methods for the quantification of bacterial sulfate reduction in coastal marine sediments: II. Estimation from chemical and bacteriological field data. Geomicrobiol. J. 1:49–64.

Jørgensen, B.B., and T. Fenchel. 1974. The sulfur cycle of a marine model system. Mar. Biol. 24:189–201.

Koehler, F.E., and S. Roberts. 1983. An evaluation of different forms of sulfur fertilizers. p. 833–842. In A.I. More (ed.) Proc. Int. Sulphur '82 Conf., Vol. 2, London. Nov. 1982. The British Sulphur Corporation Limited.

Kolthoff, I.M., and E.B. Sandell. 1963. Textbook of quantitative inorganic analysis. 3rd ed. Macmillan, New York.

Konopka, A.E., R.H. Miller, and L.E. Sommers. 1984 Microbiology of the sulfur cycle. p. 23–55. In M.A. Tabatabai (ed.) Sulfur in agriculture. ASA, Madison, WI.

Krebs, H.A. 1929. Uber der Wirkung der schwermetalle auf die Autoxydation der Alkalisulfide und des Schwefewasserstoffs. Biochim. Z. 204:344–346.

Laishley, E.J., R.D. Bryant, B.W. Kobryn, and J.B. Hyne. 1983. The effect of particle size and molecular composition of elemental sulfur on ease of microbiological oxidation. Alberta Sulphur Res. Ltd. Q. Bull. 20:33–50.

Li, P., and A.L. Caldwell. 1966. The oxidation of elemental sulfur in soil. Soil Sci. Soc. Am. Proc. 30:370–372.

London, J., and S.C. Rittenberg. 1964. Path of sulfur in sulfide and thiosulfate oxidation by thiobacilli. Proc. Natl. Acad. Sci. USA 52:1183–1190.

Luther, III, G.W., and T.M. Church. 1988. Seasonal cycling of sulfur and iron in porewaters of a Delaware salt marsh. Mar. Chem. 23:295–309.

Luther, III, G.W., E.A. Giblin, and R. Varsolona. 1985. Polarographic analysis of sulfur species in marine porewaters. Limnol. Oceanogr. 30:727–736.

Melville, G.E., J.R. Freney, and C.H. Williams. 1971. Reduction of organic sulfur compounds in soil with tin and hydrochloric acid. Soil Sci. 112:245–248.

Neptune, A.M.L., M.A. Tabatabai, and J.J. Hanway. 1975. Sulfur fractions and carbon-nitrogen-phosphorus-sulfur relationships in some Brazilian and Iowa soils. Soil Sci. Soc. Am. Proc. 39:51–55.

Nevell, W., and M. Wainwright. 1987. Influence of soil moisture on sulphur oxidation in brown earth soils exposed to atmospheric pollution. Biol. Fert. Soils 5:209–214.

Nor, Y.M., and M.A. Tabatabai. 1975. Colorimetric determination of microgram quantities of thiosulfate and tetrathionate. Anal. Lett. 8:537–547.

Nor Y.M., and M.A. Tabatabai. 1976. Extraction and colorimetric determination of thiosulfate and tetrathionate in soils. Soil Sci. 122:171–178.

Nor, Y.M., and M.A. Tabatabai. 1977. Oxidation of elemental sulfur in soils. Soil Sci. Soc. Am. J. 41:736–741.

Smittenberg, J., G.W. Harmsen, A. Quispel, and D. Otzen. 1951. Rapid methods for determining different types of sulphur compounds in soil. Plant Soil 3:353–360.

Starkey, R.L. 1956. Transformation of sulfur by microorganisms. Ind. Eng. Chem. 48:1429–1437.

Sulphur Institute. 1977. Agri-Sul starts production in Canada. Sulphur Inst. J. 12:2–3.

Tabatabai, M.A. 1974. Determination of sulphate in water samples. Sulphur Inst. J. 10:11–13.

Tabatabai, M.A. 1984. Importance of sulphur in crop production. Biogeochemistry 1:45–62.

Tabatabai, M.A. 1982. Sulfur. p. 501–538. *In* A.L. Page et al. (ed.) Methods of soil analysis. Part 2. 2nd ed. ASA and SSSA, Madison, WI.

Tabatabai, M.A. 1992. Methods of measurement of sulphur in soils, plant materials, and water. p. 307–344. *In* R.W. Howarth et al. (ed.) Sulphur cycling on the continents: Wetlands, terrestrial ecosystems and associated water bodies. SCOPE 48. John Wiley and Sons, New York.

Tabatabai, M.A., and A.A. Al-Khafaji. 1980. Comparison of nitrogen and sulfur mineralization in soils. Soil Sci. Soc. Am. J. 44:1000–1006.

Tabatabai, M.A., and J.M. Bremner. 1972a. Forms of sulfur and carbon, nitrogen and sulfur relationships in Iowa soils. Soil Sci. 114:380–386.

Tabatabai, M.A., and J.M. Bremner. 1972b. Distribution of total and available sulfur in selected soils and soil profiles. Agron. J. 64:40–44.

Trudinger P.A. 1959. The initial products of thiosulfate oxidation by *Thiobacillus X.* Biochem. Biophys. Acta 31:270–272.

Vishniac, W., and M. Santer. 1957. The thiobacilli. Bacteriol. Rev. 21:195–213.

Wainwright, M. 1984. Sulfur oxidation in soils. Adv. Agron. 37:349–396.

Waksman, S.A., and J.S. Joffe. 1922. Oxidation of sulfur in the soil. J. Bacteriol. 7:231–256.

Walker, T.W. 1964. The use of sulfur as fertilizers. Agrochimica 9:1–14.

Weir, R.G. 1975. The oxidation of elemental sulphur and sulphides in soil. p. 40–49. *In* K.D. McLachlan (ed.) Sulphur in Australasian agriculture. Sydney Univ. Press, Sydney, Australia.

Wieder, R.K., G.E. Lang, and V.A. Granus. 1985. An evaluation of wet chemical methods for quantifying sulfur fractions in freshwater wetland peat. Limnol. Oceanogr. 30:1109–1115.

Zehnder, A.J.B., and S.H. Zinder. 1980. The sulphur cycle. p. 105–145. *In* O. Hutzinger (ed.) The natural environment and the biogeochemical cycles. Springer-Verlag, Berlin.

Zhabina, N.N., and I.I. Volkov. 1978. A method for determination of various sulfur compounds in sea sediments and rocks. p. 735–746. *In* W.E. Krumbeih (ed.) Environmental biogeochemistry and geomicrobiology. Vol. 3. Ann Arbor Sci., Ann Arbor, MI.

Chapter 46

Iron and Manganese Oxidation and Reduction

WILLIAM C. GHIORSE, *Cornell University, Ithaca, New York*

Microbial oxidation and reduction of Fe and Mn are of wide-ranging importance to soil scientists (Alexander, 1977; Paul & Clark, 1989). Indeed, knowledge of the distribution, abundance, identity, and activity of Fe- and Mn-transforming microbes in soils and sediments can greatly enhance studies on such diverse agricultural and environmental problems as Fe and Mn availability to plants, metal accumulation, toxicity and mobility of metals and pesticides, and clogging in wells and wetland drainage systems. Knowledge of the biology of Fe- and Mn-transforming microorganisms may allow for future applications in which the metal mobilization and immobilization activities of these microorganisms are exploited for economic and environmental benefit (Ehrlich & Brierley, 1990). Except for the morphologically recognizable "iron bacteria," relatively little is known of the occurrence of Fe-Mn-transforming organisms in nature. Even less is known of their function in natural systems or the factors controlling their in situ activities. On the other hand, several model organisms have been isolated and characterized taxonomically (e.g., *Thiobacillus ferrooxidans, Leptothrix discophora, Shewanella putrefaciens,* and *Geobacter metallireducens* (Lovley et al., 1993)). In some cases, the biochemical mechanisms underlying their Fe- and Mn-transforming abilities have been investigated. (For reviews, see Ghiorse 1984, 1988; Ehrlich, 1987, 1990; Lovley, 1987, 1991; Nealson et al., 1988, 1989; Myers & Nealson, 1990; Ehrlich et al., 1991; Nealson & Myers, 1992).

A persistent problem has been the difficulty of distinguishing abiotic from biologically mediated (biotic) transformations, especially in environments like soil where microbial activity may alter the redox chemistry of the microenvironment, causing Fe and Mn redox changes to occur by direct or indirect mechanism (Ehrlich, 1990). These problems also apply to microbial growth media which, in some instances, may be altered by growth-induced changes in pH or E_h or metabolic products that cause chemical oxidation or reduction of Fe and Mn. These possibilities are taken into account when media are selected for enumeration, enrichment, and isolation of Fe- and Mn-transforming microorganisms. Bacteria and fungi have been implicated in both oxidation and reduction processes in soil and sediment primarily by applying cultural methods for the detection and enumeration of microorganisms that can cause Fe or Mn oxidation or reduction in selected media. Measurement of the disappearance or production of soluble forms of the metals have been used to indicate activity in soil. Inhibition of the activity by heat, metabolic inhibitors, antibiotics, and other biostatic and biocidal agents is used to establish that microbial activity is involved in the processes. The presence of Fe-Mn-transforming organisms in samples can be taken as evidence for their potential activity in the environment. Relatively few studies have addressed the problem of measuring in situ rates of microbial activity in soil because well-tested methods and standard protocols are not available.

The current literature contains ample evidence establishing the potential of soil and aquatic microorganisms to oxidize and reduce Fe and Mn, but definitive experiments showing the relative contributions of abiotic and biologically mediated reactions are lacking. It should be noted that this criticism can be made of almost any microbial process in most natural environments. Development of the required methods and protocols to overcome this problem awaits new methods for estimating activity in samples from natural environments. This may require combinations of conventional methods such as those described in this book with new in situ rate measurement techniques, as well as microscopic molecular probe techniques for distinguishing the genetic potential and expression of specific genes in individual cells. Recent advances in microbial ecology, cell biology and molecular probe technology hold much promise that such methods soon will be available. A future challenge will be to apply these powerful techniques in natural systems such as the soil and sediment that are the focus of this chapter.

The approach in this chapter is to present well-established principles and methods for analysis of organisms in soil, sediments, and drainage systems responsible for biological Fe and Mn transformations. Microscopic observations, especially combined phase contrast and epifluorescence microscopy are recommended as the first step in examining any sample that may contain active microorganisms; however, extensive description of the morphological features of iron bacteria are beyond the scope of the chapter (for recent descriptions see Hanert, 1981a,b; Mulder & Deinema, 1981; Ghiorse, 1984a; Hackett & Lehr, 1986; Jones, 1986; Ghiorse & Ehrlich, 1992). Enumeration and isolation methods form the core of the chapter. In some cases, the organisms are easily counted and isolated directly from colonies on enumeration plates. Others are readily obtained in enrichment cultures, but do not readily form colonies on solid media. Most probable number (MPN) enumeration methods are recommended for the latter.

46–1 IRON-DEPOSITING AND MANGANESE-OXIDIZING HETEROTROPHS

46–1.1 Enrichment and Isolation

46–1.1.1 Principles

Heterotrophic Fe-depositing and Mn-oxidizing bacteria and fungi occur widely in soil and aquatic sediments. Most are strict aerobes, some are microaerophiles, and a few are facultative anaerobes (Ghiorse, 1984a). Their enrichment and isolation are facilitated by the wide range of possible media for heterotrophs, but excess organic nutrients may hinder detection of Fe- and Mn-depositing activity. Care must be exercised in selecting media with regard to the concentration of organic C, phosphate, and Fe and Mn sources. Possible abiotic reactions of Fe and Mn with media components should be considered. Low organic nutrient and low phosphate media usually produce the best results. Iron and Mn sources must provide enough soluble Fe or Mn to reveal activity without hindering growth. Low solubility sources or low concentrations (mg/Kg) of Fe and Mn generally are most successful.

For enrichment and isolation media, consideration of the composition of the soil or sediment is important as well as the chemical composition of the medium. The use of a natural material extract medium, possibly supplemented with a small amount of yeast extract or peptone is recommended. The strategy is to stimulate growth of the heterotrophs while limiting abiotic reactions. Microorganisms that possess the ability to deposit (accumulate) Fe(III) oxides or oxidize Mn(II) should be encouraged to grow, while controlling the growth of other bacteria. Low incubation temperature may help to achieve the latter goal. Oxygen is required for the formation of the metal oxides, but microaerophilic conditions may favor growth of the bacteria involved. Carbon dioxide has been reported to stimulate the processes in soil enrichments under microaerophilic conditions (Uren & Leeper, 1978; Hanert, 1981b; Mulder & Deinema, 1981). Thus, if possible, it is advantageous to use low-oxygen enrichment strategies that encourage the growth of microaerophiles.

A variety of complex media have been tested for enrichment of Fe-depositing and Mn-oxidizing bacteria from soil and sediments (Bromfield & Skerman, 1950; Chapnick et al., 1982; Cullimore & McCann, 1977; Hanert, 1981b; Ghiorse, 1984a; Ehrlich, 1990). These media range from natural water or soil slurries to complex mixtures containing high concentrations of organic nutrients. Generally, pH is adjusted to the neutral range (6–8), but more acid media have been used when appropriate (Bromfield, 1978). Media in the circumneutral pH range will not reveal enzymatic Fe(II) oxidation because Fe(II) autoxidizes at pH > 3. The autoxidation occurs more rapidly in water at neutral pH under aerobic conditions. Thus, at neutral pH, the propensity of a culture to accumulate colloidal Fe(III) oxide is revealed, rather than the ability to oxidize Fe(II). Contrastingly, enzymatic Mn oxidation is detected in circumneutral pH media; though at a pH above 8, Mn oxidation will occur abiotically. Salts of Fe(II) such as $FeCl_2$, or $FeSO_4$ are used to produce Fe(III) oxide, which forms spontaneously in Fe (II)-containing neutral pH media exposed to O_2. Reduced Fe sources include metallic Fe as iron powder or iron wire, FeS, $FeCl_2$, $FeSO_4$, and various Fe(III)-organic complexes; sources of Mn(II) include $MnCO_3$, $MnCl_2$, $MnSO_4$, and organic Mn(II) complexes (Ghiorse & Hirsch, 1978). Organic nutrients are provided with adequate C, N, P, trace elements, and growth factors to support heterotrophic growth in the presence of Mn(II) or Fe(III) oxide. Iron- and Mn-oxidizing fungi often will grow on the complex media designed for heterotrophic bacteria. To detect the maximum number of Fe- and Mn-depositing bacteria, it is advisable to use both the natural material based medium and the complex medium listed below.

Detection is based largely on color of colonies. The Fe(III) oxides give a yellowish to reddish brown hue to colonies of Fe-accumulating bacteria, depending on the concentration and mineral type of oxide. The Mn(III, IV) oxides produce a yellowish brown to dark brown color in colonies. Confirmation of the presence of Fe(III) or Mn(III, IV) oxides is done with spot test reagents; Prussian blue for Fe(III) oxides, a dye reaction (e.g., leucocrystal violet) or chemical test (production of O_2 from H_2O_2) for Mn (III, IV) oxides.

46–1.1.2 Materials

1. Diluted Fe- or Mn-containing extract medium.
2. Diluted Fe- or Mn-containing PYG medium.
3. Spot test reagents for Fe(III) and Mn(III, IV) oxides. Prussian Blue reagents for Fe(III); 1 N of HCl and 1% potassium ferrocyanide. Add one drop of acid followed by a drop of ferrocyanide solution. Blue color indicates Fe(III). Leuco crystal violent reagent (Kessick et al., 1972) for Mn(III, IV) oxides. Dilute 2 mL of a 0.1% leuco crystal violet[1] in 0.1 N of perchloric acid with 5 mL 6 M acetic acid-acetate buffer (final pH = 4.0). Add a drop of test material. Violet color indicates Mn(III, IV) oxide. The dye solution oxidizes slowly in the presence of O_2 or upon exposure to UV light. Store spot test reagents in dark at 4 °C.
4. 0.01% aqueous acridine orange solution.

---

[1]Leuco crystal violet and other leuco bases of common triphenyl methane dyes such as leuco malachite green are readily oxidized by Mn(III, IV) oxides. The leuco bases are available from organic chemical supply companies.

46–1.1.3 Procedures

Natural material based enrichment media can be prepared from extract of soil, or sediment. For enrichments from soil or sediment, prepare an extract from a slurry of material mixed 1:1 with distilled water (e.g., 50 g/50 mL). The slurry is filtered (Whatman no. 1) and the clarified filtrate is filter sterilized (0.2 μm) and used to prepare 50 mL of Fe- or Mn-containing enrichment medium in 250-mL Erlenmeyer flasks. Add 0.1 g of $FeCO_3$, 0.25 to 0.5 g of Fe^o wire or Fe^o powder or 0.05 g of Mn $SO_4 \cdot H_2O$ or $MnCl_2$, or 0.1 g of $MnCO_3$ to the enrichment flasks. Addition of 0.1 g/L yeast extract is optional. It will provide small amounts of nutrients and, stimulate more rapid growth (Hanert, 1981b). The enrichment cultures are inoculated with a small amount (0.1–1 g) of soil or sediment. Cultures should be incubated in cotton-plugged flasks, stationary in the dark. Shaking may inhibit slow-growing microaerophiles. Incubation temperature should be within ±5 °C of the mean daily temperature of the environment under investigation. Lower temperature incubation with no added nutrients promotes slower growth of all bacteria in the enrichment culture. Many Fe- and Mn-depositing bacteria (e.g., prosthecate bacteria) prefer low nutrient concentrations and, therefore, grow more slowly than other heterotrophs that can rapidly exploit available nutrients. Long-term (weeks), low-temperature incubation under low-nutrient conditions may produce different results than short-term ambient temperature incubation. Therefore, it is advisable to set up enrichment cultures under various nutrient and temperature regimes.

Microaerobic or oxygen gradient enrichments can be established by adding 0.1% molten agar (Difco Noble or Difco Purified Agar) or agarose to the medium. If purified agar is not available, standard grade agar can be used if it is washed several times with distilled water to remove soluble organics. Washing is monitored as a decline in intensity of the straw-yellow color of the wash water. The medium is dispensed into 15 × 150 mm screw capped test tubes to fill them completely with a small (max. 5 mm) air space at the top. Proportional amounts of solid Fe or Mn sources are added to each tube to maintain the concentrations indicated above. Most of the O_2 can be stripped from the medium by bubbling with 0.2 μm-filtered N_2 before inoculation. Inoculate with a small amount of soil or sediment. Screw caps are left loose to allow air to enter the tube at the top. Growth of microaerophilic bacteria will occur in the oxygen gradient and form bands of growth and iron or manganese oxide beneath the surface of the tube.

Numerous enrichment media for Fe- and Mn-depositing heterotrophs are described in the literature (Ghiorse & Hirsch, 1978; Hanert, 1981b; Cullimore & McCann, 1977). The most successful of these contain a very low concentration of complex organic nutrients (0.25 g/L or less of yeast extract, peptone, or beef extract, or mixtures of these components) and mineral salts. The concentration of phosphate salts should be kept low because excess phosphate may interfere with Fe and Mn oxidation. Addition of an inorganic N source generally is not required if peptone or beef extract is used. If these complex sources of N are not used, then $(NH_4)_2SO_4$ and KNO_3 should be added. A useful general purpose medium for growth of soil and sediment consists of 0.25 g/L each of peptone, yeast extract, and glucose (PYG) plus 0.5 g/L of $MgSO_4 \cdot 7H_2O$ and 0.01 g/L of $CaCl_2$. A vitamin mixture can be added if desired (Staley, 1968). A 10- or 100-fold dilution of this medium with an Fe or Mn source added as indicated above provides a balanced low-nutrient enrichment medium in which many Fe- and Mn-depositing bacteria can grow.

Enrichment cultures should be checked periodically (at least once a day at the beginning) for signs of growth (turbidity) as well as for Fe or Mn deposition. When O_2 is low, Fe may oxidize slowly at intermediate levels, or near its source at the bottom of the flask or tube. When O_2 is high, gel-like deposits of FeOOH may form quickly. Iron-depositing bacteria usually form rusty colored films or colonies of rust-colored spots on the walls of glass containers. Manganese-oxidizing bacteria similarly form brownish films or small spots. Suspected colonies and films should be examined microscopically for the presence of bacteria and by the spot tests for Fe(III) or Mn(II) oxide.

To test for the presence of bacteria in the Fe-Mn oxide deposits, collect three colonies or fragments of film from the vessel wall with a pasteur pipette. Place one on a piece of filter paper and test with a drop of leuco crystal violet spot test reagent. Mn(III, IV) oxides turn blue. The second colony is tested with Prussian Blue reagents for Fe(III) oxide. Mount the third on a glass microscope slide in a drop of 0.1 g/L of acridine orange under a coverslip. View by phase contrast and epifluorescence microscopy at a magnification of 500 to 1000×. The phase contrast image will reveal characteristic shapes and structures of Fe- and Mn-depositing bacteria but, the diagnosis of bacteria in the aggregates is confirmed by epifluorescence. The acridine orange-DNA and -RNA complexes in the bacterial cells will shine through the Fe and Mn oxides with a bright orange, yellow, or apple-green fluorescence (see color plates in Hanert, 1981b). This direct method of detection is recommended for screening enrichment cultures.

Once diagnosed, the bacterial growth is streaked on the natural material extract medium or the 10-fold diluted PYG medium solidified with at least 1.5% agar. Higher agar concentrations can be used to suppress rapidly spreading bacteria in the sample. Isolation may be facilitated by a second passage in the enrichment medium followed by streaking on agar plates. Select colonies containing Fe or Mn oxide using the same spot tests described above. Repeat streaking and colony selection until you are satisfied that an axenic (uniform morphology) culture is obtained. Bear in mind that amorphous Fe and Mn oxides formed in the cultures may be mistaken for bacteria in the microscope; also, the oxides may concentrate nutrients that attract bacteria. Satellite bacteria are commonly associated with Fe and Mn deposits in enrichment cultures. The satellite bacteria are difficult to remove by streaking on agar; therefore, it may be advantageous to purify the culture by streaking on media without an added Fe or Mn source. Most Fe- or Mn-depositing heterotrophs obtained in such enrichments do not depend on Fe or Mn deposition for growth. Many Mn-depositing bacterial lose this property if they are transferred frequently on artificial media. Iron-depositing bacteria generally do not lose this property. Fungal and actinomycete growth will be recognized by colony morphology and microscopic examination. Fungal hyphae are much thicker (5–10 μm) than actinomycete hyphae (0.5–1.0 μm). Both can become heavily encrusted with Fe or Mn oxides in older cultures (Bromfield & Skerman, 1950), but Fe or Mn deposits may form at some distance from the hyphae in young cultures (Bromfield, 1978; Emerson et al., 1989).

46–1.2 Dilution Spread Plate Counts

46–1.2.1 Principles

The number of Fe- and Mn-depositing microorganisms in soil and sediment can be estimated by a viable cell plate count. This is achieved by serially diluting the sample in a suitable diluent and plating samples by the spread plate method. As a rule of thumb, the viable count of aerobic heterotrophic bacteria in soil or sediment will be 10% or less of the total count, and the number of viable Fe- and Mn-depositing will be a small fraction of the viable count, frequently < 10%. In principle, Fe depositors should outnumber Mn depositors, because Fe deposition can be a nonspecific Fe-oxide accumulation process. Manganese-oxidizing activity appears to be a more specific, enzyme-linked process. Thus, it is less commonly observed among heterotrophic bacteria than iron deposition (Ghiorse, 1984a). The dilutions to be plated should be selected for each sample by estimating the number of aerobic heterotrophs in the sample. This can be determined from a total direct count (see chapter 8 by Zuberer in this book) or by a preliminary aerobic heterotrophic plate count on 10-fold diluted PYG agar (46–2.1.3). The number of Fe- or Mn-depositing bacteria should be in a range of 0.1 to 10% of the aerobic heterotroph plate count or 0.01 to 1.0% of the total direct count. Three or four of the 10-fold dilution tubes should be plated in triplicate for statistical accuracy.

46–1.2.2 Materials

1. Diluted extract Fe or Mn agar.
2. Diluted Fe or Mn PYG agar.
3. Ninety-milliliter sterile dilution buffer consisting of a 1:800 dilution of 34 g of KH_2PO_4/L of distilled water, adjusted to pH 7.2 with 1 N of NaOH (Seeley & VanDemark, 1981).
4. One hundred-milliliter milk dilution bottles.
5. Pipettes, 1 and 10 mL.
6. Bent glass rod spreader.
7. Rotary spread plate table.

8. Toothpicks.
9. Spot test reagents for Fe and Mn (46–1.1.2).

46–1.2.3 Procedure

Prepare diluted extract Fe or Mn agar following procedures described in 46–1.1.3, but adding 20 g/L of agar. Insoluble powders (Fe°, $FeCO_3$ $MnCO_3$) can be dispersed evenly in the molten agar and poured into plates. Iron wire is not recommended for plate counts. Follow the usual procedures for spread plate counts of soil or sediment. Aseptically add 10 g of soil or sediment to 90 mL of sterile dilution buffer in a 125-mL milk dilution bottle. Serially dilute the sample in 3 or 4, 10-fold dilutions to arrive at the proper range as estimated above. Dispense 1.0 or 0.1 mL to each plate. Evenly spread the suspension on the plate using a bent glass rod and rotary table. Allow excess liquid to absorb into the agar before inverting. Bind the edge of the plate with parafilm to prevent drying. Incubate at 20 °C or appropriate temperature ±5 °C above mean daily temperature of the environment from which the sample was collected. Check daily at first for growth and appearance of yellow or brown colonies that test positive for Fe or Mn oxide by spot tests (46–1.1.1). Use a toothpick to remove a small piece of the colony. Immerse it in spot test reagents. An initial count can be done at 5 to 7 d, but some colonies will accumulate Fe and Mn oxides more slowly. Allow at least 21 d before final counts are done. Be careful when counting colonies that have grown out from brown or black Fe-Mn-containing particles. These may test positive simply because the original material contains (Fe(III) or Mn(III, IV) oxide.

46–2 IRON-OXIDIZING AUTOTROPHS

46–2.1 Acidophiles (Enrichment, Isolation, and Enumeration)

46–2.1.1 Principles

Iron-oxidizing autotrophic bacteria use Fe^{2+} as an energy source and CO_2 as a C source. They are divided arbitrarily into two groups based on the pH range for growth. The acidophiles grow at pH values < 5, optimally at pH 2 to 4; neutrophiles (see section 46–2.2) grow at pH above 5, and optimally at pH 6.5 to 7.5. The aqueous chemistry of Fe governs availability of the energy source, Fe^{2+}. This species becomes unstable in the presence of O_2 as the pH increases above 3. The rate of abiological Fe^{2+} oxidation in water or soil solution is slow at pH 2 to 3 or less, but at pH above 5, when O_2 is abundant, the rate increases. Thus, aerobic bacteria depending on Fe^{2+} as an energy source are either acidophilic in acidic, high O_2 environments or microaerophilic in neutral pH environments. Acidophilic bacteria occur widely in mining regions where coal and mineral deposits contribute sulfide minerals (e.g., pyrite) to the soil and sediments. These minerals are oxidized by the bacteria-producing sulfuric acid and metal ions (Fe^{2+}) in solution. The Fe^{2+} is oxidized by acidophilic Fe-oxidizing bacteria.

Thiobacillus ferrooxidans is the best-known acidophilic S- and Fe-oxidizing bacterium in soil. This organism is a mesophile ($T_{opt} = 30$–35 °C) (Alexander, 1977; Kuenen & Tuovinen, 1981; Ehrlich, 1990; Norris, 1990). Several other mesophilic Fe-oxidizing autotrophic acidophiles such as *Leptospirillum ferrooxidans* (Norris, 1990) have been described. In addition, several thermophilic Fe and S oxidizers are known (Kuenen & Tuovinen, 1981; Ehrlich, 1990; Norris, 1990). *Thiobacillus ferrooxidans* uses both reduced S and Fe as sources of energy; *L. ferrooxidans* uses only Fe. Both organisms may occur in the same enrichment cultures when Fe^{2+} is provided as the sole energy source and CO_2 is the C source.

Isolation of Fe-oxidizing autotrophs is best achieved by streaking the enrichment culture on a solidified medium. The use of highly purified agar or agarose (see Kuenen & Tuovinen, 1981; Mishra et al., 1983; Holmes & Yates, 1990) is necessary because organic compounds tend to inhibit growth of the autotrophs or promote the growth of acidophilic heterotrophs. Once isolated colonies are obtained, the culture must be checked thoroughly for the presence of satellite heterotrophs that frequently are found associated with these organisms (Ehrlich, 1990).

No standard enumeration procedure has been published for acidophilic Fe-oxidizing autotrophs in soil per se; however, methods for viable cell plate counts have been published for use in pure cultures and acid mine drainage water (Kuenen & Tuovinen, 1981; Tuovinen & Kelly, 1973; Mishra et al., 1983).

46–2.1.2 Materials

The medium of Silverman and Lundgren (1959) (9K medium) and the medium of Tuovinen and Kelly (1973) (TK medium) or Kuenen and Tuovinen (1981) (KT medium) are suitable for cultivating Fe-oxidizing thiobacilli and leptospirilla and similar organisms (Ehrlich, 1990). 9K medium contains in g/L: $FeSO_4 \cdot 7H_2O$, 44.22; $(NH_4)SO_4$, 3.0; KCl, 0.1; K_2HPO_4, 0.5; $MgSO_4 \cdot 7H_2O$, 0.5; $Ca(NO_3)_2$, 0.01. Dissolve the $FeSO_4 \cdot 7H_2O$ in 300 mL of distilled water; adjust to pH~2 with 1 mL of 10 N H_2SO_4; autoclave, cool, and mix with filtered $FeSO_4 \cdot 7H_2O$ solution. The TK medium contains in g/L: $FeSO_4 \cdot 7H_2O$, 33.3; (NH_4) So_4, 0.4; K_2HPO_4, 0.5; $MgSO_4 \cdot 7H_2O$, 0.4. Dissolve salts in 1 L of 0.1L N of H_2SO_4. Sterilize by passage through a 0.2 μm filter. KT medium contains in $g \cdot L^{-1}$: K_2HPO_4, 0.5; $(NH_4)_2SO_4$, 0.5; $MgSO_4 \cdot 7H_2O$, 0.5; and 1 N H_2SO_4, 5.0 mL (Solution I). Solution II contains in g/L: $FeSO_4 \cdot 7H_2O$, 167; 1 N H_2SO_4, 50 mL. Solutions I and II are sterilized by autoclaving. Mix 4 parts Solution I and 1 part Solution II for the final medium, pH~2. The pH may be varied by altering the amount of H_2SO_4. Lowering the pH to 1.3 will avoid formation of a ferric iron precipitate that forms at higher pH, but growth may be inhibited. Media can be solidified by adding 1% or less electrophoresis-grade agarose (Holmes & Yates, 1990; Mishra et al., 1983) or washed high-purity agar (Tuovinen & Kelly, 1973; Kuenen & Tuovinen, 1981) to the liquid media. For solid 9K medium, add 10 g of solidifying agent to the 700 mL of salts solution before autoclaving. For solid TK medium, autoclave 10 g of agar or agarose in one-half of the H_2SO_4 solution, dissolve salts in the remaining 500 mL; mix after autoclaving. For KT medium mix 4 parts Solution I, 2 parts Solution II, and 4 parts 1% agar or agarose after autoclaving.

46–2.1.3 Procedure

Enrichment cultures are prepared by adding 100-mL of sterile 9K or TK medium to 300-mL sterile Erlenmeyer flasks and inoculating with 1 g of soil or sediment. Flasks are incubated at 25 to 30 °C on a rotary shaker operated at 120 rpm. Aeration can be improved by using baffle flasks. Carbon dioxide can be supplied by bubbling glass-wool filtered 1% CO_2 air mixture in glass tubes inserted through a rubber stopper. Ferrous iron oxidation is indicated by the formation of a yellow-orange mixed precipitate consisting of $Fe_2(SO_4)_3$ that forms above pH 2.

The enrichment cultures should last at least 14 to 21 d. Ferrous Fe-oxidizing autotrophs grow slowly. Thiobacilli growing on Fe^{2+} and CO_2 may oxidize 100 mol of Fe^{2+} for each mole of CO_2 fixed (Ehrlich, 1990). Therefore, Fe precipitates may be visible in a culture long before the cells causing the precipitate are seen; the number of cells produced is very low. Acidophilic hetereotrophs may appear first. *Thiobacillus ferrooxidans* is a short gram-negative rod; *Leptospirillum ferrooxidans* is a slender vibrioid gram-negative rod. Both are motile with polar flagella. Consult original papers listed in Ehrlich (1990) and Norris (1990) for description of these and other acidophiles.

Isolation can be achieved by streaking enrichment cultures in the usual manner on solidified media. Bind the plates with parafilm and incubate for 14 to 21 d at 25 to 30 °C. Check for growth of rust-colored colonies. Streak colonies to purify them taking care to note colony and cellular morphology. Look for morphologically different satellite bacteria in the microscope. The satellite bacteria are acidophilic heterotrophs living from organic acids and alcohols excreted by the autotrophs (Ehrlich, 1990); therefore, they are never present in high numbers. These heterotrophs will grow when organic energy sources are supplied, but they will not oxidize Fe^{2+} (Norris, 1990). Conversely, true Fe^{2+} oxidizing autotrophs will grow best when supplied Fe^{2+} and CO_2 in the absence of organic energy sources.

Enumeration can be achieved with a standard MPN technique using the 9K, TK, or KT media under conditions described for enrichment cultures. Tubes that show the formation of the yellow-orange precipitate are scored positive. Viable cell plate counts may be done by the dilution spread plate technique (described above) using solid 9K, TK, or KT medium. Count Prussian Blue-positive rust-colored colonies after 14 to 21 d of incubation. Growth on agar media may be enhanced by low oxygen tension (Kuenen & Tuovinen, 1981).

46–2.2 Neutrophiles (Enrichment, Isolation, and Enumeration)

46–2.2.1 Principles

A large number of neutrophilic (pH 5–8) Fe-depositing bacteria are included in a group known classically as iron bacteria (Hanert, 1981a,b; Mulder & Deinema, 1981; Ghiorse, 1984a; Jones, 1986). This group includes many genera known to deposit Fe in morphologically distinctive structures (Pringsheim, 1949). It is questionable whether any of these bacteria are truly chemolithoautotrophic as defined originally by Winogradsky (1888) (see Ghiorse, 1984, for discussion). An exception appears to be *Gallionella ferruginea,* which has long been suspected to be a chemolithoautotrophic bacterium using Fe^{2+} oxidation as an energy source to fix CO_2 (Wolfe, 1964; Hanert, 1981a; Ghiorse, 1984). Recently, *G. ferruginea* has been shown to possess key enzymes for CO_2 fixation (Lüthers & Hanert, 1989) and carboxysome-like structures containing ribulose *bis*-phosphate carboxylase found in *T. ferrooxidans* and other obligately autotrophic bacteria. *Gallionella ferruginea* is not common in agricultural soil. It prefers a sessile life style in aqueous environments containing gradients of Fe and O_2 such as Fe seep areas, drainage ditches, water wells, and the like. It can be a major cause of well clogging and it usually is a component of Fe-depositing bacterial consortia. It also occurs in water distribution systems where low concentrations of O_2 mix with CO_2- and Fe^{2+}-rich waters encourages their chemolithoautotrophic growth (Hanert, 1981a; Ghiorse, 1986; Ghiorse & Ehrlich, 1992).

It is important to note that other classical iron bacteria such as *Leptothrix ochracea* (Mulder & Deinema, 1981) also may be capable of chemolithoautotrophic growth; however, these bacteria have so far resisted cultivation (Ghiorse & Ehrlich, 1992). Reliable enrichment, isolation, and enumeration procedures have been published for *G. ferruginea* (Hanert, 1981a).

46–2.2.2 Materials

1. FeS suspension.
2. Modified Kucera-Wolfe medium.
3. Carbon dioxide.
4. Screw capped test tubes (16 × 150 mm).
5. Low temperature incubator (15–20 °C).
6. Phase contrast microscope.

46–2.3.3 Procedures

Detailed instructions on the procedures for enrichment and isolation of *G. ferruginea* are given by Hanert (1981a). FeS suspension is prepared by mixing 78 g of $(NH_4)_2 Fe(SO_4)_2 \cdot 6H_2O$ with 44 g of $Na_2S \cdot 9H_2O$ or 140 g of $Fe(SO_4)_2 \cdot 7H_2O$ with 120 g of $Na_2S \cdot 9H_2O$ in boiled deionized water cooled to 50 °C with constant stirring in a nearly full glass-stoppered 1-L bottle. The bottle is filled with boiled deionized water and stoppered while the FeS precipitate settles. The FeS precipitate is washed by decanting and replacing the supernatant with boiled deionized water at 50 °C several (5–10) times at minimum of 4-h intervals until the pH in the FeS precipitate remains unchanged. Successive stable pH readings in the neutral range indicates that S^2-ions are no longer being released from the FeS precipitate. Precipitation of FeS may be accelerated by adding a few drops of a saturated $FeCl_3$ solution. The FeS suspension is stored in a completely filled glass-stoppered bottles to prevent oxidation. Smaller (100 mL) quantities are removed for autoclaving in 100-mL milk dilution bottles.

Modified Kucera-Wolfe mineral medium is prepared following the procedures of Kucera and Wolfe (1957) as modified by Hanert (1981a). The medium contains per liter of distilled water: 1.0 g of NH_4Cl, 0.2 g of $MgSO_4·H_2O$, 0.1 g of $CaCl_2·2H_2O$, and 0.05 g of $K_2 HPO_4·3H_2O$. A phosphate concentration of 1/10th the original Kucera-Wolfe concentration is used to prevent precipitation of Ca, Mg, and Fe phosphates during autoclaving (Hanert, 1981a).

Enrichment cultures are set up by adding 9 mL of cold (5–10 °C) mineral medium to 16 × 125 mm sterile screw-capped tubes. The medium is bubbled with cotton-filtered CO_2 for 5 s through a sterile Pasteur pipette inserted to the bottom of the tube. One milliliter of sterile FeS suspension is carefully added to the bottom of each tube. Enrichment cultures should be inoculated immediately, with material known from microscopic examination to contain stalks of *Gallionella* (see Hanert, 1981a). Incubate at 15 to 20 °C and look for growth on the sides of the tubes in 3 to 5 d or longer. Fluffy yellow-white to orange colonies will grow in a discrete zone where the proper O_2 and Fe^{2+} concentrations exist in the gradient established in the tube. The presence of *G. ferruginea* in colonies is confirmed in a phase contrast microscope at 400 to 1000×. Actively growing cultures of *G. ferruginea* are identified by the occurrence of characteristic bean-shaped cells at the end of twisted or convoluted stalks (Fig. 46–1).

Isolation may be achieved by the end-point dilution method and enumeration by the dilution MPN method using variations of the same procedure. A series of tubes is prepared to the FeS addition stage and held at low temperature. This helps to keep more CO_2 in solution until the tubes are inoculated. Care must be taken that the inocula are well dispersed so that individual cells can be diluted away from colonies or clumps of flocculent material.

46–3 IRON- AND MANGANESE-REDUCING HETEROTROPHS

46–3.1 Enumeration of Non-Enzymatic Iron- and Manganese-Reducers

46–3.1.1 Principle

Simple enumeration procedures for Fe- and Mn-reducing heterotrophic microorganisms in soil and sediment have been developed using a

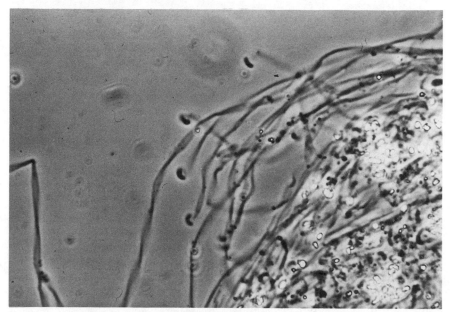

Fig. 46–1. Phase contrast photomicrograph showing four bean-shaped *Gallionella ferruginea* cells at the ends of Fe (III) oxide impregnated stalks (center). The cells are best observed at the edges of a micro-colony such as this one growing in modified Kucera-Wolfe medium. Magnification 2000×.

differential plate count procedure with agar media containing hydrous FeIII or Mn(III, IV) oxides (Ehrlich, 1990; Lovley, 1991). Non-enzymatic Mn reducers may be detected in this way under aerobic or anaerobic conditions. In principle, non-enzymatic Fe reducers could be detected in this way only under anaerobic conditions; however, because of the rapid reoxidation of Fe^{2+} at circumneutral pH in the presence of O_2. Plate counts depend on the production of clear zones of Fe- or Mn-oxide reduction that form around colonies growing on organic substrates in the medium. The zones of reduction are thought to result from reactions of extracellular excretion products of metabolism (usually acidic products of sugar metabolism) with the oxides to reduce them non-enzymatically in the vicinity of the colony. This has been termed *indirect reduction* (Ehrlich, 1990). Although Fe- and Mn-respiring bacteria have been detected in this way (Myers & Nealson, 1988; Ehrlich, 1990), as a rule neither direct enzymatic reduction for Fe or Mn respiration can be inferred from the formation of a clear zone around a colony (Ghiorse, 1988a; Lovley, 1991). Nevertheless, the Fe- and Mn-reducing bacteria and fungi detected in this way may be important members of the total Fe- and Mn-reducing community. The accumulation of metabolic products and acidic conditions, especially in microenvironments of soil and sediments, may encourage indirect reduction mechanisms.

46–3.1.2 Materials

1. Nutrient agar.
2. 20 g/L of $KMnO_4$ solution or sonicated δMnO_2 powder (Myers & Nealson, 1988).
3. Fe(III) oxide (46–3.2.3).
4. Spot test reagents for Fe an Mn (46–1.1.2).

46–3.1.3 Procedure

Prepare and sterilize nutrient agar medium according to manufacturers instructions and dispense molten agar into petri dishes (20 mL). Prepare and sterilize overlay agar tubes containing 5 mL of the same medium. Cool plates to solidify agar. Keep overlay agar molten at 50 °C. Add 0.05 mL of 20 g/Kg of K_2MnO_4 each of the molten overlay agar tubes and mix to form an even suspension of brown Mn oxide in the agar. Alternatively, 0.5 mL of a sterile concentrated suspension of δMnO_2 (Myers & Nealson, 1988) or hydrous Fe(III) oxide (Lovley, 1991) can also be mixed into the overlay agar. Pour the molten suspension over the solidified agar in petri dishes. Rotate plates to disperse the molten overlay suspension evenly. Allow top agar to solidify and dry.

Plates are inoculated and spread as described above (46–1.2) with 0.1 or 1 mL samples from standard dilution series. Incubate at 20 to 25 °C noting growth and clear zones around colonies. The clear zones may be tested for the absence of Fe(III) and Mn(III, IV) oxide with leuco crystal violet or Prussian blue spot tests (46–1.1.2).

46–3.2 Enrichment and MPN Enumeration of Iron- and Manganese-Respiring Bacteria

46–3.2.1 Principles

Recently Lovley and Phillips (1988) and Myers and Nealson (1988) showed that Fe- and Mn-respiring bacteria can be obtained from sediments by anaerobic enrichment with a nonfermentable electron donor (e.g., acetate and lactate) and hydrous Fe or Mn oxide as the sole electron acceptor. Enrichment methods have been developed for anoxic lake and river sediments, but these are applicable to soil and subsurface sediments (Lovley et al., 1990; D.R. Lovley, 1991, personal communication). Dissimilatory Fe and Mn respiration is now accepted as an important biological process in the oxidation of organic matter under anaerobic conditions; characterization of these dissimilatory Fe- and Mn-reducing bacteria is progressing rapidly (Ehrlich, 1990; Myers & Nealson, 1990; Lovley, 1991; Nealson & Myers, 1992).

Enumeration of Fe- and Mn-respiring bacteria is best achieved using a standard MPN method and strictly anaerobic cultural procedures. Tubes are scored on the basis of a color change of the medium. Bacterial Fe reduction results in a color change from reddish brown to black as amorphous Fe(III) oxide is reduced to magnetite, the principal product of Fe reduction. Reduction of brown Mn(III, IV) oxide results is the formation of a whitish precipitate of rhodochrosite ($MnCO_3$).

46–3.2.2 Materials

1. Basal salts medium.
2. Fe(III) or Mn (IV) oxide slurry.
3. Trace element solution.
4. Vitamin solution.
5. Bellco anaerobic pressure tubes with butyl rubber stoppers and aluminum crimp seals.
6. O_2-free N_2/CO_2 (80:20) gas mixture.
7. Tubes and syringes for anaerobic dilution.

46–3.2.3 Procedure

Prepare basal salts medium by the following constituents in g/900 mL of deionized water/$NaHCO_3$, 2.5; $CaCl_2 \cdot 2H_2O$, 0.1; KCl, 0.1; NH_4Cl, 1.5; $NaH_2PO_4 \cdot 6H_2O$, 0.6; a nonfermentable C-energy source, 3.0 (for discussion, see Myers & Nealson, 1990 and Lovley, 1991). Sodium salts of acetate, lactate, pyruvate, succinate or a mixture of these compounds are the best C-energy substrates for enrichment cultures. Add 10 mL each of trace element and vitamin solutions. Mix constituents and adjust pH to 6.5 to 7.0. The trace element solution contains in grams per liter: NTA, 1.5; $MgSO_4$, 0.3; $MnSO_4$, 0.5; NaCl, 1.0; $FeSO_4$, 0.1; $CaCl_2$, 0.1; ZnCl, 0.13; $CuSO_4$, 0.01; $AlK(SO)_4)_2$, 0.01; H_3BO_2, 0.01; $NaMoO_4$, 0.025; $NiCl_2 \cdot 6H_2O$, 0.024; Na_2WO_4, 0.025. The vitamin solution contains in milligram per liter: biotin, 2; folic acid, 2; pyridoxine HCl, 10; riboflavin, 5; thiamine, 5; nicotinic acid, 5; pantothenic acid, 5; vitamin B_{12}, 0.1; p-aminobenzoic acid, 5; thioctic acid, 5.

Add 1 or 2 mL of Fe(III) oxide slurry or 0.5 mL of Mn oxide slurry to the anaerobic pressure tubes containing 9 mL of the basal salts medium. Bubble the mixture with the O_2-free N_2/CO_2 gas mixture for 10 min using a stainless steel canula to reach the bottom of each tube. Cap each anaerobically under a flow of O_2-free N_2 using thick butyl rubber stoppers and seal with an aluminum crimp. Sterilize by autoclaving.

Fe(III) oxide slurry is prepared by dissolving 108 g of $FeCl_3$ in 1 L of distilled water and slowly raising the pH to 7 with drops of 10 N of NaOH. Do not allow bulk pH to go above 7 at any time during the neutralization process. Wash the resultant floc by centrifugation to remove excess salt. Store concentrated suspension in distilled water at 4 °C.

Manganese(IV) oxide slurry is prepared by dissolving 3.16 g of $KMnO_4$ and 3.2 g of NaOH in 1 L of distilled water, dissolve 5.94 g of $MnCl_2$ in another liter of water. Add the $MnCO_2$ solution slowly to the basic $KMnO_4$ solution while mixing constantly with a magnetic stir bar. Wash the resultant floc to remove salts. Store the concentrated suspension in distilled water at 4 °C.

For enrichment cultures and MPN enumeration, dilute samples anaerobically in an appropriate diluent (e.g., basal salts medium without C-energy source). Set up anaerobic MPN tubes following standard procedures. Use syringes and needles flushed with filter sterilized O_2-free N_2 or Ar gas to transfer dilution samples and inoculate tubes. Incubate at an environmentally relevant temperature or 20 to 25 °C.

Iron(III) reduction is indicated by conversion of the non-magnetic, orange-red Fe(III) oxide to a black magnetic precipitate (magnetite). Manganese(IV) reduction is indicated by conversion of dark brown Mn(IV) oxide to a whitish rhodochrosite ($MnCO_3$) precipitate.

A wide variety of electron donors including aromatic compounds and H_2 can yield successful enrichments (Lovley, 1991; Myers & Nealson, 1990); but it is not always certain whether the organisms metabolizing the added electron donors are the Fe(III) or Mn(IV)-reducers (Lovley, 1991). This determination requires isolation and study of the individual members of the Fe-Mn-reducing community. It is advisable to use only nonfermentable compounds such as those listed above for MPN determination in the enrichment cultures.

ACKNOWLEDGMENT

Derek R. Lovley kindly provided guidance for section 46–3.2. Thanks are due to Patti Lisk for help in preparing the manuscript.

REFERENCES

Alexander, M. 1977. Introduction to soil microbiology. 2nd ed. John Wiley and Sons, New York.

Bromfield, S.M. 1978. The oxidation of manganous ions under acid conditions by an acidophilous actinomycete from acid soil. Aust. J. Soil. Res. 16:91–100.

Bromfield, S.M., and V.B.D. Skerman. 1950. Biological oxidation of manganese in soils. Soil Sci. 69:337–348.

Chapnick, S.D., W.S. Moore, and K.H. Nealson. 1982. Microbially mediated manganese oxidation in a freshwater lake. Limnol. Oceanogr. 27:1004–1014.

Cullimore, D.R., and A.E. McCann. 1977. The identification, cultivation, and control of iron bacteria in ground water. p. 219–261. In F.A. Skinner and J.M. Shewan (ed.) Aquatic microbiology. Academic Press, London.

Eaglesham, B.S., R.E. Garen, and W.C. Ghiorse. 1986. Ultrastructure of Gallionella ferruginea. p. 190. In Abstract, Annu. Meet. ASM, J-11. Am. Soc. for Microbiol., Washington, DC.

Ehrlich, H.L. 1987. Manganese oxide reduction as a form of anaerobic respiration. Geomicrobiol. J. 5:423–431.

Ehrlich, H.L. 1990. Geomicrobiology. 2nd ed. Marcel Dekker, New York.

Ehrlich, H.L., and C.L. Brierley. 1990. Microbial mineral recovery. McGraw-Hill Publ. Co., New York.

Ehrlich, H.L., W.J. Ingledew, and J.C. Salerno. 1991. Iron and manganese-oxidizing bacteria. p. 147–170. In J.M. Shively and L.L. Barton (ed.) Variation in autotrophic life. Academic Press, London.

Emerson, D., R. Garen, and W.C. Ghiorse. 1989. Formation of Metallogenium-like structures by a manganese-oxidizing fungus. Arch. Microbiol. 151:223–231.

Ghiorse, W.C. 1984. Biology of iron- and manganese-depositing bacteria. Ann. Rev. Microbiol. 38:515–550.

Ghiorse, W.C. 1986. Biology of *Leptothrix, Gallionella,* and *Crenothrix* relationship to plugging. p. 97–108. *In* R. Cullimore (ed.) Proceedings of international symposium on biofouled aquifers: Prevention and restoration. Am. Water Resourc. Assoc., Bethesda, MD.

Ghiorse, W.C. 1988a. Microbial reduction of manganese and iron. p. 305–331. *In* A.J.B. Zehnder (ed.) Biology of anaerobic microorganisms. John Wiley and Sons, New York.

Ghiorse, W.C. 1988b. The biology of manganese transforming microorganisms in soil. p. 75–85. *In* R.D. Graham et al. (ed.) Manganese in soils and plants. Kluver, Dordrecht, Netherlands.

Ghiorse, W.C. 1989. Manganese and iron as physiological electron donors and acceptors in aerobic-anaerobic transition zones. p. 163–169. *In* Y. Cohen and B. Rosenberg (ed.) Microbial maps: Physiological ecology of benthic microbial communities. Am. Soc. for Microbiol., Washington, DC.

Ghiorse, W.C., and H.L. Ehrlich. 1994. Microbial biomineralization of iron and manganese. p. 75–99. *In* R.W. Fitzpatrick and H.C.W. Skinner (ed.) Iron and manganese biomineralization processes in modern and ancient environments. CATENA, West Germany.

Ghiorse, W.C., and P. Hirsch. 1978. Iron and manganese deposition by budding bacteria. p. 897–909. *In* W.E. Krumbein (ed.) Environmental biogeochemistry and geomicrobiology. Ann Arbor Science, Ann Arbor, MI.

Hackett, G., and J.H. Lehr. 1986. Iron bacteria occurrence, problems and control methods in water wells. Natl. Water Well Assoc., Worthington, OH.

Hanert, H.H. 1981a. The genus *Gallionella.* p. 509–516. *In* M.P. Starr et al. (ed.) The prokaryotes. Vol. I. Springer-Verlag, Berlin.

Hanert, H.H. 1981b. The genus *Siderocapsa* (and other iron- and manganese-oxidizing eubacteria). p. 1049–1059. *In* M.P. Starr et al. (ed.) The prokaryotes. Vol. I. Springer-Verlag, Berlin.

Holmes, D.S., and J.R. Yates. 1990. Basic principles of genetic manipulation of *Thiobacillus ferrooxidans* by biohydrometallurgical applications. p. 29–54. *In* H.L. Ehrlich and C.L. Brierley (ed.) Microbial mineral recovery. McGraw-Hill Publ. Co., New York.

Jones, J.G. 1986. Iron transformations by freshwater bacteria. Adv. Microb. Ecol. 9:149–185.

Kessick, M.A., J. Vuceta, and J.J. Morgan. 1972. Spectrophotometric determination of oxidized manganese with leuco crystal violet. Environ. Sci. Technol. 6:642–644.

Kucera, S., and R.S. Wolfe. 1957. A selective enrichment method for *Gallionella ferruginea.* J. Bacteriol. 74:344–349.

Kuenen, J.G., and O.H. Tuovinen. 1981. The genus *Thiobacillus* and *Thiomicrospira.* p. 1023–1036. *In* M.P. Starr et al. (ed.) The prokaryotes. Vol. I. Springer-Verlag, Berlin.

Lovley, D.R. 1987. Organic matter mineralization with the reduction of ferric iron: A review. Geomicrobiol. J. 5:375–399.

Lovley, D.R. 1991. Dissimilatory Fe(III) and Mn(IV) reduction. Microbiol. Rev. 55:259–287.

Lovley, D.R., F.H. Chapelle, and E.J.P. Phillips. 1990. Fe(III)-reducing bacteria in deep buried sediments of the Atlantic Coastal Plain. Geology 18:954–957.

Lovley, D.R., S.J. Giovannoni, D.C. White, J.E. Champine, E.J.P. Phillips, Y.A. Gorby, and S. Goodwin. 1993. *Geobacter metallireducens* gen. nov. sp. nov., a microorganism capable of coupling the complete oxidation of organic compounds to the reduction of iron and other metals. Arch. Microbiol. 159:336–344.

Lovley, D.R., and E.J.P. Phillips. 1988. Novel mode of microbial energy metabolism: Organic carbon oxidation coupled to dissimilatory reduction of iron and manganese. Appl. Environ. Microbiol. 54:1472–1480.

Lüthers, S., and H.H. Hanert. 1989. The ultrastructure of chemolithoautotrophic *Gallionella ferruginea* and *Thiobacillus ferrooxidans* as revealed by chemical fixation and freeze-etching. Arch. Microbiol. 151:245–251.

Mishra, A.K., P. Roy, and S.S.R. Mahapatra. 1983. Isolation of *Thiobacillus ferrooxidans* from various habitats and their growth pattern on solid medium. Curr. Microbiol. 8:147–152.

Mulder, E.G., and M.H. Deinema. 1981. The sheathed bacteria. p. 425–440. *In* M.P. Starr et al. (ed.) The prokaryotes. Vol. I. Springer-Verlag, Berlin.

Myers, C.R., and K.H. Nelson. 1988. Bacterial manganese reduction and growth with manganese oxide as the sole electron acceptor. Science 240:1319–1321.

Myers, C.R., and K.H. Nealson. 1990. Iron mineralization by bacteria: Metabolic coupling of iron reduction to cell metabolism in *Alteromonas putrefaciens* strain MR-1. p. 131–149. *In* R.B. Frankel and R.P. Blakemore (ed.) Iron biominerals. Plenum Press, New York.

Nealson, K.H., and C.R. Myers. 1992. Microbial reduction of manganese and iron: new approaches to carbon cycling. Appl. Environ. Microbiol. 58:439–443.

Nealson, K.H., R.A. Rosson, and C.R. Myers. 1989. Mechanisms of oxidation and reduction of manganese. p. 383–411. *In* R.J. Beveridge and R.J. Doyle (ed.) Metal ions and bacteria. John Wiley and Sons, New York.

Nealson, K.H., B.M. Tebo, and R.A. Rosson. 1988. Occurrence and mechanism of microbial oxidation of manganese. Adv. Appl. Microbiol. 33:279–318.

Norris, P.R. 1990. Acidophilic bacteria and their activity in mineral sulfide oxidation. p. 3–27. *In* H.L. Ehrlich and C.L. Brierley (ed.) Microbial mineral recovery. McGraw-Hill Publ. Co., New York.

Paul, E.A., and F.E. Clark. 1989. Soil microbiology and biochemistry. Academic Press, New York.

Pringsheim, E.G. 1949. Iron bacteria. Biol. Rev. 24:200–245.

Seeley, H.W., Jr., and P.J. VanDemark. 1981. Microbes in action. W.H. Freeman and Co., San Francisco.

Silverman, M.P., and D.G. Lundgren. 1959. Studies on the chemoautotrophic iron bacterium *Ferrobacillus ferrooxidans*. I. An improved medium and a harvesting procedure for securing high cell yields. J. Bacteriol. 77:642–647.

Staley, J.T. 1968. *Prosthecomicrobium* and *Ancalomicrobium:* New prosthecate freshwater bacteria. J. Bacteriol. 95:1921–1942.

Tuovinen, O.H., and D.P. Kelly. 1973. Studies on the growth of *Thiobacillus ferrooxidans*. I. Use of membrane filters and ferrous iron agar to determine viable numbers and comparison with $^{14}CO_2$ fixation and iron oxidation as measures of growth. Arch. Mikrobiol. 88:285–298.

Uren, N.C., and G.W. Leeper. 1978. Microbial oxidation of divalent manganese. Soil Biol. Biochem. 10:85–87.

Winogradsky, S. 1888. Ueber Eisenbacterien. Bot. Ztg. 46:262–270.

Wolfe, R.S. 1958. Cultivation, morphology, and classification of the iron bacteria. J. Am. Water Works Assoc. 50:1241–1249.

Wolfe, R.S. 1964. Iron and manganese bacteria. p. 82–97. *In* H. Heukelebian and N.C. Dondero (ed.) Principles and applications in aquatic microbiology: Proceedings of the Rudolfs Research Conference. John Wiley and Sons, New York.

SUBJECT INDEX